THE MAMMALS
OF
NORTH AMERICA

THE MAMMALS

OF

NORTH AMERICA

E. RAYMOND HALL, Ph.D.

RESEARCH ASSOCIATE, MUSEUM OF NATURAL HISTORY,
AND PROFESSOR OF ZOOLOGY (EMERITUS)
THE UNIVERSITY OF KANSAS

Volume I

SECOND EDITION

A WILEY-INTERSCIENCE PUBLICATION

JOHN WILEY & SONS, New York · Chichester · Brisbane · Toronto

Published April 3, 1981

Copyright © 1981 by John Wiley & Sons, Inc.

All rights reserved. Published simultaneously in Canada.

Library of Congress Cataloging in Publication Data
Hall, Eugene Raymond.
  The mammals of North America.

  "A Wiley-Interscience publication."
  Bibliography: p.
  Includes indexes.
  1. Mammals—North America.   2. Mammals—West
Indies.   I. Title.
QL715.H15   1979   599′.097   79-4109
ISBN 0-471-05443-7 (v. 1)
ISBN 0-471-05444-5 (v. 2)

Printed in the United States of America

10 9 8 7 6 5 4 3 2 1

PREFACE

The purposes of the present volume are to summarize the results of taxonomic studies of North American native mammals of the post-Columbian period (1492 through June 1977) and to show by means of maps the geographic ranges of most of the species and subspecies. Some information published after June 1977 is in the addenda to Volume II.

North America as here understood includes Greenland, all of Panamá, and the entire continent between these regions, as well as the Greater Antilles and the Lesser Antilles south to and including Grenada.

The sequence of orders and families is essentially that of Hall and Kelson (1959).[1] It followed that of Miller and Kellogg (1955), which in its main features is the sequence adopted by Simpson (1945). Genera and species are arranged in order of inferred geologic age, oldest to youngest.[2]

Subspecies are arranged alphabetically. Under each subspecies and monotypic species, reference is made to the name in the original description. To this, where necessary, are added (a) reference to the first use of the current name combination and (b) references to any junior synonyms.

For the nominate subspecies the current two-word combination (genus and species) suffices for the first use of the current name combination because the name automatically becomes a three-word combination as soon as any subspecies is recognized. An example is *Sciurus nelsoni*, proposed in 1893, which became *Sciurus nelsoni nelsoni* on June 3, 1898, when *Sciurus nelsoni hirtus* was named.

---

[1] Any linear classification of course conveys only a small fraction of the available information on phylogenetic relationships as compared with a three- or even a two-dimensional diagram. The old (1945–1959) sequence of orders and families used here is convenient because most mammalogists are familiar with it, but it fails to reflect important new information uncovered by paleontologists since 1945. Kukalova-Peck's (1973) and McKenna's (1975) linear sequences of orders living now in North America, although different from each other and from the one used by Hall and Kelson (1959), contain sequences that do reflect some of the newer information alluded to above.

Kukalova-Peck	McKenna	Hall and Kelson
Marsupialia	Marsupialia	Marsupialia
Insectivora	Cingulata	Insectivora
Primates	Pilosa	Chiroptera
Rodentia	Lagomorpha	Primates
Chiroptera	Carnivora	Edentata
Edentata	Soricomorpha	Lagomorpha
Lagomorpha	Chiroptera	Rodentia
Carnivora	Primates	Cetacea
Cetacea	Artiodactyla	Carnivora
Artiodactyla	Cetacea	Pinnipedia
Perissodactyla	Perissodactyla	Sirenia
Sirenia	Sirenia	Perissodactyla
	Rodentia	Artiodactyla

[2] "Oldest to youngest" in many instances on the following pages (although not in all) yields a sequence closely resembling a cladistic sequence, but as with the higher taxonomic categories a mere listing line by line of these lower taxa conveys far less concerning their relationships than does a diagram, even a two-dimensional one (a cladogram).

CHOICE OF TECHNICAL NAMES

In choosing names of genera, species, and subspecies, the tenth edition of *Systema Naturae* 1758 by Carolus Linnaeus has been taken as the starting place. Thereafter the law of priority has been adhered to for competing names. Against this background, questions concerning nomenclature have been decided as nearly as possible in accordance with the International Zoological Code.

USE OF COMMAS AND PARENTHESES WITH TECHNICAL NAMES

The name of the original describer of a subspecies or monotypic species (in the sense that the person coined and published the scientific name) immediately follows the scientific name without an intervening comma, semicolon, colon, or period.

Parentheses (round brackets) enclose the name of the original describer when the current scientific name (of a subspecies or monotypic species) is combined with a generic name other than the one used in the original description. Examples are *Ammospermophilus leucurus notom* (Hansen) and *Dicrostonyx hudsonius* (Pallas). Hansen used the generic name *Citellus* when he proposed the subspecific name *notom* and Pallas used the generic name *Mus* when he proposed the specific name *hudsonius*.

A comma is interposed between the scientific name and the name of any author (person) other than the original describer who uses that scientific name. For example, in the fourth line below the name *Dicrostonyx groenlandicus rubricatus* (Richardson) there is a comma between *rubricatus* and Anderson. The comma is there because Anderson did not propose (coin) the name *rubricatus;* Richardson did that.

For consistency a comma is used in the present work in the same sense as it was in Hall and Kelson (1959). See paragraph above. To identify the name of any user other than the original describer of a taxon, Article 51(i) of the 1961 International Code of Zoological Nomenclature prescribed "some distinctive manner" other than use of a comma.

Citation to place and date for any scientific (technical) name in a list of synonyms is complete enough for the reader to go directly to the original source. Consequently, most of such sources are not to be found in the terminal section of "Literature Cited." In other parts of the text, for instance in a listing of the distinctive morphological features of a particular taxon, a more abbreviated form of citation (*e.g.,* Orr, 1963:424) is used and requires the reader to go to the terminal section of "Literature Cited" to find the original source.

Where pagination overlaps in two or more publications by a given author in one year in the section of "Literature Cited," an alphabetic letter ordinarily has been added to the year in both the text and "Literature Cited" to aid the reader of the text in finding a particular publication.

DESCRIPTIVE CONTENT

The native mammals listed here are those thought to have been present in post-Columbian time, in other words in 1492 and in some or all of the time since then. In addition to the available name and junior synonyms for each genus, species, and subspecies, the listing contains (1) maps showing the geographic distribution of most species and the subspecies of each species, (2) keys in most genera by means of which specimens may be identified to species, and (3) mention of selected morphological features of species and higher taxonomic categories. Information is omitted on habits and reproduction such as was provided in "The Mammals of North America" by E. R. Hall and K. R. Kelson (2 vols., The Ronald Press Co., New York, March 31, 1959) and is to be sought there and in the subsequent literature.

The descriptions along with artificial keys were designed to permit the identification of any specimen as to species and to each taxonomic category of higher rank. The aim in constructing keys was to use qualitative characters insofar as was possible (*e.g.*, presence *vs.* absence of a certain pair of teeth); whenever this was not possible and quantitative characters were used (*e.g.*, "broader skull" *vs.* "narrower skull"), it was intended that reference be made to illustrations of skulls.

Identification of a specimen to subspecies cannot be made from the present work, but the citations to original descriptions and especially to more recent literature prefaced by the words "Revised by" (or "Reviewed by" for more abbreviated accounts) guides a person to published information, enabling him to identify a specimen to subspecies. The citations to revisions are at the beginning of an account of a genus or a species. For example, "Revised by Jackson, N. Amer. Fauna, 38:54–76, September 30, 1915," immediately follows the name of the genus (*Scapanus*) of western moles and "Revised by Lidicker . . . 1960," follows the name of the species *Dipodomys merriami*.

DESCRIPTIVE MATERIAL

Capitalized color names are those of Ridgway (Color Standards and Color Nomenclature, Washington, D.C., 1912); noncapitalized color names relate to no particular standard, except that many of them were taken *verbatim* from the early literature and some of them may refer to unidentified color charts.

Linear measurements, unless otherwise specified, are in millimeters. External measurements are arranged in this order: total length; length of tail; length of hind foot, including the longest claw; and height of ear from notch. The external measurements are those of adults. Rarely, if ever, are both the minimum and maximum measurements for a species found in one subspecies; the measurements given are intended to show the known extremes for the species as a whole. Likewise the two extremes for any genus that contains two or more species ordinarily are not to be found in any one species of that genus.

MARGINAL RECORDS

In the paragraphs of "Marginal Records" the localities of occurrence are arranged in clockwise order, beginning with the northernmost locality. If more than one locality lies on the line of latitude that is northernmost for a given kind of mammal, the westernmost one on that line is recorded first. The marginal records in roman type are represented by solid black dots on the corresponding map. Other marginal localities set in italic type are not shown on the map because symbols for them would overlap symbols for others listed in roman type. Some marginal records lack citations as to source. The source of any such record is to be found in "The Mammals of North America" by Hall and Kelson, 1959.

In a few instances a symbol on a map is in such a position that if the line separating two subspecies (*e.g., Sylvilagus audubonii arizonae* and *S. a. cedrophilus,* on Map 225) were unbroken, the line would divide the black symbol into halves. Any such instance signifies that two specimens identified as two subspecies are recorded from the same place (in this instance Flagstaff, Arizona). The assumption is that the specimens actually were obtained some distance apart. The locality (Flagstaff) is listed in the "Marginal Records" of each of the two subspecies.

Users of the maps should remember that although symbols represent actual geographic occurrences, some parts of the shaded areas represent only the author's guesses as to occurrence. Furthermore, on almost every map some parts of the shaded area are ecologically unsuited for the species concerned; consequently the species does not occur in the unsuitable areas.

A shaded arrow projecting from North America into South America signifies that the species (not necessarily the subspecies) occurs on both continents. The same is true for an arrow pointing into Asia. Wording in the text (marginal records) identifies any subspecies or monotypic species the range of which extends onto another continent.

NUMBER OF TAXA

A total of 3607 subspecies and monotypic species of native mammals are admitted here, *versus* 3209 by Hall and Kelson in 1959. The increase of 398 results from (among other things) the discovery in Central America of many species, especially of bats, previously known only from South America, and the recognition of some previously unnamed subspecies from North America.

In the same period (1959–1977) some subspecific names have been placed as synonyms of others and some species have been reduced to subspecific rank, although a few subspecies have been elevated to specific rank.

CRITERIA FOR SPECIES VERSUS SUBSPECIES

When classifying specimens as to subspecies or species, I prefer to take the following steps: (*a*) select for initial, intensive study a large series, 30 or more individuals, from one restricted locality; (*b*) segregate these by sex; (*c*) arrange specimens of each sex from oldest to youngest; (*d*) divide these into age-groups and within a given age-group, of one sex, from one locality, of what is judged to be one species, measure the amount of individual variation; (*e*) with this measurement as a "yardstick," compare individuals, and if possible series, comparable as to sex and age (and season where characteristics of the pelage are involved) from this and other localities.

If crossbreeding occurs in nature at a place or places where the geographic ranges of two kinds of mammals meet, the two kinds are to be treated as subspecies of one species. If no crossbreeding occurs, the two kinds are to be regarded as two distinct, full species. However it is difficult or impossible to apply this criterion when assessing, for example, differences between some insular populations. Frequently the populations on two islands near each other differ enough to warrant subspecific or possibly specific distinction.

A means of deciding on specific *versus* subspecific status for the two populations is to find on the adjacent mainland a continuously distributed, related kind of mammal, which there breaks up into subspecies. Ascertain the degree of difference between each pair of mainland subspecies that intergrade directly. If the maximum degree of differences between the insular kinds is greater than the difference between two subspecies on the mainland, which intergrade directly, and greater than that between either insular kind and the related population on the nearby mainland, the two insular kinds may properly be treated as full species. If the maximum degree of difference between the insular kinds is no greater than, or is less than, the difference found on the mainland between pairs of subspecies that intergrade directly, the insular kinds may properly be treated as subspecies of one species. In fine, the criterion is degree of difference with the limitation of geographic adjacency, rather than intergradation or lack of it.

Although species and subspecies seem to have the same kinds of distinguishing characters, which appear to be inherited by means of essentially the same kinds of mechanisms in the germ plasm, there are two noteworthy differences between species and subspecies. One already implied is that in a species that is continuously distributed over a given area, its characters at the boundaries of its range are sharp, definite, and precise. At any one place some of its characters comprised of size, shape, and color, are either those of one species or are instead unequivocally those of some other, whereas the characters of a subspecies, particularly at or near the place where two subspecies meet, more often than not are various combinations of those of the two subspecies, and in many individual characters there is blending (after Hall, 1943). For anyone wishing more information about my concept of species and subspecies, as well as about my procedures in making decisions, some further details are to be found on pages 23 and 24 of Hall (1951).

ACKNOWLEDGMENTS

For help since 1959 in preparing the present account of the mammals of North America, I am especially indebted to the persons named below. Of these, on the basis of time devoted to the project at the University of Kansas, Lohmann devoted the most time, and Kortlucke, Ratliff, LaVal, and Kinman follow in that order.

Ticul Alvarez	Richard K. LaVal
Ann B. Bueker	Eleanor K. Lohmann
Evelyn L. Davis	Charles A. Long
David E. Duncan	Ronald M. Nowak
Mary F. Hall	Thu Nowak
Robert W. Henderson	Fidelma (Bunker) O'Connor
Barry L. Keller	Russell M. Park
Kenneth E. Kinman	Donald B. Patrick
Sheila M. Kortlucke	Beatríz Péfaur
Margie Kuhn	Myriam (Jiménez) Post

Theresia R. Ratliff   Jeffrey A. Seims
David F. Schmidt   Barry Siler
Thomas Swearingen

Assistance with particular segments was received from other persons at the major mammal collections in the United States, including the American Museum of Natural History in New York; the Field Museum of Natural History in Chicago; the U.S. National Museum in Washington, D.C.; the Museum of Vertebrate Zoology in Berkeley, California; the Museum of Zoology in Ann Arbor, Michigan; the Museum of Natural History in Urbana, Illinois; Texas A & M University at College Station; and the Museum of Natural History in Lawrence, Kansas. Karl F. Koopman, Guy G. Musser, Joseph C. Moore, Don E. Wilson, William Z. Lidicker, Jr., M. Raymond Lee, Donald F. Hoffmeister, David J. Schmidly, and Robert S. Hoffmann of those collections contributed. I am indebted also to those who helped with "The Mammals of North America" by Hall and Kelson (1959:viii, ix).

Research grants are gratefully acknowledged from the University of Kansas General Research Fund; the University of Kansas Biological Sciences; the University of Kansas Endowment Association; the Floyd T. Amsden Fund; the Raymond F. Rice Foundation; the Dane G. Hansen Foundation; the Office of Naval Research (41-64); Biomedical Grant (RR-07031); the National Science Foundation (grants G 5633, G 12512, GE 7739, GB 2218, GB 100, GZ 378, GB 5909, GB 30969, DEB 76-21408); and the U.S. Fish and Wildlife Service (14-16-0008-2109).

A large share of the financial assistance used in organizing the following account came from the National Science Foundation. Also the Office of Biological Services of the Fish and Wildlife Service of the U.S. Department of the Interior provided support at a critical period. Without the assistance from these two agencies, the present work would not yet have been completed.

E. RAYMOND HALL

Lawrence, Kansas
November 1980

CONTENTS

Volume I

CONTENTS

Volume II

ZOOGEOGRAPHY

The Life-zones of Merriam (see end papers inside the hard covers of this volume) continue to serve usefully for tabulating numbers of kinds of mammals—fewer kinds toward the North Pole and more kinds toward the Equator. Further information on zoogeography is to be found in Hall and Kelson (1959:xxiii–xxx). Also, information on latitudinal gradients in species density as presented by John W. Wilson III (1974:124–140, and other papers cited therein) is pertinent to in-depth study of the geographic distribution of North American mammals.

Concerning the interchange of mammals between Asia and North America, a gradual increase in knowledge about the geographic extent of the land bridge, Beringia, and the times of its ice-free existence in the late Pleistocene, as well as the existing water barrier formed *ca.* 13,000 years ago, afford opportunity to more precisely date times of differentiation of mammalian stocks into subspecies and species. On this subject, one of several recent publications which includes an extensive bibliography, is "Evolutionary relationships of some Beringian mammals" by C. F. Nadler, R. S. Hoffmann, N. N. Vorontsov, and R. I. Sukernik (1976:325–336).

Concerning the interchange of mammals between North America and South America, Hershkovitz (1969:1) wrote: "The Neotropical Region, which is defined on the basis of its living mammals, is comprised of the Brazilian, Patagonian, and West Indian Subregions. The Middle American Province of the Brazilian Subregion was the primary center of origin, evolution, and dispersal for mammals now living in continental South America. The West Indies also derived its fauna from Middle America, and perhaps also from South America. Faunal interchange between these regions must have taken place since the Middle Tertiary, at the latest. By the time the Isthmian land bridge between Middle and South America was completed during the Pliocene-Pleistocene transition, nearly all modern genera of Neotropical mammals were already differentiated within their present geographic ranges." **See** addenda.

Concerning subspeciation, which is a step toward the formation of species, the southwestern quarter of the United States, and especially the region made up of Utah, Arizona, and some adjacent parts of California, Nevada, and New Mexico, has a larger number of subspecies than any other continental area of equal size in the world. Maps 304–308, 325, and 342 illustrate the large number. The reasons seem to be high degree of relief caused by alternating mountain ranges and valleys or enclosed basins, great range in elevation (below sea level in Death Valley to more than 14,000 ft. on nearby Mount Whitney), deeply cut, steep, rock-walled canyons (the Grand Canyon of the Colorado, for one), resultant wide range in temperature, sharply marked zonation of plants, and tremendous diversity of types of soil. Natural selection frequently aided by isolation caused by natural barriers has resulted in truly amazing geographic variation in animal morphology and coloration. Adequate series of specimens from closely spaced collecting stations in parts of this region reveal uniformity in size, shape, and color throughout most of the geographic range of any one of scores of subspecies, and reveal also that integradation frequently occurs in narrow belts. Well-planned investigations there will yield much additional information on the formation of species in nature as this process goes on at the twig-ends of the tree of mammalian life.

THE MAMMALS
OF
NORTH AMERICA

CLASS MAMMALIA

Beings especially notable for possessing mammary glands (mammae) that permit the female to nourish the newborn young with milk; hair present although confined to early stages of development in most of the Cetacea; mandibular ramus of lower jaw made up of single bone (dentary); lower jaw articulating directly with skull without intervention of quadrate bone; two exoccipital condyles; differing from both Aves and Reptilia in possessing diaphragm and in having nonnucleated red blood corpuscles; resembling Aves and differing from Reptilia in having "warm blood," complete double circulation, and four-chambered heart; differing from Amphibia and Pisces in presence of amnion and allantois and in absence of gills.

Mammals generally are considered by man to be the pinnacle of development of animal life—a consideration difficult to divorce entirely from self-interest, since man is a mammal. In many particulars birds also are complex and highly developed morphologically. Considered as a whole, however, the Mammalia seem to have achieved the highest point in physical organization of any of the five great classes of the Subphylum Vertebrata, and, except for the birds (Class Aves), were the last to develop on earth. Both the mammals and the birds had their beginnings in the much older Class Reptilia.

The name "mammal" and the idea that it expresses both are comparatively recent. The English naturalist, John Ray, 284 years ago in his *"Synopsis Methodica Animalium Quadrupedum et Serpentini Generis"* (1695) bracketed the terrestrial or quadrupedal mammals with the aquatic as "vivipara" in contrast to the "ovipara" or birds. Thus the idea was first expressed, but 63 years elapsed before it was adopted. This was done by the great Swedish naturalist Linnaeus, in preparing the 10th edition of his *"Systema Naturae,"* published in 1758. He caught on to John Ray's idea, removed from the fishes the cetaceans (whales and porpoises), combined them and hairy quadrupeds in a special class, and coined for it the word "Mammalia." The essential component is the Latin *mamma,* meaning breast. By analogy with *anima* (animal) and the derived technical term Animalia, Linnaeus, 221 years ago, coined the word Mammalia. From this the English-speaking peoples subsequently derived the vernacular name mammal and its plural, mammals.

SUBCLASS THERIA

Young developing to point of being recognizable as mammals in reproductive tract of female instead of developing in egg outside female as, for example, in Order Monotremata of Subclass Prototheria; members of the extinct Subclass Allotheria, comprising orders Multituberculata and Triconodonta, having, in general, platelike teeth and triconodont teeth (talon and talonid absent or poorly developed) respectively, also may have laid eggs.

INFRACLASS METATHERIA

Differs from Infraclass Pantotheria (extinct orders Pantotheria and Symmetrodonta) in presence of inflected angle on lower jaw instead of lacking angle (Order Symmetrodonta) or having noninflected angle (Order Pantotheria); differs from Infraclass Eutheria in that placenta is absent or incomplete, in presence of epipubic bones, and in other characters as are listed below for the Order Marsupialia.

1

ORDER **MARSUPIALIA**—Marsupials

Monographed by A. Cabrera, Genera Mammalium: Monotremata, Marsupialia, Mus. Nac. Cien. Nat., Madrid, pp. 1–177, 18 pls., June 23, 1919.

Nonplacental (loose allantoic placenta in some; in *Dasyurus* a yolk-sac placenta; chorionic placenta, lacking villi, in the Peramelidae); young extruded from bifid reproductive tract of female in notably undeveloped condition, undergoing further development (corresponding to terminal part of embryonic development in a placental mammal) while firmly attached for some time after birth to mammae through which milk is injected; in several genera mammae are in an abdominal pouch (= marsupium); females having two lateral canals on uterus; males having vasa deferentia that pass ureters laterally instead of mesially; long epipubic bones, or their rudiments, present in both sexes; skull with small braincase and large preorbital region; brain lacking corpus callosum; nasals expanded posteriorly; jugal bone contributing to glenoid fossa; tympanic bone small, annular, or tubular and but rarely fused with other bones of skull; auditory bulla, if present, formed entirely or principally from alisphenoid.

SUBORDER **POLYPROTODONTIA**

Incisors $\frac{4}{3}$ or more, subequal and smaller than canines; molars with sharp cusps (not bluntly tuberculated or with transversely ridged crowns).

SUPERFAMILY **DIDELPHOIDEA**

Incisor teeth $\frac{5}{4}$. All living members of this superfamily belong to the one family Didelphidae.

FAMILY **DIDELPHIDAE**—Opossums

Differs from extinct family Caroloameghiniidae (Lower Eocene of South America) in tuberculosectorial, instead of bunodont, lower molars. Other characters, not necessarily distinctive as compared with Caroloameghini-idae, are as follows: 5 subequal incisors above and 4 below; long pointed snout; tail more or less prehensile; in species of North America and Central America less than proximal half of tail furred, remainder of tail being naked; stomach simple; 5 toes on each foot; clawless hallux of hind foot opposable to other digits; marsupial pouch in *Didelphis*, *Chironectes*, and *Metachirops* and represented in other genera by rudimentary traces, such as two folds of skin; precranial part of skull long and pointed; braincase small; zygomatic arches large; hard palate with fenestrae opposite molars; upper molars tritubercular; lower molars tritubercular with well-developed trigonids; epipubic bones present; dental formula uniform throughout family, being i. $\frac{5}{4}$; c. $\frac{1}{1}$; p. $\frac{3}{3}$; m. $\frac{4}{4}$, 50 teeth in all; 3rd premolar above and below are the only teeth having milk predecessors.

The careful comparisons of the Central American genera by Miss Mary E. Works, in an unpublished manuscript (1950) at the University of Kansas, show that *Didelphis*, *Chironectes*, and *Metachirops* form a natural group of genera, whereas another natural group is made up of *Marmosa*, *Caluromys*, and *Philander*.

In this latter group the teeth are unlike in the different genera but do agree in having the 2nd post-canine tooth in the upper jaw larger than the 3rd. There is more uniformity in the teeth of the first group, composed of the three genera *Didelphis*, *Chironectes*, and *Metachirops*, in which the two mentioned teeth are subequal.

In the period 1900–1975 inclusive, four of the authors who have studied the four-eyed opossums have applied three available generic names in different ways as shown below. *Metachirops* Matschie, 1916, had not been proposed when Allen chose names in 1900. Each of the four authors presents compelling arguments for his choice of names. Their publications should be consulted by anyone who undertakes to review the history of these names. The publications by Hershkovitz (1949:11–12; 1976:295–303) and Pine (1973:391–402) list the reasons for their contrary conclusions and also contain citations to the earlier pertinent literature. Pine's 1973 application of names is tentatively adopted in this volume. The debate between Hershkovitz and Pine probably will continue beyond June 30, 1977, the cutoff date for this volume.

Authors	Gray and Black Four-eyed Opossums	Brown Four-eyed Opossums
J. A. Allen, 1900	*Metachirus*	*Metachirus*
Tate, 1939	*Metachirops*	*Metachirus*
Hershkovitz, 1949	*Philander*	*Metachirus*
Pine, 1973	*Metachirops*	*Philander*

KEY TO NORTH AMERICAN GENERA OF DIDELPHIDAE

1. Hind feet with toes fully webbed; zygomatic breadth ⅔ or more of basal length. . . . *Chironectes,*   p. 9
1′. Hind feet with toes not webbed; zygomatic breadth less than ⅔ of basal length.
  2. More than basal ⅓ of tail haired; more than 33 caudal vertebrae; upper tooth-rows prominently bowed outward. . . . .*Caluromys,*   p. 11
  2′. Less than proximal ⅓ of tail haired; fewer than 33 caudal vertebrae; upper tooth-rows not prominently bowed outward (relatively straight and convergent anteriorly).
    3. Tail less than ¾ as long as head and body. . . . .*Monodelphis,*   p. 18
    3′. Tail more than ¾ as long as head and body.
      4. Pelage of two distinct types of hair, underfur and coarse, long, white-tipped guard-hair; sternum of 8 segments. . . . .*Didelphis,*   p. 3
      4′. Pelage of one principal type of hair, the long, white-tipped guard-hairs being absent; sternum of 7 segments.
        5. Hind foot more than 33; greatest length of skull more than 50; total length more than 465.
          6. Color grayish; clearly marked light spot above each eye; more than basal 15% of tail densely haired; postorbital processes prominent; 4 vacuities in posterior part of hard palate. . . . .*Metachirops,*   p. 8
          6′. Color brownish; faintly marked light spot above each eye; less than basal 15% of tail densely haired; no postorbital processes; only 2 vacuities in posterior part of hard palate. . . . .*Philander,*   p. 17
        5′. Hind foot less than 33; greatest length of skull less than 50; total length less than 465. . . . .*Marmosa,*   p. 12

Genus **Didelphis**—Opossums

Revised by J. A. Allen, Bull. Amer. Mus. Nat. Hist., 14:149–188, June 15, 1901. See also Gardner, Special Publ. Mus. Texas Tech Univ., 4:1–81, July 3, 1973.

1758. *Didelphis* Linnaeus, Syst. nat., ed. 10, 1:54. Type, *Didelphis marsupialis* Linnaeus.
1777. *Didelphys* Schreber, Die Säugthiere . . . , fasc. 3, p. 532.
1819. *Sarigua* Muirhead, Brewster's Encyclopaedia, 13:429 (under Mazology).
1853.*Didelphus* Lapham, Trans. Wisconsin Agric. Soc., 2:337.
1914. *Leucodidelphis* Von Ihering, Rev. Mus. Paulista, 9:347, February. Type, *Didelphis paraguayensis* J. A. Allen [= *D. azarae* Temminck]. Proposed as a subgenus.

Pelage consisting of underfur and white-tipped over-hairs (which are wanting or few in other genera of the family; basal tenth or so of tail furred, remainder naked; 1st digit (hallux) on hind foot clawless and opposed to the other toes for grasping; sagittal and occipital crests high in old individuals; postorbital processes present; palate fenestrated (see Fig. 1); female having well-developed marsupium in which the mammae are situated; 8 separate bones in sternum (in *D. v.*

virginiana) as contrasted with only 7 bones in other genera examined.

KEY TO NORTH AMERICAN SPECIES OF DIDELPHIS

1. Posterior extension of lachrymal (that forms lower anterior margin of orbit) terminating in pronounced, sometimes squared, point; hair of cheek usually buff. . . . .*D. marsupialis,*   p. 3
1′. Posterior extension of lachrymal (that forms lower anterior margin of orbit) terminating in rounded edge; hair of cheek white. . . . .*D. virginiana,*   p. 5

Didelphis marsupialis
Southern Opossum

External measurements in males: up to 1017, 535, 80; in females: up to 948, 477, 70. Greatest length of skull, 88–125 (males); 84–116 (females). Tail usually longer than combined length of head and body. Head and body usually dark except for paler base of rostral vibrissae and cheek (light yellow to orange-buff); ears, lower legs, and feet black; tail black proximally (up to 50% of bare tail in southern populations). Diploid chromosome number 22, fundamental number 20; autosomes consist of 3 pairs large, 7 pairs medium-sized acrocentric chromosomes. This species is aggressive when confronted with a stressful situation and is not known to feign death.

Gardner (1973) listed under *Didelphis marsupialis caucae* all the subspecific names previously proposed for *D. marsupialis* in Central America and the Mexican state of Tabasco. This listing, he verbally explained, was to show affinity of the taxa concerned in the specific sense with *D. marsupialis* instead of with *D. virginiana*, and was not intended to indicate that the taxa (*battyi, particeps, richmondi,* and *tabascensis*) were invalid. Owing to the large amount of material of *D. marsupialis* now available from Central America, a systematic study of it could be rewarding by revealing the amount and kind of geographic variation there and the number of subspecies to be recognized.

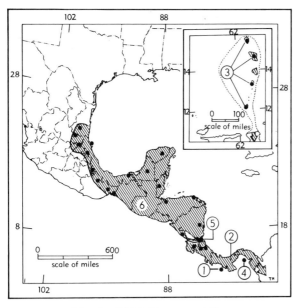

Map 1. *Didelphis marsupialis.*

1. *D. m. battyi*
2. *D. m. caucae*
3. *D. m. insularis*
4. *D. m. particeps*
5. *D. m. richmondi*
6. *D. m. tabascensis*

Didelphis marsupialis battyi Thomas

1902. *Didelphis marsupialis battyi* Thomas, Nov. Zool., 9:137, April 10, type from Coiba Island, Panamá. Known only from Coiba Island.

Didelphis marsupialis caucae J. A. Allen

1900. *Didelphis karkinophaga caucae* J. A. Allen, Bull. Amer. Mus. Nat. Hist., 13:192, October 23, type from Cali, Upper Cauca Valley, Colombia.
1902. *Didelphis marsupialis caucae* J. A. Allen, Bull. Amer. Mus. Nat. Hist., 16:257, August 18.
1902. *Didelphis marsupialis etensis* J. A. Allen, Bull. Amer. Mus. Nat. Hist., 16:262, August 18, type from Eten, Piura, Perú.

MARGINAL RECORDS.—Costa Rica: Pozo Azul (Gardner, 1973:70); San José; Santa Teresa Peralta, thence E to coast and along same into South America, and up Pacific Coast from South America to point of beginning.

Didelphis marsupialis insularis J. A. Allen

1902. *Didelphis marsupialis insularis* J. A. Allen, Bull. Amer. Mus. Nat. Hist., 16:259, August 18, type from Caparo, Trinidad.

MARGINAL RECORDS.—Island of Dominica; Martinique Island; St. Vincent Island; Grenada Island, and in the South American area on Trinidad Island.

Didelphis marsupialis particeps Goldman

1917. *Didelphis marsupialis particeps* Goldman, Proc. Biol. Soc. Washington, 30:107, May 23, type from San Miguel Island, Golfo de Panamá, Panamá. Known only from San Miguel Island.

Didelphis marsupialis richmondi J. A. Allen

1901. *Didelphis richmondi* J. A. Allen, Bull. Amer. Mus. Nat. Hist., 14:175, June 15, type from Greytown, Nicaragua.
1920. D[idelphis]. m[arsupialis]. richmondi, Goldman, Smiths. Miscl. Coll., 69(5):46, April 24.

MARGINAL RECORDS.—Nicaragua: type locality. Costa Rica: San José.

Didelphis marsupialis tabascensis J. A. Allen

1901. *Didelphis marsupialis tabascensis* J. A. Allen, Bull. Amer. Mus. Nat. Hist., 14:173, June 15, type from Teapa, Tabasco.

MARGINAL RECORDS (Gardner, 1973:69, 70, as *D. m. caucae*).—Tamaulipas: Ejido Santa Isabel, 2 km. W Pan American Highway. Veracruz: Hda. Tamiahua, Cabo Rojo. Tabasco: Frontera. Campeche: Apazote, near Yahaltuma. Yucatán: Chichén-Itzá. Belize: Central Farm, Cayo District. Honduras: 2 mi. W San Pedro Sula; Patuca. Nicaragua: Kurinwas [= Curinguas] River; Toro Rapids; 3 km. N, 4 km. W Diriamba, thence northward along Pacific Coast to Oaxaca: Tapanatepec; mts. near Santo Domingo; 10 km. S Yetla. Veracruz: 3 km. SE Orizaba. Puebla: Metlaltoyuca. San Luis Potosí: *ca.* 2 km. W Xilitla; El Salto, Río Naranjo.

Didelphis virginiana
Virginia Opossum

External measurements in males: up to 930, 446, 80; in females: up to 912, 415, 76. Greatest length of skull, 85–142 (males); 78–119 (females). Tail usually shorter than combined length of head and body. Body gray, black, reddish, or rarely white; head whitish with darker body color usually extending from dorsum to between eyes; ears either black or black with paler tips; lower legs black; in *D. v. virginiana* toes of hind feet and distal half of forefeet white, remainder of feet black; in other subspecies feet either black or black with white toes; tail black proximally (usually less than 20% of bare tail in *D. v. virginiana*, up to 30% of bare tail in *D. v. pigra*, usually more than 30% of bare tail in *D. v. californica*, from 40 to 60% of bare tail in *D. v. yucatanensis*). Diploid chromosome number 22, fundamental number 32; autosomes consist of 3 pairs large subtelocentric, 3 pairs medium-sized subtelocentric, and 4 pairs medium-sized acrocentric chromosomes. This species is less aggressive than *D. marsupialis* in stressful situations and often feigns death ("plays possum").

D. virginiana and *D. marsupialis* are here treated as two species because Gardner (1973) did so. Until at least one macromorphological feature that differentiates *all* individuals is found, the possibility remains that Gardner was dealing with a single species.

The species ranges from southern Canada to northern Costa Rica and is sympatric with *D. marsupialis* in parts of Central America and southeastern México. The Great Plains of southern Canada and the United States, together with the more southern desertlike country of southern New Mexico and northwestern México, constituted a barrier for the opossum; it did not occur on the plains or to the westward, but after the species was introduced in the first part of the twentieth century on the Pacific Coast of the United States the opossum multiplied amazingly and at this writing is well established in nearly all the coastal country from Crescent Beach, British Columbia (Cowan and Guiguet, 1960:40) south to San Diego County, California.

Didelphis virginiana californica Bennett

1816. *Did[elphys]. mes-americana* Oken, Lehrbuch der Naturgeschichte, 2(3):1152, type locality in northern México. Oken's names are non-Linnaean.
1833. *Didelphis Californica* Bennett, Proc. Zool. Soc. London, p. 40, May 17, type probably from northern, or northwestern, part of Republic of México; restricted to Sonora by Hershkovitz [Fieldiana-Zool., Field Mus. Nat. Hist., 31(47):548, July 10, 1951].
1973. *Didelphis virginiana califorica*, Gardner, Special Publ. Mus. Texas Tech Univ., 4:30, July 3.
1833. *Didelphis breviceps* Bennett, Proc. Zool. Soc. London, p. 40, May 17, type probably from northern or northwestern part of Republic of México.
1901. *Didelphis marsupialis texensis* J. A. Allen, Bull. Amer. Mus. Nat. Hist., 14:172, June 15, type from Brownsville, Cameron Co., Texas.

MARGINAL RECORDS.—Sonora: Llano; Oputo. Chihuahua: Batopilas. Durango: Chacala; Durango (Baker and Greer, 1962:62); *28 mi. S, 17 mi. W Vicente Guerrero, 8350 ft. (ibid.)*. Zacatecas: *ca.* 8 mi. S Victor Rosales (Baker, 1968:318, as *D. marsupialis* only). San Luis Potosí: Bledos (Gardner, 1973:72). Coahuila: 1 mi. SW San Pedro de las Colonias, 3700 ft.; 1 mi. N San Lorenzo, 4200 ft.; Monclova (Gardner, 1973:72); ½ mi. S Sabinas (*ibid.*). Texas (Hall and Kelson, 1959:8, as *D. m. texensis*): Pinnacle Spring; Rosillas Mts.; mouth of Pecos River; Rockport, thence southward along Gulf Coast to Campeche (Gardner, 1973:74): 1 km. SW Puerto Real, Isla del Carmen; Apazote, near Yahaltuma; 65 km. S, 128 km. E Escárcega. Honduras: Yaruca (Gardner, 1973:74). Nicaragua (Gardner, 1973:74): Jalapa; Peña Blanca; La Esperanza, 5 km. S, 3½ km. E San Carlos. Costa Rica: 6 mi. S, 2 mi. W Cañas (117435 KU). Nicaragua: Sapoá (Gardner, 1973:74), thence northward along Pacific Coast to Sinaloa: 3 mi. NE San Miguel (Gardner, 1973:72). Sonora: Hermosillo.

Didelphis virginiana pigra Bangs

1898. *Didelphis virginia pigra* Bangs, Proc. Boston Soc. Nat. Hist., 28:172, March, type from Oak Lodge, opposite Micco, Brevard Co., Florida.

MARGINAL RECORDS.—South Carolina: Hilton Head Island (Gardner, 1970 MS., p. 100), thence S and W along Atlantic Coast and Gulf of Mexico to Texas (Gardner, 1970 MS., p. 103): O'Connorsport; 10 mi. W Cuero; 5 mi. N Belleville; 9 mi. NE Sour Lake. Louisiana (Lowery, 1974:57, Map 2): Caddo Parish; West

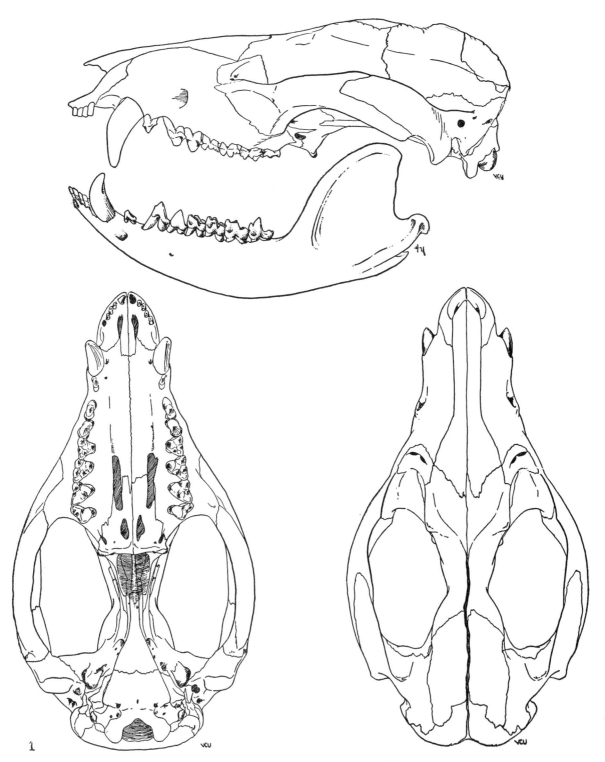

Fig. 1. *Didelphis virginiana virginiana*, 1 mi. N Lawrence, Douglas Co., Kansas, No. 3780 K.U., ♂, X 1.

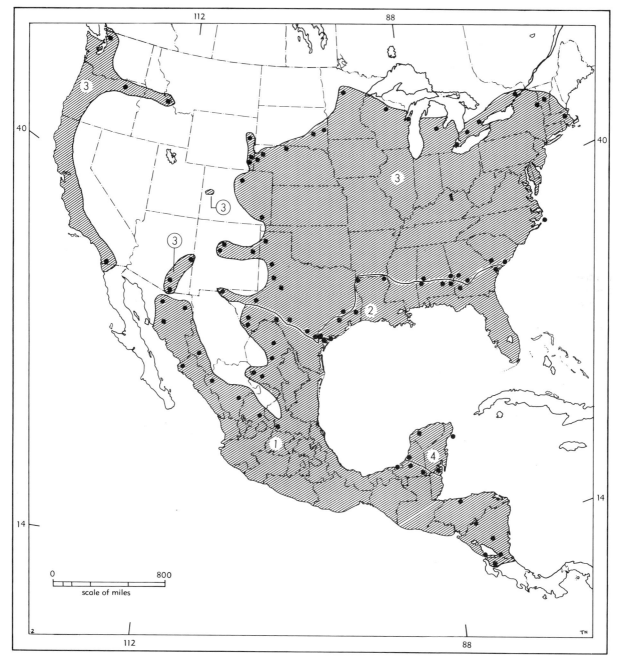

Map 2. *Didelphis virginiana.*

1. *D. v. californica*   3. *D. v. virginiana*
2. *D. v. pigra*   4. *D. v. yucatanensis*

Carroll Parish. Alabama: Myrtlewood; Catoma Creek; Seale. Georgia: Pretoria (Gardner, 1970 MS., p. 100).

Didelphis virginiana virginiana Kerr

1792. *Didelphis virginiana* Kerr, The animal kingdom . . . , p. 193. Type locality, Virginia.

1795. *D[idelphys]. pilosissima* Link, Beyträge zur Naturgeschichte, p. 67, based on Sarique à longs poils Buff [on]., V. 4. T. 29.

1795. *D[idelphys]. illinensium* Link, Beyträge zur Naturgeschichte, p. 67, based on Sarique à longs poils Buff [on]., V. 4. T. 29.

1806. *Didelphis woapink* Barton, Facts, observations, and conjectures relative to the generation of the opossum of North America, p. 2.

MARGINAL RECORDS.—Minnesota: near Hackensack (Hazard, 1963:118, as *D. marsupialis* only). Wisconsin (Long and Copes, 1968:283, 284): 8½ mi. N Tomahawk; Pierce Township, Kewaunee Co. Michigan: Isabella County. Ontario: Lot 10, Concession 2, Chatham Twp., Kent Co.; Middlesex County; Toronto; near Morrisburg. New York: Crown Point. Vermont: Rochester. Massachusetts: Newton Center, thence southward along Atlantic Coast at least to North Carolina: Hatteras. South Carolina: Christ Church Parish, Charleston Co. (Gardner, 1970 MS., p. 98); Savannah River Plant (Wood and Odum, 1964:545, as *D. marsupialis* only). Georgia: Flint River, Vienna (Gardner, 1970 MS., p. 98); 1½ mi. W Junction City. Alabama: Auburn; Greensboro. Texas (Gardner, 1970 MS., pp. 99, 100, unless otherwise noted): 17 mi. SW Huntsville; Runge; Benton; 40 mi. W Kerrville; El Paso; Monahans (V. Bailey, 1932:7, as *D. marsupialis texensis*); Colorado (Hall and Kelson, 1959:8); Double U Ranch, 4 mi. NE Draw (Packard and Garner, 1964:387); 5 mi. NE Abernathy (Bowers and Judd, 1969:277). New Mexico (Sands, 1960:393, as *D. marsupialis* only): *Belen;* Los Chavez; between towns Tijeras and San Antonio; 1 mi. S, 1 mi. E Ragland. Texas: Tascosa. Colorado: near head Caddoa Creek, 12 mi. N Gaume's Ranch; Green Mtn. Cemetery, Boulder (Armstrong, 1972:41). Wyoming: Springer Wildlife Management Unit, 20 mi. SW Torrington (Brown, 1965a:142); near Dull Center (Long, 1965a:516); 2 mi. NW Lingle (Brown, 1965a:142); *2 mi. S Torrington (ibid.).* Nebraska: 3 mi. N McGrew (Choate and Genoways, 1967:238); Alliance (Jones, 1964c:61); Cherry County, W of Crookston (*ibid.*). South Dakota: Sanborn County; Brookings County. Introduced and now established on Pacific Coast, inland to Washington: Clear Lake. Oregon: Birch Creek, below Pilot Rock. Idaho: 2 mi. W Salmon (Gardner, 1970 MS., p. 100). California: San Diego County. Introduced also in Arizona: Alpine; Nogales; Fort Lowell Road, Tucson (Cockrum, 1961:24). Introduced also in Colorado: *vic. Grand Junction.*

Didelphis virginiana yucatanensis J. A. Allen

1901. *Didelphis yucatanensis* J. A. Allen, Bull. Amer. Mus. Nat. Hist., 14:178, June 15, type from Chichén-Itzá, Yucatán. *Didelphis nelsoni* (J. A. Allen, *op. cit.*:160), a *nomen nudum,* may have been intended to apply to this subspecies.
1973. *Didelphis virginiana yucatanensis,* Gardner, Special Publ. Mus. Texas Tech Univ., 4:34, July 3.
1901. *Didelphis yucatanensis cozumelae* Merriam, Proc. Biol. Soc. Washington, 14:101, July 19, type from Cozumel Island, Yucatán (*fide* Gardner, 1970 MS.).

MARGINAL RECORDS.—Yucatán: Colegio Peninsular (Birney, *et al.,* 1974:4). Quintana Roo: Cozumel Island (Gardner, 1973:75). Belize: Orange Walk. Campeche: 5 km. S Champotón (Gardner, 1973:75).

Genus **Metachirops** Matschie
The Gray and the Black Four-eyed Opossums

For use of this name rather than *Philander* Tiedemann 1808, see Pine, Proc. Biol. Soc. Washington, 86:391, December 14, 1973.

1916. *Metachirops* Matschie, Sitzungsb. Gesell. naturforsch. Freunde, Berlin, p. 262, October. Type *Didelphis quica* Temminck.
1919. *Holothylax* Cabrera, Genera Mammalium: Monotremata, Marsupialia, Mus. Nac. Cienc. Nat., Madrid, p. 47, June 23. Type, *Didelphis opossum* Linnaeus.

External measurements: 489–610; 253–329; 35–50. Greatest length of skull, 57.9–80; color pale gray to black; white spot above each eye; tail furred at base and naked distally; marsupium present; 4 fenestrae in bony palate.

These animals are smaller and more agile than members of the genus *Didelphis;* the agility is reflected in the looser attachments of the vertebrae and in the longer neural spines than in *Didelphis.*

Metachirops opossum
Gray Four-eyed Opossum

This species ranges into South America and differs from the one other species of the genus, *Metachirops mcilhennyi* of Perú, in being gray (not black), and in having shorter fur and less extensively haired base of tail (averaging 17% instead of 27% of length of tail). After Gardner and Patton (1972:3).

Metachirops opossum fuscogriseus (J. A. Allen)

1900. *Metachirus fuscogriseus* J. A. Allen, Bull. Amer. Mus. Nat. Hist., 13:194, October 23, type from Greytown, Nicaragua (see J. A. Allen, Bull. Amer. Mus. Nat. Hist., 30:247, December 2, 1911).

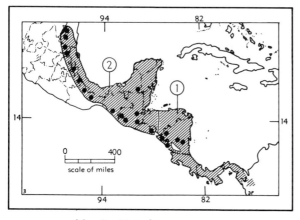

Map 3. *Metachirops opossum.*

1. *M. o. fuscogriseus*   2. *M. o. pallidus*

1924. *Metachirops opossum fuscogriseus*, Miller, Bull. U.S. Nat. Mus., 128:7, April 29.

MARGINAL RECORDS.—Nicaragua: San Juan [= San Juan Telpaneca, 3500 ft.]; *Río Coco;* Cara de Mono, 50 m. (Phillips and Jones, 1969:346, identified to species only); Escondido River, 50 mi. above Bluefields, thence along Caribbean Coast to South America, thence northward along Pacific Coast to Nicaragua: 3 km. N, 4 km. W Diriamba, 600 m. (Phillips and Jones, 1969:346, identified to species only); Volcán de Chinandega [= Volcán el Viejo].

Metachirops opossum pallidus (J. A. Allen)

1901. *Metachirus fuscogriseus pallidus* J. A. Allen, Bull. Amer. Mus. Nat. Hist., 14:215, July 3, type from Orizaba, Veracruz.

1924. *Metachirops opossum pallidus*, Miller, Bull. U.S. Nat. Mus., 128:7, April 29.

MARGINAL RECORDS.—Tamaulipas: 7 km. SW La Purísima (Alvarez, 1963:395); thence along Caribbean Coast to Campeche: 65 km. S, 128 km. E Escárcega (93191 KU). Guatemala: Finca Sepacuite. El Salvador: Lake Olomega (Burt and Stirton, 1961:20). Guatemala: Lake Atescatempa; Finca Cipres; Finca Carolina. Chiapas: 1 km. S Mapastepec. Oaxaca: Coatlán (Goodwin, 1969:31); 5 mi. W Chiltepec (*ibid.*). Veracruz: Orizaba. Puebla: Metlaltoyuca. San Luis Potosí: Xilitla; El Salto. Tamaulipas (Alvarez, 1963:395): *Rancho Pano Ayuctle;* 2 km. W El Carrizo.

Genus **Chironectes** Illiger—Water Opossum

1811. *Chironectes* Illiger, Prodromus systematis mammalium et avium . . . , p. 76. Type *Lutra minima* Zimmermann.

External measurements: 651–685; 380–386; 70–72. Greatest length of skull, 74.2–76.5. Pelage relatively short, fine, and dense; color marbled black and gray, rounded black areas being confluent along mid-line of back. Toes webbed; pisiform area of forefoot enlarged and simulating, in some respects, a 6th digit; sole of hind foot granular and without plantar tubercles; tail furred only at base; marsupium present; skull with notably broad braincase; only 2 fenestrae in posterior part of bony palate; sagittal and lambdoidal crests present in adults. Although specialized for aquatic life and consequently superficially unlike *Didelphis*, *Chironectes* agrees in deep-seated parts of its anatomy with *Didelphis* and

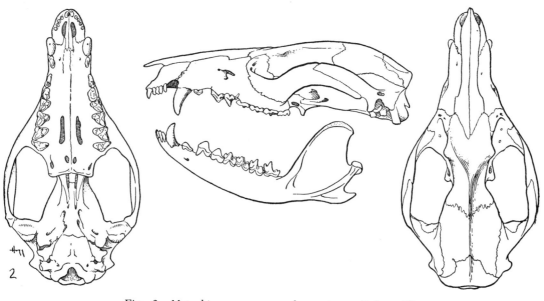

Fig. 2. *Metachirops opossum fuscogriseus,* 5 km. SE Turrialba, Costa Rica, No. 26923 K.U., ♂, X 1.

Metachirops and is closely related to those genera.

Chironectes minimus
Water Opossum

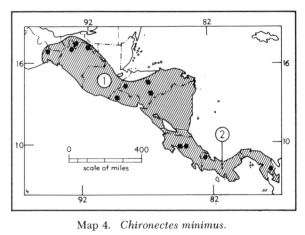

Map 4. *Chironectes minimus.*

1. *C. m. argyrodytes*   2. *C. m. panamensis*

See characters of the genus.

Chironectes minimus argyrodytes Dickey

1928. *Chironectes argyrodytes* Dickey, Proc. Biol. Soc. Wash
ington, 41:15, February 4, type caught in Río Sucio at Hda.
Zapotitán, 1500 ft., La Libertad, El Salvador.
1959. *Chironectes minimus argyrodytes,* Hall and Kelson,
Mammals of North America, Ronald Press, p. 1079, March
31.

MARGINAL RECORDS.—Tabasco: 3⅛ km. N Teapa
(Lay, 1963:374). Chiapas: Palenque (Alvarez del Toro,
1952:184). Honduras: Las Flores; Minas de Oro;
Tegucigalpa. El Salvador: type locality. Oaxaca: 30⅖
km. N Matías Romero (Schaldach, 1965:130 as *C.
minimus* only). Chiapas: Pichucalco (Alvarez del Toro,
1952:184 as *C. minimus* only).

Chironectes minimus panamensis Goldman

1914. *Chironectes panamensis* Goldman, Smiths. Miscl. Coll.,
63(5):1, March 14, type from Cana, 2000 ft., Darién,
Panamá.
1958. *Chironectes minimus panamensis,* Cabrera, Rev. Mus.
Argentino de Cienc. Nat., 4:44, March 27.

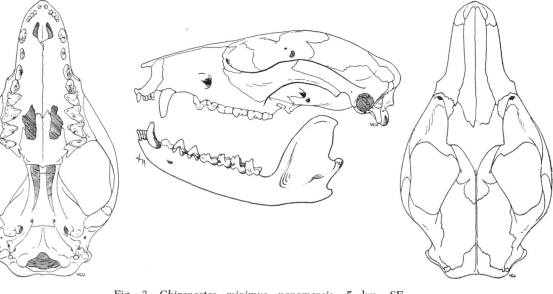

Fig. 3. *Chironectes minimus panamensis,* 5 km. SE
Turrialba, Costa Rica, No. 26928 K.U., ♂, X 1.

MARGINAL RECORDS.—Costa Rica: Vijagual; Carillo. Panamá (Handley, 1966:756): Sibube; Tacarcuna Village, thence into South America.

Genus **Caluromys** J. A. Allen—Wooly Opossums

For use of *Caluromys* J. A. Allen in place of *Philander* authors (not Tiedemann, 1808), see Hershkovitz, Proc. Biol. Soc. Washington, 62:12, March 17, 1949; and Hopwood, Proc. Zool. Soc. London, 117:533, October 30, 1947.

1856. *Philander* Burmeister, Erläuterungen zur Fauna Brasiliens . . . , Berlin, p. 74. Type, *Philander cayopollin* Burmeister [= *Didelphis philander* Linnaeus]. Not *Philander* Tiedemann, 1808.
1900. *Caluromys* J. A. Allen, Bull. Amer. Mus. Nat. Hist., 13:189, October 12. Type, *Didelphis philander* Linnaeus.
1916. *Micoureus* Matschie, Sitzungsb. Gesell. naturforsch. Freunde, Berlin, 8:259, 269, December 15, and *op. cit.*, p. 281 (September 10, 1917). Type, *Didelphis laniger* Desmarest [= *Didelphis lanata* Olfers]. Not *Micoureus* Lesson, 1842 [= *Marmosa* Gray, 1821].
1920. *Mallodelphys* Thomas, Ann. Mag. Nat. Hist., ser. 9, 5:195, February. Substitute name for *Micoureus* Matschie, 1916, preoccupied.

External measurements: 587–760; 395–490; 45–47. Greatest length of skull, 58.7–61.0. Pelage long, fine, and wooly; hair extending along almost all of, or more than, basal third of tail; color pattern ornate, including dark stripe on face and in some subspecies reddish and blackish on body; tail longer than head and body and having 37 instead of 28–30 caudal vertebrae as in the other Central American genera except possibly *Monodelphis*, in which the number is unknown to me; marsupium probably rudimentary in *C. philander* (Pine, 1973:392, 393); temporal ridges infrequently uniting to form sagittal crest; postorbital processes well developed; maxillary tooth-rows curved and converging anteriorly; M3 not larger than M1; posterior part of palate lacking fenestrae.

Caluromys derbianus
Wooly Opossum

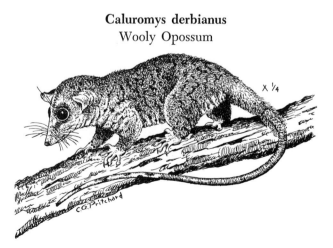

See account of the genus.

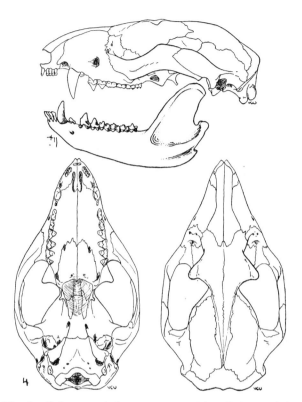

Fig. 4. *Caluromys derbianus aztecus*, 3 km. E San Andrés Tuxtla, Veracruz, No. 23367 K.U., ♀, X 1.

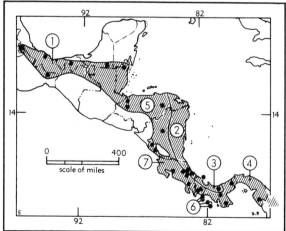

Map 5. *Caluromys derbianus.*

Guide to subspecies
1. *C. d. aztecus*
2. *C. d. canus*
3. *C. d. centralis*
4. *C. d. derbianus*
5. *C. d. fervidus*
6. *C. d. nauticus*
7. *C. d. pallidus*

Caluromys derbianus aztecus (Thomas)

1913. *Philander laniger aztecus* Thomas, Ann. Mag. Nat. Hist., ser. 8, 12:359, October, type from San Juan de la Punta, Veracruz.
1955. *Caluromys derbianus aztecus,* Miller and Kellogg, Bull. U.S. Nat. Mus., 205:10, March 3.

MARGINAL RECORDS.—Veracruz (Hall and Dalquest, 1963:201): Potrero; 3 km. E San Andrés Tuxtla, 1000 ft. Tabasco: Teapa (Thomas, 1913:359). Oaxaca: near Sarabia (Schaldach, 1965:131). Veracruz: *type locality.*

Caluromys derbianus canus (Matschie)

1917. *Micoureus canus* Matschie, Sitzungsb. Gesell. naturforsch. Freunde, Berlin, 4:284, September, type from Nicaragua. Recorded only from type locality.
1952. *Caluromys derbianus canus,* Hall and Kelson, Univ. Kansas Publ., Mus. Nat. Hist., 5:324, December 5.

Caluromys derbianus centralis (Hollister)

1914. *Philander centralis* Hollister, Proc. Biol. Soc. Washington, 27:103, May 11, type from Talamanca, Costa Rica.
1946. *Caluromys derbianus centralis,* Goodwin, Bull. Amer. Mus. Nat. Hist., 87:285, December 31.

MARGINAL RECORDS.—Costa Rica: 5 km. SE Turrialba, 1950 ft.; type locality. Panamá (Handley, 1966:754, as *C. d. fervidus*): Isla Bastimentos; Santa Fe.

Caluromys derbianus derbianus (Waterhouse)

1841. *Didelphys derbianus* Waterhouse, *in* The naturalist's library (edit. Jardine), 30 (Mammals, 11):97, type locality, Cauca Valley, Colombia, South America.
1955. *Caluromys derbianus derbianus,* Miller and Kellogg, Bull. U.S. Nat. Mus., 205:9, March 3.

MARGINAL RECORDS.—Panamá: Tabernilla; eastward into South America, and westward to Panamá: Cana; Guánico (Handley, 1966:754); Santiago (*ibid.*).

Caluromys derbianus fervidus (Thomas)

1913. *Philander laniger fervidus* Thomas, Ann. Mag. Nat. Hist., ser. 8, 12:359, October, type from "Guatemala," Lowlands of east-central Guatemala or northern Honduras according to Goodwin (Bull. Amer. Mus. Nat. Hist., 79:114, May 29, 1942).
1942. *Caluromys derbianus fervidus,* Goodwin, Bull. Amer. Mus. Nat. Hist., 79:114, May 29.

MARGINAL RECORDS.—Campeche: 65 km. S, 128 km. E Escárcega (93191 KU). Belize: Kates Lagoon, thence along Caribbean Coast at least to Honduras: San Pedro Sula; Catacamas; Ilama.

Caluromys derbianus nauticus (Thomas)

1913. *Philander laniger nauticus* Thomas, Ann. Mag. Nat. Hist., ser. 8, 12:359, October, type from Gobernadora Island, off W coast Panamá.
1955. *Caluromys derbianus nauticus,* Miller and Kellogg, Bull. U.S. Nat. Mus., 205:10, March 3.

MARGINAL RECORDS.—Panamá: Brava Island; Parida Island; type locality; Cebaco Island.

Caluromys derbianus pallidus (Thomas)

1899. *Philander laniger pallidus* Thomas, Ann. Mag. Nat. Hist., ser. 7, 4:286, October, type from Bogava [= Bugaba], 2500 m., Chiriquí, Panamá.
1946. *Caluromys derbianus pallidus,* Goodwin, Bull. Amer. Mus. Nat. Hist., 87:285, December 31.

MARGINAL RECORDS.—Nicaragua: Matagalpa. Costa Rica: Fuentes; Irazú. Panamá: Boquerón; type locality. Costa Rica: Puerto Cortés; Puntarenas. Nicaragua (Phillips and Jones, 1968:320, unless otherwise noted, as *C. derbianus* only): Hda. Calera, Naddaime (Biggers, 1967:678, as *C. derbianus* only); *3 km. N, 4 km. W Diriamba, 600 m.; 15 km. NW Masaya, ca. 100 m.;* 5 km. N Sabana Grande, 45 m.

Genus Marmosa Gray—Murine Opossums

Revised by Tate, Bull. Amer. Mus. Nat. Hist., 66:1–250, pls. 1–26, 29 figs. in text, August 10, 1933.

1821. *Marmosa* Gray, London Med. Repos., 15:308, April 1. Type, *Didelphis murina* Linnaeus.
1842. *Asagis* Gloger, Gemeinnütziges Hand- und Hilfs-buch der Naturgeschichte, 1:82. Type not designated; according to Thomas the type is *Didelphis murina* Linnaeus.
1842. *Notagogus* Golger, *loc. cit.* Type not designated; according to Thomas the type is *Didelphis murina* Linnaeus.
1842. *Micoureus* Lesson, Nouveau tableau du règne animal . . . mammifères, p. 186. Type by subsequent designation (Thomas, Catalogue of the Marsupialia and Monotremata, p. 340, 1888), *Didelphis cinerea* Temminck.
1843. *Thylamys* Gray, List of the . . . Mammalia in the . . . British Museum, p. 101. Type by monotypy, *Didelphis elegans* Waterhouse.
1854. *Grymaeomys* Burmeister, Systematische Uebersicht der Thiere Brasiliens . . . , 1:138. Type by subsequent designation (Thomas, Catalogue of the Marsupialia and Monotremata, p. 340, 1888), *Didelphis murina.*
1856. *Microdelphis* Burmeister, Erläuterungen zur Fauna Brasiliens . . . , Berlin, p. 83. Type by subsequent designation (Thomas, Catalogue of the Marsupialia and Monotremata, p. 354, 1888), *Didelphis tristriata* Kuhl.
1916. *Marmosops* Matschie, Sitzungsb. Gesell. naturforsch. Freunde, Berlin, p. 267, December 15. Type *Didelphis incana* Lund.

Maximum external measurements, 450, 281, 33. Greatest length of skull, 47. Grayish or brownish above, some individuals having black markings on face. Tail longer than head and body, underside of tip modified for grasping; 6 plantar tubercles, 2 exterior, 2 interior, 2 divided by base of 3rd digit; external pads united in some species; marsupium absent; temporal ridges not uniting into a sagittal crest; postorbital processes present or absent according to species; toothrows convergent anteriorly; M3 usually larger than M1.

KEY TO NORTH AMERICAN SPECIES OF MARMOSA

1. Fur grayish and often wooly; external anterior and posterior pads of hind foot united; 9–14 spirals of scales per centimeter of tail-length; greatest length of skull more than 44. . . . . . . *M. alstoni,* p. 13
1'. Fur rarely (in *M. canescens*) grayish and never wooly; external posterior pads of hind foot separate; 16 or more spiral rows of scales per centimeter of tail-length; greatest length of skull less than 44.
  2. Plantar pad between digits 3 and 4 smaller than that between digits 2 and 3; the 3 minute hairs accompanying each scale of the tail flattened, appressed, provided with median dorsal keel, somewhat petiolate at insertion, and often black; nasals exceeding premaxillae when skull is in norma verticalis, no pointed supraorbital processes; tympanic bulla with anteromedial process.
    3. Underparts with median area buffy to creamy white. . . . . *M. impavida,* p. 17
    3'. Underparts silvery gray. . . . . *M. invicta,* p. 17
  2'. Plantar pad between digits 3 and 4 approx. same size as that between digits 2 and 3; the 3 minute hairs accompanying each scale of the tail not flattened, not appressed, lacking a median dorsal keel, not petiolate at insertion, and rarely black; nasals not exceeding premaxillae when skull is in norma verticalis; pointed supraorbital processes present; anteromedial process of tympanic bulla lacking or represented by only a minute spine.
    4. Upper parts gray, rarely with faint wash of cinnamon; small openings between palatal slit and molars. . . . . *M. canescens,* p. 16
    4'. Upper parts brown or cinnamon; no small openings between palatal slits and molars.
      5. Supraorbital processes reduced and scarcely pointed; no trace of postorbital constriction; temporal ridges widely separated. *M. mexicana,* p. 15
      5'. Supraorbital processes large and grooved; postorbital constriction marked; temporal ridges often approximated. . . . . *M. robinsoni,* p. 14

cinerea-group

Marmosa alstoni

Alston's Opossum

slightly, if at all, constricted; supraorbital processes without dorsal grooves; palatal fenestrae smaller than in other species of approx. equal size.

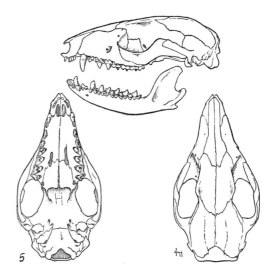

External measurements: 385–450; 195–281; 27–33. Greatest length of skull, 44.9–45.7. One of the largest species of the genus; upper parts grayish brown; underparts, and in some individuals face, cream-buff; basal 25–50 mm. of tail furred, often frizzled and longer than elsewhere on animal; braincase broad; postorbital region

Fig. 5. *Marmosa alstoni alstoni,* A. Caliente, Cartago, Costa Rica, No. 26922 K.U., ♂, X 1.

Collins (1973:62) credits R. H. Pine with suggesting that *M. a. alstoni* and *M. a. nicaraguae* are to be included in [as subspecies of ?] the earlier named *Marmosa cinerea* (Temminck 1824).

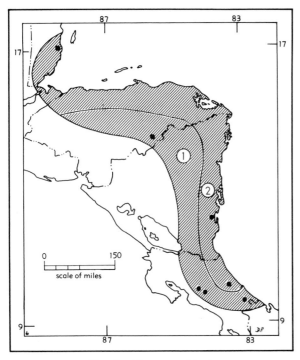

Map 6.   *Marmosa alstoni.*

1. *M. a. alstoni*   2. *M. a. nicaraguae*

Marmosa alstoni alstoni (J. A. Allen)

1900. *Caluromys alstoni* J. A. Allen, Bull. Amer. Mus. Nat. Hist., 13:189, October 12, type from Tres Ríos, Cartago, Costa Rica.
1905. [*Marmosa*] *alstoni*, Trouessart, Catalogus Mammalium . . . , Suppl., fasc., 4, p. 855.

MARGINAL   RECORDS.—Honduras: Segovia River. Costa Rica: Cubre; A. Caliente, 1300 m., Cartago; Escazú. Occurs also in Colombia, South America.

Marmosa alstoni nicaraguae Thomas

1905. *Marmosa cinerea nicaraguae* Thomas, Ann. Mag. Nat. Hist., ser. 7, 16:313, September, type from Bluefields, sea level, Nicaragua.
1933. *Marmosa alstoni nicaraguae*, Tate, Bull. Amer. Mus. Nat. Hist., 66:69, August 10.

MARGINAL RECORDS.—Belize: Double Falls. Nicaragua: type locality. Costa Rica: Siquirres, Río Pacuare.

murina-group
Marmosa robinsoni
South American Mouse-opossum

Total length up to 421, tail as long as 230, hind foot 22–29. Greatest length of skull, 33.5–43.5. Upper parts some shade of cinnamon or russet; face paler; tail densely clothed with fine hairs that almost conceal scales and give tail whitish appearance; posterior border of nasals rounded; supraorbital ridges well developed, moderately pointed and with pronounced dorsal grooves; a pronounced constriction postorbitally.

Marmosa mitis Bangs, formerly used (*e.g.*, Hall and Kelson, 1959:13) for this species, is regarded as a junior synonym of *M. robinsoni robinsoni*, which occurs only in South America.

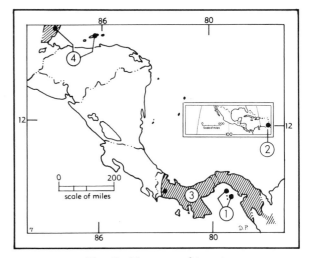

Map 7.   *Marmosa robinsoni.*

1. *M. r. fulviventer*   3. *M. r. isthmica*
2. *M. r. grenadae*   4. *M. r. ruatanica*

Marmosa robinsoni fulviventer Bangs

1901. *Marmosa fulviventer* Bangs, Amer. Nat., 35:632, August, type from San Miguel Island, Golfo de Panamá, Panamá.
1966. *M*[*armosa*]. *r*[*obinsoni*]. *fulviventer*, Handley, Ectoparasites of Panama, Field Mus. Nat. Hist., p. 755, November 22 (see Cabrera, Rev. Mus. Argentino de Cienc. Nat., 4:24, March 27, 1958, for use of *robinsoni* instead of *mitis*).

MARGINAL RECORDS.—Panamá: Saboga Island; type locality.

Marmosa robinsoni grenadae Thomas

1911. *Marmosa grenadae* Thomas, Ann. Mag. Nat. Hist., ser. 8, 7:514, type from Annandale, Grenada. G. M. Allen (Bull. Mus. Comp. Zool., 54:194, July, 1911) arranged *M. gre-*

nadae as a synonym of the earlier named *Marmosa chapmani* J. A. Allen (Bull. Amer. Mus. Nat. Hist., 13:197, October 23, 1900, type from Trinidad), but Goodwin (Amer. Mus. Novit., 2070:8, 9, December 29, 1961) showed *M. grenadae* to be a valid subspecies of *M. mitis* [= *M. robinsoni*].

MARGINAL RECORDS.—*Carriacou* (G. M. Allen, 1911:195, as *M. chapmani*, on testimony of A. H. Clark); *Isle Ronde* (*ibid.*); type locality.

Marmosa robinsoni isthmica Goldman

1912. *Marmosa isthmica* Goldman, Smiths. Miscl. Coll., 56(36):1, February 19, type from Río Indio, near Gatún, Canal Zone, Panamá.
1958. *Marmosa robinsoni isthmica*, Cabrera, Rev. Mus. Argentino de Cienc. Nat., 4:24, March 27.

MARGINAL RECORDS.—Panamá: From Boquete east over at least northern half of Panamá into South America.

Marmosa robinsoni ruatanica Goldman

1911. *Marmosa ruatanica* Goldman, Proc. Biol. Soc. Washington, 24:237, November 28, type from Ruatán Island, Caribbean Coast of Honduras.

MARGINAL RECORDS.—Belize: Silkgrass; *Bokowina*. Honduras: type locality.

Marmosa mexicana
Mexican Mouse-opossum

X ½

External measurements: 290–386; 150–190; 20–27. Greatest length of skull, 32.4–40.8. Upper

parts reddish brown; underparts yellowish to buffy; eye-rings intensely black; tail having thin growth of fine hair throughout its length and faintly bicolored; skull not constricted postorbitally, supraorbital ridges projecting laterally only slightly; no accessory fenestra between M2 and normal palatal fenestra. Comparison with *M. canescens* made in account of that species.

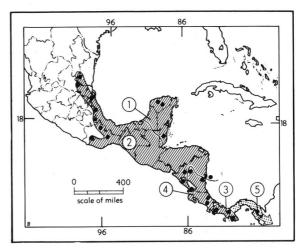

Map 8. *Marmosa mexicana* and *Marmosa impavida*.

1. *M. m. mayensis*   3. *M. m. savannarum*
2. *M. m. mexicana*   4. *M. m. zeledoni*
5. *M. impavida caucae*

Marmosa mexicana mayensis Osgood

1913. *Marmosa mayensis* Osgood, Proc. Biol. Soc. Washington, 26:176, August 8, type from Izamal, east of Mérida, Yucatán.
1917. *Marmosa mexicana mayensis*, Goldman, Proc. Biol. Soc. Washington, 30:109, May 23.

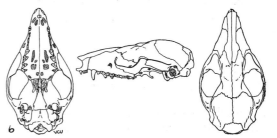

Fig. 6. *Marmosa mexicana mexicana*, 5 km. N Jalapa, 4500 ft., Veracruz, No. 19067 K.U., ♀, X 1.

MARGINAL RECORDS.—Yucatán: type locality; Chichén-Itzá. Belize: Bokowina.

Marmosa mexicana mexicana Merriam

1897. *Marmosa murina mexicana* Merriam, Proc. Biol. Soc. Washington, 11:44, March 16, type from Juquila, 1500 m., Oaxaca.

1902. *Marmosa mexicana*, Bangs, Bull. Mus. Comp. Zool., 39:19, April.

MARGINAL RECORDS.—Tamaulipas: Aserradero del Infiernillo. Veracruz: Truxpan [= Tuxpan] southward along coast to base of Yucatán Peninsula, across base of peninsula, and southward to Corn Islands. Nicaragua: Matagalpa; Managua (Tate, 1933:134); San Emilio, Lake Nicaragua (*ibid.*), thence northwestward up coast approx. to Oaxaca: type locality; Agua Zarca (Goodwin, 1969:31); Campamento Vista Hermosa (*ibid.*). Veracruz (Hall and Dalquest, 1963:199): 3 km. SE Orizaba, 5500 ft.; 4 km. W Tlapacoyan, 1700 ft. San Luis Potosí: El Salto, 1750 ft. (Jones and Alvarez, 1964:302).

Marmosa mexicana savannarum Goldman

1917. *Marmosa mexicana savannarum* Goldman, Proc. Biol. Soc. Washington, 30:108, May 23, type from Boquerón, Chiriquí, Panamá. Regarded by Tate (Bull. Amer. Mus. Nat. Hist., 64:133, August 10, 1933) as indistinguishable from *M. m. mexicana*.

MARGINAL RECORDS.—Panamá: type locality; Province of Veraguas west of Colón and Coclé; Mariato; *Bogava*.

Marmosa mexicana zeledoni Goldman

1911. *Marmosa zeledoni* Goldman, Proc. Biol. Soc. Washington, 24:238, November 28, type from Navarro, near Orosi, Caribbean slope, between 2500 and 3000 ft., Cartago, Costa Rica.
1917. *Marmosa mexicana zeledoni* Goldman, Proc. Biol. Soc. Washington, 30:109, May 23.

MARGINAL RECORDS.—Nicaragua: Río Tuma, Matagalpa, below 1000 ft.; *Escondido River* (Tate, 1933:136); Bluefields. Costa Rica: Río Pacuare; *Agua Buena*; Boruca.

Marmosa canescens
Grayish Mouse-opossum

External measurements: 261–285; 135–167; 18–21. Greatest length of skull, 31.3–36.1. Dorsally distinctly gray and in this feature unique among Mexican members of the genus, rarely with a faint wash of cinnamon; underparts yellowish-buff or cream-buff; skull with prominent, winglike supraorbital processes, which, in adults, continue posteriorly as converging temporal ridges; palate ordinarily having an accessory fenestra on each side between M2 and the normal palatal fenestra. The accessory fenestrae, the gray instead of red color, and the sharp and much produced supraorbital ridges (as opposed

to low blunt ridges, or absence of ridges) are differences from *M. mexicana*.

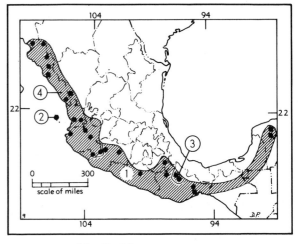

Map 9.   *Marmosa canescens*.

1. *M. c. canescens*   3. *M. c. oaxacae*
2. *M. c. insularis*   4. *M. c. sinaloae*

Marmosa canescens canescens (J. A. Allen)

1893. *Didelphis (Micoureus) canescens* J. A. Allen, Bull. Amer. Mus. Nat. Hist., 5:235, September 22, type from Santo Domingo de Guzmán, Isthmus of Tehuantepec, Oaxaca.
1897. *Marmosa canescens* J. A. Allen, Bull. Amer. Mus. Nat. Hist., 9:58, March 15.
1913. *Marmosa gaumeri* Osgood, Proc. Biol. Soc. Washington, 26:175, August 8, type from Yaxcaba, SW Chichén-Itzá, Yucatán.

MARGINAL RECORDS.—Michoacán: Los Reyes; 6 mi. S, 1 mi. E Tacámbaro, 4000 ft. Guerrero: Cueva del Cañón del Zopilote, 13 km. S Puente de Mexcala, 720 m. (Ramírez-P. and Sánchez-H., 1974:107, 108). Puebla: Tehuacán, 1700 m. Oaxaca: Tlapacingo; type locality. Yucatán: Yaxcaba; Chichén-Itzá. Oaxaca: Tehuantepec, thence up coast to point of beginning.

Marmosa canescens insularis Merriam

1898. *Marmosa insularis* Merriam, Proc. Biol. Soc. Washington, 12:14, January 27, type from María Madre Island, Tres Marías Islands, Nayarit. Known only from Tres Marías Islands.
1933. *Marmosa canescens insularis*, Tate, Bull. Amer. Mus. Nat. Hist., 66:144, August 10.

Marmosa canescens oaxacae Merriam

1897. *Marmosa oaxacae* Merriam, Proc. Biol. Soc. Washington, 11:43, March 16, type from City of Oaxaca, 4600 ft., Oaxaca.
1933. *Marmosa canescens oaxacae*, Tate, Bull. Amer. Mus. Nat. Hist., 66:143, August 10.

MARGINAL RECORDS (Goodwin, 1969:32).—
Oaxaca: Cerro San Felipe, *ca.* 7000 ft.; type locality.

Marmosa canescens sinaloae J. A. Allen

1898. *Marmosa sinaloae* J. A. Allen, Bull. Amer. Mus. Nat.
Hist., 10:143, April 12, type from Tatamales, Sinaloa.
1933. *Marmosa canescens sinaloae,* Tate, Bull. Amer. Mus.
Nat. Hist., 66:142, August 10.

MARGINAL RECORDS.—Sonora: 2 mi. SSE
Alamos, 1300 ft. (Loomis and Stephens, 1962:111).
Sinaloa (Armstrong and Jones, 1971:748): 1 mi. S El
Cajón; 44 km. ENE Sinaloa; 13 mi. ESE Badiraguato.
Durango: Ventanas (Baker and Greer, 1962:63).
Nayarit: Rancho Palo Amarillo. Jalisco (Genoways and
Jones, 1973:1, 2): 2⅔ mi. E Etzatalán, 4300 ft.; 27 mi. S,
12 mi. W Guadalajara; 8 mi. E Jilotlán de los Dolores,
2000 ft.; 14½ mi. S Pihuamo, 1100 ft.; 4 mi. SW Puerto
Vallarta, 20 ft. Nayarit: Tepic (Tate, 1933:141); 3½ *mi. E
San Blas* (Hooper, 1955:7). Sinaloa: Teacapán, Isla
Palmito del Verde (Armstrong and Jones, 1971:748);
Culiacán.—Tate (1933:141) recorded the specimen
from Tepic, Nayarit, as *M. c. canescens,* but noted that
it had some characters of *M. c. sinaloae.* Hooper
(1955:7), probably relying on Tate's subspecific identification, recorded two more specimens as *M. c. canescens* from nearby at 3½ mi. E San Blas, Nayarit. All
three of these specimens are here tentatively referred
to *M. c. sinaloae* because Genoways and Jones (1973:1,
2) refer all specimens from Jalisco to *M. c. sinaloae.*

noctivaga-group
Marmosa impavida (Tschudi)
Pale Mouse-opossum

External measurements (of Colombian specimens): 266–310; 142–182; 16–20; 17–19.
Greatest length of skull, 32.0–35.0. Upper parts
bone brown to light wood brown or Prout's
Brown; venter pale, buffy to creamy white; upper
surface of skull smoothly rounded, temporal
ridges widely separated; nasals moderately expanded at maxillofrontal suture; palate rather
narrow, usually fenestrated behind posterior
palatal openings; bullae moderately compressed.

This species belongs to the *fuscata*-section of
the *noctivaga*-group, which reaches its northern
limit of known occurrence in Panamá. See Map 8.

Marmosa impavida caucae Thomas

1900. *Marmosa caucae* Thomas, Ann. Mag. Nat. Hist., ser. 7,
5:221, February, type from Río Cauqueta [= Río Cauquita],
near Cali, 1000 m., Colombia.
1958. *Marmosa impavida caucae,* Cabrera, Rev. Mus. Argentino de Cienc. Nat., 4:16, March 27.

MARGINAL RECORDS (Handley, 1966:755).—
Panamá: Cerro Pirre, 5300 ft.; *Loma Cana,* 4900 ft.

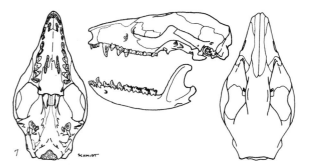

Fig. 7. *Marmosa impavida caucae,* Cerro Mali, 4700 ft.,
Darién, Panamá, No. 337956 U.S.N.M., ♂, X 1.

Marmosa invicta Goldman
Panamá Mouse-opossum

1912. *Marmosa invicta* Goldman, Smiths. Miscl. Coll.,
60(2):3, September 20, type from Cana, 2000 ft., Daríen,
Panamá.

External measurements of two specimens,
adult male and adult female, are: 248, 240; 137,
136; 19, 17.5. Greatest length of skull, 27.2, 27.0.
Upper parts bone brown; underparts silvery-
gray; tympanic bulla triangular and having process; supraorbital region smooth; palatal length
less than zygomatic breadth; pronounced median
frontal depression; palatal fenestrae small with
2nd minute pair behind 1st.

Like the previous species, *M. invicta* is a
member of the *fuscata*-section.

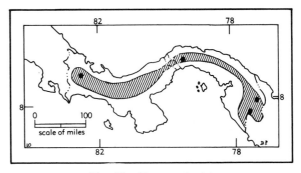

Map 10. *Marmosa invicta.*

MARGINAL RECORDS (Handley, 1966:755).—
Panamá: Cerro Azul; Tacarcuna Casita; *Tacarcuna
Laguna;* type locality; Cylindro.

Genus **Philander** Tiedemann
Brown Four-eyed Opossums

For use of this name rather than *Metachirus* Burmeister
1854, see Pine, Proc. Biol. Soc. Washington, 86:391, December 14, 1973.

1808. *Philander* Tiedemann, Zoologie . . . , 1:426. Type *P*[*hilander*]. *virginianus* Tiedemann [= *Didelphis nudicaudata* É. Geoffroy St.-Hilaire, Catalogue des mammifères du Muséum National d'Histoire Naturelle, Paris, 1803:142].

1854. *Metachirus* Burmeister, Systematische Uebersicht der Thiere Brasiliens . . . , 1:135. Type, *Didelphys myosuros* Temminck [= *D. nudicaudata* É. Geoffroy St.-Hilaire].

External measurements of holotype of *Philander nudicaudatus dentaneus*: 597, 332, 48. Greatest length of skull, 63. Pelage brown (gray to black in genus *Metachirops*); ratio of haired part of tail to naked part of tail less than in *Metachirops*; entire tail pale in contrast to white-tipped tail of *Metachirops*; no marsupium; no postorbital processes; posterior palatine vacuities only 2 (rarely indications of 2 more); interparietal rectangular; in upper jaw, 2nd postcanine tooth larger than 3rd.

Philander nudicaudatus
Brown Four-eyed Opossum

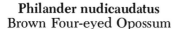

See characters of the genus. This species is widely distributed in South America.

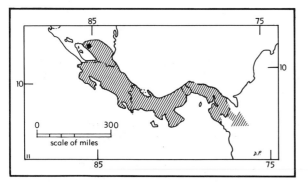

Map 11. *Philander nudicaudatus dentaneus.*

Philander nudicaudatus dentaneus (Goldman)

1912. *Metachirus nudicaudatus dentaneus* Goldman, Smiths. Miscl. Coll., 56(36):2, February 19, type from Gatún, Canal Zone, Panamá.

MARGINAL RECORDS.—Nicaragua: [district of] Chontales, thence southward at least through Panamá.

Genus **Monodelphis** Burnett
Short-tailed Opossums

1830. *Monodelphis* Burnett, Quart. Jour. Sci. Lit. Art, 28:351. Type by subsequent selection (Matschie, Sitzungsb. Gesell. naturforsch. Freunde, Berlin, 1916, p. 271), *Didelphis brevicaudata* Erxleben.

1842. *Peramys* Lesson, Nouveau tableau du règne animal . . . mammifères, p. 187. Type, *Didelphys brachyura* Schreber [= *D. brevicaudata* Erxleben].

1856. *Microdelphys* Burmeister, Erläuterungen zur Fauna Brasiliens . . . , Berlin, p. 83. Type, *Sorex americanus* Müller.

1919. *Minuania* Cabrera, Genera Mammalium: Monotremata, Marsupialia, Mus. Nat. Cienc. Nat., Madrid, p. 43, June 23. Type, *Didelphis dimidiata* Wagner.

External measurements of the type of *P. melanops* (= *M. adusta*), an adult male, are: 168, 60, 16.5. Condylobasal length of skull, 28. Pelage short, dense, and rather stiff; tail approx. half as long as head and body (always shorter than body alone), only slightly prehensile, its basal tenth haired as is body; remainder of tail thinly provided with fine hairs; hind foot having 5 toes and 5 plantar tubercles; skull resembling that of *Marmosa* but lacking supraorbital crests, although some specimens have rudiments of postorbital processes; in some species, old animals have indications of sagittal crest; palate having only one pair of fenestrae; marsupium absent.

Monodelphis adusta (Thomas)
Short-tailed Murine Opossum

1897. *Peramys adustus* Thomas, Ann. Mag. Nat. Hist., ser. 6, 20:219, August, type from "W. Cundinamarca," Colombia.

1940. *Monodelphis adusta*, Cabrera and Yepes, Mamíferos Sud-Americanos, p. 33.

1912. *Peramys melanops* Goldman, Smiths. Miscl. Coll., 60(2):2, September 20, type from Cana, 2000 ft., Darién, Panamá. Handley (1966:754–755) concluded that *P. melanops*, as well as *P. peruvianus* Osgood, Field Mus. Nat. Hist. [Fieldiana] (Zool.), Publ. 168, Vol. 10:93, May 31, 1913, was conspecific with *M. adusta*, which he considered to be monotypic.

External measurements small (see account of genus); upper parts dark brown (lacking stripes or spots that are present in some of the South American members of the genus), suffused with cinnamon on cheeks and sides of neck; face,

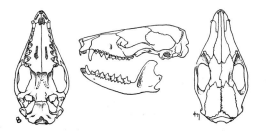

Fig. 8. *Monodelphis adusta*, Mera, Ecuador, No. 67274 A.M.N.H., ♂, X 1.

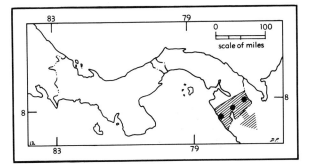

Map 12. *Monodelphis adusta*.

shoulders, and lower part of back black; throat and sides of abdomen Mouse Gray; underparts slightly grayer and with median buffy-white stripe extending from white pectoral region to buffy inguinal region; feet and tail black all around. The short tail is a diagnostic character among American opossums.

MARGINAL RECORDS.—Panamá (Handley, 1966:754): Tacarcuna Village, 1950 ft.; Cana, 2000 ft.; Guayabo, thence southwestward into South America.

Infraclass EUTHERIA

Placenta complete; epipubic bones absent; skull of most members having large braincase and small preorbital region; corpus callosum present in brain; jugal bone not forming part of glenoid fossa; tympanic annulus (in Insectivora and some Chiroptera) and tympanic bulla (in other orders) formed by tympanic bone.

Order INSECTIVORA—Insectivores

Monographed by A. Cabrera, Genera Mammalium: Insectivora, Galeopithecia, Mus. Nac. Cienc. Nat., Madrid, pp. 1–232, 19 pls., November 29, 1925.

Size generally small; feet plantigrade, pentadactyl; digits with claws; snout usually long and pointed; placenta deciduous and discoidal; uterus bicornuate; cerebral hemispheres smooth and short, exposing cerebellum and often corpora quadrigemina; form of skull primitive; braincase low; orbital and temporal fossae confluent; tympanic bone usually annular, either entirely free or enclosed in an entotympanic bulla; or incorporated into the structure of the bulla; basisphenoid and petrosal both taking part in enclosure of tympanic cavity; vertical portion of palatine excluded from orbit by maxilla; angular process of mandible typically not inflected; dental formula primitively i. $\frac{3}{3}$, c. $\frac{1}{1}$, p. $\frac{4}{4}$, m. $\frac{3}{3}$; incisors may be reduced to $\frac{2}{1}$ and enlarged or reduced; canine either caniniform or, more commonly, incisiform or premolariform, two-rooted; premolars acuminate, the posteriormost frequently sectorial; molars lophodont or bunodont; lower molars usually with 5 sharp tubercles, and the uppers tri- or quadri-tubercular; teeth always rooted, diphyodont; milk dentition usually shed early and seldom functional; clavicle present (except in *Potamogale*); interclavicle vestigial, sometimes absent; pollex and hallux not opposable; entepicondyloid foramen and centrale usually present; distal ends of tibia and fibula usually fused; glenoid fossa for humerus widely separated from sternum; prespinous fossa present; testes abdominal, inguinal, or in a prepenile scrotum; predominantly insectivorous or carnivorous; usually nocturnal; terrestrial, fossorial, or amphibious.

The order has served as a catch-all for a number of groups of mammals of doubtful relationships and is a loose, and possibly unnatural, assemblage. Many of the members show no close relationship except that they are all primitive placentals. The description of the order here given is conservatively framed to include the following Recent families: Soricidae, Talpidae, Solenodontidae, Potamogalidae, Chrysochloridae, Erinaceidae, and even the Macroscelididae.

Key to North American Familes of Insectivora

1. Paracone and metacone of each of 1st 2 upper molars forming a W-shaped loph.
   2. Zygomatic arch present. . . Talpidae, p. 66
   2'. Zygomatic arch absent.
      3. First upper incisor hooked and with a 2nd point that resembles a talon; crown surface of M3 less than half that of M2; M1 and M2 with a hypocone. . . . Soricidae, p. 24
      3'. First upper incisor simple and without talonlike cusp; crown surface of M3 half or more of the area of M2; M1 and M2 without a hypocone. . . . . Nesophontidae, p. 22
1'. Amphicone (paracone and metacone suppressed) of each of 1st 2 upper molars together with labial cusplets forming a V-shaped loph. . . . . . Solenodontidae, p. 20

Family SOLENODONTIDAE
Solenodons

External measurements: 500; 215; 65. Greatest length of skull, 90; general form of body that of a large shrew; snout elongate, tip bare, nostrils opening laterally; eyes small; ears visible above pelage; tail so sparsely haired as to seem naked; pelage composed of hair of two lengths; claws of forefeet longer than those of hind feet; axillary and inguinal odoriferous glands; one pair of inguinal mammae; cranium elongate; sagittal and especially lambdoidal crests pronounced; rostrum tubular; zygomatic arches absent, although maxillary and squamosal roots are present; I1 and i2 greatly enlarged, the latter with a hollow

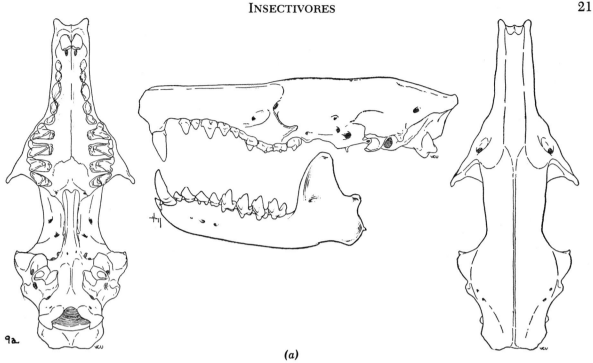

(a)

Fig. 9a. *Solenodon paradoxus*, Las Vegas, Dominican
Republic, No. 29502 K.U., sex?, X 1.

groove on medial surface; occlusal surface of
upper molars consisting of high, V-shaped amphicone and low, internal protocone and hypo-
cone. Dentition, i. $\frac{3}{3}$, c. $\frac{1}{1}$, p. $\frac{3}{3}$, m. $\frac{3}{3}$.

Members of this family are found only on Cuba
and Hispaniola.

Genus **Solenodon** Brandt—Solenodons

1833. *Solenodon* Brandt, Mém. Acad. Imp. Sci., St.
Pétersbourg, ser. 6, Sci. Math. Phys. et Nat., 2:459. Type
Solenodon paradoxus Brandt.

Characters as for the family.

KEY TO SUBGENERA OF SOLENODON

1. Claws on forefeet approx. same length as
toes; I3 in contact with C; P3 with base
oval; prenasal bone present. . *Solenodon*,   p. 21
1'. Claws on forefeet considerably longer
than toes; I3 separated from C by small
diastema; P3 with base triangular; prenasal bone absent. . . . . . *Atopogale*,   p. 22

Subgenus **Solenodon** Brandt
Hispaniolan Solenodon

1833. *Solenodon* Brandt, Mém. Acad. Imp. Sci., St.
Pétersbourg, ser. 6, Sci. Math. Phys. et Nat., 2:459. Type,
Solenodon paradoxus Brandt.

Pelage coarse; claws on forefeet as long as toes;
skull with small, rounded, prenasal bone placed
horizontally in front of premaxillae; mesopterygoid space wider anteriorly than posteriorly;
tympanic rings farther apart posteriorly than anteriorly; I3 in contact with canine; C with small
anterior cusp formed by cingulum; P3 simple
with oval base; 16 thoracic vertebrae.

Solenodon paradoxus Brandt
Hispaniolan Solenodon

1833. [*Solenodon*] *paradoxus* Brandt, Mém. Acad. Imp. Sci.,
St. Pétersbourg, ser. 6, Sci. Math. Phys. et Nat., 2:459, type
from Hispaniola. Known only from Hispaniola.

See characters of the subgenus. Still living, according to G. M. Allen (1942:12), in areas of stony forest in northeastern part of Dominican Republic, Hispaniola.

MARGINAL RECORDS.—Dominican Republic: km. 2 Site; Río San Juan; Naranjo Abajo; La Vega; vic. San José de las Matas.

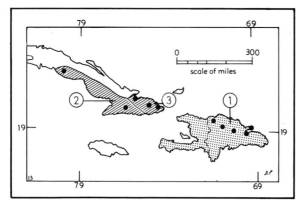

Map 13. *Solenodon*.

Guide to kinds
1. *Solenodon paradoxus*
2. *Solenodon cubanus cubanus*
3. *Solenodon cubanus poeyanus*

Subgenus **Atopogale** Cabrera—Cuban Solenodon

1925. *Atopogale* Cabrera, Genera Mammalium: Insectivora, Galeopithecia, Mus. Nac. Cienc. Nat., Madrid, p. 177, November 29. Type, *Solenodon cubanus* Peters. Used as a subgenus by Aguayo, Bol. Hist. Nat. Soc. Felipe Poey, 1:131, November 1950.

Pelage longer and finer than in *Solenodon;* claws more delicate but considerably longer than toes; no prenasal bone; tympanic rings separated more widely anteriorly than posteriorly; mesopterygoid space narrower anteriorly than posteriorly; pterygoid ending in large hamular process; teeth generally smaller than in *Solenodon;* I3 separated from canine by short diastema; C lacking small anterior cusp; P3 relatively large and with posterior prolongation giving base triangular outline; 15 thoracic vertebrae.

Solenodon cubanus
Cuban Solenodon

See characters of the subgenus. Barbour (Proc. New England Zool. Club, 23:4, 1944) suggests that the subspecies in southeastern and south-central Cuba is extinct, but Varona (1974:7) records two specimens taken in about 1956 south of Bayamo in the Sierra Maestra.

Solenodon cubanus cubanus Peters

1861. *Solenodon cubanus* Peters, Monatsb. preuss. Akad. Wiss., Berlin, 169, type from Bayamo, Cuba.
1925. *Atopogale cubana,* Cabrera, Genera Mammalium: Insectivora, Galeopithecia, Mus. Nac. Cienc. Nat., Madrid, p. 177, November 29.

MARGINAL RECORDS.—Cuba: near Trinidad; mts. east of Bayamo.

Solenodon cubanus poeyanus Barbour

1944. *Solenodon poeyanus* Barbour, Proc. New England Zool. Club, 23:6, March 7, type from vic. Nipe Bay, Cuba. Known only from mts. northeastern Cuba.
1950. *Solenodon (Atopogale) cubanus poeyanus,* Aguayo, Bol. Hist. Nat. Soc. Felipe Poey, 1(3):131, November.

MARGINAL RECORDS.—Cuba: type locality; Sierra de Toar; Nabuiabo Swamp, Finca La Caridad, between Baracoa and Duaba.—Varona (1974:7) regards *S. c poeyanus* as doubtfully separable from *S. cubanus* at the subspecific level.

FAMILY **NESOPHONTIDAE**
Nesophontid Insectivores

Greatest length of skull, 52. Lacking jugal bone and zygomatic arch; seemingly lacking auditory bulla; braincase low; rostrum elongate and tubular; sagittal and lambdoidal crests present, the latter especially prominent; occlusal outline of upper molars V-shaped; M3 no less than half as large as M2; incisors simple, smaller than canines; diastema between first upper incisors; canines daggerlike and two-rooted. Dentition, i. $\frac{3}{3}$, c. $\frac{1}{1}$, p. $\frac{3}{3}$, m. $\frac{3}{3}$. Only one genus, *Nesophontes,* is known.

Genus **Nesophontes** Anthony
Nesophontid Insectivores

1916. *Nesophontes* Anthony, Bull. Amer. Mus. Nat. Hist., 35:725, November 16. Type, *Nesophontes edithae* Anthony.

Generic characters are those of the family. The external appearance of these insectivores is unknown, since our knowledge of them is based on bones recovered from caves and kitchen middens in the Greater Antilles. One or more of the six species listed here probably became extinct after the arrival of the Spaniards. Bones of *Nesophontes* and *Rattus* have been found together (Miller, 1929a:3) and *Rattus* is thought not to have arrived before the Spaniards did. Miller (*loc. cit.*) says in relation to Hispaniola "it seems not improbable, however, that if any part of the island remains uninvaded by the roof rat, the native animal might now be found to exist there." From Cuba two additional species not listed below, *Nesophontes major* Arredondo and *N. submicrus* Arredondo, have been named from Holocene deposits of pre-Columbian age (Arredondo, 1970:130 and 137 respectively). Species not mapped.

KEY TO SPECIES OF NESOPHONTES

1. Occurring in Puerto Rico. . . *N. edithae*,   p. 23
1'. Occurring in Cuba or Hispaniola.
  2. Palatal length less than 10.7.
                  N. zamicrus,   p. 24
  2'. Palatal length more than 10.7.
    3. Occurring in Cuba.
        4. Upper premolars widely separated; 1st premolar widely separated from canine.
               N. longirostris,   p. 23
        4'. Upper premolars not widely separated from each other; 1st premolar not widely separated from canine. . . . *N. micrus*,   p. 23
    3'. Occurring in Hispaniola.
        5. Articular process to front of m1 more than 12.2; depths through coronoid process more than 7.3; combined length of m1 and m2 more than 4.2.
               N. paramicrus,   p. 23
        5'. Articular process to front of m1 less than 12.2; depth through coronoid process less than 7.3; combined length of m1 and m2 less than 4.2. *N. hypomicrus*,   p. 24

Nesophontes edithae Anthony
Puerto Rican Nesophontes

1916. *Nesophontes edithae* Anthony, Bull. Amer. Mus. Nat. Hist., 35:725, November 16, type from Cueva Catedral, near Morovis, Puerto Rico. Known only from skeletal remains from Puerto Rico.

Skulls of presumptive males are larger than those of presumptive females; the greatest length of skull is 52 in the former and 40.8 in the latter.

Nesophontes longirostris Anthony
Slender Cuban Nesophontes

1919. *Nesophontes longirostris* Anthony, Bull. Amer. Mus. Nat. Hist., 41:633, December 30, type from cave near the beach at Daiquirí, Cuba. Known only from skeletal remains from Cuba.

This species is distinguished from *N. micrus*, also of Cuba, by the longer and relatively more slender skull.

Nesophontes micrus G. M. Allen
Cuban Nesophontes

1917. *Nesophontes micrus* G. M. Allen, Bull. Mus. Comp. Zool., 61:5, January, type from Sierra de Hato-Nuevo, Province of Matanzas, Cuba. Known only from skeletal remains. Existed on Isle of Pines and throughout Cuba (see Koopman and Ruibal, Breviora Mus. Comp. Zool., 46:2, 3, June 24, 1955). Patton (Caribbean Jour. Sci., 6:181, December, 1966) records remains of *Nesophontes* sp. from Cayman Brac, British West Indies.

The differences between this species and *N. edithae* of Puerto Rico have been listed in detail by Anthony (1919:633) as well as by the describer of *N. micrus* in the original description. Some of the differences characterizing *N. micrus* are as follows: occipital region more constricted; premaxillary with a shallow concavity laterally instead of being flat; lachrymal foramen and 1st upper premolar larger.

Nesophontes paramicrus Miller
Large Hispaniolan Nesophontes

1929. *Nesophontes paramicrus* Miller, Smiths. Miscl. Coll., 81(9):3, March 30, type from cave approximately 4 mi. E St. Michel, Haiti. Known only from skeletal remains from Haiti.

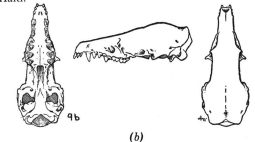

(b)

Fig. 9*b*. *Nesophontes paramicrus*, composite drawing based on three specimens in U.S.N.M., without catalogue nos.; labels bear annotations "VIP" and "IIP," X 1.

Resembles *N. micrus* of Cuba but upper molars without well-defined sulcus which, in *N. micrus*, lies between base of metacone and posterior commisure of protocone; lower molars with entoconid and metaconid obviously less nearly terete than in the Cuban species. Patterson (1962) states "*N. 'paramicrus,*' . . . agrees in size with the Cuban *N. micrus*, from which Miller separated it on molar characters. I have examined several hundred specimens of the Hispaniolan form and compared them with Cuban material. The supposed differences are not constant and I have so far been unable to find others that might validate Miller's species." Varona (1974:9) states that principally therefore he includes *N.*

paramicrus in the synonymy of *N. micrus.* Actually he did not do that. Instead he merely omitted *N. paramicrus* from his Catálogo. Consequently, the taxon in question is listed here, but only provisionally. It may merit subspecific status.

Nesophontes hypomicrus Miller
Miller's Nesophontes

1929. *Nesophontes hypomicrus* Miller, Smiths. Miscl. Coll., 81(9):4, March 30, type from cave near Atalaye Plantation, approximately 4 mi. E St. Michel, Haiti. Known only from skeletal remains from Haiti.

Resembles *N. paramicrus* but smaller.

Nesophontes zamicrus Miller
Hispaniolan Nesophontes

1929. *Nesophontes zamicrus* Miller, Smiths. Miscl. Coll. 81(9):7, March 30, type from cave approximately 4 mi. E St. Michel, Haiti. Known only from skeletal remains from Haiti.

Resembles *N. hypomicrus* but differs from it and all other known species of the genus in lesser size. Platal length, 10.6; 4 longest maxillary teeth, 5.0; 4 longest mandibular teeth, 5.6.

FAMILY **SORICIDAE**—Shrews

External measurements: 71–174; 12–80; 8.5–21.2. Condylobasal length of skull, 13–23.8. Shape generally mouselike; nose long and pointed; eyes small and often partly hidden in fur; ears provided with pinnae which may be reduced (*Blarina*) or well developed (*Notiosorex*); feet pentadactyl and normally developed. Skull triangular; zygomatic arches and auditory bullae lacking; tympanic bone annular; first upper incisors procumbent with tips curved or hooked ventrad and in addition possessing a 2nd, posterior, ventrally projecting, unicuspidlike conule which seems to serve as an additional unicuspid tooth; other incisors, canines, and all premolars except p4 and P4 simple and unicuspid; crowns of upper molars W-shaped; M3 markedly smaller than M1 or M2.

The family has been divided into four subfamilies, and all North American soricids from the Holocene are referred to the Soricinae (Repenning, 1967:27).

KEY TO NORTH AMERICAN GENERA OF SORICIDAE

In the upper jaw the 1st tooth (I1) has 2 large cusps, therefore is not counted as a unicuspid. The last 4 grinding teeth (P4–M3) of the upper jaw are not unicuspids, but the teeth intervening between the 1st tooth (I1) and the anterior grinding tooth (P4) are so termed. In some shrews certain unicuspids are not visible in lateral view. Therefore, it is necessary to examine the teeth from the crown (occlusal) surface with a hand lens or lower-power dissecting microscope to be certain of the number of unicuspids. In *Microsorex* the 3rd and 5th unicuspids are reduced to minute pegs partly hidden between the adjoining teeth; thus, at first glance, *Microsorex* seems to possess only 3 unicuspids.

1. Three unicuspid teeth in each side of upper jaw. . . . . *Notiosorex*, p. 64
1′. Four or 5 unicuspid teeth in each side of upper jaw.
  2. Four unicuspid teeth in each side of upper jaw. . . . . *Cryptotis*, p. 57
  2′. Five unicuspid teeth in each side of upper jaw.
    3. Tail less than $\frac{2}{5}$ of length of head and body; viewed from the side, 3rd and 4th unicuspids subequal and each one less than a $\frac{1}{4}$ as large as the 1st or 2nd unicuspid; lateral edge of braincase produced into a sharp, pointed angle, even in young specimens. . . . *Blarina*, p. 53
    3′. Tail more than $\frac{2}{5}$ of length of head and body; viewed from the side, 3rd and 4th unicuspids unequal in size and one or both more than $\frac{1}{4}$ as large as the 1st or 2nd unicuspid; lateral edge of braincase not produced into a sharp, pointed angle, even in old specimens.
      4. At least 4, and usually all 5, of the upper unicuspids easily visible in lateral view; ridge extending from apex medially to cingulum of 2nd and 3rd upper teeth without a distinct caudad bend near terminus when viewed occlusally; in lateral view base of lower incisor and premolar separated by space nearly equal to anteroposterior diameter of canine. . . . . *Sorex*, p. 25
      4′. Only 3 upper unicuspids easily visible in lateral view; ridge extending from apex medially to cingulum of 2nd and 3rd upper unicuspidate teeth with a distinct caudad bend near terminus when viewed occlusally; in lateral view base of lower incisor and premolar separated by space equal to less than $\frac{2}{5}$ (usually $\frac{1}{4}$) of anteroposterior diameter of canine. . . . . *Microsorex*, p. 51

Genus **Sorex** Linnaeus—Long-tailed Shrews

North American species revised by Jackson, *in* N. Amer. Fauna, 51:vi + 238, 13 pls., 24 figs., July 24, 1928.

1758. *Sorex* Linnaeus, Syst. nat., ed. 10, 1:53. Type, *Sorex araneus* Linnaeus.

1762. *Musaraneus* Brisson, Regnum animale, p. 126. Type, Musaraneus (type); included also *M. aquaticus* from Europe and *M. brasiliensis* from Brazil.

1829. *Oxyrhin* Kaup, Skizzirte Entwickelungs-Geschichte und natürliches System der europäischen Thierwelt, p. 120. Type, *Sorex tetragonurus* Hermann by subsequent designation (Miller, Catalogue of the mammals of western Europe, p. 29, November 23, 1912).

1835. *Amphiosorex* Duvernoy, Mém. Soc. Mus. d'Hist. Nat. Strasbourg, 2:23. Type, *Sorex hermanni* Duvernoy [= *Neomys fodiens* (Pennant) skull plus *Sorex araneus tetragonurus* Herman skin].

1838. *Corsira* Gray, Proc. Zool. Soc. London, for 1837, p. 123. Type, *Sorex araneus* Linnaeus.

1842. *Otisorex* De Kay, Zoology of New-York . . . , pt. 1, Mammalia, p. 22 and Pl. 5, Fig. 1. Type, *Otisorex platyrhinus* De Kay [= *Sorex cinereus* Kerr].

1848. *Hydrogale* Pomel, Arch. Sci. Phys. et Nat., Geneve, 9:248, November. Type, *Sorex fimbripes* Bachman [= *Sorex cinereus* Kerr]. Not *Hydrogale* Kaup, 1829.

1858. *Neosorex* Baird, Mammals, *in* Repts. Expl. Surv. . . . , 8(1):11, July 14. Type, *Neosorex navigator* Baird.

1884. *Atophyrax* Merriam, Trans. Linnaean Soc. New York, 2:217, August. Type, *Atophyrax bendirii* Merriam.

1890. *Homalurus* Schulze, Schriften des Naturwissenschaft. Vereins des Harzes in Wernigerode, 5:28. Type, *Sorex alpinus* Schinz.

1927. *Soricidus* Altobello, Rev. Franc. Mamm., 1:6. Type, *Soricidus monsvairani* Altobello [= *Sorex araneus tetragonurus* Hermann].

1967. *Ognevia* Dolgov and Heptner, Zool. Jour., Acad. Sci. S.S.S.R., Moscow, 46(9):1422. Type, *Sorex mirabilis* Ognev. Proposed as a subgenus.

External measurements: 71–174; 25–80; 10–21. Condylobasal length of skull, 14.2–23.8. Body slender; tail a third to more than a half of total length, hairy in young, glabrous in old adults; snout long and slender, having well-developed vibrissae; eyes minute but visible; pinnae of ears usually projecting slightly above pelage; 3 pairs of inguinal mammae; color varying from tan to black; uni-, bi-, or tricolored; skull triangular; sagittal and lambdoidal crests present on older individuals. Five upper unicuspid teeth; pigmentation varying according to species: 1st, 2nd, and 4th well developed with pigmented apices; 3rd variable in size; 5th smaller than any other, often an unpigmented peg difficult to see in lateral view; articular process of mandible having two horizontal articular facets; angular process long and slender. Dentition, i. $\frac{3}{1}$, c. $\frac{1}{1}$, p. $\frac{3}{1}$, m. $\frac{3}{3}$.

KEY TO NORTH AMERICAN SPECIES OF SOREX

1. Hind foot more than 18; pelage grayish, never distinctly brown.
    2. Rostrum short and slightly downcurved; anterior part of premaxilla scarcely shallower dorsoventrally than middle part; vertical depth of rostrum at level of 3rd unicuspid equal to approx. half the distance from anterior border of infraorbital foramen to posterior border of 1st incisor; posterior end of internal cutting edge of anterior portion of internal basal shelf of 1st and 2nd upper molars usually without cusplike lobe; hind foot distinctly fimbriated.
        3. Skull having well-developed sagittal and lambdoidal crests; condylobasal length averaging less than 19.3; known only from Point Gustavus, Glacier Bay, Alaska. . . . . *S. alaskanus*,   p. 43
        3'. Skull smooth, lacking well-developed sagittal and lambdoidal crests; condylobasal length averaging more than 19.3; not occurring at Point Gustavus, Glacier Bay, Alaska. *S. palustris*,   p. 40
    2'. Rostrum long and much curved downward; anterior part of premaxilla much shallower dorsoventrally than middle part; vertical depth of rostrum at 3rd unicuspid less than half the distance from anterior border of infraorbital formen to posterior border of 1st incisor; posterior end of internal cutting edge of anterior portion of internal basal shelf of 1st and 2nd upper molars usually with distinct cusplike lobe; hind foot slightly fimbriated. . . . . *S. bendirii*,   p. 43
1'. Hind foot less than 18, or, when more than 18, pelage brownish.
    4. Third unicuspid not smaller than 4th.
        5. Known geographic range north of United States–Mexican boundary.
            6. Posterior border of infraorbital foramen posterior to plane of space between M1 and M2.
                7. Total length more than 115; hind foot more than 13; condylobasal length more than 16.8. . . . . *S. dispar*,   p. 47
                7'. Total length less than 115; hind foot less than 13; condylobasal length less than 16.8. . . . . *S. gaspensis*,   p. 46
            6'. Posterior border of infraorbital foramen even with, or anterior to, plane of space between M1 and M2.
                8. Maxillary breadth less than 4.6.
                    9. Condylobasal length 14.7 or more; not occurring in eastern Oregon.
                        10. Known geographic range confined to Sierra Nevada of California. *S. lyelli*,   p. 30

Subgenus **Otisorex** De Kay

1842. *Otisorex* De Kay, Zoology of New-York . . . , pt. 1, Mammalia, p. 22, and Pl. 5, Fig. 1. Type, *Otisorex platyrhinus* De Kay [= *Sorex cinereus* Kerr].

Postmandibular foramen absent in most species; upper unicuspids usually having pigmented ridge, extending from apex of tooth to cingulum and uninterrupted by anteroposterior groove.

Sorex cinereus
Masked Shrew

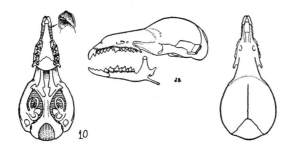

Fig. 10. *Sorex cinereus ugyunak,* Kaolak, 69° 56′ 00″, 160° 14′ 51″, 178 ft., Alaska, No. 43166 K.U., ♀, X 2.

External measurements: 71–111; 25–50; 10–14. Condylobasal length of skull, 14.6–16.9. Color grayish or tan to brownish, bicolored or tricolored in some subspecies; paler in winter, darker in summer. Skull having relatively narrow rostrum and relatively high braincase; teeth narrow in exterointernal diameter; 3rd unicuspid tooth larger than or, less commonly, equal to 4th (in some individuals of *S. c. ohionensis* 3rd unicuspid may actually be smaller than 4th); unicuspids with distinctly pigmented internal ridge extending from apex medially to cingulum, in some specimens ending in small internal cusplet.

Introduced in Newfoundland (Banfield, 1974:10). According to Hoffmann (1971:195) the species occurs in eastern Siberia "where it is sympatric with the phenetically similar *S. caecutiens* and *S. minutus*."—See addenda.

Sorex cinereus acadicus Gilpin

1867. *Sorex acadicus* Gilpin, Proc. and Trans. Nova Scotian Inst. Nat. Sci., 1(2):2. Type locality, Nova Scotia, assumed to be near Halifax, Halifax Co.
1940. *Sorex cinereus acadicus,* R. W. Smith, Amer. Midland Nat., 24:219, July 31.

MARGINAL RECORDS.—Nova Scotia: Cape North; Newport. [Possibly the geographic range of *acadicus* is more extensive than shown on the map herewith. Without mentioning record stations of occurrence, Smith (1940:219) includes "New Brunswick, and eastern Quebec" in the range of *acadicus*. Miller and Kellogg (1955:13) do likewise and include also Prince Edward Island in the range of *acadicus* as does Cameron (1959:42). Cameron·(1953:31) assigns "eastern Quebec south of the St. Lawrence River" to the range of *S. c. cinereus.*]

Sorex cinereus cinereus Kerr

1792. *Sorex arcticus cinereus* Kerr, The animal kingdom . . . , p. 206. Type locality, Fort Severn, Ontario.

1925. *Sorex cinereus cinereus*, Jackson, Jour. Mamm., 6:56, February 9.

1827. *Sorex personatus* I. Geoffroy St.-Hilaire, Mém. Mus. Hist. Nat., Paris, 15:122. Type locality, eastern United States.

1828. *Sorex forsteri* Richardson, Zool. Jour., 3:516, April, type from "Hudson's Bay countries."

1837. *Sorex cooperi* Bachman, Jour. Acad. Nat. Sci. Philadelphia, 7(2):388, type from "North Western Territory."

1837. *Sorex fimbripes* Bachman, Jour. Acad. Nat. Sci. Philadelphia, 7(2):391, type from Drury Run, Pennsylvania.

1842. *Otisorex platyrhinus* De Kay, Zoology of New-York . . . , pt. 1, Mammalia, p. 22, type from Tappan, Rockland Co., New York.

1891. *Sorex idahoensis* Merriam, N. Amer. Fauna, 5:32, July 30, type from Timber Creek, 8200 ft., Salmon River Mts. [now Lemhi Mts.], Lemhi Co., Idaho.

1926. *Sorex frankstounensis* Peterson, Ann. Carnegie Mus., 16:292, March, type from Frankstown Cave, near Holidaysburg, Blair Co., Pennsylvania.

MARGINAL RECORDS.—Mackenzie: Aklavik (Youngman, 1964:1); Anderson River Region (Fort Anderson); Lake St. Croix. Keewatin: mouth of Windy River (Harper, 1956:11). Manitoba: Churchill; York Factory. Quebec: Papps Cove, Richmond Gulf (Edwards, 1963:5, recorded only as *Sorex cinereus*); *Golfcourse Cove, Richmond Gulf (ibid.)*; Seal Lake; S Leaf Bay; Fort Chimo; George River Hudsons Bay Post; Godbout. New Brunswick: Bathurst, 15 mi. from Miramichi Road. Prince Edward Island: Alberton;

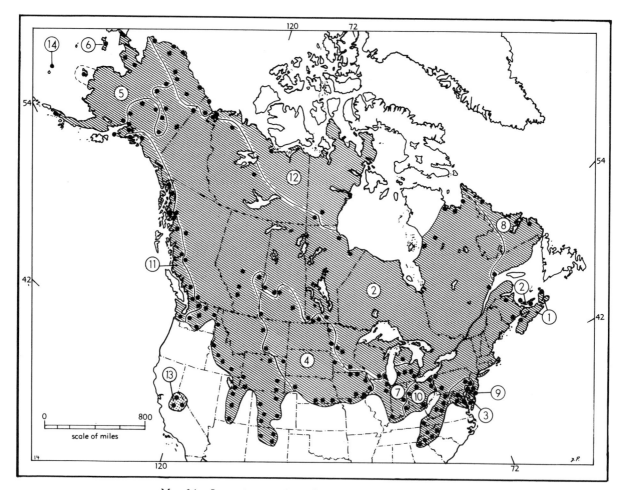

Map 14. *Sorex cinereus, Sorex lyelli,* and *Sorex hydrodromus.*

1. S. c. acadicus	5. S. c. hollisteri	9. S. c. nigriculus
2. S. c. cinereus	6. S. c. jacksoni	10. S. c. ohionensis
3. S. c. fontinalis	7. S. c. lesueurii	11. S. c. streatori
4. S. c. haydeni	8. S. c. miscix	12. S. c. ugyunak
13. S. lyelli		14. S. hydrodromus

Georgetown. New Brunswick: Hampton. New Jersey: Mauricetown. Pennsylvania: Kennett Square. West Virginia: Cheat Bridge; Cranberry Glades. North Carolina: Grandfather Mtn.; Mt. Mitchell; Highlands Plateau on main Blue Ridge of Appalachian Mts., a few mi. from both Georgia and South Carolina. Georgia: Beech Creek, near confluence with Tallulah River, Towns Co. (Wharton, 1968:158, as *S. cinereus* only). Tennessee: Buck Fork, Little Pigeon River. Kentucky: Big Black Mtn. West Virginia: Jobs Knob. Maryland: Bittinger. Pennsylvania: Westmoreland County; Erie County. Michigan: Roscommon County. Wisconsin (Jackson, 1961:32): *Milwaukee;* Milton. Minnesota: Steele County; Long Prairie; Ottertail County. Manitoba: Winnipeg; Oak Lake. Saskatchewan (Beck, 1958:8): Saskatoon; Revenue. Alberta: South Edmonton; forks of Blindmans and Red Deer rivers; Calgary. Montana: Zortman. Wyoming (Long, 1965a:519): *3 mi. WNW Monarch;* 4 mi. NNE Banner; Springhill; 2 mi. SW Pole Mtn. Colorado: *Fort Collins;* Loveland; Summit House, Pikes Peak. New Mexico: Twining; Pecos Baldy; 1 mi. N, 6 mi. E Cuba (Findley, *et al.,* 1975:10). Colorado: Hermit; Ruby Lake; Mud Springs. Wyoming: 9 mi. S Robertson (Long, 1965a:519). Utah: Wildcat Ranger Station, 8000 ft.; ½ mi. S Mt. Nebo Ranger Station, 6125 ft.; head Lambs Canyon, 9000 ft.; Beaver Creek, S fork Ogden River, 6500 ft. Wyoming: Cokeville (Long, 1965a:519). Idaho: American Falls; Ketchum; Sawtooth Lake; ½ mi. E Black Lake; Cedar Mtn. Washington: Loon Lake; Signal Peak, 4000 ft., Yakima Indian Reservation; Mt. Rainier; Lake Keechelus; Bauerman Ridge. British Columbia: Hope; Stuie (Cowan and Guiguet, 1965:45); Hazelton; Stikine River at Great Glacier. Alaska: *White Pass;* Seldovia; Hope; Tyonek; S fork Kuskokwim River, 10 mi. above mouth Post River; Mt. Sischoo; Nenana; Eagle; Fairbanks; Tanana. Yukon Territory (Youngman, 1964:1): Rampart House. [See remarks on geographic range under *S. c. acadicus.*]

Sorex cinereus fontinalis Hollister

1911. *Sorex fontinalis* Hollister, Proc. U.S. Nat. Mus., 40:378, April 17, type from Cold Spring Swamp, near Beltsville, Prince Georges Co., Maryland.
1937. *Sorex cinereus fontinalis*, Poole, Jour. Mamm., 18:96, February 11.

MARGINAL RECORDS.—Pennsylvania: Ridgewood near Reading; 3 mi. N Swarthmore. Delaware: *Keeny.* Maryland: Cambridge; Hollywood. Virginia: 6 mi. Washington, D.C. Maryland: Cabin John.—**See addenda.**

Sorex cinereus haydeni Baird

1858. *Sorex haydeni* Baird, Mammals, *in* Repts. Expl. Surv. . . ., 8(1):29, July 14, type from Fort Union, Nebraska [later Fort Buford, now Mondak, Montana, near Buford, Williams Co., North Dakota].
1925. *Sorex cinereus haydeni*, Jackson, Jour. Mamm., 6:56, February 9.—**See addenda.**

MARGINAL RECORDS.—Alberta: Islay. Saskatchewan: Osler; Indian Head. Manitoba: Killarney; Carberry. Minnesota: Kittson County; Moorhead; Madison. Iowa (Bowles, 1975:31–33): Eden Township; Hayden Prairie, 1 mi. S, 5 mi. W Lime Springs; *Ridgeway; Conover;* Dyersville; Bettendorf; Henry County; 3 mi. S, 6½ mi. E Albia. Missouri: 1 mi. W Pattonsburg (Easterla and Damman, 1977:10, as *S. cinereus* only). Nebraska: 2½ mi. S, 3 mi. E Crete (Choate and Genoways, 1967:239); 1 mi. S Saronville (*ibid.*); 1¾ mi. S Kearney (Jones, 1964c:65); Crescent Lake National Wildlife Refuge (Rickart, 1972:36, owl pellets). Wyoming (Long, 1965a:520): *1½ mi. E Buckhorn;* 3 mi. NW Sundance; *Warren Peak.* Montana: *Crow Agency;* Fort Custer. Saskatchewan: Cypress Hills (Beck, 1958:8).

Sorex cinereus hollisteri Jackson

1900. *Sorex personatus arcticus* Merriam, Proc. Washington Acad. Sci., 2:17, March 14, type from St. Michael, Alaska. Not *Sorex arcticus* Kerr, 1792.
1925. *Sorex cinereus hollisteri* Jackson, Jour. Mamm., 6:55, February 9, a renaming of *Sorex personatus arcticus* Merriam.

MARGINAL RECORDS.—Alaska: Bettles; Richardson, Tanana River; Swede Lake; Lake Clark, head Nogheling River; Kings Cove, Alaska Peninsula; Nunivak Island; Sawtooth Mts.; thence westward into Siberia, and eastward into Alaska: Cloud Lake.

Sorex cinereus jacksoni Hall and Gilmore

1932. *Sorex jacksoni* Hall and Gilmore, Univ. California Publ. Zool., 38:392, September 17, type from Sevoonga, 2 mi. E North Cape, St. Lawrence Island, Bering Sea, Alaska. Known only from St. Lawrence Island.
1967. *Sorex cinereus jacksoni*, Hoffmann and Peterson, Syst. Zool., 16:134, June 30.

Sorex cinereus lesueurii (Duvernoy)

1842. *Amphisorex lesueurii* Duvernoy, Mag. de Zool. d'Anat. Comp. et Paleont., Paris, 1842, livr. 25, p. 33, Pl. 50, type from Wabash River Valley, Indiana.
1942. *Sorex cinereus lesueurii*, Bole and Moulthrop, Sci. Publs., Cleveland Mus. Nat. Hist., 5:95, September 11.

MARGINAL RECORDS.—Michigan: Clinton County; Livingston County; Washtenaw County. Indiana: Randolph County; New Harmony. Illinois: St. Anne; Chicago. Wisconsin (Jackson, 1961:32): *Delavan; Tichigan Lake;* Racine.—**See addenda.**

Sorex cinereus miscix Bangs

1899. *Sorex personatus miscix* Bangs, Proc. New England Zool. Club, 1:15, February 28, type from Black Bay, Labrador.
1925. *Sorex cinereus miscix*, Jackson, Jour. Mamm., 6:56, February 9.

MARGINAL RECORDS.—Labrador: Okak; 20 mi. above mouth Paradise River. Quebec: Bay of Seven Islands.

Sorex cinereus nigriculus Green

1932. *Sorex cinereus nigriculus* Green, Univ. California Publ. Zool., 38:387, June 9, type from alluvial tidewater marsh on Tuckahoe River, E Tuckahoe, Cape May Co., New Jersey. Known only from type locality.

Sorex cinereus ohionensis Bole and Moulthrop

1942. *Sorex cinereus ohionensis* Bole and Moulthrop, Sci. Publs., Cleveland Mus. Nat. Hist., 5:89, September 11, type from Hunting Valley, Cuyahoga Co., Ohio.

MARGINAL RECORDS.—Ohio: Mechanicsville; Ellsworth; 5 mi. N Minford (Goodpaster and Hoffmeister, 1968:116). Indiana: Rexville (Lindsey, 1960:254). Ohio: Mercer County (Gottschang, 1965:48, as *S. cinereus* only); Maple Grove.

Sorex cinereus streatori Merriam

1895. *Sorex personatus streatori* Merriam, N. Amer. Fauna, 10:62, December 31, type from Yakutat, Alaska.
1925. *Sorex cinereus streatori*, Jackson, Jour. Mamm., 6:56, February 9.

MARGINAL RECORDS.—Alaska: Port Nell Juan (head Prince William Sound); Valdez Narrows (Prince William Sound); Skagway; Taku River; Thomas Bay; Anan Creek. British Columbia: Observatory Inlet; Kimsquit (Cowan and Guiguet, 1965:45); Alta Lake (*ibid.*); Mt. Baker Range, 49th parallel, 6000 ft. Washington: Whatcom Pass, 5200 ft.; Cedarville, thence to and up coast, including several coastal islands in British Columbia, to point of beginning.

Sorex cinereus ugyunak Anderson and Rand

1945. *Sorex cinereus ugyunak* Anderson and Rand, Canadian Field-Nat., 59:62, October 16, type from Tuktoyaktok (Tuktak), about 20 mi. SW Toker Point, on Arctic Coast near NE corner Mackenzie River Delta, Mackenzie.

MARGINAL RECORDS.—Alaska: Point Barrow; Okpilak River, near Barter Island. Yukon Territory (Youngman, 1975:44): Head Point, near Herschel Island; Driftwood Creek, 60 mi. NE Old Crow. Northwest Territories: type locality; Coronation Gulf; Chesterfield; Padley Post, 45 mi. SW. Alaska: Brooks Range, *ca.* 80 mi. W Alaska-Yukon boundary; Anaktuvuk Pass; Chandler Lake; Kaolak, 160° 14' 51", 69° 56' 00", 178 ft.; Cape Sabine (Childs, 1969:50); Wainwright.

Sorex hydrodromus Dobson
Pribilof Shrew

1889. *Sorex hydrodromus* Dobson, Ann. Mag. Nat. Hist., ser. 6, 4:373, November, type stated to be from Unalaska Island but probably from St. Paul Island (Hoffmann and Peterson,

Syst. Zool., 16:132, 134, June 30, 1967). Known only from the one island.
1895. *Sorex pribilofensis* Merriam, N. Amer. Fauna, 10:87, December 31, type from St. Paul Island, Pribilof Group (in Bering Sea), Alaska.

External measurements (Hall and Kelson, 1959:41): 92–103; 32–37; 13–14.5. Condylobasal length of skull, 15.4–16.0. Tricolored (brown, pale brown, gray) in summer pelage, bicolored (brown and gray) in winter; 3rd unicuspid larger than 4th; internal ridges of unicuspids well developed and heavily pigmented.

Sorex lyelli Merriam
Mt. Lyell Shrew

1902. *Sorex tenellus lyelli* Merriam, Proc. Biol. Soc. Washington, 15:75, March 22, type from Mount Lyell [near head Lyell Fork of Tuolumne River], Tuolumne Co., California.
1928. *Sorex lyelli*, Jackson, N. Amer. Fauna, 51:57, July 24.

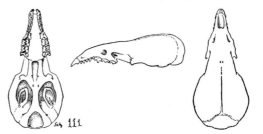

Fig. 11. *Sorex lyelli*, Vogelsang Lake, Mariposa Co., California, No. 23001 M.V.Z., ♀, X 2.

External measurements: 102–103; 39–41; 11.0–12.0. Condylobasal length of skull, 15.2–15.4. Upper parts brownish; underparts pale olive-gray or smoke-gray; tail bicolored; 3rd unicuspid tooth equal to or larger than 4th; skull similar to that of *S. cinereus* but wider interorbitally and having more nearly flat braincase.

MARGINAL RECORDS.—California: near Williams Butte, Mono Co.; Mammoth; Vogelsang Lake.

Sorex preblei Jackson
Preble's Shrew

1922. *Sorex preblei* Jackson, Jour. Washington Acad. Sci., 12:263, June 4, type from Jordan Valley, 4200 ft., Malheur Co., Oregon.

External measurements: 85–95; 35–36; 11–11. Condylobasal length of skull, 14.2–14.6. Upper parts brownish; underparts paler; tail bicolored. Skull resembling that of *S. lyelli* but smaller and tooth-row relatively shorter.

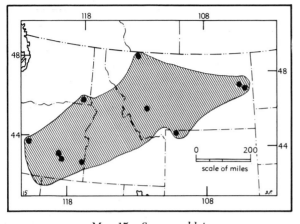

Map 15. *Sorex preblei.*

MARGINAL RECORDS.—Montana (Hoffmann, *et al.*, 1969:579, 581): St. Marys; Bloomfield; Glendive. Wyoming: Lamar Ranger Station (*ibid.*). Montana: vic. Butte (*ibid.*). Oregon: Jordon Valley, 4200 ft.; Diamond, 4300 ft.; 22 mi. S, 6 mi. E Burns (Verts, 1975:22); shore of East Lake, Paulina Mts., Deschutes Co. (275685 USNM, *fide* R. S. Hoffmann). Washington: Big Spring, 3 mi. NE Clearwater Guard Station, 5100 ft. (Armstrong, F. H., 1957:6). Not found: Washington: Woodroad 17 (*ibid.*). Specimens listed above from St. Marys, Bloomfield, and Glendive, Montana, and from Lamar Ranger Station, Wyoming, are only tentatively identified by Hoffmann, *et al.*, 1969, as *Sorex preblei.* See addenda.

Sorex milleri Jackson
Carmen Mountain Shrew

1947. *Sorex milleri* Jackson, Proc. Biol. Soc. Washington, 60:131, October 9, type from Madera Camp, 8000 ft., Sierra del Carmen, Coahuila.

External measurements: 95; 44; 11. Condylobasal length of skull, 15.5. Summer pelage unknown; upper parts brownish; underparts paler; internal ridges of unicuspids pigmented to cingula; 3rd unicuspid approx. equal to 4th.

MARGINAL RECORDS.—Coahuila: type locality; *13 mi. E San Antonio de las Alazanas, 9350 ft.* Nuevo León: Cerro Potosí, near La Jolla.

Sorex longirostris
Southeastern Shrew

External measurements: 79–108; 27–40; 10–13. Condylobasal length of skull, 13.9–16.4. Reddish brown or brownish above, paler below; 3rd unicuspid smaller than or equal to 4th; unicuspids broader than long and having internal ridge that is not pigmented; skull relatively broad and top flattened.

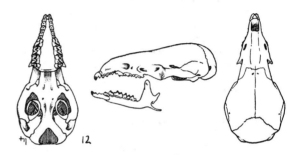

Fig. 12. *Sorex longirostris fisheri*, Dismal Swamp, Virginia, No. 75167 U.S.N.M., ♀, X 2.

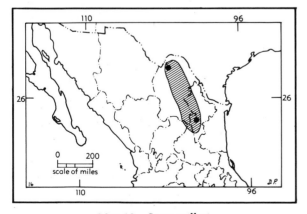

Map 16. *Sorex milleri.*

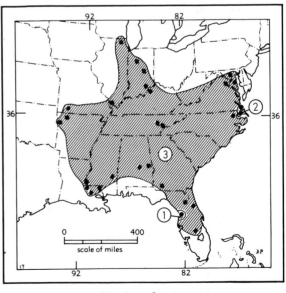

Map 17. *Sorex longirostris.*

Guide to subspecies	2. *S. l. fisheri*
1. *S. l. eionis*	3. *S. l. longirostris*

Sorex longirostris eionis Davis

1957. *Sorex longirostris eionis* Davis, Amer. Mus. Novit., 1844:3, October 10, type from Homosassa Springs, Citrus Co., Florida. Known only from type locality.

Sorex longirostris fisheri Merriam

1895. *Sorex fisheri* Merriam, N. Amer. Fauna, 10:86, December 31, type from Lake Drummond, Dismal Swamp, Virginia.
1928. *Sorex longirostris fisheri*, Jackson, N. Amer. Fauna, 51:87, July 24.

MARGINAL RECORDS.—Virginia: type locality. North Carolina: Chapanoke.

Sorex longirostris longirostris Bachman

1837. *Sorex longirostris* Bachman, Jour. Acad. Nat. Sci. Philadelphia, ser. 1, 7(2):370, type locality, Hume Plantation, swamps Santee River [= Cat Island, mouth Santee River], South Carolina.
1848. [*Musar[aneus]. (Croc[idura].)] Bachmani (longirostris junior Bachm[an])*, Pomel, Arch. Sci. Phys. et Nat., 9:249
1868. *Sorex wagneri* Fitzinger, Sitzungsb. k. Akad. Wiss., Wien, 57:512.

MARGINAL RECORDS.—Illinois: Pistakee Lake. Indiana: near Lafayette (Mumford, 1969:35); Marion County (*ibid.*); Hanover College Campus (Lindsay, 1960:254). Kentucky: Bernheim Forest. Tennessee: 10 mi. SW Knoxville (Tuttle, 1964:146, as *S. longirostris* only); C.C.C. Camp, Great Smoky Mts. National Park. Virginia: Falls Church. Maryland: Sandy Spring; *Laurel;* Chesapeake Beach. North Carolina: Bertie County. Florida: Welaka; near Davenport; 5 mi. E Gainesville. Georgia: Grady County (Golley, 1962:40). Alabama: Auburn; Bear Swamp, 4 mi. NE Autaugaville. Mississippi: *Wayne County* (Jones and Long, 1961:252); Jones County (*ibid.*). Louisiana (Lowery, 1974:75): 10 mi. E Franklinton; *ca.* 5 mi. SE Walker; *3 mi. NNE Denham Springs; Plains;* 5 mi. NE St. Francisville. Mississippi: Wilkinson County (Jones and Long, 1961:252). Arkansas: Devils Den State Park (Sealander, 1977: 149); 3 mi. N War Eagle (Sealander, 1960:525). Missouri: Ketchum Branch (Brown, 1961:527). Illinois: Olive Branch.

Sorex vagrans
Vagrant Shrew

Subspecies revised by Findley, Univ. Kansas Publ., Mus. Nat. Hist., 9:1–68, 18 figs., December 10, 1955.—**See addenda.**

External measurements: 90–153; 31–67; 11–175. Condylobasal length of skull, 16.1–23.0. Tail a little more than a third to almost a half of total length; color pattern tricolored through bicolored to almost unicolored; color reddish (Sayal- or Snuff-Brown) to grayish in summer pelage, and black to pale gray in winter; 3rd unicuspid

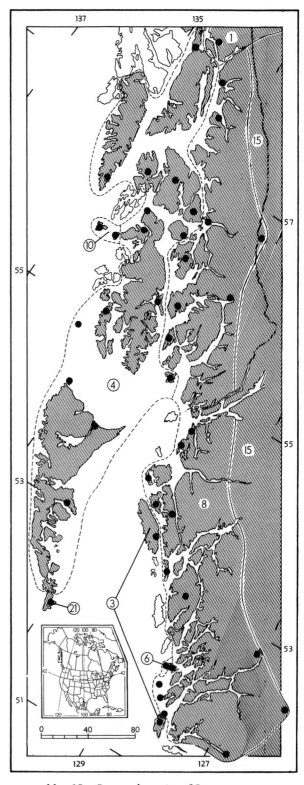

Map 18.   Some subspecies of *Sorex vagrans*.
(See also Map 19.)

1. *S. v. alascensis*	6. *S. v. insularis*	15. *S. v. obscurus*
3. *S. v. calvertensis*	8. *S. v. longicauda*	21. *S. v. prevostensis*
4. *S. v. elassodon*	10. *S. v. malitiosus*	

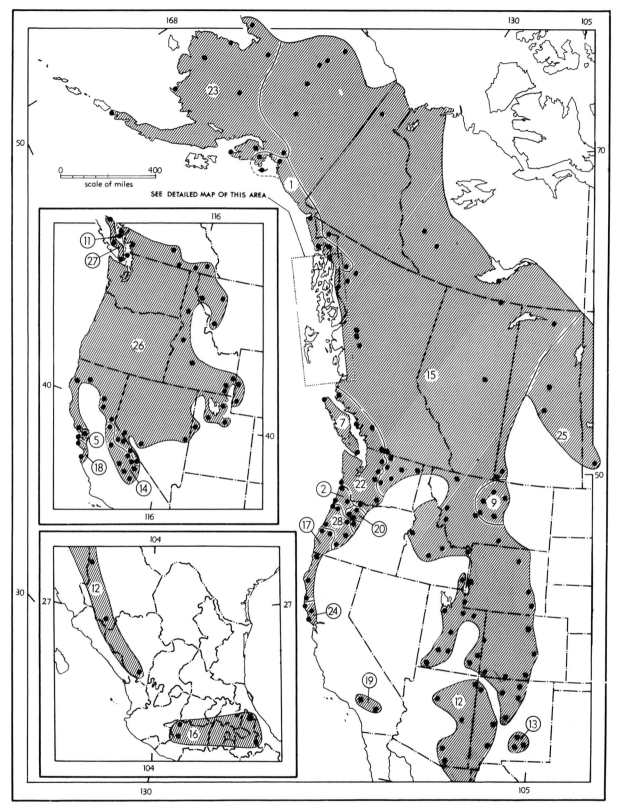

Map 19.   Some subspecies of *Sorex vagrans*. (See also Map 18.)

1. *S. v. alascensis*	11. *S. v. mixtus*	17. *S. v. pacificus*	23. *S. v. shumaginensis*
2. *S. v. bairdi*	12. *S. v. monticola*	18. *S. v. paludivagus*	24. *S. v. sonomae*
5. *S. v. halicoetes*	13. *S. v. neomexicanus*	19. *S. v. parvidens*	25. *S. v. soperi*
7. *S. v. isolatus*	14. *S. v. obscuroides*	20. *S. v. permiliensis*	26. *S. v. vagrans*
9. *S. v. longiquus*	15. *S. v. obscurus*	21. *S. v. prevostensis*	27. *S. v. vancouverensis*
10. *S. v. malitiosus*	16. *S. v. orizabae*	22. *S. v. setosus*	28. *S. v. yaquinae*

smaller than 4th; unicuspids, except 5th, having pigmented ridge extending from near apex of each tooth medially to cingulum and sometimes ending as internal cusplet. *S. vagrans* differs from members of the *ornatus*-group in less flattened skull, and in more ventrally situated foramen magnum that encroaches more on the basioccipital and less on the supraoccipital. The dental features mentioned above distinguish *S. vagrans* from *S. trowbridgii*, *S. saussurei*, *S. cinereus*, *S. merriami*, and *S. arcticus*—kinds that occur where *vagrans* occurs. The large *S. palustris* and *S. bendirii* can be distinguished from *S. vagrans* by larger size and darker color.

S. *vagrans* is an intergrading chain of subspecies, the end members of which differ so much in size and ecological requirements that two subspecies live together without crossbreeding along the Pacific Coast from southern British Columbia to the Golden Gate in California. The details of this distribution and probable reasons for it have been published by Findley (1955).

See the account of *Sorex ornatus* on page 38 concerning the possibility that it is only subspecifically distinct from the species *Sorex vagrans*. *S. v. vagrans* and *S. ornatus californicus* may intergrade.

Sorex vagrans alascensis Merriam

1895. *Sorex obscurus alascensis* Merrian, N. Amer. Fauna, 10:76, December 31, type from Yakutat, Alaska.
1955. *Sorex vagrans alascensis*, Findley, Univ. Kansas Publ., Mus. Nat. Hist., 9:41, December 10.
1900. *Sorex glacialis* Merriam, Proc. Washington Acad. Sci., 2:16, March 14, type from Point Gustavus, E side of entrance to Glacier Bay, Alaska.

MARGINAL RECORDS.—Alaska: Valdez Narrows, Prince William Sound; N shore Yakutat Bay; E side Chilkat River, 100 ft., 4 mi. N, 9 mi. W Haines. British Columbia: Sheslay River. Alaska: Juneau; Glacier Bay; Montague Island, Prince William Sound; Port Nell Juan.

Sorex vagrans bairdi Merriam

1895. *Sorex bairdi* Merriam, N. Amer. Fauna, 10:77, December 31, type from Astoria, Oregon.
1955. *Sorex vagrans bairdi*, Findley, Univ. Kansas Publ., Mus. Nat. Hist., 9:35, December 10.

MARGINAL RECORDS.—Oregon: type locality; Portland; N slope Three Sisters; Taft.

Sorex vagrans calvertensis Cowan

1941. *Sorex obscurus calvertensis* Cowan, Proc. Biol. Soc. Washington, 54:103, July 31, type from Safety Cove, Calvert Island, British Columbia.

1955. *Sorex vagrans calvertensis*, Findley, Univ. Kansas Publ., Mus. Nat. Hist., 9:39, December 10.

MARGINAL RECORDS.—British Columbia: Larson Harbor, Banks Island; type locality.

Sorex vagrans elassodon Osgood

1901. *Sorex longicauda elassodon* Osgood, N. Amer. Fauna, 21:35, September 26, type from Cumshewa Inlet near old Indian village of Clew, Moresby Island, Queen Charlotte Islands, British Columbia.
1955. *Sorex vagrans elassodon*, Findley, Univ. Kansas Publ., Mus. Nat. Hist., 9:40, December 10.

MARGINAL RECORDS.—Alaska: Hawk Inlet [Admiralty Island]; Kupreanof Island; Mitkof Island; St. John Harbor, Zarembo Island; Kasaan Bay, Prince of Wales Island; Duke Island. British Columbia: Massett, Graham Island, Queen Charlotte Islands; type locality; Langara Island, Queen Charlotte Islands. Alaska: Forester Island; Rocky Bay, Dall Island; Shakan [really on Kosciusko Island]; Point Baker; Kuiu Island; Port Conclusion, Baranof Island.

Sorex vagrans halicoetes Grinnell

1913. *Sorex halicoetes* Grinnell, Univ. California Publ. Zool., 10:183, March 20, type from salt marsh near Palo Alto, Santa Clara Co., California.
1928. *Sorex vagrans halicoetes*, Jackson, N. Amer. Fauna, 51:108, July 24.

MARGINAL RECORDS.—California: Berkeley; *Elmhurst*; *Palo Alto*; San Mateo.

Sorex vagrans insularis Cowan

1941. *Sorex obscurus insularis* Cowan, Proc. Biol. Soc. Washington, 54:103, July 31, type from Smythe Island, Bardswell Group, British Columbia.
1955. *Sorex vagrans insularis*, Findley, Univ. Kansas Publ., Mus. Nat. Hist., 9:39, December 10.

MARGINAL RECORDS.—British Columbia: Bardswell Group: *type locality*; Townsend Island; *Reginald Island*.

Sorex vagrans isolatus Jackson

1922. *Sorex obscurus isolatus* Jackson, Jour. Washington Acad. Sci., 12:263, June 4, type from mouth Millstone Creek, Nanaimo, Vancouver Island, British Columbia.
1955. *Sorex vagrans isolatus*, Findley, Univ. Kansas Publ., Mus. Nat. Hist., 9:38, December 10.

MARGINAL RECORDS.—British Columbia: Vancouver Island: Cape Scott; Victoria.

Sorex vagrans longicauda Merriam

1895. *Sorex obscurus longicauda* Merriam, N. Amer. Fauna, 10:74, December 31, type from Wrangell, Alaska.
1955. *Sorex vagrans longicauda*, Findley, Univ. Kansas Publ., Mus. Nat. Hist., 9:37, December 10.

MARGINAL RECORDS.—Alaska: Port Snettisham. British Columbia: Great Glacier, Stikine River. Alaska: Burroughs Bay. British Columbia (Cowan and Guiguet, 1965:51, unless otherwise noted): Kimsquit; Stuie; head Rivers Inlet (Hall and Kelson, 1959:32); Hecate Island; Spider Island; Goose Island; Dufferin Island; Princess Royal Island; Campania Island; Pitt Island; McCauley Island; Freeman Pass; Metlakatla (Hall and Kelson, 1959:32); Port Simpson. Alaska: Gravina Island; Helm Bay; Etolin Island; Sergief Island, mouth Stikine River; Sumdum Village. Cowan and Guiguet (1965:51) state that *longicauda* "intergrades with *setosus* between Rivers Inlet and Powell River"; indeed they list under *longicauda* material from Granville Channel and Ruth Island.

Sorex vagrans longiquus Findley

1955. *Sorex vagrans longiquus* Findley, Univ. Kansas Publ., Mus. Nat. Hist., 9:49, December 10, type from 25 mi. ESE Big Sandy, Eagle Creek, Chouteau Co., Montana.

MARGINAL RECORDS.—Montana: Bearpaw Mts.; Zortman; Big Snowy Mts.; 16 mi. N White Sulphur Springs; Highwood Mts.

Sorex vagrans malitiosus Jackson

1919. *Sorex obscurus malitiosus* Jackson, Proc. Biol. Soc. Washington, 32:23, April 11, type from E side Warren Island, Alaska. Known only from Warren and Coronation islands, Alaska.
1955. *Sorex vagrans malitiosus*, Findley, Univ. Kansas Publ., Mus. Nat. Hist., 9:40, December 10.

Sorex vagrans mixtus Hall

1938. *Sorex obscurus mixtus* Hall, Amer. Nat., 72:462, September 10, type from Vanada, Texada Island, Georgia Strait, British Columbia. Known only from type locality.
1955. *Sorex vagrans mixtus*, Findley, Univ. Kansas Publ., Mus. Nat. Hist., 9:38, December 10.

Sorex vagrans monticola Merriam

1890. *Sorex monticolus* Merriam, N. Amer. Fauna, 3:43, September 11, type from San Francisco Mtn., 11,500 ft., Coconino Co., Arizona.
1895. *Sorex vagrans monticola* Merriam, N. Amer. Fauna, 10:69, December 31.
1932. *Sorex melanogenys* Hall, Jour. Mamm., 13:260, August 9, type from Marijilda Canyon, 8600 ft., Graham Mts. [= Pinaleno Mts.], Graham Co., Arizona.

MARGINAL RECORDS.—Arizona: Tunitcha Mts. New Mexico: Chusca Mts.; Copper Canyon, Magdalena Mts.; Mimbres Mts., near Kingston. Chihuahua: Chuhuichupa (Anderson, 1972:233); Guadalupe y Calvo. Durango: Cueva, 8500 ft. (Baker and Greer, 1962:65). Arizona: Santa Rita Mts., Boulder Spring, Santa Rita Range Reserve (Lange, 1959:99); Santa Catalina Mts.; White River, Horseshoe Cienega, 8300 ft., White Mts.; type locality.

Sorex vagrans neomexicanus V. Bailey

1913. *Sorex obscurus neomexicanus* V. Bailey, Proc. Biol. Soc. Washington, 26:133, May 21, type from Cloudcroft, 9000 ft., Sacramento Mts., Otero Co., New Mexico.
1955. *Sorex vagrans neomexicanus*, Findley, Univ. Kansas Publ., Mus. Nat. Hist., 9:50, December 10.

MARGINAL RECORDS.—New Mexico: NW slope Capitan Mts.; 10 mi. NE Cloudcroft; type locality.

Sorex vagrans obscuroides Findley

1955. *Sorex vagrans obscuroides* Findley, Univ. Kansas Publ., Mus. Nat. Hist., 9:58, December 10, type from Bishop Creek, 6600 ft., Inyo Co., California.

MARGINAL RECORDS.—California: Pyramid Peak; near Mammoth; *Round Valley;* type locality; Mt. Whitney; Kern Lakes; Halstead Meadows; Horse Corral Meadows; E Fork Indian Canyon.

Sorex vagrans obscurus Merriam

1891. *Sorex vagrans similis* Merriam, N. Amer. Fauna, 5:34, July 30, type from near Timber Creek, 8200 ft., Salmon River Mts. [now Lemhi Mts.], 10 mi. W Junction [near present town of Leadore], Lemhi Co., Idaho. (Not *S. similis* Hensel, 1855 [= *Neomys similis*], type from bone deposits at Cagliari, Sardinia.)
1895. *Sorex obscurus* Merriam, N. Amer. Fauna, 10:72, December 31, a renaming of *Sorex vagrans similis* Merriam.
1955. *Sorex vagrans obscurus*, Findley, Univ. Kansas Publ., Mus. Nat. Hist., 9:43, December 10.

MARGINAL RECORDS.—Alaska: Chandler Lake, 68° 12' N, 152° 45' W; *Yukon River, 20 mi. above Circle; mts. near Eagle.* Yukon: Old Crow (Youngman, 1975:50). Mackenzie: Nahanni River Mts.; Fort Simpson; Fort Resolution, Mission Island. Alberta: Wood Buffalo Park. Saskatchewan: Stony Rapids (Beck, 1961:185, as *S. vagrans* only). Alberta: Athabaska River, 30 mi. above Athabaska Landing. Saskatchewan: Cypress Hills; Middle Creek (Beck, 1958:9). Montana: 4 mi. S Fort Logan; Pryor Mts. Wyoming: 1 mi. S, 1 mi. W Buffalo; Springhill; 1 mi. N, 5 mi. W Horse Creek P.O. Colorado: Boulder; Hunters Creek; 5 mi. S, 1 mi. W Cuchara Camps. New Mexico: 3 mi. N Red River, 10,700 ft.; Pecos Baldy; Manzano Mts.; Jemez Mts. Colorado: Navajo River; ¼ mi. N Middle Well, Prater Canyon, 7500 ft. (Anderson, 1961:37). Utah: *Abajo Mts.;* Elk Ridge; La Sal Mts., 11,000 ft. Colorado: Baxter Pass. Utah: jct. Trout and Ashley creeks, 9700 ft.; Mirror Lake, 10,000 ft.; Mt. Baldy Ranger Station; Wildcat Ranger Station; Pine Valley Mts.; Puffer Lake; South Willow Creek Canyon (Egoscue, 1965:685). Idaho: Preuss Mts. Wyoming (Long, 1965a:522); Evanston; *3 mi. W Stanley;* 10 mi. SE Afton. Idaho: 4 mi. S Trude; head Pahsimeroi River, Pahsimeroi Mts.; Perkins Lake; 1 mi. NE Heath; ½ mi. E Black Lake. Montana: Sula; 8 mi. NE Stevensville; St. Mary. Washington: head Pass Creek; Conconully; Wenatchee; Easton; Stehekin; Pasayten River. British Columbia: Second Summit, Skagit River,

5000 ft.; Babine Mts., 6 mi. N Babine Trail, 5200 ft.; Hazelton; 23 mi. N Hazelton; Flood Glacier, Stikine River; Cheonee Mts.;, Level Mtn.; W side Mt. Glave, 4000 ft., 14 mi. S, 2 mi. E Kelsall Lake. Alaska: head Toklat River; Tanana; Alatna; Bettles.

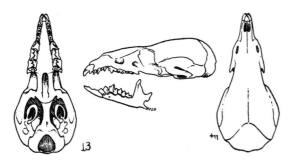

Fig. 13. *Sorex vagrans obscurus*, Stonehouse Creek, 5½ mi. W jct. Stonehouse Creek and Kelsall River, British Columbia, No. 28545 K.U., ♀, X 2.

Sorex vagrans orizabae Merriam

1895. *Sorex orizabae* Merriam, N. Amer. Fauna, 10:71, December 31, type from W slope Mt. Orizaba, 9500 ft., Puebla.
1928. *Sorex vagrans orizabae*, Jackson, N. Amer. Fauna, 51:113, July 24.

MARGINAL RECORDS.—Michoacán: Patambán. Veracruz: Cofre de Perote. Puebla: type locality. Michoacán: Mt. Tancítaro.

Sorex vagrans pacificus Coues

1877. *Sorex pacificus* Coues, Bull. U.S. Geol. and Geog. Surv. Territories, 3(3):650, May 15, type from Fort Umpqua, mouth Umpqua River, Douglas Co., Oregon.
1955. *Sorex vagrans pacificus*, Findley, Univ. Kansas Publ., Mus. Nat. Hist., 9:34, December 10.

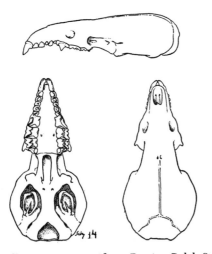

Fig. 14. *Sorex vagrans pacificus*, Russian Gulch State Park, Mendocino Co., California, No. 95645 M.V.Z., ♂, X 2.

MARGINAL RECORDS.—Oregon: Marsfield; type locality. California: Gasquet; 5 mi. S Dyerville; Mendocino.

Sorex vagrans paludivagus von Bloeker

1939. *Sorex vagrans paludivagus* von Bloeker, Proc. Biol. Soc. Washington, 52:93, June 5, type from salt marsh at mouth Elkhorn Slough, Moss Landing, Monterey Co., California.

MARGINAL RECORDS.—California: San Gregorio; *type locality; mouth Salinas River;* Seaside.

Sorex vagrans parvidens Jackson

1921. *Sorex obscurus parvidens* Jackson, Jour. Mamm., 2:161, August 19, type from spring known as Thurmans Camp, Bluff Lake, about 7500 ft., San Bernardino Mts., California.
1955. *Sorex vagrans parvidens*, Findley, Univ. Kansas Publ., Mus. Nat. Hist., 9:58, December 10.

MARGINAL RECORDS.—California: Camp Baldy, San Antonio Canyon, 4200 ft.; type locality.

Sorex vagrans permiliensis Jackson

1918. *Sorex obscurus permiliensis* Jackson, Proc. Biol. Soc. Washington, 31:128, November 29, type from Permilia Lake, W base Mt. Jefferson, Cascade Range, Marion Co., Oregon.
1955. *Sorex vagrans permiliensis*, Findley, Univ. Kansas Publ., Mus. Nat. Hist., 9:36, December 10.

MARGINAL RECORDS.—Oregon: Sand Creek, Mt. Hood; type locality; Detroit.

Sorex vagrans prevostensis Osgood

1901. *Sorex longicauda prevostensis* Osgood, N. Amer. Fauna, 21:35, September 26, type from N end Prevost Island [Kunghit Island on some maps], off coast of Houston Stewart Channel, Queen Charlotte Islands, British Columbia. Known only from type locality.
1955. *Sorex vagrans prevostensis*, Findley, Univ. Kansas Publ., Mus. Nat. Hist., 9:41, December 10.

Sorex vagrans setosus Elliot

1899. *Sorex setosus* Elliot, Field Columb. Mus., Publ. 32, Zool. Ser., 1:274, May 19, type from Happy Lake, Olympic Mts., Clallam Co., Washington.
1955. *Sorex vagrans setosus*, Findley, Univ. Kansas Publ., Mus. Nat. Hist., 9:36, December 10.

MARGINAL RECORDS.—British Columbia: Rivers Inlet; Alta Lake (Cowan and Guiguet, 1965:52); *Agassiz;* Chilliwack Lake. Washington: Barron; Lyman Lake; Mt. Stuart; Satus Pass. Oregon: 2 mi. W Parkdale, 1500 ft. Washington: Ilwaco. British Columbia: Lund, Malaspina Inlet.

Sorex vagrans shumaginensis Merriam

1900. *Sorex alascensis shumaginensis* Merriam, Proc. Washington Acad. Sci., 2:18, March 14, type from Popof Island, Shumagin Islands, Alaska.
1955. *Sorex vagrans shumaginensis*, Findley, Univ. Kansas Publ., Mus. Nat. Hist., 9:42, December 10.

MARGINAL RECORDS.—Alaska: Nome River; Nulato; Kuskokwim River, 200 mi. above Bethel, Crooked Creek; 6 mi. WSW Snowshoe Lake; Seldovia; mts. near Hope; Morhzovoi Bay; Goodnews Bay; Russian Mission; St. Michael.

Sorex vagrans sonomae Jackson

1921. *Sorex pacificus sonomae* Jackson, Jour. Mamm., 2:162, August 19, type from Sonoma Co. side Gualala River, Gualala, California.
1955. *Sorex vagrans sonomae*, Findley, Univ. Kansas Publ., Mus. Nat. Hist., 9:32, December 10.

MARGINAL RECORDS.—California: Point Arena; Monte Rio; Inverness.

Sorex vagrans soperi Anderson and Rand

1945. *Sorex obscurus soperi* Anderson and Rand, Canadian Field-Nat., 59:47, October 16, type from 2½ mi. NW Lake Audy, Riding Mtn. National Park, Manitoba.
1955. *Sorex vagrans soperi*, Findley, Univ. Kansas Publ., Mus. Nat. Hist., 9:48, December 10.

MARGINAL RECORDS.—Saskatchewan: La Ronge (Beck, 1961:185, as *S. vagrans* only); Prince Albert National Park, 1700 ft. Manitoba: type locality.

Sorex vagrans vagrans Baird

1858. *Sorex vagrans* Baird, Mammals, *in* Repts. Expl. Surv. . . . , 8(1):15, July 14, type from Shoalwater Bay [known also as Willapa Bay], Pacific Co., Washington.
1858. *Sorex suckleyi* Baird, Mammals, *in* Repts. Expl. Surv. . . . , 8(1):18, July 14, type from Steilacoom, Pierce Co., Washington.
1891. *Sorex dobsoni* Merriam, N. Amer. Fauna, 5:33, July 30, type from Alturas or Sawtooth Lake, about 7200 ft., E base Sawtooth Mts., Blaine Co., Idaho.
1895. *Sorex amoenus* Merriam, N. Amer. Fauna, 10:69, December 31, type from near Mammoth, 8000 ft., head Owens River, E slope Sierra Nevada, Mono Co., California.
1895. *Sorex nevadensis* Merriam, N. Amer. Fauna, 10:71, December 31, type from Reese River, 6000 ft., Nye–Lander County line, Nevada.
1899. *Sorex shastensis* Merriam, N. Amer. Fauna, 16:87, October 28, type from Wagon Camp, Mt. Shasta, 5700 ft., Siskiyou Co., California.

MARGINAL RECORDS.—British Columbia: Okanagan; Westbridge; Kuskonook; Cranbrook. Montana: Flathead Lake; 6 mi. E Hamilton; Prospect Creek. Idaho: Cedar Mtn.; New Meadows; Alturas Lake; 10 mi. SE Irwin. Wyoming (Long, 1965a:523): 13

mi. N, 2 mi. W Afton; *15 mi. N, 3 mi. E Sage;* 6 mi. N, 2 mi. E Sage. Idaho: 1 mi. W Bancroft; Swan Lake. Utah: Beaver Creek, South Fork, Ogden River; Midway Fish Hatchery; N end Skull Valley (Egoscue, 1961:122); west side Deep Creek Mts., Queen of Sheba Canyon, 8000 ft. Nevada: Baker Creek; Reese River; 2 mi. S Hinds Hot Springs. California: Mono Lake; near Mammoth; Alvord; Mount Conness; Donner; Buck Ranch; Warner Creek, Drake Hot Springs; Canyon Creek; Cuddeback; Novato Point, thence northward along coast to Washington: Friday Harbor, San Juan Island. British Columbia (Cowan and Guiguet, 1965: 53): Vancouver; *Bowen Island;* Powell River.

Sorex vagrans vancouverensis Merriam

1895. *Sorex vancouverensis* Merriam, N. Amer. Fauna, 10:70, December 31, type from Goldstream, Vancouver Island, British Columbia.
1928. *Sorex vagrans vancouverensis*, Jackson, N. Amer. Fauna, 51:106, July 24.

MARGINAL RECORDS.—British Columbia (Cowan and Guiguet, 1965: 53): Sayward, thence southeastward along coast, including *Saltspring Island, Saturna Island,* and *South Pender Island,* to Sooke; *Sahtlam (Cowichan Valley);* Alberni; *Sproat Lake.*

Sorex vagrans yaquinae Jackson

1918. *Sorex yaquinae* Jackson, Proc. Biol. Soc. Washington, 31:127, November 29, type from Yaquina Bay, Lincoln Co., Oregon.
1955. *Sorex vagrans yaquinae*, Findley, Univ. Kansas Publ., Mus. Nat. Hist., 9:34, December 10.

MARGINAL RECORDS.—Oregon: type locality; Philomath; McKenzie Bridge; Crescent Lake; Prospect; Gardiner.

Sorex ornatus
Ornate Shrew

External measurements: 89–108; 32–44; 12–13. Condylobasal length of skull, 15.9–17.6. Upper parts grayish brown, underparts paler; tail indistinctly bicolored. Braincase flattened on top; foramen magnum placed dorsally, encroaching

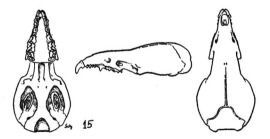

Fig. 15. *Sorex ornatus californicus,* salt marsh, 1 mi. W Avon, Contra Costa Co., California, No. 74572 M.V.Z., ♀, X 2.

more into supraoccipital and less into basioccipital; cranium relatively narrow; 3rd unicuspid smaller than 4th; unicuspids relatively narrow; metacone of M1 relatively high.

Rudd's (1955:21–34) suggestion that this species, through its subspecies *Sorex ornatus californicus*, intergrades or hybridizes with *Sorex vagrans vagrans* at Tolay Creek on the north side of San Pablo Bay, California, raises the question of whether all the subspecies of *S. ornatus* should be arranged instead as subspecies of *S. vagrans*. The latter name is the oldest of those concerned. See also under *Sorex sinuosus* beyond.

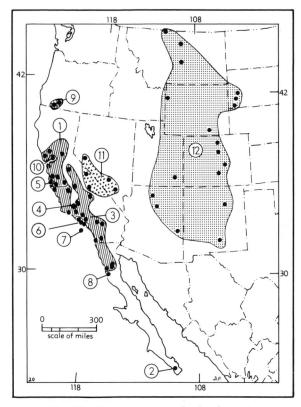

Map 20.　*Sorex ornatus* and related species.

1. S. o. californicus	5. S. o. salarius	9. S. trigonirostris
2. S. o. lagunae	6. S. o. salicornicus	10. S. sinuosus
3. S. o. ornatus	7. S. o. willetti	11. S. tenellus
4. S. o. relictus	8. S. juncensis	12. S. nanus

Sorex ornatus californicus Merriam

1895. *Sorex californicus* Merriam, N. Amer. Fauna, 10:80, December 31, type from Walnut Creek, Contra Costa Co., California.
1922. *Sorex ornatus californicus*, Jackson, Jour. Washington Acad. Sci., 12:264, June 4.

MARGINAL RECORDS.—California: Rumsey, 500 ft.; Auburn; summit Pacheco Pass; Mendota; Stonewall Creek, 6³⁄₁₀ mi. NE Soledad, 1300 ft.; Gilroy; Petaluma.

Sorex ornatus lagunae Nelson and Goldman

1909. *Sorex lagunae* Nelson and Goldman, Proc. Biol. Soc. Washington, 22:27, March 10, type from La Laguna, 5500 ft., Sierra Laguna, Baja California. Known only from type locality.
1928. *Sorex ornatus lagunae*, Jackson, N. Amer. Fauna, 51:169, July 24.

Sorex ornatus ornatus Merriam

1895. *Sorex ornatus* Merriam, N. Amer. Fauna, 10:79, December 31, type from head San Emigdio Canyon, Mt. Piños, Kern Co., California.
1903. *Sorex oreinus* Elliot, Field Columb. Mus., Publ. 74, Zool. Ser., 3:172, May 7, type from Aguaje de las Fresas, 6000 ft., Sierra San Pedro Mártir, Baja California.

MARGINAL RECORDS.—California: El Portal, 1800–2500 ft.; Little Lake; Piru Creek; Big Bear Valley; San Bernardino Mts., 7400–7500 ft.; Santa Ysabel. Baja California: Aguaje de las Fresas; La Grulla, Sierra San Pedro Mártir; San Quintín. California: San Diego; Los Angeles; Santa Barbara; Paraiso Springs; Summit Lake, 12 mi. NW Lemoore; San Emigdio Creek; Bakersfield; Orosi.

Sorex ornatus relictus Grinnell

1932. *Sorex ornatus relictus* Grinnell, Univ. California Publ. Zool., 38:389, June 9, type from Buena Vista Lake, 290 ft., Kern Co., California. Known only from type locality.

Sorex ornatus salarius von Bloeker

1939. *Sorex ornatus salarius* von Bloeker, Proc. Biol. Soc. Washington, 52:94, June 5, type from salt marsh at mouth Salinas River, Monterey Co., California.

MARGINAL RECORDS.—California: *Moss Landing; Sugarloaf Peak, 3 mi. NNE Natividad;* Carmel.

Sorex ornatus salicornicus von Bloeker

1932. *Sorex ornatus salicornicus* von Bloeker, Proc. Biol. Soc. Washington, 45:131, September 9, type from Playa del Rey, Los Angeles Co., California.

MARGINAL RECORDS.—California: Point Mugu; type locality; Nigger Slough.

Sorex ornatus willetti von Bloeker

1942. *Sorex willetti* von Bloeker, Bull. Southern California Acad. Sci., 40:163, January 31, type from Avalon Canyon, Santa Catalina Island, Los Angeles Co., California. Known only from type locality.
1967. *Sorex ornatus willetti*, von Bloeker, Proc. Symp. Biol. California Islands, Santa Barbara Bot. Garden, California, p. 247, August 18.

Sorex tenellus Merriam
Inyo Shrew

1895. *Sorex tenellus* Merriam, N. Amer. Fauna, 10:81, December 31, type from along Lone Pine Creek, at upper edge of Alabama Hills at about 5000 ft. [not "summit of Alabama Hills," as stated by Merriam in original description; see also A. B. Howell, Jour. Mamm., 4:266, November 1, 1923], near Lone Pine, Owens Valley, Inyo Co., California.

1902. *Sorex tenellus myops* Merriam, Proc. Biol. Soc. Washington, 15:76, March 22, type from Pipers Creek (Cottonwood Creek), near main peak of White Mts., 9500 ft., Mono Co., California.

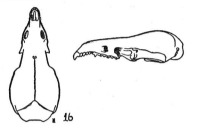

Fig. 16. *Sorex tenellus*, Rainbow Falls, 8200 ft., Kyle Canyon, Clark Co., Nevada, No. 86842 M.V.Z., ♂, X 2.

External measurements: 85–103; 36–42; 9–12.5. Condylobasal length of skull, 15.1–15.2. Upper parts drab, underparts paler; tail bicolored; skull resembling that of *S. ornatus* but smaller with relatively shorter palate and narrower braincase.

MARGINAL RECORDS.—Nevada: ¼ mi. N Wichman, 5500 ft.; Rainier Mesa (Jorgensen and Hayward, 1963:582); Kyle Canyon, 8000 ft. California: type locality; Pipers [= Cottonwood] Creek, 9500 ft.

Sorex trigonirostris Jackson
Ashland Shrew

1922. *Sorex trigonirostris* Jackson, Jour. Washington Acad. Sci., 12:264, June 4, type from Ashland, 1975 ft., Jackson Co., Oregon.

External measurements: 95, 34, 12. Condylobasal length of skull, 15.6. Color as in *S. ornatus;* skull with short angular rostrum; mastoidal region relatively angular and prominent from dorsal aspect; teeth as in *S. ornatus.*

MARGINAL RECORDS.—Oregon: W slope Grizzly Mts.; type locality.

Sorex nanus Merriam
Dwarf Shrew

1895. *Sorex tenellus nanus* Merriam, N. Amer. Fauna, 10:81, December 31, type from Estes Park, Larimer Co., Colorado.
1928. *Sorex nanus*, Jackson, N. Amer. Fauna, 51:174, July 24.

External measurements: 91–105; 39–42; 10–10.5. Condylobasal length of skull, 14.1–14.5. Upper parts brownish, underparts grayish; tail indistinctly bicolored. Skull resembles that of *S. tenellus* but smaller and relatively slightly narrower.

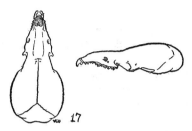

Fig. 17. *Sorex nanus*, North Fork Camp Ground, sec. 28, T. 16 N, R. 78 W, Albany Co., Wyoming, No. 184 A.B. Mickey, ♀, X 2.

MARGINAL RECORDS.—Montana (Thompson, 1977:249): Pratt Canyon, 1400 m., West Butte, Sweetgrass Hills; north face Mount Baldy [= Baldy Mtn.], Bear Paw Mts. South Dakota (Martin, 1971:835): 3½ mi. E Boxelder; 2¼ mi. N Fairburn; below Hat Creek bridge on Highway 71, *ca.* 15 mi. S Hot Springs. Wyoming: 3 mi. NW Centennial, 8480 ft. and 10,600 ft. (Brown, 1967:620). Colorado: type locality; Silver Plume; near Colorado Springs, 7000–8000 ft.; West Cliff, 8300 ft. New Mexico: 13 mi. NW Las Vegas; 2 mi. E Ruidoso (Findley and Poorbaugh, 1957:513). Arizona: near Hannagan Meadows, White Mts. (Bradshaw, 1961:96); Inner Basin, 10 mi. N Flagstaff (Marshall and Weisenberger, 1971:132); N rim Grand Canyon, 9 mi. E Swamp Point, within Grand Canyon National Park, near boundary of Kaibab National Forest. Utah: Gooseberry Ranger Station, Elk Ridge, 8560 ft. Wyoming: South Cascade Canyon, 10,050 ft., Grand Teton National Park. Montana: *Snowy Peak, Big Snowy Mts., 7800 ft., Golden Valley Co.* (Hoffmann, *et al.*, 1969:581); Yogo Peak, 8700 ft., Little Belt Mts., Judith Basin Co. (Hoffmann and Taber, 1960:232).

Sorex juncensis Nelson and Goldman
Tule Shrew

1909. *Sorex californicus juncensis* Nelson and Goldman, Proc. Biol. Soc. Washington, 22:27, March 10, type from Socorro, 15 mi. S San Quintín, Baja California. Known only from type locality.
1928. *Sorex juncensis*, Jackson, N. Amer. Fauna, 51:172, July 24.

External measurements: 101, 41, 12.5. Condylobasal length of skull, 15.6. Color as in *S. ornatus;* skull resembles that of *S. ornatus* but relatively higher and narrower.

Sorex sinuosus Grinnell
Suisun Shrew

1913. *Sorex sinuosus* Grinnell, Univ. California Publ. Zool., 10:187, March 20, type from Grizzly Island, near Suisun, Solano Co., California. Known only from Grizzly Island and adjacent tidal marshes (see Rudd, Syst. Zool., 4:23, March, 1955).

External measurements: 99, 37, 12. Condylobasal length of skull, 16.4. Almost black both dorsally and ventrally; skull and teeth resemble those of *S. ornatus.* Rudd's (1955:21–34) finding of crossbreeding between *S. sinuosus* and *S. ornatus californicus* suggests that *S. sinuosus* should be arranged as a subspecies of *S. ornatus.*

Sorex veraepacis
Verapaz Shrew

External measurements: 117–128; 45–58; 14.0–16.0. Condylobasal length of skull, 18.0–19.9. Dorsum dark brown, venter but little if any paler; tail unicolored. Cranium relatively deep, broad, and angular; relatively broad interorbitally; 3rd unicuspid smaller than 4th; unicuspids with well-defined internal ridges running from apices to cingula, these ridges more or less pigmented in unworn teeth.

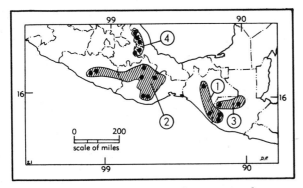

Map. 21. *Sorex veraepacis* and *Sorex macrodon.*

1. *S. v. chiapensis* 3. *S. v. veraepacis*
2. *S. v. mutabilis* 4. *S. macrodon*

Sorex veraepacis chiapensis Jackson

1925. *Sorex veraepacis chiapensis* Jackson, Proc. Biol. Soc. Washington, 38:129, November 13, type from San Cristóbal, 9500 ft., Chiapas.

MARGINAL RECORDS.—Chiapas: type locality. Guatemala: Calel; Volcano Santa Maria, Quezaltenango. Chiapas: Pinabete.

Sorex veraepacis mutabilis Merriam

1895. *Sorex saussurei caudatus* Merriam, N. Amer. Fauna, 10:84, December 31, type from Reyes (near Cuicatlán),

10,200 ft., Oaxaca. Not *S. caudatus* Hodgson, 1849 (*nomen nudum*) [= *S. caudatus* Horsfield, 1851, from Sikim and Darjeeling, India].

1898. *Sorex saussurei mutabilis* Merriam, Science (n.s.), 8:782, December 2, a renaming of *S. saussurei caudatus* Merriam.

1925. *Sorex veraepacis mutabilis,* Jackson, Proc. Biol. Soc. Washington, 38:130, November 13.

MARGINAL RECORDS.—Oaxaca: type locality; Tontontepec [= Totontepec]; *Mt. Zempoaltepec* (Goodwin, 1969:34); Ozolotepec; Río Molino, 2250 m., 3 km. SW San Miguel (Schaldach, 1966:288); Cerro San Felipe (Goodwin, 1969:34). Guerrero: near Puerto Chico (Musser, 1964:2,5); Omilteme.

Sorex veraepacis veraepacis Alston

1877. *Sorex verae-pacis* Alston, Proc. Zool. Soc. London, p. 445, October, type from Cobán, Guatemala.

1877. *C*[*orsira*]. *teculyas* [sic] Alston, Proc. Zool. Soc. London, p. 445, October (in synonymy).

MARGINAL RECORDS.—Guatemala: type locality; Todos Santos.

Sorex macrodon Merriam
Large-toothed Shrew

1895. *Sorex macrodon* Merriam, N. Amer. Fauna, 10:82, December 31, type from Orizaba, 400 ft., Veracruz.

External measurements: 128–130; 50–52; 15.0–15.5. Condylobasal length of skull, 19.2–19.6. Upper parts clove brown or sepia, underparts little if any paler; tail essentially unicolored. Skull large and massive; rostrum relatively broad anteriorly; anterior nares broad; borders of premaxillae relatively thick; teeth relatively large; 3rd unicuspid smaller than 4th; unicuspids with well-developed internal ridges.

MARGINAL RECORDS.—Puebla: 19° 52' N, 97° 20' W, $12\frac{1}{10}$ km. by road NE Tezuitlán (Heaney and Birney, 1977:543). Veracruz (Hall and Dalquest, 1963:204): Las Vigas, 8500 ft.; Xico [= Jico]; type locality; 3 km. W Acultzingo, 7000 ft.

Sorex palustris
Water Shrew

External measurements: 144–158; 63–78; 18–21. Condylobasal length of skull, 18.8–21.5. Black dorsally, white, gray, or brownish ventrally; tail markedly bicolored; hind feet relatively large and with fringe of stiff hairs. Anterior part of rostrum comparatively short, little decurved; anterior part of premaxillae approx. as deep as middle part; depth of rostrum measured at 3rd unicuspid equal to approx. half the distance between anterior border of infraorbital foramen and posterior border of I1; 3rd unicuspid smaller than 4th; protocone of M1 and M2 usually without posterior cusplike lobe.

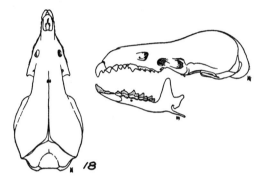

Fig. 18. *Sorex palustris navigator*, Cottonwood Creek, Mt. Grant, 7400 ft., Mineral Co., Nevada, No. 63521 M.V.Z., ♂, X 2.

Water shrews occur along the borders of ponds and streams in meadows, marshes, or woods.

Sorex palustris albibarbis (Cope)

1862. *Neosorex albibarbis* Cope, Proc. Acad. Nat. Sci. Philadelphia, 14:188, type from Profile Lake, Franconia Mts., Grafton Co., New Hampshire. [Green, A contribution to the mammalogy . . . of Pennsylvania, p. 11, March 31 1930, believes that *N. albibarbis* is preoccupied by *S. fimbripes* Bachman, 1837, which we have placed as a synonym of *Sorex cinereus cinereus*.]
1903. *Sorex palustris albibarbis*, Rhoads, The mammals of Pennsylvania and New Jersey, p. 191.

MARGINAL RECORDS.—Quebec: Mistassini Post (Cameron and Morris, 1951:124); St. Félicien (Harper, 1961:34); St. Rose, Temiscouata Dist. Maine: Lincoln; Mt. Desert Island, 20 mi. NE Penobscot Bay; Brunswick. Rhode Island: *2 mi. SW West Greenwich Center* (Layne and Shoop, 1971:215). Connecticut: *South Woodstock;* Nehantic State Forest (Wetzel and Shelar, 1964:311). New York: Beaver Dam Brook. Pennsylvania: Bushkill Creek, 7 mi. E Cresco; Mifflin County; near Carter Camp P.O., NE slope Mt. Broadhead. New York: 1 mi. S Red House Lake, Allegheny State Park; 9 mi. W Lowville (Connor, 1966:22); Tupper Lake. Ontario: North Bay.

Sorex palustris brooksi Anderson

1934. *Sorex palustris brooksi* Anderson, Canadian Field-Nat., 48:134, November 1, type from Black Creek, 150 ft., Comox District, E coast Vancouver Island, British Columbia.

MARGINAL RECORDS.—British Columbia (Vancouver Island): Quatsino (Cowan and Guiguet, 1965:54); Black Creek at Miracle Beach (*ibid.*); *type locality; Courtenay* (Cowan and Guiguet, 1965:54); *Comox;* near Victoria.

Sorex palustris gloveralleni Jackson

1915. *Neosorex palustris acadicus* G. M. Allen, Proc. Biol. Soc. Washington, 28:15, February 12, type from Digby, Digby Co., Nova Scotia. Not *Sorex acadicus* Gilpin, 1867 [= *Sorex cinereus* Kerr, 1792].
1926. *Sorex palustris gloveralleni* Jackson, Jour. Mamm., 7:57, February 15, a renaming of *Neosorex palustris acadicus* G. M. Allen, 1915.

MARGINAL RECORDS.—Quebec: Point St. Charles and Seal River; Mt. Albert. Nova Scotia: Ingonish Centre (Cameron, 1959:16). New Brunswick: Charlotte County. Quebec: Godbout.

Sorex palustris hydrobadistes Jackson

1926. *Sorex palustris hydrobadistes* Jackson, Jour. Mamm., 7:57, February 15, type from Withee, Clark Co., Wisconsin.

MARGINAL RECORDS.—Wisconsin: Basswood Lake, 10 mi. SE Iron River (Jackson, 1961:39). Ontario: *E end Lake Superior.* Michigan: Otsego County. Wisconsin: Marinette County; type locality. Minnesota: Elk River. South Dakota: Fort Sisseton. Fort Wadsworth in South Dakota not found.

Sorex palustris labradorensis Burt

1938. *Sorex palustris labradorensis* Burt, Occas. Pap. Mus. Zool., Univ. Michigan, 383:1, August 27, type from Red Bay, Strait of Belle Isle, Labrador.

MARGINAL RECORDS.—Labrador: Cartwright, on coast S Hamilton Inlet, lat. 53° 48′ N, long. 56° 59′ W; type locality; Astray Lake (Harper, 1961:33, 145 [Map 4]).

Sorex palustris navigator (Baird)

1858. *Neosorex navigator* Baird, Mammals, *in* Repts. Expl. Surv. . . . , 8(1):11, July 14, type from near head Yakima River, Cascade Mts., Washington.
1895. *Sorex (Neosorex) palustris navigator*, Merriam, N. Amer. Fauna, 10:92, December 31.

MARGINAL RECORDS.—Alaska: 20 mi. NE Anchorage. Yukon: Nisutlin River. Northwest Territories: upper end Glacier Lake, 2500 ft. (Youngman, 1968:73). British Columbia: Tupper Creek (Cowan and Guiguet, 1965:56). Alberta: Smoky Valley, 50 mi. N Jasper House; Brazeau Valley; Banff. Montana: St. Marys

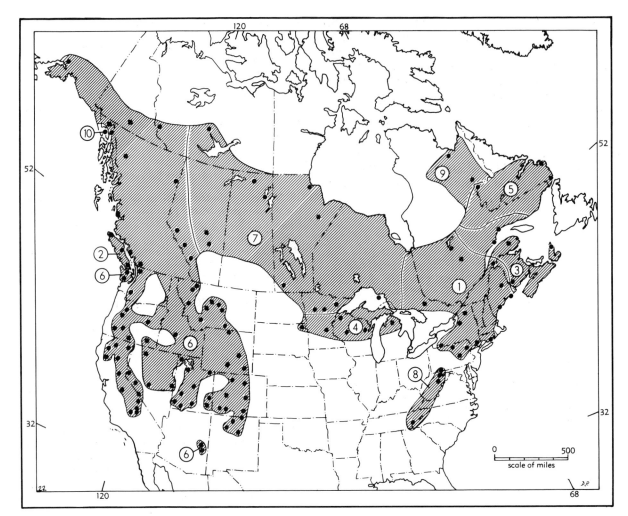

Map 22.   *Sorex palustris* and *Sorex alaskanus.*

1. *S. p. albibarbis*	4. *S. p. hydrobadistes*	8. *S. p. punctulatus*
2. *S. p. brooksi*	5. *S. p. labradorensis*	9. *S. p. turneri*
3. *S. p. gloveralleni*	6. *S. p. navigator*	10. *S. alaskanus*
	7. *S. p. palustris*	

Lake; Paola; Pattee Canyon, *ca.* 4000 ft. (Kinsella, 1967:475); Ward Peak, 6000 ft.; Sheep Creek, 16 mi. N White Sulphur Springs; Highwood Mts.; Moccasin Mts., 5 mi. NW Hilger; Tyler, 10 mi. W North Fork at Willow Creek; Pryor Mts. Wyoming (Long, 1965a:526): Eatons Ranch, Wolf; 21½ mi. S, 24½ mi. W Douglas; *Laramie Peak;* 5 mi. W Horse Creek P.O. Colorado: Boulder; Lake Moraine; 5 mi. S, 1 mi. W Cucharas Camps (Armstrong, 1972: 50). New Mexico: Costilla Pass, 9000 ft.; Willis; 10 mi. N Jemez Springs (Findley, *et al.,* 1975:17); *1 mi. N, 6 mi. E Cuba (ibid.);* 6 mi. W Hopewell, 9900 ft. Colorado: Hermit; Rico. Utah: North Creek, 7 mi. W Monticello, Abajo Mts., 8000 ft.; Warner Ranger Station, 9700 ft., La Sal Mts. Colorado: Crested Butte; Middle Park. Wyoming (Long, 1965a: 526): 2 mi. S Bridgers Peak; *8 mi. N, 20 mi. E Savery.* Utah: Carter Creek, 9000 ft.; Mammoth Ranger Sta-

tion, Manti National Forest; Kaiparowits Plateau; Springdale, 3850 ft.; Pine Valley; Beaver; South Willow Creek Canyon, 7000 ft. (Egoscue, 1965:685); Pine Canyon, 6600 ft., Raft River Mts. Nevada: Baker Creek, 11,100 ft.; 1–2 mi. E Jefferson, 7600–8000 ft.; *Jett Canyon, Toyabe Mts.;* head Big Creek, 8000 ft., Pine Forest Mts. Oregon: Steen Mts. Idaho: Sawtooth City. Oregon: Strawberry Mts.; Anna Creek, 6000 ft., Mt. Mazama; near Drews Creek. California: Parker Creek, Warner Mts. Nevada: 2 mi. W Mt. Rose summit; Cottonwood Creek, 7400 ft., Mt. Grant; Arlemont, 4900 ft. California: White Mts.; Lone Pine; Whitney Meadows, Mt. Whitney; Halstead Meadows, Sequoia National Park; Merced Grove, 5400 ft.; Blue Canyon, 4700–5000 ft.; Mill Creek, 5000 ft., Mt. Lassen; South Yolla Bolly Mtn., 6000 ft.; Canyon Creek; Upper Ash Creek, Mt. Shasta. Oregon: Prospect; McKenzie Bridge; Permilia Lake, Mt. Jefferson. Washington: Mt. St. Helens, 5500 ft.; Mt. Baker; head N fork, Quinault River, 4000 ft.; Elwha. British Columbia: Chilliwack Valley; Powell River; Neckis River (Cowan and Guiguet, 1965:56); Telegraph Creek. Alaska: Haines. British Columbia: Stonehouse Creek (Cowan and Guiguet, 1965:56). Southern segment of range: Arizona: White River, White Mts.; Prieto Plateau.

Sorex palustris palustris Richardson

1828. *Sorex palustris* Richardson, Zool. Jour., 3(12, January–Apr. 1): 517, April, type from "marshy places, from Hudson's Bay to the Rocky Mountains."

MARGINAL RECORDS.—Northwest Territories: Grandin River. Saskatchewan: Stony Rapids (Beck, 1964:166, as *S. palustris* only); Wollaston Lake (Beck, 1958:9). Manitoba: Churchill; Hill River, near Swampy Lake. Ontario: Michipicoten Island. Minnesota: Tower; Itasca County; Itasca Park. Manitoba: Aweme. Alberta: Bashaw (H. C. Smith, 1972:53, as *S. palustris* only); Edmonton.

Sorex palustris punctulatus Hooper

1942. *Sorex palustris punctulatus* Hooper, Occas. Pap. Mus. Zool., Univ. Michigan, 463:1, September 15, type from 6 mi. NW Durbin, Shavers Fork, Cheat River, 3600 ft., Randolph Co., West Virginia.

MARGINAL RECORDS.—Pennsylvania: Tumbling Cove Run, Negro Mts. (Doutt, *et al.*, 1973:34, as *S. p. albibarbis*). West Virginia: 1 mi SSE Cranesville, 2600 ft.; Blister Run, 4 mi. NNW Durbin, 3650 ft. Tennessee: W Prong, Little Pigeon River, Great Smoky Mts. National Park. West Virginia: *type locality.*

Sorex palustris turneri Johnson

1951. *Sorex palustris turneri* Johnson, Proc. Biol. Soc. Washington, 64:110, August 24, type from Fort Chimo (on eastern bank Koksoak River, lat. 58° 8′ N, long. 68° 15′ W), Ungava District, Quebec.

MARGINAL RECORDS.—Quebec: type locality; Lac Le Fer (Harper, 1961:33).

Sorex alaskanus Merriam
Glacier Bay Water Shrew

1900. *Sorex navigator alaskanus* Merriam, Proc. Washington Acad. Sci., 2:18, March 14, type from Point Gustavus, Glacier Bay, Alaska. Known only from type locality.
1926. *Sorex alaskanus*, Jackson, Jour. Mamm., 7:58, February 15.

External measurements: 145–160; 65–72; 18.5–19. Condylobasal length of skull, 18.4–19.2. Color as in *S. palustris*. Skull short, heavy, and angular; rostrum and mesopterygoid space relatively short; sagittal and lambdoidal crests much

developed; a distinct inframaxillary ridge extending above base of unicuspids. *S. alaskanus* may be only a subspecies of *S. palustris.*

Sorex bendirii
Pacific Water Shrew

External measurements: 147–174; 61–80; 18.5–21. Condylobasal length of skull, 20.8–23.8. The largest North American species of *Sorex;* upper parts dark brown, underparts scarcely paler except in subspecies *albiventer* in which they are whitish; tail unicolored; hind feet weakly fimbriate; anterior part of rostrum relatively long, distinctly curved ventrally; anterior end of premaxilla decidedly shallower than middle part; depth of rostrum measured at 3rd unicuspid less than half the distance between anterior border of infraorbital foramen and posterior border of I1; anteroposterior diameter of upper unicuspids relatively more than in subgenus *Neosorex;* protocone of M1 and M2 usually with distinct posterior cusplike lobe.

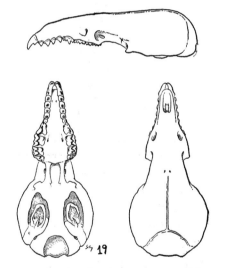

Fig. 19. *Sorex bendirii bendirii,* Russian Gulch State Park, Mendocino Co., 40 ft., California, No. 95649 M.V.Z., ♀, X 2.

Sorex bendirii albiventer Merriam

1895. *Sorex (Atophyrax) bendirii albiventer* Merriam, N. Amer. Fauna, 10:97, December 31, type from Lake Cushman, Olympic Mts., Mason Co., Washington.

MARGINAL RECORDS.—Washington: Neah Bay; Canyon Creek; Duckabush; Harstine Island; Shelton; Quinault Lake; Lapush.

Sorex bendirii bendirii (Merriam)

1884. *Atophyrax bendirii* Merriam, Trans. Linnaean Soc. New York, 2:217, August 28, type from about 1 mi.

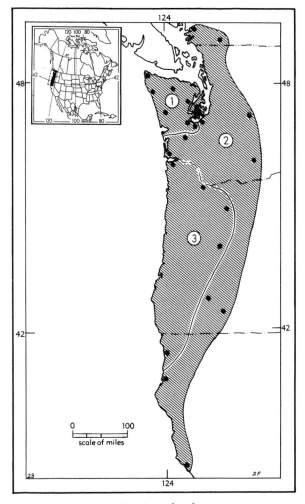

Map 23.   *Sorex bendirii.*

Guide to subspecies	2. *S. b. bendirii*
1. *S. b. albiventer*	3. *S. b. palmeri*

from Williamson River, 18 mi. SE Fort Klamath, Klamath Co., Oregon.

1890. *Sorex bendirii,* Dobson, A monograph of the Insectivora . . . , pt. 3, fasc. 1, Pl. 23, Fig. 17 and explanation.

MARGINAL RECORDS.—British Columbia: Port Moody; Chilliwack. Washington: Easton; Signal Peak, 4000 ft. Oregon: type locality. California: Gualala; Carson Camp, Mad River, Humboldt Bay. Oregon: Prospect. Washington: Ilwaco; Oakville; Steilacoom.

Sorex bendirii palmeri Merriam

1895. *Sorex (Atophyrax) bendirii palmeri* Merriam, N. Amer. Fauna, 10:97, December 31, type from Astoria, Clatsop Co., Oregon.

MARGINAL RECORDS.—Oregon: type locality; Portland; Camas Prairie, E base Cascade Mts.; McKenzie Bridge. California: Requa, thence up coast to point of beginning.

Subgenus Sorex Linnaeus

1758. *Sorex* Linnaeus, Syst. nat., ed. 10, 1:53. Type, *Sorex araneus* Linnaeus.

Postmandibular foramen usually present and well developed; each upper unicuspid lacking a pigmented ridge from the apex to the cingulum.

Sorex fumeus
Smoky Shrew

External measurements: 111–127; 45–52; 13–15. Condylobasal length of skull, 17.8–19.0. Pelage grayish or blackish in winter, brownish in summer, underparts slightly paler than upper parts but general effect unicolored, tail usually bicolored. Unicuspids with ridge extending medially from apex approx. halfway to internal cingulum; this ridge pigmented near apex of tooth; unicuspids broader than long; 3rd unicuspid larger than 4th.

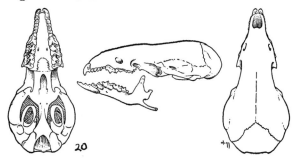

Fig. 20.   *Sorex fumeus fumeus,* 1½ mi. W Camden on Gauley, Webster Co., West Virginia, No. 39056 K.U., ♂, X 2.

Sorex fumeus fumeus Miller

1895. *Sorex fumeus* Miller, N. Amer. Fauna, 10:50, December 31, type from Peterboro, Madison Co., New York.

MARGINAL RECORDS.—Ontario: Fraserdale. Quebec: Lake Edward; St. Joachim. New York: Tupper Lake. Vermont: Mt. Mansfield. New Hampshire: Mossy Brook, Mt. Monadnock; Intervale; Ossipee. Rhode Island: 2 mi. SW West Greenwich Center (Layne and Shoop, 1971:215). New Jersey: Greenwood Lake. Pennsylvania: Chester County. Virginia: Paris; Madison County. North Carolina: Magnetic City, Roan Mtn. South Carolina: *Jones Gap;* Jocassee. Georgia: Brasstown Bald, 4700 ft. Kentucky: Mammoth Cave. Ohio: North Chagrin Metropolitan Park, Cuyahoga Co. Ontario: Elgin County; Pancake Bay; Schreiber; Thunder Bay. Also recorded from Racine, Wisconsin.

Sorex fumeus umbrosus Jackson

1917. *Sorex fumeus umbrosus* Jackson, Proc. Biol. Soc. Washington, 30:149, July 27, type from James River, Antigonish Co., Nova Scotia.

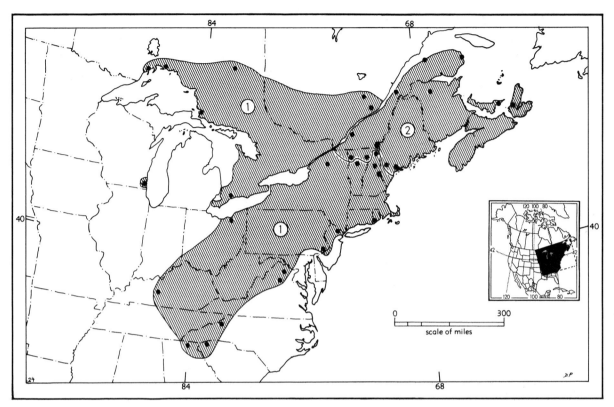

Map 24. *Sorex fumeus.*

1. *S. f. fumeus*   2. *S. f. umbrosus*

MARGINAL RECORDS.—Quebec (Manville, 1961:108): Ste. Félicité; Cap des Rosiers. New Brunswick: *ca.* 3½ mi. SW summit Mt. Carleton (Peterson and Symansky, 1963:278, as *S. fumeus* only). Prince Edward Island: Fortune (Cameron, 1959:43). Nova Scotia: West Mabou (Cameron, 1959: 16). Maine: Brunswick; "Lynchtown," Twp. 5, R. 4, Oxford Co. New Hampshire: First Connecticut Lake, 7 mi. N Pittsburg. Vermont: Brighton; St. Albans. Quebec: Mont St. Hilaire (Wrigley, 1969:204); Rivière-du-Loup.

Sorex arcticus
Arctic Shrew

External measurements: 101–117; 30–45; 12.5–15. Condylobasal length of skull, 17.8–19.1. Tricolored in most pelages, dorsal region darkest (grayish to brownish), distinctly set off from brownish (or tan) flanks, and still paler venter; *S. a. tundrensis* bicolored in some pelages, color of sides being scarcely differentiated from that of belly; 3rd unicuspid larger than 4th; ridges extending from apices of unicuspids medially toward cingula but incomplete, weakly pigmented, and not ending in internal cusplets.

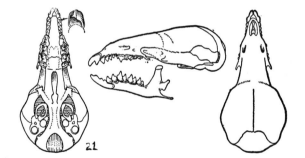

Fig. 21. *Sorex arcticus tundrensis,* 1½ mi. W and ¾ mi. N Umiat, 69° 22′ 18″, 152° 08′ 10″, 370 ft., Alaska, No. 43212 K.U., ♂, X 2.

This shrew is often taken in bogs in the southern part of its range and occurs on the tundra in the Far North. According to Hoffmann (1971:194, 199) *S. arcticus* occurs also in eastern Siberia. Karyotypes reported by Meylan and Hausser (1973:143–158) suggest that study of specimens from between the known geographic ranges of *S. tundrensis* and *S. arcticus* would clarify the subspecific versus specific status of those two taxa and some related American taxa.

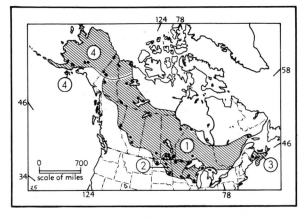

Map 25. *Sorex arcticus.*

1. *S. a. arcticus*   3. *S. a. maritimensis*
2. *S. a. laricorum*   4. *S. a. tundrensis*

Sorex arcticus arcticus Kerr

1792. *Sorex arcticus* Kerr, The animal kingdom . . . , p. 206. Type locality, settlement on Severn River, Hudson Bay, now known as Fort Severn, mouth of Severn River, Ontario.
1837. *Sorex richardsonii* Bachman, Jour. Acad. Nat. Sci. Philadelphia, 7:383. Type locality, probably plains of Saskatchewan.
1877. *Sorex sphagnicola* Coues, Bull. U.S. Geol. and Geog. Surv. Territories, 3:650, May 15, type from Fort Liard, Mackenzie.
1892. *Sorex belli* Merriam, Proc. Biol. Soc. Washington, 7:25, April 13. *Nomen nudum*, see Jackson, N. Amer. Fauna, 51:69, July 24, 1928, and Youngman, Nat. Mus. Canada, Natural Sci., Zool., 7:1, 1972 [= February 5, 1973].

MARGINAL RECORDS.—Northwest Territories: Fort Norman; Fort Rae; Great Slave Lake. Manitoba: Shamattawa River, tributary Hayes River. Ontario: type locality. Quebec: Saguenay County. Ontario: Macdiarmid, Lake Nipigon. Manitoba: Whitemouth River, S of Reynolds; *Red River Settlement;* S end Lake Manitoba. North Dakota: 6 mi. N Lostwood. Saskatchewan: Indian Head. Alberta: Blindmans and Red Deer rivers; Island Lake, 15 mi. W Lake St. Ann. British Columbia: Tupper Creek, Peace River District; *Charlie Lake, Peace River District* (Cowan and Guiguet, 1965:43). Northwest Territories: Fort Liard. Yukon: vic. Yukon Crossing (Youngman, 1975:44).

Sorex arcticus laricorum Jackson

1858. *Sorex pachyurus* Baird, Mammals, *in* Repts. Expl. Surv. . . . , 8(1):20, July 14, type from Pembina, North Dakota [not Minnesota, as stated by Baird]. Not [*S.*] *Pachyurus* Küster, 1835, type from Cagliari, Sardinia, *qui Pachyura etrusca* Savi, 1822, type from Pisa, Italy.
1925. *Sorex arcticus laricorum* Jackson, Proc. Biol. Soc. Washington, 38:127, November 13, type from Elk River, Sherburne Co., Minnesota.

MARGINAL RECORDS.—Manitoba: Carberry; Emerson. Minnesota: Lake of the Woods County;

Grand Portage (Timm, 1975:10). Michigan: Chippewa County. Wisconsin: SE McFarland, near Lake Kegonsa (Nelson, 1934:252); Withee. Minnesota: 1 mi. N, 2 mi. W Victoria (Heaney and Birney, 1975:29). South Dakota: Fort Sisseton. North Dakota: Fort Totten. Manitoba: Max Lake, Turtle Mts.; *Aweme.*

Sorex arcticus maritimensis R. W. Smith

1939. *Sorex arcticus maritimensis* R. W. Smith, Jour. Mamm., 20:244, May 15, type from Wolfville, Kings Co., Nova Scotia.

MARGINAL RECORDS.—Nova Scotia and west in New Brunswick to Maugerville and Hampton.

Sorex arcticus tundrensis Merriam

1900. *Sorex tundrensis* Merriam, Proc. Washington Acad. Sci., 2:16, March 14, type from St. Michael, Alaska.
1956. *Sorex arcticus tundrensis*, Bee and Hall, Univ. Kansas Mus. Nat. Hist., Miscl. Publ., 8:22, March 10.

MARGINAL RECORDS.—Mackenzie: Anderson River, near Liverpool Bay; Peel River, Mackenzie Delta. Yukon (Youngman, 1975:48): 20 mi. S Chapman Lake; Fortymile. Alaska: *near Eagle;* Savage River; Mt. McKinley; Kodiak Island; Nushagak, thence north and east along coast to Yukon (Youngman, 1964:2): Firth River, 15 mi. S mouth Joe Creek, 1560 ft.
Youngman (1975:45, 48) lists three characters in which *S. tundrensis* differs from *S. arcticus* and judges the two taxa to be species instead of subspecies. Specimens from Yukon between Yukon Crossing (*arcticus*) and Chapman Lake (*tundrensis*) are needed to learn whether the two intergrade and therefore are subspecies.

Sorex gaspensis Anthony and Goodwin
Gaspé Shrew

1924. *Sorex gaspensis* Anthony and Goodwin, Amer. Mus. Novit., 109:1, March 10, type from Mt. Albert, 2000 ft., Gaspé Peninsula, Quebec.

External measurements: 95–115; 47–55; 12.0–12.5. Condylobasal length of skull, 15.8–16.3. Mouse-gray or neutral gray, underparts slightly paler; tail bicolored; skull similar to but smaller than that of *S. dispar.*

Most individuals of this species have been captured in vic. streams in coniferous forests. Goodwin (1929:242) suggests that the habits of the Gaspé Shrew are similar to those of *Sorex palustris.*

MARGINAL RECORDS.—Quebec: type locality; Cascapedia Valley at Red Camp, 8 mi. inland. New Brunswick: *ca.* 3½ mi. SW summit Mt. Carleton (Peterson and Symansky, 1963:278). Quebec: Big Berry Mtn., 35 mi. inland.—See addenda.

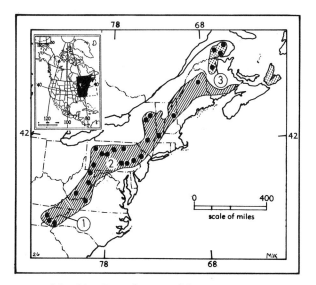

Map 26. *Sorex dispar* and *Sorex gaspensis*.

   1. *S. d. blitchi*   2. *S. d. dispar*
          3. *S. gaspensis*

Sorex dispar
Long-tailed Shrew

x ½

External measurements: 103–139; 48–66; 13–16. Condylobasal length of skull, 17.3–18.5. Mouse-gray dorsally, nearly the same ventrally; tail ordinarily unicolored; orbital region of skull relatively elongate and depressed; unicuspids relatively narrow, 1st and 2nd subequal, 3rd and 4th smaller than 1st and 2nd, 3rd equal to or slightly smaller than 4th, 5th relatively large but smaller than 3rd.

Sorex dispar blitchi Schwartz

1956. *Sorex dispar blitchi* Schwartz, Jour. Elisha Mitchell Sci. Soc., 72(1):26, May 24, type from 2 mi. NE Wagon Road Gap, 4525 ft., Haywood Co., North Carolina.

MARGINAL RECORDS.—Tennessee: Walker Prong, Great Smoky National Park, 4400–4500 ft.; *between Highway 71 and West Prong, Little Pigeon*

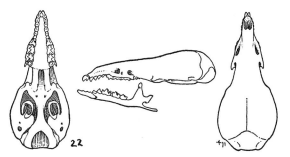

Fig. 22. *Sorex dispar dispar*, ½ mi. SE Dellslow, Monongalia Co., West Virginia, No. 39050 K.U., ♂, X 2.

River, 3400 ft. North Carolina: type locality; *talus above Highway 107, 4400 ft.;* Clingmans Dome, 6400–6642 ft.

Sorex dispar dispar Batchelder

1896. *Sorex macrurus* Batchelder. Proc. Biol. Soc. Washington, 10:133, December 8, type from Beedes (⅜ mi. S, ½ mi. E Saint Huberts, lat. 44° 09′, long. 73° 46′, see Martin, 1966:131), Essex Co., New York. Not *S. macrourus* Lehmann, 1822.

1911. *Sorex dispar* Batchelder, Proc. Biol. Soc. Washington, 24:97, May 15, a renaming of *S. macrurus* Batchelder.

MARGINAL RECORDS.—Maine: SW edge Lower South Branch Pond. New Hampshire: Tuckermans Ravine, Mt. Washington. Massachusetts: Mt. Graylock. New Jersey: Stillwater Twp., Sussex Co. Pennsylvania: Northampton County; Berks County; Perry County. West Virginia: Spruce. Virginia: 4⅘ mi. NNE Mt. Lake. West Virginia: Winding Gulf, 4 mi. SW Pemberton, 2000 ft.; ½ mi. SE Dellslow. Pennsylvania: 2 mi. SSE Rector; Jefferson County; Venango County; Clearfield County; Lycoming County; Lake Leigh. New York: Hunter Mtn.; Mt. Marcy; type locality.—See addenda.

Sorex trowbridgii
Trowbridge's Shrew

External measurements: 110–132; 48–62; 12–15. Condylobasal length of skull, 16.4–18.8. Color dark gray or blackish, sometimes with brownish hue, above and below; tail sharply

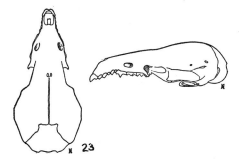

Fig. 23. *Sorex trowbridgii mariposae*, ½ mi. NE Dutch Flat, Placer Co., California, No. 88148 M.V.Z., ♀, X 2.

bicolored; 3rd unicuspid smaller than 4th; internal ridge of unicuspids weakly pigmented, never pigmented to cingulum, not ending in internal cusplet, separated from cingulum by anteroposterior groove.

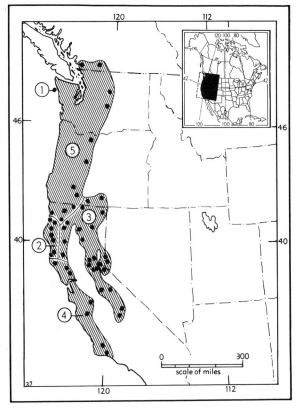

Map 27. *Sorex trowbridgii.*

Guide to subspecies
1. *S. t. destructioni*
2. *S. t. humboldtensis*
3. *S. t. mariposae*
4. *S. t. montereyensis*
5. *S. t. trowbridgii*

Sorex trowbridgii destructioni Scheffer and Dalquest

1942. *Sorex trowbridgii destructioni* Scheffer and Dalquest, Jour. Mamm., 23:334, August 13, type from Destruction Island, Jefferson Co., Washington. Known only from Destruction Island.

Sorex trowbridgii humboldtensis Jackson

1922. *Sorex trowbridgii humboldtensis* Jackson, Jour. Washington Acad. Sci., 12:264, June 4, type from Carsons Camp, Mad River, Humboldt Bay, Humboldt Co., California.

MARGINAL RECORDS.—California: Orick; Hoopa Valley; type locality; Dyerville; Briceland; Sherwood; Mendocino, thence up coast to point of beginning.

Sorex trowbridgii mariposae Grinnell

1913. *Sorex montereyensis mariposae* Grinnell, Univ. California Publ. Zool., 10:189, March 20, type from Yosemite Valley, 4000 ft., Mariposa Co., California.
1923. *Sorex trowbridgii mariposae* Grinnell, Univ. California Publ. Zool., 21:314, January 27.

MARGINAL RECORDS.—Oregon: Lakeview. California: Parker Creek, Warner Mts.; Hayden Hill. Nevada: Verdi; 3 mi. S Mt. Rose, 8500 ft. California: Myers; Mt. Tallac; Cisco, 6000 ft.; Fyffe, 3600 ft.; type locality; Giant Forest, Sequoia National Park; Kaweah River, E fork; Sweetwater Creek, 3800 ft., 2 mi. E Feliciana Mtn.; Placerville; middle fork American River; Dutch Flat, 3400 ft.; S base Mt. Lassen, Mill Creek, 5000 ft.; Carberrys Ranch; South Yolla Bolly Mtn.; 8 mi. E Hearst; Canyon Creek, 4600 ft.; Jackson Lake; Beswick. Oregon: Swan Lake Valley.

Sorex trowbridgii montereyensis Merriam

1895. *Sorex montereyensis* Merriam, N. Amer. Fauna, 10:79, December 31, type from Monterey, Monterey Co., California.
1922. *Sorex t[rowbridgii]. montereyensis,* Jackson, Jour. Washington Acad. Sci., 12:264, June 4.

MARGINAL RECORDS.—California: Gualala; Mt. St. Helena; Mt. Veeder; summit Pacheco Peak; Tassajara Creek, 6 mi. below Tassajara Springs; Peachtree River, San Rafael Mts.; Santa Barbara Botanic Gardens, Mission Canyon, thence up coast to point of beginning.

Sorex trowbridgii trowbridgii Baird

1858. *Sorex trowbridgii* Baird, Mammals, *in* Repts. Expl. Surv. . . . , 8(1):13, July 14, type from Astoria, mouth Columbia River, Clatsop Co., Oregon.

MARGINAL RECORDS.—British Columbia: Hope. Washington: Stehekin; 2 mi. S Blewett Pass. Oregon: 2 mi. W Parkdale; Three Sisters; Drew; Prospect. California: Stud Horse Canyon, Siskiyou Mts.; Happy Camp, Klamath River; Forest Glen, Trinity Mts.; mouth Klamath River, thence northward along coast to British Columbia: Vancouver (Cowan and Guiguet, 1965:56).

Sorex merriami
Merriam's Shrew

External measurements: 85–107; 33–41; 11–13. Condylobasal length of skull, 15.1–16.9. Grayish above; feet and underparts whitish; 3rd unicuspid larger than 4th; unicuspid row crowded; unicuspids higher (dorsoventrally) than long (anteroposteriorly), lacking heavily pigmented internal ridge.—See addenda.

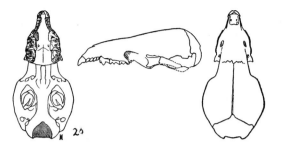

Fig. 24. *Sorex merriami leucogenys*, Chiatovich Creek, 8200 ft., Esmeralda Co., Nevada, No. 38398 M.V.Z., ♀, X 2.

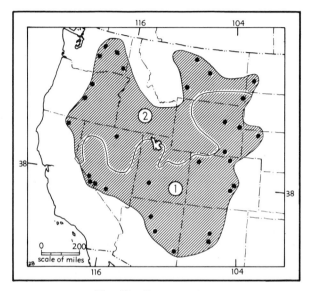

Map 28. *Sorex merriami*.

1. *S. m. leucogenys* 2. *S. m. merriami*

Most specimens of this rare shrew have come from open arid places.

Sorex merriami leucogenys Osgood

1909. *Sorex leucogenys* Osgood, Proc. Biol. Soc. Washington, 22:52, April 17, type from mouth of canyon of Beaver River, about 3 mi. E Beaver, Beaver Co., Utah.
1939. *Sorex merriami leucogenys*, Benson and Bond, Jour. Mamm., 20:348, August 14.

MARGINAL RECORDS.—Montana: 5 mi. N, 3½ mi. W of the South Dakota town of Camp Crook (Pefaur and Hoffmann, 1971:247). Wyoming (Long, 1965a:528): 42 mi. S, 13 mi. W Gillette; 10 mi. N Hat Creek P.O. Nebraska: Rush Creek, 3 mi. S Rushville (McDaniel, 1967:493). Colorado: Owl Canyon; Black Forest (Armstrong, 1972:51); Eightmile Creek, 6365 ft. (*ibid.*). New Mexico: Tree Springs; Red Canyon, ½ mi. S, 5 mi. W Manzano (Jones, 1961:399, listed as *S. merriami*). Arizona: Rose Peak; Sawmill Springs, 8 mi. SE Mormon Lake; Buggelin Tank, 5½ mi. S, 10 mi. E

Grand Canyon Village (Hoffmeister, 1971:164); *Long Jim Canyon, 3 mi. S, 4 mi. E Grand Canyon Village* (*ibid.*). Nevada: Rainier Mesa (Jorgensen and Hayward, 1963:582); ½ mi. SE Indian Spring, 7700 ft., Mt. Magruder. California: Cottonwood Creek, 9500 ft. Nevada: Chiatovich Creek, 8200 ft. Utah: type locality. Colorado: ⅖ mi. S U.S. 40, beside road to Juniper Hot Springs. Wyoming (Long, 1965a:528): 3⅗ mi. SW Laramie; *5 mi. N, 2 mi. E Laramie*.

Sorex merriami merriami Dobson

1890. *Sorex merriami* Dobson, A monograph of the Insectivora. . . , pt. 3 (fasc. 1; Pl. 23, Fig. 6), May, type from Little Bighorn River, about 1 mi. above Fort Custer, Bighorn Co., Montana.

MARGINAL RECORDS.—Montana: Eagle Creek, Madison Ranch, 6 mi. NNW Warrick, Bearpaw Mts.; 12 mi. E Winnet (Hoffmann, *et al.*, 1969:581). North Dakota: Medora. Nevada: Desert Ranch, 100 mi. NE Golconda. California: Indian Well Cave, Lava Beds National Monument. Oregon: *ca.* 22 mi. SE Bend (Gashwiler, 1976:13); 7 mi. SE Antelope. Washington: 17 mi. E Ellensburg; near East Wall, Grand Coulee, 16 mi. N Coulee City; 5 mi. SE Creston; Cloverland Grade, about 3½ mi. SW Asotin. Montana: Cayuse Hills (Hoffmann, *et al.*, 1969:581).

Sorex saussurei
Saussure's Shrew

External measurements: 104–128; 41–60; 13.5–15.0. Condylobasal length of skull, 17.4–18.5. Upper parts fuscous to clove brown, underparts paler or much the same depending upon subspecies; tail indistinctly bicolored. Skull flattened; braincase rounded laterally; mesopterygoid space relatively narrow; 3rd unicuspid usually equal to 4th in size, sometimes slightly larger or smaller; unicuspids with poorly defined unpigmented internal ridge running from apex of tooth to cingulum; cingula narrow, sloping, and indistinct.

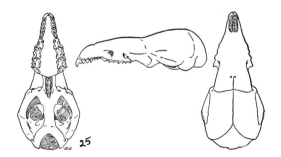

Fig. 25. *Sorex saussurei veraecrucis*, 6 km. SSE Altotonga, 9000 ft., Veracruz, No. 19093 K.U., ♀, X 2.

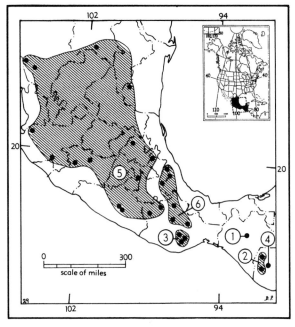

Map 29. *Sorex saussurei.*

1. *S. s. cristobalensis*   4. *S. s. salvini*
2. *S. s. godmani*   5. *S. s. saussurei*
3. *S. s. oaxacae*   6. *S. s. veraecrucis*

Sorex saussurei cristobalensis Jackson

1925. *Sorex saussurei cristobalensis* Jackson, Proc. Biol. Soc Washington, 38:129, November 13, type from San Cristóbal, 8400 ft., Chiapas. Known only from type locality.

Sorex saussurei godmani Merriam

1897. *Sorex godmani* Merriam, Proc. Biol. Soc. Washington, 11:229, July 15, type from Volcán Santa María, 9000 ft., Quezaltenango, Guatemala.
1928. *Sorex saussurei godmani,* Jackson, N. Amer. Fauna, 51:158, July 24.

MARGINAL RECORDS.—Guatemala: Todos Santos; type locality.

Sorex saussurei oaxacae Jackson

1925. *Sorex saussurei oaxacae* Jackson, Proc. Biol. Soc. Washington, 38:128, November 13, type from mountains near Ozolotepec, 10,000 ft., Oaxaca.

MARGINAL RECORDS.—Oaxaca: San Lorenzo Mixtepec (Goodwin, 1969:39); type locality; Río Molino, 2250 m., 3 km. SW San Miguel Suchixtepec (Schaldach, 1966:288).

Sorex saussurei salvini Merriam

1897. *Sorex salvini* Merriam, Proc. Biol. Soc. Washington, 11:229, July 15, type from Calel, 10,200 ft., Quezaltenango, Guatemala. Known only from type locality.

1928. *Sorex saussurei salvini,* Jackson, N. Amer. Fauna, 51:159, July 24.

Sorex saussurei saussurei Merriam

1892. *Sorex saussurei* Merriam, Proc. Biol. Soc. Washington, 7:173, September 29, type from N slope Sierra Nevada de Colima, approximately 8000 ft., Jalisco.
1925. *Sorex durangae* Jackson, Proc. Biol. Soc. Washington, 38:127, November 13, type from El Salto, Durango (regarded as indistinguishable from *Sorex saussurei saussurei* by Findley, Univ. Kansas Publ., Mus. Nat. Hist., 7:617, June 10, 1955).

MARGINAL RECORDS.—Coahuila: Sierra Guadalupe. Tamaulipas: Miquihuana. Hidalgo: Encarnación. Puebla: Huauchinango. México: Mt. Popocatépetl. Oaxaca: Tamazulapan; Tlapancingo. Guerrero: Chilpancingo; Omilteme. Michoacán: Pátzcuaro; Mt. Tancítaro. Jalisco: type locality; San Sebastián. Durango (Baker and Greer, 1962:64): El Salto; 1½ mi. W San Luis.

Sorex saussurei veraecrucis Jackson

1925. *Sorex saussurei veraecrucis* Jackson, Proc. Biol. Soc. Washington, 38:128, November 13, type from Xico [= Jico], 6000 ft., Veracruz.

MARGINAL RECORDS.—Veracruz: 6 km. SSE Altotonga, 9000 ft.; *2 km. E Las Vigas, 8000 ft.* (Hall and Dalquest, 1963:205); type locality. Oaxaca: Mt. Zempoaltepec; Cerro San Felipe (Goodwin, 1969:38); [Pápalo Santos] Reyes, near Cuicatlán. Puebla: Mt. Orizaba.

Sorex oreopolus
Mexican Long-tailed Shrew

External measurements: 88–112; 36–46; 12.5–14. Condylobasal length of skull, 16.4–17.8. Upper parts clove brown, sepia, or between sepia and bister; underparts paler, grayish in some; tail bicolored. Skull relatively deep; molars relatively small; 3rd unicuspid smaller than 4th except in *S. o. emarginatus* where reverse is true. Resembling *Sorex saussurei* but smaller.
See addenda for *Sorex arizonae.*

Sorex oreopolus emarginatus Jackson

1925. *Sorex emarginatus* Jackson, Proc. Biol. Soc. Washington, 38:129, November 13, type from Sierra Madre, near Bolaños, 7600 ft., Jalisco.
1955. *Sorex oreopolus emarginatus,* Findley, Univ. Kansas Publ., Mus. Nat. Hist., 7:616, June 10.

MARGINAL RECORDS.—Durango: 7 mi. SW Las Adjuntas, 8900 ft. (Baker and Greer, 1962:63); *2 mi. N Pueblo Nuevo, 6000 ft. (ibid.).* Zacatecas: Plateado, 7600–8500 ft. Jalisco: type locality.

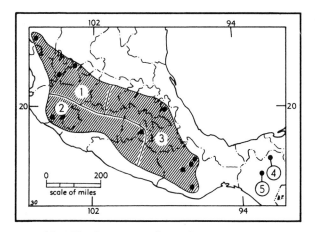

Map 30. *Sorex oreopolus* and related species.

Guide to kinds
1. *S. o. emarginatus*
2. *S. o. oreopolus*
3. *S. o. ventralis*
4. *S. sclateri*
5. *S. stizodon*

Sorex oreopolus oreopolus Merriam

1892. *Sorex oreopolus* Merriam, Proc. Biol. Soc. Washington, 7:173, September 29, type from N slope Sierra Nevada de Colima, approx. 10,000 ft., Jalisco.

MARGINAL RECORDS.—Jalisco: 8200 ft., SE Autlán; type locality. Morelos: Laguna Seca, 2860 m. (Ramírez-P., 1971:266). Jalisco: *Volcano de Nieve*.

Sorex oreopolus ventralis Merriam

1895. *Sorex obscurus ventralis* Merriam, N. Amer. Fauna, 10:75, December 31, type from Cerro San Felipe, 10,000 ft., Oaxaca.
1955. *Sorex oreopolus ventralis*, Findley, Univ. Kansas Publ. Mus. Nat. Hist., 7:617, June 10.

MARGINAL RECORDS.—Puebla: Huauchinango. Oaxaca: Llano de los Flores, 9200 ft. (Hooper, 1961:120); near Cajonos; *mts. near Ozolotepec*; Arroyo Agua Fría (Goodwin, 1969:39); 15 mi. W Oaxaca.

Sorex sclateri Merriam
Sclater's Shrew

1897. *Sorex sclateri* Merriam, Proc. Biol. Soc. Washington, 11:228, July 15, type from Tumbalá, 5000 ft., Chiapas. Known only from type locality.

External measurements: 125–126; 52; 16. Condylobasal length of skull, 19.6–19.9. Winter pelage everywhere brown or blackish brown; summer pelage unknown; skull relatively narrow; interorbital region elongate; 3rd unicuspid equal to or larger than 4th; unicuspids without internal pigmented ridges or cusplets.

Sorex stizodon Merriam
San Cristóbal Shrew

1895. *Sorex stizodon* Merriam, N. Amer. Fauna, 10:98, December 31, type from San Cristóbal, 9000 ft., Chiapas. Known only from type locality.

External measurements: 107; 41; 13.5. Condylobasal length of skull, 17.5. Upper parts bister or a little darker, underparts slightly paler. Skull broad and flattened; rostrum relatively short and wide; teeth with little pigment; 3rd unicuspid approx. equal to 4th in size; width of M1 more than anteroposterior length.

Genus Microsorex Coues—Pygmy Shrew

Revised by Jackson, N. Amer. Fauna, 51:200–210, July 24, 1928, and by Long, Trans. Kansas Acad. Sci., 74: 181–196, April 7, 1972.—See addenda.

1877. *Microsorex* Coues, Bull. U.S. Geol. and Geog. Surv. Territories, 3:646, May 15. Type, *Sorex hoyi* Baird.

External measurements: 78–98; 27–35; 8.5–12. Condylobasal length of skull, 13.0–15.8. Brownish above, paler below, tail indistinctly bicolored. Skull resembles that of *Sorex* but relatively narrower and more flattened; rostrum relatively short and broad; infraorbital foramen relatively small; mandible short and heavy; 1st and 2nd unicuspids with distinct internal ridge terminating in pronounced internal cusp; apices of unicuspids curved posteriorly; 3rd unicuspid disklike, much compressed anteroposteriorly; 4th unicuspid normal; 5th unicuspid minute; molariform teeth resembling those of *Sorex*; bases of lower incisor and premolar separated by space equal to approx. a fourth anteroposterior diameter of canine.

Microsorex hoyi
Pygmy Shrew

See characters of the genus. Long (1972), in a classification characterized by him as tentative (*op. cit.*:181), recognized two species instead of one. He chose to arrange *S. thompsoni* Baird and *M. winnemana* Preble as two subspecies of the species *M. thompsoni*. He arranged the remaining names applied to *Microsorex* as applying to five subspecies of *M. hoyi*. Long's classification may be correct, but he did not make it clear how the two species could be distinguished. All the valid taxa are here arranged as subspecies of the species *M. hoyi*.

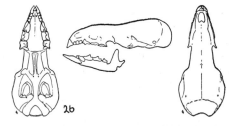

Fig. 26. *Microsorex hoyi washingtoni*, Thompson River, 7 mi. E Thompson Falls, 2250 ft., Sanders Co., Montana, No. 80762 M.V.Z., ♀, X 2.

Microsorex hoyi alnorum (Preble)

1902. *Sorex (Microsorex) alnorum* Preble, N. Amer. Fauna, 22:72, October 31, type from Robinson Portage, about 35 mi. SW Oxford Lake, Manitoba.

1925. *Microsorex hoyi alnorum*, Jackson, Proc. Biol. Soc. Washington, 38:126, November 13.

MARGINAL RECORDS.—Quebec (Harper, 1961: 34): Fort Chimo; NW Bay of Attikamagen Lake. Ontario: Favourable Lake. Manitoba: type locality; 3 mi. N Gillam (Long, 1972:190); 3 mi. S Fort Churchill (Wrigley, 1974:204).

Microsorex hoyi eximius (Osgood)

1901. *Sorex (Microsorex) eximius* Osgood, N. Amer. Fauna, 21:71, September 26, type from Tyonek, Cook Inlet, Alaska.

1925. *Microsorex hoyi eximius*, Jackson, Proc. Biol. Soc. Washington, 38:125, November 13.

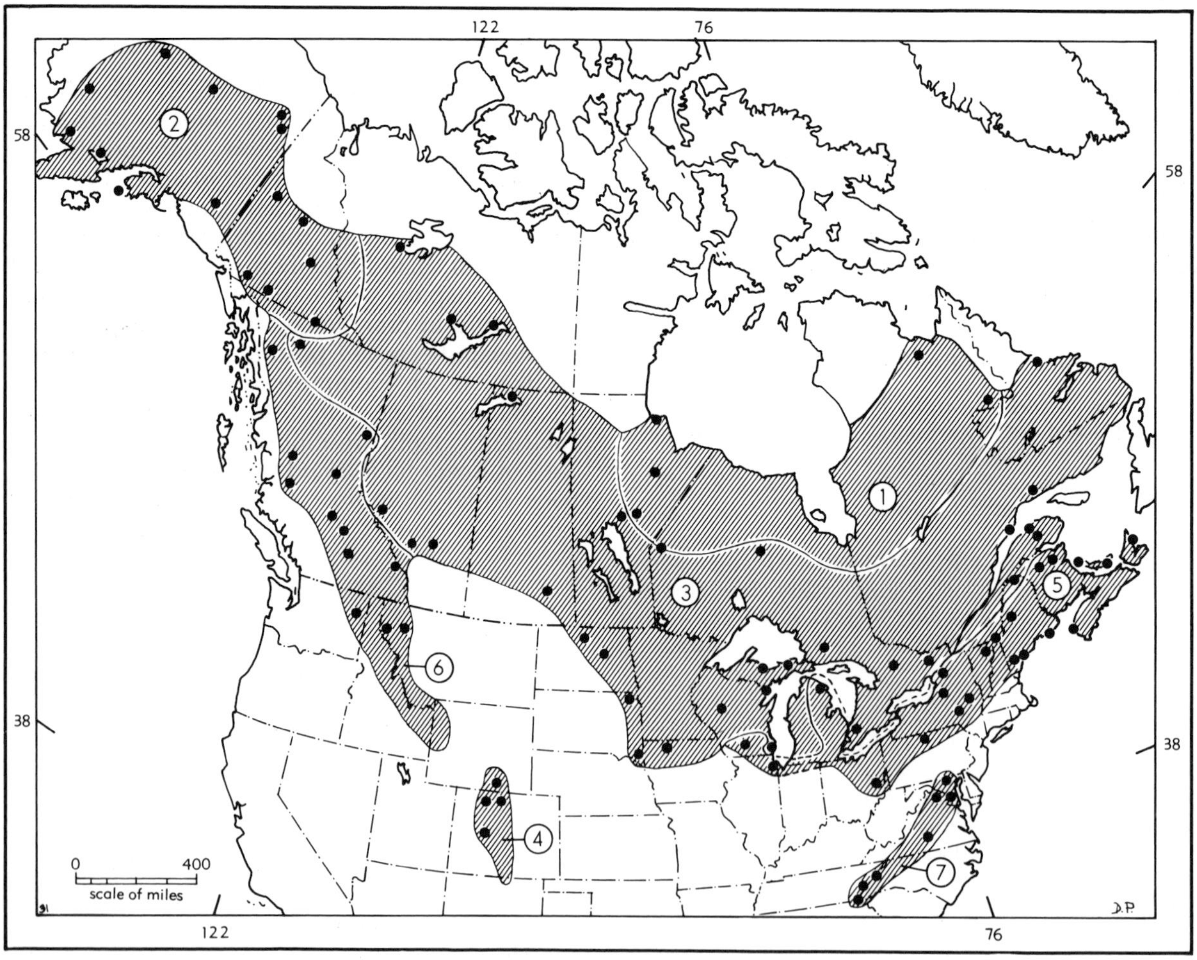

Map 31. *Microsorex hoyi*.

1. *M. h. alnorum*   3. *M. h. hoyi*   6. *M. h. washingtoni*
2. *M. h. eximius*   4. *M. h. montanus*   7. *M. h. winnemana*
                            5. *M. h. thompsoni*

MARGINAL RECORDS.—Alaska (Rausch, R. L., 1967:9, 10): upper Sheenjek River region [Small Lake, 67° 22′ N, 143° 48′ W]; Lower Rampart, Porcupine River. Yukon (Long, 1972: 189, unless otherwise noted): 14 mi. E Dawson City, 1300 ft.; 6 mi. N Mayo, 1900 ft.; Sheldon Lake; Liard River Valley, 2100 ft.; Haines Road (Hall and Kelson, 1959:51, as *M. h. intervectus*); Dezadeash Lake. Alaska (Rausch, R. L., 1971:9, 10): jct. Taylor and Alaska highways; Barabori; Lake Clark, near Nondalton; 80 mi. up Kakwok River; Nyac; Nulato; Tanana.

Microsorex hoyi hoyi (Baird)

1858. *Sorex hoyi* Baird, Mammals, *in* Repts. Expl. Surv. . . . , 8(1):32, July 14, type from Racine, Wisconsin.
1901. [*Microsorex*] *hoyi*, Elliot, Field Columb. Mus., Publ. 45, Zool. Ser., 2:377, March 6.
1925. *Microsorex hoyi intervectus* Jackson, Proc. Biol. Soc. Washington, 38:125, November 13, type from Lakewood, Oconto Co., Wisconsin.

MARGINAL RECORDS.—Mackenzie: Fort Franklin, Great Bear Lake; Fort Rae; Big Island, Great Slave Lake. Saskatchewan: Fond-du-lac (Beck, 1958:9). Manitoba: Echimamish River. Ontario: Favourable Lake; Attawapiscat Lake. Labrador: Hopedale. Quebec: Point St. Charles (C. F. Jackson, 1938:432); Godbout; Ste. Anne des Monts; Grand Cascapedia. Maine: Allagash (Long, 1972:188). Ontario: near Leitrim; Algonquin Park; Coldstream; Biscotasing (Long, 1972:188). Michigan: 12 mi. NW Newberry (Long, 1972:188); Marquette County; Menominee County. Wisconsin: type locality; Hewett. Iowa: Deweys Pasture (Bowles, 1975:34). South Dakota: ¼ mi. N, ½ mi. E Vermillion; Ft. Sisseton. North Dakota (Long, 1972:188): 2 mi. W Fort Totten, 1400 ft.; Lower Souris Refuge. Saskatchewan: Indian Head (Beck, 1958:9). Alberta: forks Blindman and Red Deer rivers; Sundre, Bearberry Creek (Long, 1972:186); Entrance. British Columbia: McDame Post (*ibid.*).

Microsorex hoyi montanus Brown

1966. *Microsorex hoyi montanus* Brown, Proc. Biol. Soc. Washington, 79:50, May 23, type from edge Trails Divide Pond, ¼ mi. S Univ. Wyoming Summer Science Camp Lodge, *ca.* 10,000 ft., *ca.* 6¼ mi. W Centennial, Albany Co., Wyoming (pers. comm., 10 May 1968).

MARGINAL RECORDS.—Wyoming: type locality; 7 *mi. W Centennial* (Brown, 1967:620). Colorado: 36 mi. W Fort Collins [9700 ft.] (Spencer and Pettus, 1966:677, as *M. hoyi* only); 2⁷⁄₁₀ mi. NW Gothic, 9800 ft. (DeMott, 1975:417, as *M. hoyi* only); 3 mi. SW Rabbit Ears Pass, 9900 ft. (Vaughan, 1969:53).

Microsorex hoyi thompsoni (Baird)

1858. *Sorex thompsoni* Baird, Mammals, *in* Repts. Expl. Surv. . . . , 8(1):34, July 14, type from Burlington, Chittenden Co., Vermont.

1925. *Microsorex hoyi thompsoni*, Jackson, Proc. Biol. Soc. Washington, 38:126, November 13.

MARGINAL RECORDS.—New Brunswick: Bathurst, 15 mi. from Miramichi Road. Prince Edward Island: Alberton; Georgetown. Nova Scotia: Ingonish Centre, Cape Breton Island; Little River, Digby Neck. Maine: Eagle Lake, Mt. Desert Island; Brunswick; Norway (Long, 1972:192). New York (Connor, 1960:22, 23): Petersburg Mtn., near Cobleskill; 2 mi. N Oneonta. Pennsylvania: Potter County (Doutt, *et al.*, 1973:40). Ohio: Zanesville. Illinois: Chicago (Long, 1972:193). Wisconsin (Long, 1972:192, as *M. thompsoni thompsoni*): Madison; *Milwaukee*. Michigan: Barnhart Lakes (*ibid.*). New York: Locust Grove; Canton. Vermont: Brighton. New Hampshire: First Connecticut Lake. Maine: Brassua Lake. New Brunswick: Trousers Lake. Schorger (1973:76) reported a specimen from Cherokee Marsh (short distance N Madison) as *M. h. hoyi.*

Microsorex hoyi washingtoni Jackson

1925. *Microsorex hoyi washingtoni* Jackson, Proc. Biol. Soc. Washington, 38:125, November 13, type from Loon Lake, Stevens Co., Washington.

MARGINAL RECORDS (Long, 1972:191, unless otherwise noted).—British Columbia: Telegraph Creek; Fort St. John; Willow River; Kootenay National Park, Vermillion Crossing. Montana: S fork Flathead River, 20 mi. S Hungry Horse Dam (Hall and Kelson, 1959:51); 7 mi. E Thompson Falls. Washington: type locality. British Columbia: Pass Lake; Chimney Lake; Williams Lake; Anahim Lake (Cowan and Guiguet, 1965:58); Ootsa Lake.

Microsorex hoyi winnemana Preble

1910. *Microsorex winnemana* Preble, Proc. Biol. Soc. Washington, 23:101, June 24, type from bank of Potomac River near Stubblefield Falls, 4 mi. below Great Falls of Potomac River, Fairfax Co., Virginia.
1925. *Microsorex hoyi winnemana*, Jackson, Proc. Biol. Soc. Washington, 38:126, November 13.

MARGINAL RECORDS.—Maryland: near Prettyboy Reservoir (Lee, 1974:60, as *M. hoyi* only); Berwyn. Virginia: Altavista. North Carolina: Pisgah National Forest. Georgia: Beech Creek near confluence with Tallulah River, Towns Co. (Wharton, 1968:158, as *M. hoyi* only). North Carolina: Newfound Gap, Great Smoky Mts. National Park (Hoffmeister, 1968:331). Virginia: type locality.

Genus Blarina Gray—Short-tailed Shrews

Revised by Merriam, N. Amer. Fauna, 10:9–16, December 31, 1895. See also Bole and Moulthrop, Sci. Publs., Cleveland Mus. Nat. Hist., 5(6):99–113, September 11, 1942.

1838. *Blarina* Gray, Proc. Zool. Soc. London for 1837, p. 124, June 14. Type, *Sorex talpoides* Gapper.

External measurements: 95–134; 17–30; 11.5–17. Condylobasal length of skull, 20.2–24.7. External ear not apparent; eyes minute; tail always less than half length of head and body; color grayish-black, sometimes with silvery or brownish cast; 1st and 2nd upper unicuspids large, 3rd and 4th smaller and subequal; 5th minute. Dentition, i. $\frac{3}{2}$, c. $\frac{1}{0}$, p. $\frac{2}{1}$, m. $\frac{3}{3}$.

KEY TO SPECIES OF BLARINA

1. Occurring in Dismal Swamp, Virginia, S to heart of Albemarle-Pamlico Peninsula, North Carolina . . . . . *B. telmalestes*, p. 57
1'. Not occurring in Dismal Swamp, Virginia, S to heart of Albemarle-Pamlico Peninsula, North Carolina *B. brevicauda*, p. 54

Blarina brevicauda
Short-tailed Shrew

For diagnosis see account of genus. In Nebraska in "zones of contact" between *Blarina brevicauda brevicauda* and *B. b. carolinensis*, Genoways and Choate (1972: 106–116) found the two taxa reacting to each other as two species except that one of 66 individuals was identified as either an intergrade or a hybrid.

Blarina brevicauda aloga Bangs

1902. *Blarina brevicauda aloga* Bangs, Proc. New England Zool. Club, 3:76, March 31, type from West Tisbury, Marthas Vineyard Island, Dukes Co., Massachusetts. Known only from Marthas Vineyard Island.

Blarina brevicauda angusta Anderson

1943. *Blarina brevicauda angusta* Anderson, Ann. Rep. Provancher Soc. Nat. Hist., Quebec, p. 52, September 7, type from Kellys Camp, Berry Mountain Brook, near head Grand Cascapedia River, Gaspé Co., Quebec.

MARGINAL RECORDS.—Quebec: Ste. Anne des Monts; 2 mi. W Coin du Banc (Manville, 1961:108). New Brunswick: St. Leonard; Baker Lake.

Blarina brevicauda brevicauda (Say)

1823. *Sorex brevicaudus* Say, *in* Long, Account of an exped. . . . to the Rocky Mts., 1:164. Type locality, W bank

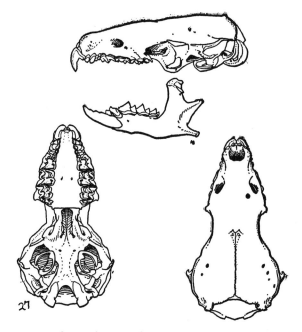

Fig. 27. *Blarina brevicauda kirtlandi*, 1 mi. W Highland Park, Lake Co., Illinois, No. 81749 M.V.Z., ♀, X 2.

of Missouri River, near Blair, formerly Engineer Cantonment, Washington Co., Nebraska.

1858. *Blarina brevicauda*, Baird, Mammals, *in* Repts. Expl. Surv. . . . , 8(1):42, July 14.

1891. *Blarina costaricensis* J. A. Allen, Bull. Amer. Mus. Nat. Hist., 3:205, April 17, type probably from upper Mississippi Valley, although alleged by describer to be from La Carpintera, Costa Rica.

1943. *Blarina fossilis* Hibbard, Univ. Kansas Sci. Bull., 29 (pt. 2):238, October 15, type from Pleistocene, Locality No. 5, Lincoln Co., Kansas. Regarded as a synonym of *B. b. brevicauda* by Graham and Semken, Jour. Mamm., 57:437, August 27, 1976.

MARGINAL RECORDS.—Ontario: Rat Portage, Lake of the Woods; Quetico Park. Minnesota: 2 mi. N, 4 mi. E Grand Portage (Timm, 1975:13). Wisconsin (Jackson, 1961:55): Ogema; T. 18 N, R. 3 E, Juneau Co. Illinois: Dekalb. Missouri: Kimmswick. Iowa (Bowles, 1975:36): ½ mi. N, 2 mi. E Mt. Ayr; ½ mi. S, 3 mi. W Henderson. Nebraska (Jones, 1964c:68): Crete; Saronville; Kearney; 5 mi. S Gothenburg; Perch; Valentine. North Dakota: Fort Berthold; Turtle Mts.; Pembina.

Blarina brevicauda carolinensis (Bachman)

1837. *Sorex carolinensis* Bachman, Jour. Acad. Nat. Sci. Philadelphia, 7(2):366. Type locality, eastern South Carolina.

1895. *Blarina brevicauda carolinensis*, Merriam, N. Amer. Fauna, 10:13, December 31.

1899. *Blarina brevicauda hulophaga* Elliot, Field Columb. Mus., Publ. 38, Zool. Ser., 1:287, May 25, type from Dougherty, Murray Co., Oklahoma. Arranged as a synonym

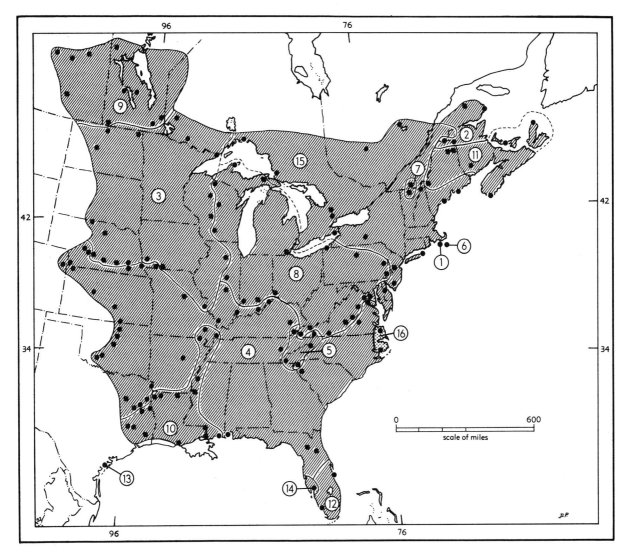

Map 32. *Blarina brevicauda* and *Blarina telmalestes*.

1. *B. b. aloga*	5. *B. b. churchi*	9. *B. b. manitobensis*	13. *B. b. plumbea*
2. *B. b. angusta*	6. *B. b. compacta*	10. *B. b. minima*	14. *B. b. shermani*
3. *B. b. brevicauda*	7. *B. b. hooperi*	11. *B. b. pallida*	15. *B. b. talpoides*
4. *B. b. carolinensis*	8. *B. b. kirtlandi*	12. *B. b. peninsulae*	16. *B. telmalestes*

of *B. b. carolinensis* by Jones and Glass, Southwestern Nat., 5:136, November 1, 1960.

MARGINAL RECORDS.—Nebraska (Jones, 1964c:70): 2 mi. N, 5 mi. W North Platte, 2720 ft.; *2 mi. N North Platte*; 5 mi. S, 2½ mi. W Brady; 1½ mi. W DeWitt; 1 mi. SE Nebraska City. Iowa (Bowles, 1975:37, 38): *3½ mi. S Sidney*; 2 mi. S, ¾ mi. W Bedford. Missouri: Boone Co. (Easterla, 1968d:448). Illinois: Alto Pass; White Heath. Indiana: New Harmony; Corydon; Bascom (Mumford, 1969:36, would assign specimens from southern Indiana to *B. b. kirtlandi*). Kentucky: Quicksand. Tennessee: Harriman. South Carolina: Greenville. Virginia: Amelia County. Maryland: Cambridge, thence southward along coast (excluding range of *Blarina telmalestes*) to Florida: Welaka; Gainesville. Alabama: Alabama Port. Arkansas: Beebe. Texas: 10 mi. S Texarkana; Harrison County (Davis, 1966:37, as *B. brevicauda* only); Rusk County (*ibid.*, as *B. brevicauda* only); Cherokee County; Henderson County (Davis, 1966:37, as *B. brevicauda* only). Oklahoma (Jones and Glass, 1960:140, 141): Wichita Mtn. Wildlife Refuge; Kiowa Agency, 17 mi. SE Old Fort Cobb; 3 mi. N Edmond; (SE ¼, sec. 18, T. 19 N, R. 1 E) 9 mi. W Stillwater; mouth Salt Fork River; *3 mi. E Arkansas River bridge at Ponca City*. Kansas: 3 mi. SE Arkansas City; ½ mi. N, 1 mi. E Halstead; ¾ mi. S, 3½ mi. W Hays; 11½ mi. N, 15 mi. W St. Francis. Colorado: 1 mi. E Laird. Nebraska: 5 mi. N, 2 mi. W Parks.

Blarina brevicauda churchi Bole and Moulthrop

1942. *Blarina brevicauda churchi* Bole and Moulthrop, Sci. Publs., Cleveland Mus. Nat. Hist., 5:109, September 11, type from Roan Mtn., Mitchell Co., North Carolina.

MARGINAL RECORDS.—Kentucky: Black Mtn. Virginia: Whitetop Mtn. North Carolina: type locality. South Carolina: 8 mi. E Caesars Head. North Carolina: 9 mi. SW Murphy.

Blarina brevicauda compacta Bangs

1902. *Blarina brevicauda compacta* Bangs, Proc. New England Zool. Club, 3:77, March 31, type from Nantucket Island, Nantucket Co., Massachusetts. Known only from type locality.

Blarina brevicauda hooperi Bole and Moulthrop

1942. *Blarina brevicauda hooperi* Bole and Moulthrop, Sci. Publs., Cleveland Mus. Nat. Hist., 5:110, September 11, type from Lyndon, Caledonia Co., Vermont.

MARGINAL RECORDS.—Quebec: North Hatley. Vermont: type locality.

Blarina brevicauda kirtlandi Bole and Moulthrop

1942. *Blarina brevicauda kirtlandi* Bole and Moulthrop, Sci. Publs., Cleveland Mus. Nat. Hist., 5:99, September 11, type from Holden Arboretum, Kirtland Twp., Lake Co., and Chadron Twp., Geauga Co. (county line bisects type locality), Ohio.

MARGINAL RECORDS.—Michigan: Keweenaw County; Luce County; Monroe County, thence eastward along southern shore of Lake Erie to Pennsylvania: McKean; Drury Run; Nazareth; Philadelphia. Maryland: Oxon Hill. Virginia: Fairfax County; Spotsylvania County; Cumberland County; Campbell County; Patrick County; Taswell County; Scott County. Ohio: Cincinnati. Indiana: Kurtz; Bicknell (Mumford, 1969:36, would assign specimens from southern Indiana to *B. b. kirtlandi*). Illinois: Bloomington. Wisconsin (Jackson, 1961:55): Mazomanie; Rib Hill; Mercer.

Blarina brevicauda manitobensis Anderson

1947. *Blarina brevicauda manitobensis* R. M. Anderson, Bull. Nat. Mus. Canada, 102:23, January 24, type from Max Lake, Manitoba.

MARGINAL RECORDS.—Manitoba: The Pas; Overflowing River, Lake Winnipegosis; Lake St. Martin Reserve; Telford, near Whiteshell Forest Reserve, 10 mi. W Ingoman; Sandilands Forest Reserve, SW Marchand; Max Lake, Turtle Mts., 2100 ft. Saskatchewan: Regina; Keatley (Nero, 1960:41, as *B. brevicauda* only); Wakaw (Beck, 1958:10); Nipawin (Riome, 1968:201).

Blarina brevicauda minima Lowery

1943. *Blarina brevicauda minima* Lowery, Occas. Pap. Mus. Zool., Louisiana State Univ., 13:218, November 22, type from Comite River, 13 mi. NE Baton Rouge, East Baton Rouge Parish, Louisiana.

MARGINAL RECORDS.—Missouri (Easterla, 1968d:448): Duck Creek Wildlife Area, 6 mi. N Puxico; Big Oak Tree State Park. Mississippi: Rosedale; Washington. Louisiana: 2 mi. N Angie (Lowery, 1974:80). Mississippi: Biloxi. Louisiana (*ibid.*): 1 mi. N Slidell; Avery Island. Texas: 7 mi. NE Sour Lake; San Jacinto County (Davis, 1966:37, as *B. brevicauda* only); Walker County (*ibid.*, as *B. brevicauda* only); Nacogdoches County (*ibid.*, as *B. brevicauda* only); Joaquin. Louisiana (Lowery, 1974:80): *3 mi. S, 2 mi. W Blanchard;* 8 mi. NW Shreveport; Evergreen; Haile. Arkansas: 30 ft. S Missouri–Arkansas state line, 4⅘ mi. SW Naylor, *in* Missouri (Easterla, 1968d:448).

Blarina brevicauda pallida R. W. Smith

1940. *Blarina brevicauda pallida* R. W. Smith, Amer. Midland Nat., 24:223, July 31, type from Wolfville, Kings Co., Nova Scotia.

MARGINAL RECORDS.—Nova Scotia: Cape North; Barrington Passage. New Brunswick: Scotch Lake. Maine: Mooselookmeguntic Lake; Ashland; Caribou. Prince Edward Island.

Blarina brevicauda peninsulae Merriam

1895. *Blarina carolinensis peninsulae* Merriam, N. Amer. Fauna, 10:14, December 31, type from Miami River, Dade Co., Florida.
1897. [*Blarina brevicauda*] *peninsulae*, Trouessart, Catalogus Mammalium . . . , fasc. 1, p. 188.

MARGINAL RECORDS.—Florida: Oak Lodge; Everglade.

Blarina brevicauda plumbea Davis

1941. *Blarina brevicauda plumbea* Davis, Jour. Mamm., 22:317, August 14, type from ½ mi. W Marano Mill, Aransas Co., Texas. Known only from Aransas National Wildlife Refuge, Aransas Co., Texas.

Blarina brevicauda shermani Hamilton

1955. *Blarina brevicauda shermani* Hamilton, Proc. Biol. Soc. Washington, 68:37, May 20, type from 2 mi. N Fort Myers, Lee Co., Florida. Known only from type locality.

Blarina brevicauda talpoides (Gapper)

1830. *Sorex talpoides* Gapper, Zool. Jour., 5:202. Type locality, between York and Lake Simcoe, Ontario.
1902. *Blarina brevicauda talpoides,* Bangs, Proc. New England Zool. Club, 3:75, March 31.
1858. *Blarina angusticeps* Baird, Mammals, *in* Repts. Expl. Surv. . . . , 8(1):34, July 14, type from Burlington,

Chittenden Co., Vermont. Regarded by Merriam, N. Amer. Fauna, 10:10, December 31, 1895, as based on a deformed skull. See also Bole and Moulthrop, Sci. Publs., Cleveland Mus. Nat. Hist., 5(6):111, September 11, 1942.

MARGINAL RECORDS.—Quebec: St. Méthode; *Val Jalbert* (Harper, 1961:35). New Hampshire: Pittsburg. Maine: Mt. Desert Island; Small Point Beach. New York: eastern end Long Island at Orient. New Jersey: Mays Landing. Delaware: Smyrna. New Jersey: Princeton. New York: Bath; *Elba*. Ontario: Point Abino; near Cedardale Mill; Lake Simcoe; Pancake Bay; Schrieber. Quebec: 47° 50′ N, 75° 35′ W, S of Clova (MacLeod and Cameron, 1961:281).

Blarina telmalestes Merriam
Swamp Short-tailed Shrew

1895. *Blarina telmalestes* Merriam, N. Amer. Fauna, 10:15, December 31, type from Lake Drummond, Dismal Swamp, Norfolk Co., Virginia.

External measurements:118–128; 26.4–28; 16–17. Greatest length of skull, 24. Skull relatively slender; dentition little pigmented. Only slightly differentiated from *Blarina brevicauda talpoides*. The range of *B. telmalestes* is surrounded by that of the smaller, darker *B. b. carolinensis*. If the range of *telmalestes* were adjacent to that of the northern *B. b. talpoides*, the two kinds probably would be considered as subspecies of a single species.

MARGINAL RECORDS.—Virginia: type locality. North Carolina: 7 mi. NW Swanquarter, Hyde Co. (Paul, 1965:496).

Genus **Cryptotis** Pomel—Small-eared Shrews

Revised by Choate, Univ. Kansas Publ., Mus. Nat. Hist., 19:195–317, December 30, 1970.

1848. *Cryptotis* Pomel, Arch. Sci. Phys. et Nat., Genève, 9:249, November. Type, *Sorex cinereus* Bachman [= *Sorex parvus* Say].

External measurements: total length, 69–135; tail, 19–52 per cent of head and body; hind foot, 9.0–17.5. Condylobasal length, 14.4–23.7; maxillary breadth, 4.7–7.8; maxillary tooth-row, 5.2–9.0; cranial breadth, 7.4–12.4. Upper parts brownish or blackish, underparts same or paler; eyes minute; pinnae of ears inconspicuous, snout pointed; skull approx. conical; braincase low, laterally angular; rostrum wedge-shaped; unicuspids 4, never in 2 pairs, 4th always smaller than 3rd, usually minute. Dentition, i. $\frac{3}{1}$, c. $\frac{1}{1}$, p. $\frac{2}{1}$, m. $\frac{3}{3}$ = 30.

KEY TO RECENT NORTH- AND MIDDLE-AMERICAN SPECIES OF CRYPTOTIS

After Choate (1970:223, 224)

1. Tail elongate, more than 45% of length of head and body; rostrum elongate; postcentrocrista and metacone usually present on M3, metacrista sometimes present; entoconid present on m3.
   2. Dentition bulbous; posterior surfaces of P4–M2 never recessed; rostrum broad.
      3. Size large (total length, 123–135; condylobasal length, 22.5–23.7); occurs only in southern México. . . . .*C. magna*,   p. 64
      3′. Size medium (total length, 109; condylobasal length, 20.4); occurs only in southern Central America. . . . .*C. endersi*,   p. 64
   2′. Dentition not bulbous; posterior surfaces of P4–M2 sometimes slightly recessed; rostrum slender. . . . .*C. gracilis*,   p. 63
1′. Tail not elongate, less than 45% of length of head and body; rostrum not markedly elongate; postcentrocrista, metacone, and metacrista usually reduced or lacking on M3; entoconid reduced or lacking on m3 in certain taxa.
   4. Dentition bulbous. . . . .*C. nigrescens*,   p. 62
   4′. Dentition not bulbous.
      5. Forefeet enlarged; claws conspicuously long and broad.
         6. Size large (total length, 103–128; condylobasal length, 20.4–21.9; cranial breadth, 10.6–11.8); talonid of m3 reduced, consisting only of hypoconid, and shortened anteroposteriorly; winter pelage usually almost black; upper surfaces of feet usually black. . . . .*C. goodwini*,   p. 60
         6′. Size medium to large (in region of potential geographic sympatry with *C. goodwini*: total length, 101–111; condylobasal length, 19.3–20.5; cranial breadth, 10.1–10.8); talonid of m3 only moderately reduced, usually consisting only of hypoconid but vestigial entoconid sometimes present, not shortened anteroposteriorly; winter pelage dark brown, usually with slight olive cast, upper surfaces of feet usually pale.
                           C. goldmani.   p. 59

5'. Forefeet small; claws short and slender.
    7. Size medium (total length, 83–112; condylobasal length, 17.5–20.2); talonid of m3 almost always consisting of both hypoconid and well-developed entoconid; posterior surfaces of P4–M2 only slightly if at all recessed; color of venter usually dark, only slightly paler than dorsum.
                                            C. mexicana,   p. 58
    7'. Size small (in region of potential geographic sympatry with *C. mexicana*: total length, 69–99; condylobasal length, 15.3–18.4); talonid of m3 consisting only of hypoconid; posterior surfaces of P4–M2 moderately to considerably recessed; color of venter whitish, considerably paler than dorsum. . . . .*C. parva,*   p. 60

Cryptotis mexicana
Mexican Small-eared Shrew

External measurements: total length, 83–112; tail averaging, according to subspecies, 33–42 per cent of head and body; hind foot, 11–17. Condylobasal length, 17.5–20.2; maxillary breadth, 5.4–6.8; maxillary tooth-row, 5.8–7.6; cranial breadth, 8.8–10.7. Upper parts in winter near Bister or Clove in life; underparts only slightly paler (never white); summer pelage seldom distinctively colored, blackish gray. "Cranial characteristics: rostrum relatively short; braincase angular; anterior limit of zygomatic plate varying from slightly anterior to metastyle of M1 to above juncture of M1 and M2; posterior limit of zygomatic plate at level of or posterior to maxillary process, above M3; dentition not bulbous; anterior element of ectoloph of M1 not reduced relative to posterior element; posterior surfaces of P4–M2 negligibly or only slightly recessed; protoconal basin of M1 not reduced relative to hypoconal basin; M3 consisting of paracrista, precentrocrista, postcentrocrista, and metacone (vestigial metacrista, hypocone, or cingular cusplet sometimes present); talonid of m3 consisting of well-developed hypoconid and entoconid, . . . latter infrequently reduced." (Choate, 1970:224.)

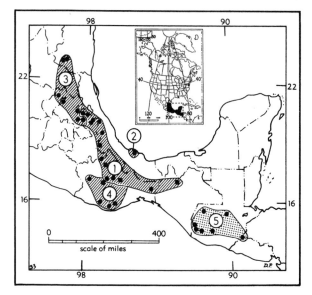

Map 33.   *Cryptotis mexicana* and *Cryptotis goodwini.*

  1. *C. m. mexicana*   3. *C. m. obscura*
  2. *C. m. nelsoni*   4. *C. m. peregrina*
              5. *C. goodwini*

Cryptotis mexicana mexicana (Coues)

1877. *Blarina (Soriciscus) mexicana* Coues, Bull. U.S. Geol. and Geog. Surv. Territories, 3:652, May 15, type from Jalapa, Veracruz.
1911. *Cryptotis mexicana,* Miller, Proc. Biol. Soc. Washington, 24:221, October 31.

MARGINAL RECORDS (Choate, 1970:234).— Hidalgo: Tenango de Doria. Puebla: 6 km. N Villa Juárez; Xocoyolo. Veracruz: 4 km. W Tlapacoyan; *2⁶/₁₀ mi. W Banderilla*; type locality. Chiapas: 3 mi. E Pueblo Nuevo Solistahuacán. Oaxaca: 8 km. NW "Colonia Rudolfo Figuroa"; Cerro Zempoaltepec; near San Pedro Cajonos; Cerro San Felipe; Papalo Santos Reyes; Teotitlán del Camino. Veracruz: Orizaba. Puebla: 2 mi. NW Zacapoaxtla; *Huauchinango*.

Cryptotis mexicana nelsoni (Merriam)

1895. *Blarina nelsoni* Merriam, N. Amer. Fauna, 10:26, December 31, type from Volcán Tuxtla, 4800 ft., Veracruz. Known only from type locality.
1970. *Cryptotis mexicana nelsoni,* Choate, Univ. Kansas Publ., Mus. Nat. Hist., 19:234, December 30.

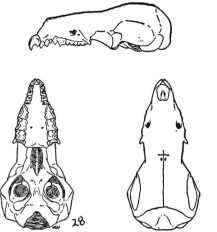

Fig. 28.   *Cryptotis mexicana mexicana,* 4 km. W Tlapacoyan, 1700 ft., Veracruz, No. 23413 K.U., ♀, X 2.

Cryptotis mexicana obscura (Merriam)

1895. *Blarina obscura* Merriam, N. Amer. Fauna, 10:23, December 31, type from Tulancingo, 8500 ft., Hidalgo.
1970. *Cryptotis mexicana obscura*, Choate, Univ. Kansas Publ., Mus. Nat. Hist., 19:235, December 30.
1954. *Cryptotis mexicana madrea* Goodwin, Amer. Mus. Novit., 1670:1, June 28, type from 5 mi. NW Gómez Farías, 3500 ft., Tamaulipas.

MARGINAL RECORDS (Choate, 1970:237).—Tamaulipas: Rancho del Cielo. Veracruz: Zacualpan. Puebla: *Honey;* Lago Tejocotal, 11 km. E Acaxochitlán (in Hidalgo). Hidalgo: type locality; Encarnación. Querétaro: Pinal de Amoles. Tamaulipas: Asserradero del Infiernillo, cave 11 mi. W Gómez Farías.

Cryptotis mexicana peregrina (Merriam)

1895. *Blarina mexicana peregrina* Merriam, N. Amer. Fauna, 10:24, December 31, type from mts. 15 mi. SW Oaxaca de Juárez, 9500 ft., Oaxaca (see Goldman, Smiths. Miscl. Coll., 115:218, July 31, 1951, for corrected type locality).
1911. *C[ryptotis]. mexicana peregrina*, Miller, Proc. Biol. Soc. Washington, 24:222, October 31.
1966. *Notiosorex (Xenosorex) phillipsii* Schaldach, Säugetierkund. Mitteil., 14:289, October, type from Río Molino, 3 km. SW San Miguel Suchixtepec, 2250 m., Oaxaca.

MARGINAL RECORDS (Choate, 1970:239).—Oaxaca: La Muralla, Cerro Yucumino, 8 mi. S Tlaxiaco; type locality; Lovené, Río Jalatengo, S of San Miguel Suchixtepec; Finca Sinai, 10 km. E Santos Reyes Nopala.

Cryptotis goldmani
Goldman's Small-eared Shrew

External measurements: total length, 90–117; tail averaging, according to subspecies, 34–40 per cent of head and body; hind foot, 11–16. Condylobasal length, 18.2–21.0; maxillary breadth, 5.7–7.2; maxillary tooth-row, 6.8–7.8; cranial breadth, 9.2–11.0. Upper parts in winter varying from Olive-Brown to Bister with distinctive olive sheen in life; underparts considerably paler owing to pale buffy or whitish tips on hairs. Upper parts in summer varying from Clove Brown to Mummy Brown (slight olivaceous sheen) in life; underparts buffy, more grayish than in winter. "Cranial characteristics: rostrum relatively long; braincase angular; anterior limit of zygomatic plate above metastyle or between mesostyle and metastyle of M1; posterior limit of zygomatic plate at level of, or posterior to, maxillary process, above M3 or metastyle of M2; dentition not bulbous; anterior element of ectoloph of M1 slightly reduced relative to posterior element; posterior surfaces of P4–M2 decidedly recessed throughout most of geographic range (less

so in Oaxaca); protoconal basin of M1 reduced relative to hypoconal basin; M3 usually consisting only of paracrista and precentrocrista (precentrocrista frequently vestigial in *C. g. alticola*), although postcentrocrista and rudimentary metacone sometimes present; talonid of m3 consisting only of hypoconid in *C. g. alticola,* but vestigial entoconid frequently present in *C. g. goldmani.*" (Choate, 1970:240.)

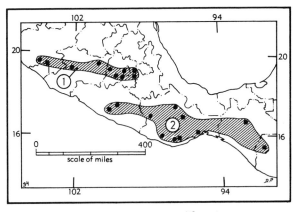

Map 34. *Cryptotis goldmani.*

1. *C. g. alticola*   2. *C. g. goldmani*

Cryptotis goldmani alticola (Merriam)

1895. *Blarina alticola* Merriam, N. Amer. Fauna, 10:27, December 31, type from Mt. Popocatépetl, 11,500 ft., México.
1970. *Cryptotis goldmani alticola*, Choate, Univ. Kansas Publ., Mus. Nat. Hist., 19:245, December 30.
1967. *Cryptotis euryrhynchis* Genoways and Choate, Proc. Biol. Soc. Washington, 80:203, December 1, type from Volcán de Fuego [also called Volcán de Colima], 9800 ft., Jalisco.

MARGINAL RECORDS (Choate, 1970:246, 247, unless otherwise noted).—Michoacán: 12 mi. W Ciudad Hidalgo. México: N edge Refugio San Cayetano, 3 mi. S Bosenchere; *Salazar.* Distrito Federal: Cerro de Santa Rosa. México: Monte Río Frío, 45 km. ESE Mexico City (Hall and Kelson, 1959: 60); *12 km. ESE Amecameca;* type locality; Lagunas de Zempoala, 10 mi. NNW Cuernavaca, Morelos; N slope Nevado de Toluca. Michoacán: Cerro de Tancítaro. Jalisco: Volcán de Fuego; 20 mi. SE Autlán; *12 mi. SW Ciudad Guzmán.*

Cryptotis goldmani goldmani (Merriam)

1895. *Blarina mexicana goldmani* Merriam, N. Amer. Fauna, 10:25, December 31, type from mountains near Chilpancingo, 10,000 ft. (9600 ft., see Choate, Univ. Kansas Publ., Mus. Nat. Hist., 19:247, December 30, 1970), Guerrero.
1970. *Cryptotis goldmani goldmani*, Choate, Univ. Kansas Publ., Mus. Nat. Hist., 19:247, December 30.

1895. *Blarina fossor* Merriam, N. Amer. Fauna, 10:28, December 31, type from Cerro Zempoaltepec, 10,500 ft., Oaxaca.

1895. *Blarina mexicana machetes* Merriam, N. Amer. Fauna, 10:26, December 31, type from mountains near Santa María Ozolotepec, 10,000 ft., Oaxaca.

1911. *Cryptotis frontalis* Miller, Proc. Biol. Soc. Washington, 24:222, October 31, type from "near Tehuantepec City," Oaxaca.

1933. *Cryptotis griseoventris* Jackson, Proc. Biol. Soc. Washington, 46:80, April 27, type from San Cristóbal de las Casas, 9500 ft., Chiapas.

1933. *Cryptotis guerrerensis* Jackson, Proc. Biol. Soc. Washington, 46:80, April 27, type from Omilteme, *ca.* 8000 ft., Guerrero.

MARGINAL RECORDS (Choate, 1970:249).— Oaxaca: 11 mi. NE Llano de las Flores; Cerro Zempoaltepec. Chiapas: San Cristóbal de las Casas; *6 mi. SE San Cristóbal de las Casas.* Guatemala: Todos Santos Cuchumatán. Oaxaca: "near the City of Tehuantepec"; San Juan Ozolotepec; *San Miguel Suchixtepec;* 3 km. SW San Miguel Suchixtepec; Lachao; 2 km. NE San Andrés Chicahuaxtla. Guerrero: S slope Cerro Teotepec; 3 mi. NW Omilteme; *type locality.*

Cryptotis goodwini Jackson
Goodwin's Small-eared Shrew

1933. *Cryptotis goodwini* Jackson, Proc. Biol. Soc. Washington, 46:81, April 27, type from Calel, 10,200 ft., Guatemala.

External measurements: total length, 103–128; tail averaging 35 per cent of head and body; hind foot, 14–17. Condylobasal length, 20.4–21.9; maxillary breadth, 6.5–7.3; maxillary tooth-row, 7.0–8.4; cranial breadth, 10.6–11.8. Upper parts in winter Clove Brown in life; underparts paler. Upper parts in summer Bister in life; underparts only slightly paler. "Cranial characteristics: rostrum relatively long, slender; braincase not especially angular; anterior limit of zygomatic plate above metastyle of M1; posterior limit of zygomatic plate at level of or posterior to maxillary process, above M3; dentition not bulbous; anterior element of ectoloph of M1 reduced relative to posterior element; posterior surfaces of P4–M2 decidedly recessed; protoconal basin of M1 reduced relative to hypoconal basin; M3 consisting primarily of paracrista, precentrocrista usually vestigial and frequently absent; talonid of m3 reduced, short, consisting only of hypoconid [hypoconid frequently vestigial]." (Choate, 1970:250.)

MARGINAL RECORDS (Choate, 1970:251).— Guatemala: 3½ mi. SW San Juan Ixcoy; Finca Xicacao. El Salvador: Hda. Montecristo, Santa Ana. Guatemala: Tecpán; Volcán Santa María; Finca La Paz; S slope Volcán Tajamulco. See Map 33.

Cryptotis parva
Least Shrew

External measurements: total length, 69–103; tail averaging, according to subspecies, 19–37 per cent of head and body; hind foot, 9–13. Condylobasal length, 14.4–18.4; maxillary breadth, 4.7–6.0; maxillary tooth-row, 5.2–6.8; cranial breadth, 7.4–8.9. Upper parts in winter near Clove Brown to blackish in life; underparts paler (nearly white in adults); tail bicolored, especially in northern subspecies; summer pelage paler. Smaller than all other species except *C. nigrescens,* which is dark colored ventrally at all ages. "Cranial characteristics: rostrum short; braincase not especially angular, almost pentagonal . . . ; anterior limit of zygomatic plate varying from slightly anterior to mesostyle of M1 to above metastyle of that tooth; posterior limit of zygomatic plate above, or slightly anterior to, maxillary process; dentition not bulbous; anterior element of ectoloph of M1 reduced relative to posterior element; posterior surfaces of P4–M2 decidedly recessed (less so in southern México and Guatemala); protoconal basin of M1 reduced relative to hypoconal basin; M3 usually consisting only of paracrista and precentrocrista, . . . latter sometimes reduced or absent (vestigial postcentrocrista or metacone present in a few specimens); talonid of m3 consisting only of hypoconid." (Choate, 1970:252, 253.)

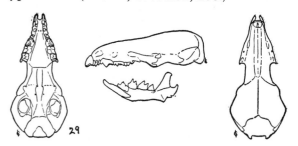

Fig. 29. *Cryptotis parva parva,* Monroe, Ouachita Parish, Louisiana, No. 70505 M.V.Z., ♀, X 2.

Cryptotis parva berlandieri (Baird)

1858. *Blarina berlandieri* Baird, Mammals, *in* Repts. Expl. Surv. . . . , 8(1):53, July 14, lectotype from Matamoros, Tamaulipas.

1941. *Cryptotis parva berlandieri,* Davis, Jour. Mamm., 22:413, November 13.

1903. *Blarina pergracilis* Elliot, Field Columb. Mus. Publ. 71, Zool. Ser., 3:149, March 20, type from Ocotlán, Jalisco.

1911. *Cryptotis pergracilis macer* Miller, Proc. Biol. Soc. Washington, 24:223, October 31, type from near Guanajuato, Guanajuato.

1933. *Cryptotis pergracilis nayaritensis* Jackson, Proc. Biol. Soc. Washington, 46:79, April 27, type from Tepic, 3000 ft., Nayarit.

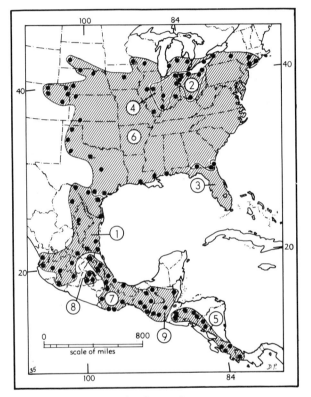

Map 35. *Cryptotis parva.*

Guide to subspecies
1. *C. p. berlandieri*
2. *C. p. elasson*
3. *C. p. floridana*
4. *C. p. harlani*
5. *C. p. orophila*
6. *C. p. parva*
7. *C. p. pueblensis*
8. *C. p. soricina*
9. *C. p. tropicalis*

MARGINAL RECORDS (Choate, 1970:262, unless otherwise noted).—Texas: Del Rio; Dilley (Hall and Kelson, 1959:57); W shore Lake Corpus Christi (Raun, 1960:195, as *C. parva* only); Brownsville. Tamaulipas: *type locality;* 2 mi. S, 10 mi. W Altamira. San Luis Potosí: 10 km. E Platanito; Alvarez. Guanajuato: Guanajuato. Michoacán: 12 km. W Morelia; *3 mi. E Pátzcuaro; Colonia Ibarra, Pátzcuaro* (Hall and Kelson, 1959:58); "1 hr. 20 min. [by mule] NE Rancho Barolosa"; La Palma. Jalisco: 21 mi. SW Guadalajara; 3½ mi. N Mascota. Nayarit: Tepic. Nuevo León: 7⅘ km. N, 3¾ km. W Monterrey (Hoffmeister, 1977:150). Coahuila: 10 mi. NE Melchor Múquiz.

Cryptotis parva elasson Bole and Moulthrop

1942. *Cryptotis parva elasson* Bole and Moulthrop, Sci. Publs., Cleveland Mus. Nat. Hist., 5:97, September 11, type from Bettsville, Seneca Co., Ohio.

MARGINAL RECORDS.—Ohio: Austinburg; Aurora Pond; Smoky Creek in Green Township; Antwerp; Evansport; Bay Point.

Cryptotis parva floridana (Merriam)

1895. *Blarina floridana* Merriam, N. Amer. Fauna, 10:19, December 31, type from Chester Shoal, 11 mi. N Cape Canaveral, Brevard Co., Florida.

1927. *Cryptotis parva floridana*, Harper, Proc. Boston Soc. Nat. Hist., 38:270, March.

MARGINAL RECORDS.—Georgia: St. Marys; thence throughout *peninsular Florida.* Georgia: 10 mi. SSW Thomasville, approx. 7 mi. Florida boundary.

Cryptotis parva harlani (Duvernoy)

1842. *Brachysorex harlani* Duvernoy, Mag. de Zool. d'Anat. Comp. et Paleont., Paris, livr. 25, p. 40, Pl. 53. Type locality, New Harmony, Indiana.

1942. *Cryptotis parva harlani*, Bole and Moulthrop, Sci. Publs., Cleveland Mus. Nat. Hist., 5:97, September 11.

MARGINAL RECORDS.—Indiana: Orland; Pleasant Mills; Terre Haute; type locality. Illinois: Bloomington.

Cryptotis parva orophila (J. A. Allen)

1895. *Blarina (Soriciscus) orophila* J. A. Allen, Bull. Amer. Mus. Nat. Hist., 7:340, November 8, type from Volcán Irazú [= Irazú Range], Prov. Cartago, Costa Rica.

1970. *Cryptotis parva orophila*, Choate, Univ. Kansas Publ., Mus. Nat. Hist., 19:262, December 30.

1908. *Blarina olivaceus* J. A. Allen, Bull. Amer. Mus. Nat. Hist., 24:669, October 13, type from San Rafael del Norte, 5000 ft., Nicaragua.

MARGINAL RECORDS (Choate, 1970:264).—Honduras: Lago de Yojóa; Cerro Cantoral; Montserrat Cloud Forest, near Yuscarán. Nicaragua: San Raphael del Norte; Santa María de Ostuma. Costa Rica: Monte Verde. Panamá: Cerro Punta–Boquete trail; *Santa Clara.* Costa Rica: *Río Navarro, El Muñeco;* Cerro Tablazo; Zarcero. El Salvador: Cerro Montecristo. Honduras: Belén.

Cryptotis parva parva (Say)

1823. *Sorex parvus* Say, *in* Long, Account of an exped. . . . to the Rocky Mts., 1:163, type from W bank Missouri River, near Blair, formerly Engineer Cantonment, Washington Co., Nebraska.

1912. *Cryptotis parva*, Miller, Bull. U.S. Nat. Mus., 79:24, December 31.

1837. *Sorex cinereus* Bachman, Jour. Acad. Nat. Sci. Philadelphia, 7:373, type from "Goose Creek about twenty-two miles from Charleston," South Carolina. Not *Sorex arcticus cinereus* Kerr, 1792.

1858. *Blarina exilipes* Baird, Mammals, *in* Repts. Expl. Surv.
. . . , 8(1):51, July 14, type from Washington, Mississippi.
1858. *B*[*larina*]. *eximius* Baird, Mammals, *in* Repts. Expl.
Surv. . . . , 8(1):52, July 14. Based on two specimens: one from DeKalb Co., Ill., and one from St. Louis, Missouri.

MARGINAL RECORDS.—Minnesota: Homer. Wisconsin: Prairie du Sac, thence along shores S ¼ Lake Michigan into Michigan: Allegan County; Clinton County; Lake Angelus NE of Pontiac, thence east, presumably, to W end Lake Ontario and along S shore of same to Niagara River, along W bank Niagara River to Lake Erie and west along N shore of same (as at Long Point in Ontario), northern boundary of Ohio, E half N boundary of Indiana, and southwest to Illinois: Mason County; near Carbondale (Pfeiffer and Gass, 1963:427, as *C. parva* only). Indiana: Ripley County (Lindsay, 1960:256). Thence along southern margins of geographic ranges of *C. p. harlani* and *C. p. elasson* to S shore Lake Erie and east along same to New York: Depew; 3 mi. E Ithaca. Connecticut: Salisbury (Goodwin, 1932:39, but see Jarrell, 1965:671); Westbrook (Jarrell, 1965:671); *Darien;* Stamford (Linsley, 1842:346). Thence south along Atlantic Coast to approx. 31°. Georgia: Chessers Island, Okefinokee Swamp, and along coast of western Florida to Alabama: Alabama Port. Louisiana (Lowery, 1974:86): 1 mi. N Slidell; Avery Island, thence to and west along Gulf Coast to Texas: Victoria; 9 mi. E Pleasanton; Waco; Young County (Davis, 1966:40, as *C. parva* only); Yellowhouse Canyon, 4½ mi. N Slaton (Packard and Judd, 1968:535); vic. Buffalo Lake Wildlife Refuge (*ibid.*); Stinnett. Kansas: 17 mi. SW Meade; 2 mi. S Ludell. Colorado: Dry Willow Creek; 4 mi. E Flagler (Armstrong, 1972:53); 3 mi. S Boulder (*ibid.*); 1 mi. W Livermore (Marti, 1972:448); 5 mi. W Crook. Nebraska (Jones, 1964c:73): 4 mi. NNW Keystone; *near North Platte;* Neligh; W Crookston. South Dakota: Cottonwood Range Experimental Station, 1½ mi. S Cottonwood. Iowa: Ames (Bowles, 1975:39).

Cryptotis parva pueblensis Jackson

1933. *Cryptotis pergracilis pueblensis* Jackson, Proc. Biol.
Soc. Washington, 46:79, April 27, type from Huauchinango, 5000 ft., Puebla.
1970. *Cryptotis parva pueblensis*, Choate, Univ. Kansas Publ., Mus. Nat. Hist., 19:264, December 30.
1956. *Cryptotis celatus* Goodwin, Amer. Mus. Novit., 1791: 1, September 28, type from Las Cuevas, Santiago Lachiguiri, District of Tehuantepec, Oaxaca.

MARGINAL RECORDS (Choate, 1970:266).—San Luis Potosí: 15 km. NE Xilitla. Puebla: Metlaltoyuca. Veracruz: 7 km. W El Brinco; *5 km. N Jalapa; 7 km. NNW Cerro Gordo;* Boca del Río; Catemaco. Chiapas: Yajalón; Volcán Kagchiná, 3½ km. N Las Margaritas; *Cueva Los Llanos, 9 km. S Las Margaritas;* Huixtla; *Finca Esperanza, 4 mi. NE Esquintla;* Prusia; Villa Flores. Oaxaca: Santiago Lachiguiri; Pluma Hidalgo;

ca. 6 mi. N Puerto Escondido (km. 212); *4 km. W San Gabriel Mixtepec; Santa Catarina Juguila; ca. 5 mi. NE Putla; Teotitlán del Camino.* Veracruz: Orizaba. Puebla: type locality. San Luis Potosí: *Xilitla; 2 km. SW Huichihuayán.* Not found: Chiapas: Finca Prusia.

Cryptotis parva soricina (Merriam)

1895. *Blarina soricina* Merriam, N. Amer. Fauna, 10:22, December 31, type from Tlalpam, 10 mi. S Mexico City, 7600 ft., Distrito Federal.
1970. *Cryptotis parva soricina,* Choate, Univ. Kansas Publ., Mus. Nat. Hist., 19:267, December 30.

MARGINAL RECORDS (Choate, 1970:268).—México: 1 km. S San Juan Zitlaltepec; Tlapacoyán. Distrito Federal: type locality; *Bosque Chapultepec.*

Cryptotis parva tropicalis (Merriam)

1895. *Blarina tropicalis* Merriam, N. Amer. Fauna, 10:21, December 31, type from Cobán, Guatemala. A renaming of the preoccupied *Sorex micrurus* Tomes, Proc. Zool. Soc. London, 1861:279, 1862, lectotype (informally designated by Thomas—see Handley and Choate, Proc. Biol. Soc. Washington, 83:195–201, May 27, 1970) from Cobán, *ca.* 4400 ft., Alta Verapaz, Guatemala. Not *G*[*alemys*]. *micrurus* Pomel, Arch. Sci. Phys. et Nat., Genève, 9:249, 1848 [= *Sorex talpoides* Gapper].
1970. *Cryptotis parva tropicalis,* Choate, Univ. Kansas Publ., Mus. Nat. Hist., 19:268, December 30.

MARGINAL RECORDS (Choate, 1970:269, 270).—Chiapas: La Libertad. Belize: Mountain Pine Ridge, 12 mi. (by road) S Cayo. Guatemala: type locality; *La Primavera;* Panajachel; Finca La Paz. Chiapas: *Unión Juárez.*

Cryptotis nigrescens
Blackish Small-eared Shrew

External measurements: total length, 82–108; tail averaging, according to subspecies, 37–41 per cent of head and body; hind foot, 10.5–14.0. Condylobasal length, 17.8–19.8; maxillary breadth, 5.6–6.6; maxillary tooth-row, 6.1–7.7; cranial breadth, 8.3–10.4. Winter pelage "salt-and-pepper" brown or brownish black in life; summer pelage "salt-and-pepper" gray in life. "Cranial characteristics: rostrum relatively short; braincase not especially angular; anterior limit of zygomatic plate varying from slightly anterior to mesostyle of M1 to slightly posterior of that point; posterior limit of zygomatic plate usually at level of, or posterior to, maxillary process, and above or slightly anterior to M3; dentition decidedly bulbous; anterior element of ectoloph of M1 not reduced relative to posterior element; posterior surfaces of P4–M2 not at all recessed; protoconal basin of M1 not reduced relative to hypoconal basin; M3 consisting only of paracrista and pre-centrocrista; talonid of m3 elongate and unspecialized, but consisting only of hypoconid." (Choate, 1970:270.)

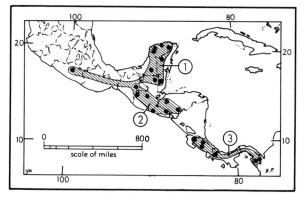

Map 36. *Cryptotis nigrescens.*

1. *C. n. mayensis*   2. *C. n. merriami*
3. *C. n. nigrescens*

Cryptotis nigrescens mayensis (Merriam)

1901. *Blarina mayensis* Merriam, Proc. Washington Acad. Sci., 3:559, November 29, type from Maya ruin at Chichén-Itzá, Yucatán.
1970. *Cryptotis nigrescens mayensis*, Choate, Univ. Kansas Publ., Mus. Nat. Hist., 19:275, December 30.

MARGINAL RECORDS (Choate, 1970:277).—Yucatán: Buctzotz; *Calotmul;* Nabalam. Quintana Roo: 2 km. SE Laguna de Chichancanab. Belize: Baking Pot. Guatemala: Uaxactún. Guerrero: Cueva del Cañón del Zopila[o]te, 13 km. S Puente de Mexcala. Campeche: La Tuxpeña. Yucatán: *Uxmal;* Actun Spukil; 6 km. S Mérida; *Temax.* Not found: Belize: "Central Farm." Yucatán: Senotillo.

Cryptotis nigrescens merriami Choate

1970. *Cryptotis nigrescens merriami* Choate, Univ. Kansas Publ., Mus. Nat. Hist., 19:277, December 30, type from Jacaltenango, 5400 ft., Huehuetenango, Gutemala.

MARGINAL RECORDS (Choate, 1970:278, 279).—Chiapas: Volcán Kagchiná. Guatemala: La Primavera. Honduras: San José, E Llama; Montserrat Cloud Forest, near Yuscarán. El Salvador: Cerro Cacaquatique; 2 mi. NW Apaneca. Guatemala: Jacaltenango. Chiapas: *Cueva Los Llanos.*

Cryptotis nigrescens nigrescens (J. A. Allen)

1895. *Blarina (Soriciscus) nigrescens* J. A. Allen, Bull. Amer. Mus. Nat. Hist., 7:339, November 8, type from "San Isidro (San José)," Costa Rica.
1911. C[ryptotis]. nigrescens, Miller, Proc. Biol. Soc. Washington, 24:222, October 31.
1912. *Cryptotis merus* Goldman, Smiths. Miscl. Coll.,

60(2):17, September 20, type from near head Río Limón, 4500 ft., Mt. Pirre, eastern Panamá.
1950. *Cryptotis zeteki* Setzer, Jour. Washington Acad. Sci., 40:299, September 29, type from Cerro Punta, 6500 ft., Chiriquí, Panamá.
1954. *Cryptotis tersus* Goodwin, Amer. Mus. Novit., 1677:1, June 28, type from Santa Clara, 4200 ft., Chiriquí, Panamá.

MARGINAL RECORDS (Choate, 1970:281).—Costa Rica: 4½ km. NE Tilarán; 14 mi. N San Isidro de El General. Panamá: Cerro Punta; *Cerro Tacarcuna;* Cerro Malí; E slope Cerro Pirre, near head Río Limón; *Río Candela, Volcán de Chiriquí; Santa Clara.* Costa Rica: Cerro Tablazo.

Cryptotis gracilis Miller
Talamancan Small-eared Shrew

1911. *Cryptotis gracilis* Miller, Proc. Biol. Soc. Washington, 24:221, October 31, type from head Lari River, 6000 ft., near base of Pico Blanco, Limón (formerly Talamanca), Costa Rica.
1944. *Cryptotis jacksoni* Goodwin, Amer. Mus. Novit., 1267:1, December 10, type from Volcán Irazú, Costa Rica.

External measurements: total length, 91–116; tail averaging 52 per cent of head and body; hind foot, 13–15. Condylobasal length, 18.6–20.7; maxillary breadth, 5.4–6.2; maxillary tooth-row, 6.8–7.7; cranial breadth, 9.1–10.3. Upper parts in winter and summer near Clove Brown to almost black in life; underparts only slightly paler. "Cranial characteristics: rostrum elongate, slender; braincase almost circular in dorsal outline; anterior limit of zygomatic plate above metastyle or between mesostyle and metastyle of M1; posterior limit of zygomatic plate at level of, or slightly posterior to, maxillary process, and above M3 or juncture of M2 and M3; dentition not bulb-

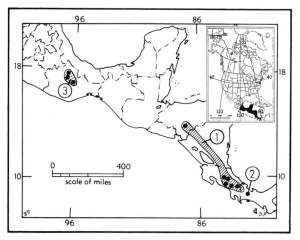

Map 37. Three species of *Cryptotis.*

1. *Cryptotis gracilis*   2. *Cryptotis endersi*
3. *Cryptotis magna*

ous; anterior element of ectoloph of M1 slightly reduced relative to posterior element; posterior surfaces of P4–M2 negligibly or only slightly recessed; protoconal basin of M1 slightly reduced relative to hypoconal basin; M3 consisting of paracrista, precentrocrista, postcentrocrista, vestigial metacrista, and well-developed metacone; talonid of m3 consisting of well-developed hypoconid and entoconid (specimen from Honduras lacks entoconid)." (Choate, 1970:282.)

MARGINAL RECORDS (Choate, 1970:285).— Honduras: W slope Cerro Uyuca, 12 km. WNW El Zamorano. Costa Rica: Hda. El Retiró, Volcán Turrialba; type locality. Panamá: *Cerro Punta–Boquete trail.* Costa Rica: Cerro Estaquero; *La Piedra; N slope Cerro de la Muerte;* Cerro Las Vueltas; 11 km. NNE Heredia.

Cryptotis endersi Setzer
Enders' Small-eared Shrew

1950. *Cryptotis endersi* Setzer, Jour. Washington Acad. Sci., 40:300, September 29, type from Bocas del Toro, Chiriquí, Panamá. Known only from type specimen, a juvenile.

External measurements: total length, 109; tail, 36 [49% of head and body]; hind foot, 13 (12 in orig. descr.). Condylobasal length, 20.4; maxillary tooth-row, 7.5; cranial breadth, 9.9. Juvenal pelage as in *C. n. nigrescens.* "Cranial characteristics: rostrum exceptionally elongate, broad; braincase not angular; anterior limit of zygomatic plate above metastyle of M1; posterior limit of zygomatic plate above M3; dentition bulbous; anterior element of ectolph of M1 not reduced relative to posterior element; posterior surfaces of P4–M2 not recessed; protoconal basin of M1 not reduced relative to hypoconal basin; M3 consisting of paracrista, precentrocrista, vestigial postcentrocrista, and well-developed metacone; talonid of m3 consisting of hypoconid and entoconid, . . . latter not well developed." (Choate, 1970:285, 286.)

Cryptotis magna (Merriam)
Big Small-eared Shrew

1895. *Blarina magna* Merriam, N. Amer. Fauna, 10:28, December 31, type from Totontepec, 6800 ft., Oaxaca.
1912. *Cryptotis magna,* Miller, Bull. U.S. Nat. Mus., 79:28, December 31.

External measurements: total length, 123–135; tail averaging 52 per cent of head and body; hind foot, 16.0–17.5. Condylobasal length, 22.5–23.7; maxillary breadth, 7.3–7.8; maxillary tooth-row, 8.2–9.0; cranial breadth, 11.4–12.4. Upper parts in winter Clove Brown in life; underparts almost as dark but slightly paler; upper parts in summer Fuscous-Black to Chaetura Black in life; underparts slightly paler. "Cranial characteristics: skull massive; rostrum elongate, broad; braincase angular; anterior limit of zygomatic plate above metastyle of M1; posterior limit of zygomatic plate at level of, or posterior to, maxillary process and above M3; dentition moderately bulbous; anterior element of ectoloph of M1 slightly reduced relative to posterior element; posterior surfaces of P4–M2 not recessed; protoconal basin of M1 not reduced relative to hypoconal basin; M3 consisting of paracrista, precentrocrista, and postcentrocrista; talonid of m3 consisting of both hypoconid and entoconid." (Choate, 1970:288.)

MARGINAL RECORDS (Choate, 1970:289, unless otherwise noted).—Oaxaca: San Isidro; *12 mi. NE Llano de las Flores; type locality;* Cerro Zempoaltepec; 12 mi. SSW Vista Hermosa; *S slope Cerro Pelon, 9200 ft.* (Musser, 1964:6).

Genus Notiosorex Coues—Desert Shrews

Revised by Merriam, N. Amer. Fauna, 10:31–34, December 31, 1895.

1877. *Notiosorex* Coues, Bull. U.S. Geol. and Geog. Surv. Territories, 3:646, May 15. Type, *Sorex (Notiosorex) crawfordi* Coues.
1950. *Megasorex* Hibbard, Contrib. Mus. Paleo., Univ. Michigan, 8:127, June 29. Type, *Notiosorex gigas* Merriam.

Tail relatively short; 3 unicuspids forming a uniform series, 3rd more than half as large as 2nd and never minute; unicuspids narrow-based without an internal cusplet. Dentition: i. $\frac{3}{2}$, c. $\frac{1}{0}$, p. $\frac{1}{1}$, m. $\frac{3}{3}$.

KEY TO NORTH AMERICAN SPECIES OF NOTIOSOREX

1. Posterior borders of P4, M1, and M2 nearly straight when viewed from occlusal surface (see Fig. 31). . . . . .*N. gigas,* p. 65
1'. Posterior borders of P4, M1, and M2 "excavated" (deeply concave) when viewed from occlusal surface (see Fig. 30).
N. crawfordi, p. 64

Notiosorex crawfordi
Crawford's Desert Shrew

External measurements: 87–101; 23–39; 11–13. Greatest length of skull, 18.1 (Baker, 1962:283). Grayish to plumbeous above, the same or paler below; cranial characters those of the genus; differs from *N. gigas* in concave posterior borders of P4, M1, and M2.

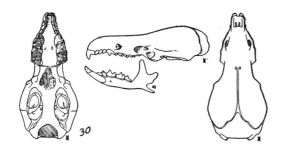

Fig. 30. *Notiosorex crawfordi crawfordi*, 5 mi. E and 1 mi. N Grapevine Peak, 5500 ft., Nye Co., Nevada, No. 92391 M.V.Z., ♂, X 2.

This shrew has most often been taken in desert areas and a number have been captured in apiaries. Little is known of its habits.

Notiosorex crawfordi crawfordi (Coues)

1877. *Sorex (Notiosorex) crawfordi* Coues, Bull. U.S. Geol. and Geog. Surv. Territories, 3:651, May 15, type from near old Fort Bliss, approximately 2 mi. above El Paso, El Paso Co., Texas.
1895. *Notiosorex crawfordi*, Merriam, N. Amer. Fauna, 10:32, December 31.

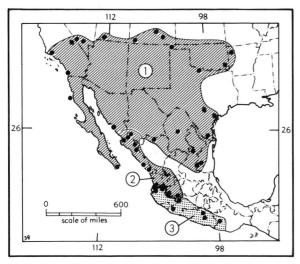

Map 38. *Notiosorex crawfordi* and *Notiosorex gigas*.

1. *N. c. crawfordi*   2. *N. c. evotis*
3. *N. gigas*

MARGINAL RECORDS.—Colorado: Eightmile Creek, 6780 ft. (Armstrong, 1972:53); 3 mi. NW Higbee. Oklahoma: Tesquite Canyon. Arkansas: 5 mi. S West Fork, 10 mi. NE Natural Dam (Preston and Sealander, 1969:641). Oklahoma: northeastern Pushmataha County. Texas: 5 mi. W Wichita Falls (Dalquest, 1968:15, owl pellet specimen only); San Antonio; Refugio County (Davis, 1966:41, as *N. crawfordi* only); *Corpus Christi; Riviera* (Davis, 1941:319, as *N. crawfordi* only); 2 mi. S Faysville (Davis, 1941:320, as

N. crawfordi only). Tamaulipas: Jaumave; Palmilla [= Palmillas]. Nuevo León: 3 mi. SE Galeana, 5100 ft. (Dalby and Baker, 1967:195, as *N. crawfordi* only). Coahuila: 3 mi. NW Cuatro Ciénegas. Durango: 7 mi. NNE Boquilla, 6400 ft. (Baker, 1966:345). Sonora: 1 mi. W Maytorena (Cockrum and Bradshaw, 1963:3). Baja California: Santa Anita; San Martín Island (Schulz, *et al.*, 1970:148). California: San Diego; El Rincon Creek; Grapevine Canyon, Saline Valley (Coulombe and Banta, 1964:277). Nevada: 1 mi. N, 5 mi. E Grapevine Peak; Ranger Mtn. (Jorgensen and Hayward, 1963:582). Utah: Oak Creek Canyon, Zion National Park, 4100 ft. (Wauer, 1965:496, as *N. crawfordi* only). Colorado: Mesa Verde National Park (Douglas, 1967:322).

Notiosorex crawfordi evotis (Coues)

1877. *Sorex (Notiosorex) evotis* Coues, Bull. U.S. Geol. and Geog. Surv. Territories, 3:652, May 15, type from Mazatlán, Sinaloa.
1895. *Notiosorex crawfordi evotis*, Merriam, N. Amer. Fauna, 10:34, December 31; also Armstrong and Jones, Jour. Mamm., 52:750, December 16, 1971.

MARGINAL RECORDS.—Sinaloa (Armstrong and Jones, 1971:750): 1 mi. S El Cajón, 1800 ft.; 44 km. ENE Sinaloa, 600 ft.; 20 km. N, 5 km. E Badiraguato, 1800 ft. Michoacán: 2 mi. E La Palma. Jalisco: 21 mi. SW Guadalajara. Nayarit: 5 mi. W Tepic (Schlitter, 1973:423). Sinaloa (Armstrong and Jones, 1971: 750): 15 mi. SE Escuinapa, 125 ft.; Mazatlán; 5 mi. WNW El Carrizo; El Fuerte.—See addenda.

Notiosorex gigas Merriam
Merriam's Desert Shrew

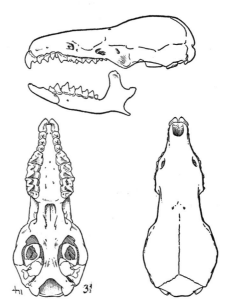

Fig. 31. *Notiosorex gigas*, Los Reyes, Michoacán, No. 125896 U.S.N.M., ♂, X 2.

1897. *Notiosorex gigas* Merriam, Proc. Biol. Soc. Washington, 11:227, July 15, type from mountains at Milpillas, near San Sebastián, Jalisco.

External measurements: 127–130; 39–43; 16–17. Condylobasal length, 22.2–22.8; breadth of braincase, 10.5. Pelage plumbeous, barely paler on underparts than on upper parts. Differs cranially from *N. crawfordi* in straight posterior borders of P4, M1, M2.

MARGINAL RECORDS.—Nayarit: 7$\frac{3}{10}$ km. ESE Amatlán de Cañas, 5000 ft. (Jones, 1966:249). Michoacán: Los Reyes. Guerrero: Cueva del Cañón del Zopilote, 13 km. S Puente de Mezcala, 720 m. (Ramírez-P. and Sánchez-H., 1974:107, 108). Oaxaca: San Vicente (Goodwin, 1969:43). Guerrero: 14 mi. S Chilpancingo (Jones, 1966:249). Michoacán: 7$\frac{1}{2}$ mi. by road E Dos Aguas, 5600 ft. (Winkelmann, 1962:109). Jalisco: type locality.—See addenda.

FAMILY TALPIDAE—Moles

The American members of the family have been revised by Jackson (N. Amer. Fauna, 38:1–100, illustrated, September 30, 1915).

External measurements: 100–237; 18–83.5; 13.0–30. Greatest length of skull, 21.1–44.6. Pelage resembling velvet; snout long; opening for eye in skin minute or wanting; forefeet much broader horizontally than thick (vertically); zygomatic arch complete but delicate; crowns of upper molars W-shaped.

From the American shrews, all of which belong to the family Soricidae, the American moles differ as follows: pinna of ear absent, instead of present (small and inconspicuous in *Blarina* and *Cryptotis*); clavicle short and broad instead of long and slender; humerus less, instead of more, than twice as long as wide; pelvis 3, instead of less than 3, times, as long as wide; os falciforme present, instead of absent, on forefoot; terminal phalanges of forefeet bifurcate, instead of simple; zygomata present rather than absent; external pterygoid region rounded and much inflated instead of angular and not inflated; first upper incisor flat and without elongated crown instead of not flat and with elongated crown.

The American members of the family Talpidae all are highly modified for burrowing and for a life underground; one example of this modification is the enlargement of the foreleg and pectoral girdle. All the species are primarily insectivorous but some (*e.g.*, *Scapanus townsendii*) eat some plant material. None of the American genera is truly aquatic, but *Condylura* is semiaquatic in that it enters water through burrows of its own making which enter the sides and bottoms of streams under water.

The family has been divided into five subfamilies (see Cabrera, 1925:79), and only two, Condylurinae and Scalopinae, occur in the New World.

KEY TO NORTH AMERICAN GENERA OF TALPIDAE

1. Tail less than $\frac{1}{4}$ of total length; width of palm equaling or exceeding length.
  2. Tail naked or but scantily haired; nostrils superior; auditory bullae complete; interior basal projection of upper molars narrow, simple.
    3. Tail slender, essentially naked; foretoes webbed; 2 lower incisors; geographic range east of Rocky Mts. . . . .*Scalopus*,   p. 71
    3'. Tail fleshy and scantily haired; foretoes not webbed; 3 lower incisors; geographic range west of Rocky Mts. . . . . . .*Scapanus*,   p. 67
  2'. Tail densely covered with hair; nostrils lateral; auditory bullae incomplete; interior basal projection of upper molars broad, lobed.
    Parascalops,   p. 71
1'. Tail more than $\frac{1}{4}$ of total length; width of palm less than length.
  4. Anterior end of snout with circular fringe of fleshy processes; nostrils circular to oval, and anterior; interior basal projection of 1st and 2nd upper molars trilobed; premolars $\frac{4}{4}$.
    Condylura,   p. 75
  4'. Anterior end of snout without circular fringe of fleshy processes; nostrils crescentic and lateral; interior basal projection of 1st and 2nd upper molars bilobed; premolars $\frac{3}{3}$. *Neurotrichus*,   p. 66

SUBFAMILY SCALOPINAE

Anterior nasal openings of skull directed anteriorly; snout lacking fleshy appendages.

Genus **Neurotrichus** Günther—Shrew-mole

Revised by Jackson, N. Amer. Fauna, 38:92–98, September 30, 1915.

1880. *Neurotrichus* Günther, Proc. Zool. Soc. London, p. 441, October. Type, *Urotrichus gibbsii* Baird.

Smallest of the American moles; external measurements: 107–126; 33–42; 15.7–17.0. Greatest length of skull, 21.5–24.2. Color dark mouse-gray or blackish mouse-gray. Tail approx. half as long as head and body, moderately fleshy, constricted at base, scaled, annulated, provided with coarse hairs; palms of forefeet longer than broad; toes not webbed; 6 tubercles on sole of hind foot; auditory bullae incomplete; braincase broad and skull scarcely constricted interorbitally; zygomata short; upper molars with bicuspidate internal edge; 1st and 2nd subequal and 3rd smaller. Dentition, i. $\frac{3}{3}$, c. $\frac{1}{1}$, p. $\frac{2}{2}$, m. $\frac{3}{3}$.

Neurotrichus gibbsii
Shrew-mole

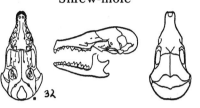

Fig. 32. *Neurotrichus gibbsii hyacinthinus*, 5 mi. NNE Point Reyes Lighthouse, Marin Co., California, No. 96374 M.V.Z., ♀, X 1.

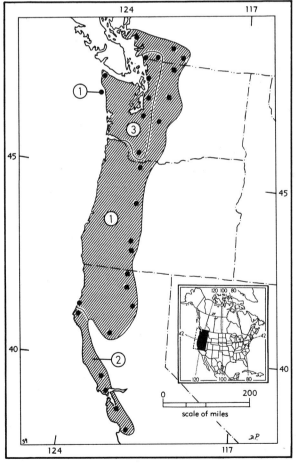

Map 39. *Neurotrichus gibbsii.*

1. *N. g. gibbsii*   2. *N. g. hyacinthinus*
3. *N. g. minor*

See characters of the genus.

Neurotrichus gibbsii gibbsii (Baird)

1858. *Urotrichus gibbsii* Baird, Mammals, *in* Repts. Expl. Surv. . . . , 8(1):76, July 14, type from Naches Pass, 4500 ft., Pierce Co., Washington (*fide* Dalquest and Burgner, 1941:12).
1880. *Neurotrichus gibbsi,* Günther, Proc. Zool. Soc. London, Pl. 42, October.
1899. *Neurotrichus gibbsi major* Merriam, N. Amer. Fauna, 16:88, October 28, type from Carberry Ranch, 4100 ft., between Mt. Shasta and Mt. Lassen, Shasta Co., California.

MARGINAL RECORDS.—British Columbia: Roberts Creek; Hope; Allison Pass. Washington: Baker Lake; Stevens Pass; type locality. Oregon: Multnomah Falls; McKenzie Bridge; Crater Lake; Fort Klamath. California: Beswick; Mt. Shasta; Carberry Ranch, Shasta Co.; South Yolla Bolly Mtn.; Eureka. Washington: Destruction Island.

Neurotrichus gibbsii hyacinthinus Bangs

1897. *Neurotrichus gibbsi hyacinthinus* Bangs, Amer. Nat., 31:240, March, type from Nicasio, Marin Co., California.

MARGINAL RECORDS.—California: Cuddeback; Guerneville; type locality; Palo Alto; Fremont Peak.

Neurotrichus gibbsii minor Dalquest and Burgner

1941. *Neurotrichus gibbsii minor* Dalquest and Burgner, Murrelet, 22:12, April 30, type from Univ. Washington Campus, Seattle, King Co., Washington.

MARGINAL RECORDS.—British Columbia: *lowland S side Fraser River* (Cowan and Guiguet, 1965:63); Vedder Crossing; *Cultus Lake.* Washington: Mt. Vernon; Cottage Lake; Spanaway; Yacolt; Neah Bay British Columbia: Crescent Beach.

Genus Scapanus Pomel—Western Moles

Revised by Jackson, N. Amer. Fauna, 38:54–76, September 30, 1915.

1848. *Scapanus* Pomel, Arch. Sci. Phys. et Nat., Genève, 9:247, November. Type, *Scalops townsendii* Bachman.

External measurements: 132–237; 21–51; 18–28. Greatest length of skull, 29.7–44.6. Body robust; tail short and thick, tapered toward tip, slightly constricted proximally, indistinctly annulated, bearing coarse hairs; snout shorter and less truncate than in *Scalopus,* naked anterior to nostrils; palms of forefeet as broad as long; soles of hind feet bearing 1–3 (usually 2) distinct tubercles; neither forefeet nor hind feet webbed; 4 pairs of mammae, 2 pectoral, 1 abdominal, 1 inguinal; skull conoidal, flattened; braincase relatively broad; interparietal large, somewhat rectangular; auditory bullae complete, depressed;

infraorbital foramen relatively small, but larger than in *Scalopus;* I1 long and broad; C simple and conical, approx. two-thirds as large as I1; M1 and M2 subequal, M3 much smaller than either of those teeth. Functional dentition, i. $\frac{3}{3}$, c. $\frac{1}{1}$, p. $\frac{4}{4}$ (in some specimens $\frac{2}{3}$ and in others $\frac{3}{3}$), m. $\frac{3}{3}$.

KEY TO SPECIES OF SCAPANUS

1. Unicuspid teeth evenly spaced and not crowded; rostrum long and narrow; color dark, almost black (except in some populations of the small *S. orarius*).
    2. Total length averaging more than 200; greatest length of skull more than 40; sublachrymal-maxillary ridge distinct. . . . . *S. townsendii,*   p. 68
    2′. Total length averaging less than 200; greatest length of skull less than 40; sublachrymal-maxillary ridge little developed. . . . . . . *S. orarius,*   p. 68
1′. Unicuspid teeth unevenly spaced and usually crowded; rostrum short and broad; color usually brown or gray, seldom almost black. . . . . *S. latimanus,*   p. 69

Scapanus townsendii (Bachman)
Townsend's Mole

1839. *Scalops Townsendii* Bachman, Jour. Acad. Nat. Sci., Philadelphia, 8:58, November, type from vic. Vancouver, Clark Co., Washington. (See True, Proc. U.S. Nat. Mus., 19:63, December 21, 1896.)
1848. *Scapanus Tow[n]sendii,* Pomel, Arch. Sci. Phys. et Nat., Genève, 9:247, November.

External measurements: 195–237; 34–51; 24–28. Greatest length of skull, 41.2–44.6. Upper parts blackish brown to almost black; sublachrymal ridge well developed.

MARGINAL RECORDS.—British Columbia: Huntingdon. Washington: Sauk; Skykomish; Spray Park, Mt. Rainier National Park, 5500 ft.; Yacolt. Oregon:

Oregon City; Drain; Grants Pass. California: Smith River; Ferndale.

Scapanus orarius
Coast Mole

External measurements: 162–175; 26–37; 21.5–23. Greatest length of skull, 32.8–39.0. Resembles *S. townsendii* but smaller; feet and claws relatively smaller; sublachrymal-maxillary ridge not much developed. Differs from *S. latimanus* in evenly spaced and uncrowded unicuspid teeth, narrower rostrum, undeveloped and indistinct sublachrymal-maxillary ridge, "weaker" mandible, and smaller incisors.

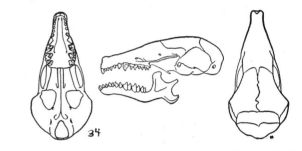

Fig. 34. *Scapanus orarius orarius,* Eureka, Humboldt Co., California, No. 19188 M.V.Z., ♂, X 1.

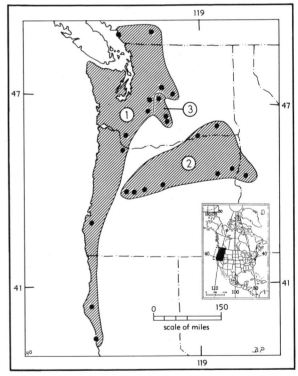

Map 40. *Scapanus orarius.*

1. *S. o. orarius*   2. *S. o. schefferi*
            3. *S. o. yakimensis*

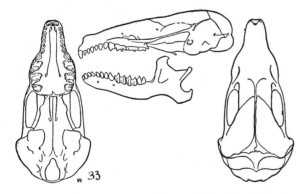

Fig. 33. *Scapanus townsendii,* Ferndale, Humboldt Co., California, No. 19115 M.V.Z., ♂, X 1.

Scapanus orarius orarius True

1896. *Scapanus orarius* True, Proc. U.S. Nat. Mus., 19:52, December 21, type from Shoalwater Bay (= Willapa Bay), Pacific Co., Washington.

MARGINAL RECORDS.—British Columbia: Hope. Washington: Wenatchee; Merritt; Lester; Owyhigh Lakes, 5100 ft., Mt. Rainier National Park; Yacolt. Oregon: Portland; Myrtle Point. California: Cuddeback; Mendocino. British Columbia: Vancouver (Cowan and Guiguet, 1965:61).

Scapanus orarius schefferi Jackson

1915. *Scapanus orarius schefferi* Jackson, N. Amer. Fauna, 38:63, September 30, type from Walla Walla, Walla Walla Co., Washington.

MARGINAL RECORDS.—Washington: Dayton. Idaho: Cambridge. Oregon: Halfway; near Baker; "Blue Mountains Country near Prineville"; N base Three Sisters, 5000 and 5500 ft.; McKenzie Bridge; Vida. Washington: Walla Walla.

Scapanus orarius yakimensis Dalquest and Scheffer

1944. *Scapanus orarius yakimensis* Dalquest and Scheffer, Murrelet, 25:27, September 19, type from ¾ mi. N Union Gap, Yakima Co., Washington.

MARGINAL RECORDS.—Washington: Easton; Selah; type locality.

Scapanus latimanus
Broad-footed Mole

Revised by F. G. Palmer, Jour. Mamm., 18:280–314, August 14, 1937.

External measurements: 132–192; 21–45; 18–25. Greatest length of skull, 29.7–37.4. Smaller and less blackish than *S. townsendii;* differences from *S. orarius* are listed in the account of that species.

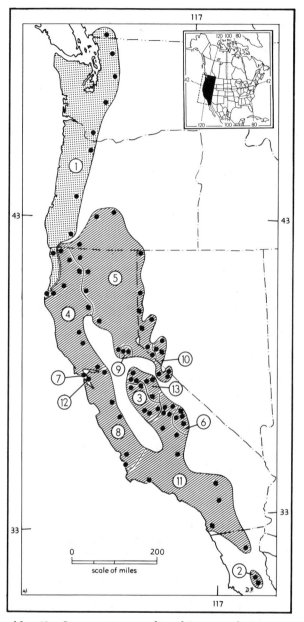

Map 41. *Scapanus townsendii* and *Scapanus latimanus.*

1. *S. townsendii*

2. *S. l. anthonyi*	8. *S. l. latimanus*
3. *S. l. campi*	9. *S. l. minusculus*
4. *S. l. caurinus*	10. *S. l. monoensis*
5. *S. l. dilatus*	11. *S. l. occultus*
6. *S. l. grinnelli*	12. *S. l. parvus*
7. *S. l. insularis*	13. *S. l. sericatus*

Scapanus latimanus anthonyi J. A. Allen

1893. *Scapanus anthonyi* J. A. Allen, Bull. Amer. Mus. Nat. Hist., 5:200, August 18, type from Sierra San Pedro Mártir, 7000 ft., Baja California.

1937. *Scapanus latimanus anthonyi*, F. G. Palmer, Jour. Mamm., 18:312, August 14.

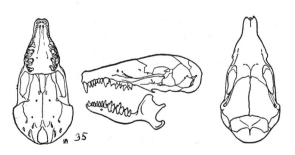

Fig. 35. *Scapanus latimanus monoensis*, East Walker River, 2 mi. NW Morgan's Ranch, Nevada, No. 63520 M.V.Z., ♂, X 1.

MARGINAL RECORDS.—Baja California: Vallecitos, 7500 ft.; La Grulla, 7200 ft.

Scapanus latimanus campi Grinnell and Storer

1916. *Scapanus latimanus campi* Grinnell and Storer, Univ. California Publ. Zool., 17:1, August 23, type from Snelling, 250 ft., Merced Co., California.

MARGINAL RECORDS.—California: El Portal, 2000 ft.; Dunlap, 2000 ft.; Minkler, 200 ft.; 3 mi. N Sanger; type locality.

Scapanus latimanus caurinus F. G. Palmer

1937. *Scapanus latimanus caurinus* F. G. Palmer, Jour. Mamm., 18:290, August 14, type from Laytonville, Mendocino Co., California.

MARGINAL RECORDS.—California: E fork Illinois River, ¼ mi. S Oregon line; Poker Flat, 5800 ft., 12 mi. NW Happy Camp; Scott River, 6 mi. NW Callahan; Red Bluff; Snow Mtn.; Long Valley; Lake Chabot; Cuddeback; Horse Mtn., 4700–5000 ft.

Scapanus latimanus dilatus True

1894. *Scapanus dilatus* True, Diagnoses of new North American mammals, p. 2, April 26 (preprint of Proc. U.S. Nat. Mus., 17:242, November 15, 1894), type from Fort Klamath, 4200 ft., Klamath Co., Oregon.
1913. *Scapanus latimanus dilatus*, Grinnell, Proc. California Acad. Sci., 3:269, August 28.
1897. *Scapanus truei* Merriam, Proc. Biol. Soc. Washington, 11:102, April 26, type from Lake City, Modoc Co., California.
1897. *Scapanus alpinus* Merriam, Proc. Biol. Soc. Washington, 11:102, April 26, type from Crater Lake, 7000 ft., Klamath Co., Oregon.

MARGINAL RECORDS.—Oregon: Fremont. Nevada: Twelve Mile Creek, 1½ mi. E California boundary, 5300 ft.; 1 mi. W Hausen; N side State-line Peak; 3 mi. W Reno; 4 mi. W Nixon, S end Pyramid Lake; 5 mi. S, 3¾ mi. E Minden; Holbrook. California: Auburn; Chico; Tower House; Sisson; head Doggett Creek [= Deer Camp], 5800 ft. Oregon: 6 mi. S Medford, 1600 ft.; Crater Lake.

Scapanus latimanus grinnelli Jackson

1914. *Scapanus latimanus grinnelli* Jackson, Proc. Biol. Soc. Washington, 27:56, March 20, type from site of old Fort Independence (on Carl Walters ranch), 2 mi. N Independence, 3900 ft., Inyo Co., California.

MARGINAL RECORDS.—California: type locality; Lone Pine; Olancha, Owens Lake; Upper Funston Meadow, 6700 ft., Kern River.

Scapanus latimanus insularis F. G. Palmer

1937. *Scapanus latimanus insularis* F. G. Palmer, Jour. Mamm., 18:297, August 14, type from Angel Island, San Francisco Bay, Marin Co., California. Known only from Angel Island.

Scapanus latimanus latimanus (Bachman)

1842. *Scalops latimanus* Bachman, Boston Jour. Nat. Hist., 4:34, type probably from Santa Clara, Santa Clara Co., California (*fide* Osgood, Proc. Biol. Soc. Washington, 20:52, 1907).
1912. *Scapanus latimanus latimanus*, Grinnell and Swarth, Univ. California Publ. Zool., 10:131, April 13.
1856. *Scalops californicus* Ayres, Proc. California Acad. Sci., 1:54. Based on specimens from San Francisco, California.

MARGINAL RECORDS.—California: Pinole; Bells Station; 1 mi. SE Summit San Benito Mtn., 4400 ft.; Santa Margarita; beach, between Santa Maria River and Oso Flaco Lake (McCully, 1967:480, as *S. latimanus* only).

Scapanus latimanus minusculus Bangs

1899. *Scapanus californicus minusculus* Bangs, Proc. New England Zool. Club, 1:70, July 31, type from Fyffe, 3500 ft., Eldorado Co., California.
1912. *Scapanus latimanus minusculus*, Grinnell and Swarth, Univ. California Publ. Zool., 10:133, April 13.

MARGINAL RECORDS.—California: Placerville; type locality.

Scapanus latimanus monoensis Grinnell

1918. *Scapanus latimanus monoensis* Grinnell, Univ. California Publ. Zool., 17:423, April 25, type from Taylor Ranch, 5300 ft., 2 mi. S Benton Station, Mono Co., California.

MARGINAL RECORDS.—Nevada: in bend Walker River, 4½ mi. E Wabuska; East Walker River, 2 mi. NW Morgans Ranch, 5050 ft. California: Pellisier Ranch, 5 mi. N Benton Station; type locality; Farrington's (near Williams Butte), 6800 ft., Mono Lake. Nevada: West Walker River, 10½ mi. S Yerington, 4500 ft.

Scapanus latimanus occultus Grinnell and Swarth

1912. *Scapanus latimanus occultus* Grinnell and Swarth, Univ. California Publ. Zool., 10:131, April 13, type from Santa Ana Canyon, 400 ft., 12 mi. NE Santa Ana, Orange Co., California.

MARGINAL RECORDS.—California: Kings River Canyon, 5000 ft.; Twin Lakes, 9800 ft., Sequoia National Park; Parker Meadow, 6400 ft.; Tehachapi; San Bernardino Peak; Strawberry Valley, 6000 ft., San Jacinto Mts. Baja California: Laguna Hanson, Sierra Juárez. California: Lakeside; Santa Barbara; 6 mi. WNW Porterville, 380 ft.; Big Meadow, 7660 ft.; Hume, 5300 ft.

Scapanus latimanus parvus F. G. Palmer

1937. *Scapanus latimanus parvus* F. G. Palmer, Jour. Mamm., 18:300, August 14, type from Alameda, Alameda Co., California. Known only from Alameda Island, Alameda Co., California.

Scapanus latimanus sericatus Jackson

1914. *Scapanus latimanus sericatus* Jackson, Proc. Biol. Soc. Washington, 27:55, March 20, type from Yosemite, Yosemite Valley, Mariposa Co., California.

MARGINAL RECORDS.—California: Twain Harte P.O., 4000 ft.; Tuolumne Meadows, 8600 ft.; Shaver Ranger Station, 5300 ft.; type locality; Merced Grove Big Trees, 5400 ft.; 3 mi. NE Coulterville, 3200 ft.

Genus Parascalops True—Hairy-tailed Mole

Revised by Jackson, N. Amer. Fauna, 38:77–82, September 30, 1915.

1894. *Parascalops* True, Diagnoses of new North American mammals, p. 2, April 26 (preprint of Proc. U.S. Nat. Mus., 17:242, November 15, 1894). Type, *Scalops breweri* Bachman.

External measurements: 139–153; 23–36; 18–20. Greatest length of skull, 31.0–33.8. Color blackish; tail thick, fleshy, constricted at base, annulated, and densely covered with long coarse hairs; snout shorter than in *Scalopus* or *Scapanus* and with median longitudinal groove on its anterior half; nostrils lateral, crescentic, with concavities directed upward; palms of forefeet as broad as long; toes not webbed; auditory bullae incomplete; M1 and M2 having trilobed internal basal shelf; corresponding part of M3 bilobed. Dentition, i. $\frac{3}{3}$, c. $\frac{1}{1}$, p. $\frac{4}{4}$, m. $\frac{3}{3}$.

Parascalops breweri (Bachman)
Hairy-tailed Mole

1842. *Scalops breweri* Bachman, Boston Jour. Nat. Hist., 4:32, type locality in eastern North America; type supposed by Bachman to have been taken on the island of Marthas Vineyard, Massachusetts, a locality where the animal probably does not occur.

1895. *Parascalops breweri*, True, Science, n. s., 1:101, January 25.

See characters of the genus.

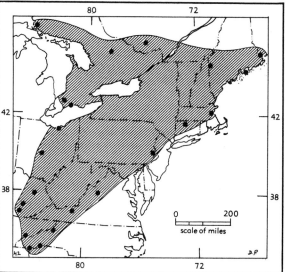

Fig. 36. *Parascalops breweri*, 2 mi. S Center Ossipee, Carroll Co., New Hampshire, No. 11335 K.U., sex ?, X 1.

MARGINAL RECORDS.—Ontario: Pancake Bay; Rock Lake, Algonquin Park. Quebec: Meaches Lake. Maine: Lake Umbagog. New Brunswick: Charlotte County. Maine: Mount Desert Island. Massachusetts: Harvard. Connecticut: West Winstead. Pennsylvania: Brownsburg. Virginia: Camp Todd, near Mt. Solon; Mountain Lake. North Carolina (Gordon and Bailey, 1963:580, 581, unless otherwise noted): Avery County; Transylvania County; just off U.S. Highway 28, 1⅝ mi. from intersection U.S. Highway 28 and U.S. Highway 64. Tennessee: Chapman Prong, 3200 ft. Kentucky: near Plato (Fassler, 1974:38); ½ mi. SW Morrill (Wallace and Houp, 1969:9); Triplet Creek, near Clearfield. Ohio: Franklin County; Rocky River Metropolitan Park. Ontario: Port Bruce on Lake Erie; 15–20 mi. W London.

Map 42. *Parascalops breweri*.

Genus Scalopus É. Geoffroy St.-Hilaire
Eastern Mole

Revised by Jackson, N. Amer. Fauna, 38:27–54, September 30, 1915.

1803. *Scalopus* É. Geoffroy St.-Hilaire, Catalogue des mammifères du Muséum National d'Histoire Naturelle, Paris, p.

77. Type, *Sorex aquaticus* Linnaeus. The Catalogue of 1803 by É. Geoffroy St.-Hilaire containing generic and also specific names has been regarded as not meeting the requirements of the International Code of Zoological Nomenclature by some authors, who therefore use the next available names. See Sherborn (1922, Index Animalium, Vol. A-B, lviii), Pocock (1939, Fauna of British India . . . , Vol. 1, p. 364, footnote), Ellerman and Morrison-Scott (1951, Checklist of Palaearctic and Indian Mammals 1758–1946, British Museum, London, p. 282, lines 16–23), Laurie and Hill (List of land mammals of New Guinea, Celebes, and adjacent islands 1758–1952, p. 100, footnote), and Harrison (1964, The mammals of Arabia, Vol. 1, p. 19, column 1, lines 23–26). Other authors, myself included, regard those names of 1803 proposed by É. Geoffroy St.-Hilaire as meeting the requirements of the Code and therefore use them. See, for example, Setzer (1952, Proc. U.S. Nat. Mus., 102:343), and Hershkovitz (1955, Proc. Biol. Soc. Washington, 68:185–191).

1811. *Scalops* Illiger, Prodromus Systematis mammalium et avium . . . , p. 126. Type, *Sorex aquaticus* Linnaeus.

1827. *Talpasorex* Lesson, Manuel de mammalogie . . . , p. 124. Type, *Scalops pennsylvanica* Harlan. Not *Talpasorex* Schinz.

External measurements: 128–208; 18–38; 15–22. Greatest length of skull, 29.3–39.5. Tail less than a fourth of total length, terete, indistinctly annulated and nearly naked; nose a distinct snout, naked anterior to nostrils; palms of forefeet broader than long; toes webbed on hind feet and on forefeet; interparietal short and narrow; frontal sinuses swollen; anterior nares of skull directed forward; auditory bullae complete; external pterygoid region much inflated posteriorly and slightly inflated anteriorly; I1 long and broad; C approx. two-thirds as large as first incisor and simple; M1 and M2 subequal; no persistent lower canine; lower premolars increasing in size posteriorly; lower molars successively decreasing in size posteriorly. Functional dentition, i. $\frac{3}{2}$, c. $\frac{1}{0}$, p. $\frac{3}{3}$, m. $\frac{3}{3}$.

Scalopus aquaticus montanus and *Scalopus aquaticus* from northern México and from Texas, respectively, seem now to be relict populations of the once more widely distributed species *Scalopus aquaticus*. Teeth probably will be found, if looked for in Pleistocene strata of western Texas.

Scalopus aquaticus
Eastern Mole

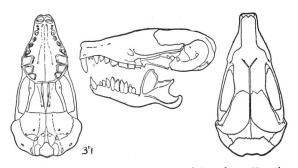

Fig. 37. *Scalopus aquaticus machrinoides*, Hamilton, Greenwood Co., Kansas, No. 95767 M.V.Z., ♂, X 1.

For characters see account of the genus.

Scalopus aquaticus aereus (Bangs)

1896. *Scalops texanus aereus* Bangs, Proc. Biol. Soc. Washington, 10:138, December 28, type from Stilwell, Adair Co., Oklahoma.

1912. *Scalopus aquaticus aereus*, Miller, Bull. U.S. Nat. Mus., 79:8, December 31 (*fide* Hall and Kelson, Univ. Kansas Publ., Mus. Nat. Hist., 5:326, December 5, 1952).

1899. *Scalops machrinus intermedius* Elliot, Field Columb. Mus., Publ. 37, Zool. Ser., 1:280, May 15, type from Alva, Woods Co., Oklahoma. Regarded as inseparable from *aereus* by Yates and Schmidly, Occas. Pap. Mus. Texas Tech Univ., 45:25, June 3, 1977.

1914. *Scalopus aquaticus pulcher* Jackson, Proc. Biol. Soc. Washington, 27:20, February 2, type from Delight, Pike Co., Arkansas.

MARGINAL RECORDS.—Kansas (as *S. a. intermedius*): Little Salt Marsh; 4 mi. NE Harper. Oklahoma: Garnett (as *S. a. intermedius*); Scraper. Arkansas (Yates and Schmidly, 1977: 26): Fayetteville; Lake City; Wilmot. Louisiana (Lowery, 1974:93): 9 mi. S Delhi; White Castle; Avery Island; Holmwood. Texas (Yates and Schmidly, 1977:27): Sour Lake; $8\frac{2}{3}$ mi. W Jasper; 3 mi. W Lufkin; 3 mi. SE Hallsville P.O.; Texarkana. Oklahoma: Dougherty (as *S. a. intermedius*). Texas (as *S. a. intermedius*): Belknap; Paducah; 6 mi. S, $2\frac{1}{2}$ mi. W Mobeetie; Stinnett. Colorado: near Cimarron River, 16 mi. S Stonington (Vaughan, 1961:172, as *S. aquaticus* only). Kansas: 9 mi. N, 3 mi. E Elkhart (as *S. a. intermedius*).

Scalopus aquaticus alleni Baker

1951. *Scalopus aquaticus alleni* Baker, Univ. Kansas Publ., Mus. Nat. Hist., 5:22, February 28, type from Rockport, Aransas Co., Texas.

MARGINAL RECORDS.—Texas (Yates and Schmidly, 1977:28): 3 mi. NE Gause; Eagle Lake; Aransas Wildlife Refuge; Padre Island; 7 mi. SW Somerset; Mason.

Scalopus aquaticus anastasae (Bangs)

1898. *Scalops anastasae* Bangs, Proc. Boston Soc. Nat. Hist.,

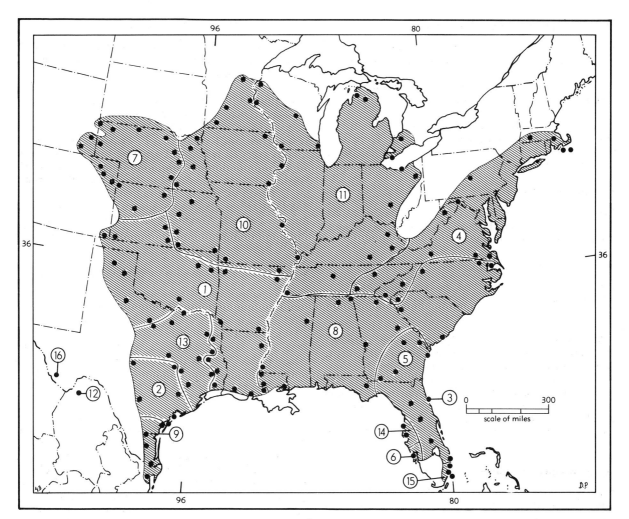

Map 43. *Scalopus aquaticus.*

1. *S. a. aereus*	7. *S. a. caryi*	12. *S. a. montanus*
2. *S. a. alleni*	8. *S. a. howelli*	13. *S. a. nanus*
3. *S. a. anastasae*	9. *S. a. inflatus*	14. *S. a. parvus*
4. *S. a. aquaticus*	10. *S. a. machrinoides*	15. *S. a. porteri*
5. *S. a. australis*	11. *S. a. machrinus*	16. *S. a. texanus*
6. *S. a. bassi*		

28:212, March, type from Point Romo, Anastasia Island, St. John Co., Florida. Known only from Anastasia Island, Florida.

1915. *Scalopus aquaticus anastasae*, Jackson, N. Amer. Fauna, 38:39, September 30.

Scalopus aquaticus aquaticus (Linnaeus)

1758. [*Sorex*] *aquaticus* Linnaeus, Syst. nat., ed. 10, 1:53. Type locality, eastern United States (Philadelphia, Pennsylvania; fixed by Jackson, N. Amer. Fauna, 38:33, September 30, 1915).

1905. *Scalopus aquaticus*, Oberholser, Mammals and summer birds of western North Carolina, Biltmore Forest School, Biltmore, North Carolina, p. 3, June 30.

1803. *Scalopus virginianus* É. Geoffroy St.-Hilaire, Catalogue des mammifères du Muséum National d'Histoire Naturelle, Paris, p. 78. Type locality, Virginia (?).

1814. *Talpa cupreata* Rafinesque, Précis des découvertes et travaux somiologiques . . . , p. 14. Type locality, "Atlantic States."

1825. *Scalops pennsylvanica* Harlan, Fauna Americana, p. 33, type from an unknown locality.

MARGINAL RECORDS.—Massachusetts: Holyoke; Middleboro; Nantucket; West Tisbury. Virginia: Dismal Swamp; Brunswick County; Montgomery County. North Carolina: Buncombe County. South Carolina: Greenville. North Carolina: Highlands. Tennessee: Briceville; Walden Ridge near Rathburn. West Virginia: Moorefield; Berkeley Springs. Pennsylvania: Mifflintown.

Scalopus aquaticus australis (Chapman)

1893. *Scalops aquaticus australis* Chapman, Bull. Amer. Mus. Nat. Hist., 5:339, December 22, type from Gainesville, Alachua Co., Florida.
1905. *Scalopus aquaticus australis,* Elliot, Field Columb. Mus., Publ. 105, Zool. Ser., 6:470.

MARGINAL RECORDS.—Georgia: Pinetucky; Hursmans Lake; Montgomery. Florida: Hypoluxo; Orange Hammock; Eustis; Levy Lake. Georgia: 10 mi. SSW Thomasville; Nashville.

Scalopus aquaticus bassi Howell

1939. *Scalopus aquaticus bassi* Howell, Jour. Mamm., 20:363, August 14, type from Englewood, Sarasota Co., Florida. Known only from type locality.

Scalopus aquaticus caryi Jackson

1914. *Scalopus aquaticus caryi* Jackson, Proc. Biol. Soc. Washington, 27:20, February 2, type from Neligh, Antelope Co., Nebraska.

MARGINAL RECORDS.—Nebraska (Jones, 1964c: 76): Fort Niobrara Wildlife Refuge; 10 mi. NE Stuart; type locality; 3 mi. S Columbus; Inland. Kansas: Smith Center; Logan County; 23 mi. (by road) NW St. Francis. Colorado: Wray; Akron; Merino. Wyoming (Long, 1965a:528): Horse Creek, 3 mi. W Meriden; Lingle. Nebraska (Jones, 1964c:76): 7 mi. S Gering; 5½ mi. N, 2½ mi. W Harrison; Monroe Canyon; 8 mi. N Harrison; Warbonnet Canyon; Chadron.

Scalopus aquaticus howelli Jackson

1914. *Scalopus aquaticus howelli* Jackson, Proc. Biol. Soc. Washington, 27:19, February 2, type from Autaugaville, Autauga Co., Alabama.

MARGINAL RECORDS.—North Carolina: Jackson; Moran. South Carolina: Charleston. Georgia: Crawfordsville; Columbus. Florida: Rock Bluff. Louisiana: 3 mi. N Lacombe (Lowery, 1974:92); *Madisonville* (ibid.); Baton Rouge; *St. Francisville;* 38 mi. NNW Baton Rouge (loc. cit.). Mississippi: Washington; Cedarbluff. Alabama: Huntsville; Sand Mtn., near Carpenter. Georgia: Young Harris. South Carolina: Abbeville. North Carolina: Wilkesboro.

Scalopus aquaticus inflatus Jackson

1914. *Scalopus inflatus* Jackson, Proc. Biol. Soc. Washington, 27:21, February 2, type from Tamaulipas, 45 miles from Brownsville, Texas.
1977. *Scalopus aquaticus inflatus,* Yates and Schmidly, Occas. Pap. Mus. Texas Tech Univ., 45:28, June 3.

MARGINAL RECORDS (Yates and Schmidly, 1977:29).—Texas: Corpus Christi; Brownsville. Tamaulipas: type locality. Texas: 1 mi. S Linn; S Falfurrias.

Scalopus aquaticus machrinoides Jackson

1914. *Scalopus aquaticus machrinoides* Jackson, Proc. Biol. Soc. Washington, 27:19, February 2, type from Manhattan, Riley Co., Kansas.

MARGINAL RECORDS.—Minnesota: 10 mi. SW Onamia (Heaney and Birney, 1975:30); Ramsey County. Iowa (Bowles, 1975:41, 42): Decorah; *SE ¼ of SE ¼ of sec. 18, T. 85 N, R. 3 E;* Princeton Twp., Scott County; sec. 12 Jefferson Twp., Lee County. Missouri: St. Louis; Charleston. Arkansas: Greenway. Missouri: Washburn (Yates and Schmidly, 1977:29). Kansas: Shole Creek; 6 mi. NE Wellington; Halstead; 6 mi. S Solomon; Blue Rapids. Nebraska (Jones, 1964c:78): Angus; Everett; Wayne. South Dakota: Vermillion. Minnesota (Heaney and Birney, 1975:30): 1½ mi. S, 5½ mi. W Kinbrae; 1½ mi. N Morton.

Scalopus aquaticus machrinus (Rafinesque)

1832. *Talpa machrina* Rafinesque, Atlantic Jour., 1:61. Type locality, near Lexington, Fayette Co., Kentucky.
1905. *Scalopus aquaticus machrinus,* Elliot, Field Columb. Mus., Publ. 105, Zool. Ser., 6:470.
1832. *Talpa sericea* Rafinesque, Atlantic Jour., 1:62, type locality near Nicholasville and Harrodsburg, Kentucky.
1842. *Scalops argentatus* Audubon and Bachman, Jour. Acad. Nat. Sci. Philadelphia, 8:292, type locality in southern Michigan.

MARGINAL RECORDS.—Wisconsin (Jackson, 1961:62, 67): Meenon Twp., Burnett Co.; Plainfield; Racine. Michigan: Charlevoix County; Valentine Lake. Ontario: Shathroy; Point Pelee. Ohio: Cleveland; Salem; Fairfield County. Kentucky: Rowan County; Quicksand; Pulaski County (Fassler, 1974:38, as *S. aquaticus* only). Tennessee: Nashville; Ellendale. Wisconsin: White City Resort, 1½ mi. N Dubuque, Iowa; Prescott.

Scalopus aquaticus montanus Baker

1951. *Scalopus montanus* Baker, Univ. Kansas Publ., Mus. Nat. Hist., 5:19, February 28, type from Club Sierra del Carmen, 2 mi. N, 6 mi. W Piedra Blanca, Coahuila. Known only from type locality.
1977. *Scalopus aquaticus montanus,* Yates and Schmidly, Occas. Pap. Mus. Texas Tech Univ., 45:29, June 3.

Scalopus aquaticus nanus Davis

1942. *Scalopus aquaticus nanus* Davis, Amer. Midland Nat., 27:383, March, type from 13 mi. E Centerville, Leon Co., Texas.
1942. *Scalopus aquaticus cryptus* Davis, Amer. Midland Nat., 27:384, March, type from College Station, Brazos Co., Texas. Regarded as inseparable from *nanus* by Yates and Schmidly, Occas. Pap. Mus. Texas Tech Univ., 45:30, June 3, 1977.

MARGINAL RECORDS.—Texas (Yates and Schmidly, 1977:30, 31, unless otherwise noted): Denison (Davis, 1942:383, as *S. a. aereus*); Gilmer; 10 mi. NE Nacogdoches; 3 mi. W Ratcliff; $9\frac{7}{10}$ mi. N Spurger; $2\frac{1}{2}$ mi. N Hockley (Davis, 1942:385, as *S. a. cryptus*); 3 mi. SE College Station; *Waco;* 7 mi. E Laguna Park; Possum Kingdom Lake; Denton (Davis, 1942:383, as *S. a. aereus*).

Scalopus aquaticus parvus (Rhoads)

1894. *Scalops parvus* Rhoads, Proc. Acad. Nat. Sci. Philadelphia, 46:157, type from Tarpon Springs, Pinellas Co., Florida.
1915. *Scalopus aquaticus parvus,* Jackson, N. Amer. Fauna, 38:41, September 30.

MARGINAL RECORDS.—Florida: Port Richey; Port Tampa City.

Scalopus aquaticus porteri Schwartz

1952. *Scalopus aquaticus porteri* Schwartz, Jour. Mamm., 33:381, August 19, type from Uleta, Dade Co., Florida.

MARGINAL RECORDS.—Florida: type locality; Biscayne Gardens; Lemon City.

Scalopus aquaticus texanus (J. A. Allen)

1891. *Scalops argentatus texanus* J. A. Allen, Bull. Amer. Mus. Nat. Hist., 3:221, April 29, type from Presidio County, Texas. Known only from type locality which, as Baker [see below] points out, might be in present-day Brewster, Jeff Davis, or Presidio counties, Texas.
1951. *Scalopus aquaticus texanus* Baker, Univ. Kansas Publ., Mus. Nat. Hist., 5:21, February 28. Not of V. Bailey, 1905.

SUBFAMILY **CONDYLURINAE**
Star-nosed Mole

Anterior nares of skull opening obliquely upward; snout provided with ring of fleshy appendages.

Genus **Condylura** Illiger—Star-nosed Mole

Revised by Jackson, N. Amer. Fauna, 38:82–91, September 30, 1915.

1811. *Condylura* Illiger, Prodromus systematis mammalium et avium . . . , p. 125. Type, *Sorex cristatus* Linnaeus.

External measurements: 183–211; 65–83.5; 26–30. Greatest length of skull, 33.9–35.2. Blackish brown to nearly black. Nose having ring of 22 fleshy appendages; tail approx. as long as body, constricted at base, scaled, annulated; palms of forefeet as broad as long; premaxillae much extended beyond nasals anteriorly; first upper incisors large, incurved, and projecting anteriorly.

Condylura cristata
Star-nosed Mole

C.G. Pritchard

Characters as for the genus.

Condylura cristata cristata (Linnaeus)

1758. [*Sorex*] *cristatus* Linnaeus, Syst. nat., ed. 10, 1:53. Type locality, Pennsylvania.
1819. *Condylura cristata,* Desmarest, Jour. de Phys., Chim., Hist. Nat., et des Arts, 89:230, September.

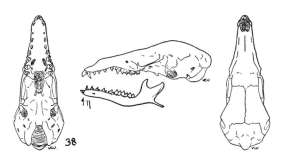

Fig. 38. *Condylura cristata cristata,* 2 mi. S Center Ossipee, Carroll Co., New Hampshire, No. 11336 K.U., ♂, X 1.

MARGINAL RECORDS.—Quebec: Papps Cove, Richmond Gulf (Edwards, 1963:5, as *C. cristata* only). Labrador: Hamilton Inlet; 8 mi. N Pinware (Threlfall, 1969:198). New Brunswick: Hampton. Virginia: Richmond County; Fairfax County. West Virginia: Lake Terra Alta. Pennsylvania: New Lexington. Ohio: Ellsworth. Indiana: Dearborn County; Bartholomew County (Rust, 1966:538); Worden Twp., Clark County. consin: Milwaukee County (Jackson, 1961:69); Wood County (Rus, 1966:538); Worden Twp., Clark County. Minnesota: *Carlos Avery Game Management Area, 5½ mi. N, 1 mi. W Lino Lakes* (Heaney and Birney, 1975:31); Elk River; Fort Ripley; Detroit Lakes. North Dakota: Tower. Manitoba: Riding Mtn.; Pine Falls; Island Lake (Beck and Wilson, 1969:93). Ontario: Moose Factory. Quebec: East Main River. For Illinois, Hoffmeister (1954:1) indicates there are no authentic records. The alleged occurrence in Olmstead County, Minnesota (Gunderson and Beer, 1953:36), is not shown on Map 44 because Heaney and Birney (1975:31) suspect the specimen was misidentified.

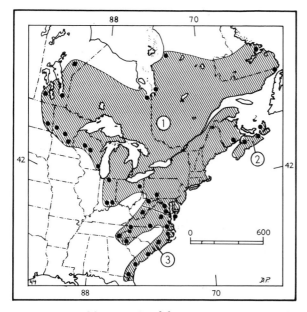

Map 44. *Condylura cristata.*

1. *C. c. cristata*   2. *C. c. nigra*
            3. *C. c. parva*

Condylura cristata nigra Smith

1940. *Condylura cristata nigra* R. W. Smith, Amer. Midland Nat., 24:218, July, type from Wolfville, Kings Co., Nova Scotia.

MARGINAL RECORDS.—Nova Scotia: Frizzleton; James River; Newport.

Condylura cristata parva Paradiso

1959. *Condylura cristata parva* Paradiso, Proc. Biol. Soc. Washington, 72:103, July 24, type from 5 mi. NW Stuart, Patrick Co., Virginia.

MARGINAL RECORDS (Paradiso, 1959:106, unless otherwise noted).—Virginia: $2\frac{3}{10}$ mi. E Wattsville, thence southward along coast to South Carolina: $2\frac{1}{2}$ mi. S Georgetown. Georgia: Okefenokee Swamp, Mixons Ferry; Marlow. North Carolina: Garland; Wenona. Virginia: Dismal Swamp; Richmond (Paradiso, 1959:103); type locality. South Carolina: "Upper South Carolina" (Hall and Kelson, 1959:76). North Carolina: Hayesville; Roan Mtn., Magnetic City. Tennessee: Shady Valley. West Virginia: $4\frac{1}{8}$ mi. NE Richwood.

Order **CHIROPTERA**—Bats

Families and genera revised by Miller, Bull. U.S. Nat. Mus., 57:17 + 282, 13 pls., 49 figs. in text, June 29, 1907.

Placental mammals unique in being modified for flight; bones of fingers greatly lengthened and supporting wing-membrane made of relatively naked double layer of skin; membrane extends to and usually also between hind legs (interfemoral- or tail-membrane much reduced in some species) and enclosing tail; tail usually present, variable in length and sometimes extending beyond membrane; pectoral region much developed and sternum usually keeled; clavicle well developed; ulna much reduced; calcar present and often of extreme development; thumb free of wing membrane and clawed; knee joint directed outward and posteriorly; ears well developed, often enormously so; eyes small in many taxa; dentition tuberculosectorial, molars in some genera broadened and having reduced cusps, in some other genera (*e.g.*, *Leptonycteris*) much elongated.

The keys to families and genera contained in this section are based mainly on those of Miller (1907). Where other published sources have been employed, individual mention of the source is given. Length of upper tooth-row of bats is taken from the posteriormost point of the last tooth in the jaw to the anterior face of the canine on the same side.

Suborder **MICROCHIROPTERA**

All New World bats belong to this suborder. Second finger closely bound to 3rd; humerus with trochiter and trochin large, the former usually articulating with the scapula; margin of external ear not forming complete ring.

Key to North American Families of Microchiroptera

1. Premaxillaries usually free, always incomplete, their boundaries never obliterated. Emballonuridae, p. 77
1'. Premaxillaries always fused with surrounding parts, complete or incomplete, their boundaries obliterated early in life.
  2. Ischia fused beneath posterior extremity of sacrum. . . . . Noctilionidae, p. 85
  2'. Ischia not fused beneath sacrum.
    3. Fibula robust, its diameter usually about half that of tibia and contributing much to strength of short, stout leg. . . . . Molossidae, p. 238
    3'. Fibula slender or rudimentary, not contributing much to strength of long, slender leg.
      4. Third phalanx of middle finger cartilaginous except at extreme base.
        5. Humerus with trochiter much longer than trochin, projecting conspicuously beyond head, and forming a complete secondary articulation with scapula. Vespertilionidae, p. 182
        5'. Humerus with trochiter scarcely longer than trochin, not projecting conspicuously beyond head, and its articulation with scapula frequently slight or absent.
          6. Presternum not broadened anteriorly; keel of mesosternum high; claw of thumb well developed. . . . . Natalidae, p. 176
          6'. Presternum greatly broadened anteriorly; keel of mesosternum a low ridge; claw of thumb rudimentary. . . . . Furipteridae, p. 179
    4'. Third phalanx of middle finger bony.
      7. Canine teeth shearlike; molars much reduced, without trace of crushing surface. Desmodontidae, p. 173
      7'. Canine teeth not shearlike or specially modified; molars well developed with at least some trace of crushing surface.
        8. Toes with two phalanges each; thumb and foot with suction cups; no nose-leaf, flaps, or folds of skin on face. . . . . Thyropteridae, p. 180
        8'. Toes, except hallux, with 3 phalanges each; no suction cups; nose-leaf, flaps, or folds of skin on face. . . . . Phyllostomatidae, p. 88

Family **EMBALLONURIDAE**
Sac-winged Bats

Skull having well-developed postorbital processes (partly obscured in *Diclidurus*); premaxillaries incomplete, possessing nasal portions only, never fused with maxillaries or with each other; auditory bullae emarginate on inner side. Humerus having well-developed trochiter; capitellum nearly in line with shaft; 2nd digit of manus having metacarpal but no phalanges; 7th cervical vertebra not fused to pectoral girdle; fibula complete, slender. Hind leg slender; tail perforating upper surface—not free edge—of interfemoral membrane; muzzle lacking leaflike excrescences.

The emballonurids, second only to the Old World Rhinopomidae, seem to combine the greatest number of primitive characters with the least degree of specialization according to Miller (1907:84).

KEY TO NORTH AMERICAN GENERA OF EMBALLONURIDAE

1. Postorbital process broad, almost obliterated by markedly widened supraorbital ridge; pelage white. . . . . *Diclidurus,*   p. 85
1'. Postorbital process slender; pelage not white.
  2. Clavicle markedly expanded, maximum width approx. ⅓ of length. . . . . *Cyttarops,*   p. 84
  2'. Clavicle normal (not markedly expanded), maximum width approx. ⅛ of length.
    3. Basisphenoidal pit divided by median septum.
      4. Anterior upper premolars tricuspidate.
        5. Sagittal crest pronounced, extending onto postorbital process of frontal.   *Cormura,*   p. 80
        5'. Sagittal crest indistinct, not extending onto postorbital process of frontal.
                                                  Centronycteris,   p. 82
      4'. Anterior upper premolars simple spicules, not tricuspidate.
        6. Distal part of rostrum with dorsal inflations. . . . . *Balantiopteryx,*   p. 83
        6'. Distal part of rostrum without dorsal inflations. . . . . *Saccopteryx,*   p. 79
    3'. Basisphenoidal pit not divided by median septum.
      7. Angle in profile between facial and cranial regions of skull less than 140°.
        8. Distal part of rostrum with dorsal inflations and with evident median depression between inflations. . . . . *Balantiopteryx,*   p. 83
        8'. Distal part of rostrum without dorsal inflations. . . . . *Peropteryx,*   p. 81
      7'. Angle in profile between facial and cranial regions of skull more than 160°.
        9. Sagittal crest pronounced. . . . . *Cormura,*   p. 80
        9'. Sagittal crest absent. . . . . *Rhynchonycteris,*   p. 78

Genus **Rhynchonycteris** Peters
Long-nosed Bats

1867. *Rhynchonycteris* Peters, Monatsb. preuss. Akad. Wiss., Berlin, p. 477, July 25. Type, *Vespertilio naso* Wied-Neuwied.
1823. *Proboscidea* Spix, Simiarum et vespertilionum Brasiliensium . . . , p. 61. Type, *Proboscidea saxatilis* Spix [= *Vespertilio naso* Wied-Neuwied]. Not *Proboscidea* Brugière, 1791, a nematode.
1907. *Rhynchiscus* Miller, Proc. Biol. Soc. Washington, 20:65, June 12. Type, *Vespertilio naso* Wied-Neuwied, a renaming of *Rhynchonycteris* Peters, erroneously assumed to be preoccupied by *Rhinchonycteris* Tschudi, 1844–1846 [= *Anoura* Gray, 1838].

No wing sacs; forearm dotted with tufts of fur; interfemoral membrane haired to exsertion of tail; muzzle greatly elongated; no angle between rostrum and forehead; first upper premolar large, triangular, possessing cingulum provided with a small anterior and a small posterior cusp (after Sanborn, 1937:325). Dentition, i. ⅔, c. ⅓, p. ⅔, m. ⅗. The genus is monotypic.

Rhynchonycteris naso (Wied-Neuwied)
Brazilian Long-nosed Bat

1820. *Vespertilio naso* Wied-Neuwied, Reise nach Brasilien . . . , 1:251, footnote, type from bank of Rio Mucuri near Morro d'Arara, Minas Geraes, Brazil.
1867. *Rhynchonycteris naso,* Peters, Monatsb. preuss. Akad. Wiss., Berlin, p. 478, July 25.
1823. *Proboscidea rivalis et saxatilis* Spix, Simiarum et vespertilionum Brasiliensium . . . , p. 62. Types from Rio Amazon and Rio San Francisco, Brazil.
1835–1841. *Emballonura lineata* Temminck, Monographies de mammalogie . . . , 2:297, type from Dutch Guiana.
1855. *Proboscidea villosa* Gervais, Mammiferes *in* [Castelnan] Expéd. dans les parties centrales de l'Amér. du Sud . . . , p. 68, pl. 11, type from Prov. Goyaz, Brazil.
1914. *Rhynchiscus naso priscus* G. M. Allen, Proc. Biol. Soc. Washington, 27:109, July 10, type from Xcopen, Quintana Roo. (Regarded as a synonym of *R. naso* by Sanborn, Field Mus. Nat. Hist., Publ. 399, Zool. Ser., 20:325, 326, 328, December 28, 1937.)

Length of head and body approx. 40; length of forearm, 35.3–40.7; greatest length of skull,

Fig. 39. *Rhynchonycteris naso*, 15 km. W Piedras Negras, 300 ft., Veracruz, No. 19096 K.U., ♀, X 2.

11.2–12.6; interorbital breadth, 2.3–2.8; zygomatic breadth, 6.7–7.3; length of upper tooth-row, 4.3–4.7. Color of upper parts depending greatly on amount of wear; brownish hairs tipped with gray in fresh pelage, gray wanting in worn pelage; much the same below, but gray more pronounced; 2 curved, gray or white, lines usually apparent on lower back and rump. See also generic description.

MARGINAL RECORDS.—Veracruz: 5 km. SW Boca del Río; *14 km. SW Coatzacoalcos, 100 ft.* (Hall and Dalquest, 1963:208). Chiapas: 21 km. WSW Teapa (Carter, *et al.*, 1966:488), thence across base of Peninsula of Yucatán to Quintana Roo: Chetumal. Costa Rica (Starrett and Casebeer, 1968:2): 4 mi. W Puerto Viejo de Sarapiquí, 92 m.; Río Madre de Dios, near Finca La Lola, 50 m. Panamá (Handley, 1966:757): Sibube; Fort Gulick; Armila, and southward into South America; along Pacific Coast northward to Panamá (*ibid.*): Río Jesucito; San Félix. Costa Rica (Starrett and Casebeer, 1968:2): 6½ mi. N, 2 mi. W Puntarenas, 30 m.; Río Diria at Santa Cruz, 60 m.; 5 km. N, 4 km. W Liberia, 50 m. Guatemala: Ocas (Jones, 1966:442). Oaxaca: 25 km. NE Matías Romero (Schaldach, 1965:131). Veracruz: *Río Blanco, 20 km. W Piedras Negras, 400 ft.* (Hall and Dalquest, 1963:208).

Genus **Saccopteryx** Illiger—White-lined Bats

1811. *Saccopteryx* Illiger, Prodromus systematis mammalium et avium . . . , p. 121. Type, *Vespertilio lepturus* Schreber.
1838–39. *Urocryptus* Temminck, Tijdschr. natuurl. Gesch. Phys., 5:31. Type, *Urocryptus bilineatus* Temminck.

Length of head and body approx. 49. Large glandular sac close to forearm near elbow and opening on upper surface of antebrachial membrane; wing-membrane from tarsus. Skull with slight angle between rostrum and forehead; premaxillae large, ending on upper surface of rostrum; postorbital processes long and relatively broad; sagittal crest on braincase; basisphenoidal pits large, separated by median septum; 1st upper premolar (P2) a small spicule (after Sanborn, 1937:328). Dentition, i. $\frac{1}{3}$, c. $\frac{1}{1}$, p. $\frac{2}{2}$, m. $\frac{3}{3}$.

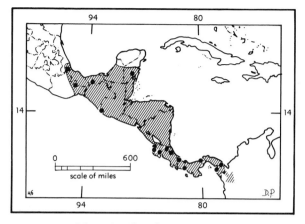

Map 45. *Rhynchonycteris naso.*

KEY TO NORTH AMERICAN SPECIES OF SACCOPTERYX

1. Upper tooth-row more than 5.8.
   S. bilineata, p. 79
1′. Upper tooth-row less than 5.8.
   S. leptura, p. 80

Saccopteryx bilineata (Temminck)
Greater White-lined Bat

1838–39. *Urocryptus bilineatus* Temminck, Tijdschr. natuurl. Gesch. Phys., 5:33, type from Surinam.
1867. *Saccopteryx bilineata*, Peters, Monatsb. preuss. Akad. Wiss., Berlin, p. 471, after July 29.
1855. *Saccopteryx insignis* Wagner, *in* Schreber, Die Säugthiere . . . Suppl., 5:695, type from Rio de Janeiro, Brazil.
1899. *Saccopteryx perspicillifer* Miller, Bull. Amer. Mus. Nat. Hist., 12:176, October 20, type from Caura, Trinidad.
1904. *Saccopteryx bilineata centralis* Thomas, Ann. Mag. Nat. Hist., ser. 7, 13:251, April, type from Teapa, Tabasco. See Alvarez, Rev. Soc. Mexicana Hist. Nat., 19:22–23, December, 1968, who suggests validation of *S. b. centralis*.

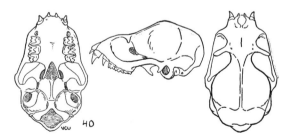

Fig. 40. *Saccopteryx bilineata*, 14 km. SW Coatzacoalcos, 100 ft., Veracruz, No. 19141 K.U., ♀, X 2.

Length of forearm, 41.7–51.7; condylobasal length, 13.2–15.7; length of upper tooth-row, 5.8–7.4. Upper parts black in fresh pelage (deep brown when worn), having 2 wavy lines of yellowish to whitish passing down back onto rump; underparts brownish or grayish in overall tone. Interfemoral membrane thinly haired to exsertion of tail.

MARGINAL RECORDS.—Jalisco (Watkins, *et al.*, 1972:7): 5 mi. NW Cuitzamala; *2 km. NNW Barro de Navidad.* Colima: Hda. Magdalena. Veracruz: Plantación Tres Encinos, approx. 30 km. E Córdoba (Villa-R., 1967:142); Boca del Río. Tabasco: Teapa. Campeche: 1 km. N, 13 km. W Escárcega (91466 KU); 46 km. S Campotón (93196 KU); 65 km. S, 128 km. E Escárcega (93203 KU). Guatemala: Uaxactún. Belize: Kates Lagoon. Panamá (Handley, 1966:757): Almirante; Armila; Cana, thence southward into South America and northward along Pacific Coast to point of beginning.

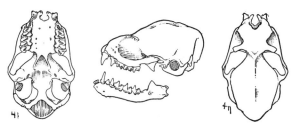

Fig. 41. *Saccopteryx leptura*, Porto Velho, Brazil, No. 21662 F.M.N.H., ♀, X 2.

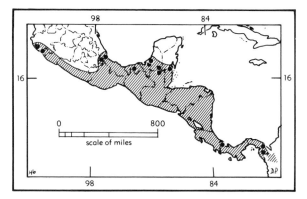

Map 46. *Saccopteryx bilineata.*

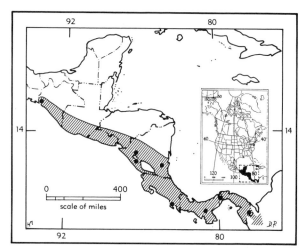

Map 47. *Saccopteryx leptura.*

Saccopteryx leptura (Schreber)
Lesser White-lined Bat

1774. *Vespertilio lepturus* Schreber, Die Säugthiere . . . , Thiel 1, Heft 8, pl. 57, type from Dutch Guiana.
1811. *Saccopteryx lepturus,* Illiger, Prodromus systematis mammalium et avium . . . , p. 121.

Length of forearm, 37.4–42.3; greatest length of skull, 13.1–14.4; interorbital breadth, 3.2–3.8; zygomatic breadth, 8.4–9.1; length of upper tooth-row, 5.1–5.5. Upper parts uniform brown having 2 whitish longitudinal lines extending down back onto rump; underparts slightly paler shade of brown.

MARGINAL RECORDS.—Chiapas: 14 km. SE Tonalá, 100 ft. (Carter, *et al.*, 1966:489). Nicaragua: 2 mi. SE Dario, 1500 ft. (Davis, *et al.*, 1964:375); 6 mi. W Rama, 50 ft. (*ibid.*). Panamá, Canal Zone: Fort Randolph; Paya Village (Handley, 1966:757), thence southeastward into South America, thence northwestward along coast to Panamá: 2 mi. S San Francisco, 200 ft. (Davis, *et al.*, 1964:375). Costa Rica: Rincón de Osa (Starrett and Casebeer, 1968:3). Nicaragua (Jones, 1964d:506): 3 km. N, 4 km. E Sabana Grande, 50 m.; *4 km. N Sabana Grande,* thence along coast to point of beginning.

Genus **Cormura** Peters
Wagner's Sac-winged Bat

1867. *Cormura* Peters, Monatsb. preuss. Akad. Wiss., Berlin, p. 475, after July 29. Type *Emballonura brevirostris* Wagner.

Length of head and body approx. 60. Wing sac in center of antebrachial membrane, opening of sac directed distally and extending from anterior border almost to elbow; base of interfemoral membrane almost naked; wing-membrane from metatarsus. Rostrum short and broad; rim of orbit and zygoma broad; no angle between rostrum and forehead; first upper premolar provided with distinct anterior and posterior cusps (after Sanborn, 1937:349). Dentition, i. $\frac{3}{3}$, c. $\frac{1}{1}$, p. $\frac{2}{2}$, m. $\frac{3}{3}$. This little-known genus is monotypic.

Cormura brevirostris (Wagner)
Wagner's Sac-winged Bat

1843. *Emballonura brevirostris* Wagner, Wiegmann's Arch. für Naturgesch., Jahrg. 9, 1:367, type from Marabitanas, Rio Negro, Amazonas, Brazil.
1867. *Cormura brevirostris,* Peters, Monastb. preuss. Akad. Wiss., Berlin, p. 475, after July 29.

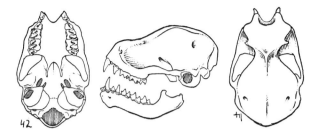

Fig. 42. *Cormura brevirostris*, Boca Curaray, Ecuador, No. 71637 A.M.N.H., ♀, X 2.

Length of forearm, 43.3–49.8; greatest length of skull, 15–16.7; zygomatic breadth, 9.4–10.1; length of upper tooth-row, 6.1–6.8. Upper parts either deep blackish brown or reddish brown; underparts paler.

MARGINAL RECORDS.—Nicaragua: Prinzapolka. Panamá (Handley, 1966:757): Almirante; Armila; Capetí, thence southeastward into South America, and northwestward along coast to Nicaragua: Peña Blanca.

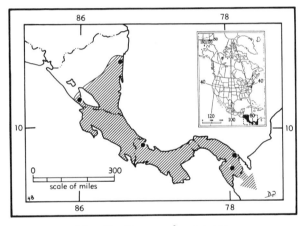

Map 48. *Cormura brevirostris*.

Genus **Peropteryx** Peters—Doglike Bats

1867. *Peropteryx* Peters, Monatsb. preuss. Akad. Wiss., Berlin, p. 472, after July 29. Type, *Vespertilio caninus* Wied-Neuwied [= *Emballonura macrotis* Wagner].

Length of head and body approx. 46. Angle of approx. 115° between expanded rostrum and forehead; basisphenoidal pit not divided. Wing sac near upper edge of antebrachial membrane; opening of sac directed distally (after Sanborn, 1937:339); ears not joined at bases; dental formula as in *Saccopteryx*.

KEY TO NORTH AMERICAN SPECIES OF PEROPTERYX

1. Greatest length of skull more than 16; forearm more than 45. . . . *P. kappleri*, p. 82

1′. Greatest length of skull less than 15.5; forearm less than 48.2. . . . *P. macrotis*, p. 81

Peropteryx macrotis
Lesser Doglike Bat

Length of forearm, 38.3–48.2; greatest length of skull, 12–15; interorbital breadth, 2.3–3.3; zygomatic breadth, 7.6–8.9; length of upper tooth-row, 4.6–6.2. Upper parts varying much in color: buffy brown, reddish brown, grayish brown, blackish brown, or dull sepia; underparts somewhat paler; ears and membranes dark brown.

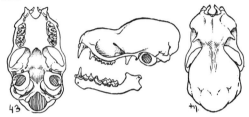

Fig. 43. *Peropteryx macrotis macrotis*, 35 km. SE Jesús Carranza, 500 ft., Veracruz, No. 23454 K.U., ♂, X 2.

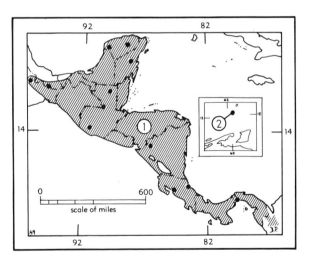

Map 49. *Peropteryx macrotis*.

1. *P. m. macrotis*   2. *P. m. phaea*

Peropteryx macrotis macrotis (Wagner)

1821. *Vesp[ertilio]. caninus* Wied-Neuwied, *in* Schinz, Das Thierreich . . . , 1:179, type from E coast Brazil. Not *Vespertilio caninus* Blumenbach, 1797.
1843. *Emballonura macrotis* Wagner, Wiegmann's Arch. für Naturgesch., Jahrg. 9, 1:367, type from Mato Grosso, Brazil.
1935. *Peropteryx macrotis macrotis*, G. M. Allen, Jour. Mamm., 16:227, August 12.

MARGINAL RECORDS.—Yucatán: Chichén-Itzá (Ingles, 1959:381). Quintana Roo: Felipe Carillo

Puerto (91477 KU). Guatemala: Tikal (Rick, 1968:517); 12 km. NNW Chinajá (Jones, 1966:443). Costa Rica: 1 mi. NW Limón, Las Cuevas, sea level (Starrett and Casebeer, 1968:3). Panamá: Balboa (Handley, 1966:758), thence southeastward along Caribbean Coast into South America, thence northwestward along Pacific Coast to Costa Rica: 8⅛ mi. W Atenas on road to San Matéo, 671 m. (Starrett and Casebeer, 1968:3). Nicaragua: 5 km. N, 9 km. E Condega, 800 m. (Jones, *et al.*, 1971:3). Guatemala: Patalul. Oaxaca: 5 mi. W Chiltepec (Goodwin, 1969:45). Veracruz: 35 km. SE Jesús Carranza, 5000 ft. Yucatán: Calcehtok.

Peropteryx macrotis phaea G. M. Allen

1911. *Peropteryx canina phaea* G. M. Allen, Bull. Mus. Comp. Zool., 54:222, July, type from Point Saline, Grenada, Lesser Antilles.
1937. *Peropteryx macrotis phaea*, Sanborn, Field Mus. Nat. Hist., Zool. Ser., 20:342, December 28.

MARGINAL RECORDS.—Grenada, Lesser Antilles: *Mt. Pleasant Estate;* type locality.

Peropteryx kappleri
Greater Doglike Bat

Length of forearm, 45.0–53.6; greatest length of skull, 16–17.8; interorbital breadth, 2.6–3.5; zygomatic breadth, 9.5–10.9; length of upper tooth-row, 6.8–7.8. According to Sanborn (1937:343), two color phases, one close to Mummy Brown and the other slightly darker than Prout's Brown. In each phase, underparts paler than upper parts.

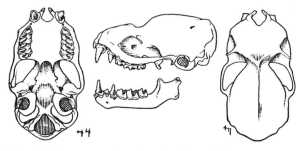

Fig. 44. *Peropteryx kappleri kappleri,* 38 km. SE Jesús Carranza, 500 ft., Veracruz, No. 23440 K.U., ♂, X 2.

Peropteryx kappleri kappleri Peters

1867. *Peropteryx kappleri* Peters, Monatsb. preuss. Akad. Wiss., Berlin, p. 473, after July 29, type from Surinam.

MARGINAL RECORDS.—Veracruz: 38 km. SE Jesús Carranza. Tabasco: Cueva de Escorpión, 5 km. NNE Teapa (Villa-R., 1967:149). Guatemala: Escobas. Honduras: 5 km. N Talanga, 750 m. (Valdez and LaVal, 1971:247). Nicaragua: Muy Muy (Allen, 1910:110). Panamá (Handley, 1966:757): Almirante; Yavisa, thence into South America.

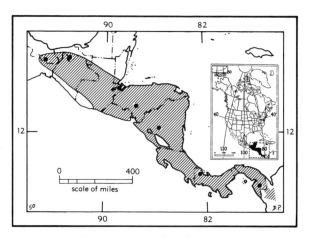

Map 50.   *Peropteryx kappleri kappleri.*

Genus Centronycteris Gray

1838. *Centronycteris* Gray, Mag. Zool. Bot., 2:499, February. Type, *Vespertilio calcaratus* Wied-Neuwied [= *V. maximiliani* Fischer].

Angle wide (Fig. 45) between rostrum and forehead; lower edge of orbit so little expanded that edge of tooth-row can be seen from above; 1st upper premolar (P2) having one distinct anterior cusp and one distinct posterior cusp; basisphenoidal pit divided by median septum. No wing sac known; wing-membranes from metatarsus; fur long and soft; back without lines; dental formula as in *Saccopteryx* (after Sanborn, 1937:336).

Centronycteris maximiliani
Thomas' Bat

Selected measurements of the type, an adult male, are: length of head and body, 52; length of forearm, 45; greatest length of skull, 15; zygomatic breadth, 10; length of upper tooth-row, 6.1. Upper parts near raw umber; paler below; hairs on interfemoral membrane reddish.

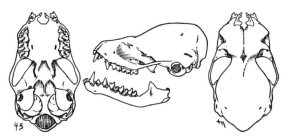

Fig. 45. *Centronycteris maximiliani centralis,* 35 km. SW Jesús Carranza, Veracruz, No. 32088 K.U., sex ?, X 2.

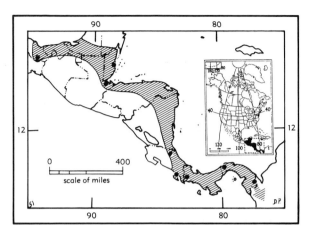

Map 51. *Centronycteris maximiliani centralis.*

Centronycteris maximiliani centralis Thomas

1912. *Centronycteris centralis* Thomas, Ann. Mag. Nat. Hist., ser. 8, 10:638, December, type from Bogava, Chiriquí, Panamá.

1936. *Centronycteris maximiliani centralis*, Sanborn, Field Mus. Nat. Hist., Zool. Ser., 20:94, August 15.

MARGINAL RECORDS.—Veracruz: 35 km. SE Jesús Carranza, 350 ft. Belize: Double Falls. Guatemala: Escobas, near San Tomás. Panamá (Handley, 1966:758): Barro Colorado Island; Cerro Malí, thence southeastward into South America, and northwestward along coast to Panamá: Bugaba (Handley, 1966:758). Costa Rica: Puntarenas, Rincón de Osa, Osa Productos Forestales, Camp Seattle, 35 m. (Starrett and Casebeer, 1968:4); La Selva (LaVal, 1977:78).

Genus **Balantiopteryx** Peters

1867. *Balantiopteryx* Peters, Monatsb. preuss. Akad. Wiss., Berlin, p. 476, after July 29. Type, *Balantiopteryx plicata* Peters.

Wing sac near center of antebrachial membrane and with opening directed proximally; rostrum greatly inflated; basisphenoidal pit divided or not by median septum (after Sanborn, 1937:351).

KEY TO NORTH AMERICAN SPECIES OF
BALANTIOPTERYX

1. White line present on edge of alar membrane; interpterygoid fossa narrow, V-shaped; forearm more than 38.2.
   B. plicata, p. 83
1'. White line on edge of alar membrane wanting; interpterygoid fossa broadly U-shaped; forearm less than 38.9. *B. io,* p. 84

Balantiopteryx plicata
Peters' Bat

Length of head and body approx. 48; length of forearm, 38.3–46.2; greatest length of skull, 11.5–14.8; zygomatic breadth, 8.3–9.3; length of upper tooth-row, 5–5.6. Interpterygoid fossa narrow. Upper parts dark gray to rich brown; underparts dark gray anteriorly, paler on lower abdomen.

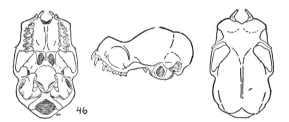

Fig. 46. *Balantiopteryx plicata plicata,* Puente Nacionál, Veracruz, No. 19157 K.U., ♂, X 2.

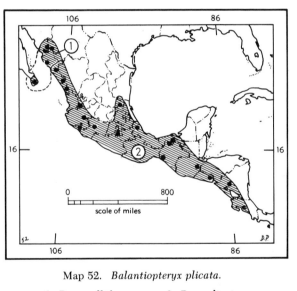

Map 52. *Balantiopteryx plicata.*

1. *B. p. pallida* 2. *B. p. plicata*

Balantiopteryx plicata pallida Burt

1948. *Balantiopteryx plicata pallida* Burt, Occas. Pap. Mus. Zool., Univ. Michigan, 515:1, October 30, type from San Bernardo, Río Mayo, Sonora.

MARGINAL RECORDS.—Sonora: type locality; *Camoa, Río Mayo.* Chihuahua: 40 km. N, 6 km. W Choix, Sinaloa (Anderson, 1972:235). Sinaloa: 10 mi. NNE Los Mochis (Jones, *et al.,* 1962:151). Baja California: *Santa Anita;* San José del Cabo. Sonora: Chinobampo.

Balantiopteryx plicata plicata Peters

1867. *Balantiopteryx plicata* Peters, Monatsb. preuss. Akad. Wiss., Berlin, p. 476, after July 29, type from Puntarenas, Puntarenas, Costa Rica.

1938. *Balantiopteryx ochoterenai* Martínez and Villa, Anal. Inst. Biol., Univ. Nac. Autó. México, 9(3–4):339, November 14, type from Cuautla, Morelos. Regarded as identical with *plicata* by Burt and Hooper, Occas. Pap. Mus. Zool., Univ. Michigan, 430:2, May 27, 1941.

MARGINAL RECORDS.—Sinaloa (Jones, *et al.*, 1962:151, 152): 4 mi. N Terrero; ½ mi. E Piaxtla; 5 mi. SSE Rosario; *10 mi. SE Escuinapa.* Nayarit: *San Blas.* Jalisco (Watkins, *et al.*, 1972:7): 2 mi. E Bolaños, 3550 ft.; 6 mi. N, 2 mi. E Atoyac, 4400 ft. Michoacán: Apatzingán. Morelos: Cuernavaca. San Luis Potosí: Cueva Sabinas, near Valles. Veracruz: Puente Nacionál. Tabasco (Villa-R., 1967:153): Cueva de Coconá, Teapa; Rancho El Tumbo, 4 km. E Estación El Zapote, Macuspana. Guatemala: El Rancho, Zacapa. Nicaragua: Río Viejo, 7 mi. WNW Dario (Jones, 1964d:506). Costa Rica: *near Hda. La Pacífica* (3170 LaVal at KU); type locality. El Salvador: Río Goáscoran. Chiapas: Finca Ocuilapa, 14 km. NNE Tonalá (Villa-R., 1967:153). Colima: Colima.

Balantiopteryx io Thomas
Thomas' Sac-winged Bat

1904. *Balantiopteryx io* Thomas, Ann. Mag. Nat. Hist., ser. 7, 13:252, April, type from Río Dolores, near Cobán, Alta Verapaz, Guatemala.

Length of head and body averaging 50; length of forearm, 35.6–38.8; greatest length of skull, 12.4–12.9; zygomatic breadth, 8.2–8.3; length of upper tooth-row, 4.5–4.8. Interpterygoid fossa broadly U-shaped. Upper parts dark brown; underparts paler.

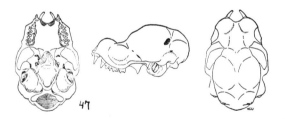

Fig. 47. *Balantiopteryx io*, 38 km. SE Jesús Carranza, 500 ft. Veracruz, No. 23456 K.U., ♂, X 2.

MARGINAL RECORDS.—Veracruz: Grutas Atoyac, 2 km. E Atoyac (Hall and Dalquest, 1963:215); *5 km. S Potrero, 1700 ft.*; 38 km. SE Jesús Carranza. Tabasco (Villa-R., 1967:157): Cueva La Murcielaguera, 5³⁄₁₀ km. NE Teapa; *Cueva de Don Louis, 3³⁄₁₀ km. NE Teapa.* Belize: Lucy Cave, Glenwood Farm, Sibun River, 35 m. (Kirkpatrick, *et al.*, 1975:330). Guatemala (Jones, 1966:444): *Lanquin Cave;* Escobas; Chimoxan; type locality. Oaxaca: 1 mi. S Tollosa (Baker and Greer,

1960:414); *Pequeña Cueva cerca de Montebello, 24 km. N Matías Romero* (Villa-R., 1967:157); 5 mi. W Chiltepec (Goodwin, 1969:48).

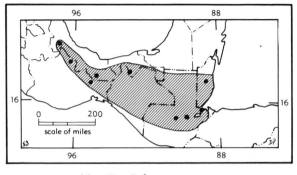

Map 53. *Balantiopteryx io.*

Genus **Cyttarops** Thomas—Short-eared Bat

1913. *Cyttarops* Thomas, Ann. Mag. Nat. Hist., ser. 8, 11:134, January. Type, *Cyttarops alecto* Thomas.

Frontal cup pronounced; postorbital processes long and curved posteroventrally; supraorbital notch deep; length of lower premolars less than a third length of lower molars; hard palate ending opposite last molar; tibia grooved on plantar surface; clavicle wide but relatively less so than in *Diclidurus;* tail simple (not recurved at tip); uropatagium lacking glandularlike structures; pelage on back approx. 6 mm. long, and silky.

Cyttarops alecto Thomas
Short-eared Bat

1913. *Cyttarops alecto* Thomas, Ann. Mag. Nat. Hist., ser. 8, 11:135, January. Type from Mocajatuba, state of Para, Brazil.

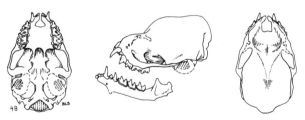

Fig. 48. *Cyttarops alecto*, Puerto Viejo, Costa Rica, No. 26625 L.A.C.M., ♂, X 2.

Length of head and body, 46–47.2; tail, 20; hind foot, 8.0; ear, 10; greatest length of skull, 12.6–13.6; zygomatic breadth, 8.0; upper tooth-row (C–M3), 5.3–5.5. Lambdoidal crest absent; when viewed dorsally, braincase obscures exoccipital condyles; tragus 2.8 mm. long on inner edge, lower half of outer margin forming an enormous angular lobe; broad antebrachial membrane extending to base of distal phalanx of thumb; color above and below dull smoky gray, bases of hairs slightly paler than tips on foreback, darker on hind back.

MARGINAL RECORDS.—Nicaragua: 4½ km. NW Rama (Baker and Jones, 1975:2). Costa Rica: Puerto Viejo de Sarapique, 100 m. (Starrett and Casebeer, 1968:6); Los Diamantes, 300 m. (Starrett and de la Torre, 1964:54), and eastward into South America.

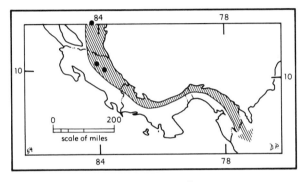

Map 54. *Cyttarops alecto.*

Genus **Diclidurus** Wied-Neuwied—Ghost Bats

1820. *Diclidurus* Wied-Neuwied, Isis von Oken, 1819, p. 1629. Type, *Diclidurus albus* Wied-Neuwied.

Braincase large, flattened anteriorly and descending abruptly to rostrum; rostrum broad, bearing lateral ridges; wing sac absent; nose simple; tail shorter than and perforating interfemoral membrane. Dental formula as in *Saccopteryx*. Ojasti and Linares (1972:422–427) discussed the status and distribution of the four species of *Diclidurus*, and suggested that *Depanycteris* Thomas, 1920, is congeneric with *Diclidurus*.

Diclidurus virgo Thomas
Northern Ghost Bat

1903. *Diclidurus virgo* Thomas, Ann. Mag. Nat. Hist., ser. 7, 11:377, April, type from Escazú, San José, Costa Rica.

Selected measurements of the type, an adult female, are: length of forearm, 66; greatest length of skull, 18; length of upper tooth-row, 8.1. Color white. For additional characters see generic account.

MARGINAL RECORDS.—Nayarit: Playa Novillero, 24 km. W Tecuala (Villa-R. and Ramírez-P., 1971:155). Oaxaca: Paso Real, 14 km. NNE Tuxtepec (Villa-R., 1967:159). Veracruz: Sontecomapan (*ibid.*) [locality listed as Catemaco on page 163]. Guatemala: 4 km. S Chinajá (Jones, 1966:445). Honduras: 3 mi. N Gracias

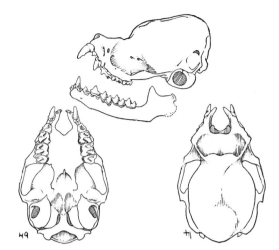

Fig. 49. *Diclidurus virgo,* Costa Rica, No. 7947 A.M.N.H., sex ?, X 2.

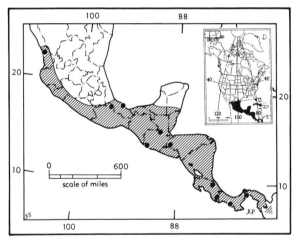

Map 55. *Diclidurus virgo.*

on Río Grande. Costa Rica: La Palma. Panamá: Fort Gulick (Handley, 1966:758), thence into South America, and northwestward along coast to Panamá: Albrook Field (*ibid.*); Pueblo Nuevo. Costa Rica: 2⁷⁄₁₀ mi. NW Villa Neilly, 55 m. (Starrett and Casebeer, 1968:5). Guatemala: Champerico.

FAMILY **NOCTILIONIDAE**—Bulldog Bats

Skull without distinct postorbital processes; premaxillae with both nasal and palatal branches, nasal branches markedly elongate; premaxillae, in adult, fused to each other and to maxillae; palate complete, closed anteriorly. Humerus having trochiter smaller than trochin, poorly articulated with scapula; capitellum slightly offset from axis of humeral shaft; 2nd digit of manus having metacarpal as long as that of 3rd; 7th cervical vertebra not fused with 1st thoracic; fibula threadlike, extending to head of tibia, but cartilaginous proximally; ischia fused together and to ventral side of sacrum. Muzzle and nose without excrescences; lips full and appearing swollen; internal cheek pouches present.

The family contains only one genus.

Genus **Noctilio** Linnaeus—Bulldog Bats

1766. *Noctilio* Linnaeus, Syst. nat., ed. 12, 1:88. Type, *Noctilio americanus* Linnaeus [= *Vespertilio leporinus* Linnaeus].
1808. *Noctileo* Tiedemann, Zoologie . . . , 1:536. Type, [*Vespertilio*] *leporinus* Linnaeus; a *lapsus?*
1821. *Celaeno* Leach, Trans. Linnean Soc. London, 13:69. Type, *Celaeno brooksiana* Leach [= *V. leporinus*].
1906. *Dirias* Miller, Proc. Biol. Soc. Washington, 19:84, June 4. Type, *Noctilio albiventer* Spix. Regarded as synonymous with *Noctilio* by Osgood, Field Mus. Nat. Hist., Publ. 149, Zool. Ser., 10:31, 32, October 20, 1910.

Length of head and body approx. 98. Braincase deep, oval in outline; mastoidal region conspicuously flaring, shelflike; sagittal crest distinct; rostrum strongly arched; nares opening anteriorly, somewhat tubular; palate concave transversely, almost flat anteroposteriorly, extending posteriorly beyond tooth-rows; auditory bullae small. Dentition: i. $\frac{2}{1}$, c. $\frac{1}{1}$, p. $\frac{1}{2}$, m. $\frac{3}{3}$. Ears separate, slender, pointed; tragus well developed; muzzle pointed with pad strongly projecting; chin having well-developed transverse ridges; tail well developed, more than half as long as femur, extending approx. to middle of interfemoral membrane; tip of tail free on dorsal surface of interfemoral membrane. Hair short; feet robust; calcar greatly elongated.

Most bats have a musky smell but in *Noctilio* this odor is unusually strong and penetrating.

KEY TO SPECIES OF NOCTILIO

1. Combined length of tibia and foot more than 45; foot more than 23; forearm more than 73; weight more than 45 grams.
　　　　　　　　　　N. leporinus,　p. 86
1'. Combined length of tibia and foot less than 45; foot less than 23; forearm less than 73; weight less than 45 grams.
　　　　　　　　　　N. albiventris,　p. 87

Noctilio leporinus
Greater Bulldog Bat

Revised by W. B. Davis, Jour. Mamm., 54:862–874, December 14, 1973.

Measurements of an adult male, the type of *N. l. mexicanus*, are: length of forearm, 83.2; greatest

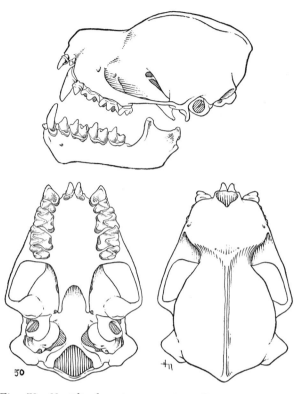

Fig. 50. *Noctilio leporinus mastivus*, Papayo, Guerrero, No. 1681 K.U., ♀, X 2.

length of skull, 28.5; interorbital breadth, 7.4; zygomatic breadth, 19.8; length of upper tooth-row, 10.7. Upper parts rich dark ochraceous-tawny in males, grayer in females, usually with a paler line mid-dorsally; underparts paler in both sexes. See key for additional characters. Males average slightly larger than females. In adults: sagittal crest high in males, poorly developed in females; ramus of lower mandible 30 per cent deeper in males than in females.

Noctilio leporinus mastivus (Vahl)

1797. *Vespertilio mastivus* Vahl, Skrivter af Naturh.-Selsk. Kjøbenhavn, 4:132, type from St. Croix, Virgin Islands, West Indies.
1884. *Noctilio leporinus mastivus*, True, Proc. U.S. Nat. Mus., 7(App. Circ. 29):603, November 29.
1915. *Noctilio leporinus mexicanus* Goldman, Proc. Biol. Soc. Washington, 28:136, June 29, type from Papayo, Guerrero.

MARGINAL RECORDS (Davis, 1973:870, 871, unless otherwise noted).—México and Central America: Sinaloa: San Benito, 400 ft.; 1½ mi. N Badiraguato (Jones, *et al.*, 1972:5). Veracruz: Río San Juan, 21 mi. W Santiago Tuxtla; 1 mi. E Jaltipan, 50 ft. Tabasco: 5 mi. SW Teapa (Hall and Kelson, 1959:88). Campeche: 1 mi. SW Puerto Real. Yucatán: 19 km. E Progreso (Birney, *et al.*, 1974:5), thence along coast to Guatemala: 25 km.

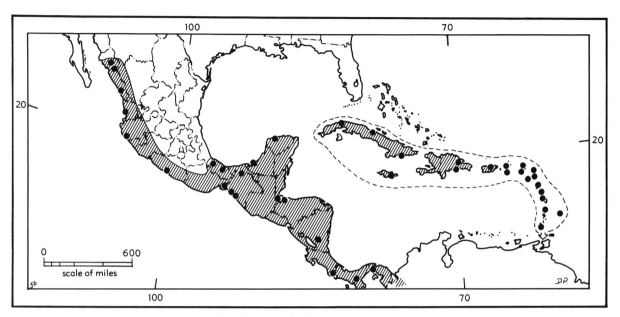

Map 56. *Noctilio leporinus mastivus.*

SSW Puerto Barrios, 30 ft. Honduras: 7 mi. NW San Pedro Sula, 100 ft.; *2 mi. W San Pedro Sula,* thence along coast to Nicaragua: Cacao [one of several inland localities], thence along coast into South America and northward through Panamá: 18 km. WSW Chepo, 200 ft.; 2 mi. S San Francisco, 200 ft.; *Coiba Island, Hermosa Bay.* Costa Rica: 9 mi. ENE Puerto Golfito, thence up coast to Chiapas: 5 km. SE Pijijiapan, 100 ft.; 12 km. SSE Tonalá. Oaxaca: 4 mi. E Tapanatepec, 800 ft., thence up coast to Guerrero: Papayo, 25 ft. Jalisco: Cuitzamala, 25 ft. Nayarit: San Blas. Sinaloa: Chele (Koopman, 1961:536). Greater and Lesser Antilles: Cuba: Habana; Cueva Grande, Caguana, Mayajigua (Dusbábek, 1969:323); cave in eastern Cuba (Anthony, 1919:636). Jamaica: Kingston. Dominican Republic (Armstrong and Johnson, 1969:133, as *N. leporinus* only): Río Yuna, 6 mi. E Arenosa; Hda. Nigua. Puerto Rico: Old Loiza. Virgin Islands (Hall and Kelson, 1959:88): Botany Bay, St. Thomas; type locality. Antillean Islands (Hall and Kelson, 1959: 88, unless otherwise noted): St. Martin (Koopman, 1968:2); Barbuda (*ibid.*); St. Kitts; Antigua (Koopman, 1968:2); Monserrat; Guadeloupe (Jones and Phillips, 1970:137, 139, as *N. leporinus* only); Dominica; Martinique (Koopman, 1968:2); Santa Lucia; *St. Vincent;* Barbados; Grenada, thence into South America.

Noctilio albiventris
Lesser Bulldog Bat

Revised by W. B. Davis, Jour. Mamm., 57:687–707, December 10, 1976.

Selected measurements of an adult female, the type of *N. labialis minor,* are: length of head and body, 67; length of forearm, 58.4; greatest length

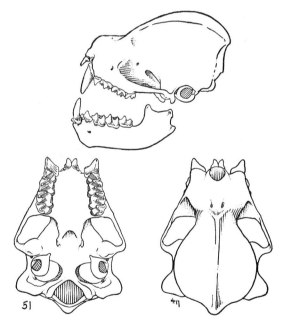

Fig. 51. *Noctilio albiventris minor,* Summit, Panamá, No. 261402 U.S.N.M., ♂, X 2.

of skull, 17.2; zygomatic breadth, 14.6; length upper tooth-row, 7.5. Upper parts grayish brown to yellowish or bright rufous, with narrow whitish median stripe from shoulder region to rump; underparts paler. Many males are bright rufous and many females are dull brown to drab. Ear laid forward extending slightly beyond muzzle; chin with small plications.

Noctilio albiventris minor Osgood

1910. *Noctilio minor* Osgood, Field Mus. Nat. Hist., Publ. 149, Zool. Ser., 10:30, October 20, type from Encontrados, Zulia, Venezuela.
1912. *Noctilio albiventer minor* Osgood, Field Mus. Nat. Hist., Publ. 155, Zool. Ser., 10:62, January 10.
1975. *Noctilio albiventris minor*, Hershkovitz, Jour. Mamm., 56:244, February 20.

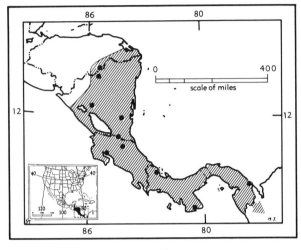

Map 57. *Noctilio albiventris minor.*

MARGINAL RECORDS.—Honduras: Río Coco, 78 mi. ENE Danlí, 900 ft. (Davis, *et al.*, 1964:376). Nicaragua: 2 mi. W Rama, 50 ft. (*ibid.*); El Toro Rapids, Lake Nicaragua. Costa Rica: 4 mi. W Puerto Viejo (*ibid.*). Panamá (Handley, 1966:758): Almirante; Armila, thence southeastward into South America, and

northwestward along Pacific Coast to Panamá: Guánico (Handley, 1966:758). Costa Rica: Río Higuerón, near sea level, ½ mi. E Finca Jinénez [*sic*] (Brown, 1968:756); *Río Guacimal, 1320 m. (ibid.).* Nicaragua: 4 km. W Teustepe, 140 m. (Jones, *et al.*, 1971:3); 55 mi. NNE Jinotega, 1100 ft. (Davis, *et al.*, 1964:376).

FAMILY PHYLLOSTOMATIDAE
American Leaf-nosed Bats

Skull without postorbital processes; premaxillae complete, fused to each other and to maxillae, the palatal branches isolating two lateral palatal foramina; teeth variable according to genus. Tragus present, variable; nose-leaf usually present. Humerus with well-developed trochiter, but smaller than trochin; capitellum distinctly offset from axis of shaft; 2nd digit of manus with well-developed metacarpal and small phalanx, 3rd finger with 3 complete bony phalanges; 7th cervical vertebra free from anterior thoracic vertebra; fibula present, cartilaginous proximally; pelvis normal; sacrum forming flattened, narrow urostyle posteriorly.

This family, the American leaf-nosed bats, comprises a diverse assemblage of genera a few of which lack the nose-leaf. Except in *Lonchorina* and *Centurio*, the leaf is never so large or complex, respectively, as in some Old World nose-leafed families. Phyllostomatids are tropical and subtropical in distribution.

The species *Pygoderma bilabiatum*, listed in Hall and Kelson (1959:145), seems not to occur in North America (Koopman, 1958:584).

KEY TO SUBFAMILIES OF PHYLLOSTOMATIDAE

1. Trochiter not impinging on scapula; nose-leaf absent (a trace in *Aello*). . . . . .Chilonycterinae, p. 89
1'. Trochiter impinging on scapula; nose-leaf present but sometimes small.
    2. Upper molars essentially normal, the configuration never obliterating the fundamental W-shaped pattern. . . . . .Phyllostomatinae, p. 97
          and some Glossophaginae, p. 118
   2'. Upper molars modified and fundamental W-shaped pattern absent or much obscured.
      3. Upper molars with trace of commissures and styles; lower molars with 5 cusps.
          some Glossophaginae, p. 118
     3'. Upper molars without commissures and styles; cusps of lower molars (when present) strictly lateral.
        4. Crowns of both upper and lower molars trenchant. . . . . .Carollinae, p. 134
        4'. Crowns of both upper and lower molars grooved or flattened.
          5. Crowns of molars with well-developed cusps rising from a flattened crushing surface.
            6. Tail absent; calcar well developed. . . . . .Stenoderminae, p. 141
            6'. Tail present (but short); calcar poorly developed. . . . .some Brachyphyllinae, p. 168
         5'. Crowns of molars with distinct longitudinal groove, cusps (when present) strictly lateral.
          7. Crowns of lower molars with distinct cusps on both margins of groove.
           Sturnirinae, p. 138
         7'. Crowns of lower molars without distinct cusps on both margins of groove.
           some Brachyphyllinae, p. 168

SUBFAMILY CHILONYCTERINAE
Mustached Bats and Allies

Revised by Smith, Univ. Kansas Mus. Nat. Hist., Miscl. Publ., 56:1–132, March 10, 1972.

Trochiter not impinging on scapula and much reduced relative to the trochiter in other members of the family; nose-leaf wanting; lips expanded and ornamented with flaps and folds, forming "funnel" into mouth when open; "funnel" surrounded by short, bristlelike hairs; upper lip, with nostrils and various bumps and ridges, forming labio-nasal plate; pinnae funnel-shaped, lower edge confluent with ridge extending along lower lip; tragus complex; eyes small, inconspicuous; calcar long, usually flexed forward against tibia; distal part of tail protruding dorsally from uropatagium; both greater and lesser tuberosities of humerus equal head of humerus in proximal extension; no special relationship between humerus and scapula; supraglenoid fossa well developed, forming effective locking mechanism in conjunction with supraglenoid process; upper molars with unmodified dilambdodont cusp pattern; last upper molar only slightly molariform; dentition, i. $\frac{2}{2}$, c. $\frac{1}{1}$, p. $\frac{2}{3}$, m. $\frac{3}{3}$. Color varies according to age of pelage and stage of molt.

For bats of this subfamily, Dalquest and Werner (1954:159) recommended family rank as also did Smith (1972:41); reasons for the recommendation have considerable weight, but the grade of differentiation from related groups seems to be about the same as that between some other groups of mammals currently treated as subfamilies. Therefore the chilonycterines here are accorded subfamily rank.

Smith (op. cit.: 10–15) has recommended also that the family group name be changed from Chilonycterinae—again for reasons of considerable weight. Nevertheless, the rules for family group names are less precise than for names of lesser taxa and do not prohibit the continued use of a family group name based on a generic name subsequently reduced to subgeneric rank. Therefore Chilonycterinae, long in use, is retained here as the more readily recognized subfamily name.

KEY TO GENERA OF CHILONYCTERINAE

1. Rostrum elevated, producing abruptly rising forehead; p2 not noticeably smaller than p1; ears united. . . . .*Aello*, p. 96
1'. Rostrum not, or only slightly, elevated more or less in same plane as braincase; p2 much smaller than p1; ears separated.
Pteronotus, p. 89

Genus Pteronotus Gray
Mustached Bats and Naked-backed Bats

1838. *Pteronotus* Gray, Mag. Zool. Bot., 2:500, February. Type, *Pteronotus davyi* Gray. Not *Pteronotus* Rafinesque, 1815, a *nomen nudum*.
1839. *Chilonycteris* Gray, Ann. Nat. Hist., 4:4, September. Type, *Chilonycteris macleayii* Gray. Valid as a subgenus.
1840. *Lobostoma* Gundlach, Wiegmann's Arch. für Naturgesch. Jahrg., 4:357. Type, presumably *Lobostoma quadridens* Gundlach (= *Chilonycteris macleayii*).
1843. *Phyllodia* Gray, Proc. Zool. Soc. London, p. 50, October. Type, *Phyllodia parnellii* Gray. Valid as a subgenus.
1901. *Dermonotus* Gill, Proc. Biol. Soc. Washington, 14:177, September 25. Replacement name for *Pteronotus* Gray, 1838, thought to be preoccupied by *Pteronotus* Rafinesque, 1815, a *nomen nudum*.

Rostrum and braincase more or less in same plane; zygomatic arches complete, well developed, and in same horizontal plane as rostrum and braincase; lambdoidal ridges strongly developed; presphenoid elevated above level of basisphenoid; p2 reduced to small peglike, unicuspid tooth excluded lingually from tooth-row; pelage short and stiff; pinnae pointed, separate (weakly connected dorsally by low ridge on rostrum in *parnelli*, *macleayii*, *fuliginosus*, and *personatus*); tragus lanceolate with small secondary fold on cranial edge, or spatulate with marked secondary fold.

KEY TO SPECIES OF PTERONOTUS

1. Wing membranes fused in center of back.
   2. Forearm more than 49; maxillary tooth-row more than 7.1.
      P. gymnonotus, p. 95
   2'. Forearm less than 50; maxillary tooth-row less than 7.1. . . . . .*P. davyi*, p. 94
1'. Wing membranes not fused in center of back.
   3. Wing membranes attached low on sides; basioccipital narrowly constricted between auditory bullae; forearm more than 50. *P. parnellii*, p. 90
   3'. Wing membranes attached high on sides; basioccipital not constricted between auditory bullae; forearm less than 50.
      4. Labio-nasal plate simple, lacking lateral spikes; uropatagium and wing membrane attaching to ankle by short ligament. *P. personatus*, p. 93
      4'. Labio-nasal plate complex, having prominent lateral spikes; uropatagium and wing membrane attaching to ankle by long ligament.
         5. Condylobasal length more than

14.2 (Jamaica), 13.8 (Cuba); absent in Hispaniola and Puerto Rico. . . *P. macleayii*,   p. 92
5'. Condylobasal length less than 14.2
(Jamaica), 13.8 (Cuba); present in Hispaniola and Puerto Rico. *P. quadridens*,   p. 92

Pteronotus parnellii
Parnell's Mustached Bat

Length of head and body, 60–67; total length,
73–102; length of forearm, 48.9–65.4; condylobasal length, 17.0–21.9; zygomatic breadth,
10.0–13.4; maxillary tooth-row, 7.3–9.9; wing
membrane attached low on side of body and attached to ankle by short, tightly bound ligament;
back covered by wide band of short, stiff hairs;
labio-nasal plate simple; margin above each nostril with several irregular wartlike tubercles;
deep emargination between nostrils; area lateral
to nostrils lacking spikelike projection; tragus
simple, lanceolate, secondary fold small; anteromedial edge of pinna smooth, lanceolate portion broad; ears connected by two low, inconspicuous ridges fusing on top of muzzle forming
prominent rostral tubercle; profile of skull flattened, rostrum not elevated, forehead sloping
gradually onto long, broad braincase; upper incisors robust and peglike, inner pair distinctly
bifurcate having broad, rounded heel; lower incisors also robust, inner pair trilobate, outer pair
bilobate; tympanic rings smaller than in other
species, covering approx. a third of auditory bullae; basioccipital region narrowly constricted between bullae having two narrow, deep longitudinal furrows into which M. longus capitis inserts.

Pteronotus parnellii gonavensis (Koopman)

1955. *Chilonycteris parnellii gonavensis* Koopman, Jour.
Mamm., 36:110, February 25, type from cave near En Café,
La Gonave Island, Haiti. Known only from type locality.
1972. *Pteronotus parnellii gonavensis*, Smith, Univ. Kansas
Mus. Nat. Hist., Miscl. Publ., 56:68, March 10.

Pteronotus parnellii mesoamericanus Smith

1972. *Pteronotus parnellii mesoamericanus* Smith, Univ.
Kansas Mus. Nat. Hist., Miscl. Publ., 56:71, March 10, type
from 1 mi. S, ¾ mi. E Yepocapa, 4280 ft., Chimaltenango,
Guatemala.

MARGINAL RECORDS (Smith, 1972:74, 75, unless
otherwise noted).—Quintana Roo: Aeropuerto, 4 km.
WSW Puerto Juárez, 5 m. Belize: Stann Creek Valley
(Hall and Kelson, 1959:92). Honduras: Balfate.
Nicaragua: 2 mi. SE Darío, 1500 ft.; 1 km. S El Castillo,
130 m. Costa Rica: *Comelco Ranch* (3002 LaVal at KU);
1 mi. W La Irma (2951 LaVal at KU); 9 mi. ENE Puerto

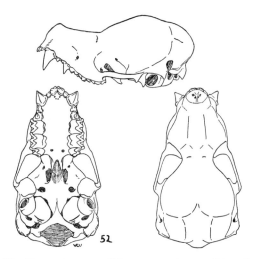

Fig. 52. *Pteronotus parnellii mesoamericanus*, 3 km. E San
Andrés Tuxtla, 1000 ft., Veracruz, No. 23517 K.U., ♂, X 2.

Golfito, 100 ft. Panamá: Guabalá, 50 ft.; Guánico;
Penonomé. El Salvador: Comacarán, 700 ft.
Guatemala: type locality. Chiapas (Baker, 1967:426):
Kilometer 184, Hwy. 200, N Huixtla; 42 km. W Cintalapa. Veracruz: 2 mi. WNW San Andrés Tuxtla, 1000
ft. Chiapas: 21 km. WSW Teapa, Tabasco, 200 ft. Campeche: 5 km. S Champotón. Yucatán: Gruta de Balen-
kanche, 5 km. E Chichén-Itzá.

Pteronotus parnellii mexicanus (Miller)

1902. *Chilonycteris mexicana* Miller, Proc. Acad. Nat. Sci.
Philadelphia, 54:401, September 12, type from San Blas,
Nayarit.
1963. *Pteronotus parnellii mexicana*, Hall and Dalquest,
Univ. Kansas Publ., Mus. Nat. Hist., 14:217, May
20.

MARGINAL RECORDS (Smith, 1972:70, 71, unless
otherwise noted).—Sonora: Cueva de la Tigre, 14⁹⁄₁₀ mi.
SSE Carbo (Cockrum and Bradshaw, 1963:3).
Chihuahua: Carimeche, Río Mayo. Durango: Chacala
(Hall and Kelson, 1959:93). Zacatecas (Matson and Pat-
ten, 1975:2): Monte Escobedo, 2225 m.; 6⅔ km. S Jalpa,
1311 m. Jalisco: ½ km. N, 3 km. W Jamay, 1650 m.
(Watkins, *et al.*, 1972:9). Michoacán: S shore Lake
Chapala (Hall and Kelson, 1959:93); 5 km. SW Turan-
deo, 1900 m. Distrito Federal: Tacubaya. San Luis
Potosí (Hall and Kelson, 1959:93): 1 km. W
Huichihuayán; 3 km. N Taninul. Tamaulipas: 10 km.
N, 8 km. W El Encino, 400 ft.; 7½ km. NNW Ciudad
Victoria (Villa-R., 1967:176); 2 mi. S, 10 mi. W Piedra,
1200 ft. Veracruz: 8 km. NW Potrero, 1700 ft.; 38 km.
SE Jesús Carranza, 500 ft. Oaxaca: Tehuantepec; km.
183, 36½ km. N San Gabriel Mixtepec (Schaldach,
1966:290). Guerrero: 12 km. N Zacatula. Sinaloa: ½ mi.
S Concepción; 12 mi. NE San Benito, 1000 ft. Sonora:
15 mi. NW Guaymas (Hall and Kelson, 1959:93).

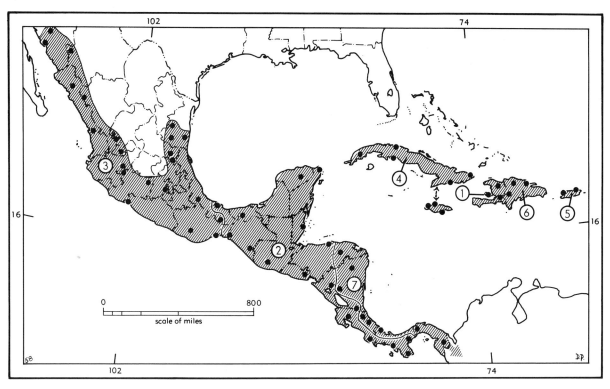

Map 58. *Pteronotus parnellii.*

Guide to subspecies	2. *P. p. mesoamericanus*	5. *P. p. portoricensis*
	3. *P. p. mexicanus*	6. *P. p. pusillus*
1. *P. p. gonavensis*	4. *P. p. parnellii*	7. *P. p. rubiginosus*

Pteronotus parnellii parnellii (Gray)

1843. *Phyllodia Parnellii* Gray, Proc. Zool. Soc. London, p. 50, October, type from Jamaica.
1972. *Pteronotus parnellii parnellii,* Smith, Univ. Kansas Mus. Nat. Hist., Miscl. Publ., 56:63, March 10.
1861. *Chilonycteris osburni* Tomes, Proc. Zool. Soc. London, p. 66, May, types from Sportsman Hall and Oxford Cave, Manchester, Jamaica.
1861. *Chilonycteris boothi* Gundlach, Monatsb. preuss. Akad. Wiss., Berlin, p. 154, type from Fundador or Güines, Matanzas and Habana provinces, respectively, Cuba.

MARGINAL RECORDS (Smith, 1972:67).—Cuba: Cueva del Agua, Sagua La Grande; Baracoa; Cueva de Las Majaes, Caney; Cueva de Los Masones, Trinidad; San Vicente. Jamaica: Windsor Cave; Healthshire Hills Caves; Lucea.

Pteronotus parnellii portoricensis (Miller)

1902. *Chilonycteris portoricensis* Miller, Proc. Acad. Nat. Sci. Philadelphia, 54:400, September 12, type from Cueva de Fari, near Pueblo Viejo, Puerto Rico.
1972. *Pteronotus parnellii portoricensis,* Smith, Univ. Kansas Mus. Nat. Hist., Miscl. Publ., 56:68, March 10.

MARGINAL RECORDS (Smith, 1972:69).—Puerto Rico: Pueblo Viejo; Cueva de Trujillo Alto.

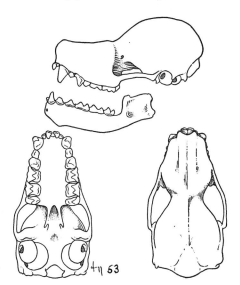

Fig. 53. *Pteronotus parnellii portoricensis,* Trujillo Alto, Puerto Rico, from Anthony (1918:343), No. 39370 A.M.N.H., ♂, X 2.

Pteronotus parnellii pusillus (G. M. Allen)

1917. *Chilonycteris parnellii pusillus* G. M. Allen, Proc. Biol. Soc. Washington, 30:168, October 23, type from Arroyo Salado, Dominican Republic.
1972. *Pteronotus parnellii pusillus*, Smith, Univ. Kansas Mus. Nat. Hist., Miscl. Publ., 56:67, March 10.

MARGINAL RECORDS (Smith, 1972:68).—Dominican Republic: type locality; Rancho La Guardia. Haiti: Diquini; St. Michel. Dominican Republic: Cueva Durán, 4 km. SW Monción.

Pteronotus parnellii rubiginosus (Wagner)

1843. *Chilonycteris rubiginosus* Wagner, Wiegmann's Arch. für Naturgesch., Jahrg., 9:361, December 31, type from Caicara, Mato Grosso, Brazil.
1972. *Pteronotus parnellii rubiginosus*, Smith, Univ. Kansas Mus. Nat. Hist., Miscl. Publ., 56:75, March 10.

MARGINAL RECORDS (Smith, 1972:76).—Honduras: San José Río Tinto, 340 m. Nicaragua: Bonanza, 850 ft.; Los Cocos, 14 km. S Boaco, 220 m. Costa Rica: La Selva (3108 LaVal at KU); 3 km. S Moravia. Panamá: Almirante; Bas Obispo; Río Chucunaque, thence into South America.

Pteronotus macleayii
Macleay's Mustached Bat

Length of head and body, 46–53; total length, 61–78; length of forearm, 41.2–45.1; condylobasal length, 14.1–15.7; zygomatic breadth, 7.6–8.4; maxillary tooth-row, 6.4–7.0; wing membrane and uropatagium attached to ankle by long ligament tightly bound to distal half of tibia; tragus long and spatulate, secondary fold prominent, sparsely clothed in short bristlelike hairs, with longer hairs on cranial and distal edges; labionasal plate distinct, squared lappet above each

nostril and prominent lateral spikes to either side of nostrils; one or two (occasionally three) toothlike serrations on anteromedial edge of pinna; ears connected by low, inconspicuous ridge on top of nose, rostral tubercle not prominent; clitoris unusually long; pelage tricolored dorsally; profile of skull flattened, rostrum slightly elevated, forehead rising abruptly onto round, high braincase; rostrum longer than braincase; marked diastema between outer upper incisor and canine; inner incisors distinctly bifurcate; lower incisors small and trilobate having short, rounded heels.

Pteronotus macleayii griseus (Gosse)

1851. *Chilonycteris grisea* Gosse, A naturalist's sojourn in Jamaica, p. 326, type from Phoenix Park, near Savanna la Mar, Westmoreland Parish, Jamaica.
1972. *Pteronotus macleayii grisea*, Smith, Univ. Kansas Mus. Nat. Hist., Miscl. Publ., 56:82, March 10.

MARGINAL RECORDS.—Jamaica: Lucea; type locality; Kingston.

Pteronotus macleayii macleayii (Gray)

1839. *Chilonycteris Macleayii* Gray, Ann. Nat. Hist., 4:5, September, type from Cuba.
1970. *P[teronotus]. macleayii*, Vaughan and Bateman, Jour. Mamm., 51:218, May 20.

MARGINAL RECORDS (Smith, 1972:82).—Cuba: Cueva del Río, San Vicente; Cueva de Mudo, Güines; Soledad, Harvard Botanical Gardens; Baracoa; Cueva de Pedernales, Isla de Pinos.

Pteronotus quadridens
Sooty Mustached Bat

Length of head and body, 39–43; total length, 48–64; length of forearm, 35.9–39.5; condylobasal length, 12.8–13.8; zygomatic breadth, 7.1–7.8; maxillary tooth-row, 5.6–6.2; wing membrane and uropatagium attached to ankle by long ligament tightly bound to distal half of tibia; tragus long and spatulate, secondary fold prominent, sparsely covered with short bristlelike hairs, with longer hairs on cranial and distal edges; labionasal plate moderately complex, margin above each nostril with three or four wartlike tubercles, nostrils separated by slight emargination, prominent lateral spike on either side of nostrils; three or four toothlike serrations on anteromedial edge of long, lanceolate portion of ear; ears connected by two low, inconspicuous ridges fusing on top of nose, rostral tubercle not prominent; pelage distinctly tricolored dorsally, hairs of back having three well-marked bands, pelage bicolored vent-

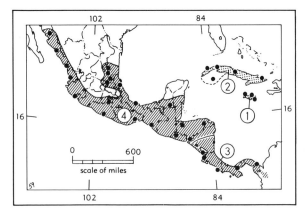

Map 59. *Pteronotus macleayii* and *Pteronotus personatus*.

| 1. *P. m. griseus* | 3. *P. p. personatus* |
| 2. *P. m. macleayii* | 4. *P. p. psilotis* |

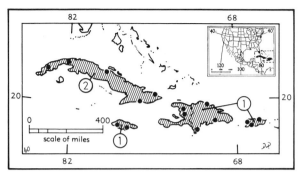

Map 60. *Pteronotus quadridens.*

1. *P. q. fuliginosus*   2. *P. q. quadridens*

rally; profile of skull flattened, rostrum slightly elevated, forehead rising abruptly onto round, high braincase; rostrum approx. same length as braincase; marked diastema between outer upper incisor and canine; inner incisors distinctly bifurcate; lower incisors small, and trilobed having short, rounded heels.

Pteronotus quadridens fuliginosus (Gray)

1843. *Chilonycteris fuliginosa* Gray, Proc. Zool. Soc. London, p. 20, July, type from Port-au-Prince, Haiti.
1976. *Pteronotus quadridens fuliginosus,* Silva-Taboada, Poeyana, No. 153:7, November 30.
1904. *Chilonycteris macleayii inflata* Rehn, Proc. Acad. Nat. Sci. Philadelphia, 56:190, March 26, type from Cueva di Fari, near Pueblo Viejo, Puerto Rico.

MARGINAL RECORDS (Smith, 1972:86).—Puerto Rico: Cueva de Trujillo Alto; Cueva di Fari, near Pueblo Viejo; Mayaguez. Dominican Republic: Sosúa; Caña Honda, Santo Domingo; 6 mi. NE Oviedo. Haiti: 4 mi. SE Cerca La Source; St. Michel. Jamaica: Oxford Cave, Balaclava; St. Clair Cave.

Pteronotus quadridens quadridens (Gundlach)

1840. *L[obostoma]. quadridens* Gundlach, Wiegmann's Arch. für Naturgesch., Jahrg. 6:357, type from Cafetal San Antonio El Fundador, Canímar, Matanzas, Cuba.
1976. *Pteronotus quadridens quadridens,* Silva-Taboada, Poeyana, No. 153:7, November 30.
1916. *Chilonycteris torrei* G. M. Allen, Proc. New England Zool. Club, 6:4, February 8, type from La Cueva de la Majana, Baracoa, Prov. Oriente, Cuba, a synonym of *P. q. quadridens* according to Silva-T. (Poeyana, No. 153:7, November 30, 1976).

MARGINAL RECORDS (Smith, 1972:87).—Cuba: Cueva de Mudo, Güines; Cueva de Colón, Yaguajay, Punta Caguanes; Baracoa; Banabacóa, Dos Caminos; Cueva del Río, San Vicente.

Pteronotus personatus
Wagner's Mustached Bat

Length of head and body, 49–55; total length, 64–74; length of forearm, 40.8–47.4; condylobasal length, 13.2–15.0; zygomatic breadth, 7.9–9.3; maxillary tooth-row, 5.7–6.5; uropatagium and wing membranes attach to ankle by short ligament tightly bound to tibia; tragus spatulate, secondary fold moderately well developed and shelflike, clothed with short hairs; labio-nasal plate simple, margin above nostrils smooth or with several wartlike tubercles, lacking lateral spikes on either side of nostrils; nostrils not separated by deep emargination; anteromedial edge of pinna having three or four (occasionally as many as six) toothlike serrations; ears united by two low, inconspicuous ridges fusing on top of muzzle, forming prominent rostral tubercle; profile of skull flattened, rostrum slightly elevated, forehead sloping gradually onto long, oval braincase; upper incisors reduced, slight diastema between outer incisor and canine; inner incisors distinctly bifurcate, with narrow, rounded heel; lower incisors reduced and having short rounded heel, inner pair distinctly trilobate, outer pair weakly trilobate; small, shelflike cuspule on anterior lingual surface of postcentrocrista on M1 and M2 (this cuspule, found in no other chilonycterid, in a position roughly approximating that of metaconule); anterior opening of infraorbital canal slightly anterior to maxillary root of zygomatic arch. See Map 59.

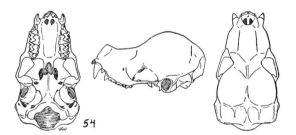

Fig. 54. *Pteronotus personatus psilotus,* 3 km. E San Andrés Tuxtla, 1000 ft., Veracruz, No. 23575 K.U., ♂, X 2.

Pteronotus personatus personatus (Wagner)

1843. *Chilonycteris personata* Wagner, Wiegmann's Arch. für Naturgesch., Jahrg., 9:367, December 31, type from Mato Grosso, Brazil.
1970. *P[teronotus]. personatus,* Vaughan and Bateman, Jour. Mamm., 51:218, May 20.

MARGINAL RECORDS (Smith, 1972:92, unless otherwise noted).—Nicaragua: $6\frac{9}{10}$ mi. E San Juan del Sur (Baker and Jones, 1975:2). Costa Rica: $4\frac{2}{3}$ mi. S

Liberia; Rincón de Osa. Panamá: Penonomé Caves; Armila, Quebrada Venado, thence into South America.

Pteronotus personatus psilotis (Dobson)

1878. *Chilonycteris psilotis* Dobson, Catalogue of the Chiroptera in the . . . British Museum, p. 451, type locality unknown; fixed by de la Torre (Fieldiana-Zool., Field Mus. Nat. Hist., 37:696, June 19, 1955) as Tehuantepec, Oaxaca.
1972. *Pteronotus personatus psilotis*, Smith, Univ. Kansas Mus. Nat. Hist., Miscl. Publ., 56:92, March 10.
1938. *Chilonycteris torrei continentis* Sanborn, Occas. Pap. Mus. Zool., Univ. Michigan, 373:1, 26 May, type from Laguna de Zotz, Petén, Guatemala.

MARGINAL RECORDS (Smith, 1972:93, unless otherwise noted).—Sonora: Río Cuchujaqui, 8 mi. by road SSE Alamos. Sinaloa: Cueva Chinacaterra, 4½ mi. S, 8 mi. W Pericos, 800 ft.; 1 mi. W Matatán. Nayarit: San Blas. Colima: ½ mi. S, 7 mi. W Santiago. Morelos: Alpuyeca. Querétaro: 5⁹⁄₁₀ km. NW Jalpan, 750 m. (Spenrath and LaVal, 1970:396). San Luis Potosí: 16 mi. S Valles (Jones and Alvarez, 1964:302); 1 km. S El Salto. Tamaulipas: 7 mi. W Ocampo, 2400 ft. Veracruz: near Tuxpan (Hall and Kelson, 1959:91); 3 km. ENE San Andrés Tuxtla, 1000 ft. Campeche: 65 km. S, 128 km. E Escárcega. Honduras: 5 km. NE Ilama, 120 m.; San José Río Tinto, 340 m.; Río Laure, 10 km. E San Lorenzo, 25 ft. El Salvador: 2 km. W Suchitoto, 390 m. Chiapas: 20 km. SE Pijijiapan. Oaxaca: San Gerónimo. Guerrero: 2 mi. NW Acapulco.

Pteronotus davyi
Davy's Naked-backed Bat

Length of head and body, 53–59; total length, 71–83; length of forearm, 40.6–49.2; condylobasal length, 13.7–15.3; zygomatic breadth, 8.3–9.5; maxillary tooth-row, 6.0–6.9; wing membranes fused dorsally at mid-line, resulting in a naked-backed appearance; wing membrane and uropatagium attached to ankle by long ligament (nearly half as long as tibia) tightly bound to tibia; labio-nasal plate moderately complex in structure, margin above each nostril ornamented by series or irregularly shaped, wartlike tubercles, lateral spikes on either side of nostrils; tragus moderately complex, spatulate, with prominent

secondary fold, distal tip curled craniad, and moderately long hairs ornamenting distal and cranial edges; anteromedial edge of pinna smooth (lacking toothlike serrations); ears not noticeably connected by ridge, rostral tubercle only weakly developed; pelage unicolored or only slightly bicolored dorsally (hair beneath dorsally fused wing membranes noticeably longer than that exposed in scapular, nape, and head regions); profile of skull flattened, but rostrum short and noticeably upturned; forehead relatively abruptly elevated onto elongated braincase; proximal nasal root distinctly "scooped out" forming pronounced basin; rostrum extremely broad in contrast to length as exemplified by relatively short maxillary tooth-row; upper incisors small; inner pair bifurcate and with narrow rounded heel directed toward canine, inner and outer incisors compressed together, forming two groups in contact only at inside tips of inner two incisors; diastema between outer incisor and canine; lower incisors small, trilobate, with short rounded heel; tympanic ring large, nearly covering auditory bulla; basioccipital region between bullae extremely broad and cup-shaped having two shallow, oval pits in which M. longus capitus inserts.

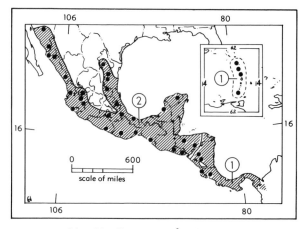

Map 61. *Pteronotus davyi.*

1. *P. d. davyi*   2. *P. d. fulvus*

Pteronotus davyi davyi Gray

1838. *Pteronotus davyi* Gray, Mag. Zool. Bot., 2:500, February, type from island of Trinidad.

MARGINAL RECORDS (Smith, 1972:100).—Nicaragua: 3 mi. E San Ramón; Cuapa. Costa Rica: Playa del Coco; 4¾ mi. W Atenas, thence into South America. Also, Lesser Antilles: Marie Galante Island (near Guadeloupe); Dominica; Martinique; Grenada.

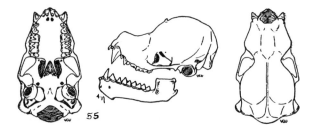

Fig 55. *Pteronotus davyi fulvus*, 3 km. E San Andrés Tuxtla, 1000 ft., Veracruz, No. 23578 K.U., ♂, X 2.

Pteronotus davyi fulvus (Thomas)

1892. *Ch[ilonycteris]*. *Davyi fulvus* Thomas, Ann. Mag. Nat. Hist., ser. 6, 10:410, November, type from Las Peñas, Jalisco, México.
1912. *Pteronotus davyi fulvus*, Miller, Bull. U.S. Nat. Mus., 79:33, December 31.
1958. *Pteronotus suapurensis calvus* Goodwin, Amer. Mus. Novit., 1871:1, February 26, type from Tehuantepec, Oaxaca, México.

MARGINAL RECORDS (Smith, 1972:102, unless otherwise noted).—Sonora: 13 mi. S Carbó, 1200 ft.; La Aduana (Cockrum and Bradshaw, 1963:4). Sinaloa: 1 mi. S, 6 mi. E El Carrizo; 1 mi. E Sinaloa; Pánuco. Nayarit: 2 mi. S Compostela. Zacatecas: 6⅔ km. S Jalpa (Matson and Patten, 1975:2). Jalisco: 9 mi. NNE Guadalajara (Watkins, *et al.*, 1972:10). Michoacán: Lake Chapala. Morelos: 2 km. W Tequesquitengo. Veracruz: Mirador (Hall and Kelson, 1959:94). Puebla: 2 mi. W Villa Avila Camacho (LaVal, 1972:449). San Luis Potosí: El Salto, 7 mi. N Naranjo. Tamaulipas: Rancho Pano Ayuctle, 8 mi. N Gómez Farías, 300 ft. (Hall and Kelson, 1959:94). Nuevo León: Cueva La Boca, 4 km. SE Santiago, 445 m. (Jiménez-G., 1971:134). Tamaulipas: Rancho Santa Rosa, 25 km. N, 13 km. W Ciudad Victoria. Veracruz: 3 km. ENE San Andrés Tuxtla, 1000 ft. Campeche: 5 km. S Champotón. Yucatán: Chichén-Itzá. Honduras: Subirana; Chichicaste. El Salvador: Tabancó; Finca El Marne, 8 km. SW Santa Ana, 960 m. Chiapas: 11 km. NW Escuintla, 100 ft. Oaxaca: Tehuantepec; 18 mi. NW Sola de Vega. Guerrero: 2 mi. NW Acapulco, 50 ft. Colima: ½ mi. S, 7 mi. W Santiago.

Pteronotus gymnonotus (Natterer, in Wagner)
Big Naked-backed Bat

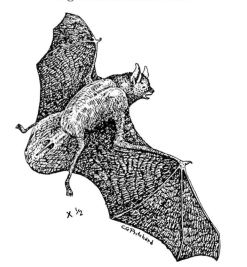

1843. *Chilonycteris gymnonotus* Natterer, *in* Wagner, Wiegmann's Arch. für Naturgesch., Jahrg. 9:367, December 31, type from Cuyaba, Mato Grosso, Brazil.
1977. *Pteronotus gymnonotus*, J. D. Smith, Jour. Mamm., 58:246, May 31.

1904. *Dermonotus suapurensis* J. A. Allen, Bull. Amer. Mus. Nat. Hist., 20:229, June 29, type from Suapuré, Bolívar, Venezuela.
1942. *Pteronotus suapurensis centralis* Goodwin, Jour. Mamm., 28:88, February 16, type from Matagalpa, 3000 ft., Matagalpa, Nicaragua.

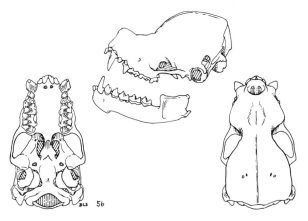

Fig. 56. *Pteronotus gymnonotus*, Cara de Mono, 50 m., Zelaya, Nicaragua, No. 114770 K.U., ♂, X 2.

Length of head and body, 58–67; total length, 81–96; length of forearm, 49.0–56.0; condylobasal length, 15.2–16.9; zygomatic breadth, 9.5–10.8; maxillary tooth-row, 6.8–7.8; differs from *P. davyi* mainly in larger size.

MARGINAL RECORDS (Smith, 1972:106, as *P. suapurensis*).—Veracruz: 3 km. ENE San Andrés Tuxtla, 1000 ft. Guatemala: Lanquin Cave, 1022 ft. Honduras: San José Río Tinto, 340 m. Nicaragua: 10 km. W Rama. Panamá: Madden Dam; Armila, Quebrada Venado, thence southeastward into South America and northwestward to Panamá: Penonomé. Costa Rica: Rincón de Osa. Nicaragua: Matagalpa. El Salvador: Finca El Marne, 8 km. S Santa Ana.

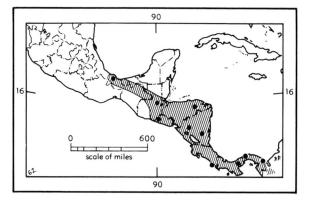

Map 62. *Pteronotus gymnonotus*.

Genus **Aello** Leach
Ghost-faced Bats

1821. *Aello* Leach, Trans. Linnean Soc. London, 13:69. Type, *Aello cuvieri* Leach.

1821. *Mormoops* Leach, Trans. Linnean Soc. London, 13:76. Type, *Mormoops blainvillii* Leach [= *Aello cuvieri* Leach]. Authors have used *Mormoops* in place of *Aello*, owing to Opinion 462 (16[pt. 1]: 1–12, April 2, 1957) of the International Comm. on Zool. Nomenclature, although *Aello* has page priority.

Long, lax pelage; short rounded ears that are strongly connected dorsally on rostrum by two conspicuous ridges; lower edge of pinna confluent with lower lip; ornamentation of lower lip complex, with many intricate folds and scallops, forming shieldlike plate (bearing many wartlike tubercles) in front of narrow central tubercle; secondary transverse ridge below lower lip further complicates structure; labio-nasal plate complex with nostrils surrounded by separate pads and separated by long, wartlike ridge; margin above and between nostrils with several long, irregularly shaped tubercles; tragus complex, secondary fold prominent; pubescence restricted to several long hairs and numerous short bristles on cranial edge; wing membranes attaching relatively high on sides of body but not on mid-line as in *Pteronotus;* uropatagium and wing membrane attaching to ankle by short ligament not bound to tibia; rostrum strongly upturned; forehead rising abruptly almost at right angle to rostral part of skull; zygomatic arches angled as result of rostro-braincase flexion; mastoid flanges nearly absent and mastoid bone little more than short spicules above and behind auditory bullae; parietals noticeably inflated in area that encases cerebelli; infraorbital foramen at maxillary root of zygomatic arch; incisive foramina large; basioccipital region relatively broad between bullae and basisphenoid with cuplike depression, into which M. longus capitus inserts; molars dilambdodont; wide diastemas in upper tooth-row between outer incisor and canine and between 1st and 2nd premolars; upper incisors thin and bladelike, heel on inner pair absent; upper canine deeply grooved on anteromedial surface; lower incisors small and delicate, trilobate, heels absent; 2nd lower premolar long and narrow; ramus markedly curved posteriorly because of cranial flexion.

KEY TO SPECIES OF AELLO

1. Forearm longer than 50; mainland.
   A. megalophylla,   p. 97

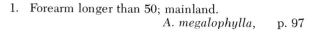
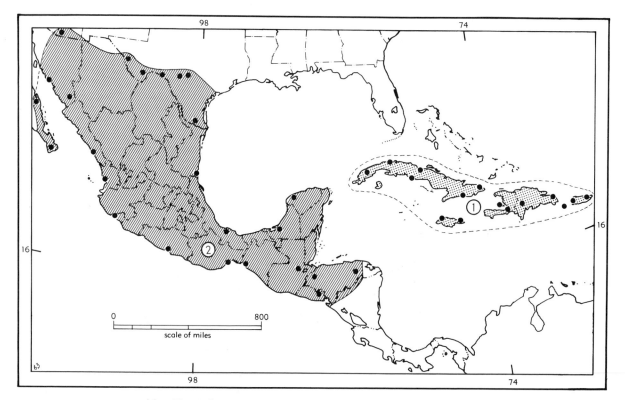

Map 63.   *Aello cuvieri* (1) and *Aello megalophylla megalophylla* (2).

1'. Forearm shorter than 50; Greater Antilles.
A. cuvieri, p. 97

Aello cuvieri Leach
Antillean Ghost-faced Bat

1821. *Aello Cuvieri* Leach, Trans. Linnean Soc., 13:71, type from unknown locality, restricted by Smith, Univ. Kansas Mus. Nat. Hist., Miscl. Publ., 56:111, March 10, 1972, to Jamaica.
1821. *Mormoops Blainvillii* Leach, Trans. Linnean Soc., 13:77, type from Jamaica.
1840. *L[obostoma]. cinnamomeum* Gundlach, Wiegmann's Arch. für Naturgesch., Jahrg., 6:357, type from Cafetal St. [San] Antonio El Fundador, Matanzas, Cuba.

Length of head and body approx. 51; total length, 80–86; length of forearm, 44.8–49.0; condylobasal length, 12.6–13.5; zygomatic breadth, 8.0–8.9; maxillary tooth-row, 7.2–7.8; differs from *A. megalophylla* primarily in being smaller.

MARGINAL RECORDS (Smith, 1972:111, 112).—Cuba: 4 mi. [from] San José de Las Lajas; Cueva Grande, Yaguajay; Baracoa; Santiago de Cuba; Cueva de los Masones, Trinidad; Cueva del Indio, San Vicente. Haiti: La Gonave Island (Hall and Kelson, 1959:95); Diquini. Dominican Republic: Rancho La Guardia; Cueva Las Lagunas de Nisibon. Puerto Rico: Mona Island; Mayaguez; Cueva Trujillo Alto. Jamaica: Lucea; Port Antonio.

Aello megalophylla
Peters' Ghost-faced Bat

Length of head and body, 59–69; total length, 85–97; length of forearm, 51.0–58.8; condylobasal length, 13.6–14.9; zygomatic breadth, 9.0–10.1; maxillary tooth-row, 7.5–8.3; differs from *A. cuvieri* primarily in being larger.

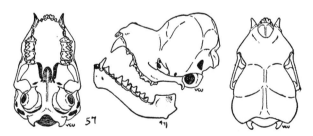

Fig. 57. *Aello megalophylla megalophylla*, 3 km. E San Andrés Tuxtla, 1000 ft., Veracruz, No. 23636 K.U., ♂, X 2.

Aello megalophylla megalophylla Peters

1864. *Mormops megalophylla* Peters, Monatsb. preuss. Akad. Wiss., Berlin, p. 381, type from México [regarded as Parrás, Coahuila, by Smith, Univ. Kansas Mus. Nat. Hist., Miscl. Publ., 56:116, March 10, 1972].
1902. *Mormoops megalophylla senicula* Rehn, Proc. Acad.

Nat. Sci. Philadelphia, 54:169, June 11, type from Fort Clark, Kinney Co., Texas.
1962. *Mormoops megalophylla rufescens* Davis and Carter, Southwestern Nat., 7:65, June 1, type from 5 mi. W Alamos, Sonora.

MARGINAL RECORDS (Smith, 1972:117, 118, unless otherwise noted).—Arizona: 5 mi. N, 2 mi. W Patagonia, 4450 ft. (Cockrum, 1961:31). Texas: Presidio Co., ZH Canyon, ca. 9 mi. W Valentine (Mollhagen, 1973:428); Giant Dagger Yucca Flats, Big Bend National Park (Easterla, 1968a:516); 9 mi. SW Comstock (Davis and Carter, 1962:67); Frio Cave (Jameson, 1959:62, as *M. m. senicula*); Nye [Ney] Cave; Edinburg. Tamaulipas: Tampico. Veracruz: 3 km. E San Andrés Tuxtla, 1000 ft. Campeche: 1 km. N, 13 km. W Escárcega, 65 m. Yucatán: Mérida. Guatemala: Lanquin Cave, 1022 ft. Honduras: 2 km. S San Nicolás, 660 m.; San José Río Tinto, 340 m. El Salvador: Encuentros Mine, 3 mi. W Divisadero, 700 ft. Chiapas: 5 mi. N Arriaga, 800 ft. Oaxaca: 8 km. NW Salina Cruz. Guerrero: 2 mi. NW Acapulco, 50 ft. Colima: ½ mi. S, 7 mi. W Santiago, sea level. Nayarit: San Blas. Sinaloa: Mazatlán. Sonora: La Aduana; Rancho San José, 15 mi. NW Guaymas. Baja California: Santa Anita (Davis and Carter, 1962:66); ¼ mi. S Mulegé. Other subspecies occur in South America.

SUBFAMILY PHYLLOSTOMATINAE

Teeth essentially normal, although in some genera mesostyle and commissures of M1 and M2, and paraconid and metaconid of m1 and m2 are reduced; humerus with definite secondary impingement (not articulation) with scapula; epitrochlea large, having slightly developed spinous process; nose-leaf present; lower lip without platelike outgrowths (after Miller, 1907:122).

KEY TO NORTH AMERICAN GENERA
OF PHYLLOSTOMATINAE

(Adapted from Miller, 1907:122–123)

6. Rostrum as long as braincase; molars
wider than palate, their W-shaped pattern
much modified. . . . . . . *Vampyrum,* p. 117
6'. Rostrum shorter than braincase; molars
narrower than palate, their W-shaped pattern essentially normal.
  7. Second and 3rd lower premolars not
differing greatly in size.
    8. Auditory bullae large, their great-
est diameter much exceeding the
distance between them. *Macrotus,* p. 104
    8'. Auditory bullae small, their great-
est diameter approx. equal to the
distance between them
                Micronycteris, p. 98
  7'. Middle lower premolar much smaller
than 3rd.
    9. First lower premolar in contact or
nearly so with 3rd, the 2nd displaced inward from tooth-row.
      10. Length of rostrum much less
than breadth of braincase.
          Macrophyllum, p. 107
      10'. Length of rostrum approx.
equal to breadth of braincase.
          Trachops, p. 115
  9'. First lower premolar distant from
3rd, the 2nd being in normal position in tooth-row.
    11. Dorsal profile of rostrum
strongly convex, a deep depression present between
orbits. . . . . *Lonchorhina,* p. 106
    11'. Dorsal profile of rostrum not
convex; no depression be-
tween orbits. *Phylloderma,* p. 114

Genus **Micronycteris** Gray
Large-eared Bats

Revised by Andersen, Ann. Mag. Nat. Hist., ser. 7, 18:
50–58, July, 1906; and by Sanborn, Fieldiana-Zool., Chicago
Nat. Hist. Mus., 31:215–233, April 29, 1949.

1866. *Micronycteris* Gray, Proc. Zool. Soc. London, p. 113,
May. Type, *Phyllophora megalotis* Gray.

Skull slender, lightly constructed; rostrum narrow, tapering, more than half as long as, but
shorter than, braincase; frontal region rising from
axis of rostrum at angle of approx. 45° (but varying
according to species); auditory bullae small,
greatest diameter of one approximating interbul-
lar space; middle lower premolar approx. as large
as 3rd lower premolar. Dentition, i. $\frac{2}{2}$, c. $\frac{1}{1}$, p. $\frac{2}{3}$, m.
$\frac{3}{3}$. Tail extending to middle of interfemoral
membrane.

KEY TO NORTH AMERICAN SUBGENERA AND SPECIES OF MICRONYCTERIS

1. Ears connected by high, notched band; P3
about same size as P4.
          subgenus *Micronycteris,* p. 98
  2. Interauricular band slightly notched
medially; venter brown. *M. megalotis,* p. 99
  2'. Interauricular band deeply notched
medially; venter pale, almost white.
    3. Calcar shorter than foot, about 6;
p3 small, compared to p2.
          M. minuta, p. 99
    3'. Calcar longer than foot, about 10;
p3 only slightly smaller than p2.
          M. schmidtorum, p. 100
1'. Ears connected by low unnotched band, or
none; P3 and P4 usually differing somewhat in size.
  4. Interauricular band present, but low
and unnotched; 3rd metacarpal shortest, 5th longest; greatest length of
skull more than 22.
    subgenus *Xenoctenes, M. hirsuta,* p. 100
  4'. Interauricular band absent; 4th or 5th
metacarpal shortest, 3rd or 5th longest;
greatest length of skull less than 22.
    5. Fifth metacarpal shortest, 3rd
longest; P4 straight; I2 bifid, medial cusp elongated.
      subgenus *Lampronycteris,*
          M. brachyotis, p. 102
    5'. Fourth metacarpal shortest, 3rd or
5th longest; P4 slightly recurved;
I2 minute, unicuspid.
      6. Fifth metacarpal longest; P3
larger than P4.
        subgenus *Glyphonycteris,*
          M. sylvestris, p. 102
      6'. Third metacarpal longest; P3
smaller than P4.
        subgenus *Trinycteris,*
          M. nicefori, p. 101

Subgenus **Micronycteris** Gray

1866. *Micronycteris* Gray, Proc. Zool. Soc. London, p. 113,
May. Type, *Phyllophora megalotis* Gray.
1856. *Schizostoma* Gervais, Mammifères *in* [Castelnau]
Expéd. dans les parties centrales de l'Amér. du Sud . . . ,
pt. 7, p. 38. Type, *Schizostoma minuta* Gervais. Not
Schizostoma Bronn, 1835, a mollusk.
1862. *Schizastoma* Gray, Catalogue of the bones of Mammalia
in the . . . British Museum, p. 38, a *lapsus?*

Ears connected by high, notched band; 3rd
metacarpal shortest, 5th longest; forearm, 31–38;

P3 large, about equal to P4; greatest length of skull, 17–20.

Micronycteris megalotis
Brazilian Large-eared Bat

Length of head and body approx. 48; interorbital breadth, 3.8–5.5; zygomatic breadth, 8.2–9.8; length upper tooth-row, 6.8–7.8. Upper parts medium darkish brown tinged with russet or dark brown without russet tinge; underparts paler brown; intermediate colors present on some specimens. Nose lancet approx. 1½ times higher than breadth of its base; calcar longer than foot and more than half as long as lower leg.

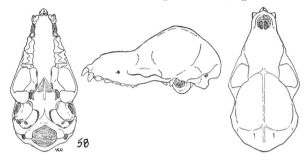

Fig. 58. *Micronycteris megalotis mexicana*, 4 km. W Peso de San Juan, 250 ft., Veracruz, No. 23639 K.U., ♂, X 2.

Micronycteris megalotis megalotis (Gray)

1842. *Phyllophora megalotis* Gray, Ann. Mag. Nat. Hist., ser. 1, 10:257, type from Brazil.
1866. *Micronycteris megalotis*, Gray, Proc. Zool. Soc. London, p. 113, May.

MARGINAL RECORDS.—Lesser Antilles: Grenada (Jones and Phillips, 1970:132), thence southward into South America.

Micronycteris megalotis mexicana Miller

1898. *Micronycteris megalotis mexicanus* Miller, Proc. Acad. Nat. Sci. Philadelphia, 50:329, August 2, type from Platanar, Jalisco.
1904. *Macrotus pygmaeus* Rehn, Proc. Acad. Nat. Sci. Philadelphia, 56:444, June 30, type from Izamal, Yucatán. Goodwin (Bull. Amer. Mus. Nat. Hist., 102:246, August 31, 1953) regards *M. pygmaeus* as indistinguishable from *Micronycteris megalotis mexicana*.

MARGINAL RECORDS.—Tamaulipas: Pano Ayuctle. Veracruz (Hall and Dalquest, 1963:221): Plan del Río, 1000 ft.; 14 km. SW Coatzacoalcos, 100 ft. Yucatán: 10 km. NE Mérida (93348 KU). Quintana Roo:

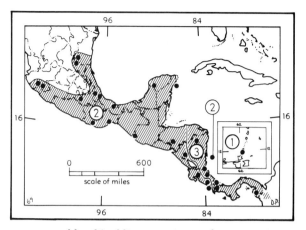

Map 64. *Micronycteris megalotis.*

1. *M. m. megalotis*   2. *M. m. mexicana*
3. *M. m. microtis*

Cozumel Island (Jones and Lawlor, 1965:412). Honduras: Sabana Grande, Tegucigalpa. Nicaragua: 11 mi. SE Dario (Jones, *et al.*, 1971:5). Costa Rica (Gardner, *et al.*, 1970:715): 5 mi. E Tilarán, 700 m.; Fila la Maquina, *ca.* 7½ km. E Canaán, *ca.* 8700 ft.; 2 km. W Rincón de Osa. Chiapas: Kilometer 184, Hwy. 200, N Huixtla (Baker, 1967:426, as *M. megalotis* only). Oaxaca (Goodwin, 1969:57): San Antonio; Puerto Escondido. Jalisco: Tolimán, 2200 ft. (Watkins, *et al.*, 1972:11); type locality; *Ciudad Guzmán*. Morelos: Cuautla. Veracruz: Cuesta de Don Lino, near Jalapa. San Luis Potosí: 8 mi. (by road) E Santa Barbarita. Island population: Nicaragua: Isla del Maíz Grande (Jones, *et al.*, 1971:5).

Micronycteris megalotis microtis Miller

1898. *Micronycteris microtis* Miller, Proc. Acad. Nat. Sci. Philadelphia, 50:328, July 12, type from Greytown [= San Juan del Norte], Comarca de San Juan del Norte, Nicaragua.
1949. *Micronycteris (Micronycteris) megalotis microtis*, Sanborn, Fieldiana-Zool., Chicago Nat. Hist. Mus., 31:219, April 29.

MARGINAL RECORDS.—Nicaragua (Jones, *et al.*, 1971:6): *Río Coco*; Bonanza; type locality: Cariari (Gardner, *et al.*, 1970:715). Panamá (Handley, 1966:760): Almirante; Boquete; Río Jesucito, thence southeastward into South America.

Micronycteris minuta (Gervais)
Gervais' Large-eared Bat

1856 [1855]. *Schizostoma minutum* Gervais, Mammifères, *in* [Castelnau] Expéd. dans les parties centrales de l'Amér. du Sud . . . , pt. 7, p. 50, pl. 7, fig. 1, type from Capella Nova, Brazil.
1901. *Micronycteris minuta*, Thomas, Ann. Mag. Nat. Hist., ser. 7, 8:191, September.

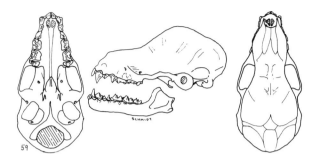

Fig. 59. *Micronycteris minuta*, 2 mi. W Guabala, Chiriqui, Panamá, No. 331112 U.S.N.M., ♂, X 2.

Selected measurements of five specimens from Nicaragua and Costa Rica: forearm, 33.2–35.7; greatest length of skull, 17.7–19.0; postorbital breadth, 4.0–4.1; mandibular tooth-row, 5.7–7.4. Tips of dorsal hairs pale brown, bases white; venter near (e) Pallid Neutral Gray. Differs from *M. schmidtorum* in shorter calcar and smaller p3.

MARGINAL RECORDS.—Nicaragua: 6 km. N Tuma, 500 m. (Valdez and LaVal, 1971:248). Costa Rica: Turrialba, IICA, 600 m. (Gardner, *et al.*, 1970: 716). Panamá: Boca de Río Paya (Handley, 1966:760), thence southeastward into South America, and northwestward along coast to Panamá: Cerro Hoya, 1800 ft. (*ibid.*). Costa Rica (Gardner, *et al.*, 1970:716): Río Corrogres, *ca.* 2 km. NW Santa Ana; Finca La Pacífica, 4 km. NW Cañas, 45 m.

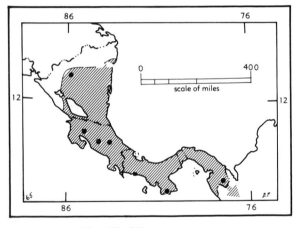

Map 65. *Micronycteris minuta*.

Micronycteris schmidtorum Sanborn
Schmidt's Large-eared Bat

1935. *Micronycteris schmidtorum* Sanborn, Field Mus. Nat. Hist., Publ. 340, Zool. Ser., 20:81, May 15, type from Bobós, Izabal, Guatemala.

Selected measurements of the type, an adult male: length of head and body, approx. 47; total length, 64; length of forearm, 35.3; greatest length of skull, 20.5; interorbital breadth, 4.3; zygomatic breadth, 9.1; length upper tooth-row, 7.9. Upper parts medium brown; underparts paler, almost white.

MARGINAL RECORDS.—Guatemala: type locality. Nicaragua: Cacao, 22 km. by road W Muelle de las Bueyes, 400 ft. (Davis, *et al.*, 1964:378). Panamá: Guánico (Handley, 1966:760). Costa Rica (Starrett and Casebeer, 1968:10): 3 km. S Playas del Coco; 9 km. N Liberia along Interamerican Highway, 4 km. E Interamerican Highway, 150 m. Honduras: Copán.—A juvenile from Isla Cozumel, Quintana Roo, if correctly identified as this species (Jones, *et al.*, 1973:9), extends the known geographic range 350 mi. farther northward than shown on Map 66.

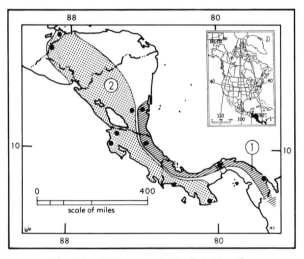

Map 66. *Micronycteris nicefori* (1) and *Micronycteris schmidtorum* (2).

Subgenus Xenoctenes Miller

1907. *Xenoctenes* Miller, Bull. U.S. Nat. Mus., 57:124, June 29. Type, *Schizostoma hirsutum* Peters.

Ears connected by low band, not notched; 3rd metacarpal shortest, 5th metacarpal longest; P3 large, about equal to P4.

Micronycteris hirsuta (Peters)
Hairy Large-eared Bat

1869. *Schizostoma hirsutum* Peters, Monatsb. preuss. Akad. Wiss., Berlin, p. 396. Type from an unknown locality; tentatively fixed at Pozo Azul, San José, Costa Rica, by Goodwin, Bull. Amer. Mus. Nat. Hist., 87:302, December 31, 1946.
1906. *Micronycteris hirsuta*, Andersen, Ann. Mag. Nat. Hist., ser. 7, 18:57, July.

Range of measurements for six specimens, two each from Honduras, Nicaragua, and Costa Rica: forearm, 40.0–43.5; 3rd metacarpal, 34.3–36.5;

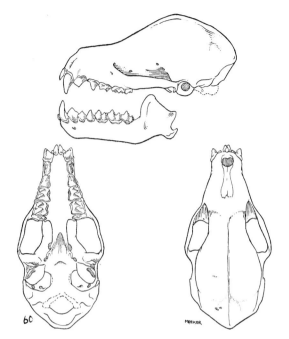

Fig. 60. *Micronycteris hirsuta*, 5 km. SE Turrialba, 1950 ft., Prov. Cartago, Costa Rica, No. 26933 K.U., ♀, X 2.

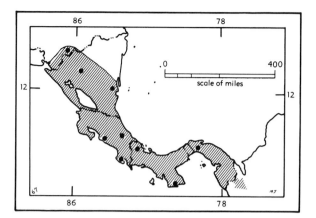

Map 67. *Micronycteris hirsuta*.

greatest length of skull, 22.6–24.4; postorbital width, 4.6–5.0; zygomatic width, 10.7–11.7; maxillary tooth-row, 8.7–9.4. Pelage long, about 11 mm. in center of back; dorsal fur grayish-brown distally, white at base; ventrally, gray with silvery gray tips.

MARGINAL RECORDS.—Honduras: 7 km. E Danlí (LaVal, 1969:819). Nicaragua: 6 km. N Tuma, 500 m. (Valdez and LaVal, 1971:247); 9 mi. E Rama (Baker and Jones, 1975:3). Costa Rica: 5 km. SE Turrialba, 1950 ft. (26933 KU). Panamá: upper Río Changena, 4800 ft. (Handley, 1966:759), thence southwestward along coast to South America, and northwestward to Panamá (*ibid.*): Cerro Azul; Guánico. Costa Rica: 2½ mi. SW Rincón, 15 m. (Gardner, *et al.*, 1970:716); Pozo Azul.

Subgenus **Trinycteris** Sanborn

1949. *Trinycteris* Sanborn, Fieldiana-Zool., Chicago Nat. Hist. Mus., 31:228, April 29. Type, *Micronycteris (Trinycteris) nicefori* Sanborn.

Ears not connected by band; 4th metacarpal shortest, 3rd longest; nose-leaf narrow, tip pointed; rostrum elongated as in *Micronycteris (Glyphonycteris) sylvestris*; P3 and p3 small, almost flat, anterior cusp small; P4 recurved as in *Glyphonycteris*; upper outer incisors small, not filling space between canines and inner incisors; lower incisors small, not crowded, faintly trifid.

Micronycteris nicefori Sanborn
Nicéforo's Large-eared Bat

1949. *Micronycteris nicefori* Sanborn, Fieldiana-Zool., Chicago Nat. Hist. Mus., 31:230, April 29, type from Cúcuta, Colombia.

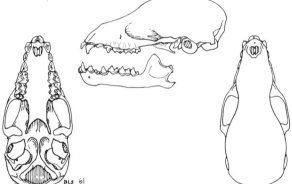

Fig. 61. *Micronycteris nicefori*, Ft. Gulick, Canal Zone, Panamá, No. 312927 U.S.N.M., ♀, X 2.

Head and body of specimen from Venezuela, 50. Selected measurements of five specimens from Colombia: forearm, 35.0–38.6; 3rd metacarpal, 33.9–35.7; 5th metacarpal, 32.6–34.6. Skull of holotype: greatest length, 20.5; interorbital breadth, 4.3; upper tooth-row, 7.3. Upper parts olive brown to walnut brown; underparts paler; faint light gray line on lower back of most specimens; calcar less than half length of foot.

MARGINAL RECORDS.—Nicaragua: 3 km. NW Rama (Baker and Jones, 1975:3). Costa Rica: La Selva (3223 LaVal at KU). Panamá (Handley, 1966:760): Fort Gulick; Armila, thence southeastward into South America. See Map 66.

Subgenus **Lampronycteris** Sanborn

1949. *Lampronycteris* Sanborn, Fieldiana-Zool., Chicago Nat. Hist. Mus., 31:223, April 29. Type, *Micronycteris (Lampronycteris) platyceps* Sanborn.

Ears not connected by band; 5th metacarpal shortest, 3rd metacarpal longest; calcar shorter than foot with claws; braincase lower than in subgenera *Micronycteris* or *Xenoctenes;* teeth essentially as in *Micronycteris,* but P4 longer and narrower with its inner border more nearly straight and internal ledge more nearly horizontal.

Micronycteris brachyotis (Dobson)
Dobson's Large-eared Bat

1878. *Schizostoma brachyotis* Dobson, Proc. Zool. Soc. London, p. 880, type from Cayenne, French Guiana.
1961. *Micronycteris brachyotis,* Goodwin and Greenhall, Bull. Amer. Mus. Nat. Hist., 122:231, June 26.
1949. *Micronycteris (Lampronycteris) platyceps* Sanborn, Fieldiana-Zool., Chicago Nat. Hist. Mus., 31:224, April 29, type from Guanapo, Trinidad.

Selected measurements of the type of *M. platyceps,* an adult female, and two topotypes are, respectively: total length, 68, 70, 70; length of forearm, 39.8, 39.6, 40.0; greatest length of skull, 21.2, 21.3, 21.6; interorbital breadth, 4.9, 5.2, 5.2; zygomatic breadth, 10.3, 10.4, 10.6; length upper tooth-row, 8.0, 8.2, 8.2. Upper parts olive brown; chest and belly tawny-olive; throat yellowish.

MARGINAL RECORDS.—Guatemala: Chuntuquí (Jones, 1966:450); Tikal (Rick, 1968:517), thence southward along the Caribbean Coast to Panamá: Fort

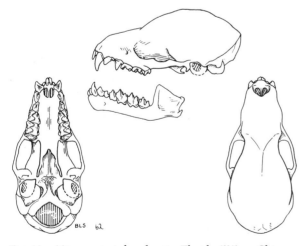

Fig. 62. *Micronycteris brachyotis,* Florida, 525 m., Chiapas, México, No. 9102 T.C.W.C., ♀, X 2.

San Lorenzo (Handley, 1966:759), and into South America, then northwestward to Panamá: Guánico (*ibid.*). Nicaragua: Volcán de Chinandega. Oaxaca: 29 km. S Matías Romero (Schaldach, 1965:131); *Mazahuito, 18 mi. S Matías Romero* (Goodwin, 1969:61, 260). Chiapas: Florida (Jones, 1966:450).

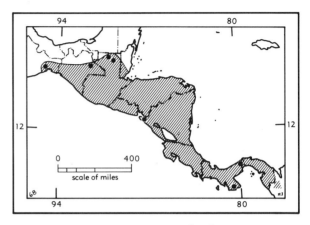

Map 68. *Micronycteris brachyotis.*

Subgenus **Glyphonycteris** Thomas

1896. *Glyphonycteris* Thomas, Ann. Mag. Nat. Hist., ser. 6, 18:302, October. Type, *Glyphonycteris sylvestris* Thomas.

Ears not connected by a band; 4th metacarpal shortest, 5th metacarpal longest; P3 larger than P4; P4 slightly recurved; braincase expanded less than in *Micronycteris* but more than in *Lampronycteris.*

Micronycteris sylvestris (Thomas)
Large-eared Forest Bat

1896. *Glyphonycteris sylvestris* Thomas, Ann. Mag. Nat. Hist., ser. 6, 18:303, October, type from between 1400 and 2000 ft., near Hda. Miravalles, Guanacaste, Costa Rica.
1949. *Micronycteris (Glyphonycteris) sylvestris,* Sanborn, Fieldiana-Zool., Chicago Nat. Hist. Mus., 31:231, April 29.

Measurements of 11 specimens from Veracruz: head and body, 51–64; greatest length of skull, 20.0–21.4; interorbital breadth, 4.7–5.0; zygomatic breadth, 10.1–10.8; maxillary tooth-row, 7.8–8.3. Upper parts three-banded, with whitish middle portion, and blackish tips and bases; underparts with dark gray bases and pale gray tips. Externally, resembles *Carollia brevicauda,* but feet smaller, forearm bare, and dorsal fur grayer.

MARGINAL RECORDS.—Nayarit: 2 mi. E El Venado (Jones, 1964e:509). Jalisco: 10 km. NW Soyatlán del Oro. Veracruz: 9 km. E Totutla, 2500 ft. (Hall and Dalquest, 1963:222); *15 km. ENE Tlacotepec,* thence southeastward along Caribbean Coast to Panamá: Ar-

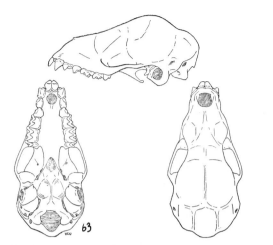

Fig. 63. *Micronycteris sylvestris*, 15 km. ENE Tlacotepec, 1500 ft., Veracruz, No. 23645 K.U., ♀, X 2.

mila (Handley, 1966: 760) and into South America, thence northwestward along Pacific Coast to Costa Rica: 2 mi. S, 11 mi. W Las Cañas, 10 ft. (Davis and Carter, 1962:67); *type locality.*

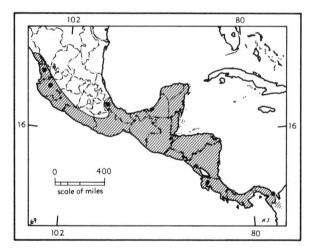

Map 69. *Micronycteris sylvestris.*

Genus **Barticonycteris** Hill

1964. *Barticonycteris* Hill, Mammalia, 28:556, December. Type, *Barticonycteris daviesi* Hill.

Closely allied to *Micronycteris*, except only one pair upper incisors, nearly matching upper canines in bulk and height; lower incisors elongated, laterally compressed, bases extending posteriorly beyond anterior faces of lower canines and forming scooplike structure; premolars large, compressed in tooth-row; massive shieldlike cingula on upper incisors, canines, and premolars. Dentition, i. $\frac{1}{2}$, c. $\frac{1}{1}$, p. $\frac{2}{3}$, m. $\frac{3}{3}$.

Barticonycteris daviesi Hill
Davies' Large-eared Bat

1964. *Barticonycteris daviesi* Hill, Mammalia, 28:557, December, type from forest reserve "24 miles from Bartica, along the Potaro" Road, Guyana.

Measurements of holotype and three specimens from Perú: head and body, 69, 83, 77, 84;

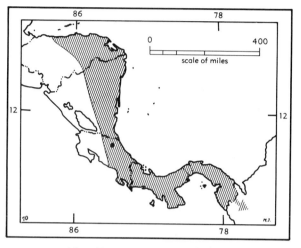

Map 70. *Barticonycteris daviesi.*

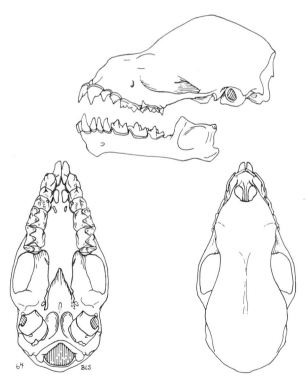

Fig. 64. *Barticonycteris daviesi*, San Blas Armila, Panamá, No. 335105 U.S.N.M., ♀ (details of area of anterior palatine foramina from No. 3108 R. K. LaVal at K.U., ♂, from La Selva, Costa Rica), X 2.

forearm, 57.5, 53.8, 58.1, *ca*. 58. Skull of holotype: greatest length, 27.2; interorbital constriction, 6.5; zygomatic width, 13.3; width of braincase, 10.9; maxillary tooth-row, 11.1. Ears not connected by band or ridge; dorsal pelage long, dark brownish-gray, ventral pelage pale gray.

MARGINAL RECORDS.—Costa Rica: La Selva (3108 LaVal at KU). Panamá: *San Blas Armila* (335105 USNM), thence into South America.

Genus Macrotus Gray—Leaf-nosed Bats

Revised by Anderson and Nelson, Amer. Mus. Novit., 2212:1–39, March 17, 1965, and by Rehn, Proc. Acad. Nat. Sci. Philadelphia, 56:427–446, June 27, 30, 1904. See also remarks on *M. waterhousii* below.

1843. *Macrotus* Gray, Proc. Zool. Soc. London, p. 21, July. Type, *Macrotus waterhousii* Gray.
1891. *Otopterus* Lydekker, *in* Flower and Lydekker, An introduction to . . . mammals living and extinct, p. 673. Type, *Macrotus waterhousii*, a renaming of *Macrotus* Gray, erroneously assumed to be preoccupied by *Macrotis* Reid, 1837, a marsupial.

Rostrum moderately long, considerably lower than braincase; medial upper incisors chisel-shaped, long; lateral upper incisors weak. Head long; muzzle conical; nose-leaf simple, erect, lanceolate; nose-pad rounded; nostrils elongate, distinct; lower lip with triangular pad bearing longitudinal groove; ears large, united; tragus lanceolate; uropatagium large; tail long, projecting somewhat beyond posterior margin of uropatagium, which envelops tail except its free apex; calcanea short and stout (after Rehn, 1904:428). Dentition, i. $\frac{2}{2}$, c. $\frac{1}{1}$, p. $\frac{2}{3}$, m. $\frac{3}{3}$.

Macrotus waterhousii
Waterhouse's Leaf-nosed Bat

Total length, 77.0–108; tail, 25.0–42.0; length of forearm, 44.7–58.0; greatest length of skull, 22.1–26.8; interorbital breadth, 3.2–4.9; zygomatic breadth, 10.5–12.5. Upper parts varying from buffy gray to dark brown; underparts pale drab to buffy brown, usually with silvery wash. Ears large and subovate; teeth robust to weak; rostrum robust to slender.

By means of a chromosomal and sophisticated mathematical analysis of mainland populations of *Macrotus* only, Davis and Baker (1974) listed *Macrotus californicus* as a species instead of as a subspecies of *M. waterhousii*. By studying only insular populations of *Macrotus*, Buden (1975:758) found statistically significant differences between the five previously named insular subspecies and noted (*op. cit.*:768) that "On

morphological grounds alone five subspecies might be recognized" there. Nevertheless, two sentences later he proposed by "correlating morphology, geography, and clinal trends" to recognize two instead of five subspecies in the West Indies! Because he did not list the valid names and synonyms, his proposal was not realized and he was taxonomically inconclusive. Nagorsen and Peterson (1975:2) recorded for *M. w. waterhousii*, an insular subspecies, a chromosome number of "2N = 46, FN = 60." For the time being the systematic classification below, by Anderson and Nelson (1965), can stand, and probably will be altered little if any when the somewhat contradictory suggestions in the literature cited immediately above have been reassessed.

Macrotus waterhousii bulleri H. Allen

1890. *Macrotus bulleri* H. Allen, Proc. Amer. Philos. Soc., 28:73, May 10, type from Bolaños, Jalisco.
1965. *Macrotus waterhousii bulleri*, Anderson and Nelson, Amer. Mus. Novit., 2212:28, March 17.

MARGINAL RECORDS (Anderson and Nelson, 1965:30, 31, unless otherwise noted).—Durango: Chacala; Santa Ana, 12 km. E Cosalá (Sinaloa), 1300 ft. Sinaloa: 6 mi. W Santa Lucia, 5650 ft.; *Copala*. Jalisco: type locality; NW side Río Verde, 12 mi. S, 4 mi. E Yahualico, 5200 ft. Hidalgo: Jacala. Michoacán: S shore Lake Chapala (Hall and Kelson, 1959:102). Jalisco: Ameca (*ibid.*). Nayarit: María Madre, Tres Marías Islands. Sinaloa: 3 mi. SE Camino Real, 500 ft.

Macrotus waterhousii californicus Baird

1858. *Macrotus californicus* Baird, Proc. Acad. Nat. Sci. Philadelphia, 10:116, type from Old Fort Yuma, Imperial Co., California, on right bank Colorado River, opposite present town of Yuma, Arizona.—See addenda.
1965. *Macrotus waterhousii californicus*, Anderson and Nelson, Amer. Mus. Novit., 2212:31, March 17.

MARGINAL RECORDS.—Arizona: Virgin Narrows, NE Littlefield; Golden Keyes Mine, 10 mi. W Bagdad (Irwin and Baker, 1967:195); Superior; 5 mi. N Clifton (Cockrum, 1961:33); Tombstone. Sonora: Santa Maria Mine, El Tigre Mts. (Anderson and Nelson, 1965:34). Chihuahua (Anderson and Nelson, 1965:34): Barranca de Cobre; La Bufa. Tamaulipas: *Conrado Castillo* (Choate and Clifton, 1970:359); Jaumave (Anderson and Nelson, 1965:34). San Luis Potosí: 2 km. W El Custodio, 1100 m. (Alvarez and Ramírez-P., 1972: 168). Sinaloa: 12 mi. N, 3 mi. W Los Mochis (*ibid.*). Baja California: 1 km. S Las Cuevas (Jones, *et al.*, 1965:53); Cape St. [= San] Lucas (Anderson and Nelson, 1965:34); San Antonio (Jones, *et al.*, 1965:53). California: De Luz; 2 mi. N Owensmouth; Indian Wells; Riverside Mts., 35 mi. N Blythe (Anderson and Nelson,

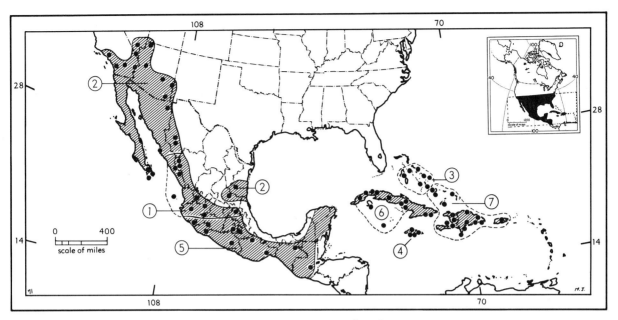

Map 71. *Macrotus waterhousii.*

Guide to subspecies	2. *M. w. californicus*	5. *M. w. mexicanus*
	3. *M. w. compressus*	6. *M. w. minor*
1. *M. w. bulleri*	4. *M. w. jamaicensis*	7. *M. w. waterhousii*

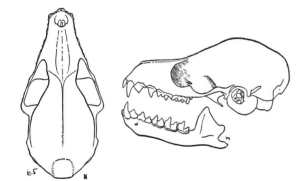

Fig. 65. *Macrotus waterhousii californicus,* 14 mi. E Searchlight, Nevada, No. 61421 M.V.Z., ♂, X 2.

1965:33). Nevada: 14 mi. E Searchlight, Colorado River; Frenchmans Mine, 7 mi. E Las Vegas.

Macrotus waterhousii compressus Rehn

1904. *Macrotus waterhousii compressus* Rehn, Proc. Acad. Nat. Sci. Philadelphia, 56:434, June 30, type from Eleuthera, Bahamas.

MARGINAL RECORDS (Anderson and Nelson, 1965:21).—Bahama Islands: 4 mi. S Georgetown, Eleuthera Island; Orange Creek, Cat Island; San Salvador (Buden, 1975:768); Hamiltons Cave, Long Island; Pigeon Cay, Great Exuma Island; *Exuma Cays, Darby Island;* Conch Sound, Andros Island; Nassau, New Providence Island.

Macrotus waterhousii jamaicensis Rehn

1904. *Macrotus waterhousii jamaicensis* Rehn, Proc. Acad. Nat. Sci. Philadelphia, 56:432, June 27, type from Spanishtown, Jamaica.

MARGINAL RECORDS (Anderson and Nelson, 1965:21, unless otherwise noted).—Jamaica: Windsor; Port Antonio; Kingston; *type locality; Portland Point Lighthouse;* Bluefields (Hall and Kelson, 1959:101).

Macrotus waterhousii mexicanus Saussure

1860. *Macrotus mexicanus* Saussure, Revue et Mag. Zool., Paris, ser. 2, 12:486, November, type from Yautepec, near Cuautla, Morelos.
1965. *Macrotus waterhousii mexicanus,* Anderson and Nelson, Amer. Mus. Novit., 2212:25, March 17.
1876. *Macrotus bocourtianus* Dobson, Ann. Mag. Nat. Hist., ser. 4, 18:436, type from Vera Paz, Guatemala.

MARGINAL RECORDS (Anderson and Nelson, 1965:28, unless otherwise noted).—Colima: Colima. Michoacán: La Salada. Morelos: Cuernavaca; *type locality; Cuautla* (Hall and Kelson, 1959:102). Puebla: 2 mi. SE Izúcar de Matamoros; *1 mi. E Raboso, 4350 ft.* Oaxaca: Reyes [= Pápalo Santos Reyes]. Chiapas: Palenque. Guatemala: Vera Paz (Jones, 1966:451). Oaxaca: Tehuantepec; *3 mi. NW Tehuantepec.* Guerrero: Chilpancingo. Michoacán: 45 km. S, 15 km. E Nueva Italia, 250 m. (Alvarez, 1968:25).—Anderson and Nelson (1965:27) and Jones, *et al.* (1973:10), on the authority of Gaumer (1917:292, 293), think this bat was found at several localities in Yucatán.

Macrotus waterhousii minor Gundlach

1864. *Macrotus minor* Gundlach, Monatsb. preuss. Akad. Wiss., Berlin, p. 382, type from western Cuba.
1904. *Macrotus waterhousii minor,* Rehn, Proc. Acad. Nat. Sci. Philadelphia, 56:435, June 30.

MARGINAL RECORDS (Anderson and Nelson, 1965:24, 25).—Cuba: Guanajay; *ca.* 4 mi. S San José de las Lajas; Hormiguero, Cantabria Cave; Cubitas (Cueva del Indio, Sierra de); Daiquirí; *Siboney;* Santiago; El Cobre (mine); Cumanayagua, Mine Carlotta; *Soledad, Vilches Cave; 2 mi. SE Atkins Gardens; Guabairo Cave, Guabairo;* Cienfuegos. Grand Cayman Island. Isle of Pines: Caballos Mts., Kennan Caves; *Casas Mts.; Nueva Gerona.* Cuba: Luis Lazo.—Varona (1974:16) questions the subspecific identification of specimens from Grand Cayman Island.

Macrotus waterhousii waterhousii Gray

1843. *Macrotus waterhousii* Gray, Proc. Zool. Soc. London, p. 21, July, type from Haiti.
1931. *Macrotus waterhousii heberfolium* Shamel, Jour. Washington Acad. Sci., 21:252, June 4, type from Kingston, Providenciale Island, Bahama Islands. Known only from the type locality. Inseparable from *M. w. waterhousii* according to Anderson and Nelson, Amer. Mus. Novit., 2212:14, March 17, 1965, but significantly different from the four other insular subspecies according to Buden, Jour. Mamm., 56:758, November 18, 1975.

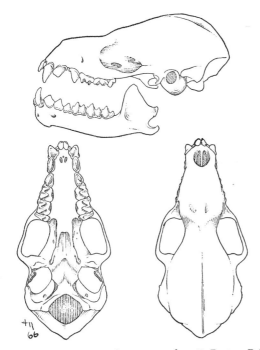

Fig. 66. *Macrotus waterhousii waterhousii,* Port-au-Prince, Haiti, No. 30764 F.N.M.H., ♂, X 2.

MARGINAL RECORDS (Anderson and Nelson, 1965:18, 19, unless otherwise noted).—Crooked Island: Gordon. *East Plana Cay* (Buden, 1975:768); North Caicos (*ibid.*). Dominican Republic: Sosua; San Lorenzo, Samana Bay; *Caña Honda.* Puerto Rico: Cueva de Clara, *ca.* ½ mi. N, 3 mi. W Morovis (Choate and Birney, 1968:404). Dominican Republic: La Romana (caves near River Chavon, E of); San Domingo City; Bohechio, 4 km. SW Padre Las Casas; *San Juan River (cave at Laguna, near the); Cabral, 10 km. S El Firme;* Los Patos. Beata Island. Dominican Republic: *Pedernales (7 km. ESE of).* Haiti: Port-au-Prince; La Gonave Island (Hall and Kelson, 1959:102); St. Michel, Cave I; l'Acul (cave 3 mi. W of). Great Inagua Island: Salt Pond Hill Cave. Acklins Island: Salt Point, Jamaica Bay.

Genus **Lonchorhina** Tomes
Tomes' Long-eared Bat

1863. *Lonchorhina* Tomes, Proc. Zool. Soc. London, p. 81, May. Type, *Lonchorhina aurita* Tomes.

Skull having a distinct concavity at base of rostrum between orbits; middle of braincase low, rising little above occiput; auditory bullae small; basisphenoid pits wide and deep; teeth much as in *Micronycteris,* but crowns of incisors wider (width approximately equal to height); 2nd lower premolar smaller than either of others. Dental formula as in *Macrotus.* Ears large, separate; nose-leaf nearly as long as ear; tail longer than femur, extending to edge of membrane.

Lonchorhina aurita
Tomes' Long-eared Bat

Selected measurements of 22 individuals from Chiapas, Guatemala, Costa Rica and Venezuela are: head and body, 54–67; tail, 42–55; foot, 12–16; ear, 28–32; forearm, 45.0–49.5; greatest length of skull, 20.3–21.2; zygomatic breadth,

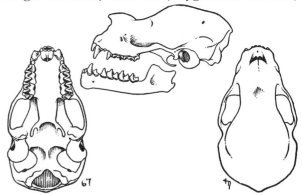

Fig. 67. *Lonchorhina aurita aurita,* Chilibrillo River, Panamá, No. 173849 U.S.N.M., sex ?, X 2.

9.7–11.3; interorbital breadth, 4.5–5.1; maxillary tooth-row, 6.2–6.7.

The subspecies *L. a. occidentalis* Anthony (1923:13), with type locality in Ecuador, is extralimital.

Lonchorhina aurita aurita Tomes

1863. *Lonchorhina aurita* Tomes, Proc. Zool. Soc. London, p. 83, May, type from "West Indies." [According to P. Hershkovitz, *in Litt.* September 8, 1958, the type locality is Trinidad on basis of Thomas, Jour. Trinidad Field Nat. Club, 1:5, 1893.]

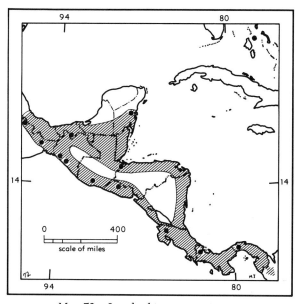

Map 72. *Lonchorhina aurita aurita.*

MARGINAL RECORDS.—Quintana Roo: 2 km. N Felipe Carrillo Puerto (91540 KU). Guatemala: Quebradas (Jones, 1966:451). Panamá: Río Changena, 4800 ft. (Handley, 1966:760); Chilibrillo Cave, near Ahlajuela, thence southeastward into South America, and northwestward along coast to Costa Rica: Comelco Ranch (3048 LaVal at KU). El Salvador: *Cueva Hedionda; Santa Ana;* Suchitoto. Guatemala: 1 mi. S, ¾ mi. E Yepocapa (Jones, 1966:451). Chiapas (Carter, *et al.*, 1966:490): 5 km. SE Pijijiapam, 100 ft.; 15 km. SE Tonalá, 100 ft. Oaxaca: National Route 185, 29 km. N Matías Romero (Schaldach, 1965:131); Cerro Piñón, 12 km. S Acatlán de Perez Figueroa (Constantine, 1966:126). Tabasco: 5 mi. W Teapa. Also: Bahama Islands: Nassau Harbour, New Providence.

Genus **Macrophyllum** Gray—Long-legged Bat

1838. *Macrophyllum* Gray, Mag. Zool. Bot., 2:489, February. Type, *Macrophyllum nieuwiedii* Gray [= *Phyllostoma macrophyllum* Schinz].

1891. *Dolichophyllum* Lydekker, *in* Flower and Lydekker, An introduction to . . . mammals living and extinct, p. 673. Type, *Macrophyllum nieuwiedii* Gray, a renaming of *Macrophyllum* Gray, which Lydekker considered to be a homonym of *Macrophylla* Hope, 1837, a beetle.

Resembling *Micronycteris*, but: ears separate; tail longer than femur and continued to outer edge of uropatagium; tibia, uropatagium, feet, and claws elongated, much as in *Noctilio* and *Myotis vivesi;* posterior edge of uropatagium having papillae. Skull with short rostrum, its length less than breadth of braincase; nares emarginate laterally and above, leaving noticeable flattened area over roots of incisors; basioccipital pits obsolete; auditory bullae small, not covering half of cochleae. Teeth essentially like those of *Micronycteris*, except that middle lower premolar (p3) minute and so crowded inward that 1st and 3rd are almost in contact, 1st upper premolar (P3) not much larger than outer incisor, middle upper incisors project more conspicuously, and crowns of lower incisors relatively wider (after Miller, 1907:128).

Macrophyllum macrophyllum (Schinz)
Long-legged Bat

1821. *Phyllost[oma]. macrophyllum* Schinz, Das Thierreich . . ., 1:163, type from Rio Mucurí, Minas Geraes, Brazil.
1912. *Macrophyllum macrophyllum*, Nelson, Proc. Biol. Soc. Washington, 25:93, May 4.
1838. *Macrophyllum nieuwiedii* [*sic*] Gray, Mag. Zool. Bot., 2:489. Based on *Ph[yllostoma]. macrophyllum* Wied-Neuwied, Beiträge zur Naturgesch. Brasil., 2:188, 1826, type from Brazil.

Measurements of 13 specimens from Central America: head and body, 43–46; tail, 42–45; foot, 11–14; ear, 15–19; forearm, 34–35.3; greatest length of skull, 16.3–17.2; zygomatic breadth, 9.0–9.7; postorbital constriction, 3.0–3.4; maxillary tooth-row, 5.2–5.6. Dorsum medium brown (hairs paler basally); venter same or barely paler.

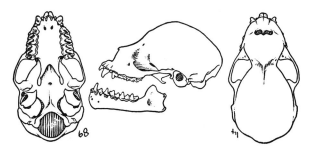

Fig. 68. *Macrophyllum macrophyllum*, Old Panamá, Panamá, No. 178236 U.S.N.M., ♂, X 2.

MARGINAL RECORDS.—Tabasco: $4\frac{2}{5}$ km. SE Teapa (Lay, 1963:374). Honduras: Lancetilla, 40 m. (Valdez and LaVal, 1971:248); near Río Sicre. Nicaragua: Cacao, 22 km. by road W Muelle de las Bueyes, 400 ft. (Davis, et al., 1964:378). Costa Rica: Finca La Lola, Río Lola, Río Madre de Dios, 50 m. (Starrett and Casebeer, 1968:10). Panamá (Handley, 1966:760): Almirante; Panamá Viejo; Boca de Río Paya, thence southeastward into South America, and northwestward along coast to Panamá: Cerro Hoya, 1800 ft. (ibid.). Costa Rica: Río Colorado, 9 km. N Liberia along Interamerican Highway and 4 km. E Interamerican Highway, 150 m. (Starrett and Casebeer, 1968:10). El Salvador: Cueva Hedionda.

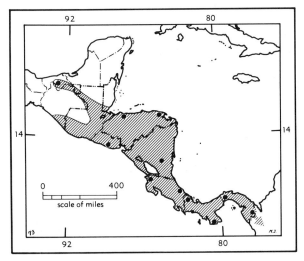

Map 73. *Macrophyllum macrophyllum*.

Genus **Tonatia** Gray—Round-eared Bats

Species reviewed by Goodwin, Jour. Mamm., 23:204–209, June 3, 1942.

1827. *Tonatia* Gray, in Griffith, The animal kingdom . . . by Baron Cuvier . . . , 5:71. Type, *Vampyrus bidens* Spix.
1836. *Lophostoma* D'Orbigny and Gervais, Voy. dans l'Amér. Mérid., Atlas Zool., 4(2; Mamm.):11. Type, *Lophostoma sylvicolum* D'Orbigny.

Length of head and body, 42.5–77. Skull resembling that of *Micronycteris* but more robust; zygoma abruptly expanded near anterior base; no basisphenoidal pits; auditory bullae small, covering less than half of cochleae; sagittal crest present, not divided anteriorly; ears large, rounded, separate or conjoined; tail shorter than femur, reaching approx. to middle of membrane; upper canines relatively large and nearly touching medial incisors, lateral incisors being displaced out of normal position in tooth-row; lower incisors reduced to 2; middle lower premolar much reduced. Dentition, i. $\frac{2}{1}$, c. $\frac{1}{1}$, p. $\frac{2}{3}$, m. $\frac{3}{3}$.

KEY TO NORTH AMERICAN SPECIES OF TONATIA

(W. B. Davis, MS)

1. Forearm less than 40; greatest length of skull less than 22; weight 8–12 grams. *T. nicaraguae*, p. 109
1'. Forearm more than 40; greatest length of skull more than 23; weight more than 15 grams.
  2. Width of cranium at postorbital constriction greater than width across upper canines; ear from notch usually less than 27; forearm 55–60; faint medial whitish stripe on head. *T. bidens*, p. 108
  2'. Width of cranium at postorbital constriction less than width across upper canines; ear from notch more than 30; no medial white stripe on head.
    3. Sagittal crest accentuated, about 2 mm. high; greatest length of skull 27–30; forearm 51–57; maxillary tooth-row 9.5 or more; weight 30–40 grams. . . . . *T. silvicola*, p. 110
    3'. Sagittal crest moderate, about 1 mm. high; greatest length of skull 25–27; forearm 47–53; maxillary tooth-row 9 or less; weight 18–25 grams. . . . . . . *T.* sp. novum, p. 110

Tonatia bidens (Spix)
Spix's Round-eared Bat

1823. *Vampyrus bidens* Spix, Simiarum et vespertilionum Brasiliensium . . . , p. 65, type from bank of Rio São Francisco, Bahia, Brazil.
1840. [*Tonatia*] *bidens*, Gray, in Griffith, The animal kingdom . . . by the Baron Cuvier . . . , 5:69.
1838. *Phyllostoma childreni* Gray, Mag. Zool. Bot., 2:488. Type from South America.

Measurements of 12 specimens from Central America: forearm, 55.4–60.5; greatest length of skull, 26.2–29.1; zygomatic breadth, 13.3–14.8; postorbital constriction, 5.4–5.8; maxillary tooth-row, 9.2–10.2. Upper parts varying from approx. ochraceous to blackish brown; dorsal fur of head bisected by longitudinal white stripe; underparts paler, grayer, washed with buff. "Skull large and massive, rostrum broad, flat and not constricted in orbital region; superior outline evenly elevated from front of nasals and without depression in orbital region; palate narrow, tooth-rows only slightly converging anteriorly; saggital crest present in adult specimens but undeveloped." (Goodwin, 1946:304.)

MARGINAL RECORDS.—Honduras: Lancetilla (Valdez and LaVal, 1971:248). Panamá (Handley, 1966:761): Almirante; Cerro Azul, 2000 ft., thence

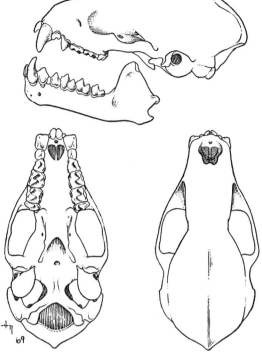

Fig. 69. *Tonatia bidens*, Palmar (Pacific), Costa Rica, No. 139439 A.M.N.H., ♂, X 2.

southeastward into South America and northwestward along coast to Costa Rica: Palmar; Río Damitas, 14½ mi. N Quepos (Gardner, *et al.*, 1970:716). Guatemala: Izabal (Carter, *et al.*, 1966:490).

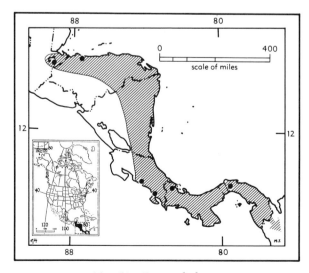

Map 74. *Tonatia bidens*.

Tonatia nicaraguae Goodwin
Pygmy Round-eared Bat

1942. *Tonatia nicaraguae* Goodwin, Jour. Mamm., 23:205, June 3, type from Kanawa Creek, 100 ft., near Cukra, Zelaya, Nicaragua.

1942. *Tonatia minuta* Goodwin, Jour. Mamm., 23:206, June 3, type from Boca Curaray, Ecuador [now, 1978, in Perú]. (Regarded as synonymous with *T. nicaraguae* by Handley, Ectoparasites of Panamá, Field Mus. Nat. Hist., p. 761, November 22, 1966. Handley (*ibid.*) used the name *T. minuta*, but *T. nicaraguae* has page priority.)

Measurements of 12 specimens from Central America: forearm, 32–36.4; greatest length of skull, 19.4–20.4; zygomatic breadth, 9.0–9.3; postorbital constriction, 3.0–3.2; maxillary tooth-row, 6.6–7.0. Upper parts dark brown, almost

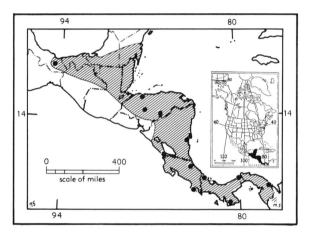

Map 75. *Tonatia nicaraguae*.

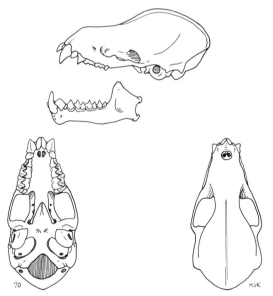

Fig. 70. *Tonatia nicaraguae*, Santa Rosa, 17 km. N, 15 km. E Boaco, 300 m., Boaco, Nicaragua, No. 110696 K.U., ♂, X 2.

black; underparts dull drab. "Skull small, evenly elevated behind nasals, rostrum parallel-sided, lower incisors small, faintly bifid; middle lower premolar minute." (Goodwin, 1946:305.)

MARGINAL RECORDS.—Veracruz: near Tenochititlán (Lackey, 1970:384). Honduras: Lancetilla (Valdez and LaVal, 1971:248). Nicaragua: type locality. Costa Rica: Cariari (Gardner, *et al.*, 1970:716). Panamá (Handley, 1966:761): Subube; Fort Sherman; *Armila;* Puerto Obaldía, thence into South America, thence northwestward to Panamá: Cerro Hoya (Handley, 1966:761). Costa Rica: Osa Península, 2½ mi. SW Rincón, 15 m. (Gardner, *et al.*, 1970:716). Nicaragua: 12½ mi. S, 13 mi. E Rivas, 125 ft. (Davis and Carter, 1962:67). Honduras (LaVal, 1969:820): Chichicaste; Comayagua.

Tonatia silvicola
D'Orbigny's Round-eared Bat

Maxillary tooth-row more than 9.2 mm.; weight more than 27 grams; see key for additional means of identification.

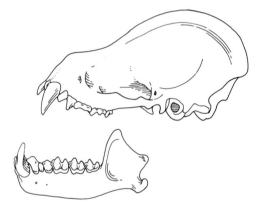

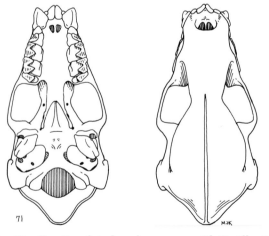

Fig. 71. *Tonatia silvicola* subsp. novum, El Castillo, Río San Juan, Nicaragua, No. 18773 T.C.W.C., ♂, X 2.

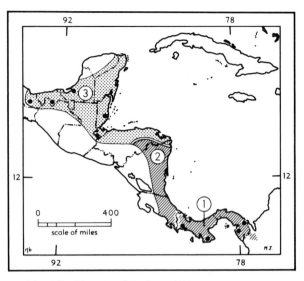

Map 76. *Tonatia silvicola* and *Tonatia* sp. novum.

1. *T. s. silvicola*   2. *T. silvicola* subsp. novum
                    3. *Tonatia* sp. novum

Tonatia silvicola silvicola (D'Orbigny)

1836. *Lophostoma sylvicolum* D'Orbigny, Voy. dans l'Amér. Mérid., Atlas Zool., Pl. 6, described in Vol. 4 (pt. 2):11, 1847, by D'Orbigny and Gervais. Type locality, Yungas de Bolivia between Río Securé and Río Isibara.

MARGINAL RECORDS.—Panamá (Handley, 1966: 761): Bugaba; Armila, thence into South America, and northwestward to Panamá (*ibid.*): Río Esnápe; Guánico.

Tonatia silvicola subsp. novum—**See** Addenda.

RANGE.—Caribbean versants of Honduras, Nicaragua, and Costa Rica.

Tonatia sp. novum—See Addenda
Davis' Round-eared Bat

Maxillary tooth-row less than 9.2 mm.; weight less than 27 grams; see key for additional means of identification.

MARGINAL RECORDS.—Campeche: 12 km. W Escárcega (Jones, 1964e:509). Belize: Freetown, Sittee River. Guatemala: Izabal (Carter, *et al.*, 1966:490). Tabasco: 1 mi. E Teapa (Lay, 1963:375). Veracruz: near Tenochtitlán (Lackey, 1970:384).

Genus **Mimon** Gray—Spear-nosed Bats

1847. *Mimon* Gray, Proc. Zool. Soc. London, p. 14, April 13. Type, *Phyllostoma bennettii* Gray.
1855. *Tylostoma* Gervais, Mammifères, *in* [Castelnau] Expéd. dans les parties centrales de l'Amér. du Sud . . . , p. 49.

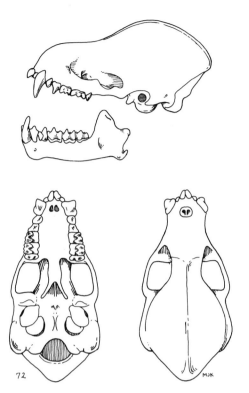

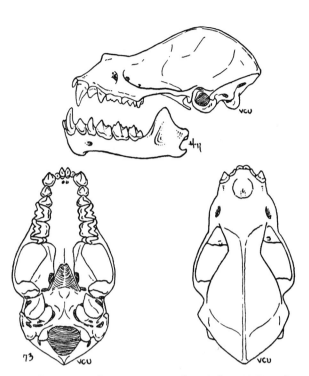

Fig. 72. *Tonatia* sp. novum, 12 km. W Escárcega, Campeche, México, No. 93346 K.U., ♂, X 2.

Fig. 73. *Mimon bennettii cozumelae*, 3 km. N Presidio, Veracruz, No. 19169 K.U., ♀, X 2.

Type, *Phyllostoma crenulatum* É. Geoffroy St.-Hilaire. Not *Tylostoma* Sharpe, 1849, a genus of Mollusca.
1891. *Anthorhina* Lydekker, *in* Flower and Lydekker, An introduction to . . . mammals living and extinct, p. 674, a renaming of *Tylostoma* Gervais.

Skull slender; rostrum relatively broadly arched; zygoma without expansion either anteriorly or posteriorly; basisphenoidal pits broad and shallow, separated by a median septum; auditory bullae small, covering about half of cochleae, unusually narrow; ears large, not conjoined; tail approx. as long as femur, terminating in middle of membrane. Dentition, i. $\frac{2}{1}$, c. $\frac{1}{1}$, p. $\frac{2}{2}$, m. $\frac{3}{3}$.

KEY TO NORTH AMERICAN SPECIES OF MIMON

1. Wing membrane from side of foot near base of outer toe; white line from crown to tail; tympanic bullae large. *M. crenulatum*, p. 112
1'. Wing membrane from side of ankle; no white line on back; tympanic bullae small . . . . *M. bennettii*, p. 111

Mimon bennettii
Bennett's Spear-nosed Bat

Length of head and body, 65–75; tail, 15–26; ear, 32–38. Length of forearm in North America: 54.6–60.7; greatest length of skull, 25.3–27.4; zygomatic breadth, 13.3–14.4; maxillary tooth-row, 8.9–9.8. Upper parts fulvous brown; slightly

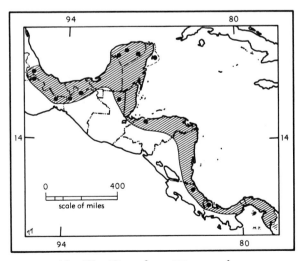

Map 77. *Mimon bennettii cozumelae*.

paler ventrally. Skull and especially tympanic ring larger than in *M. crenulatum*.

Mimon bennettii cozumelae Goldman

1914. *Mimon cozumelae* Goldman, Proc. Biol. Soc. Washington, 27:75, May 11, type from Cozumel Island, Quintana Roo.
1965. *Mimon bennettii cozumelae*, Schaldach, W. J., Jr., Anal. Inst. Biol., Univ. Nac. Autó. México, 35:132, August 25. Carter, Pine, and Davis, Southwestern Nat., 11:491, December 31, 1966, treated *M. cozumelae* and *M. bennettii* as two species because integradation had not been demonstrated.

MARGINAL RECORDS.—Yucatán: Izamel [= Izamal] (Handley, 1960:462); Tekom. Quintana Roo: type locality. Honduras: Lancetilla (Valdez and LaVal, 1971:248). Costa Rica: Finca La Selva, 1 mi. SW Puerto Viejo, 100 m. (Gardner, *et al.*, 1970:717). Panamá: 20 km. SSW Changuinola (Bocas del Toro), banks of the Río Changuinola (Handley, 1966:761), thence southeastward into South America and northward to Guatemala: Tikal, El Petén (Jones, 1966:451). Tabasco: 1 mi. E Teapa (Winkelmann, 1962:112, as *M. cozumelae*). Chiapas: 11 mi. W Mal Paso, 400 ft. (Carter, *et al.*, 1966:491). Oaxaca: Río Sarabia, 29 km. N Matías Romero (Schaldach, 1965:132); 1 mi. NW Los Limones (Constantine, 1966:125). Veracruz: 3 km. N Presidio, 1500 ft. Yucatán: Calcehtok, Actun Tuz-ic.

Mimon crenulatum (É. Geoffroy) Striped Spear-nosed Bat

Selected measurements are: length of head and body, 66 (in female from Campeche); length of

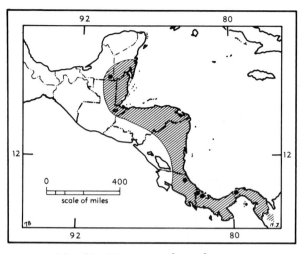

Map 78.   *Mimon crenulatum keenani.*

forearm, 45–51.9; greatest length of skull, 20.9–23.3; zygomatic breadth, 12.2–13.0; length of upper tooth-row, 7.8–8.3. Upper parts dark brown to blackish with indistinct white line from crown to tail; slightly paler ventrally; nose-leaf hairy.

Mimon crenulatum keenani Handley

1960. *Mimon crenulatum keenani* Handley, Proc. U.S. Nat. Mus., 112:460, October 6, type from Fort Gulick, Panamá Canal Zone.

MARGINAL RECORDS.—Campeche: 65 km. S, 128 km. E Escárcega at Laguna Alvarado (Jones, 1964e:510). Costa Rica: Cariari (Gardner, *et al.*, 1970:717). Panamá (Handley, 1966:761): Sibube; Almirante; type locality, thence southeastward into South America.

Genus **Phyllostomus** Lacépède
Spear-nosed Bats

1799. *Phyllostomus* Lacépède, Tableau des divisions, sous-divisions, ordres et genres des mammifères, p. 16 (published as a supplement to Discours d'ouverture et de clôture du cours d'histoire naturelle. . . . Type, *Vespertilio hastatus* Pallas.
1866. *Alectops* Gray, Proc. Zool. Soc. London, p. 114, May. Type, *Alectops ater* Gray [= *P. elongatus* É. Geoffroy St.-Hilaire, 1810].

Skull robust; rostrum broad, low, flattened; sagittal crest high; paroccipital process shelflike; zygomata heavy, slightly expanded anteriorly and posteriorly; basisphenoidal pits present but faintly expressed; basicranial plane forming angle with roof of posterior nares; auditory bullae small, flattened; ears widely separated; nose-leaf

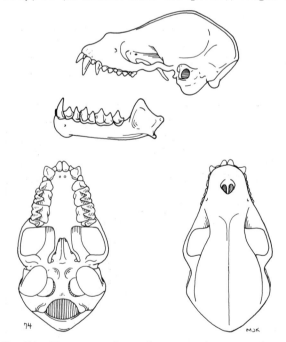

Fig. 74. *Mimon crenulatum keenani*, 65 km. S, 128 km. E Escárcega, at Laguna Alvarado, Campeche, No. 93512 K.U., ♀, X 2.

well developed; tail less than half length of femur, projecting from base of membrane; teeth essentially as in *Micronycteris*. Dentition, i. $\frac{2}{2}$, c. $\frac{1}{1}$, p. $\frac{2}{2}$, m. $\frac{3}{3}$.

KEY TO NORTH AMERICAN SPECIES OF PHYLLOSTOMUS

1. Length of forearm less than 75; zygomatic breadth less than 18. . . . . *P. discolor*, p. 113
1'. Length of forearm more than 75; zygomatic breadth more than 18. *P. hastatus*, p. 113

Phyllostomus hastatus
Spear-nosed Bat

Selected measurements of an adult male and female from Costa Rica are, respectively: length of head and body, 128, 109; ear, 30, 33; forearm, 92.5, 88; condylobasal length, 35.8, 34.2; zygomatic breadth, 23, 20.8; length of upper tooth-row, 14.5, 14. Upper parts seal brown, grayer on shoulders and sides of head; underparts paler, washed with gray; some individuals golden throughout. Skull robust; sagittal crest well developed.

Phyllostomus hastatus panamensis J. A. Allen

1904. *Phyllostomus hastatus panamensis* J. A. Allen, Bull. Amer. Mus. Nat. Hist., 20:233, June 29, type from Boquerón, Chiriquí, Panamá.
1904. *Phyllostomus hastatus caurae* J. A. Allen, Bull. Amer. Mus. Nat. Hist., 20:234, June 29, type from Cali, upper Cauca Valley, Colombia.

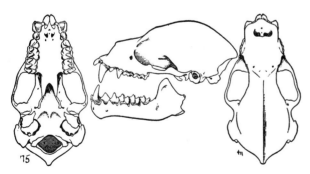

Fig. 75. *Phyllostomus hastatus panamensis*, Chilibrillo Cave, 10 mi. N Pedro Miguel, Panamá, No. 45063 K.U., ♀, X 1.

MARGINAL RECORDS.—Honduras: Patuca. Nicaragua (Jones, *et al.*, 1971:8): Bonanza, 850 ft.; El Recreo, 25 m. Panamá (Handley, 1966:762): Almirante; Boca de Río Paya, thence into South America, and northward to Panamá: Guánico (*ibid.*). Costa Rica: 27 de Abril. Nicaragua: Metagalpa.

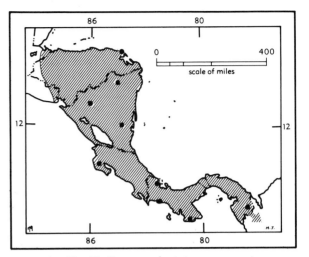

Map 79. *Phyllostomus hastatus panamensis*.

Phyllostomus discolor
Pale Spear-nosed Bat

Extreme measurements (cranial, 47 individuals; external, 402 individuals) of specimens from El Salvador (Felten, 1956:188) are: head and body, 75–91; tail, 9–22; foot, 12–16.5; forearm, 57–66; greatest length of skull, 28.2–31.9; zygomatic breadth, 14.6–16.3; interorbital breadth, 6.7–7.4; maxillary tooth-row, 9.2–10.4. Upper parts dark brown posteriorly, becoming ochraceous tawny on head and shoulder; underparts cinnamon buff. Skull relatively slender; sagittal crest weakly developed.

Davis and Carter (1962:69) refer all specimens from Central America to the subspecies *P. d. verrucosus*. Handley (1966:761) refers specimens from Panamá to the subspecies *P. d. discolor* without comment on specimens from farther north.

Figure 67, page 108 of Hall and Kelson (1959), incorrectly identified as *Phyllostomus discolor verrucosus*, actually is *Trachops cirrhosus*.

Phyllostomus discolor verrucosus Elliot

1905. *Phyllostoma verrucossum* [*sic*] Elliot, Proc. Biol. Soc. Washington, 18:236, December 9, type from Niltepec, Oaxaca.
1936. *Phyllostomus discolor verrucosus*, Sanborn, Field Mus. Nat. Hist., Publ. 361, Zool. Ser., 20:97, August 15.

MARGINAL RECORDS.—Veracruz: Orizaba; Laguna Asmolapan, 1½ km. E Catemaco (Villa-R., 1967:219). Guatemala: Escobas, Izabal. Honduras: Las Flores. Panamá: Barro Colorado Island (Hall and Kelson, 1959:108, as *P. d. discolor*), thence southward into South America, and northward to Costa Rica: Río

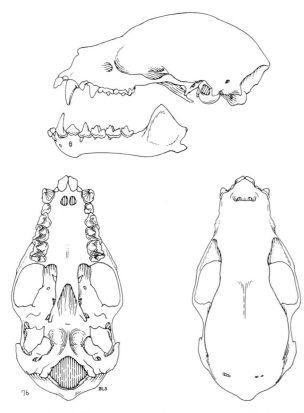

Fig. 76. *Phyllostomus discolor verrucosus*, 14 km. S Boaco, 220 m., Nicaragua, No. 97473 K.U., ♀, X 2.

Tenorio, 3 mi. S, 10 mi. W Las Cañas, 10 ft. (Davis and Carter, 1962:68). El Salvador: Tabanco (Burt and Stirton, 1961:29). Guatemala: Patulul. Chiapas: 10 km. SE Tonalá (Villa-R., 1967:219). Oaxaca: type locality.

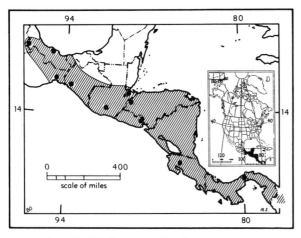

Map 80. *Phyllostomus discolor verrucosus*.

Genus **Phylloderma** Peters—Spear-nosed Bats

1865. *Phylloderma* Peters, Monatsb. preuss. Akad. Wiss., Berlin, p. 513. Type, *Phyllostoma stenops* Peters.

1866. *Guandira* Gray, Proc. Zool. Soc. London, p. 114, May. Type, *Guandira cayanensis* Gray. Not *Guandira* Gray, 1843, a *nomen nudum*.

Closely resembling *Phyllostomus* but differing in having bifid medial incisors, narrower-crowned molariform teeth, and a minute p3. Dentition, i. $\frac{2}{2}$, c. $\frac{1}{1}$, p. $\frac{2}{3}$, m. $\frac{3}{3}$.

Phylloderma stenops
Northern Spear-nosed Bat

Selected measurements of the type of *P. septentrionalis* followed by those of two paratypes from Honduras, are: head and body, 121, 110, 112; tail, 16, 20, 20; forearm, 80.0, 82.5, 81.1; greatest length of skull, 34.5, 35.5, 33.1; postorbital width, 9.6, 9.2, 9.6; zygomatic breadth, 18.0, 17.6, 18.0; length of upper tooth-row, 11, 11, 11. Upper parts dark brown, hairs buffy at base; underparts buffy.

Specimens from Honduras and Chiapas differ from *Phyllostomus discolor* in their short, wooly fur and white wing tips, as well as in dental characters. Because specimens from Panamá are intermediate between *Phylloderma septentrionalis* and *P. stenops*, Handley (1966:762) concluded that the two nominal species were only subspecifically different.

Phylloderma stenops septentrionalis Goodwin

1940. *Phylloderma septentrionalis* Goodwin, Amer. Mus. Novit., 1075:1, June 27, type from Las Pilas, about 4000 ft., 6 mi. N Marcala, La Paz, Honduras.

MARGINAL RECORDS.—Chiapas: 10 mi. W Mal Paso (Carter, *et al.*, 1966:491). Honduras: Las Flores, Tegucigalpa; type locality. Costa Rica: La Selva (3258 LaVal at KU).

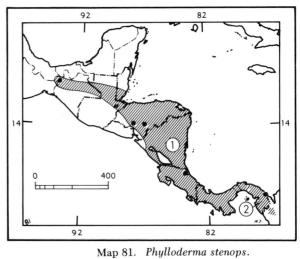

Map 81. *Phylloderma stenops*.

1. *P. s. septentrionalis*   2. *P. s. stenops*

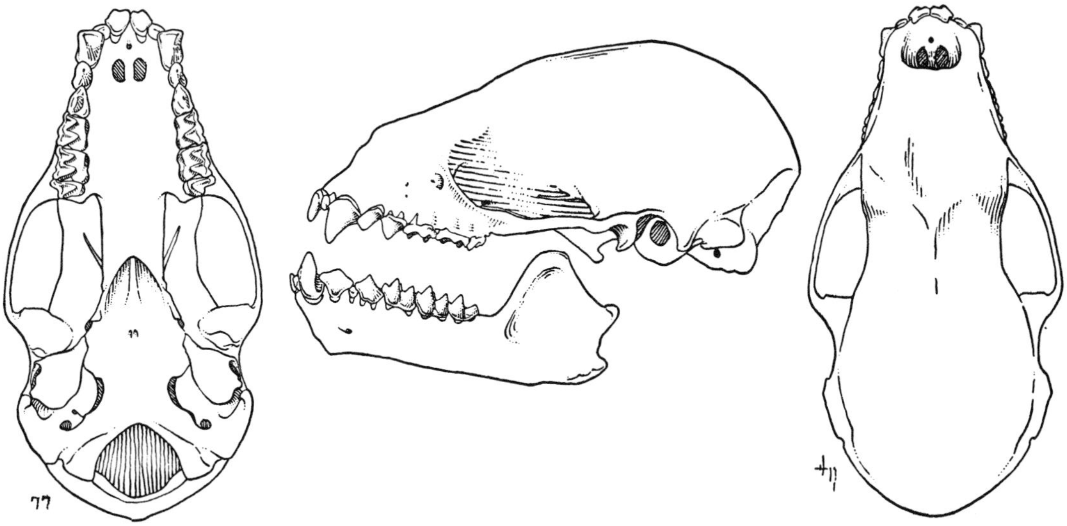

Fig. 77. *Phylloderma stenops septentrionalis*, Las Pilas, La Paz, Honduras, No. 126867 A.M.N.H., sex ?, X 2.

Phylloderma stenops stenops Peters

1865. *Phylloderma stenops* Peters, Monatsb. preuss. Akad. Wiss., Berlin, p. 648, type from Cayenne, French Guiana.
1966. *P[hylloderma]. s[tenops]. stenops*, Handley, Ectoparasites of Panama, Field Mus. Nat. Hist., p. 762, November 22.

MARGINAL RECORDS.—Panamá: Armila (Handley, 1966:762), thence southward into South America.

Genus **Trachops** Gray—Fringe-lipped Bats

1847. *Trachops* Gray, Proc. Zool. Soc. London, p. 14, April 13. Type, *Trachops fuliginosus* Gray [= *Vampyrus cirrhosus* Spix].
1825. *Istiophorus* Gray, Zool. Jour., 2:242. Type, *Vampyrus cirrhosus* Spix, or possibly *V. soricinus* Spix. *Not Istiophorus* Lacépède, 1802, a fish.
1846. *Histiophorus* Agassiz, Nomen. Zool., Index Univ., p. 183, an emendation of *Istiophorus* Gray.
1865. *Trachyops* Peters, Monatsb. preuss. Akad. Wiss., Berlin, p. 512, an emendation of *Trachops?*

Skull relatively elongated and rounded; interorbital region somewhat depressed; zygomata expanded anteriorly and posteriorly, the anterior expansion but faintly expressed; auditory bullae covering approx. half of cochleae and as high as wide; maxillary tooth-row resembling that of *Phyllostomus* but lateral incisor greatly reduced and cheek-teeth relatively larger; p3 present and lower molars relatively narrower; ear longer than head; tail much shorter than femur and projecting from upper surface of membrane; front and sides of lips and chin bearing numerous small warty excrescences. Dentition, i. $\frac{2}{2}$, c. $\frac{1}{1}$, p. $\frac{2}{3}$, m. $\frac{3}{3}$.

Trachops cirrhosus
Fringe-lipped Bat

Length of head and body, 72–87; forearm, 55.9–61.7; greatest length of skull, 27.7–29.2; zygomatic breadth, 13.3–14.2; interorbital constriction, 4.8–5.3; maxillary tooth-row, 9.7–11.2. Upper parts approx. cinnamon brown or, in some specimens, a trifle darker; underparts dull brownish washed with gray.

Trachops cirrhosus cirrhosus (Spix)

1823. *Vampyrus cirrhosus* Spix, Simiarum et vespertilionum Brasiliensium . . . , p. 64, type from Brazil.
1878. *Trachyops* [sic] *cirrhosus*, Dobson, Catalogue of the Chiroptera in the . . . British Museum, p. 481.
1865. *Trachops fuliginosus* Gray, Proc. Zool. Soc. London, p. 14, June, type from Pernambuco, Brazil.

MARGINAL RECORDS.—Costa Rica: Finca La Selva, 1 mi. SW Puerto Viejo, 100 m. (Gardner, *et al.*, 1970:717, as *T. cirrhosus* only). Panamá (Handley, 1966:762): Isla Bastimentos; Armila, thence into South America, and northwestward along coast to Panamá: Guánico (*ibid.*). Costa Rica: 5 km. N Villa Neilly on Interamerican Highway, 150 m. (Starrett and Casebeer, 1968:11, as *T. cirrhosus* only); Río Tenorio, 3 mi. S, 10 mi. W Las Cañas, 10 ft. (Davis and Carter, 1962:69, as *T. cirrhosus* only).

Trachops cirrhosus coffini Goldman

1925. *Trachops coffini* Goldman, Proc. Biol. Soc. Washington, 38:23, March 12, type from El Gallo, 8 mi. W Yaxha, on the Remate-El Cayo trail, Petén, Guatemala (see de la Torre, Proc. Biol. Soc. Washington, 69:189, December 31, 1956).
1956. *Trachops cirrhosus coffini*, Felten, Senckenbergiana Biol., 37:189, April 15.

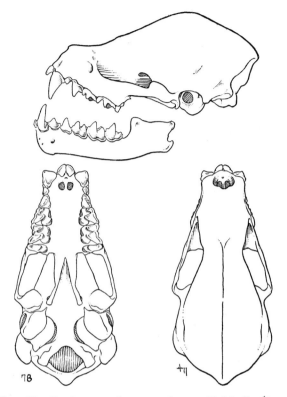

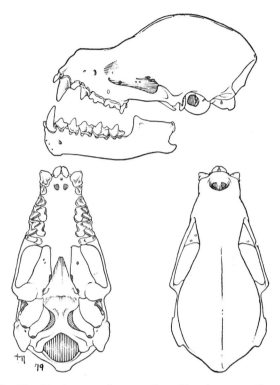

Fig. 78. *Trachops cirrhosus cirrhosus*, Chilibrillo Cave, Panamá, No. 174884 U.S.N.M., sex ?, X 2.

Fig. 79. *Trachops cirrhosus coffini*, Chinaltenango, Guatemala, No. 64634 F.N.M.H., ♂, X 2.

MARGINAL RECORDS.—Oaxaca: Cerro Piñón, 12 km. S Acatlán de Perez Figueroa (Constantine, 1966:126). Veracruz: 35 km. SE Jesús Carranza, 350 ft. (Hall and Dalquest, 1963:224). Chiapas: 12 mi. W Mal Paso, 400 ft. (Carter, *et al.*, 1966:491, as *T. cirrhosus* only). Campeche: 65 km. S, 128 km. W Escárcega (93381 KU). Guatemala: Tikal (Jones, 1966:452).

Belize: Belize. Honduras: Lancetilla (Valdez and LaVal, 1971:249, as *T. cirrhosus* only). Nicaragua: Río Coco, 64 mi. NNE Jinotega, 1000 ft. (Carter, *et al.*, 1966:491, as *T. cirrhosus* only); Cara de Mono, 50 m. (Jones, *et al.*, 1971:8). El Salvador (Burt and Stirton, 1961:30): Lake Olomega; Cueva Hedionda. Chiapas (Carter, *et al.*, 1966:491, as *T. cirrhosus* only): 18 km. SE Tonalá, 100 ft.; 5 mi. N Arriaga. Oaxaca (Goodwin, 1969:63, 262): Río Ostuta, 4 mi. W Zanatepec; 6 mi. N Matías Romero.

Genus **Chrotopterus** Peters—False Vampire Bats

1865. *Chrotopterus* Peters, Monatsb. preuss. Akad. Wiss., Berlin, p. 505. Type, *Vampyrus auritus* Peters.

Skull resembling that of *Phyllostomus* in general; rostrum and interorbital region subcylindrical; paroccipital expansion small but distinct; tympanic covering less than half cochlea; diameter of tympanic nearly twice its height at inner edge. Ears separate, simple and large (middle of conch, when laid forward, extending to nostril); tail barely perceptible in base of wide interfemoral membrane; lips and chin nearly smooth; fur unusually long and soft (after Miller, 1907:134). Dentition, i. $\frac{2}{1}$, c. $\frac{1}{1}$, p. $\frac{2}{3}$, m. $\frac{3}{3}$.

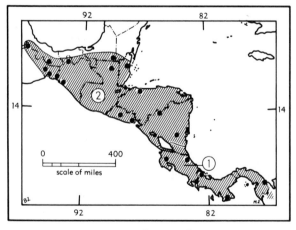

Map 82. *Trachops cirrhosus*.

1. *T. c. cirrhosus*   2. *T. c. coffini*

Chrotopterus auritus
Peters' False Vampire Bat

Length of head and body, 103–122; tail, 7–17; foot, 22–30; ear, 42–48; forearm, 77.2–83.5; greatest length of skull, 35.3–38.5; zygomatic breadth, 18.1–20.4; interorbital breadth, 6.0–6.8; maxillary tooth-row, 13.0–13.9. Upper parts dull brown; underparts slightly paler and washed with gray; ears and membranes dark brown but edged with whitish.

Chrotopterus auritus auritus (Peters)

1856. *Vampyrus auritus* Peters, Monatsb. preuss. Akad. Wiss., Berlin, p. 415, type from México.
1865. *Chrotopterus auritus* Peters, Monatsb. preuss. Akad. Wiss., Berlin, p. 505.

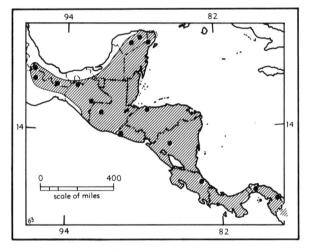

Map 83. *Chrotopterus auritus auritus*.

MARGINAL RECORDS.—Yucatán: 6 km. N Tizimin (93382 KU). Quintana Roo: 1½ km. S, 1 km. E Pueblo Nuevo X-Can (91549 KU). Honduras: Lancetilla, 40 m. (Valdez and LaVal, 1971:249, as *C. auritus* only). Costa Rica: Finca La Lola, 50 m. (Starrett and Casebeer, 1968:12). Panamá: Armila (Handley, 1966:762, as *C. auritus* only), thence into South America, and northwestward to Panamá: Cerro Azul (*ibid.*). Nicaragua: Santa María de Ostuma, 1250 m. (Jones, *et al.*, 1971:9). El Salvador: Barra de Santiago (Burt and Stirton, 1961:30). Guatemala: 1 km. WNW Sacapulas, 1200 m. (Carter, *et al.*, 1966:492). Chiapas: Zapaluta Cave, 1³⁄₁₀ mi. SSE Zapaluta, 5700 ft. (Davis, *et al.*, 1964:379, as *C. auritus* only). Oaxaca: Campamento Vista Hermosa (Goodwin, 1969:64); 3 mi. W Los Limones (Constantine, 1966:125). Veracruz: 38 km. SE Jesús Carranza. Chiapas: 5 mi. SW Teapa, Tabasco (Goodwin, 1969:64). Yucatán: Yaxcach.

Genus **Vampyrum** Rafinesque

1815. *Vampyrum* Rafinesque, Analyse de la nature, p. 54. Type, *Vespertilio spectrum* Linnaeus. (For use of this name in place of *Vampyrus* Leach, Trans. Linnean Soc. London, 13:79, 1821, and for selection of type see Andersen, Ann. Mag. Nat. Hist., ser. 8, 1:433, May, 1908.)

Skull elongated; breadth of braincase less than one-third greatest length; sagittal crest well developed, especially posteriorly; paroccipital expansions distinct; rostrum subcylindrical; zygomata slightly expanded anteriorly and posteriorly; tympanic small, covering less than half cochlear surface; height of inner edge of tympanic less than diameter. Externally resembles *Phyllostomus*, but chin smooth as in *Chrotopterus*; muzzle much elongated; tail absent; inter-

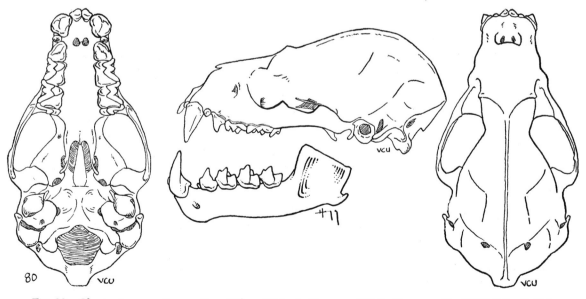

Fig. 80. *Chrotopterus auritus auritus*, 38 km. SE Jesús Carranza, 500 ft., Veracruz, No. 23661 K.U., ♀, X 2.

femoral membrane wide; ear extending to extremity of muzzle; fur normal (after Miller, 1907:135). Molars notably narrow and heightened, their basic W-shaped pattern discernible but much distorted; lower canines markedly enlarged, their bases nearly in contact posteromedially. Dentition, i. $\frac{2}{2}$, c. $\frac{1}{1}$, p. $\frac{2}{3}$, m. $\frac{3}{3}$.

Vampyrum spectrum (Linnaeus)
Linnaeus' False Vampire Bat

1758. *Vespertilio spectrum* Linnaeus, Syst. nat., 10th ed., 1:31, January 1. Habitat, South America. Thomas, Proc. Zool. Soc. London, 1911:130, suggested Surinam as the type locality, March, 1911.

1949. *Vampyrum spectrum,* Sanborn, Jour. Mamm., 30:280, August 17.

1917. *Vampyrus spectrum nelsoni* Goldman, Proc. Biol. Soc. Washington, 30:115, May 23, type from Coatzacoalcos, Veracruz. (Regarded as inseparable from *V. spectrum* by Husson, Zool. Verhand., Rijksmuseum Nat. Hist. Leiden, 58:122, November 13, 1962.)

1949. Sanborn (Jour. Mamm., 30:281, August 17) stated that the names *Vespertilio guianensis* Lacépède, 1799, *V. nasutus* Shaw, 1800, and *V. maximus* Geoffroy, 1806 were all based on La Grande Sérotine de Guiane of Buffon, 1789, and he listed them as synonyms of *Vampyrum spectrum* (Linnaeus), 1758. The description and drawing in Buffon, 1789 (Hist. Nat., vol. 8, p. 289, Pl. 73 facing p. 289, in copy in Univ. Kansas Library) pertain to a species other than

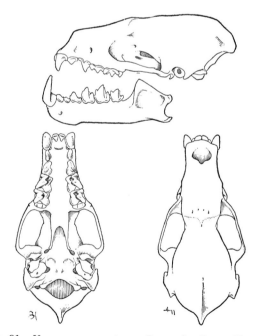

Fig. 81. *Vampyrum spectrum,* Boquerón, Prov. Chiriquí, Panamá. No. 18707 A.M.N.H., ♂, X 1.

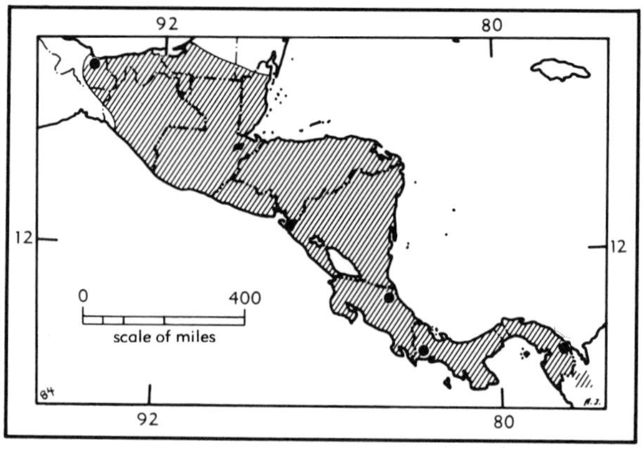

Map 84. *Vampyrum spectrum.*

Vespertilio spectrum Linnaeus, 1758. Consequently none of the three names is here listed as a synonym of *Vampyrum spectrum.*

Length of head and body approx. 130. Length of forearm, 100–107; greatest length of skull, 49.2–52.2; zygomatic breadth, 23.2–24.9; length of upper tooth-row (C–M3), 20.3–21.4. Upper parts dark reddish brown; underparts slightly paler. See account of genus for additional characters.

MARGINAL RECORDS.—Veracruz: Coatzacoalcos. Nicaragua: Chinandega (Goldman, 1917:116). Costa Rica: Finca La Selva, near Puerto Viejo (Casebeer, *et al.,* 1963:188). Panamá (Handley, 1966:762): Boquerón; Armila, thence southward into South America. The specimen from Jamaica (Koopman and Williams, 1951:19) probably was an accidental occurrence.

SUBFAMILY GLOSSOPHAGINAE
Long-tongued Bats

External characters summarized by Sanborn, Field Mus. Nat. Hist., Zool. Ser., 24:271–277, 1 fig., January 6, 1943.

Upper molars with styles reduced and closely approximated to paracone and metacone, this in connection with obsolescence of commissures nearly obliterating W-pattern. Lower molars having the five typical cusps, but reduced in height, especially paraconid; commissures rudimentary. All cheek-teeth elongated; rostrum much produced. Tongue long and highly extensible, its surface armed with conspicuous bristlelike papillae. Nose-leaf present, well developed, though never large (after Miller, 1907:136).

KEY TO NORTH AMERICAN GENERA OF GLOSSOPHAGINAE

1. Two upper and 2 lower molars on each side.
  2. Lower incisors present. . . . .*Leptonycteris,* p. 131
  2'. Lower incisors absent. . . . .*Lichonycteris,* p. 134
1'. Three upper and 3 lower molars on each side.
  3. Upper premolars 3–3. . . . .*Anoura,* p. 126
  3'. Upper premolars 2–2.
    4. Lower incisors well developed in adult.
      5. Upper incisors approx. equal in size; crowns of lower incisors broad, flat; zygomatic arch complete. . . . .*Glossophaga,* p. 119
      5'. Upper incisors differing markedly in size; crowns of lower incisors narrow, trenchant; zygomatic arch incomplete.
        6. Premolars flattened laterally, lacking cingula; fur on dorsum pale at base; uropatagium not furred. . . . .*Lonchophylla,* p. 122
        6'. Premolars not flattened, having well-developed cingula; fur on dorsum dark at base; uropatagium sparsely furred. . . . .*Lionycteris,* p. 124
    4'. Lower incisors minute or absent in adult.
      7. Zygomatic arch complete; lower incisors usually present in adult, but minute.
                                     Monophyllus, p. 124
      7'. Zygomatic arch incomplete; lower incisors absent in adult.
        8. Pterygoid processes convex on inner sides, the hamular processes not in contact with auditory bullae. . . . .*Hylonycteris,* p. 131
        8'. Pterygoids deeply concave on inner sides, the hamular processes in contact with auditory bullae.
          9. Rostrum greatly elongated; middle cusp of lower premolars longest.
                                    Choeronycteris, p. 128
          9'. Rostrum of normal proportions; cusps of lower premolars subequal. *Choeroniscus,* p. 130

Genus Glossophaga É. Geoffroy St.-Hilaire
Long-tongued Bats

Revised by Miller, Proc. U.S. Nat. Mus., 46:413–429, December 31, 1913.

1818. *Glossophaga* É. Geoffroy St.-Hilaire, Mém. Mus. Hist. Nat., Paris, 4:418. Type, *Vespertilio soricinus* Pallas.
1838. *Phyllophora* Gray, Mag. Zool. Bot., 2:489. Type, *Phyllophora amplexicaudata* Gray [= *Glossophaga amplexicaudata* Spix].
1847. *Nicon* Gray, Proc. Zool. Soc. London, p. 15, April 13. Type, *Nicon caudifer* Gray [= *Vespertilio soricinus* Pallas].

Skull with large, elongate, low, rounded, smooth braincase; rostrum shorter than braincase; basisphenoidal pits shallow; tympanic ring covering less than half surface of cochleae; tail shorter than tibia and not extending beyond middle of interfemoral membrane; zygoma complete; incisor series above and below unbroken; diastema between I1 and C less than width of outer incisor; crown of lower incisors about as broad as long. Dentition, i. $\frac{2}{2}$, c. $\frac{1}{1}$, p. $\frac{2}{3}$, m. $\frac{3}{3}$.

In the four known species, dorsal hairs two-banded, with wide, pale bases and narrow, dark tips; tip color varies from pale brown to reddish brown to almost black; in Central America *G. commissarisi* is normally darker than the other two species. Because the North American species are highly variable in the distinguishing features,

a combination of characters must be used for correct identification, and specimens with missing incisors or much worn teeth may be nearly impossible to identify.

KEY TO SPECIES OF GLOSSOPHAGA

1. Tip of I1 protrudes well past tip of I2; lower incisors crowded. *G. soricina,* p. 119
1'. Tip of I1 even with tip of I2; lower incisors not crowded.
  2. Rostrum approx. as long as braincase; lower incisors relatively large. *G. longirostris,* p. 122
  2'. Rostrum shorter than braincase; lower incisors relatively small.
    3. Postpalatal ridge of relatively uniform height; lower incisors with wide gap in center, narrower gap between i1 and i2. *G. alticola,* p. 120
    3'. Postpalatal ridge much lower along posterior half than along anterior half; lower incisors all separated by wide gaps. *G. commissarisi,* p. 121

Glossophaga soricina
Pallas' Long-tongued Bat

In 30 specimens from México, Central America, and Trinidad: length of head and body,

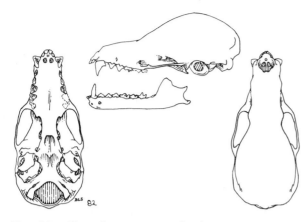

Fig. 82. *Glossophaga soricina leachii*, Finca Tepeyac, 10½ km. N, 9 km. E Matagalpa, 960 m., Nicaragua, No. 97569 K.U., ♂, X 2.

49–64; length of forearm, 32.3–38.9; condylobasal length, 19.8–21.5; breadth of braincase, 8.1–8.9; length of maxillary tooth-row, including C, 6.6–7.4. Rostrum approx. as long as braincase; braincase relatively low; slope of forehead moderate; postpalatal ridge relatively uniform in height; protruding I1, with tips extending well past tips of I2; lower incisors relatively large, crowded.

Glossophaga soricina antillarum Rehn

1902. *Glossophaga soricina antillarum* Rehn, Proc. Acad. Nat. Sci. Philadelphia, 54:37, April 23, type from Port Antonio, Jamaica.

MARGINAL RECORDS.—Bahama Islands. Jamaica: Sewell Cave, Montego Bay (Goodwin, 1970:574); type locality.

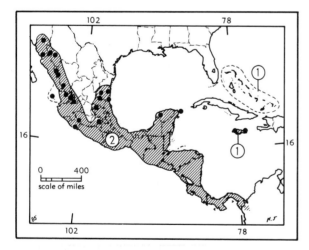

Map 85. *Glossophaga soricina.*

1. *G. s. antillarum*　　　　2. *G. s. leachii*

Glossophaga soricina leachii (Gray)

1844. *Monophyllus leachii* Gray, *in* The zoology of the voyage of H.M.S. Sulphur . . . , 1(1, Mamm.):18, April, type from Realejo, Chinandega, Nicaragua.
1913. *Glossophaga soricina leachii*, Miller, Proc. U.S. Nat. Mus., 46:419, December 31.
1898. *Glossophaga mutica* Merriam, Proc. Biol. Soc. Washington, 12:18, January 27, type from María Madre Island, Tres Marías Islands, Nayarit.
1938. *Glossophaga morenoi* Martínez and Villa, Anal. Inst. Biol., Univ. Nac. Autó. México, 9(3–4):347, November 14, type from Xiutepec, Morelos. According to Davis and Russell (Jour. Mamm., 35:68, February 10, 1954) indistinguishable from *G. s. leachii*.

MARGINAL RECORDS.—Sonora: below Rebeico Dam, 28 mi. by road E Mazatán (Findley and Jones, 1965:330, as *G. soricina* only). Chihuahua: Río Septentrión, 1½ mi. SW Tocuina (Anderson, 1960:7); La Bufa, 3500 ft. (Anderson, 1972:238). Durango: Chacala; Santa Ana (Jones, 1964a:751); Pueblo Nuevo, 5000 ft. (Baker and Greer, 1962:67). Nayarit: Santiago. Jalisco: 2 mi. E Bolaños, 3550 ft. (Watkins, *et al.*, 1972:12). Zacatecas: 5½ mi. S Moyahua, 4000 ft. (Genoways and Jones, 1968:743). Michoacán: Hda. El Sabino. Distrito Federal: Chicomostoc, Cerro Teutli, 2620 m. Puebla: *Tuchitan.* Querétaro: 2 mi. SSE Conca (Schmidly and Martin, 1973:91). San Luis Potosí: El Salto. Tamaulipas: 10 km. N, 8 km. W El Encino (Alvarez, 1963:400); 3 mi. S, 16 mi. W Piedra; Altamira. Yucatán: Mérida (69067 KU). Quintana Roo: Rancho del Pirata, Isla Mujeres (91563 KU), thence southeastward into South America and thence northwestward along Pacific Coast to Michoacán: 3 km. S Melchor Ocampo (Alvarez, 1968:25). Nayarit: María Madre Island, Tres Marías Islands. Sonora: *Río Alamos, 8 mi. by road S Alamos* (Baker and Christianson, 1966:310, as *G. soricina* only); 1 km. SW La Aduana, *ca.* 1600 ft. (Loomis and Davis, 1965:497).

Glossophaga alticola Davis
Davis' Long-tongued Bat

1944. *Glossophaga soricina alticola* Davis, Jour. Mamm., 25:377, December 12, type from 13 km. NE Tlaxcala, 7800 ft., Tlaxcala.
1967. *G[lossophaga]. alticola,* Baker, Southwestern Nat., 12:411, December 31.

Length of head and body, 56–65; hind foot, 10–12; ear, 12–16; forearm, 34.9–38.7; condylobasal length, 17.8–19.8; depth of braincase, 6.8–7.6; breadth of braincase, 8.5–9.0; maxillary tooth-row, 6.5–7.2. Rostrum shorter than braincase; braincase relatively high, domed; slope of forehead steep; postpalatal ridge relatively uniform in height; tips of upper incisors even with each other; lower incisors small, well-spaced, usually with larger space in middle than between i1 and i2.

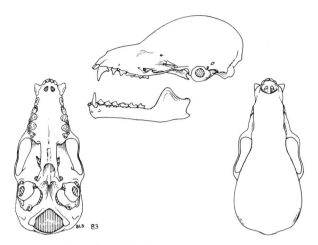

Fig. 83. *Glossophaga alticola*, 3 mi. SW Managua, Managua, Nicaragua, No. 70621 K.U., ♂, X 2.

MARGINAL RECORDS.—Tlaxcala: type locality. Oaxaca: 5 mi. W Chiltepec (Goodwin, 1969:70). Chiapas: 4 mi. S, 17 mi. W Las Cruces (60941 KU). Guatemala: 2¼ mi. N, 2½ mi. W San Cristóbal, 2900 ft. (64890 KU). Costa Rica: 1 km. SSE Bagaces, 75 m. (88038 KU). Nicaragua: 3 mi. SW Managua (70621 KU). El Salvador: San Salvador, 680 m. (88040 KU). Oaxaca (Goodwin, 1969:70): Tapanatepec: Santiago Lachiguirí. Guerrero: Kilometer 231, Hwy. 95 (Baker, 1967:427); 8 mi. N, 1 mi. W Teloloapan, 3600 ft. (66358 KU); *Cueva de Tia Juana, 1¹/₂ km. SSW Yerbabuena, 1840 m.* (28396 KU). Morelos: Santa Clara.

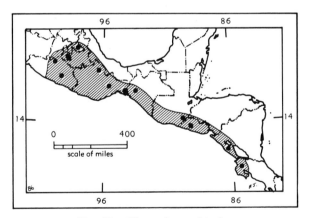

Map 86. *Glossophaga alticola*.

Glossophaga commissarisi Gardner
Commissaris' Long-tongued Bat

1962. *Glossophaga commissarisi* Gardner, Los Angeles Co. Mus. Contrib. Sci., 54:1, May 11, type from 10 km. SE Tonalá, Chiapas.

Length of head and body, 43–60; hind foot, 9–13; ear, 11–16; forearm, 33.0–35.7; condylobasal length, 18.1–19.4; depth of braincase,

6.3–7.3; maxillary tooth-row, 6.5–7.3. Rostrum shorter than braincase; braincase relatively low; slope of forehead moderate; postpalatal ridge much lower (sometimes absent) along posterior half than along anterior half; tips of upper incisors even with each other; lower incisors small, widely spaced, and usually evenly spaced.

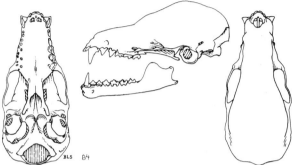

Fig. 84. *Glossophaga commissarisi*, Bonanza, 850 ft., Nicaragua, No. 96264 K.U., ♂, X 2.

MARGINAL RECORDS.—Sinaloa: 20 km. N, 5 km. E Badiraguato, 1800 ft. (96461 KU); Santa Lucia, 3600 ft. (94744 KU). Durango (Baker and Greer, 1962:67): 2 mi. N Pueblo Nuevo, 6000 ft.; *6 mi. S Pueblo Neuvo, 3000 ft.* Jalisco: 14 mi. WSW Ameca, 5000 ft. (105465 KU); 6 mi. E Limón, 2700 ft. (103408 KU). Oaxaca (Goodwin, 1969:69): 4 mi. W Chiltepec; Sierra Madre, N Zanatepec. Chiapas: 13 km. ENE Pichucalco, 60 m. (Baker, *et al.*, 1973:78, 85); Sabana de San Quintín, 215 m. (102359 KU). Honduras: Lancetilla (103259 KU); Río Coco, 78 mi. ENE Danlí, 900 ft. (*ibid.*). Nicaragua: Bonanza (Jones, 1964d:507); 6 mi. W Rama, 50 ft. (Davis, *et al.*, 1964:380). Costa Rica: La Selva (3130 LaVal at KU). Panamá (Handley, 1966:763): Almirante; Armila, thence into South America, and northwestward along coast to Costa Rica: 4 km. SW Rincón (LaVal,

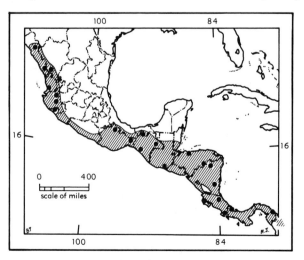

Map 87. *Glossophaga commissarisi*.

1970:3); 7 mi. SW Filadelfia (60457 KU). Guatemala: *2¼ mi. N, 2½ mi. W San Cristóbal* (Jones, 1966:453); Astillero, 25 ft. (64893 KU). Chiapas: 20 km. SE Pijijiapam (Gardner, 1962:2). Colima: Cerro Chino (*ibid.*). Nayarit: 8 mi. E San Blas (*ibid.*).

Glossophaga longirostris
Miller's Long-tongued Bat

Length of head and body, 56–65; length of forearm, 35.4–39.4; condylobasal length, 21.0–22.4; breadth of braincase, 8.8–9.4; length of upper tooth-row, 7.8–8.8. Rostrum approx. equal to braincase in length; braincase relatively low, slope of forehead relatively slight; postpalatal ridge variable, usually uniform in height but sometimes lower posteriorly; tips of upper incisors even with each other, or sometimes tips of I1 extend slightly past tips of I2; lower incisors large, as in *soricina*, but less crowded.

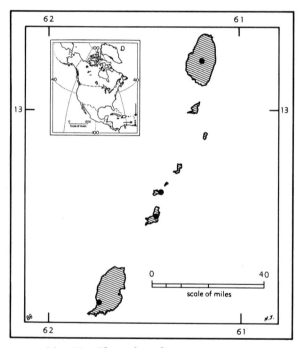

Map 88. *Glossophaga longirostris rostrata.*

Glossophaga longirostris rostrata Miller

1913. *Glossophaga rostrata* Miller, Proc. Biol. Soc. Washington, 26:32, February 8, type from Westerhall estate, Grenada, Lesser Antilles.
1913. *Glossophaga longirostris rostrata* Miller, Proc. U.S. Nat. Mus., 46:423, December 31.

MARGINAL RECORDS.—Lesser Antilles: St. Vincent (Koopman, 1968:2); Union Island; Carriacou; Old Fort, St. Georges, Grenada. Two specimens from

Dominica may belong to this species, but are not positively identifiable. Jones and Phillips (1970:137) also list this species from Dominica, but do not state the origin of the record.

Genus Lonchophylla Thomas
Long-tongued Bats

1903. *Lonchophylla* Thomas, Ann. Mag. Nat. Hist., ser. 7, 12:458, October. Type, *Lonchophylla mordax* Thomas.

"Like *Glossophaga*, but zygomatic arch incomplete; inner upper incisor higher than wide and more than double the bulk of outer tooth, which stands by itself near middle of space between large incisor and canine; lower incisors having narrow trifid cutting edges, the outer separated from canine by a space nearly equal to the length of its crown; and last upper molar nearly as large as either of the others, its parastyle short." (Miller, 1907:139.)
Dentition, i. $\frac{2}{2}$, c. $\frac{1}{1}$, p. $\frac{2}{3}$, m. $\frac{3}{3}$.

KEY TO NORTH AMERICAN SPECIES OF
LONCHOPHYLLA

(Adapted from R. H. Pine, *in Litt.*)

1. Last upper premolars spindle shaped in cross section, lingual cusp poorly developed or absent; posterior cusp on anterior lower premolar lacking, or if present, poorly developed and not hooklike.
   L. mordax, p. 122
1'. Last upper premolars triangular or T-shaped in cross section, lingual cusp prominent; hooklike posterior cusp present on anteriormost lower premolar.
   2. Last upper premolar T-shaped (lingual cusp forming stem of T); posterior palatal emargination V-shaped; forearm less than 35 . . . . *L. thomasi,* p. 123
   2'. Last upper premolar triangular; posterior palatal emargination U-shaped; forearm more than 39. *L. robusta,* p. 123

Lonchophylla mordax
Brazilian Long-tongued Bat

Length of head and body, 55–58; selected measurements of type (of *L. concava*) plus five specimens from Costa Rica: length of forearm, 32.7–33.9; greatest length of skull, 22.4–23.4; interorbital breadth, 4.1–4.7; maxillary tooth-row, 7.2–8.0. Upper parts near (*n*) Mummy Brown; underparts slightly paler; rostrum longer than in *L. thomasi*.

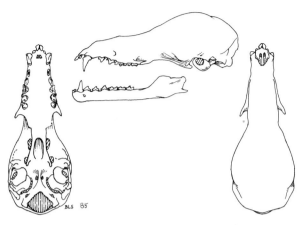

Fig. 85. *Lonchophylla mordax concava*, 20 km. SW San Isidro de General, 865 m., San José, Costa Rica, No. 88035 K. U., ♀, X 2.

Lonchophylla mordax concava Goldman

1914. *Lonchophylla concava* Goldman, Smiths. Miscl. Coll., 63(5):2, March 14, type from Cana, 2000 ft., Darién, Panamá.
1966. *L[onchophylla]*. *m[ordax]*. *concava*, Handley, Ectoparasites of Panama, Field Mus. Nat. Hist., p. 763, November 22.

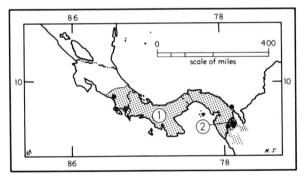

Map 89. *Lonchophylla mordax concava* (1) and *Lonchophylla thomasi* (2).

MARGINAL RECORDS.—Costa Rica: 20 km. SW San Isidro del General (Armstrong, 1969:809). Panamá (Handley, 1966:763): *Armila*; Puerto Obaldía, thence into South America, thence northwestward to Panamá: type locality. Costa Rica: 9 mi. ENE Puerto Golfito (Davis, *et al.*, 1964:380); 2 mi. W Rincón (Armstrong, 1969:808).

Lonchophylla thomasi J. A. Allen
Thomas' Long-tongued Bat

1904. *Lonchophylla thomasi* J. A. Allen, Bull. Amer. Mus. Nat. Hist., 20:230, June 29, type from Cuidad [sic] Bolívar, Venezuela.

Selected measurements of holotype and three specimens from Surinam: forearm, 30.0–32.3; 3rd

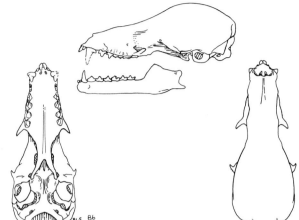

Fig. 86. *Lonchophylla thomasi*, San Blas, Armila (Quebrada Venado), Panamá, No. 335180 U.S.N.M., ♀, X 2.

metacarpal, 32–34; 5th metacarpal, 29–31; greatest length of skull, 19.6–21.0; breadth of braincase, 8.0–8.5. As compared to *L. m. concava*, smaller and darker; venter little if any paler than dorsum; skull smaller, with rostrum shorter, broader, and less tapering; braincase more convex and higher.

MARGINAL RECORD.—Panamá: Boca de Río Paya (Handley, 1966:763), thence into South America.

Lonchophylla robusta Miller
Panamá Long-tongued Bat

1912. *Lonchophylla robusta* Miller, Proc. U.S. Nat. Mus., 42:23, March 6, type from cave on Chilibrillo River, near Alhajuela, Panamá.

Selected measurements of the type, an adult male, and a female topotype, are, respectively: length of head and body, 56, 60; length of forearm, 43.6, 43; condylobasal length, 25.2, 25.4; interorbital breadth, 5.2, 5.4; length of upper tooth-row, 9.8, 10. Upper parts dark brown; underparts slightly paler, having a still paler wash. Skull relatively broad and robust; braincase well inflated; 2nd upper premolar broad, its inner lobe conspicuously developed.

MARGINAL RECORDS.—Nicaragua: 3 km. NW Rama (Baker and Jones, 1975:4). Costa Rica: Finca Los Diamantes (Armstrong, 1969: 809); 4 km. SW Hda. de Moravia (Walton, 1963:87). Panamá (Handley, 1966:763): Almirante; Isla Bastimentos; Barro Colorado Island; Buena Vista; type locality; Cerro Azul; *Armila*; Puerto Obaldía; Tacarcuna Village; Cana, thence into South America.

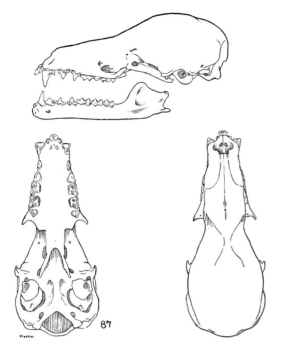

Fig. 87. *Lonchophylla robusta*, Chilibrillo Cave, 10 mi. N Pedro Miguel, Panamá, No. 45075 K.U., ♀, X 2.

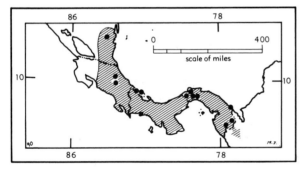

Map 90. *Lonchophylla robusta*.

Genus **Lionycteris** Thomas
Long-tongued Bats

1913. *Lionycteris* Thomas, Ann. Mag. Nat. Hist., ser. 8, 12:270, July. Type, *Lionycteris spurrelli* Thomas.

Skull much as in *Glossophaga;* premolars high-crowned, lacking characteristic lateral compression and anteroposterior elongation of other glossophagines; upper incisors as in *Lonchophylla;* diastema between C1 and P1 as in *Lonchophylla;* zygomatic arch lacking. Dentition, i. $\frac{2}{2}$, c. $\frac{1}{1}$, p. $\frac{2}{3}$, m. $\frac{3}{3}$.

Uropatagium well developed and sparsely furred, tail reaching halfway to posterior edge; hairs on dorsum darker at bases than at tips; color dark brown.

Lionycteris spurrelli Thomas
Little Long-tongued Bat

1913. *Lionycteris spurrelli* Thomas, Ann. Mag. Nat. Hist., ser. 8, 12:271, July, type from Condoto, Chocó, Colombia, 300 ft.

Selected measurements of four specimens from Brazil, Colombia, Guyana, and, in parentheses, the "immature" holotype: Head and body, (49); forearm, 34.9–36 (33); 3rd metacarpal, 34.2–36.2; 4th metacarpal, 31.8–32.6; greatest length of skull, 19.6, 19.7 (18.7); interorbital width, 3.7, 4.1 (3.7); maxillary tooth-row, 6.4, 6.2 (6.1). Description as for genus.

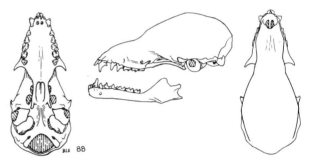

Fig. 88. *Lionycteris spurrelli*, San Blas, Armila (Quebrada Venado), Panamá, No. 335186 U.S.N.M., ♀, X 2.

MARGINAL RECORDS.—Panamá (Handley, 1966:763): Armila; Cana, thence southward into South America.

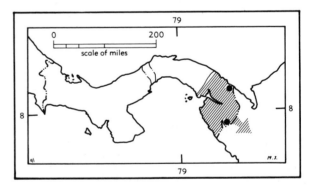

Map 91. *Lionycteris spurrelli*.

Genus **Monophyllus** Leach—Long-tongued Bats

Revised by Schwartz and Jones, Proc. U.S. Nat. Mus., 124(3635):1–20, December, 1967. Varona (1974:18) treats *Monophyllus* as a subgenus of *Glossophaga*.

1821. *Monophyllus* Leach, Trans. Linnean Soc. London, 13:75. Type, *Monophyllus redmani* Leach.

"In general like *Glossophaga;* zygomatic arch complete; tail about half as long as femur, projecting beyond edge of . . . narrow interfemoral membrane. Teeth essentially as in *Glossophaga,* but upper incisors much smaller and of equal length, the outer sharp pointed, the inner with flat cutting edge, neither of the incisors in contact with the other or with canine; lower incisors . . . minute, with roundish flat crowns, the four teeth standing as two pairs, one pair on each side of a broad median space; upper and lower premolars with conspicuous styles; upper molars with inner margin obliquely truncate." (Miller, 1907:139.)

Dentition, i. $\frac{2}{2}$, c. $\frac{1}{1}$, p. $\frac{2}{3}$, m. $\frac{3}{3}$. Pelage dark brown to pale brown (underparts grayish in two of the specimens of *M. p. luciae*).

KEY TO SPECIES OF MONOPHYLLUS

1. Upper premolars separated by diastema half or more of length of 1st premolar.
   M. redmani, p. 125
1'. Upper premolars separated by diastema less than half length of 1st premolar.
   M. plethodon, p. 126

Monophyllus redmani
Leach's Long-tongued Bat

Selected measurements: total length, 59–80; tail, 7–11; length of forearm, 35.5–42.8; hind foot, 9–14; greatest length of skull, 19.0–23.9; condylobasal length, 17.9–22.6; interorbital breadth, 3.8–4.6; zygomatic breadth, 8.2–10.4; alveolar length of upper tooth-row (C–M3), 6.8–8.9. See key for difference from *M. plethodon*.

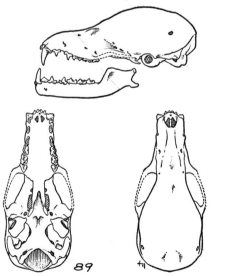

Fig. 89. *Monophyllus redmani clinedaphus*, Baracoa, eastern Cuba, No. 113689 U.S.N.M., sex ?, X 2.

Monophyllus redmani clinedaphus Miller

1900. *Monophyllus clinedaphus* Miller, Proc. Washington Acad. Sci., 2:36, March 30, type locality restricted to "vicinity of Baracoa, Oriente Province, Cuba" by Schwartz and Jones, Proc. U.S. Nat. Mus., 1967:6, December.
1967. *Monophyllus redmani clinedaphus,* Schwartz and Jones, Proc. U.S. Nat. Mus., 124(3635):6, December.
1902. *Monophyllus cubanus* Miller, Proc. Acad. Nat. Sci. Philadelphia, 54:410, September 12, type from Baracoa, Oriente, Cuba.
1918. *Monophyllus cubanus ferreus* Miller, Proc. Biol. Soc. Washington, 31:40, May 16, type from cave 8 mi. WSW Jérémie, Haiti.

MARGINAL RECORDS (Schwartz and Jones, 1967:11, unless otherwise noted).—Cuba: Cueva de la Numancia, Aguacate; type locality. Bahamas (Buden, 1975:376): Crooked Island; Acklins Island; *North Caicos,* as species only; Middle Caicos, as species only. Haiti: La Gonave Island (Koopman, 1955:110); St. Michel de l'Atalaye; *cave at Diquini* (Miller, 1929a:8); 8 mi. WSW Jérémie. Cuba: *Guisa [20° 15′ N, 76° 32′ W]* (Schwartz and Jones, 1967:10); Pinar del Río (Koopman and Ruibal, 1955:3); 9 km. SW San José de las Lajas. Not found: Rangel in the Sierra del Rosario in Pinar del Río Province (Schwartz and Jones, 1967:10).

Monophyllus redmani portoricensis Miller

1900. *Monophyllus portoricensis* Miller, Proc. Washington Acad. Sci., 2:34, March 30, type from cave near Bayamón, Puerto Rico.
1967. *Monophyllus redmani portoricensis,* Schwartz and Jones, Proc. U.S. Nat. Mus., 124(3635):11, December.

MARGINAL RECORDS (Schwartz and Jones, 1967:12).—Puerto Rico: *Cueva de Fari, Pueblo Viejo [18° 26′ N, 66° 07′ W];* Cueva de Trujillo Alto, Trujillo Alto; *ca. 1 km. NE Cidra;* 7½ km. E Guánico; $17\frac{7}{10}$ km. NE Utuado.

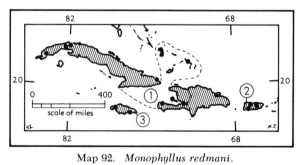

Map 92. *Monophyllus redmani.*

1. *M. r. clinedaphus*
2. *M. r. portoricensis*
3. *M. r. redmani*

Monophyllus redmani redmani Leach

1821. *Monophyllus redmani* Leach, Trans. Linnean Soc. London, 13:76, type from Jamaica.

MARGINAL RECORDS (Schwartz and Jones, 1967:6, unless otherwise noted).—Jamaica: Windsor, Trelawny Parish; Kingston (Miller, 1902:410); *Oxford Cave, Balaclava.*

Monophyllus plethodon
Insular Long-tongued Bat

Selected measurements: total length, 67–80; tail, 8–16; length of forearm, 38–45.7; hind foot, 12–15; greatest length of skull, 21.4–24.2; condylobasal length, 19.5–22.6; interorbital breadth, 4.4–5.0; zygomatic breadth, 9.4–11.0; alveolar length of upper tooth-row (C–M3), 6.8–8.5. See key for difference from *M. redmani.*

Monophyllus plethodon frater Anthony

1917. *Monophyllus frater* Anthony, Bull. Amer. Mus. Nat. Hist., 37:565, September, type from cave near Morovis, Puerto Rico. Known only from type locality.
1967. *Monophyllus plethodon frater,* Schwartz and Jones, Proc. U.S. Nat. Mus., 124(3635):15, December.

Monophyllus plethodon luciae Miller

1902. *Monophyllus luciae* Miller, Proc. Acad. Nat. Sci. Philadelphia, 54:411, September 12, type from St. Lucia, Lesser Antilles.
1967. *Monophyllus plethodon luciae,* Schwartz and Jones, Proc. U.S. Nat. Mus., 124(3635):13, December.

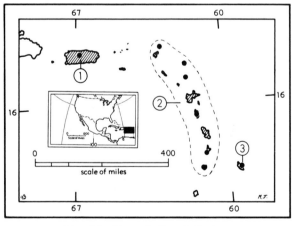

Map 93. *Monophyllus plethodon.*

1. *M. p. frater*   2. *M. p. luciae*
3. *M. p. plethodon*

MARGINAL RECORDS (Schwartz and Jones, 1967:15).—Lesser Antilles: Anguilla; Barbuda; Antigua; Dominica; type locality; Clifton Hill, 400 ft., St. George Parish, St. Vincent (110088 KU).—See addenda.

Monophyllus plethodon plethodon Miller

1900. *Monophyllus plethodon* Miller, Proc. Washington Acad. Sci., 2:35, March 30, type from St. Michael Parish, Barbados, Lesser Antilles.

MARGINAL RECORDS.—Barbados: *Jack-in-the-box Gully, St. Thomas Parish* (Schwartz and Jones, 1967:13); type locality.

Genus **Anoura** Gray—Tailless Bats

Revised by Sanborn, Field Mus. Nat. Hist., Publ. 323, Zool. Ser., 20:23–27, December 11, 1933.

1838. *Anoura* Gray, Mag. Zool. Bot., 2:490, February. Type, *Anoura geoffroyi* Gray.
1844–46. *Rhinchonycteris* Tschudi, Untersuchungen über die fauna Peruana . . . , p. 71. Type, *Rhinchonycteris peruana* Tschudi [= *Anoura geoffroyi* Gray].
1846. *Anura* Agassiz, Nomen. Zool., Index Univ., p. 27, an emendation?
1868. *Lonchoglossa* Peters, Monatsb. preuss. Akad. Wiss., Berlin, p. 364. Type, *Glossophaga caudifera* E. Geoffroy St.-Hilaire.
1868. *Glossonycteris* Peters, Monatsb. preuss. Akad. Wiss., Berlin, p. 365. Type, *Glossonycteris lasiopyga* Peters.

This American genus contains *A. geoffroyi, A. cultrata,* and *A. werckleae* of North America. *A. brevirostrum* (Carter, 1968:427) and *A. caudifer* (Tamsitt and Valdivieso, 1966:230) are limited to South America. Tail absent or rudimentary. Dentition: i. $\frac{2}{0}$, c. $\frac{1}{1}$, p. $\frac{3}{3}$, m. $\frac{3}{3}$.

KEY TO SPECIES OF ANOURA

1. Limited to South America. *A. caudifer* and *A. brevirostrum.*
1'. Not limited to South America.
   2. Tail absent, zygomata rudimentary.
                               A. geoffroyi,   p. 127
   2'. Tail rudimentary, zygomata complete.
      3. Pits in ventral surface of basisphenoid only slightly expanded anteriorly and without raised shelf anteriorly; pelage blackish.
                         A. cultrata,   p. 126
      3'. Pits in ventral surface of basisphenoid expanded anterolaterally beneath alisphenoids and partly covered by raised shelf anteriorly; pelage orange brown.
                       A. werckleae,   p. 128

Anoura cultrata Handley
Handley's Tailless Bat

1960. *Anoura cultrata* Handley, Proc. U.S. Nat. Mus., 112:463, October 6, type from Tacarcuna Village, 3200 ft., Río Purco, Darién, Panamá.

Measurements (holotype): total length, 94; tail vertebrae, 6; hind foot, 14; ear from notch, 16; forearm, 43.2; greatest length of skull, 26.3; zygomatic breadth, 10.7; interorbital breadth, 4.7; maxillary tooth-row, 9.0; postpalatal length, 9.4 (after Handley, 1960:464). Skull resembling that of *A. geoffroyi*, but differing as follows: larger; braincase more tapering anteriorly; rostrum thickened; zygomata complete. Coloration shiny blackish (between Blackish Brown-3 and Black).

MARGINAL RECORDS.—Costa Rica: 18 mi. NE Naranjo, 3100 ft. (Carter, *et al.*, 1966:492). Panamá (Handley, 1966:764): upper Río Changena, 2400 ft.; Tacarcuna Village, 1950 ft.; Cerro Punta, 5300 ft. Costa Rica: Monteverde (LaVal, 1977:79). Occurs also in South America.

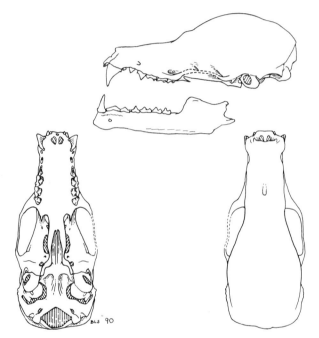

Fig. 90. *Anoura cultrata*, Cerro Tacarcuna, 4100 ft., Darién, Panamá, No. 337991 U.S.N.M., ♂, X 2.

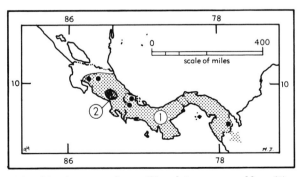

Map 94. *Anoura cultrata* (1) and *Anoura werckleae* (2).

Anoura geoffroyi
Geoffroy's Tailless Bat

Length of head and body approx. 60. Selected measurements are: length of forearm, 40–47.3; greatest length of skull, 24.5–27.0. Additional measurements of a female from Texolo, Veracruz, are: total length, 64; interorbital breadth, 4.4; mastoidal breadth, 10.0. Upper parts dull brown usually becoming silvery gray over sides of neck and shoulders; underparts grayish brown.

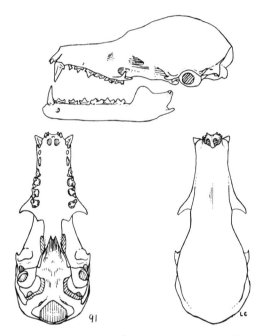

Fig. 91. *Anoura geoffroyi lasiopyga*, 2 mi. WSW Teopisca, Chiapas, No. 61037 K.U., ♂, X 2.

Anoura geoffroyi geoffroyi Gray

1838. *Anoura geoffroyi* Gray, Mag. Zool. Bot., 2:490, February, type from Brazil.

MARGINAL RECORDS.—Within the scope of this work known only from the Lesser Antilles: Grenada (Jones and Phillips, 1970:132, 133), thence southward into South America.

Anoura geoffroyi lasiopyga (Peters)

1868. *Glossonycteris lasiopyga* Peters, Monatsb. preuss. Akad. Wiss., Berlin, p. 365, type from southern México.
1933. *Anoura geoffroyi lasiopyga*, Sanborn, Field Mus. Nat. Hist., Publ. 323, Zool. Ser., 20:27, December 11.

MARGINAL RECORDS.—Sinaloa: 12 mi. NE San Benito, 1000 ft. (Jones, *et al.*, 1972:9). Durango: 1½ mi. W San Luis, 7500 ft. (Baker and Greer, 1962:67). Zacatecas: Santa Rosa, 1219 m. (Matson and Patten, 1975:3). Jalisco: 9 mi. N Guadalajara, 4000 ft. (Watkins,

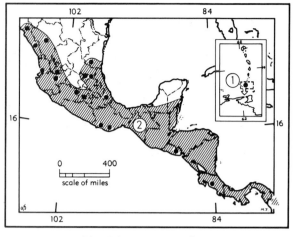

Map 95. *Anoura geoffroyi.*

1. *A. g. geoffroyi*   2. *A. g. lasiopyga*

et al., 1972:13). Michoacán: 3 km. SW Turundeo, 1900 m. (Alvarez and Ramírez-P., 1972:169). México: Barranca de los Idolos, 32 km. SW México, D.F., 3300 m. (Villa-R., 1967:240). San Luis Potosí: 5⅔ km. NW Xilitla (Spenrath and LaVal, 1970:396). Querétaro: Hda. X-Conca (Schmidly and Martin, 1973:91). Tamaulipas: 5 mi. W Gómez Farías (Baker and Lopez, 1968:361, as *A. geoffroyi* only). Veracruz: Texolo [probably Teocelo] (Sanborn, 1933:27), thence southward along coast to South America, then northward along Pacific Coast to Costa Rica: San Rafael de Tarrazi, 30 km. S San José (39249 KU). El Salvador: Mt. Cacaguatique (Sanborn, 1933:27); *Volcán de San Vicente* (Felten, 1956:196). Oaxaca: Río Guajolote (Schaldach, 1966:290); *San Gabriel Mixtepec (ibid.); ca.* 11 mi. E Juquila, 6100 ft. (Baker and Womochel, 1966:306). Nayarit: 2 mi. SE Jalcocotán, 3000 ft. (Anderson, 1956:350). Sinaloa (Jones, 1964e:510): *1½ km. S Santa Lucia;* Santa Lucia; *1 km. NE Santa Lucia.* Not found: Jalisco: San Sebastián (Sanborn, 1933:27).

Anoura werckleae Starrett
Starrett's Tailless Bat

1969. *Anoura werckleae* Starrett, Los Angeles Co. Mus. Contrib. Sci., 157:1, February 27, type from 6⅗ mi. S restaurant "La Georgina," 2500 m., Cerro de la Muerte massif, Prov. San José, Costa Rica. Known only from type locality.

Length of head and body approx. 65; hind foot, 11.9–12.2; ear from notch, 14; length of forearm, 40.7–43.2; greatest length of skull, 25.8–26.2; zygomatic breadth, 10.6–10.8; maxillary toothrow, 9.0–9.2. Upper parts orange-brown, somewhat paler ventrally. *A. werckleae* appears to be found only in association with the tree *Wercklea lutea* Rolfe. See Map 94.

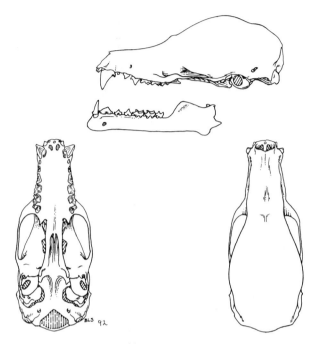

Fig. 92. *Anoura werckleae,* 6⅗ mi. S La Georgina, San José Prov., Costa Rica, No. 25438 L.A.C.M., ♀, X 2.

Genus **Choeronycteris** Tschudi
Long-tongued Bats

1844. *Choeronycteris* Tschudi, Untersuchungen über die Fauna Peruana . . . , p. 70. Type, *Choeronycteris mexicana* Tschudi.

Rostrum elongated, comprising 40 to about 55 per cent of length of skull; zygomata incomplete; pterygoid processes strongly concave on inner sides; hamulae in contact with auditory bullae; tail approx. half length of femur and not extending to mid-point of interfemoral membrane; calcar weakly developed; lower incisors absent in adults; W-shaped pattern of upper molars nearly obliterated. Dentition, i. $\frac{2}{0}$, c. $\frac{1}{1}$, p. $\frac{2}{3}$, m. $\frac{3}{3}$.

KEY TO SPECIES OF CHOERONYCTERIS

1. Length of rostrum less than half length of skull. . . . .*C. mexicana,* p. 128
1'. Length of rostrum more than half length of skull. . . . .*C. harrisoni,* p. 129

Choeronycteris mexicana Tschudi
Mexican Long-tongued Bat

1844. *Choeronycteris mexicana* Tschudi, Untersuchungen über die Fauna Peruana . . . , p. 72, type from México.

Selected measurements: head and body, 60–86; tail, 7–16; forearm, 42.0–46.7; greatest

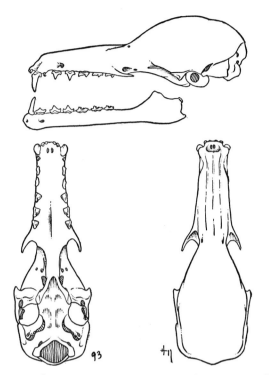

Fig. 93. *Choeronycteris mexicana*, 1 mi. S, 4 mi. W Bella, Unión, 1000 ft., Coahuila, No. 34550 K.U., ♂, X 2.

length of skull, 29.3–31.1; length of rostrum, 11.5–12.7; interorbital breadth, 3.5–4.2; maxillary tooth-row, 10.9–11.8.

MARGINAL RECORDS.—California: Hermosa Way, San Diego. Arizona: Alamo Canyon Tunnel, 10 mi. N Tucson (Lange, 1960:439); tunnel at narrows of Sabina Canyon, 20 mi. NE Tucson (Cockrum, 1961:34); *vic. Tucson, Pima Co.* (Schaldach and McLaughlin,

1960:8); *1 mi. E Lime Peak* (Constantine, 1961:404); 1 mi. N Paradise (*ibid.*); *Guadalupe Canyon* (Mumford and Zimmerman, 1962:101). New Mexico: 10 mi. S Rodeo (Findley, 1957:513). Coahuila: 4 mi. S, 9 mi. W San Buenaventura, 1800 ft.; cave NE Hermanas (Axtell, 1962:76). Texas: Santa Ana National Wildlife Refuge (LaVal and Shifflett, 1972:40). Tamaulipas: 4 km. N Joya Verde, 4000 ft. Hidalgo: Río Tasquillo, 26 km. E Zimapán. Tlaxcala: 13 km. NE Tlaxcala. Honduras: 12 mi. N Tegucigalpa, 2800 ft. (Davis, *et al.*, 1964:380). Guatemala (Jones, 1966:455): Dueñas; Hda. California. Oaxaca: *ca.* 11 mi. E Juquila, 6100 ft. (Baker and Womochel, 1966:306). Michoacán: 2 mi. W Pátzcuaro. Jalisco: Los Masos. Nayarit: Tres Marías Islands. Sinaloa: 12 mi. N, 4 mi. W Los Mochis (Jones, *et al.*, 1972:9). Baja California: 8 mi. N Santa Catarina.

Choeronycteris harrisoni (Schaldach and McLaughlin)
Trumpet-nosed Bat

1960. *Musonycteris harrisoni* Schaldach and McLaughlin, Los Angeles Co. Mus., Contrib. Sci., 37:3, May 19, type from 2 km. SE Pueblo Juárez (formerly Hda. La Magdalena) Colima, México.
1966. *Choeronycteris harrisoni*, Handley, Proc. Biol. Soc. Washington, 79:86, May 23. Based on comparison of the species of five genera of glossophagine bats, Handley (*ibid.*)

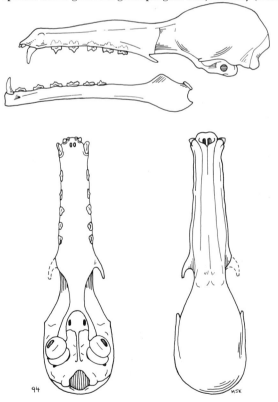

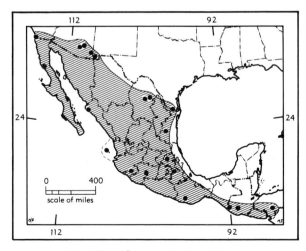

Map 96. *Choeronycteris mexicana*.

Fig. 94. *Choeronycteris harrisoni*, Las Juntas, approx. 5 km. E Pueblo Juárez, Colima, México, No. 98874 K.U., ♂, X 2.

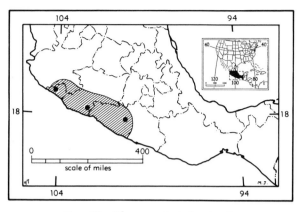

Map 97. *Choeronycteris harrisoni.*

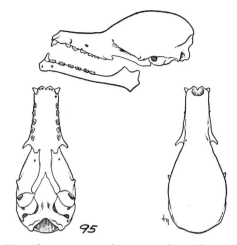

Fig. 95. *Choeroniscus godmani*, La Flor Archaga, Honduras, No. 127597 A.M.N.H., ♂, X 2.

concluded that *Choeronycteris* and *Musonycteris* are congeneric.

Selected measurements: total length, 80–89; tail, 8–12; forearm, 41.6–42.9; greatest length of skull, 32.0–35.2; length of rostrum, 15.8–18.3; interorbital breadth, 3.6–4.1; length of maxillary tooth-row, 12.0–14.1. Posterior dorsal region between Mummy Brown and Clove Brown, lightening on middle back and shoulder region toward brownish light drab; underparts like shoulder region (after Schaldach and McLaughlin, 1960:3).

MARGINAL RECORDS (Schaldach and McLaughlin, 1960:8, unless otherwise noted).—Colima: type locality; *4 km. S Cerro de Ortega.* Guerrero: Cañón de Zopilote, 14½ mi. by road N Zumpango, 2000 ft. (Winkelmann, 1962:108). Michoacán: Arroyo del Chivo, 20 km. N Infiernillo, Guerrero (Villa-R., 1967:257, 258).

Genus **Choeroniscus** Thomas—Long-tailed Bats

1928. *Choeroniscus* Thomas, Ann. Mag. Nat. Hist., ser. 10, 1:122, January. Type, *Choeronycteris minor* Peters.

Resembling *Choeronycteris* in most features; rostrum much elongated and slender, less than half length of skull; mandible long, slender; nose-leaf small, triangular; cusps of lower premolars subequal; upper incisors minute. Dentition, i. $\frac{2}{0}$, c. $\frac{1}{1}$, p. $\frac{2}{3}$, m. $\frac{3}{3}$.

Choeroniscus godmani (Thomas)
Godman's Bat

1903. *Choeronycteris Godmani* Thomas, Ann. Mag. Nat. Hist., ser. 7, 11:288, March, type from Guatemala.
1928. [*Choeroniscus*] *godmani* Thomas, Ann. Mag. Nat. Hist., ser. 10, 1:122, January.

Measurements for 20 specimens from Central America: head and body, 53–55; forearm, 31.6–

34.4; greatest length of skull for 10 males, 19.2–20.0; for 10 females, 20.5–21.2; interorbital breadth, 3.1–3.9; length of maxillary tooth-row, 6.6–7.5. Dorsum uniform dark brown in females, grayish brown in males; venter barely paler.

MARGINAL RECORDS.—Sinaloa: San Ignacio (Jones, 1964e:510). Nayarit: 8 mi. E San Blas (Gardner, 1962:102, 103). Jalisco: 2 km. NW Emiliano Zapata, 20 m., (Watkins, *et al.*, 1972:14). Oaxaca: Río Sarabia (Goodwin, 1969:71). Veracruz: 2 km. SE Sontecomapan [= Zontecomapán] (Carter, *et al.*, 1966:492). Guatemala: 25 km. SSW Puerto Barrios, 300 ft. (*ibid.*). Honduras: Lancetilla (Valdez and LaVal, 1971:249); 5 km. NE Jesús de Otoro (LaVal, 1969:820); *Cantoral*; La Flor Archaga; 6 km. E Danlí (LaVal, 1969:820). Nicaragua: Santa Rosa, 300 m. (Jones, *et al.*, 1971:9). Costa Rica: San José; 4 km. NW Cañas (Gardner, *et al.*, 1970:719), thence northwestward along coast to Chiapas: Pijijiapan. Oaxaca: Kilometer 183, 36½ km. N San Gabriel Mixtepec (Schaldach, 1966:291, but see

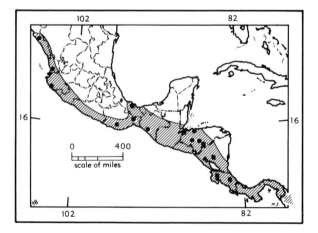

Map 98. *Choeroniscus godmani.*

Marginal Records for *Hylonycteris underwoodi minor*), thence northwestward along coast to point of beginning. Also occurs in South America (Handley, 1976:22).

Genus **Hylonycteris** Thomas

1903. *Hylonycteris* Thomas, Ann. Mag. Nat. Hist., ser. 7, 11:286, March. Type, *Hylonycteris underwoodi* Thomas.

Resembling *Choeroniscus* in all respects except that pterygoids are normal, convex medially rather than concave, not inflated, and not in contact with bullae. The genus in monotypic.

Hylonycteris underwoodi Thomas
Underwood's Long-tongued Bat

External measurements, taken on specimens in alcohol are: head and body, six males 49–66, seven females 51–72; tail, 6–9, 7–11 (including type of *H. u. minor*). Other selected measurements: forearm, 17 males 31.4–33.7, seven females 31.5–35.9; greatest length of skull, 20.2–22.2, 20.6–23.0 (including type of *H. underwoodi*); least interorbital breadth, 3.3–4.3, 3.4–4.3 (including type of *H. underwoodi*); breadth of braincase, 8.0–8.9, 8.1–8.9; mastoid breadth, 8.0–8.7, 8.0–8.8; length of maxillary tooth-row, 6.8–7.8, 7.1–8.4.

Hylonycteris underwoodi minor Phillips and Jones

1971. *Hylonycteris underwoodi minor* Phillips and Jones, Jour. Mamm., 52:77, February 26, type from 10 mi. SE Tuxpán, 4200 ft., Jalisco.

MARGINAL RECORDS (Phillips and Jones, 1971:79, unless otherwise noted).—Jalisco: 2 mi. N

Milpillas, 3000 ft.; 14 mi. WSW Ameca, 5000 ft.; type locality. Oaxaca: 8 mi. SSW Juchatengo, 6300 ft.; Yautepec (Goodwin, 1969:72); Chiltepec (*ibid.*); *San José Chacalapa, ca. 1500 ft.* [The Oaxacan locality *36½ km. N San Gabriel Mixtepec* (Villa-R., 1967:264) listed as *36 km. N San Gabriel Mixtepec* as a locality for *H. underwoodi* by Villa-R., 1967:265 and Phillips and Jones, 1971:79, is a Schaldach (1966:287) collecting locality. Schaldach (1966:291) there obtained specimens of bats he identified as *Choeroniscus godmani*, which approximate descriptions of specimens identified as *Hylonycteris underwoodi* by Villa-R. (1967:266). This locality would be slightly more marginal than 8 mi. SSW Juchatengo if these bats are in fact *H. underwoodi*. The locality is listed above in the **Marginal Records** for *Choeroniscus godmani*.]

Hylonycteris underwoodi underwoodi Thomas

1903. *Hylonycteris underwoodi* Thomas, Ann. Mag. Nat. Hist., ser. 7, 11:286, March, type from Rancho Redondo, San José, Costa Rica.

MARGINAL RECORDS (Phillips and Jones, 1971:79, unless otherwise noted).—Veracruz: 15 km. ENE Tlacotepec, 1500 ft. Oaxaca: Tuxtepec (Goodwin, 1969:72). Tabasco: Cueva de Don Luis, 3 km. E Teapa, 50 m. (Villa-R., 1967:265). Guatemala: 25 km. SSW Puerto Barrios, 75 m. Costa Rica (Gardner, *et al.*, 1970:719): Cariari; Río Chitaría, above highway, *ca.* 2500 ft. Panamá: Santa Clara (Handley, 1966:764). Costa Rica (Gardner, *et al.*, 1970:719): 2 km. S San Vito, Finca Las Cruces, 1300 m.; Fila la Maquina, *ca.* 3 km. E Canaán, *ca.* 6700 ft. Oaxaca: Ixtlán (Goodwin, 1969:72); *600 m. S Vista Hermosa; Vista Hermosa.* Veracruz: Metlác (Hall and Kelson, 1959:121).

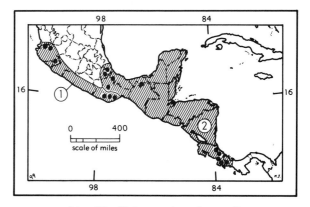

Map 99. *Hylonycteris underwoodi.*

1. *H. u. minor*   2. *H. u. underwoodi*

Genus **Leptonycteris** Lydekker—Long-nosed Bat

1860. *Ischnoglossa* Saussure, Revue et Mag. Zool., Paris, ser. 2, 12:491, November. Type, *Ischnoglossa nivalis* Saussure. Not *Ischnoglossa* Kraatz, 1856, a beetle.

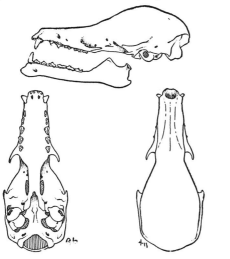

Fig. 96. *Hylonycteris underwoodi minor*, 15 km. ENE Tlacotepec, Veracruz, No. 23709 K.U., ♂, X 2.

1891. *Leptonycteris* Lydekker, *in* Flower and Lydekker, An introduction to . . . mammals living and extinct, p. 674. Type, *Ischnoglossa nivalis* Saussure.

Skull of the usual glossophagine type; zygomata slender but complete; tail absent; interfemoral membrane much reduced; calcar small; lower incisors normally present but often lost presumably by traumatic means; molars elongated and W-shaped pattern nearly lost. Dentition, i. $\frac{2}{2}$, c. $\frac{1}{1}$, p. $\frac{2}{3}$, m. $\frac{2}{2}$.

The genera *Leptonycteris* and *Lichonycteris* are unique among bats in lacking the 3rd molars. Normally these two genera are easily separated on the basis of the presence of lower incisors in *Leptonycteris* and their absence in *Lichonycteris*. *Leptonycteris*, however, often lacks lower incisors in which case identification depends primarily on the upper incisors, which are evenly and widely spaced between canines in *Lichonycteris* but form an almost continuous line (or are separated by a median space into two pairs) in *Leptonycteris*.

KEY TO SPECIES OF LEPTONYCTERIS

1. Fur long, lax; uropatagium moderately hairy and with conspicuous fringe of hairs 3–4 mm. long; length of 3rd finger more than 105. . . . .*L. nivalis*, p. 132
1′. Fur short and dense; uropatagium sparsely haired and with slight fringe; length of 3rd finger less than 105. *L. yerbabuenae*, p. 133

Leptonycteris nivalis (Saussure)
Big Long-nosed Bat

1860. *M* [= *Ischnoglossa*]. *nivalis* Saussure, Revue et Mag. Zool., Paris, ser. 2, 12:492, November, type from near snow line of Mt. Orizaba, Veracruz.
1900. *Leptonycteris nivalis*, Miller, Proc. Biol. Soc. Washington, 13:126, April 6.
1957. *Leptonycteris nivalis longala* Stains, Univ. Kansas Publ., Mus. Nat. Hist., 9:355, January 21, type from 12 mi. S and 2 mi. E Arteaga, 7500 ft., Coahuila. (Regarded as indistinguishable from *L. nivalis* by Davis and Carter, Proc. Biol. Soc. Washington, 75:194, August 28, 1962.)

Length of head and body, 76–78; condylobasal length, 26.2–28.3; interorbital breadth, 4.3–5.4; zygomatic breadth, 10.5–12.0; length of maxillary tooth-row, 8.2–9.6. Upper parts medium brown posteriorly, paler over shoulders; underparts paler than posterior part of back, approx. same as on shoulder region. Length of three phalanges of 3rd finger more or less than length of 3rd metacarpal.

MARGINAL RECORDS (Baker and Cockrum, 1966:330, 331, unless otherwise noted).—Texas: Pinto

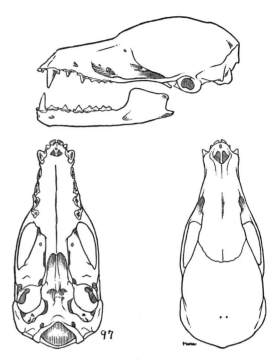

Fig. 97. *Leptonycteris nivalis*, 12 mi. S, 2 mi. E Arteago, 7500 ft., Coahuila, No. 33077 K.U., ♂, X 2.

Canyon, *ca.* 14 mi. E Ruidosa (Mollhagen, 1973:428); Emory Peak. Coahuila: 12 mi. S, 2 mi. E Arteaga, 7500 ft. Tamaulipas: 6½ mi. N, 13 mi. W Jiménez. Veracruz: *Veracruz* (Hall and Dalquest, 1963:229); near Boca del Río (Novick, 1963:50, as *L. nivalis* only). Guatemala (Jones, 1966: 455, and identified by J. E. Hill August 22, 1974, *in Litt.* to E. R. Hall, on basis of key above, as

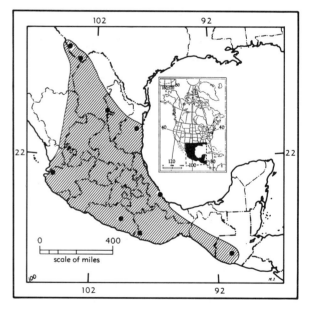

Map 100. *Leptonycteris nivalis*.

L. nivalis and not *L. yerbabuenae*): *Dueñas*; Ciudad Vieja. Oaxaca: 6 km. SSW Cacahuatepec, 300 m. (Webb and Baker, 1971:143). Guerrero: 12 mi. S Mexcala (Hoffmeister, 1957: 459). Jalisco: San Sebastián.

Leptonycteris yerbabuenae Martínez and Villa-R.
Little Long-nosed Bat

Treated by Ramírez-P. and Alvarez, Southwestern Nat., 16: 249–259, February 18, 1972.

According to Watkins, *et al.*, Special Publ. Mus. Texas Tech Univ., 1:16, December 8, 1972, Ramírez-P. and Alvarez should not have designated a lectotype, because the Code, they claim, does not provide for designation of a lectotype when the holotype has been lost (see Articles 73, 74, and 75 of the International Code of Zoological Nomenclature, November 6, 1961). But Article 17(2) does permit the use of a name proposed for one of a pair of species when the other species, *L. nivalis* (Saussure) in this instance, of that pair can clearly be identified as previously named. Therefore, the earlier specific name *yerbabuenae* 1940 instead of the later specific name *sanborni* 1957 is employed here.

1940. *Leptonycteris nivalis yerbabuenae* Martínez and Villa-R., Anal. Inst. Biol., Univ. Nac. Autó. México, 11:313, August, type from Yerbabuena, Guerrero.

1967. *Leptonycteris yerbabuenae*, Villa-R., Los murciélagos de México, Anal. Inst. Biol., Univ. Nac. Autó. México, p. 252, February 6.

1957. *Leptonycteris nivalis sanborni* Hoffmeister, Jour. Mamm., 38:456, November 20, type from mouth Miller Canyon, Huachuca Mts., 10 mi. SSE Fort Huachuca, Cochise Co., Arizona.

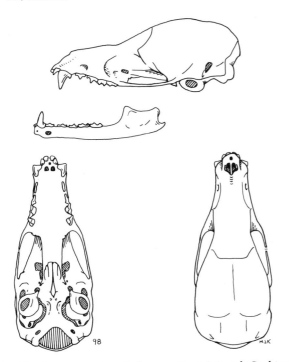

Fig. 98. *Leptonycteris yerbabuenae*, 2 mi. S Portal, Cochise Co., Arizona, No. 102085 K.U., ♂, X 2.

Length of head and body, 69–84; length of forearm, 51.3–54.3; condylobasal length, 25.1–26.5; zygomatic breadth, 10.1–11.3; interorbital width, 4.4–4.9; maxillary tooth-row, 8.5–9.2. Upper parts usually reddish brown; underparts heavily washed with brown or cinnamon. Length of three phalanges of 3rd finger not greater than length of 3rd metacarpal.

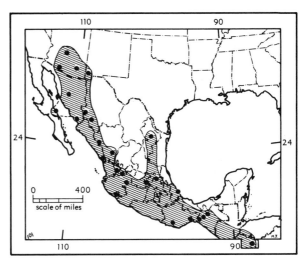

Map 101. *Leptonycteris yerbabuenae*.

MARGINAL RECORDS (Davis and Carter, 1962:197, unless otherwise noted).—Arizona: Glendale (Constantine, 1966:126); *Phoenix* (*ibid.*); Redfield Canyon, sec. 35, T. 11 S, R. 19 E (Cockrum, 1961:34); *Buckelew Cave, W end Blue Mtn., 17 mi. S San Simon* (Cockrum and Ordway, 1959:9, as *L. n. nivalis*); *1 mi. N Paradise* (*ibid.*). New Mexico: 17 mi. NNE Rodeo (Baker and Cockrum, 1966:331). Chihuahua: Carimechi, Río Mayo; Batopilas (Anderson, 1972:239). Durango: Santa Ana (Jones, 1964a:751); Agua Caliente (Crossin, *et al.*, 1973:197). Jalisco: 2 mi. E Bolaños, 3550 ft. (Watkins, *et al.*, 1972:15). Zacatecas: Santa Rosa (Matson and Patten, 1975:4). Jalisco: 500 m. N, 3 km. W Jamay, 1650 m. (Ramírez-P. and Alvarez, 1972:258). Querétaro (Schmidly and Martin, 1973: 91): 1 mi. NE Peña Blanca; *7 mi. ENE Pinal de Amoles.* Nuevo León: 8 km. NNW Los Ramones (*ibid.*). Hidalgo: Jacala (Ramírez-P. and Alvarez, 1972:258); *6 km. NW Tasquillo, 5000 ft.* Veracruz: 3 km. W Boco del Río. Chiapas: 4 km. N Tuxtla Gutiérrez (Ramírez-P. and Alvarez, 1972:258). El Salvador: 3 1/10 mi. W San Miguel (Jones and Bleier, 1974:144). Chiapas: 7 mi. WSW Ocozocoautla, 2500 ft. (Davis, *et al.*, 1967:380); 42 km. W Cintalapa (Baker, 1967:427). Oaxaca: 6 mi. NW Mixteguilla; *ca.* 11 mi. E Juquila, 6100 ft. (Baker and Womochel, 1966:306). Colima: Pueblo La Jola (Baker and Cockrum, 1966:331). Sinaloa: Matatán (Jones, *et al.*, 1972:10). Sonora: 1 km. SW La Aduana, 1600 ft. (Loomis and Davis, 1965:497). Baja California: 2 mi. W Santa Rosalía, Cerro del Elote

(Baker and Cockrum, 1966:331). Sonora: 25 mi. N Hermosillo, 1500 ft. Arizona: 2 mi. W Tonoga Well (Constantine, 1961:404).

Genus **Lichonycteris** Thomas

1895. *Lichonycteris* Thomas, Ann. Mag. Nat. Hist., ser. 6, 16:55, July. Type, *Lichonycteris obscura* Thomas.

Closely resembling *Leptonycteris;* differing in having more or less evenly spaced upper incisors; no lower incisors; "upper incisors . . . [with] crowns narrow though scarcely trenchant, longer than high, that of inner tooth distinctly emarginate on cutting edge, so that it appears bilobed when viewed from in front, that of outer tooth with sharp, backward-directed cusp near inner edge, and a flattish outer projection." (Miller, 1907:143.)

Lichonycteris obscura Thomas
Brown Long-nosed Bat

1895. *Lichonycteris obscura* Thomas, Ann. Mag. Nat. Hist., ser. 6, 16:56, July, type from Managua, Managua, Nicaragua.

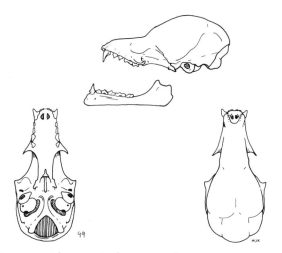

Fig. 99. *Lichonycteris obscura*, S side Río Mico, El Recreo, 25 m., Zelaya, Nicaragua, No. 106026 K.U., ♂, X 2.

Measurements: head and body, 46–55; forearm, 30.0–33.9; greatest length of skull, 17.8–19.9; interorbital breadth, 3.3–4.4; maxillary tooth-row, 5.8–6.4. Upper parts uniform dark brown; underparts slightly darker.

MARGINAL RECORDS.—Guatemala: 25 km. SSW Puerto Barrios, 300 ft. (Carter, *et al.*, 1966:493). Nicaragua: 6 mi. W Rama, 50 ft. (Davis, *et al.*, 1964:380). Panamá (Handley, 1966:764): Almirante; Armila, thence southeastward into South America and

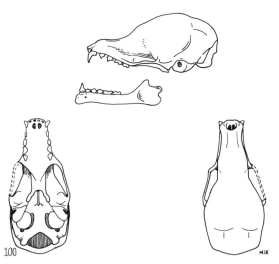

Fig. 100. *Lichonycteris obscura*, S side Río Mico, El Recreo, 25 m., Zelaya, Nicaragua, No. 106025 K.U., ♂, X 2.

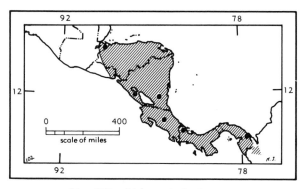

Map 102. *Lichonycteris obscura*.

northwestward along coast to Costa Rica: San José. Nicaragua: type locality.

SUBFAMILY **CAROLLIINAE**

Trochiter impinging on scapula; nose-leaf present; upper molars so modified that the W-shaped pattern is much altered or almost obliterated; upper molars lacking commissures and styles; lower molars with cusps strictly lateral; crowns of upper and lower molars trenchant.

One genus, *Carollia*, occurs in the area of our study.

Genus **Carollia** Gray—Short-tailed Bats

Pine's (Tech. Monog., Texas A&M Univ., Texas Agric. Exp. Station, 8:1–125, 3 figs., 8 tables, 7 maps, September 26, 1972) arrangement of the North American species is followed here. He did not study or identify subspecies.

1838. *Carollia* Gray, Mag. Zool. Bot., 2:488, February. Type, *Carollia Braziliensis* Gray [= *Vespertilio perspicillatus*

Linnaeus]. *Carollia* is not a homonym of *Carolia* Cantraine, 1838, and is available.

1855. *Hemiderma* Gervais, Mammifères, *in* [Castelnau] Expéd. dans les parties centrales de l'Amér. du Sud . . . , p. 43. Type *Phyllostoma brevicaudum* Schinz.

1866. *Rhinops* Gray, Proc. Zool. Soc. London, p. 115, May. Type, *Rhinops minor* Gray [= *Phyllostoma brevicaudum* Schinz].

Skull relatively robust; rostrum approx. two-thirds as long as braincase; braincase rising, but not abruptly so, well above frontal region; zygomata incomplete; auditory bullae small, covering less than half cochlear surface; ears small, separate; tail about half as long as femur, extending to middle of interfemoral membrane; bony palate prolonged posteriorly considerably beyond last molar, forming posterior projection of nasal cavity; lingual cusps of molars reduced or absent; lower molars distinctly different in form from lower premolars. Dentition, i. $\frac{2}{2}$, c. $\frac{1}{1}$, p. $\frac{2}{2}$, m. $\frac{3}{3}$.

KEY TO SPECIES OF CAROLLIA

1. When viewed from above, i2 obscured by cingulum of canine, and/or upper tooth-row straight; lower jaw V-shaped.
   . . . . *C. perspicillata*, p. 137
1′. When viewed from above, i2 easily visible; upper tooth-row bowed lingually or with notch or "step" in labial outline; lower jaw tending to be U-shaped.
   2. Labial outline of upper tooth-row with notch or "step"; p2 twice height of m1; occlusal surface of m1 with straight profile. . . . . *C. castanea*, p. 135
   2′. Labial outline of upper tooth-row curved, without notch or "step"; p2 less than twice height of m1; occlusal surface of m1 with cusp or cusps, resulting in jagged profile.
      3. Pelage long, thick, and fine; forearm hairy; hair on nape of neck with broad dark basal band contrasting strongly with and sharply demarcated from broad whitish band distal to it. *C. brevicauda*, p. 136
      3′. Pelage short, sparse, and coarse; forearm naked or sparsely furred; hair on nape of neck with narrow, generally indistinct basal band, not strongly demarcated from paler band distal to it. . . . *C. subrufa*, p. 136

Carollia castanea H. Allen
Allen's Short-tailed Bat

1890. *Carollia castanea* H. Allen, Proc. Amer. Philos. Soc., 28:19, February 25, type from Angostura, Costa Rica.

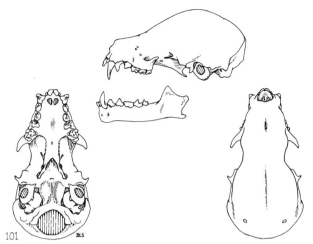

Fig. 101. *Carollia castanea*, 1 km. NE Esquipulas, 420 m., Matagalpa, Nicaragua, No. 114869 K.U., ♂, X 2.

Overall length, 59–69; hind foot, 11–12; ear, 14–18; forearm, 35.1–38.3; greatest length of skull, 18.8–21.3; postorbital width, 4.9–5.7; mandibular tooth-row, 6.8–7.9. Upper parts range from dull, dark gray-brown to rich chestnut or light tan; hair indistinctly three-banded; dark band at base, paler band in middle and dark band at tip; underparts with middle pale band missing; forearms naked; skull with more globular braincase, lower sagittal crest, and proportionately smaller rostrum than in other species; P2 lingual to labial border of M1, resulting in "notch" in tooth-row; viewed from above, i2 not obscured by cingulum of c1; crown of i1 ovoid, not triangular; p2 longer than p1, and twice height of m1; m1 with straight profile, lacking well-developed cusps.

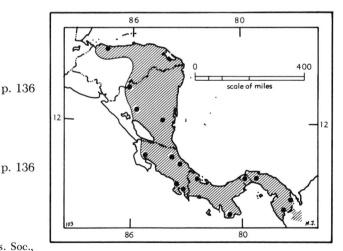

Map 103. *Carollia castanea*.

MARGINAL RECORDS (Pine, 1972:21, unless otherwise noted).—Honduras: Lancetilla. Nicaragua: Cara de Mono, 50 m. (Jones, *et al.*, 1971:10); *Cacao, 22 km. W Muelle de los Bueyes, 400 ft.* Costa Rica: Río Sarapiquí, Puerto Viejo, 300 ft.; 2½ mi. E Turrialba, 520 m. Panamá (Handley, 1966:764): Almirante; Barro Colorado Island; Tacarcuna Village, 3200 ft., thence into South America, and northwestward to Panamá (*ibid.*): Cana, 2000 ft.; Cerro Azul, 2000 ft.; Cerro Hoya, 3000 ft. Costa Rica: 9 mi. ENE Puerto Golfito, 100 ft.; 4 mi. NE Palmar, 300 ft.; La Pacífica (Fleming, *et al.*, 1972:569). Nicaragua (Jones, *et al.*, 1971:10): 1 km. NE Esquipulas, 420 m.; 7 km. N, 4 km. E Jalapa, 660 m. Honduras: *Río Coco, 78 mi. ENE Danlí, 900 ft.*

Carollia subrufa (Hahn)
Hahn's Short-tailed Bat

1905. *Hemiderma subrufum* Hahn, Proc. Biol. Soc. Washington, 18:247, December 9, type from Santa Efigenia, Oaxaca.
1924. *Carollia subrufa*, G. S. Miller, Bull. U.S. Nat. Mus., 128:54, April 29.

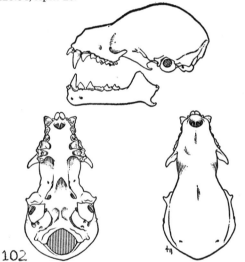

Fig. 102. *Carollia subrufa*, Sololá, Moca, Guatemala, No. 41866 F.M.N.H., ♂, X 2.

Overall length, 68–73; hind foot, 12–15; ear, 17–21; forearm, 37.6–40.6; greatest length of skull, 20.4–22.8; postorbital width, 4.8–5.6; mandibular tooth-row, 7.6–8.6. Upper parts dull, dark gray-brown to reddish-tan to pale gray or yellow; hairs indistinctly four-banded with narrow, indistinct dark basal band, followed by a whitish band, then darker band, and narrow paler tip; hairs of underparts lacking the whitish band; pelage coarser and sparser than in *brevicauda;* forearm usually naked; as compared to *brevicauda,* rostrum less robust and shorter, braincase more globular; tubular rear extension of palate shorter than in *brevicauda;* upper tooth-row forms smooth, concave curve from C1 to M3, without distinct notch; occlusal surface of i2 visible from above; occlusal surface of i1 more nearly triangular than in *castanea* but less nearly triangular than in *brevicauda;* upper edge of m1 jagged.

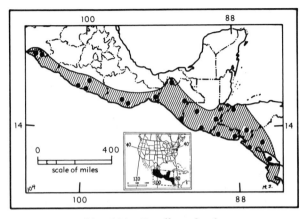

Map 104.   *Carollia subrufa.*

MARGINAL RECORDS (Pine, 1972:27, 28, unless otherwise noted).—Colima: Hda. Magdalena; Río Naranjo, Puente San Vicente, S Cerro de Ortega. Jalisco: *14¹/₂ mi. S Pihuamo, 1100 ft.* (Watkins, *et al.*, 1972:16). Guerrero: 3 mi. N Colotlipa. Oaxaca: San Gerónimo; type locality. Chiapas: 13 km. ENE Pichucalco, 60 m. (Baker, *et al.*, 1973:78, 85); 10 mi. S Zapaluta, 3000 ft. Guatemala: 1 km. WNW Sacapulas, 1200 m.; Jocotán, near Chiquimula, 1350 ft. Honduras: 2 mi. W San Pedro Sula; *4 mi. W La Lima;* 3 mi. S Sabana Grande, 1500 ft. Nicaragua: 2 mi. SE Dario, 1500 ft.; 6 km. E Moyogalpa, NW end Isla Ometepe, 400 m.; San Antonio, 15 m. El Salvador: Río San Miguel, 13° 25' N, 225 ft.; 20 mi. W La Libertad. Guatemala: 8 km. NW Puerto de San José, 50 ft. Chiapas: 4 km. NW Tapachula. Oaxaca: 1 mi. WNW Chacalapa. Guerrero: 2 mi. NW Acapulco, 50 ft.; El Papayo, 25 ft.

Carollia brevicauda (Schinz)
Silky Short-tailed Bat

1821. *Phyllost*[*oma*]. *bernicaudum* Schinz, Das Thierreich . . . , I. Säugethiere und Vögel, p. 164, type from "*Fazenda von Coroaba* in den grossen Wäldern an den Ufern des kleinen Flusses *Jucú* unweit des Río do Espirito Santo," Brazil (from Pine, Tech. Monog., Texas A&M Univ., Texas Agric. Exp. Station, 8:29, September 26, 1972). Pine concluded that "*bernicaudum*" was a misprint for "*brevicaudum*".
1866. *C*[*arollia*]. *brevicauda*, Peters, Monatsb. preuss. Akad. Wiss., Berlin, p. 519.
1838. *Phyllostoma Grayi* [*sic*] Waterhouse, Mammalia, *in* Darwin, The zoology of the voyage of H.M.S. Beagle . . . , p. 3, type from Pernambuco, Brazil.
1840. *Ph*[*yllostoma*]. *bicolor* Wagner, Supplementband, 1st Abtheilung: Die Affen und Flederthiere, *in* Schreber, Die Säugethiere . . . , p. 400, type from Brazil.
1843. *Phyllostoma lanceolatum* Gray, List of the . . . Mam-

malia in the . . . British Museum, p. 20, type from South America. Treated as *nomen nudum* by Pine (1972:32).

1866. *R[hinops]. minor* Gray, Proc. Zool. Soc. London, p. 155, type from Bahia, Brazil.

Overall length, 59–77; hind foot, 10–16; ear, 17–22; forearm, 37.4–42.2; greatest length of skull, 21.0–24.1; postorbital width, 5.0–5.8; mandibular tooth-row, 7.7–8.8. Upper parts dark gray to dull gray-brown to dull, light chestnut brown; hairs four-banded with broad, dark basal band, contrasting sharply with broad whitish band, followed by darker subterminal band and narrow buffy or whitish tip; underparts with bands indistinct; fur denser, longer, finer and silkier than in *subrufa* or *perspicillata*; forearm and toes hairier than in most *subrufa* and *perspicillata*; rostrum generally more elongate and robust than in *subrufa*, with braincase less globular; rear extension of palate longer than in *subrufa*, but shorter than in *perspicillata*; upper tooth-row more curved and more divergent posteriorly than in *perspicillata*, and upper teeth less crowded; lower jaw U-shaped; occlusal surface of i2 visible from above; occlusal surface of i1 more nearly triangular than in *castanea* or *subrufa*; upper edge of m1 jagged.

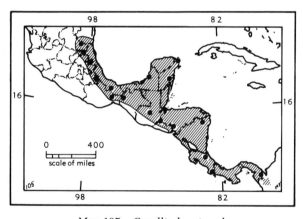

Map 105. *Carollia brevicauda*.

MARGINAL RECORDS (Pine, 1972:41–44, unless otherwise noted).—San Luis Potosí: El Salto, thence along Gulf of Mexico to Campeche: 5 km. S Champotón. Quintana Roo: 4 km. WSW Puerto Juárez, 5 m.; 6 km. N, 16 km. E Chetumal. Guatemala: Escobas, 2000 ft. Nicaragua: Bonanza, 850 ft. Costa Rica: 5½ mi. S, 1 mi. E Puerto Viejo, 400 ft. Panamá: Tacarcuna Village Camp, 3200 ft., thence into South America, and northwestward to Costa Rica: 9 mi. ENE Puerto Golfito, 100 ft. Honduras: 6 km. E Amatillo, 60 m.; 1 km. W Nueva Ocotepeque, 840 m. Guatemala: San Cristóbal Verapaz, 1380 m. Chiapas: 12 mi. W Mal Paso, 400 ft. Oaxaca: Río Grande, 6 mi. S Matías Romero. Veracruz: Achotal; 4 km. WNW Fortín, 3200 ft. Puebla: 2 mi. W

Villa Avila Camacho, 250 m. (LaVal, 1972:450, as *Carollia "subrufa"*). San Luis Potosí: 10 km. NNW Xilitla.

Carollia perspicillata (Linnaeus)
Seba's Short-tailed Bat

Overall length, 67–95; hind foot, 12–17; ear, 17–22; forearm, 37.5–44.8; greatest length of skull, 21.4–25.2; postorbital width, 5.0–6.1; mandibular tooth-row, 8.1–9.7. Upper parts nearly black, to various browns, grays and bright orange; hairs three-banded with dark basal band, lighter middle band, and darker terminal band; bands not contrasting sharply; underparts with bands very indistinct or lacking; fur shorter, coarser, and sparser than in *brevicauda*; forearms and toes less hairy than in *brevicauda*; rostrum more elongate and robust than in other species of *Carollia*; rear extension of palate averages longer than in any other species of *Carollia*; upper tooth-row relatively straight, and upper teeth more crowded; lower tooth-row straight, and lower jaw V-shaped; viewed from above, i2 obscured by cingulum of c1; upper edge of m1 jagged; lower tooth-row longer and m3 larger than in *brevicauda*.

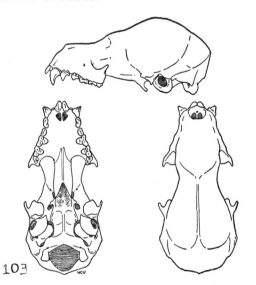

Fig. 103. *Carollia perspicillata azteca*, Mirador, 3500 ft., Veracruz, No. 23710 K.U., ♀, X 2.

Carollia perspicillata azteca Saussure

1860. *Carollia azteca* Saussure, Revue et Mag. Zool., Paris, ser. 2, 12:480, November, type locality restricted to Peréz, Veracruz, by Dalquest (Louisiana State Univ. Studies, Biol. Ser., 1:29, December 28, 1953).

1924. *Carollia perspicillata azteca*, Miller, Bull. U.S. Nat. Mus., 128:54, April 29.

MARGINAL RECORDS (Pine, 1972:70–72, unless otherwise noted).—Quintana Roo: 4 km. WSW Puerto Juárez, 5 m. Belize: Augustine, 1550 ft. Honduras: Lancetilla. Nicaragua: Bonanza, 850 ft. Costa Rica: 5½ mi. S, 1 mi. E Puerto Viejo, 400 ft. Panamá: Tacarcuna Village Camp, 3200 ft., thence into South America, and northwestward to Panamá: 2 mi. W Soná. Costa Rica: Los Huecos. Nicaragua: 4 km. S San Antonio, 15 m. El Salvador: Potosí Mine, Comacarán, Guatemala: Río San Símon, *ca.* 6 km. NE Raxrujá. Chiapas: Cerro Hueco Cave, 2 mi. SE Tuxtla Gutiérrez, *ca.* 2600 ft. Oaxaca: 8 mi. N Matías Romero. Puebla: Villa Avila Camacho, 250 m. (LaVal, 1972: 450). Veracruz: 3 km. E San Andrés Tuxtla, 1000 ft.; 1 mi. E Jaltipan. Campeche: 7½ km. W Escárcega, 65 m. Yucatán: 13 km. W Peto. Also recorded from *Jamaica* by G. M. Allen (1911:232, subspecific identity uncertain), but according to R. E. Goodwin (1970:578) it is unlikely that *C. perspicillata* is, or ever has been, in Jamaica. Also occurs in South America.

Carollia perspicillata perspicillata (Linnaeus)

1758. [*Vespertilio*] *perspicillatus* Linnaeus, Syst. nat., ed. 10, 1:31, type from Surinam (according to Thomas, Proc. Zool. Soc. London, p. 130, March 22, 1911).
1924. *Carollia perspicillata perspicillata*, Miller, Bull. U.S. Nat. Mus., 128:53, April 29.
1818. *Glossophaga amplexicauda* É. Geoffroy St.-Hilaire, Mém. Mus. Hist. Nat., Paris, 4:418, type from vicinity of Rio de Janeiro, Brazil.
1821. *Phyllost[oma]. brachyotos* Schinz, Das Thierreich . . . I. Säugethiere und Vögel, p. 164, type from "In den dichten Waldungen am Mucuri in Brasilien."
1838. *Carollia Braziliensis* Gray, Mag. Zool. Bot., 2:488. Probably a substitute name for *Phyllostoma brachyotos* Schinz, 1821.
1843. *Phyllostoma calcaratum* Wagner, Wiegmann's Arch. für Naturgesch., Jahrg 1:366, type from Brazil.
1844. *Carollia verrucata* Gray, *in* The zoology of the voyage of H.M.S. Sulphur . . . , 1(1:Mamm.):20. Type locality unknown.

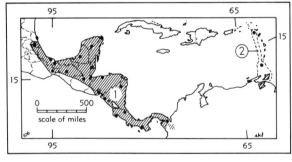

Map 106. *Carollia perspicillata.*

1. *C. p. azteca*   2. *C. p. perspicillata*

MARGINAL RECORDS (recorded as *Hemiderma perspicillatum,* and not identified to subspecies).— Lesser Antilles: Redonda (Hahn, 1907:110), but

Koopman (1968:3), Jones and Phillips (1970:132), and Pine (1972:69) regard the occurrence of *Carollia perspicillata* on Redonda as doubtful; Grenada (G. M. Allen, 1911:232). Also occurs on Trinidad and South American mainland.

Subfamily STURNIRINAE

Trochiter impinging on scapula; nose-leaf present; upper molars so modified that W-shaped pattern is at least partly obliterated, cusps and commissures much reduced; crowns of molars having distinct longitudinal groove, cusps being laterally placed; crowns of lower molars having distinct cusps on both margins of groove.

Genus **Sturnira** Gray

1842. *Sturnira* Gray, Ann. Mag. Nat. Hist., 10:257, December. Type, *Sturnira spectrum* Gray [= *Phyllostoma lilium* É. Geoffroy St.-Hilaire].
1849. *Nyctiplanus* Gray, Proc. Zool. Soc. London, p. 58, December 20. Type, *Nyctiplanus rotundatus* Gray [= *Phyllostoma lilium* É. Geoffroy St.-Hilaire].
1938. *Sturnirops* Goodwin, Amer. Mus. Novit., 976:1, May 4, type *Sturnirops mordax* from El Sauce Peralta, Cartago, Costa Rica, regarded as generically inseparable by Cabrera, Rev. Mus. Argentino de Cienc. Nat., 4:78, March 27, 1958.

Braincase moderately high with moderately developed sagittal crest; rostrum more than half as long as braincase; greatest interorbital breadth slightly more than depth in same region, and about equal to distance from incipient postorbital process to canine (after Miller, 1907:149); calcar small, tail absent. Dentition, i. $\frac{2}{2}$, c. $\frac{1}{1}$, p. $\frac{2}{2}$, m. $\frac{3}{2-3}$.

KEY TO SPECIES OF STURNIRA

1. Lower incisors trilobate.
   2. Length of forearm more than 45; upper tooth-rows almost parallel. *S. thomasi,* p. 138
   2'. Length of forearm 45 or less; upper tooth-rows evenly curved. *S. lilium,* p. 139
1'. Lower incisors bilobate.
   3. I1 unicuspid; hind feet heavily furred. *S. ludovici,* p. 140
   3'. I1 bicuspid; hind feet sparsely furred. *S. mordax,* p. 140

Sturnira thomasi de la Torre and Schwartz
Sofaian Bat

1966. *Sturnira thomasi* de la Torre and Schwartz, Proc. Biol. Soc. Washington, 79:299, December 1, type from Sofaia, 1200 ft., Guadeloupe, French Leeward Islands, Lesser Antilles. Known only from the island of Guadeloupe.

Length of head and body (male holotype and two females), 80, 80, 81; forearm, 48.1, 46.1, 47.7; greatest length of skull, 26.2, 24.9, 25.1; zygomatic breadth, 12.7, 12.2, 12.5; maxillary tooth-row, 7.7, 6.9, 6.9 (after Genoways and Jones, 1975:925). Rostrum long, tubular; upper tooth-rows almost parallel; i1 trilobate; lower molars having well-developed lingual cusps; m3 absent. Upper parts dark golden brown, hairs having four bands. Band sequence from tip to base: dark yellowish-brown, yellowish-buff, pale grayish-brown, white. Underparts yellowish buff; pelage thick, silky.

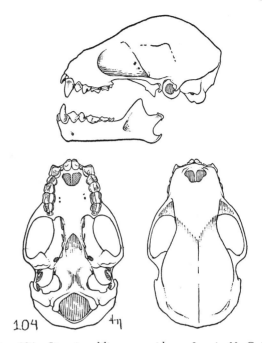

Sturnira lilium
Yellow-shouldered Bat

Length of head and body, 62–65; forearm, 36.6–44.8. Lower incisors trilobate; upper tooth-row evenly curved. Upper parts pinkish buff to pale grayish-brown overlaid with brown; shoulder marked by reddish or straw-colored epaulettes. Underparts like back but lacking dark overlay in some populations.

Sturnira lilium angeli de la Torre

1966. *Sturnira angeli* de la Torre, Proc. Biol. Soc. Washington, 79:271, December 1, type from 6 mi. NE Roseau, 1000 ft., St. Paul Parish, Dominica, Windward Islands, West Indies. Known only from the island of Dominica.
1976. *Sturnira lilium angeli*, Jones and Phillips, Occas. Pap. Mus. Texas Tech Univ., 40:10, April 16.

Sturnira lilium luciae Jones and Phillips

1976. *Sturnira lilium luciae* Jones and Phillips, Occas. Pap. Mus. Texas Tech Univ., 40:11, April 16, type from ½ mi. SE Bogius, Dauphin Parish, St. Lucia, Lesser Antilles, 100 ft. Known only from island of St. Lucia.

Sturnira lilium parvidens Goldman

1917. *Sturnira lilium parvidens* Goldman, Proc. Biol. Soc. Washington, 30:116, May 23, type from Papayo, about 25 mi. NW Acapulco, Guerrero.

MARGINAL RECORDS.—Sonora: E side Río Jaqui, *ca.* 1 mi. S El Novillo (Findley and Jones, 1965:330, as *S. lilium* only). Chihuahua: 1½ mi. SW Tocuina (Anderson, 1960:7). Durango: Santa Ana (Jones, 1964a:751); 6 mi. N Pueblo Nuevo, 3000 ft. (Baker and Greer, 1962:69). Jalisco: 4½ mi. W Villa Guerrero, 5200 ft. (Watkins, *et al.*, 1972:16). Zacatecas: Río Juchipila, 1⅔ km. N Santa Rosa, 1100 m. (Baker, *et al.*, 1967:225). Jalisco: Atotonilco el Alto, 5000 ft. (Watkins, *et al.*, 1972:16). Morelos: Oaxtepec (Baker, 1967:427, as *S. lilium* only). Oaxaca: Chiltepec (Goodwin, 1969:79). Veracruz: Ojo de Agua del Río Atoyac (Baker, 1967:427, as *S. lilium* only); Mirador. Querétaro: Río Galindo (Schmidly and Martin, 1973:91); 7 mi. ENE *Pinal de Amoles (ibid.)*; 17¾ km. NW Jalpan (Spenrath and LaVal, 1970:396). Tamaulipas: Pano Ayuctle, near Gómez Farías; *Rancho del Cielo* (Baker and Gomez,

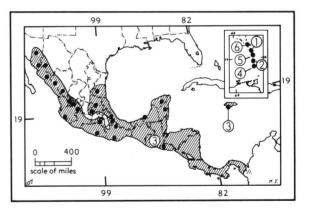

Fig. 104. *Sturnira lilium parvidens*, 2 mi. N Cuidad Guzmán, Jalisco, No. 31866 K.U., ♂, X 2.

Map 107. *Sturnira lilium* and *Sturnira thomasi*.

1. *S. l. angeli*
2. *S. l. luciae*
3. *S. l. parvidens*
4. *S. l. paulsoni*
5. *S. l. zygomaticus*
6. *S. thomasi*

1968:361, as *S. lilium* only). Campeche: 1 km. SW Puerto Real, Isla del Carmen (91631 KU). Yucatán: 3½ km. N Piste (Birney, *et al.*, 1974:6). Quintana Roo: 4 km. NNE Felipe Carrillo Puerto (91629 KU), thence southeastward along Caribbean Coast to South America, thence northwestward along Pacific Coast to El Salvador: San Salvador. Oaxaca: San Gabriel Mixtepec, 800 m. (Schaldach, 1966:291). Guerrero: type locality. Jalisco: 5 mi. NW Barro de Navidad, 200 ft. (Watkins, *et al.*, 1972:17). Sinaloa: 12 mi. NE San Benito, 1000 ft. (Jones, *et al.*, 1972:10). Sonora: *Río Alamos, 8 mi. by road S Alamos* (Baker and Christianson, 1966:310, as *S. lilium* only); 1 km. SW La Aduana, *ca.* 1600 ft. (Loomis and Davis, 1965:497). The species has also been recorded as occurring on Jamaica.

Sturnira lilium paulsoni de la Torre and Schwartz

1966. *Sturnira paulsoni* de la Torre and Schwartz, Proc. Biol Soc. Washington, 79:301, December 1, type from Lowrt, 1000 ft., St. Andrew Parish, Saint Vincent, British Windward Islands, Lesser Antilles. Known only from St. Vincent Island.
1976. *Sturnira lilium paulsoni*, Jones and Phillips, Occas. Pap. Mus. Texas Tech Univ., 40:11, April 16.

Sturnira lilium zygomaticus Jones and Phillips

1976. *Sturnira lilium zygomaticus* Jones and Phillips, Occas. Pap. Mus. Texas Tech Univ., 40:11, April 16, type from Balata, Martinique, Lesser Antilles. Known only from island of Martinique.

Sturnira ludovici
Anthony's Bat

Length of head and body, 65–75. Resembling *S. lilium* externally but averaging larger [forearm 40.3–46.5]; rostrum seems longer; lower incisors deeply bilobate in young individuals but simple or weakly bilobate and often with a minute lobe in fully adult specimens; 2nd upper molar turned inward and not in line with 1st molar (after Hershkovitz, 1949:442).

Sturnira ludovici ludovici Anthony

1924. *Sturnira ludovici* Anthony, Amer. Mus. Novit., 139:8, October 20, type from near Gualea, about 4000 ft., northwestern Ecuador.
1927. *Sturnira lilium bogotensis* Shamel, Proc. Biol. Soc. Washington, 40:129, September 26, type from Bogotá, Colombia. Regarded as a synonym by Hershkovitz, Proc. U.S. Nat. Mus., 99:441, May 10, 1949.
1940. *Sturnira hondurensis* Goodwin, Amer. Mus. Novit.,

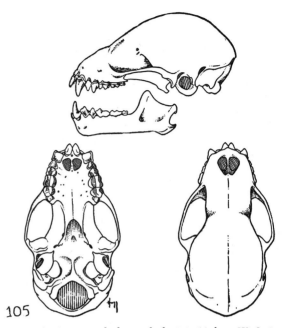

Fig. 105. *Sturnira ludovici ludovici*, 11 km. W Quiroga, Michoacán, No. 95704 U.M.M.Z., ♀, X 2.

1075:1, June 27, type from La Cruz Grande, near San José, about 3000 ft., La Paz, Honduras. Regarded as a synonym by Hershkovitz, Proc. U.S. Nat. Mus., 99:441, May 10, 1949.

MARGINAL RECORDS.—Tamaulipas: 3 mi. W El Carrizo, 1500 ft. (Musser, 1964:7). San Luis Potosí: 3 km. W Xilitla (Dalquest, 1953:32). Veracruz: Mirador (de la Torre, 1952:1), thence southward in Central America into South America, and north to Oaxaca: Río Molino, 35 km. SW San Miguel Suchixtepec (Schaldach, 1966:291); *Río Guajolote (ibid.)*; *ca.* 11 mi. E Juquila, 6100 ft. (Baker and Womochel, 1966:306). Colima: Pueblo Juárez (Villa-R., 1967:275). Michoacán: 11 km. W Quiroga (de la Torre, 1952:1). Querétaro: 2 mi. NW Conca (Schmidly and Martin, 1973:91).

Sturnira ludovici occidentalis Jones and Phillips

1964. *Sturnira ludovici occidentalis* Jones and Phillips, Univ. Kansas Publ., Mus. Nat. Hist., 14:477, March 2, type from Plumosas, 2500 ft., Sinaloa.

MARGINAL RECORDS (Jones and Phillips, 1964:481, unless otherwise noted).—Sinaloa: 5 km. SW Palmito (Jones, *et al.*, 1972:11). Durango: ½ mi. W Revolcaderos, 6600 ft.; *6 mi. S Pueblo Nuevo, 3000 ft.* Sinaloa: type locality. Jalisco: 1 km. E Soyatlán del Oro, 4600 ft. (Watkins, *et al.*, 1972:17); 10 mi. SE Tuxpan, 4200 ft. *(ibid.)*; 20 km. WNW Purificación, 1400 ft.; 4 km. N Durazno.

Sturnira mordax (Goodwin)
Talamancan Bat

1938. *Sturnirops mordax* Goodwin, Amer. Mus. Novit., 976:1, May 4, type from El Sauce Peralta, Cartago, Costa Rica.
1964. *Sturnira mordax*, Davis, Carter and Pine, Jour. Mamm., 45:381, September 15.

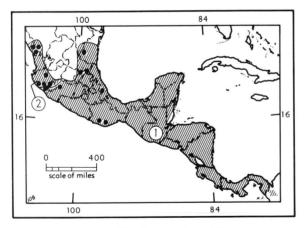

Map 108. *Sturnira ludovici.*

1. *S. l. ludovici*　　　　2. *S. l. occidentalis*

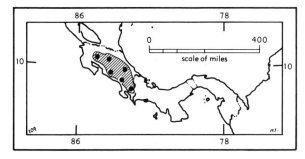

Map 109. *Sturnira mordax*.

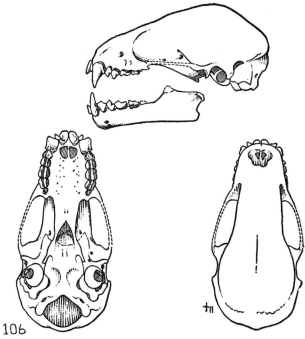

106

Fig. 106. *Sturnira mordax*, El Sauce Peralta, Costa Rica, No. 250310 U.S.N.M., from photo of holotype, ♂, X 2.

Measurements of 20 specimens: head and body, 60 (holotype only); forearm, 44.6–48.0; greatest length of skull, 23.8–26.6; zygomatic breadth, 12.4–13.0; interorbital breadth, 5.5–6.2; maxillary tooth-row, 6.5–7.2.

MARGINAL RECORDS.—Costa Rica: Monte Verde (3056 LaVal at KU); Cariblanco, 18 mi. NE Naranjo,

2900 ft. (Davis, *et al.*, 1964:381); type locality; Finca Las Cruces (Armstrong, 1969:809); San Gerardo, *ca.* 4500 ft. (Gardner, *et al.*, 1970:720); Colorado (*ibid.*).

SUBFAMILY STENODERMINAE

Trochiter impinging on scapula; nose-leaf usually present, but sometimes much reduced or absent; upper molars with cusps and commissures so reduced that W-shaped pattern is obliterated; upper molars lacking styles; crowns of upper and lower molars much flattened and with well-developed cusps arising from crushing surface.

Differs from Brachyphyllinae as follows: posterior palatal emargination usually U-shaped; less than 50 per cent of length of tongue free; pectoral ridge of humerus expanded proximally, forming triangular ridge; 4 lumbar vertebrae; tail absent; calcar well developed; 3rd phalanx of 3rd digit in resting position nearly parallel with metacarpal (modified from Silva-T. and Pine, 1969:10–12).

KEY TO NORTH AMERICAN GENERA OF STENODERMINAE

1. Rostrum less than half as long as braincase. . . . . *Centurio*, p. 167
1'. Rostrum more than half as long as braincase.
  2. Nasal region occupied by a narrow emargination extending back from nares to orbital level. . . . . *Chiroderma*, p. 149
  2'. Nasal region without emargination.
    3. Interpterygoid space extended forward as a deep palatal emargination.
      4. Rostrum strongly depressed between high supraorbital ridges; nares extending halfway from front of premaxillae to point of juncture of supraorbital ridges. . . . *Stenoderma*, p. 166
      4'. Rostrum rising above level of low supraorbital ridges; nares extending much less than halfway from front of premaxillae to juncture of supraorbital ridges.
        5. Borders of palatal emargination strongly converging anteriorly; inner upper incisor with crown slender, noticeably higher than long. . . . . *Phyllops*, p. 165
        5'. Borders of palatal emargination not strongly converging anteriorly; inner upper incisor with crown short and thick, scarcely or not higher than long.
          6. Upper molars 3–3. . . . . *Ardops*, p. 164
          6'. Upper molars 2–2. . . . . *Ariteus*, p. 166
    3'. Interpterygoid space not extended forward as a deep palatal emargination.
      7. Inner upper incisor slightly higher than outer, but not twice as large, the two teeth usually not conspicuously different in form or size.
        8. Length of rostrum fully ¾ that of braincase, depth of rostrum at front of 2nd premolar more than half depth of braincase. . . . . *Uroderma*, p. 142

8'. Length of rostrum slightly more than half that of braincase; depth of rostrum at front of 2nd premolar less than half that of braincase.

 9. Inner upper incisor bifid, M3 and m3 present or absent, but when present so reduced that they have no effect on the form of surrounding bone. . . . . *Artibeus,*   p. 153

  9'. Inner upper incisor entire, M3 and m3 well developed and affecting form of surrounding bone. . . . . *Enchisthenes,*   p. 163

7'. Inner upper incisor much higher than outer, usually at least twice as large, the two teeth conspicuously different in form and size.

 10. First lower molar with distinct posterointernal cusp, crown notably different in form from that of last premolar.

  11. Upper molars 3–3, the 2nd with large metacone. . . . . *Platyrrhinus,*   p. 144

  11'. Upper molars 2–2, the 2nd with metacone obsolete. . . . . *Vampyrodes,*   p. 146

 10'. First lower molar without posterointernal cusp, the crown resembling that of last premolar.

  12. Inner cusps of 2nd lower molar large, their height about half width of crown. *Vampyressa,*   p. 147

  12'. Inner cusps of m2 obsolete or absent.

   13. Leaflet behind nose-leaf; basal lappet of ear small and round; m3 present. *Mesophylla,*   p. 151

   13'. Leaflet behind nose-leaf absent; basal lappet of ear absent; m3 absent. *Ectophylla,*   p. 152

Genus **Uroderma** Peters
Tent-making Bat and Davis' Bat

Revised by Davis, Jour. Mamm., 49:676–698, November 26, 1968.

1865. *Uroderma* Peters, Monatsb. preuss. Akad. Wiss., Berlin, p. 588. Type, *Phyllostoma personatum* Peters [= *Uroderma bilobatum* Peters].

Rostrum at least three-fourths as long, and half as deep, as braincase; incisors resembling those of *Artibeus* but lateral upper one distinctly and evenly bilobed; incisors subequal; interpterygoid space not extended forward as a deep palatal emargination; nose-leaf well developed, lanceolate; facial stripes (when present) pale, one pair from nose-leaf to between ears and second pair from corners of mouth to bases of ears. Dentition, i. $\frac{2}{2}$, c. $\frac{1}{1}$, p. $\frac{2}{2}$, m. $\frac{3}{3}$.

KEY TO SPECIES OF URODERMA

1. Conch of ear concolor; facial stripes faint to absent; mesethmoid in cross section cross-shaped, and in frontal view shieldlike. . . . . . . *U. magnirostrum,*   p. 142

1'. Conch of ear edged with yellowish white (in museum specimens) and usually bright yellow (in life); facial stripes well developed; mesethmoid in cross section rod-like, and in frontal view strap-shaped.
      U. bilobatum,   p. 143

Uroderma magnirostrum Davis
Davis' Bat

1968. *Uroderma magnirostrum* Davis, Jour. Mamm., 49:679, November 26, type from 10 km. E San Lorenzo, 25 ft., Dept. de Valle, Honduras.

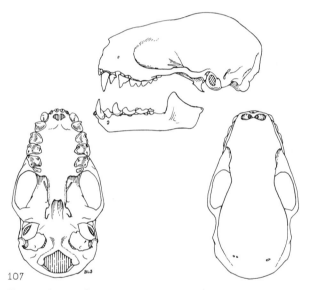

Fig. 107. *Uroderma magnirostrum,* 3 km. N, 4 km. W Sapoa, 40 m., Nicaragua, No. 97697 K.U., ♀, X 2.

Selected measurements: length of head and body, 58–65; length of forearm, 36.0–46.6; greatest length of skull, 21.9–24.9; zygomatic breadth, 12.0–13.5; upper tooth-row C–M3, 7.5–8.5. Upper parts grayish brown, tending to be paler (more yellowish) than in *bilobatum;* ear conch concolored; facial stripes faint to absent; rostrum deep, dorsal profile of skull nearly straight from tip of snout to crown; mesethmoid in cross section cross-shaped, and in frontal view shieldlike.

MARGINAL RECORDS (Davis, 1968:680, unless otherwise noted).—Guerrero: 20 km. N Tecpan (*fide* J.

Ramírez-P.). Oaxaca: 20 mi. NW La Ventosa. Chiapas: 21 km. SE Tonalá, 100 ft. El Salvador: 13 km. W La Libertad, 15 m. Honduras: type locality. Nicaragua: San Antonio, 35 m. (Jones, et al., 1971:11); 3 km. N, 4 km. W Sapoá, 40 m. Panamá: Isla Cébaco. Also occurs in South America.

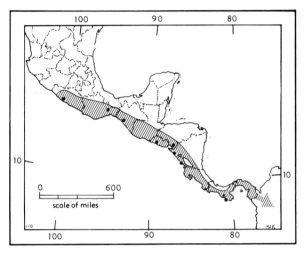

Map 110. *Uroderma magnirostrum.*

Uroderma bilobatum
Tent-making Bat

Selected measurements: head and body, 54–61 (two adult males); length of forearm, 38.9–45.7; greatest length of skull, 21.1–24.5; zygomatic

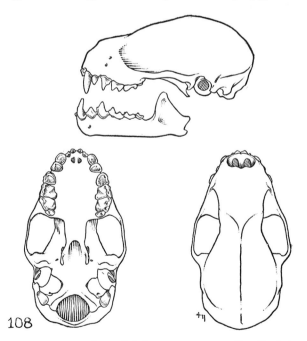

Fig. 108. *Uroderma bilobatum convexum,* San Gerónimo, Costa Rica, No. 256485 U.S.N.M., ♂, X 2.

breadth, 12.0–14.3; upper tooth-row C–M3, 7.2–8.6. Upper parts grayish brown, tending to be darker (less yellowish) than in *magnirostrum;* ear conch edged with yellowish white (in museum specimens) and usually bright yellow (in life); facial stripes prominent; rostrum shallow, dorsal profile of skull having distinct step between rostrum and braincase; mesethmoid in cross section rodlike, and in frontal view strap-shaped. In México and Central America has been taken only at elevations of less than 5010 ft. according to Davis (1968:697).

Uroderma bilobatum convexum Lyon

1902. *Uroderma convexum* Lyon, Proc. Biol. Soc. Washington, 15:83, April 25, type from Colón, Panamá.
1968. *Uroderma bilobatum convexum,* Davis, Jour. Mamm., 49:693, November 26.

MARGINAL RECORDS (Davis, 1968:695, 696, unless otherwise noted).—Nicaragua: ½ mi. S Chinandega (Baker and McDaniel, 1972:4); 3 km. N, 4 km. E Sabana Grande. Costa Rica: Los Chiles; La Irma (2926 LaVal at KU); 9 mi. ENE Golfito. Panamá: Olá, to N coast, E into South America, thence up Pacific Coast, including Panamanian Islands of San Miguel and *San José* (Kellogg, 1946:2), to point of beginning.

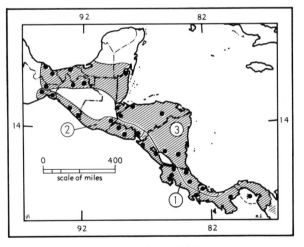

Map 111. *Uroderma bilobatum.*
1. *U. b. convexum*   2. *U. b. davisi*
3. *U. b. molaris*

Uroderma bilobatum davisi Baker and McDaniel

1972. *Uroderma bilobatum davisi* Baker and McDaniel, Occas. Pap. Mus. Texas Tech Univ., 7:1, November 3, type from 3 mi. NW La Herradura, La Paz, El Salvador, 20 m.

MARGINAL RECORDS (Baker and McDaniel, 1972:3, 4, unless otherwise noted).—Oaxaca (Davis, 1968:695, as *U. b. convexum*): 20 mi. NW La Ventosa; Tapanatepec. Chiapas: 11 9/10 mi. SE Tres Picos; 6⅘ mi. N

Tapachula, Rancho San Jorge. Guatemala: 20 km. SSE Chiquimula (Davis, 1968:695, as *U. b. convexum*). El Salvador: *1 mi. W Ilopango Airport;* 1⅕ mi. W Suchitoto; type locality. Honduras: 6 km. E Amatillo, 60 m. (Davis, 1968:695, as *U. b. convexum*).

Uroderma bilobatum molaris Davis

1968. *Uroderma bilobatum molaris* Davis, Jour. Mamm., 49:696, November 26, type from 16 mi. NW Palenque, 100 ft., Chiapas.

MARGINAL RECORDS (Davis, 1968:697, 698, unless otherwise noted).—Veracruz: 4⅕ mi. N Santiago Tuxtla (Baker and McDaniel, 1972:4); 1 mi. E Jaltipan. Tabasco: 13⅜ mi. N Villahermosa (Baker and McDaniel, 1972:4). Belize: Rockstone Pond, *ca.* 35 km. NNW Belice. Honduras (Baker and McDaniel, 1972:4): 23 mi. N San Pedro Sula; 10³⁄₁₀ mi. by road SSW Dulce Nombre de Culmi. Nicaragua: vic. Rama (*ibid.*). Costa Rica: 7³⁄₁₀ mi. SE Puerto Viejo (Baker and McDaniel, 1972:4); Peralta. Panamá: *Isla Bastimentos; Almirante; 7 km. SSW Changuinola.* Nicaragua: San Francisco. Honduras: Copán. Chiapas: Palenque; 8 km. S Solusuchiapa. Oaxaca: near Matías Romero.

Genus **Platyrrhinus** Saussure
Broad-nosed Bats

1860. *Platyrrhinus* Saussure, Revue et Mag. Zool., Paris, ser. 2, 12:429, October. Type, *Phyllostoma lineatum* É. Geoffroy St.-Hilaire (not preoccupied, because of a one-letter difference, by *Platyrhinus* Clairville, 1798, proposed for a beetle). Drs. de la Torre and Starrett, Nat. Hist. Miscellanea, Chicago Acad. Sci. No. 167:1, February 13, 1959, regard *Platyrrhinus* Fabricius, 1801, having a double r, as an emendation instead of an incorrect spelling, and maintain that Fabricius' name of 1801 and Saussure's name of 1860 are homonyms and that *Vampyrops* Peters, 1865, is the first available generic name for the neotropical broad-nosed bats. Because the *Platyrrhinus* of Fabricius, 1801, was listed only in a synonymy and because Fabricius in no way indicated that he intended it to be an emendation, I think the double r was an incorrect subsequent spelling, not an emendation.

1865. *Vampyrops* Peters, Monatsb. preuss. Akad. Wiss., Berlin, p. 356. Type, *Phyllostoma lineatum* É. Geoffroy St.-Hilaire.

Resembling *Uroderma,* but incisors markedly unequal, inner pair being at least twice as high as outer pair; cutting edges entire; interfemoral membrane much narrower, and conspicuously fringed with hairs 2–6 mm. long.

Following Alston (1879:48) many authors applied the name *Vampyrops lineatus* to Central American specimens on the now seemingly erroneous assumption that *Vampyrops helleri* Peters, 1865, of México, was the same as *Phyllostoma lineatum* É. Geoffroy St.-Hilaire, 1810, from South America.

KEY TO NORTH AMERICAN SPECIES OF PLATYRRHINUS

1. Forearm usually less than 42; greatest length of skull less than 24; color pale buffy brown. . . . .*P. helleri,* p. 144
1'. Forearm usually more than 42; greatest length of skull more than 24; color dark blackish-brown or bright blackish-brown.
  2. Forearm more than 55; greatest length of skull more than 31; color dark blackish-brown; facial stripes buffy.
        P. vittatus, p. 145
  2'. Forearm less than 52; greatest length of skull less than 30; color either dark blackish-brown or bright blackish-brown; facial stripes buffy or white.
        P. dorsalis, p. 145

Platyrrhinus helleri (Peters)
Heller's Broad-nosed Bat

1866. *Vampyrops helleri* Peters, Monatsb. preuss. Akad. Wiss., Berlin, p. 392, type from México.
1959. *Platyrrhinus helleri,* Hall and Kelson, Mammals of North America, Ronald Press, p. 131, March 31.
1891. *Vampyrops zarhinus* H. Allen, Proc. Acad. Nat. Sci. Philadelphia, 43:400, type from "Brazil"; actually Bas Obispo, Canal Zone, Panamá; see Goldman, Smiths. Misc. Coll., 69(5):200, April 26, 1920.

Measurements of six specimens from Central America: head and body, 55–66; forearm, 37.5–40.0; greatest length of skull, 21.7–22.9; zygomatic breadth, 12.5–12.6; maxillary tooth-row, 7.9. Four facial stripes and dorsal stripe white, distinct; color warm buffy-brown.

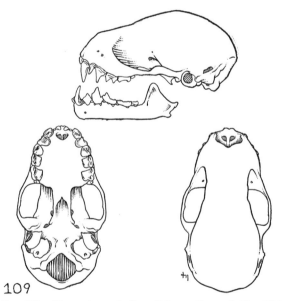

109

Fig. 109. *Platyrrhinus helleri,* Cabima, Panamá, No. 173833 U.S.N.M., sex ?, X 2.

MARGINAL RECORDS (Carter, *et al.*, 1966:494, unless otherwise noted).—Veracruz: Río Quezalapam, 2 mi. E Lago Catemaco, *ca.* 2000 ft. Guatemala: Tikal (Rick, 1968:518); 10 mi. N Sebol (Jones, 1966:457); 25 km. SSW Puerto Barrios, 75 m. Honduras: 4 mi. SW Tela, 25 ft.; 6 km. E Danlí (LaVal, 1969:820). Nicaragua: El Recreo (Jones, *et al.*, 1971:12). Panamá: Bas Obispo, Canal Zone (H. Allen, 1891:400, as *Vampyrops zarhinus*), thence southeastward into South America, and northwestward along Pacific Coast to El Salvador: 3½ km. E La Libertad (LaVal, 1969:820). Guatemala: 8 mi. E Coatepeque, 2200 ft. Chiapas (Baker, 1967:427): *Kilometer 184, Hwy. 200, N Huixtla;* 45 km. W Cintalapa. Oaxaca: Fulta [= Tutla] (Villa-R., 1967:285).

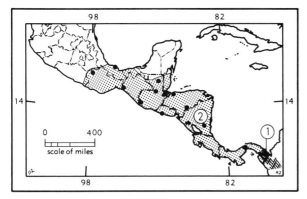

Map 112. *Platyrrhinus dorsalis* (1) and *Platyrrhinus helleri* (2).

Platyrrhinus vittatus (Peters)
Greater Broad-nosed Bat

1859. *Artibeus vittatus* Peters, Monatsb. preuss. Akad. Wiss., Berlin, p. 225, type from Puerto Cabello, Carabobo, Venezuela.
1959. *Platyrrhinus vittatus*, Hall and Kelson, Mammals of North America, Ronald Press, p. 132, March 31.
1865. V [*ampyrops*]. *vittatus* Peters, Monatsb. preuss. Akad. Wiss., Berlin, p. 356.

Head and body of adult from Colombia: 95. Measurements of 15 specimens from Vara Blanca, Costa Rica: forearm, 59.2–63.7; greatest length of skull, 31.6–33.7; zygomatic breadth, 18.2–19.9; postorbital constriction, 7.4–8.0; maxillary toothrow, 12.8–13.7.

Differs from *P. helleri* in much larger size, dark blackish-brown color, buffy dorsal and facial stripes, and poorly developed lower facial stripes.

MARGINAL RECORDS (as *Vampyrops vittatus*).—Costa Rica: Monte Verde (2944 LaVal at KU); Cariblanco, 18 mi. NE Naranjo, 3000 ft. (Davis, *et al.*, 1964:383). Panamá (Handley, 1966:766): Río Changena; Tacarcuna Village, 1950 ft., thence south-

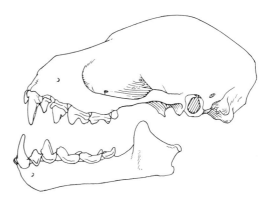

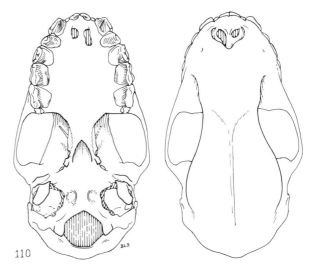

110

Fig. 110. *Platyrrhinus vittatus*, 4 mi. W of top Cerro Mali, 4800 ft., Darién, Panamá, No. 99335 K.U., ♂, X 2.

eastward into South America, and northwestward to Costa Rica: Finca Las Cruces, 2 km. S San Vito, 1300 m. (Gardner, *et al.*, 1970:720); San José (Davis, *et al.*, 1964:383).

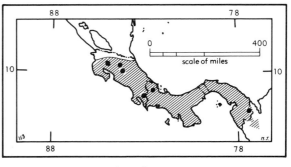

Map 113. *Platyrrhinus vittatus*.

Platyrrhinus dorsalis (Thomas)
Thomas' Broad-nosed Bat

1900. *Vampyrops dorsalis* Thomas, Ann. Mag. Nat. Hist., ser. 7, 5:269, March, type from Paramba, Ecuador, 1100 m.

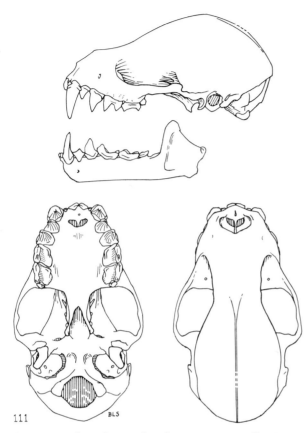

Fig. 111. *Platyrrhinus dorsalis*, Tacarcuna Vill. Camp, Darién, Panamá, No. 309611 U.S.N.M., ♀, X 2.

1902. *Vampyrops umbratus* Lyon, Proc. Biol. Soc. Washington, 15:151, June 20, type from San Miguel, Colombia.
1914. *Vampyrops oratus* Thomas, Ann. Mag. Nat. Hist., ser. 8, 14:411, November, type from Galifari, 6500 ft., Sierra del Avila, Venezuela.
1972. *Vampyrops aquilus* Handley and Ferris, Proc. Biol. Soc. Washington, 84:521, February 29, type from head Río Pucro, 4100 ft., Cerro Malí, Darién, Panamá. Regarded as a junior synonym of *V. dorsalis* by Carter and Rouk, Jour. Mamm., 54:976, December 14, 1973.

Selected measurements of nine specimens from Perú: head and body, 69–80; forearm, 44.6–50.1 (up to 51.3 in Panamá); greatest length of skull, 26.5–28.7; zygomatic breadth, 15.3–16.9; breadth of braincase, 11.2–12.0; maxillary toothrow, 10.2–11.4. Intermediate in size between *P. helleri* and *P. vittatus*. Color, including dorsal and facial stripes, as in *P. vittatus*.

MARGINAL RECORDS.—Panamá: head Río Pucro, 4100 ft., Cerro Malí (Handley and Ferris, 1972:521, as *Vampyrops aquilus*); *Tacarcuna Village, 1950 ft.* (Handley, 1966:766). Occurs in South America.

Also occurs in Costa Rica according to Baker, *et al.*, 1976:22. See Map 112.

Genus **Vampyrodes** Thomas

1900. *Vampyrodes* Thomas, Ann. Mag. Nat. Hist., ser. 7, 5:270, March. Type, *Vampyrops Caracciolae* Thomas [= *Vampyrodes caraccioli* (Thomas)].

"Similar to . . . [*Platyrrhinus*], but with only 2–2 upper molars, and these conspicuously differing from each other in form, owing to the reduction of the metacone in the second to a mere trace." (Miller, 1907:156.)
Only one species, *V. major*, is recorded from our area.

Vampyrodes major G. M. Allen
San Pablo Bat

1908. *Vampyrodes major* G. M. Allen, Bull. Mus. Comp. Zool., 52:38, July, type from San Pablo, Panamá.
1924. *Vampyrodes ornatus* Thomas, Ann. Mag. Nat. Hist., ser. 9, 13:532, May, type from San Lorenzo, Río Marañón, 500 ft., Perú.
1966. *V*[*ampyrodes*]. *c*[*araccioli*]. *major*, Handley, Ectoparasites of Panama, Field Mus. Nat. Hist., p. 766, November 22. Goodwin and Greenhall (Bull. Amer. Mus. Nat. Hist., 122:257, June 26, 1961) emended the spelling of *caraccioli* to *caraccioloi*. Starrett and Casebeer (Los Angeles Co. Mus. Contrib. Sci., 148:13, June 28, 1968) concluded that *Vampyrodes major* G. M. Allen and *Vampyrops caracciolae* Thomas [= *Vampyrodes caraccioloi* are two species.

Length of head and body approx. 76. Selected measurements of 12 specimens from Costa Rica: forearm, 52.2–54.4; greatest length of skull, 27.4–28.3; postorbital constriction, 6.6–7.0; zygomatic breadth, 16.7–17.8; maxillary toothrow, 9.8–10.1. Pair of broad white facial stripes extending from nose backward, one on each side, over eye to above ear; white line extending from top of head down middle of back; another white mark extending from near corner of mouth to ear. These stripes are present in *Uroderma* and *Platyrrhinus*, but only *P. vittatus* overlaps *Vampyrodes* in size. *P. vittatus* can be distinguished by its darker color, dusky facial stripes, and greater number of molars.

MARGINAL RECORDS (Davis, *et al.*, 1964:384, unless otherwise noted).—Oaxaca: near Valle Nacionál. Chiapas: Solosuchiapa [= Solusuchiapa] (Jones, 1964e:511); Florida, 50 km. E Altimirano, *ca.* 525 m. Guatemala: Escobas. Honduras: Lancetilla (Valdez and LaVal, 1971:247); Río Coco, 78 mi. ENE Danlí, 900 ft. Nicaragua: Santa Rosa, 17 km. N, 15 km. E Boaco, 300 m. (Jones, *et al.*, 1971:12). Costa Rica:

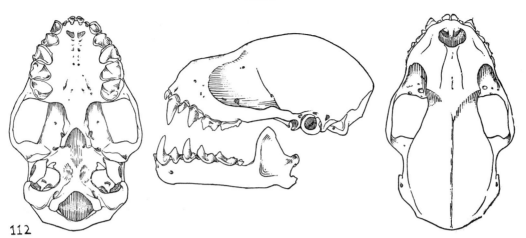

Fig. 112. *Vampyrodes major*, Barro Colorado Island, Canal Zone, Panamá, No. 45085 K.U., ♂, X 2.

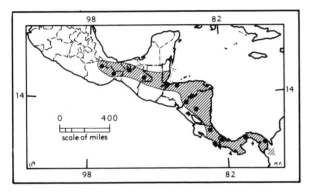

Map 114. *Vampyrodes major*.

Puerto Viejo de Sarapiquí, 100 m. (Starrett and Casebeer, 1968:12). Panamá (Handley, 1966:766): Almirante; Barro Colorado Island; Armila, thence southeastward into South America and northwestward along coast to Costa Rica: 2½ mi. SW Rincón, 15 m. (Gardner, *et al.*, 1970:720). Nicaragua: Vijagual [= Bijagual]. Oaxaca: Río Sarabia, 30⅖ km. N Matías Romero (Schaldach, 1965:135, as *V. carracioloi major*).

Genus **Vampyressa** Thomas—Yellow-eared Bats

Revised by Goodwin, Amer. Mus. Novit., 2125:1–24, April 5, 1963, and Peterson, Royal Ontario Mus., Life Sci. Contrib., 73:1–17, September 20, 1968.

1900. *Vampyrops* (*Vampyressa*) Thomas, Ann. Mag. Nat. Hist., ser. 7, 5:270, March. Type, *Phyllostoma pusillum* Wagner.
1968. *Metavampyressa* Peterson, Royal Ontario Mus., Life Sci. Contrib., 73:13, September 20. Type, *Vampyressa nymphaea* Thomas. Proposed as a subgenus on basis of a combination of characters (listed by Peterson *supra cit.*, on pages 12 and 13) not duplicated in other species of the genus.

In general like *Vampyrops* but molars ordinarily $\frac{2}{2}$ (always $\frac{2}{2}$ in N. Amer. species), their surface sculpture much more distinct; middle upper incisor faintly bifid; metacone of M2 so reduced that tooth is irregularly pyriform in outline; cusp posterior to protocone on M1 absent on M2; m1 without cusps on inner side, and resembling last premolar (adapted from Miller, 1907:156). Two pairs of narrow whitish facial stripes present; no trace of external tail; interfemoral membrane narrow; nose-leaf well developed, with pointed tip and well defined median rib.

KEY TO NORTH AMERICAN SPECIES OF VAMPYRESSA

1. Upper parts smoke-gray, facial stripes conspicuous; greatest length of skull more than 20. . . . . *V. nymphaea*, p. 148
1'. Upper parts whitish brown anteriorly, uniform pale brown posteriorly, facial markings reduced and inconspicuous; greatest length of skull less than 20. *V. pusilla*, p. 147

Vampyressa pusilla
Little Yellow-eared Bat

Length of head and body, 49–52. Selected measurements of a pregnant female from Belize: total length, 52; hind foot, 8; ear from notch, 13. Measurements of nine specimens: forearm, 30.3–33.6; greatest length of skull, 18.5–19.0; zygomatic breadth, 10.5–11.1; length of upper tooth-row, 5.8–6.2. Upper parts whitish brown anteriorly, uniform pale brown posteriorly; facial stripes reduced. *V. p. thyone* is smaller and differently colored than *V. nymphaea*.

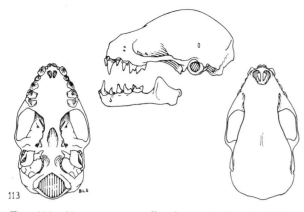

Fig. 113. *Vampyressa pusilla thyone*, 1 km. N, 2½ km. W Villa Somoza, 330 m., Chontales, Nicaragua, No. 111043 K.U., ♀, X 2.

Vampyressa pusilla thyone Thomas

1909. *Vampyressa thyone* Thomas, Ann. Mag. Nat. Hist., ser. 8, 4:231, type from Chimbo, 1000 ft., near Guayaquil, Bolívar Prov., Ecuador, September.
1963. *Vampyressa pusilla thyone*, Goodwin, Amer. Mus. Novit., 2125:14, April 5.
1912. *Vampyressa minuta* Miller, Proc. U.S. Nat. Mus., 42:25, March 6, type from Cabima, Panamá, Panamá.

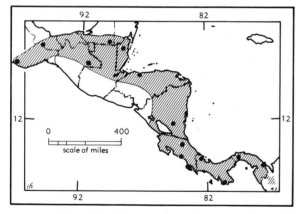

Map 115. *Vampyressa pusilla thyone*.

MARGINAL RECORDS.—Campeche: 65 km. S, 128 km. E Escárcega (93503 KU). Belize: Rockstone Pond, 30 mi. NW Belize (Peterson, 1965:676). Honduras: Lancetilla (Valdez and LaVal, 1971:249). Nicaragua: 1 km. N, 2½ km. W Villa Somoza (Jones, *et al.*, 1971:13). Costa Rica: La Selva (3126 LaVal at KU). Panamá (Handley, 1966:767): Sibube; Armila, thence southeastward to South America and northwestward to Panamá (*ibid.*); Cabima; Cerro Hoya. Costa Rica: 6 mi. W Rincón (Armstrong, 1969:809); Hda. Quebrada de Azul, Quepos (Gardner, *et al.*, 1970:721). Chiapas: Florida, 50 km. E Altamirano (*ibid.*). Oaxaca: Finca Sinai, 2200 m., 10 km. E Nopala (Arnold and Schonewald, 1972:171). Veracruz: 2 km. W Suchilapa (Rick,

1968:518). Not found: Nicaragua (Dept. Managua): Hda. San José (Jones, *et al.*, 1971:13).

Vampyressa nymphaea Thomas
Big Yellow-eared Bat

1909. *Vampyressa nymphaea* Thomas, Ann. Mag. Nat. Hist., ser. 8, 4:230, type from Novita, 150 ft., Río San Juan, Chocó, Colombia.

Selected measurements of five specimens from Costa Rica and one from Nicaragua: total length, 55–60; forearm, 36.2–39.0; greatest length of skull, 21.1–21.8; zygomatic breadth, 12.1–12.8; maxillary tooth-row, 7.2–7.8. Upper parts

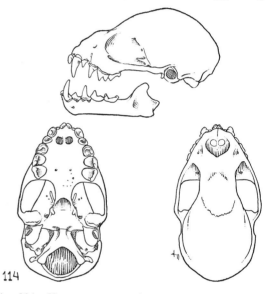

Fig. 114. *Vampyressa nymphaea*, Barro Colorado Island, Canal Zone, Panamá, No. 52455 K.U., ♀, X 2.

smoke-gray; underparts slightly paler; facial stripes conspicuous, upper pair extending posteriorly beyond ears; faint mid-dorsal white stripe present in most specimens.

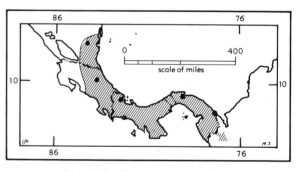

Map 116. *Vampyressa nymphaea*.

MARGINAL RECORDS.—Nicaragua: Río Mico, El Recreo, 25 m. (Jones, *et al.*, 1971:13). Costa Rica: Cariari, Río Tortuguero, 50 m. (Gardner, *et al.*, 1970:721). Panamá (Handley, 1966:767): Almirante; Mandinga; Armila. Also occurs in South America.

Genus **Chiroderma** Peters

1860. *Chiroderma* Peters, Monatsb. preuss. Akad. Wiss., Berlin, p. 747. Type, *Chiroderma villosum* Peters.
1866. *Mimetops* Gray, Proc. Zool. Soc. London, p. 117, May (cited in synonymy as manuscript name). Included two species: *Chiroderma villosum* Peters; *C. pictum* Gray.

Skull resembling that of *Platyrrhinus* but lacking nasal bones; cusps of molars much thickened and encroaching to considerable degree onto crushing surface of crown; externally much resembling *Platyrrhinus* but nose-leaf broader and forearm and uropatagium more heavily furred. Dentition, i. $\frac{2}{2}$, c. $\frac{1}{1}$, p. $\frac{2}{2}$, m. $\frac{2}{2}$.

KEY TO NORTH AMERICAN SPECIES OF CHIRODERMA

1. Forearm more than 41; anterior lower premolar relatively small, with anterior cusp small or absent.
    2. Forearm more than 53 (57.5 in one known specimen). *C. improvisum*, p. 151
    2'. Forearm less than 53.
        3. Dorsal and facial stripes prominent. *C. salvini*, p. 149
        3'. Dorsal and facial stripes absent or faint. . . . . *C. villosum*, p. 150
1'. Forearm less than 41; anterior lower premolar relatively large, with anterior cusp large. . . . . *C. trinitatum*, p. 150

Chiroderma salvini
Salvin's White-lined Bat

Measurements of 13 specimens from Central America: length of head and body, 67–77; length of forearm, 43.9–51.5; greatest length of skull, 24.2–27; interorbital breadth, 5.5–6.8; zygomatic breadth, 15.5–17.5; length of maxillary tooth-row, 8.9–10. Upper parts dark brown, white median stripe pronounced, extending from nape to uropatagium; facial markings distinct; underparts paler, more grayish, than upper parts. Skull robust; molariform teeth robust; inner pair of upper incisors high, slender; outer pair tiny; lower incisors small, subequal.

Chiroderma salvini salvini Dobson

1878. *Chiroderma salvini* Dobson, Catalogue of the Chiroptera in the . . . British Museum, p. 532, type from Costa Rica.

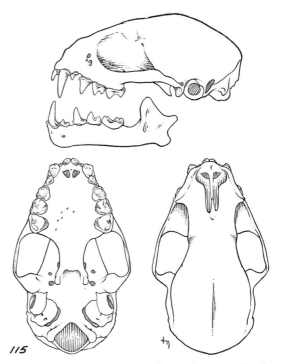

Fig. 115. *Chiroderma salvini salvini*, Tapasuna, Honduras, No. 47640 F.M.N.H., ♂, × 2.

MARGINAL RECORDS.—Veracruz: ½ mi. NE Las Minas (Handley, 1966:298). Honduras: Subirana, 2800 ft. (*ibid.*). Costa Rica: Angostura (Goodwin, 1946:322). Panamá (Handley, 1966:767): Cerro Azul, 2000 ft.; Cana, 2000 ft., thence southeastward into South America, thence northwestward to Honduras: Comayagua (LaVal, 1969:820, as *C. salvini* only); Marcala (*ibid.*); Ruinas de Copán, 1900 ft. (Carter, *et al.*, 1966:495, as *C. salvini* only). Guatemala: Villalobos, 13 km. S Guatemala City, 4200 ft. (*ibid.*). Puebla: 1½ km. S Atlixco (Alvarez and Ramírez-P., 1972:169).

Chiroderma salvini scopaeum Handley

1966. *Chiroderma salvini scopaeum* Handley, Anal. Inst. Biol., Univ. Nac. Autó. México, 36:297, June 20, type from Pueblo Juárez, Colima.

MARGINAL RECORDS (Handley, 1966:298, unless otherwise noted).—Chihuahua: 1½ mi. SW Tocuina, Río Septentrión, 1500 ft. Sinaloa: 1 mi. S Santa Lucia, 5650 ft. Durango: Paso de Sihuacori (Crossin, *et al.*, 1973:199). Nayarit: 7 mi. N Acaponeta. Zacatecas: 5½ mi. S Moyahua (Genoways and Jones, 1968:744). Jalisco (Watkins, *et al.*, 1972:18): El Salto, 24 mi. W Guadalajara, 4500 ft.; 8 mi. E Jilotlán de los Dolores, 2000 ft. Guerrero: Acahuizotla, 2800 ft. Oaxaca: 20 mi. S, 5 mi. E Sola de Vega, 4800 ft. (99702 KU). Colima: type locality; *3 km. S Pueblo Nuevo*. Jalisco: 17 km. SE Talpa de Allende, 5200 ft. Nayarit: 8 mi. E San Blas.

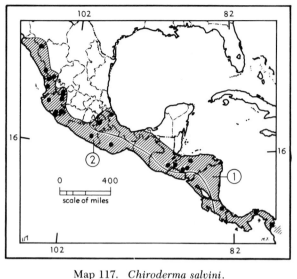

Map 117. *Chiroderma salvini.*

1. *C. s. salvini*   2. *C. s. scopaeum*

Chiroderma trinitatum
Goodwin's Bat

Smallest species in genus *Chiroderma*; forearm less than 41; canines short; p4 low crowned; p1 larger than in *villosum* and *salvini*; supraorbital ridge absent. Measurements of type and two specimens from Panamá: head and body, 57.5, 56, 57; ear from meatus, 14.2, 17, 18; hind foot, 12.5, 10, 11; forearm, 40.5, 38.5, 37.6. Greatest length of skull, 22.5, 20.9, 20.7; zygomatic breadth, 13.7, 12.8, 13.1; interorbital breadth, 5.7, 5.4, 5.6; maxillary tooth-row, 7.7, 7.3, 7.3. Distinct white line down full length of back.

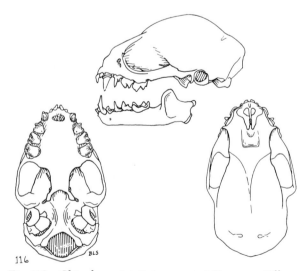

Fig. 116. *Chiroderma trinitatum gorgasi,* Tacarcuna Village Camp, Panamá, No. 309901 U.S.N.M., ♀, X 2.

Chiroderma trinitatum gorgasi Handley

1960. *Chiroderma gorgasi* Handley, Proc. U.S. Nat. Mus., 112:464, October 6, type from Tacarcuna Village, 3200 ft., Río Pucro, Darién, Panamá.
1965. *Chiroderma trinitatum gorgasi,* Barriga-Bonilla, Caldasia, 9:246, July 3.

MARGINAL RECORDS (Handley, 1966:767, unless otherwise noted).—Panamá: upper Río Changena, 2400 ft.; Armila; Tacarcuna Village, 1950 ft., thence into South America.

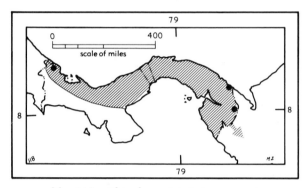

Map 118. *Chiroderma trinitatum gorgasi.*

Chiroderma villosum
Shaggy-haired Bat

Selected measurements of 16 specimens are: length of head and body, 65–82; length of ear, 14–20; length of forearm, 42.7–50.3; greatest length of skull, 24.0–27.2; interorbital breadth, 5.6–6.2; zygomatic breadth, 14.9–17.2; length of maxillary tooth-row, 8.4–9.7. Upper parts olivaceous brown (whitish median line present in some individuals); underparts paler, somewhat buffy. Smaller than *C. salvini* and readily distinguishable from it by absence of, or only faint, facial markings.

Chiroderma villosum jesupi J. A. Allen

1900. *Chiroderma jesupi* J. A. Allen, Bull. Amer. Mus. Nat. Hist., 13:88, May 12, type from Cacagualito, Santa Marta, Colombia.
1960. *Chiroderma villosum jesupi,* Handley, Proc. U.S. Nat. Mus., 112:466, October 6.
1912. *Chiroderma isthmicum* Miller, Proc. U.S. Nat. Mus., 42:25, March 6, type from Cabima, Panamá.

MARGINAL RECORDS.—Quintana Roo: 8 km. N, 5½ km. E Playa del Carmen (Birney, *et al.,* 1974:6). Costa Rica: Puerto Viejo (Armstrong, 1969:809). Panamá (Handley, 1966:767): Almirante; Armila, thence southeastward into South America and northwestward to Panamá: Cerro Hoya (*ibid.*). Costa Rica: 2 mi. W Rincón (Armstrong, 1969:809). Nicaragua (Jones, *et al.,* 1971:14): 2 km. N, 3 km. E Mérida, 200 m., Isla

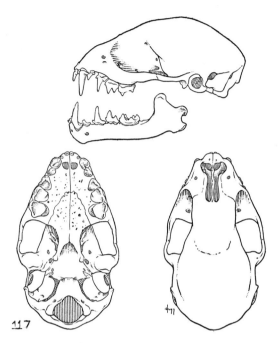

Fig. 117. *Chiroderma villosum jesupi*, Barro Colorado Island, Canal Zone, Panamá, No. 45096 K.U., ♂, X 2.

de Ometepe; 4½ km. N Cosiguina, 15 m. Guatemala: 20 km. SSE Chiquimula, 550 m. (Carter, *et al.*, 1966:494). Chiapas: 12 km. SSE Tonalá, 100 m. (Handley, 1966:298). Oaxaca: 18 mi. NW Sola de Vega (Goodwin, 1969:85). Veracruz: Presidio; Achotal. Campeche: 103 km. SE Escárcega (93437 KU).—**See** addenda.

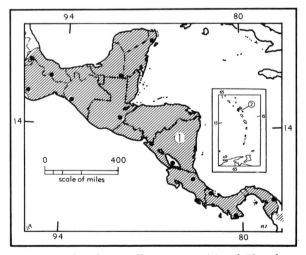

Map 119. *Chiroderma villosum jesupi* (1) and *Chiroderma improvisum* (2).

Chiroderma improvisum Baker and Genoways
Guadeloupe White-lined Bat

1976. *Chiroderma improvisum* Baker and Genoways, Occas. Pap. Mus. Texas Tech Univ., 39:1, April 16, type from

Basse-Terre, 2 km. S, 2 km. E Baie-Mahault, Guadeloupe, Lesser Antilles. Known only from type locality.—**See** addenda.

Length of head and body, 87; forearm, 57.5; greatest length of skull, 29.9; interorbital breadth, 6.5; zygomatic breadth, 18.9; length of upper tooth-row, 10.7. Upper parts grayish brown with distinct white line down center of posterior half of back; underparts gray, tips of hairs there with white band producing a "frosted" effect; indistinct white line above and below each eye.

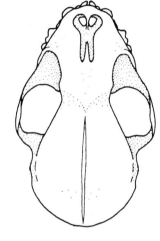

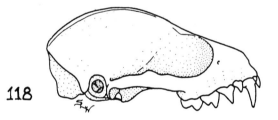

Fig. 118. *Chiroderma improvisum*, Basse-Terre, 2 km. S, 2 km. E Baie-Mahault, Guadeloupe, Lesser Antilles, No. 19900 T.T.U., holotype (after Baker and Genoways, 1976: Fig. 4), ♂, X 2.

Genus **Mesophylla** Thomas

1901. *Mesophylla* Thomas, Ann. Mag. Nat. Hist., ser. 7, 8:143, August. Type *Mesophylla macconnelli* Thomas.

Externally resembling *Vampyressa pusilla*; occlusal surfaces of m2 and M2 lacking longitudinal ridge seen in *Ectophylla*; m3 present but minute; m2 larger than m1; 1st and 2nd molars not separated by space; basisphenoid pits relatively shallow; tail absent. Dentition, i. $\frac{2}{2}$, c. $\frac{1}{1}$, p. $\frac{2}{2}$, m. $\frac{2}{3}$. Some zoologists regard *Mesophylla* as a subgenus of *Ectophylla*.

Mesophylla macconnelli Thomas
McConnell's Bat

1901. *Mesophylla macconnelli* Thomas, Ann. Mag. Nat. Hist.,
ser. 7, 8:145, August. Type from Kanuku Mts., 2000 ft.,
about 59° W and 3° N, Guyana.

1962. *Ectophylla macconnelli flavescens* Goodwin and
Greenhall, Amer. Mus. Novit., 2080:2, April 24. Type from
Talpara, Trinidad. Arranged as a synonym of *M. macconnelli* by Starrett and Casebeer, Los Angeles Co. Mus. Contrib. Sci., 148:14, June 28, 1968.

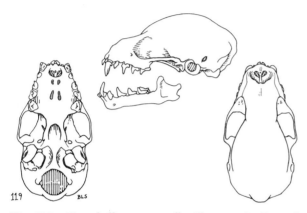

Fig. 119. *Mesophylla macconnelli*, Changuinola, Panamá,
No. 315564 U.S.N.M., ♂, X 2.

Length of head and body, 45–49; length of
forearm, 29.5–33.0; greatest length of skull,
16.8–18.8; zygomatic breadth, 9.2–10.7; interorbital breadth, 4.0–4.7; maxillary tooth-row, 5.5–
6.3; nose-leaf not crenulated; supplementary
leaflet immediately behind nose-leaf. According
to Thomas (1901:145) "fur close and thick, hairs
about 5 millim. long on the back . . . head and
anterior back dull brownish white, darkening
posteriorly to a brown very near Ridgway's 'wood
brown'." Specimens from Trinidad, according to
Goodwin and Greenhall (1962:2), are pale
grayish buff; ears, nose-leaf, thumb, and second
and third metacarpals bright yellow in life.

Goodwin and Greenhall (1962:2) regarded
Mesophylla as no more than a subgenus of *Ec-*

tophylla and proposed the subspecific name *Ectophylla macconnelli flavescens* for specimens
from Trinidad. Handley (1966:768) implied that
E. m. flavescens was not worthy of recognition,
since "Individual variation among the Panamanian specimens exceeds the variation described
by Goodwin and Greenhall . . . as geographic."
Later Starrett and Casebeer (1968:14) concluded
that *Mesophylla* was a valid genus but followed
Handley in treating this bat as a monotypic
species.

MARGINAL RECORDS.—Costa Rica: Moravia de
Chirripó, 1116 m. (Starrett and Casebeer, 1968:14).
Panamá (Handley, 1966:768): Almirante; Tacarcuna
Village, 1950 ft. Also occurs in South America.

Genus Ectophylla H. Allen

1892. *Ectophylla* H. Allen, Proc. U.S. Nat. Mus., 15:441.
Type, *Ectophylla alba* H. Allen.

Externally resembling a small whitish *Vampyressa*. In dental structure the most aberrant
genus of the stenodermine bats. Basin-shaped
crowns of 2nd upper and lower molars crossed by
distinct longitudinal ridge; 1st upper molar with
low protocone; 2nd lower molar with only one
cusp, a well-developed paracone; 2nd lower
molar markedly basined, broadly oval, wider
than mandibular ramus; tail absent; tip of calcar
free.

The genus is monotypic.

Ectophylla alba H. Allen
Honduran White Bat

1892. *Ectophylla alba* H. Allen, Proc. U.S. Nat. Mus., 15:442,
October 26, type from Segovia River, eastern Honduras [=
Comarca de El Cabo, northern Nicaragua, according to Miller and Kellogg, Bull. U.S. Nat. Mus., 205:77, March 3,
1955].

Fig. 120. *Ectophylla alba*, Puerto Viejo, 100 m., Heredia,
Costa Rica, No. 88025 K.U., ♀, X 2.

Selected measurements of the type: length of
head and body "(from crown of head to base of
tail)" (H. Allen, 1892:442), 36; length of forearm,

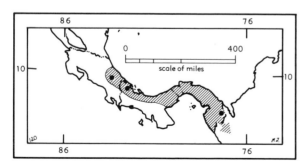

Map 120. *Mesophylla macconnelli*.

25. Measurements of nine specimens from Costa Rica: forearm, 26.4–29.6; greatest length of skull, 16.1–17.0; zygomatic breadth, 9.7–10.8; postorbital constriction, 4.1–4.5; maxillary tooth-row, 5.9–6.3. Upper parts dull whitish paler over posterior parts; sides fawn colored; underparts dull white. Dentition, i. $\frac{2}{2}$, c. $\frac{1}{1}$, p. $\frac{2}{2}$, m. $\frac{2}{2}$.

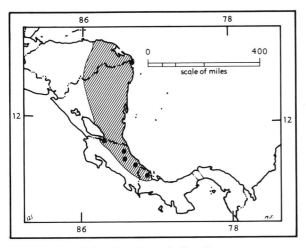

Map 121. *Ectophylla alba*.

MARGINAL RECORDS.—Nicaragua: type locality. Costa Rica (Gardner, *et al.*, 1970:722, unless otherwise noted): San Emilio, Lake Nic-Nac (Casebeer, *et al.*, 1963:186); Cariari; Turrialba, 600 m.; Vesta. Panamá (Handley, 1966:768): Almirante; *Sibube*.

Genus **Artibeus** Leach—Fruit-eating Bats

1821. *Artibeus* Leach, Trans. Linnean Soc. London, 13:75. Type, *Artibeus jamaicensis* Leach.

1821. *Madataeus* Leach, Trans. Linnean Soc. London, 13:81. Type, *Madataeus lewisii* Leach [= *Artibeus jamaicensis* Leach].
1827. *Medateus* Gray *in* Griffith, The animal kingdom . . . by the Baron Cuvier, 5:74, an emendation?
1838. *Arctibeus* Gray, Mag. Zool. Bot., 2:487, an emendation?
1843. *Medateus* Gray, List of the . . . Mammalia in the . . . British Museum, p. xviii, an emendation?
1847. *Arctibius* Bonaparte, Proc. Zool. Soc. London, p. 115, November 10, an emendation?
1856. *Pteroderma* Gervais, Mammifères, *in* [Castelnau] Expéd. dans les parties centrales de l'Amér. du Sud . . . , pt. 7, p. 34. Type, *"perspicillatum."*
1856. *Artibaeus* Gervais (*loc. cit.*), an emendation?
1856. *Dermanura* Gervais, *op. cit.*:36. Type, *Dermanura cinereum* Gervais.
1892. *Artobius* Winge, Jordf. og Nulevende Flagermus fra Lagoa Santa, Minas Geraes, Brasilien, p. 10, an emendation?

Skull with moderately wide, slightly elevated braincase; zygomata widespreading and short; rostrum low and about equal to lachrymal breadth; median depth in lachrymal region less than half lachrymal breadth; palate moderately wide, distance between 2nd upper premolars approx. equal to that from incisor to hypocone of 1st molar. Ears separate; nose-leaf well developed; no external tail; interfemoral membrane narrow; calcar short but distinct. (After Miller, 1907:161.) Upper incisors small, crowded, the inner distinctly bilobed, the outer much smaller and entire; lower incisors smaller than uppers, closely crowded and faintly bilobed; molars robust, crowns of crushing type and finely corrugated; molar teeth $\frac{2-2}{2-2}$, $\frac{2-2}{3-3}$ or $\frac{3-3}{3-3}$, sometimes according to species and sometimes individually. Usually paired facial stripes present but median dorsal stripe always absent.

KEY TO NORTH AMERICAN SPECIES OF ARTIBEUS

1. Length of forearm less than 48; greatest length of skull less than 24.
  2. Interfemoral membrane deeply incised (width less than 7.2); posterior edge conspicuously hairy.
    3. Forearm usually less than 42.0; skull usually shorter than 21.0; maxillary tooth-row usually less than 7.0. . . . .*A. toltecus*, p. 160
    3'. Forearm usually more than 42.0; skull usually longer than 21.0; maxillary tooth-row usually more than 7.0. . . . .*A. aztecus*, p. 159
  2'. Interfemoral membrane less deeply incised (width more than 7.2); posterior edge nearly bare.
    4. Narrow talon on M1; molars normally $\frac{2}{3}$. . . . .*A. watsoni*, p. 161
    4'. Wide or narrow talon on M1; molars normally $\frac{2}{2}$.
      5. In geographic area here covered, occurs only on Grenada; narrow talon on M1.
          A. cinereus, p. 161
      5'. In geographic area here covered, occurs only on mainland; wide talon on M1. *A. phaeotis*, p. 162
1'. Length of forearm more than 48; greatest length of skull more than 24.
  6. Greatest length of skull usually more than 30; length of forearm usually more than 64; preorbital and postorbital processes well developed. . . . .*A. lituratus*, p. 157
  6'. Greatest length of skull usually less than 29; length of forearm 66 or less; preorbital and postorbital processes poorly developed or absent.

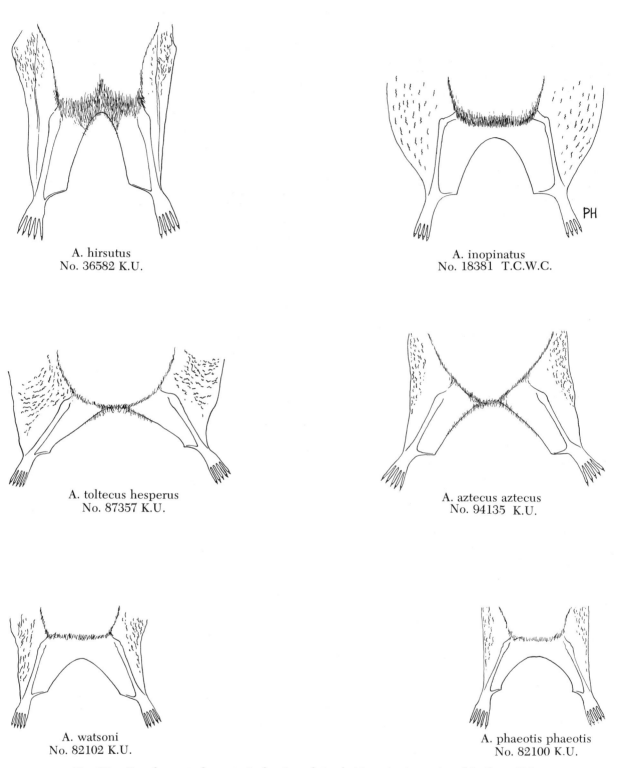

A. hirsutus
No. 36582 K.U.

A. inopinatus
No. 18381 T.C.W.C.

PH

A. toltecus hesperus
No. 87357 K.U.

A. aztecus aztecus
No. 94135 K.U.

A. watsoni
No. 82102 K.U.

A. phaeotis phaeotis
No. 82100 K.U.

Fig. 121. Dorsal aspect of uropatagia showing relative hairiness in six species of *Artibeus*, X 1.

7. Interfemoral membrane densely furred or with noticeable fringe of hair on free edge; color silvery gray.
    8. Interfemoral membrane moderately furred (but with noticeable fringe of hairs on its free edge); maxillary tooth-row 9.2 or less; posterior border of palate evenly concave; greatest length of skull including canines less than 26.1. . . . .*A. inopinatus,* p. 159
    8'. Interfemoral membrane densely furred; maxillary tooth-row 9.8 or more; posterior border of palate having broad posterior spine; greatest length of skull including canines more than 26.1.
                                           A. hirsutus, p. 158
7'. Interfemoral membrane nearly bare. . . . .*A. jamaicensis,* p. 155

Artibeus jamaicensis
Jamaican Fruit-eating Bat

Length of head and body approx. 75; length of forearm, 52.0–67.4; greatest length of skull, 26.2–31.6; zygomatic breadth, 16.2–20.6; length of upper tooth-row, 9.8–11.5. Upper parts variable individually, but usually some shade of brown, sometimes smoky brown; facial markings present or absent, and when present of varying degrees of development; underparts often grayish and usually paler than upper parts. Skull short, broad, robust; braincase variable in degree of inflation both individually and geographically; sagittal crest moderately developed; rostrum short, low, wide; basisphenoidal pits absent. Number of molars variable.

Artibeus jamaicensis jamaicensis Leach

1821. *Artibeus Jamaicensis* Leach, Trans. Linnean Soc. London, 13:75, type from Jamaica.
1821. *Madataeus Lewisii* Leach, Trans. Linnean Soc. London, 13:81, type from Jamaica.
1851. *Artibeus carpolegus* Gosse, A naturalist's sojourn in Jamaica, p. 271, type from Content, Jamaica.
1889. *Dermanura eva* Cope, Amer. Nat., 23:130, February, type from St. Martins, Lesser Antilles.
1890. *Artibeus coryi* J. A. Allen, Bull. Amer. Mus. Nat. Hist., 3:173, November 14, type from St. Andrews, Caribbean Sea.
1904. *Artibeus insularis* J. A. Allen, Bull. Amer. Mus. Nat. Hist., 20:231, June 29, type from St. Kitts, Lesser Antilles.
1906. *Artibeus jamaicensis praeceps* Andersen, Ann. Mag. Nat. Hist., ser. 7, 18:421, December, type from Guadeloupe, Lesser Antilles.

MARGINAL RECORDS.—Haiti: cave near St. Michel. Dominican Republic: *Río Yuna, 6 mi. E Arenosa* (Armstrong and Johnson, 1969:133, as *A. jamaicensis* only); San Gabriel Cave. Puerto Rico: Old Loiza. Anegada Island. Lesser Antilles (Jones and Phillips, 1970:137): Anguilla; Barbuda; Barbados; St. Vincent (as unnamed subspecies, See addenda); St. Lucia; Martinique; Dominica; Guadeloupe; Monserrat; St. Kitts; Saba. St. Andrews Island. Old Providence Island. Jamaica: Content; Constant Springs. Haiti: La Gonave Island; cave near Daiquini.

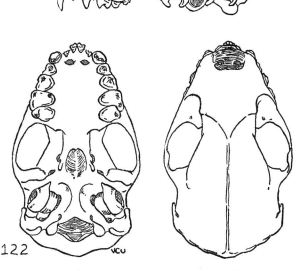

Fig. 122. *Artibeus jamaicensis yucatanicus,* 3 km. W Boca del Río, 10 ft., Veracruz, No. 23727 K.U., ♀, X 2.

Artibeus jamaicensis parvipes Rehn

1902. *Artibeus parvipes* Rehn, Proc. Acad. Nat. Sci. Philadelphia, 54:639, December 8, type from Santiago de Cuba, Oriente, Cuba.
1908. *Artibeus jamaicensis parvipes,* Andersen, Proc. Zool. Soc. London, 2:261, September 7.

MARGINAL RECORDS.—Bahama Islands: Abrahams Hill, Mariguana Island; Great Inagua Island. Cuba: Baracoa. Isle of Pines: Nueva Gerona. Cuba: Pinar del Río. Note: G. M. Allen (1911:235) states that the supposed occurrence of this bat at Key West, Florida, should be disregarded. In making this statement, G. M. Allen (*loc. cit.*) probably had in mind the account of Maynard (Quart. Jour. Boston Zool. Soc., 2(2):22, April, 1883).

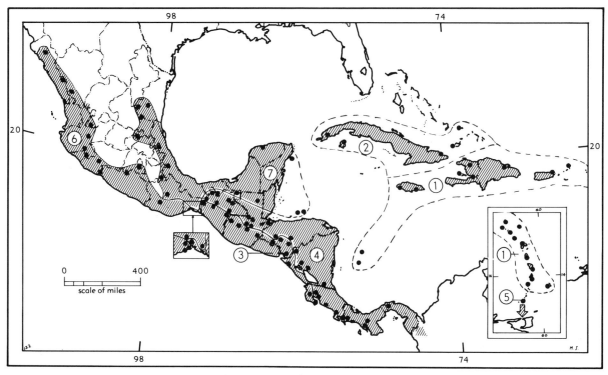

Map 122. *Artibeus jamaicensis.*

Guide to   2. *A. j. parvipes*   5. *A. j. trinitatis*
subspecies   3. *A. j. paulus*   6. *A. j. triomylus*
   1. *A. j. jamaicensis*   4. *A. j. richardsoni*   7. *A. j. yucatanicus*

Artibeus jamaicensis paulus Davis

1970. *Artibeus jamaicensis paulus* Davis, Jour. Mamm.,
   51:119, February 20, type from 7½ km. WNW La Libertad,
   ca. 500 ft., Dept. of La Libertad, El Salvador.

MARGINAL RECORDS (Davis, 1970:120, 121).—
Chiapas: 7 mi. NW Cintalapa, 2600 ft.; 4 mi. N Chiapa
de Corzo, K 1110, 3000 ft.; *4 mi. NE Chiapa de Corzo,
2000 ft.;* 1³⁄₁₀ mi. SE Zapaluta [= La Trinitaria], 5700 ft.
Guatemala: 4 km. WSW San Miguel Acatán, 5850 ft.; 1
km. WNW Sacapulas, 3900 ft.; San Miguel Chicaj, 3100
ft.; *2 km. SE Salamá, 3100 ft.;* 20 km. SSE Chiquimula,
1800 ft. Honduras: La Esperanza, 5200 ft.; *3 km. W
Comayagua, 1890 ft.;* Comayagua, 1890 ft.; 10 mi. SE
Tegucigalpa, 5135 ft.; 10 km. E San Lorenzo, 25 ft.
Nicaragua: San Antonio, 50 ft. Costa Rica: 3 mi. S, 10
mi. W Las Cañas, 10 ft.; 3 km. N Tambor, sea level,
thence up coast into Oaxaca at 95° W long. and inland
to La Ventosa.

Artibeus jamaicensis richardsoni J. A. Allen

1908. *Artibeus jamaicensis richardsoni* J. A. Allen, Bull.
   Amer. Mus. Nat. Hist., 24:669, October 13, type from
   Matagalpa, Matagalpa, Nicaragua.

MARGINAL RECORDS (Davis, 1970:115–117).—
Tabasco: 1–3 mi. E Teapa. Chiapas: Yaxoquintela, 37

km. NE Altamirano, 1900 ft.; Florida, 50 km. E Al-
tamirano, 1700 ft. Guatemala: Chinaja, 550 ft.; Río San
Simón, 6 km. NE Raxrujá, 450 ft.; 25 km. SSW Puerto
Barrios, 300 ft., thence eastward along coast into South
America and northward to Panamá: 18 km. WSW
Chepó, 200 ft.; 2 mi. S San Francisco, 200 ft.; 2 mi. W
Soná, 200 ft.; 11 mi. W Concepción, 200 ft. Costa Rica:
9 mi. ENE Puerto Golfito, 100 ft.; 1–4 mi. NE Palmar,
300 ft.; Julieta, near sea level; 6½ mi. N, 2 mi. W Pun-
tarenas, 100 ft.; 5 mi. E Tilarán, 2300 ft. Nicaragua: El
Castillo, 130 ft.; San Francisco, at K 92, 400 ft.; 2 mi. SE
Dario, 1500 ft.; Yalaguina, 2300 ft. Honduras: 10 mi.
NE Talanga, 3400 ft.; 5 km. NE Jesús de Otoro, 2100 ft.
Guatemala: Finca Los Alpes (*ca.* 20 km. S Languin),
3250 ft.; 1 km. W Cobán, *ca.* 3500 ft. Chiapas: La Sol-
edad, 16 mi. NE Las Margaritas, 3600 ft.; 3 mi. SSE
Soyaló, 3000 ft.; 10 mi. W Mal Paso, 400 ft.; 21 km.
WSW Teapa (Tabasco), 200 ft.

Artibeus jamaicensis trinitatis Andersen

1906. *Artibeus planirostris trinitatis* Andersen, Ann. Mag.
   Nat. Hist., ser. 7, 18:420, December, type from St. Anns,
   Trinidad.
1949. *A[rtibeus]. j[amaicensis]. trinitatis*, Hershkovitz, Proc.
   U.S. Nat. Mus., 99:447, May 10.
1906. *Artibeus planirostris grenadensis* Andersen, Ann. Mag.
   Nat. Hist., ser. 7, 18:420, December, type from Grenada.

MARGINAL RECORD.—Lesser Antilles: Grenada, thence into South America.

Artibeus jamaicensis triomylus Handley

1966. *Artibeus jamaicensis triomylus* Handley, Anal. Inst. Biol., Univ. Nac. Autó. México, 36:299, June 20, type from Pápayo, Guerrero.

MARGINAL RECORDS (Villa-R., 1967:296, unless otherwise noted).—Sinaloa: Sinaloa de Leyva; 2 mi. E Aguacaliente (Jones, *et al.*, 1972:13). Durango: near Pueblo Nuevo (Davis, 1970:119). Zacatecas: 5½ mi. S Moyahua (Genoways and Jones, 1968:744). Jalisco: El Zapote; 2 mi. N Ciudad Guzmán (Handley, 1966:300); Pihuamo. Michoacán: El Guayabo (Handley, 1966:300). Guerrero: 12 km. NNW Teloloapan; *Acuitlapán, 1830 m.* Morelos: *Coatetelco, 993 m.; Alpuyeca, 3500 ft.;* Hda. de Atliuayan (Novick, 1960:508); *Cueva del Salitre (ibid.).* Oaxaca (Goodwin, 1969:87): 18 mi. NW Sola de Vega; Las Cuevas (10 mi. NW Tehuantepec); Salina Cruz. Guerrero: Zacatula (Alvarez, 1968:27), thence up coast to point of beginning.

Artibeus jamaicensis yucatanicus J. A. Allen

1904. *Artibeus yucatanicus* J. A. Allen, Bull. Amer. Mus. Nat. Hist., 20:232, June 29, type from Chichén-Itzá, Yucatán.
1908. *Artibeus jamaicensis yucatanicus,* Andersen, Proc. Zool. Soc. London, 2:263, September 7.

MARGINAL RECORDS (Davis, 1970:118, unless otherwise noted).—Tamaulipas: 30 mi. NNW El Mante, 975 ft., thence along coast to Yucatán: 19 km. E Progresso (Birney, *et al.*, 1974:7). Quintana Roo: Cozumel Island (Jones and Lawlor, 1965:412). Honduras: Guanaja Island; Roatán Island. Guatemala: Tikal (Rick, 1968:518). Tabasco: ½ mi. S Banancán. Oaxaca (Goodwin, 1969:86): Tapanatepec; *Zanatepec;* 20 mi. NW La Ventosa; Cerro San Felipe. Puebla: 10 km. S Ajalpan, 1600 m. (Alvarez and Ramírez-P.,

1972:171). Veracruz: Mirador (Handley, 1966:300). Puebla: 2 mi. W Villa Avila Camacho, 250 m. (LaVal, 1972:450). Querétaro: 11 mi. W Jalpan, 2500 ft. San Luis Potosí: El Salto, 2000 ft. Tamaulipas: Cueva del Abra, 2 mi. W El Abra, 1000 ft. *A. j. yucatanicus* has been recorded (Baker, *et al.*, 1973:78, 85) from *10 km. S Solusuchiapa, 395 m., Chiapas.*

Artibeus lituratus
Big Fruit-eating Bat

Length of head and body, 87–100; length of forearm, 63.0–75.8; greatest length of skull, 29.7–34.0; zygomatic breadth, 17.1–20.2; length of upper tooth-row, 10.3–12.2. Upper parts as in *A. jamaicensis* but averaging browner and darker; underparts seal brown, hairs lacking silvery tips; "skull . . . with prominent postorbital and preorbital processes, these united with a beading that traverses the frontal region obliquely and meets the prominent sagittal crest" (Davis, 1970:107).

Artibeus lituratus intermedius J. A. Allen

1897. *Artibeus intermedius* J. A. Allen, Bull. Amer. Mus. Nat. Hist., 9:33, March 11, type from San José, Costa Rica.
1969. *Artibeus lituratus intermedius,* Goodwin, Bull. Amer. Mus. Nat. Hist., 141:87, April 30.

MARGINAL RECORDS (recorded as *A. lituratus* or *A. l. palmarum*).—Sinaloa: 12 mi. NE San Benito (Jones, *et al.*, 1972:14). Durango: Santa Ana (Jones, 1964a:751). Sinaloa: Santa Lucia (Jones, 1964e:512). Durango: Tecomates (Crossin, *et al.*, 1973:197). Nayarit: Huajimic. Zacatecas: 5½ mi. S Moyahua (Genoways and Jones, 1968:744). Morelos: Cuernavaca. Querétaro: 17¾ km. NW Jalpan (Spenrath and LaVal, 1970:396). San Luis Potosí: El Salto.

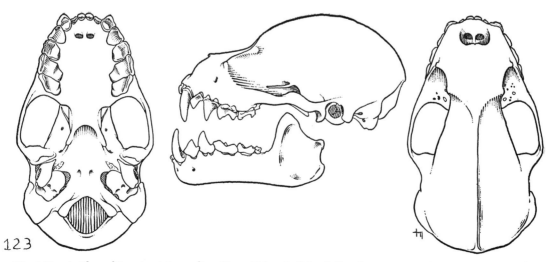

Fig. 123. *Artibeus lituratus intermedius,* Barro Colorado Island, Canal Zone, Panamá, No. 45086 K.U., ♀, X 2.

Tamaulipas: 10 km. N, 8 km. W El Encino (Alvarez, 1963:402). Campeche: 5 km. S Champotón (91784 KU). Yucatán: Pisté (91779 KU). Quintana Roo: Cozumel Island (Jones and Lawlor, 1965:412), thence southward along Caribbean Coast to South America, and then northward along Pacific Coast to Costa Rica: type locality. El Salvador: Corinto. Guerrero: Zacatula (Alvarez, 1968:27). Sinaloa: 1 mi. S El Dorado (Jones, *et al.*, 1962:153).

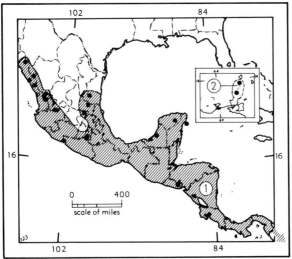

Map 123. *Artibeus lituratus*.

1. *A. l. intermedius*   2. *A. l. palmarum*

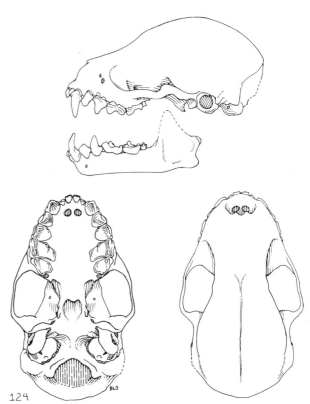

Fig. 124. *Artibeus hirsutus*, 2 mi. ENE Tala, 4500 ft., Jalisco, No. 36582 K.U., ♂, X 2.

Artibeus lituratus palmarum J. A. Allen and Chapman

1897. *Artibeus palmarum* J. A. Allen and Chapman, Bull. Amer. Mus. Nat. Hist., 9:16, February 26, type from Botanical Gardens at Port of Spain, Trinidad.
1949. *A*[*rtibeus*]. *l*[*ituratus*]. *palmarum*, Hershkovitz, Proc. U.S. Nat. Mus., 99:447, May 10.
1899. *Artibeus femurvillosum* Bangs, Proc. New England Zool. Club, 1:73, November 24, type from Santa Marta, Colombia.

MARGINAL RECORDS.—Lesser Antilles: St. Vincent Island; Douglaston, Grenada.

Artibeus hirsutus Andersen
Hairy Fruit-eating Bat

1906. *Artibeus hirsutus* Andersen, Ann. Mag. Nat. Hist., ser. 7, 18:420, December, type from La Salada, Michoacán.

Length of head and body, 79–86. Resembling *A. jamaicensis* but averaging smaller; tibia and interfemoral membrane densely haired; color of fur of upper side of body in adults drab with silvery tinge. Length of maxillary tooth-row, 9.5–10.4; forearm, 52.0–59.7. (After Andersen, 1906:420.)

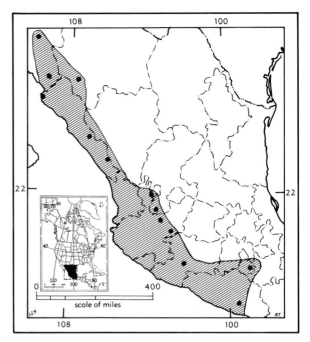

Map 124. *Artibeus hirsutus*.

MARGINAL RECORDS.—Sonora: tributary, W side Río Yaqui, *ca*. 1 mi. S El Novillo (Findley and Jones, 1965:331). Chihuahua: Río Batopilas, near La Bufa (Anderson, 1960:5). Sinaloa: 6 km. E Cosalá (Jones, *et al.*, 1972:13); Santa Lucía (Jones, 1964e:512). Jalisco: 4 mi. W Villa Guerrero, 5500 ft. (Watkins, *et al.*, 1972:18). Zacatecas: 5½ mi. S Moyahua (Genoways and Jones, 1968:744). Jalisco (Watkins, *et al.*, 1972:18): 9 mi. N Guadalajara, 4000 ft.; 4 mi. SSW Ocotlán. Michoacán: type locality. Morelos: Jonacatepec. Guerrero: 4 mi. N Colotlipa, 3000 ft.; thence up coast to Sinaloa: 1 mi. N Zaragoza (Jones, *et al.*, 1972:13). Sonora: *1 km. SW La Aduana, 1600 ft.* (Loomis and Davis, 1965:497); ¼ mi. W Aduana.

Artibeus inopinatus Davis and Carter
Honduran Fruit-eating Bat

1964. *Artibeus inopinatus* Davis and Carter, Proc. Biol. Soc. Washington, 77:119, June 26, type from Choluteca, 10 ft., Dept. Choluteca, Honduras.

Length of forearm, 48–53; greatest length of skull to front of canines, 25.0–26.0; zygomatic breadth, 15.4–16.3; length of upper tooth-row, 8.8–9.2. Resembling *A. hirsutus* but distinguishable by smaller size and absence of spine on posterior border of palate, which is evenly concave.

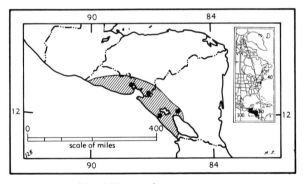

Map 125. *Artibeus inopinatus*.

MARGINAL RECORDS.—El Salvador: Divisidero (Davis and Carter, 1964:121). Honduras: type locality. Nicaragua (Baker and Jones, 1975:4): San Francisco, 400 ft., Boaco; 25 mi. WNW Managua, in León.

Artibeus aztecus
Highland Fruit-eating Bat

Length of head and body, 61–71; length of forearm, 40.0–48.3; greatest length of skull, 21.0–23.8; zygomatic breadth, 12.0–14.4; length of upper tooth-row, 6.8–7.9; interfemoral membrane deeply incised (width less than 7.2) and conspicuously hairy; coloration intensely black to pale brown. Resembling *A. toltecus* but distin-

guishable by larger size. This species occurs mainly in moutainous areas above 3000 ft.

Artibeus aztecus aztecus Andersen

1906. *Artibeus aztecus* Andersen, Ann. Mag. Nat. Hist., ser. 7, 18:422, December, type from Tetela del Volcán, Morelos.

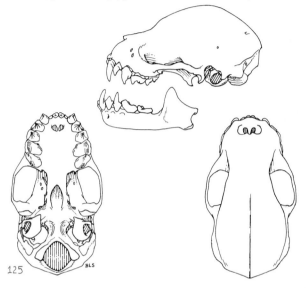

Fig. 125. *Artibeus aztecus aztecus*, Santa Lucia, 3600 ft., Sinaloa, No. 94135 K.U., ♂, X 2.

MARGINAL RECORDS (Davis, 1969:21, 22, unless otherwise noted).—Nuevo León: Rancho La Laguna, 2⅖ km. N Zaragoza, 1371 m. (Jiménez-G., 1971:135). Tamaulipas: Rancho del Cielo, 5 mi. NW Gómez Farías, 3300 ft. Querétaro: 20 mi. E Landa, 5400 ft. Oaxaca: Santo Tomás Teipan, 7000 ft. (Goodwin, 1969:90); Kilometer 183, 1700 m., 36½ km. N San Gabriel Mixtepec (Schaldach, 1966:291, as *A. aztecus* only). Guerrero: 2 mi. W Omiltemi, 7900 ft. Jalisco: N slope Nevada de Colima, 8000 ft. Sinaloa: *near Santa Lucía;* Rancho Batel, 5 mi. NE Santa Lucía, 5200 ft. (Koopman, 1961:536, as *A. aztecus* only). Durango: 2 mi. N Pueblo Nuevo, 6000 ft. (Baker and Greer, 1962:70, as *A. aztecus* only). Nayarit: El Maguey, E Huajicori, 6400 ft. Michoacán: 3 km. SW Turundeo, 1900 m. (Alvarez and Ramírez-P., 1972:170). Querétaro: 12⅖ mi. WSW San Joaquín (Schmidly and Martin, 1973:91). San Luis Potosí: Cerro Campanario. Not found: Nayarit: 2 km. NE Tomates, 5 km. W Río Chihuacora.

Artibeus aztecus major Davis

1969. *Artibeus aztecus major* Davis, Southwestern Nat., 14:22, May 16, type from 7½ km. E Canaán, *ca.* 8000 ft., Province of San José, Costa Rica.

MARGINAL RECORDS (Davis, 1969:23, unless otherwise noted).—Costa Rica: Monte Verde (2946

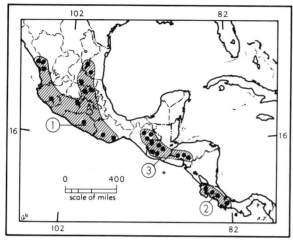

Map 126.　*Artibeus aztecus.*

1. *A. a. aztecus*　　　　　　　2. *A. a. major*
　　　　　3. *A. a. minor*

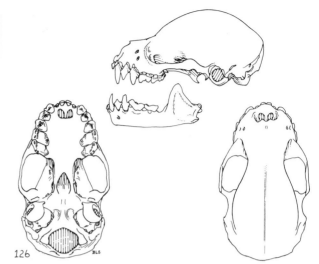

Fig. 126.　*Artibeus toltecus hesperus*, 15 km. NW Cihuatlán, Jalisco, No. 87357 K.U., ♀, X 2.

LaVal at KU); Vara Blanca, 6500–6800 ft.; Villa Mills, 10,000 ft.; *type locality.* Panamá: Cerro Punta, 5300 ft. Costa Rica: *3 km. E Canaán, ca. 6000 ft.; San Gerardo, 4500 ft.*

Artibeus aztecus minor Davis

1969. *Artibeus aztecus minor* Davis, Southwestern Nat., 14:22, May 16, type from San Cristóbal Verapaz, 4485 ft., Alta Verapaz, Guatemala.

MARGINAL RECORDS (Davis, 1969:22).—Chiapas: 12 mi. SE San Cristóbal de las Casas, 6700 ft.; Finca Patihuitz, 33 mi. NE Las Margaritas, *ca.* 6000 ft. Guatemala: type locality; 1 km. SE San Jerónimo, 3250 ft. Honduras: La Esperanza, 5400 ft.; 12$\frac{7}{10}$ km. S Comayagua, 1885 ft.; 10 mi. SE Tegucigalpa, *ca.* 5200 ft. Guatemala: 2 mi. S Chimaltenango, 5700 ft.; Verdenango, Atitlan Lake, 5173 ft.; San Pedro Solomá, 7400 ft. Chiapas: 1$\frac{1}{10}$ mi. SSE Zapaluta (= La Trinitaria), 5700 ft.; *2 km. W Teopisca, 7000 ft.*

Artibeus toltecus
Lowland Fruit-eating Bat

Length of head and body, 51–63; length of forearm, 36.3–43.0 (normally 37–40, rarely 41 and in only four of 209 specimens exceeding 42); greatest length of skull, 19.0–21.0; zygomatic breadth, 11.0–12.5; length of upper tooth-row, 6.1–7.0; interfemoral membrane deeply incised (width less than 7.2) and conspicuously hairy; coloration as in *A. aztecus*, but distinguishable by smaller size (after Davis, 1969:23–28).

Artibeus toltecus hesperus Davis

1969. *Artibeus toltecus hesperus* Davis, Southwestern Nat., 14:25, May 16, type from Agua del Obispo, 3300 ft., Guerrero, México.

MARGINAL RECORDS (Davis, 1969:26, unless otherwise noted).—Sinaloa: 12 mi. NE San Benito (Jones, *et al.*, 1972:15). Durango: Ventanas (Andersen, 1908:300, as *A. t. toltecus*). Sinaloa: Santa Lucía (94947 KU); 2 mi. NW Palmito. Durango: Paso de Sihuacori (Crossin, *et al.*, 1973:199). Jalisco: 4$\frac{1}{2}$ mi. W Villa Guerrero (Watkins, *et al.*, 1972:20). Zacatecas: 5$\frac{1}{2}$ mi. S Moyahua (Genoways and Jones, 1968:744, as *A. toltecus* only). Jalisco (Watkins, *et al.*, 1972:20): 9 mi. NE Guadalajara, 4000 ft.; 8 mi. E Jilotlán de los Dolores, 2000 ft. Morelos: 6 km. W Yautepec (Alvarez and Ramírez-P., 1972:170). Oaxaca: 12 mi. NW Sola de Vega (Goodwin, 1969:89, as *A. toltecus* only); 8 km. NW Tapanatepec, *ca.* 100 ft.; *4 mi. E Tapanatepec, ca. 500 ft.* Chiapas: 8 km. N Berriozábal, 1065 m. (Baker, *et al.*, 1973:78, 85); 38 km. N Huixtla, *ca.* 100 ft. Guatemala: Villalobos, 13 km. S Guatemala City, 4200 ft. El Salvador: San Salvador (Felten, 1956:351, as *A. cinereus toltecus*); Volcán de San Miguel (Burt and Stirton, 1961:35, as *A. cinereus*, but *A. toltecus hesperus*, W. B. Davis, *in Litt.*, October 12, 1970); 20 km. W La Libertad, 800 ft., thence up coast to point of beginning. Not found: Chiapas: Rancho San Miguel, Municipio de Cintalapa, *ca.* 2000 ft.—The southernmost record of occurrence of subspecies *A. t. hesperus* is in Nicaragua: Isla de Ometepe, Rivas (Jones, Smith, and Turner, 1971:15), 220 mi. SE Volcán de San Miguel, El Salvador. The Nicaraguan occurrence is not shown on Map 127.

Artibeus toltecus toltecus (Saussure)

1860. *Stenoderma tolteca* Saussure, Revue et Mag. Zool., Paris, ser. 2, 12:427, October, type from México; restricted to Mirador, Veracruz, by Hershkovitz, Proc. U.S. Nat. Mus., 99:449, May 10, 1949.
1958. *Artibeus toltecus*, Davis, Proc. Biol. Soc. Washington, 71:165, December 31.

MARGINAL RECORDS (Davis, 1969:27, 28, unless otherwise noted).—Nuevo León: Cueva La Boca, 4 km.

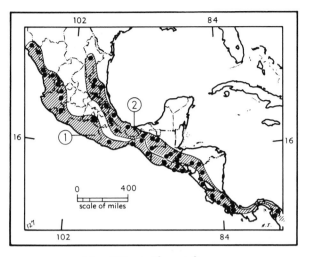

Map 127. *Artibeus toltecus.*

1. *A. t. hesperus*   2. *A. t. toltecus*

SE Santiago, 445 m. (Jiménez-G., 1971:136). Tamaulipas: Rancho Pano Ayuctle, 300 ft. (Jones, *et al.*, 1962:153); *Rancho del Cielo, 3300 ft.; Río Sabinas, ca. 300 ft., near Gómez Farías.* San Luis Potosí: 19 km. SE Ebano, *ca.* 100 ft. Veracruz: 1 mi. SE Plan del Río, 1200 ft.; 2 mi. E Lago Catemaco, *ca.* 3000 ft. Chiapas: 8 km. S Solusuchiapa, 400 ft.; 37 km. NE Altamirano, 1920 ft.; Florida, *ca.* 525 m. (Davis, *et al.*, 1964:385). Guatemala: Finca Los Alpes, 1000 m. (Jones, 1966:461). Honduras: Copán, 2150 ft.; La Flor Archaga, 4500 ft.; 6 km. E Danlí, 2200 ft. Nicaragua: Jinotega (Hall and Kelson, 1959: 141); *17 km. by road N Matagalpa, 4875 ft.* Costa Rica: Los Diamantes, *ca.* 1000 ft. Panamá (Handley, 1966:768, as *A. toltecus*): upper Río Changena, 2400–4800 ft.; Tacarcuna Village, 1950 ft., thence southeastward into South America, thence northwestward to Panamá (Handley, 1966:768, as *A. toltecus* only): Cana, 2000 ft.; Cerro Punta, 5300–5700 ft. Costa Rica: 15 km. E Potrero Grande, *ca.* 3000 ft.; 3 km. NE Canaán, *ca.* 6000 ft.; 2 km. ESE Bajos de Jorco, Río Jorco, 3000 ft.; 5 mi. E Tilarán, 2275 ft. Nicaragua: Volcán Mombacho, 4000 ft.; *Santa María de Ostuma, 4062 ft.* Honduras: Nueva Ocotepeque, 2800 ft. Guatemala: 1 km. SE San Jerónimo, 3250 ft.; *2 km. SE Salamá, ca. 3100 ft.* Chiapas: 14 mi. SSE Zapaluta [= La Trinitaria], 2700 ft.; *6½ km. SE Rayón* (Baker, *et al.*, 1973:78, 85). Oaxaca: Vista Hermosa, 1600 m. (99696 KU). Veracruz: Ojo de Agua de Atoyac, Potrero, 1500 ft. Puebla: 2 mi. W Villa Avila Camácho, 250 m. (LaVal, 1972:450). San Luis Potosí: 1 km. W Huichihuayán, *ca.* 400 ft.; El Salto, 30 km. NE Ciudad del Maíz, *ca.* 2000 ft. Not found: Veracruz: 5 km. NNW San Juan Diaz Covarrubia, 520 ft. (*sic*).

Artibeus cinereus
Gervais' Fruit-eating Bat

Selected measurements are: length of forearm, 37.2–41.8; greatest length of skull, 19.7–21.6; zygomatic breadth, 11.3–12.5; length of maxillary tooth-row, 6.2–7.2. Larger than *A. watsoni*, especially in length of skull and palate. Resembling *A. watsoni* but less specialized; m3 usually lacking; dorsolateral and lateral swelling of frontal bone in orbital region lacking.

Only one subspecies, *A. c. cinereus*, is pertinent to the present work.

Artibeus cinereus cinereus (Gervais)

1856. *Dermanura cinereum* Gervais, Mammifères, *in* [Castelnau] Expéd. dans les parties centrales de l'Amér. du Sud . . . , p. 36, Pl. 8, Fig. 4, Pl. 9, Figs. 4 and 4a, and Pl. 11, Fig. 3 (for 1855), type from Brazil.
1901. *Artibeus cinereus* Thomas, Ann. Mag. Nat. Hist., 8:143, August.

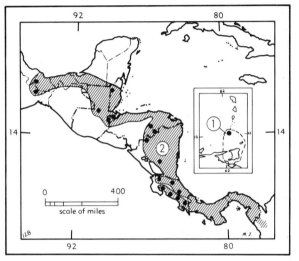

Map 128. *Artibeus cinereus cinereus* (1) and *Artibeus watsoni* (2).

MARGINAL RECORDS.—Lesser Antilles: Grenada (Jones and Phillips, 1970:132).

Artibeus watsoni Thomas
Thomas' Fruit-eating Bat

1901. *Artibeus Watsoni* Thomas, Ann. Mag. Nat. Hist., ser. 7, 7:542, June, type from Bogava [= Bugaba], Chiriquí, Panamá.
1906. *Dermanura jucundum* Elliot, Proc. Biol. Soc. Washington, 19:50, May 1, type from Achotal, Veracruz.

Length of head and body approx. 58; length of forearm, 35.0–46.8; greatest length of skull, 18.7–21.2; zygomatic breadth, 10.1–13.0; length of upper tooth-row, 6.0–7.2. Upper parts light brown or tan; underparts paler; two pairs of facial stripes present. Ears light-rimmed.

Occupies humid lowlands and mid-elevations (rarely to 5000 ft.). Appears to be absent from dry tropical sections of Middle America.

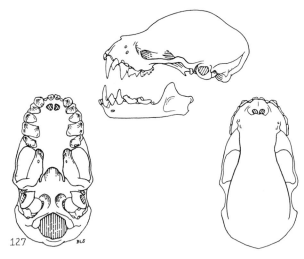

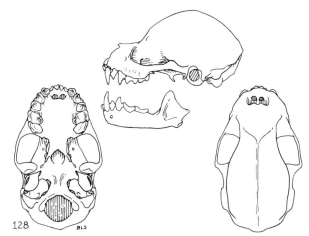

Fig. 127. *Artibeus watsoni*, 7 km. N, 4 km. E Jalapa, 660 m., Nuevo Segovia, Nicaragua, No. 111161 K.U., ♂, X 2.

Fig. 128. *Artibeus phaeotis palatinus*, 6½ km. N, 1 km. E Cosigüina, 10 m., Chinandega, Nicaragua, No. 115074 K.U., ♀, X 2.

MARGINAL RECORDS (Davis, 1970:393, 394, unless otherwise noted).—Veracruz: Achotal, *ca.* 500 ft. Belize: Gallon Jug. Guatemala: 25 km. SSW Puerto Barrios, 300 ft., thence along coast to South America, and up Pacific Coast to Costa Rica: 6½ mi. N, 2 mi. W Puntarenas, 100 ft.; 2 km. ESE Bajos de Jorco, Río Jorco, *ca.* 1800 ft.; 9½ mi. N San Isidro General, *ca.* 5000 ft.; *2 km. SW Ujarras, Río Ceibo, ca. 1500 ft.*; 14 km. NE Potrero Grande, *ca.* 3500 ft.; Turrialba, *ca.* 1700 ft.; Puerto Viejo, *ca.* 200 ft.; Santa Clara, *ca.* 600 ft. Nicaragua (Jones, *et al.*, 1971:15): 1 km. N, 1½ km. W Villa Somoza; 7 km. N, 4 km. E Jalapa. Honduras: Río Coco, 78 mi. ENE Danlí, 900 ft.; 40 km. E Cataco[a]mas, *ca.* 1600 ft. Guatemala: 16 km. N, 4 km. E Los Amates, 50 ft.; 10 mi. N Sebol (= Embarcadero), 900 ft. Oaxaca: 18 mi. N Matías Romero; *24 mi. N Matías Romero.*

Artibeus phaeotis
Pygmy Fruit-eating Bat

Length of head and body, 51–60; length of forearm, 35.2–41.8; greatest length of skull, 17.5–20.5; zygomatic breadth, 10.4–12.3; length of upper tooth-row, 5.5–6.7. Two pairs of whitish facial stripes present. Ears usually rimmed with yellowish white. Distinguishable from *A. watsoni* by smaller molars, wider talon on M1, and absence of m3 (normally present in *watsoni*).

Occurs at lower and middle elevations (sea level to 4000 ft.).

Artibeus phaeotis nanus Andersen

1906. *Artibeus nanus* Andersen, Ann. Mag. Nat. Hist., ser. 7, 18:423, December, type from Tierra Colorada, Sierra Madre del Sur, Guerrero.
1965. *Artibeus phaeotis nanus*, Jones and Lawlor, Univ. Kansas Publ., Mus. Nat. Hist., 16:412, April 13.

MARGINAL RECORDS (Davis, 1970:400, unless otherwise noted).—Sinaloa: San Ignacio (Jones, *et al.*, 1972:14); Puerta de Cañoa, 11 mi. N, 2½ mi. E Mazatlán; Presidio (Davis, 1958:164). Nayarit: 40 mi. E Acaponeta. Jalisco (Watkins, *et al.*, 1972:21): 2 mi. N Milpillas, 3000 ft.; 8 mi. E Jilotán de los Dolores, 2000 ft. Guerrero: type locality; Tres Palos; Zacatula (Alvarez, 1968:27, as *A. turpis nanus*). Michoacán: *3 km. S Melchor Ocampo (ibid.)*, thence up coast to point of beginning.

Artibeus phaeotis palatinus Davis

1970. *Artibeus phaeotis palatinus* Davis, Southwestern Nat., 14:400, February 16, type from 15 km. SW Retalhuleu, 240 ft., Retalhuleu, Guatemala.

MARGINAL RECORDS (Davis, 1970:401).—Oaxaca: 8 mi. NW Tapanatepec, *ca.* 100 ft. Chiapas: near Tonalá, *ca.* 100 ft.; 20 km. SE Pijijiapan, *ca.* 100 ft.; *2 km. N Sesecapa (= 12 km. ESE Mapastepec), 100 ft.*; near Escuintla, 100 ft. Guatemala: El Carmen, 1600 ft. El Salvador: 20 km. W Chalatenango, 800 ft. Nicaragua: 1 mi. SE Yalaguina, 2200 ft.; *Daraili, 5 km. N, 15 km. E Condega, ca. 3000 ft.*; 4 mi. E Matagalpa, 2600 ft.; San Antonio, 50 ft.; 11 km. S, 3 km. E Rivas, 160 ft. Costa Rica: 1⅘ mi. NE Cañas, *ca.* 300 ft.; 2 km. ESE Bajos de Jorco, Río Jorco, *ca.* 3000 ft., thence up Pacific Coast to point of beginning.

Artibeus phaeotis phaeotis (Miller)

1902. *Dermanura phaeotis* Miller, Proc. Acad. Nat. Sci. Philadelphia, 54:405, September 12, type from Chichén-Itzá, Yucatán.
1965. *Artibeus phaeotis phaeotis*, Jones and Lawlor, Univ. Kansas Publ., Mus. Nat. Hist., 16:412, April 13.
1906. *Artibeus turpis* Andersen, Ann. Mag. Nat. Hist., ser. 7, 18:422, December, type from Teapa, Tabasco.

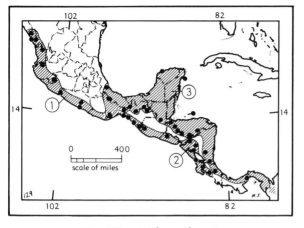

Map 129. *Artibeus phaeotis.*

1. *A. p. nanus*   2. *A. p. palatinus*

3. *A. p. phaeotis*

Enchisthenes hartii (Thomas)
Little Fruit-eating Bat

1892. *Artibeus hartii* Thomas, Ann. Mag. Nat. Hist., ser. 6, 10:409, November, type from Trinidad, Lesser Antilles.
1908. *Enchisthenes harti*, Andersen, Proc. Zool. Soc. London, 2:224, September 7.

Length of head and body of an adult male from Nicaragua, 60. Selected measurements of 13 specimens from Costa Rica: length of forearm, 37.0–42.6; greatest length of skull, 20.2–21.9; zygomatic breadth, 11.7–12.9; length of upper tooth-row, including canine, 6.7–7.2. Upper parts dark brown, almost blackish on head and shoulders; underparts paler than back, darkest anteriorly; facial stripes narrow, buffy (after Goodwin, 1942:137). Color Mummy Brown in a Oaxacan specimen (Goodwin, 1969:90).

MARGINAL RECORDS (Davis, 1970:398, 399, unless otherwise noted).—Yucatán: 10 mi. W Progreso, *ca.* 50 ft. Quintana Roo: Cozumel Island (Jones and Lawlor, 1965:412). Honduras: Mangrove Bay, Guanaja Id., thence southeastward along coast into South America. Thence up coast to Panamá: La Chorrera (Handley, 1966:769). Costa Rica: 2½ mi. E Turrialba, 1600 ft. Nicaragua: El Castillo, 130 ft.; Cacao, 22 km. W Muelle de las Bueyes, 400 ft. Honduras: Río Coco, 76–78 mi. ENE Danlí, *ca.* 900 ft.; 21 km. E Danlí, 1500 ft.; *7–9 km. E Danlí, 2000 ft.;* Comayagua, 1900 ft.; 5 km. NE Jesús de Otero [= Otoro], 2100 ft.; 7 km. N Santa Bárbara, 400 ft. Guatemala: 25 km. SW Puerto Barrios, 300 ft.; Río San Simón, 6 km. NE Raxrujá, 450 ft. Chiapas: Florida, 50 km. E Altamirano, 1700 ft.; Yaxoquintela, 37 km. NE Altamirano, 1920 ft.; 2 km. S Ixtapa, 3900 ft.; *8 km. N Berriozábal, 1065 m.* (Baker, *et al.*, 1973:78, 85); vic. Mal Paso, 400 ft. Oaxaca (Goodwin, 1969:89): Sierra Madre, N of Zanatepec; San Bartolo Yautepec; Chiltepec. Veracruz: Plan del Río (Davis, 1958:163); *1 mi. SE Plan del Río, 1000 ft.,* thence along coast to beginning.

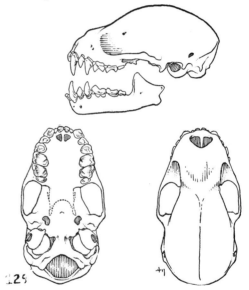

Fig. 129. *Enchisthenes hartii,* La Flor Archaga, Honduras, No. 26239 A.M.N.H., ♂, X 2.

Genus **Enchisthenes** Andersen

1906. *Enchisthenes* Andersen, Ann. Mag. Nat. Hist., ser. 7, 18:419, December. Type, *Artibeus hartii* Thomas.

Resembling *Artibeus,* but inner upper incisor not bifid, and 3rd molar both above and below well developed and affecting form of surrounding bone (after Miller, 1907:162). Goodwin (1969:90) regarded the two characters mentioned above as insufficient for assigning this animal to a genus other than *Artibeus.* The genus contains only one known species.

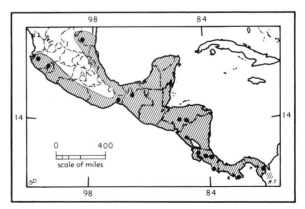

Map 130. *Enchisthenes hartii.*

MARGINAL RECORDS.—Tamaulipas: Aserradero del Infiernillo; *Rancho del Cielo* (Baker and Lopez, 1968:361). Chiapas: 8 km. N Berriozábal, 1065 m. (Baker, *et al.*, 1973:78, 85). Honduras: La Flor Archaga; Chichicaste (LaVal, 1969:821). Costa Rica (Gardner, *et al.*, 1970:722): Vara Blanca; Río Pacuare, Pacuare. Panamá: Tacarcuna Village (Handley, 1966:768), thence southeastward into South America, and northwestward to Panamá: Cerro Hoya, 2600 ft. (*ibid.*). Costa Rica: 2 km. SW Rincón, 15 m. (Gardner, *et al.*, 1970:722); Monte Verde (3114 LaVal at KU). Oaxaca: Sierra Madre, N Zanatepec (Goodwin, 1969:90). Jalisco: 2 mi. N Ciudad Guzmán; 10 mi. SE Talpa de Allende, 5350 ft. (Jones, 1964e:512). Also reported from Arizona: *Tucson Mts.* (Irwin and Baker, 1967:195).

Genus **Ardops** Miller—Tree Bats

Revised by Jones and Schwartz, Proc. U.S. Nat. Mus., 124(3634):1–13, December, 1967.

1906. *Ardops* Miller, Proc. Biol. Soc. Washington, 19:84, June 4. Type *Stenoderma nichollsi* Thomas.

Skull resembling that of *Artibeus*, but relatively broader; rostrum moderately flattened; supraorbital ridges thickened, rounded; interpterygoid space extending anteriorly to level of 1st molar; borders of palatal emargination almost parallel; inner upper incisors short and thick; M3 small, peglike; 1st lower molar lacks metaconule. Dentition, i. $\frac{2}{2}$, c. $\frac{1}{1}$, p. $\frac{2}{2}$, m. $\frac{3}{3}$.

The genus is known only from the Lesser Antilles. "*Ardops* is related to three other endemic Antillean genera, *Ariteus* of Jamaica, *Phyllops* of Cuba and Hispaniola, and *Stenoderma* of Puerto Rico and the Virgin Islands. Of these, *Ardops* may be related most closely to *Ariteus*, from which it differs principally in having a broader rostrum, narrower mesopterygoid fossa, distinctive sphenoid-basioccipital region, . . . and the absence of any trace of a metaconule on the first lower molar." (Jones and Schwartz, 1967:1,2.) *Ariteus* lacks the small, peglike M3 found in *Ardops*.

Ardops nichollsi
Tree Bat

Selected measurements: length of head and body, 53–73; length of forearm, 42.5–51.9; greatest length of skull, 20.4–24.0; zygomatic breadth, 13.4–16.0; length of maxillary tooth-row (C-M3), 6.2–8.4; length of mandibular tooth-row (c-m3), 6.3–8.1. Upper parts Prout's Brown or Bister to Buffy Brown; tricolored appearance when pelage is parted; faint grayish-brown patch over shoulder; white spot at jct. wing with body in all

subspecies, better developed in females than in males. Underparts rich brownish tinged with grayish white, not tricolored. Bats of this species roost in arboreal vegetation.

Ardops nichollsi annectens Miller

1913. *Ardops annectens* Miller, Proc. Biol. Soc. Washington, 26:33, February 8, type from Guadeloupe, Lesser Antilles. Known only from type locality.—**See** addenda.
1967. *Ardops nichollsi annectens*, Jones and Schwartz, Proc. U.S. Nat. Mus., 124(3634):10, December.

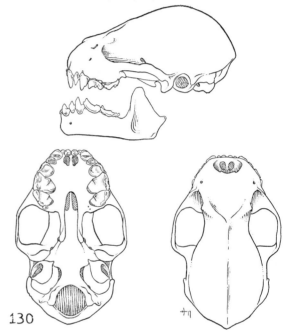

Fig. 130. *Ardops nichollsi luciae*, St. Lucia, West Indies, No. 110918 U.S.N.M., sex ?, X 2.

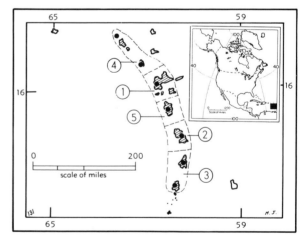

Map 131. *Ardops nichollsi.*

1. *A. n. annectens*   3. *A. n. luciae*
2. *A. n. koopmani*   4. *A. n. montserratensis*
          5. *A. n. nichollsi*

Ardops nichollsi koopmani Jones and Schwartz

1967. *Ardops nichollsi koopmani* Jones and Schwartz, Proc.
U.S. Nat. Mus., 124(3634):11, December, type from near
Balata, Martinique, Lesser Antilles. Known only from type
locality.

Ardops nichollsi luciae (Miller)

1902. *Stenoderma luciae* Miller, Proc. Acad. Nat. Sci.
Philadelphia, 54:407, September 12, type from St. Lucia,
Lesser Antilles.
1967. *Ardops nichollsi luciae,* Jones and Schwartz, Proc. U.S.
Nat. Mus., 124(3634):9, December.

MARGINAL RECORDS (Jones and Phillips,
1970:138).—Lesser Antilles: St. Lucia; St. Vincent.

Ardops nichollsi montserratensis (Thomas)

1894. *Stenoderma montserratense* Thomas, Proc. Zool. Soc.
London, p. 133, June, type from Montserrat, Lesser
Antilles.—**See** addenda.
1967. *Ardops nichollsi montserratensis,* Jones and Schwartz,
Proc. U.S. Nat. Mus., 124(3634):7, December.

MARGINAL RECORDS (Jones and Phillips,
1970:138).—Lesser Antilles: St. Eustatius; Montserrat.

Ardops nichollsi nichollsi (Thomas)

1891. *Stenoderma nichollsi* Thomas, Ann. Mag. Nat. Hist., ser.
6, 7:529, June, type from Dominica, Lesser Antilles. Known
only from Dominica.
1906. *Ardops nichollsi,* Miller, Proc. Biol. Soc. Washington,
19:84, June 4.

Genus **Phyllops** Peters

1865. *Phyllops* Peters, Monatsb. preuss. Akad. Wiss., Berlin,
p. 356. Type, *Phyllostoma albomaculatum* Gundlach [=
Arctibeus falcatus Gray].

"Like *Ardops,* but inner upper incisor with
crown higher than long, and without distinct secondary cusp, first and second upper molars with
hypocone much lower than protocone, first lower
molar with well-developed metaconid connected
with ridge on inner side of protoconid, and
palatal emargination with sides strongly converging, in continuation of the divergent pterygoids."
(Miller, 1907:164, 165.)

Phyllops vetus, known only as a fossil, is not
here included because Anthony (1917:336) states
that "judging from the condition of the specimens
[*P. vetus*] . . . has not been frequenting this region since the more recent animal remains, bats
of the present day, were deposited."

KEY TO SPECIES OF PHYLLOPS

1. Palatal emargination U-shaped; length of
   forearm more than 40. . . . . .*P. falcatus,*   p. 165
1′. Palatal emargination V-shaped; length of
    forearm less than 40. . . . . .*P. haitiensis,*   p. 165

Phyllops falcatus (Gray)
Cuban Fig-eating Bat

1839. *Arctibeus falcatus* Gray, Ann. Nat. Hist., ser. 1, 4:1,
September, type from Cuba.
1907. *Phyllops falcatus,* Miller, Bull. U.S. Nat. Mus., 57:165,
June 29.

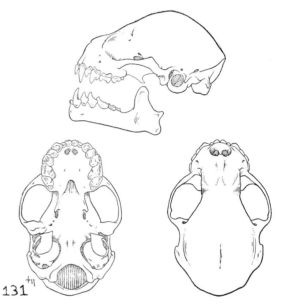

131

Fig. 131. *Phyllops falcatus,* Santiago, Cuba, No. 143844
U.S.N.M., ♀, X 2.

Length of head and body approx. 48; length of
forearm of the type, an immature female, 41.9.
Frontal area between supraorbital ridges flattened; palatal emargination U-shaped, extending
anteriorly approx. to level of middle of 2nd upper
molar.

MARGINAL RECORDS.—Cuba: Matanzas; Daiquirí; near Cienfuegos.

Phyllops haitiensis (J. A. Allen)
Dominican Fig-eating Bat

1908. *Ardops haitiensis* J. A. Allen, Bull. Amer. Mus. Nat.
Hist., 24:581, September 11, type from Caña Honda,
Dominican Republic.
1917. *Phyllops haitiensis,* Anthony, Bull. Amer. Mus. Nat.
Hist., 37:337, May 28.

Differing from *P. falcatus* mainly in smaller
size (forearm, 39), and V-shaped palatal emargi-

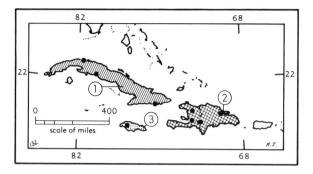

Map 132. *Phyllops falcatus* (1), *Phyllops haitiensis* (2), and *Ariteus flavescens* (3).

nation that extends anteriorly approx. to level of anterior edge of 2nd upper molar.

MARGINAL RECORDS.—Haiti: cave near St. Michel. Dominican Republic: Río Yuna, 6 mi. E Arenosa (Armstrong and Johnson, 1969:133); type locality. Haiti: Port-au-Prince.

Genus **Ariteus** Gray—Jamaican Fig-eating Bat

1838. *Ariteus* Gray, Mag. Zool. Bot., 2:491, February. Type, *Istiophorus flavescens* Gray.
1876. *Peltorhinus* Peters, Monatsb. preuss. Akad. Wiss., Berlin, p. 433. Type, *Artibeus achradophilus* Gosse.

Length of head and body approx. 51. "Like *Ardops*, but without the small upper molar; first lower molar with minute though evident metaconid." (Miller, 1907:165.) Dentition, i. $\frac{2}{2}$, c. $\frac{1}{1}$, p. $\frac{2}{2}$, m. $\frac{2}{3}$.

The genus is monotypic.

Ariteus flavescens (Gray)
Jamaican Fig-eating Bat

1831. *Istiophorus flavescens* Gray, Zool. Misc., No. 1, p. 37, February, type from an unknown locality. Known definitely only from Jamaica.
1838. *Ariteus flavescens* Gray, Mag. Zool. Bot., 2:491, February.
1851. *Artibeus achradophilus* Gosse, A naturalist's sojourn in Jamaica, p. 271, type from Content, Jamaica.

Characters as for the genus.

Genus **Stenoderma** É. Geoffroy St.-Hilaire

1818. *Stenoderma* É. Geoffroy St.-Hilaire, Description de l'Egypte . . . , 2:114. Type, "le sténoderme roux" [= *Stenoderma rufa* Desmarest, 1820]. Not *Stenoderma* Oken, 1816; Oken's names are not available.
1869. *Histiops* Peters, Monatsb. preuss. Akad. Wiss., Berlin, p. 399. Type, *Artibaeus undatus* Gervais [= *Stenoderma rufa* Desmarest].

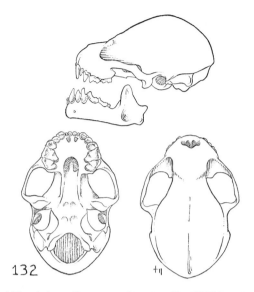

132

Fig. 132. *Ariteus flavescens*, Jamaica, No. 252771 U.S.N.M., ♀, X 2.

"In general like *Ardops*, but skull with nasal region much depressed between high supraorbital ridges; anterior nares directed chiefly upward and extending fully halfway from front of premaxillaries to point of juncture of supraorbital ridges which are not angulated at middle, but extend in a nearly straight line from front of orbit to sagittal crest; incisive foramina separated from roots of incisors by space equal to their greatest diameter; inner upper incisor with high slender crown, as in *Phyllops*; first and second upper molars with low but distinct metaconule on surface of crown between hypocone and metacone." (Miller, 1907:166.) Dentition, i. $\frac{2}{2}$, c. $\frac{1}{1}$, p. $\frac{2}{2}$, m. $\frac{3}{3}$.

Much can be said in support of the classification of the Microchiroptera recently proposed by Varona (1974) in his catalogue of mammals of the Antilles. His division of the family Phyllostomatidae into seven subfamilies instead of the four used here emphasizes the diversity in this group. His recognition of only two genera, in the Antilles, in subfamily Stenodermatinae emphasizes undoubted relationships between several species. This reduction in number of genera results from placing the generic names *Phyllops* and *Ardops* as junior synonyms of *Ariteus* and reducing *Ariteus* to subgeneric rank. However testing Varona's classification on mainland animals may recommend retention of more features of the older arrangement used here than he retains in listing only the kinds known from the Antilles.

Stenoderma rufum
Desmarest's Fig-eating Bat

Selected measurements are: head and body, 53–71; forearm, 46.7–49.5; hind foot, 10–16; greatest length of skull including incisors, 22.0–23.9; condylobasal length, 18.2–19.6; length of upper tooth-row (C–M3), 6.5–7.2; postorbital breadth, 5.5–6.0. Bone fragments from Puerto Rico represent a large, extinct subspecies. Upper parts Buckthorn Brown to Dresden Brown; underparts less reddish to pale reddish-gray; pinnae of ears light to dark brown.

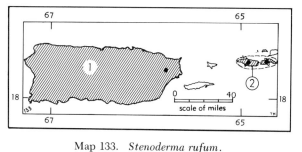

Map 133. *Stenoderma rufum*.
1. *S. r. darioi*   2. *S. r. rufum*

MARGINAL RECORDS.—Virgin Islands (Hall and Tamsitt, 1968:1): St. Thomas Island; St. John Island (in dry arborescent vegetation).

Genus Centurio Gray—Wrinkle-faced Bat

Revised by Rehn, Proc. Acad. Nat. Sci. Philadelphia, 53:295–302, June 8, 1901.

1842. *Centurio* Gray, Ann. Mag. Nat. Hist., ser. 1, 10:259, December. Type, *Centurio senex* Gray.
1861. *Trichocoryes* H. Allen, Proc. Acad. Nat. Sci. Philadelphia, p. 359. Type, *Centurio mcmurtrii* H. Allen [= *Centurio senex* Gray]. Proposed as a subgenus of *Centurio*.
1866. *Trichocorytes* Gray, Proc. Zool. Soc. London, p. 118, May, an emendation. (Raised to generic rank.)
1897. *Trichocoryctes* Trouessart, Catalogus Mammalium . . . , fasc. 1, p. 164, a *lapsus*.

Length of head and body approx. 61. Skull with high, rounded braincase and exceedingly short rostrum; palate short, approx. half as long as wide; upper canines having anterior basal cavity. No true nose-leaf but face completely covered with wrinkled growths.

The genus is probably monotypic.

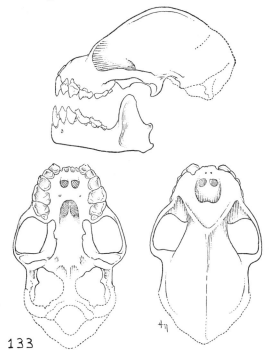

133

Fig. 133. *Stenoderma rufum darioi*, Cueva Catedral, Arecibo Prov., Puerto Rico, No. 40951 A.M.N.H., sex ?, X 2.

Stenoderma rufum darioi Hall and Tamsitt

1968. *Stenoderma rufum darioi* Hall and Tamsitt, Royal Ontario Mus., Life Sci. Occas. Pap., 11:1, August 4, type from 1 mi. NW El Yunque Peak, 355 m., Puerto Rico. Known only from rain forests of Puerto Rico.

Stenoderma rufum rufum Desmarest

1820. *Stenoderma rufa* Desmarest, Mammalogie . . . , p. 117, type from an unknown locality. Not *St[enoderma]. rufus* Oken, 1816; Oken's names are unavailable.
1855. *Artibaeus undatus* Gervais, Mammifères, *in* [Castelnau] Expéd. dans les parties centrales de l'Amér. du Sud . . . , p. 35, Pl. 9, Fig. 3, type probably is the skull of the holotype of *Stenoderma rufum* (see Anthony, Mem. Amer. Mus. Nat. Hist., n.s., 2(2):354, October 12, 1918).

Centurio senex
Wrinkle-faced Bat

Reviewed by Paradiso, Mammalia, 31:595–604, December, 1967.

Selected measurements: length of forearm, 41.0–46.5; greatest length of skull, 17.3–18.9; condylobasal length, 14.4–15.8; zygomatic breadth, 14.4–15.7; interorbital breadth, 4.7–5.6; length of maxillary tooth-row, 4.8–5.6. Upper parts dark or medium brown; underparts appreciably paler. Only one subspecies occurs in the area here treated. A second subspecies occurs on Trinidad.

Centurio senex senex Gray

1842. *Centurio senex* Gray, Ann. Mag. Nat. Hist., ser. 1, 10:259, December, type locality erroneously given as Am-

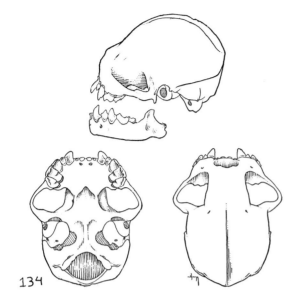

Fig. 134. *Centurio senex senex*, Sololá, Moca, Guatemala, No. 41802 F.M.N.H., ♂, X 2.

boyna, East Indies; Realejo, Chinandega, Nicaragua, subsequently suggested by Goodwin (Bull. Amer. Mus. Nat. Hist., 87:327, December 31, 1946).

1854. *Centurio flavogularis* Lichtenstein and Peters, Monatsb. preuss. Akad. Wiss., Berlin, p. 335, type from "Cuba."

1860. *Centurio mexicanus* Saussure, Revue et Mag. Zool., Paris, ser. 2, 12:381, type from "Les regions chaudes du México."

1861. *Centurio mcmurtrii* H. Allen, Proc. Acad. Nat. Sci. Philadelphia, 13:360, November 26, type from Mirador, Veracruz.

1891. *Centurio minor* Ward, Amer. Nat., 25:750, August, type from Cerro de los Pájaros, Las Vigas, Veracruz.

MARGINAL RECORDS.—Sinaloa: 12 mi. NE San Benito (Jones, *et al.*, 1972:16); 5 mi. WSW Plomosas (Jones, 1964e:513). Durango: Paso de Sihuacori (Cros-

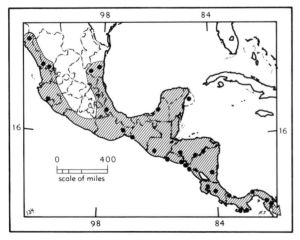

Map 134. *Centurio senex senex*.

sin, *et al.*, 1973:199). Tamaulipas: *Aserradero del Infiernillo;* Pano Ayuctle; 3 mi. S, 14 mi. W Piedra. Veracruz: Orizaba (Hall and Dalquest, 1963:239). Campeche: 5 km. S Champotón (91803 KU). Quintana Roo: Cozumel Island. Honduras: La Esperanza, 1660 m. (LaVal, 1969:821); Chichicaste, 480 m. (*ibid.*); Río Coco, 78 mi. ENE Danlí, 900 ft. (Davis, *et al.*, 1964:386). Nicaragua: S side Río Mico, El Recreo (Jones, *et al.*, 1971:16). Costa Rica: Cartago. Panamá (Paradiso, 1967:601): *Sibube;* Almirante; Cristóbal; El Real; Cerro Malí, thence into South America, thence northwestward to Panamá (*ibid.*): *Tacarcuna Village;* Cerro Hoya; Isla Cebaco. Costa Rica: Finca Taboga (Armstrong, 1969:809). Nicaragua (Jones, *et al.*, 1971:16): San Antonio; 4½ km. N Cosigüina. El Salvador: Lake Olomega (Paradiso, 1967:601); 20 mi. W La Libertad, 10 ft. (Davis, *et al.*, 1964:386). Guatemala (Jones, 1966:461): *Yepocapa;* Mocá. Chiapas: 45 km. W Cintalapa (Baker, 1967:428). Oaxaca: Matías Romero (Paradiso, 1967:600). Jalisco: 4 km. N Durazno (Jones, 1964e:513).

SUBFAMILY **BRACHYPHYLLINAE**

If the "Brachyphyllina" of H. Allen 1898 (Trans. Amer. Philos. Soc., 19:258) is a family-group name, it has priority over Phyllonycterinae Miller 1907 (Bull. U.S. Nat. Mus., 57:171, June 29). Also, *Brachyphylla* Gray 1834 is older than *Phyllonycteris* Gundlach 1861 and therefore should be the nominal type-genus of the subfamily. Article 39 of the 1961 Code deals with changes in family-group names.

Rostrum slightly to moderately elongated; posterior palatal emargination V-shaped; nose-leaf always present but not extending free; approx. 50 per cent of length of tongue free; pectoral ridge of humerus not expanded proximally; 5 lumbar vertebrae; tail present; calcar poorly developed or lacking; 3rd phalanx of 3rd digit in resting position nearly at right angle with metacarpal (modified from Silva-T. and Pine, 1969:10–15). Dentition, i. $\frac{2}{2}$, c. $\frac{1}{1}$, p. $\frac{2}{2}$, m. $\frac{3}{3}$.

KEY TO GENERA OF BRACHYPHYLLINAE

1. Forearm more than 55; length of maxillary tooth-row nearly equals width across 2nd upper molars. . . . . . . *Brachyphylla*,   p. 169
1'. Forearm less than 55; length of maxillary tooth-row greatly exceeds width across 2nd upper molars.
    2. Second and 3rd lower molars distinctly cuspidate; zygomatic arches complete, although weak. *Erophylla*,   p. 170
    2'. Second and 3rd lower molars not distinctly cuspidate; zygomatic arches incomplete. . . . . . *Phyllonycteris*,   p. 171

Genus **Brachyphylla** Gray—Fruit-eating Bats

1834. *Brachyphylla* Gray, Proc. Zool. Soc. London, p. 122, March 12. Type, *Brachyphylla cavernarum* Gray.

Length of head and body (78–118) essentially equal to total length owing to vestigial tail. Skull relatively long, narrow; upper incisors markedly different in size and shape, inner one large, higher than long, recurved, outer one rounded, minute, flat-crowned; shaft of upper canine with large secondary cusp extending nearly to middle of posterior edge; anterior upper premolar minute; posterior upper premolar high and short; crowns of upper and lower molars heavily wrinkled; 1st lower molar with distinct posterointernal cusp, differing markedly from last premolar; interpterygoid space not extended forward as a palatal emargination; nasal region without emargination; ears small, separate; nose-leaf rudimentary; tail approx. a fourth length of femur, wholly enclosed by interfemoral membrane.

Brachyphylla cavernarum
Antillean Fruit-eating Bat

Selected measurements: total length, 91–118; length of forearm, 60–66.4; greatest length of skull, 29.5–32.3; interorbital breadth, 6.0–6.8; zygomatic breadth, 14.6–17.8; length of maxillary tooth-row (excluding canine), 8.4–9.2.

Upper parts ivory yellow, hairs tipped with sepia, patches on neck, shoulders, and sides appearing paler because dark wash is absent; underparts brown. Skull robust; rostrum little longer than broad, slightly flattened above; interorbital constriction slight; braincase broadest posteriorly, with well-developed sagittal crest;

zygomatic arches widely and evenly bowed; interpterygoid notch not extending anteriorly to level of anterior root of zygoma; tympanic completely covering cochleae. First upper premolar small; 2nd upper premolar with greatly heightened cusp on cutting edge (much the highest cusp in molariform series); cusps progressively lower in posterior teeth.

Brachyphylla cavernarum cavernarum Gray

1834. *Brachyphylla cavernarum* Gray, Proc. Zool. Soc. London, p. 123, March 12, type from St. Vincent, Lesser Antilles.

MARGINAL RECORDS.—Puerto Rico: Corozal. Virgin Islands (Varona, 1974:27): *St. John; St. Croix*. Lesser Antilles (Jones and Phillips, 1970:137, 138): Anguilla; St. Martin; Barbuda; Antigua; Guadeloupe; Dominica; Martinique; St. Lucia; St. Vincent; Monserrat; St. Eustatius; Saba.—**See** addenda.

Brachyphylla cavernarum minor Miller

1913. *Brachyphylla minor* Miller, Proc. Biol. Soc. Washington, 26:32, February 8, type from Coles Cave, St. Thomas Parish, Barbados, Lesser Antilles. Known only from Barbados.
1970. *Brachyphylla cavernarum minor*, Jones and Phillips, Studies on the fauna of Curaçao and other Caribbean islands, 32:138, March.

Brachyphylla cavernarum nana Miller

1902. *Brachyphylla nana* Miller, Proc. Acad. Nat. Sci. Philadelphia, 54:409, September 12, type from El Guamá, Pinar del Río, Cuba.
1974. *Brachyphylla cavernarum nana*, Varona, Catálogo Mam. viv. y exting. de las Antillas, Acad. Cienc. Cuba, p. 27.

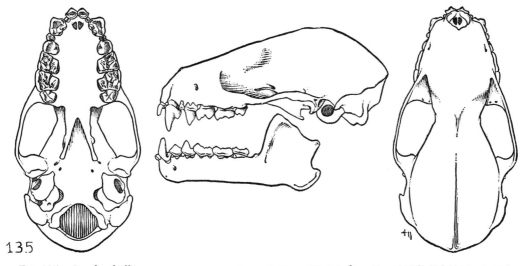

135

Fig. 135. *Brachyphylla cavernarum cavernarum*, Antigua, West Indies, No. 123270 U.S.N.M., ♂, X 2.

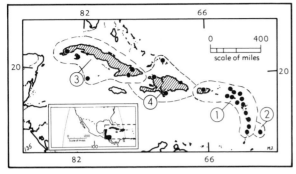

Map 135. *Brachyphylla cavernarum.*

1. *B. c. cavernarum*　　　3. *B. c. nana*
2. *B. c. minor*　　　　　　4. *B. c. pumila*

MARGINAL RECORDS.—Cuba: type locality; Munición (Stubbe, 1970:397); *La Fangosa* (*ibid.*); Santiago. Grand Cayman, British West Indies (Johnston, 1974:171). Cuba: Isle of Pines.—See addenda.

Brachyphylla cavernarum pumila Miller

1918. *Brachyphylla pumila* Miller, Proc. Biol. Soc. Washington, 31:39, May 16, type from cave near Port-de-Paix, Haiti.
1974. *Brachyphylla cavernarum pumila*, Varona, Catálogo Mam. viv. y exting. de las Antillas, Acad. Cienc. Cuba, p. 27.

MARGINAL RECORDS.—Caicos Islands: Conch Bar, Middle Caicos (Grand Caicos on some maps) (Buden, 1977:221). Dominican Republic: Los Patos. Haiti: type locality.—See addenda.

Genus **Erophylla** Miller

1906. *Erophylla* Miller, Proc. Biol. Soc. Washington, 19:84, June 4. Type, *Phyllonycteris bombifrons* Miller.

Length of head and body approx. 65. Calcar short but distinct; interfemoral membrane narrow; nose-leaf bifid distally; zygomatic arches usually complete; lower molars distinctly cuspidate and with trenchant edge.

Taxa below are arranged according to G. M. Allen (1917:166) instead of according to Buden (1976:1–15), even though the latter records some data on geographic variation in *Erophylla* not recorded by the former.

KEY TO SPECIES OF EROPHYLLA

1. Skull with high, rounded braincase rising from plane of rostrum at a distinct angle; rostrum relatively short, tapered; occurring west of Windward Channel, Greater Antilles. . . . .*E. bombifrons,* p. 170
1'. Skull with low, flattened braincase not rising above rostral plane at a distinct angle; rostrum relatively long and less narrowed; occurring east of Windward Channel, Greater Antilles. . . . . .*E. sezekorni,* p. 171

Erophylla bombifrons
Brown Flower Bat

Selected measurements of the type of *E. b. bombifrons*, a male: length of forearm, 48.4; greatest length of skull, 24.4; interorbital breadth, 5; zygomatic breadth, 12; length of upper toothrow, 8. Upper parts dark brown; slightly paler below. Braincase expanded, high and rounded, rising from rostral plane at definite angle; rostrum relatively short and narrow.

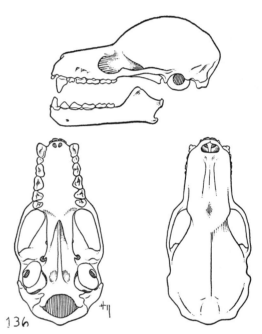

Fig. 136. *Erophylla bombifrons bombifrons,* Cueva Catedral, Puerto Rico, No. 39339 A.M.N.H., sex ?, X 2, from Anthony (1918:358).

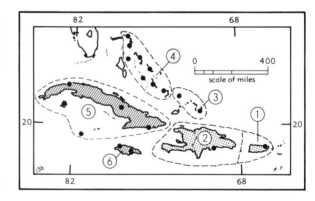

Map 136. *Erophylla bombifrons* and *Erophylla sezekorni.*

1. *E. b. bombifrons*　　　　4. *E. s. planifrons*
2. *E. b. santacristobalensis*　5. *E. s. sezekorni*
3. *E. s. mariguanensis*　　　6. *E. s. syops*

Erophylla bombifrons bombifrons (Miller)

1899. *Phyllonycteris bombifrons* Miller, Proc. Biol. Soc. Washington, 13:36, May 29, type from cave near Bayamón, Puerto Rico.
1906. *Erophylla bombifrons,* Miller, Proc. Biol. Soc. Washington, 19:84, June 4.

MARGINAL RECORDS.—Puerto Rico: *Cueva de Fari, Pueblo Viejo;* type locality.

Erophylla bombifrons santacristobalensis (Elliot)

1905. *Phyllonycteris santa-cristobalensis* Elliot, Proc. Biol. Soc. Washington, 18:236, December 9, type from San Cristóbal, Dominican Republic. Known only from the type locality.
1906. *E[rophylla]. santacristobalensis,* Miller, Proc. Biol. Soc. Washington, 19:84, June 4.—See addenda.

Erophylla sezekorni
Buffy Flower Bat

Selected measurements: length of forearm, 47–49; greatest length of skull, 24.2–26.4; zygomatic breadth, 11–12; length of upper tooth-row, 7.6–8.0. Upper parts pale yellowish brown or buffy; underparts paler in varying degree. Braincase low, somewhat flattened, not rising above rostral plane at distinct angle; rostrum relatively long, and broader than in *E. bombifrons.*

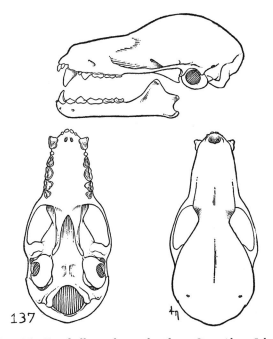

Fig. 137. *Erophylla sezekorni planifrons,* Great Abaco Island, Bahama Islands, No. 5179 K.U., ♀, X 2.

Erophylla sezekorni mariguanensis Shamel

1931. *Erophylla planifrons mariguanensis* Shamel, Jour. Washington Acad. Sci., 21:252, June 4, type from Abrahams Hill, Mariguana [= Mayaguana] Island, Bahamas.

MARGINAL RECORDS.—Bahama Islands: type locality; Stubbs Guano Cave, East Caicos.

Erophylla sezekorni planifrons (Miller)

1899. *Phyllonycteris planifrons* Miller, Proc. Biol. Soc. Washington, 13:34, May 29, type from Nassau, New Providence Island, Bahama Islands.
1917. *E[rophylla]. s[ezekorni]. planifrons,* G. M. Allen, Proc. Biol. Soc. Washington, 30:167, October 23.

MARGINAL RECORDS.—Bahama Islands: cave near Israel's Point, Great Abaco; Marsh Harbor, Great Abaco; Eleuthera Island; Cat Island (Varona, 1974:30); Long Island (*ibid.*); Crooked Island (*ibid.*); *Pequeña Exuma* (*ibid.*); Gran Exuma (*ibid.*); *ca.* 6 mi. from Nassau, New Providence.

Erophylla sezekorni sezekorni (Gundlach)

1861. *Phyllonycteris sezekorni* Gundlach, Monatsb. preuss. Akad. Wiss., Berlin (for 1860), p. 818, type from Cuba.
1906. *E[rophylla]. sezekorni,* Miller, Proc. Biol. Soc. Washington, 19:84, June 4.

MARGINAL RECORDS.—Cuba: Habana; *Las Villas;* 7 km. W Banao; Siboney. Gran Caimán (Varona, 1974:30, as species only); Isla de Pinos (Varona, 1974:30).

Erophylla sezekorni syops G. M. Allen

1917. *Erophylla sezekorni syops* G. M. Allen, Proc. Biol. Soc. Washington, 30:167, October 23, type from Montego Bay, Jamaica.

MARGINAL RECORDS.—Jamaica: type locality; *Mount Plenty Cave, 1 mi. N Goshen* (Goodwin, 1970:575); St. Clair Cave, 3 mi. S Ewarton (*ibid.*).

Genus **Phyllonycteris** Gundlach

1861. *Phyllonycteris* Gundlach, Monatsb. preuss. Akad. Wiss., Berlin (for 1860), p. 817. Type, *Phyllonycteris poeyi* Gundlach.
1898. *Reithronycteris* Miller, Proc. Acad. Nat. Sci. Philadelphia, 50:333, August 2. Type, *Reithronycteris aphylla* Miller. Valid as a subgenus (Koopman, Jour. Mamm., 33:255, May 14, 1952).
1904. *Rhithronycteris* Elliot, Field Columb. Mus., Publ. 95, Zool. Ser., 4:687, July, an invalid emendation.

Rostrum deep; zygomatic arches incomplete; auditory bullae covering more than half surface of cochleae; upper incisors small, inner pair twice size of lateral pair; 1st upper molar small, low; upper molars longer than broad, basin-shaped; ears moderately large, separate; nose-leaf rudimentary; calcar absent; interfemoral membrane narrow and extending to middle of tibia; tail approx. half length of femur. Dentition, i. $\frac{2}{2}$, c. $\frac{1}{1}$, p. $\frac{2}{2}$, m. $\frac{3}{3}$.

KEY TO SPECIES OF PHYLLONYCTERIS

1. Basicranial region anterior to basioccipital deeply grooved. . . . .*P. aphylla*, p. 173
1'. Basicranial region normal, not deeply grooved.
   2. Greatest length of skull more than 26.0. . . . .*P. major*, p. 172
   2'. Greatest length of skull less than 26.0.
      3. Distance from anterior edge of premaxilla to posterior lip of incisive foramen more than 70% of palatal width taken immediately anterior to canine. . . .*P. poeyi*, p. 172
      3'. Distance from anterior edge of premaxilla to posterior lip of incisive foramen less than 70% of palatal width taken immediately anterior to canine.   *P. obtusa*, p. 172

Phyllonycteris major Anthony
Puerto Rican Flower Bat

1917. *Phyllonycteris major* Anthony, Bull. Amer. Mus. Nat. Hist., 37:567, September 7, type from Cueva Catedral, near Morovis, Puerto Rico. Known only from type locality.

Selected cranial measurements: greatest length, 26.7–28.1; interorbital breadth, 5.5–5.9; length of upper molar series, 6.7–6.8. Skull with long, deep, somewhat tubular rostrum; interorbital constriction slight; braincase high, rounded, but not rising abruptly from rostral plane; palate long, narrowed anteriorly; interpterygoid notch V-shaped, not reaching plane of last molars. Known from skeletal remains only.

Phyllonycteris obtusa Miller
Haitian Flower Bat

1929. *Phyllonycteris obtusa* Miller, Smiths. Miscl. Coll. 81(9):10, March 30, type from Crooked Cave, near Atalaye Plantation, about 4 mi. E St. Michel, Haiti.

Selected measurements of the type and an additional specimen are: greatest length of skull, ___, 22.2; interorbital breadth, 5.6, 5.4; alveolar length of upper molar series, 7.0, 7.2. Resembling *P. major* but distinctive in the shorter relative length of that part of skull anterior to the palatine foramina. Known from skeletal remains only.

MARGINAL RECORDS.—Haiti: cave near Port-de-Paix; type locality; cave at Diquini.

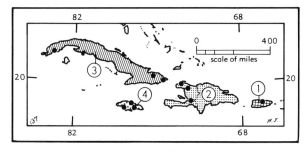

Map 137. *Phyllonycteris.*

1. *P. major*   3. *P. poeyi*
2. *P. obtusa*   4. *P. aphylla*

Phyllonycteris poeyi Gundlach
Cuban Flower Bat

1861. *Ph[yllonycteris]. poeyi* Gundlach, Monatsb. preuss. Akad. Wiss., Berlin, p. 817, type from Cuba.

Selected measurements of two specimens are: greatest length of skull, 25.7, 24.7; interorbital breadth, 7.3, 6.9; length of upper molar series, 6.1, 5.7. Similar to *P. major* but smaller, narrower braincase, relatively weaker dentition, especially 2nd upper premolar. Fur grayish white, hairs of crown and back distinctly washed with clay color at tip, fur of underparts washed with pale cream-buff apically; hairs of silky texture that produces silvery reflections in certain lights; membranes and ears light brown; outermost phalanges and adjacent part of membrane whitish (after Miller, 1904:345).

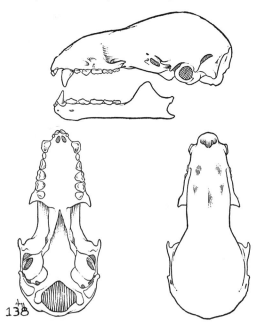

Fig. 138. *Phyllonycteris poeyi*, Guanajay, Cuba, No. 103555 U.S.N.M., ♀, X 2.

MARGINAL RECORDS.—Cuba: El Guama; La Munición (Stubbe, 1970:397); Baracoa.

Phyllonycteris aphylla (Miller)
Jamaican Flower Bat

1898. *Reithronycteris aphylla* Miller, Proc. Acad. Nat. Sci. Philadelphia, 50:334, August 2, type from Jamaica.
1952. *Phyllonycteris (Reithronycteris) aphylla*, Koopman, Jour. Mamm., 33:257, May 14.

Selected measurements of the type, an adult male: length of head and body, approx. 76; total length, 88; length of forearm, 48; greatest length of skull, 26; interorbital breadth, 5.4; length of upper tooth-row, 8. Skull resembling those of the other species of genus *Phyllonycteris* but basisphenoidal-presphenoidal region much elevated and deep longitudinal dorsal groove proceeds posteriorly from roof of internal nares.

MARGINAL RECORDS.—Jamaica: Dairy Cave, St. Ann Parish; Riverhead Cave and St. Clair Cave, St. Catherine Paris (Henson and Novick, 1966:351); Wallingford Cave, St. Elizabeth Parish.

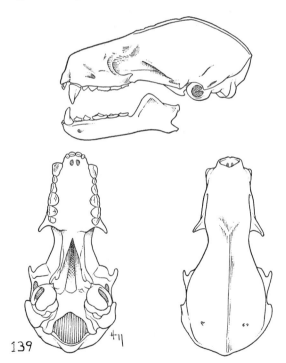

Fig. 139. *Phyllonycteris (Reithronycteris) aphylla*, ♂, type (after Miller, 1907:174) X 2.

FAMILY DESMODONTIDAE
Vampire Bats

Like Phyllostomatidae in respect to wing, pectoral girdle, and pelvis, except that tuberosities of humerus more nearly equal in size, and both more distinctly exceed head. Fibula large, extending to head of tibia. All long bones of leg and wing deeply grooved for accommodating muscles, this especially noticeable in tibia, fibula, and femur. Teeth highly specialized for cutting, all trace of crushing surface being absent, and cheek-teeth so reduced that length of entire upper row less than 1.7 times length of canine along alveolus. Stomach a slender caecumlike structure. Nostrils surrounded by dermal outgrowths forming rudimentary nose-leaf (after Miller, 1907:176).

The close resemblance of the three genera of vampires to phyllostomatids, as evidenced by comparisons of serum proteins, karyotypes, sperm morphology, and ectoparasite host specificity by Forman, *et al.* (1968:417–425) and Machado-Allison (1967:225) suggests that these bats are more closely related than previously was realized. As a result of these findings Koopman and Jones (1970:25) reduced the vampires (*Diphylla, Diaemus,* and *Desmodus*) to subfamily rank within the family Phyllostomatidae. The sanguinivorous habits and corresponding morphological adaptations (see Park and Hall, 1951:64–72, and Rouk and Glass, 1970:455–472), as well as other obvious differences, contrast so sharply with those of other bats that the vampires are here accorded family rank.

KEY TO GENERA OF DESMODONTIDAE

1. Inner lower incisors 4-lobed (inner) or 7-lobed (outer); lower cheek-teeth 4; legs thickly furred. . . . . *Diphylla*, p. 175
1'. Lower incisors bilobed or notched or entire; lower cheek-teeth 3; legs sparsely furred.
   2. Upper cheek-teeth usually 3; preorbital process present; lower incisors subequally bilobed or notched, or entire; wing tips white. . . . *Diaemus*, p. 174
   2'. Upper cheek-teeth 2; preorbital process absent; lower inner incisors bilobed, outer subequally bilobed; wing tips not white. . . . *Desmodus*, p. 173

Genus Desmodus Wied-Neuwied—Vampire Bat

1826. *Desmodus* Wied-Neuwied, Beiträge zur Naturgeschichte von Brasilien, 2:231. Type, *Desmodus rufus* Wied-Neuwied [= *Phyllostoma rotundum* É. Geoffroy St.-Hilaire].
1834–36. *Edostoma* D'Orbigny, Voy. dans l'Amér. Mérid., p. viii. Type, *Edostoma cinerea* D'Orbigny.
1905. *Desmodon* Elliot, Field Columb. Mus., Publ. 105, Zool. Ser., 6:530, an invalid emendation.

Braincase broad posteriorly, markedly narrower anteriorly; rostrum extremely reduced, little more than a support for the large incisors and canines; palate deeply concave transversely; auditory bullae covering more than half cochlear surface; upper incisors greatly enlarged, with long, curved, sharp edge; lower incisors minute, bilobed; canines large, sharp-edged posteriorly; cheek-teeth minute, probably nonfunctional. Ears relatively small, separate, pointed; thumb long, having distinct basal pad; calcar reduced to a small nubbin; tail absent. Dentition, i. $\frac{1}{2}$, c. $\frac{1}{1}$, p. $\frac{1}{2}$, m. $\frac{1}{1}$. The genus is monotypic.

Desmodus rotundus
Vampire Bat

Selected measurements of a male and female from Costa Rica are, respectively: length of head and body, 86, 80; length of forearm, 59.5, 63.0; greatest length of skull, 25.8, 25.0; interorbital breadth, 6.0, 5.3; zygomatic breadth, 12.7, 11.5. Upper parts dark grayish brown, palest anteriorly; underparts paler, sometimes with faint buffy wash.

Desmodus rotundus murinus Wagner

1840. D[esmodus]. murinus Wagner, in Schreber, Die Säugthiere . . . , Suppl., 1:377, type from México.
1912. Desmodus rotundus murinus, Osgood, Field Mus. Nat. Hist., Publ. 155, Zool. Ser., 10:63, January 10.

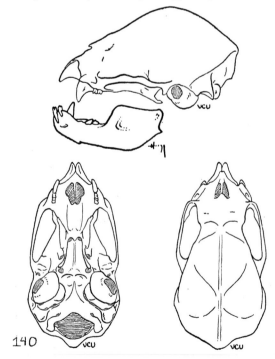

Fig. 140. Desmodus rotundus murinus, 3 km. W Boca del Río, 10 ft., Veracruz, No. 23755 K.U., ♀, X 2.

MARGINAL RECORDS.—Sonora: Pótam; 1 km. SW La Aduana, ca. 1600 ft. (Loomis and Davis, 1965:497). Chihuahua: Urique, 1700 ft. (82299 KU). Tamaulipas (Alvarez, 1963:406): 6½ mi. N, 3 mi. W Jiménez, 1250 ft.; Sierra de Tamaulipas, 3 mi. S, 16 mi. W Piedra, 1400 ft., thence along Caribbean Coast to South America and then northward on Pacific Coast to Guerrero: Zacatula (Alvarez, 1968:28). Nayarit?: Rancho Palo Amarillo, near Amatlán de Cañas. Sinaloa: 6 km. E Cosalá (Jones, et al., 1972:16).—See addenda.

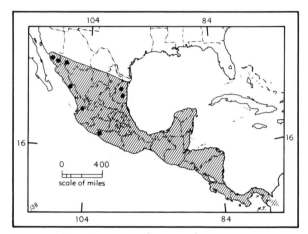

Map 138. Desmodus rotundus murinus.

Genus Diaemus Miller

1906. Diaemus Miller, Proc. Biol. Soc. Washington, 19:84, June 4. Type, Desmodus youngi Jentink.

Externally resembles Desmodus, but differs as follows: thumb shorter; wing tips white; hairs on foot extending ca. 5 mm. beyond tips of toes; insertion of wing membrane on back halfway between lateral line and dorsal mid-line; 3rd metacarpal, 4th metacarpal, forearm subequal; rostrum larger in relation to braincase; bullae more inflated; preorbital process present; minute 2nd upper molars usually present; cutting edges of lower incisors entire or notched, not bilobed; coronoid process higher than mandibular condyle. Dentition, i. $\frac{1}{2}$, c. $\frac{1}{1}$, p. $\frac{1}{2}$, m. $\frac{2}{1}$. The genus in monotypic.

Diaemus youngi
White-winged Vampire Bat

Selected measurements of one specimen from Tamaulipas, two from Tabasco, and one from Costa Rica: head and body, 98, 91; forearm, 55.2, 56.0; 3rd metacarpal, 55.8; greatest length of skull, 25.6, 26.4, 26.2, 26.6; zygomatic breadth, 14.6; postorbital constriction, 7.1, 6.8; breadth of braincase, 14.0, 14.2, 14.0, 13.7. Pelage of upper parts Prout's Brown, with gray or white bases;

pelage of underparts paler, with gray or pale yellow tips.

Diaemus youngi cypselinus Thomas

1928. *Diaemus youngi cypselinus* Thomas, Ann. Mag. Nat. Hist., ser. 10, 2:288, September, type from Pebas, Perú.

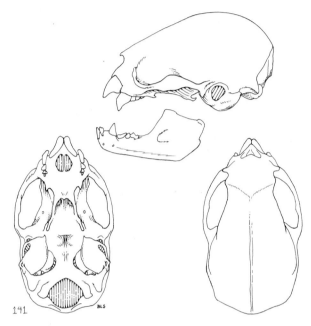

Fig. 141. *Diaemus youngi*, El Valle, Nueva Esparta, Venezuela, No. 119066 K.U., ♂, X 2.

MARGINAL RECORDS.—Tamaulipas: 3 km. W Kilómetro 619 carretera México-Laredo, near El Encino (Villa-R., 1967:339). Tabasco: 2¾ mi. SE Teapa (Lay, 1963:375). Nicaragua: 5 mi. N, 1 mi. W San Juan del Sur, Rivas (Baker and Jones, 1975:4). Costa Rica: Finca La Pacífica, 4 km. NW Cañas, 45 m. (Gardner, *et al.*, 1970:723). Panamá (Handley, 1966:769): Isla Bastimentos; Armila, thence into South America.

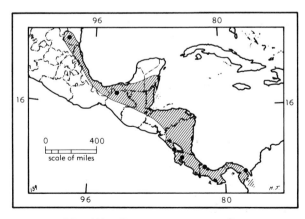

Map 139. *Diaemus youngi cypselinus*.

Genus **Diphylla** Spix—Hairy-legged Vampire Bat

1823. *Diphylla* Spix, Simiarum et vespertilionum Brasiliensium . . . , p. 68. Type, *Diphylla ecaudata* Spix.
1896. *Haematonycteris* H. Allen, Proc. U.S. Nat. Mus., 18:777. "Name based on a probably abnormal specimen of *Diphylla*" (Miller, Bull. U.S. Nat. Mus., 57:179, June 29, 1907).

Externally resembling *Desmodus* but usually smaller; ears shorter and rounded; thumb short and lacking basal pad; calcar well formed but small; interorbital region higher, broader; two pairs upper incisors, outer pair being minute; a 2nd pair of molars present above, but much reduced; lower incisors relatively large, inner pair having 4 lobes, and outer pair 7. Dentition, i. $\frac{2}{2}$, c. $\frac{1}{1}$, p. $\frac{1}{2}$, m. $\frac{2}{2}$.

Diphylla ecaudata Spix
Hairy-legged Vampire Bat

1823. *Diphylla ecaudata* Spix, Simiarum et vespertilionum Brasiliensium . . . , p. 68, Pl. 36, Fig. 7, type locality Brazil (restricted to Rio San Francisco, Bahia, by Cabrera, Rev. Mus. Argentino de Cienc. Nat., Cienc. Zool., 4(no. 1):94, March 27, 1958).
1903. *Diphylla centralis* Thomas, Ann. Mag. Nat. Hist., ser. 7, 11:378, April, type from Boquete, 4500 ft., Chiriquí, Panamá. Regarded as indistinguishable from *D. ecaudata* by Burt and Stirton, Miscl. Publ. Mus. Zool., Univ. Michigan, 117:37, September 22, 1961. Ojasti and Linares, Acta Biol. Venezuelica, 7:431, May, 1972, regard *D. e. centralis* as a subspecies occurring in Central America, with South American specimens being assigned to *D. e. ecaudata*.

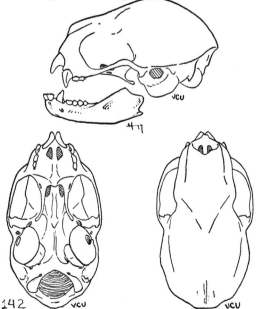

Fig. 142. *Diphylla ecaudata*, 7 km. NW Potrero, 1700 ft., Veracruz, No. 32111 K.U., ♂, X 2.

Villa-R., Los Murciélagos de México, Anal. Inst. Biol., Univ. Nac. Autó. México, p. 343, February 6, 1967, lists measurements for 20 specimens from México that support the conclusion of Burt and Stirton (*ante*) that *D. e. centralis* is indistinguishable from *D. e. ecaudata*.

Selected measurements of a male from Panamá (type of *D. centralis* Thomas) and another male from Honduras are, respectively: length of head and body, 87, 84; length of forearm, 54.0, 55.4; greatest length of skull, 22.8, 23.1; zygomatic breadth, 12.6, 12.9. Dark brown, somewhat paler on shoulders, nape, and underparts. The genus contains only one species.

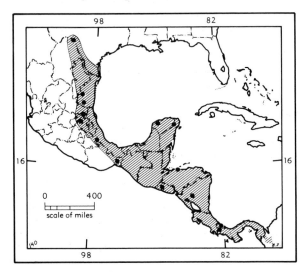

Map 140. *Diphylla ecaudata.*

MARGINAL RECORDS.—Texas: 12 mi. W Comstock (Reddell, 1968:769). Tamaulipas: 7½ km. NNW Ciudad Victoria. Yucatán: 10 km. NE Mérida (93530 KU). Quintana Roo: 1½ km. S, 1 km. E Pueblo Nuevo X-Can (91845 KU). Honduras: Lancetilla, 40 m. (Valdez and LaVal, 1971:249). Panamá: Almirante (Handley, 1966:769), thence southeastward along coast to South America, and northwestward to Panamá: Boquete (*ibid.*). Nicaragua: 2 km. N, 6 km. E Esquipulas (Jones, *et al.*, 1971:17). El Salvador: Talnique. Oaxaca: Santa Efingenia. Veracruz: Orizaba. Hidalgo: Jacala. San Luis Potosí: 2 km. WSW Ahuactlán. Tamaulipas: Cueva el Pachón.

Family **NATALIDAE**—Funnel-eared Bats

Taxa in this family arranged as proposed by Varona (Catálogo Mam. viv. y exting. de las Antillas, Acad. Cienc. Cuba, pp. 31–33, 1974).

Thumb well developed, base enclosed in membrane; width of presternum equals combined length of presternum and mesosternum; keel of presternum projecting posteriorly; meso-

sternum with low keel, higher posteriorly; palatal branches of premaxillae bony; posterior orifice of antorbital canal in funnel-shaped concavity. Fur soft and long; tail enclosed in uropatagium; ears large and funnellike; large glandular structure (natalid organ) on face of male. Dentition, i. $\frac{2}{3}$, c. $\frac{1}{1}$, p. $\frac{3}{3}$, m. $\frac{3}{3}$.

Genus **Natalus** Gray—Funnel-eared Bats

1838. *Natalus* Gray, Mag. Zool. Bot., 2:496, December. Type, *Natalus stramineus* Gray.
1855. *Spectrellum* Gervais, Mammifères, *in* [Castelnau] Expéd. dans les parties centrales de l'Amér. du Sud . . . , pt. 7, p. 51. Type, *Spectrellum macrourum* Gervais.
1855. *Nyctiellus* Gervais, Mammifères, *in* [Castelnau] Expéd. dans les parties centrales de l'Amér. du Sud . . . , pt. 7, p. 84. Type, *Vespertilio lepidus* Gervais. Valid as a subgenus.
1898. *Chilonatalus* Miller, Proc. Acad. Nat. Sci. Philadelphia, 50:326, July 12. Type, *Natalus micropus* Dobson, from Jamaica.
1906. *Phodotes* Miller, Proc. Biol. Soc. Washington, 19:85, June 4. Type, *Natalus tumidirostris* Miller, from Island of Curaçao.

Characters as for the family.

Key to Subgenera of Natalus

1. Braincase rounded; lower lip grooved or cleft. . . . . *Natalus*, p. 176
1'. Braincase flattened; lower lip entire. *Nyctiellus*, p. 179

Subgenus **Natalus** Gray

1838. *Natalus* Gray, Mag. Zool. Bot., 2:406, December. Type, *Natalus stramineus* Gray.
1898. *Chilonatalus* Miller, Proc. Acad. Nat. Sci. Philadelphia, 50:326, July 12. Type, *Natalus micropus* Dobson.

Braincase rounded; lower lip grooved or cleft; tibia ordinarily between 50 and 59 per cent of length of forearm.

Key to North American Species of Subgenus Natalus

1. Rostrum tipped downward; lower lip with shallow invagination (a cleft) length of tibia about 59% length of forearm. *N. stramineus*, p. 176
1'. Rostrum tipped upward; lower lip deeply grooved; length of tibia about 50% length of forearm. . . . . *N. micropus*, p. 178

Natalus stramineus
Mexican Funnel-eared Bat

Length of head and body approx. 50; total length, 93–110; length of forearm, 35.0–45.5;

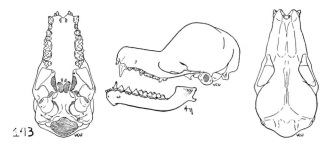

Fig. 143. *Natalus stramineus saturatus*, 3 km. E San Andrés, Tuxtla, 1000 ft., Veracruz, No. 23815 K.U., ♂, X 2.

greatest length of skull, 15.0–18.3; interorbital breadth, 2.9–3.6; zygomatic breadth, 6.5–9.6; length of upper tooth-row, 6.5–8.1. The mainland subspecies are smaller than those in the Antilles. In México and Central America two color phases are known; in light phase upper parts buffy to pinkish-cinnamon, in dark phase rich yellowish- or reddish-brown; underparts paler. Color in the Antilles has been described (Goodwin, 1959c:8, 9) as: upper parts Light Cinnamon Buff to Tawny Olive; underparts Pinkish Buff. Skull long, nar-

row; braincase rounded, rising abruptly from rostrum.

Natalus stramineus jamaicensis Goodwin

1959. *Natalus major jamaicensis* Goodwin, Amer. Mus. Novit., 1977:9, December 22, type from St. Clair, St. Catherine Parish, Jamaica, British West Indies.
1974. *Natalus stramineus jamaicensis*, Varona, Catálogo Mam. viv. y exting. de las Antillas, Acad. Cienc. Cuba, p. 32.

MARGINAL RECORDS (Goodwin, 1959c:10).— Jamaica: Wallingford Cave, Balaclava; type locality.

Natalus stramineus major Miller

1902. *Natalus major* Miller, Proc. Acad. Nat. Sci. Philadelphia, 54:389, September 12, type from near Savaneta, Dominican Republic.
1974. *Natalus stramineus major*, Varona, Catálogo Mam. viv. y exting. de las Antillas, Acad. Cienc. Cuba, p. 32.

MARGINAL RECORDS (Goodwin, 1959c:8).— Dominican Republic: type locality; Los Patos; Maniel Viejo. Haiti: Port-au-Prince.

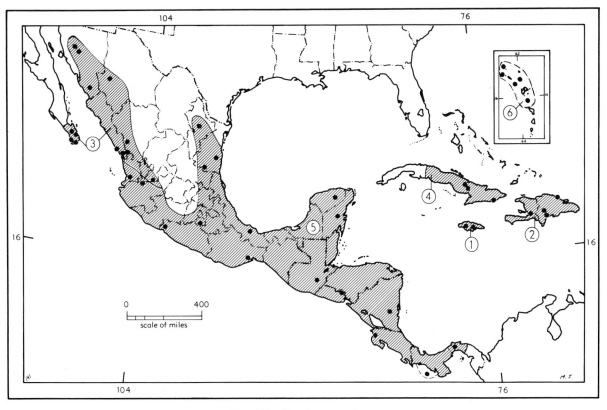

Map 141. *Natalus stramineus*.

1. *N. s. jamaicensis*   3. *N. s. mexicanus*   5. *N. s. saturatus*
2. *N. s. major*   4. *N. s. primus*   6. *N. s. stramineus*

Natalus stramineus mexicanus Miller

1902. *Natalus mexicanus* Miller, Proc. Acad. Nat. Sci. Philadelphia, 54:399, September 12, type from Santa Anita, Baja California.

1959. *Natalus stramineus mexicanus*, Goodwin, Amer. Mus. Novit., 1977:6, December 22.

MARGINAL RECORDS.—Sonora: 14$\frac{9}{10}$ mi. SSE Carbó (Cockrum and Bradshaw, 1963:6); 13 mi. SW Ures (Broadbooks, 1961:403, as *N. mexicanus*). Chihuahua: Mojaráchic. Durango: Ventanas (Baker and Greer, 1962:72). Zacatecas: near Santa Rosa, on the Río Juchipila (Matson and Patten, 1975:6). Sinaloa: Plumosas, 22 km. E Matatan, 2500 ft. (92907 KU); ½ mi. W Rosario, 100 ft. (39626 KU); 5 mi. NW Mazatlán (85681 KU). Sonora: Tesia (Goodwin, 1959c:6). Also, Baja California (Jones, *et al.*, 1965:53, unless otherwise noted): 2½ km. N San Antonio; Las Cuevas; *1 km. S Las Cuevas*; type locality; *NW Santa Anita* (Goodwin, 1959c:6); 5 km. SE Pescadero.

Natalus stramineus primus Anthony

1919. *Natalus primus* Anthony, Bull. Amer. Mus. Nat. Hist., 41:642, December 30, type from Cueva de los Indios, Daiquirí, Cuba.

1974. *Natalus stramineus primus*, Varona, Catálogo Mam. viv. y exting. de las Antillas, Acad. Cienc. Cuba, p. 33.

MARGINAL RECORDS.—Cuba: *Las Villas* (Goodwin, 1959c:10); 7 km. W Banao; Camaguey (Goodwin, 1959c:10); type locality. This subspecies is known only from skeletal material, and is thought to be extinct.

Natalus stramineus saturatus Dalquest and Hall

1949. *Natalus mexicanus saturatus* Dalquest and Hall, Proc. Biol. Soc. Washington, 62:153, August 23, type from 3 km. E San Andrés Tuxtla, 1000 ft., Veracruz.

1959. *Natalus stramineus saturatus*, Goodwin, Amer. Mus. Novit., 1977:7, December 22.

MARGINAL RECORDS.—Nuevo León: Cueva la Boca, 2 mi. E Santiago, 1700 ft. (Davis and Carter, 1962:71). Tamaulipas: 3 mi. S, 14 mi. W Piedra. Veracruz: type locality. Yucatán: Gruta de Balankanche, 5 km. E Chichén-Itzá (91847 KU). Quintana Roo: 2 km. N Felipe Carrillo Puerto (91903 KU). Nicaragua: S side Río Mico, El Recreo, 25 m. (Jones, *et al.*, 1971:17). Panamá: Chilibrillo Caves; Coiba Island. Costa Rica: Curiol de Santa Rosa (Starrett and Casebeer, 1968:15). El Salvador: Hda. Santa Rosa. Guatemala: Progresso. Oaxaca: Bisilana (Goodwin, 1959c:7). Guerrero: 12 km. N Zacatula (Alvarez, 1968:28). Nayarit: 6 mi. SSE Las Varas. Jalisco: Itztlán [= Etzatlán]. Morelos: Tequesquitengo (Goodwin, 1959c:7). Tamaulipas: Antigua [= Antigo] Morelos (*ibid.*); *El Pachón*.

Natalus stramineus stramineus Gray

1838. *Natalus stramineus* Gray, Mag. Zool. Bot., 2:496, December, type locality unknown, probably Antigua, British

West Indies (Goodwin, Amer. Mus. Novit., 1977:2, December 22, 1959).

1928. *Natalus dominicensis* Shamel, Proc. Biol. Soc. Washington, 41:67, March 16, type from Dominica, Lesser Antilles. Regarded as identical with *stramineus* by Goodwin, Amer. Mus. Novit., 1977:4, December 22, 1959.

MARGINAL RECORDS (Jones and Phillips, 1970:137).—Lesser Antilles: Anguilla; Antigua; Dominica; Montserrat; Saba.—**See** addenda.

Natalus micropus
Cuban Funnel-eared Bat

Length of head and body approx. 42; length of forearm, 31–34; interorbital breadth, 2.6–2.8; rostrum tipped upward; natalid organ of medium size, rounded, situated at base of muzzle; lower lip deeply grooved; ears large, truncated to 30° angle beyond median lobe; tibia *ca.* half length of forearm.

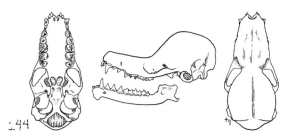

Fig. 144. *Natalus micropus tumidifrons*, Watling Island, Bahama Islands, No. 122021 U.S.N.M., ♂, X 2.

Natalus micropus brevimanus Miller

1898. *Natalus (Chilonatalus) brevimanus* Miller, Proc. Acad. Nat. Sci. Philadelphia, 50:328, July 12, type from Old Providence Island, off Caribbean Coast of Nicaragua. Known only from type locality.

1974. *Natalus micropus brevimanus*, Varona, Catálogo Mam. viv. y exting. de las Antillas, Acad. Cienc. Cuba, p. 31.

Natalus micropus macer (Miller)

1914. *Chilonatalus macer* Miller, Proc. Biol. Soc. Washington, 27:225, December 29, type from Baracoa, Oriente, Cuba.

1974. *Natalus micropus macer*, Varona, Catálogo Mam. viv. y exting. de las Antillas, Acad. Cienc. Cuba, p. 32.

MARGINAL RECORDS.—Cuba: type locality; Guantánamo.

Natalus micropus micropus Dobson

1880. *Natalus micropus* Dobson, Proc. Zool. Soc. London, p. 443, October, type from Kingston, Jamaica.

MARGINAL RECORDS.—Jamaica: St. Clair Cave, 3 mi. S Ewarton (Goodwin, 1970:577); type locality.

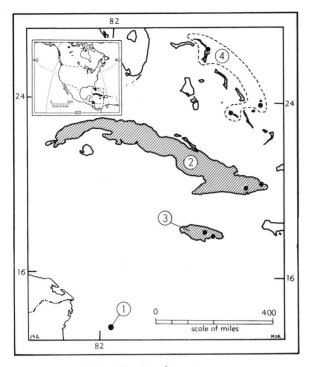

Map 142. *Natalus micropus.*

1. *N. m. brevimanus*   3. *N. m. micropus*
2. *N. m. macer*   4. *N. m. tumidifrons*

Natalus micropus tumidifrons (Miller)

1903. *Chilonatalus tumidifrons* Miller, Proc. Biol. Soc. Washington, 16:119, September 30, type from Watling Island, Bahamas.
1974. *Natalus micropus tumidifrons,* Varona, Catálogo Mam. viv. y exting. de las Antillas, Acad. Cienc. Cuba, p. 32.

MARGINAL RECORDS.—Bahama Islands: Great Abaco Island; type locality; Gran Exuma (Varona, 1974:32).

Subgenus **Nyctiellus** Gervais

1855. *Nyctiellus* Gervais, Mammifères, *in* [Castelnau] Expéd. dans les parties centrales de l'Amér. du Sud . . . , pt. 7, p. 84. Type, *Vespertilio lepidus* Gervais.

Braincase flattened; rostrum tipped downward; 1st premolars and canines reduced in size; duct area of pararhinal glands forming broad, low swelling on muzzle distally; natalid organ small, rounded, on median part of muzzle; lower lip entire; ears relatively small, constricted beyond median lobe; tibia approx. 47 per cent of length of forearm.
The subgenus is monotypic.

Natalus lepidus (Gervais)
Gervais' Funnel-eared Bat

1837. *Vespertilio lepidus* Gervais, L'Institut, Paris, 5(218):253, August, type from Cuba.

1884. *Natalus lepidus,* True, Proc. U.S. Nat. Mus., 7(App. Circ. 29):603, November 29.

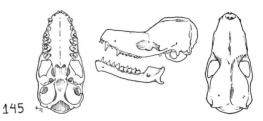

Fig. 145. *Natalus lepidus,* Long Island, Bahama Islands, No. 44535 F.M.N.H., ♂, X 2.

Selected measurements are: length of forearm, 27.3–30.4; greatest length of skull, 13.0–13.5; interorbital breadth, 2.5–2.8; zygomatic breadth, 6.2–6.5; length of upper tooth-row, 5.1–5.4. Pelage buffy or yellowish washed above with brown. Other characters as for the subgenus.

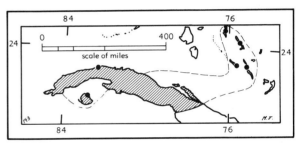

Map 143. *Natalus lepidus.*

MARGINAL RECORDS.—Bahama Islands: Eleuthera Island (Varona, 1974:33); Sheep Hill Cave, Cat Island; Miller's Cave and Mortimer's, Long Island. Isle of Pines: Nueva Gerona. Cuba: *near Havana;* Almendarez, Havana. Pequeña Exuma Island (Varona, 1974:33).

FAMILY **FURIPTERIDAE**—Smoky Bats

Resembling Natalidae, but differing as follows: thumb small, included in membrane to base of minute claw; presternum slightly wider than long, and keel projecting anteriorly; mesosternum with low ridge, no keel; palatal branches of premaxillae mere cartilaginous filaments; posterior orifice of antorbital canal not situated in a funnel-shaped concavity; teeth as in *Natalus* except canines reduced to height of premolars.

Genus **Furipterus** Bonaparte—Smoky Bats

1828. *Furia* F. Cuvier, Mém. Mus. d'Hist. Nat., Paris, 16:149–155. Preoccupied by *Furia* Linnaeus 1758, a genus of Nematodes.

1837. *Furipterus* Bonaparte, Iconografia della fauna Italica, I, fasc. 21 (under *Plecotus auritus*, p. 3). Type, *Furia horrens* F. Cuvier.

Dark slaty blue, somewhat paler below; height of braincase about two-thirds its length. Otherwise as in description of family. The genus is monotypic.

Furipterus horrens (F. Cuvier)
Eastern Smoky Bat

1828. *Furia horrens* F. Cuvier, Mém. Mus. d'Hist. Nat., Paris, 16:155, type from Mana River, French Guiana.
1856. *Furipterus horrens*, Tomes, Proc. Zool. Soc. London, pt. 24, p. 176, lam. 42.

Selected measurements of two specimens from Surinam: forearm, 35.3, 35.5; greatest length of skull, 11.8, 11.9; zygomatic breadth, 7.7, 7.9; breadth of braincase, 5.9, 6.2; interorbital constriction, 3.1, 3.0; maxillary tooth-row, 4.7, 4.9.

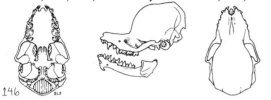

Fig. 146. *Furipterus horrens*, 20 km. SSW Changuinola, Panamá, No. 315734 U.S.N.M., ♂, X 2.

MARGINAL RECORDS.—Costa Rica: La Selva (2971 R. K. LaVal). Panamá: Río Changuinola, 20 km. SSW Changuinola (Handley, 1966:770). Also in South America.

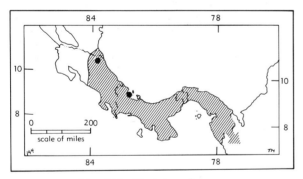

Map 144. *Furipterus horrens*.

FAMILY **THYROPTERIDAE**
Disk-winged Bats

"Shoulder joint and wing as in the Natalidae, except that trochiter is distinctly larger than trochin, second finger is reduced to a rudimentary

metacarpal less than half as long as that of third finger, there are three bony phalanges in third digit, and first phalanx of thumb bears a large sucking disk, . . . foot abnormal, the toes with only two phalanges each, the third and fourth digits, together with their claws, anchylosed together from base to tip; fibula reduced to a minute osseous thread closely applied to the tibia and disappearing about midway between heel and knee; sole with well-developed sucking disk attached to metatarsals; pelvis not essentially abnormal, but with very small pectineal process, and with obturator foramen much reduced by bony outgrowth from its sides . . . , ischia wide apart posteriorly, a symphysis pubis in males, sacrum with the posterior two vertebrae distinct, the others fused; . . . skull without postorbital processes, much as in the Natalidae, the braincase large, smooth, and rounded, the rostrum slender and weak; premaxillaries complete, the very slender and easily broken palatal branches isolating two foramina; teeth normal, not essentially different from those of the Natalidae; ear, tragus, and muzzle as in *Natalus*." (Miller, 1907:191.)

The family contains but a single genus.

Genus **Thyroptera** Spix—Disk-winged Bats

1823. *Thyroptera* Spix, Simiarum et vespertilionum Brasiliensium . . . , p. 61. Type, *Thyroptera tricolor* Spix.
1854. *Hyonycteris* Lichtenstein and Peters, Monatsb. preuss. Akad. Wiss., Berlin, p. 335. Type, *Hyonycteris discifera* Lichtenstein and Peters.

Braincase approx. 1½ times the length of the rostrum, rising at angle of about 50° from dorsal plane of rostrum; rostrum reduced; interorbital constriction pronounced; palate abruptly narrowed behind tooth-rows; auditory bullae small, covering less than half surface of cochleae; resembling *Natalus* externally but legs and tail not elongated, adhesive disks present on thumb and sole, and muzzle with small wartlike excrescence above nostrils. Dentition, i. $\frac{2}{3}$, c. $\frac{1}{1}$, p. $\frac{3}{3}$, m. $\frac{3}{3}$.

KEY TO SPECIES OF THYROPTERA

1. Underparts white or pale yellowish; ears blackish; calcar with two cartilaginous projections extending into posterolateral border of membrane. . . . . *T. tricolor*, p. 181
1'. Underparts brown; ears yellowish; calcar with one cartilaginous projection extending into posterolateral border of membrane. . . . . *T. discifera*, p. 181

Thyroptera discifera
Peters' Disk-winged Bat

Selected measurements (after Wilson, 1976:309): total length, 61–80; tail, 24–33; forearm, 31.1–35.4; condylocanine length, 12.1–12.9; mastoid breadth, 6.7–7.0. In each instance the minimum measurement is from an adult female and the maximum measurement is from immature animals (a male and a female). Diagnostic features, additional to those mentioned in the key above, are: pelage yellowish brown, slightly paler ventrally; two terminal vertebrae projecting *ca.* 2 mm. beyond uropatagium but in *T. tricolor* the tail projects 5–8 mm. beyond uropatagium. *T. discifera* so far has been found clinging to the under surface of banana leaves. *T. bicolor* roosts in rolled leaves of *Heliconia*.

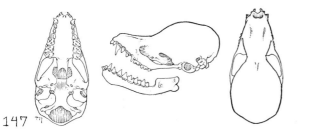

Fig. 147. *Thyroptera discifera discifera*, Loreto, Cumeria, Perú, No. 46160 F.M.N.H., ♂, X 2.

Thyroptera discifera abdita Wilson

1976. *Thyroptera discifera abdita* Wilson, Proc. Biol. Soc. Washington, 89:307, October 12, type from Escondido River, 50 mi. E Bluefields, Nicaragua. Known only from type locality. The trinomen *Thyroptera discifera discifera*, Lichtenstein and Peters, 1875, long was erroneously applied to specimens from Central America because the

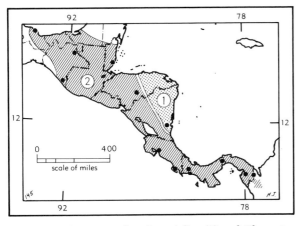

Map 145. *Thyroptera discifera abdita* (1) and *Thyroptera tricolor albiventer* (2).

type locality of *T. d. discifera* was mistakenly thought to be in Honduras, whereas it actually was in Venezuela of South America.

Thyroptera tricolor
Spix's Disk-winged Bat

Length of head and body, approx. 34–52; length of forearm, 32.0–38.0; greatest length of skull, 14.4–15.1; interorbital breadth, 2.6–2.9; zygomatic breadth, 7.1–7.6; length of upper tooth-row, 5.6–6.7. Upper parts reddish brown or somewhat darker; underparts white, but lateral extent of white area variable. Ears smaller than in *T. discifera;* calcar having two cartilaginous projections rather than one; braincase higher, more globose; teeth larger.

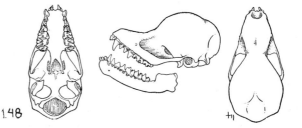

Fig. 148. *Thyroptera tricolor albiventer*, San José, Cauca, Colombia, No. 31678 A.M.N.H., ♀, X 2.

Thyroptera tricolor albiventer (Tomes)

1856. *Hyonycteris albiventer* Tomes, Proc. Zool. Soc. London, pt. 24, p. 176, type from Río Napo, near Quito, Ecuador.
1931. *Thyroptera tricolor albiventer*, Dunn, Jour. Mamm., 12:430, November 11 (*lapsus* for *T. t. albigula*, as noted under *T. t. albiventer* by Cabrera, Rev. Mus. Argentino de Cienc. Nat., 4:97, March 27, 1958).
1923. *Thyroptera tricolor albigula* G. M. Allen, Proc. New England Zool. Club, 9:1, December 10, type from Gutiérrez, in mountains about 25 mi. inland from Chiriquito [= Chiriquicito], on trail from Chiriquí Lagoon to Boquete, Chiriquí, Panamá. Regarded as inseparable from *T. t. albiventer* by Cabrera, *supra cit.*

MARGINAL RECORDS.—Veracruz: 1 km. E Sontecomapan [= Zontecomapan], 50 m. (Villa-R., 1967:353). Chiapas: Ruinas de Palenque (102026 KU). Belize: South Stann Creek, 15 mi. W All Pines. Honduras: San Marcos de Guaymaca. Panamá (Handley, 1966:770): Barro Colorado Island; Tacarcuna Village, thence southeastward into South America and northwestward to Panamá (*ibid.*): Río Jesucito; on trail from Chiriquicito to Boquete. Costa Rica: 2½ mi. SW Rincón (Gardner, *et al.*, 1970:723); Esparta. Chiapas: 11 km. NW Escuintla (Gardner, 1963:41).

FAMILY
VESPERTILIONIDAE
Vespertilionid Bats

American members of the genus revised by Miller, N. Amer. Fauna, 13:1–135, 3 pls., 40 figs. in text, October 16, 1897.

"Humerus with trochiter . . . noticeably larger than trochin and projecting distinctly beyond head, its surface of . . . [impingement] on scapula decidedly more than half as large as glenoid fossa, . . . capitellum scarcely out of line with shaft; ulna usually fused with radius at its head, the shaft reduced to a scarcely ossified fibrous strand; second finger with fully developed metacarpal and one small bony phalanx; . . . shoulder girdle strictly normal in its general structure, . . . foot normal; fibula thread-like, complete or with upper extremity cartilaginous, extending to head of tibia; . . . skull without postorbital processes; premaxillaries without palatal branches, the palate widely emarginate anteriorly; palate abruptly narrowed behind toothrows, the sides of its posterior extension parallel or nearly so; teeth usually normal, though in a few genera showing a tendency to reduction of the cusps; ears usually though not invariably separate, the anterior border with distinct basal lobe (except in Tomopeatinae); tragus usually well developed, simple; . . . tail well developed, extending to edge of wide interfemoral membrane." (Miller, 1907:195, 196.)

Most bats found north of México are vespertilionids; hence the name "common bats." Vespertilionids are not, however, the most common bats of México and Central America. This family, nevertheless, is almost worldwide in distribution, being absent only from Arctic and Antarctic regions and from certain oceanic islands.

Vespertilionids are insect-eaters. Locally they are at times as numerous as birds. Their nocturnal feeding complements the diurnal feeding of most birds.

KEY TO NORTH AMERICAN SUBFAMILIES OF VESPERTILIONIDAE

1. Nostrils opening forward beneath a conspicuous horseshoe-shaped ridge or low nose-leaf. . . . . Nyctophilinae, p. 236
1'. Nostrils opening laterally or vertically, the muzzle occasionally with warty elevations, but never with horseshoe-shaped ridge or low nose-leaf. Vespertilioninae, p. 182

SUBFAMILY VESPERTILIONINAE

Sternum slender, its entire length considerably more than twice greatest width of presternum; median lobe much smaller than body of presternum; 6 ribs connected with sternum; 7th cervical vertebra not fused with 1st dorsal; scapula with coracoid curved outward; nostrils simple, sometimes tubularly elongated, but never margined by special outgrowths; lower incisors in all known genera 3–3 (after Miller, 1907:197).

This subfamily contains the vast majority of kinds of vespertilionids.

KEY TO NORTH AMERICAN GENERA OF VESPERTILIONINAE

1. Cheek-teeth $\frac{6-6}{}$ [M. fortidens, a few specimens of M. riparius, and a larger percentage of M. l. occultus have a reduced number of premolars (fewer than ⅜)]. . . . . Myotis, p. 183
1'. Cheek-teeth fewer than $\frac{6-6}{6}$.
  2. Upper premolars 2–2.
    3. Upper incisors 1–1; metacarpal of 3rd, 4th, and 5th fingers successively much shortened.
                             Nycteris, p. 219
    3'. Upper incisors 2–2; metacarpal of 3rd, 4th, and 5th fingers subequal in length.
      4. Lower premolars 3–3.
        5. Auditory bullae not especially enlarged; rostrum broad, concave on each side above; ear shorter than head. . . . . Lasionycteris, p. 209
        5'. Auditory bullae much enlarged; rostrum narrow, evenly convex above; ear much longer than head. . . . . Plecotus, p. 232
      4'. Lower premolars 2–2.
        6. Lower canine small, its tip unequally bifid; diameter of auditory bulla equal to length of tooth-row exclusive of incisors; ear much longer than head. . . . . . Euderma, p. 231
        6'. Lower canine well developed, its tip not bifid; diameter of auditory bulla much less than length of tooth-row; ear not greatly enlarged.
          7. Upper outer incisor approx. equal in size to upper inner incisor; outer upper incisor having a flat or convex surface directed toward canine. . . Pipistrellus, p. 210

7'. Upper outer incisor distinctly larger than upper inner incisor; outer upper incisor
    having a well-developed concave surface directed toward canine. . . . . . *Myotis*, p. 183
2'. Upper premolars 1–1.
    8. Upper incisors 2–2. . . . . *Eptesicus*, p. 213
    8'. Upper incisors 1–1.
       9. Skull short, deep, depth of braincase including bullae approx. half greatest length; meta-
          carpal of 3rd, 4th, and 5th fingers successively much shortened. . . . . *Nycteris*, p. 219
       9'. Skull normal, depth of braincase including bullae much less than half greatest length; meta-
           carpal of 3rd, 4th, and 5th fingers subequal in length.
           10. Third lower incisor noticeably smaller than 1st or 2nd. . . . . *Rhogeessa*, p. 228
           10'. Third lower incisor as large as 1st or 2nd. . . . . *Nycticeius*, p. 226

Genus **Myotis** Kaup—Mouse-eared Bats

North American species revised by Miller and G. M. Allen, Bull. U.S. Nat. Mus., 144:1–209, May 25, 1928.

Neotropical species revised by LaVal, Nat. Hist. Mus. Los Angeles Co. Sci. Bull., 15:1–54, 23 figs., 14 tables, February 14, 1973.

1829. *Myotis* Kaup, Skizzirte Entwickelungs-Geschichte und natürliches System der Europäischen Thierwelt, 1:106. Type, *Vespertilio myotis* Borkhausen.

1829. *Nystactes* Kaup, Skizzirte Entwickelungs-Geschichte und natürliches System der Europäischen Thierwelt, 1:108. Type, *Vespertilio bechsteinii* Kuhl. Not *Nystactes* Gloger, 1827, an avian genus.

1830. *Leuconoë* Boie, Isis, p. 256. Type, *Vespertilio daubentonii* Kuhl. Valid as a subgenus.

1841. *Selysius* Bonaparte, Iconografia della fauna Italica . . ., 1:(introd.)3. Type, *Vespertilio mystacinus* Kuhl. Valid as a subgenus.

1841. *Capaccinius* Bonaparte, Iconografia della fauna Italica . . ., 1:1. Type, *Vespertilio capaccinii* Bonaparte.

1842. *Trilatitus* Gray, Ann. Mag. Nat. Hist., ser. 1, 10:258. Included 3 species: *hasseltii* Temminck; *macellus* Temminck; *blepotis* Temminck [= *Miniopterus schreibersii blepotis*].

1849. *Tralatitus* Gervais, Dict. Univ. Hist. Nat., 13:213, a *lapsus* for *Trilatitus* Gray.

1856. *Brachyotis* Kolenati, Allgem. deutsch. Naturh. Zeit. Dresden, neue Folge, 2:131. Type, *Vespertilio mystacinus* Kuhl. Not *Brachyotis* Gould, 1837, an avian genus.

1856. *Isotus* Kolenati, Allgem. deutsch. Naturh. Zeit. Dresden, neue Folge, 2:131. Type, *Vespertilio nattereri* Kuhl. Valid as a subgenus.

1866. *Tralatitius* Gray, Ann. Mag. Nat. Hist., ser. 3, 17:90, a *lapsus* for *Trilatitus* Gray.

1867. *Pternopterus* Peters, Monatsb. preuss. Akad. Wiss., Berlin, p. 706. Type, *Vespertilio (Pternopterus) lobipes* Peters (?) [= *Vespertilio muricola* Gray].

1870. *Exochurus* Fitzinger, Sitzungsb. k. Akad. Wiss., Wien, 62:75. Included 3 species: *macrodactylus* Temminck; *horsfieldii* Temminck; *macrotarsus* Waterhouse.

1870. *Aeorestes* Fitzinger, Sitzungsb. k. Akad. Wiss., Wien, 62:427. Included 3 species: *villosissimus* É. Geoffroy St.-Hilaire; *albescens* É. Geoffroy St.-Hilaire; *nigricans* Wied-Neuwied.

1870. *Comastes* Fitzinger, Sitzungsb. k. Akad. Wiss., Wien, 62:565. Included 2 species: *Vespertilio capaccinii* Bonaparte; *Vespertilio dasycneme* Boie.

1899. *Euvespertilio* Acloque, Fauna France, Mamm., p. 38. Included *emarginatus* and *murinus* in a composite sense.

1906. *Pizonyx* Miller, Proc. Biol. Soc. Washington, 19:85. Type, *Myotis vivesi* Menegaux. Valid as a subgenus (see Patton and Findley, Los Angeles Co. Mus. Contrib. Sci., 183:7, April 17, 1970).

1910. *Chrysopteron* Jentink, Notes Leiden Mus., 32:74. Type, *Kerivoula weberi* Jentink. Valid as a subgenus.

1916. *Rickettia* Bianchi, Ann. Mus. Zool., Acad. St. Pétersbourg, 21:77. Type, *Vespertilio (Leuconoe) rickettii* Thomas.

1916. *Dichromyotis* Bianchi, Ann. Mus. Zool., Acad. St. Pétersbourg, 21:78. Type, *Vespertilio formosus* Hodgson.

1916. *Paramyotis* Bianchi, Ann. Mus. Zool., Acad. St. Pétersbourg, 21:79, a renaming of *Nystactes* Kaup, 1829.

1958. *Hesperomyotis* Cabrera, Rev. Mus. Argentino de Cienc. Nat., Zool., 4:103, March 27. Type, *Myotis simus* Thomas. Valid as a subgenus.

Upper incisors well developed, closely crowded, outer distinctly larger than inner, and with crowns higher than long and subterete; inner incisor having distinct posterior secondary cusp; outer incisor, in many species, having well-developed concave surface directed toward canine; outer incisor separated from canine by space not quite equal to diameter of both incisors together. "Lower incisors with crowns about equal in length, forming a continuous, strongly convex row between canines. . . . Canines well developed, simple, with distinct though rather small cingulum. . . . Skull slender and lightly built, without special peculiarities of form, the rostrum nearly as long as braincase. . . . Ear well developed, slender, occasionally rather large; tragus slender and nearly or quite straight." (Miller, 1907:201, 202.) Number of upper premolars varies somewhat, both individually and in different species (notably in *M. lucifugus* and *M. fortidens*); consequently caution is advisable when identifying any given specimen that might be of this genus. Normal dentition, i. $\frac{2}{3}$, c. $\frac{1}{1}$, p. $\frac{3}{3}$, m. $\frac{3}{3}$.

KEY TO NORTH AMERICAN SPECIES OF MYOTIS

A. Occurring East of Great Plains of Canada and United States

1. Ear more than 16; extending more than 2 mm. beyond nose when laid forward. . . . . *M. keenii*, p. 204
1'. Ear less than 16; not extending 2 mm. beyond nose when laid forward.
  *2. Calcar keeled.
    3. Black mask across face; greatest length of skull usually less than 14.5. . . . *M. subulatus*, p. 187
    3'. No black mask; greatest length of skull usually more than 14.5. . . . . *M. sodalis*, p. 194
  *2'. Calcar not keeled.
    4. Dorsal hairs not banded; wing attached to foot at ankle. . . . . *M. grisescens*, p. 197
    4'. Dorsal hairs banded; wing attached to foot at base of toe.
     *5. Sagittal crest absent; fur glossy. . . . . *M. lucifugus*, p. 191
     *5'. Sagittal crest present; fur dull.
      6. Greatest length of skull normally more than 15; forehead rising gradually (see Fig. 156); central parts of Kansas, Oklahoma, and Texas, west of 96° 30'. . . *M. velifer*, p. 195
      6'. Greatest length of skull normally less than 15; forehead rising abruptly (see Fig. 155); southeastern United States, east of 96° 30'. . . . . *M. austroriparius*, p. 195

B. Occurring in or West of Great Plains, or in México, Central America, or Lesser Antilles

1. Toes greatly elongated, nearly equal to tibia in length; forearm more than 56. . . . . . *M. vivesi*, p. 208
1'. Toes much shorter than tibia; forearm less than 56.
  2. Restricted to Lesser Antilles.
    3. Forearm usually less than 35; greatest length of skull less than 13.5. . . . *M. dominicensis*, p. 200
    3'. Forearm usually more than 35; greatest length of skull more than 14.0. *M. martiniquensis*, p. 202
  2'. Not occurring in Lesser Antilles.
    *4. Uropatagium having fringe of hairs.
     5. Dorsal fur with black bases, white tips; tropical lowlands only. . . . . *M. albescens*, p. 200
     5'. Dorsal fur not black with white tips; not in tropical lowlands.
      6. Forearm usually more than 40; greatest length of skull usually more than 16.4; fringe usually well developed. . . . . *M. thysanodes*, p. 203
      6'. Forearm usually less than 40; greatest length of skull usually less than 16.4; fringe poorly developed. . . . . *M. evotis*, p. 207
    *4'. Uropatagium not having fringe of hairs.
     7. Forearm less than 28; skull much flattened. . . . . *M. planiceps*, p. 208
     7'. Forearm more than 29; skull not much flattened.
      8. Dorsal fur with black bases, white tips, resulting in "silver-tipped" appearance; tropical lowlands only. *M. albescens*, p. 200
      8'. Dorsal fur not as above.
       9. Occurring primarily in tropical lowlands, or in highlands of Central America.
        10. Premolars 2/2; restricted to México. . . . . *M. fortidens*, p. 193
        *10'. Premolars usually 3/3; not restricted to México.
         *11. Sagittal crest usually present; distance across C1–C1 usually more than postorbital constriction.
          12. Dorsal fur on uropatagium extends on tibia at least halfway from knee to foot and usually to foot; upper premolars in tooth-row. . . . . *M. keaysi*, p. 201
          12'. Dorsal fur on uropatagium does not extend to knee; P3 usually crowded to lingual side of tooth-row. . . . . . . *M. riparius*, p. 202
         *11'. Sagittal crest absent; distance across C1–C1 usually less than postorbital constriction.
          13. Forearm usually more than 38; greatest length of skull usually more than 14; restricted to mts. of Costa Rica and western Panamá. . . . . *M. oxyotus*, p. 202
          13'. Forearm usually less than 39; greatest length of skull usually less than 14; widespread.
           14. Occiput raised above braincase; greatest length of skull usually 13.0 or less. . . . . *M. elegans*, p. 200
           14'. Occiput not raised above braincase; greatest length of skull usually more than 13.0. . . . . *M. nigricans*, p. 199
       9'. Occurring in Canada, United States, Mexican highlands, Baja California, or highlands of northern Central America.

15. Ear more than 16; extends more than 2 mm. beyond nose when laid forward.
  16. Ears black.
    17. Maxillary tooth-row 6.0 or less; known only from Sierra San Pedro Mártir, Baja California. . . . .*M. milleri,* p. 206
    17'. Maxillary tooth-row 6.0 or more; widespread, but not recorded from Sierra San Pedro Mártir. . . . .*M. evotis,* p. 207
  16'. Ears dark, but not black.
    18. Ear 17–19, extending about 4 mm. beyond tip of nose; greatest length of skull 15.6 or less. . . . .*M. keenii,* p. 204
    18'. Ear 20–22, extending about 8 mm. beyond tip of nose; greatest length of skull 15.7 or more. . . . .*M. auriculus,* p. 205
15'. Ear less than 16; extends less than 2 mm. beyond nose when laid forward.
  *19. Calcar keeled.
    20. Foot more than 8.5; forearm usually more than 35. . . . .*M. volans,* p. 197
    20'. Foot less than 8.5; forearm usually less than 35.
      21. Frontal area of skull rising abruptly from rostrum; dorsal fur dull, not glossy; black mask usually absent. . . . .*M. californicus,* p. 185
      21'. Frontal area of skull rising gently from rostrum; dorsal fur glossy; black mask usually present. . . . .*M. subulatus,* p. 187
  *19'. Calcar not keeled.
    *22. Sagittal crest present.
      23. Forearm usually more than 38; greatest length of skull usually more than 15; not occurring in Baja California. . . . .*M. cobanensis,* p. 197 and *M. velifer,* p. 195
      23'. Forearm usually less than 40; greatest length of skull usually less than 15.5; restricted to southern tip of Baja California. . . . .*M. peninsularis,* p. 196
    *22'. Sagittal crest absent.
      24. Greatest length of skull usually less than 14; fur usually dull, not glossy.
                                               M. yumanensis, p. 189
      24'. Greatest length of skull usually more than 14; fur usually glossy. *M. lucifugus,* p. 191

---

    * This key to species represents the thinking of several specialists. One thinks that all calcars are "keeled," that "calcar not keeled" means *keel on calcar not visible to the naked eye,* and that "uropatagium not having fringe of hairs" also should be qualified by adding *visible to the naked eye;* another specialist thinks that whoever uses the character "sagittal crest absent" or "sagittal crest present" should realize that the development of the crest is individually variable in a few species and in two of them may be a secondary sexual character. Also, a few specimens of *M. riparius* have a reduced number of premolars (fewer than ⅜) as does a larger percentage of the subspecies *M. l. occultus.*

Myotis californicus
California Myotis

Length of head and body, 35.0–44.8; ear, 11.2–14.6; length of forearm, 29.0–36.2; greatest length of skull, 12.6–14.2; zygomatic breadth, 7.4–8.6; breadth of braincase, 5.8–7.0; length of upper tooth-row, 4.8–5.2. Upper parts brown to distinctly yellowish; underparts usually paler; hairs extending sparingly onto upper side of uropatagium to a line connecting knees, extending half as far on ventral side of membrane. Skull delicate and slender; rostrum relatively long and tapering; braincase rising abruptly from rostral level, flat-topped; sagittal crest obsolete or absent. Ear extending beyond muzzle when laid forward; calcar keeled. For distinguishing *M.*

californicus from *M. subulatus* (= *M. leibii* of Bogan 1974) in southwestern North America by means of a bivariate scattergram, see Bogan (1974:fig. 1).

Myotis californicus californicus (Audubon and Bachman)

1842. *Vespertilio californicus* Audubon and Bachman, Jour. Acad. Nat. Sci. Philadelphia, ser. 1, pt. 2, 8:285, type from "California"; subsequently restricted to Monterey, Monterey Co., California (see Miller and G. M. Allen, Bull. U.S. Nat. Mus., 144:153, May 25, 1928).
1897. *Myotis californicus,* Miller, N. Amer. Fauna, 13:69, October 16.
1862. *Vespertilio nitidus* H. Allen, Proc. Acad. Nat. Sci. Philadelphia, p. 247, type from Monterey, Monterey Co., California.

1864. *Vespertilio oregonensis* H. Allen, Smiths. Miscl. Coll., 165:61, June, based on specimens from Old Fort Yuma, California, and Cabo San Lucas, Baja California.

1866. *Vespertilio exilis* H. Allen, Proc. Acad. Nat. Sci. Philadelphia, 18:283, type from Cabo San Lucas, Baja California.

1866. *Vespertilio tenuidorsalis* H. Allen, Proc. Acad. Nat. Sci. Philadelphia, 18:283, type from Cabo San Lucas, Baja California.

1914. *Myotis californicus quercinus* H. W. Grinnell, Univ. California Publ. Zool., 12:317, December 4, type from Seven Oaks, 5000 ft., San Bernardino Co., California.

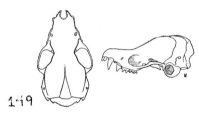

Fig. 149. *Myotis californicus stephensi*, Cave Spring, Esmeralda Co., Nevada, No. 40519 M.V.Z., ♂, X 2.

MARGINAL RECORDS.—British Columbia: Hemp Creek (Cowan and Guiguet, 1965:78); Selkirk College, Castlegar (van Zyll de Jong, et al., 1975:322). Washington (Miller and Allen, 1928:156, doubtfully as *M. c. caurinus*): Colville; Blue Creek. Montana (Hoffmann, et al., 1969:739, as *M. californicus* only): Kalispell; Grantsdale; *Hughes Creek*. Utah: Aspen Grove, Mt. Timpanogos. California: Yosemite Valley; Mt. Whitney; Kenworthy; Julian, thence southward, W of Sierra San Pedro Mártir, Baja California, to Sonora: Bahía Kino (Cockrum and Bradshaw, 1963:6, as *M. californicus*). Arizona: 12 mi. ENE Nogales (*ibid.*); Oracle; Yarnell; 30 mi. E Heber (Cockrum and Bradshaw, 1963:6, as *M. californicus*). New Mexico: Yeso Tank, 12 mi. S Canjilon (120060 KU); Jemez Springs (120064 KU); Carlsbad Cave. Texas: Giant Dagger Yucca Flats (Easterla, 1968a:516, as *M. californicus*). Coahuila: El Río Alamos, Rancho Las Margaritas; *La Gacha*; Guadalupe. Zacatecas: 8 mi. S Majoma (Genoways and Jones, 1968:744). Baja California: Cape St. Lucas; Miraflores (Jones, et al., 1965:54), thence northward, W of Sierra San Pedro Mártir, Baja California, to California: San Clemente Island (von Bloeker, 1967:249); San Nicolas Island (*ibid.*); Santa Catalina Island (*ibid.*); Santa Barbara (Miller and Allen, 1928:154); Monterey (*ibid.*); Vacaville; Trinity Mts. E of Hoopa; Mt. Shasta. Oregon: Blue River; Mt. Hood. Washington: Lyle; Orondo. British Columbia: Hedley.

Myotis californicus caurinus Miller

1897. *Myotis californicus caurinus* Miller, N. Amer. Fauna, 13:72, October 16, type from Massett, Graham Island, Queen Charlotte Islands, British Columbia.

MARGINAL RECORDS.—Alaska: Howkan, Long Island. British Columbia: Stuie; Lillooet; Okanagan Landing. Washington: Chelan; Mt. Rainier; White Salmon. Oregon: Marmot; Holly; McKenzie Bridge; Fish Lake. California: Mt. Sanhedrin; Howell Mtn.; Walnut Creek; Menlo Park, thence northward along coast, including coastal islands and Langara Island (Cowan and Guiguet, 1965:78) to point of beginning. Also from California (von Bloeker, 1967:249): Santa Cruz Island; Santa Rosa Island.

Myotis californicus mexicanus (Saussure)

1860. *V[espertilio]. mexicanus* Saussure, Revue et Mag. Zool., Paris, ser. 2, 12:282, July, type from an unknown locality. (Dalquest, Louisiana State Univ. Studies, Biol. Ser., 1:49, December 28, 1953, gives the type locality as "The desert (warmer part) of the state of México, México.")

1897. *Myotis californicus mexicanus*, Miller, N. Amer. Fauna, 13:73, October 16.

1866. *Vespertilio agilis* H. Allen, Proc. Acad. Nat. Sci. Philadelphia, 18:282, type from Mirador, Veracruz.

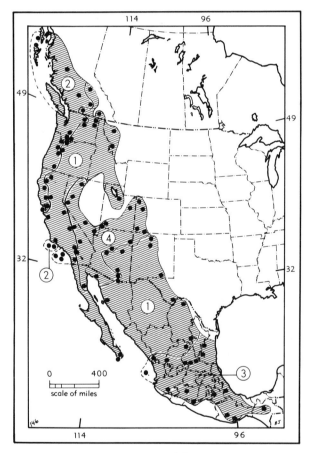

Map 146. *Myotis californicus.*

1. *M. c. californicus*   3. *M. c. mexicanus*
2. *M. c. caurinus*   4. *M. c. stephensi*

MARGINAL RECORDS.—Tamaulipas: San José. Veracruz: Mirador. Chiapas: Tenejapa, 6700 ft. (Carter, et al., 1966:496). Oaxaca (Goodwin, 1969:96): Chontecomatlán; Cerro San Felipe. Tlaxcala: 13 km. NE Tlaxcala. Guerrero: 4³⁄₁₀ km. N Teloloapan, 1480 m. (Villa-R., 1967:381). Michoacán: Pátzcuaro. Jalisco: Los Masos. Nayarit: Tres Marías Islands (See addenda). Sinaloa: 5 mi. E Plomosas, 5500 ft. (Jones, et al., 1971:410). Zacatecas: San Juan Capistrano. San Luis Potosí: Cerro Peñón Blanco; Hda. La Parada; 10 km. NW Villar. Tamaulipas: Miquihuana.

Myotis californicus stephensi Dalquest

1900. Myotis californicus pallidus Stephens, Proc. Biol. Soc. Washington, 13:153, June 13, type from Vallecito, San Diego Co., California. Not K[erivoula]. pallida Blyth, 1863 [= Myotis pallidus = Myotis formosus formosus Hodgson, 1835, from Nepal], from Chaibassa, Orissa, India.

1946. Myotis californicus stephensi Dalquest, Proc. Biol. Soc. Washington, 59:67, March 11. Dalquest (op. cit.) named a new type, also from Vallecito, California. However, if the International Code of Zoological Nomenclature (Art. 72d) is to be retroactive, the type of M. c. stephensi Dalquest must be the original type of M. c. pallidus Stephens, and the type named by Dalquest has no status in nomenclature.

MARGINAL RECORDS.—Nevada: Cottonwood Range; Calico Spring, Red Rock Canyon (O'Farrell and Bradley, 1969:128, as M. californicus only); Nevada Test Site (Jorgensen and Hayward, 1965:3); White Spot Spring, Desert Game Range (O'Farrell, et al., 1967:164, as M. californicus only). Utah: St. George; campground Zion National Park; Sunnyside. Colorado (Armstrong, 1972:64, 66, unless otherwise noted): Echo Park, 10 mi. N Artesia (as M. leibii melanorhinus); Rifle; Allison (Bogan, 1975:25, as M. californicus only); Rock Springs, Mesa Verde National Park. Arizona: Fort Defiance; Fort Verde; Santa Catalina Mills. Sonora: Sierra del Pinacate. Baja California: Sierra San Pedro Mártir. California: La Puerta Valley; Lavic; Lone Pine Creek. Nevada: 9 mi. E, 2 mi. S Yerington; Fallon.

Myotis subulatus
Small-footed Myotis

Length of head and body, 34.4–48.0; ear, 12.2–15.0; length of forearm, 29.6–36.0; greatest length of skull, 13.1–14.7; zygomatic breadth, 8–9; breadth of braincase, 6.2–7.1; length of upper tooth-row, 4.8–5.5. Pelage long, silky, the tips frequently glossy; ears and face black; upper parts light buff to golden brown; underparts buffy to almost white. Skull small, delicate, resembling that of M. californicus; braincase not rising abruptly from rostral level but instead sloping upward gradually; low sagittal crest sometimes present; braincase conspicuously flattened. Ear, when laid forward, barely exceeding muzzle; fur

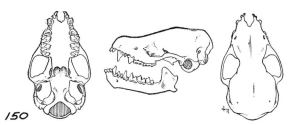

Fig. 150. Myotis subulatus subulatus, 5 mi. W Elkader, Logan Co., Kansas, No. 5561 K.U., ♂, X 2.

extending slightly onto membranes below; foot small, delicate; calcar long, slender, terminating in a minute lobule, keeled. For distinguishing M. subulatus from M. californicus in southwestern North America by means of a bivariate scattergram, see Bogan (1974:fig. 1).

In much of the earlier literature (prior to 1928), the name M. subulatus was applied to the bat now known as M. keenii septentrionalis. Glass and Baker (1965:205) proposed that the name Vespertilio subulatus Say be suppressed and later (1968:259) proposed that this name be considered "nonassignable." To obtain stability in nomenclature, Glass and Baker should have assigned subulatus to the species we now call M. yumanensis or to the species they call M. leibii. Because they did neither, I choose to follow the last reviser (Miller and Allen, 1928).

As pointed out by Glass and Ward (1959:201), specimens from western Oklahoma are intermediate between M. s. melanorhinus and M. s. leibii and overlap both those subspecies in color and size. Since there are no large series from eastern Oklahoma, Arkansas, or Missouri with which to compare the specimens from western Oklahoma, and since they do not differ greatly from melanorhinus from New Mexico and Colorado, they are here assigned to that subspecies. The specimen from eastern Oklahoma is assigned to M. s. leibii on geographic grounds. Probably there is some gene flow across central Oklahoma, and certainly there are no differences between the populations from the eastern United States and those from the western United States that warrant the recognition of two species. Actually, M. s. subulatus seems to differ more from M. s. melanorhinus than does M. s. leibii.

Also, since the most distinctive character of M. s. subulatus seems to be its pale, blond color, and since Texas specimens are not that color, but are more nearly like those from New Mexico, the Texas specimens are here assigned to M. s. melanorhinus, the color of which varies much.

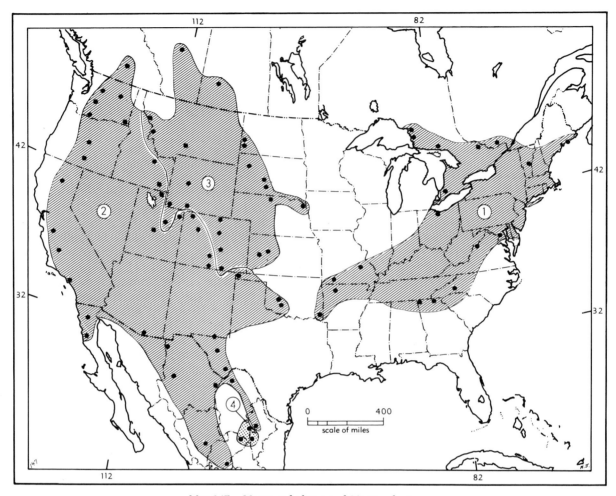

Map 147.   *Myotis subulatus* and *Myotis planiceps.*

1. *Myotis s. leibii*   3. *Myotis s. subulatus*
2. *Myotis s. melanorhinus*   4. *Myotis planiceps*

Myotis subulatus leibii (Audubon and Bachman)

1842. *Vespertilio leibii* Audubon and Bachman, Jour. Acad.
Nat. Sci. Philadelphia, ser. 1, 8:284, type from Erie County,
Ohio.

1928. *Myotis subulatus leibii*, Miller and G. M. Allen, Bull.
U.S. Nat. Mus., 144:171, May 25.

1913. *Myotis winnemana* Nelson, Proc. Biol. Soc. Washington, 26:183, August 8, type from Plummers Island,
Montgomery Co., Maryland.

MARGINAL RECORDS.—Ontario (Fenton, 1972:1,
2, unless otherwise noted): Lake Superior Provincial
Park; mine near Alona Bay; mine near Webbwood;
Fourth Chute (Hall and Kelson, 1959:175). Quebec: La
Fliche Cave, near Wakefield. Vermont: Vershire.
Maine: Otter Point, Mt. Desert Id., thence southward
along coast to Maryland: Plummers Island. Virginia:
Millboro Cave. North Carolina: Bat Cave. Georgia
(Baker, 1967:142): Toccoa Experimental Station,
Union Co.; Howard Waterfall Cave, Trenton. Okla-
homa: 250 yd. W Mtn. Fork R., McCurtain Co. Game
Reserve (8066 Oklahoma State Univ.). Arkansas: Salt

Peter Cave, near Boxley (Sealander, 1967:666). Missouri: Marvel Cave (Gunier and Elder, 1973:489); 3
mi. S Graniteville. Ohio: type locality. Ontario: Mt.
Brydges.

Myotis subulatus melanorhinus (Merriam)

1890. *Vespertilio melanorhinus* Merriam, N. Amer. Fauna,
3:46, September 11, type from Little Spring, 8250 ft., N
base San Francisco Mtn., Coconino Co., Arizona.

1928. *Myotis subulatus melanorhinus*, Miller and G. M. Allen, Bull. U.S. Nat. Mus., 144:169, May 25.

1894. *V[espertilio]. nitidus henshawii* H. Allen, Bull. U.S.
Nat. Mus., 43:103, March 14, type from near Wingate,
McKinley Co., New Mexico.

1903. *Myotis orinomus* Elliot, Field Columb. Mus., Publ. 79,
Zool. Ser., 3:228, August 15, type from La Grulla, 8000 ft.,
Sierra San Pedro Mártir, Baja California.

MARGINAL   RECORDS.—British   Columbia:
Oyama. Washington: 5 mi. S Grand Coulee Dam; Bly.
Idaho: Double Springs; ½ mi. E Portneuf. Utah: 15 mi.
N Logan; *Salt Lake City*; Ferron; 25 mi. E Vernal. Col-

orado (Armstrong, 1972:66): Little Snake River, S of Sunny Peak; Dry Fork, White River, 6200 ft.; 1½ mi. N Crestone, 8050 ft.; 7 mi. E Antonito. Oklahoma (Glass and Ward, 1959:201): 3 mi. N Kenton; Windmill Cave, Kiowa Co.; Narrows, Wichita Mts. National Wildlife Refuge. New Mexico: Guadalupe Canyon [probably in Texas]. Texas (Hall and Kelson, 1959, as *M. s. subulatus*): Terlingua Creek; Fort Davis. Coahuila: Fronteriza Mts. (Easterla and Baccus, 1973:426). Nuevo León: Monterrey (Villa-R., 1967:382). Chihuahua: 2 mi. SW Hechicero (Anderson, 1972:246). Zacatecas: 9⁷⁄₁₀ mi. NW Cuauhtemoc, 7100 ft. (Best, *et al.*, 1972:97). Durango: 1 mi. N Chorro (Baker and Greer, 1962:72). Chihuahua (Anderson, 1972:246): Rancho San Ignacio, 4 mi. S, 1 mi. W Santo Tomás; Llano de las Carretas, 27 mi. W Cuervo. Arizona: Huachuca Mts. Baja California: Laguna Hanson; Santa Eulalia. California: Santa Ysabel; 1 mi. SW Cholame; 7½ mi. ESE Panoche; Petes Valley. Oregon: Warner Valley; Twelve Mile Creek. Washington: Lyle; Selah; Wenatchee.

Myotis subulatus subulatus (Say)

1823. V[*espertilio*]. *subulatus* Say, *in* Long, Account of an exped. . . . to the Rocky Mts. . . . , 2:65 (footnote), type from vic. jct. Arkansas River and Apishapa Creek, Colorado (see Glass and Baker, Proc. Biol. Soc. Washington, 81:257, August 30, 1968).

1897. *Myotis subulatus*, Miller, N. Amer. Fauna, 13:75, October 16. Note: Although the referred specimens in Miller, p. 76, are *M. keenii septentrionalis*, his name combination *Myotis subulatus* was intended to apply to *V. subulatus* (Say).

1886. *Vespertilio ciliolabrum* Merriam, Proc. Biol. Soc. Washington, 4:2, December 17, type from a bluff on Hackberry Creek, about 1 mi. from Castle Rock, near Banner, Trego Co., Kansas.

MARGINAL RECORDS.—Alberta: Red Deer River near Rumsey. Saskatchewan: South Saskatchewan River Valley N of Stewart Valley (Beck, 1958:12). North Dakota: 1 mi. S, 1 mi. W Medora (Jones and Genoways, 1966:88); Little Missouri River, 8 mi. N, 7½ mi. W Amidon (Jones and Stanley, 1962:263). South Dakota (Jones and Genoways, 1967:186): 10 mi. S, 5 mi. W Reva; Philip; 14 mi. N Longvalley. Nebraska: 1 mi. S, 18 mi. E Valentine (Farney and Jones, 1975:327); Crystal Lake (Jones, 1976:89—specimen lost). Kansas: Banner; 5 mi. W Elkader (Cockrum, 1952:63). Colorado (Armstrong, 1972:66): near Wooten, 7500 ft.; Colorado Springs; Boulder; 27 mi. NNW Fort Collins. Wyoming: Kinney Ranch. Utah: Soldier Canyon. Wyoming: Fort Bridger; Bull Lake. Montana (Hoffmann, *et al.*, 1969:739, unless otherwise noted): Big Timber (Hall and Kelson, 1959:176); *Sleeping Child*; Grantsdale; Amador Mines on Cedar Creek; *Frenchtown*. Jones (1964c:85), among others, reported a record from Crystal Lake in eastern Nebraska based on a specimen of questionable identity that does not now seem to be extant.

Myotis yumanensis
Yuma Myotis

Length of head and body, 37.8–49.0; ear, 11.0–14.5; length of forearm, 32.0–38.0; greatest length of skull, 13.0–14.2; zygomatic breadth, 7.8–8.8; breadth of braincase, 6.6–7.5; length of maxillary tooth-row, 4.6–5.2. Upper parts tawny, buffy, or even brown; darker subspecies often with buffy wash; underparts paler, buffy to yellowish white; membranes pale brownish; fur dull, lacking brassy sheen of *lucifugus* except in some specimens of *M. y. saturatus*. Braincase rising abruptly from level of rostrum; sagittal crest usually absent; foot relatively large, robust; tail barely extending beyond membrane.

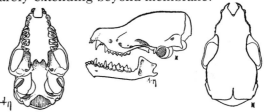

Fig. 151. *Myotis yumanensis yumanensis*, ½ mi. S Pyramid Lake, Nevada, No. 88052 M.V.Z., ♂, X 2.

Myotis yumanensis lambi Benson

1947. *Myotis yumanensis lambi* Benson, Proc. Biol. Soc. Washington, 60:45, May 19, type from San Ignacio, lat. 27° 17′ N, Baja California. Known only from type locality.

Myotis yumanensis lutosus Miller and G. M. Allen

1928. *Myotis yumanensis lutosus* Miller and G. M. Allen, Bull. U.S. Nat. Mus., 144:72, May 25, type from Pátzcuaro, Michoacán.

MARGINAL RECORDS.—Sinaloa: ½ mi. SE Vaca, 650 ft. (Jones, *et al.*, 1971:416). Zacatecas: San Juan Capistrano. San Luis Potosí: Ahualulco. Hidalgo: Río Tasquillo. Morelos: 1 km. S Oaxtepec (Ramírez-P., 1971:124). México: Valle de Bravo, 1820 m. (Alvarez and Ramírez-P., 1972:173). Michoacán: type locality; El Molino. Jalisco (Jones, *et al.*, 1971:416): El Salto, 24 mi. W Guadalajara, 1280 m.; 2 mi. S La Cuesta, 1500 ft. Sinaloa (*loc. cit.*): Cosalá, 1300 ft.; 2 mi. E Aguacaliente, 800 ft.; 6 km. NE El Fuerte, 150 m. See addenda.

Myotis yumanensis oxalis Dalquest

1947. *Myotis yumanensis oxalis* Dalquest, Amer. Midland Nat., 38:228, July, type from Oxalis, San Joaquin Valley, Fresno Co., California.

MARGINAL RECORDS.—California: Davis; type locality; Berkeley.

Myotis yumanensis saturatus Miller

1897. *Myotis yumanensis saturatus* Miller, N. Amer. Fauna, 13:68, October 16, type from Hamilton, Skagit Co., Washington.

MARGINAL RECORDS.—British Columbia: Princess Royal Island; Kimsquit; Shuswap; Kamloops; Okanagan; *Skagit;* Cultus Lake (Cowan and Guiguet, 1965:88). Washington: Goldendale; Hamilton. Oregon: Crooked River. California: Baird; ½ mi. S Oroville; Lake Alta; Yosemite Valley; San Simeon, thence northward along coast and coastal islands to point of beginning.

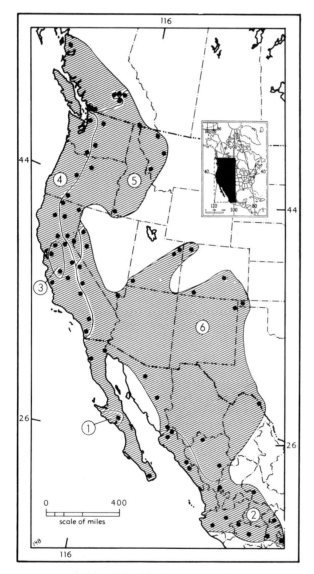

Map 148. *Myotis yumanensis.*

1. *M. y. lambi*   4. *M. y. saturatus*
2. *M. y. lutosus*   5. *M. y. sociabilis*
3. *M. y. oxalis*   6. *M. y. yumanensis*

Myotis yumanensis sociabilis H. W. Grinnell

1914. *Myotis yumanensis sociabilis* H. W. Grinnell, Univ. California Publ. Zool., 12:318, December 4, type from Old Fort Tejon, Tehachapi Mts., Kern Co., California.

MARGINAL RECORDS.—British Columbia: Sicamous; Creston. Montana: Belton; Seeley Lake, Missoula Co. (Hoffmann, *et al.*, 1969:737, as *M. yumanensis* only); Conner (*ibid.*). Idaho: South Fork, Owyhee River. California: Alturus Lake. Nevada: Calneva. California: Weldon; San Bernardino Mts.; Lake Hodges, near Escondido; Fresno; *Tahoe City;* Nevada City; Dale's; Beswick. Oregon: Klamath Falls; Lone Rock. Washington: Selah; Stehekin. British Columbia: Lehman; *Okanagan Landing;* Kamloops.

Myotis yumanensis yumanensis (H. Allen)

1864. *Vespertilio yumanensis* H. Allen, Smiths. Miscl. Coll., 7(Publ. 165):58, June, type from Old Fort Yuma, Imperial Co., California, on right bank of Colorado River, opposite present town of Yuma, Arizona.
1897. *Myotis yumanensis*, Miller, N. Amer. Fauna, 13:66, October 16.
1866. *Vespertilio obscurus* H. Allen, Proc. Acad. Nat. Sci. Philadelphia, p. 281, type from Baja California.
1866. *Vespertilio macropus* H. Allen, Proc. Acad. Nat. Sci. Philadelphia, p. 288, type from Fort Mohave, Colorado River, Arizona. Not *Myotis macropus* Gould, 1854, from Australia.
1903. *Myotis californicus durangae* J. A. Allen, Bull. Amer. Mus. Nat. Hist., 19:612, November 12, type from San Gabriel, Río Sestín, Durango.
1928. *Myotis lucifugus phasma* Miller and G. M. Allen, Bull. U.S. Nat. Mus., 144:53, May 25, type from Snake River, S of Sunny Peak, Routt [now Moffatt] Co., Colorado.

MARGINAL RECORDS.—Colorado: Snake River S of Sunny Peak (Harris and Findley, 1962:199); Allison (*ibid.*); Fort Carson (Constantine, 1966:126). Oklahoma: Pigeon Cave, 1 mi. N, 3 mi. E Kenton (Glass and Ward, 1959:197). New Mexico: Apache Canyon, near Clayton. Texas: Del Rio. Durango: San Gabriel; Arroyo de Bucy. Sonora (Harris, 1974:606, unless otherwise noted): 8 mi. (by road) E Alamos; Rebeico Dam, 28 mi. by road E Mazatán (Findley and Jones, 1965:331, as *M. yumanensis* only); 1 mi. (by road) E Santa Ana; Colonia Lerdo. Baja California: Cape San Lucas; Rancho San Antonio. California: Mt. Whitney; jct. U.S. Hwy. 395 with McGee Creek, near Crowley Lake (Harris, 1974:606). Nevada: 3 mi. W Sutcliffe; *E shore Pyramid Lake, 15 mi.* [by road?] *N Nixon;* Fallon; Colorado River. Utah (Harris, 1974:606): Pine Valley, 6500 ft. (Stock, 1970:429); along Green River, 15 mi. SW Ouray, 4500 ft.; 1½ mi. W Jensen P.O.—In the light of Harris' (1974) identification of specimens from Sonora and Chihuahua, as *M. y. yumanensis*, specimens previously listed under other subspecific names from Pilares, Sonora (Jones, *et al.*, 1971:416), Guirocoba, Sonora (*ibid.*), and Río Piedras Verdes, 8 mi. NW Colonia Juárez, Chihuahua (Anderson,

1972:249), need restudy to make certain of their sub-specific identity.

Myotis lucifugus
Little Brown Myotis

×1

Length of head and body, 41–54; ear, 11.0–15.5; length of forearm, 33.0–41.0; greatest length of skull, 14.0–16.0; zygomatic breadth, 8.1–9.8; breadth of braincase, 7.0–7.8; length of upper tooth-row, 5.0–6.6. Upper parts cinnamon-buff to dark brown; underparts buffy to pale gray, often with lighter wash; buffy shoulder spot sometimes present; pelage long and silky, individual hairs shiny (almost having metallic sheen) at tips. Braincase rising gradually from rostrum. Ear when laid forward reaching approx. to nostril; tragus approx. half as high as ear.

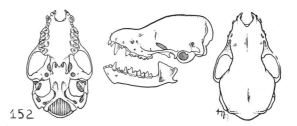

Fig. 152. *Myotis lucifugus lucifugus*, 1½ mi. E, ¼ mi. N Blue Rapids, Kansas, No. 44694 K.U., ♂, X 2.

Myotis lucifugus alascensis Miller

1897. *Myotis lucifugus alascensis* Miller, N. Amer. Fauna, 13:63, October 16, type from Sitka, Alaska.

MARGINAL RECORDS.—Alaska: Mole Harbor, Admiralty Island. British Columbia: Tetana Lake (Cowan and Guiguet, 1965:84); Fort St. James; Assiniboine. Montana: Corvallis. Oregon: Cornucopia. Washington: Godman Springs. Idaho: Felton Mills. British Columbia: Okanagan. Washington: Whatcom Pass. Oregon: Bend. California: Castle Lake; Chews Ridge, Santa Lucia Mts., thence along coast and coastal islands to Alaska: type locality.

Myotis lucifugus carissima Thomas

1904. *Myotis* (*Leuconoë*) *carissima* Thomas, Ann. Mag. Nat. Hist., ser. 7, 13:383, May, type from Lake Hotel, Yellowstone National Park, Wyoming.
1917. *Myotis lucifugus carissima*, Cary, N. Amer. Fauna, 42:43, October 3.
1916. *Myotis yumanensis altipetens* H. W. Grinnell, Univ. California Publ. Zool., 17:9, August 23, type from 1 mi. E Merced Lake, 7500 ft., Yosemite National Park, California.
1919. *Myotis albicinctus* G. M. Allen, Jour. Mamm., 1:2, November 28, type from upper limit of timber, 11,000 ft., Mt. Whitney, Tulare Co., California.

MARGINAL RECORDS.—British Columbia (Cowan and Guiguet, 1965:84): Cache Creek; Vernon; Osoyoos. Washington: Newport. Oregon: Baker County. Montana: Corvallis; Cut Bank; Glasgow. North Dakota: 8 mi. N Towner; Cannon Ball. Nebraska (Jones, 1964c:83): 6 mi. S Chadron; Agate. Colorado (Armstrong, 1972:59): 1 mi. S La Salle; Fort Carson. New Mexico: Sierra Grande; 4 mi. W Eagle Nest (Barbour and Davis, 1970:150); Montezuma (Studier, *et al.*, 1967:565, as *M. lucifugus* only). Colorado: Conejos

River (Harris and Findley, 1962:199, as *M. lucifugus* only). Utah: Donkey Lake. California: Bluff Lake; *Bear Lake; Mt. Whitney, 11,000 ft.* (G. M. Allen, 1919:2, as *M. albicinctus*); Yosemite National Park; Gilmore Lake; Castle Lake. Oregon: Paulina Lake. Washington: Vantage; Stehekin.

Myotis lucifugus lucifugus (Le Conte)

1831. *V[espertilio]. lucifugus* Le Conte, *in* McMurtrie, The animal kingdom . . . by the Baron Cuvier, 1:(App.) 431, type from Georgia; probably Le Conte Plantation, near Riceboro, Liberty Co. (but see Davis and Rippy, 1968:113).
1897. *Myotis lucifugus,* Miller, N. Amer. Fauna, 13:59, October 16.

1832. *Vespertilio gryphus* F. Cuvier, Nouv. Ann. Mus. d'Hist. Nat. Paris, 1:15, type from New York.
1832. ? *Vespertilio salarii* F. Cuvier, Nouv. Ann. Mus. d'Hist. Nat. Paris, 1:16, type from New York.
1832. ? *Vespertilio crassus* F. Cuvier, Nouv. Ann. Mus. d'Hist. Nat. Paris, 1:18, type from New York.
1832. *Vespertilio domesticus* Green *in* Doughty, Cabinet of natural history, 2:290, type from western Pennsylvania.
1839. ? *Vespertilio lanceolatus* Wied-Neuwied, Reise in das innere Nord-America . . . , 1:364, footnote, type from Bethlehem, Pennsylvania.
1840. *Vespertilio carolii* Temminck, Monographies de mammalogie . . . , 2:237, based on specimens from Philadelphia and New York.
1841. *Vespertilio Virginianus* Audubon and Bachman, Proc.

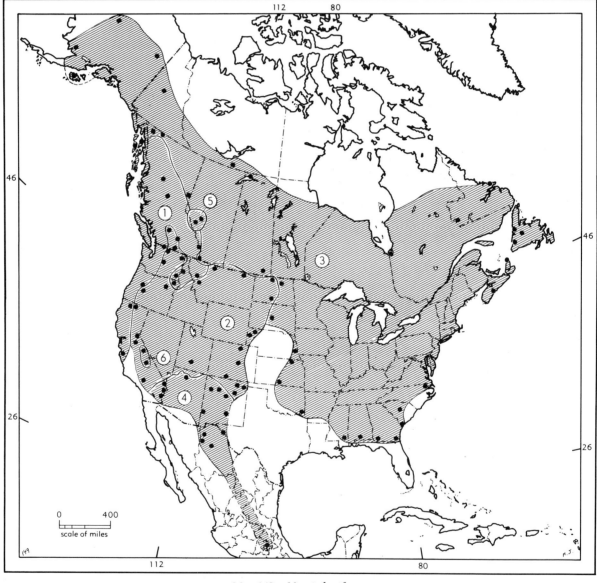

Map 149. *Myotis lucifugus.*

1. *M. l. alascensis*	3. *M. l. lucifugus*	5. *M. l. pernox*
2. *M. l. carissima*	4. *M. l. occultus*	6. *M. l. relictus*

Acad. Nat. Sci. Philadelphia, 1:93, October, type from mountains of Virginia.

1860 [= 1862]. V[espertilio]. brevirostris Wied-Neuwied, Verzeichniss der auf senier Resise in Nord-Amerika beobachteten Säugethiere, p. 19, type from Freiburg, Pennsylvania.

1864. Vespertilio affinis H. Allen, Smiths. Miscl. Coll., 165:53, figs. 48–50, June, type from Fort Smith, Arkansas.

MARGINAL RECORDS.—Saskatchewan: Crackingstone Point, Lake Athabasca (Beck, 1958:12). Quebec: Rupert House. Labrador: Ashuanipi Lake (Harper, 1961:35); Makkovik. Newfoundland: Nicholsville; South Brook (Cameron, 1959:74); Bay St. George. Nova Scotia: Cape North; Halifax. North Carolina: Bertie County (Davis and Rippy, 1968:116). South Carolina (Davis and Rippy, 1968:116): Columbia (sight record only); Beaufort. Georgia (Golley, 1962:58): Charlton County; Decatur County. Alabama: Sander's Cave (LaVal, 1967:645). Mississippi: Pitt's Cave (ibid.). Oklahoma: Beaver's Bend State Park (Glass and Ward, 1959:197). Kansas (Jones, et al., 1967:11): Double-entrance "S" Cave; Harvard Cave, $2^{1}/_{2}$ mi. SW Sun City; $\frac{1}{4}$–$\frac{1}{2}$ mi. N, $1\frac{1}{2}$ mi. W Blue Rapids. Nebraska: $\frac{1}{2}$–1 mi. W Meadow (Jones, 1964c:84). South Dakota: Fort Pierre. North Dakota: Devils Lake. Saskatchewan: Estevan (Beck, 1958:12). British Columbia: Akamina Pass (Cowan and Guiguet, 1965:85). Not found: South Dakota: Cedar Island (Miller and G. M. Allen, 1928:46).

Myotis lucifugus occultus Hollister

1909. Myotis occultus Hollister, Proc. Biol. Soc. Washington, 22:43, March 10, type from W side Colorado River, 10 mi. above Needles, San Bernardino Co., California.

1967. Myotis lucifugus occultus, Findley and Jones, Jour. Mamm., 48:443, August 21. (See also Barbour and Davis, Jour. Mamm., 51:150, 151, February 20, 1970.)

1909. Myotis baileyi Hollister, Proc. Biol. Soc. Washington, 22:44, March 10, type from base White Mts., 7500 ft., near Ruidosa, Lincoln Co., New Mexico.

MARGINAL RECORDS (Findley and Jones, 1967:443, unless otherwise noted).—Arizona: South Rim (Hoffmeister, 1971:164). New Mexico: Bear Ridge, Zuni Mts.; Corrales; Sandia Park, Sandia Mts.; $2\frac{1}{2}$ mi. N, 7 mi. E Cloudcroft. Texas: Fort Hancock. México: 5 km. NW Texcoco, 7600 ft. Distrito Federal: Coapa, 2240 m. (Alvarez and Ramírez-P., 1972:172). Chihuahua (Anderson, 1972:246): 1 mi. S, $\frac{1}{2}$ mi. E Santa Clara, 6100 ft.; Río Gavilán, 7 mi. SW Pacheco, 5700 ft.; Río Casas Grandes, 10 mi. N Nueva Casas Grandes. New Mexico: 1 mi. N Redrock. California: Ripley, 5 mi. S Blythe; Blythe; Riverside Mts.; type locality. Arizona: Mohave Desert (no precise locality).

Myotis lucifugus pernox Hollister

1911. Myotis pernox Hollister, Smiths. Miscl. Coll., 56(26):4, December 5, type from Henry House, Alberta.

1943. Myotis lucifugus pernox, Crowe, Bull. Amer. Mus. Nat. Hist., 80:395, February 4.

MARGINAL RECORDS.—Alaska: Fort Yukon. Yukon: Mayo Landing. Mackenzie: Salt River. Alberta: type locality; Entrance. British Columbia: Tupper Creek; Screw Creek, 10 mi. S, 50 mi. E Teslin Lake, Yukon (Youngman, 1975:53); Atlin. Alaska: Kodiak Island; Bristol; Nulato. [See Youngman, 1975:53, 54, for diagnostic features of this subspecies and for its geographic range as outlined by marginal records immediately above.]

Myotis lucifugus relictus Harris

1974. Myotis lucifugus relictus Harris, Jour. Mamm., 55:598, August 20, type from Keeler, 3600 ft., Inyo Co., California.

MARGINAL RECORDS (Harris, 1974:606).—California: Long Valley Resort, W side Crowley Lake; jct. U.S. Hwy. 395 with McGee Creek, near Crowley Lake; 3 mi. W Lone Pine, Lone Pine Creek; type locality.

Myotis fortidens
Cinnamon Myotis

Length of head and body, 46.0–53.6; length of forearm, 35.6–38.8; greatest length of skull, 14.8–15.5; zygomatic breadth, 9.4–9.7; breadth of braincase, 6.8–7.4; length of upper tooth-row, 5.4–5.8. Upper parts cinnamon-brown, the hairs black basally; underparts but little paler. In general closely resembling M. lucifugus but molariform teeth markedly larger, sagittal crest well developed, zygomatic breadth averaging greater, premolars $\frac{2-2}{2-2}$ ($\frac{3-3}{3-3}$ in M. lucifugus).

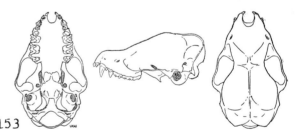

Fig. 153. Myotis fortidens fortidens, 20 km. ENE Jesús Carranza, 200 ft., Veracruz, No. 32112 K.U., ♂, X 2.

Myotis fortidens fortidens Miller and G. M. Allen

1928. Myotis lucifugus fortidens Miller and G. M. Allen, Bull. U.S. Nat. Mus., 144:54, May 25, type from Teapa, Tabasco.

1950. Myotis fortidens, Hall and Dalquest, Univ. Kansas Publ., Mus. Nat. Hist., 1:586, January 20.

1967. Myotis fortidens fortidens, Findley and Jones, Jour. Mamm., 48:442, August 21.

1902. Pipistrellus cinnamomeus Miller, Proc. Acad. Nat. Sci. Philadelphia, 54:390, September 12, type from Montecristo, Tabasco. Not Vespertilio cinnamomeus Wagner, 1855, a renaming of Vespertilio ruber É. Geoffroy St.-Hilaire, 1806 [= Myotis ruber É. Geoffroy St.-Hilaire, from Paraguay].

MARGINAL RECORDS (Findley and Jones, 1967:442, unless otherwise noted).—Sinaloa: 6 km. E Cosalá; Escuinapa. Nayarit: 8 mi. E San Blas. Jalisco: El Tabaco, 200 ft. (Jones, *et al.*, 1971:411). Colima: 6 km. N Agua Zarca. Guerrero: Pápayo. Oaxaca (Goodwin, 1969:96): San Bartolo Yautepec; Tuxtepec. Veracruz: Río Blanco, 20 km. WNW Piedras Negras. Tabasco: ½ mi. W Miramar; Monte Cristo. Chiapas: 10 mi. S Zapaluta; 1 mi. SE Puerto Madero. Oaxaca: Tequixistlán.

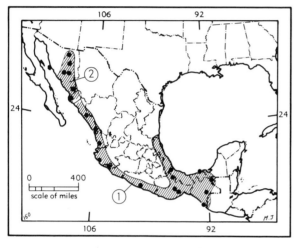

Map 150. *Myotis fortidens.*

1. *M. f. fortidens*   2. *M. f. sonoriensis*

Myotis fortidens sonoriensis Findley and Jones

1967. *Myotis fortidens sonoriensis* Findley and Jones, Jour. Mamm., 48:441, August 21, type from east bank of Río Yaquí, *ca*. 1 mi. S El Novillo, Sonora.

MARGINAL RECORDS (Findley and Jones, 1967:442, unless otherwise noted).—Sonora: *ca.* 9 mi. N Nacozari; 2 mi. by road S Moctezuma; type locality; 10 mi. by road E Alamos. Sinaloa: Río Fuerte, 1 mi. N, ½ mi. E San Miguel. Sonora: *La Aduana;* 3 mi. E Mazatán; *La Estancia, 6 mi. N Nacori.*

Myotis sodalis Miller and G. M. Allen
Indiana Myotis

1928. *Myotis sodalis* Miller and G. M. Allen, Bull. U.S. Nat. Mus., 144:130, May 25, type from Wyandotte Cave, Crawford Co., Indiana.

Length of head and body, 41.4–49.0; ear, 10.4–14.8; length of forearm, 36.0–40.6; greatest length of skull, 14.2–15.0; zygomatic breadth, 8.3–9.3; breadth of braincase, 6.8–7.2; length of upper tooth-row, 5.2–5.6. Pelage unusually fine and fluffy: upper parts dull grayish chestnut, each hair slightly glossy at tip, basal two-thirds blackish; underparts having general effect of pinkish

white; membranes and ears blackish brown. Skull resembling that of *M. lucifugus* but with smaller, narrower, and lower braincase; delicate but complete sagittal crest usually present in adults. Calcar obviously keeled.

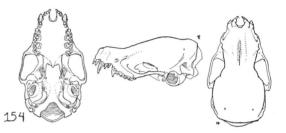

Fig. 154. *Myotis sodalis*, White Rock Camp, ½ mi. E Fiftysix, Stone Co., Arkansas, No. 47581 M.V.Z., ♀, X 2.

MARGINAL RECORDS.—Vermont: Vershire. Massachusetts: Worcester. Connecticut: Roxbury. New Jersey: Hibernia (Davis, 1965:151). Virginia: Madden's Cave; Nellie's Hole. South Carolina: *upper western tip of state.* Florida: Old Indian Cave, 3 mi. N Marianna. Alabama: Anniston. Mississippi: Chalk Mine, 4 mi. E Iuka (LaVal, 1967:647). Arkansas: near Cushman; 25 mi SW Fayetteville. Oklahoma: Bower's Trail Cave (4½ mi. NW Honobia) (Glass and Ward, 1959:200); 5 mi. S Kansas. Missouri: Rocheport Cave; *2 mi. SE Arkoe* (Easterla and Watkins, 1969:373); 6 mi. S Maryville (*ibid.*); *4 mi. SE Maryville* (*ibid.*). Iowa: near Colfax (Bowles, 1975:47); Dubuque (Muir and Polder, 1960:603). Wisconsin: 2¼ mi. W Beetown. Illinois: 8 mi. SE Galena. Michigan: Grosse Isle. New York: Brownville Township, S shore Black River, opposite Brownville (Fenton, 1966:526). Vermont: Brandon.

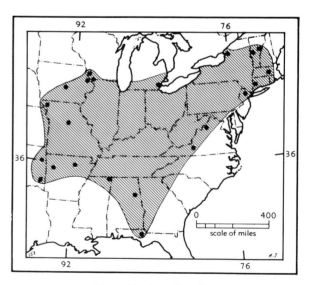

Map 151. *Myotis sodalis.*

Myotis austroriparius (Rhoads)
Southeastern Myotis

Revised by LaVal, Jour. Mamm., 51:542–552, August 28, 1970.

1897. *Vespertilio lucifugus austroriparius* Rhoads, Proc. Acad. Nat. Sci. Philadelphia, 49:227, May 22, type from Tarpon Springs, Pinellas Co., Florida.

1928. *Myotis austroriparius*, Miller and G. M. Allen, Bull. U.S. Nat. Mus., 144:76, May 25.

1943. *Myotis austroriparius gatesi* Lowery, Occas. Pap. Mus. Zool., Louisiana State Univ., 13:219, November 22, type from Louisiana State Univ. Campus, Baton Rouge.

1955. *Myotis austroriparius mumfordi* Rice, Quart. Jour. Florida Acad. Sci., 18:67, May 17, type from Bronsons Cave, Spring Mill State Park, 3 mi. E Mitchell, Lawrence Co., Indiana.

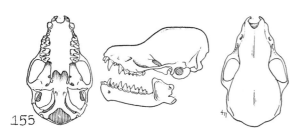

Fig. 155. *Myotis austroriparius*, Alachua Co., Florida, No. 9523 K.U., ♀, X 2.

Total length, 78–93; forearm, 34.0–41.3; 4th metacarpal, 31.0–37.8; greatest length of skull, 13.5–15.2; occipital depth, 6.1–7.2. Fur thick, wooly; overhairs sparse; fur gray to bright orange-brown, females normally more brightly colored than males; color highly variable, probably a function of molt and age of pelage. Skull with abruptly rising forehead and highly inflated braincase; sagittal crest usually present.

MARGINAL RECORDS.—Indiana: Ray's Cave, Greene Co.; Bronsons Cave, 3 mi. E Mitchell; Saltpeter Cave, Crawford County. Kentucky: Dixon Cave, Mammoth Cave National Park (J. S. Hall, 1961:400). Tennessee: Decatur County (from map in LaVal, 1970:544). Mississippi (LaVal, 1967:646): 4 mi. E Luka [= Iuka]; Pitt's Cave, 7 mi. N Waynesboro. Alabama: Sander's Cave, 1 mi. N, 3 mi. W Brooklyn (LaVal, 1967:645). Georgia (LaVal, 1967:645): Colquitt; Athens. North Carolina: Cape Fear River, *ca.* 30 mi. above Wilmington (Davis and Rippy, 1968:116). Georgia: Sapelo Island (LaVal, 1967:645). Florida (Rice, 1957:16–29): Orange County; Manatee County; Devil's Den, Levy Co.; St. Marks; Old Indian Cave, Jackson Co. Louisiana: Belle Chasse (LaVal, 1967:646); Avery Island (*ibid.*); Vowells Mill (Lowery, 1974:105). Texas: 8 mi. SW Gary (Michael, *et al.*, 1970:620); New Boston (Packard, 1966:128). Oklahoma: 5½ mi. ENE Cloudy (LaVal, 1967:646); Honobia (Glass and Ward,

1959:198); 8 mi. E Smithville (*ibid.*). Arkansas: 12 mi. NW Hot Springs; 1 mi. S Arkansas A&M College (Baker and Ward, 1967:130). Tennessee (from map in LaVal, 1970:544): Louderdale County; Obion County. Illinois: Union County (LaVal, 1970:551); Cave Spring Cave, 2½ mi. S Eichorn.

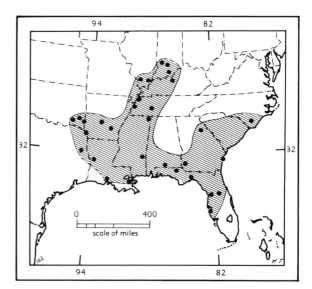

Map 152. *Myotis austroriparius.*

Myotis velifer
Cave Myotis

Length of head and body, 44.2–55.0; ear, 13.0–16.6; length of forearm, 36.5–47.0; greatest length of skull, 14.2–17.6; zygomatic breadth, 9.0–11.6; breadth of braincase, 7.0–8.2; length of upper tooth-row, 6.0–7.0. Pelage of back dull, of moderate length; upper parts dull sepia to drab; underparts paler. Skull large, robust; rostrum borad, its area when viewed from above but little less than that of braincase; sagittal crest well developed; breadth of maxillary teeth exceeding that in any other North American species of *Myotis*, except *M. lucifugus* (subspecies *occultus*); calcar well developed, terminating in a minute lobule, but not keeled.

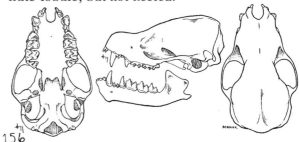

Fig. 156. *Myotis velifer grandis*, Havard Cave, 4½ mi. SW Sun City, Barber Co., Kansas, No. 9713 K.U., ♂, X 2.

Myotis velifer brevis Vaughan

1954. *Myotis velifer brevis* Vaughan, Univ. Kansas Publ., Mus. Nat. Hist., 7:509, July 23, type from Madera Canyon, 5000 ft., Santa Rita Mts., Pima Co., Arizona. Regarded as a synonym of *M. v. velifer* by Hayward (WRI-SCI, Western New Mexico Univ., 1:4–5, February 15, 1970) because specimens assigned to *M. v. brevis* were one end of a cline, and therefore did not deserve nomenclatural recognition.

MARGINAL RECORDS.—Nevada: mine, Jackass Flats, 4½ mi. N Davis Dam, ¾ mi. W Lake Mohave, *ca.* 800 ft. (Cockrum and Musgrove, 1964:636, as *M. velifer* only). Arizona: Big Sandy Creek; Camp Verde; Nantan Plateau (Hall and Kelson, 1959:166, as *M. v. velifer*); Snow Flat, Graham Mts. New Mexico: E side Big Hatchet Mts. (Hawyward, 1972, *in Litt.*). Sonora: Tajitos (Baker and Patton, 1967:284, as *M. velifer* only). Arizona: Ehrenburg. California: Riverside Mts., 35 mi. N Blythe; Needles.

Myotis velifer grandis Hayward

1970. *Myotis velifer grandis* Hayward, WRI-SCI, Western New Mexico Univ., 1:5, February 15, type from Harvard [= Havard] Cave, 4½ mi. SW Sun City, Kansas.

MARGINAL RECORDS.—Kansas: Pratt; Harper. Oklahoma: Enid; Fort Reno; Cache Creek. Texas: Vernon (Dalquest, 1968:15); 7⅖ mi. NW, 6³⁄₁₀ mi. W Hamlin (Hayward, 1970:11); 27 mi. SE Amarillo (*ibid.*); Shamrock (*ibid.*). Kansas: 5 mi. N, 6 mi. W Fowler (Jones, *et al.*, 1967:12); 6 mi. E Ford (Birney and Rising, 1968:520).

Myotis velifer incautus (J. A. Allen)

1896. *Vespertilio incautus* J. A. Allen, Bull. Amer. Mus. Nat. Hist., 8:239, November 21, type from San Antonio, Bexar Co., Texas.
1928. *Myotis velifer incautus*, Miller and G. M. Allen, Bull. U.S. Nat. Mus., 144:92, May 25.

MARGINAL RECORDS.—New Mexico: Carlsbad. Texas: Lampasas; Williamson County (Davis, 1966:49, as *M. velifer* only); Travis County (*ibid.*, as *M. velifer* only); New Braunfels; Santa Ana National Wildlife Refuge (Hayward, 1972, *in Litt.*). Tamaulipas: Charco Escondido, 20 mi. S Reynosa (*ibid.*); San Fernando, 180 ft. (Alvarez, 1963:408); *Soto la Marina;* Sierra de Tamaulipas, 2 mi. S, 10 mi. W Piedra, 1200 ft. (Alvarez, 1963:408). Durango (Baker and Greer, 1962:73): Hda. Atotonilco; San Gabriel; Río Sestín. Chihuahua (Anderson, 1972:248): 5 mi. E Parral, 5700 ft.; Rancho San Ignacio, 4 mi. S, 1 mi. W Santo Tomás; Ramos, 4800 ft.

Myotis velifer velifer (J. A. Allen)

1890. *Vespertilio velifer* J. A. Allen, Bull. Amer. Mus. Nat. Hist., 3:177, December 10, type from Santa Cruz del Valle, Guadalajara, Jalisco.
1897. *Myotis velifer*, Miller, N. Amer. Fauna, 13:56, October 16.

1901. *Myotis californicus jaliscensis* Menegaux, Bull. Mus. Hist. Nat. Paris, 7:321, type from near Lake Zacoalco, Jalisco.

MARGINAL RECORDS.—Sonora: Río Yaqui, *ca.* 1 mi. S El Novillo (Findley and Jones, 1965:331, as *M. velifer* only). Chihuahua: Urique, 1700 ft. (Anderson, 1972:248). San Luis Potosí: Ahualulco; Río Verde. Veracruz: 5 km. N Jalapa. Tabasco: Macuspana (Hayward, 1972, *in Litt.*). Guatemala: Santa Clara (Jones, 1966:463). Honduras: La Esperanza (LaVal, 1969:821, as *M. velifer* only); 2 mi. S Zamorano, 2800 ft. (Davis, *et al.*, 1964:386). Guatemala: Ciudad Vieja (Jones, 1966:463). Chiapas: 1 km. S Trinitaria (Hayward, 1972, *in Litt.*). Oaxaca: Tehuantepec; *ca.* 11 mi. E Juquila, 6100 ft. (Baker and Womochel, 1966:306). Michoacán: Rancho Reparto, 6000 ft. (Hooper, 1961:121). Colima: 4 mi. N Colima (Hayward, 1972, *in Litt.*). Jalisco: San Marcos. Durango: Huasamota; Santa Ana (Jones, 1964a:751). Sinaloa (Jones, *et al.*, 1962:154): 1 mi. N, ½ mi. E San Miguel; El Fuerte.

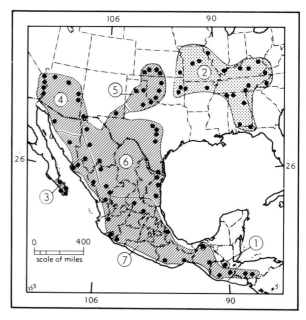

Map 153. *Myotis velifer* and related species.

1. *M. cobanensis*
2. *M. grisescens*
3. *M. peninsularis*
4. *M. velifer brevis*
5. *M. velifer grandis*
6. *M. velifer incautus*
7. *M. velifer velifer*

Myotis peninsularis Miller
Peninsular Myotis

1898. *Myotis peninsularis* Miller, Ann. Mag. Nat. Hist., ser. 7, 2:124, August, type from San José del Cabo, Baja California. Regarded as a subspecies of *M. velifer* by Miller and Allen (Bull. U.S. Nat. Mus., 144:93, May 25, 1928) but as a full species by Hayward (WRI-SCI, Western New Mexico Univ., 1:5, February 15, 1970).

Length of head and body, 44.8–55.0; forearm, 37.0–40.6; greatest length of skull, 14.2–15.6; zygomatic breadth, 9.0–10.2; breadth of braincase, 7.0–7.2; length of upper tooth-row, 6.0–6.4. Differs little from *M. velifer* except in much smaller size.

MARGINAL RECORDS.—Baja California: La Paz; 1 km. S Las Cuevas (Jones, *et al.*, 1965:54); type locality; 5 km. SE Pescadero (Jones, *et al.*, 1965:54). See Map 153.

Myotis cobanensis Goodwin
Guatemalan Myotis

1955. *Myotis velifer cobanensis* Goodwin, Amer. Mus. Novit., 1744:2, August 12, type from Cobán, 1305 m., Alta Verapaz, Guatemala. Known only from type locality.
1958. *Myotis cobanensis*, de la Torre, Proc. Biol. Soc. Washington, 71:167–170, December 31.

Length of head and body, 34.5; ear, 12.5; length of forearm, 41.1; greatest length of skull, 15.4; zygomatic breadth, 9.3; breadth of braincase, 7.2; length of maxillary tooth-row, 6.2. Pelage "about Sepia" according to Goodwin (1955b:2) and darker than in many individuals of *M. v. velifer*. Differences from *M. v. velifer* according to de la Torre (1958:169, 170) are smaller overall size, shorter rostrum, and more nearly globular braincase that is more elevated above rostrum. The species is known only from the holotype. See Map 153.

Myotis grisescens A. H. Howell
Gray Myotis

1909. *Myotis grisescens* A. H. Howell, Proc. Biol. Soc. Washington, 22:46, March 10, type from Nickajack Cave, near Shellmound, Marion Co., Tennessee.

Length of head and body, 45.4–53.0; ear, 13.0–16.5; length of forearm, 40.6–45.8; greatest length of skull, 15.5–16.4; zygomatic breadth, 9.4–10.2; breadth of braincase, 7.4–8.0; length of upper tooth-row, 5.8–6.2. Hairs of upper parts not markedly darkened basally; upper parts either dusky or russet, but underparts paler and washed with whitish or pale buffy. Skull large; sagittal and lambdoidal crests conspicuous. Wing inserts on tarsus instead of side of foot as in other North American species.

MARGINAL RECORDS.—Illinois: 4 mi. E Nebo; Cave Spring, ½ mi. N, 5½ mi. E Elizabethtown. Indiana: *southern Indiana.* Kentucky (Hall and Wilson, 1966:318–320): Morgan's Cave; Chrisman's Cave; Carter Caves. Virginia: Grigsby Cave (Holsinger, 1964:151). North Carolina: Asheville (Tuttle and

Robertson, 1969:370). Tennessee: ⅖ mi. SW Washington (*ibid.*); type locality. Alabama: Anniston. Florida: Marianna. Alabama: Sander's Cave, 1 mi. N, 3 mi. W Brooklyn (LaVal, 1967:646); Saltpetre Cave, near Six Mile (Chermock and White, 1953:24). Mississippi: Chalk Mine, 4 mi. E Iuka (LaVal, 1967:646). Arkansas: near Marcella; 25 mi. SW Fayetteville. Oklahoma: Flower Creek Cave (3 mi. N, 3 mi. E Ft. Gibson) (Glass and Ward, 1959:199); Scraper. Kansas: Lightning Creek, 15½ mi. W Pittsburg (Jones, *et al.*, 1967:6). Missouri: Columbus; Rocheport Cave. See Map 153.

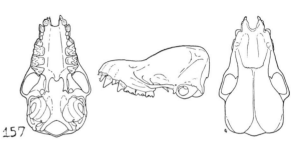

Fig. 157. *Myotis grisescens*, Hunter's Cave (= Kelly Cave), Boone Co., Missouri, No. 63095 M.V.Z., ♂, X 2.

Myotis volans
Long-legged Myotis

Length of head and body, 43–54; ear, 11–14; length of forearm, 35.2–41.2; greatest length of skull, 12.2–15.0; zygomatic breadth, 8–9; breadth of braincase, 6.7–7.6; length of upper tooth-row, 4.6–5.6. Pelage long, soft, extending distally on interfemoral membrane for a distance approx. equal to length of femur, extending onto wing below to a line joining elbow and knee; upper parts varying from ochraceous tawny to dark smoky brown; underparts smoky brown to dull yellowish white washed with buffy; tips of hairs above slightly burnished. Skull small, delicate; rostrum short; braincase abruptly elevated from rostral level; occipital region somewhat inflated; sagittal crest low, poorly defined; palatal region resembling that of *M. lucifugus*. Ears low, rounded, barely reaching rostrum when laid forward; foot small; calcar distinctly keeled.

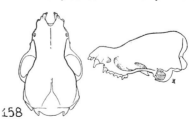

Fig. 158. *Myotis volans interior*, Smiths Creek, 5800 ft., Lander Co., Nevada, No. 63535 M.V.Z., ♂, X 2.

Myotis volans amotus Miller

1914. *Myotis longicrus amotus* Miller, Proc. Biol. Soc. Washington, 27:212, October 31, type from Cofre de Perote, 12,500 ft., Veracruz.
1928. *Myotis volans amotus,* Miller and G. M. Allen, Bull. U.S. Nat. Mus., 144:145, May 25.

MARGINAL RECORDS.—Jalisco: 15 mi. S, 9 mi. E Talpa de Allende, 6900 ft. (Jones, *et al.,* 1971:415); Los Masos; *SE slope El Nevado de Colima, 9100 ft.* (Baker and Phillips, 1965:691). Veracruz: type locality.

Myotis volans interior Miller

1914. *Myotis longicrus interior* Miller, Proc. Biol. Soc. Washington, 27:211, October 31, type from 5 mi. S Twining, 11,300 ft., Taos Co., New Mexico.
1928. *Myotis volans interior,* Miller and G. M. Allen, Bull. U.S. Nat. Mus., 144:142, May 25.

MARGINAL RECORDS.—Montana (Hoffmann, *et al.,* 1969:739, as *M. volans* only): Mt. Aeneas, Swan Range; 3 mi. N Maiden. North Dakota (Genoways, 1967:355): Granville; 1 mi. S, 1 mi. W Medora. South Dakota: 10 mi. S, 5 mi. W Reva (Andersen and Jones, 1971:371); Davenport Cave, 3 mi. S, ½ mi. W Sturgis, 4400 ft. (Turner and Jones, 1968:445); *2 mi. W Nemo, 4700 ft. (ibid.);* Wind Cave National Park headquarters (Jones and Genoways, 1967:190). Nebraska (Jones, 1964c:87): *1 mi. N, 10 mi. W Crawford;* 3 mi. S Glen. Colorado (Armstrong, 1972:63): Ft. Collins; Glen Cove, 11,450 ft. New Mexico: Raton Range. Texas: Nichols' Pasture, 3 mi. N Vera (Mollhagen and Baker, 1972:97); 5 mi. E Mt. Livermore; Brewster County (Davis, 1966:50, as *M. volans* only). Coahuila: Club Sierra del Carmen, 4950 ft., 2 mi. N, 6 mi. W Piedra Blanca; *Fronteriza Mts.* (Easterla and Baccus, 1973:426). Durango: 6 mi. SW El Salto (Gardner, 1965:104, as *M. volans* only). Chihuahua: Colonia Garcia. Arizona: Santa Rita Mts. California: Dulzura; San Emigdio; Dudley; Nevada City; Mt. Shasta. Oregon: Fremont. Washington: Walla Walla. Idaho: Mission.

Myotis volans longicrus (True)

1886. *Vespertilio longicrus* True, Science, 8:588, December 24, type from vicinity of Puget Sound, Washington.
1911. *Myotis altifrons* Hollister, Smiths. Miscl. Coll., 56(26):3, December 5, type from Henry House, Alberta.
1928. *Myotis volans longicrus,* Miller and G. M. Allen, Bull. U.S. Nat. Mus., 144:140, May 25.
1938. *Myotis ruddi* Silliman and von Bloeker, Proc. Biol. Soc. Washington, 51:167, August 23, type from Lime Kiln Creek, 250 ft., southwestern Santa Lucia Mts., Monterey Co., California.

MARGINAL RECORDS.—British Columbia: Atlin (Cowan and Guiguet, 1965:86); *S end Atlin Lake; Kispiox;* Hazelton. Alberta: Henry House; Dried Meat

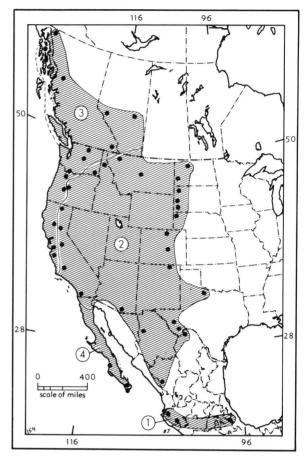

Map 154. *Myotis volans.*

1. *M. v. amotus*	3. *M. v. longicrus*
2. *M. v. interior*	4. *M. v. volans*

Lake, near Camrose. British Columbia: Cranbrook. Washington: Oroville; Entiat; Carson. Oregon: Estacada; E of Mt. Thielson. California: Hurleton; Pacheco Pass; Priest Valley, thence northward along coast, including coastal islands, to Alaska: Mole Harbor, Admiralty Id.

Myotis volans volans (H. Allen)

1866. *V[espertilio]. volans* H. Allen, Proc. Acad. Nat. Sci. Philadelphia, 18:282, type from Cabo San Lucas, Baja California.
1914. *Myotis volans,* Goldman, Proc. Biol. Soc. Washington, 27:102, May 11.
1909. *Myotis capitaneus* Nelson and Goldman, Proc. Biol. Soc. Washington, 22:28, March 10, type from San Jorge, 30 mi. SW Comondú, Baja California.

MARGINAL RECORDS.—Baja California: San Jorge; Miraflores (Jones, *et al.,* 1965:54); type locality.

Myotis nigricans
Black Myotis

Length of head and body, 37.6–49.0; ear, 10.2–13.2. Following measurements based on 444 North American specimens: forearm, 31.0–39.6; greatest length of skull, 12.8–14.5; postorbital width, 3.2–3.9; mastoid width, 6.4–7.4; width across upper canines, 3.1–3.8; maxillary toothrow, 4.6–5.3; 55 bacula from North American specimens averaged .68 long, .27 deep, and .33 wide. Fur usually between 4 and 4.5; tibia/forearm ratio *ca.* .40/1; upper parts monotone, or tips contrasting with bases; tips Bone Brown to Cinnamon; underparts white through buff and yellow to dark brown; fur silky, rarely wooly; tips often slightly glossy; fur on dorsum of uropatagium rarely extends as far as knees; sagittal crest absent; forehead moderately steep, and braincase moderately inflated.—**See** addenda.

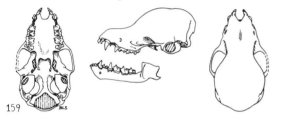

Fig. 159. *Myotis nigricans nigricans*, 3 km. E San Andrés Tuxtla, 1000 ft., Veracruz, No. 23839 K.U., ♂, X 2.

Myotis nigricans carteri LaVal

1973. *Myotis nigricans carteri* LaVal, Nat. Hist. Mus. Los Angeles Co. Sci. Bull., 15:13, February 14, type from 16 mi. NE Tamazula, Jalisco.

MARGINAL RECORDS (LaVal, 1973:14).— Nayarit: 8 mi. E San Blas. Jalisco: type locality. Colima: Cerro Chino, *ca.* 1500 m. Jalisco: 15 km. NW Cihuatlan, *ca.* 100 m.; Talpe de Allende (Río de Talpa), *ca.* 600 m. Nayarit: 10 mi. WSW Tepic, *ca.* 700 m.—**See** addenda.

Myotis nigricans nigricans (Schinz)

1821. *Vesp*[*ertilio*]. *nigricans* Schinz, Das Thierreich . . . , I, Säugethiere und Vögel, pp. 179, type from Fazenda do Agá, near Rio Iritiba, Espírito Santo, Brazil.
1897. *Myotis nigricans*, Miller, N. Amer. Fauna, 13:74, October 16.
1837. *Vespertilio parvulus* Temminck, Monographies de Mammalogie. . . . , 2:246, type from Brazil.
1866. *Vespertilio concinnus* H. Allen, Proc. Acad. Nat. Sci. Philadelphia, 18:280, type from San Salvador, El Salvador.

1866. *Vespertilio mundus* H. Allen, Proc. Acad. Nat. Sci. Philadelphia, 18:280, type from Maracaibo, Venezuela.
1904. *Myotis chiriquensis* J. A. Allen, Bull. Amer. Mus. Nat. Hist., 20:77, February 29, type from Boquerón, Chiriquí, Panamá.
1914. *Myotis bondae* J. A. Allen, Bull. Amer. Mus. Nat. Hist., 33:384, July 9, type from Bonda, Santa Marta, Colombia.
1914. *Myotis maripensis* J. A. Allen, Bull. Amer. Mus. Nat. Hist., 33:385, July 9, type from Maripa, Venezuela.
1928. *Myotis nigricans extremus* Miller and G. M. Allen, Bull. U.S. Nat. Mus., 144:181, May 25, type from Huehuetán, 300 ft., Chiapas.—**See** addenda.
1961. *Myotis nigricans dalquesti* Hall and Alvarez, Univ. Kansas Publ., Mus. Nat. Hist., 14:71, December 29, type from 3 km. E San Andrés Tuxtla, 1000 ft., Veracruz.

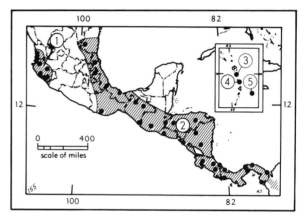

Map 155. *Myotis nigricans, Myotis dominicensis,* and *Myotis martiniquensis.*

1. *M. n. carteri*   3. *M. dominicensis*
2. *M. n. nigricans*   4. *M. m. martiniquensis*
                  5. *M. m. nyctor*

MARGINAL RECORDS (LaVal, 1973:10).— Tamaulipas: 10 km. N, 8 km. W El Encino, 120 m. Veracruz: 13 km. WNW Potrero, 600 m. Puebla: 2 mi. W Villa Avila Camacho, 250 m. Veracruz: near Paraje Nuevo; Catemaco, Hotel Playa Azul, 400 m. Chiapas: 8 km. S Solosuchiapa (on Río Teapa), 120 m.; Ocosingo, *ca.* 800 m. Honduras: Copán; 5 km. NE Ilama, 120 m.; 40 km. E Catacamas, 500 m. Nicaragua: 6 km. N Tuma, 550 m.; Cacao, 22 km. W Muelle de las Bueyes, 120 m. Costa Rica: Río Chitaría, 750 m.; Pandora, *ca.* 100 m. Panamá: Sibube, sea level; Fort Sherman, sea level; Mandinga, *ca.* 100 m.; Armila, Quebrado Venado, *ca.* 100 m., thence into South America, and northward to Panamá: Jaqué; Fort Clayton, sea level; Boquerón, 380 m. Costa Rica: La Piedra, *ca.* 4 km. SW Cerro Chirripo, 3150 m.; Monteverde, *ca.* 1400 m. Nicaragua: San Antonio, 35 m. El Salvador: 2{7/10} mi. E Colón, 610 m. Guatemala: Finca Los Arcos, Cueva do los Ladrones. Chiapas: Huehuetán, sea level. Oaxaca: 36½ km. N San Gabriel Mixtepec, Juquila.

Myotis elegans Hall
Elegant Myotis

1962. *Myotis elegans* Hall, Univ. Kansas Publ., Mus. Nat. Hist., 14:163, May 21, type from 12½ mi. N Tihuatlán, 300 ft., Veracruz.

External measurements of three females: head and body, 45, 40, 39. Measurements of 24 specimens: forearm, 31.9–34.1; greatest length of skull, 12.5–13.4; postorbital width, 3.2–3.5; mastoid width, 6.5–7.1; width across upper canines, 3.1–3.5; maxillary tooth-row, 4.5–4.8; 3 bacula measured .50, .58, .66 long; .22, .14, .13 deep; .36, .29, .33 wide; fur 3–5. Tips of dorsal hairs Bister to Cinnamon, with darker, contrasting bases; underparts like upper parts, or brighter; fur on dorsum of uropatagium extends to knees; sagittal crest absent; forehead gently sloping; braincase

Fig. 160. *Myotis elegans*, 12½ mi. N Tihuatlan, 300 ft., Veracruz, No. 88398 K.U., holotype, ♀, X 2.

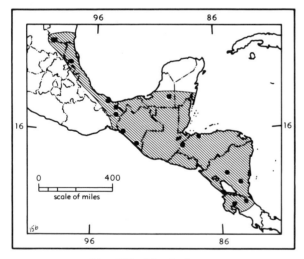

Map 156. *Myotis elegans.*

flattened, not inflated; braincase gently sloping upward from front to rear, except having conspicuously raised portion above occipital region.

MARGINAL RECORDS (LaVal, 1973:15, unless otherwise noted).—San Luis Potosí: 8 km. W El Naranjo. Veracruz: type locality; 1 km. NE Sontecomapan, 40 m.; 4½10 km. S, 2³⁄10 km. W Tenochitilan, 22 m.

Campeche: 65 km. S, 128 km. E Escárcega, *ca.* 100 m. Honduras: Lancetilla, 40 m. Nicaragua: 17 km. N, 15 km. E Boaco, 300 m.; 10 mi. W Rama, 40 m. Costa Rica: 1 mi. E Puerto Viejo, 120 m.; *La Selva* (3247 LaVal at KU); Hda. La Pacífica (2958 LaVal at KU). Nicaragua: 1 km. NW Sapoa, 40 m. Honduras: 5 km. E Santa Rita, 750 m. Chiapas: Carretera Arriaga a Tapachula, Puente Bado Ancho, *ca.* 100 m.; Finca Ocuilapa, 14 km. NNE Tonalá, 100 m. Veracruz: 2 km. W Suchilapa, *ca.* 100 m.

Myotis dominicensis Miller
Dominican Myotis

1902. *Myotis dominicensis* Miller, Proc. Biol. Soc. Washington, 15:243, December 16, type from Dominica, Lesser Antilles. Known only from Dominica.

Measurements of 27 specimens: head and body, 42–54; ear, 12–13; forearm, 33.0–35.0; greatest length of skull, 12.6–13.2; postorbital width, 3.1–3.5; mastoid width, 6.4–6.8; width across upper canines, 3.2–3.4; maxillary tooth-row, 4.7–4.9; 6 bacula measured .58–.65 long, .22 deep, and .29–.36 wide; fur 4–5; tibia/forearm ratio about .35/1. Upper parts monotone, Bister to Sudan Brown; underparts paler, tips contrasting slightly with bases; fur wooly in texture; fur on dorsum of uropatagium extends halfway from knee to foot; sagittal crest absent; profile of skull as in *M. nigricans*.

Specimens of *Myotis* from the islands of St. Martin and Grenada in the Lesser Antilles are referred to by LaVal (1973:17, 18) as inadequate in number for him to identify taxonomically. He notes that *Myotis nesopolus* Miller (1900:123), described from Curaçao, is a name that may eventually prove applicable to one or more of the island populations in the Lesser Antilles. Varona (1974:34) lists Montserrat as habitat for *Myotis* unidentified to subspecies. See Map 155.—See addenda.

Myotis albescens (É. Geoffroy)
Silver-tipped Myotis

1806. *Vespertilio albescens* É. Geoffroy St.-Hilaire, Ann. Mus. d'Hist. Nat. Paris, 8:204, type from Paraguay; restricted to Yaguaron, Paraguay, by LaVal, Nat. Hist. Mus. Los Angeles Co. Sci. Bull., 15:26, February 14, 1973.
1900. *Myotis albescens*, Thomas, Ann. Mus. Civ. Storia Nat. Genova, 40:546, July 4.
1826. *Vespertilio leucogaster* Wied-Neuwied, Beiträge zur Naturgeschichte von Brasilien, Weimar, 2:271, type from Moucouri River, Brazil.—See addenda.
1840. *Vespertilio arsinoë* Temminck, Monographies de Mammalogie . . . , 2:247, type from Surinam.
1947. *Myotis argentatus* Dalquest and Hall, Univ. Kansas Publ., Mus. Nat. Hist., 1:239, December 10, type from 14 km. SW Coatzacoalcos, 100 ft., Veracruz.

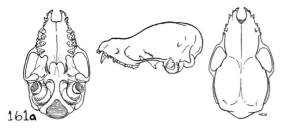

Fig. 161a. *Myotis albescens*, Tacuaral, Paraguay, No. 105664 U.S.N.M., ♀, X 2.

Length of head and body, 44.2–50.0; ear, 11.2–14.6. Measurements of 54 North American specimens: forearm, 33.0–38.4; greatest length of skull, 13.7–15.1; postorbital width, 3.7–4.3; mastoid width, 6.8–7.6; width across upper canines, 3.6–3.9; maxillary tooth-row, 4.9–5.4; 5 bacula are .86–.94 long, .29–.33 deep, and .29–.42 wide; fur, 3–5. Hairs of upper parts white at tips, grayish brown to black at bases; hair of underparts tipped with buffy, dark basally; upper side of uropatagium furred about halfway to knees; sagittal crest absent, forehead gently sloping; postorbital region notably wide compared to rostrum and braincase.

Study of geographic variation probably will reveal that at least some of the four names listed above as synonyms are instead the names of recognizable subspecies.

MARGINAL RECORDS (LaVal, 1973:29, 30, unless otherwise noted).—Veracruz: 14 km. SW Coatzacoalcos, 30 m. Chiapas: 8 km. WNW Mal Paso, 120 m. Honduras: eastern Honduras [= 10⅘ mi. by road SSW Dulce Nombre de Culmi]. Nicaragua: 6 km. N Tuma, 550 m.; 17 km. N, 15 km. E Boaco, 300 m.; Río Sequia, *ca.* 50 m. Costa Rica: Cariari, 50 m.; *La Selva* (LaVal, 1977:81). Panamá: 7 km. SSW Changuinola, sea level; 2 mi. W Soná, 60 m.; Barro Colorado Island; Tacarcuna Village Camp, 1000 m.; mouth Río Paya, *ca.* 300 m., thence into South America.

Myotis keaysi
Hairy-legged Myotis

Length of head and body, 36–47; ear, 12–14. Measurements of 311 North American specimens: forearm, 31.2–41.2; greatest length of skull, 12.6–14.7; postorbital width, 3.0–3.6; mastoid width, 6.3–7.5; width across upper canines, 3.1–4.0; maxillary tooth-row, 4.5–5.6; 55 bacula from North American specimens averaged .53 long, .15 deep, and .24 wide; fur 4–6. Upper parts usually Bister to Sudan Brown, tips contrasting slightly with bases; underparts warm gray to buff to yellow to orange, with strongly contrasting dark bases; fur wooly or straight, but not silky; fur on dorsum of uropatagium usually extends along tibia to feet; sagittal crest present; forehead steeply sloping and braincase well inflated.

Myotis keaysi pilosatibialis LaVal

1973. *Myotis keaysi pilosatibialis* LaVal, Nat. Hist. Mus. Los Angeles Co. Sci. Bull., 15:24, February 14, type from 1 km. W Talanga, Francisco Morazán, 750 m., Honduras.

MARGINAL RECORDS (LaVal, 1973:25, unless otherwise noted).—Tamaulipas: Rancho del Cielo, 1350 m.; 8 km. SW Chamal, Canyon of Río Boquilla, 240 m. Veracruz: 9 km. N Jalapa, Cueva de Zatiopan, 1700 m.; Quezalapam, 2 mi. E Lago Catemaco. Tabasco: 3 3/10 km. NE Teapa, *ca.* 100 m. Campeche: La Tuxpeña. Yucatán: Hunucmá, *ca.* 50 m. Quintana Roo: Pueblo Nuevo X-Can, 10 m.; Cozumel Island, San Miguel, sea level; 2 km. N Felipe Carrillo Puerto, 30 m. Belize: Augustine, *ca.* 400 m. Guatemala: 25 km.

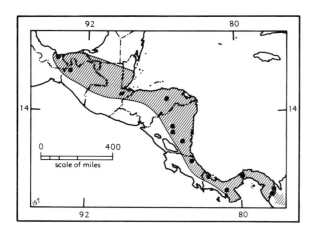

Map 157. *Myotis albescens.*

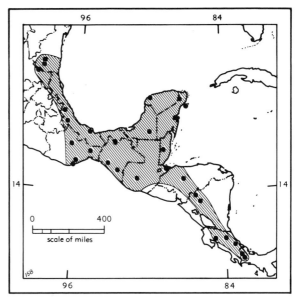

Map 158. *Myotis keaysi pilosatibialis.*

SSW Puerto Barrios, 90 m. Honduras: type locality. Nicaragua: Daraili, 5 km. N, 14 km. E Condega, 940 m.; 17 km. N Matagalpa, 1465 m. Costa Rica: *La Selva* (3164 LaVal at KU); Volcán Barba, 3⅕ km. S Los Cartagos, *ca.* 1800 m.; 4 km. SW Moravia de Turrialba, 1200 m. Panamá: 6 mi. NE El Volcán, 2040 m.; 36 km. N Concepción, 2000 m. Costa Rica: Monteverde, *ca.* 1400 m. Guatemala: 4 mi. NW Patzún, Chocoyos, 2180 m. Chiapas: Finca Prusia, 1160 m.; 9 mi. SE, 8 mi. NE Tonalá, *ca.* 500 m. Oaxaca: 24 mi. N Matías Romero, Montebello, 100 m.; ½ mi. W Chiltepec; 16 km. ENE Piedra Blanca; Vista Hermosa, 1600 m. Veracruz: 4 km. WNW Fortín, 960 m. San Luis Potosí: El Salto Falls, 600 m. Also occurs in South America.

Myotis martiniquensis
Schwartz' Myotis

External measurements of a male from Barbados: head and body, 40; tail, 37; ear, 14. Measurements of 20 specimens from Martinique and Barbados: forearm, 34.7–38.5; greatest length of skull, 14.3–14.8; postorbital width, 3.2–3.7; mastoid width, 7.1–7.6; width across upper canines, 3.6–3.9; maxillary tooth-row, 5.3–5.6; fur usually 5 mm. long, wooly. Hairs of upper parts medium brown, with little contrast between tips and bases; pelage of underparts buffy distally, and dark brown basally; fur on dorsum of uropatagium extending well past knees, but not onto tibiae; sagittal crest usually present. See Map 155.

Myotis martiniquensis martiniquensis LaVal

1973. *Myotis martiniquensis* LaVal, Nat. Hist. Mus. Los Angeles Co. Sci. Bull., 15:35, February 14, type from *ca.* 6 km. E La Trinité, Tartane, Martinique, Lesser Antilles. Known only from Martinique.

Myotis martiniquensis nyctor LaVal and Schwartz

1975. *Myotis martiniquensis nyctor* LaVal and Schwartz, Caribbean Jour. Sci., 14:189 (for December, 1974), type from Cole's Cave, St. Thomas Parish, Barbados, Lesser Antilles. Known only from type locality.

Myotis riparius Handley
Riparian Myotis

1960. *Myotis simus riparius* Handley, Proc. U.S. Nat. Mus., 112:466, type from Tacarcuna Village, 3200 ft., Río Pucro, Darién, Panamá.

External measurements of two specimens from Panamá: head and body, 49, 50; ear, 14, 13. Measurements of 20 specimens from Central America: forearm, 31.8–38.6; greatest length of skull,

13.1–14.5; postorbital width, 3.3–3.7; mastoid width, 6.8–7.5; width across upper canines, 3.6–3.9; maxillary tooth-row, 4.7–5.3; 11 bacula from Central and South America, .65–.86 long, .22–.29 deep, and .36–.50 wide; fur 3–4, wooly. Pelage of upper parts monotone, or slightly darker at base, varying from dark gray to Cinnamon distally; hairs on underparts with tips buff to yellow to medium brown, bases dark; fur sparse on uropatagium, extending barely halfway to knees; sagittal crest present; small upper premolars crowded, P3 usually displaced to lingual side of tooth-row and one-fourth or less height of P1.

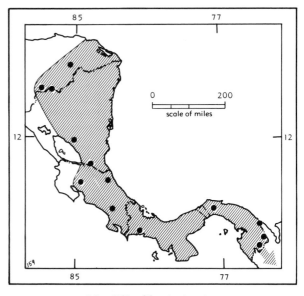

Map 159. *Myotis riparius.*

MARGINAL RECORDS (LaVal, 1973:34, unless otherwise noted).—Honduras: 40 km. E Catacamas, 500 m. Costa Rica: Cariari, 50 m.; *La Selva* (LaVal, 1977:81). Panamá: Cerro Azul; Armila; type locality; Cerro Malí, 1230 m. [into South America]; 3 mi. W David, 45 m. Costa Rica: Fila La Maquina, *ca.* 3 km. E Canaán, 2000 m.; Monteverde (LaVal, 1977:81). Nicaragua: 1 km. S El Castillo, 130 m.; 1 km. N, 2½ km. W Villa Somoza, 330 m.; 4½ km. N, 2 km. E Jalapa, 680 m. Honduras: *7 km. SE Danlí, 620 m.;* 1 km. SE Danlí, 780 m.

Myotis oxyotus
Montane Myotis

Measurements of 33 specimens from Costa Rica and Panamá: forearm, 38.0–43.5; greatest length of skull, 13.8–14.9; postorbital width,

3.6–4.0; mastoid width, 6.8–7.5; width across upper canines, 3.5–3.9; maxillary tooth-row, 5.1–5.6. Thirteen bacula from North American specimens: .58–.65 long, .14–.29 deep, and .29–.42 wide. Fur, silky and 4.5–5 on back. Hairs of upper parts Snuff Brown at tips, darker at bases; hairs of underparts grayish buff to medium warm brown at tips, black at bases; fur on dorsum of uropatagium usually extends slightly past knees; sagittal crest absent; forehead moderately to steeply sloping; braincase globular; rostrum long and low. Where *oxyotus* (24 specimens) and *nigricans* (63 specimens) occur at elevations above 1800 m. in Costa Rica, mandibular tooth-row almost always longer, 6.8 (6.7–7.0) in *oxyotus*, than in *nigricans*, 6.3 (6.0–6.7), with only one specimen of *oxyotus* as short as 6.7 and only two specimens of *nigricans* as long as 6.7.

LaVal (1973:40, 41) recognized the possibility that *V. oxyotus* Peters 1867 may intergrade with the more southern *V. chiloensis* Waterhouse 1838, in which event the latter name will replace *oxyotus* as the name for the species.

Myotis oxyotus gardneri LaVal

1973. *Myotis oxyotus gardneri* LaVal, Nat. Hist. Mus. Los Angeles Co. Sci. Bull., 15:42, February 14, type from Fila

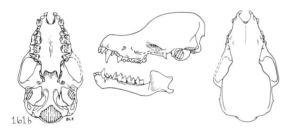

Fig. 161b. *Myotis oxyotus gardneri*, Cerro Punta, Panamá, No. 318872 U.S.N.M., sex ?, X 2.

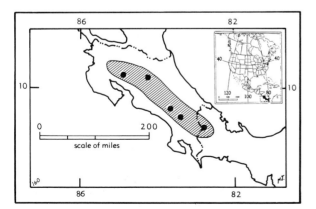

Map 160. *Myotis oxyotus gardneri*.

La Maquina, *ca.* 7½ km. E Canaán, San José, 2610 m., Costa Rica.

MARGINAL RECORDS (LaVal, 1973:42, unless otherwise noted).—Costa Rica: Monte Verde (3208 LaVal at KU); Vara Blanca, 1800 m.; 1 km. NW Villa Mills, 3060 m.; type locality; *La Piedra, ca. 4 km. SW Cerro Chirripo, 3150 m.* Panamá: Casa Tilley, Cerro Punta, 1600 m.; *Finca Lara, Cerro Punta, 1740 m.*

Myotis thysanodes
Fringed Myotis

Length of head and body, 43–52; ear, 16–19; length of forearm, 39.8–46.0; greatest length of skull, 16.2–17.2; zygomatic breadth, 9.2–10.8; breadth of braincase, 7.2–8.2; length of upper tooth-row, 6.0–6.6. Upper parts yellowish brown to darker olivaceous tones; underparts barely, if any, paler. Skull resembling that of *M. evotis* but larger, more robust, broader; sagittal crest well developed; length of upper tooth-row exceeded by greatest breadth of palate including molars; fringe of short, stiff hairs along free edge of interfemoral membrane well developed; ear large, projecting 3–5 mm. beyond muzzle when laid forward.

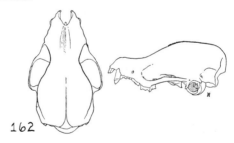

Fig. 162. *Myotis thysanodes thysanodes*, Horse Spring, 4750 ft., San Bernardino Co., California, No. 86119 M.V.Z., ♂, X 2.

Myotis thysanodes aztecus Miller and G. M. Allen

1928. *Myotis thysanodes aztecus* Miller and G. M. Allen, Bull. U.S. Nat. Mus., 144:128, May 25, type from San Antonio, Oaxaca.

MARGINAL RECORDS.—Veracruz: 3 mi. ESE Las Vigas (Davis and Carter, 1962:72). Chiapas: 7 mi. SE San Cristóbal de las Casas, 8000 ft. (Carter, *et al.*, 1966:495). Oaxaca (Goodwin, 1969:96, 101): *type locality;* Hda. de Cinco Señores; Río Molino. México: 1 km. N, 8½ km. W Río Frío, 3450 m. (Alvarez and Ramírez-P., 1972:173).

Myotis thysanodes pahasapensis Jones and Genoways

1967. *Myotis thysanodes pahasapensis* Jones and Genoways, Jour. Mamm., 48:231, May 20, type from 6 mi. N Newcastle, 6000 ft., Weston County, Wyoming.

MARGINAL RECORDS (Jones and Genoways, 1967:235, unless otherwise noted).—Wyoming: 1½ mi. E Buckhorn. South Dakota: Cliff Shelf, 2 mi. N, 2½ mi. E Interior (124920 KU). Nebraska: 10 mi. S, 2½ mi. E Gering (Farney and Jones, 1975:328). South Dakota: Custer State Park; *3 mi. N Hot Springs; ½ mi. S, 1½ mi. W Minnekahta, 4200 ft.* (Turner and Jones, 1968:445). Wyoming: *type locality.*

Myotis thysanodes thysanodes Miller

1897. *Myotis thysanodes* Miller, N. Amer. Fauna, 13:80, October 16, type from Old Fort Tejon, Tehachapi Mts., Kern Co., California.

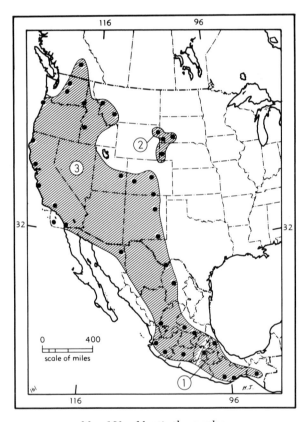

Map 161. *Myotis thysanodes.*

1. *M. t. aztecus*   2. *M. t. pahasapensis*
3. *M. t. thysanodes*

MARGINAL RECORDS.—British Columbia: Vernon. Washington: Anatone. Montana (Hoffmann, *et al.*, 1969:739, as *M. thysanodes* only): Missoula; Lewis and

Clark Cavern, Jefferson Co. Idaho: Karney Lake (Jones and Genoways, 1967:235). Utah: 4 mi. N Thompsons (Krutzsch and Heppenstall, 1955:126). Colorado (Armstrong, 1972:62): Grizzly Gulch, Black Canyon of Gunnison National Monument; 6 mi. N, 1 mi. W Colorado Springs; near Wooton, 7500 ft. New Mexico: Montezuma (Studier, 1968:362); Carlsbad. Texas: SE slope Mariscal Mtn. Coahuila: 4⅕ mi. W Cuatro Ciénegas, 2900 ft. (Villa-R., 1967:376). San Luis Potosí: Hda. La Parada; Hda. Capulín. Michoacán: Pátzcuaro. Jalisco: Los Masos; La Laguna. Zacatecas: Hda. San Juan Capistrano. Sonora: near El Tigre. California: Dulzura; San Clemente Island (von Bloeker, 1967:248); Lebec; Stonewall Creek near Soledad; Howell Mtn.; Willow Creek. Oregon: Tillamook. Washington: 1½ mi. NW Moses Coulee (Williams, 1968:26); Carleton (Johnson, 1962:44).

Myotis keenii
Keen's Myotis

Length of head and body, 40.4–55.0; ear, 14.2–18.6; length of forearm, 34.6–38.8; greatest length of skull, 14.6–15.6; zygomatic breadth, 8.2–9.2; breadth of braincase, 6.8–7.6; length of maxillary tooth-row, 5.4–6.0. Pelage long, silky, dull; color essentially as in *M. lucifugus*. Skull relatively lightly built, slender, sagittal crest sometimes present; length of upper tooth-row slightly exceeds greatest palatal breadth including molars. Metacarpals subequal (not of graded sizes as in *M. lucifugus*); ears relatively long (when laid forward surpassing muzzle by *ca.* 4 mm.) and slender.

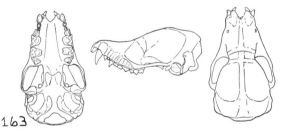

Fig. 163. *Myotis keenii septentrionalis*, Crystal Cave, 5 mi. N Bentonville, Benton Co., Arkansas, No. 83535 M.V.Z., ♀, X 2.

In Wisconsin, Long (1976:80) differentiates *M. keenii* from *M. sodalis* and *M. lucifugus* by the 11 or fewer elastic bands in the uropatagium. Each of the other two species has more than 11 bands.

M. keenii differs from the allopatric *M. auriculus* of the Southwest in shorter ear (especially in *M. k. septentrionalis*), shorter forearm, 3rd, 4th, and 5th metacarpals subequal (instead of 3rd longest and 4th shortest), smaller skull, shorter maxillary tooth-row, sagittal crest lacking or weakly developed (instead of always strongly developed). M3 longer anteroposteriorly in rela-

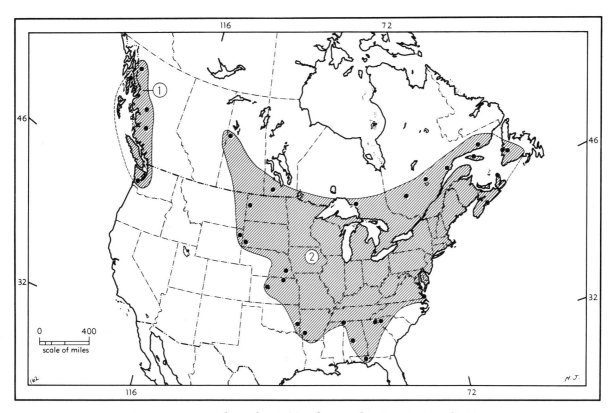

Map 162. *Myotis keenii keenii* (1) and *Myotis keenii septentrionalis* (2).

tion to combined length of 3 upper molars, with metaconule and its loph present instead of absent and with re-entrant angle between parastyle and mesostyle poorly developed instead of trenchant. Baculum more complex. *M. k. keenii* and *M. k. septentrionalis* differ much from each other, and their geographic ranges are widely separated. They may be species instead of subspecies.

Myotis keenii keenii (Merriam)

1895. *Vespertilio subulatus keenii* Merriam, Amer. Nat., 29:860, September, type from Massett, Graham Island, Queen Charlotte Islands, British Columbia.

1928. *Myotis keenii keenii*, Miller and G. M. Allen, Bull. U.S. Nat. Mus., 144:104, May 25.

MARGINAL RECORDS.—British Columbia: Telegraph Creek; Telkwa; Stuie. Washington: Lake Cushman, thence up coast, including coastal islands, to Alaska: Wrangell.

Myotis keenii septentrionalis (Trouessart)

1897. [*Vespertilio gryphus*] var. *septentrionalis* Trouessart, Catalogus mammalium . . . , fasc. 1, p. 131. Type locality, Halifax, Nova Scotia.

1928. *Myotis keenii septentrionalis*, Miller and G. M. Allen, Bull. U.S. Nat. Mus., 144:105, May 25.

MARGINAL RECORDS.—Saskatchewan: Buffalo Narrows, Churchill Lake. Manitoba: Souris. Ontario: Michipicoten. Quebec: 26 mi. S Clova (MacLeod and Cameron, 1961:281); Lake Edward; Godbout; Anticosti Island; Natashquan (Harper, 1961:36). Newfoundland: Lewis Hills; Spruce Brook. Nova Scotia: Cape North (Cameron, 1959:18); Halifax, thence southward along Atlantic Coast to Virginia: Norfolk County. South Carolina: Pickens County (Golley, 1966:55). Georgia: Young Harris. Florida: 3 mi. N Marianna. Alabama: Saltpetre Cave, near Six Mile (Chermock and White, 1953:24). Mississippi: 4 mi. E Iuka (LaVal, 1967:647). Arkansas: Delight (Dellinger and Black, 1940:188). Oklahoma: Kiamichi Mts. Cave (Glass and Ward, 1959:200). Kansas (Jones, *et al.*, 1967:8): Hays; *3 mi. S, 1 mi. E Codell;* 11 mi. N Morrowville. Nebraska: Lincoln (Jones, 1964c:81). South Dakota: Elk Mtn. Wyoming: ½ mi. E Buckhorn (Findley, 1960:20). North Dakota: Fort Buford.—See addenda.

Myotis auriculus
Mexican Long-eared Myotis

Length of head and body, 41–52; ear, 18–22; forearm, 35.7–40.4; greatest length of skull, 15.7–17.0; zygomatic breadth, 8.7–10.0; breadth of braincase, 7.2–7.9; length of maxillary toothrow, 6.3–6.8; sagittal crest present in all specimens examined. Resembles *M. keenii* in size and color, but forearm considerably longer and ear larger; 3rd metacarpal longest and 4th shortest (3rd, 4th, and 5th subequal in *M. keenii*). Dorsal pelage of *M. a. auriculus* varying from Brussels Brown to Raw Umber with ears and wings dark brown; *M. a. apache* much paler (varying from Tawny-Olive to Light Brownish Olive) with ears and membranes pale brownish. Molars markedly larger than in *M. keenii* (length plus width of M2 3.15 mm. or more instead of 3.10 mm. or less—see Genoways and Jones, 1969: Fig. 2). Baculum smaller and less complex than in *M. keenii* and *M. evotis* (see Fig. 165). For further comparison, see accounts of those two species.

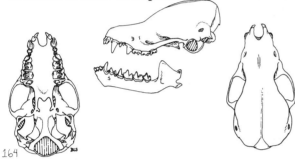

Fig. 164. *Myotis auriculus auriculus*, 2 mi. S, 10 mi. W Piedra, 1200 ft., Tamaulipas, No. 55110 K.U., holotype, ♀, X 2.

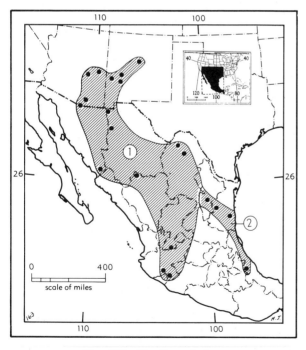

Map 163. *Myotis auriculus.*

1. *M. a. apache*   2. *M. a. auriculus*

Myotis auriculus apache Hoffmeister and Krutzsch

1955. *Myotis evotis apache* Hoffmeister and Krutzsch, Nat. Hist. Miscl., Chicago Acad. Sci., 151:1, December 28, type from Snow Flat, 8750 ft., Graham Mts., Graham Co., Arizona.
1969. *Myotis auriculus apache*, Genoways and Jones, Southwestern Nat., 14:11, May 16.

MARGINAL RECORDS (Genoways and Jones, 1969:11, 12, unless otherwise noted).—New Mexico: Embuda Cave, Sandia Mts.; Springtime Canyon, San Matea Mts., 38 mi. S, 6 mi. W Magdalena. Chihuahua: 5 mi. N, 2½ mi. W San Francisco; Pacheco, 1900 m. (Anderson, 1972:242). Coahuila: Fronteriza Mts. (Easterla and Baccus, 1973:426); 4 mi. W Hda. La Mariposa, 2300 ft. Jalisco: 10 mi. NNE Pihuamo, 3500 ft.; *Los Masos;* 4 km. E Venustiano Carranza, 2160 m. (Jones, *et al.*, 1971:410). Zacatecas: 16 km. NW Yahualica (Jalisco), 2164 m. (Matson and Patten, 1975:6). Durango: Navarro. Sonora: Mina Aduana, Alamos. Arizona: 1 mi. E Pena Blanca Spring (Cockrum, 1961:41); McClearys, 30 mi. SE Tucson; *Santa Catalina Mills* (Miller and Allen, 1928:117, as *M. evotis chrysonotus*); Wilbank's Ranch, Sierra Ancha, 7200 ft.; Cooley's Ranch (Genoways and Jones, 1969:8). New Mexico: 3 mi. NE Aragon; North Fork Water Canyon, Magdalena Mts.

Myotis auriculus auriculus Baker and Stains

1955. *Myotis evotis auriculus* Baker and Stains, Univ. Kansas Publ., Mus. Nat. Hist., 9:83, December 10, type from 2 mi. S, 10 mi. W Piedra, 1200 ft., Sierra de Tamaulipas, Tamaulipas.
1969. *Myotis auriculus auriculus*, Genoways and Jones, Southwestern Nat., 14:10, May 16.

MARGINAL RECORDS (Genoways and Jones, 1969:10).—Nuevo León: Iturbide, Sierra Madre Oriental, 5000 ft. Tamaulipas: Ranchero Guayabos, 1500 ft.; type locality. Veracruz: Perote.

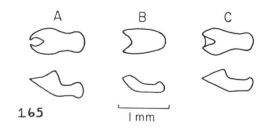

Fig. 165. Dorsal and lateral views of bacula of three species of *Myotis*. A, *Myotis evotis evotis*, No. 86115 K.U.; B, *Myotis auriculus auriculus*, No. 55108 K.U.; C, *Myotis keenii septentrionalis*, No. 79031 K.U.

Myotis milleri Elliot
Miller's Myotis

1903. *Myotis milleri* Elliot, Field Columb. Mus., Publ. 74, Zool. Ser., 3:172, May 7, type from LaGrulla, Sierra San

Pedro Mártir, Baja California. Known definitely only from type locality.

Length of head and body, 41.2–46.6; ear, 19–20; length of forearm, 34–37; greatest length of skull, 14.8–15.2; zygomatic breadth, 8.8–8.9; breadth of braincase, 7.0–7.2; length of upper tooth-row, 5.4–6.0. Externally resembling *M. evotis* but smaller; braincase less elevated, noticeably flat-topped, sagittal crest absent in each of six specimens examined; teeth small, crown area of upper molars equal to that of upper molars of *M. keenii;* color as in *M. e. evotis.*

Myotis evotis
Long-eared Myotis

Length of head and body, 41.6–56.0; ear, 18.0–22.4; length of forearm, 35.5–41.0; greatest length of skull, 15.0–16.4; zygomatic breadth,

8.6–10.1; breadth of braincase, 7.0–8.2; length of maxillary tooth-row, 6.0–6.8. Upper parts light to medium brown, ears conspicuously darker. Skull closely resembling that of *M. keenii;* upper profile curving gradually from rostrum to summit of braincase; sagittal crest often present but never large; braincase viewed from above oval and bulging posteriorly beyond lambdoidal ridges. Ear and tragus large; when laid forward ear extends about 7 mm. beyond muzzle; tail membrane usually edged with sparse and inconspicuous fringe of hairs.

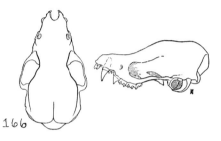

Fig. 166. *Myotis evotis evotis,* Burned Corral Canyon, Nevada, No. 57366 M.V.Z., ♂, X 2.

M. e. evotis occurs together with *M. auriculus apache* in central Arizona and central New Mexico, where *M. evotis* has longer ears, darker ears that contrast more with pelage of body, black instead of brownish bases on hairs of back, more hair (forming slight fringe) on edge of uropatagium, smaller skull, forehead rising less abruptly, palatal spine shorter and pointed (instead of broader and rounded), less flaring angular process on mandible, mandible shorter (less instead of more than 82.5%) in relation to condylobasal length; M3 longer anteroposteriorly, and baculum more complex (see Fig. 165).

Myotis evotis evotis (H. Allen)

1864. *Vespertilio evotis* H. Allen, Smiths. Miscl. Coll., 7(Publ. 165):48, June. Type locality, by subsequent restriction, Monterey, California (see Dalquest, Proc. Biol. Soc. Washington, 56:2, February 25, 1943).

1896. *Vespertilio chrysonotus* J. A. Allen, Bull. Amer. Mus. Nat. Hist., 8:240, November 21, type from Kinney Ranch, Bitter Creek, Sweetwater Co., Wyoming.

1897. *Myotis evotis,* Miller, N. Amer. Fauna, 13:77, October 16.

1909. *Myotis micronyx* Nelson and Goldman, Proc. Biol. Soc. Washington, 22:28, March 10, type from Comondú, Baja California.

MARGINAL RECORDS.—British Columbia: Summit Lake, near Prince George (Cowan and Guiguet, 1965:81). Alberta: Red Deer River near Rumsey. Saskatchewan: sec. 14, T. 20, R. 13, W of 3rd merid. (Maher, 1972:236, as *M. evotis* only). North Dakota:

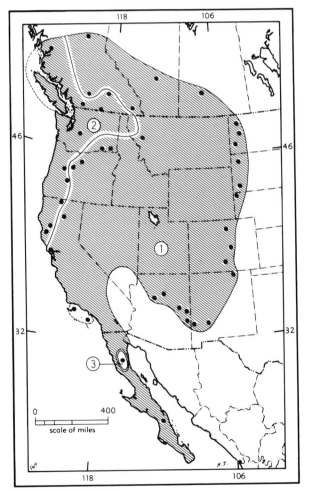

Map 164. *Myotis evotis* and *Myotis milleri.*

1. *M. evotis evotis* 2. *M. evotis pacificus*
3. *M. milleri*

Beaver Creek, 4 mi. W Grinnell; 5 mi. N, 6½ mi. W Killdeer (Genoways and Jones, 1972:6); 1 mi. S, 1 mi. W Medora (Jones and Genoways, 1966:88). South Dakota: 10 mi. S, 5 mi. W Reva (Andersen and Jones, 1971:369); Corral Draw. Nebraska: Warbonnet Canyon (Jones, 1964c:80). Colorado: Loveland; 3 mi. N Colorado Springs (Armstrong, 1972:61); 12½ mi. W Gardner, 6000 ft. (*ibid.*). New Mexico: Vermejo River; Springtime Canyon, 38 mi. S, 6 mi. W Magdalena (Findley, 1960:16); 9 mi. E Mogollon (Baker and Patton, 1967:284, as *M. evotis* only). Arizona: 4 mi. S Hannagan Meadow (Hoffmeister and Krutzsch, 1955:3); 25 mi. NE White River; 30 mi. E Heber (Baker and Patton, 1967:284, as *M. evotis* only); Sunset Crater; 30 mi. W Flagstaff (Baker and Patton, 1967:284, as *M. evotis* only). Baja California: Comondú, thence northward along Pacific Coast to California: Santa Catalina Island (von Bloeker, 1967:248); Santa Cruz Island (*ibid.*); San Rafael; Antelope Creek. Oregon: Sisters; Fossil. Washington: South Touchet; Godman Springs. Montana: National Bison Range (Hoffmann, *et al.*, 1969:737, as *M. evotis* only). British Columbia (Cowan and Guiguet, 1965:81): Cranbrook; Merritt.—**See addenda.**

Myotis evotis pacificus Dalquest

1943. *Myotis evotis pacificus* Dalquest, Proc. Biol. Soc. Washington, 56:2, February 25, type from 5 mi. N, 3½ mi. E Yacolt, 500 ft., Clark Co., Washington.

MARGINAL RECORDS.—British Columbia: Kimsquit; Hope (Cowan and Guiguet, 1965:81); Sugar Lake (*ibid.*); Rock Creek. Washington: Easton. Oregon: McKenzie Bridge; Fremont. California: Beswick; South Yolla Bolly Mtn.; Mt. Sanhedrin, thence northward along Pacific Coast, including coastal islands, to point of beginning.

Myotis planiceps Baker
Flat-headed Myotis

1955. *Myotis planiceps* Baker, Proc. Biol. Soc. Washington, 68:165, December 31, type from 7 mi. S, 4 mi. E Bella Unión, 7200 ft., Coahuila.

Measurements of the holotype and a specimen from Nuevo León: length of head and body, 51,—; length of forearm, 25.6, 27.5; 3rd metacarpal, 24.7, 25.6; condylobasal length, 13.3, 14.1; width across upper molars, 4.9, 4.7; maxillary tooth-row, 4.7, 5.0; length of fur at middle of back, 8.2, 9.9. In external measurements the smallest of New World *Myotis*; skull more flattened than that of any other American species of the genus but larger in most measurements than several other small species; fur with wide black bases, tipped with Cinnamon-Brown on upper parts and buff on underparts; occlusal surface of 2nd premolar only slightly smaller than that of 1st premolar, in both jaws; upper incisors half length of canines.

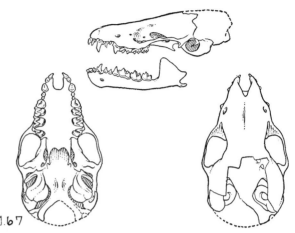

Fig. 167. *Myotis planiceps*, 7 mi. S and 4 mi. E Bella Unión, 7200 ft., Coahuila, No. 48242 K.U., ♂, X 3.

MARGINAL RECORDS.—Coahuila: type locality. Nuevo León: Cerro Potosí, 2800 m. (Jiménez-G., 1971:136). Zacatecas: 16 km. SW Concepción del Oro, 2316 m. (Matson and Patten, 1975:6). See Map 147.

Myotis vivesi Menegaux
Fish-eating Bat

1901. *Myotis vivesi* Menegaux, Bull. Mus. Hist. Nat. Paris, 7:323, type from islet of Cardonal or Islo, Archipelago of Salsipuedes, off San Rafael Bay, Baja California (probably Isla Partida, 28° 53′ N lat., 113° 04′ W long.; see Reeder and Norris, Jour. Mamm., 35:83–85, February 10, 1954).

Measurements of the two female cotypes: total length, 145, 140; tibia, 24.0, 24.6; foot, 23.0, 23.8. Additional measurements of these and other specimens are: length of head and body, 71–76.2; length of forearm, 59–62.2; greatest length of

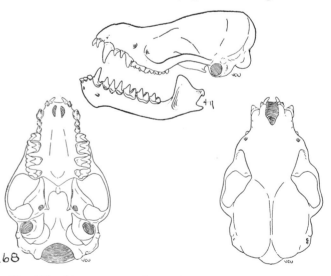

Fig. 168. *Myotis vivesi*, Isla Partida, 28° 53′ N, 113° 04′ W, Gulf of California, Baja California, No. 9254 K.U., ♂, X 2.

skull, 21.0–22.0; zygomatic breadth, 14.0–14.6; breadth of braincase, 10.0–10.8; length of upper tooth-row, 8.8–9.4; largest species of New World *Myotis.* Upper parts dark buff or pale tan; underparts whitish; feet and legs greatly elongated; claws much enlarged and laterally compressed; glandular mass present near middle of forearm; ventral hairs of distal half of uropatagium grow anteriorly.

This curious species of bat seems to feet primarily on small marine fish, but also consumes crustaceans and insects. The characteristic long feet with their long, compressed claws assist in the capture of the prey. By day the bats secrete themselves in crevices among rocks or under rocks, or in holes excavated by sea birds.

MARGINAL RECORDS.—Sonora: Isla San Jorge; Isla Alcatraz, Bahía Kino (Cockrum and Bradshaw, 1963:6); *Bahía de San Carlos;* Guaymas (Baker and Patton, 1967:285); *Isla Blanca [27° 55′ N, 110° 59′ W]* (Orr, 1965:497). Baja California: small island S end Bahía Rosario (24° 15′ N, 110° 09′ W), *ca.* 7 mi. SE Punta Coyote (Patten and Findley, 1970:2); Cayo Island, off SW end San José Island (Banks, 1964:489); Puerto San Bartolomé; Punta Malarrimo; Isla Encantada.

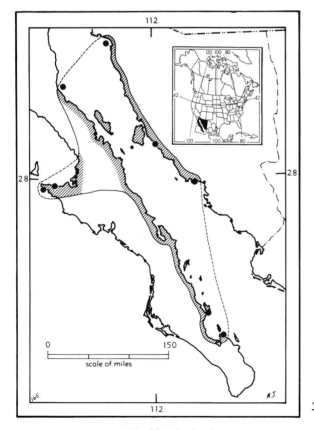

Map 165. *Myotis vivesi.*

Genus **Lasionycteris** Peters—Silver-haired Bat

1866. *Lasionycteris* Peters, Monatsb. preuss. Akad. Wiss., Berlin, 1865, p. 648. Type, *Vespertilio noctivagans* Le Conte.
1875. *Vesperides* Coues, *in* Coues and Yarrow, Report . . . mammals . . . Nevada, Utah, California, New Mexico, and Arizona . . . , Repts. . . . Expl. Surv. West of . . . Hundredth meridian . . . , 5:83. Proposed as a subgenus. Type, *Vespertilio (Vesperides) noctivagans* Le Conte.

Length of head and body approx. 60. Structure of teeth as in *Myotis;* inner incisor strongly bicuspidate and outer one simple; P3 absent; hypocone distinctly indicated in M1 and M2; M3 with more than half the crown area of M1; in M3 metacone nearly as large as paracone, and its three commissures well developed. Skull flattened; rostrum broad; depth of braincase including auditory bullae approx. three-fourths of mastoid breadth; sagittal crest obsolete; interorbital region wide, flattish; upper edge of orbit with low "bead"; central part of bead forming angle suggesting incipient postorbital process; upper surface of rostrum distinctly concave on each side between lachrymal region and nares; other features of skull essentially as in *Myotis,* except for general tendency toward broadening and shortening. Ear short, nearly as broad as long; interfemoral membrane furred on basal half above (after Miller, 1907:203, 204). Dentition, i. $\frac{2}{3}$, c. $\frac{1}{1}$, p. $\frac{2}{3}$, m. $\frac{3}{3}$.

This genus is monotypic.

Lasionycteris noctivagans (Le Conte)
Silver-haired Bat

1831. *V[espertilio]. noctivagans* Le Conte, *in* McMurtrie, The animal kingdom . . . by the Baron Cuvier . . . , 1:(app.)431, type from eastern United States.
1866. *Lasionycteris noctivagans,* Peters, Monatsb. preuss. Akad. Wiss., Berlin, 1865, p. 648.

Upper parts dark brownish black, strongly washed with silver; underparts slightly lighter, silvery wash less pronounced. Other characters as for the genus.

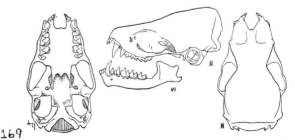

Fig. 169. *Lasionycteris noctivagans,* 4 mi. W Fallon, Nevada, No. 88058 M.V.Z., ♂, X 2.

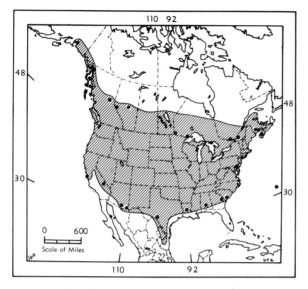

Map 166. *Lasionycteris noctivagans.*

MARGINAL RECORDS.—Alaska: vic. Prince William Sound (Manville and Young, 1965:13); Canyon Island, Taku River, E of Juneau (Barbour and Davis, 1969:109). British Columbia: Charlie Lake. Alberta: Henry House; Edmonton (E. T. Jones, 1974:244, as silver-haired bat, only). Saskatchewan: [near?] *Frobisher Lake.* Ontario: Northern Light Lake. Quebec: Hull (Rand, 1945:115); Mont St. Hilaire (Wrigley, 1969:205). New Brunswick: Rusigonis (Gorham and Johnston, 1963:228); St. John County. Nova Scotia: near Lake Kedjimkujik (Bleakney, 1965:155), thence southward along Atlantic Coast to South Carolina: Charleston County (Golley, 1966:56). Georgia: Jefferson County (Golley, 1962:64). Alabama: Autaugaville. Louisiana: 3 mi. W Tullos (LaVal, 1967:647). Texas: *ca.* 18 mi. W Bandera (Blair, 1952:95). Arizona: Fly Park, Chiricahua Mts.; Tucson (Cockrum, 1961:47). California: Death Valley; Pacific Grove, thence northward along Pacific Coast and larger coastal islands to British Columbia: Skidegate, Moresby Id. Also, Bermuda (Van Gelder and Wingate, 1961:1).

Genus **Pipistrellus** Kaup—Pipistrelles

American representatives revised by Hall and Dalquest, Univ. Kansas Publ., Mus. Nat. Hist., 1:591–602, January 20, 1950.

1829. *Pipistrellus* Kaup, Skizzirte Enwickelungs-Geschichte und natürliches System der Europäischen Thierwelt, 1:98. Type, *Vespertilio pipistrellus* Schreber.
1838. *Romicia* Gray, Mag. Zool. Bot., 2:495, February. Type, *Romicia calcarata* Gray [= *Vespertilio kuhlii* Kuhl].

1856. *Hypsugo* Kolenati, Allgem. deutsch. Naturh. Zeit. Dresden, neue Folge, 2:131. Included two species: *Vesperugo maurus* Blasius; *Vesperugo krascheninikowii* Eversmann.
1856. *Nannugo* Kolenati, Allgem. deutsch. Naturh. Zeit. Dresden, neue Folge, 2:131. Included three species: *Vesperugo nathusii* Keyserling and Blasius; *Vespertilio pipistrellus* Daubenton; *Vespertilio kuhlii* Kuhl.

Length of the head and body are approx. 41. Teeth essentially as in *Myotis* and *Lasionycteris* except for fewer premolar teeth; upper incisors subequal, outer one extending considerably beyond its cingulum; inner upper incisor simple, or more often with well-developed secondary cusp; anterior upper premolar barely or not in tooth-row; other teeth with no special peculiarities; skull essentially as in *Myotis*, though with tendency to greater breadth. External characters not essentially different from those of *Myotis*, but ear usually shorter and broader, and tragus less acutely pointed; in one species, tragus bent forward at tip. Dentition, i. $\frac{2}{3}$, c. $\frac{1}{1}$, p. $\frac{2}{2}$, m. $\frac{3}{3}$.

The genus is widely distributed in the Eastern Hemisphere, and in the Western Hemisphere occurs southward to Honduras.

KEY TO NORTH AMERICAN SPECIES OF PIPISTRELLUS

1. Foot less than half as long as tibia; skull nearly straight in dorsal profile; palate extending far behind molars.   *P. hesperus,*   p. 210
1'. Foot more than half as long as tibia; skull concave in dorsal profile; palate extending little behind molars. . . . . *P. subflavus,*   p. 212

Pipistrellus hesperus
Western Pipistrelle

Subspecies revised by Findley and Traut, Jour. Mamm., 51:741–765, November 30, 1970.

Smoke Gray to Buff Brown dorsally; total length, 60–86; foot less than half as long as tibia; tragus blunt, its terminal part bent forward; skull nearly straight in dorsal profile; inner upper incisors unicuspidate; outer upper incisor with accessory cusp on anterointernal surface; P1, viewed from occlusal surface, less than a seventh of area of canine, and from labial aspect concealed by canine and 4th premolar; lower, 3rd premolar lower than anterior cusp on canine; lower premolars crowded, distance between canine and 1st molar less than length of 2nd lower molar.

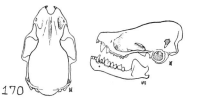

Fig. 170. *Pipistrellus hesperus hesperus*, Crystal Spring, Lincoln Co., Nevada, No. 52203 M.V.Z., ♂, X 2.

Pipistrellus hesperus hesperus (H. Allen)

1864. *Scotophilus hesperus* H. Allen, Smiths. Miscl. Coll., 7(Publ. 165):43, June, type from Old Fort Yuma, Imperial Co., California, on right bank Colorado River, opposite present town of Yuma, Arizona.

1897. *Pipistrellus hesperus*, Miller, N. Amer. Fauna, 13:88, October 16.

1866. *Vesperugo merriami* Dobson, Ann. Mag. Nat. Hist., ser. 5, 18:124, August, type from Red Bluff, Tehama Co., California.

1897. *Pipistrellus hesperus australis* Miller, N. Amer. Fauna, 13:90, October 16, type from Barranca Ibarra, Jalisco.

1904. *Pipistrellus hesperus apus* Elliot, Field Columb. Mus., Publ. 90, Zool. Ser., 3:269, March 7, type from Providencia Mines, Sonora.

MARGINAL RECORDS (Findley and Traut, 1970:756–761, unless otherwise noted).—Washington: Vantage; Almota. Oregon: Riverside; Watson. Idaho: Salmon Creek, 8 mi. W Rogerson (Davis, 1939:120). Nevada: Golconda; Middlegate, 5 mi. W Eastgate; 3 mi. S Schurz; *Fletcher; Arlemont area; Cave Spring;* Fish Lake. California: Furnace Creek, Death Valley. Nevada: Mercury; Middle Stormy Spring, 11 mi. S Lock's Ranch, 5000 ft.; *Crystal Spring;* Ash Spring. Utah: St. George; Springerdale [*sic*]; Belknap Ranger Station, Tushar Mts.; Skull Valley; *4 mi. S [of] Hwy. 40 on road to Dugway; Ogden City;* 3 mi. E Ogden (Hall and Kelson, 1959:181); Richfield; Willow Tank Spring; 10 mi. E Escalante; Fruita; *Notom;* 23 mi. N Hanksville; *Buckham Wash, San Rafael River;* 14 mi. N Green River; 35 mi. N Green River, mouth Florence Creek Canyon, 4300 ft. (Hall and Kelson, 1959:181); *near Green River, 15 mi. SW Ouray* (*ibid.*); Desert Springs, 10 mi. SW Ouray (*ibid.*); 14 mi. N Dragon. Colorado: Rifle; Grand Junction; Tabaguache Creek, 8 mi. NW Nucla. New Mexico: sec. 20, T. 31 N, R. 7 W, Pine River Canyon; Chaco Wash, W Durham; *Glenwood* (Jones and Suttkus, 1972:262, as *P. hesperus* only); 14 mi. S, 6½ mi. W Glenwood; *5 mi. N Redrock;* Redrock; Old Maloney Dam, sec. 9, T. 29 S, R. 20 W; Pasco Ranch, Guadalupe Mts., sec. 22, T. 34 S, R. 21 W. Sonora: Pilares (Burt, 1938:24); E bank Río Jaqui, 1 mi. S El Novillo. Chihuahua: 25 mi. S, 1½ mi. E Creel, Barranca de Cobre; *La Bufa, 2800 ft.* Zacatecas: Hda.

San Juan Capistrano. Jalisco: 2 mi. E Bolaños, 3550 ft. Zacatecas: 3 mi. N Moyahua, 4500 ft. Jalisco: *Barranca Ibarra;* vic. Guadalajara. Morelos: 1 km. S Oaxtepec (Ramírez-P., 1971:125). Guerrero: $4\frac{3}{10}$ km. N Teloloapan (*ibid.*). Michoacán: Río Marqués, 7 mi. S Lombardía, 1500 ft. Jalisco: 6 mi. E El Limón, 2700 ft. (103680 KU). Sinaloa: 1 mi. E Santa Lucía, 5650 ft.; 10 mi. NNW Los Mochis. Baja California: Cerralvo Island; Santa Anita; Todos Santos (Jones, *et al.*, 1965:55); Margarita Island; Comondú area; San Fernando; Agua Caliente, thence northward along coast to opposite California: Clear Lake Lodge; Stillwater; Dale's, on Paines Creek; Marysville Buttes; Tracy area; *Panoche area;* 1 mi. S New Idria, 3700 ft.; Friant Dam, 12 mi. NE Fresno; 3 mi. above Bayden Cave, Hwy. 180, Kings Canyon; *9 mi. W Bishop;* Benton Station area; Coleville. Nevada: 2 mi. N, 9 mi. E Yerington; 12 mi. SE Fallon; Lovelock; Deep Hole, 4000 ft. (Hall, 1946:151); *12 mi. N, 2 mi. E Gerlach* (*ibid.*); mouth Little High Rock Canyon (*ibid.*). Oregon: Trout Creek;

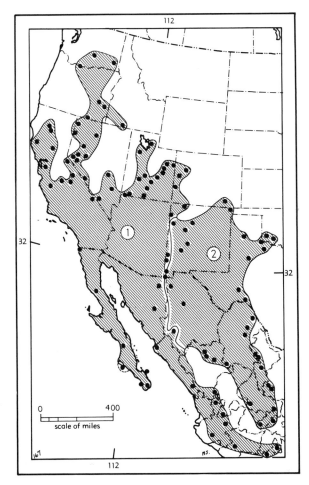

Map 167. *Pipistrellus hesperus.*

1. *P. h. hesperus*   2. *P. h. maximus*

mouth Deschutes River. Washington: Maryhill (Dalquest, 1948:165).

Pipistrellus hesperus maximus Hatfield

1936. *Pipistrellus hesperus maximus* Hatfield, Jour. Mamm., 17:261, August 17, type from Dog Spring, Hidalgo Co., New Mexico.

1936. *Pipistrellus hesperus santarosae* Hatfield, Jour. Mamm., 17:261, type from Santa Rosa, Guadalupe Co., New Mexico.

1951. *Pipistrellus hesperus potosinus* Dalquest, Proc. Biol. Soc. Washington, 64:105, August 24, type from Presa de Guadalupe, San Luis Potosí.

1959. *Pipistrellus hesperus oklahomae* Glass and Morse, Jour. Mamm., 40:531, November 20, type from north side Granite Mts., Granite, Greer Co., Oklahoma.

MARGINAL RECORDS (Findley and Traut, 1970:761–763, unless otherwise noted).—Colorado: Rock Crossing, 30 mi. S La Junta. New Mexico: 3 mi. W Kenton, Oklahoma. Texas: Palo Duro Canyon Cow Camp; Panther Cave, 22 mi. SE Childress. Oklahoma: 1 mi. N Granite (Glass and Morse, 1959:533); *Lugert area; Svoboda Pond, 1/2 mi. N, 1 mi. W Mountain Park;* W Cache Creek at S Refuge boundary. Texas: Nichols' Pasture, 3 mi. N Vera (Baker, 1964:205); Justiceburg; Sheffield [area]; mouth Pecos River. Coahuila: Las Margaritas; 4 mi. W Hda. La Mariposa, 2300 ft.; 4 mi. S, 9 mi. W San Buenaventura, 1800 ft. Nuevo León: near Cemetery, Villa de García; Rancho Rodeo, 7$\frac{3}{10}$ mi. S Santa Catarina, 2400 ft.; Iturbide. Tamaulipas: Joya Verde, 35 km. SW Ciudad Victoria, 3800 ft.; La Joya Salos, 20 mi. SE Jaumave. San Luis Potosí: Hda. Capulín. Hidalgo: Río Tasquillo, 26 km. E Zimapán. Querétaro: 1 mi. NE Peña Blanca (Schmidly and Martin, 1973:92). San Luis Potosí: San Luis Potosí area. Coahuila: W foot Pico de Jimulco, 5000 ft. Durango: 1½ mi. NW Nazas, 4100 ft.; 2 mi. S El Palmito, 4800 ft.; *Indé; 3 mi. E Las Nieves, 5400 ft.; Navarro, 75 km. by road from Hda. de Parral* [sic]. Chihuahua: 5 mi. N Cerro Campaña; *30 mi. SW Gallego, 7000 ft.;* San Francisco. New Mexico: Long [= Lang] Ranch, Animas Valley; tank on Double Adobe Creek, sec. 3, T. 31 S, R. 19 W; *NE 1/4 sec. 18, T. 30 S, R. 18 W;* Garcia Falls Canyon, 6 mi. N Monticello; *Springtime Canyon, 32 mi. S, 28 mi. W Socorro, San Mateo Mts.;* Socorro; Laguna; Fort Wingate; Jemez Springs area. Colorado: *jct. Plum [= Plumb] and Chacuaco creeks.*

Pipistrellus subflavus
Eastern Pipistrelle

Silvery gray to darker than Mummy Brown, dorsally; total length, 73–89; foot more than half as long as tibia; tragus tapering and straight; dorsal profile of skull convex in interorbital region; inner upper incisor bicuspidate; outer upper incisor unicuspidate (lacking accessory cusp on anterointernal surface); P1 viewed from occlusal surface more than a seventh of area of canine and visible from labial aspect; lower, 3rd premolar as

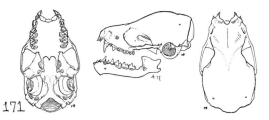

Fig. 171. *Pipistrellus subflavus subflavus*, Bat Cave, 2 mi. N War Eagle, Benton Co., Arkansas, No. 81387 M.V.Z., ♀, X 2, but lower jaw from cave along Missouri River 1 mi. SE Leavenworth Co., Kansas, No. 38797 K.U., ♂, X 2.

high as anterior cusp of canine; lower premolars less crowded than in *P. hesperus* and distance between canine and 1st molar less than length of 2nd lower molar.

Pipistrellus subflavus clarus Baker

1954. *Pipistrellus subflavus clarus* Baker, Univ. Kansas Publ., Mus. Nat. Hist., 7:585, November 15, type from 2 mi. W Jiménez, 850 ft., Coahuila.

MARGINAL RECORDS.—Texas: *12 mi. W Comstock* (Reddell, 1968:769); *Comstock; Devils River;* Del Rio. Coahuila: type locality; 2 mi. S, 3 mi. E San Juan de Sabinas, 1160 ft.

Pipistrellus subflavus floridanus Davis

1957. *Pipistrellus subflavus floridanus* Davis, Proc. Biol. Soc. Washington, 70:213, December 31, type from Homosassa Springs, head Homosassa River, Citrus Co., Florida.

MARGINAL RECORDS (Davis, 1959:529, 530, unless otherwise noted).—Georgia: 5 mi. W Ludowici. Florida: Lake Underhill; Bassenger; Sugarloaf Key (Layne, 1974:389); Tarpon Springs (Hall and Dalquest, 1950:599); Taylor County.

Pipistrellus subflavus subflavus (F. Cuvier)

1832. V[espertilio]. *subflavus* F. Cuvier, Nouv. Ann. Mus. Hist. Nat. Paris, 1:17, type from eastern United States, probably Georgia. Type locality restricted by Davis, Jour. Mamm., 40:522, November 20, 1959, to LeConte Plantation, 3 mi. SW Riceboro, Liberty Co., Georgia.

1897. *Pipistrellus subflavus*, Miller, N. Amer. Fauna, 13:90, October 16.

1835–1841. *Vespertilio erythrodactylus* Temminck, Monographies de mammalogie . . . , 2:238. Type locality, vicinity of Philadelphia, Pennsylvania.

1841. (?) *Vespertilio monticola* Audubon and Bachman, Proc. Acad. Nat. Sci. Philadelphia, 1:92, October, based on specimens from Grey Sulphur Springs, Virginia.

1864. *Scotophilus georgianus* H. Allen, Smiths. Miscl. Coll., 7(Publ. 165):35, June. Allen attributed the name georgianus to Cuvier 1832 but Cuvier's name may apply to some other species.

1894. *Vesperugo carolinensis* H. Allen, Bull. U.S. Nat. Mus.,

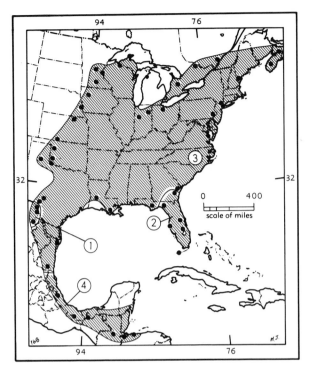

Map 168. *Pipistrellus subflavus.*

1. *P. s. clarus*   3. *P. s. subflavus*
2. *P. s. floridanus*   4. *P. s. veraecrucis*

43:121, March 14. Allen attributed the name *carolinensis* to É. Geoffroy St.-Hilaire, 1806 [= *Eptesicus fuscus*], type locality, "Carolina."

1897. *Pipistrellus subflavus obscurus* Miller, N. Amer. Fauna, 13:93, October 16, type from Lake George, Warren Co., New York.

MARGINAL RECORDS.—Michigan: 1 mi. N Mohawk, 47° 20′ N, 88° 22′ W, *ca.* 1200 ft. (Stones and Haber, 1965:688). Indiana (Mumford, 1969:47): Tippecanoe County; Wells County. Ohio: Fremont. Ontario: Rockwood, near Guelph; Renfrew County. Quebec: Joliette. Nova Scotia (Bleakney, 1965:528, 529): Five Mile River Cave near Maitland; *near Gays River*; Grafton Lake Federal Fish Hatchery. Maine: Windham. New York: Hastings-on-Hudson. Pennsylvania: Germantown, thence southward along Atlantic Coast to Georgia: type locality. Florida: 18 mi. above mouth Appalachicola River (Davis, 1959:527), thence westward along Gulf Coast to Louisiana (Lowery, 1974:110): *Tulane Univ. Riverside Campus, Belle Chasse*; Houma; 11 mi. W Forest Hill on Calcasieu River, thence to and southwestward along coast to Tamaulipas: Matamoros; near headwaters Río Sabinas, 10 km. N, 8 km. W El Encino (Davis, 1959:527); *Rancho Pano Ayuctle, 6 mi. N Gómez Farías (ibid.)*. Texas: Rocksprings; Kerr County (Davis, 1966:54, as *P. subflavus* only); Hardeman County (Dalquest, 1968:15); Fissure Cave (Milstead and Tinkle, 1959:138, as *P. subflavus* only). Oklahoma: 1 mi. S, 3 mi. W Reed; *3 mi.*

N *Jester*; Alabaster Cavern, Woodward Co. Kansas (Jones, *et al.*, 1967:15, unless otherwise noted): Double-entrance "S" Cave [11½ mi. S, 16 mi. E Coldwater]; *5½ mi. SW Sun City* (Hall, 1955:35); *Havard Cave (2½ mi. SW Sun City)*; ½–½ mi. N, 1½ mi. W Blue Rapids. Nebraska (Jones, 1964c:90): *1–1½ mi. NE Louisville*; ½–1 mi. W Meadow. Iowa: Storm Lake (Kunz and Schlitter, 1968:170). Minnesota: St. Peter; St. Cloud; Marine. Wisconsin: Hurley.

Pipistrellus subflavus veraecrucis (Ward)

1891. *Vesperugo veraecrucis* Ward, Amer. Nat., 25:745, August, type from Las Vigas, Canton of Jalapa, Veracruz.

1950. *Pipistrellus subflavus veraecrucis [sic]*, Hall and Dalquest, Univ. Kansas Publ., Mus. Nat. Hist., 1:601, January 20.

MARGINAL RECORDS.—Veracruz: type locality. Chiapas: 11 mi. W Mal Paso, 400 ft. (Carter, *et al.*, 1966:496). Honduras: Jilamo Farm, Tela District. Guatemala (Carter, *et al.*, 1966:496): 25 km. SSW Puerto Barrios, 75 m.; Sebol, 900 ft. Veracruz: 30 km. SSE Jesús Carranza.

Genus **Eptesicus** Rafinesque—Big Brown Bats

North American species revised under name *Vespertilio* by Miller, N. Amer. Fauna, 13:95–104, October 16, 1897. See also Davis, Jour. Mamm., 46:229–240, May 20, 1965, and Southwestern Nat., 11:245–274, July 20, 1966.

1820. *Eptesicus* Rafinesque, Annals of nature . . . , p. 2. Type, *Eptesicus melanops* Rafinesque [= *Vespertilio fuscus* Palisot de Beauvois].

1829. *Cnephaeus* Kaup, Skizzirte Entwickelungs-Geschichte und natürliches System der Europäischen Thierwelt, 1:103. Type, *Vespertilio serotinus* Schreber.

1837. *Noctula* Bonaparte, Iconografia della fauna Italica . . . , fasc. 21, vol. 1. Type, *Noctula serotina* Bonaparte.

1856. *Cateorus* Kolenati, Allgem. deutsch. Naturh. Zeit. Dresden, neue Folge, 2:131. Type, *Vespertilio serotinus* Schreber.

1858. *Amblyotis* Kolenati, Sitzungsb. k. Akad. Wiss., Wien, 29:252. Type, *Amblyotis atratus* Kolenati [= *Vespertilio nilssonii* Keyserling and Blasius].

1866. *Pachyomus* Gray, Ann. Mag. Nat. Hist., ser. 3, 17:90. Type, *Scotophilus pachyomus* Tomes.

1870. *Nyctiptennis* Fitzinger, Sitzungsb. k. Akad. Wiss., Wien, 62:424. Type, *Vespertilio smithii* Wagner.

1891. *Adelonycteris* H. Allen, Proc. Acad. Nat. Sci. Philadelphia, p. 466, a renaming of *Vesperus* Keyserling and Blasius, 1839, which included parts of both *Vespertilio* and *Eptesicus* and which is a homonym of *Vesperus* Latreille, 1829, an insect.

1916. *Pareptesicus* Bianchi, Ann. Mus. Zool., Acad. St. Pétersbourg, 21:76. Type, *Vesperugo pachyotis* Dobson.

1916. *Rhyneptesicus* Bianchi, Ann. Mus. Zool., Acad. St. Pétersbourg, 21:76. Type, *Vesperugo nasutus* Dobson.

1926. *Neoromicia* Roberts, Ann. Transvaal Mus., 11:245. Type, *Eptesicus zuluensis* Roberts.

1934. *Vespadelus* Iredale and Troughton, Mem. Australian Mus., 6:95.

Upper incisors well developed, inner larger than outer and usually having distinct secondary cusp; outer incisor separated from canine by space equal to greatest diameter of incisor; lower incisors subequal, trifid, closely crowded and distinctly imbricated, forming strongly convex row between canines; crown on 3rd lower incisor wider than on 1st or 2nd; hypocone on M1 and M2 always indicated and in some species well developed; distinct concavity between hypocone and protocone; M3 variable in form, usually with well-developed metacone and three commissures in smaller species, but having metacone and 3rd commissure obsolete in larger forms; skull essentially as in *Pipistrellus*; rostrum flattish or more often rounded off above; palatal emargination at least as deep as wide (after Miller, 1907:208, 209). Ears small; tragus short, straight; interfemoral membrane bearing sprinkling of hairs above on basal fourth.

The names *Vesperus* (*Marsipolaemus*) *albigularis* and *Vesperus propinquus*, both proposed in 1872 by Peters, as new species from "Mexico" and ". . . Guatemala," respectively, have appeared in previous lists of North American mammals as taxa of the genus *Eptesicus*. These two names are omitted from the present list because Davis (1965:229) deduced that Peters' specimens on which the two names were based came from the Old World (not the Americas) and are junior synonyms (or subspecies) of the Old World *Vespertilio murinus* Linnaeus, 1758, and *Eptesicus nilssoni* (Keyserling and Blasius, 1839), respectively. Neither of these species occurs in the New World.

Key to North American Species of Eptesicus

Principally after Davis (1965:239)

1. M2 more than 2.4 × 1.8; forearm averaging more than 45*; greatest length of skull averaging more than 17.2.
    2. Greatest length of skull more than 22; total length more than 128.   *E. guadeloupensis,*   p. 217
    2'. Greatest length of skull less than 22; total length less than 128.   *E. fuscus,*   p. 214
1'. M2 less than 2.4 × 1.9; forearm averaging less than 45; greatest length of skull averaging less than 17.2.

* Forearm recorded as less than 45 in some *E. fuscus* from Antillean region.

3. M2 averaging more than 1.8 × 1.6; pelage on back more than 8; forearm usually more than 42.5.   *E. andinus,*   p. 218
3'. M2 averaging less than 1.8 × 1.6; pelage on back less than 8; forearm usually less than 42.5. . . . .*E. furinalis,*   p. 217

See account of *E. lynni,* from Jamaica, a fifth North American species.

Eptesicus fuscus
Big Brown Bat

Subspecies revised by G. M. Allen, Canadian Field-Nat., 47:31, 32, February, 1932; western subspecies revised by Engels, Amer. Midland Nat., 17:653–660, May, 1936. See also Davis, Jour. Mamm., 46:229–240, May 20, 1965; Southwestern Nat., 11:245–274, July 20, 1966.

Total length, 87–138; tail, 34–56; length of forearm, 39–53.6; greatest length of skull, 15.1–23.0; zygomatic breadth, 11.1–14.2; breadth of braincase, 7.5–9.6; maxillary tooth-row, 7.1–9.8. Along the Pacific Coast, geographic variation is considerable, total length, for instance, being a fifth more in central California than in Baja California del Sur. Apex of 2nd triangle of M3 less than half height of anterior side of 1st triangle. Upper parts brown, usually dark, but pale in some subspecies, and sometimes reddish brown; underparts paler than upper parts, sometimes cinnamon or even buffy.

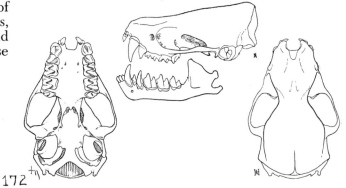

172

Fig. 172. *Eptesicus fuscus pallidus,* E slope Irish Mtn., 6900 ft., Lincoln Co., Nevada, No. 47851 M.V.Z., ♂, X 2.

Eptesicus fuscus bahamensis (Miller)

1897. *Vespertilio fuscus bahamensis* Miller, N. Amer. Fauna, 13:101, October 16, type from Nassau, New Providence, Bahamas. Known only from the type locality.
1945. *Eptesicus fuscus bahamensis,* Shamel, Proc. Biol. Soc. Washington, 58:108, July 18.

Eptesicus fuscus bernardinus Rhoads

1902. *Eptesicus fuscus bernardinus* Rhoads, Proc. Acad. Nat. Sci. Philadelphia, 53:619, February 6, type from near San Bernardino, San Bernardino Co., California.

1904. *Eptesicus fuscus melanopterus* Rehn, Proc. Acad. Nat. Sci. Philadelphia, 46:590, October 18, type from Mt. Tallac, Eldorado Co., California. Not *Vesperus melanopterus* Jentink, July 15, 1904 [= *Eptesicus melanopterus* from Dutch Guiana].

MARGINAL RECORDS.—British Columbia (Cowan and Guiguet, 1965:70, unless otherwise noted): Stuie; Williams Lake; *Chimney Lake;* Kamloops; *Horseshoe Lake;* Vernon; Anarchist Mtn.; Newgate (Hall and Kelson, 1959:185). Washington: Newport; Grand Ronde River. California: 8 mi. SW Ravendale. Nevada: Wadsworth. California: Robert's Ranch, Weyman Creek, White Mts.; Kenworthy; Trabuco Canyon; Santa Cruz Island (von Bloeker, 1967:249, 250), thence northward along Pacific Coast, including larger coastal islands, to British Columbia: Esquimalt (Cowan and Guiguet, 1965:70); *ca.* 15 mi. SE Campbell River (Campbell, 1972:12).

Eptesicus fuscus dutertreus (P. Gervais)

1837. *Vespertilio dutertreus* P. Gervais, Ann. Sci. Nat., Paris, ser. 2; Zool. 8:61, July, type from Cuba.
1955. *Eptesicus fuscus dutertreus,* Miller and Kellogg, Bull. U.S. Nat. Mus., 205:103, March 3.
1839. *Scotophilus cubensis* Gray, Ann. Nat. Hist., 4:7, September, type from Cuba.

MARGINAL RECORDS.—Andros Island. Watling Island, Cockburntown. Long Island: Clarencetown. Cuba: El Cobre. Gran Caimán Island (Varona, 1974:35). Cuba: Pinar del Río. Not found: McKinnon's Cave and Mortimer's, Long Island.

Eptesicus fuscus fuscus (Palisot de Beauvois)

1796. *Vespertilio fuscus* Palisot de Beauvois, Catalogue raisonné du muséum de Mr. C. W. Peale, Philadelphia, p. 18 (p. 14 of English edition by Peale and Beauvois), type from Philadelphia, Pennsylvania.
1900. *Eptesicus fuscus,* Méhely, Magyarország denevéreinek monographiája (Monographia Chiropterorum Hungariae), pp. 206, 338.
1806. *Vespertilio carolinensis* É. Geoffroy St.-Hilaire, Ann. Mus. Hist. Nat. Paris, 8:193. Type locality, "Carolina."
1818. *Vespertilio phaiops* Rafinesque, Amer. Month. Mag., 3:445. Type locality, Kentucky.
1820. *Vespertilio melanops* Rafinesque, Annals of nature . . . , p. 2, a renaming of *phaiops* Rafinesque.
1823. *Vespertilio arquatus* Say, *in* Long, Account of an exped. . . . to the Rocky Mts. . . . , 1:167, type from Engineer Cantonment, *ca.* 12 mi. SE Blair, Washington Co., Nebraska.
1835–1841. *Vespertilio ursinus* Temminck, Monographies de mammalogie, . . . , 2:235.
1843. *Scotophilus greenii* Gray, List of the . . . Mammalia in the . . . British Museum, p. 30, a *nomen nudum* (*fide* Miller, N. Amer. Fauna, 13:27, 96, October 16, 1897).

MARGINAL RECORDS.—Quebec: *central Quebec.* New Brunswick: *St. Andrews* (Gorham and Johnston, 1963:228). Maine: Eastport, thence southward along Atlantic Coast to Georgia: *ca.* 10 mi. SSW Thomasville, thence westward along Gulf Coast to Louisiana (Lowery, 1974:113): New Orleans; Merryville, and down coast and inland to Nuevo León: 3 mi. SW La Escondida, 6300 ft. (Genoways and Jones, 1967:478); Río Ramos, 20 km. NW Montemorelos. Texas: McLennan County (Davis, 1966:56, as *E. fuscus* only); Acme (Dalquest, 1968:15). Kansas (Jones, *et al.*, 1967:16): Double-entrance "S" Cave, 11½ mi. S, 17 mi. E Coldwater; McCracken; Hays. Nebraska: *4¹/₂ mi. E Red Cloud* (Jones, 1964c:92); *Red Cloud (ibid.);* Hastings; 1 mi. W Niobrara. Minnesota (Long and Severson, 1969:623): Fairmont; Brown County; Itasca Park. Manitoba: Lake Winnipeg; Whiteshell Forest Reserve, *ca.* 80 mi. E Winnipeg (Buckner, 1961:529). Ontario: Cavern Lake, 30 mi. NE Port Arthur.

Eptesicus fuscus hispaniolae Miller

1918. *Eptesicus hispaniolae* Miller, Proc. Biol. Soc. Washington, 31:39, May 16, type from Constanza, Dominican Republic.
1945. *Eptesicus fuscus hispanolae* [*sic*], Shamel, Proc. Biol. Soc. Washington, 58:108, July 18.

MARGINAL RECORDS.—Haiti: Port-de-Paix. Dominican Republic: San Gabriel Cave. Haiti: near St. Michel. Jamaica: Sherwood Forest (Sanborn, 1941:383; the three specimens concerned should be reexamined to ascertain if they are *E. lynni* named 4 years later from Jamaica by Shamel, Proc. Biol. Soc. Washington, 58:107, July 18, 1945).

Eptesicus fuscus miradorensis (H. Allen)

1866. S[*cotophilus*]. *miradorensis* H. Allen, Proc. Acad. Nat. Sci. Philadelphia, 18:287, type from Mirador, Veracruz.
1912. *Eptesicus fuscus miradorensis,* Miller, Bull. U.S. Nat. Mus., 79:62, December 31.
1920. *Eptesicus fuscus pelliceus* Thomas, Ann. Mag. Nat. Hist., ser. 9, 5:361, April, type from La Culata, 4000 m., near Mérida, Venezuela.

MARGINAL RECORDS.—Durango (Baker and Greer, 1962:74): Arroyo de Bucy; Ciudad [= Ciudad Durango]. Tamaulipas (Alvarez, 1963:411): *Nicolás, 56 km. NW Tula, 5500 ft.; Miquihuana, 6200 ft.;* 14 mi. N, 6 mi. W Palmillas, 5500 ft.; *Joya Verde, 35 km. SW Ciudad Victoria, 3800 ft.;* Sierra de Tamaulipas, 2 mi. S, 10 mi. W Piedra, 1200 ft. Veracruz: *6 km. SSE Altotonga, 9000 ft.* (Hall and Dalquest, 1963:250); *3 km. E Las Vigas;* 5 km. N Jalapa. Guatemala: Flores; Cobán (Jones, 1966:464). Honduras: El Manteado. Costa Rica: San José. Panamá (Handley, 1966:771): *upper Río Changena, 4800 ft.;* El Valle, thence into South America, and northwestward to Panamá: Boquete (*ibid.*). Honduras: *El Pedrero.* El Salvador: Los Esesmiles (Burt and Stirton, 1961:39). Guatemala: 5 mi. S Guatemala [City] (Jones, 1966:464); Zunil. Oaxaca: 42 km. W Cintalapa, Chiapas (Baker and Patton, 1967:285, as *E. fuscus* only); Chontecomatlán (Good-

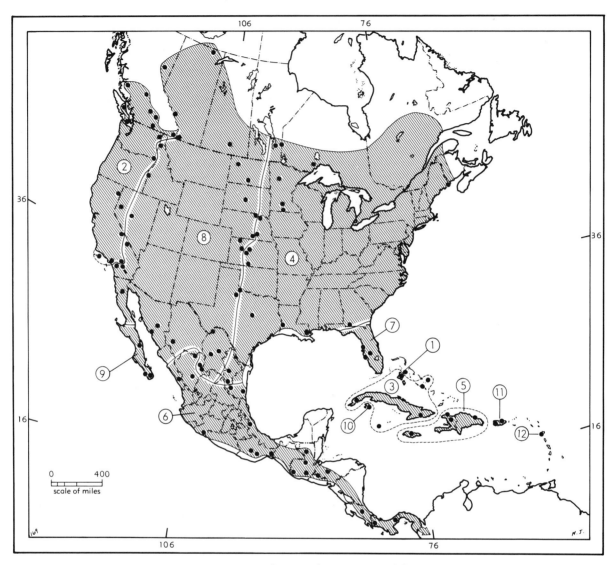

Map 169. *Eptesicus fuscus* and *Eptesicus guadeloupensis*.

1. *E. f. bahamensis*   7. *E. f. osceola*
2. *E. f. bernardinus*   8. *E. f. pallidus*
3. *E. f. dutertreus*   9. *E. f. peninsulae*
4. *E. f. fuscus*   10. *E. f. petersoni*
5. *E. f. hispaniolae*   11. *E. f. wetmorei*
6. *E. f. miradorensis*   12. *E. gaudeloupensis*

win, 1969:103); Oaxaca. Michoacán: 22 km. S Arteaga (Alvarez, 1968:30). Sinaloa: 2 km. W Palmitos, Hwy. 40, 6000 ft. (Irwin and Baker, 1967:195, as *E. fuscus* only). Villa-R. (1967:401) reported the subspecies *E. f. pallidus* from two localities in the southern half of Zacatecas.

Eptesicus fuscus osceola Rhoads

1902. *Eptesicus fuscus osceola* Rhoads, Proc. Acad. Nat. Sci. Philadelphia, 53:618, February 6, type from Tarpon Springs, Pinnellas Co., Florida.

MARGINAL RECORDS.—Florida: type locality; southern Highlands County; Englewood.

Eptesicus fuscus pallidus Young

1908. *Eptesicus pallidus* Young, Proc. Acad. Nat. Sci. Philadelphia, 60:408, October 14, type from Boulder, Colorado.

1912. *Eptesicus fuscus pallidus*, Miller, Bull. U.S. Nat. Mus., 79:62, December 31.

MARGINAL RECORDS.—Alberta: Pine Lake, Wood Buffalo Park. Manitoba: *Whiteshell Forest Reserve, ca. 80 mi.* E Winnipeg (Buckner, 1961:529, 1 *E. f. pallidus* with 15 specimens of *E. f. fuscus*). Saskatchewan: Regina. North Dakota: 4 mi. W Grinnell (Long and Severson, 1969:623); Cannon Ball. South Dakota:

Fort Pierre (Jones and Genoways, 1967:192). Nebraska: 4 mi. N, 2½ mi. W Spencer (Jones, 1964c:93); Funk (Kunz, 1965:201); *Alma* (*ibid.*). Kansas (Jones, *et al.*, 1967:17): Herndon; Castle Rock. Texas: Walkup Cave (Milstead and Tinkle, 1959:139); River Styx Cave, King Co. (Packard and Judd, 1968:536). Coahuila: 1 mi. N Boquillas, 700 ft.; 4 mi. S, 9 mi. W San Buenaventura, 2000 ft.; Acatita, 3600 ft. Durango (Baker and Greer, 1962:74): 10 mi. WSW Lerdo; 12 mi. SSW Mapimí. Chihuahua: 15 mi. S, 6 mi. E Creel, 7300 ft. (Anderson, 1972:252). Sonora: Río Yaqui, *ca.* 1 mi. S El Novillo (Findley and Jones, 1965:331, as *E. fuscus* only); Guaymas. Baja California: La Grulla. California: Escondido; Julian; Hanapah Canyon, Panamint Mts. Nevada: Peterson Creek, Shoshone Mts. Idaho: Payette Valley; Fort Sherman. Alberta: Waterton Lakes (Hall and Kelson, 1959:186); Johnson's Canyon (Banfield, 1958:10). British Columbia: Charlie Lake near Fort St. John (Cowan and Guiguet, 1965:70). According to Reeder (1965:332), a specimen in alcohol from 64° 15′ N, 145° 50′ W, in Alaska, may have been brought to the place where found by "vehicular traffic along the Alaska Highway." Locality not shown on Map 169.

Eptesicus fuscus peninsulae (Thomas)

1898. *Vespertilio fuscus peninsulae* Thomas, Ann. Mag. Nat. Hist., ser. 7, 1:43, January, type from Sierra Laguna, Baja California.
1912. *Eptesicus fuscus peninsulae*, Miller, Bull. U.S. Nat. Mus., 79:63, December 31.

MARGINAL RECORDS.—Baja California: Commondú; *type locality; Agua Caliente;* Miraflores (Jones, *et al.*, 1965:55); Todos Santos (*ibid.*).

Eptesicus fuscus petersoni Silva-T.

1974. *Eptesicus fuscus petersoni* Silva-T., Poeyana, 128:1, June 18, type from "Cueva de los Lagos, Cerro de las Guanábanas, Isla de Pinos." Known only from Isla de Pinos.

Eptesicus fuscus wetmorei Jackson

1916. *Eptesicus wetmorei* Jackson, Proc. Biol. Soc. Washington, 29:37, February 24, type from Maricao, Puerto Rico.
1945. *Eptesicus fuscus wetmorei*, Shamel, Proc. Biol. Soc. Washington, 58:108, July 18.

MARGINAL RECORDS.—Puerto Rico: Cueva de Farid (Pueblo Viejo); San Germán; *type locality.*

Eptesicus guadeloupensis Genoways and Baker
Guadeloupe Big Brown Bat

1975. *Eptesicus guadeloupensis* Genoways and Baker, Occas. Pap. Mus. Texas Tech Univ., 34:1, July 18, type from 2 km. S, 2 km. E Baie-Mahault, Basse-Terre, Guadeloupe. Known only from type locality.

Selected measurements of the type (a male), and two females, are: length of head and body, 73, 75, 72; length of forearm, 49.6, 51.5, 51.1; length of ear from notch, 23.0, 22.5, 24.0; length of tibia, 24.4, 24.8, 25.7; greatest length of skull, 22.5, 22.7, 23.1; breadth of braincase, 9.4, 9.2, 9.5; length of maxillary tooth-row, 8.1, 8.1, 8.3. Membranes black; pelage black basally, both dorsally and ventrally; hairs of upper parts tipped with a dark chocolate brown; underparts pale, hairs tipped with dark buff to almost white in some areas. Skull resembling that of *E. fuscus* but proportionally longer and narrower; largest New World member of the genus.

Genoways and Baker (1975:6) remark that Dobson's (1878:194) record of *Eptesicus fuscus* from Barbados, generally "discounted by recent authors . . . should be re-evaluated." See Map 169.

Eptesicus lynni Shamel
Lynn's Brown Bat

1945. *Eptesicus lynni* Shamel, Proc. Biol. Soc. Washington, 58:107, July 18, type from cave 3 mi. E Montego Bay, Jamaica. Known only from type locality.

Selected measurements of the type, an adult female, are: length of head and body, 55.0; length of forearm, 44.5; greatest length of skull, 16.8; breadth of braincase, 8.0; length of upper tooth-row, 6.0. Dichromatic; "typical" phase reddish-brown; alternate phase paler sometimes almost white (only specimens in alcohol described); underparts paler than back in each phase.

See statement under *Eptesicus fuscus hispaniolae* on need to compare three specimens from Sherwood Forest, Jamaica, recorded by Sanborn (1941:383) as *E. f. hispaniolae* to ascertain whether they are correctly identified or are instead *E. lynni.* If the two names apply to the same subspecific taxon, *E. lynni* falls as a synonym of *E. f. hispaniolae.*

Eptesicus furinalis
Argentine Brown Bat

Selected measurements of specimens from Costa Rica and specimens recorded by Davis (1965:236, 237; 1966:266) are: total length, 96–105; length of forearm, 37.4–42.5; greatest length of skull, 15.0–17.1; zygomatic breadth, 9.8–11.6; breadth of braincase, 6.8–8.0; length of upper tooth-row, 5.4–6.1. Upper parts blackish brown or sometimes a little paler, often frosted with buff; underparts dark brown, strongly washed with some shade of buff.

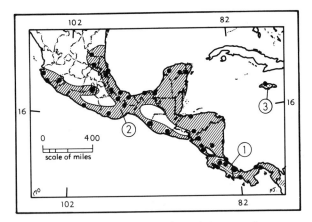

Map 170. *Eptesicus furinalis* and *Eptesicus lynni.*

1. *E. furinalis carteri*   2. *E. furinalis gaumeri*
                3. *E. lynni*

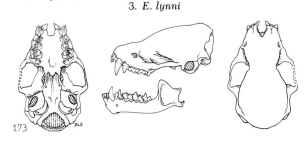

Fig. 173. *Eptesicus furinalis furinalis*, Village No. 8, Friesland Colony, Itacurubi del Rosario, Paraguay, No. 92660 K.U., ♂, X 2.

Eptesicus furinalis carteri Davis

1965. *Eptesicus gaumeri carteri* Davis, Jour. Mamm., 46:233, May 20, type from Turrialba, 2600 ft., Costa Rica.

MARGINAL RECORDS (Davis, 1965:233).—Costa Rica: Santa Teresa Peralta; La Iberia; *type locality; 3 mi. SE Turrialba.* Panamá: *7 km. SSW Changuinola;* Sibube. Costa Rica: *Villa Quesada; 18 mi. NE Naranjo.*

Eptesicus furinalis gaumeri J. A. Allen

1897. *Adelonycteris gaumeri* J. A. Allen, Bull. Amer. Mus. Nat. Hist., 9:231, September 28, type from Izamal, Yucatán.
1966. *Eptesicus furinalis gaumeri*, Davis, Southwestern Nat., 11:268, July 20.

MARGINAL RECORDS (Davis, 1965:234, 235, unless otherwise noted).—San Luis Potosí: El Salto. Veracruz: *4 km. ENE Tuxpan;* Tuxpan; 25 mi. W Santiago Tuxtla; 1 mi. E Jaltipan. Tabasco: Miramar. Campeche: Balchacaj, Laguna de Terminos. Yucatán: 2½ km. NW Dzitya (Birney, *et al.,* 1974:8); Piste. Quintana Roo: X-Can. Honduras: Lancetilla. Nicaragua: 1 km. N, 2½ km. W Villa Somoza, 330 m. (Jones, *et al.,* 1971:20). Costa Rica: Tortuguero (Starrett and Casebeer, 1968:16). Panamá: San Pablo, Canal Zone; Tocumen, thence southward into South America, thence northwestward along Pacific Coast to Panamá: Tapia. Costa

Rica: Boca de Barranca (Starrett and Casebeer, 1968:16). Nicaragua: 3 mi. NNW Diriamba; 1 mi. SE Yalaguina. Honduras: Comayaguela. Guatemala: 2 km. E Taxisco. Chiapas: Kilometer 184 on Hwy. 200 E of Huixtla (Baker and Patton, 1967:285). Oaxaca: 6⅔ km. N San Gabriel Mixtepec, 700 m. (Arnold and Schonewald, 1972:172). Guerrero: Supongo. Jalisco: 15 km. NW Cihuatlán; Sierra de Cuale; 10 mi. NNE Pihuamo, 3500 ft. (Watkins, *et al.,* 1972:27). Morelos: Cueva de Salitre. Oaxaca: Tolocita (= Tolosa); 5 mi. W Chiltepec (Goodwin, 1969:103). Veracruz: 2 km. N Motzorango; *Río Blanco, 20 km. N Piedras Negras;* 12½ mi. N Tihuatlán. San Luis Potosí: 1 km. W Huichihuayán.

Eptesicus andinus J. A. Allen
Andean Brown Bat

1914. *Eptesicus andinus* J. A. Allen, Bull. Amer. Mus. Nat. Hist., 33:382, July 9, type from Valle de las Papas, 10,000 ft., Colombia.

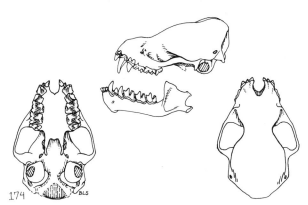

Fig. 174. *Eptesicus andinus*, Finca El Paraíso, 4050 ft., Chiapas, No. 66516 K.U., ♂, X 2.

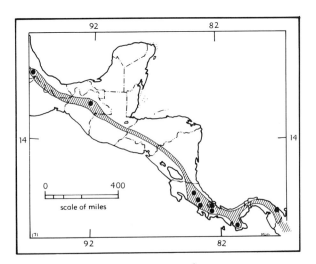

Map 171. *Eptesicus andinus.*

1920. *Eptesicus chiriquinus* Thomas, Ann. Mag. Nat. Hist., ser. 9, 5:362, April, type from Boquete, Chiriquí, Panamá.

1920. *Eptesicus inca* Thomas, Ann. Mag. Nat. Hist., ser. 9, 5:363, April, type from Chanchamayo, Cuzco, Perú.

Selected measurements are: total length, 99–120; length of forearm, 42.5–48.0; greatest length of skull, 15.9–18.1; zygomatic breadth, 10.7–12.0; breadth of braincase, 7.2–8.2; length of upper tooth-row, 6.0–6.9. Pelage nearly black, long (9–10 mm.), and underparts nearly as dark as upper parts.

MARGINAL RECORDS (Davis 1966:252, unless otherwise noted).—Veracruz: Ojo de Agua de Ata[o]yac (Baker, R. J., and Patton, J. L., 1967:285). Chiapas: Finca El Paraíso, thence southward to Costa Rica: San José, campus Univ. Costa Rica, 1200 m. (Starrett and Casebeer, 1968:17); 4⅕–6¾ mi. S La Georgina, 8200–9000 ft. Panamá: Río Changuina, 2400 ft.; Armila (Quebrada Venado), thence southeastward into South America, and northwestward through mts. to Panamá: Cerro Hoya; San Vicente, 1800 ft. Costa Rica: San Isidro General.

Genus **Nycteris** Borkhausen—Hairy-tailed Bats

North American species revised under the name *Lasiurus* by Miller, N. Amer. Fauna, 13:105–115, October 16, 1897; and by Hall and Jones, Univ. Kansas Publ., Mus. Nat. Hist., 14:73–98, December 29, 1961.

1797. *Nycteris* B[orkhause]n, Der Zoologe (Compendiose Bibliothek gemeinnützigsten Kenntnisse für alle Stände, pt. 21), Heft 4–7, p. 66. Type, *Vespertilio borealis* Müller [= *Nycteris borealis*].

Nycteris Borkhausen 1797, by the law of priority, was correctly applied to American bats for two decades (1909–1929). See Miller (Proc. Biol. Soc. Washington, 22:90, April 17, 1909). For part of those two decades, as well as at certain earlier times, the name *Nycteris* Geoffroy and Cuvier 1795 had been incorrectly (because the 1795 name was a *nomen nudum*) applied to Old World bats of a different genus (*Petalia* Gray 1838). In 1913 Old World zoologists, anxious to avoid changing the generic name of the Old World bats concerned from *Nycteris* to *Petalia*, outnumbered other zoologists on the recently re-formed International Commission on Zoological Nomenclature, and voted to suspend the rules in this case. This required a change in generic name for the New World bats from *Nycteris* to *Lasiurus!* Stability of nomenclature is the ultimate aim, of course, and experience convinces the writer that it is to be attained by adhering to the rules instead of by suspending them. Therefore *Nycteris* is the generic name used here. See Hall and Kelson, Mammals of North America (p. 188, March 31, 1959) for an earlier explanation of the Commission's suspension of the rules, which resulted in use of the generic name *Lasiurus* for some of the American species.

1831. *Lasiurus* Gray, Zool. Miscl., No. 1, p. 38. Type, *Vespertilio borealis* Müller.

1871. *Atalapha* Peters, Monatsb. preuss. Akad. Wiss., Berlin, p. 907, and other authors [*nec Atalapha* Rafinesque, 1814].

Interfemoral membrane large and most of its upper surface furred; mammae 4; 3rd, 4th, and 5th fingers progressively shortened; ear short and rounded; skull short and broad; nares and palatal emargination wide and shallow (width transversely exceeding length anteroposteriorly); sternum prominently keeled; i. $\frac{1}{3}$, c. $\frac{1}{1}$, p. $\frac{1}{2}$ or $\frac{2}{2}$, m. $\frac{3}{3}$; when 2 upper premolars present, anterior one minute, peglike, and displaced lingually; M3 much reduced, area of its crown less than a third that of M1.

KEY TO NORTH AMERICAN SPECIES OF NYCTERIS

Nycteris intermedia
Northern Yellow Bat

External measurements: total length, 121–164; tail, 51–77; foot, 8–13; forearm, 45.2–62.8. Skull: condylocanine length, 16.9–21.5; zygomatic breadth, 12.6–15.6; interorbital breadth, 4.7–5.5; length of mandibular tooth-row, 7.8–9.7. Upper parts yellowish orange, or yellowish brown, or brownish gray faintly washed with black to pale yellowish gray.

Nycteris intermedia floridana (Miller)

1902. *Dasypterus floridanus* Miller, Proc. Acad. Nat. Sci. Philadelphia, 54:392, September 12, type from Lake Kissimmee, Osceola Co., Florida.

MARGINAL RECORDS (Hall and Jones, 1961:85, unless otherwise noted).—New Jersey: Westfield (Koopman, 1965:695). Virginia: Willoughby Beach. South Carolina: 5 mi. NW Charleston. Florida: 5 mi. W Jacksonville; Bunnell; 2 mi. SW Deland; type locality; Palm Beach; Miami; *Fort Myers* (Layne, 1974:389); 1 mi. NE Punta Gorda; Seven Oaks [near present town of Safety Harbor]; head Chassahowitzka River; Old Town; Aucilla River, 15 mi. S Waukenna. Alabama: Chickasaw (Linzey and Linzey, 1969: 845). Mississippi: Hancock County, thence westward along Gulf Coast to Texas: Houston; Eagle Lake; Bexar County (Davis, 1966:60, as *L. intermedius* only); Austin. Louisiana (Lowery, 1974:126): 18 mi. E Clarence; 4 mi. W Ferriday. Georgia: W edge Camilla; Bibb County (Golley, 1962:73). South Carolina: Barnwell County (Golley, 1966:63).

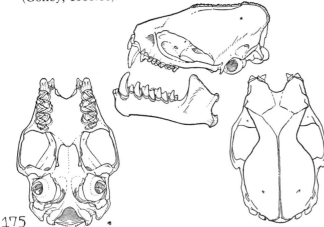

175

Fig. 175. *Nycteris intermedia floridana*, Houston, Harris Co., Texas, No. 84218 M.V.Z., ♀, X 2.

Nycteris intermedia insularis (Hall and Jones)

1961. *Lasiurus intermedius insularis* Hall and Jones, Univ. Kansas Publ., Mus. Nat. Hist., 14:85, December 29, type from Cienfuegos, Las Villas Prov., Cuba. Silva-T. (1976:10–15) has presented weighty evidence that this taxon should stand as a species instead of as a subspecies of *N. intermedia*. The degree of difference between *N. i.*

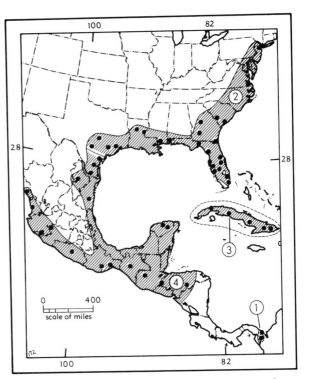

Map 172.   *Nycteris castanea* and *Nycteris intermedia*.

1. *N. castanea*   3. *N. i. insularis*
2. *N. i. floridana*   4. *N. i. intermedia*

floridana and *N. i. insularis* is great, but is less between *N. i. insularis* and *N. i. intermedia*. Furthermore, specimens of *N. i. intermedia* seem to increase in size southward. When more specimens from northern Central America become available, the degree of difference between *N. i. insularis* and *N. i. intermedia* may prove to be less than now appears to be the case. Provisionally therefore, the name combination *N. i. insularis* is here retained, although more specimens from Honduras may well show that specific rank for *insularis* is required.

MARGINAL RECORDS (Hall and Jones, 1961:87).—Cuba: Laguna La Deseada, San Cristóbal; type locality; San Blas; San Germán; Bayate, Guantánamo.

Nycteris intermedia intermedia (H. Allen)

1862. *Lasiurus intermedius* H. Allen, Proc. Acad. Nat. Sci. Philadelphia, 14:246, "April" (between May 27 and August 1), type from Matamoros, Tamaulipas.

MARGINAL RECORDS (Hall and Jones, 1961:84, unless otherwise noted).—Texas: San Patricio County (Davis, 1966:60, as *L. intermedius* only); *Padre Island*, thence along coast to Yucatán: Tekom; Izamal. Honduras: Río Yeguare, between Tegucigalpa and Danlí; 1 km. NW Nueva Ocotepeque, 840 m. (Carter, *et al.*, 1966:497). Guatemala: 1 km. NE Aguacatán, 1620 m. (*ibid.*). Chiapas: San Bartolomé. Oaxaca: Tehuantepec; Chontecomatlán (Goodwin, 1969:105). Guerrero: Cañón de Zopilote, 14½ mi. N Zumpango (Winkelmann, 1962:108, as *L. intermedius* only). Colima: Pueblo Juárez [= La Magdalena] (Gardner, 1962:103). Nayarit: 8 mi. E San Blas (*ibid.*). Sinaloa: 3 mi. N Mazatlán (Loomis and Jones, 1964:32). Michoacán: Briseñas, 1536 m. (Villa-R., 1967:411). Tamaulipas: Sierra de Tamaulipas, 1200 ft., 2 mi. S, 10 mi. W Piedra. Nuevo León: Santiago (Jiménez-G., 1971:141). Texas: 5⅝ mi. N Mission; Kleberg County (Davis, 1966:60, as *L. intermedius* only).—See addenda.

Nycteris ega
Southern Yellow Bat

External measurements: total length, 109–124; tail, 45–58; forearm, 42.7–52.2. Cranial measurements: greatest length, 14.7–16.2; interorbital width, 4.3–5.0; length of upper tooth-row, 5.0–5.9. Upper parts yellowish brown (much as in *Nycteris intermedia floridana* from Louisiana) having overlay of grayish or blackish anterior to shoulders; hair on basal half of interfemoral membrane more yellowish than elsewhere. The species occurs also in South America.

Nycteris ega panamensis (Thomas)

1901. *Dasypterus ega panamensis* Thomas, Ann. Mag. Nat. Hist., ser. 7, 8:246, September, type from Bogava [= Bugaba], Chiriquí, Panamá.

MARGINAL RECORDS.—Costa Rica: Interamerican Highway, near bridge over Río Puerto Nuevo, 90 m. (Starrett and Casebeer, 1968:18). Panamá (Handley, 1966:771): type locality; Cerro Hoya, 1800 ft.; Fort Clayton, thence southeastward into South America.

Nycteris ega xanthina (Thomas)

1897. *Dasypterus ega xanthinus* Thomas, Ann. Mag. Nat. Hist., ser. 6, 20:544, December, type from Sierra Laguna, Baja California.

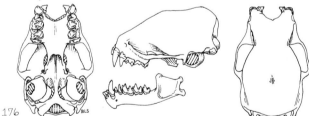

Fig. 176. *Nycteris ega xanthina*, 2 mi. S, 10 mi. W Piedra, 1200 ft., Tamaulipas, No. 55316 K.U., ♂, X 2.

MARGINAL RECORDS (Hall and Jones, 1961:91, unless otherwise noted).—California: Pomona (Stewart, 1969:194); Palm Springs; *Joshua Tree National Monument* (Constantine, 1966:126). Arizona: Phoenix; *Scottsdale* (Constantine, 1966:126); Tucson. New Mexico: Guadalupe Canyon (Mumford, *et al.*, 1964:45, as *L. ega* only). Durango: Aguajequiroz, 12 mi. SSW Mapimí, 5000 ft. Coahuila: 4 mi. W Hda. La Mariposa, 2300 ft. Texas: 5 mi. SE Brownsville (Baker,

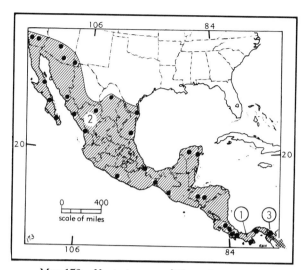

Map 173. *Nycteris ega* and *Nycteris egregia*.

1. *N. ega panamensis*   2. *N. ega xanthina*
3. *N. egregia*

et al., 1971:849). Tamaulipas: Sierra de Tamaulipas, 3 mi. S, 10 mi. W Piedra. Veracruz: Achotal. Yucatán: Yaxcach. Quintana Roo: 7 mi. N, 37 mi. E Puerto de Morelos. Honduras: La Esperanza (LaVal, 1969:821, as *L. ega* only); Tegucigalpa. Costa Rica: San José; Lajas Villa Quesada. Chiapas: Kilometer 184 on Hwy. 200 E Huixtla (Baker and Patton, 1967:285, as *L. ega* only). Oaxaca: 42 km. W Cintalapa, Chiapas (*ibid.*, as *L. ega* only). Guerrero: El Pápayo (Hall and Kelson, 1959:194). Sinaloa: Esquinapa; 1 mi. S Pericos. Sonora: Río Alamos, 8 mi. by road S Alamos (Baker and Christianson, 1966:310, as *L. ega* only). Baja California: Miraflores; *type locality;* Comondú; Santa Ana. Not found: New Mexico: Granite Gap, Peloncillo Mts. (Jones and Findley, 1963:175, as *L. ega* only).

Nycteris pfeifferi (Gundlach)
Pfeiffer's Hairy-tailed Bat

1861. *Atalapha pfeifferi* Gundlach, Monatsb. preuss. Akad. Wiss., Berlin, p. 152, type from Cuba. Known only from Trinidad, Cuba (Miller, Jour. Mamm., 12:410, November 11).

Head and body, approx. 55. Differing from red bat (*N. borealis borealis*) of eastern United States in being larger (forearm 42 mm. or more instead of 42 mm. or less; tibia about 21 mm. instead of 18.6–20 mm.), and in marked reduction of pale tips of hairs on both upper and lower surfaces of body. Skull equal in size to that of largest individuals of *N. borealis borealis* (after Miller, 1931:409, 410). See Map 174.

Nycteris degelida (Miller)
Jamaican Hairy-tailed Bat

1931. *Lasiurus degelidus* Miller, Jour. Mamm., 12:410, November 11, type from Sutton's, District of Vere, Jamaica.

Head and body, approx. 55. Skull in two females larger than in two males of *Nycteris pfeifferi*, and cheek-teeth (in crown view) conspicuously larger. Color of females about as in brightest males of *Nycteris borealis borealis* but with no trace of grayish "frosting" on upper parts; underparts darker than in mainland animal, and, in two of four specimens, lacking whitish tips on hairs of chest; white shoulder spots thus thrown into strong relief (after Miller, 1931:410).

MARGINAL RECORDS.—Jamaica: type locality; Spanishtown (Miller, 1931:410). See Map 174.

Nycteris minor (Miller)
Small Hairy-tailed Bat

1931. *Lasiurus minor* Miller, Jour. Mamm., 12:410, November 11, type from Voûte l'Église, 1350 ft., a cave near Jacmel road a few km. N Trouin, Haiti. (Regarded as a subspecies of *N. borealis* by Koopman, *et al.*, Jour. Mamm., 38:168, May 27, 1957.)

Skull noticeably smaller than in Cuban or Jamaican species (greatest length of skull of two specimens, 13 and 12), and about the same as in *N. b. borealis* from Virginia; braincase, as compared with that of the mainland animal, more rounded when viewed from above, and more nearly flat-topped when viewed from behind; lachrymal ridge and tubercle poorly developed; upper cheek-teeth essentially like those of *N. borealis borealis* except that P4 smaller (after Miller, 1931:410).

MARGINAL RECORDS.—Bahama Islands: New Providence Island; Orange Creek, Cat Island (G. M. Allen and Sanborn, 1937:228; identification stated to be uncertain). Puerto Rico: near Moca (Starrett and Rolle, 1963:264, as *L. borealis minor*). Haiti: type locality. Bahama Islands: Andros Island.—Although not marginal, occurrences of this bat on Long, Gran [= Great] Inagua, and Caicos islands may be noted (Varona, 1974:37, as *L. b. minor*). See Map 174.

Nycteris borealis
Red Bat

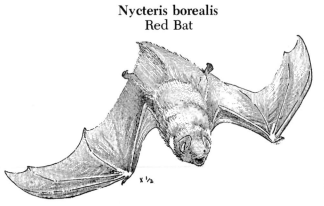

x ½

Length of head and body, approx. 55; total length, 91–112; length of forearm, 37.1–43.6; greatest length of skull (excluding incisors), 11.5–13.5; maxillary tooth-row, 3.8–4.9. Upper parts brick red to rusty red washed with white (males more brightly colored than females); underparts usually slightly paler; anterior part of shoulder with buffy white patch. Ears low, broad, rounded, naked inside, densely furred outside on basal two-thirds; tragus triangular. Uropatagium densely furred above, sparsely furred below.

Nycteris borealis borealis (Müller)

1776. *Vespertilio borealis* Müller, Des Ritters Carl von Linné . . . vollständiges Natursystem . . . , Suppl., p. 20, type from New York.
1910. *Nycteris borealis* Hollister, Bull. Wisconsin Nat. Hist. Soc., 8:30, May.
1777. [*Vespertilio*] *noveboracensis* Erxleben, Systema regni animalis . . . , 1:155. Based on "the New York bat of Pennant (Synop. Quad., p. 367), 'Die nordamerikanische Fledermaus' of Schreber (Säugthiere, I, p. 176), and 'Der

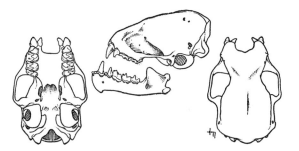

Fig. 177. *Nycteris borealis borealis*, 1½ mi. S Galena, Cherokee Co., Kansas, No. 38813 K.U., ♀, X 2.

Neujorker' of Müller" (*antea*) (Miller, N. Amer. Fauna, 13:32, October 16, 1897).
1781. *Vespertilio lasiurus* Schreber, Die Säugthiere . . . , Abth. 1, p. 62. Type locality, North America.
1796. *Vespertilio rubellus* Palisot de Beauvois, Catalogue raisonné du muséum de Mr. C. W. Peale, Philadelphia, p. 204. Type locality unknown.
1815. *Vespertilio rubra* Ord, *in* Guthrie, A new geog., hist., coml. grammar . . . , Philadelphia, Amer. ed. 2, 2:291. Based on the red bat of Wilson, Amer. Ornith., 6:60.
1818. *Vespertilio tesselatus* Rafinesque, Amer. Month. Mag., 3:445. Type locality unknown.
1818. *Vespertilio monachus* Rafinesque, Amer. Month. Mag., 3:445. Type locality unknown.
1820. *Vespertilio rufus* Warden, Desc. États-Unis de l'Amér. Septentrionale, 5:606. Based on the red bat of Wilson, Amer. Ornith., 6:60.
1870. *Lasiurus funebris* Fitzinger, Sitzungsb. k. Akad. Wiss., Wien, 62:46. Based on *Nycticejus noveboracensis* Temminck, 1840, type locality Tennessee.
1930. *Myotis quebecensis* Yourans, Naturaliste Canadien, ser. 3, 57(vol. 1:65, March, type from Anse-à-Wolfe, Quebec. For

status see "La Direction" [= Georges Maheux], Naturaliste Canadien, ser. 3, 57(vol. 1):185, 186, October, 1930.

MARGINAL RECORDS.—Saskatchewan: Reindeer Lake (Beck, 1958:14). Manitoba: Winnipeg. Ontario: North Bay. Quebec: Ottawa. New Brunswick: vic. Long Lake, Tobique Valley; Grand Maman Island. Maine: Mount Desert Island (Manville, 1960: 416, as *L. borealis* only), thence southward along Atlantic Coast to Florida: Gainesville; Old Town, thence westward along Gulf Coast to Tamaulipas: Matamoros. Texas: Edinburg. Tamaulipas: 1½ mi. NW Tinaja (Schmidly, *et al.*, 1974:88). Coahuila: Fortín. Texas: Mt. Emory, Chisos Mts. (Anderson, 1972:254). Chihuahua (Anderson, 1972:253): Ojo de Galeana, 4³⁄₁₀ mi. SE Galeana; Colonia Juárez. Texas: 5 mi. S, 8 mi. E El Paso City Hall (Jones and Lee, 1962:77). New Mexico: 6½ mi. N, 2 mi. E Las Cruces (Findley, *et al.*,

1975:55); Rattlesnake Spring, 5 mi. SSW Carlsbad Cavern (Jones, 1961:539); *Carlsbad Cavern (ibid.)*. Texas: Littlefield (Packard and Garner, 1964:388); 9 mi. E Stinnett (Milstead and Tinkle, 1959:139). Kansas: 1 mi. E Coolidge. Colorado: NW of Littleton; Greeley. South Dakota: Moon (Turner and Davis, 1970:363). North Dakota: Yellowstone River, probably Buford. Alberta: Calgary. Also, *Coral Harbour, Southampton Island* (Banfield, 1961:264); 42° 42' N, 62° 58' W; Bermuda.

Nycteris borealis frantzii (Peters)

1871. *Atalapha frantzii* Peters, Monatsb. preuss. Akad. Wiss., Berlin, p. 908, type from Costa Rica.

MARGINAL RECORDS.—Quintana Roo: Cozumel Island (Jones and Lawlor, 1965:417—closely resembles *N. b. frantzii* but reported as *L. b. teliotus* and

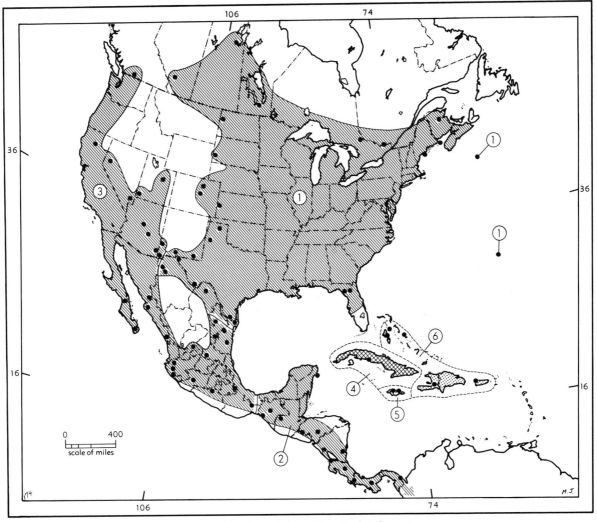

Map 174. *Nycteris borealis* and related species.

1. *N. borealis borealis*	3. *N. borealis teliotis*	5. *N. degelida*
2. *N. borealis frantzii*	4. *N. pfeifferi*	6. *N. minor*

locality questioned). Honduras: La Flor Archaga. Nicaragua: 3 km. NW Rama, Zelaya (Baker and Jones, 1975:4). Costa Rica: San José. Panamá: Armila (Handley, 1966:771), thence southeastward to South America, and northwestward to Panamá (ibid.): Colobre; Boquete, 3500 ft. Nicaragua: 12½ mi. S, 13 mi. E Rivas, 125 ft. (Carter, et al., 1966:496). El Salvador: Los Esesmiles (Burt and Stirton, 1961:39, as L. b. "probably . . . ornatus" [= teliotis according to Handley, 1960:472]). Guatemala: Barrillos. Chiapas (Carter, et al., 1966:496): 15 km. SE Tonalá, 100 ft.; Tenejapa, 6700 ft.

Nycteris borealis teliotis (H. Allen)

1891. *Atalapha teliotis* H. Allen, Proc. Amer. Philos. Soc., 29:5, April 10, type from an unknown locality, probably some part of California.
1912. *Nycteris borealis teliotis*, Miller, Bull. U.S. Nat. Mus., 79:64, December 31.
1951. *Lasiurus borealis ornatus* Hall, Univ. Kansas Publ., Mus. Nat. Hist., 5:226, December 15, type from Peñuela, Veracruz. Equals *Nycteris borealis mexicana* (Saussure), Auct., but *mexicana* Saussure is properly assignable to *Nycteris cinerea*. Indistinguishable from *N. b. teliotis* according to Handley (Proc. U.S. Nat. Mus., 112:472, 1960).

MARGINAL RECORDS.—British Columbia: Skagit. California: Dale's. Nevada: 5 mi. SW Fallon; Overton. Utah: St. George; *La Verkin Cave*; Kenilworth Mine. Arizona: Montezuma Well; pond near Aztec Peak, Sierra Ancha Mts. (Johnson and Johnson, 1964:322). New Mexico: Glenwood (Jones, 1961: 539). Arizona: Cave Creek, near Portal. New Mexico: Guadalupe Canyon (Mumford, et al., 1964:44). Zacatecas: 9⁷⁄₁₀ mi. NW Cuauhtemoc, 7100 ft. (Best, et al., 1972:97). San Luis Potosí: Bledos. Nuevo León: 3 mi. SW La Escondida (Genoways and Jones, 1967:478); 1 km. W Apodaca (Jiménez-G., 1971:141). Tamaulipas: *3 mi. S, 14 mi. W Piedra* (ibid.); 2 mi. S, 10 mi. W Piedra (ibid.). Veracruz: Peñuela. Oaxaca: Guichicovi. Guerrero: 3⅝ km. N Teloloapan, 1480 m. (Alvarez and Ramírez-P., 1972:174). Michoacán: Nuevo San Juan, 5 mi. SW Uruapan. Jalisco (Watkins, et al., 1972:28): Cihuatlán, 15 ft.; 2 mi. S La Cuesta,

1500 ft.; 17 km. SE Talpa de Allende, 5200 ft. Sinaloa: 2 km. W Palmitos, Hwy. 40, 6000 ft. (Irwin and Baker, 1967:195, as *L. borealis* only); 10 mi. NNW Los Mochis (Jones, et al., 1962:155). Sonora: *Guirocoba*; Río Cuchujaqui, 8 mi. by road SSE Alamos (Baker and Christianson, 1966:311). Baja California: San José del Cabo (Jones, 1970:361); Comondú, thence northward along Pacific Coast to point of beginning.

Nycteris seminola (Rhoads)
Seminole Bat

1895. *Atalapha borealis seminola* Rhoads, Proc. Acad. Nat. Sci. Philadelphia, 47:32, March 19, type from Tarpon Springs, Pinellas Co., Florida.
1896. *Atalapha borealis peninsularis* [Coues], The Nation, 62:404, May 21. Type locality, Florida. Described by Cory, Hunting and fishing in Florida . . . , pp. 115, 116, 1896.

Resembling *N. borealis* and differing mainly in color. Upper parts rich mahogany brown lightly frosted with grayish white; posterior part of underparts slightly paler than back; throat and chest whitish. Lachrymal not developed (present in *N. borealis*).

MARGINAL RECORDS.—New York: Ithaca. Pennsylvania: Hopewell; shore of Susquehanna River near mouth Fishing Creek. North Carolina (Barkalow and Funderburg, 1960:394): *Wrightsboro*; 10 mi. S Wilmington. South Carolina: Charleston, thence southward along coast to Florida: Micco; Miami (Layne, 1974:389); vic. Fort Myers (ibid.); Seven Oaks, thence westward along Gulf Coast to Texas: San Jacinto River, on Farm Road 945, NE Cleveland (Baker and Mascarello, 1969:250); Rusk County (Davis, 1966:60). The alleged occurrence at Brownsville, Texas, shown on Map 175 needs verification. Oklahoma: Little River, 6 mi. S Eagletown (Glass,

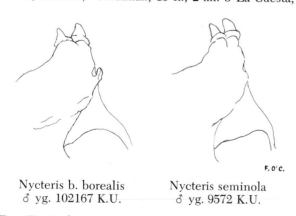

Nycteris b. borealis
♂ yg. 102167 K.U.

Nycteris seminola
♂ yg. 9572 K.U.

F. O'C.

Fig. 178. Right anterior part of skulls showing difference in lachrymal region, X approx. 6½. *N. b. borealis* from Lawrence, Douglas Co., Kansas. *N. seminola* from Chesser's Island, Charlton Co., Georgia.

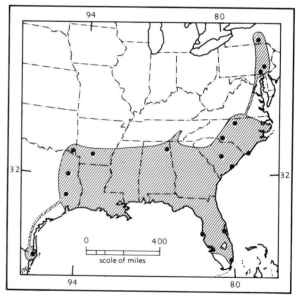

Map 175. *Nycteris seminola*.

1958:587). Arkansas: Newport Landing. Alabama: Fort Payne. South Carolina: 9 mi. N Orangeburg. North Carolina: Camp Steere; Raleigh. Villa-R. (1967:408) stated that the only Mexican record of this species came from near Tecolutla, Veracruz, and was based on a badly deteriorated specimen that he did not see. Identity of the specimen from Brownsville, Texas, verified by R. H. Pine, *in litt.*, August 15, 1972. Both the Brownsville record and those from New York and Pennsylvania should be considered extralimital. Also, *Bermuda* (Van Gelder and Wingate, 1961:1).

Nycteris castanea (Handley)
Tacarcuna Bat

1960. *Lasiurus castaneus* Handley, Proc. U.S. Nat. Mus., 112:468, October 6, type from Tacarcuna Village, 3200 ft., Río Pucro, Darién, Panamá.

Selected measurements of holotype: total length, 112; length of tail vertebrae, 48; length of hind foot, 11; length of forearm, 44.8; greatest length of skull, 13.0; zygomatic breadth, 9.9; interorbital breadth, 4.2; length of maxillary tooth-row, 4.7. Upper parts deep chestnut (between Morocco Red and Chestnut), face and muzzle entirely black; underparts blackish brown with only scattered buff-tipped hairs except on collar; ears, wings, membranes, and lips entirely blackish. M3 much reduced, 2nd commissure shorter than 1st; hypocone much reduced on M1 and M2; P4 double-rooted. Rostrum broad and deep; lachrymal ridge not developed; braincase narrow, deep, and tilted up away from plane of palate. Otherwise similar to *Nycteris borealis* (after Handley, 1960:468, 469).

Fig. 179. *Nycteris castanea*, San Blas Armila (Quebrada Venado), Panamá, No. 335422 U.S.N.M., ♀, X 2.

MARGINAL RECORDS.—Panamá (Handley, 1966:771): Armila; type locality. See Map 172.

Nycteris egregia (Peters)
Big Red Bat

1871. *Atalapha egregia* Peters, Monatsb. preuss. Akad. Wiss., Berlin, p. 912, type from Santa Catarina, Brazil.

Measurements of holotype (taken by G. S. Miller, *fide* C. O. Handley): total length, 127; forearm, 48. Measurements of specimen from

Panamá: total length, 130; forearm, 50.0; condylobasal length, 15.8; zygomatic breadth, 12.1; maxillary tooth-row, 6.3. Color as in *N. borealis frantzii* except underparts same as upper parts, not paler. Size exceeds that of all other reddish species of *Nycteris*.

MARGINAL RECORDS.—Panamá: Armila (Handley, 1966:772), thence into South America. See Map 173.

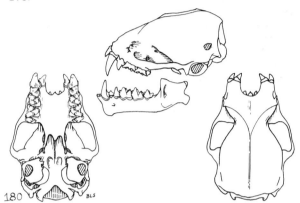

Fig. 180. *Nycteris egregia*, San Blas Armila (Quebrada Venado), Panamá, No. 335424 U.S.N.M., ♀, X 2.

Nycteris cinerea
Hoary Bat

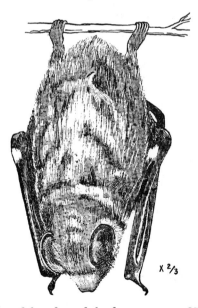

X 2/3

Length of head and body, approx. 85; total length, 134–140; length of forearm, 46–55; greatest length of skull, 17.0–18.5; maxillary tooth-row, 5.3–6.5. Upper parts varying considerably in color but not geographically, yellowish brown to mahogany brown strongly frosted with silver; underparts whitish on belly, pale brown on chest, yellowish on throat; ears rimmed with black. Skull robust; rostrum broad, short; zygomatic arches widespread.

Nycteris cinerea cinerea (Palisot de Beauvois)

1796. *Vespertilio cinereus* (misspelled *linereus*) Palisot de Beauvois, Catalogue raisonné du muséum de Mr. C. W. Peale, Philadelphia, p. 18, type from Philadelphia, Pennsylvania.

1912. *Nycteris cinerea*, Miller, Bull. U.S. Nat. Mus., 79:64, December 31.

1823. *Vespertilio pruinosus* Say, *in* Long, Account of an exped. . . . to the Rocky Mts. . . . , 1:167, type from Engineer Cantonment, Washington Co., Nebraska.

1861. *A*[*talapha*]. *mexicana* Saussure, Revue et Mag. Zool., Paris, ser. 2, 13:97, March, type from an unknown locality, probably from Veracruz, Puebla, or Oaxaca.

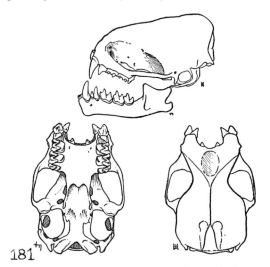

Fig. 181. *Nycteris cinerea cinerea*, Camp Verde, Yavapai Co., Arizona, No. 71588 M.V.Z., ♀, X 2.

MARGINAL RECORDS.—Keewatin: Bear Island, Southampton Island. Nova Scotia: Halifax, thence southward along Atlantic Coast to Florida: between Tavares and Apopka, thence westward along Gulf Coast to Tamaulipas: Matamoros. Veracruz: Jalapa (Hall and Dalquest, 1963:251). Guatemala (Carter, *et al.*, 1966:497): *San Pedro Soloma, 2270 m.;* 1 km. NE Aguacatán, 1620 m. Oaxaca: 10 mi. NE Cerro San Felipe (Goodwin, 1969:105). Michoacán: Barrancá Seca. Sinaloa: 2 km. W Palmitos, Hwy. 40, 6000 ft. (Irwin and Baker, 1967:195). Chihuahua: Mojarachic. Sonora: Río Yaqui, *ca.* 1 mi. S El Novillo (Findley and Jones, 1965:331, as *L. cinereus* only); Isla Datil (Turner's Island) (Baker and Christianson, 1966:311). Baja California (Banks, 1967c:225): Bahía de los Angeles; La Laguna; San Antonio; *Valladares*; Laguna Hanson. California: Chula Vista; South Farallon Island (Tenaza, 1966:533). British Columbia: Victoria (Cowan and Guiguet, 1965:75); Vancouver; Alta Lake. Alberta: Edmonton (E. T. Jones, 1974:244). Mackenzie: Resolution. Also, Bermuda (Vand Gelder and Wingate, 1961:1, as *Lasiurus cinereus* only). This species occurs

in Hawaii and South America where different subspecific names have been proposed.

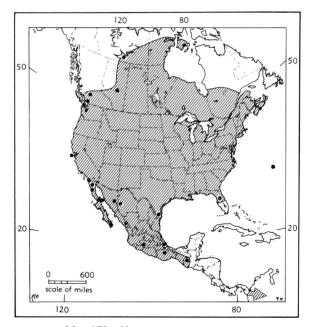

Map 176. *Nycteris cinerea cinerea*.

Genus Nycticeius Rafinesque—Evening Bats

Revised by Miller, N. Amer. Fauna, 13:118–121, October 16, 1897. See also Watkins, Mammalian species, 23:1–4, November 29, 1972.

1819. *Nycticeius* Rafinesque, Jour. Phys. Chim. Hist. Nat. et des Arts, Paris, 88:417, June. Type, *Vespertilio humeralis* Rafinesque.

1827. *Nycticeus* Lesson, Manuel de mammalogie . . . , p. 98, a variant spelling of *Nycticeius* Rafinesque.

1827. *Nycticejus* Temminck, Monographies de mammalogie . . . , 1:xvii, a variant spelling of *Nycticeius* Rafinesque.

1830. *Nycticeyx* Wagler, Naturliches System der Amphibien . . . , p. 13, a variant spelling of *Nycticeius* Rafinesque.

1831. *Nycticea* Le Conte, *in* McMurtrie, The animal kingdom . . . by the Baron Cuvier . . . , 1(App.):432, a variant spelling of *Nycticeius* Rafinesque.

1947. *Scoteanax* Troughton, Furred animals of Australia, p. 353. Type, *Nycticejus ruepellii* Peters. Valid as a subgenus.

1947. *Scotorepens* Troughton, Furred animals of Australia, p. 354. Type, *Scoteinus orion* Troughton. Valid as a subgenus.

Lower teeth as in *Eptesicus;* I1 simple, unicuspid, nearly half as high as C and in contact with C or nearly so; upper premolar without cusp on inner side; M3 with area of crown more than half that of M1 or M2; mesostyle, metacone, and three commissures well developed on M3. Dentition, i. $\frac{1}{3}$, c. $\frac{1}{1}$, p. $\frac{1}{2}$, m. $\frac{3}{3}$.

KEY TO SPECIES OF NYCTICEIUS

1. Total length less than 85; occurring in Cuba. . . . .*N. cubanus*,   p. 227
1'. Total length usually more than 85; occurring on mainland. . . . . *N. humeralis*,   p. 227

Nycticeius humeralis
Evening Bat

Length of head and body, approx. 56; total length, 85–99; length of forearm, 33.6–39; greatest length of skull, 14.0–14.7. Skull short, low, broad, and robust; dorsal profile from nares to occiput almost straight; teeth small for size of skull. Pelage short, sparse; upper parts medium to dark brown, hairs plumbeous basally; underparts paler, sometimes as pale as tawny; membranes naked.

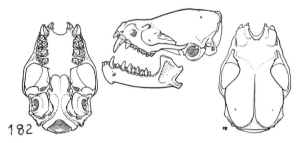

Fig. 182. *Nycticeius humeralis humeralis*, St. Andrews Parish, Charleston Co., South Carolina, No. 97176 M.V.Z., ♂, X 2.

Nycticeius humeralis humeralis (Rafinesque)

1818. *Vespertilio humeralis* Rafinesque, Amer. Month. Mag., 3(6):445, October, type from Kentucky.
1819. *N[ycticeius]. humeralis* Rafinesque, Jour. Phys. Chim. Hist. Nat. et des Arts, Paris, 88:417, June.
1831. *Nycticea crepuscularis* Le Conte, *in* McMurtrie, The animal kingdom . . . by the Baron Cuvier . . . , 1(App.):432. Type locality not stated.

MARGINAL RECORDS.—Iowa: S fork Iowa River, 2½ mi. S Eldora (Bowles, 1975:55). Illinois: Sugar Mound. Michigan: Climax. Ontario: Point Pelee. Pennsylvania: Carlisle; Buckingham (Doutt, *et al.*, 1973:87). District of Columbia: Washington, thence southward along Atlantic Coast to Florida: Kenansville; Jupiter (Layne, 1974:389); Indian Key, Tampa Bay, thence westward along Gulf Coast to Texas: Bee County (Davis, 1966:62, as *N. humeralis* only); 20 mi. SW Hunt. Oklahoma: 3 mi. W Willis (119430 KU). Kansas: ½ mi. S, 3 mi. E Wilmore (120200 KU); 6 mi. E Ford (Birney and Rising, 1968:522); Burdett (121793 KU); sec. 1, T. 1 S, R. 6 W (Jones, *et al.*, 1967:22). Nebraska: 2 mi. N, 2 mi. E Bellwood (Kunz, 1965:201).—See addenda.

Nycticeius humeralis mexicanus Davis

1944. *Nycticeius humeralis mexicanus* Davis, Jour. Mamm., 25:380, December 12, type from Río Ramos, 1000 ft., 20 km. NW Montemorelos, Nuevo León.

MARGINAL RECORDS.—Coahuila: 2 mi. W Jiménez, 850 ft. Texas (Davis, 1966:62, as *N. humeralis* only): Hidalgo County; Cameron County. Tamaulipas: 7 mi. W La Pesca (Baker and Webb, 1967:186). Veracruz: 4 km. NE Tuxpan (Hall and Dalquest, 1963:252). San Luis Potosí: 19 km. SW Ebano. Nuevo León: Linares; type locality. Coahuila: 8 mi. N, 4 mi. W Musquiz, 1800 ft.

Nycticeius humeralis subtropicalis Schwartz

1951. *Nycticeius humeralis subtropicalis* Schwartz, Jour. Mamm., 32:233, May 21, type from 2½ mi. W Monroe Station, Collier Co., Florida.

MARGINAL RECORDS (Layne, 1974:389).—Florida: vic. Ochopee; *type locality;* 6 mi. W Peters.

Nycticeius cubanus (Gundlach)
Cuban Evening Bat

1861. *Vesperus cubanus* Gundlach, Monatsb. preuss. Akad. Wiss., Berlin, p. 150, type from near Cárdenas, Matanzas, Cuba.
1904. *Nycticeius cubanus*, Miller, Proc. U.S. Nat. Mus., 27:338, January 23.

Length of head and body, approx. 52. Closely resembling *N. humeralis;* differs mainly in smaller size (total length, 73–83; length of forearm, 28.6–32.4).

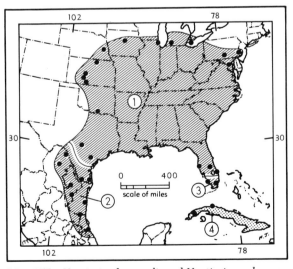

Map 177. *Nycticeius humeralis* and *Nycticeius cubanus*.

1. *N. h. humeralis*   3. *N. h. subtropicalis*
2. *N. h. mexicanus*   4. *N. cubanus*

MARGINAL RECORDS.—Cuba: type locality; Pinar del Río.

Genus **Rhogeessa** H. Allen—Little Yellow Bats

Revised by LaVal, Univ. Kansas Mus. Nat. Hist., Occas. Pap., 19:1–47, November 9, 1973.

1866. *Rhogeessa* H. Allen, Proc. Acad. Nat. Sci. Philadelphia, 18:285. Type, *Rhogeessa tumida* H. Allen.
1873. *Rhogoessa* Marschall, Nomenclator Zool., Mamm., p. 11, a variant spelling.
1906. *Baeodon* Miller, Proc. Biol. Soc. Washington, 19:85, June 4. Type, *Rhogeessa alleni* Thomas.

Length of head and body, 36–53; tail, 26–44; forearm, 24.5–34.4; length of skull including incisors, 11.0–15.4. Upper parts brownish; underparts paler. Resembling *Nycticeius* but skull narrower relative to length; also, i3 lacking inner lobe and therefore much smaller than i2 or i1 (i3 in *R. alleni* less than half as large as i2 and in many specimens a mere spicule nearly concealed beneath cingulum of C and having a cross-sectional area one-sixth or less of that of i2). Compared with *Nycticeius humeralis:* upper tooth-rows divergent posteriorly (instead of nearly parallel); basisphenoidal fossae ordinarily absent (instead of ordinarily present); M3 lacking (instead of having) 3rd commissure and mesostyle. Females of North American populations average 4 per cent larger than males in linear measurements.

KEY TO NORTH AMERICAN SPECIES OF RHOGEESSA

1. Greatest length of skull including incisors more than 14.5; i3 unicuspid; 3rd metacarpal averaging 2.2 times as long as 1st phalanx of 3rd digit. . . . .*R. alleni,*   p. 230
1′. Greatest length of skull including incisors less than 14.5; i3 often bicuspid; 3rd metacarpal averaging more than 2.2 times as long as 1st phalanx of 3rd digit.
   2. Ear from notch usually more than 16.5; pelage of dorsum 3-banded, darker basally than distally. . . . .*R. gracilis,*   p. 230
   2′. Ear from notch usually less than 16.5; pelage of dorsum 2-banded, paler basally than distally.
      3. Greatest length of skull including incisors averaging less than 11.6; lingual cingulum of C1 smooth (lacks cusps). . . . . . .*R. mira,*   p. 229
      3′. Greatest length of skull including incisors averaging more than 11.6; lingual cingulum of C1 not smooth (usually has cusps).
         4. Uropatagium sparsely to heavily furred from base to line half

way between knees and feet; i3 usually less than ¼ size of i2. . . . .*R. parvula,*   p. 229
4′. Uropatagium furred only at base; i3 usually more than ¼ size of i2. . .*R. tumida,*   p. 228

Rhogeessa tumida H. Allen
Central American Yellow Bat

1866. *R[hogeessa]. tumida* H. Allen, Proc. Acad. Nat. Sci. Philadelphia, 18:286, type from Mirador, Veracruz.
1903. *Rhogeessa io* Thomas, Ann. Mag. Nat. Hist., (7)11:382, April, type from Valencia, Carabobo, Venezuela.
1903. *Rhogeessa velilla* Thomas, Ann. Mag. Nat. Hist., (7)11:383, April, type from Puná, 10 m., Isla Puná, Gulf of Guyaquil, Ecuador.
1913. *Rhogeessa bombyx* Thomas, Ann. Mag. Nat. Hist., (8)12:569, December, type from Condoto, 300 ft., Chocó, Colombia.
1958. *Rhogeessa tumida riparia* Goodwin, Amer. Mus. Novit., 1923:5, December 31, type from Cumanacoa, 700 ft., Sucre, Venezuela.
1958. *Rhogeessa parvula aeneus* Goodwin, Amer. Mus. Novit., 1923:6, December 31, type from Chichén-Itzá, *ca.* 10 m., Yucatán.

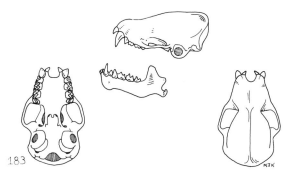

Fig. 183. *Rhogeessa tumida,* 3 mi. N La Pesca, Tamaulipas, No. 55208 K.U., ♂, X 2.

Length of head and body, 41–49; ear from notch, 11–14; forearm, 25.0–33.2; length of skull including incisors, 11.0–14.1. Hairs on dorsum of Central American specimens bicolored: distal third Fuscous-Black to Pinkish Cinnamon, basally buffy gray to buffy yellow. Ventrally, hairs of uniform color or tips Buffy Brown to Light Ochraceous-Buff and paler at bases. Upper side of uropatagium sparsely haired rarely as far as knees. Sagittal crest present as often as absent; i3 same size as i2 or smaller, but cusps smaller than on i2; lingual cingulum of C having two small cusps or mere traces thereof. In general, nearly bare uropatagium and larger i3 separate *R. tumida* from *R. parvula,* and on the Isthmus of Tehuantepec tips of hairs on dorsum of *R. tumida* are darker.

LaVal's (1973b:31) statement that there is "a bewildering amount and kind of geographic variation" in this taxon suggests that most of the names he places as synonyms of *R. tumida* are names of valid subspecies, or that he has misidentified some *R. parvula* from Chiapas as *R. tumida*.

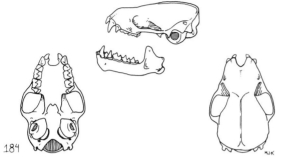

Fig. 184. *Rhogeessa tumida*, Finca San Salvador, 15 km. SE San Clemente, 1000 m., Chiapas, No. 102619 K.U., ♀, X 2.

MARGINAL RECORDS (LaVal, 1973b:43, 44).—Tamaulipas: 4 mi. N La Pesca. Veracruz: type locality. Tabasco: Rancho El Tumbo, 4 km. E F.F.C.C. El Zapote, Macuspana. Campeche: 5 km. S Champotón. Yucatán: 10 mi. W Progreso. Quintana Roo: Isla Cozumel, 4 km. N San Miguel. Belize: Turneffe Id., Calabash Cay. Honduras: 23 mi. N San Pedro Sula, thence down coast to Panamá: Fort Sherman, thence along coast into South America, and back up Pacific Coast to Panamá: La Palma de Darién; Guanico Arriba, *ca.* 200 m. Costa Rica: "Pacific Coast." Nicaragua: $6\frac{9}{10}$ mi. E San Juan del Sur. El Salvador: Puerto del Triunfo. Guatemala: Astillero. Chiapas: Rancho San Miguel, 32 mi. SW Cintalapa. Oaxaca: 3 km. W Estación Vicente, Municipio de Acatlán, 60 m. Veracruz: $12\frac{1}{2}$ mi. N Tihuatlán, 300 ft. San Luis Potosí: 3 km. N Taninul, 650 ft. Tamaulipas: 30 mi. N El Mante, Río Cielito; Santa María, 870 m.

Rhogeessa parvula H. Allen
Little Yellow Bat

1866. *R*[*hogeessa*]. *parvula* H. Allen, Proc. Acad. Nat. Sci. Philadelphia, 18:285, type from Tres Marías Islands, Nayarit.
1958. *Rhogeessa tumida major* Goodwin, Amer. Mus. Novit., 1923:4, December 31, type from San Bartolo Yautepec, 800 m., Oaxaca.

Length of head and body, 36–49; ear from notch, 11–14; forearm, 25.8–32.8; length of skull including incisors, 11.4–14.1. Hairs on dorsum bicolored: distal third Hair Brown to Warm Buff; remainder, which may or may not contrast with distal third, pale grayish to buff to pale yellow. Ventrally, hairs of uniform color or tips Cartridge Buff to Light Ochraceous-Buff and bases slightly contrasting with tips. Upper side of uropatagium

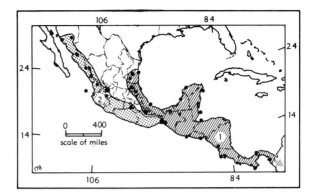

Map 178. *Rhogeessa tumida* (1) and *Rhogeessa parvula* (2).

sparsely to thickly furred usually to line half way between knees and feet. Third metacarpal 2.7 times as long as 1st phalanx of 3rd digit where *alleni*, *gracilis*, and *parvula* occur in same area. Sagittal crest absent in 115 of 131 skulls; i3 varying in size from nearly equal to i2 to minute but never so small as in *alleni*; lingual cingulum of C well developed, bearing two cusps of variable size; postorbital region narrow relative to length of skull.

Fig. 185. *Rhogeessa parvula*, W side Alamos, Sonora, No. 24854 K. U., ♀, X 2.

MARGINAL RECORDS (LaVal, 1973b:42, 43, unless otherwise noted).—Sonora: 28 mi. E Mazatán, 500 m. Sinaloa: *1 mi. S El Cajón, 1800 ft.*; 16 km. NNE Choix, 1700 ft.; 7 mi. ESE Sanalona, 600 ft. Durango: Santa Ana (12 km. E Cosalá of Sinaloa), 1300 ft. Sinaloa: 2 mi. E Palmito; *5 mi. WSW Plomosas, 800 ft.* Nayarit: *ca.* 40 mi. E Acaponeta. Zacatecas: Santa Rosa (Matson and Patten, 1975:8). Jalisco: 9 mi. N Guadalajara, 4000 ft. Morelos: Río Oaxtepec, 1 km. S Oaxtepec, 890 m. Oaxaca: *Río Ostuta, 4 mi. W Zanatepec, sea level*; 20 mi. W Tapanatepec, sea level, thence up coast including type locality (Tres Marías Islands of Nayarit), to Sonora: Estero Tastiota, sea level.

Rhogeessa mira LaVal
Least Yellow Bat

1973. *Rhogeessa mira* LaVal, Univ. Kansas Mus. Nat. Hist., Occas. Pap., 19:26, November 9, type from 20 km. N El Infiernillo, elevation 125 m., Michoacán.

Length of head and body, 37–39; ear from notch, 10–12; forearm, 24.5–26.9; length of skull including incisors, 11.0–11.6; smallest species of the genus. Hairs on dorsum bicolored: distal third Buckthorn Brown to Buffy Brown, basal two-thirds more buffy than tips but not contrasting strongly with tips. Hairs on venter same color as basal two-thirds of hairs on dorsum. Upper side of uropatagium sparsely furred to or past knees. Sagittal crest absent; i3 only slightly smaller than i2; lingual cingulum of C smooth, lacking slightest suggestion of cusps.

MARGINAL RECORDS (LaVal, 1973b:43).— Michoacán: *type locality; 7 km. N El Infiernillo.*

Rhogeessa gracilis Miller
Slender Yellow Bat

1897. *Rhogeessa gracilis* Miller, N. Amer. Fauna, 13:126, October 16, type from Piaxtla, Puebla.

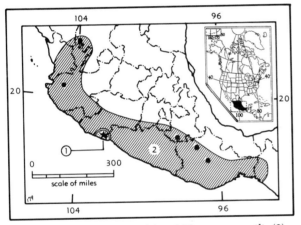

Map 179.   *Rhogeessa mira* (1) and *Rhogeessa gracilis* (2).

Length of head and body, 46–48; ear from notch, 17.5–18.0 (longest in genus); forearm, 30.2–33.4; length of skull including incisors, 12.9–13.8. Hairs on dorsum tricolored: distal fourth Light Ochraceous-Buff, succeeding fourth pale buff, basal half dark grayish brown. Hairs on venter bicolored: distally Pinkish Buff, basally dark grayish brown. Upper side of uropatagium sparsely furred almost to knees. Third metacarpal 2.6 times as long as 1st phalanx of 3rd digit. Sagittal crest present in seven of eight skulls, but poorly developed; i3 nearly as large as i2; lingual cingulum of C low, smooth, lacking cusps; skull narrow relative to length; shaft of baculum more nearly straight than in any other species of genus.

MARGINAL RECORDS (LaVal, 1973b:42).— Jalisco: 5 mi. NE Huejuquilla, 6200 ft.; 10 mi. SE Talpa

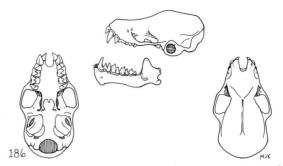

Fig. 186.   *Rhogeessa gracilis*, 5 mi. NE Huejuquilla, 6200 ft., Jalisco, No. 108976 K.U., ♂, X 2.

de Allende, 5350 ft.; *17 km. SE Talpa de Allende, 5200 ft.* Puebla: type locality. Oaxaca: Valero Taujano, 2 mi. W Tomallim [= Tomellin?]; Cerro San Felipe, San Felipe del Agua, *ca.* 1700 m.; *Isthmus of Tehuantepec.*

Rhogeessa alleni Thomas
Allen's Yellow Bat

1892. *Rhogeessa alleni* Thomas, Ann. Mag. Nat. Hist., ser. 6, 10:477, December, type from Santa Rosalía, near Autlán, Jalisco. (From 1906–1973 known as *Baeodon alleni*.)

Length of head and body, 46–53; ear from notch, 14–16; forearm, 30.8–34.4; length of skull including incisors, 14.7–15.4; largest species of the genus. Hairs on dorsum tricolored: distal fourth Dresden Brown, middle half buffy, basal fourth gray. Hairs on venter bicolored: distally Light Ochraceous-Buff, basally gray. Upper side of uropatagium almost bare. Third metacarpal 2.2 times as long as 1st phalanx of 3rd digit. Sagittal crest in eight of nine specimens examined; sagittal and lambdoidal crests forming "helmet" above occiput; i3 mere spicule nearly concealed beneath cingulum of c; 1 or 2 cusps on lingual side of C; only 3.1–3.3 across postorbital constriction.

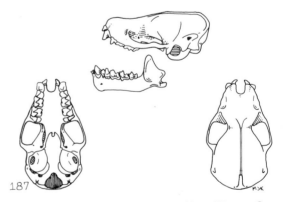

Fig. 187.   *Rhogeessa alleni*, Cuicatalán, 590 m., Oaxaca, No. 29439 K.U., ♂, X 2.

MARGINAL RECORDS (LaVal, 1973b:42, unless otherwise noted).—Zacatecas (Matson and Patten, 1975:8): 16 km. W Jalpa; 6⅖ km. S Jalpa. Puebla: 10 km. W Acatlán, 6000 ft. Oaxaca: 2 mi. NNW Tamazulapan, 1990 m.; Cuicatlán, 590 m.; 2 mi. N, 6 mi. W Nejapa. Michoacán: *20 km. N El Infiernillo*; 7 km. N Infiernillo, *ca.* 175 m. Jalisco: type locality; Piedra Gorda, *ca.* 8 km. NW Soyatlán del Oro, 1600 m.—**See** addenda.

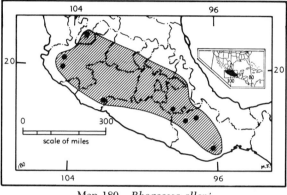

Map 180. *Rhogeessa alleni.*

Genus **Euderma** H. Allen—Spotted Bat

Revised by Handley, Proc. U.S. Nat. Mus., 110:95–246, September 3, 1959.

1892. *Euderma* H. Allen, Proc. Acad. Nat. Sci. Philadelphia, 43:467, January 19. Type, *Histiotus maculatus* J. A. Allen.

Skull with sharply ridged supraorbital region; exceptionally elongated braincase; zygoma relatively heavy, expanding in middle third of arch; auditory bullae roughly elliptical, greatest diameter of each approx. equaling length of upper tooth-row exclusive of incisors; lower canine relatively smallest in subfamily, and when viewed from slightly anterior to a lateral view, appearing unequally bilobed. Dentition, i. $\frac{2}{3}$, c. $\frac{1}{1}$, p. $\frac{2}{2}$, m. $\frac{3}{3}$.

The genus is represented by one apparently relict species from a restricted habitat.

Euderma maculatum (J. A. Allen)
Spotted Bat

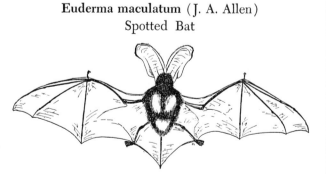

1891. *Histiotus maculatus* J. A. Allen, Bull. Amer. Mus. Nat. Hist., 3:195, February 20, type from near Piru, Ventura Co., California; probably at mouth Castac Creek, Santa Clara Valley, Los Angeles Co., 8 mi. E Piru.
1894. *Euderma maculata*, H. Allen, Bull. U.S. Nat. Mus., 43:61, March 14.

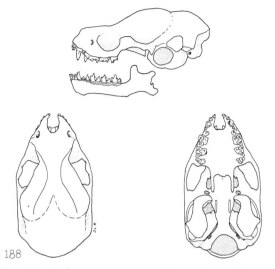

Fig. 188. *Euderma maculatum*, Weir Tank, 1½ mi. E Springtime Campground, 7200 ft., Socorro Co., New Mexico, No. 25000 M.S.B., ♂, X 2.

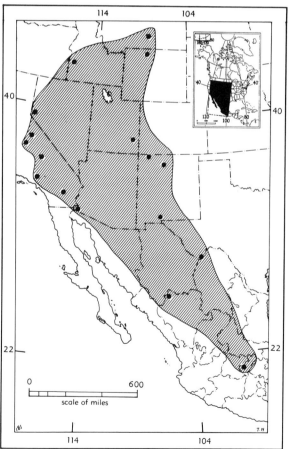

Map 181. *Euderma maculatum.*

Length of head and body of an adult female from Utah approx. 65. Extreme measurements (Handley, 1959:224) of both sexes are: total length, 107–115; length of ear from notch, 37–47; length of forearm, 48–51. In two males and two females (Handley, 1959:218): greatest length of skull, 18.4–19.0; interorbital breadth, 3.9–4.3; zygomatic breadth, 9.9–10.9; length of maxillary tooth-row (one male, two females), 7.3–7.7. Upper parts blackish with two large white, roughly circular spots on the shoulders, one at the base of the tail, and patches at the posterior base of each auricle; underparts white tipped with incompletely concealed black bases. Membranes pale grayish brown; auricles and tragi yellowish brown.

MARGINAL RECORDS (Handley, 1959:127, unless otherwise noted).—Montana: Billings. Wyoming (Mickey, 1961:401): Byron. Utah: 5 mi. NW Monticello. New Mexico: 2 mi. N Aztec (C. J. Jones, 1961:539); Ghost Ranch (*ibid.*); Mesilla Park. Texas: Big Bend National Park (Easterla, 1970:306). Querétaro: 1 mi. NE Peña Blanca (Schmidly and Martin, 1973:92). Durango: Navarro (Gardner, 1965:105). Arizona: 4 mi. S Yuma. California: Mecca; type locality; Red Rock Canyon; Friant (Medeiros and Heckmann, 1971:858); Yosemite Valley. Nevada: Reno. Idaho: 15 mi. SW Caldwell.

Genus **Plecotus** É. Geoffroy Saint-Hilaire
Big-eared Bats

Revised by Handley, Proc. U.S. Nat. Mus., 110:95–246, September 3, 1959.

1818. *Plecotus* É. Geoffroy Saint-Hilaire, Description des Mammifères qui se trouvent en Égypte. *In* France. Commission des monuments d'Égypte. Description de l'Égypte . . . , Hist. Nat., 2:112, 118.

Skull slender, highly arched; rostrum relatively greatly reduced; lachrymal region smoothly rounded. Ears much enlarged, joined basally across forehead. Muzzle having dorsolateral glandular masses, in some species rising above muzzle as prominent lumps. Dentition, i. $\frac{2}{3}$, c. $\frac{1}{1}$, p. $\frac{2}{3}$, m. $\frac{3}{3}$.

KEY TO NORTH AMERICAN SUBGENERA AND SPECIES
OF PLECOTUS

(Adapted from Handley, 1959:137)

1. Nostril not elongated posteriorly; glands on each side of muzzle between eye and nostril not conspicuous; accessory basal lobe of auricle developed into projecting lappet; calcar having well-developed keel; breadth of braincase amounting to more than half of greatest length of skull. subgenus *Idionycteris*, *P. phyllotis*, p. 232

1'. Nostril elongated posteriorly; muzzle bearing conspicuous glandular mass on each side between eye and nostril; accessory basal lobe of auricle absent; calcar not keeled; breadth of braincase amounting to less than half of greatest length of skull. . . . . . subgenus *Corynorhinus*, p. 233
  2. White or whitish tips of hair on venter contrasting sharply with black or blackish bases of hair. . . *P. rafinesquii*, p. 234
  2'. Brown or brownish tips of hair on venter not contrasting sharply with gray or brownish bases of hair.
    3. Greatest length of skull usually less than 15.5; tragus usually less than 13; known only from México. . . . . . *P. mexicanus*, p. 233
    3'. Greatest length of skull usually more than 15.5; tragus usually more than 13; known from Appalachians, Ozarks, western United States, southwestern Canada, and México.
          P. townsendii, p. 234

Subgenus **Idionycteris** Anthony

1923. *Idionycteris* Anthony, Amer. Mus. Novit., 54:1, January 17. Type, *Idionycteris mexicanus* Anthony.—See addenda.

Nostril unspecialized; accessory basal lobe of auricle developed into projecting lappet; calcar having well-developed keel; breadth of braincase more than half of greatest length of skull.

Plecotus phyllotis (G. M. Allen)
Allen's Big-eared Bat

1916. *Corynorhinus phyllotis* G. M. Allen, Bull. Mus. Comp. Zool., 60:352, April, type from San Luis Potosí, probably near city of same name.
1953. *Plecotus phyllotis*, Dalquest, Louisiana State Univ. Studies, Biol. Sci. Ser., 1:63, December 28.
1923. *Idionycteris mexicanus* Anthony, Amer. Mus. Novit., 54:1, January 17, type from Miquihuana, Tamaulipas.

Head and body approx. 65; in 14 males from 1 mi. NE Aztec Peak in Arizona (Hayward and Johnson, 1961:402), external measurements: total length, 115 (103–118); tail, 50 (46–55); hind foot, 10.8 (9–12); ear from notch, 40 (38–43). Greatest length of skull in holotype, a female, 17.2; length of upper tooth-row of same female, 6.3. Upper parts tawny olive, hairs blackish brown basally; underparts slightly paler.

MARGINAL RECORDS (Handley, 1959:131, unless otherwise noted).—Utah: Squaw Spring, sec. 25, T. 30 S, R. 19 E, 5100 ft. (Armstrong, 1974:115); 5 mi. N Blanding, 6000 ft. (Black, 1970:190). New Mexico: Weir Tank (Findley, *et al.*, 1975:61). Nuevo León: Huasteca Canyon, 17 mi. SW Monterrey, 4500 ft. (Car-

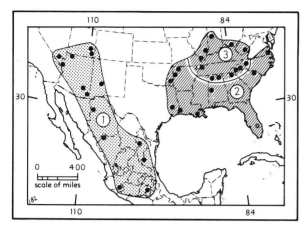

Map 182. *Plecotus phyllotis* and *Plecotus rafinesquii*

1. *P. phyllotis*   2. *P. rafinesquii macrotis*
3. *P. rafinesquii rafinesquii*

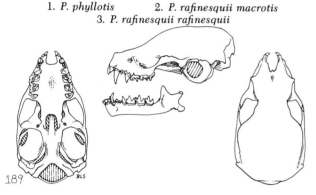

Fig. 189. *Plecotus phyllotis*, Wilbank's Ranch, Sierra Ancha, 7200 ft., Gila Co., Arizona, No. 83649 K.U., ♂, X 2.

ter, *et al.*, 1966:497). Tamaulipas: Miquihuana. Distrito Federal: Ciudad Universitaria, 2250 m. (Villa-R., 1967:427). Jalisco: Volcán de Fuego [= Volcán de Colima], 9800 ft. (Genoways and Jones, 1967:479). Durango: Navarro, 6100 ft. (*ibid.*). Chihuahua: 11$\frac{1}{10}$ mi. SE Nueva Casas Grandes (Bogan and Williams, 1970:131, 132). Arizona: 5 mi. WSW Portal, 5400 ft.; Oak Grove Canyon (Hayward and Johnson, 1961:402); 2 mi. W Union Pass, Black Mts. (Baker and Mascarello, 1969:251). Nevada: Calico Spring (O'Farrell and Bradley, 1969:128). Arizona: vic. Pipe Springs National Monument, 5000 ft. (Genoways and Jones, 1967:480).

Subgenus **Corynorhinus** H. Allen

1865. *Corynorhinus* H. Allen, Proc. Acad. Nat. Sci. Philadelphia, 17:173. Type, *Plecotus macrotis* Le Conte.
1865. *Corynorhynchus* Peters, Monatsb. preuss. Akad. Wiss., Berlin, p. 524, October, a *lapsus*?
1875. *Corinorhinus* Dobson, Ann. Mag. Nat. Hist., 16:348, November.

·Nostril elongated posteriorly; accessory basal lobe of auricle absent; calcar not keeled; breadth of braincase less than half of greatest length of skull.

Plecotus mexicanus (G. M. Allen)
Mexican Big-eared Bat

1916. *Corynorhinus megalotis mexicanus* G. M. Allen, Bull. Mus. Comp. Zool., 60:347, April, type from near Pacheco, Chihuahua.
1959. *Plecotus mexicanus*, Handley, Proc. U.S. Nat. Mus., 110:141, September 3.

Length of head and body approx. 50; extreme measurements (Handley, 1959:226) of both sexes (13 males, 57 females) are: total length, 90–103; length of forearm, 39.3–45.2. In 16 males and 68 females (Handley, 1959:219): greatest length of skull, 14.7–15.9; interorbital breadth, 3.2–3.6; zygomatic breadth, 7.6–8.6; length of maxillary tooth-row, 4.6–5.0. Upper parts vary from Verona Brown to Fuscous; underparts Bone Brown to Fuscous-Black. Differs from Mexican *Plecotus townsendii* usually in having fewer cross-ribs on the interfemoral membrane, a smaller tragus on the average, and a smaller skull.

MARGINAL RECORDS (Handley, 1959:151, unless otherwise noted).—Sonora: Santa María Mine. Chihuahua: near Pacheco [Sierra de Breña, 8000 ft.]; Mojarachic [= Mafuarachic]; Barranca del Cobre, 23 mi. S, 1$\frac{1}{2}$ mi. E Creel, *ca.* 1000 m. (Anderson, 1972:255). Zacatecas: Sierra de Valparaíso [13 mi. W Valparaíso, 8200 ft.]. Guanajuato: Santa Rosa (9500 ft.). México: Monte Río Frío, 10,500 ft., 28 mi. ESE Ciudad México. Veracruz: 6 mi. WSW Zacualpilla, 6500 ft. Nuevo León: 22 mi. SSE Monterrey. Tamaulipas: Cueva Chica de la Perra, *ca.* 8 mi. NW Gómez Farías, 7000 ft. (Mollhagen, 1971:21). Veracruz: 2$\frac{1}{2}$ mi. E Las Vigas, 8500 ft.; *Jico (5500 ft.)*. Puebla: 2 mi. NW Esperanza. Morelos: no exact locality. Michoacán: 2 mi. N Pátzcuaro, 7100 ft. Jalisco (Watkins, *et al.*, 1972:30): 12 mi. S Tolimán, 7700 ft.; 15 mi. S, 9 mi. E Talpa de Allende.—Also, Yucatán: 8 km. by highway from Tixpehual on highway to Tixkokob (Koopman, 1974:873). Quintana Roo: Cozumel Island (*ibid.*).

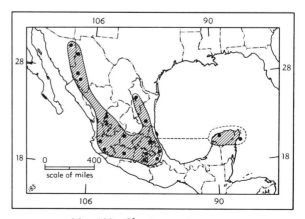

Map 183. *Plecotus mexicanus*.

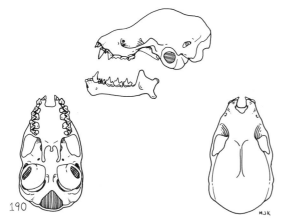

Fig. 190. *Plecotus mexicanus*, 4 km. E Las Vigas, 8500 ft., Veracruz, No. 29849 K.U., ♀, X 2.

Plecotus rafinesquii
Rafinesque's Big-eared Bat

External measurements: total length, 91–106; forearm, 40.4–45.8. Cranial measurements: greatest length, 15.3–16.7; zygomatic breadth, 8.2–9.2; breadth of braincase, 7.4–8.2; maxillary tooth-row, 5.0–5.6. Upper parts gray at tips, black at bases; underparts white at tips, black at bases; 2 large fleshy lumps on snout; I1 bicuspid; long hairs on toes projecting beyond toenails. (See Map 182.)

Plecotus rafinesquii macrotis Le Conte

1831. *Plec[otus]. macrotis* Le Conte, *in* McMurtrie, The animal kingdom . . . by the Baron Cuvier . . . , 1(App.):431, type from Georgia (see Miller, N. Amer. Fauna, 45:51, October 16, 1897), probably vic. Le Conte Plantation, 5 mi. S Riceboro, Liberty Co. (see Handley, Proc. U.S. Nat. Mus., 110:162, September 3, 1959).
1959. *Plecotus rafinesquii macrotis* Handley, Proc. U.S. Nat. Mus., 110:161, September 3.
1837. *Plecotus lecontii* Cooper, Ann. Lyc. Nat. Hist. New York, 4:72, a renaming of *macrotis* Le Conte, which Cooper considered as "nonwise distinctive of the species" (*loc. cit.*).

MARGINAL RECORDS (Handley, 1959:163, unless otherwise noted).—Virginia: Lake Drummond, thence to Atlantic Coast and southward and westward to Louisiana: 9 mi. SE Houma; Starks Road (Lowery, 1974:133). Texas: Polk County (Michael and Birch, 1967:672). Oklahoma: 2½ mi. W Smithville. Arkansas: Mulberry; Osage River (Black, 1936:30). Alabama: Eutaw. South Carolina: Society Hill. Taylor (1964:301) recorded from ½ mi. W Large Cave, 3 mi. SW Reed, western Oklahoma, three individuals as *Corynorhinus rafinesqui*. The locality is not shown on the distribution map for *Plecotus rafinesquii* because the specimens were not saved (pers. comm.) and probably were *Plecotus townsendii* as defined by Handley (1959:165, 186, 191, 197).

Plecotus rafinesquii rafinesquii Lesson

1818. *Vespertilio megalotis* Rafinesque, Amer. Month. Mag., 3(6):446, October. Type locality, lower part Ohio River, probably in southern Indiana, Illinois, or western Kentucky in area between Wabash and Green rivers. Not *Vespertilio megalotis* Bechstein, 1800.
1827. *Plecotus rafinesquii* Lesson, Manuel de mammalogie . . . , p. 96, a renaming of *Vespertilio megalotis* Rafinesque.

MARGINAL RECORDS (Handley, 1959:165, unless otherwise noted).—Indiana: 1 3/10 mi. N West Lafayette (Wilson, 1960:500). Ohio: Waggoner Ripple Cave. West Virginia: Collison Cave. North Carolina: 10 mi. NW Taylorville; head of S fork Mills River, 3300 ft.; Highlands, 3850 ft. Georgia: Young Harris. Alabama: Monte Sano, 1600 ft.; 5 mi. E Leighton. Tennessee: 1 mi. NE Reelfoot Lake (Hoffmeister and Goodpaster, 1962:87); *10 mi. NE Tiptonville*. Illinois: ½ mi. N Olive Branch; *3 mi. WSW Mill Creek* (Layne, 1958:232); *4 mi. NW Mill Creek (ibid.)*; Mount Carmel.

Plecotus townsendii
Townsend's Big-eared Bat

X ½

External measurements: total length, 90–112; length of forearm, 39.2–47.6. Cranial measurements: greatest length, 15.2–17.2; zygomatic breadth, 8.2–9.6; breadth of braincase, 7.4–8.4; length of maxillary tooth-row, 4.8–5.6. Upper parts and underparts brown (pale brown to nearly black) at tips, black at bases, contrasting little with tips; two large fleshy lumps on snout; I1 unicuspid; hairs on toes not projecting beyond toenails.

Plecotus townsendii australis (Handley)

1955. *Corynorhinus townsendii australis* Handley, Jour. Washington Acad. Sci., 45:147, May 23, type from 2 mi. W Jacala, Hidalgo.
1959. *Plecotus townsendii australis* (Handley), Proc. U.S. Nat. Mus., 110:185, September 3.

MARGINAL RECORDS (Handley, 1959:189, unless

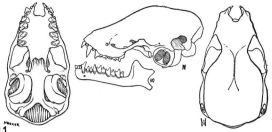

Fig. 191. *Plecotus townsendii pallescens*, 7 mi. S Cleveland Ranch, White Pine Co., Nevada, No. 45899 M.V.Z., ♂, X 2.

otherwise noted).—Coahuila: 4 mi. W Hda. La Mariposa, 2300 ft.; 4 mi. S, 9 mi. W San Buenaventura, 2000 ft.; ½ mi. N Muralla, 4500 ft.; 1 mi. S, 4 mi. W Bella Unión, 7000 ft. Tamaulipas: 5 km. S Miquihuana, 2150 m. (Alvarez and Ramírez-P., 1972:174). Hidalgo: type locality; Grutas Xoxafi, 6⅖ mi. SE Yoltepec [= Yolotepec?, 6600 ft.]. Oaxaca (Goodwin, 1969:107): 3 mi. SW Oaxaca de Juárez; Mitla; Tehuantepec. Guerrero: 1 mi. SSE Almolonga, 5600 ft. (Davis and Carter, 1962:73). Jalisco: San Pedro, Guadalajara; 17 km. NNW Soyatlán del Oro (Villa-R., 1967:430); San Andrés, 4900 ft., 10 mi. W Magdalena. Zacatecas: Sierra de Valparaíso [13 mi. W Valparaíso, 8200 ft.]. Durango: San Juan, 3800 ft., 12 mi. W Lerdo; 7 mi. N Campana, 3750 ft. (Baker and Greer, 1962:75). Chihuahua: 1 mi. N, 1 mi. W Salaices (Anderson, 1972:256).

Plecotus townsendii ingens (Handley)

1955. *Corynorhinus townsendii ingens* Handley, Jour. Washington Acad. Sci., 45:148, May 23, type from Hewlitt Cave, 12 mi. W Fayetteville, Washington Co., Arkansas.
1959. *Plecotus townsendii ingens* (Handley), Proc. U.S. Nat. Mus., 110:189, September 3.

MARGINAL RECORDS (Handley, 1959:190, unless otherwise noted).—Missouri: Stone County. Arkansas: Basset Cave near Hicks; Devils Icebox, 25 mi. SW Fayetteville. Oklahoma: Cave Springs Cave, Adair Co. (Glass, 1961:200). Arkansas: type locality.

Plecotus townsendii pallescens (Miller)

1897. *Corynorhinus macrotis pallescens* Miller, N. Amer. Fauna, 13:52, October 16, type from Keam Canyon, Navajo Co., Arizona.
1959. *Plecotus townsendii pallescens,* Handley, Proc. U.S. Nat. Mus., 110:190, September 3.
1914. *Corynorhinus macrotis intermedius* H. W. Grinnell, Univ. California Publ. Zool., 12:320, December 4, type from Auburn, 1300 ft., Placer Co., California.

MARGINAL RECORDS (Handley, 1959:194–196, unless otherwise noted).—British Columbia: Williams Lake (Cowan and Guiguet, 1956:67); Adams River, NW Shuswap Lake; Creston, Kootenay River. Montana (Hoffmann, et al., 1969:741, as *P. townsendii* only): Kalispell; 9 mi. W Loma; Judith Mts., 7 mi. N, 9 mi. E Lewistown. South Dakota: Ludlow Cave Hills; White River; mouth Spring Creek, Cheyenne River; 1 mi. S, 2 mi. E Hot Springs, 3400 ft. (Turner and Jones,

1968:447); 5½ mi. E Minnekahta, 4000 ft. (ibid.). Colorado (Armstrong, 1972:73): Ft. Collins; Colorado Springs; Trinidad; 2 mi. N, 19 mi. W Campo. Oklahoma; *Tesequite Canyon, 1⁷/₁₀ mi. SE Kenton.* Kansas: 6 mi. NW Aetna; Sun City. Oklahoma: *Marehew Cave, ¹/₂ mi. S Oklahoma–Kansas boundary;* Alva; 3 mi. SE Southard; Wichita Mts. National Wildlife Refuge. Texas: Hardeman County (Jameson, 1959:63); *Panther Cave, 10 mi. SE Quanah; 13 mi. NE Seymour* (Baccus, 1971:180); Kimble County (Davis, 1966:63, as *P. townsendii* only); Devil's Sink-hole, 6½ mi. NE Rocksprings; East Painted Cave, NW Del Rio; Langtry; Viviani Mines, Mariscal Mts. Chihuahua: Tinaja de Ponce, 2600 ft.; Casas Grandes; La República, 3900 ft. (Anderson, 1972:256). Baja California: Santa Catalina Island (Orr and Banks, 1964: 209). Sonora: Santa María Mine; Sáric. California: Potholes; Dulzura; San Clemente Island (von Bloeker, 1967:250); Santa Catalina Island (ibid.); San Nicolas Island (ibid.); Santa Cruz Island (ibid.); Old Fort Tejon, near Lebec; 4 mi. SE Porterville, 550 ft.; Auburn, 1300 ft.; mouth Battle Creek, near Bloody Island, 350 ft.; Lava Beds National Monument; *5 mi. SW Tule Lake.* Washington: Selah. British Columbia: Keremeos, Similkameen Valley; Riske Creek. Not found: California: Fort Creek; Old Station, 4300 ft.; Pioneer Cave.

Plecotus townsendii townsendii Cooper

1837. *Plecotus townsendii* Cooper, Ann. Lyc. Nat. Hist. New York, 4:73, November, type from Columbia River, restricted to Fort Vancouver, Clark Co., Washington, by Handley, Proc. U.S. Nat. Mus., 110:197, September 3, 1959.

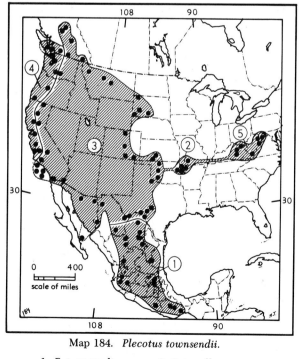

Map 184. *Plecotus townsendii.*

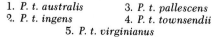

1. *P. t. australis*   3. *P. t. pallescens*
2. *P. t. ingens*   4. *P. t. townsendii*
       5. *P. t. virginianus*

MARGINAL RECORDS (Handley, 1959:200, 201, unless otherwise noted).—British Columbia: Comox, Vancouver Island; Newcastle Island; Hope (Hall and Kelson, 1959:200). Washington: Blakely Island; Seattle; Boulder Cave, Naches River, 43 mi. NW Yakima. Oregon: McKenzie Bridge; Siskiyou Mts., Oregon Caves National Monument. California: *Happy Camp;* Pope Creek, 8 mi. NW Monticello; 4 mi. ESE Mission San Jose; Hernandez; 6½ mi. SE Shandon; 5 mi. SSW Adelaida, thence westward to Pacific Coast and northward along coast to point of beginning.

Plecotus townsendii virginianus (Handley)

1955. *Corynorhinus townsendii virginianus* Handley, Jour. Washington Acad. Sci., 45:148, May 23, type from Schoolhouse Cave, 4⅗ mi. NE Riverton, 2205 ft., Pendleton Co., West Virginia.
1959. *Plecotus townsendii virginianus* (Handley), Proc. U.S. Nat. Mus., 110:201, September 3.

MARGINAL RECORDS (Handley, 1959:203, unless otherwise noted).—West Virginia: Preston County; Bakers Cave, near Durgon; *Tory's Cave, 10 mi. S Franklin.* Virginia: Burkes Garden, 3200 ft. Kentucky (Rippy and Harvey, 1965:499): Old Landing, Lee Co.; *northwestern Lee County;* Natural Bridge State Park; Bat Cave, Carter Caves State Park, Carter Co. West Virginia: The Sinks, Cave No. 1, 3500 ft., ½ mi. W Osceola.

SUBFAMILY NYCTOPHILINAE

Differ from Vespertilioninae in abruptly truncate muzzle, on anterior surface of which nostrils open forward beneath distinct horseshoe-shaped ridge or small nose-leaf (after Miller, 1907:234). Lower incisors 2–2 or 3–3.

Genus Antrozous H. Allen—Antrozoine Bats

1862. *Antrozous* H. Allen, Proc. Acad. Nat. Sci. Philadelphia, 14:248. Type, *Vespertilio pallidus* Le Conte.
1959. *Bauerus* Van Gelder, Amer. Mus. Novit., 1973:1, November 19. Type, *Antrozous (Bauerus) dubiaquercus* Van Gelder. Valid as a subgenus (see Pine, *et al.*, Jour. Mamm., 52:664, December 16, 1971).

Braincase high; rostrum more than half as long as braincase; basisphenoid pits absent; tympanic ring large, covering most of cochlea. Snout blunt, piglike; low horseshoe-shaped ridge on muzzle instead of evident nose-leaf possessed by non-American Recent genera; ears separate, large, and extending well beyond muzzle when laid forward. Dentition, i. $\frac{1}{2-3}$, c. $\frac{1}{1}$, p. $\frac{1}{2}$, m. $\frac{3}{3}$.

KEY TO SPECIES OF ANTROZOUS

1. Greatest length of tibia more than 27.6; Cuba. . . . .*A. koopmani,* p. 238

1'. Greatest length of tibia less than 27.6; mainland.
  2. Lower incisors 2; forehead slightly convex along mid-line; hairs paler at bases than at tips. . . . .*A. pallidus,* p. 236
  2'. Lower incisors usually 3; forehead flat along mid-line; hairs darker at bases than at tips. . . . .*A. dubiaquercus,* p. 238

Antrozous pallidus
Pallid Bat

External measurements: total length, 92–135; ear, 23–37; length of forearm, 48.0–60.2. Greatest length of skull, 18.6–23.6. Upper parts creamy, yellowish, even light brown; underparts paler, sometimes almost whitish. Other characters as for the genus.

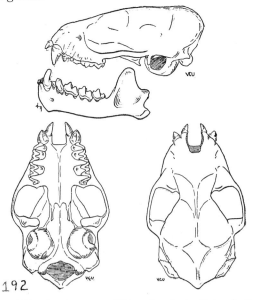

Fig. 192. *Antrozous pallidus bunkeri,* Natural Bridge, Barber Co., Kansas, No. 9302 K.U., ♀, X 2.

Antrozous pallidus bunkeri Hibbard

1934. *Antrozous bunkeri* Hibbard, Jour. Mamm., 15:227, type from 7 mi. S Sun City, Barber Co., Kansas, "in the tunnel by the Natural Bridge on the south fork of Bear Creek."
1960. *Antrozous pallidus bunkeri*, Morse and Glass, Jour. Mamm., 41:15, February 20.

MARGINAL RECORDS (Jones, *et al.*, 1967:25, unless otherwise noted).—Kansas: Natural Bridge, 4½ mi. S, ¼ mi. E Sun City; Aetna; *May Cave, 4 mi. S Aetna.* Oklahoma: Woodward County (Morse and Glass, 1960:11); Greer County, W end Wichita Mts. (Morse and Glass, 1960:15). Texas: Los Lingos Canyon, Briscoe Co. (Packard and Judd, 1968:536). Kansas: *1 mi. SW Aetna; "5½ mi. S Sun City" [ca. 4 mi. S Sun City].*

Antrozous pallidus cantwelli V. Bailey

1936. *Antrozous pallidus cantwelli* V. Bailey, N. Amer. Fauna, 55:391, August 29, type from Rogersburg, Asotin Co., Washington.

MARGINAL RECORDS.—British Columbia: Okanagan Landing. Washington: type locality. Oregon: Home. Nevada: Quinn River Crossing. Oregon: Catlow Cave, 110 mi. S Burns; The Dalles. Washington: Wenatchee. British Columbia: between Oliver and Okanagan Falls.

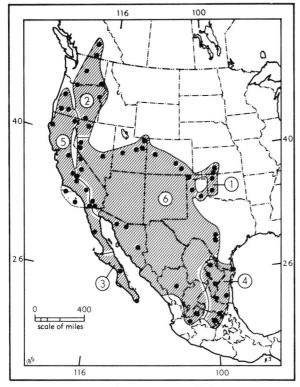

Map 185. *Antrozous pallidus.*

1. *A. p. bunkeri*
2. *A. p. cantwelli*
3. *A. p. minor*
4. *A. p. obscurus*
5. *A. p. pacificus*
6. *A. p. pallidus*

Antrozous pallidus minor Miller

1902. *Antrozous minor* Miller, Proc. Acad. Nat. Sci. Philadelphia, 54:389, September 3, type from Comondú, Baja California.
1951. *Antrozous pallidus minor*, Goldman, Smiths. Miscl. Coll., 115:356, July 31.

MARGINAL RECORDS.—Baja California: type locality southward to Cape St. Lucas.

Antrozous pallidus obscurus Baker

1967. *Antrozous pallidus obscurus* Baker, Southwestern Nat., 12:329, October 31, type from Acuña, 800 m., Tamaulipas.

MARGINAL RECORDS (Baker, 1967:330, unless otherwise noted).—Nuevo León: Ojo de Agua, 2½ mi. SW Sabinas Hidalgo, 1500 ft. Texas: 5 mi. SE Brownsville (Baker, *et al.*, 1971:850). Tamaulipas: type locality. Querétaro: Jalpan; Cadereyta, 2100 m. Tamaulipas: Tula. Nuevo León: Maguayes, Río Pilón; Huasteca Canyon, 17 mi. SW Monterrey, 4500 ft.

Antrozous pallidus pacificus Merriam

1897. *Antrozous pallidus pacificus* Merriam, Proc. Biol. Soc. Washington, 11:180, July 1, type from Old Fort Tejon, 3200 ft., Tehachapi Mts., Kern Co., California.

MARGINAL RECORDS.—Oregon: Eugene; Fort Klamath. California: Goose Lake; Snelling; White River; Kelso Valley; Bear Valley, San Bernardino Mts.; Campo. Baja California: San Fernando, thence northward along Pacific Coast, to California: coastal islands (Santa Catalina and Santa Cruz, von Bloeker, 1967:251), thence up coast to Ferndale. Oregon: ridge between Salt and Evans creeks, *ca.* 40 mi. N Rogue River P.O.

Antrozous pallidus pallidus (Le Conte)

1856. *V[espertilio]. pallidus* Le Conte, Proc. Acad. Nat. Sci. Philadelphia, 7:437, type from El Paso, El Paso Co., Texas.
1864. *Antrozous pallidus*, H. Allen, Smiths. Miscl. Coll., 7(Publ. 165):68, June.

MARGINAL RECORDS.—Utah: 6 mi. N Jensen. Colorado: 7 mi. W Rifle; Pueblo; 3 mi. NW Higbee. Oklahoma: 6½ mi. N Kenton. Texas: Tascosa; Kerrville; 16 mi. S Kerrville. Jalisco (Watkins, *et al.*, 1972:31, 32): 12 mi. W Encarnación de Díaz, 5600 ft.; Santa Cruz del Valle, Guadalajara; *1½ mi. WNW Amatitán, 4100 ft.* Durango: 1 mi. N Chorro, 6450 ft. Sonora: Río Yaqui, *ca.* 1 mi. S El Novillo (Findley and Jones, 1965:331, as *A. pallidus* only); Sáric; Tajitos (Baker and Patton, 1967:285, as *A. pallidus* only). California: Jacumba; Vallecitos; Walker Basin; Independence; 5 mi. N Benton Station. Nevada: 2 mi. N, 9 mi. E Yerington; 3 mi. WSW Lahontan Dam; Burned Corral Canyon, Quinn Canyon Mts., 6700 ft. Utah: volcanic caves, 10 mi. W Meadow; Price; Willow Creek, 25 mi. S Ouray, 5250 ft.

Antrozous koopmani Orr and Silva-T.
Cuban Bat

1960. *Antrozous koopmani* Orr and Silva-T., Proc. Biol. Soc. Washington, 73:84, August 10, Cueva del Hoyo García, Municipio de San Juan y Martínez, Provincia de Pinar del Río, Cuba. Holotype, a subfossil skull lacking lower mandibles.

As now known, closely resembles the species *A. pallidus* except that lower extremities, especially the tibia, actually and relatively, longer. The longer tail of *A. koopmani*, almost the same length as in *Antrozous (Bauerus) dubiaquercus*, reduces the earlier supposed degree of difference between the subgenera *Bauerus* and *Antrozous* according to Silva-T. (1976:19). He records fossil and subfossil remains (crania and/or lower jaws) from six localities additional to the type locality, and records two entire specimens in liquid preservative from Bayate. Reasons for arranging *A. koopmani* as a species, instead of as a subspecies of *Antrozous pallidus*, are not given by Silva-T. (*op. cit.*).

MARGINAL RECORDS.—Cuba: type locality; Bayate, Guantánamo, Provincia de Oriente (Silva-T., 1976:20).

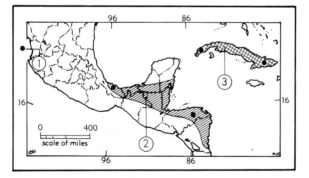

Map 186. *Antrozous dubiaquercus* and *Antrozous koopmani.*

1. *A. d. dubiaquercus*   2. *A. d. meyeri*
3. *A. koopmani*

Antrozous dubiaquercus
Van Gelder's Bat

External measurements: total length, 115–123; tail, 47–58; forearm, 48.0–55.9; greatest length of skull, 19.9–21.5; maxillary tooth-row, 6.9–7.5. Upper parts golden brown to dark brown; ears shorter than those of *A. pallidus;* differs in phallic morphology and bacular shape from *A. pallidus;* lower incisors 3 except in holotype, which has 2.

Antrozous dubiaquercus dubiaquercus Van Gelder

1959. *Antrozous (Bauerus) dubiaquercus* Van Gelder, Amer. Mus. Novit., 1973:2, November 19, type from María Magdalena Island, Tres Marías Islands, Nayarit. Known only from María Magdalena Island.

Antrozous dubiaquercus meyeri (Pine)

1966. *Baeodon meyeri* Pine, Southwestern Nat., 11:308, June 30, type from Río Quezalapam, 2 mi. E Lago Catemaco, *ca.* 610 m., Los Tuxtlas, Veracruz.
1971. *Antrozous dubiaquercus meyeri,* Pine, *et al.,* Jour. Mamm., 52:665, December 16.
1967. *Antrozous (Bauerus) meyeri* (Pine), Southwestern Nat., 12:485, December 31.

MARGINAL RECORDS.—Veracruz: type locality. Honduras: 40 km. E Catacamas, 500 m. (Pine, *et al.,* 1971:665).

FAMILY MOLOSSIDAE—Free-tailed Bats

"Humerus with trochiter much larger than trochin, the discrepancy in size usually more noticeable than in the Vespertilionidae, trochin articulating with scapula by a surface nearly as large as glenoid fossa, . . . capitellum almost directly in line with nearly straight shaft; ulna less reduced than in the Vespertilionidae, the very slender shaft usually about half as long as radius; second finger with well-developed metacarpal and one rudimentary phalanx; . . . shoulder girdle normal . . . , except that seventh cervical vertebra is fused with first dorsal; foot short and broad, but of normal structure; fibula complete, bowed outward from tibia . . . ; skull without postorbital processes; premaxillaries with nasal branches present or absent, when present forming two palatal foramina, when absent allowing the formation of one; . . . teeth normal; ears variable in size and form, sometimes joined across forehead, the tragus much reduced, the antitragus usually very large . . . ; muzzle obliquely truncate, usually sprinkled with short, modified hairs with spoon-shaped tips, the nostrils usually opening on a special pad . . . ; membranes thick and leathery, the uropatagium short, the tail projecting conspicuously beyond its free edge." (Miller, 1907:242, 243.)

This family is widely distributed through the warmer parts of the Old World and New World.

KEY TO NORTH AMERICAN GENERA OF MOLOSSIDAE

1. Bony palate with conspicuous median emargination extending back of roots of incisors.
   2. Upper premolars 1–1. *Mormopterus,*   p. 246

2'. Upper premolars 2–2. . . . *Tadarida*, p. 240
1'. Bony palate without conspicuous median emargination, but a small notch may be present that never extends back of roots of incisors.
   3. Upper incisor with length along cingulum equal to or greater than height of shaft. . . . . *Molossus*, p. 252
   3'. Upper incisor with length along cingulum decidedly less than height of shaft.
      4. Palate conspicuously dome-shaped. . . . . *Promops*, p. 251
      4'. Palate arched transversely but little, if any, longitudinally.
         5. Rostrum noticeably flattened, its length about equal to lachrymal breadth. *Molossops*, p. 239
         5'. Rostrum subcylindrical, its length considerably greater than lachrymal breadth.
           Eumops, p. 246

Genus **Molossops** Peters—Dog-faced Bats

1865. *Molossops* Peters, Monatsb. preuss. Akad. Wiss., Berlin, p. 575. Type, *Dysopes temminckii* Burmeister.
1920. *Cynomops* Thomas, Ann. Mag. Nat. Hist., ser. 9, 5:189, February. Type, *Molossus cerastes* Thomas.

Small to medium-sized, free-tailed bats; ears not connected to each other; nasal branches of premaxillaries joined anteriorly; one pair or two pairs of lower incisors present (after Goodwin and Greenhall, 1961:282).

KEY TO SPECIES OF MOLOSSOPS

1. Greatest length of head and body usually less than 62; membranes pale brown; underparts yellowish white. *M. planirostris*, p. 239
1'. Greatest length of head and body usually more than 62; membranes dark brown; underparts dull grayish brown.
         M. greenhalli, p. 239

Molossops planirostris
Southern Dog-faced Bat

Length of head and body, approx. 60. Skull with high, broad, flattened rostrum with conspicuous laterally projecting lachrymal ridges. Other characters as for the genus.

Molossops planirostris planirostris (Peters)

1865. *M[olossus]. planirostris* Peters, Monatsb. preuss. Akad. Wiss., Berlin, p. 575, type from Guyana.—See addenda.
1907. *Molossops planirostris*, Miller, Bull. U.S. Nat. Mus., 57:248, fig. 39, June 29.

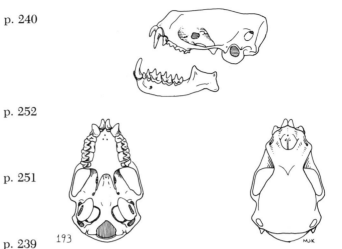

Fig. 193. *Molossops planirostris planirostris*, 2 km. NE Maripa, Bolívar, Venezuela, No. 119091 K.U., ♂, X 2.

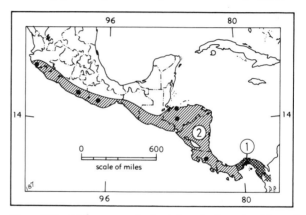

Map 187. *Molossops planirostris planirostris* (1) and *Molossops greenhalli mexicanus* (2).

MARGINAL RECORDS.—Panamá (Handley, 1966:772): Fort Clayton; *Panamá; Pacora*. Also occurs in South America.

Molossops greenhalli
Greenhall's Dog-faced Bat

External measurements: head and body, 90–111; tail, 28–37; ear from notch, 13.8–18; forearm, 33.3–38.2; weight, 10.8–16.5. Skull: greatest length, 17.0–20.6; zygomatic breadth, 11.2–13.5; length of maxillary tooth-row, 6.2–7.6.

Color of upper parts dark brown to reddish brown; base of hairs light to white; underparts paler than back; nose, membranes, ears, feet and tail dark brown to blackish; narrow strip of fur extending along outer side of forearm to base of metacarpals. Skull short, broad, and low, with relatively high flattened rostrum; conspicuous laterally projecting lachrymal ridges; braincase relatively smooth and flattened on top, with or without low sagittal crest; upper incisors long, in contact with each other, and projecting forward; inner lower incisors crowded forward from tooth-row, cutting edges deeply bifid; outer lower incisors, when present, smaller than inner, faintly trifid and crowded against canines; 1st lower premolar smaller than 2nd; upper canines with broad groove on front surface of shaft (mostly after Goodwin and Greenhall, 1961:283).

Molossops greenhalli mexicanus Jones and Genoways

1967. *Molossops greenhalli mexicanus* Jones and Genoways, Proc. Biol. Soc. Washington, 80:207, December 1, type from 7½ mi. SE Tecomate, 1500 ft., Jalisco.

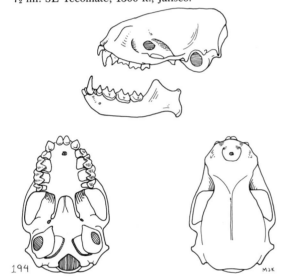

Fig. 194. *Molossops greenhalli mexicanus*, 3 km. N Agua del Obispo, Chilpancingo, Guerrero, No. 99741 K.U., ♀, X 2.

MARGINAL RECORDS (Jones and Genoways, 1967:209, unless otherwise noted).—Jalisco: type locality. Guerrero: 3 km. N Agua del Obispo, 3180 ft. Oaxaca: 20 mi. S, 5 mi. E Sola de Vega, 4800 ft. Honduras: Lancetilla (Valdez and LaVal, 1971:250, as *M. greenhalli* only); Comayagua (LaVal, 1969:821, as *M. greenhalli* only). Costa Rica: *ca.* 2 km. NW Santa Ana (Finca Lornessa) (Gardner, *et al.*, 1970:725, as *M. greenhalli* only), thence southeastward into South America.

Genus **Tadarida** Rafinesque—Free-tailed Bats

Revised by Shamel, Proc. U.S. Nat. Mus., 78:1–27, May 6, 1931.

1814. *Tadarida* Rafinesque, Précis des découvertes et travaux somiologiques . . . , p. 55. Type, *Cephalotes teniotis* Rafinesque.
1902. *Nyctinomops* Miller, Proc. Acad. Nat. Sci. Philadelphia, 54:393, September 12. Type, *Nyctinomus femorosaccus* Merriam.

Differs from other North American molossids, except *Mormopterus*, in deep vertical grooves or wrinkles on upper lip, Z-shaped structure of M3, and separation of premaxillae between upper incisors. Dentition, i. $\frac{1}{3 or 2}$, c. $\frac{1}{1}$, p. $\frac{2}{2}$, m. $\frac{3}{3}$.

KEY TO NORTH AMERICAN SPECIES OF TADARIDA

1. Second phalanx of 4th digit more than 5.0; ears not extending appreciably beyond muzzle when laid forward, inner bases not conjoined; breadth of anterior part of rostrum markedly greater than interorbital breadth. . . . .*T. brasiliensis*, p. 240
1'. Second phalanx of 4th digit less than 5.0; ears extending appreciably beyond muzzle when laid forward, inner bases conjoined; breadth of anterior part of rostrum barely greater than interorbital breadth.
2. Length of forearm more than 55.0. *T. macrotis*, p. 245
2'. Length of forearm less than 55.0.
3. M1, viewed from crown surface, nearly square; posterior margin of palate behind a line connecting posterior margins of M3–M3. *T. femorosacca*, p. 243
3'. M1, viewed from crown surface, broader posteriorly than anteriorly owing to expansion posteromedially of hypocone which forms a broad heel; posterior margin of palate almost on a line connecting posterior margins of M3–M3.
4. Forearm more than 45.0; greatest length of skull more than 20.0. . . . . .*T. aurispinosa*, p. 243
4'. Forearm less than 45.0; greatest length of skull less than 19.0. *T. laticaudata*, p. 244

Tadarida brasiliensis
Brazilian Free-tailed Bat

Subspecies reviewed by Schwartz, Jour. Mamm., 36:106–109, February 28, 1955.

Length of head and body, 46.6–65.2; length of forearm, 36.6–46.4; greatest length of skull, 14.6–18.4; zygomatic breadth, 8.4–10.8; length of

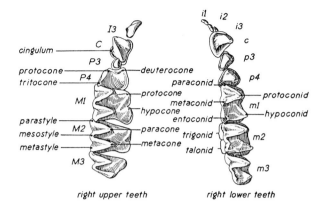

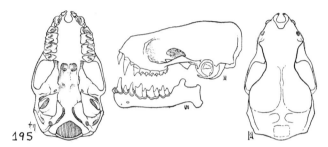

Fig. 195. *Tadarida brasiliensis mexicana*, Greenmonster Canyon, Monitor Range, Nye Co., Nevada, No. 57472 M.V.Z., ♂, X 2. Occlusal view of teeth, X 6.

upper tooth-row, 5.0–6.6. Upper parts dark brown, bases of hairs whitish; underparts slightly paler. For other characters see key to the species.

Tadarida brasiliensis antillularum (Miller)

1902. *Nyctinomus antillularum* Miller, Proc. Acad. Nat. Sci. Philadelphia, 54:398, September 12, type from Roseau, Dominica, West Indies.
1955. *Tadarida brasiliensis antillularum*, Schwartz, Jour. Mamm., 36:108, February 28.

MARGINAL RECORDS.—Puerto Rico: *no specific locality.* Lameshur on St. John Island (Koopman, 1975:3, 4). Lesser Antilles (Jones and Phillips, 1970:137): St. Martin; *St. Bartholemew;* Barbuda; Antigua; Guadeloupe; Dominica; Martinique; St. Lucia; Montserrat; St. Kitts; St. Eustatius.

Tadarida brasiliensis bahamensis (Rehn)

1902. *Nyctinomus bahamensis* Rehn, Proc. Acad. Nat. Sci. Philadelphia, 54:641, December 12, type from Governors Harbor, Eleuthera, Bahama Islands.
1955. *Tadarida brasiliensis bahamensis*, Schwartz, Jour. Mamm., 36:108, February 28.

MARGINAL RECORDS.—Bahama Islands: Little Abaco Island; Marsh Harbour, Great Abaco Island; type locality; Long Island.

Tadarida brasiliensis brasiliensis (I. Geof. St.-Hilaire)

1824. *Nyctinomus brasiliensis* I. Geoffroy St.-Hilaire, Ann. Sci. Nat., 1:343, type from Brazil, by subsequent restriction (Shamel, Proc. U.S. Nat. Mus., 78:4, May 6, 1931), Curityba, Paraná.
1920. *Tadarida brasiliensis*, Thomas, Proc. U.S. Nat. Mus., 58:222, November 10.
1827. *Dysopes nasutus* Temminck, Monographies de mammalogie . . . , 1:233. Type locality, Brazil.
1837. *Molossus rugosus* D'Orbigny, Voy. dans l'Amér. Mérid., Atlas Zool., Pl. 10, Figs. 3–5. Type locality [probably Tucumán], Corrientes, Argentina.
1840. *Dysopes naso* Wagner, *in* Schreber, Die Säugthiere . . . , Suppl., 1:475. Type locality, Brazil.
1861. *Dysopes multispinosus* Burmeister, Reise dürch die La Plata-Staaten . . . Argentinischen Republik . . . , 2:391. Type locality, Tucumán, Argentina.

MARGINAL RECORDS.—Costa Rica: San José. Panamá: Cerro Punta, 5700 ft. (Handley, 1966:772), thence eastward into South America.

Tadarida brasiliensis constanzae Shamel

1931. *Tadarida constanzae* Shamel, Proc. U.S. Nat. Mus., 78:10, May 6, type from Constanza, Dominican Republic.
1955. *Tadarida brasiliensis constanzae*, Schwartz, Jour. Mamm., 36:108, February 28.

MARGINAL RECORDS.—Haiti: Bombardopolis; St. Michel. Dominican Republic: type locality.

Tadarida brasiliensis cynocephala (Le Conte)

1831. *Nyct[icea]. cynocephala* Le Conte, *in* McMurtrie, The animal kingdom . . . by the Baron Cuvier . . . , 1[App.]:432, type from Georgia, probably in neighborhood Le Conte Plantation, Liberty Co.
1955. *Tadarida brasiliensis cynocephala*, Schwartz, Jour. Mamm., 36:108, February 28.
1837. *Molossus fuliginosus* Cooper, Ann. Lyc. Nat. Hist. New York, 4:67, November, type from Milledgeville, Georgia.

MARGINAL RECORDS.—Arkansas: Little Rock (Sealander and Price, 1964:152). Alabama: Sauta Cave, near Scottsboro. South Carolina: Columbia; Georgetown County (Golley, 1966:68). Georgia: Savannah. Florida: Enterprise; Kissimmee; Miami (Layne, 1974:389); *Everglades National Park (ibid.);* Marco Island, thence westward along Gulf Coast to Louisiana (Lowery, 1974:141): Port Sulphur; *Houma; Morgan City;* Merryville. Texas (Davis, 1966:70, as *T. cynocephala*): Anderson County; Nacogdoches County; *Sabine County.* Louisiana: Marthaville (Lowery, 1974:141); Ruston. Also Ohio: Vandalia (Mills, 1971:479); 10 mi. NW Portsmouth (Smith and Goodpaster, 1960:117).—See addenda.

Tadarida brasiliensis intermedia Shamel

1931. *Tadarida intermedia* Shamel, Proc. U.S. Nat. Mus., 78:7, May 6, type from Valley of Comitán, Chiapas.

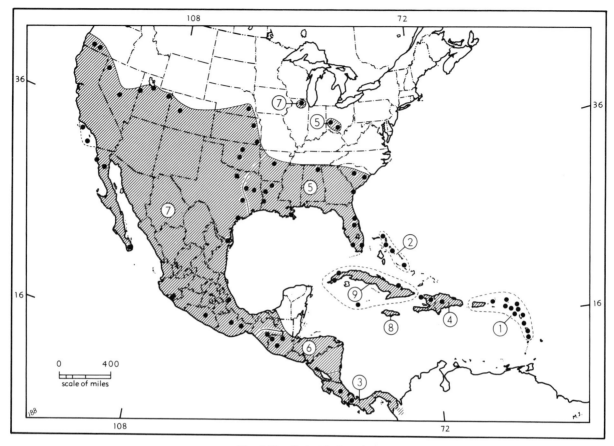

Map 188. *Tadarida brasiliensis.*

1. *T. b. antillularum*	4. *T. b. constanzae*	7. *T. b. mexicana*
2. *T. b. bahamensis*	5. *T. b. cynocephala*	8. *T. b. murina*
3. *T. b. brasiliensis*	6. *T. b. intermedia*	9. *T. b. muscula*

1955. *Tadarida brasiliensis intermedia,* Schwartz, Jour. Mamm., 36:108, February 28.

MARGINAL RECORDS.—Chiapas: type locality. Guatemala: Cobán; Chimaltenango; Jacatlenango.

Tadarida brasiliensis mexicana (Saussure)

1860. *Molossus mexicanus* Saussure, Revue et Mag. Zool., Paris, ser. 2, 12:283, July, type from Cofre de Perote, 13,000 ft., Veracruz.
1955. *Tadarida brasiliensis mexicana,* Schwartz, Jour. Mamm., 36:108, February 28.
1889. *Nyctinomus mohavensis* Merriam, N. Amer. Fauna, 2:25, October 30, type from Fort Mojave, Arizona.
1894. *Nyctinomus brasiliensis californicus* H. Allen, Bull. U.S. Nat. Mus., 43:166, March 14, type from California.
1942. *Tadarida texana* Stager, Bull. Southern California Acad. Sci., 41(pt. 1):49, May 31, type from Ney Cave, 20 mi. N Hondo, Medina Co., Texas (*fide* Blair, Texas Jour. Sci., 4:96, March 30, 1952).

MARGINAL RECORDS.—Oregon: Medford; Kingsley Air Force Base, Klamath Falls (Constantine,

1961:404). Nevada: mouth Little High Rock Canyon; Greenmonster Canyon, Monitor Range. Utah: cave at Salt Springs, near Utah–Nevada boundary; Salt Lake City, 4250 ft.; Jensen. Colorado: Newcastle. Nebraska: Lincoln. Kansas: Lawrence; Galena. Oklahoma: Stillwater; Norman. Texas: 3 mi. E Roanoke; Grimes County (Davis, 1966:69, as *T. mexicana*); Aransas County (*ibid.*, as *T. mexicana*); Brownsville, thence southward along Gulf Coast to Veracruz: 5 km. N Jalapa. Oaxaca (Goodwin, 1969:112): Mazatlán; Cerro San Felipe. Guerrero: Mexicapan (Villa–R. and Cockrum, 1962:47); *Cuetzala del Progreso* (*ibid.*). Colima: Colima, thence along coast to Baja California: Miraflores (Jones, *et al.*, 1965:56); San Pedro Mártir; San Telmo. California (von Bloeker, 1967:252): Santa Cruz Island; San Clemente Island. Also, Illinois: DeKalb (Walley, 1970:113).—See addenda.

Tadarida brasiliensis murina (Gray)

1827. *Nyctinomus murinus* Gray, in Griffith, The animal kingdom . . . by the Baron Cuvier . . . , 5:66, type from Jamaica. Known only from Jamaica.

1955. *Tadarida brasiliensis murina*, Schwartz, Jour. Mamm., 36:108, February 28.

Tadarida brasiliensis muscula (Gundlach)

1861. *Nyctinomus musculus* Gundlach, Monatsb. preuss. Akad. Wiss., Berlin, p. 149, paratypes from Cuba.
1955. *Tadarida brasiliensis muscula*, Schwartz, Jour. Mamm., 36:108, February 28.

MARGINAL RECORDS.—Cuba: Cabañas; Cueva de los Americanos, ½ mi. S Gibara. Grand Cayman [= Gran Caimán Isla (Varona, 1974:40)]. Cuba: Pinar del Río.

Tadarida aurispinosa (Peale)
Peale's Free-tailed Bat

1848. *Dysopes aurispinosus* Peale, U.S. Expl. Exp., 8:21, type obtained on board U.S.S. Peacock at sea, approx. 100 mi. S Cape San Roque, Brazil.
1931. *Tadarida aurispinosa*, Shamel, Proc. U.S. Nat. Mus., 78:11, May 6.
1941. *Tadarida similis* Sanborn, Field Mus. Nat. Hist., Zool. Ser., 27:386, December 8, type from Bogotá, Colombia. Considered to be a synonym of *T. aurispinosa* by Carter and Davis, Proc. Biol. Soc. Washington, 74:164, August 11, 1961.

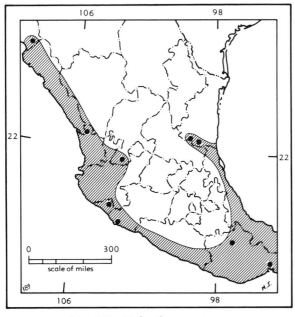

Map 189. *Tadarida aurispinosa.*

Length of hind foot, 9.9–11.0; length of forearm, 47.8–51.4; greatest length of skull, 20.5–21.6; interorbital breadth, 3.6–4.0; zygomatic breadth, 11.6–12.0; length of upper tooth-row, 7.8–8.2 (Carter and Davis, 1961: 162, 163). Upper parts wood brown to russet; underparts wood brown (Shamel, 1931:12).

MARGINAL RECORDS.—Sonora: 8 mi. S Alamos (Baker and Jones, 1972:308). Nayarit: Huajicori (Gardner, 1963:42). Zacatecas: Río Juchipila, 1¾ km. N Santa Rosa, 1100 m. (Baker, *et al.*, 1967:225). Colima: near Pueblo Juárez (Jones and Alvarez, 1964:303); 9 *km. W Pueblo Nuevo* (Alvarez and Aviña, 1964:246). Michoacán: 48 km. S [SE?] Coahuayana (*ibid.*). Oaxaca: Chicatlán (*ibid.*). San Luis Potosí: El Salto, 1750 ft. (Jones and Alvarez, 1964:303, recorded from partial skull on cave floor). Tamaulipas: Cueva del Abra, 6 mi. NNE Antiguo Morelos (Alvarez, 1963:416). Oaxaca: Juchitán (Alvarez and Aviña, 1964:246). Also occurs in South America.

Tadarida femorosacca (Merriam)
Pocketed Free-tailed Bat

1889. *Nyctinomus femorosaccus* Merriam, N. Amer. Fauna, 2:23, October 30, type from Agua Caliente [= Palm Springs], Riverside Co., California.
1924. *Tadarida femorosacca*, Miller, Bull. U.S. Nat. Mus., 128:86, April 29.

Length of head and body, 54.4–65.2; length of forearm, 45.5–49.2; greatest length of skull, 18.4–20.3; zygomatic breadth, 9.6–11.4; length of upper tooth-row, 7.0–7.7. Upper parts Vandyke Brown, sometimes distinctly reddish; underparts slightly paler, sometimes with buffy wash. Other characters as for the species-group (see artificial key).

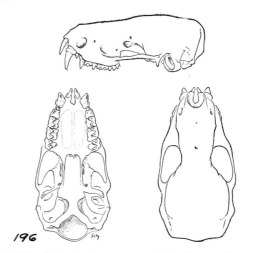

Fig. 196. *Tadarida femorosacca*, Palm Canyon, Borego Valley, San Diego Co., California, No. 94702 M.V.Z., sex ?, X 2.

MARGINAL RECORDS.—Arizona: near Alamo Crossing, Bill Williams River, *ca.* 1000 ft. (Cockrum and Musgrove, 1965:509); 8 mi. NW Roosevelt (*ibid.*); Eden (Davis, W. H., 1966:28). New Mexico: sec. 1, T. 34 S, R. 22 W (Findley, *et al.*, 1975:69). Sonora: Rebeico Dam, 28 mi. by road E Mazatán (Findley and Jones, 1965:331); 1 mi. NW Alamos. New Mexico: Carlsbad Caverns (Easterla, 1968a:516). Texas: Giant

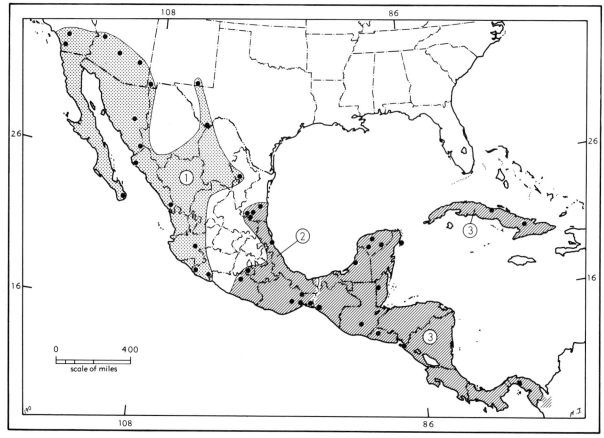

Map 190. *Tadarida femorosacca* and *Tadarida laticaudata*.

1. *T. femorosacca*   2. *T. laticaudata ferruginea*   3. *T. laticaudata yucatanica*

Dagger Yucca Flats, Big Bend National Park (Easterla, 1968a:515). Nuevo León: Monterrey (Jiménez-G., 1971:142). Jalisco: Zacoalco. Michoacán (Alvarez and Aviña, 1964:246): 30 km. N Infiernillo; 48 km. S Cuahuayana [= 48 km. SE Coahuayana]. Sinaloa (Jones, *et al.*, 1972:22): Rosario; 1½ mi. NW Topolobambo. Baja California: Santa Anita. California: 2 mi. E Suncrest Store; type locality.

Tadarida laticaudata
Broad-tailed Bat

Total length, 92–112; length of tail, 34–45; length of hind foot, 9–11; length of forearm, 40.8–45; greatest length of skull, 16.7–18.5; zygomatic breadth, 9.3–10.6; length of upper tooth-row, 6.1–6.9. Upper parts Vandyke Brown, hairs extensively white at base; underparts bister (white at base except on abdomen where also basal part of hair is bister) but hairs tipped with pinkish buff. This bat is smaller than *T. aurispinosa*. From *T. femorosacca*, *laticaudata* differs in shape of M1 (see key).

Molossus caecus Rengger, 1830, Naturgeschichte der Säugethiere von Paraguay, Basel,

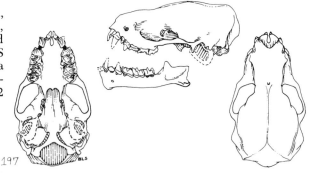

Fig. 197. *Tadarida laticaudata ferruginea*, 2 mi. S, 10 mi. W Piedra, 1200 ft., Tamaulipas, No. 55385 K.U., ♀, X 2.

with type locality at Asunción, Paraguay, was included in the synonymy of *Tadarida molossa* by Hall and Kelson (1959:209). Measurements given in the original description are too small for *T. macrotis* [formerly *T. molossa*, Hall and Kelson, 1959]. Rengger's statement that the ears of his *Molossus caecus* are joined suggests that it was the earlier named *Molossus laticaudata* Geoffroy 1805. If so, *Molossus caecus* Rengger 1830 is a synonym of *Tadarida laticaudata laticaudata* (É.

Geoffroy 1805). This subspecies does not occur in North America. Consequently the name *Molossus caecus* can be omitted from the North American list. I am indebted to Karl F. Koopman (*in Litt.*) for information on this matter.

Tadarida laticaudata ferruginea Goodwin

1954. *Tadarida laticaudata ferruginea* Goodwin, Amer. Mus. Novit., 1670:2, June 28, type from 8 mi. N Antiguo Morelos, Tamaulipas.

MARGINAL RECORDS (Alvarez, 1963:416, unless otherwise noted).—Tamaulipas: Sierra de Tamaulipas, 2 mi. S, 10 mi. W Piedra, 1200 ft. Veracruz: Tuxpan (Alvarez and Aviña, 1964:244). Oaxaca: Magoñé (Goodwin, 1969:112, as *T. femorosacca*); Tapanatepec (*ibid.*); Juchitán [de Zaragoza] (Alvarez and Aviña, 1964:244); Tequisistlán (*ibid.*). Guerrero: 7 km. N Balsas, 700 m. (Alvarez and Ramírez-P., 1972:175). Morelos: 2 km. SE Alpuyeca (*ibid.*). Tamaulipas: 8 mi. N Antiguo Morelos; 20 mi. SW El Mante; 5 mi. S El Mante.

Tadarida laticaudata yucatanica (Miller)

1902. *Nyctinomops yucatanicus* Miller, Proc. Acad. Nat. Sci. Philadelphia, 54:393, September 12, type from Chichén-Itzá, Yucatán.
1962. *Tadarida laticaudata yucatanica*, Jones and Alvarez, Univ. Kansas Publ., Mus. Nat. Hist., 14:132, March 7.

MARGINAL RECORDS (Jones and Alvarez, 1962:132,133, unless otherwise noted).—Yucatán: Mérida; Chichén-Itzá. Quintana Roo: Cozmuel Island. Belize: El Cayo, thence southward along Caribbean Coast to Panamá: Pacora, thence southeastward into South America and northwestward along Pacific Coast to Nicaragua: Potosí (Jones, *et al.*, 1971:21). El Salvador: San Salvador. Guatemala: Dueñas. Chiapas: Río Ocuilapa, 12 km. SSE Tonalá (Alvarez and Aviña, 1964:245). Campeche: San José Carpizo. Yucatán: Uxmal. Also reported from Cuba: Yaguajay (Dusbábek, 1969:322); 5 km. NW Omaja (Silva-T. and Koopman, 1964:1).

Tadarida macrotis (Gray)
Big Free-tailed Bat

X.⅓

1839. *Nyctinomus macrotis* Gray, Ann. Nat. Hist., 4:5, September, type from Cuba.
1924. *Tadarida macrotis*, Miller, Bull. U.S. Nat. Mus., 128:86, April 29.
1843. *Dysopes auritus* Wagner, Wiegmann's Arch. für Naturgesch., Jahrg. 9, 1:368, type from Cuyaba, Brazil.
1876. *Nyctinomus megalotis* Dobson, Proc. Zool. Soc. London, p. 728, November 7, type from Surinam.
1891. *Nyctinomus depressus* Ward, Amer. Nat., 25:747, August, type from Tacubaya, Distrito Federal, México.
1894. *Nyctinomus macrotis nevadensis* H. Allen, Bull. U.S. Nat. Mus., 43:171, March 17, type from California.
1900. *Promops affinis* J. A. Allen, Bull. Amer. Mus. Nat. Hist., 13:91, December, type from Taguaga [= Taganga], Colombia.
1914. *Nyctinomus aequatorialis* J. A. Allen, Bull. Amer. Mus. Nat. Hist., 33:386, July 9, type from Ecuador.

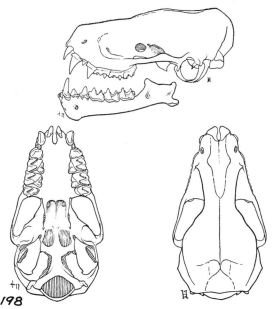

198

Fig. 198. *Tadarida macrotis*, Pine Canyon, Chisos Mts., 6000 ft., Brewster Co., Texas, No. 81683 M.V.Z., ♀, X 2.

Length of head and body, 66.6–79.0; length of forearm, 58.0–63.8; greatest length of skull, 22.2–24.0; zygomatic breadth, 10.2–13.0; length of upper tooth-row, 8.2–9.5. Upper parts dark (mummy) brown; underparts slightly paler. Skull large, robust, and with relatively long rostrum.

MARGINAL RECORDS.—Colorado: Grand Junction; Cheyenne Cañon, Colorado Springs (Armstrong, 1972:75); Rocky Ford (Constantine, 1961:405). Kansas: 9 mi. N Elkhart. Oklahoma: 4 mi. S Elkhart, Kansas. Texas (Axtell, 1961:52, unless otherwise noted): Petersborg (Packard and Judd, 1968:536); Lubbock; Alpine; Welder Wildlife Refuge Center. Nuevo León: Monterrey (Jiménez, 1971:142). Veracruz: Casona en Paso de Ovejas (Villa-R., 1967:448). Distrito Federal: Multifamiliar Miguel Alemán, Ciudad de México (*ibid.*). Guerrero: Cañón de los Sabinos, 20 km. E Teloloapan (*ibid.*). Michoacán: 48 km. S [SE?] Coahuayana (*ibid.*). Sinaloa: 5 mi. WSW Plomosas, 800 ft. (Jones, *et al.*, 1972:23). California: San Diego. Nevada: Henderson (Bradley, *et al.*, 1965:220, as *T. molossa*). Utah: Pine Valley, Desert Range Experimental Station, U.S. Forest Service, sec. 33, T. 25 S, R. 17 W, Salt Lake Meridian. Also known from Cuba: 7 km. W Banao. Dominican Republic: *no specific locality.* Jamaica: *no specific locality.* Extramarginal occurrences: *British Columbia: Essondale, near Westminster.* Iowa: *Marshalltown; Cedar Rapids.* Kansas: *Pittsburg* (Hays and Ireland, 1967:196). Also known from South America.

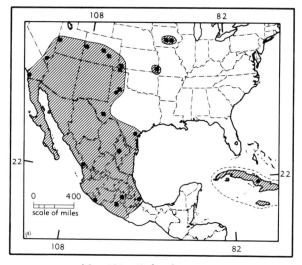

Map 191. *Tadarida macrotis.*

Genus **Mormopterus** Peters—Goblin Bats

1865. *Mormopterus* Peters, Monatsb. preuss. Akad. Wiss., Berlin, p. 258. Type, *Nyctinomus jugularis* Peters.

Closely resembling *Tadarida;* differing from *Tadarida* primarily in possessing only one upper premolar and in slightly smaller ears that are joined basally. Dentition, i. $\frac{1}{3 \text{ or } 2}$, c. $\frac{1}{1}$, p. $\frac{1}{2}$, m. $\frac{3}{3}$.

Mormopterus minutus (Miller)
Little Goblin Bat

1899. *Nyctinomus minutus* Miller, Bull. Amer. Mus. Nat. Hist., 12:173, October 20, type from Trinidad, Santa Clara, Cuba.
1907. *M[ormopterus]. minutus* Miller, Bull. U.S. Nat. Mus., 57:254, June 29.

Measurements of the type, an adult male, are: length of head and body, approx. 46; total length, 74; length of forearm, 29. Cranial measurements of a female topotype are: greatest length, 13.4;

zygomatic breadth, 8.6; breadth of braincase, 7; length of upper tooth-row, 5. Skull flattened and lightly constructed; dorsal profile nearly straight; rostrum short and broad. Other characters as for the genus.

MARGINAL RECORDS.—Cuba: type locality; Omaja; a few mi. S Guaro, near Preston.

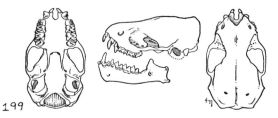

Fig. 199. *Mormopterus minutus*, Oriente, Omaja, Cuba. Skull is of No. 72994 F.M.N.H., ♀. Mandible is of No. 72977 F.M.N.H., ♂. X 2.

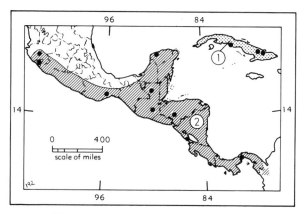

Map 192. *Mormopterus minutus* (1) and *Promops centralis centralis* (2).

Genus **Eumops** Miller—Mastiff Bats

Revised by Eger, Royal Ontario Mus., Life Sci. Contrib., 110:1–69, June 16, 1977.

1906. *Eumops* Miller, Proc. Biol. Soc. Washington, 19:85, June 4. Type, *Molossus californicus* Merriam.
1932. *Molossides* G. M. Allen, Jour. Mamm., 13:257, August 9. Type, *Molossides floridanus* G. M. Allen.

Skull slender but robust; rostrum well developed; dorsal profile of skull almost straight; premaxillaries wholly lacking palatal branches; 1st upper premolar small but usually well formed, and usually crowded from normal position in dental arcade. External form slender, ears large, rounded or angular, usually conjoined basally, and extending slightly beyond nostril when laid forward. Dentition, i. $\frac{1}{2}$, c. $\frac{1}{1}$, p. $\frac{2}{2 \text{ or } 1}$, m. $\frac{3}{3}$.

KEY TO NORTH AMERICAN SPECIES OF EUMOPS

(Adapted from Sanborn, 1932:348, 349)

1. Length of forearm more than 52.
    2. Length of forearm more than 73.
                    E. perotis,   p. 247
    2'. Length of forearm less than 73.
        3. Length of forearm averaging more than 64; length of upper tooth-row more than 11. . . *E. underwoodi,*   p. 249
        3'. Length of forearm averaging less than 64; length of upper tooth-row less than 11.
            4. Lachrymal ridge slight; tragus small (3.5 from posterior notch), pointed.
                    E. auripendulus,   p. 248
            4'. Lachrymal ridge absent; tragus large (4–5 from posterior notch), quadrate. *E. glaucinus,*   p. 249
1'. Length of forearm less than 52.
    5. Palate extending behind last molars; 3rd commissure of M3 same length as 2nd; lower incisors not crowded; basisphenoid pits large and deep.
                    E. hansae,   p. 250
    5'. Palate not extending behind last molars; 3rd commissure of M3 shorter than 2nd; lower incisors bunched; basisphenoid pits small and shallow.
                E. nanus,   p. 251

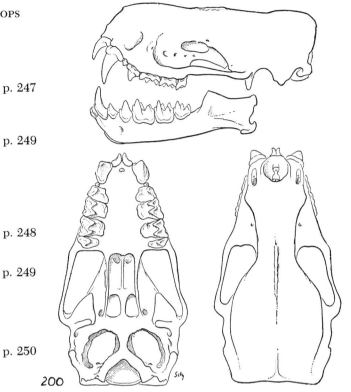

200

Fig. 200. *Eumops perotis californicus,* 1½ mi. N Barrett Junction, San Diego Co., California, No. 94706 M.V.Z., ♀, X 2.

Eumops perotis
Greater Mastiff Bat

Length of head and body, approx. 105; length of forearm, 73–80; greatest length of skull, 30.3–32.9; zygomatic breadth, 17.2–19.8; length of upper tooth-row, 11.9–13.6. Upper parts sooty brown; slightly paler below.

Eumops perotis californicus (Merriam)

1890. *Molossus californicus* Merriam, N. Amer. Fauna, 4:31, October 8, type from Alhambra, Los Angeles Co., California.

1932. *Eumops perotis californicus,* Sanborn, Jour. Mamm., 13:351, November 2.

MARGINAL RECORDS.—California: Yosemite Valley (Vaughan, 1959:8); Fresno; Traver. Nevada: Las Vegas (Bradley and O'Farrell, 1967:672, as *E. perotis* only). Arizona: Secret Pass, 5 mi. S, 15 mi. W Kingman (Cox, 1965:687); Lower Ruins, Tonto National Monument (Johnson and Johnson, 1964:322); ¼ mi. N Eagle Creek Pump Station, 10 mi. WSW Morenci (Cox, 1965:687). New Mexico: 32 mi. S, ½ mi. E Rodeo, 4500 ft. (Rowlett, 1972:640, as *E. perotis* only). Texas: Langtry. Coahuila: cave near Cuatro Ciénegas. Zacatecas: Santa Rosa (Matson and Patten, 1975:10, as *E. perotis* only). Sonora: Pilares. Sinaloa: Kilometer

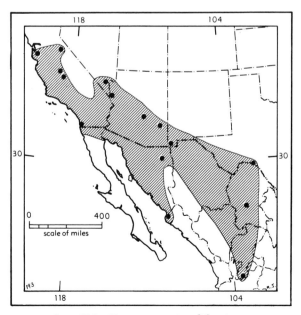

Map 193. *Eumops perotis californicus.*

1665, Hwy. 15, near Sonoran border (Jones, *et al.*, 1972:24). California: Otay; Hayward.

Eumops perotis gigas (Peters)

1864. *Dysopes (Molossus) gigas* Peters, Monatsb. preuss. Akad. Wiss., Berlin, p. 383, type from "in dem Gebirge Taburete, District Callajabas auf Cuba." See Cabrera, Cat. Mam. America del Sur, Rev. Mus. Argentino de Cienc. Nat., 4:127, March 27, 1958, who regards *D. gigas* as a subspecies of *Eumops perotis*, as also does Varona, Catálogo Mam. viv. y exting. de las Antillas, Acad. Cienc. Cuba, 1974:41, who defines the type locality as "Loma Taburete, en el partido de Cayajabos, jurisdicción del Mariel, Pinar del Río, Cuba," and speculates that this subspecies may be extinct. Known only from the type. Not mapped.—**See** addenda.

Eumops auripendulus
Shaw's Mastiff Bat

Length of head and body, approx. 81; length of forearm, 56–63; greatest length of skull, 23.0–25.7; zygomatic breadth, 14.3–15.3; length of upper tooth-row, 9.4–10.6. Upper parts dark reddish brown, hairs buffy white basally; underparts slightly paler. Lachrymal ridge slight; lambdoidal crest well developed and extending posteriorly beyond occipital area; 3rd commissure on M3 less developed than in other kinds of *Eumops*; posterior half of outer edge of M3 shorter than in *E. glaucinus*.

Relying on Cabrera (1958:123, 124), Goodwin (1960:5), and Eger (1974:1–8), the synonymy for

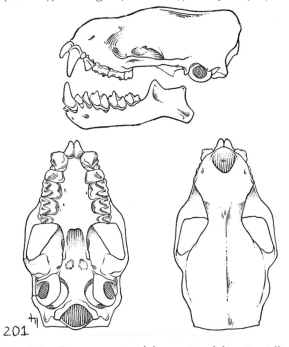

Fig. 201. *Eumops auripendulus auripendulus*, Turrialba, Cartago Prov., Costa Rica, No. 26952 K.U., ♂, X 2.

Eumops a. auripendulus would be about as follows:

Eumops auripendulus auripendulus (Shaw)

1800. *Vespertilio auripendulus* Shaw, General Zoology, vol. 1, pt. 1, p. 137, based on the "Slouch-eared Bat" of Pennant, History of quadrupeds, 3rd ed., 2:313, 1793.
1960. *Eumops auripendulus*, Goodwin, Amer. Mus. Novit., 1994:5, March 8.
1805. *Molossus amplexi-caudatus* É. Geoffroy St.-Hilaire, Ann. Mus. Hist. Nat. Paris, 6:156, based on Buffon's illustration, Hist. Nat. Générale et Particulière, Paris, suppl. 7:Pl. 75, 1789.
1843. *Dysopes longimanus* Wagner, Wiegmann's Arch. für Naturgesch., Jahrg. 9(1ª parte):367.
1843. *Dysopes leucopleura* Wagner, Wiegmann's Arch. für Naturgesch., Jahrg. 9(1ª parte):367.
1900. *Promops milleri* J. A. Allen, Bull. Amer. Mus. Nat. Hist., 13:92, May 12, type from Guayabamba, Perú.
1904. *Promops barbatus* J. A. Allen, Bull. Amer. Mus. Nat. Hist., 20:228, June 27, type from La Union, Venezuela.
1956. *Eumops abrasus oaxacensis* Goodwin, Amer. Mus. Novit., 1757:2, March 8, type from Mazatlán, *ca.* 3000 ft., Oaxaca.

MARGINAL RECORDS.—Quintana Roo: 10 mi. E Yucatán boundary on road to Valladolid from Puerto Morelos (Ingles, 1959:385—incorrectly reported by Villa-R., 1967:451, as *Eumops maurus*). Jamaica: Kingston (Eger, 1974:6). Belize: Rockstone Pond (*op. cit.*:5). Honduras: 7 km. N Santa Bárbara (*op. cit.*:6). Costa Rica: Turrialba, 600 m. (26952 KU). Panamá (Handley, 1966:773): 7 km. SSW Changuinola; Bohio, thence into South America and northward to Panamá: Fort Kobbe (*ibid.*). Costa Rica: Villa Neily (Eger, 1974:5). Nicaragua: Hda. Mecatepe (Jones, *et al.*, 1971:22). El Salvador: Chilata (Eger, 1974:6). Guatemala: Finca Cipres. Oaxaca: Mazatlán, *ca.* 3000 ft. Tabasco: Teapa (*op. cit.*:7).

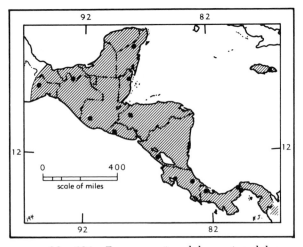

Map 194. *Eumops auripendulus auripendulus*.

Eumops underwoodi
Underwood's Mastiff Bat

Length of head and body, approx. 130; length of forearm, 65.3–71.5; zygomatic breadth, 16.7–18.8; length of upper tooth-row, 11.1–12.3. Upper parts rich ochraceous brown to dark brown, hairs whitish basally. Skull short, wide, and strongly ridged; interorbital region hourglass-shaped.

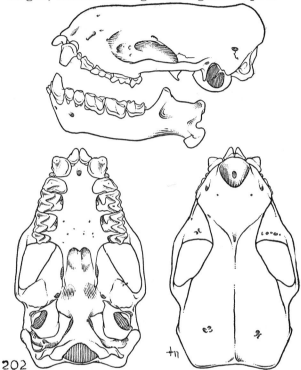

Fig 202. *Eumops underwoodi underwoodi*, 2 mi. N Apo, Tancítaro Mtn., Michoacán, No. 89461 U.M.M.Z., ♀, X 2.

Eumops underwoodi sonoriensis Benson

1947. *Eumops sonoriensis* Benson, Proc. Biol. Soc. Washington, 60:133, December 31, type from Rancho de Costa Rica, 270 ft., Río Sonora, Sonora.
1949. *Eumops underwoodi sonoriensis*, Hall and Villa, Univ. Kansas Publ., Mus. Nat. Hist., 1:446, December 27.

MARGINAL RECORDS.—Arizona: 1 mi. N, 11½ mi. E Topawa (Hoffmeister, 1959:16); *3 mi. WNW Buenos Aires* (Constantine, 1961:405); *2 mi. E Sasabe*. Sonora: type locality; 10 mi. NW Noche Buena.

Eumops underwoodi underwoodi Goodwin

1940. *Eumops underwoodi* Goodwin, Amer. Mus. Novit., 1075:2, June 27, type from El Pedrero, 6 km. N Chinaela [= Chinacla], La Paz, Honduras.

MARGINAL RECORDS.—Chihuahua: Naranjo (Eger, 1977:55). Jalisco: Juntas del Salitre, *ca.* 6 km. N

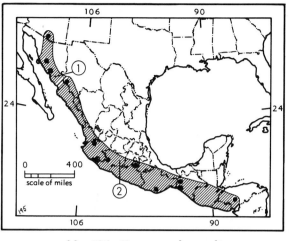

Map 195. *Eumops underwoodi*.

1. *E. u. sonoriensis*   2. *E. u. underwoodi*

Soyatlan del Oro (Mitchell, 1965:100, as *E. underwoodi* only). Morelos: Palo Bolero (Alvarez and Aviña, 1964:247, as *E. u. sonoriensis* [= *E. u. underwoodi*, Carter, *et al.*, 1966:498]). Chiapas: 3 mi. E Cintalapa, 1700 ft. (Carter, *et al.*, 1966:498). Honduras: type locality. Chiapas: 10 km. SE Tonalá (Alvarez and Aviña, 1964:248). Oaxaca: Río Jalatengo, Kilometer 178 on Puerto Angel road, 1300 m. (Arnold and Schonewald, 1972:173). Michoacán: Rancho Escondido, 2 mi. N Apo, Tancítaro Mtn. Colima: 26 km. W Pueblo Juárez (Alvarez and Aviña, 1964:247, as *E. u. sonoriensis* [= *E. u. underwoodi*, Carter, *et al.*, 1966:498]).

Eumops glaucinus
Wagner's Mastiff Bat

Length of head and body, 80–85; length of forearm, 57.7–66.0; condylobasal length of skull, 21.0–25.3; zygomatic breadth, 13.9–17.2; length of maxillary tooth-row, 8.8–10.5. Upper parts dark cinnamon brown, slightly paler below. Supraoccipital region extending posteriorly beyond lambdoidal crest; lachrymal ridges lacking. Tragus broad and truncate distally.

Eumops glaucinus floridanus (G. M. Allen)

1932. *Molossides floridanus* G. M. Allen, Jour. Mamm., 13:257, August 9, type from Melbourne, Brevard Co., Florida, possibly from stratum 2 of a Pleistocene deposit.
1971. *Eumops glaucinus floridanus*, Koopman, Amer. Mus. Novit., 2478:5, December 14.

MARGINAL RECORDS (Layne, 1974:389, 390).—Florida: type locality ("Pleistocene or early post-Pleistocene age"); Miami; *Coral Gables; Coconut Grove.*

Eumops glaucinus glaucinus (Wagner)

1843. *Dysopes glaucinus* Wagner, Wiegmann's Arch. für Naturgesch., Jahrg. 9, 1:368, type purportedly from Cuyaba, Mato Grosso, Brazil.
1906. *E*[*umops*]. *glaucinus*, Miller, Proc. Biol. Soc. Washington, 19:85, June 4.
1861. *Molossus ferox* Gundlach, Monatsb. preuss. Akad., Wiss., Berlin, p. 149, type from Cuba.
1889. *Nyctinomus orthotis* H. Allen, Proc. Amer. Philos., Soc., 26:561, December 18, type from Spanishtown, Jamaica.

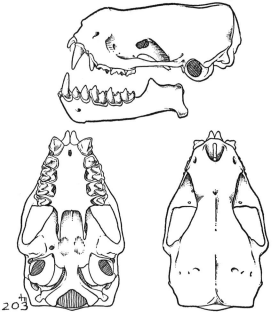

Fig. 203. *Eumops glaucinus glaucinus*, Jesús Carranza, Veracruz, No. 19234 K.U., ♀, X 2.

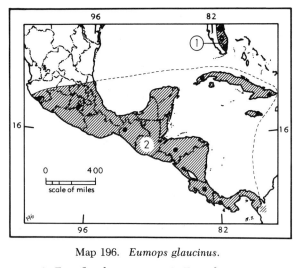

Map 196. *Eumops glaucinus*.

1. *E. g. floridanus*   2. *E. g. glaucinus*

MARGINAL RECORDS.—Cuba: Pinar del Río; Guantánamo. Jamaica: St. Andrews; *Spanishtown,*

thence southward into South America and northward to Panamá: David, 150 ft. (Handley, 1966:773). Costa Rica: Finca Lornessa, *ca.* 2 km. NW Santa Ana (Gardner, *et al.*, 1970:725). Honduras: Comayagua, 580 m. (LaVal, 1969:821). Chiapas: 3 mi. E Cintalapa, 1700 ft. (Carter, *et al.*, 1966:499). Morelos. Colima: near Pueblo Juárez [= La Magdalena] (Gardner, 1962:103).

Eumops hansae Sanborn
Sanborn's Mastiff Bat

1932. *Eumops hansae* Sanborn, Jour. Mamm., 13:356, November 2, type from Colonia Hansa, near Joinville, Santa Catherina, Brazil.
1955. *Eumops amazonicus* Handley, Proc. Biol. Soc. Washington, 68:177, December 31, type from Manáos [= Manaus], Amazonas, Brazil. Regarded as inseparable from *hansae* by Gardner, *et al.*, Jour. Mamm., 51:726, 727), November 30, 1970.

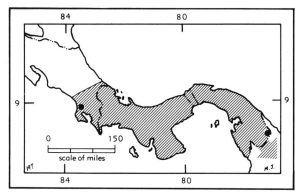

Map 197. *Eumops hansae*.

Measurements of the male holotype of *E. hansae*, a male from Costa Rica, female from Brazil (holotype of *E. amazonicus*), and female from Panamá: forearm, 41.6, 40.2, 36.9, 37.4; greatest length of skull, 21.7, 20.7, 18.4, 19.4; zygomatic breadth, 12.8, 11.7, 10.7, 11.0; postorbital constriction, 4.3, 4.2, 4.0, 4.1; maxillary tooth-row, 8.1, 7.5, 6.8, 7.2. Resembles *E. bonariensis* but

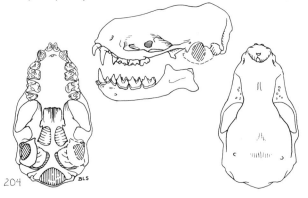

Fig. 204. *Eumops hansae*, Tacarcuna Village Camp, Darién, Panamá, No. 310278 U.S.N.M., ♀, X 2.

differs as follows: color darker; tips of upper incisors widely separated; lower incisors in a straight line, not crowded or bunched; palate extends behind last molars; basisphenoid pits large, deep, and well defined. Like many other molossids, this species shows a high degree of sexual dimorphism.

MARGINAL RECORDS.—Costa Rica: 10 mi. S Palmar Sur (Gardner, *et al.*, 1970:726). Panamá: Tacarcuna Village, 1950 ft. (Handley, 1966:773), thence into South America.

Eumops nanus (Miller)
Dwarf Mastiff Bat

1900. *Promops nanus* Miller, Ann. Mag. Nat. Hist., ser. 7, 6:470, November, type from Bogava [= Bugaba], 800 ft., Chiriquí, Panamá.
1906. *Eumops nanus* Miller, Proc. Biol. Soc. Washington, 19:85, June 4.

Length of head and body, approx. 42; length of forearm, 39–49; greatest length of skull, 16.4–20.1; zygomatic breadth, 10.4–12.4; length of upper tooth-row, 6.6–8.1. Upper parts reddish brown to blackish brown; underparts drab to gray. Skull resembling that of *E. glaucinus* but smaller, relatively broader across rostrum, and with crests less developed. Differs from *E. hansae* as outlined in the account of that species.

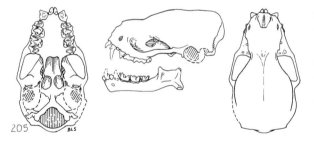

Fig. 205. *Eumops nanus*, Tole, Panamá, No. 331971 U.S.N.M., ♀, X 2.

Although both Eger (1977:35) and Sanborn (1932:356) regarded *E. nanus* as a subspecies of *E. bonariensis* (Peters), *E. nanus* is here retained as a species (see Handley, 1976:40) because actual intergradation has not yet been shown to occur between it and *Eumops bonariensis delticus* Thomas.

MARGINAL RECORDS.—Yucatán: 2½ km. NW Dzitya (Birney, *et al.*, 1974:11), thence along Caribbean Coast to Panamá: Boquerón; *type locality*, and into South America. Thence northwestward along coast to Nicaragua: Potosí, 5 m., Chinandega (114142 KU). Veracruz: 3–4 mi. ENE Tlacotalpan (Alvarez and

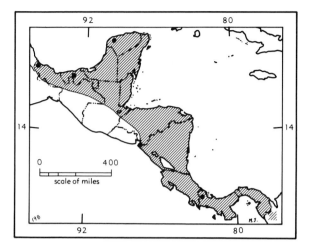

Map 198. *Eumops nanus*.

Aviña, 1964:248). Tabasco: Huastecas, 18 mi. N, 2 mi. E Teapa (*ibid.*).

Genus Promops Gervais
Mastiff Bats

1855. *Promops* Gervais, Mammifères, *in* [Castelnau] Expéd. dans les parties centrales de l'Amér. du Sud . . . , pt. 7, p. 58. Type, *Promops ursinus* Gervais [= *Molossus nasutus* Spix].

Skull short, broad (somewhat rounded when viewed from above), robust; sagittal crest pronounced; rostrum markedly short and deep; palate strongly dome-shaped; tips of upper incisors diverging. Dentition, i. $\frac{1}{2}$, c. $\frac{1}{1}$, p. $\frac{2}{2}$, m. $\frac{3}{3}$. Ears short and rounded, extending barely to nostril when laid forward; pad of muzzle small but distinct and without processes.

Promops centralis
Thomas' Mastiff Bat

Selected measurements of the type, an adult female, are: length of forearm, 54; greatest length of skull (occiput to base of incisors), 20.2; length of upper tooth-row, 8.3. Upper parts dark brown to glossy black, the hairs whitish basally; underparts slightly paler; fur long, approx. 5 mm. in center of back. Other characters as for the genus.

Promops centralis centralis Thomas

1915. *Promops centralis* Thomas, Ann. Mag. Nat. Hist., ser. 8, 16:62, July, type from northern Yucatán. Ojasti and Linares, Acta Biol. Venezuelica, 7:433, May, 1972, concluded that *P. centralis* consisted of two subspecies, *P. c. centralis* and *P. c. davisoni*.

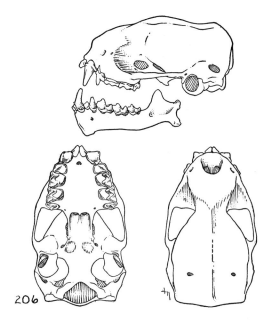

Fig. 206. *Promops centralis centralis*, Baja Verapaz, Guatemala, No. 42088 F.M.N.H., ♀, X 2.

MARGINAL RECORDS.—Yucatán: Colegio Peninsular (Birney, *et al.*, 1974:12). Guatemala: Libertad. Panamá (Handley, 1966:773): Fort Clayton; *Corozal; Fort Amador*, thence into South America, and northward along coast to Nicaragua: 5 mi. N, 1 mi. W San Juan del Sur, Rivas (Baker and Jones, 1975:4). Honduras: *La Esperanza* (LaVal, 1969:821); El Pedrero. Guatemala: Salamá. Oaxaca: Tehuantepec (Goodwin, 1969:114). Colima: 26 km. W Pueblo Juárez (Alvarez and Aviña, 1964:249). Jalisco: Tenamastlán [= Tenamaxtlán] (*ibid.*); *9 mi. NW Cuautla, 5900 ft.* (Watkins, *et al.*, 1972:33). See Map 192.

Genus **Molossus** É. Geoffroy St.-Hilaire
Mastiff Bats

1805. *Molossus* É. Geoffroy St.-Hilaire, Ann. Mus. Hist. Nat. Paris, 6:151. Type, *Vespertilio molossus* Pallas. See Miller, Proc. U.S. Nat. Mus., 46:85–92, August 23, 1913, and Husson, Zool. Verhand., Rijksmuseum Nat. Hist. Leiden, 58:251–264, November 13, 1962.
1811. *Dysopes* Illiger, Prodromus systematis mammalium et avium . . . , p. 76, a renaming of *Molossus* É. Geoffroy St.-Hilaire.

Skull resembling that of *Promops* but sagittal and lambdoidal crests better developed, especially in males; palate arched anteroposteriorly but not strongly domed; basisphenoid pits well developed; height of upper incisor less than width of crown through heel, tips converging. Dentition, i. $\frac{1}{1}$, c. $\frac{1}{1}$, p. $\frac{1}{2}$, m. $\frac{3}{3}$. Tragus minute; fur short, usually 3 mm. or less on center of back. Males distinctly larger than females.

KEY TO NORTH AMERICAN MAINLAND SPECIES OF MOLOSSUS

1. Forearm more than 47 (males) or 46.5 (females).
   2. Floor of basisphenoid pits 1 mm. or longer; dorsal hairs with contrasting white or gray bases.   *M. sinaloae,*   p. 254
   2'. Floor of basisphenoid pits less than 1 mm. long; dorsal hairs with little contrast between tips and bases. *M. ater,*   p. 252
1'. Forearm less than 46.5 (males) or 45.5 (females).
   3. Fur approx. 3 mm. on center of back; pale basal band on dorsal hairs.
                  M. molossus,   p. 255
   3'. Fur approx. 2 mm. on center of back; little contrast between tips and bases of dorsal hairs.
     4. Forearm more than 43 (males) or 41 (females); greatest length of skull more than 20.5 (males) or 19.5 (females); floor of basisphenoid pits more than .5 mm. long.
              M. pretiosus,   p. 253
     4'. Forearm less than 43 (males) or 41.5 (females); greatest length of skull less than 20 (males) or 19 (females); floor of basisphenoid pits less than .5 mm. long.
              M. bondae,   p. 254

Molossus ater
Black Mastiff Bat

Selected measurements: total length, 120–141; tail, 35–55; ear, 14–20; forearm, 47.1–51.2; greatest length of skull, 21.5–23.5; zygomatic breadth, 13.3–14.3; maxillary tooth-row, 7.6–8.3. Upper parts rich russet or blackish; underparts slightly paler. Skull robust; sagittal crest and lambdoidal crest well developed.

Molossus ater nigricans Miller

1902. *Molossus nigricans* Miller, Proc. Acad. Nat. Sci. Philadelphia, 54:395, September 12, type from Acaponeta, Nayarit.
1962. *Molossus ater nigricans*, Jones, Alvarez, and Lee, Univ. Kansas Publ., Mus. Nat. Hist., 14:155, May 18. (See also Goodwin, Amer. Mus. Novit., 1994:4, March 8, 1960.)
1955. *Cynomops malagai* Villa-R., Acta Zool. Mexicana, 1(4):2, September 15, type from Tuxpan de Rodríguez Cano, 20° 57' N, 97° 24' W, 4 m., Veracruz. (See Jones, Proc. Biol. Soc. Washington, 78:93, July 21, 1965, for status of *C. malagai*.)

MARGINAL RECORDS.—Sinaloa:1 mi. S Pericos (Jones, *et al.*, 1962:155). Durango: Santa Ana (Jones, 1964a:752). Nayarit: type locality. Jalisco: El Zapote

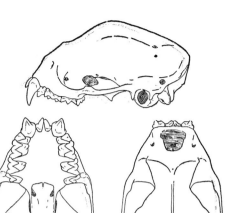

Fig. 207. *Molossus ater nigricans*, Río Atoyac, 8 km. NW Potrero, Veracruz, No. 17857 K.U., ♀, X 2.

(Watkins, *et al.*, 1972:34). Morelos: Hda. Cocoyotla. Querétaro: 17⅗ km. NW Jalpan (Spenrath and LaVal, 1970:396). San Luis Potosí: El Salto. Tamaulipas (Alvarez, 1963:417): Rancho Santa Rosa, 25 km. N, 13 km. W Ciudad Victoria, 260 m.; *3 mi. NE Guemes*; Altamira. Yucatán: Pisté (92038 KU). Quintana Roo: Felipe Carrillo Puerto (92053 KU). Guatemala: Toocog, 15 km. SE La Libertad, 540 ft. (Jones, 1966:466). Nicaragua: Corozo, 15 km. NNE Jalapa, 660 m. (Jones, *et al.*, 1971:22). Panamá: 2 mi. W El Volcán, 4100 ft. (Handley, 1966:773), thence southeastward into South America, and northwestward to Panamá: La Concepción, 800 ft. (*ibid.*). Costa Rica: Boruca. Nicaragua: Potosí, 5 m. (Jones, *et al.*, 1971:22). Guatemala: Astillero (Jones, 1966:466). Oaxaca: Tehuantepec (Goodwin, 1969:114).

Molossus pretiosus
Miller's Mastiff Bat

Selected measurements of 59 adults from Nicaragua: length of head and body, 63–85; total length, 105–130; length of forearm, 41.0–46.1 (49.8 in one from Oaxaca); greatest length of skull, 19.5–22.2; zygomatic breadth, 12.0–13.7; maxillary tooth-row, 7.1–7.9. Pelage uniform Seal Brown, but variation from black to rufous not uncommon; skull short and broad with relatively low and weak sagittal and lambdoidal crests. Because *M. pretiosus* has been widely confused with *M. ater*, the former may be more common than indicated by the few records in the literature.

Molossus pretiosus macdougalli Goodwin

1956. *Molossus pretiosus macdougalli* Goodwin, Amer. Mus. Novit., 1757:3, March 8, type from San Blas, 3 km. SE Tehuantepec, approx. 100 ft., Oaxaca. The name combina-

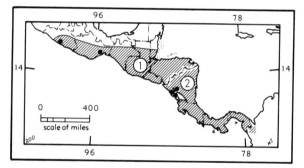

Map 200. *Molossus pretiosus.*

1. *M. p. macdougalli*   2. *M. p. pretiosus*

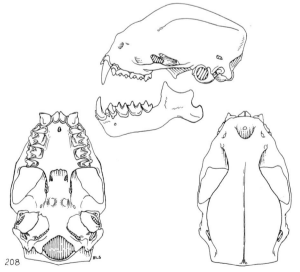

Fig. 208. *Molossus pretiosus pretiosus*, 3 km. N, 4 km. W Diriamba, 600 m., Carazo, Nicaragua, No. 111264 K.U., ♂, X 2.

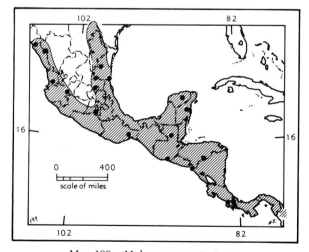

Map 199. *Molossus ater nigricans.*

tion *Molossus macdougalli* is used by Goodwin, Bull. Amer. Mus. Nat. Hist., 141 (Art. 1):114, April 30, 1969.

MARGINAL RECORDS (Goodwin, 1969:114).—Guerrero: Acapulco. Oaxaca: Tehuantepec; *type locality.*

Molossus pretiosus pretiosus Miller

1902. *Molossus pretiosus* Miller, Proc. Acad. Nat. Sci. Philadelphia, 54:396, September 12, type from La Guaira, Venezuela.

MARGINAL RECORDS (Jones, *et al.*, 1971:25).—Nicaragua: Los Cocos, 14 km. S Boaco, 220 m.; *San Francisco, 19 km. S, 2 km. E Boaco, 200 m.*, thence southward along Caribbean Coast into South America, thence northward along Pacific Coast to Nicaragua: 3 km. N, 4 km. W Diriamba; 6 mi. WSW Managua.

Molossus sinaloae J. A. Allen
Allen's Mastiff Bat

Selected measurements: total length, 115–148; tail, 38–54; ear, 12–17; forearm, 46.2–48.6; greatest length of skull, 20.0–21.5; zygomatic breadth, 11.5–12.6; maxillary tooth-row, 7.1–7.9.

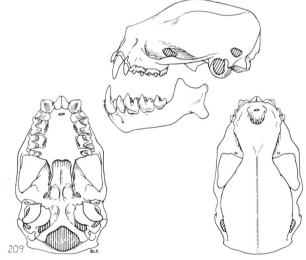

Fig. 209. *Molossus sinaloae sinaloae*, Santa Rosa, 17 km. N, 15 km. E Boaco, 300 m., Boaco, Nicaragua, No. 111265 K.U., ♂, X 2.

Molossus sinaloae sinaloae J. A. Allen

1906. *Molossus sinaloae* J. A. Allen, Bull. Amer. Mus. Nat. Hist., 22:236, July 25, type from Escuinapa, Sinaloa.

MARGINAL RECORDS.—Sinaloa: type locality. Jalisco: Teuchitlán (Watkins, *et al.*, 1972:34). Morelos (Alvarez and Aviña, 1964:249, 250): Jintepec; *Oaxtepec.* Puebla: 2 mi. E Izúcar de Matamoras (61289 KU). Yucatán: Calcehtok. Guatemala: Bobós, 200 ft. (Jones, 1966:466). Nicaragua: S side Río Mico, El Recreo, 25 m. (Jones, *et al.*, 1971:27). Costa Rica: Puerto

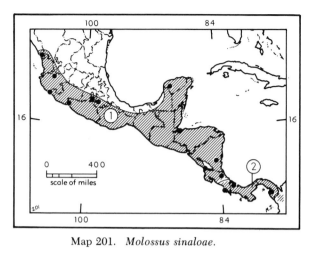

Map 201. *Molossus sinaloae.*

1. *M. s. sinaloae* 2. *M. s. trinitatus*

Viejo (127975 KU). Nicaragua: San Juan del Sur, 20 m. (Jones, *et al.*, 1971:27). Guerrero: Zacatula (Alvarez, 1968:30). Colima: 2 km. E Santiago (87421 KU).

Molossus sinaloae trinitatus Goodwin

1959. *Molossus trinitatus* Goodwin, Amer. Mus. Novit., 1967:1, October 29, type from Belmont, Port of Spain, Trinidad.
1966. *M[olossus]. s[inaloae]. trinitatus*, Handley, Ectoparasites of Panama, Field Mus. Nat. Hist., p. 773, November 22.

MARGINAL RECORDS.—Panamá: 7 km. SSW Changuinola (Handley, 1966:773), thence along coast to South America, and then northward to Panamá: El Real. Costa Rica: 2 mi. W Rincón (Armstrong, 1969:809, as *M. sinaloae* only).

Molossus bondae J. A. Allen
Bonda Mastiff Bat

1904. *Molossus bondae* J. A. Allen, Bull. Amer. Mus. Nat. Hist., 20:228, June 29, type from Bonda, Santa Marta, Colombia.

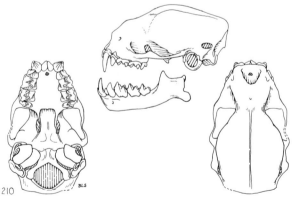

Fig. 210. *Molossus bondae*, Turrialba, Costa Rica, No. 57151 K.U., ♀, X 2.

Selected measurements of five females from Costa Rica: forearm, 39.1–41.4; greatest length of skull, 18.6–19.0; zygomatic breadth, 11.4–11.9; maxillary tooth-row, 6.4–6.8. Upper parts either dark brown 'or approx. rich russet; underparts paler than back.

MARGINAL RECORDS.—Nicaragua: Greytown (Miller, 1913:89). Costa Rica (Gardner, et al., 1970:727): Cariari, Río Tortuguero; Turrialba, 1950 ft. Panamá (Handley, 1966:772): Almirante; Fort Sherman; Tacarcuna Village, thence southeastward into South America—Alvarez and Ramírez-P. (1972:175) perhaps correctly identify as this species a specimen from 6 km. NE San Miguel, Isla Cozumel, Quintana Roo, as well as Honduran specimens from Los Encuentros and El Manteado (see Goodwin, 1942:145), although the unfinished studies of Jones, et al. (1971:25) suggest that the specimens from all three places represent a subspecies of M. molossus.

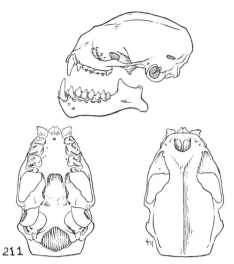

Fig. 211. *Molossus molossus coibensis*, Barro Colorado Island, Canal Zone, Panamá, No. 45092 K.U., ♂, X 2.

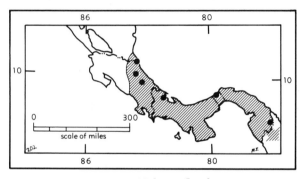

Map 202. *Molossus bondae*.

Molossus molossus
Pallas' Mastiff Bat

Selected measurements: length of head and body, 56.6–58.0 (66.0 in *M. m. fortis*); tail, 35–40; forearm, 37.8–40.5; tibia and hind foot, 20.4–24; greatest length of skull, 15.4–17.5; condylobasal length, 14.0–16.2; maxillary tooth-row, 5.4–6.4 (7.2 in *M. m. fortis*); mandibular tooth-row (c–m3), 5.0–7.3. Upper parts dark brown to grayish brown; underparts paler. Gular sac area whitish; basal part of hairs whitish to brownish.

Molossus molossus aztecus Saussure

1860. *M[olossus]. aztecus* Saussure, Revue et Mag. Zool., Paris, ser. 2, 12:285, July, type from Amecameca, México. Considered as a subspecies of *M. major* [= *M. molossus*] by Hershkovitz, Proc. U.S. Nat. Mus., 99:454, May 10, 1949.

MARGINAL RECORDS (Gardner, 1966:5, unless otherwise noted).—Sinaloa: Alisos, *ca*. 50 km. (by road) NNE Badiraguato. Jalisco: Los Masos. México: type

locality. San Luis Potosí: Río Verde (Dalquest, 1953:69). Tamaulipas: 3 km. N El Limón (Alvarez and Aviña, 1964:250). Belize: Mountain Pine Ridge (Murie, 1935:20). Nicaragua: Santa Rosa, 17 km. N, 15 km. E Boaco, 300 m. (Jones, *et al.*, 1971:23). Costa Rica: La Irma (2927 LaVal at KU). Nicaragua (Jones, *et al.*, 1971:23): Rivas, 60 m.; Potosí, 5 m. El Salvador: San Salvador (Felten, 1957:13, as *M. major aztecus* and *M. tropidorhynchus coibensis*). Chiapas: Huehuetan (Miller, 1913:91). Oaxaca: Salina Cruz (Goodwin, 1969:115, as *M. aztecus aztecus*); 5 mi. N Juchatengo. Colima: Tlapeixtes, 4 km. ENE Manzanillo (Gardner, 1966:4, 5). Jalisco: Río de Talpa, Talpa de Allende.

Molossus molossus coibensis J. A. Allen

1904. *Molossus coibensis* J. A. Allen, Bull. Amer. Mus. Nat. Hist., 20:227, June 29, type from Coiba Island, Panamá. Considered as a subspecies of *M. major* [= *M. molossus*] by Hershkovitz, Proc. U.S. Nat. Mus., 99:454, May 10, 1949, but arranged as a subspecies of *M. tropidorhynchus* by Felten, Senckenbergiana Biol., 38:14, January 15, 1957.

MARGINAL RECORDS.—Panamá: Fort Sherman (Handley, 1966:772, as *M. coibensis*), thence southeastward into South America, and northwestward to Panamá (*ibid.*): Pacora; type locality; Boquerón.

Molossus molossus debilis Miller

1913. *Molossus debilis* Miller, Proc. U.S. Nat. Hist. Mus., 46:90, August 23, type from St. Kitts, Lesser Antilles.
1970. *Molossus molossus debilis*, Jones and Phillips, Studies on the fauna of Curaçao and other Caribbean islands, Utrecht, 32:143, March.

MARGINAL RECORDS (Jones and Phillips, 1970:137, unless otherwise noted).—Anguilla; Barbuda; Antigua; Montserrat; Nevis; *type locality*; St. Eustatius; St. Croix (Koopman, 1975:4); *St. Martin*.

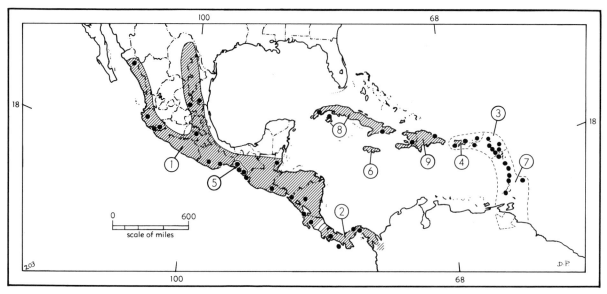

Map 203. *Molossus molossus.*

1. *M. m. aztecus*	4. *M. m. fortis*	7. *M. m. molossus*
2. *M. m. coibensis*	5. *M. m. lambi*	8. *M. m. tropidorhynchus*
3. *M. m. debilis*	6. *M. m. milleri*	9. *M. m. verrilli*

Molossus molossus fortis Miller

1913. *Molossus fortis* Miller, Proc. U.S. Nat. Mus., 46:89, August 23, type from Luquillo, Puerto Rico.
1975. *M(olossus). m(olossus). fortis,* Koopman, Amer. Mus. Novit., 2581:4, June 20.

MARGINAL RECORDS (Koopman, 1975:3, 4, unless otherwise noted).—Virgin Gorda; *St. John; St. Thomas; Culebra.* Puerto Rico: San Germán (Anthony, 1918:363); type locality.

Molossus molossus lambi Gardner

1966. *Molossus aztecus lambi* Gardner, Los Angeles Co. Mus. Contrib. Sci., 111:1, November 9, type from 11 km. NW Escuintla, Chiapas, México.

MARGINAL RECORDS (Gardner, 1966:3).—Chiapas: 15 mi. SW Las Cruces; type locality; 20 km. SE Pijijiapan. Not found: Chiapas: 12 mi. E Ortiz Rubio on Villa Flores Road.

Molossus molossus milleri Johnson

1952. *Molossus milleri* Johnson, Proc. Biol. Soc. Washington, 65:197, November 5.
1974. *Molossus molossus milleri,* Varona, Catálogo Mam. viv. y exting. de las Antillas, Acad. Cienc. Cuba, p. 42.
1838. *Molossus fuliginosus* Gray, Mag. Zool. Bot., 2:501, February, based on material from an unknown locality; lectotype from Jamaica. Not *Molossus fuliginosus* Cooper, 1837. [= *Tadarida cynocephala* (Le Conte)]. Known precisely only from Jamaica.

"Like *Molossus fortis* . . . but rostral part of skull obviously shortened . . ." (Miller, 1913:90). Varona (1974:43) "records *Molossus molossus* ssp." from Gran Caimán Isla [= Grand Cayman]. Conceivably the bat(s) are the Jamaican *M. m. milleri.*

Molossus molossus molossus (Pallas)

1766. *V[espertilio]. molossus* Pallas, Miscellanea Zool. . . . , p. 49, Spicilegia zoologica, . . . , fasc. 3, p. 8, 1767. Lectotype from Martinique (Husson, Zool. Verhand., Rijksmuseum Nat. Hist. Leiden, 58:257, November 13, 1962).
1962. *Molossus molossus,* Husson, Zool. Verhand., Rijksmuseum Nat. Hist. Leiden, 58:251, November 13.
1792. *V[espertilio]. mol[ossus]. major* Kerr, The animal kingdom . . . , p. 97. Type locality, Martinique.
1792. *V[espertilio]. mol[ossus]. minor* Kerr (*loc. cit.*). Type locality, Martinique, Lesser Antilles.
1805. *Molossus obscurus* É. Geoffroy St.-Hilaire, Ann. Mus. d'Hist. Nat. Paris, 6:155, type locality restricted to Martinique by Husson, Zool. Verhand., Rijksmuseum Nat. Hist. Leiden, 58:258, November 13, 1962.
1805. *Molossus longicaudatus* É. Geoffroy St.-Hilaire, Ann. Mus. d'Hist. Nat. Paris, 6:155 (see Miller, Proc. U.S. Nat. Mus., 46:90, August 23, 1913, who regards the type locality as Martinique).
1805. *Molossus fusciventer* É. Geoffroy St.-Hilaire, Ann. Mus. d'Hist. Nat. Paris, 6:155 (see Miller, Proc. U.S. Nat. Mus., 46:90, August 23, 1913, who thinks "Probably from Martinique").

MARGINAL RECORDS (Jones and Phillips, 1970:137).—Lesser Antilles: Guadeloupe southward (excluding the Grenadines) into South America.

Molossus molossus tropidorhynchus Gray

1839. *Molossus tropidorhynchus* Gray, Ann. Nat. Hist., 4:6, September, type from Cuba.
1974. *Molossus molossus tropidorhynchus*, Varona, Catálogo Mam. viv. y exting. de las Antillas, Acad. Cienc. Cuba, p. 42.

Head and body *ca.* 55. Resembles *M. m. pygmaeus* of Curaçao, West Indies, and *M. m. coibensis* but paler (usually between Raw Umber and Wood Brown), and without evident trace of drab.

MARGINAL RECORDS.—Cuba: Pinar del Río; El Cobre; Nueva Genora [= Gerona], Isla de Pinos (Dusbábek, 1969:323).—Varona (1974:43) records "*Molossus molossus* ssp." from Gran Caimán Isla [= Grand Cayman]. Conceivably the bat(s) are *M. m. tropidorhynchus*.

Molossus molossus verrilli J. A. Allen

1908. *Mollossus* [*sic*] *verrilli* J. A. Allen, Bull. Amer. Mus. Nat. Hist., 24:581, September 11, type from Samaná, Dominican Republic.
1951. *Molossus major verrilli*, Hershkovitz, Fieldiana-Zool., Chicago Nat. Hist. Mus., 31:558, July 10. Husson, Zool. Verhand., Rijksmuseum Nat. Hist. Leiden, 58:257, November 13, 1962, considered *major* to be a synonym of *molossus*.

MARGINAL RECORDS.—Dominican Republic: type locality. Haiti: Pétionville.

ORDER **PRIMATES**—Man, Monkeys, Lemurs, and Allies

The Recent (living) members may be characterized as follows: plantigrade; pentadactyl; ethmoturbinals 4–3; clavicle present, well developed; teeth tuberculosectorial; 3rd incisor and 1st premolar absent above and below; orbit directed more or less anteriorly and separated from temporal fossa by bony bar or partition.

SUBORDER **ANTHROPOIDEA** Man, Monkeys, and Allies

Incisors chisel-shaped; lower canine pointed (caniniform); molars bunodont; auditory bullae little inflated; ethmoturbinals 3; tibia and fibula always separate; penis pendulous, with or without baculum.

The Suborder Anthropoidea as used here excludes primates belonging to the Suborder Prosimii.

Key to Families and Genera of North American Primates

1. Premolars $\frac{2-2}{2-2}$.
   2. Tail absent (externally); forelimbs not modified for walking; C less than twice as long as I2.
        family Hominidae, p. 273
        Homo, p. 273
   2'. Tail present (approx. as long as head and body); forelimbs modified for walking; C more than twice as long as I2 (introduced from Africa). . . . . family Cercopithecidae, p. 271
        Cercopithecus, p. 271
1'. Premolars $\frac{3-3}{3-3}$.
   3. Nape chestnut; top of head between ears white; claws (true nails absent) projecting beyond fleshy ends of digits for a distance more than width of base of claw; molars $\frac{2-2}{2-2}$ (32 teeth) in adults. . . . . family Callithricidae, p. 270
        Saguinus, p. 270
   3'. Nape not chestnut; top of head between ears not white; nails (true claws absent) not projecting beyond fleshy ends of digits for a distance as much as width of base of nail; molars $\frac{3-3}{3-3}$ (36 teeth) in adults. . . . . family Cebidae, p. 259
      4. Back uniformly gray; hair of face with 3 narrow black lines (one from outer corner of each eye to nape and a 3rd on center of face and head; these lines border whitish areas, one above each eye); greatest breadth of skull is across orbits. *Aotus*, p. 259
      4'. Back reddish or black; hair of face lacking 3 black lines; greatest breadth of skull not across orbits.
        5. Forearms distinctly reddish or yellow; overall length of upper dental arcade less than 25. . . . . *Saimiri*, p. 265
        5'. Forearms black; overall length of upper dental arcade more than 25.
           6. Topknot absent and face black; less than 34% of length of skull (gnathion to inion) lies behind glenoid fossa; less than ⅛ of skull projects behind lower jaw (when skull, with lower mandible in place, rests on horizontal surface on inferior margin of lower jaw); height of vertical ramus of lower jaw (as measured from notch between coronoid and articular processes) more than distance between external auditory meatus and infraorbital foramen. . . . . *Alouatta*, p. 260
           6'. Topknot (tuft of hair directed upward and forward) present (*Ateles*) or face, chest, and shoulders white (*Cebus*); more than 34% of skull lies behind glenoid fossa; more than ⅛ of skull lies behind lower jaw; height of vertical ramus of lower jaw less than distance between external auditory meatus and infraorbital canal.
              7. Pelage black or reddish (without white face, chest, and shoulders); topknot present; tail averaging more than 1½ times as long as head and body; distal ¼ or more of underside of tail bare; thumbs normally absent; occlusal area of M3 not smaller than that of P2; occlusal area of m3 larger than that of p3; inferior margin of lower mandible approx. straight. . . . . *Ateles*, p. 266
              7'. Pelage black except for white face, chest, and shoulders; topknot absent; tail averaging less than 1½ times as long as head and body; underside of tail haired for entire length; occlusal area of M3 less than that of P2; occlusal area of m3 smaller than that of p3; inferior margin of lower mandible with angular part produced downward. . . . . *Cebus*, p. 263

SUPERFAMILY CEBOIDEA
New World Monkeys

Premolars $\frac{3=3}{3=3}$; nose broad in most (not all) species. This superfamily includes all the New World monkeys. All Old World monkeys have premolars $\frac{2=2}{2=2}$ and are narrow-nosed.

FAMILY CEBIDAE—Capuchins, Howlers, Spider Monkeys, and Allies

Dentition, i. $\frac{2}{2}$, c. $\frac{1}{1}$, p. $\frac{3}{3}$, m. $\frac{3}{3}$; nose usually broad and nostrils opening laterally; tail present and well developed; bare ischial callosities and cheek pouches absent; no bony external auditory meatal tube. Arboreal in habit.

Possibly one or more kinds of monkeys of this family lived in the West Indies in post-Columbian time. Williams and Koopman (1952:1–16) describe: (1) remains of *Montaneia anthropomorpha* Ameghino 1910 [= *Ateles fusciceps* Gray 1866] from Cuba (probably of pre-Columbian age); (2) the distal end of a tibia of a small primate from Hispaniola (probably of pre-Columbian age); and (3) *Xenothrix mcgregori*, new genus and new species from Jamaica, p. 12 (probably of pre-Columbian age). Rímoli (1977:8) described *Saimiri bernensis*, new species from Hispaniola (probably of pre-Columbian age).

If the Cuban *Ateles fusciceps*, based on 16 teeth of one animal found in a burial cave, was imported from the mainland of South America or Panamá as Williams and Koopman think likely, the currently used subspecific name "*Ateles fusciceps robustus* J. A. Allen" is to be replaced by *Ateles fusciceps anthropomorphus* (F. Ameghino), and the citation to the original description will be:

1910. *Montaneia anthropomorpha* F. Ameghino, An. Mus. Nac. Buenos Aires, ser. 3, vol. 13:317, September 16, type from human burial cave near Sancti Spíritus, Cuba.

Genus **Aotus** Illiger—Douroucouli Monkeys

1811. *Aotus* Illiger, Prodromus systematis mammalium et avium, p. 71. Type, *Simia trivirgata* Humboldt, 1809.
1823. *Nyctipithecus* Spix, Simiarum et vespertilionum Brasiliensium . . . , p. 24. Included *Nyctipithecus felinus* Spix and *N. vociferans* Spix.
1824. *Nocthora* F. Cuvier, Hist. Nat. Mamm., 5, livr. XLIII, pl., August. Type, *Simia trivirgata* Humboldt.

Size small (see measurements of species); muzzle conical and truncated; tail not prehensile; eyes enormously enlarged; laryngeal pouch absent; thoracolumbar vertebrae 22; brain small, the external surface of cerebral hemispheres rela-

tively smooth and almost devoid of convolutions; occipital region of braincase not produced posteriorly; premaxillaries small and distinctly separated from each other; nails small and slightly compressed laterally.

There seems to be more individual variation in color in this genus than in some other genera of Central American primates. The differences in color have been used in several instances to separate some of the nominal species. Possibly there is only one species having several subspecies.

KEY TO NORTH AMERICAN SPECIES OF AOTUS

1. Occuring on Azuero Peninsula, Panamá.
   A. bipunctatus, p. 259
1'. Occurring in Canal Zone, Panamá, and eastward into South America.
   A. trivirgatus, p. 260

Aotus bipunctatus Bole
Bole's Douroucouli

1937. *Aotus bipunctatus* Bole, Sci. Publs., Cleveland Mus. Nat. Hist., 7:152, August 31, type from Paracoté, 3 mi. E Montijo Bay, and 1½ mi. S mouth Río Angulo, Veraguas, Panamá. Known only from type locality.

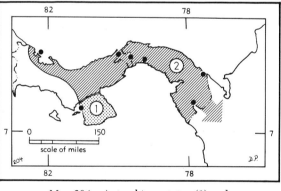

Map 204. *Aotus bipunctatus* (1) and *Aotus trivirgatus griseimembra* (2).

External measurements of a male and a female (the type) are, respectively: 839, 806; 430, 413; 101, 95. Greatest length of skull,—, 61.5. Back and sides pale wood brown lightly washed with ochraceous-tawny along mid-dorsal region; underparts light ochraceous-buff to slightly paler, slightly darker in axillary and inguinal regions; tail suffused with russet basally, becoming darker (blackish) distally, distal half black; supraorbital spots white. Skull larger than in *A. trivirgatus. A. bipunctatus* almost surely is no more than a subspecies of *A. trivirgatus.*

Aotus trivirgatus
Three-banded Douroucouli

Total length of the type of *A. t. griseimembra,* 1047; occipitonasal length, 58.2. Average external measurements of five adults (Boca de Cupe, 2; Gatún, 1; Cana, 1; Río Indio, 1; all in Panamá): 657 (620–685); 360 (325–390); 88 (83–90). Occipitonasal length, 57.2 (56.5–59.8). Pelage short (25), semiwooly, brownish gray, changing to black on distal two-thirds of tail and backs of hands and feet; underparts of body ochraceous white; face bare; top of head blackish but with triangular white area above each eye, the two white areas separated by a narrow (3 mm. wide) anterior extension of median line from black area on top of head (description based on specimens from Panamá).

These monkeys are nocturnal, and several together have been found in hollow trees. The species has a wide geographic range in South America.

A specimen of *Aotus vociferans* (Spix) is said to have been taken in the forest of Quindín, Costa Rica (Sclater, Proc. Zool. Soc. London, 1872, p. 3), but this record is almost certainly erroneous [Elliot, Rev. Primates, vol. 2, pp. 14–15 (1912), June, 1913]. Hershkovitz (Proc. U.S. Nat. Mus., 98:408, May 10, 1949) regards the specimen as *Aotus trivirgatus lemurinus* (I. Geoffroy St.-Hilaire) that occurs naturally only in Colombia, South America.

A specimen of *Aotus* received alive by the Zoological Society of London from San Juan del Norte, Nicaragua, was named by Sclater (*loc. cit.*) as a new species, *Nyctipithecus rufipes.* As pointed out by Hershkovitz (1949:405), the "original description and color plate indicate that the type most probably originated in Brazil and was transported as a pet to Nicaragua. The monkey cannot be identified with *griseimembra.* . . . So far, there is not one authenticated record of occurrence of the genus in Central America outside of Panamá."

Aotus trivirgatus griseimembra Elliot

1912. *Aotus griseimembra* Elliot, Bull. Amer. Mus. Nat. Hist., 31:33, March 4, type from Hda. Cincinnati, Santa Marta, Colombia.
1949. *Aotus trivirgatus griseimembra,* Hershkovitz, Proc. U.S. Nat. Mus., 98:402, May 10.
1914. *Aotus zonalis* Goldman, Smiths. Miscl. Coll., 63(5):6, March 14, type from Gatún, 100 ft., Canal Zone, Panamá. Regarded as a synonym of *A. t. griseimembra* by Hershkovitz (*supra cit.*).

MARGINAL RECORDS.—Panamá (Handley, 1966:774): Isla Bastimentos; Gatún; Alhajuela; Pacora; Armila; Cana, thence into South America.

Genus **Alouatta** Lacépède—Howler Monkeys

A. palliata and *A. pigra* treated by Smith, Jour. Mamm., 51:358–369, May 20, 1970, but nomenclatural changes provided by him apply only to populations north of Honduras, except that he concluded that the name *Mycetes villosus* Gray 1845 was based on a specimen from Brazil. Neither of the two North American species occurs in Brazil (see Cabrera, Rev. Mus. Argentino de Cienc. Nat., 4:155, 156, 1958—and for *A. pigra* see Smith, Jour. Mamm., 51:362, Fig. 3, 1970). Therefore, the name *villosus* used in the specific sense by Sclater (Proc. Zool. Soc. London, p. 5, 1872) and other authors including Hall and Kelson (Mammals of North America, Ronald Press, pp. 221–223, 1959) is not applicable to North American taxa. For taxa south of Honduras the nomenclature that follows is that of Lawrence (Bull. Mus. Comp. Zool., 75:315–354, November, 1933).

1799. *Alouatta* Lacépède, Tableau des divisions, sousdivisions, ordres et genres des mammifères, p. 4. (Published as supplement to Discours d'ouverture et de clôture du

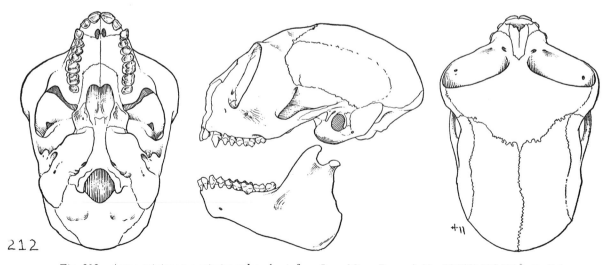

212

Fig. 212. *Aotus trivirgatus griseimembra,* ½ mi. from Juan Mina, Panamá, No. 284776 U.S.N.M., ♀, X 1.

cours d'histoire naturelle. . . .) Type, *Simia belzebub* Linnaeus.

1811. *Mycetes* Illiger, Prodromus systematis mammalium et avium . . . , p. 70. Type species not designated; species included were *Simia belzebub* Linnaeus and *S. seniculus* Linnaeus.

1812. *Stentor* É. Geoffroy St.-Hilaire, Ann. Mus. Hist. Nat. Paris, 19:107, October 12. Type species not designated; species included were *Stentor seniculus* (Linnaeus), *S. ursinus* Humboldt, *S. stramineus* É. Geoffroy St.-Hilaire, *S. fuscus* É. Geoffroy St.-Hilaire, *S. flavicaudatus* É. Geoffroy St.-Hilaire, and *S. niger* É. Geoffroy St.-Hilaire.

Total length, 1000–1500. Muzzle short, somewhat flattened; tail prehensile and naked on under surface distally; thumb well developed; hyoid bone enlarged into large capsule, forming resonance chamber; angular region of mandible greatly enlarged; much enlarged laryngeal pouch present; face bare and black; beard present on males; M3 smaller than M2 or M1; premolars smaller than M1 or M2; p2 larger than p3.

KEY TO NORTH AMERICAN SPECIES OF ALOUATTA

1. Interorbital breadth usually more than 12; condylobasal length usually more than 103; occurring in Yucatán Peninsula, Guatemala, Belize, eastern Tabasco, and northeastern Chiapas. . . . . .*A. pigra*, p. 263
1'. Interorbital breadth usually less than 12; condylobasal length usually less than 103; widespread in tropical lowlands except in area outlined above. . . . .*A. palliata*, p. 261

Alouatta palliata
Mantled Howler Monkey

Average external measurements of four males and six females from Veracruz, México, are, respectively: 1145, 1104; 621, 605; 144, 140. Extremes of cranial measurements of 19 specimens from Nicaragua: condylobasal length, 94.4–100.9; zygomatic breadth, 76.1–86.0; interorbital breadth, 9.4–12.5; mastoid breadth, 55.2–58.9; breadth across canines, 28.1–31.8; maxillary tooth-row, 37.7–41.9. Metacone of M2 small and displaced lingually, as compared with metacone on M1; metacone absent on M3; stylar region of M1 poorly developed; stylar shelf narrow and mesostyle absent on M2; M3 lacking stylar shelf; pterygoid wings narrow, sharply pointed; supraoccipital region forming 90° angle with inferior margin of mandible, when mandible articulated with cranium; post-glenoid process narrow, with sharp point; pelage black dorsally, grading laterally into yellowish mantle extending from axial region to flank; pelage thinner and coarser than in *A. pigra*.

Howler monkeys live in bands of 6 to 50 individuals. The loud, resounding roar of the male carries long distances, and in early morning the cries of different individuals from different places in the jungle produce a constant roar. Howlers can walk on the ground but seldom do so. They spend almost all of their lives in trees. When one member of a band is shot, others sometimes throw small missiles with remarkably poor aim at the hunter. So far as we know, however, they never actually physically attack humans.

Alouatta palliata aequatorialis Festa

1903. *Alouata* [*sic*] *aequatorialis* Festa, Boll. Mus. Zool., Anat. Comp., Univ. Torino, 18(435):3, February 11, type from Vinces, Ecuador.
1933. *Alouatta palliata aequatorialis*, Lawrence, Bull. Mus. Comp. Zool., 75:322, November.
1913. *Alouatta palliata inconsonans* Goldman, Smiths. Miscl. Coll., 60(22):17, February 28, type from Cerro Azul, 2500 ft., near headwaters Chagres River, Panamá.

MARGINAL RECORDS.—Panamá: Bocas del Toro, thence eastward along Caribbean Coast to South America, thence westward along Pacific Coast but north of range of *A. v. trabeata*, but including Insoleta Island; Sevilla Island. Costa Rica: Puerto Cortez.

Alouatta palliata coibensis Thomas

1902. *Alouatta palliata coibensis* Thomas, Nov. Zool., 9:135, April 10, type from Coiba Island, Panamá. Known only from Coiba Island.

Alouatta palliata mexicana Merriam

1902. *Alouatta palliata mexicana* Merriam, Proc. Biol. Soc. Washington, 15:67, March 22, type from Minatitlán, Veracruz.

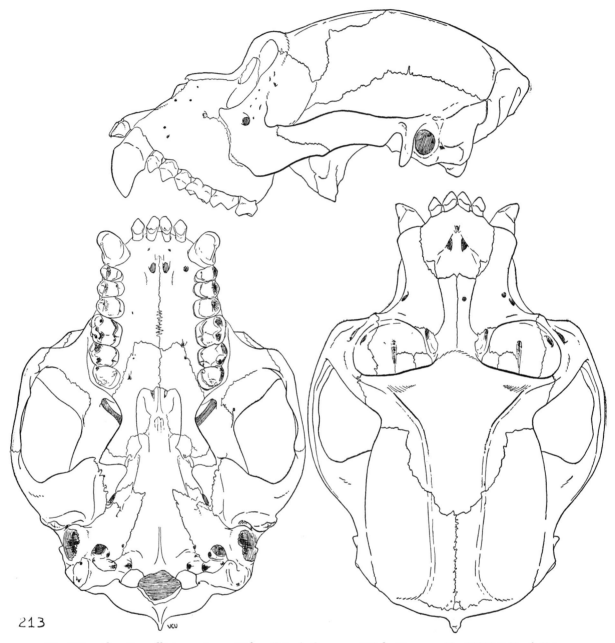

Fig. 213. *Alouatta palliata mexicana*, 20 km. E Jesús Carranza, 300 ft., Veracruz, No. 23932 K.U., ♂, X 1.

MARGINAL RECORDS.—Veracruz: 10 km. NW Minatitlán. Tabasco: 6 mi. S Cárdenas (66239 KU); 5 mi. SE Macuspana (Smith, 1970:368). "Chiapas" (*ibid*.). Oaxaca: Ubero (Goodwin, 1969:117). Veracruz: Achotal; Pasa Nueva (Hall and Dalquest, 1963:258).

Alouatta palliata palliata (Gray)

1849. *Mycetes palliatus* Gray, Proc. Zool. Soc. London, p. 138, June 1, type from Lake Nicaragua, not Caracas, Venezuela; see Sclater, Proc. Zool. Soc. London, pp. 7–8, June, 1872.

1908. *Alouatta palliata matagalpae* J. A. Allen, Bull. Amer. Mus. Nat. Hist., 24:670, October 13, type from Lavala, Matagalpa, Nicaragua. For status see J. A. Allen, Bull. Amer. Mus. Nat. Hist., 28:114, April 30, 1910.

MARGINAL RECORDS.—Honduras: Chamelecón. Nigaragua: Río Mahogany, 20 km. above jct. with Río Escondido (106305 KU). Costa Rica: Cuabré; Guayábo. Nicaragua: 3 km. N, 4 km. W Sapoá (97869 KU). Guatemala: *along Pacific Coast*. Honduras: Copán.

Alouatta palliata trabeata Lawrence

1933. *Alouatta palliata trabeata* Lawrence, Bull. Mus. Comp.
Zool., 75:328, November, type from Capina, Herrera,
Panamá. (According to Barbara Lawrence, *in Litt.*, January
7, 1957, Capina is approx. 10 mi. SW Chitre and between
the Río de la Villa and the Río Parita.)

MARGINAL RECORDS.—Panamá: *type locality;*
Parita.

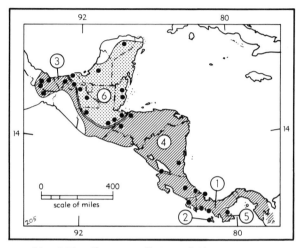

Map 205. *Alouatta palliata* and *Alouatta pigra.*

1. *A. palliata aequatorialis*
2. *A. palliata coibensis*
3. *A. palliata mexicana*
4. *A. palliata palliata*
5. *A. palliata trabeata*
6. *A. pigra*

Alouatta pigra Lawrence
Lawrence's Howler Monkey

1933. *Alouatta palliata pigra* Lawrence, Bull. Mus. Comp.
Zool., 75:333, November, type from Uaxactún, Petén,
Guatemala.
1970. *Alouatta pigra*, Smith, Jour. Mamm., 51:358, May 20.
1933. *Alouatta palliata luctuosa* Lawrence, Bull. Mus. Comp.
Zool., 75:337, November, type from Mountain Cow, Belize.

External measurements differ little from those
of *A. palliata.* Cranial measurements for 14
specimens from Petén, Guatemala, are: condylobasal length, 106.5–121.0; zygomatic
breadth, 72.2–90.7; interorbital breadth, 11.3–
17.1; mastoid breadth, 52.8–63.6; breadth across
canines, 27.0–36.8; maxillary tooth-row, 38.5–
44.5.

Differs from *A. palliata* as follows: metacone
and paracone of M2 well developed and in line;
stylar shelf and cusps well developed on M1 and
M2, shelf reduced and cusps absent on M3;
pterygoid wings large, blunt-tipped and widely
flared; in males, supraoccipital region forming 50°
angle with ventral margin of mandible, when
mandible is articulated with cranium; post-

glenoid process with blunt, irregularly shaped
point; pelage black and comparatively dense and
soft.

MARGINAL RECORDS (Smith, 1970:368, unless
otherwise noted).—Yucatán: Dzitás. Belize: Rockstone
Pond; Hummingbird Highway. Guatemala: lower
Motagua (Hall and Kelson, 1959:223, as *A. v. villosa*);
Quirigua; Zacapa; San Mateo Ixtatán, *ca.* 11,000 ft.
Chiapas: Laguna Ocotal; 10 km. SW Palenque.
Tabasco: 5 mi. SE Macuspana; Frontera. Campeche:
Apozote.

Genus **Cebus** Erxleben—Capuchin Monkeys

1777. *Cebus* Erxleben, Systema regni animalis . . . , 1:44.
Type, *Simia capucina* Linnaeus.
1792. *Sapajus* Kerr, The animal kingdom . . . ,
1(Mamm.):74–79. Type not designated.
1862. *Calyptrocebus* Reichenbach, Die vollständigste
Naturgeschichte der Affen, p. 55. Type not designated.
1862. *Pseudocebus* Reichenbach, Die vollständigste
Naturgeschichte der Affen, p. 55. Type not designated.
1862. *Otocebus* Reichenbach, Die vollständigste Naturgeschichte der Affen, pp. 55–56. Type not designated.
1862. *Eucebus* Reichenbach, Die vollständigste Naturgeschichte der Affen, p. 56. Type not designated.

Size small (see measurements under species);
pelage black except that face, chest, and shoulders white or pale; head crested or noncrested;
ectopterygoid fossa present; thumb well developed; mandible not modified; laryngeal pouch
absent; brain large, elongated hemispheres overlaying cerebellum, braincase produced posteriorly; tail prehensile in some degree but not naked
on under surface; facial part of skull short.

Cebus capucinus
Capuchin

Average external measurements of three adults
from Panamá: 947; 503; 135. Occipitonasal
length, 86. Pelage black except on face, shoulders, and sides of neck, which are white; head not
crested; hind foot less than 150; tail averaging
less than 1½ times length of head and body. Angular part of mandible produced downward below
inferior margin of horizontal ramus; approx. 39
per cent of length of skull (gnathion to inion) behind glenoid fossa when skull, with lower jaw in
place, rests on inferior points of lower mandible
on horizontal surface; upper cheek-teeth decreasing in size posteriorly (P2–M3); occlusal area of
last lower molar less than that of 2nd lower
cheek-tooth (p3).

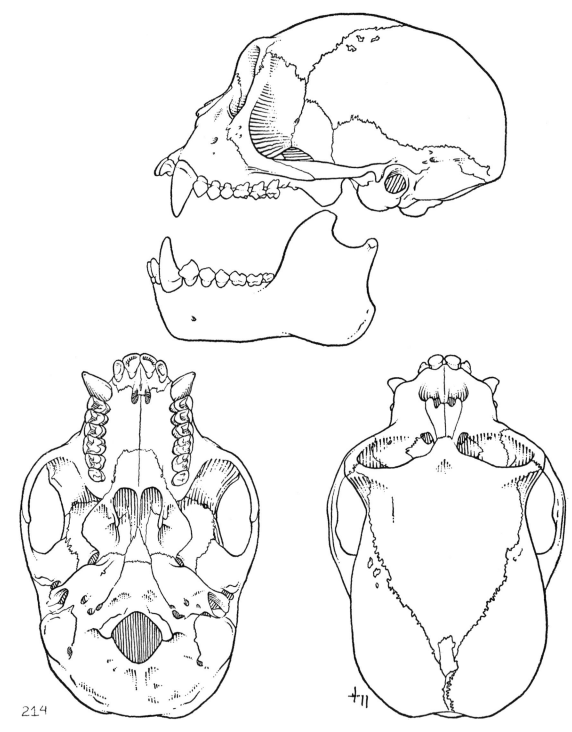

Fig. 214. *Cebus capucinus imitator*, Barro Colorado Island, Canal Zone, Panamá, No. 45098 K.U., ♂, X 1.

Cebus capucinus capucinus (Linnaeus)

1758. [*Simia*] *capucina* Linnaeus, Syst. nat., ed. 10, 1:29. Type locality unknown; subsequently fixed as northern Colombia by Goldman, Proc. Biol. Soc. Washington, 27:99, May 11, 1914.

1909. *C*[*ebus*]. *capucinus*, Elliot, Bull. Amer. Mus. Nat. Hist., 26:229, April 17.

MARGINAL RECORDS.—Panamá: Cerro Bruja; Cerro Azul, thence southward into South America.

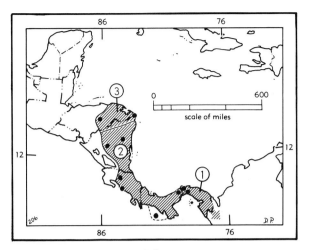

Map 206. *Cebus capucinus.*

1. *C. c. capucinus*   2. *C. c. imitator*
3. *C. c. limitaneus*

Cebus capucinus imitator Thomas

1903. *Cebus imitator* Thomas, Ann. Mag. Nat. Hist., ser. 7, 11:376, April, type from Boquete, Chiriquí, Panamá.
1914. *Cebus capucinus imitator*, Goldman, Proc. Biol. Soc. Washington, 27:99, May 11.

MARGINAL RECORDS.—Nicaragua: Ocotal, thence down the Caribbean Coast to Panamá: Gatún; *Barro Colorado Island*; Coiba Island, thence up Pacific Coast to Costa Rica: Pozo Azul; Cataratos, San Carlos. Nicaragua: Muy Muy.

Cebus capucinus limitaneus Hollister

1914. *Cebus capucinus limitaneus* Hollister, Proc. Biol. Soc. Washington, 27:105, May 11, type from Río Segovia, eastern Honduras, subsequently restricted to Cabo Gracias a Dios at the mouth of the Río Segovia, eastern border between Honduras and Nicaragua by Hershkovitz, Proc. U.S. Nat. Mus., 98:347, May 10, 1949.

MARGINAL RECORDS.—Honduras: type locality; Catacamas.

Genus Saimiri Voigt—Squirrel Marmosets

1831. *Saimiri* Voigt, *in* G. Leopold v. Cuvier, Das Thierreich . . . , 1:95. Type, *Simia sciurea* Linnaeus. For use of this name in place of *Chrysothrix* Kaup see Palmer, Proc. Biol. Soc. Washington, 11:174, June 9, 1897.
1835. *Chrysothrix* Kaup, Das Thierreich . . . , 1:50. Type, *Simia sciurea* Linnaeus.
1840. *Pithesciurus* Lesson, Nouveau tableau de règne animal . . . mammifères, p. 7. Type, *Pithesciurus saimiri* Lesson.

Size small (see measurements under species); pelage short, thick, soft, and brilliantly colored; eyes large and placed close together; ears large and shaped approx. as in man; thumbs short; tail nonprehensile, long, covered with short hair, slightly tufted at tip; ectopterygoid fossa present; thumb well developed; mandible not modified; laryngeal pouch absent; brain large, elongated hemispheres overlaying cerebellum, braincase produced posteriorly; facial part of skull small relative to cranial part; foramen magnum in horizontal plane.

Saimiri oerstedii
Titi Monkey

External measurements: 612–695; 350–415; 80–90. Occipitonasal length, 54–56. Skin of lips including area around nostrils black and nearly devoid of hair; hair around eyes, ears, on throat and sides of neck white; top of head black to grayish (depending on subspecies); back, hind feet, forefeet, and forearms reddish or yellow; shoulders and hind feet suffused with gray; underparts whitish or light ochraceous; tail bicolored like body except that distal fourth is black.

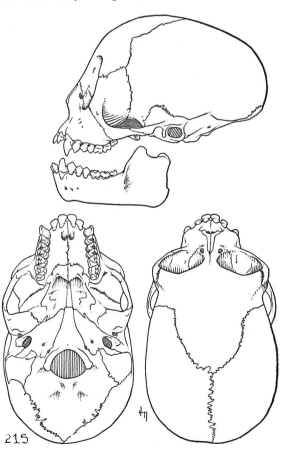

Fig. 215. *Saimiri oerstedii oerstedii*, Burica Península, Panamá, No. 291050 U.S.N.M., ♀, X 1.

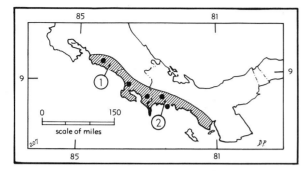

Map 207. *Saimiri oerstedii.*

1. *S. o. citrinellus*   2. *S. o. oerstedii*

Saimiri oerstedii citrinellus Thomas

1904. *Saimiri oerstedi citrinellus* Thomas, Ann. Mag. Nat. Hist., ser. 7, 13:250, April, type from Pozo Azul, 9 mi. upstream from mouth Río Pirris, San José, Costa Rica. Known only from the type locality.

Saimiri oerstedii oerstedii (Reinhardt)

1872. *Chrysothrix örstedii* Reinhardt, Vidensk. Meddel. nat. For. Kjöbennavn, ser. 3, 4(nos. 6–9):157, type from vicinity of David, Chiriquí, Panamá.
1901. *Saimiri oerstedii,* Miller and Rehn, Proc. Boston Soc. Nat. Hist., 30:297, December 27.

MARGINAL RECORDS.—Costa Rica: Palmar; Coto. Panamá: Boquerón; *Veragua;* Sevilla Island.

Genus Ateles É. Geoffroy St.-Hilaire
Spider Monkeys

Revised by Kellogg and Goldman, Proc. U.S. Nat. Mus., 96:1–45, November 2, 1944.

1799. *Sapajou* Lacépède, Tableau-des divisions, sous divisions, ordres et genres des mammifères, p. 4. Type, *Sapajou paniscus* Lacépède [= *Simia paniscus* Linnaeus]. For use of *Ateles,* in place of *Sapajou* that has 6 years priority, see Opinion 91 of the International Commission on Zoological Nomenclature.
1806. *Ateles* É. Geoffroy St.-Hilaire, Ann. Mus. d'Hist. Nat. Paris, 7:262. Type, *Simia paniscus* Linnaeus.
1806. *Atelocheirus* É. Geoffroy St.-Hilaire, Ann. Mus. d'Hist. Nat. Paris, 7:272. Type, *Ateles belzebuth* É. Geoffroy St.-Hilaire.
1815. *Paniscus* Rafinesque, Analyse de la nature, p. 53. Type *Simia paniscus* Linnaeus. Not *Paniscus* Schrank, 1802, an insect.
1910. *Montaneia* Ameghino, Anal. Mus. Nac. Hist. Nat. Buenos Aires, ser. 3, 13:317. Type, *Montaneia anthropomorpha* Ameghino, from an Indian grave in a cave near Sancti Spíritus, Cuba.

Ectopterygoid fossa absent; tail prehensile; thumb absent or poorly developed and appressed to 2nd digit; mandible not modified; laryngeal pouch absent; brain and braincase essentially as in *Cebus;* legs and tail exceptionally long relative to length of body; head small; muzzle prominent;

molars small, each with 4 pronounced cusps; inferior margin of lower mandible approx. straight.

KEY TO NORTH AMERICAN SPECIES OF ATELES

1. General coloration of entire back chiefly black. . . . .*A. fusciceps,* p. 266
1'. General coloration of entire back not chiefly (jet) black. . . . . .*A. geoffroyi,* p. 266

Ateles fusciceps
Brown-headed Spider Monkey

Differs from *A. geoffroyi* only in nearly uniform black coloration. Additional collecting in geographically critical areas is required to show certainly that this monkey is specifically distinct from *A. geoffroyi.*

Ateles fusciceps robustus J. A. Allen

1914. *Ateles robustus* J. A. Allen, Bull. Amer. Mus. Nat. Hist., 33:652, December 14, type from Gallera, Department of Cauca, 5000 ft., Colombia.
1944. *Ateles fusciceps robustus,* Kellogg and Goldman, Proc. U.S. Nat. Mus., 96:29, November 2.
1915. *Ateles dariensis* Goldman, Proc. Biol. Soc. Washington, 28:101, April 13, type from near head Río Limón, 5200 ft., Mt. Pirre, Panamá.
See terminal paragraph under family Cebidae for name change possibly required for taxon *Ateles fusciceps robustus.*

MARGINAL RECORDS.—Panamá: Río Bayano; Armila (Handley, 1966:774); Cituro; near head Río Limón, 5200 ft.; thence into South America.

Ateles geoffroyi
Geoffroy's Spider Monkey

X ¹⁄₁₀

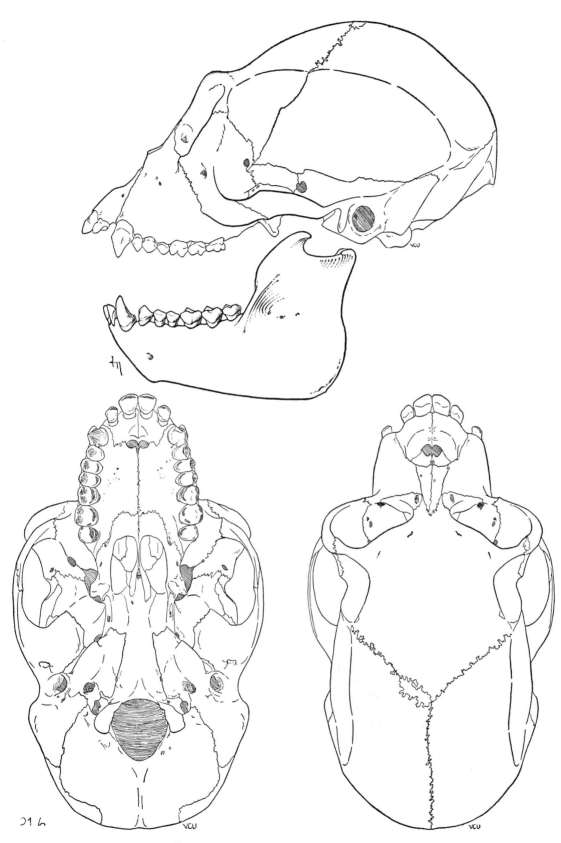

Fig. 216. *Ateles geoffroyi vellerosus*, 20 km. E Jesús Carranza, 300 ft., Veracruz, No. 23916 K.U., ♀, X 1.

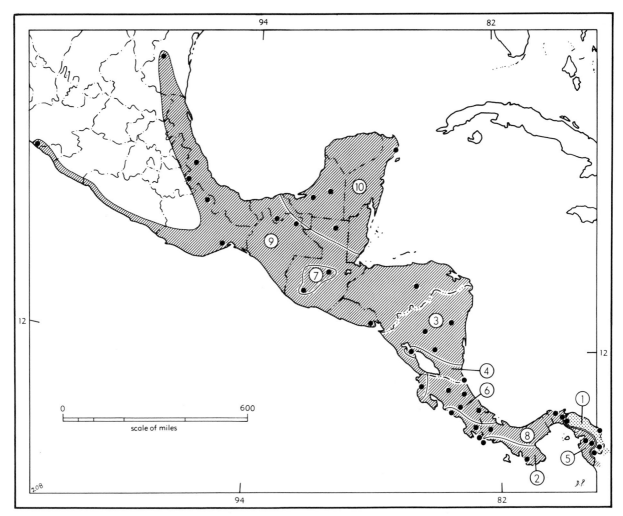

Map 208. *Ateles fusciceps* and *Ateles geoffroyi.*

Guide to	2. *A. g. azuerensis*	5. *A. g. grisescens*	8. *A. g. panamensis*
kinds:	3. *A. g. frontatus*	6. *A. g. ornatus*	9. *A. g. vellerosus*
1. *A. f. robustus*	4. *A. g. geoffroyi*	7. *A. g. pan*	10. *A. g. yucatanensis*

External measurements: 1129–1280; 698–840; 159–194. Occipitonasal length, 85–100. Pelage of back with some buff or rufescent; hair of head directed upward and backward forming more or less of a topknot; tail more than 1½ times length of head and body and distal fourth (usually more) of under surface bare; upper cheek-teeth of relatively uniform width except the 1st premolar and last molar, which are narrower (and shorter).

Ateles geoffroyi azuerensis Bole

1937. *Ateles azuerensis* Bole, Sci. Publs., Cleveland Mus. Nat. Hist., 7:149, August 31, type from Altos Negritos, 1500 ft.,

10 mi. E Montijo Bay, Mariato Suay Lands, Azuero Peninsula, Veraguas, Panamá.
1944. *Ateles geoffroyi azuerensis*, Kellogg and Goldman, Proc. U.S. Nat. Mus., 96:41, November 2.

MARGINAL RECORDS.—Panamá: Río La Vaca, near Puerto Armuelles, Burica Península; type locality.

Ateles geoffroyi frontatus (Gray)

1842. *Eriodes frontatus* Gray, Ann. Mag. Nat. Hist., ser. 1, 10:256, December, type from "South America"; actually from harbor of Culebra, Bahía de Culebra, Guanacaste, Costa Rica (see Kellogg and Goldman, Proc. U.S. Nat. Mus., 96:37, November 2, 1944).

1944. *Ateles geoffroyi frontatus*, Kellogg and Goldman, Proc. U.S. Nat. Mus., 96:37, November 2.

MARGINAL RECORDS.—Nicaragua: Río Yoya, tributary of Río Princapolca; Peña Blanca. Costa Rica: type locality. Nicaragua: Tuma.

Ateles geoffroyi geoffroyi Kuhl

1820. *Atele*[s] *geoffroyi* Kuhl, Beiträge zur Zoologie und vergleichenden Anatomie, 1:26, type from an unknown locality; fixed at San Juan del Norte, Nicaragua, by Kellogg and Goldman, Proc. U.S. Nat. Mus., 96:31, November 2.
1820. *Ateles melanochir* Desmarest, Mammalogie, . . . , pt. 1, p. 76, *in* Encyclopédie méthodique . . . , type from an unknown locality.

MARGINAL RECORDS.—Nicaragua: Monagua; type locality.

Ateles geoffroyi griescens Gray

1866. *Ateles griescens* Gray, Proc. Zool. Soc. London, p. 732 (for 1865), April, type from an unknown locality, presumably valley of Río Tuyra, Darién, Panamá (see Kellogg and Goldman, Proc. U.S. Nat. Mus., 96:43, November 2).
1944. *Ateles geoffroyi griescens*, Kellogg and Goldman, Proc. U.S. Nat. Mus., 96:43, November 2.
1866. *Ateles cucullatus* Gray, Proc. Zool. Soc. London, p. 733 (for 1865), April, type from an unknown locality.
1872. *Ateles rufiventris* Sclater, Proc. Zool. Soc. London, p. 688, November, type from Río Atrato, Darién, Colombia. Regarded as a synonym of *Ateles geoffroyi grisescens* by Hershkovitz (Proc. U.S. Nat. Mus., 98:381, May 10, 1949), who states: "Type an immature; head, tail, upper parts of trunk and limbs black, chest, belly, inner sides of upper arms and thighs rufous (*ex* type, British Museum)."

MARGINAL RECORDS.—Panamá: Chepigana; type locality, thence into South America.

Ateles geoffroyi ornatus Gray

1870. *Ateles ornatus* Gray, Catalogue of monkeys, lemurs, and fruit-eating bats in the . . . British Museum, p. 44, type from an unknown locality; fixed at Cuabre, Limón, Costa Rica, by Kellogg and Goldman, Proc. U.S. Nat. Mus., 96:39, November 2.
1944. *Ateles geoffroyi ornatus*, Kellogg and Goldman, Proc. U.S. Nat. Mus., 96:39, November 2.

MARGINAL RECORDS.—Costa Rica: Cataratos, San Carlos; Guápiles; type locality; Santa María.

Ateles geoffroyi pan Schlegel

1876. *Ateles pan* Schlegel, Mus. Hist. Nat. Pays-Bas, Leiden, 7(pt. 12; Monogr. 40 Simiae):180, cotypes from Cobán, Alta Vera Paz, Guatemala.
1944. *Ateles geoffroyi pan*, Kellogg and Goldman, Proc. U.S. Nat. Mus., 96:36, November 2.

MARGINAL RECORDS.—Guatemala: type locality; Volcán Atitlán.

Ateles geoffroyi panamensis Kellogg and Goldman

1944. *Ateles geoffroyi panamensis* Kellogg and Goldman, Proc. U.S. Nat. Mus., 96:40, November 2, type from Cerro Bruja, about 15 mi. SE Portobello, Colón, Panamá.
1862. *Ateles Beelzebuth* Geoff., Varietas *triangulifera* Weinland, Zool. Gart., a M., Jahrg. 3 (No. 3):206, 207, fig. Type locality, unknown. This name may apply here; if so, it will replace *Ateles geoffroyi panamensis* because of 82 years priority. According to Hershkovitz (Proc. U.S. Nat. Mus., 98:380, footnote, May 10, 1949), "The name is based on a menagerie individual of unknown origin. Judged by the description, the type is most probably a representative of one of the Central American subspecies of *Ateles geoffroyi*. For the present, there is no good reason for giving priority to *triangulifera* over any of the later named taxa recognized by Kellogg and Goldman."

MARGINAL RECORDS.—Costa Rica: Pozo Azul; Agua Buena. Panamá: Chiriquí; type locality; Cerro Azul, near head Chagres River. Costa Rica: Coto.

Ateles geoffroyi vellerosus Gray

1866. *Ateles vellerosus* Gray, Proc. Zool. Soc. London, p. 733 (for 1865), April, type from "Brazil?"; fixed at Mirador, 2000 ft., about 15 mi. NE Huatusco, Veracruz, by Kellogg and Goldman, Proc. U.S. Nat. Mus., 96:33, November 2, 1944.
1944. *Ateles geoffroyi vellerosus*, Kellogg and Goldman, Proc. U.S. Nat. Mus., 96:32, November 2.
1873. *Ateles neglectus* Reinhardt, Vidensk. Meddel. nat. For. Kjöbenhavn, ser. 3, 4(6–9):150, type from Mirador, Veracruz.
1914. *Ateles tricolor* Hollister, Proc. Biol. Soc. Washington, 27:141, July 10, type from Hda. Santa Efigenia, 8 mi. N Tapanatepec, Oaxaca.

MARGINAL RECORDS.—Tamaulipas: 25 km. NNW Ciudad Victoria (Villa-R., 1958:347, sight record only). Veracruz: Barranca de Boca, Cantón de Jalapa. Tabasco: Teapa. Chiapas: 6 mi. SE Palenque. Guatemala: *lowland forests of . . .* [*Caribbean Coast*]. Honduras: Catacamas. El Salvador: Laguna Lomego, San Miguel. Guatemala: *lowland forests of . . .* [*Pacific Coast*]. Oaxaca: Tehuantepec. Jalisco: 10 km. NNW Cihuatlán (Villa-R., 1958:345, sight record only). Oaxaca: Tuxtepec. Veracruz: Volcán de Orizaba.

Ateles geoffroyi yucatanensis Kellogg and Goldman

1944. *Ateles geoffroyi yucatanensis* Kellogg and Goldman, Proc. U.S. Nat. Mus., 96:35, November 2, type from Puerto Morelos, 100 ft., Quintana Roo.

MARGINAL RECORDS.—Quintana Roo: type locality. Guatemala: Uaxactún. Campeche: 7½ km. W Escárcega (92076 KU); Apazote.

FAMILY **CALLITHRICIDAE**—Marmosets

Reviewed by Hershkovitz, Proc. U.S. Nat. Mus., 98:409–424, May 10, 1949.

Total length not exceeding 750; molars $\frac{2}{2}$–$\frac{2}{2}$ (dentition, i. $\frac{2}{2}$, c. $\frac{1}{1}$, p. $\frac{3}{3}$, m. $\frac{2}{2}$; total number of teeth 32, therefore agreeing with that in man and Old World monkeys, both of which, however, have only 2 premolars and 3 molars); face naked in adults; ears large; hind limbs longer than forelimbs; claws (not nails) on all digits except hallux, which has flattened nail; thumb elongated, parallel to other digits and not opposable; ischial callosities and cheek pouches absent.

Genus **Saguinus** Hoffmannsegg—Tamarins

1807. *Saguinus* Hoffmannsegg, Mag. Gesell. naturforsch. Freunde, Berlin, 1:102. Type, *Saguinus ursula* Hoffmannsegg [= *Saguinus tamarin* Link].
1840. *Marikina* Lesson, Species des mammifères, bimanes et quadrumanes; . . . , p. 199 (listed under synonymy of *Oedipus titi* Lesson [= *Simia oedipus* Linnaeus] in an erroneous combination with [*Midas*] *bicolor* Spix and the bibliographic references thereto). Type, *Marikina bicolor* Lesson.
1763. *Cercopithecus* Gronov, Zoophylacium Gronovianum, fasc. 1, p. 5. Type, *Simia midas* Linnaeus, designated by Elliot, Bull. Amer. Mus. Nat. Hist., 30:341, December 21, 1911; eliminated from availability by Opinion 89 of the International Commission on Zoological Nomenclature.

"Hands normal; palm broad, digits not markedly elongated; first phalanges of middle digits usually free, webbing, if present, extremely narrow; length of longest finger (with claw) less than twice width of palm; sides of crown and sides of face completely covered with hair or nearly bare; ears entirely or partially exposed [instead of completely covered by mane]. Sphenoidal pits or vacuities obsolete or absent." (Hershkovitz, 1949:410.)

Subgenus **Oedipomidas** Reichenbach
Crested Bare-faced Tamarin

1840. *Oedipus* Lesson, Species des mammifères, bimanes et quadrumanes; . . . , p. 184. Type, *Oedipus titi*. Not *Oedipus* Tschudi, 1838, an amphibian.
1862. *Oedipomidas* Reichenbach, Die vollständigste Naturgeschichte der Affen, p. 5. Type, *Simia oedipus* Linnaeus, a renaming of *Oedipus* Lesson. This subgenus is monotypic.

Saguinus oedipus
Crested Bare-faced Tamarin

Conspicuous median crest of long white hairs on forehead and crown; ears small, lamina of lower posterior margin of pinna deeply emarginate or obsolete (after Hershkovitz, 1949:411).

External measurements (seven adults from Panamá): 620–685; 325–390; 83–90. Occipitonasal length, 56.5–59.8. In Panamanian specimens, face so sparsely haired as to appear bare except for narrow wedge of white pelage extending from between eyes onto crown, and narrow (1 mm. wide) line of white hairs extending from posterior angle of ear to below ear; upper parts black marbled with light ochraceous (marbling a result of an ochraceous subapical band on otherwise black hairs); nape hazel, individual hairs as long as 29 mm. and forming a partial mane and mantle; underparts, postorbital area, medial sides of legs and arms, lateral side of forearms, hands, and feet yellowish white; basal fourth of under surface of tail hazel, remainder of tail black.

In specimens from Panamá, and perhaps in all species of *Saguinus*, the long curved claws, obviously adapted for a scansorial habit, are much compressed laterally, the bases of the claws being narrower than the distance to which the claws project beyond the fleshy ends of the digits.

Saguinus oedipus geoffroyi (Pucheran)

1823. *Midas Oedipus* (varietas) Spix, Simiarum et vespertilionum Brasiliensium . . . , p. 30, pl. 23, "*habitat, ut opinor, in provincia* Guiana." Not *Simia oedipus* Linnaeus.

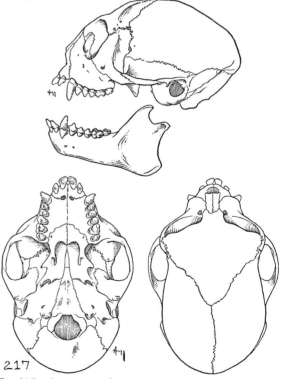

Fig. 217. *Saguinus oedipus geoffroyi*, La Cascades, Panamá, No. 257313 U.S.N.M., ♀, X 1.

1845. *Hapale Geoffroyi* Pucheran, Revue et Mag. Zool., Paris, 8(ar. 2):336, September, type from Panamá, subsequently restricted to Canal Zone by Hershkovitz, Proc. U.S. Nat. Mus., 98:417, May 10, 1949.

1862. *J[acchus]. spixii* Reichenbach, Die vollständigste Naturgeschichte der Affen, p. 1, pl. 1, fig. 2. (Copied, with slight modification, from Spix, pl. 23.)

MARGINAL RECORDS.—Panamá: Río Indio, near Gatún. Thence southward into South America and on west coast of Panamá: La Chorrera. Costa Rica: La Vaca River (Carpenter, 1935:171).—According to Hershkovitz (1966:394), *geoffroyi* is a subspecies of *S. oedipus*.

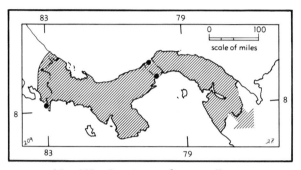

Map 209. *Saguinus oedipus geoffroyi.*

SUPERFAMILY CERCOPITHECOIDEA
Macaquelike Primates

Premolars, $\frac{2-2}{2-2}$; nose narrow. Two species have been introduced in the Lesser Antilles from the Old World.

FAMILY CERCOPITHECIDAE
Guenons, Guerzas, and Allies

Dentition, i. $\frac{2}{2}$, c. $\frac{1}{1}$, p. $\frac{2}{2}$, m. $\frac{3}{3}$; tail present and well developed; bare ischial callosities present and often brightly colored; face more or less naked. Arboreal in habitat.

SUBFAMILY CERCOPITHECINAE
Guenons

Arms and legs of approx. equal length; tail relatively long; skull usually with supraorbital ridges; muzzle somewhat extended; thumb normal; stomach simple.

Genus Cercopithecus Linnaeus—Guenons

1758. *Cercopithecus* Linnaeus, Syst. nat., ed. 10, 1:26. Type, *Simia diana* Linnaeus. For use of *Cercopithecus* rather than *Lasiopyga* Illiger, 1811, see Opinion 104, International Commission of Zoological Nomenclature.

Head rounded; muzzle truncate; cheek pouches large; tail long; ischial callosities small; m3 quadrituberculate.

KEY TO INTRODUCED SPECIES OF CERCOPITHECUS

1. Muzzle much reduced (facial angle approx. 70°); upper parts speckled reddish and black. . . . .*C. mona,* p. 271
1'. Muzzle more pronounced (facial angle approx. 50°); upper parts yellowish with strong greenish cast. . . . .*C. sabaeus,* p. 271

Cercopithecus sabaeus (Linnaeus)
Green Guenon

External measurements: 1308; 762; 152. Occipitonasal length, 92. Upper parts bright gold-green; face black; forearms and forelegs gray; underparts, cheeks, sides of neck white; tail grayish green on basal two-thirds, yellowish distally. Facial part of skull prolonged, facial angle being approx. 50°; supraorbital ridges pronounced.

1766. [*Simia*] *Sabaea* Linnaeus, Syst. nat., ed. 12, 1:38. Type locality, Cape Verde Islands; probably from Senegal (see J. A. Allen, Bull. Amer. Mus. Nat. Hist., 47:352, February 6, 1925).

1851. *C[ercopithecus]. sabaeus,* I. Geoffroy St.-Hilaire, Compte Rendu des Séances de l'Acad. des Sci. Paris, 31:874.

1851. *Cercopithecus werneri* I. Geoffroy St.-Hilaire, Compte Rendu des Séances de l'Acad. des Sci. Paris, 31:874. Type from Africa.

1851. *Cercopithecus callitrichus* I. Geoffroy St.-Hilaire, Catalogue méthodique de la collection des mammifères . . . du muséum . . . Paris, p. 23, type from an unknown locality [West Africa].

MARGINAL RECORDS.—Introduced and established in the New World on Lesser Antillean islands of St. Kitts; Barbados. Not mapped.

Cercopithecus mona (Schreber)
Mona Guenon

External measurements: 1295; 785; 145. Occipitonasal length, 77.9. Upper parts speckled reddish and black, darkest toward rump; lateral surfaces of hands and arms black; lateral surfaces of legs black speckled with small red spots; underparts and medial surfaces of limbs grayish white; patch beneath tail to hip white; upper and lower surfaces of tail colored like respective surfaces of body on basal third, black distally. Facial part of skull truncate, facial angle approx. 70°; supraorbital ridges reduced.

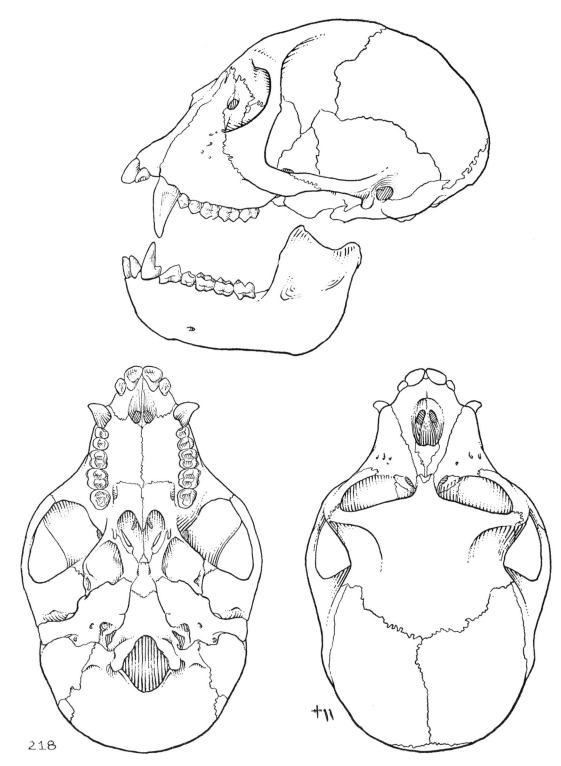

218

Fig. 218. *Cercopithecus mona*, Mpivia, French Congo, Africa, No. 220358 U.S.N.M., ♂, X 1.

1774. *Simia mona* Schreber, Die Säugthiere . . . , 2(pt. 1):pl. 15(7:97 contains the vernacular name and description). Type locality, "Barbary."

1777. *Cercopithecus mona*, Erxleben, Systema regni animalis . . . , 1:30.

MARGINAL RECORDS.—Introduced and established in the New World on Lesser Antillean islands of St. Kitts; Grenada. Not mapped.

SUPERFAMILY **HOMINOIDEA**
Manlike Primates

Radius and ulna capable of complete rotation; thumb opposable and articular face of trapezium rounded; cheek pouches absent; dentaries firmly fused at mental symphysis; tail absent (represented by coccyx). Dentition, i. $\frac{2}{2}$, c. $\frac{1}{1}$, p. $\frac{2}{2}$, m. $\frac{3}{3}$.

FAMILY **HOMINIDAE**—Men and Apes

Tail absent (externally); forelimbs not modified for walking; other characters as given for the superfamily.

Genus **Homo** Linnaeus—Man

1758. *Homo* Linnaeus, Syst. nat., ed. 10, 1:20. Type, *Homo sapiens* Linnaeus.

Hind limbs longer than forelimbs and modified for upright stance; femur longer than humerus; foramen magnum situated near center of ventral face of skull; diastema absent between i2 and c; tendencies present toward (1) formation of pronounced chin; (2) loss of supraorbital prominences; (3) reduction in prognathism; (4) reduction of height of canine and, in all but earliest representatives of genus, canine not notably higher than other teeth; (5) increase in absolute and relative size of brain (approx. 1400 grams in adult males); and (6) decrease in thickness of cranial bones.

Homo sapiens
Modern Man

X $\frac{1}{12}$

Characterized by having achieved the greatest degree of development of the generic characters. The species *H. sapiens* is the only species still existing.

Homo sapiens afer Linnaeus

1758. [*Homo sapiens*] *afer* Linnaeus, Syst. nat., ed. 10, 1:22. Type locality, Africa. Introduced and widely established in North America.

Homo sapiens americanus Linnaeus

1758. [*Homo sapiens*] *americanus* Linnaeus, Syst. nat., ed. 10, 1:20. Type locality, eastern North America. Known from North, Central, and South America.

Homo sapiens asiaticus Linnaeus

1758. [*Homo sapiens*] *asiaticus* Linnaeus, Syst. nat., ed. 10, 1:21. Type locality, Asia. Introduced and now widely established in North America.

Homo sapiens sapiens Linnaeus

1758. [*Homo*] *sapiens* Linnaeus, Syst. nat., ed. 10, 1:20. Type locality, Uppsala, Sweden. Cosmopolitan.

219

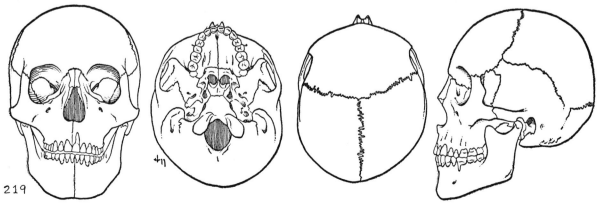

Fig. 219. *Homo sapiens americanus*, Prairie Dog Creek at State Line, Phillips Co., Kansas, No. 14PH4-1436 U.S.N.M., ♂, X ¼.

ORDER **EDENTATA**
Sloths, Anteaters, and Armadillos

Teeth incomplete in one sense or another; teeth lacking enamel in many Recent members of order, and absent in Myrmecophagidae; deciduous dentition absent except in armadillos; teeth single-rooted and absent from anteriormost parts of jaws; processes on basicranial region absent or much reduced; ilium everted and strongly crested.

SUBORDER **XENARTHRA**

Differs from Suborder Palaeanodonta (extinct) in (1) xenarthrous vertebrae (secondary articulations between vertebrae of the lumbar series), (2) ischiocaudal symphysis present (incipient in some palaeanodonts), and (3) canines, when present, lacking enamel and never sectorial.

KEY TO SUPERFAMILIES AND FAMILIES OF XENARTHRA

1. Teeth absent; mouth tubular with small terminal opening.
   superfamily Myrmecophagoidea, p. 275
   family Myrmecophagidae, p. 275
1'. Teeth present; mouth neither tubular nor with small terminal opening.
   2. Top and sides of body covered with horny scutes; teeth more than $\frac{6}{8}$.
      superfamily Dasypodoidea, p. 281
      family Dasypodidae, p. 281
   2'. Top and sides of body covered with hair, horny scutes absent; teeth fewer than $\frac{6}{8}$.
      3. Fewer than 4 claws on each foot; diameter of largest upper cheek-tooth (excepting anteriormost, caniniform) less than 8.
         superfamily Bradypodoidea, p. 279
         family Bradypodidae, p. 279
      3'. Four or more claws on some or all feet; diameter of largest upper cheek-tooth (excepting anteriormost, caniniform) more than 8.
         superfamily Megalonychoidea, p. 274
         family Megalonychidae, p. 274

INFRAORDER **PILOSA**

Exoskeleton absent or present as irregular separate plaques embedded in skin; teeth $\frac{5}{4}$ or fewer, prismatic, lacking enamel and with roots

open; all vertebrae separate; intermaxillary usually small to rudimentary; lachrymal large or small, with facial lachrymal-foramen; zygomatic arch more or less incomplete; coracoid process forming coracoscapular foramen.

SUPERFAMILY **MEGALONYCHOIDEA**

Teeth not more than $\frac{5}{4}$; skull elongate or short; zygomatic arch complete; jugal with upward-directed process; lachrymal nipple-shaped and with large lachrymal-foramen; dentaries fused and mental region prolonged as a distinct projection; thoracolumbar vertebrae, 19–25; manus pentadactyl with 5th digit somewhat reduced.

FAMILY **MEGALONYCHIDAE**
Ground Sloths

Cheek-teeth prismatic, quadrangular to transversely elliptical with 2 transverse ridges; anterior cheek-tooth caniniform and separated from others by diastema; last molar smallest; alveolar canal opening either anterior to base of ascending tubercle or on lateral side (seldom on medial side) near tubercle.

SUBFAMILY **Ortotheriinae**

KEY TO NORTH AMERICAN (RECENT?) GENERA OF ORTOTHERIINAE

1. Lesser trochanter absent from femur; humerus approx. 200 long. *Parocnus*, p. 275
1'. Lesser trochanter present on femur; humerus approx. 135 long.
   2. Walls in postorbital region having lateral constriction; tubercle on intertrochanteric ridge below lesser trochanter inconspicuous.
      Acratocnus, p. 274
   2'. Walls in postorbital region lacking lateral constriction; tubercle on intertrochanteric ridge below lesser trochanter conspicuous. *Synocnus*, p. 275

Genus **Acratocnus** Anthony
Puerto Rican Ground Sloth

1916. *Acratocnus* Anthony, Ann. New York Acad. Sci., 27:195, August 9. Type, *Acratocnus odontrigonus* Anthony.

Canines trigonal; sagittal crest present; preorbital fossa large; limbs slender; humerus approx.

135 mm. long; spines low on dorsal vertebrae; caudal vertebrae wide; lesser trochanter present on femur; entepicondylar foramen present; teeth $\frac{5}{4}$. The genus is monotypic and is extinct. It is listed here because it may have been exterminated by early man. Not mapped.

Acratocnus odontrigonus Anthony
Puerto Rican Ground Sloth

1916. *Acratocnus odontrigonus* Anthony, Ann. New York Acad. Sci., 27:195, August 9, type from Cueva de la Ceiba, Hda. Jobo, near Utuado, Puerto Rico. Known only from skeletal remains from Puerto Rico.
1918. *Acratocnus major* Anthony, Mem. Amer. Mus. Nat. Hist., n.s., 2:412, October 12, type from cave on property of Don Gervacio Torano, near Utuado, Puerto Rico. A synonym of *A. odontrigonus* according to Paula Couto, Amer. Mus. Novit., 2304:33, October 20, 1967.

Genus Synocnus Paula Couto
Lesser Haitian Ground Sloth

1967. *Synocnus* Paula Couto, Amer. Mus. Novit., 2304:35, October 20. Type, *Acratocnus comes* Miller.

Resembles *Acratocnus* from Puerto Rico but having conspicuous tubercle at middle of shaft of femur, and neck of femur shorter and less bowed outward. Weight of animal probably 50 pounds. This monotypic genus is included here because it, like *Acratocnus*, seems to have been a member of the recently man-exterminated fauna. Not mapped.

Synocnus comes (Miller)
Lesser Haitian Ground Sloth

1929. *Acratocnus* (?) *comes* Miller, Smiths. Miscl. Coll., 81(9):26, March 30, type from large cave near St. Michel, Haiti. Known only from skeletal remains, from Haiti (see Miller, *antea*, and Smiths. Miscl. Coll., 82(5):11, December 11, 1929) and La Gonave Island (see Miller, Smiths. Miscl. Coll., 82(15):5, December 24, 1930).
1967. *Synocnus comes*, Paula Couto, Amer. Mus. Novit., 2304:36, October 20.

Genus Parocnus Miller
Greater Haitian Ground Sloth

1929. *Parocnus* Miller, Smiths. Miscl. Coll., 81(9):28, March 30. Type, *Parocnus serus* Miller.

Femur with lesser trochanter absent and upper half of shaft wide and flattened. These are differences from *Acratocnus* and the latter is a point of resemblance to the extinct genus *Nothrotherium*.

The genus is monotypic. It is included here because of Miller's (1929c:11) statement that there seems to be no "doubt that a ground sloth was a member of the recently man-exterminated fauna of Hispaniola."

Parocnus serus Miller
Greater Haitian Ground Sloth

1929. *Parocnus serus* Miller, Smiths. Miscl. Coll., 81(9):29, March 30, type from large cave near St. Michel, Haiti. Known only from skeletal remains from Haiti.

Femur approx. 145; humerus approx. 200. Weight of animal in life probably 150 pounds. Not mapped.

SUPERFAMILY **MYRMECOPHAGOIDEA**
Anteaters

Clothed with hair (not scutes); teeth absent; head elongated; mouth tubular; clavicle rudimentary; 3rd digit of manus large and provided with strong claw; other toes of manus suppressed or smaller; pes with 4 or 5 subequal digits.

FAMILY **MYRMECOPHAGIDAE**
Anteaters

Characters as for superfamily. Mouth cavity tubular with small terminal aperture through which long, slender tongue, covered with sticky secretion of enormously enlarged submaxillary glands, is rapidly protruded in feeding and withdrawn with adhering particles of food. Each of the three living genera occurs in South America and Central America.

KEY TO GENERA OF MYRMECOPHAGIDAE

1. Combined length of head and body less than 300; 2 claws on forefoot; pterygoids not meeting below posterior nares.
    Cyclopes, p. 278
1'. Combined length of head and body more than 300; 4 claws on forefoot; pterygoids meeting below posterior nares.
    2. Length of head and body less than 900; tail naked distally; longest hair on tail less than $\frac{1}{3}$ as long as hind foot; prelachrymal region of skull shorter than postlachrymal region.
        Tamandua, p. 276
    2'. Length of head and body more than 900; all of tail clothed with hair; longest hair on tail more than twice length of hind foot; prelachrymal region of skull more than twice as long as postlachrymal region.
        Myrmecophaga, p. 276

Genus **Myrmecophaga** Linnaeus
Giant Anteaters

1758. *Myrmecophaga* Linnaeus, Syst. nat., ed. 10, 1:35. Type, by subsequent selection (Fleming, The philosophy of zoology . . . , 2:194, May or June, 1822), *Myrmecophaga jubata* Linnaeus [= *M. tridactyla* Linnaeus].
1900. *Falcifer* Rehn, Amer. Nat., 34:576, July. Type, *Myrmecophaga jubata* Linnaeus [= *M. tridactyla* L.].

External measurements (after Goodwin, 1946:356): 1860; 650; 135. Greatest length of skull, 365. Four claws on forefoot, 5 on hind foot; 3rd digit much the largest on forefoot; hair (285 mm. in a Costa Rican specimen) on tail more than twice as long as hind foot; pterygoids joined below narial passage and prolonging it for almost entire length of skull; prelachrymal region of skull more than twice as long as postlachrymal region; zygoma incomplete and jugal attached to maxilla; one bulla on each side of base of skull (tympanic and alisphenoidal parts not separate as in *Tamandua*); alisphenoidal part twice as large as tympanic part.

Myrmecophaga tridactyla
Giant Anteater

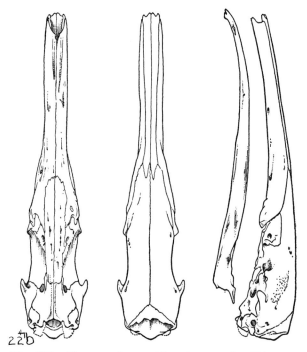

Fig. 220. *Myrmecophaga tridactyla centralis*, Talamanca, Costa Rica, No. 14107 U.S.N.M., sex ?, X ¼.

In a Costa Rican specimen: upper parts grizzled, hairs being banded alternately with black and light buff; black stripe 2.5–6.3 mm. wide extending from throat over shoulder and ending on lower back, forming with its opposite a collar; forelimbs yellowish white; head yellowish white but hairs sparse and bristly; small area above toes and band above ankles black; hind limbs, tail, and most of underparts blackish.

Myrmecophaga tridactyla centralis Lyon

1906. *Myrmecophaga centralis* Lyon, Proc. U.S. Nat. Mus., 31:570, November 14, type from Pacuare, Costa Rica.
1920. *Myrmecophaga tridactyla centralis*, Goldman, Smiths. Misc. Coll., 69(5):64–65, April 26.

MARGINAL RECORDS.—Belize: near Punta Gorda, thence southward along Caribbean Coast to South America and thence up Pacific Coast to

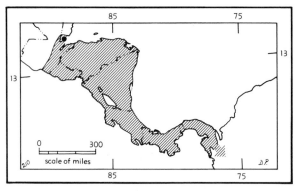

Map 210. *Myrmecophaga tridactyla centralis*.

Guatemala: *Pacific Coast region between San José and border of El Salvador.*

Genus **Tamandua** Gray—Tamandua

Revised by J. A. Allen, Bull. Amer. Mus. Nat. Hist., 20:385–398, October 29, 1904. See also Reeve, Proc. Zool. Soc. London, 111:279–302, February 17, 1942. Wetzel, Proc. Biol. Soc. Washington, 88:95–112, April 23, 1975, shows that *Tamandua mexicana* is the specific name for the subspecies in northwestern South America and all of Central America, and that *Tamandua tetradactyla* is the specific name for subspecies in remainder of South America.

1825. *Tamandua* Gray, Ann. Philos., 10(n.s.):343, November. Type, *Myrmecophaga tamandua* Cuvier [= *M. tetradactyla* Linnaeus]; see Gray, London Med. Repos., 15(88):305, April 1, 1821. Tamandua Gray (*ibid.*) is a vernacular name and therefore is not available. Not *Tamandua* Frisch, 1775; Frisch names are non-Linnaean and unavailable. Not *Tamandua* Rafinesque, 1815, which was not differentiated from *Myrmecophaga* and included no species.

1830. *Uroleptes* Wagler, Natürliches System der Amphibien . . ., p. 36. Type, *Myrmecophaga tetradactyla* Linnaeus.

1841. *Dryoryx* Gloger, Gemeinnütziges Hand- und Hilfsbuch der Naturgeschichte, 1:xxxi, 112. Type, the tamandua.

1882. *Uropeltes* Alston, Biologia Centrali-Americana, p. 191 (misprint for *Uroleptes* Wagler).

Total length approx. 1200. Three large claws and one small claw on forefoot, 5 claws on hind foot; pelage short and coarse; hair on tail less than one-third as long as hind foot; tail naked at tip and on all of underside; tail prehensile; pterygoids united below narial passage and prolonging passage for almost entire length of skull; prelachrymal region of skull shorter than postlachrymal region; zygomatic arch incomplete, jugal attached to maxilla; 3 bullae (2 sphenoidal and 1 tympanic) on each side of basicranial region.

Tamandua mexicana
Tamandua

X 1/12

Average external measurements of three females from Veracruz, México: 1132, 553; 97. Height of ear from notch, 41; greatest length of skull of other specimens, 123–128. In specimens from Veracruz: body black from immediately an-

terior to hind limbs to immediately behind forelimbs, a black stripe continuing over shoulder to axilla; otherwise tan, and tan vertebral stripe extends almost to end of black area; naked, scaly, distal part (two-thirds in specimens from Veracruz) of tail tan with irregular black markings.

Tamandua mexicana chiriquensis J. A. Allen

1904. *Tamandua tetradactyla chiriquensis* J. A. Allen, Bull. Amer. Mus. Nat. Hist., 20:395, October 29, type from Boquerón, Chiriquí, Panamá.

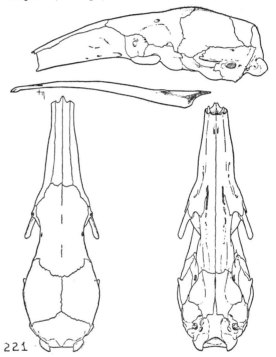

221

Fig. 221. *Tamandua mexicana mexicana*, 6 km. NW Paso Nuevo, 100 ft., Veracruz, No. 19906 K.U., ♀, X ½.

MARGINAL RECORDS.—Nicaragua: Ocotal, thence southward into South America.

Tamandua mexicana hesperia Davis

1955. *Tamandua tetradactyla hesperia* Davis, Jour. Mamm., 36:588, December 14, type from near Acahuizotla, 2800 ft., Guerrero.

MARGINAL RECORDS.—Guerrero: type locality; *Chapalapa*.

Tamandua mexicana mexicana (Saussure)

1860. "*Myrmecophaga tamandua* (?), Desm. (Var. *Mexicana* Sauss.)" Saussure, Revue et Mag. Zool., Paris, ser. 2, 12:9, January, type from Tabasco.

1889. *Myrmecophaga sellata* Cope, Amer. Nat., 23:133, February, type from Honduras. Regarded as tenable by

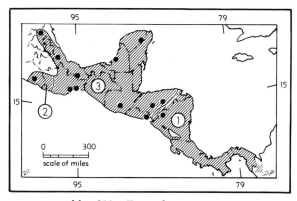

Map 211. *Tamandua mexicana.*

1. *T. m. chiriquensis*   2. *T. m. hesperia*
3. *T. m. mexicana*

Lönnberg, Arkiv för Zool., 29A(19):27, August 9, 1937, and probably synonymous with *T. t. mexicana* by Reeve, Proc. Zool. Soc. London, 111(ser. A; pts. 3–4):301, February 17, 1942.

1904. *Tamandua tetradactyla tenuirostris* J. A. Allen, Bull. Amer. Mus. Nat. Hist., 20:394, October 29, type from Pasa Nueva, Veracruz.

MARGINAL RECORDS.—San Luis Potosí: 5 km. S Tamazunchale. Veracruz: Mirador; 20 km. ENE Jesús Carranza, 200 ft. (Hall and Dalquest, 1963:262). Campeche: 10 km. SSW Champotón (92085 KU). Yucatán: 8 km. ESE Ticuch (Birney, *et al.*, 1974:12), thence southward along Caribbean Coast to Honduras: Catacamas; La Cueva Archaga. El Salvador: El Carmen (Burt and Stirton, 1961:42). Guatemala: Dueñas. Oaxaca: Huilotepec; Zapotitlán (Goodwin, 1969:117).

Genus **Cyclopes** Gray—Two-toed Anteaters

1821. *Cyclopes* Gray, London Med. Repos., 15:305. Type, *Myrmecophaga didactyla* Linnaeus. For use of *Cyclopes* rather than *Cyclothurus* Lesson, 1842, see Thomas, Ann. Mag. Nat. Hist., ser. 6, 15:191, February, 1895, and Palmer, Proc. Biol. Soc. Washington, 13:72, September 28, 1899.

1830. *Myrmedon* Wagler, Natürliches System der Amphibien, . . . , p. 36. Type, *Myrmecophaga didactyla* Linnaeus.

1836. *Myrmecolichnus* Reichenbach, k. Sächische Naturh. Mus., Dresden, ein Leitfaden, p. 51. Type, *Myrmecophaga didactyla* Linnaeus.

Size small (head and body approx. 200); forefoot with 2 digits (II and III), lateral being much the larger; dorsal profile of skull strongly convex; pterygoids and posterior part of palatines not meeting below posterior nares and thus forming channel rather than tube; ribs unusually broad and flattened; zygomatic arch absent and jugal absent in specimens seen by me.

Cyclopes didactylus
Two-toed Anteater

External measurements: 398–422; ——227; 30–36. Greatest length of skull, 50.2. Color golden brown marbled with silvery and in some specimens with dark brownish or blackish mid-dorsal and mid-ventral areas. Pelage long (15 mm.) and semiwooly.

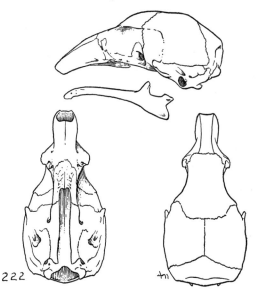

Fig. 222. *Cyclopes didactylus dorsalis*, Panamá, No. 283876 U.S.N.M., ♀, X 1.

Cyclopes didactylus dorsalis (Gray)

1865. *Cyclothurus dorsalis* Gray, Proc. Zool. Soc. London, p. 385, pl. 19, October, type from Costa Rica, subsequently restricted to Orosi, near Cartago, Costa Rica (Goodwin, Bull. Amer. Mus. Nat. Hist., 87:354, December 31, 1946).

1900. *C*[*yclopes*]. *d*[*idactylus*]. *dorsalis*, Thomas, Ann. Mag. Nat. Hist., ser. 7, 6:302, September.

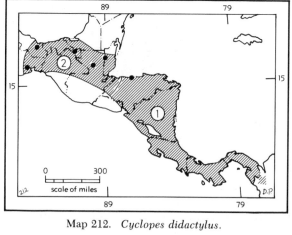

Map 212. *Cyclopes didactylus.*

1. *C. d. dorsalis*   2. *C. d. mexicanus*

MARGINAL RECORDS.—Honduras: Ceiba, thence southward into South America.

Cyclopes didactylus mexicanus Hollister

1914. *Cyclopes mexicanus* Hollister, Proc. Biol. Soc. Washington, 27:210, October 31, type from Tehuantepec, Oaxaca.
1952. *Cyclopes didactylus mexicanus*, Hall and Kelson, Univ. Kansas Publ., Mus. Nat. Hist., 5:316, November 21.

MARGINAL RECORDS.—Veracruz: Minatitlán. Tabasco: Montecristo. Belize: Santa Familia, near El Cayo de San Ignacio (Jones and Carter, 1972:535). Guatemala: Libertad. Oaxaca: type locality.

SUPERFAMILY BRADYPODOIDEA—Sloths

Pelage long, crisp, and strawlike, usually supporting growth of algae that imparts greenish tinge in wet seasons; teeth $\frac{5}{4-5}$, subcylindrical, persistently growing; clavicle present; forelimbs longer than hind limbs; 2 or 3 claws on foot; tail rudimentary or less than third of total length; stomach complex; caecum absent; uterus simple and globular.

FAMILY BRADYPODIDAE—Sloths

Characters as for the superfamily.

KEY TO GENERA OF BRADYPODIDAE

1. Tail present; 3 claws on forefoot; anterior tooth in upper jaw smaller than succeeding tooth and space between them less than length of crown of 2nd tooth.
   Bradypus, p. 279
1'. Tail absent; 2 claws on forefoot; anterior tooth in upper jaw larger than succeeding tooth and space between them several times longer than length of crown of 2nd tooth. . . . . *Choloepus*, p. 280

Genus Bradypus Linnaeus—Three-toed Sloth

1758. *Bradypus* Linnaeus, Syst. nat., ed. 10, 1:34. Type, *Bradypus tridactylus* Linnaeus.
1779. *Ignavus* Blumenbach, Handbuch der Naturgeschichte, 1:70. Type, *Ignavus tridactylus* [= *Bradypus tridactylus* Linnaeus].
1850. *Arctopithecus* Gray, Proc. Zool. Soc. London, for 1849, p. 65, January–June. Based on *Bradypus gularis* Rüppell, *Arctopithecus marmoratus* Gray, *A. blainvillii* Gray, *A. flaccidus* Gray, and *A. problematicus* Gray. Not *Arctopithecus* Virey, 1819, a primate.
1864. *Scaeopus* Peters, Monatsb. preuss. Akad. Wiss., Berlin, p. 678. Type, *Bradypus torquatus* Illiger.
1906. *Hemibradypus* R. Anthony, Compt. Rend. Acad. Sci., Paris, 142 (fasc. 5):292. Type, *Bradypus torquatus* Illiger.
1942. *Eubradypus* Lönnberg, Arkiv för Zool., 34A(9):5. Type,

Bradypus (Eubradypus) tocantinus Lönnberg [= *Arctopithecus marmoratus* Gray].
1942. *Neobradypus* Lönnberg, *loc. cit.*:15. Type, *Arctopithecus marmoratus* Gray.

Forefoot with 3 claws; tail comprising approx. one-eighth of total length; anterior tooth in upper jaw smaller than any succeeding tooth; anterior tooth in lower jaw broader than succeeding teeth but compressed anteroposteriorly; space between 1st and 2nd tooth, both above and below, shorter than length of crown of 2nd; tympanic bullae present.

Bradypus variegatus
Brown-throated Sloth

External measurements: 570–660; 66–70; 120–130. Greatest length of skull, 70–80. Pelage of two distinct types: long (as long as 110) over-hairs of large diameter, and short (as long as 35) underfur of fine texture; arms to shoulders, top of head, neck, and chin drab; face whitish having brown stripe on each side, enclosing eye; forehead having ruff of brownish hair; chest, top of

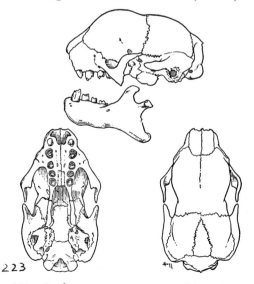

Fig. 223. *Bradypus variegatus griseus*, Monte Lirio, Canal Zone, Panamá, No. 256178 U.S.N.M., ♀, X ½.

neck, and shoulders slightly paler brown than ruff; remainder of body soiled yellowish; females having median dorsal stripe; males having dorsal speculum where hair is short and yellow, with narrow (10–20) black or brownish black mid-dorsal stripe.

The correct specific name for the Central American sloth is *Brad[ypus]. variegatus* Schinz (1825:510) according to Wetzel and Kock (1973:25–34, supplemented *in Litt.*, Wetzel, November 22, 1973).

Bradypus variegatus castaneiceps (Gray)

1871. *Arctopithecus castaneiceps* Gray, Proc. Zool. Soc. London, p. 444, August, type from woods surrounding Javali gold mine, 2000 ft., Chontales District, Nicaragua.

MARGINAL RECORDS.—Honduras: Patuca. Costa Rica: Jiménez. Nicaragua: *type locality;* Río Coco.

Bradypus variegatus ephippiger Philippi

1870. *Bradypus ephippiger* Philippi, Arch. Naturg., 1:267, pl. 3, type from northwestern South America, restricted to the Río Atrato, Colombia, by Cabrera, Rev. Mus. Argentino de Cienc. Nat., 4:209, March 27, 1958.
1913. *Bradypus ignavus* Goldman, Smiths. Miscl. Coll., 60(22):1, February 28, type from Marraganti, *ca.* 2 mi. above Real de Santa María, near head of tidewater on Río Tuyra, Darién, Panamá.

MARGINAL RECORDS.—Panamá: Real de Santa María; Tapalisa; Cituro, thence into South America.

Bradypus variegatus griseus (Gray)

1871. *Arctopithecus griseus* Gray, Ann. Mag. Nat. Hist., ser. 4, 7:302, April, type from Cordillera de Chucu, Veraguas,

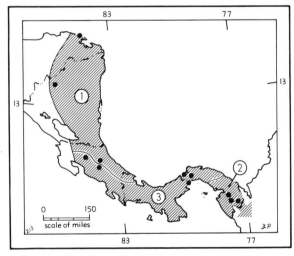

Map 213. *Bradypus variegatus.*

1. *B. v. castaneiceps*   2. *B. v. ephippiger*
3. *B. v. griseus*

Panamá (see Alston, Mamm., pp. 183, 184, December 1880, *in* Godman and Salvin, Biologia Centrali-Americana; . . .).

MARGINAL RECORDS.—Costa Rica: Vijagual; Juan Viñas. Panamá: Gatún; vic. Frijoles; La Chorrera, thence westward along Pacific Coast to point of beginning.

Genus **Choloepus** Illiger—Two-toed Sloth

1811. *Choloepus* Illiger, Prodromus systematis mammalium et avium . . . , p. 108. Type, *Bradypus didactylus* Linnaeus.

Forefoot with 2 claws; tail absent or vestigial; anterior tooth, above and below, caniniform, wearing to sharp beveled edge against opposing tooth, upper tooth anterior to lower tooth when mouth closed; anterior tooth, above and below, much larger than any succeeding tooth, and separated from succeeding tooth by diastema several times longer than crown of succeeding tooth; tympanic annulus present; pterygoid and alisphenoid inflated and forming large bulla with openings in posterior nares; dorsal profile of skull convex with highest point approx. at middle of skull; only part of occipital condyle posterior to paroccipital process.

Choloepus hoffmanni
Two-toed Sloth

External measurements, 600–640; no tail; 110–120. Greatest length of skull, 106–109. Pelage including fur and long guard-hairs but also hairs of intermediate sort; longest hairs up to 170.

In a considerable series of specimens from Costa Rica, Goodwin (1946:354) noted great differences in pallor and darkness among different individuals and described the average coloration as follows: Hair on brow and saddle white or buffy white to roots; forelimbs, shoulders, hind limbs and rump Mummy Brown or Cinnamon-

Brown, the hair more or less broadly tipped with buffy white; throat brownish white; underparts Mummy Brown, washed with brownish white; neck and shoulders, and in some individuals entire pelage, tinged with bright green. Very young specimens are uniform pale Cinnamon-Brown.

Choloepus hoffmanni hoffmanni Peters

1858. *Choloepus hoffmanni* Peters, Monatsb. preuss. Akad. Wiss., Berlin, p. 128, type from Costa Rica, subsequently restricted to Escazú by Goodwin, Bull. Amer. Mus. Nat. Hist., 87:353, December 31, 1946.

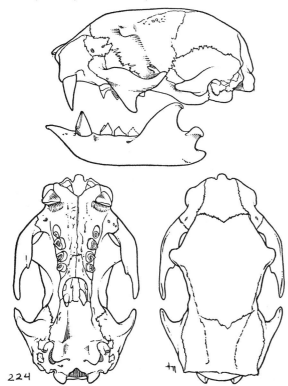

Fig. 224. *Choloepus hoffmanni hoffmanni*, La Palma, Costa Rica, No. 15954 U.S.N.M., ♀, X ½.

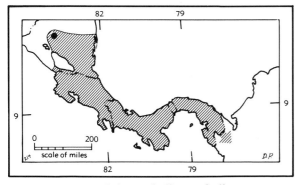

Map 214. *Choloepus hoffmanni hoffmanni*.

MARGINAL RECORDS.—Nicaragua: Matagalpa, thence southward into South America.

INFRAORDER CINGULATA

Major part of skin ossified; ossified part consisting of regularly arranged bony scutes forming an armor; scutes covered by horny epidermal plates; teeth 7 or more, simple, single-rooted; in some species 1 or 2 teeth in premaxilla, in other species all teeth in maxilla; at least 2nd and 3rd cervical vertebrae fused; zygomatic arch complete; tibia and fibula fused proximally and distally.

SUPERFAMILY DASYPODOIDEA
Armadillos

The Dasypodoidea differ from the Glyptodontoidea, the other recognized superfamily, in lacking ventrally projecting process on zygoma; dentary longer than high, ascending ramus extending obliquely, rather than steeply, upward; clavicle present; skull shallow rather than deep.

FAMILY DASYPODIDAE—Armadillos

Horns absent from head-shield; symphyseal part of mandible V-shaped; mandibular condyle flat or concave.

SUBFAMILY DASYPODINAE

Differs from Stegotheriinae in much broader and more flattened skull, occipital condyles placed relatively higher, and less reduced jaws and dentition; differs from Chlamytheriinae in lacking anteroposteriorly lengthened teeth, which are partially dissected into prisms by vertical grooves; differs from Chlamyphorinae in having skull elongated, tapering rather than short, pointed.

KEY TO TRIBES OF DASYPODINAE

1. Bands on back, 11–13; scales on tail not segmentally arranged; 3rd digit unusually enlarged. . . . . Tribe Priodontini, p. 281
1'. Bands on back, 7–11; scales on tail segmentally arranged (in rings); 3rd digit not unusually enlarged. . Tribe Dasypodini, p. 282

TRIBE PRIODONTINI

Anterior and posterior shields present; bands on back, 11–13; bands on neck, 3–4; scales on tail not segmentally arranged; forefoot with 5 digits, each bearing claw; 3rd digit and its claw much enlarged.

Genus **Cabassous** McMurtrie
Eleven-banded Armadillos

1831. *Cabassous* McMurtrie, The animal kingdom . . . by the Baron Cuvier . . . , 1:164. Type, *Dasypus unicinctus* Linnaeus.

1830. *Xenurus* Wagler, Natürliches System der Amphibien . . . , p. 36, August. Type, *Dasypus gymnurus* Wied-Neuwied [= *Dasypus unicinctus* Linnaeus]. Not *Xenurus* Boie, 1826, a genus of birds.

1841. *Arizostus* Gloger, Gemeinnütziges Hand- und Hilfsbuch der Naturgeschichte, 1:xxxii, 114. Type, *Dasypus gymnurus* [= *Dasypus unicinctus* Linnaeus].

1865. *Tatoua* Gray, Proc. Zool. Soc. London, p. 378, October. Type, *Dasypus unicinctus* Linnaeus.

1873. *Ziphila* Gray, Hand-list of the edentate, thick-skinned and ruminant mammals in the British Museum, p. 22. Type, *Ziphila lugubris* [= *Dasypus tatouay* Desmarest].

1891. *Lysiurus* Ameghino, Rev. Argentina Hist. Nat., 1(entr. 4a):254, renaming of *Xenurus* Wagler, 1830.

Teeth $\frac{8}{7}$ or $\frac{10}{8}$; teeth absent from premaxilla; scales on tail small; claws 3 and 4 largest.

Cabassous centralis (Miller)
Central American Five-toed Armadillo

1899. *Tatoua (Ziphila) centralis* Miller, Proc. Biol. Soc. Washington, 13:4, January 31, type from Chamelicón [= Chamelecón], Honduras.

1899. *C[abassous]. centralis* Palmer, Proc. Biol. Soc. Washington, 13:72, September 28.

External measurements of holotype (from dried specimen): 450; 150; 58. Greatest length of skull, 80.0. Crown shields, 37–39, pentagonal or hexagonal, arranged in bilaterally symmetrical pattern; less than 12 irregularly arranged scales on cheek; distance between bases of ears approx. $1\frac{1}{2}$ times length of ear from notch; scapular shield of 7 or 8 rows of plates of which longest row contains approx. 28 plates; 2 or 3 rows of scales on

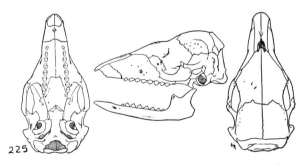

Fig. 225. *Cabassous centralis*, El Muñeco, Costa Rica, No. 67548 U.M.M.Z., ♂, X ⅓.

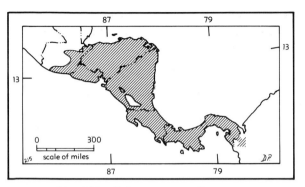

Map 215. *Cabassous centralis*.

neck in front of scapular shield; dorsal rings, 10; pelvic shield with 10 rows of plates; rostrum short; anterior opening of infraorbital canal approx. midway between anterior tip of nasals and posterior tip of pterygoid; mastoidal breadth more than 40 per cent (47% in holotype) of length of skull.

The tremendously enlarged, flattened and ventrally concave claws are indicative of the exceptional fossorial powers of this species. In the holotype the claw of the third digit is 38 mm. long and 13 mm. wide and, although rounded at the tip, is nearly uniform in width.

MARGINAL RECORDS.—Guatemala: *mts. of Quiché* and *Pacific coastal plain*, thence southward into South America.

TRIBE **DASYPODINI**

Teeth $\frac{7-8}{7-8}$, small, subcylindrical, allegedly lacking enamel, 1st two laterally compressed and last half or less (in occlusal area) as large as others; tooth-row short, terminating anterior to root of zygoma; snout long; pterygoids meeting below posterior nares; ears close together on occiput; 7–11 movable bands in carapace between anterior and posterior bucklers; pattern of sculpture on scutes V-shaped; tail scutes arranged in rings, 3rd digit not unusually enlarged; polyembryonic young.

Genus **Dasypus** Linnaeus
Nine-banded Armadillo

1758. *Dasypus* Linnaeus, Syst. nat., ed. 10, 1:50. Type by tautonomy, *Dasypus novemcinctus* Linnaeus.

1762. *Cataphractus* Storr, Prodromus methodi mamm., p. 40, Table B. Based on *Armadillo, Armadillo orientalis, A. indicus, A. mexicanus, A. brasilianus, A. guianensis,* and *A. africanus*.

1765. *Tatus* Fermin, Histoire naturelle de la Hollande Equinoxiale . . . , Amsterdam, p. 3. Type, *Tatus cucurbitalis*.

1779. *Tatu* Blumenbach, Handbuch der Naturgeschichte, 1:74. Type, *Dasypus novemcinctus* Linnaeus.

1804. *Loricatus* Desmarest, Nouv. Dict. Hist. Nat., 24(pr. 6; Tabl. method mamm.):28. No type designated; included *Dasypus giganteus* É. Geoffroy St.-Hilaire, *Loricatus flavimanus* Desmarest, *L. tatouay* Desmarest, *L. villosus* Desmarest, *L. niger* Desmarest, *L. hybridus* Desmarest, *L. pichiy* Desmarest, and *L. matacus* Desmarest.

1827. *Tatusia* Lesson [F. Cuvier *in* Lesson of *Aucts.*], Manual de mammalogie, . . . , p. 309. No type designated; included *Tatusia apar* Desmarest, *T. quadricincta* Lesson, *T. peba* Desmarest, *T. hybridus* Desmarest, *T. tatouay* Desmarest, *T. villosa* Desmarest, and *T. minuta* Desmarest.

1831. *Cachicamus* McMurtrie, The animal kingdom . . . by the Baron Cuvier . . . , 1:163. Included *Dasypus novemcinctus* Linnaeus and *D. septemcinctus* Schreber.

1835. *Cachicama* I. Geoffroy St.-Hilaire, Résumé leçons mamm., 1:53, an invalid emendation of *Cachicamus* McMurtrie.

1841. *Zonoplites* Gloger, Gemeinnütziges Hand- und Hilfsbuch der Naturgeschichte, 1:114. Based on the armadillos having 4 toes on the forefoot, the 2 middle being larger than the others.

1854. *Praopus* Burmeister, Systematische Uebersicht der Thiere Brasiliens . . . , pt. 1, p. 295. Type, *Dasypus longicaudus* Wied-Neuwied.

1856. *Cryptophractus* Fitzinger, Versamml. deutsch. Naturf. Aertz., p. 123. Type, *Cryptophractus pilosus* Fitzinger.

1864. *Hyperoambon* Peters, Monatsb. preuss. Akad. Wiss., Berlin, p. 179. Included *Dasypus pentadactylus* Peters [= *Dasypus kappleri* Krauss] and *D. peba* Desmarest [= *Dasypus novemcinctus* Linnaeus].

1874. *Muletia* Gray, Proc. Zool. Soc. London, p. 244, August. Type, *Dasypus septemcinctus* Linnaeus.

On anterior and posterior shields, hairs arise from pits occurring at intersection of furrow surrounding large plaques with radial furrows that separate secondary (peripheral) plaques, which in turn surround each large plaque. Other characters as for the tribe.

Dasypus novemcinctus
Nine-banded Armadillo

x ¹⁄₉

External measurements: 615–800; 245–370; 75–100. Greatest length of skull, 85.5–100. Weight of adults in Texas up to 17 pounds. Ordinarily approx. two-thirds that weight. Four toes on forefoot and 5 on hind foot; all digits clawed; scapular shield and pelvic shield each of 18–20 rows of ossified scutes; bases of ears touching, at least in dried skins; rostrum long; anterior open-ing of infraorbital canal nearer posterior ends of pterygoids than anterior ends of nasals; mastoidal width less than 40 per cent (usually less than 30%) of greatest length of skull.

Dasypus novemcinctus davisi Russell

1953. *Dasypus novemcinctus davisi* Russell, Proc. Biol. Soc. Washington, 66:21, March 30, type from Huitzilac, 8500 ft., Morelos.

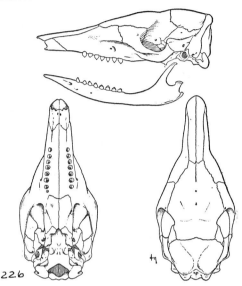

226

Fig. 226. *Dasypus novemcinctus mexicanus*, 20 km. ENE Jesús Carranza, 200 ft., Veracruz, No. 32118 K.U., ♂, X ½.

MARGINAL RECORDS.—Morelos: *5 km. N Tres Cumbres;* Tlacotepec; type locality. Guerrero (Davis and Lukens, 1958:362): Agua del Obispo, 3300 ft.; Acahuizotla, 2800 ft.

Dasypus novemcinctus fenestratus Peters

1864. *Dasypus fenestratus* Peters, Monatsb. preuss. Akad. Wiss., Berlin, p. 180, type from Costa Rica.
1911. *Dasypus novemcinctus fenestratus*, G. M. Allen, Bull. Mus. Comp. Zool., 54:199, July.

MARGINAL RECORDS.—El Salvador: Mt. Cacaguatique (Burt and Stirton, 1961:42). Nicaragua: Río Coco (J. A. Allen, 1910:94), thence southward into South America, and up coast to El Salvador: Barra de Santiago (Burt and Stirton, 1961:42).

Dasypus novemcinctus hoplites G. M. Allen

1911. *Dasypus novemcinctus hoplites* G. M. Allen, Bull. Mus. Comp. Zool., 54:195, July, type from hills back of Gouyave, Grenada, Lesser Antilles. Known only from type locality.

Dasypus novemcinctus mexicanus Peters

1864. *Dasypus novemcinctus* var. *mexicanus* Peters, Monatsb. preuss. Akad. Wiss., Berlin, p. 180, type from Matamoros,

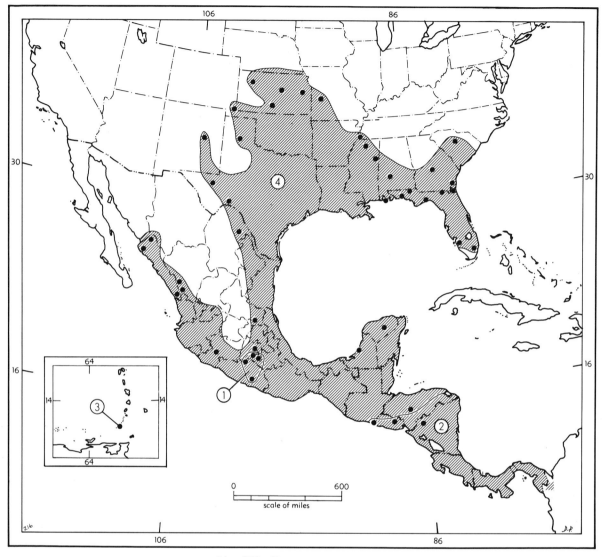

Map 216. *Dasypus novemcinctus.*

1. *D. n. davisi*   2. *D. n. fenestratus*   3. *D. n. hoplites*   4. *D. n. mexicanus*

Tamaulipas (see Hollister, Jour. Mamm., 6:60, February 9, 1925).

1920. *D*[*asypus*]. *novemcinctus mexicanus*, Goldman, Smiths Miscl. Coll., 69(5):66, April 24.

1905. *Tatu novemcinctum texanum* V. Bailey, N. Amer. Fauna, 25:52, October 24, type from Brownsville, Cameron Co., Texas. Regarded as a synonym of *D. n. mexicanus* by Hollister (*antea*).

1917. *Dasypus cucurbitinus* Gaumer, Mongrafía de los mamíferos de Yucatán, p. 21. (In synonymy of *Dasypus novemcinctus*.)

MARGINAL RECORDS.—Kansas: 5 mi. S, 5 mi. W Hoxie (Smith and Lawlor, 1964:48); Salina (Smith and Lawlor, 1964:49); Osage County. Missouri: Brown-

ington. Tennessee: Memphis (Galyon, 1968:14). Mississippi (Kennedy, *et al.*, 1974:9, 10): 10 mi. SE Waterford; Lowndes County. Alabama: 12 mi. NE Thomasville (Wolfe, 1968:210). Georgia: Bibb County (Golley, 1962:76). South Carolina: Fairfield County (Golley, 1966:15). Georgia: Camden County (Cleveland, 1970:91). Florida (Wolfe, 1968:209, 210, unless otherwise noted): Holmes Co., Rt. 81, 2 mi. N Rt. 90; Hamilton County (from map); Nassau County (Fitch, *et al.*, 1952:26); Broward County (*ibid.*); near Fort Myers Beach (*ibid.*); Wakulla County; 7 mi. SW Milton. Alabama: near Fort Morgan (Wolfe, 1968:210). Campeche: 7 km. S Champotón (92080 KU). Yucatán: Pisté (92078 KU). Honduras: La Cueva Archaga, thence northward along Pacific Coast to Sinaloa: Escuinapa; 3 mi. NE San Miguel (Jones, *et al.*, 1962:155); 16 km.

NNE Choix, 1700 ft. (Armstrong and Jones, 1971:752). Durango (Baker and Greer, 1962:77): Ventanas; 2 mi. S Pueblo Nuevo, 3000 ft. *Guanajuato.* Michoacán: Tancítaro. Distrito Federal: Cerro Zacayuca, 2 mi. N Tlalpam. San Luis Potosí: Xilitla. Coahuila: Sabinas. Texas (Cleveland, 1970:91): Terrell County; Reeves County.

New Mexico: *southeastern corner;* 9¾ mi. W Santa Rosa (Hendricks, 1963:581, as *D. novemcinctus* only); *9 mi. W Santa Rosa* (*ibid.,* as *D. novemcinctus* only). Texas: Armstrong County (Davis, 1966:251, as *D. novemcinctus* only). Colorado: *ca.* 20 mi. S Walsh (Hahn, 1966:303, as *D. novemcinctus* only). Kansas: Iuka.

Order **LAGOMORPHA**
Hares, Rabbits, and Pikas

Families and genera revised by Lyon, Smiths. Miscl. Coll., 45:321–447, June 15, 1904. For taxonomic status of group see Gidley, Science, n.s., 36:285–286, August 30, 1912.

The Order Lagomorpha is old in the geological sense; fossilized bones and teeth of both pikas and rabbits are known from deposits of Oligocene age, and even at that early time the structural features distinguishing these animals from other orders were well developed.

KEY TO NORTH AMERICAN FAMILIES AND GENERA OF LAGOMORPHA

1. Hind legs scarcely longer than forelegs; hind foot less than 40; nasals widest anteriorly; no supraorbital process on frontal; 5 cheek-teeth on each side above.
   Family Ochotonidae, p. 286
   Ochotona, p. 286
1'. Hind legs notably longer than forelegs; hind foot more than 40; nasals widest posteriorly; supraorbital process on frontal; 6 cheek-teeth on each side above.
   Family Leporidae, p. 292
   2. Interparietal fused with parietals (see Figs. 239–245); hind foot usually more than 105. . . . . *Lepus*, p. 314
   2'. Interparietal not fused with parietals (see Figs. 228–238); hind foot usually less than 105. . . '. . . . *Romerolagus*, p. 293
   Sylvilagus, p. 294

FAMILY **OCHOTONIDAE**—Pikas

Certain characters in which this family differs from the Leporidae (hares and rabbits) are: hind legs scarcely longer than forelegs; ears short, approx. as wide as high; no postorbital process on frontal; rostrum slender; nasals widest anteriorly; maxilla having a single, large fenestra (instead of fenestrae); jugal long and projecting far posteriorly to zygomatic arm of squamosal; no pubic symphysis; one less cheek-tooth above, the dental formula being i. $\frac{2}{1}$, c. $\frac{0}{0}$, p. $\frac{3}{2}$, m. $\frac{2}{3}$; 2nd upper maxillary tooth unlike 3rd in form; last lower molar simple (not double) or absent (in the extinct genus *Oreolagus*); cutting edge of 1st upper incisor V-shaped; mental foramen situated under last lower molar.

Genus **Ochotona** Link—Pikas

North American species revised by A. H. Howell, N. Amer. Fauna, 47:1–57, August 21, 1924. For a synopsis see Hall, Univ. Kansas Publ., Mus. Nat. Hist., 5:125–133, December 15, 1951.

1795. *Ochotona* Link, Beytrage zur Naturgeschichte, 1 (pt. 2):74. Type, *Lepus ogotona* Pallas.

Five cheek-teeth in lower jaw; 1st cheek-tooth (p3) with more than one reentrant angle; columns of lower molars angular internally; transverse width of any one column of a lower molariform tooth more than double the width of the neck connecting it to the other column.

Subgenus **Pika** Lacépède

1799. *Pika* Lacépède, Tableau des divisions, sous-divisions, ordres et genres des mammifères, p. 9. Type, *Lepus alpinus* Pallas.

Skull flattened; interorbital region wide; maxillary orifice roundly triangular; palatal foramina separate from anterior palatine foramina.

All living members of family Ochotonidae belong to this genus. American pikas all belong to subgenus *Pika*, which occurs also in Eurasia.

KEY TO NOMINAL AMERICAN SPECIES OF OCHOTONA

1. Occurring north of 58° N lat.; underparts creamy white, without buffy wash; an indistinct grayish "collar" on shoulders.
   O. collaris, p. 286
1'. Occurring south of 58° N lat.; underparts washed with buff; no grayish "collar" on shoulders. . . . . *O. princeps*, p. 287

Ochotona collaris (Nelson)
Collared Pika

1893. *Lagomys collaris* Nelson, Proc. Biol. Soc. Washington, 8:117, December 21, type from near head Tanana River, Alaska.
1897. [*Ochotona*] *collaris*, Trouessart, Catalogus mammalium . . . , p. 648.

Total length, 189; hind foot, 30. Upper parts drab to light drab; underparts creamy white; grayish patch on nape and shoulders; skull broad; tympanic bullae large.

MARGINAL RECORDS.—Mackenzie: Richardson Mts., 67° 54′ N, 136° 08′ W (Feist and McCourt, 1973:318); *Horn Lake, 37 mi. NW Fort McPherson,*

1000 ft. (Youngman, 1964:2); mile 63E on Little Keel River, Canol Road; Nahanni River, above Gates, 4500 ft. (Youngman, 1968:74). Yukon: Ida Lake (= McPherson Lake) (*ibid.*); vic. Teslin Lake. British Columbia (Cowan and Guiguet, 1965:93): Teslin Lake; Atlin Lake. Alaska: near Skagway (Manville and Young, 1965:14, from map). British Columbia: Stonehouse Creek, 5½ mi. W jct. Stonehouse Creek and Kelsall River. Alaska: Tanana River; 61° 15′ N, 145° 45′ W (Rausch, 1962:22, fig. 1); 61° N, 149° W (*ibid.*); Chigmit Mts. (*ibid.*); near Fort Yukon (Manville and Young, 1965:14, from map).

Ochotona princeps
Pika

Total length, 162–216; hind foot, 25–35; weight of *O. p. tutelata*: average of 6 males, 121 (108–128); two females, 121 and 129 grams. Upper parts varying from grayish to Cinnamon-Buff depending on the subspecies; underparts with wash of buff. Eight Nevadan females had an average of 3.1 (2–4) embryos. Mode was 3.

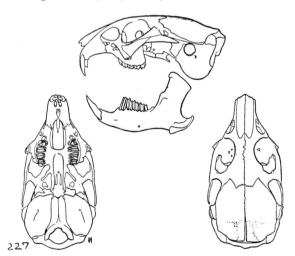

Fig. 227. *Ochotona princeps tutelata*, Greenmonster Canyon, 8150 ft., Nevada, No. 38519 M.V.Z., ♂, X 1.

Ochotona princeps albata Grinnell

1912. *Ochotona albatus* Grinnell, Univ. California Publ. Zool., 10:125, January 31, type from 11,000 ft., near Cottonwood Lakes, Sierra Nevada, Inyo Co., California.
1951. *Ochotona princeps albata*, Hall, Univ. Kansas Publ., Mus. Nat. Hist., 5:127, December 15.

MARGINAL RECORDS.—California: Bullfrog Lake; 10,000 ft., Independence Creek; type locality; Mineral King, E fork Kaweah River.

Ochotona princeps barnesi Durrant and Lee

1955. *Ochotona princeps barnesi* Durrant and Lee, Proc. Biol. Soc. Washington, 68:6, May 20, type from Johnson's Reservoir, 8800 ft., 15 mi. by road N Loa (Fishlake Plateau), Sevier Co., Utah.

MARGINAL RECORDS.—Utah: *1 mi. NW Mt. Marvine (Seven Mile Valley), 9200 ft.;* type locality; *Seven Mile Canyon, 4 mi. N Johnson's Reservoir, 8800 ft.*

Ochotona princeps brooksi A. H. Howell

1924. *Ochotona princeps brooksi* A. H. Howell, N. Amer. Fauna, 47:30, September 23, type from Sicamous, British Columbia.

MARGINAL RECORDS (Cowan and Guiguet, 1965:95).—British Columbia: type locality; *Tappen; Robbins Range, 20 mi. SE Kamloops; Shuswap.*

Ochotona princeps brunnescens A. H. Howell

1919. *Ochotona fenisex brunnescens* A. H. Howell, Proc. Biol. Soc. Washington, 32:108, May 20, type from Keechelus, Kittitas Co., Washington.
1924. *Ochotona princeps brunnescens* A. H. Howell, N. Amer. Fauna, 47:31, September 23.

MARGINAL RECORDS.—British Columbia: Alta Lake; Hope, Lake House. Washington: *Whatcom Pass;* Stevens Pass; *Cowlitz Pass.* Oregon: Mt. Hood; ½ mi. W Salt Creek Falls, 4000 ft.; Crater Lake; Mt. McLoughlin; Diamond Lake. Washington: Tumtum Mtn.; Mt. Index; Mt. Baker, 4800 ft. (Broadbooks, 1965:301, as *O. princeps* only). British Columbia: Sumas (Cowan and Guiguet, 1965:95); North Vancouver.

Ochotona princeps cinnamomea J. A. Allen

1905. *Ochotona cinnamomea* J. A. Allen, Mus. Brooklyn Inst. Arts and Sci., Sci. Bull., 1:121, March 31, type from 11,000 ft., Brigg's [= Britt's] Meadows, Beaver Range, Beaver Co., Utah (5 mi. by road W Puffer Lake, according to Hardy, Jour. Mamm., 26:432, February 12, 1946). Known from type locality only.
1934. *Ochotona princeps cinnamomea*, Hall, Proc. Biol. Soc. Washington, 47:103, June 13.

Ochotona princeps clamosa Hall and Bowlus

1938. *Ochotona princeps clamosa* Hall and Bowlus, Univ. California Publ. Zool., 42:335, October 12, type from 8400 ft., N rim Copenhagen Basin, Bear Lake Co., Idaho.

MARGINAL RECORDS.—Idaho: type locality; *Deep Lake, Bear River Mts. 2 mi. E Strawberry Creek Ranger Station, Wasatch Mts.*

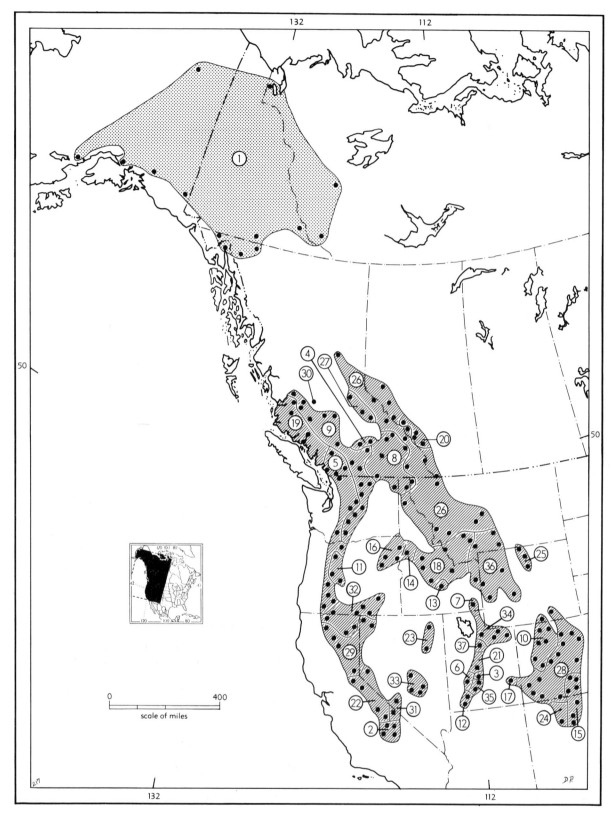

Map 217. *Ochotona collaris* and *Ochotona princeps.*

Guide to kinds				
1. *O. collaris*	6. *O. p. cinnamomea*	12. *O. p. fuscipes*	19. *O. p. littoralis*	26. *O. p. princeps*	32. *O. p. taylori*
2. *O. p. albata*	7. *O. p. clamosa*	13. *O. p. goldmani*	20. *O. p. lutescens*	27. *O. p. saturata*	33. *O. p. tutelata*
3. *O. p. barnesi*	8. *O. p. cuppes*	14. *O. p. howelli*	21. *O. p. moorei*	28. *O. p. saxatilis*	34. *O. p. uinta*
4. *O. p. brooksi*	9. *O. p. fenisex*	15. *O. p. incana*	22. *O. p. muiri*	29. *O. p. schisticeps*	35. *O. p. utahensis*
5. *O. p. brunnescens*	10. *O. p. figginsi*	16. *O. p. jewetti*	23. *O. p. nevadensis*	30. *O. p. septentrionalis*	36. *O. p. ventorum*
	11. *O. p. fumosa*	17. *O. p. lasalensis*	24. *O. p. nigrescens*	31. *O. p. sheltoni*	37. *O. p. wasatchensis*
		18. *O. p. lemhi*	25. *O. p. obscura*		

Ochotona princeps cuppes Bangs

1899. *Ochotona cuppes* Bangs, Proc. New England Zool. Club, 1:40, June 5, type from 4000 ft., Monashee Divide, Gold Range, British Columbia.
1924. *Ochotona princeps cuppes*, A. H. Howell, N. Amer. Fauna, 47:27, September 23.

MARGINAL RECORDS.—British Columbia: Glacier; Toby Creek, 18 mi. W Invermere. Montana: W Fork Yaak River (Hoffmann, *et al.*, 1969:582, as *O. princeps* only). Idaho: Cabinet Mts. Washington: Sullivan Lake. British Columbia: Kettle Valley; type locality; Mt. Revelstoke.

Ochotona princeps fenisex Osgood

1863. *Lagomys minimus* Lord, Proc. Zool. Soc. London, p. 98, lectotype from 7000 ft., Ptarmigan Hill, near head Ashnola River, Cascade Range, British Columbia. *Nec* Schinz, 1821.
1913. *Ochotona fenisex* Osgood, Proc. Biol. Soc. Washington, 26:80, March 22, renaming of *L. minimus* Lord.
1924. *Ochotona princeps fenisex*, A. H. Howell, N. Amer. Fauna, 47:28, September 23.

MARGINAL RECORDS.—British Columbia (Cowan and Guiguet, 1965:96, unless otherwise noted): Kimsquit; Rainbow Mts.; Redstone; Hanceville, 24 mi. W Williams Lake; MacGillivray Creek, Lillooet; Okanagan (Hall and Kelson, 1959:249); Hedley; *Ashnola River.* Washington: Horseshoe Basin, near Mt. Chopaka; mts. near Wenatchee; Steamboat Mtn.; Bethel Ridge, 30 mi. ESE Mt. Rainier, 3 mi. N Tieton Reservoir, 6000 ft. (Broadbooks, 1965:301, as *O. princeps* only); Easton; Lyman Lake; Barron. British Columbia (Cowan and Guiguet, 1965:96): Tulameen; *Texas Creek;* Kleena Kleene; Cariboo Mts. S of Stuie.

Ochotona princeps figginsi J. A. Allen

1912. *Ochotona figginsi* J. A. Allen, Bull. Amer. Mus. Nat. Hist., 31:103, May 28, type from Pagoda Peak, Rio Blanco Co., Colorado.
1924. *Ochotona princeps figginsi*, A. H. Howell, N. Amer. Fauna, 47:21, September 23.

MARGINAL RECORDS.—Wyoming (Long, 1965a:540): Bridger Peak; *W slope Sierra Madre Mts.* Colorado: Mt. Zirkel; Trappers Lake; *Crested Butte;* Irwin; 6 mi. E Skyway (Anderson, 1959:409); type locality; Sand Mtn., 9 mi. SW Hahns Peak P.O. Wyoming: *7½ mi. N, 18½ mi. E Savery* (Long, 1965a:540).

Ochotona princeps fumosa A. H. Howell

1919. *Ochotona fenisex fumosa* A. H. Howell, Proc. Biol. Soc. Washington, 32:109, May 20, type from Permilia [= Pamelia?] Lake, W base Mt. Jefferson [Linn Co. ?], Oregon.
1924. *Ochotona princeps fumosa* A. H. Howell, N. Amer. Fauna, 47:33, September 23.

MARGINAL RECORDS.—Oregon: *ca.* 900 ft., 15 mi. above Estacada; Lost Lake in Newberry Crater; *Paulina Lake; Three Sisters; Lava Butte;* Lost Creek Ranger Station, 10 mi. SE McKenzie Bridge.

Ochotona princeps fuscipes A. H. Howell

1919. *Ochotona schisticeps fuscipes* A. H. Howell, Proc. Biol. Soc. Washington, 32:110, May 20, type from Brian Head, Parowan Mts., Iron Co., Utah.
1941. *O[chotona]. p[rinceps]. fuscipes*, Hall and Hayward, The Great Basin Naturalist, 2:108, July 20.

MARGINAL RECORDS.—Utah: type locality; 9000 ft., Duck Creek, Kane Co. (Durrant, 1952:Fig. 18 on p. 68, and p. 71); *Kolob Reservoir,* 8094 ft. (Stock, 1970:430); Lava Point, 7890 ft. (*ibid.*).

Ochotona princeps goldmani A. H. Howell

1924. *Ochotona schisticeps goldmani* A. H. Howell, N. Amer. Fauna, 47:40, September 23, type from Echo Crater, Snake River Desert, 20 mi. SW Arco, Butte Co., Idaho.
1938. *Ochotona princeps goldmani*, Hall and Bowlus, Univ. California Publ. Zool., 42:337, October 12.

MARGINAL RECORDS.—Idaho: *S base Grassy Cone;* type locality; *Fissure Crater; Great Owl Cavern.*

Ochotona princeps howelli Borell

1931. *Ochotona princeps howelli* Borell, Jour. Mamm., 12:306, August 24, type from 7500 ft., near head Bear Creek, summit Smith Mtn., S end Seven Devils Mts., Adams Co., Idaho.

MARGINAL RECORDS.—Idaho: *½ mi. E Black Lake;* type locality.

Ochotona princeps incana A. H. Howell

1919. *Ochotona saxatilis incana* A. H. Howell, Proc. Biol. Soc. Washington, 32:107, May 20, type from 12,000 ft., Pecos Baldy, Santa Fe Co., New Mexico.
1924. *Ochotona princeps incana* A. H. Howell, N. Amer. Fauna, 47:25, September 23.

MARGINAL RECORDS.—Colorado (Armstrong, 1972:79, unless otherwise noted): North Crestone Trail, 11,500 ft.; Medano Creek (Hall and Kelson, 1959:249); West Spanish Peak, 11,900 ft. New Mexico: Wheeler Peak; type locality; Santa Fe Ski Area (Findley, *et al.*, 1975:82). Colorado: near Fort Garland (Armstrong, 1972:79).

Ochotona princeps jewetti A. H. Howell

1919. *Ochotona schisticeps jewetti* A. H. Howell, Proc. Biol. Soc. Washington, 32:109, May 20, type from head Pine Creek, near Cornucopia, S slope Wallowa Mts., Baker Co., Oregon.
1951. *Ochotona princeps jewetti*, Hall, Univ. Kansas Publ., Mus. Nat. Hist., 5:130, December 15.

MARGINAL RECORDS.—Oregon: Wallowa Lake; Cornucopia, near head East Pine Creek; *Anthony;* Strawberry Butte; Austin.

Ochotona princeps lasalensis Durrant and Lee

1955. *Ochotona princeps lasalensis* Durrant and Lee, Proc. Biol. Soc. Washington, 68:4, May 20, type from Warner Ranger Station, 9750 ft., La Sal Mts., Grand Co., Utah.

MARGINAL RECORDS.—Utah: ½ mi. N Warner Ranger Station, 9000 ft., La Sal Mts.; *Geyser Pass; Mt. Mellithin* [= *Mellenthin*], *12,280 ft.; ½ mi. S Warner Ranger Station, 9700 ft.*

Ochotona princeps lemhi A. H. Howell

1919. *Ochotona uinta lemhi* A. H. Howell, Proc. Biol. Soc. Washington, 32:106, May 20, type from Lemhi Mountains, 10 mi. W Junction, Lemhi Co., Idaho.
1924. *Ochotona princeps lemhi* A. H. Howell, N. Amer. Fauna, 47:16, September 23.

MARGINAL RECORDS.—Montana (Hoffmann, *et al.,* 1969:582, 583, as *O. princeps* only): Table Mtn., Silver Bow Co.; Anchor Lake, Madison Co.; *Mt. Bradley, Madison Co.* Idaho: mts. E Birch Creek; Ketchum; *Stanley Lake;* 5 mi. W Cape Horn; Elk Summit, *ca.* 15 mi. SE Warren. Montana: Upper Miner Lake, near Jackson (Hoffmann, *et al.,* 1969:582, as *O. princeps* only).

Ochotona princeps littoralis Cowan

1955. *Ochotona princeps littoralis* Cowan, Murrelet, 35:22, August 27, type from Hagensborg, British Columbia.

MARGINAL RECORDS.—British Columbia: type locality; Purcell Point; *Fawn Bluff;* Arran Rapids; head Rivers Inlet.

Ochotona princeps lutescens A. H. Howell

1919. *Ochotona princeps lutescens* A. H. Howell, Proc. Biol. Soc. Washington, 32:105, May 20, type from approx. 8000 ft., Mt. Inglesmaldie, near Banff, Alberta.

MARGINAL RECORDS.—Alberta: Mistaya Creek [= River], Banff–Jasper Highway; *head Dorimer River;* type locality; *Canmore;* Mt. Forget-me-not; head Brewster Creek.

Ochotona princeps moorei Gardner

1950. *Ochotona princeips moorei* Gardner, Jour. Washington Acad. Sci., 40:344, October 23, 1950, type from 10,000 ft., 1 mi. NE Baldy Ranger Station, Manti National Forest, Sanpete Co., Utah. Known from type locality only.

Ochotona princeps muiri Grinnell and Storer

1916. *Ochotona schisticeps muiri* Grinnell and Storer, Univ. California Publ. Zool., 17:6, August 23, type from 9300 ft.,

near Ten Lakes, Yosemite National Park, Tuolumne Co., California.
1934. *Ochotona princeps muiri,* Hall, Proc. Biol. Soc. Washington, 47:103, June 13.

MARGINAL RECORDS.—Nevada: 8500 ft., 3 mi. S Mt. Rose. California: Markleeville; mts. W Bishop Creek; Washburn Lake; lat. 39° N, summit of Sierra.

Ochotona princeps nevadensis A. H. Howell

1919. *Ochotona uinta nevadensis* A. H. Howell, Proc. Biol. Soc. Washington, 32:107, May 20, type from 10,500 ft., Ruby Mts., SW Ruby Valley P.O., Elko Co., Nevada.
1924. *Ochotona princeps nevadensis* A. H. Howell, N. Amer. Fauna, 47:21, September 23.

MARGINAL RECORDS.—Nevada: 7830 ft., Long Creek; type locality.

Ochotona princeps nigrescens V. Bailey

1913. *Ochotona nigrescens* V. Bailey, Proc. Biol. Soc. Washington, 26:133, May 21, type from Goat Peak, head Santa Clara Creek, 10,000 ft., Jemez Mountains, Sandoval Co., New Mexico. Known only from type locality.
1924. *Ochotona princeps nigrescens,* A. H. Howell, N. Amer. Fauna, 47:26, September 23.

Ochotona princeps obscura Long

1965. *Ochotona princeps obscura* Long, Univ. Kansas Publ., Mus. Nat. Hist., 14:538, July 6, type from Medicine Wheel Ranch, 28 mi. E Lovell, 9000 ft., Big Horn Co., Wyoming.

MARGINAL RECORDS (Long, 1965a:539).—Wyoming: type locality; *4½ mi. S, 19 mi. E Shell;* head Trappers Creek; Powder River Pass, Big Horn Mts. (Broadbooks, 1965:303, as *O. princeps* only).

Ochotona princeps princeps (Richardson)

1828. *Lepus (Lagomys) princeps* Richardson, Zool. Jour., 3:520, type from headwaters Athabaska River, near Athabaska Pass, Alberta.
1897. [*Ochotona*] *princeps,* Trouessart, Catalogus mammalium . . . , p. 648.
1912. *Ochotona levis* Hollister, Proc. Biol. Soc. Washington, 25:57, April 13, type from Chief Mountain [= Waterton] Lake, Glacier Co., Montana. (Regarded as subspecifically inseparable from *O. p. princeps* by Cowan, Murrelet, 35:20, August 27, 1955.)

MARGINAL RECORDS.—British Columbia: headwaters South Pine River. Alberta: Muskeg Creek "about" 60 mi. N Jasper House; Medicine Lake; Sunwapta Pass. British Columbia: *Thompson Pass* (Cowan and Guiguet, 1965:97). Alberta: *Pipestone River;* Baker Lake. British Columbia (Cowan and Guiguet, 1965:97): *Assiniboine;* Tornado Pass. Alberta: Waterton Lake. Montana: Chief Mountain Lake; Little Belt Mts.; Belt Mts.; Cutaway Pass, Granite Co. (Hoffmann, *et al.,* 1969:582, as *O. princeps* only); Lake Como, Bitterroot Mts. Idaho: Coeur d'Alene National Forest. British Columbia: Mt. Evans (Cowan and Guiguet, 1965:97); Spillamacheen River; Kinbasket Lake; Mt. Robson (Cowan and Guiguet, 1965:97).

Ochotona princeps saturata Cowan

1955. *Ochotona princeps saturatus* Cowan, Murrelet, 35:23, August 27, type from Mt. Huntley in Wells Gray National Park, British Columbia.

MARGINAL RECORDS.—British Columbia: Indianpoint Lake (Cowan and Guiguet, 1965:98); type locality; Murtle Lake.

Ochotona princeps saxatilis Bangs

1899. *Ochotona saxatilis* Bangs, Proc. New England Zool. Club, 1:41, June 5, type from Montgomery, "near" Mt. Lincoln, Park Co., Colorado.
1924. *Ochotona princeps saxatilis*, A. H. Howell, N. Amer. Fauna, 47:23, September 23.

MARGINAL RECORDS.—Wyoming (Long, 1965a:540): 3 mi. E Browns Peak; $^{1}/_{2}$ *mi. E Medicine Bow Peak*. Colorado: Estes Park; Niwot Ridge, 12,300 ft. (Johnson, 1967:311); Pikes Peak; Osier (Armstrong, 1972:80); Horse Spring Mtn., near Dyke (*ibid.*); Lime Creek, $39\frac{7}{10}$ mi. N Durango (*ibid.*); Lone Cone Peak (*ibid.*); Crystal Lake, 5 mi. W Lake City; Middle Brush Creek; Ten Mile Creek; Berthoud Pass; "Rocky Cut," 12,100 ft., Rocky Mountain National Park (Broadbooks, 1965:303, as *O. princeps* only). Not found: Colorado: *Irwin Lakes*.

Ochotona princeps schisticeps (Merriam)

1889. *Lagomys schisticeps* Merriam, N. Amer. Fauna, 2:11, October 30, type from Donner [= Summit], Placer Co., California.
1936. *Ochotona princeps schisticeps*, A. H. Miller, Jour. Mamm., 17:175, May 18 (*princeps* and *schisticeps* regarded as conspecific by Borell, Jour. Mamm., 12:307–308, August 24, 1931).

MARGINAL RECORDS.—Nevada: 3 mi. N, 12 mi. E Fort Bidwell, 5700 ft.; 8400–8600 ft., Duffer Peak, Pine Forest Mts. California: Tahoe; *Donner Pass*; 12 mi. NE Prattville; Lassen Peak; Mt. Shasta.

Ochotona princeps septentrionalis Cowan and Racey

1947. *Ochotona princeps septentrionalis* Cowan and Racey, Canadian Field-Nat., 60:102, April 22, type from 6500 ft., Itcha Mountains, lat. 52° 45' N, long. 125° W, British Columbia. Known from type locality only.

Ochotona princeps sheltoni Grinnell

1918. *Ochotona schisticeps sheltoni* Grinnell, Univ. California Publ. Zool., 17:429, April 25, type from 11,000 ft., "near" Big Prospector Meadow, White Mountains, Mono Co., California.

1946. *Ochotona princeps sheltoni*, Hall, Mammals of Nevada, p. 592, July 1.

MARGINAL RECORDS.—Nevada: 8700 ft., Pinchot Creek. California: type locality.

Ochotona princeps taylori Grinnell

1912. *Ochotona taylori* Grinnell, Proc. Biol. Soc. Washington, 25:129, July 31, type from 9000 ft., Warren Peak, Warner Mts., Modoc Co., California.
1951. *Ochotona princeps taylori*, Hall, Univ. Kansas Publ., Mus. Nat. Hist., 5:133, December 15.

MARGINAL RECORDS.—Oregon: N end Steens Mts.; Guano Valley; Jack Lake, 20 mi. NE Adel; Adel. California: type locality; 5400 ft., near Termo, Madeline Plains; near head Little Shasta River. Oregon: Lower Klamath Lake.

Ochotona princeps tutelata Hall

1934. *Ochotona princeps tutelata* Hall, Proc. Biol. Soc. Washington, 47:103, June 13, type from 8150 ft., Greenmonster Canyon, Monitor Mts., Nye Co., Nevada.

MARGINAL RECORDS.—Nevada: 7500 ft., Smiths Creek, Desatoya Mts.; type locality; 8700–11,000 ft., SW and W slopes Mt. Jefferson, Toquima Range; South Twin River; *Arc Dome*.

Ochotona princeps uinta Hollister

1912. *Ochotona uinta* Hollister, Proc. Biol. Soc. Washington, 25:58, April 13, type from near head E fork Bear River, Uinta Mts., Summit Co., Utah.
1924. *Ochotona princeps uinta*, A. H. Howell, N. Amer. Fauna, 47:19, September 23.

MARGINAL RECORDS.—Utah: type locality; Elk Park; *11,000–11,500 ft., The Nipple*; 10,500 ft., SW slope Bald Mtn.; 8500 ft., Morehouse Canyon, 5 mi. above Weber River; *Spirit Lake* not found.

Ochotona princeps utahensis Hall and Hayward

1941. *Ochotona princeps utahensis* Hall and Hayward, Great Basin Nat., 2:107, July 20, type from 2 mi. W Deer Lake, Garfield Co., Utah.

MARGINAL RECORDS.—Utah: 9000 ft., Donkey Lake, Boulder Mtn.; type locality.

Ochotona princeps ventorum A. H. Howell

1919. *Ochotona uinta ventorum* A. H. Howell, Proc. Biol. Soc. Washington, 32:106, May 20, type from Fremont Peak, 11,500 ft., Wind River Mts., Fremont Co., Wyoming.
1924. *Ochotona princeps ventorum* A. H. Howell, N. Amer. Fauna, 47:18, September 23.

MARGINAL RECORDS.—Montana (Hoffmann, *et al.*, 1969:583, as *O. princeps* only): head of Big Timber Creek; 1 mi. S Hellroaring Lakes. Wyoming (Long,

1965a:541): *30 mi. N, 18 mi. W Cody; Pahaska;* Needle Mtn.; *14 mi. S, 8½ mi. W Lander;* 17 mi. S, 6½ mi. W Lander; 19 mi. W, 2 mi. S Big Piney; *5 mi. S, 2 mi. W Fremont Peak; Middle Piney Lake; 2 mi. N, 8 mi. E Alpine;* Teton Pass. Idaho: Teton Canyon. Montana: South Cottonwood Canyon (Hoffmann, *et al.,* 1969:583, as *O. princeps* only).

Ochotona princeps wasatchensis Durrant and Lee

1955. *Ochotona princeps wasatchensis* Durrant and Lee, Proc. Biol. Soc. Washington, 68:2, May 20, type from 10 mi. above lower powerhouse, road to Cardiff Mine, Big Cottonwood Canyon, Salt Lake Co., Utah.

MARGINAL RECORDS.—Utah: type locality; *near Lake Solitude, 9000 ft., Silver Lake P.O. (Brighton); Mt. Timpanogos; Big Willow Canyon, 7000 ft.; Little Cottonwood Canyon, 6 mi. above Wasatch Blvd.*

Family **LEPORIDAE**—Rabbits and Hares

Hind legs longer than forelegs; ears longer than wide; frontal bone carrying supraorbital process consisting always of posterior arm and often also of anterior arm; rostrum wide; nasals not wider anteriorly than posteriorly; maxillae having numerous fenestrae; jugal projecting less than halfway from zygomatic root of squamosal to external auditory meatus (except in *Romerolagus*); pubic symphysis well marked; dental formula, i. $\frac{2}{1}$, c. $\frac{0}{0}$, p. $\frac{3}{2}$, m. $\frac{3}{3}$ (but m. $\frac{2}{3}$ in *Pentalagus* of Liu Kiu [= Ryukyu] Islands south of Japan); 2nd upper maxillary tooth like 3rd in form; last lower molar double; cutting edge of 1st upper incisor straight; mental foramen of mandible situated under 1st lower cheek-tooth. Females average larger than males in all members of this family. (See Orr, 1940:20.) The reverse is true in most other families of mammals.

"Hare" is a name applied to any lagomorph whose young are born fully haired, with eyes open, and able to run about a few minutes after birth. The young are born in the open, not in a nest. All the species of genus *Lepus* are hares. The species of leporids of all genera other than *Lepus*, in North America at least, are rabbits. Their young are born naked, blind, and helpless, in a nest especially built for them and lined with fur. Considering the degree of development of the young at birth, the gestation periods are about what would be expected: 26–30 days in *Sylvilagus* and 36–47 days in *Lepus* (see Severaid, 1950:356–357). Vernacular names are misleading because the names "jack rabbit" and "snowshoe rabbit" are applied to hares; also, "Belgian hare" is a name applied to a rabbit (genus *Oryctolagus*) that is commonly bred in captivity.

Key to Species of Sylvilagus and Romerolagus

1. Antorbital extension of supraorbital process more than half length of posterior extension; 1st upper cheek-tooth with only 1 reentrant angle on anterior face; reentrant angle of 2nd upper cheek-tooth not crenate. . . . . *Sylvilagus idahoensis,*   p. 294
1'. Antorbital extension of supraorbital process less than half of posterior extension or entirely absent; 1st upper cheek-tooth with more than 1 (usually 3) reentrant angles on anterior face; reentrant angle of 2nd upper cheek-tooth crenate.
    2. Anterior extension of supraorbital process absent (or if a point is barely indicated, then $\frac{5}{8}$ or all of posterior process fused to braincase).
       3. Tympanic bulla smaller than foramen magnum; hind foot more than 74; geographic range wholly in United States.
          4. Ear more than 58 from notch in dried skin; basilar length of skull more than 63.
                                Sylvilagus aquaticus,   p. 311
         4'. Ear less than 58 from notch in dried skin; basilar length of skull less than 63.
           5. Underside of tail white; posterior extension of supraorbital process tapering to a slender point, this point free of braincase or barely touching it and leaving a slit or long foramen. . . . . *Sylvilagus transitionalis,*   p. 305
           5'. Underside of tail brown or gray; posterior extension of supraorbital process always fused to skull, usually for entire length but in occasional specimens there is small foramen at middle of posterior extension of supraorbital process.
                                Sylvilagus palustris,   p. 299
       3'. Tympanic bulla as large as foramen magnum; hind foot less than 74; geographic range limited to southern edge of Mexican tableland at high elevations. . . *Romerolagus diazi,*   p. 293
    2'. Anterior extension of supraorbital process present, and posterior extension free of braincase or leaving a slit between the process and braincase.
       6. Tympanic bullae large (see Fig. 236). . . . . *Sylvilagus audubonii,*   p. 307
       6'. Tympanic bullae small (see Figs. 231, 233, and 235).
          7. Restricted to Pacific coastal strip from Columbia River south to tip of Baja California, west of Sierra Nevada–Cascade mt. chain; hind foot less than 81.
           Sylvilagus bachmani, p. 296 and *S. mansuetus,*   p. 299

7'. East of Pacific coastal strip mentioned in 7; hind foot usually more than 81.
  8. North of United States–Mexican boundary.
    9. In Arizona, New Mexico, and southern Colorado posterior extension of supraorbital process free of braincase, and supraoccipital shield posteriorly pointed; from central Colorado north into Canada, diameter of external auditory meatus more than crown length of last 3 cheek-teeth. . . . . *Sylvilagus nuttallii,*   p. 306
    9'. In Arizona, New Mexico, and southeastern Colorado posterior extension of supraorbital process of frontal with its tip against, or fused to, braincase, and supraoccipital shield posteriorly truncate or notched; from central Colorado north into Canada, diameter of external auditory meatus less than crown length of last 3 cheek-teeth. . . . . *Sylvilagus floridanus,*   p. 300
  8'. South of United States–Mexican boundary.
    10. Geographic range restricted to Tres Marías Islands. . . . . *Sylvilagus graysoni,*   p. 313
    10'. Geographic range not including Tres Marías Islands.
      11. Underside of tail dingy gray or buffy (not white).
        12. Tail short (less than 30) and brown like rump; ear from notch (dry) less than 53; interorbital breadth less than 16. . . . . *Sylvilagus brasiliensis,*   p. 295
        12'. Tail of moderate length (more than 30) and dingy gray; ear from notch (dry) more than 53; interorbital breadth more than 16. . . . . *Sylvilagus insonus,*   p. 312
      11'. Underside of tail distinctly white.
        13. Total length more than 476; ear from notch (dry) more than 64; interorbital breadth usually more than 19.3; geographic range, southwestern México north of the Isthmus of Tehuantepec. . . . . *Sylvilagus cunicularius,*   p. 312
        13'. Total length less than 476; ear from notch (dry) less than 64; interorbital breadth usually less than 19.3; geographic range, Canada to Panamá.

                                                  Sylvilagus floridanus,   p. 300

Genus **Romerolagus** Merriam—Volcano Rabbit

1896. *Romerolagus* Merriam, Proc. Biol. Soc. Washington, 10:173, December 29. Type, *Romerolagus nelsoni* Merriam [= *Lepus diazi* Diaz].

Total length, 300–311; tail rudimentary; hind foot, 52; ear from notch (dry), 36; upper parts grizzled buffy brown or dull cinnamon brown; underparts dingy gray; anterior projection of supraorbital process absent; jugal projecting posteriorly past squamosal root of zygomatic arch more than halfway to external auditory meatus. The two cranial characters mentioned are resemblances to pikas, although the skull otherwise resembles that of the true rabbits. The genus contains only the one living species.

Romerolagus diazi (Diaz)
Volcano Rabbit

1893. *Lepus diazi* Diaz, Catálogo, Comisión Geográfico-Exploradora de la República Méxicana Exposición Internacional Colombina de Chicago . . . , pl. 42, March, type from eastern slope of Mount Ixtaccíhuatl, Puebla. (Villa, Anal. Inst. Biol., Univ. Nac. Autó. México, 23:353, May 20, 1953, cites the name *Lepus diazi* from Ferrari-Pérez *in* Diaz.)
1911. *Romerolagus diazi* Miller, Proc. Biol. Soc. Washington, 24:228, October 31.
1896. *Romerolagus nelsoni* Merriam, Proc. Biol. Soc. Wash-

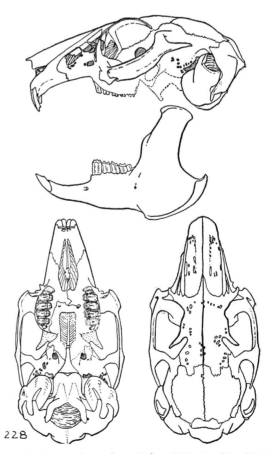

Fig. 228. *Romerolagus diazi,* 31 km. S Mexico City, Distrito Federal, No. 30815 K.U., ♀, X 1.

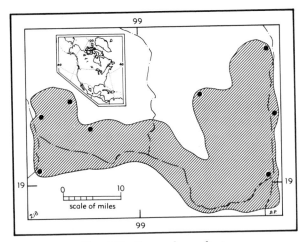

Map 218. *Romerolagus diazi.*

Sylvilagus idahoensis (Merriam)
Pygmy Rabbit

1891. *Lepus idahoensis* Merriam, N. Amer. Fauna, 5:76, July 30, type from head of Pahsimeroi Valley, near Goldburg, Custer Co., Idaho (Davis, The Recent mammals of Idaho, Caxton Printers, Caldwell, Idaho, p. 363, April 5, 1939).
1930. *Sylvilagus idahoensis*, Grinnell, Dixon, and Linsdale, Univ. California Publ. Zool., 35:553, October 10.

Total length, 250–290; tail, 20–30; hind foot, 65–72; ear from notch (dry), 36–48; weight, average of six males, 409(375–435), average of nine females, 398(246–458) grams. Upper parts pinkish to blackish or dark grayish depending on amount of wear.

ington, 10:173, December 29, type from west slope Mount Popocatépetl, 11,000 ft., México.

MARGINAL RECORDS.—México: Monte Río Frío, 45 km. ESE Mexico City. Puebla: type locality. México: Mt. Popocatépetl. Morelos: Laguna de Zempoala (Ramírez-P., 1971:268). Distrito Federal: 16 km. N Huitzilac (*op. cit.*:269); Ajusco; 31 km. S Mexico City. México: Llano Grande, 3 km. W Tlalmanalco.

Genus Sylvilagus Gray—Cottontails and Allies

Revised by Nelson, N. Amer. Fauna, 29:159–278, August 31, 1909. For a synopsis see Hall, Univ. Kansas Publ., Mus. Nat. Hist., 5:138–170, December 15, 1951.

1867. *Sylvilagus* Gray, Ann. Mag. Nat. Hist., ser. 3, 20:221, September. Type, *Lepus sylvaticus* Bachman [= *Lepus nuttalli mallurus* Thomas].

Total length, 291–538; tail, 18–73; hind foot, 71–111; ear from notch (dry), 41–74. Grayish to dark brownish above and paler below; sutures of interparietal bone distinct throughout life; 2nd to 4th cervical vertebrae broader than long with dorsal surface flattened and without carination.

The genus *Sylvilagus* is restricted to the New World; the two species *Sylvilagus brasiliensis* and S. *floridanus* are the only two that occur in South America and they occur also in North America.

Subgenus Brachylagus Miller—Pygmy Rabbit

1900. *Brachylagus* Miller, Proc. Biol. Soc. Washington, 13:157, June 13. Type, *Lepus idahoensis* Merriam.

For characters see subgenus *Sylvilagus.*

MARGINAL RECORDS.—In southeastern Washington: *ca.* 5 mi. W Lower Coulee; Ritzville; Lind; Warden. In remainder of range: Montana: Bannack; Tash Ranch, near Dillon (Hoffmann, *et al.*, 1969:583); Centennial Valley (*ibid.*). Idaho: Trail Creek near Pocatello. Utah: 3 mi. NE Clarkson; W side Utah Lake; 20 mi. W Parowan; 10 mi. SW Cedar City. Nevada: 8½ mi. NE Sharp; Fallon. California: Bodie; Crowley Lake area, 60 mi. SE Bodie; 3 mi. S Ravendale, 5000 ft. Oregon: Klamath Falls (Olterman and Verts, 1972:25); Fremont; Redmond; 10 mi. N Baker. Idaho: type locality; Junction.

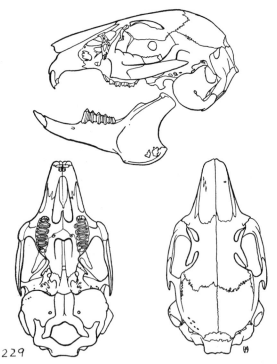

229

Fig. 229. *Sylvilagus idahoensis*, Millett P.O., Nevada, No. 37275 M.V.Z., ♂, X 1.

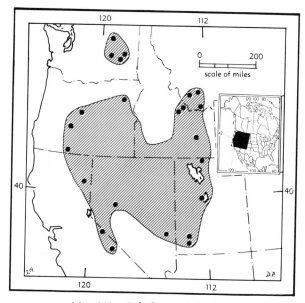

Map 219. *Sylvilagus idahoensis.*

Subgenus **Sylvilagus** Gray
Cottontails and Allies

1867. *Sylvilagus* Gray, Ann. Mag. Nat. Hist., ser. 3, 20:221, September. Type, *Lepus sylvaticus* Bachman [= *Lepus nuttalli mallurus* Thomas].

1867. *Tapeti* Gray, Ann. Mag. Nat. Hist., ser. 3, 20:224, September. Type, *Lepus brasiliensis* Linnaeus.

1897. *Microlagus* Trouessart, Catalogus mammalium . . . , p. 660. Type, *Lepus cinerascens* J. A. Allen.

1897. *Limnolagus* Mearns, Science, n.s., 5:393, March 5. Type, *Lepus aquaticus* Bachman.

1950. *Paludilagus* Hershkovitz, Proc. U.S. Nat. Mus., 100:333, May 26. Type, *Lepus palustris* Bachman.

Characters of subgeneric worth, in contrast to those of the subgenus *Brachylagus*, are: 1st premolar, in upper jaw and in lower jaw, with more than one fold in the enamel; infolded enamel, which divides each molar tooth into two parts, crenate.

Differences between some of the species are of lesser magnitude than was supposed to be the case when the five names for genera or subgenera listed immediately above were proposed. Some species can be placed in each of two subgenera with almost equal propriety. If used, four of the five subgeneric names mentioned above would contain only one species each. It seems that no useful purpose is served by attempting to fit the several species of the genus *Sylvilagus* into more than the two subgenera *Brachylagus* and *Sylvilagus*. The other names, *Tapeti* Gray, *Microlagus* Trouessart, *Limnolagus* Mearns, and *Paludilagus* Hershkovitz, are here arranged as

synonyms of the subgeneric name *Sylvilagus* Gray.

Sylvilagus brasiliensis
Forest Rabbit

Total length, 380–420; tail, 20–21; hind foot, 77–80; ear from notch (dry), 39–46. The principal characters of this species are small size, dark color, short tail, and dingy buffy (not white) undersurface of tail.

Sylvilagus brasiliensis consobrinus Anthony

1917. *Sylvilagus gabbi consobrinus* Anthony, Bull. Amer. Mus. Nat. Hist., 37:335, May 28, type from Old Panamá, Panamá. Known from type locality only.

1950. *Sylvilagus brasiliensis consobrinus,* Hershkovitz, Proc. U.S. Nat. Mus., 100:353, May 26.

Sylvilagus brasiliensis dicei Harris

1932. *Sylvilagus dicei* Harris, Occas. Pap. Mus. Zool., Univ. Michigan, 248:1, August 4, type from 6000 ft., El Copey de Dota, in Cordillera de Talamanca, Costa Rica.

1950. *Sylvilagus brasiliensis dicei,* Hershkovitz, Proc. U.S. Nat. Mus., 100:352, May 26.

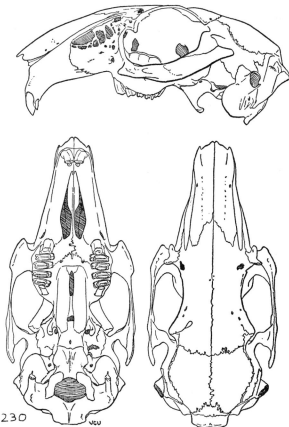

Fig. 230. *Sylvilagus brasiliensis truei,* 30 km. SSE Jesús Carranza, Veracruz, No. 32128 K.U., ♂, X 1.

MARGINAL RECORDS.—Costa Rica: Rancho de Río Jiménez; Juan Viñas; type locality; *San José.*

Sylvilagus brasiliensis gabbi (J. A. Allen)

1877. *Lepus brasiliensis* var. *gabbi* J. A. Allen, *in* Coues and Allen, Monog. N. Amer. Rodentia, p. 349, August. Name based on a specimen from Chiriquí, Panamá, and another from Talamanca [= Sipurio, Río Sixaola, near Caribbean Coast], Costa Rica. Nelson, N. Amer. Fauna, 29:259, August 31, 1909, selected the latter as the holotype.
1950. *Sylvilagus brasiliensis gabbi,* Hershkovitz, Proc. U.S. Nat. Mus., 100:351, May 26.
1908. *Lepus gabbi tumacus* J. A. Allen, Bull. Amer. Mus. Nat. Hist., 24:649, October 13, type from Tuma, Nicaragua.

MARGINAL RECORDS.—Honduras: San Pedro Sula; to Gulf Coast and southward along coast to Panamá Canal, Panamá: Gatún; Corozal; Gobernador Island; Divala; *Chiriquí;* northward east of the range of *S. b. dicei,* thence westward in Costa Rica: Vijagual, San Carlos. Nicaragua: Matagalpa; Ocotal. Honduras: San José, Santa Bárbara.

Sylvilagus brasiliensis incitatus (Bangs)

1901. *Lepus* (*Tapeti*) *incitatus* Bangs, Amer. Nat., 35:633, August, type from San Miguel Island, Bay of Panamá. Known from type locality only.
1950. *Sylvilagus brasiliensis incitatus,* Hershkovitz, Proc. U.S. Nat. Mus., 100:352, May 26.

Sylvilagus brasiliensis messorius Goldman

1912. *Sylvilagus gabbi messorius* Goldman, Smiths. Miscl. Coll., 60(2):13, September 20, type from Cana, 1800 ft., mts. of eastern Panamá.
1950. *Sylvilagus brasiliensis messorius,* Hershkovitz, Proc. U.S. Nat. Mus., 100:352, May 26.

MARGINAL RECORDS.—Panamá: Boca de Cupe; *Tacarcuna; Tapalisa;* type locality. Also occurs in South America.

Sylvilagus brasiliensis truei (J. A. Allen)

1890. *Lepus truei* J. A. Allen, Bull. Amer. Mus. Nat. Hist., 3:192, December 10, type from Mirador, Veracruz.
1950. *Sylvilagus brasiliensis truei,* Hershkovitz, Proc. U.S. Nat. Mus., 100:351, May 26.

MARGINAL RECORDS.—Tamaulipas: Rancho del Cielo, thence down coast to Tabasco: Teapa. Chiapas: Huehuetan. Oaxaca: Santo Domingo. Veracruz: Buena Vista; Motzorongo; *Mirador* (Hall and Dalquest, 1963:266). Puebla: Metlaltoyuca. San Luis Potosí: Rancho Apetsco, Xilitla.

Sylvilagus bachmani
Brush Rabbit

Size small. Total length, 300–375; tail, 20–43; hind foot, 64–81; ear from notch (dry), 50–64; weight (topotypes of *S. b. macrorhinus*) average of 16 males, 679(561–832), average of 22 females, 707(517–843) grams. Body uniformly dark brown or brownish gray, but tail whitish beneath; hair on mid-ventral part of body gray at base; only a slight crenulation on the ridge of enamel that separates an individual molariform tooth into anterior and posterior sections. From *Sylvilagus audubonii,* the only other species of *Sylvilagus* in the same geographic area, *S. bachmani* differs in smaller size, less white on underparts (the hairs on the mid-ventral part of the body being gray instead of white at base), shorter ears and legs, and a less crenulated ridge of enamel separating the anterior and posterior parts of a molariform tooth.

Sylvilagus bachmani bachmani (Waterhouse)

1839. *Lepus bachmani* Waterhouse, Proc. Zool. Soc. London, p. 103 (for 1838), February 7, type from California, probably between Monterey and Santa Barbara.
1904. *Sylvilagus* (*Microlagus*) *bachmani,* Lyon, Smiths. Miscl. Coll., 45:336, June 15.

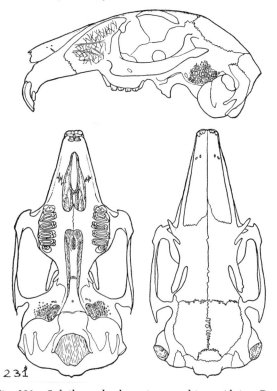

Fig. 231. *Sylvilagus bachmani macrorhinus,* Alpine Creek Ranch, 1700 ft., San Mateo Co., California, No. 53382 M.V.Z., ♀, × 1

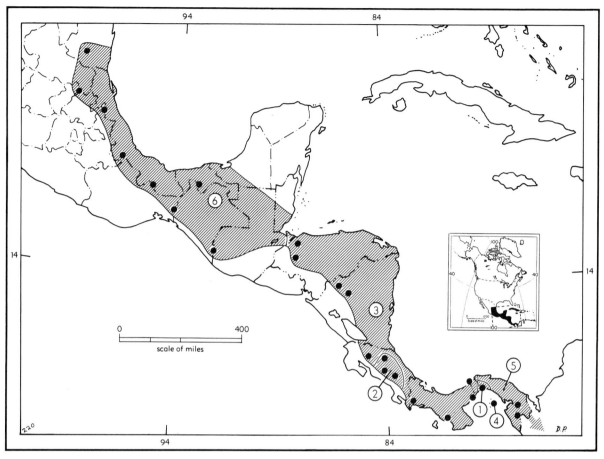

Map 220. *Sylvilagus brasiliensis*.

1. *S. b. consobrinus*	3. *S. b. gabbi*	5. *S. b. messorius*
2. *S. b. dicei*	4. *S. b. incitatus*	6. *S. b. truei*

1855. *Lepus trowbridgii* Baird, Proc. Acad. Nat. Sci. Philadelphia, 7:333, April, type from Monterey County, California.

MARGINAL RECORDS.—California: 2 mi. S mouth Salinas River; near Morro.

Sylvilagus bachmani cerrosensis (J. A. Allen)

1898. *Lepus cerrosensis* J. A. Allen, Bull. Amer. Mus. Nat. Hist., 10:145, April 12, type from Cerros [= Cedros] Island, Baja California. Known from Cerros Island only.
1909. *Sylvilagus bachmani cerrosensis*, Nelson, N. Amer. Fauna, 29:225, August 31.

Sylvilagus bachmani cinerascens (J. A. Allen)

1890. *Lepus cinerascens* J. A. Allen, Bull. Amer. Mus. Nat. Hist., 3:159, October 8, type from San Fernando, Los Angeles Co., California.
1907. *Sylvilagus bachmani cinerascens*, Nelson, Proc. Biol. Soc. Washington, 20:84, July 22.

MARGINAL RECORDS.—California: 5700 ft., San Emigdio Canyon; Reche Canyon; 3500 ft., Dos Palmas Springs, Santa Rosa Mts. Baja California: La Huerta, thence northward up coast to point of beginning.

Sylvilagus bachmani exiguus Nelson

1907. *Sylvilagus bachmani exiguus* Nelson, Proc. Biol. Soc. Washington, 20:84, July 22, type from Yubay, central Baja California.

MARGINAL RECORDS.—Baja California: Agua Dulce; Santana.

Sylvilagus bachmani howelli Huey

1927. *Sylvilagus bachmani howelli* Huey, Trans. San Diego Soc. Nat. Hist., 5:67, July 6, type from 10 mi. SE Alamo, Baja California, lat. 31° 35′ N, long. 116° 03′ W.

MARGINAL RECORDS.—Baja California: Laguna Hanson, Sierra· Juárez; Valle de la Trinidad; *Ojos Negros*.

Sylvilagus bachmani macrorhinus Orr

1935. *Sylvilagus bachmani macrorhinus* Orr, Proc. Biol. Soc. Washington, 48:28, February 6, type from Alpine Creek Ranch, 3½ mi. S and 2¼ mi. E Portola, 1700 ft., San Mateo Co., California.

MARGINAL RECORDS.—California: 10 mi. SW Suisun; W side Mt. Diablo; Summit Station, Santa Cruz Mts., thence north along coast to *Golden Gate*.

Sylvilagus bachmani mariposae Grinnell and Storer

1916. *Sylvilagus bachmani mariposae* Grinnell and Storer, Univ. California Publ. Zool., 17:7, August 23, type from McCauley Trail, 4000 ft., near El Portal, Mariposa Co., California.

MARGINAL RECORDS.—California: Carbondale; French Gulch, 6700 ft., Piute Mtn.

Sylvilagus bachmani peninsularis (J. A. Allen)

1898. *Lepus peninsularis* J. A. Allen, Bull. Amer. Mus. Nat. Hist., 10:144, April 12, type from Santa Anita, Baja California.
1909. *Sylvilagus bachmani peninsularis*, Nelson, N. Amer. Fauna, 29:255, August 31.

MARGINAL RECORDS.—Baja California: type locality; Cape San Lucas.

Sylvilagus bachmani riparius Orr

1935. *Sylvilagus bachmani riparius* Orr, Proc. Biol. Soc. Washington, 48:29, February 6, type from W side San Joaquin River, 2 mi. NE Vernalis, Stanislaus Co., California. Known from type locality only.

Sylvilagus bachmani rosaphagus Huey

1940. *Sylvilagus bachmani rosaphagus* Huey, Trans. San Diego Soc. Nat. Hist., 9:221, July 31, type from 2 mi. W Santo Domingo Mission, Baja California, lat. 30° 45′ N, long. 115° 58′ W, near huge red cliff marking entrance of Santo Domingo River Cañon from coastal plain.

MARGINAL RECORDS.—Baja California: San Quintín; El Rosario.

Sylvilagus bachmani tehamae Orr

1935. *Sylvilagus bachmani tehamae* Orr, Proc. Biol. Soc. Washington, 48:27, February 6, type from Dale's, on Paine's Creek, 600 ft., Tehama Co., California.

MARGINAL RECORDS.—Oregon: Prospect. California: Auburn; 7 mi. W, 14 mi. S Chico; Rumsey; Castle Springs; 3 mi. S Covelo; Mad River Bridge, S Fork Mtn.

Sylvilagus bachmani ubericolor (Miller)

1899. *Lepus bachmani ubericolor* Miller, Proc. Acad. Nat. Sci. Philadelphia, 51:383, September 29, type from Beaverton, Washington Co., Oregon.

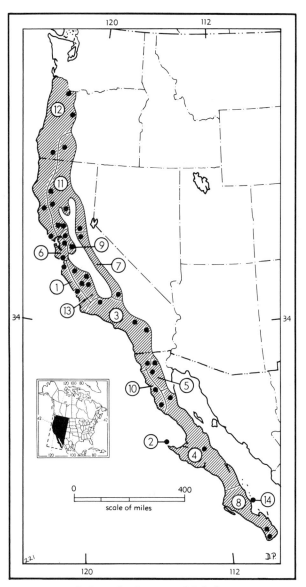

Map 221. *Sylvilagus bachmani* and *Sylvilagus mansuetus*.

1. *S. b. bachmani*	8. *S. b. peninsularis*
2. *S. b. cerrosensis*	9. *S. b. riparius*
3. *S. b. cinerascens*	10. *S. b. rosaphagus*
4. *S. b. exiguus*	11. *S. b. tehamae*
5. *S. b. howelli*	12. *S. b. ubericolor*
6. *S. b. macrorhinus*	13. *S. b. virgulti*
7. *S. b. mariposae*	14. *S. mansuetus*

1904. *Sylvilagus (Microlagus) bachmani ubericolor*, Lyon, Smiths. Miscl. Coll., 45:337, June 15.

MARGINAL RECORDS.—Oregon: Portland; Mackenzie Bridge; above Grants Pass. California: Laytonville; Maillard [= 4 mi. E Lagunitas]; from *San Francisco Bay* up coast to *mouth Columbia River* in *Oregon* and up *S bank* of that river to point of beginning.

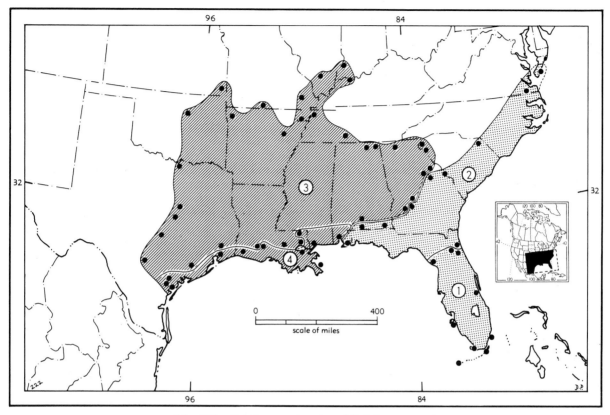

Map 222. *Sylvilagus palustris* and *Sylvilagus aquaticus*.

1. *S. p. paludicola*   2. *S. p. palustris*   3. *S. a. aquaticus*   4. *S. a. littoralis*

Sylvilagus bachmani virgulti Dice

1926. *Sylvilagus bachmani virgulti* Dice, Occas. Pap. Mus. Zool., Univ. Michigan, 166:24, February 11, type from Soledad, Monterey Co., California.

MARGINAL RECORDS.—California: The Pinnacles; Waltham Creek, 4½ mi. SE Priest Valley; 2 mi. S San Miguel; Bryson.

Sylvilagus mansuetus Nelson
San José Brush Rabbit

1907. *Sylvilagus mansuetus* Nelson, Proc. Biol. Soc. Washington, 20:83, July 22, type from San José Island, Gulf of California, Baja California. Known from San José Island only.

This insular species is closely related to *Sylvilagus bachmani* and is distinguished by paleness, proportionately longer and narrower skull, fusion to skull of anterior arm of supraorbital process, and larger jugal.

Sylvilagus palustris
Marsh Rabbit

Total length, 425–440; tail, 33–39; hind foot, 88–91; ear from notch (dry), 45–52. Upper parts blackish brown or reddish brown; underside of tail brownish or dingy gray (not white); ears, tail, and hind feet short; posterior and anterior extensions of supraorbital processes joined to skull along most (or all) of their extent. Lack of white on underside of tail is ready means of distinguishing this species from other species of the genus that occur within its geographic range.

Sylvilagus palustris paludicola (Miller and Bangs)

1894. *Lepus paludicola* Miller and Bangs, Proc. Biol. Soc. Washington, 9:105, June 9, type from Fort Island, near Crystal River, Citrus Co., Florida.
1909. *Sylvilagus palustris paludicola*, Nelson, N. Amer. Fauna, 29:269, August 31.

MARGINAL RECORDS.—Florida: Hibernia; San Mateo; Micco; Key Biscayne (Layne, 1974:390—and throughout the mainland and on nearshore islands; on the Florida Keys, including *Elliott*, Largo, *Plantation*, *Lower Matecumbe*, Long, *Upper Matecumbe*, *Big Pine*, and Key West); Cape Sable; Sanibel Island (Layne, 1974:390); Suwannee River.

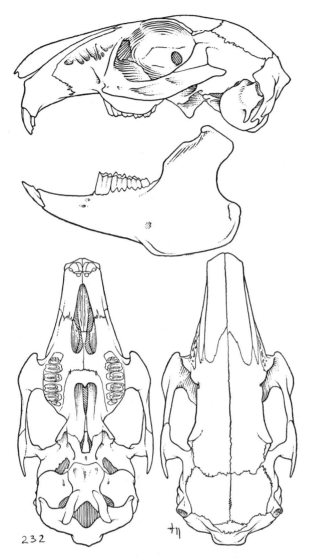

Fig. 232. *Sylvilagus palustris palustris*, Riceboro, Georgia, No. 45502 U.S.N.M., ♀, X 1.

Sylvilagus palustris palustris (Bachman)

1837. *Lepus palustris* Bachman, Jour. Acad. Nat. Sci. Philadelphia, 7:194. Type locality, eastern South Carolina.
1909. *Sylvilagus palustris*, Nelson, N. Amer. Fauna, 29:266, August 31.

MARGINAL RECORDS.—Virginia (Payne, 1975:77, 78): Hog Island [possibly introduced by man or natural rafting.—E.R.H.], southward along Atlantic Coast to northern Florida: Anastasia Island. West to Gulf Coast and along coast to Alabama: Bon Secour; Flomaton; Dothan. Georgia: Americus; Glascock County (Lowe, 1958:118, fig. 1, as *S. palustris* only); near Augusta (Caldwell, 1966:527). South Carolina: Society Hill. Virginia: Nansemond County.

Sylvilagus floridanus
Eastern Cottontail

Total length, 375–463; tail, 39–65; hind foot, 87–104; ear from notch (dry), 49–68; upper parts brownish or grayish; underside of tail white; skull with transversely thick posterior extension of supraorbital process of frontal. The geographic range is larger than that of any other North American species of the genus *Sylvilagus;* from Canada the species occurs south at least to Costa Rica and it may occur in Panamá for the species is recorded also from South America.

In the western part of the Great Plains this species is confined to the riparian growth along streams, and *Sylvilagus audubonii* occupies the remainder of the terrain. In New Mexico and southwestern Texas *S. floridanus* is confined to the boreal life-zones where timber provides denser cover than is found in the lower life-zones. The zonal range is from the Canadian Life-zone into the Tropical Life-zone. It is not surprising, therefore, that there is much geographic variation in the shape and size of the skull. There is so much geographic variation in the form of the skull that it is impossible, at this writing at least,

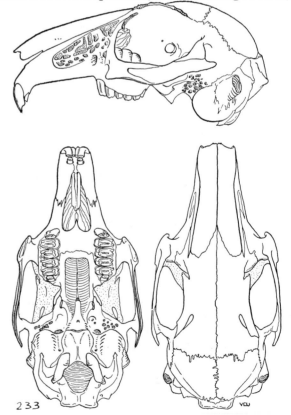

Fig. 233. *Sylvilagus floridanus mearnsi*, 4 mi. NE Lawrence, Douglas Co., Kansas, No. 3774 K.U., ♂, X 1.

to frame a description that will enable the reader to distinguish the skull from those of all other species of the genus. In any given area, however, it is possible, easily and with certainty, to distinguish the skulls of *S. floridanus* from those of the other species that occur in that area.

Genoways and Jones (1972:11) thought that *S. floridanus similis* in the preceding half-century in southwestern North Dakota had displaced the more western *S. nuttalli grangeri.*

This species has been introduced and has become established in Washington (see Dalquest, 1948:389). Wild-taken specimens have been saved from Huntingdon, British Columbia (No. 2317, adult female, taken on August 12, 1952) and Langley Prairie, British Columbia (No. 2300, adult male, taken on March 20, 1952), both in the collection of Kenneth Racey. In Oregon an adult was taken on February 1, 1951, at Mt. Scott (51652 KU), and Trethewey and Verts (1971:463) report established populations in the Willamette Valley. Chapman and Morgan (1973) have described effects of massive introductions of Middlewestern subspecies into Maryland and West Virginia.—See addenda.

Sylvilagus floridanus alacer (Bangs)

1896. *Lepus sylvaticus alacer* Bangs, Proc. Biol. Soc. Washington, 10:136, December 28, type from Stilwell, Boston Mountains, Adair Co., Oklahoma.
1904. *Sylvilagus (Sylvilagus) floridanus alacer,* Lyon, Smiths. Misc. Coll., 45:336, June 15.

MARGINAL RECORDS.—Missouri: Columbia; St. Louis. Illinois: Ozark. Tennessee: Samburg; Raleigh. Mississippi: Michigan City; Bay St. Louis. Louisiana (Lowery, 1974:162, 163): Chandeleur Islands, opposite North Island; Cheniere au Tigre. Texas: Port Lavaca; Bell County (Davis, 1966:222, as *S. floridanus* only); Brazos; Henrietta. Oklahoma: Norman. Kansas: *8 mi. NE Harper; Rago; Halstead; 4 mi. S, 14 mi. W Hamilton;* 3 mi. N Chanute.

Sylvilagus floridanus ammophilus A. H. Howell

1939. *Sylvilagus floridanus ammophilus* A. H. Howell, Jour. Mamm., 20:365, August 14, type from "Oak Lodge," E of Micco, Florida. Known from type locality only.

Sylvilagus floridanus aztecus (J. A. Allen)

1890. *Lepus sylvaticus aztecus* J. A. Allen, Bull. Amer. Mus. Nat. Hist., 3:188, December 10, type from Tehuantepec [City], Oaxaca.
1904. *Sylvilagus (Sylvilagus) floridanus aztecus,* Lyon, Smiths. Misc. Coll., 45:336, June 15.

MARGINAL RECORDS.—Oaxaca: Santa María Petapa; Santa Efigenia. Chiapas: Tonalá, 50 m. Oaxaca: Salina Cruz; *type locality;* Río Molino, 35 km. SW San Miguel Suchixtepec (Schaldach, 1966:292, as *S. floridanus* ssp. only).

Sylvilagus floridanus chapmani (J. A. Allen)

1899. *Lepus floridanus chapmani* J. A. Allen, Bull. Amer. Mus. Nat. Hist., 12:12, March 4, type from Corpus Christi, Nueces Co., Texas.
1904. *Sylvilagus (Sylvilagus) floridanus chapmani,* Lyon, Smiths. Misc. Coll., 45:336, June 15.
1899. *Lepus floridanus caniclunis* Miller, Proc. Acad. Nat. Sci. Philadelphia, 51:388, October 5, type from Fort Clark, Kinney Co., Texas.
1902. *Lepus simplicicanus* Miller, Proc. Biol. Soc. Washington, 15:81, April 25, type from Brownsville, Texas.

MARGINAL RECORDS.—Texas (Davis, 1966:222, as *S. floridanus* only, unless otherwise noted): Clyde (Hall and Kelson, 1959:260); Llano County; Hays County; Dewitt County; Victoria County; *Refugio County; Rockport* (Hall and Kelson, 1959:260). Tamaulipas: La Pesca (Alvarez, 1963:419); Ejido Eslabones, 2 mi. S, 10 mi. W Piedra, 1200 ft. (*ibid.*); Jaumave. Coahuila: Monclova; *Nadadores;* 1 mi. W Hda. La Mariposa; *12 mi. S Hda. Las Margaritas;* Rancho Las Margaritas. Texas: Comstock; Terrell County (Davis, 1966:222, as *S. floridanus* only); Stanton; Glasscock County (Davis, 1966:222, as *S. floridanus* only); Coke County (*ibid.,* as *S. floridanus* only).—See addenda.

Sylvilagus floridanus chiapensis (Nelson)

1904. *Lepus floridanus chiapensis* Nelson, Proc. Biol. Soc. Washington, 17:106, May 18, type from San Cristóbal, Chiapas.
1909. *Sylvilagus floridanus chiapensis,* Lyon and Osgood, Bull. U.S. Nat. Mus., 62:32, January 28.

MARGINAL RECORDS.—Chiapas: type locality; Comitán. Guatemala: Hda. Chabcol near Zacopa; Panajachel. Chiapas: Tuxtla.

Sylvilagus floridanus cognatus Nelson

1907. *Sylvilagus cognatus* Nelson, Proc. Biol. Soc. Washington, 20:82, July 22, type from Tajique near summit of Manzano Mts., Torrance Co., New Mexico.
1951. *Sylvilagus floridanus cognatus,* Hall and Kelson, Univ. Kansas Publ., Mus. Nat. Hist., 5:55, October 1.

MARGINAL RECORDS.—New Mexico: Santa Rosa, 35 mi. N on Conchas River; Capitan Mts.; 1 mi. N Cloudcroft (Findley, *et al.,* 1975:87, as *S. floridanus* only); Bear Trap Canyon, 20 mi. S, 19 mi. W Magdalena (Jones, *et al.,* 1960:275, as *S. floridanus* only); Datil Mts.; W slope Mt. Taylor (Findley, *et al.,* 1975:88); type locality.

Sylvilagus floridanus connectens (Nelson)

1904. *Lepus floridanus connectens* Nelson, Proc. Biol. Soc. Washington, 17:105, May 18, type from Chichicaxtle, central Veracruz.

1909. *Sylvilagus floridanus connectens*, Lyon and Osgood, Bull. U.S. Nat. Mus., 62:32, January 28.

MARGINAL RECORDS.—Tamaulipas: 2 km. W El Carrizo (Alvarez, 1963:420); Altamira. Veracruz: 3 km. W Boca del Río (Hall and Dalquest, 1963:268). Oaxaca: Mt. Zempoaltepec; 5 mi. SE Oaxaca de Juárez (Goodwin, 1969:121). Veracruz: Orizaba; *Jico*. Puebla: Metlaltoyuca. Querétaro: Pinal de Amoles. San Luis Potosí: Valles. Tamaulipas (Alvarez, 1963:420): 9 mi. SW Tula, 5200 ft.; *La Joya de Salas; 10 km. N, 8 km. W El Encino, 400 ft.*

Sylvilagus floridanus costaricensis Harris

1933. *Sylvilagus floridanus costaricensis* Harris, Occas. Pap. Mus. Zool., Univ. Michigan, 266:3, June 28, type from Hda. Santa María, 22 mi. NE Liberia, Prov. Guanacaste, 3200 ft., Costa Rica.

MARGINAL RECORDS.—Costa Rica: El Pelón; *type locality*; Tenorio.

Sylvilagus floridanus floridanus (J. A. Allen)

1890. *Lepus sylvaticus floridanus* J. A. Allen, Bull. Amer. Mus. Nat. Hist., 3:160, October 8, type from Sebastian River, Brevard Co., Florida.
1904. *Sylvilagus floridanus*, Lyon, Smiths. Miscl. Coll., 45:322, June 15.

MARGINAL RECORDS.—Florida: St. Augustine (Nelson, 1909:164); Merritt Island (Schwartz, 1956:150); Lantana Road and Military Trail (*ibid.*); *area S of Lake Okeechobee* (Layne, 1974:390); Pinecrest (*ibid.*); Naples (Schwartz, 1956:150); Blitches Ferry (Nelson, 1909:165).

Sylvilagus floridanus hesperius Hoffmeister and Lee

1963. *Sylvilagus floridanus hesperius* Hoffmeister and Lee, Amer. Midland Nat., 70:140, July 10, type from 5¼ mi. SE Kingman, Hualpai Mts., Mohave Co., Arizona.

MARGINAL RECORDS (Hoffmeister and Lee, 1963:143).—Arizona: Pine Springs, 15 mi. S Colorado Canyon; 6 mi. NW Camp Verde; Reynolds Creek R. S., Sierra Ancha; *Peterson Ranch, 6900 ft.; near Carr's Ranch; 1 mi. S Parker Creek Station, 4900 ft.; 2 mi. S Parker Creek Station, 4800 ft.*; Mayer; 5 mi. N Skull Valley; type locality; Hackberry.

Sylvilagus floridanus hitchensi Mearns

1911. *Sylvilagus floridanus hitchensi* Mearns, Proc. U.S. Nat. Mus., 39:227, January 9, type from Smiths Island, Northampton Co., Virginia.

MARGINAL RECORDS.—Virginia: type locality; *Fishermans Island.*

Sylvilagus floridanus holzneri (Mearns)

1896. *Lepus sylvaticus holzneri* Mearns, Proc. U.S. Nat. Mus., 18:554, June 24, type from Douglas spruce zone, near summit of Huachuca Mts., Cochise Co., Arizona.
1904. *Sylvilagus (Sylvilagus) floridanus holzneri*, Lyon, Smiths. Miscl. Coll., 45:336, June 15.
1896. [*Lepus sylvaticus*] subspecies *rigidus* Mearns, Proc. U.S. Nat. Mus., 18:555, June 24, type from Carrizalillo Mts., near Monument 31, Mexican boundary, New Mexico.
1903. *Lepus (Sylvilagus) durangae* J. A. Allen, Bull. Amer. Mus. Nat. Hist., 19:609, November 12, type from Rancho Bailón, northwestern Durango.

MARGINAL RECORDS.—Arizona (Hoffmeister and Lee, 1963:146): 8 mi. S Whiteriver, 6100 ft.; Prieto Plateau, 32 mi. N, 2 mi. W Clifton, S end Blue Range, 8000 ft. New Mexico: Silver City; Carrizalillo Mts., near Monument 31, Mexican Boundary Line. Chihuahua (Anderson, 1972:264): 3½ mi. ESE Los Lamentos; 40 mi. E Gallego, 5000 ft.; 4 mi. NW San Francisco de Borja, 5700 ft. Durango (Baker and Greer, 1962:80): Indé; Río Nazas, 6 mi. NW Rodeo, 4200 ft.; 4 mi. W La Pila. Zacatecas: Valparaíso; Plateado. Sinaloa (Armstrong and Jones, 1971:754): Plomosas, 2500 ft.; San Ignacio, 700 ft.; 1½ km. N Badiraguato, 750 ft. Sonora (Hoffmeister and Lee, 1963:120): 7 mi. WNW Alamos; *8 mi. WNW Alamos*. Arizona (Hoffmeister and Lee, 1963:146): 2 mi. E Baboquivari Peak, Thomas Canyon; Roskruge Mts.; 3 mi. SW Oracle, ½ mi. S Oracle R. S.; Fish Creek, Tonto National Forest, 2000 ft.; *Tonto National Monument, 2800 ft.*; Sawmill, 27 mi. NE Globe, 5600 ft.—Anderson and Ogilvie (1957:35) list, to species, a fragment from *3½ mi. ESE Los Lamentos, Chihuahua.*

Sylvilagus floridanus hondurensis Goldman

1932. *Sylvilagus floridanus hondurensis* Goldman, Proc. Biol. Soc. Washington, 45:122, July 30, type from Monte Redondo, 5100 ft., Honduras.

MARGINAL RECORDS.—Honduras: Santa Bárbara; Cedros. Nicaragua: Jinotega; Chontales ["District" of]; León. El Salvador (Burt and Stirton, 1961:65): Lake Olomega; Puerto El Triunfo; Barra de Santiago. Honduras: Ocotepeque.

Sylvilagus floridanus llanensis Blair

1938. *Sylvilagus floridanus llanensis* Blair, Occas. Pap. Mus. Zool., Univ. Michigan, 380:1, June 21, type from Old "F" Ranch headquarters, Quitaque, Texas.

MARGINAL RECORDS.—Kansas: 15 mi. N, 3 mi. E Stafford; 1 mi. NE Aetna. Oklahoma: 3 mi. SE Southard; *Fort Cobb*; Mt. Scott. Texas: Wilbarger County (Davis, 1966:222, as *S. floridanus* only); 6 mi. E Coahoma; Dawson County (Davis, 1966:222, as *S. floridanus* only); 6 mi. SW Muleshoe. Colorado: Two Buttes (Armstrong, 1972:81). Kansas: Coolidge.

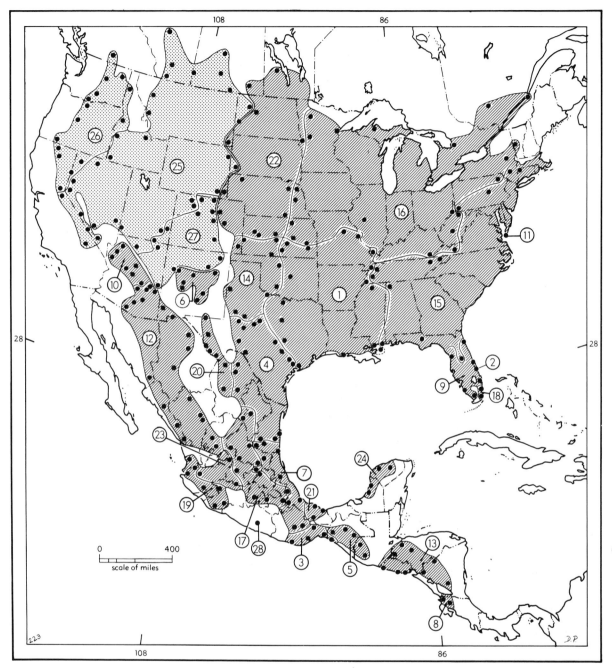

Map 223. *Sylvilagus floridanus, Sylvilagus nuttallii,* and *Sylvilagus insonus.*

Guide to
kinds
1. *S. f. alacer*
2. *S. f. ammophilus*
3. *S. f. aztecus*
4. *S. f. chapmani*
5. *S. f. chiapensis*
6. *S. f. cognatus*
7. *S. f. connectens*
8. *S. f. costaricensis*

9. *S. f. floridanus*
10. *S. f. hesperius*
11. *S. f. hitchensi*
12. *S. f. holzneri*
13. *S. f. hondurensis*
14. *S. f. llanensis*
15. *S. f. mallurus*
16. *S. f. mearnsii*
17. *S. f. orizabae*
18. *S. f. paulsoni*

19. *S. f. restrictus*
20. *S. f. robustus*
21. *S. f. russatus*
22. *S. f. similis*
23. *S. f. subcinctus*
24. *S. f. yucatanicus*
25. *S. n. grangeri*
26. *S. n. nuttallii*
27. *S. n. pinetis*
28. *S. insonus*

303

Sylvilagus floridanus mallurus (Thomas)

1898. *L[epus]. n[uttalli]. mallurus* Thomas, Ann. Mag. Nat. Hist., ser. 7, 2:320, October, type from Raleigh, Wake Co., North Carolina.
1904. *Sylvilagus floridanus mallurus,* Lyon, Smiths. Miscl. Coll., 45:323, June 15.
1837. *Lepus sylvaticus* Bachman, Jour. Acad. Nat. Sci. Philadelphia, 7:403, no type or type locality. Name given to "common gray rabbit" of eastern U.S., probably of South Carolina. Not *Lepus borealis sylvaticus* Nilsson, 1832.

MARGINAL RECORDS.—Connecticut: Bear Mountain, S along coast to Florida: Lake Julian; *Leesburg; Gulf Hammock;* Rock Bluff. Mississippi: Horn Island (Richmond, 1962:99). Alabama: Leighton. Tennessee: Arlington; Hornbeak; Highcliff; Watauga Valley. West Virginia: *Ernshaw.* Pennsylvania: Waynesburg; Potts Grove. New York: Palenville.

Sylvilagus floridanus mearnsii (J. A. Allen)

1894. *Lepus sylvaticus mearnsii* J. A. Allen, Bull. Amer. Mus. Nat. Hist., 6:171, May 31, type from Fort Snelling, Hennepin Co., Minnesota.
1904. *Sylvilagus (Sylvilagus) floridanus mearnsi,* Lyon, Smiths. Miscl. Coll., 45:336, June 15.

MARGINAL RECORDS.—Minnesota: Fertile; Duluth. Michigan: Marquette County. Ontario: Lake Simcoe. Quebec: Mt. Tremblant Province Park (Pirlot, 1962:132); St. Lambert (Wrigley, 1969:205, as *S. floridanus* only, sight record only). Vermont: 4 mi. SE Middlebury (Smith, P. B., 1971:624, as *S. floridanus* only). New York: "eastern New York." Pennsylvania: Lopez. Ohio: Oglebay. West Virginia: 7 mi. E Philippi; Gilboa. Virginia: Smith County. Kentucky: Big Black Mtn. Illinois: Sangamon. Kansas: Neosho Falls; 1 mi. N, ½ mi. E Lincolnville; *6 mi. SW Clay Center;* Strawberry. Nebraska (Jones, 1964c:107): Columbus; 4 mi. SE Carroll. Minnesota: Otter Tail County.

Sylvilagus flordanus orizabae (Merriam)

1893. *Lepus orizabae* Merriam, Proc. Biol. Soc. Washington, 8:143, December 29, type from Mt. Orizaba, 9500 ft., Puebla.
1909. *Sylvilagus floridanus orizabae,* Nelson, N. Amer. Fauna, 29:183, August 31.
1903. *Lepus floridanus persultator* Elliott, Field Columb. Mus., Publ. 71, Zool. Ser., 3:147, March 20, type from Puebla, Puebla.

MARGINAL RECORDS.—Coahuila (Baker, 1956:201): 13 mi. E San Antonio de las Alazanas. San Luis Potosí: Presa de Guadalupe; Hda. Capulín. Hidalgo: Encarnación. Veracruz: Las Vigas; *Mt. Orizaba.* Puebla: Chalchicomula. México: Mt. Popocatépetl. Morelos: *Alrededores de Huitzilac* (Ramírez-P., 1971:269). México: Volcán de Toluca. Guanajuato: Santa Rosa. San Luis Potosí: Bledos; Cerro Peñón Blanco. Coahuila: Sierra Encarnación.

Sylvilagus floridanus paulsoni Schwartz

1956. *Sylvilagus floridanus paulsoni* Schwartz, Proc. Biol. Soc. Washington, 69:147, September 12, type from 6 mi. N Homestead, Dade Co., Florida.

MARGINAL RECORDS.—Florida: Fort Lauderdale; type locality; [*Everglades National*] *Park Headquarters* (Layne, 1974:390, as *S. floridanus* only).

Sylvilagus floridanus restrictus Nelson

1907. *Sylvilagus floridanus restrictus* Nelson, Proc. Biol. Soc. Washington, 20:82, July 22, type from Zapotlán, Jalisco.

MARGINAL RECORDS.—Nayarit: Tepic; Ojo de Agua. Jalisco: *La Ciénega; Atenfuillo.* Michoacán: Mt. Tancítaro; Pátzcuaro; Dos Aguas, 7000 ft. (Hooper, 1961:122). Jalisco: *type locality;* Las Canoas; La Laguna.—See addenda.

Sylvilagus floridanus robustus (V. Bailey)

1905. *Lepus pinetis robustus* V. Bailey, N. Amer. Fauna, 25:159, October 24, type from 6000 ft., Davis Mts., Jeff Davis Co., Texas.
1951. *Sylvilagus floridanus robustus,* Hall and Kelson, Univ. Kansas Publ., Mus. Nat. Hist., 5:56, October 1.
1955. *Sylvilagus floridanus nelsoni* Baker, Univ. Kansas Publ., Mus. Nat. Hist., 7:611, April 8, type from 22 mi. S, 5 mi. W Ocampo, 5925 ft., Coahuila. Regarded as inseparable from *S. f. robustus* by Raun, 1965:521.

MARGINAL RECORDS.—Texas: The Bowl, Guadalupe Mts.; Davis Mts. (Davis, 1966:223, as *S. robustus* only). Coahuila (Raun, 1965:519, 520): Mesa del Hillcoat, Sierra del Carmen, 85 mi. NW Múzquiz; *20 mi. S, 4 mi. W Ocampo;* Sierra de la Madera, 22 mi. S, 5 mi. W Ocampo. Texas: Chisos Mts.; 35 mi. S Marfa; Chinati Mts. (Davis, 1966:223, as *S. robustus* only).

Sylvilagus floridanus russatus (J. A. Allen)

1904. *Lepus (Sylvilagus) russatus* J. A. Allen, Bull. Amer. Mus. Nat. Hist., 20:31, February 29, type from Pasa Nueva, southern Veracruz.
1909. *Sylvilagus floridanus russatus,* Nelson, N. Amer. Fauna, 29:186, August 31.

MARGINAL RECORDS.—Veracruz: Catemaco; Coatzacoalcos; *Minatitlán;* type locality; *Jimba; 7 km. NW Paso Nuevo, 100 ft.* (Hall and Dalquest, 1963:268).

Sylvilagus floridanus similis Nelson

1907. *Sylvilagus floridanus similis* Nelson, Proc. Biol. Soc. Washington, 20:82, July 22, type from Valentine, Cherry Co., Nebraska.

MARGINAL RECORDS.—Manitoba: Dauphin. Minnesota: Ten Mile Lake. Nebraska (Jones, 1964c:108): Neligh; Davenport. Kansas: *Long Island;* 3 mi. N, 2 mi. W Hoisington; Lane County; Elkader. Colorado: 3 mi. S Elbert (Armstrong, 1972:82); Arvada.

Wyoming (Long, 1965a:543): *3 mi. E Horse Creek P.O.;* 3½ *mi.* W Lagrange. Nebraska: 8 mi. E Chadron. South Dakota: *Headquarters, Wind Cave National Park, 4100 ft.* (Turner, 1974:61). Montana: *Little Missouri River, 7 mi. NE Albion;* Box Elder Creek, 25 mi. SW Sykes. North Dakota (Genoways and Jones, 1972:10): 1 mi. S, 1 mi. W Medora; 7 mi. N, 9 mi. W Killdeer. Saskatchewan: Estevan (Beck, 1958:24).

Sylvilagus floridanus subcinctus (Miller)

1899. *Lepus floridanus subcinctus* Miller, Proc. Acad. Nat. Sci. Philadelphia, 51:386, October 5, type from Hda. El Molino, near Negrete, Michoacán.—**See** addenda.
1904. *Sylvilagus (Sylvilagus) floridanus subcinctus,* Lyon, Smiths. Miscl. Coll., 45:336, June 15.

MARGINAL RECORDS.—Jalisco: Lagos. Guanajuato: Acámbaro. Michoacán: *Querendaro.* Jalisco: *Ameca;* Etzatlán.

Sylvilagus floridanus yucatanicus (Miller)

1899. *Lepus floridanus yucatanicus* Miller, Proc. Acad. Nat. Sci. Philadelphia, 51:384, September 29, type from Mérida, Yucatán.
1904. *Sylvilagus (Sylvilagus) floridanus yucatanicus,* Lyon, Smiths. Miscl. Coll., 45:336, June 15.

MARGINAL RECORDS.—Yucatán: *Progreso;* type locality; Chichén-Itzá. Campeche: Campeche; Champotón (92090 KU).

Sylvilagus transitionalis (Bangs)
New England Cottontail

1895. *Lepus sylvaticus transitionalis* Bangs, Proc. Boston Soc. Nat. Hist., 26:405, January 31, type from Liberty Hill, New London Co., Connecticut.
1909. *Sylvilagus transitionalis,* Nelson, N. Amer. Fauna, 29:195, August 31.

Total length, 388; tail, 39; hind foot, 95; ear from notch (dry), 52. Upper parts almost pinkish buff, varying to almost ochraceous buff; back overlaid by distinct black wash giving penciled effect; anterior extension of supraorbital process obsolete or short, closely appresed to orbital rim; tympanic bulla small, smaller than in any subspecies of *S. floridanus* in the United States. *S. transitionalis* is a forest-inhabiting species—more so than is *S. floridanus.*

MARGINAL RECORDS.—Vermont: west side at Canadian boundary; Montpelier. Maine: Sagadahoc County; *Androscoggin County.* New York: Miller

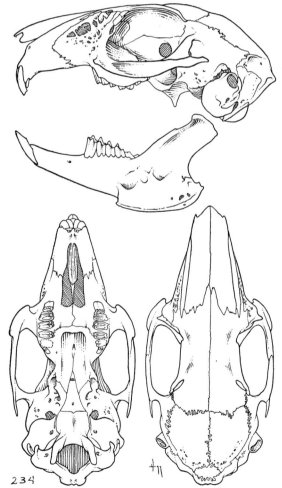

Fig. 234. *Sylvilagus transitionalis,* Exeter, Rhode Island, No. 125529 U.S.N.M., ♀, X 1.

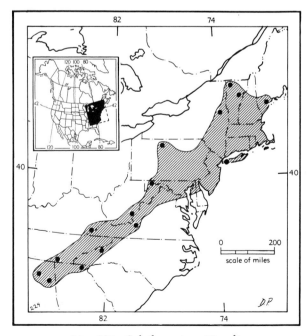

Map 224. *Sylvilagus transitionalis.*

Place. Virginia: Roanoke County. North Carolina: Roan Mtn. Georgia: Brasstown Bald Mtn. Alabama: Erin; Ardell. Tennessee: Walden Ridge, near Soddy. Kentucky: Big Black Mtn. West Virginia: Ronceverte. Maryland: 3 mi. S Piney Grove (Chapman and Paradiso, 1972:149). Pennsylvania: Renovo. New York: Lake George.

Sylvilagus nuttallii
Nuttall's Cottontail

Total length, 350–390; tail, 44–50; hind foot, 88–100; ear from notch (dry), 55–56; weight in Nevada, male, 678, three females, 928(868–1032) grams. Hind feet densely covered with long hair; ears short; tympanic bullae of moderate size. In the northern part of its range S. nuttallii occurs principally in the areas of sagebrush but occurs also in timbered areas of the Transition Life-zone and almost exclusively in timbered areas in the southern part of its range. From S. floridanus, S. nuttallii along the eastern margin of its range differs in more slender rostrum, and larger external auditory meatus. In New Mexico and Arizona, S. nuttallii differs from S. floridanus in having the posterior border of the supraoccipital shield pointed or rounded and unnotched instead of truncate or notched, and by a combination of other cranial features (see Hoffmeister and Lee, 1963:139) no one of which alone suffices for identifying every skull to species. S. nuttallii differs from S. audubonii in shorter ears, smaller tympanic bullae, and smaller hind legs. S. nuttallii usually occurs at higher elevations, or where the 2 occur at approx. the same elevation S. nuttallii occurs in wooded or brushy areas and S. audubonii lives on the plains or in relatively open country.

Genoways and Jones (1972:11) thought that in the preceding half-century S. nuttallii grangeri in southwestern North Dakota had yielded range there to the more eastern S. floridanus similis. See Map 223.

Sylvilagus nuttallii grangeri (J. A. Allen)

1895. Lepus sylvaticus grangeri J. A. Allen, Bull. Amer. Mus. Nat. Hist., 7:264, August 21, type from Hill City, Black Hills, Pennington Co., South Dakota.
1909. Sylvilagus nuttalli grangeri, Nelson, N. Amer. Fauna, 29:204, August 31.
1904. Lepus l[aticinctus]. perplicatus Elliott, Field Columb. Mus., Publ. 87, Zool. Ser., 3:255, January 7, type from Hannopee [= Hannaupah] Canyon, Panamint Mts., Inyo Co., California.

MARGINAL RECORDS.—Alberta: Delbourne (Soper, 1965:121). Saskatchewan: Cypress Hills; Dun-

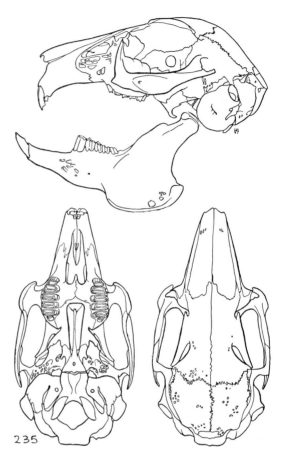

Fig. 235. Sylvilagus nuttallii grangeri, ½ mi. E Jefferson, Nevada, No. 58527 M.V.Z., ♀, X 1.

durn (Beck, 1958:24); Johnston Lake; Big Muddy Lake. North Dakota: Goodall; Medora (Bailey, 1927:137). Wyoming (Long, 1965a:544); Devils Tower; Sundance. South Dakota: Custer. Wyoming (Long, 1965:544): Wheatland; 2 mi. W Horse Creek P.O.; Sherman. Colorado (Armstrong, 1972:84): Lay; Meeker. Utah: Mt. Ellen. Arizona (Hoffmeister and Lee, 1963:148): VT[= De Motte] Park, Kaibab National Forest, 8500 ft.; Cape Royal (Hoffmeister, 1971:90, sight record only); Shiva Temple, Grand Canyon; Big Spring, Kaibab. Utah (Stock, 1970:430): Pine Valley; Enterprise Reservoir, 5760 ft. Nevada: ¼ mi. W Utah–Nevada boundary, 38° 17′ N, 7300 ft.; S end Belted Range, 5 mi. NW Whiterock Spring, 7200 ft.; Chiatovich Creek, 7000 ft.; 1 mi. S, 2½ mi. E Grapevine Peak, 6700 ft.; Clark Canyon (Deacon, et al., 1964:402); Charleston Park, Kyle Canyon, 8000 ft. California: Johnson Canyon, 6500 ft.; near Woodfords, 5500 ft. Nevada: Calvada; Hardscrabble Canyon; Paradise Valley. Idaho: S Fork Owyhee River, 12 mi. N Nevada line; Crane Creek, 15 mi. E Midvale; Lemhi. Montana: 4 mi. W Hamilton; 2 mi. N Moise Lake. Alberta: Cardston; Steveville.

Sylvilagus nuttallii nuttallii (Bachman)

1837. *Lepus nuttallii* Bachman, Jour. Acad. Nat. Sci. Philadelphia, 7:345. Type locality, probably eastern Oregon near mouth Malheur River.
1904. *Sylvilagus nuttallii*, Lyon, Smiths, Miscl. Coll., 45:323, June 15.

MARGINAL RECORDS.—British Columbia: Penticton; *Anarchist Mtn.* (Cowan and Guiguet, 1965:107). Washington: Kettle Falls. Idaho: Couer d'Alene; *Lewiston;* Fiddle Creek. Nevada: Quinn River Crossing, 5800 ft.; *¹/₂ mi. S Granite Creek, Granite Mts.; Smoke Creek, 9 mi. E California line;* 4½ mi. S Flanigan. California: Truckee; *Beckwith;* Weed; Yreka. Oregon: near Ashland; Bend; The Dalles. Washington: Grand Dalles; Yakima Valley; Douglas. British Columbia: *Keremeos* (Cowan and Guiguet, 1965:107, sight record only).

Sylvilagus nuttallii pinetis (J. A. Allen)

1894. *Lepus sylvaticus pinetis* J. A. Allen, Bull. Amer. Mus. Nat. Hist., 6:348, December 7, type from White Mts., south of Mt. Ord, Apache Co., Arizona, according to Warren (Mammals of Colorado, p. 270, 1942).
1909. *Sylvilagus nuttali pinetis*, Nelson, N. Amer. Fauna, 29:207, August 31.

MARGINAL RECORDS.—Colorado: 15 mi. E Virginia Dale (Armstrong, 1972:84); Arkins; Golden; Greenhorn Mts. New Mexico: Sierra Grande; Willis; Zuni Mts. Arizona: type locality; *Horse Cienega, 1 mi. NNE Hannagan Meadows* (Hoffmeister and Lee, 1963:147); *head Fish Creek, 9500 ft. (ibid.);* Bat Woman Cave, 16 mi. W Kayenta, 6900 ft. *(ibid.).* Utah: Navajo Mtn.; Block Canyon, 19 mi. SE Moab, 5400 ft.; *5 mi. NE La Sal P.O., 8000 ft.* Colorado (Armstrong, 1972:84): Glenwood Springs; Craig.

Sylvilagus audubonii
Desert Cottontail

×¹/₆

+11

Total length, 350–420; tail, 45–75; hind foot, 75–100; ear from notch (dry), 55–70; weight of *S. a. vallicola,* average of seven males, 912(835–988), two females, 1096 and 1191 grams. Long

hind legs, long ears, sparseness of hair on the ears, shortness of hair on feet, prominent (upturned) supraorbital process of skull and much inflated tympanic bullae are characters of this widespread species.

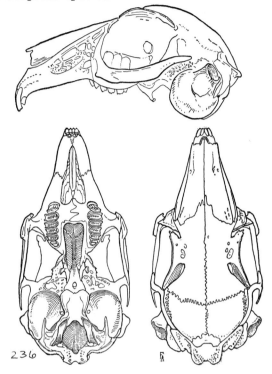

236

Fig. 236. *Sylvilagus audubonii minor,* Neville Spring, 3290 ft., Grapevine Mts., Big Bend, Brewster Co., Texas, No. 80519 M.V.Z., ♂, X 1.

Sylvilagus audubonii arizonae (J. A. Allen)

1877. *[Lepus sylvaticus]* var. *arizonae* J. A. Allen, *in* Coues and Allen, Monog. N. Amer. Rodentia, p. 332, August, type from Beals Spring, Yavapai Co. [= Beales Spring, Mohave Co.; see also Grinnell, Univ. California Publ. Zool., 40:203, September 26, 1933], Arizona.
1909. *Sylvilagus auduboni arizonae*, Nelson, N. Amer. Fauna, 29:222, August 31.
1896. *Lepus arizonae major* Mearns, Proc. U.S. Nat. Mus., 18:557, June 24, type from Calabasas, Santa Cruz Co., Arizona.
1904. *Lepus laticinctus* Elliot, Field Columb. Mus., Publ. 87, Zool. Ser., 3:254, January 7, type from Oro Grande, Mohave Desert, San Bernardino Co., California.
1904. *Lepus l[aticinctus]. rufipes* Elliot, Field Columb. Mus., Publ. 87, Zool. Ser., 3:254, January 7, type from Furnace Creek, Inyo Co., California.

MARGINAL RECORDS.—Utah: N end Newfoundland Mts. (Egoscue, 1965:685); 2 mi. SW Fish Springs; Holden; 7 mi. SW Tropic. Arizona (Hoffmeister and Lee, 1963:514, unless otherwise noted): 4 mi. S Fredonia; *32 mi. SW Fredonia; 39 mi. SW Fredonia; 40 mi.*

SW *Fredonia; foot Broad Canyon, 5050 ft.;* ½ mi. E Vulcan's Throne, Grand Canyon National Monument; Peach Springs, Hualpai Indian Reservation; Big Sandy Creek; Flagstaff (Cockrum, 1961:74); 5 mi. S, 1 mi. E New River; 5 mi. NW Bartlett Dam; Long Gulch, 5 mi. N, 1 mi. E Roosevelt; Superior (Hall and Kelson, 1959:265); *6 mi. S Superior; Willow Spring Ranch, 40 mi. NNE Tucson; Oracle;* 1½ mi. S Oracle, 4500 ft.; *Oracle Ranger Station;* Indian Cave, 4 mi. N, ⁹⁄₁₀ mi. W Cortaro; *13³⁄₅ mi. NW Tucson;* 23 mi. N, 1 mi. E Sasabe; *Figaroa Ranch, 8 mi. NW Arivaca; Arivaca; 1¹⁄₂ mi. S Arivaca;* Sapo Tank, 1½ mi. S, ¼ mi. E Arivaca. Sonora (Hoffmeister and Lee, 1963:514, unless otherwise noted): 5 mi. N Cornelio; *61 mi. N Hermosillo;* Carbó; Tecoripa (Hall and Kelson, 1959:265); Ortiz; Batamotal. Baja California: San Matías Pass. California: Vallecito; Fairmont, Antelope Valley; Little Lake, 3300 ft.; near Benton, 5300–5639 ft. Nevada: Arlemont; 4 mi. E Smith Creek Cave. Utah: N end Stansbury Mts., *ca.* 200 ft. S Hwy. 40, 4246 ft. (Egoscue, 1961:123, as *S. auduboni* only).

Sylvilagus audubonii audubonii (Baird)

1858. *Lepus audubonii* Baird, Mammals, *in* Repts. Expl. Surv. . . . , 8(1):608, July 14, type from San Francisco, San Francisco Co., California.
1909. *Sylvilagus auduboni,* Nelson, N. Amer. Fauna, 29:214, August 31.

MARGINAL RECORDS.—California: Paines Creek, 600 ft.; Rackerby; Pleasant Valley; Snelling; 2 mi. S mouth Salinas River, northward not reaching coast again except at San Francisco, thence around shores San Francisco Bay to mouth Carquinez Straits and northward along western side Sacramento Valley to Winslow, 5 mi. W Fruto.

Sylvilagus audubonii baileyi (Merriam)

1897. *Lepus baileyi* Merriam, Proc. Biol. Soc. Washington, 11:148, June 9, type from Spring Creek, E side Bighorn Basin, Bighorn Co., Wyoming.
1908. *Sylvilagus auduboni baileyi,* Lantz, Trans. Kansas Acad. Sci., 22:336.

MARGINAL RECORDS.—Montana (Hoffmann, *et al.,* 1969:583, as *S. audubonii* only): 2 mi. N Conrad; Darvall Ranch, Missouri River S of Glasgow. North Dakota: 1 mi. S Medora (Genoways and Jones, 1972:10); Wade on Cannonball River. South Dakota: Corral Draw. Nebraska (Jones, 1964c:103–104): Valentine; Mullen; 3 mi. S Imperial. Kansas: 12½ mi. S, 4 mi. W Oberlin; WaKeeney. Colorado: Monon; Regnier (Armstrong, 1972:87); Trinchera (*ibid.*); Quenda [= Querida]; Salida; Minnehaha (Armstrong, 1972:86); Evergreen (*ibid.*). Wyoming: 2 mi. N Colorado line, Hwy. 281 (Long, 1965a:545). Colorado: Fortification Creek, near Craig (Armstrong, 1972:86); White Rock [2 mi. above Meeker, 6400 ft.]; 20 mi. SW Rangely. Utah: 8 mi. S Myton; 6 mi. NW Duchesne; 10 mi. E Mountain Home. Wyoming (Long, 1965a:546): *Henrys Fork;*

Fort Bridger; Opal; Big Piney; Circle; Mammoth Hot Springs. Montana: Stillwater; Freezeout (Greenfields) Lake (Hoffmann, *et al.,* 1969:583, as *S. audubonii* only). Not found: Montana: Philips Creek.

Sylvilagus audubonii cedrophilus Nelson

1907. *Sylvilagus auduboni cedrophilus* Nelson, Proc. Biol. Soc. Washington, 20:83, July 22, type from Cactus Flat, 20 mi. N Cliff, Grant Co., New Mexico.
1907. *Sylvilagus auduboni warreni* Nelson, Proc. Biol. Soc. Washington, 20:83, July 22, type from Coventry, Montrose Co., Colorado. Subspecifically inseparable from *S. a. cedrophilus* according to Hoffmeister and Lee (Jour. Mamm., 44:515, December 15, 1963).

Although *cedrophilus* has priority by page position over *S. a. warreni,* Hoffmeister and Lee (*op. cit.*) used *warreni* in the belief that they were "first reviser(s)" and that Article 24 of the International Code of Zool. Nomenclature (November 17, 1961) therefore permitted them to choose whichever of the two names more nearly "ensured stability and universality of nomenclature." It seems that Nelson (N. Amer. Fauna, 29, April 17, 1909), not Hoffmeister and Lee, was first reviser. However that may be, it seems to be that Hoffmeister and Lee (*op. cit.*) have not shown that use of the name *warreni* will ensure stability of nomenclature any bettter than will use of the name *cedrophilus.* Consequently *cedrophilus* is here retained for use owing to its precedence in position in the work in which both names were originally proposed, which is in accordance with Recommendation 24A of the International Code (*op. cit.*).

MARGINAL RECORDS.—Utah: Willow Creek, 5250 ft. Colorado: Rifle; Coventry (Hoffmeister and Lee, 1963:516); head Prater Canyon (Anderson, 1961:40); 2½ mi. S, 1 mi. W Chimney Rock (Armstrong, 1972:87); Rio Grande County (*ibid.*); 3 mi. E Villa Grove (*ibid.*); N of entrance station, Great Sand Dunes National Monument (*ibid.*). New Mexico (Hoffmeister and Lee, 1963:516, unless otherwise noted): Hondo Canyon (Hall and Kelson, 1959:268); Cieneguilla (*ibid.*); Santa Rosa (*ibid.,* as *S. a. cedrophilus*); *1 mi. S, 3 mi. E Pastura;* Capitan (Hall and Kelson, 1959:268, as *S. a. cedrophilus*); 41⅓ mi. by road N Arrey, U.S. 85; *Big Rosa Canyon; 2 mi. S, 2¹⁄₂ mi. E Madre Mt.;* Redrock; *Cactus Flat, 20 mi. N* [= 14 mi. NW] *Cliff; Beaver Lake, Gila National Forest.* Arizona (Hoffmeister and Lee, 1963:516, unless otherwise noted): 7 mi. S Springerville; Holbrook; Winslow; Forte Verde; *Camp Verde; 6 mi. NW Camp Verde;* Flagstaff (Cockrum, 1961:76); Cataract Canyon, sec. 4, T. 22 N, R. 2 E, (= 4 mi. N Williams); Seligman; 7 mi. E Hilltop; *Pasture Wash Ranger Station, Grand Canyon National Park; Grand Canyon Village; 1 mi. SW Moran Point, Grand Canyon National Park;* Desert View (Hoffmeister, 1971:167); *11 mi. S Page;* 4 mi. S Page. Utah: Canesville; Wellington.

Sylvilagus audubonii confinis (J. A. Allen)

1898. *Lepus arizonae confinis* J. A. Allen, Bull. Amer. Mus. Nat. Hist., 10:146, April 12, type from Playa María, Baja California.

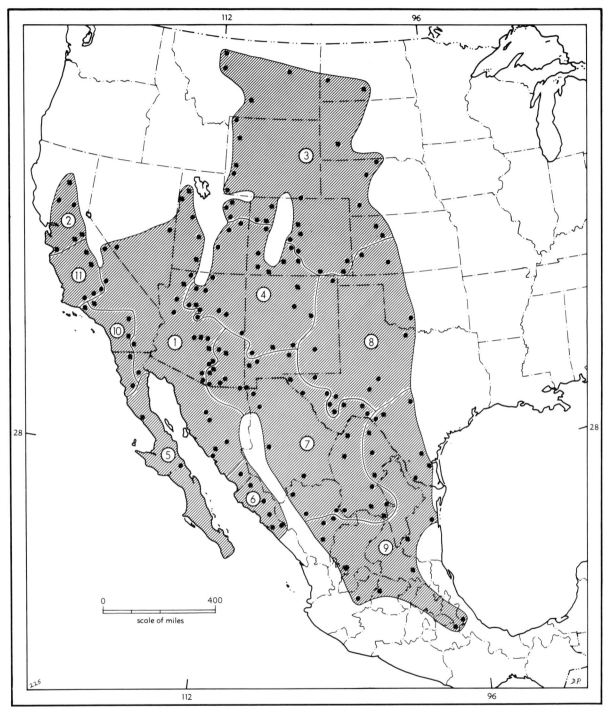

Map 225. *Sylvilagus audubonii.*

Guide to subspecies
1. *S. a. arizonae*
2. *S. a. audubonii*
3. *S. a. baileyi*
4. *S. a. cedrophilus*
5. *S. a. confinis*
6. *S. a. goldmani*
7. *S. a. minor*
8. *S. a. neomexicanus*
9. *S. a. parvulus*
10. *S. a. sanctidiegi*
11. *S. a. vallicola*

1909. *Sylvilagus auduboni confinis*, Nelson, N. Amer. Fauna, 29:220, August 31.

MARGINAL RECORDS.—Baja California: type locality; San Bruno, thence southward on peninsula to *Cape San Lucas.*

Sylvilagus audubonii goldmani (Nelson)

1904. *Lepus arizonae goldmani* Nelson, Proc. Biol. Soc. Washington, 17:107, May 18, type from Culicán, Sinaloa.
1909. *Sylvilagus auduboni goldmani* (Nelson), N. Amer. Fauna, 29:225, August 31.

MARGINAL RECORDS (Hoffmeister and Lee, 1963:518, unless otherwise noted).—Sonora: Camoa. Sinaloa: 7 mi. NE El Fuerte; Bacubirito; 12 mi. N Culicán; 2 mi. E Aguacaliente (Armstrong and Jones, 1971:754); *type locality; 6 mi. N El Dorado;* 6 mi. N, 1½ mi. E El Dorado.

Sylvilagus audubonii minor (Mearns)

1896. *Lepus arizonae minor* Mearns, Proc. U.S. Nat. Mus., 18:557, June 24, type from El Paso, El Paso Co., Texas.
1907. S[*ylvilagus*]. a[*uduboni*]. *minor*, Nelson, Proc. Biol. Soc. Washington, 20:83, July 22.

MARGINAL RECORDS.—Arizona (Hoffmeister and Lee, 1963:511): Cutter, 6 mi. E Globe; *Rice;* Ash Creek, Graham Mts., 3200 ft. New Mexico (Hoffmeister and Lee, 1963:511, unless otherwise noted): Lordsburg; *Burro Mts.;* Silver City; 12 mi. N Tularosa (Hall and Kelson, 1959:266). Texas: 3 mi. S, 14 mi. E El Paso City Hall; Hudspeth County (Davis, 1966:225, as *S. auduboni* only); Kent; Haymond; Terrell County (Davis, 1966:225, as *S. auduboni* only); Langtry. Coahuila (Baker, 1956:198, unless otherwise noted): 10 mi. S, 5 mi. E Boquillas, 1500 ft.; 13 mi. N, 11 mi. W Tanque Alvarez, 4250 ft.; 1 mi. S San Lázaro; *Saltillo; Diamante, 4 mi. S, 6 mi. E Saltillo, 7500 ft.; 7 mi. S, 4 mi. E Bella Unión, 7200 ft.;* 2 mi. N, 2 mi. E San Antonio de las Alazanas, 8700 ft.; *7 mi. S, 1 mi. E Gómez Farías, 6500 ft.;* La Ventura; N foot Sierra de Guadalupe, 10 mi. S, 5 mi. W General Cepeda, 6200 ft.; W foot Pico de Jimulco, 5000 ft. Durango: San Juan, 10 mi. WSW Lerdo, 3800 ft. (Baker and Greer, 1962:79); 4 mi. NNE Boquilla, 6300 ft. (*ibid.*); Matalotes (*ibid.*). Chihuahua (Hoffmeister and Lee, 1963:511, unless otherwise noted): 5 mi. E Parral; 2 mi. W Miñaca, 6900 ft.; 12 mi. N Dublán; plains Playas Valley, E of San Luis Mts. New Mexico: Guadalupe M:s., sec. 22, T. 34 S, R. 21 W (Hoffmeister and Lee, 1963:511). Arizona (Hoffmeister and Lee, 1963:511, unless otherwise noted): *San Bernardino Ranch;* 8 mi. E Douglas; 1½ mi. N Tombstone; Huachuca Mts.; 2½ mi. N, 1½ mi. W Nogales; *35 mi. S Tucson, Santa Rita Experimental Range; 30 mi. S Tucson;* 2 mi. SW Helmet Peak; *various localities within 8 mi. radius Tucson;* 18 mi. NE Tucson, Mt. Lemmon highway.

Sylvilagus audubonii neomexicanus Nelson

1907. *Sylvilagus auduboni neomexicanus* Nelson, Proc. Biol. Soc. Washington, 20:83, July 22, type from Fort Sumner, De Baca Co., New Mexico.

MARGINAL RECORDS.—Kansas: 1 mi. E Coolidge; Rezeau Ranch, 5 mi. N Belvidere. Oklahoma: Wichita Mts. Wildlife Refuge (Glass and Halloran, 1961:237). Texas: Wichita Falls; San Angelo; Adam [= 15 mi. E Adams]; 28 mi. S Alpine; *15 mi. S Alpine;* 7 mi. NE Marfa; Toyahvale [= 10 mi. S of]; McKittrick Canyon. New Mexico: Roswell; Emory Peak.

Sylvilagus audubonii parvulus (J. A. Allen)

1904. *Lepus (Sylvilagus) parvulus* J. A. Allen, Bull. Amer. Mus. Nat. Hist., 20:34, February 29, type from Apam, Hidalgo.
1909. *Sylvilagus auduboni parvulus*, Nelson, N. Amer. Fauna, 29:236, August 31.

MARGINAL RECORDS.—Texas: Llano; San Diego; Kleberg County; Rio Grande City. Tamaulipas: El Mulato; Miquihuana. San Luis Potosí: Río Verde. Veracruz: Perote. Puebla: Chalchicomula. Guanajuato: Silao. Jalisco (Genoways and Jones, 1973:2): 3 mi. SW Tepatitlán; *3 mi. S Yahualica, 5900 ft.;* 3 mi. S Huejúcar, 5900 ft. Durango: Durango City; 6 mi. NW La Pila, 6150 ft. (Baker and Greer, 1962:79); 6 mi. N Chocolate, 4100 ft. (*ibid.*). Coahuila (Baker, 1956:199): Monclova; La Gacha, 1600 ft.; 11 mi. W Hda. San Miguel, 2200 ft. Texas: Comstock; Edwards County (Davis, 1966:225, as *S. auduboni* only).

Sylvilagus audubonii sanctidiegi (Miller)

1899. *Lepus floridanus sanctidiegi* Miller, Proc. Acad. Nat. Sci. Philadelphia, 51:389, October 5, type from Mexican Boundary Monument No. 258, shore Pacific Ocean, California.
1909. *Sylvilagus auduboni sanctidiegi*, Nelson, N. Amer. Fauna, 29:218, August 31.

MARGINAL RECORDS.—California: Sespe; Reche Canyon near Colton; San Felipe Canyon. Baja California: Nachogüero Valley; Santo Tomás, thence northward along coast to point of beginning.

Sylvilagus audubonii vallicola Nelson

1907. *Sylvilagus auduboni vallicola* Nelson, Proc. Biol. Soc. Washington, 20:82, July 22, type from San Emigdio Ranch, Kern Co., California.

MARGINAL RECORDS.—California: Fresno Flat; Badger; 2750 ft.; Onyx; Tehachapi; Mt. Pinos, northwesterly, seldom actually reaching coast, to central Monterey County, thence easterly to point of beginning.

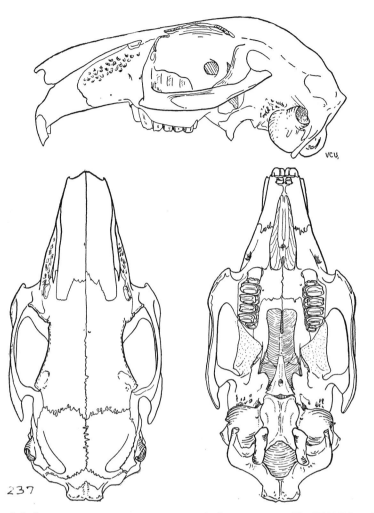

Fig. 237. *Sylvilagus aquaticus aquaticus*, Crawford Co., Kansas, No. 8544 K.U., ♂, X 1.

Sylvilagus aquaticus
Swamp Rabbit

Total length, 530–540; tail, 67–71; hind foot, 105–110; length of ear from notch (dry), 63–67. Upper parts blackish brown or reddish brown; underparts, with some white; skull robust; posterior extensions of supraorbital processes joined for their entire length with side of braincase or, in some specimens, with a small foramen between the braincase and the base of the posterior extension of the supraorbital process. This big rabbit is a stronger runner than the smaller marsh rabbit and is white instead of brownish or grayish on underside of tail. (See Map 222, p. 299.)

Lowery (1974:166, 167) presents convincing evidence that the features distinguishing *S. a. littoralis* from *S. a. aquaticus* result from adventitious coloration, at least in Louisiana. If specimens heretofore assigned to *littoralis* from Texas, Mississippi, and southwestern Alabama are likewise adventitiously colored, *S. a. littoralis* should be placed as a synonym of S. *aquaticus*, making the latter a monotypic species.

Sylvilagus aquaticus aquaticus (Bachman)

1837. *Lepus aquaticus* Bachman, Jour. Acad. Nat. Sci. Philadelphia, 7:319. Type locality, western Alabama.
1909. *Sylvilagus aquaticus*, Nelson, N. Amer. Fauna, 29:270, August 31.
1895. *Lepus aquaticus attwateri* J. A. Allen, Bull. Amer. Mus. Nat. Hist., 7:327, November 8, type from Medina River, 18 mi. S San Antonio, Bexar Co., Texas.
1899. *Lepus telmalemonus* Elliot, Field Columb. Mus., Publ. 38, Zool. Ser., 1:285, May 25, type from Washita River, near Dougherty, Murray Co., Oklahoma.

MARGINAL RECORDS.—Indiana: *ca.* 18 km. S Vincennes (Terrel, 1972:283, as *S. aquaticus* only); *near Yankeetown* (Mumford, 1969:52); sec. 17, T. 8 S, R. 6 W, Spencer Co. (Kirkpatrick, 1961:99, as *S. aquaticus* only). Tennessee: 5 mi. W Hornbeak; Henryville. Alabama: Huntsville; Big Crow Creek near Stevenson. Georgia: Murray County (Lowe, 1958:118, Fig. 1, as *S. aquaticus* only). South Carolina: *ca.* 3 mi. SE Westminster; *ca.* 5 mi. W Iva. Georgia (Lowe, 1958:118, Fig. 1, as *S. aquaticus* only): Wilkes County; Hancock County; Macon County; Randolph County. Alabama: Castleberry. Louisiana: Covington; Kleinpeter; Egan (Lowery, 1943:251). Texas: Sourlake; Richmond; Medina River, 18 mi. SW San Antonio; Travis County (Davis, 1966:226, as *S. aquaticus* only); Gurley; Limestone County (Davis, 1966:226, as *S. aquaticus* only); Montague County (Dalquest, 1968:16). Oklahoma: 7 mi. NW Stillwater. Kansas: Crawford County. Arkansas: along White River near Springdale. Missouri: 3 mi. SW Udall. Arkansas: White River near Augusta. Missouri: St. Francis River, W of Senath; Duck Creek Wildlife Area (Toll, *et al.*, 1960:398). Illinois: 6 mi. N Sesser.

Sylvilagus aquaticus littoralis Nelson

1909. *Sylvilagus aquaticus littoralis* Nelson, N. Amer. Fauna, 29:273, August 31, type from Houma, Terrebonne Parish, Louisiana.

MARGINAL RECORDS.—Louisiana (Crain and Packard, 1966:324): 1 mi. N Franklington [*sic*]; *10 mi. S Angie*. Alabama: Blakely Island, opposite Mobile. Mississippi: Bay St. Louis, thence along coast, including Breton Island, Louisiana (Lowery, 1974:167, as *S. aquaticus* only), to Texas (Davis, 1966:226, as *S. aquaticus* only): Aransas County; Refugio County; Victoria County; Chambers County. Louisiana: Hackberry; Rayne; New Orleans (Lowery, 1936:32); *Lake Catherine* (*ibid.*).

Sylvilagus insonus (Nelson)
Omilteme Cottontail

1904. *Lepus insonus* Nelson, Proc. Biol. Soc. Washington, 17:103, May 18, type from Omilteme, Guerrero. Known from type locality only.

1909. *Sylvilagus insonus*, Lyon and Osgood, Bull. U.S. Nat. Mus., 62:34, January 28 (see Hershkovitz, Proc. U.S. Nat. Mus., 100:335, May 26, 1950, for allocation of *S. insonus* to subgenus *Sylvilagus* instead of to subgenus *Tapeti*).

Total length, 435; tail, 42.5; hind foot, 95; ear from notch (dry), 61. Color grayish brown above and dingy (not white) below; tail dingy buffy below and dull rusty brown above. The collectors thought that the species was restricted to the forested parts of the Sierra Madre del Sur between 7000 and 10,000 ft. altitude in Guerrero, México. (See Map 223.)

Sylvilagus cunicularius
Mexican Cottontail

Total length, 485–515; tail, 54–68; hind foot, 108–111; ear from notch (dry), 60–63. Pelage coarse; upper parts brownish gray; skull massive; posterior extensions of supraorbital processes varying from those that project free to those that have the tips, or tips and a considerable part of the processes, attached to the braincase.

Sylvilagus cunicularius cunicularius (Waterhouse)

1848. *Lepus cunicularius* Waterhouse, A natural history of the Mammalia, 2:132, type from Zacualpan (probably in state of México).
1909. *Sylvilagus cunicularius*, Nelson, N. Amer. Fauna, 29:239, August 31.
1890. *Lepus verae-crucis* Thomas, Proc. Zool. Soc. London, p. 74, June, type from Las Vigas, Veracruz.

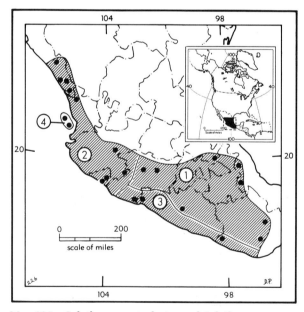

Map 226. *Sylvilagus cunicularius* and *Sylvilagus graysoni.*

1. *S. c. cunicularius*   3. *S. c. pacificus*
2. *S. c. insolitus*   4. *S. graysoni*

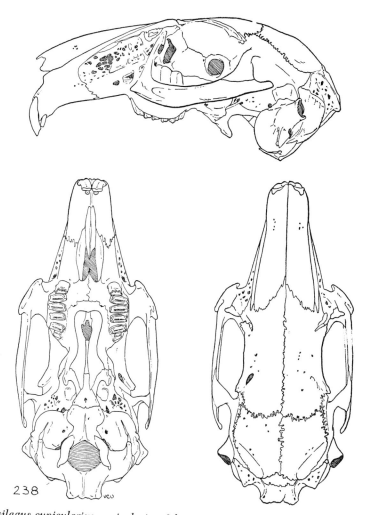

Fig. 238. *Sylvilagus cunicularius cunicularius*, 3 km. W Acultzingo, Veracruz, No. 30749 K.U., ♂, X 1.

MARGINAL RECORDS.—Hidalgo: Tulancingo. Veracruz: Las Vigas; Orizaba. Oaxaca: Mt. Zempoaltepec; Suchixtepec. Guerrero: Chilpancingo. Michoacán: Tancítaro; Pátzcuaro.

Sylvilagus cunicularius insolitus (J. A. Allen)

1890. *Lepus insolitus* J. A. Allen, Bull. Amer. Mus. Nat. Hist., 3:189, December 10, type from plains of Colima, Jalisco.
1909. *Sylvilagus cunicularius insolitus*, Nelson, N. Amer. Fauna, 29:243, August 31.

MARGINAL RECORDS.—Sinaloa: San Ignacio (Armstrong and Jones, 1971:755); Mazatlán; Rosario; Esquinapa. Nayarit: Acaponeta. Jalisco: Ajijic (Ingles, 1959:387, as *S. cunicularius* only); 8 mi. E Jilotlán de los Dolores, 2000 ft. (Genoways and Jones, 1973:4). Colima: Colima; Armeria, thence northward along Pacific Coast.—**See** addenda.

Sylvilagus cunicularius pacificus (Nelson)

1904. *Lepus veraecrucis pacificus* Nelson, Proc. Biol. Soc. Washington, 17:104, May 18, type from Acapulco, Guerrero.
1909. *Sylvilagus cunicularius pacificus*, Lyon and Osgood, Bull. U.S. Nat. Mus., 62:35, January 28.

MARGINAL RECORDS.—Guerrero: El Limón. Oaxaca: Llano Grande, thence westward along Pacific Coast to Michoacán: 14 km. N Melchor Ocampo (Alvarez, 1968:30).

Sylvilagus graysoni (J. A. Allen)
Tres Marías Cottontail

1877. *Lepus graysoni* J. A. Allen, *in* Coues and Allen, Monog. N. Amer. Rodentia, p. 347, August, type from Tres Marías Islands, Nayarit; probably María Madre Island. (See Nelson, N. Amer. Fauna, 14:16, April 29, 1899).
1904. *Sylvilagus (Sylvilagus) graysoni*, Lyon, Smiths. Miscl. Coll., 45:336, June 15.

Total length, 480; tail, 51; hind foot, 99; ear from notch (dry), 57. This insular species is closely related to *S. cunicularius* of the adjacent mainland but has notably shorter ears and is more reddish on the upper parts, sides, and legs; the skull is slenderer, especially in the rostral region. The posterior extensions of the supraorbital process are united to the braincase throughout most of their length as in *S. palustris*. The species seems to have a narrow vertical range, occurring from sea level up only to *ca.* 1000 ft. (V. E. Diersing, *in Litt.*).—**See** addenda.

MARGINAL RECORDS.—María Madre Island; María Magdalena Island.

Genus **Lepus** Linnaeus—Hares and Jack Rabbits

American species revised by Nelson, N. Amer. Fauna, 29:59–158, August 31, 1909. Concerning Shamel's (Proc. Biol. Soc. Washington, 55:25, May 12, 1942) proposed changes of names for several species, see Hall, Univ. Kansas Publ., Mus. Nat. Hist., 5:45, October 1, 1951. For synopsis, see Hall, Univ. Kansas Publ., Mus. Nat. Hist., 5:170–195, December 15, 1951.

1758. *Lepus* Linnaeus, Syst. nat., ed. 10, 1:57. Type, *Lepus timidus* Linnaeus.
1895. *Macrotolagus* Mearns, Science, n.s., 1:698, June 21. Type, *Lepus alleni* Mearns. (See Mearns, Proc. U.S. Nat. Mus., 18:552, June 24, 1896.)
1904. *Poecilolagus* Lyon, Smiths. Miscl. Coll., 45:395, June 15. Type, *Lepus americanus* Erxleben.
1904. *Lagos* Palmer, N. Amer. Fauna, 23:361, January 23. Type, *Lepus arcticus* Ross. *Lagos* J. Brooks, A catalogue of the anatomical and zoological museum, pt. 1, p. 54, July, 1828, appears to be a *nomen nudum*.
1911. *Boreolepus* Barrett-Hamilton, History of the British

Mammalia, pt. 9, p. 160, November 17. Type, *Lepus groenlandicus* Rhoads. (For status see Sutton and Hamilton, Mem. Carnegie Mus., 12(pt. 2, sec. 1):78, August 4, 1932; also A. H. Howell, Jour. Mamm., 17:331, November 16, 1936.)

Total length, 363–664; tail, 25–112; hind foot, 112–189; ear from notch (dry), 62–144. Upper parts grayish, brownish, or black; interparietal bone fused to surrounding bones; cervical vertebrae long, 2nd and 3rd being longer than wide; transverse processes of lumbar vertebrae long, the longest one equal to the length of the centrum, to which it is attached, plus half the length of preceding centrum; free extremity of transverse process of lumbar vertebra considerably expanded; distance from anterior edge of acetabulum to extreme anterior point of ilium less than distance from former point to most distant point of ischium; ulna reduced in size along middle part of shaft, and, excepting the lower extremity, placed almost entirely behind radius.

All members of genus *Lepus* are technically hares, according to the definition in the preceding account of family Leporidae. The largest members of Order Lagomorpha are members of Genus *Lepus*.

In the past it has been customary to recognize two or more subgenera of genus *Lepus*. The species are a less diverse lot than those in some other genera, however, and it seems that no useful purpose is served by recognizing subgenera. Accordingly, the several names proposed for this purpose are arranged here as synonyms of the generic name *Lepus* Linnaeus.

KEY TO NORTH AMERICAN SPECIES OF LEPUS

1. North of 34° N latitude.
  2. All white pelage (tips of ears sometimes black).
    3. North of line from Port Simpson, British Columbia, to Halifax, Nova Scotia.
      4. Basilar length of skull more than 67; ear from notch usually more than 73 dry (77 fresh); 1st upper incisors inscribing an arc of a circle the radius of which is more than 9.6 mm.
        5. Geographic range east of Mackenzie River. . . . . *Lepus arcticus,* p. 318
        5'. Geographic range west of Mackenzie River. . . . . *Lepus othus,* p. 318
      4'. Basilar length of skull less than 67; ear from notch usually less than 73 dry (77 fresh); 1st upper incisors inscribing an arc of a circle the radius of which is less than 9.6 mm.
                Lepus americanus, p. 315
    3'. South of a line from Port Simpson, British Columbia, to Halifax, Nova Scotia.
      6. Ear from notch more than 82 dry (87 fresh); least interorbital breadth more than 26.
                Lepus townsendii, p. 322
      6'. Ear from notch less than 82 dry (87 fresh); least interorbital breadth less than 26.
                Lepus americanus, p. 315
  2'. Brownish or grayish pelage.
    7. Tail blackish or brownish all around (in specimens not having completed molt on tail, white winter pelage may be present); basilar length less than 67 mm. . . . . *Lepus americanus,* p. 315
    7'. Tail partly or wholly white.
      8. Tail black on upper surface.
        9. Upper sides of hind feet without a trace of white; upper parts tawny. *Lepus capensis,* p. 332

9'. Upper sides of hind feet with more or less white or whitish; upper parts grayish or
brownish. . . . . *Lepus californicus,* p. 324
8'. Tail all white or (in some *Lepus townsendii*) with faint buffy or dusky median line on top
but this line not extending on to rump (as in *L. californicus*).
    10. Geographic range north of a line from Port Simpson, British Columbia, to Halifax,
        Nova Scotia. . . . . *Lepus townsendii,* p. 322
           11. Geographic range east of Mackenzie River. . . . . . *Lepus arcticus,* p. 318
           11'. Geographic range west of Mackenzie River. . . . . *Lepus othus,* p. 318
    10'. Geographic range south of a line from Port Simpson, British Columbia, to
        Halifax, Nova Scotia. . . . . *Lepus townsendii,* p. 322
1'. South of 34° N latitude.
    12. In state of Tamaulipas, México. . . . . *Lepus californicus,* p. 324
    12'. Range outside Tamaulipas, México.
        13. Ears with terminal black patch (on outside) *Lepus californicus,* p. 324 and *Lepus insularis,* p. 328
        13'. Ears without terminal black patch.
           14. Ear from notch, dry more than 130 (137 fresh). . . . . *Lepus alleni,* p. 331
           14'. Ear from notch, dry less than 130 (137 fresh).
               15. Ears yellow; range Pacific Coastal region of Isthmus of Tehuantepec in southern
                  Oaxaca and Chiapas. . . . . *Lepus flavigularis,* p. 330
               15'. Ears dark buff, grayish, white and black; range north of Isthmus of Tehuantepec.
                                   Lepus callotis, p. 328

Lepus americanus
Snowshoe Rabbit

Total length, 363–520; tail, 25–55; hind foot, 112–150; ear from notch (dry), 62–70. Upper parts brownish or dusky grayish; hind feet brownish or white depending on subspecies; winter pelage white except in certain populations along Pacific Coast; basilar length less than 67; 1st upper incisors inscribing an arc of a circle the radius of which is less than 9.6 mm.

Lepus americanus americanus Erxleben

1777. [*Lepus*] *americanus* Erxleben, Systema regni animalis . . ., 1:330. Type locality, Hudson Bay, Canada. Restricted to Fort Severn, Ontario, by V. Bailey, N. Amer. Fauna, 49:138, January 8, 1927.
1778. *Lepus hudsonius* Pallas, Novae species quadrupedum e glirium ordine . . ., p. 30, type locality not stated.
1790. *Lepus nanus* Schreber, Die Säugthiere . . ., 4:880–885, pl. 234B, a composite of *Lepus americanus* and *Sylvilagus floridanus.* No type or type locality designated. Range given as from Hudson Bay to Florida.
1899. *Lepus bishopi* J. A. Allen, Bull. Amer. Mus. Nat. Hist., 12:11, March 4, type from Mill Lake, Turtle Mts., North Dakota (inseparable from *L. a. americanus* according to V. Bailey, N. Amer. Fauna, 49:138, January 8, 1927 [not December, 1926]).

MARGINAL RECORDS.—Keewatin: 40 mi. NE Windy River (Harper, 1956:16); Hudson Bay. Ontario: Fort Severn; around shore of Hudson Bay to approx. 56° N, thence to Ungava: Fort Chimo. Labrador: Hamilton Inlet. Ontario: North Bay of Lake Nipissing; Michipicoten Island. Michigan: Isle Royale. Manitoba: Dog Lake. Saskatchewan: Indian Head. North Dakota: Mill Lake, Turtle Mts.; Grafton; near Fargo; Elbowoods; 7 mi. N, 9 mi. W Killdeer (Geno-

ways and Jones, 1972:9); Buford. Montana (Hoffmann, *et al.,* 1969:584, as *L. americanus* only); *lower Yellowstone River, near Sidney;* Milk River, Glasgow. Saskatchewan: Battle Creek. Alberta: Red Deer; 50 mi. N Edmonton; Fort Chipewyan; Government Hay Camp, Slave River.

Lepus americanus bairdii Hayden

1869. *Lepus bairdii* Hayden, Amer. Nat., 3:115, May. Type locality, Columbia Valley, Wind River Mts., Fremont Co., Wyoming.
1875. [*Lepus americanus*] var. *Bairdii,* J. A. Allen, Proc. Boston Soc. Nat. Hist., 17:431, February 17.

MARGINAL RECORDS.—British Columbia: St. Mary Lake (Cowan and Guiguet, 1965:102); Elko. Alberta: Waterton Lake National Park. Montana: Fort Benton; Big Snowy Mts. Wyoming (Long, 1965a:548): Whirlwind Peak; *Bull Lake;* 3 mi. E Independence Rock; 3 mi. ESE Browns Peak. Colorado (Armstrong, 1972:88): Coal Creek Canyon; head Cucharas River. New Mexico (Findley, *et al.,* 1975:92): Agua Fria Mtn., 10,500 ft.; on a high mtn. W of Las Vegas and Santa Fe, on a flat ridge E of Pecos River; *Jemez Mts.;* Gallinas Mts.; Chama. Utah: 18 mi. SE Manila; *30 mi. N Fort Duchesne;* 28 mi. N Fruitland; 21 mi. N Escalante; 10 mi. E Marysvale; City Creek Canyon, Salt Lake City. Idaho: Pocatello; *Payette;* Cuddy Mtn.; *Weippe;* Bitterroot Valley.

Lepus americanus cascadensis Nelson

1907. *Lepus bairdi cascadensis* Nelson, Proc. Biol. Soc. Washington, 20:87, December 11, type from Roabs Ranch, near Hope, British Columbia.
1936. *Lepus americanus cascadensis,* Racey and Cowan, Report Prov. Mus. British Columbia, 1936:H18.

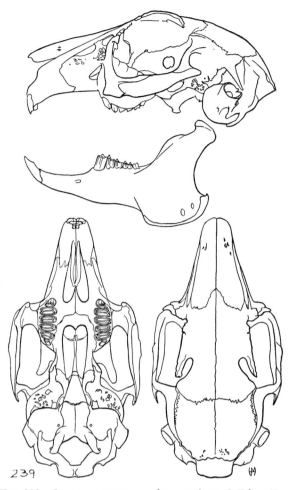

Fig. 239. *Lepus americanus tahoensis,* ½ mi. S Tahoe Tavern, Lake Tahoe, Placer Co., California, No. 37522 M.V.Z., ♂, X 1.

MARGINAL RECORDS.—British Columbia: Lillooet; Fairview-Keremeos summit. Washington: Lake Chelan; Trout Lake; *Vance; Mt. Rainier;* Entiat River, 20 mi. from mouth. British Columbia: *Chilliwack Lake; type locality; North Vancouver* (Cowan and Guiguet, 1965:104); *Whytecliffe; Sechelt Inlet; Brackendale;* Alta Lake.

Lepus americanus columbiensis Rhoads

1895. *Lepus americanus columbiensis* Rhoads, Proc. Acad. Nat. Sci. Philadelphia, 47:242, July 2, type from Vernon, British Columbia.

MARGINAL RECORDS.—British Columbia: Indianpoint Lake. Alberta: Jasper National Park; Banff National Park. British Columbia (Cowan and Guiguet, 1965:104): Radium Hot Springs; Rossland. Washington: Republic; *Moulson.* British Columbia (Cowan and Guiguet, 1965:104): *Bridesville summit; Inkaneep Creek; Vaseux Lake;* type locality; Sicamous.

Lepus americanus dalli Merriam

1900. *Lepus americanus dalli* Merriam, Proc. Washington Acad. Sci., 2:29, March 14, type from Nulato, Alaska.
1900. *Lepus americanus macfarlani* Merriam, Proc. Washington Acad. Sci., 2:30, March 14, type from Fort Anderson, near mouth Anderson River, Mackenzie. Inseparable from *L. a. dalli* according to Youngman (Nat. Mus. Canada Publ. Zool., 10:57, September 4, 1975).
1900. *Lepus saliens* Osgood, N. Amer. Fauna, 19:39, October 6, type from Caribou Crossing, between Lake Bennett and Lake Tagish, Yukon.
1907. ? *Lepus niediecki* Matschie, Niedieck's Kreuzfahrten im Beringmeer, p. 240, type locality Kasilof Lake, Kenai Peninsula, Alaska.

MARGINAL RECORDS.—Mackenzie: Fort Anderson, near mouth Anderson River; Fort Franklin; Fort Rae; Fort Resolution; Fort Smith. British Columbia (Cowan and Guiguet, 1965:105): Tupper Creek; Tetana Lake; Glenora on Stikine River; Atlin; Stonehouse Creek near jct. with Kelsall River. Alaska: Cordova; Mills Creek; Lake Clark; Naknek Lake (Schiller and Rausch, 1956:196); Bethel (Youngman, 1975:58); Yukon Delta; Anvik; Koyukuk; Noatak River; Upper John River; Arctic Village. Yukon: Old Crow River, at Timber Creek (Youngman, 1975:58).

Lepus americanus klamathensis Merriam

1899. *Lepus klamathensis* Merriam, N. Amer. Fauna, 16:100, October 28, type from head Wood River, near Fort Klamath, Klamath Co., Oregon.
1936. *Lepus americanus klamathensis,* V. Bailey, N. Amer. Fauna, 55:95, August 29.

MARGINAL RECORDS.—Oregon: Mt. Hood; mouth Davis Creek. California: vic. Fort Bidwell; 3000 ft., Rush Creek, 12 mi. N Weaverville. Oregon: *Estacada.*

Lepus americanus oregonus Orr

1934. *Lepus bairdii oregonus* Orr, Jour. Mamm., 15:152, May 15, type from 12 mi. S Canyon City, Oregon.
1942. *Lepus americanus oregonus,* Dalquest, Jour. Mamm., 23:179, June 3.

MARGINAL RECORDS.—Oregon: 22 mi. N Enterprise; *Wallowa Lake; summit Blue Mts.;* Ochoco National Forest, Harney Co.

Lepus americanus pallidus Cowan

1938. *Lepus americanus pallidus* Cowan, Jour. Mamm., 19:242, May 12, type from Chezacut Lake, Chilcotin River, British Columbia.

MARGINAL RECORDS.—British Columbia: 23 mi. N Hazelton; Nukko Lake (Cowan and Guiguet, 1965:106); Indianpoint Lake; Sicamous; *Falkland;* Bonaparte River, 5 days N Ashcroft; Hagensborg; Kimsquit; *Hazelton.*

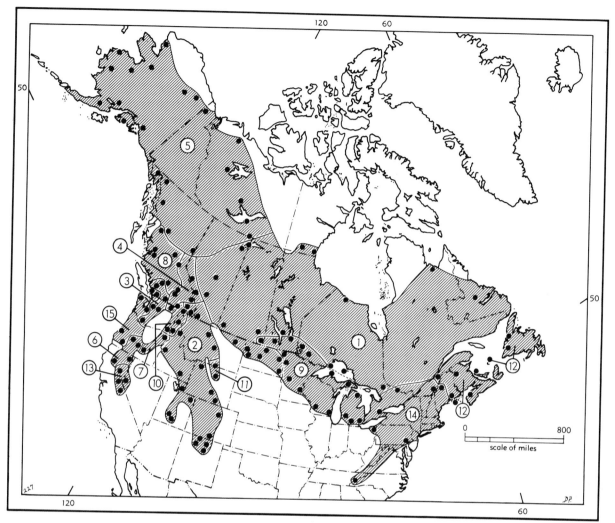

Map 227. *Lepus americanus.*

Guide to subspecies
1. *L. a. americanus*
2. *L. a. bairdii*
3. *L. a. cascadensis*
4. *L. a. columbiensis*
5. *L. a. dalli*
6. *L. a. klamathensis*
7. *L. a. oregonus*
8. *L. a. pallidus*
9. *L. a. phaeonotus*
10. *L. a. pineus*
11. *L. a. seclusus*
12. *L. a. struthopus*
13. *L. a. tahoensis*
14. *L. a. virginianus*
15. *L. a. washingtonii*

Lepus americanus phaeonotus J. A. Allen

1899. *Lepus americanus phaeonotus* J. A. Allen, Bull. Amer.
Mus. Nat. Hist., 12:11, March 4, type from Hallock, Kittson
Co., Minnesota.

MARGINAL RECORDS.—Manitoba: Selkirk Set-
tlement. Ontario: Lake of the Woods; Rainy Lake.
Michigan: Houghton; Chippewa County; Presque Isle
County; Wayne County; Jackson County; Allegan
County. Wisconsin (Jackson, 1961:110, Map 24, pre-
1900): Milwaukee County; Richland County. Min-
nesota: Blaine (Heaney and Birney, 1975:31); Moores
Lake; Warren; St. Vincent. Saskatchewan: Glen Ewen.
Manitoba: Carberry.

Lepus americanus pineus Dalquest

1942. *Lepus americanus pineus* Dalquest, Jour. Mamm.,
23:178, June 3, type from Cedar Mtn., Latah Co., Idaho.

MARGINAL RECORDS.—British Columbia: Trail;
Creston. Idaho: 5 mi. W Cocolalla; Troy. Washington:
Blue Mts., Columbia Co.; *Marcus.*

Lepus americanus seclusus Baker and Hankins

1950. *Lepus americanus seclusus* Baker and Hankins, Proc.
Biol. Soc. Washington, 63:63, May 25, type from 12 mi. E, 2
mi. N Shell, 7900 ft., Bighorn Mts., Big Horn Co., Wyoming.
1959. *Lepus americanus setzeri* Baker, Jour. Mann., 40:145,
February 20. [Proposed as a substitute name on the assump-

tion that *Lepus a. seclusus* Baker and Hankins 1950 was preoccupied by *Lepus timidus seclusus* Degerbøl 1940. Actually, Degerbøl's name was expressly for a variety, therefore an infrasubspecific name. Such a name does not preoccupy a species-group name (See Art. 45, International Code Zool. Nomenclature, November 6, 1961).]

MARGINAL RECORDS.—Wyoming (Long, 1965a:549): Medicine Wheel Ranch, 28 mi. E Lovell; *type locality; head Trappers Creek;* 9 mi. N, 9 mi. E Tensleep.

Lepus americanus struthopus Bangs

1898. *Lepus americanus struthopus* Bangs, Proc. Biol. Soc. Washington, 12:81, March 24, type from Digby, Nova Scotia.

MARGINAL RECORDS.—Newfoundland: Bay of Islands; Bay of St. George. Nova Scotia: type locality. Maine: Bucksport. Quebec: S of St. Lawrence River. New Brunswick: Andover. Prince Edward Island: Alberton. Quebec: Grosse Isle, Magdalen Islands.

Lepus americanus tahoensis Orr

1933. *Lepus washingtonii tahoensis* Orr, Jour. Mamm., 14:54, February 14, type from ½ mi. S Tahoe Tavern, Placer Co., California.
1942. [*Lepus americanus*] *tahoensis,* Dalquest, Jour. Mamm., 23:176, June 3.

MARGINAL RECORDS.—California: vicinity Mineral. Nevada: 350 yards NE jct. Nevada state line and N shore Lake Tahoe. California: Niagara Creek; Cisco.

Lepus americanus virginianus Harlan

1825. *Lepus virginianus* Harlan, Fauna Americana, p. 196. Type locality, Blue Mountains, northeast of Harrisburg, Pennsylvania.
1875. [*Lepus americanus*] var. *virginianus,* J. A. Allen, Proc. Boston Soc. Nat. Hist., 17:431, February 17.
1825. *Lepus wardii* Schinz, Das Thierreich . . . , 4:428, based on the snowshoe rabbit of the southern part of the United States (Warden, D. B., *in* Statistical, political, and historical account of the United States . . . , 1:233, 1819).
1845. *Lepus borealis* Schinz, Synopsis mammalium, 2:286–287. No type or type locality mentioned. "Habita in Montibus Virginia et Montibus Alleghanis."

MARGINAL RECORDS.—Quebec: Megantic County. Maine: *Greenville;* Sebec Lake; Mt. Desert Island. *Massachusetts: Concord; Middleboro. Rhode Island: Washington County.* Pennsylvania: type locality. Tennessee: White Rock. Ohio: Ashtabula County. Ontario: Holland River; Ottawa River.

Lepus americanus washingtonii Baird

1855. *Lepus washingtonii* Baird, Proc. Acad. Nat. Sci. Philadelphia, 7:333, April, type from Steilacoom, Washington.
1875. [*Lepus americanus*] var. *Washingtoni,* J. A. Allen, Proc. Boston Soc. Nat. Hist., 17:431, February 17.

MARGINAL RECORDS.—British Columbia: Point Grey, thence eastward along S bank Fraser River to Chilliwack. Washington: *Mt. Vernon;* Lake Kapowsin; *White Salmon.* Oregon: Drew; Florence; *Tillamook.* Washington: Sekiu River.

Lepus othus
Alaskan Hare

Total length, 565–690; tail, 53–104; hind foot, 147–189; ear from notch (dry), 75–78. Color brownish in summer, white in winter, but tips of ears always black. I have not seen specimens of *Lepus* from northeastern Alaska or areas in eastern Asia that would afford basis for deciding on specific versus subspecific status of the three nominal species, *L. timidus* of Eurasia, *L. arcticus* of Canada and Greenland, and *L. othus* of Alaska.

Lepus othus othus Merriam

1900. *Lepus othus* Merriam, Proc. Washington Acad. Sci., 2:28, March 14, type from St. Michael, Norton Sound, Alaska.

MARGINAL RECORDS.—Alaska: Kuparuk River; Killik River; Kotzebue Sound; mts. NW Nulato River; Akiak; 75 mi. below Bethel; Wales.—See addenda.

Lepus othus poadromus Merriam

1900. *Lepus poadromus* Merriam, Proc. Washington Acad. Sci., 2:29, March 14, type from Stepovak Bay, Alaska Peninsula, Alaska.
1936. *Lepus othus poadromus,* A. H. Howell, Jour. Mamm., 17:334, November 16.

MARGINAL RECORDS.—Alaska: Nushagak; Kawatna Bay, Shelikof Strait; *Cold Bay; Chiknik; type locality; Sand Point;* 15 mi. W Pavlof Mtn.

Lepus arcticus
Arctic Hare

Revised by A. H. Howell, Jour. Mamm., 17:315–337, November 16, 1936. For taxonomic status of technical names *arcticus* and *glacialis* see Rhoads, Amer. Nat., 30:234–235, March, 1896; Merriam, Science, n.s., 3:564–565, April 10, 1896; Rhoads, Science, n.s., 3:843–845, June 5, 1896; Merriam, Science, n.s., 3:845, June 5, 1896.

Total length, 480–678; tail, 34–80; hind foot, 132–174; ear from notch (dry), 70–84. Upper parts gray in summer in southern subspecies; in others white; in winter white in all subspecies, except black tips of ears. Weights of lean individuals reach 12 pounds.

Lepus arcticus andersoni Nelson

1934. *Lepus arcticus andersoni* Nelson, Proc. Biol. Soc. Washington, 47:85, March 8, type from Cape Barrow, Coronation Gulf, Mackenzie.

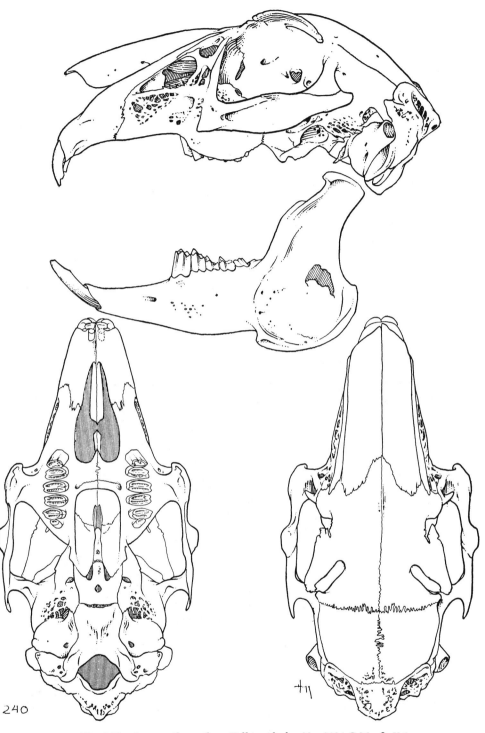

Fig. 240. *Lepus othus othus*, Teller, Alaska, No. 3194 C.M., ♀, X 1.

MARGINAL RECORDS.—Franklin: Cambridge Bay, Victoria Island. Keewatin: Back River, 125 mi. NW Baker Lake; Windy River Post (Harper, 1956: 13). Mackenzie: Lake Hanbury; Fort Rae; Fort Anderson.

Lepus arcticus arcticus Ross

1819. *Lepus arcticus* Ross, A voyage of discovery . . . , ed. 2, vol. 2, App. 4, p. 151, type locality Possession Bay, Bylot Island, lat. 73° 37′ N.

1819. *Lepus glacialis* Leach, *in* Ross, A voyage of discovery . . . , ed. 2, vol. 2, App. 4, p. 170, type locality same as for *Lepus arcticus* Ross.

MARGINAL RECORDS.—Franklin: type locality; *Egukjuak, 8 mi. E Pond Inlet, Baffin Island;* west coast Baffin Island, 67° 30'; Winter Island, Melville Peninsula; Repulse Bay, Melville Peninsula.

Lepus arcticus bangsii Rhoads

1896. *Lepus arcticus bangsii* Rhoads, Amer. Nat., 30:253 [= 236 of March issue], author's separates (preprints) published February 20, 1896, type from Codroy, Newfoundland.

MARGINAL RECORDS.—Labrador: Hopedale; Makkovik. Newfoundland: Saint Johns; type locality; Mt. St. Gregory.

Lepus arcticus banksicola Manning and Macpherson

1958. *Lepus arcticus banksicola* Manning and Macpherson, Arctic Inst. N. Amer., Tech. Pap. No. 2, p. 8, type locality on Banks Island at 71° 35' N, 123° 30' W.

MARGINAL RECORDS (Manning and Macpherson, 1958:19, 20).—Franklin (Banks Island): Cape Vesey Hamilton: 71° 13' N, 122° 35' W; 71° 30' N, 123° W; *71° 35' N, 123° 30' W; Sachs Harbour;* Cape Kellet; 10 mi. up Bernard River.

Lepus arcticus groenlandicus Rhoads

1896. *Lepus groenlandicus* Rhoads, Amer. Nat., 30:254 [= 237 of March issue], author's separates (preprints) issued February 20, 1896, type from Robertson Bay, NW Greenland.
1934. [*Lepus arcticus*] *groenlandicus*, Nelson, Proc. Biol. Soc. Washington, 47:83, March 8.
1930. *Lepus variabilis hyperboreus* Pedersen, Medd. om Grønland, 77:363, no type or type locality designated but name applied to hares of E Greenland in general vic. Scoresby Sound. Not *Lepus hyperboreus* Pallas, Zoogeographica Rosso-Asiatica, 1:152, 1811, a species of *Ochotona*.
1934. *Lepus arcticus persimilis* Nelson, Proc. Biol. Soc. Washington, 47:84, March 8, type from S side Clavering Island, east Greenland.

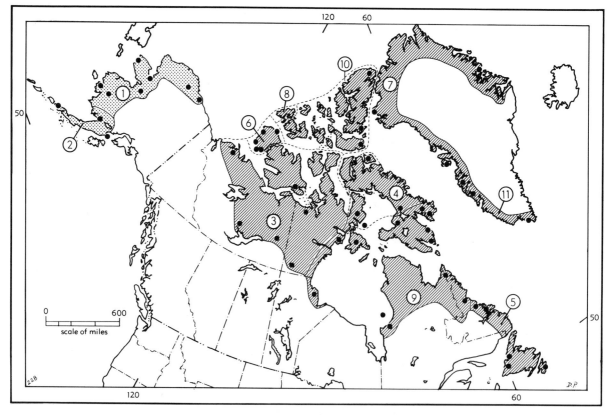

Map 228. *Lepus othus* and *Lepus arcticus.*

Guide to kinds	3. *L. a. andersoni*	6. *L. a. banksicola*	9. *L. a. labradorius*
1. *L. o. othus*	4. *L. a. arcticus*	7. *L. a. groenlandicus*	10. *L. a. monstrabilis*
2. *L. o. poadromus*	5. *L. a. bangsii*	8. *L. a. hubbardi*	11. *L. a. porsildi*

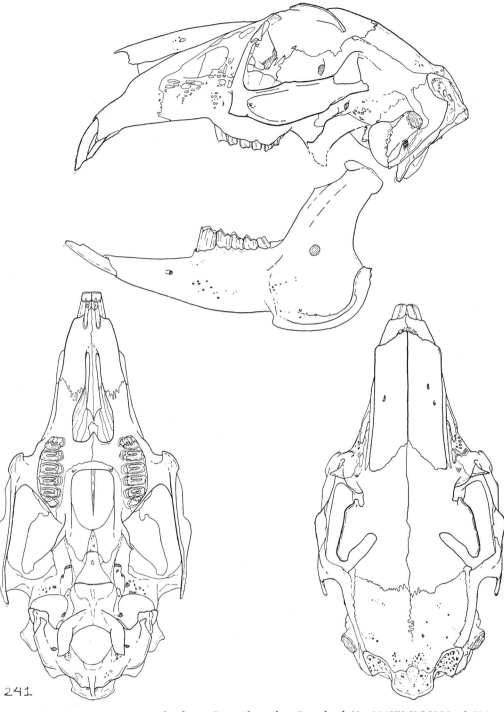

Fig. 241. *Lepus arcticus groenlandicus*, Cape Alexander, Greenland, No. 114850 U.S.N.M., ♂, X 1.

MARGINAL RECORDS.—Greenland: Cape Alexander; on E coast to Francis Joseph Fiord; on W coast to *Nugsuak Peninsula*; Disko Island; Holsteinsborg.

Bay, 5 mi. NE Mould Bay Station, Prince Patrick Island, Franklin. Known only from Prince Patrick Island.

Lepus arcticus hubbardi Handley

1952. *Lepus arcticus hubbardi* Handley, Proc. Biol. Soc. Washington, 65:199, November 5, type from near Cherie

Lepus arcticus labradorius Miller

1899. *Lepus labradorius* Miller, Proc. Biol. Soc. Washington, 13:39, May 29, type from Fort Chimo, Quebec.

1924. *Lepus arcticus labradorius,* G. M. Allen and Copeland, Jour. Mamm., 5:12, February 9.
1902. *Lepus arcticus canus* Preble, N. Amer. Fauna, 22:59, October 31, type from Hubbart Point, W coast Hudson Bay, Keewatin.

MARGINAL RECORDS.—Franklin: Pangnirtung Fiord; Nunata, Kingua Fiord; Cumberland Sound, Blacklead Island; Weddell Harbor, Frobisher Bay. Labrador: Ramah; Solomons Island, near Davis Inlet. Quebec: *type locality;* Great Whale River, Hudson Bay; Belcher Islands. Manitoba: Fort Churchill; *Hubbart Point.* Keewatin: Cape Fullerton; Southampton Island. Franklin: *Cape Dorset;* Camp Kungovik, west coast Baffin Island, 65° 35′ N lat.; *Nettilling Fiord.*

Lepus arcticus monstrabilis Nelson

1934. *Lepus arcticus monstrabilis* Nelson, Proc. Biol. Soc. Washington, 47:85, March 8, type from Buchanan Bay, Ellesmere Island, Franklin.

MARGINAL RECORDS.—Franklin: Cape Sheridan; Craig Harbor; Dundas Harbor, Devon Island.

Lepus arcticus porsildi Nelson

1934. *Lepus arcticus porsildi* Nelson, Proc. Biol. Soc. Washington, 47:83, March 8, type from near Julianehaab, lat. 61° 20′ N, Greenland.

MARGINAL RECORDS.—Greenland: Sukkertoppen; *Neria, lat. 61° 36′ N;* lat. 60° 42′ N.

Lepus townsendii
White-tailed Jack Rabbit

X ⅓

Total length, 565–655; tail, 66–112; hind foot, 145–172; ear from notch (dry), 96–113. Upper

parts grayish brown; tail all white or with dusky or buffy mid-dorsal stripe, which does not extend onto back; white in winter in northern parts of its range. Two adult males weighed 6½ and 5½ pounds (Orr, 1940:43).

This species has been introduced by modern man into Wisconsin and is now well established in appropriate habitat throughout that state except in the counties, or parts of some of those counties, bordering Lake Superior and the Upper Peninsula of Michigan (see Jackson, 1961:104–108).

Lepus townsendii campanius Hollister

1837. *Lepus campestris* Bachman, Jour. Acad. Nat. Sci. Philadelphia, 7:349, Type locality, plains of the Saskatchewan, probably near Carlton House. Not *Lepus cuniculus campestris* Meyer, 1790.
1915. *Lepus townsendii campanius* Hollister, Proc. Biol. Soc. Washington, 28:70, March 12, a renaming of *L. campestris* Bachman.

MARGINAL RECORDS.—Saskatchewan (Beck, 1958:23): Prince Albert National Park; Dafoe. Manitoba (Soper, 1961:185): Gilbert Plains; Dog Lake; Whitemouth. Ontario: Rainy River. Minnesota: Polk County; Otter Tail County; Sherburne County; Washington County. Illinois: Blanding, 6 mi. WNW Hanover. Iowa: Mediapolis (Bowles, 1975:59). Missouri: near Marshall (Enders, 1932:120, sight record only). Nebraska: Weeping Water (Jones, 1964c:114). Kansas: Red Fork, 60 mi. W Fort Riley; Greensburg. Colorado (Armstrong, 1972:91, 92): Boyero; 20 mi. W La Veta. New Mexico: near Taos; Hopewell. Colorado (Armstrong, 1972:91, 92): Antonito; U.S. Hwy. 160 to Beaver Creek Reservoir; Cochetopa Park, 26 mi. SE Gunnison; Salida; 6 mi. S, 4 mi. W Leadville; Mill City; near utility entrance, Rocky Mountain National Park; Coalmont. Wyoming (Long, 1965a:551, unless otherwise noted): Spring Creek (Hall and Kelson, 1959:280); Farson; Big Piney; head Glen Creek, Yellowstone National Park (Hall and Kelson, 1959:280). Montana (Hoffmann, *et al.,* 1969:584, as *L. townsendii* only); 23 mi. S Dillon; *Bannack; 5 mi. S Stevensville;* Springer Gulch (near Hamilton); NW edge Missoula. Alberta: Great Plains region.

Lepus townsendii townsendii Bachman

1839. *Lepus townsendii* Bachman, Jour. Acad. Nat. Sci. Philadelphia, 8(pt. 1):90, pl. 2, type from Fort Walla Walla, near present town of Wallula, Walla Walla Co., Washington.
1904. *Lepus campestris sierrae* Merriam, Proc. Biol. Soc. Washington, 17:132, July 14, type from 7800 ft., Hope Valley, Alpine Co., California. Regarded as inseparable from *L. t. townsendii* by Orr, Occas. Pap., California Acad. Sci., 19:42, May 25, 1940.

MARGINAL RECORDS.—British Columbia: Okanagan Falls. Idaho: Rathdrum Prairie; Lemhi

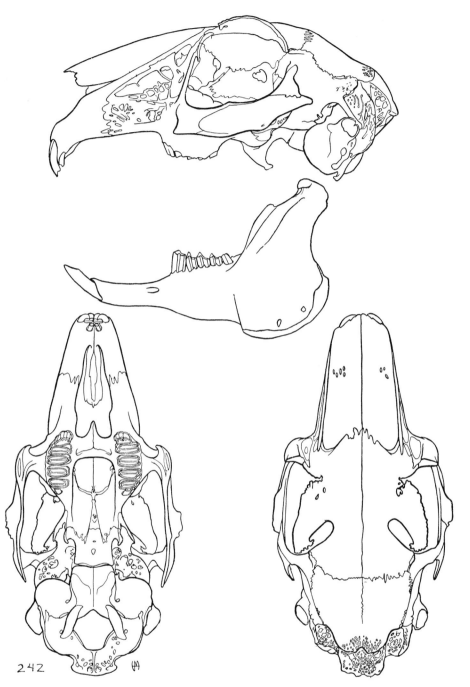

Fig. 242. *Lepus townsendii townsendii*, N end Ruby Valley, E base Ruby Mts., Elko Co., Nevada, No. 45746 K.U., ♀, X 1.

River; Teton Basin. Wyoming (Long, 1965a:552): *"near" Kelley;* Jackson; Hams Fork; Henrys Fork. Colorado (Armstrong, 1972:92, unless otherwise noted): 23 mi. N, 34 mi. W Maybell; near Hayden; Grand Lake; Dillon; Crested Butte; 2 mi. SW Doyleville (Sparks, 1968:325, as *L. townsendii* only); Coventry. Utah: Kanab; 1 mi. S Kolob Reservoir, 8050 ft. (Stock, 1970:430). Nevada: Hamilton; Desatoya Mts.; Santa Rosa Mts. California: Parker Creek, 6300 ft., Warner Mts. Nevada: 3 mi. S Mt. Rose, 8600 ft.; Lapon Canyon, 8900 ft.; Mt. Grant; Mt. Magruder. California: Tuolumne Meadows; Woodfords; Tahoe City. Oregon: Upper Klamath Lake (Olterman and Verts, 1972:25, as *L. townsendii* only); Antelope. Washington: Manson. British Columbia: *White Lake* (Cowan and Guiguet, 1965:100).

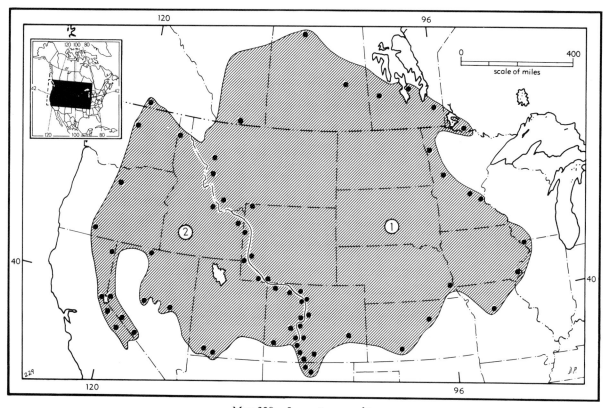

Map 229. *Lepus townsendii.*

1. *L. t. campanius*   2. *L. t. townsendii*

Lepus californicus
Black-tailed Jack Rabbit

×⅕

Total length, 465–630; tail, 50–112; hind foot, 112–145; ear from notch (dry), 99–131. Upper

parts gray to blackish; tail with black mid-dorsal stripe extending onto back. On the tableland of México and in the southwestern United States where this species occurs together with the white-sided jack rabbits, *L. californicus* can be recognized by the terminal black patch on the outside of each ear and by the less extensive area of white on the flank. To the eastward, in Tamaulipas, where only the black-tailed jack rabbit occurs, it too has extensively white flanks, and some individuals lack the terminal black patch on the ear.

A certain means for distinguishing the skulls of the black-tailed jack rabbit from those of all the white-sided jack rabbits has not yet been discovered. The same is true of the skulls of the white-tailed jack rabbit and the black-tailed jack rabbit in the Great Basin region of Nevada. The skulls, at least of adults, of these two species in the region east of the Rocky Mountains can be readily distinguished by the pattern of infolding of the enamel on the front of the 1st upper incisor teeth; *L. townsendii* has a simple groove on the anterior face of the tooth and *L. californicus*, east of the Rocky Mountains, has a bifurcation, or even trifurcation, of the infold that can readily be seen by examining the occlusal surface of the incisor.

Lepus californicus altamirae Nelson

1904. *Lepus merriami altamirae* Nelson, Proc. Biol. Soc. Washington, 17:109, May 18, type from Alta Mira, Tamaulipas.
1951. *Lepus californicus altamirae*, Hall, Univ. Kansas Publ., Mus. Nat. Hist., 5:45, October 1.

MARGINAL RECORDS.—Tamaulipas: 3 mi. N Soto la Marina (Alvarez, 1963:420); type locality; *2 mi. NW Soto la Marina* (Alvarez, 1963:420).

Lepus californicus asellus Miller

1899. *Lepus asellus* Miller, Proc. Acad. Nat. Sci. Philadelphia, 51:380, September 29, type from San Luis Potosí, San Luis Potosí.

1909. *Lepus californicus asellus*, Nelson, N. Amer. Fauna, 29:150, August 31.

MARGINAL RECORDS.—Coahuila: 3 mi. S, 3 mi. E Muralla, 3800 ft.; Jaral; *15 mi. N, 41 mi. W Saltillo.* Nuevo León: Miquihuana. San Luis Potosí: Ciudad del Maíz; Río Verde. Aguascalientes: Chicalote. Zacatecas: Valparaíso.

Lepus californicus bennettii Gray

1843. *Lepus bennettii* Gray, The zoology of the voyage of H.M.S. *Sulphur* . . . , p. 35, pl. 14, April, type from San Diego, San Diego Co., California.
1909. *Lepus californicus bennetti*, Nelson, N. Amer. Fauna, 29:136, August 31.

MARGINAL RECORDS.—California: Mt. Piños; Arroyo Seco, Pasadena; San Felipe Valley; Jacumba.

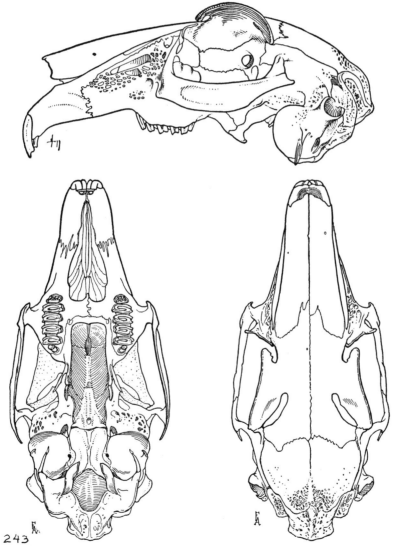

Fig. 243. *Lepus californicus texianus*, E base Burro Mesa, 3500 ft., Big Bend, Brewster Co., Texas, No. 81694 M.V.Z., ♂, X 1.

Baja California: San Quintín. Northward along coast at least to California: Montalvo.

Lepus californicus californicus Gray

1837. *Lepus californica* Gray, Charlesworth's Mag. Nat. Hist., 1:586, type from "St. Antoine," California (probably on coastal slope of mts. near Mission of San Antonio, Jolon, Monterey Co.).
1926. *Lepus californicus vigilax* Dice, Occas. Pap. Mus. Zool., Univ. Michigan, 166:11, February 11, type from Balls Ferry, Shasta Co., California. Regarded as identical with *californicus* by Orr, Occas. Pap. California Acad. Sci., 19:67, May 25, 1940.

MARGINAL RECORDS.—Oregon: Drain; Grants Pass. California: Callahan, Scott River; 3300 ft., Lymans, NW of Lyonsville; Dry Creek, Oroville–Chico Road; Snelling; Hernandez; Morro; *Carmel Point; Bolinas Bay; Freestone; Sherwood; Ferndale;* 3 mi. W Arcata. Oregon: Rogue River Valley.

Lepus californicus curti Hall

1951. *Lepus californicus curti* Hall, Univ. Kansas Publ., Mus. Nat. Hist., 5:42, October 1, type from island 88 mi. S, 10 mi. W Matamoros, Tamaulipas.

MARGINAL RECORDS.—Tamaulipas: type locality; *90 mi. S, 10 mi. W Matamoros* (Alvarez, 1963:420).

Lepus californicus deserticola Mearns

1896. *Lepus texianus deserticola* Mearns, Proc. U.S. Nat. Mus., 18:564, June 24, type from western edge Colorado Desert, base Coast Range Mts., Imperial Co., California.
1909. *Lepus californicus deserticola*, Nelson, N. Amer. Fauna, 29:137, August 31.
1932. *Lepus californicus depressus* Hall and Witlow, Proc. Biol. Soc. Washington, 45:71, April 2, type from ½ mi. S Pocatello, Bannock Co., Idaho. (Regarded as inseparable from *L. c. deserticola* by Davis, The Recent mammals of Idaho, Caxton Printers, Caldwell, Idaho, p. 359, April 5, 1939.)

MARGINAL RECORDS.—Montana (Hoffmann, *et al.*, 1969:585, as *L. californicus* only): Grasshopper Creek at Bannack; Sweetwater Creek, 21 mi. SE Dillon. Idaho: Blackfoot. Utah: Ogden; Provo; Loa. Arizona: San Francisco Mtn.; Fort Whipple; Phoenix; Rancho Bonito, Abra Valley. Sonora: El Doctor. Baja California: Calamahué [= Calamajue]; Esperanza Canyon. California: Coyote Wells; Kenworthy; Victorville; Farrington Ranch; 5 mi. SW Lone Pine; head Silver Canyon, 10,000 ft.; Mono Mills; near Woodfords, 5600 ft. Nevada: Sutcliffe; ¾ mi. S Sulphur. Idaho: 6 mi. S Murphy; Boise River; Sawtooth National Forest; Arco.

Lepus californicus eremicus J. A. Allen

1894. *Lepus texianus eremicus* J. A. Allen, Bull. Amer. Mus. Nat. Hist., 6:347, December 7, type from Fairbank, Cochise Co., Arizona.

1909. *Lepus californicus eremicus,* Nelson, N. Amer. Fauna, 29:140, August 31.

MARGINAL RECORDS.—Arizona: Casa Grande; Fort Bowie; 2 mi. E Portal. Chihuahua (Anderson, 1972:265): 2 mi. N San Francisco, 5100 ft.; Vuelta de Alamos, 4100 ft.; Colonia García. Sonora: Hermosillo; La Libertad; Agua Dulce (of Sonora, not of Arizona).

Lepus californicus festinus Nelson

1904. *Lepus festinus* Nelson, Proc. Biol. Soc. Washington, 17:108, May 18, type from Irolo, Hidalgo.
1909. *Lepus californicus festinus* Nelson, N. Amer. Fauna, 29:151, August 31.

MARGINAL RECORDS.—Hidalgo: Zimapán; *Tulancingo;* type locality. Querétaro: Tequisquiapam.

Lepus californicus magdalenae Nelson

1907. *Lepus californicus magdalenae* Nelson, Proc. Biol. Soc. Washington, 20:81, July 22, type from Magdalena Island, Baja California.

MARGINAL RECORDS.—Baja California: type locality; Margarita Island.

Lepus californicus martirensis Stowell

1895. *Lepus martirensis* Stowell, Proc. California Acad. Sci., ser. 2, 5:51, May 28, type from San Pedro Mártir Mts., Baja California.
1909. *Lepus californicus martirensis,* Nelson, N. Amer. Fauna, 29:152, August 31.

MARGINAL RECORDS.—Baja California: La Huerta; Calamahué [= Calamajue]; San Bruno; Rancho San José; San Simón.

Lepus californicus melanotis Mearns

1890. *Lepus melanotis* Mearns, Bull. Amer. Mus. Nat. Hist., 2:297, February 21, type from Independence, Kansas.
1909. *Lepus californicus melanotis,* Nelson, N. Amer. Fauna, 29:146, August 31.

MARGINAL RECORDS.—South Dakota: Lyman County; T. 101 N, R. 62 E, in Davison Co. Nebraska: Oakland. Kansas: near Doniphan Lake. Missouri: Saline County; 5 mi. E Rockbridge. Arkansas: Lonsdale. Texas: near Boston (Packard, 1963:107); 3 mi. W Carthage (*op. cit.*:108); 5 mi. SW Center (*ibid.*); 6 mi. W Newton (*ibid.*, sight record only); Brazos County; Golinda; Washburn. New Mexico: Santa Rosa; vic. Cimarron. Colorado (Armstrong, 1972:93, 94): 4 mi. W La Veta; Custer County; Semper; 2 mi. N, ¼ mi. W Fort Collins. Wyoming (Long, 1965a:552): 1½ mi. W Horse Creek; *18 mi. N* [? *Laramie*]; 1 mi. N, 3 mi. E Orin. South Dakota: 20 mi. S, 25 mi. E Rapid City.

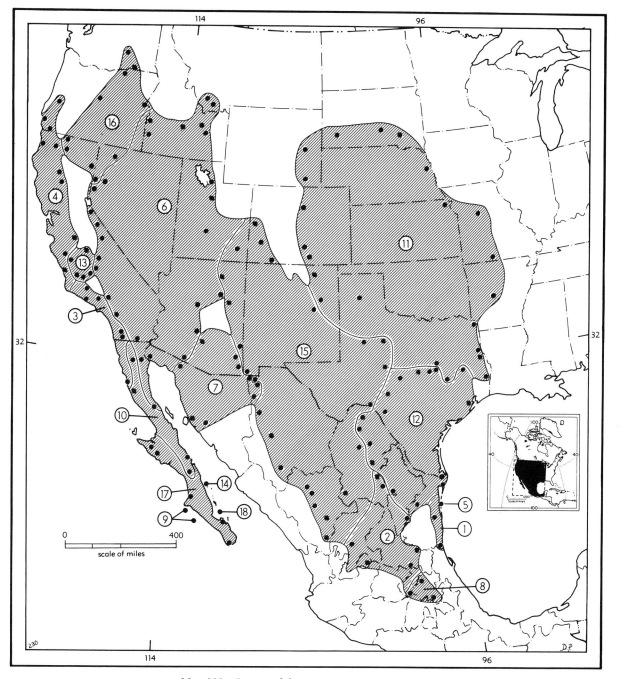

Map 230. *Lepus californicus* and *Lepus insularis*.

Guide to kinds	4. *L. c. californicus*	9. *L. c. magdalenae*	14. *L. c. sheldoni*
1. *L. c. altamirae*	5. *L. c. curti*	10. *L. c. martirensis*	15. *L. c. texianus*
2. *L. c. asellus*	6. *L. c. deserticola*	11. *L. c. melanotis*	16. *L. c. wallawalla*
3. *L. c. bennettii*	7. *L. c. eremicus*	12. *L. c. merriami*	17. *L. c. xanti*
	8. *L. c. festinus*	13. *L. c. richardsonii*	18. *L. insularis*

Lepus californicus merriami Mearns

1896. *Lepus merriami* Mearns, Preliminary diagnoses of new mammals from the Mexican border of the United States, p. 2, March 25 (preprint of Proc. U.S. Nat. Mus., 18:444, May 23, 1896), type from Fort Clark, Kinney Co., Texas.

1909. *Lepus californicus merriami*, Nelson, N. Amer. Fauna, 29:148, August 31.

MARGINAL RECORDS.—Texas: Lott; Antioch; Houston. Tamaulipas: Matamoros; 33 mi. S Washington Beach (Alvarez, 1963:421); 12 mi. NW San Carlos, 1300 ft. (*ibid.*). Nuevo León: Santa Catarina. Coahuila: Monclova; Sabinas; 25 mi. S, 8 mi. E Piedra Blanca; 16 mi. N, 21 mi. E Piedra Blanca. Texas (Davis, 1966:221, as *L. californicus* only, unless otherwise noted): Kinney County; Edwards County; Mason (Hall and Kelson, 1959:284); Burnet County; Bell County.

Lepus californicus richardsonii Bachman

1839. *Lepus richardsonii* Bachman, Jour. Acad. Nat. Sci. Philadelphia, 8(pt. 1):88, type from California (exact locality unknown, but probably on interior slope of mts. near Jolon, Monterey Co.).

1909. *Lepus californicus richardsoni*, Nelson, N. Amer. Fauna, 29:133, August 31.

1904. *Lepus tularensis* Merriam, Proc. Biol. Soc. Washington, 17:136, July 14, type from Alila [= Earlimart], Tulare Co., California.

MARGINAL RECORDS.—California: Minkler; Thompson Valley, Walker Basin; Kern Lake Basin; Carrizo Plains, 7 mi. SE Simmler; *2 mi. E Bryson*; Jolon.

Lepus californicus sheldoni Burt

1933. *Lepus californicus sheldoni* Burt, Proc. Biol. Soc. Washington, 46:37, February 20, type from Carmen Island [lat. 26° N, long. 111° 12' W, Gulf of California], Baja California. Known from type locality only.

Lepus californicus texianus Waterhouse

1848. *Lepus texianus* Waterhouse, A natural history of the Mammalia, 2:136, type locality unknown, but probably in western Texas.

1909. *Lepus californicus texianus*, Nelson, N. Amer. Fauna, 29:142, August 31.

1896. *Lepus texianus griseus* Mearns, Proc. U.S. Nat. Mus., 18:562, June 24, type from Fort Hancock, Texas.

1903. *Lepus (Macrotolagus) texianus micropus* J. A. Allen, Bull. Amer. Mus. Nat. Hist., 19:605, November 12, type from Río de las Bocas, northwestern Durango.

MARGINAL RECORDS.—Colorado (Armstrong, 1972:94): 2 mi. E Rim Rock Drive, Colorado National Monument; Norwood; Bayfield. New Mexico: Roswell. Texas (Davis, 1966:221, as *L. californicus* only, unless otherwise noted): Borden County; Fisher County; Tom Green County; Comstock (Hall and Kelson, 1959:285). Coahuila: 18 mi. S, 14 mi. E Tanque Alvarez; 8 mi. SE San Pedro de los Colonias, 3700 ft.; *3 mi. SE Torreón*. Durango: 2 km. SW Mezquital, 1450 m. (Crossin, *et al.*, 1973:197, as *L. californicus* only); 1 mi. N Chorro, 6450 ft. (Baker and Greer, 1962:77); Hda. Atotonilco, 6880 ft. (*ibid.*); Río Sestín; Río del Bocas (Nelson, 1909:146). Chihuahua (Anderson, 1972:266): La Unión, 10 km. N Guachochic, 8400 ft.; 2 mi. W Miñaca; 4 mi. SE Las Varas, 7800 ft.; White Water. New Mexico: Guadalupe Ranch. Arizona (Cockrum, 1961:71): York, 3500 ft.; Holbrook; *20 mi. W Winslow*; Painted Desert, Little Colorado River. Utah: Abajo (Blue Mts.).

Lepus californicus wallawalla Merriam

1904. *Lepus texianus wallawalla* Merriam, Proc. Biol. Soc. Washington, 17:137, July 14, type from Touchet, Plains of the Columbia, Walla Walla Co., Washington.

1909. *Lepus californicus wallawalla*, Nelson, N. Amer. Fauna, 29:132, August 31.

MARGINAL RECORDS.—Washington: Moses Coulee; type locality. Oregon: Ontario. Nevada: 4100 ft., Quinn River Crossing; 4200 ft., 4½ mi. W Flanigan. California: 5000 ft., 7 mi. E Ravendale; 3600 ft., 1 mi. SE Weed; Hornbrook. Oregon: Hay Creek; Willow Junction.

Lepus californicus xanti Thomas

1898. *Lepus californicus Xanti* Thomas, Ann. Mag. Nat. Hist., ser. 7, 1:45, January, type from Santa Anita, Baja California.

MARGINAL RECORDS.—Baja California (southern part of Peninsula): Santa Clara Mts., southward around range of *L. c. martirensis* to and down east coast; La Paz; Cape St. Lucas; San Jorgé; 20 mi. W San Ignacio.

Lepus insularis Bryant
Black Jack Rabbit

1891. *Lepus insularis* Bryant, Proc. California Acad. Sci., ser. 2, 3:92, April 23, type from Espíritu Santo Island, Gulf of California, Baja California. Known from Espíritu Santo Island only.

1895. *Lepus edwardsi* St.-Loup, Bull. Mus. Hist. Nat., Paris, 1:5, type from Espíritu Santo Island, Gulf of California, Baja California.

Total length, 574; tail, 96; hind foot, 121; ear from notch (dry), 105. This insular species, clearly a close relative of *L. californicus* of the adjacent peninsula of Baja California, is mainly glossy black on upper parts but grizzled and suffused on sides of back and body, and in some specimens on head, with dark buffy or reddish brown; underparts dark cinnamon buffy or dusky brown; ears and sides of head grayish dusky; jugals heavier than in *L. californicus* of the adjacent peninsula of Baja California.

Lepus callotis
White-sided Jack Rabbit

Total length, 432–598; tail, 47–92; length of hind foot, 118–141; ear from notch (dry), 108–149. Upper parts dark, slightly pinkish, buff heavily washed with black; backs of ears mainly white without terminal patch of black; flanks white; rump iron gray.

Although no single measurement or pair of measurements used as a ratio definitely distinguishes all specimens of *L. callotis* from all

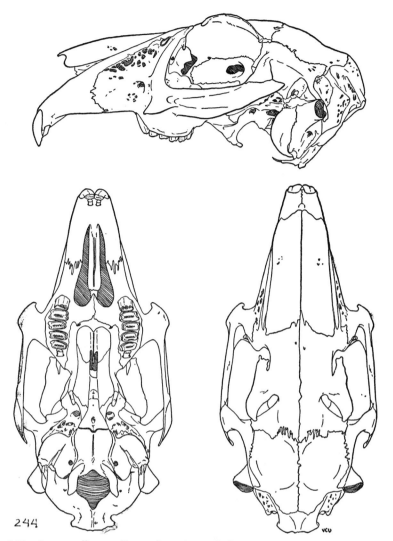

Fig. 244. *Lepus callotis callotis*, 3½ mi. S Tecolotlán, Jalisco, No. 31842 K.U., ♀, X 1.

specimens of *L. californicus*, a number of differences have been noted, and skulls can be identified to species by considering all the differences listed below. In comparison with *L. californicus*, *L. callotis* has a higher nasal aperture, smaller, more inclined supraorbital surface, more ventral placement of posteriormost point of skull, and consequently more inclined parietal, lesser breadth across auditory bullae, less compact appearance of skull in posterior view, more prominent supraorbital ridges in posterior view, smaller auditory meatus, deeper rostrum, smaller bullae, and less constriction in basioccipital (Anderson and Gaunt, 1962:2, 3).

It is suspected that the black-tailed jack rabbit, *L. californicus*, is replacing *L. callotis* over the areas of its range that have been overgrazed or otherwise altered by ranching practices (Baker and Greer, 1962:78).

Lepus callotis callotis Wagler

1830. *Lepus callotis* Wagler, Natürliches System der Amphibien . . . , p. 23, August, type from southern end of Mexican Tableland.
1830. *Lepus mexicanus* Lichtenstein, Abh. k. Akad. Wiss., Berlin, p. 101, type from Mexico (southern end of Mexican Tableland).
1833. *Lepus nigricaudatus* Bennett, Proc. Zool. Soc. London, p. 41, May 17, type from "that part of California which adjoins to Mexico" (probably southwestern part of Mexican Tableland).

MARGINAL RECORDS (Anderson and Gaunt, 1962:10–12, unless otherwise noted).—Durango: SE end of Laguna de Santiaguilla, "Santa Cruz." San Luis

Potosí: 4½ mi. SW Herradura, 7200 ft.; Arenal, 7000 ft. Hidalgo: Tulancingo, 7200 ft. Puebla: Tehuacán, 5400 ft. Oaxaca: Oaxaca City, 5200 ft.; Tlapancingo, 5200 ft. Guerrero: 6 mi. W Colotlipa, 2700 ft.; 5 mi. E Omiltemi, 6800 ft. Michoacán: ½ mi. S Las Cruces [= 24 mi. S Apatzingan], 1150 ft. Jalisco: Artenkiki [= Atenquiqui]; 11 mi. NW Ayutla; Reyes (Hall and Kelson, 1959:285); Eztatlán, 3500 ft. Durango: Durango City; 2 mi. S Sauz, 6200 ft.

Lepus callotis gaillardi Mearns

1896. *Lepus gaillardi* Mearns, Proc. U.S. Nat. Mus., 18:560, June 24, type from West Fork of Playas Valley, near Monument 63, Mexican boundary line.
1962. *Lepus callotis gaillardi*, Anderson and Gaunt, Amer. Mus. Novit., 2088:5, May 24.
1903. *Lepus (Microtolagus* [*sic*]) *gaillardi battyi* J. A. Allen, Bull. Amer. Mus. Nat. Hist., 19:607, November 12, type from Rancho Santuario, northwestern Durango.

MARGINAL RECORDS (Anderson and Gaunt, 1962:8, 9, unless otherwise noted).—New Mexico (Bo-

gan and Jones, 1975:47): ½ km. N Cloverdale; S end of W side Playas Valley, 4600 ft. Chihuahua: Arroyo del Nido, 30 mi. SW Gallego. Durango: 7½ mi. SE Torreón de Canas; Rancho Santuario; Río Campo. Chihuahua: 2 mi. W Miñaca, 6900 ft.; Rancho San Ignacio, 4 mi. S, 1 mi. W Santo Tomás; Colonia Juárez, 5000 ft.

Lepus flavigularis Wagner
Tehuantepec Jack Rabbit

1844. *Lepus callotis* var. γ *flavigularis* Wagner, *in* Schreber, Die Säugthiere . . . , Suppl., 4:106, type from México (probably near Tehuantepec City, Oaxaca).
1909. *Lepus flavigularis*, Nelson, N. Amer. Fauna, 29:125, August 31.

Total length, 595; tail, 77; hind foot, 133; ear from notch (dry), 112. Upper parts bright ochraceous buff strongly washed with black; ears entirely buff; nape with black stripe extending back from base of each ear and median stripe of buff; flanks and underparts of body white; rump iron

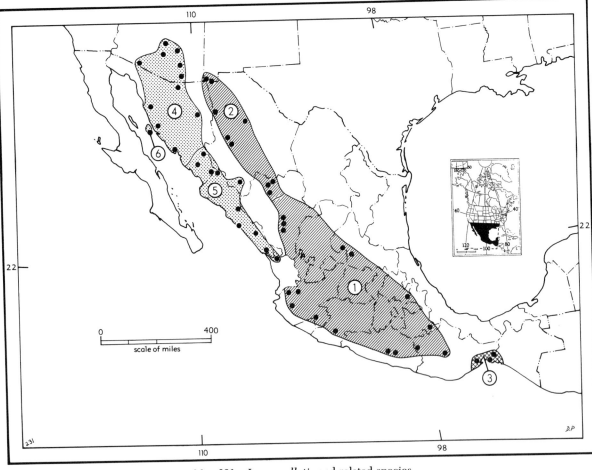

Map 231. *Lepus callotis* and related species.

1. *L. callotis callotis*   3. *L. flavigularis*   5. *L. alleni palitans*
2. *L. callotis gaillardi*   4. *L. alleni alleni*   6. *L. alleni tiburonensis*

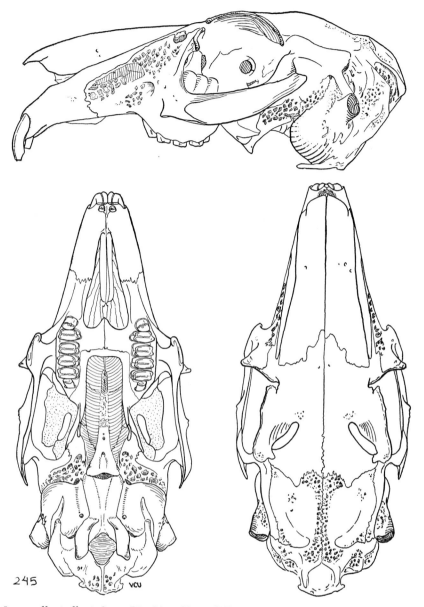

Fig. 245. *Lepus alleni alleni*, Santa Rita Mts., 30 mi. S Tucson, Pima Co., Arizona, No. 8621 K.U., ♂, X 1.

gray; tympanic bullae smaller than in any other *Lepus* of México.

MARGINAL RECORDS.—Oaxaca: Santa Efigenia; San Mateo del Mar; Huilotepec.

Lepus alleni
Antelope Jack Rabbit

Total length, 553–670; tail, 48–76; hind foot, 127–150; ear from notch, in flesh, 138–173. Top and sides of head creamy buff, slightly washed on top with black; tail white except for mid-dorsal line of black extending onto rump; sides of shoulders, flanks, sides of abdomen, rump, and outside of hind legs uniform iron gray. The average weight of 61 adult males from Arizona was 8.2 pounds.

Lepus alleni alleni Mearns

1890. *Lepus alleni* Mearns, Bull. Amer. Mus. Nat. Hist., 2:294, February 21, type from Rillito, on Southern Pacific Railroad, Pima Co., Arizona.

MARGINAL RECORDS.—Arizona: 6½ mi. S Apache Junction (Woolsey, 1959:250); 35 mi. E Florence (Cockrum, 1961:67); Cascabel; *7 mi. N Fort Huachuca* (Cockrum, 1961:68); near Fort Huachuca (*ibid.*). Sonora: Cerro Blanco; Oputo; Batamotal; La Libertad Ranch; Picu Pass. Arizona: Pinta Sands, sec. 19, T. 15 S, R. 11 W (Simmons, 1966:122); Casa Grande; *Queen Creek* (Cockrum, 1961:67).

Lepus alleni palitans Bangs

1900. *Lepus (Macrotolagus) alleni palitans* Bangs, Proc. New England Zool. Club, 1:85, February 23, type from Aguacaliente [*sic*], about 40 mi. SE Mazatlán, Sinaloa.

MARGINAL RECORDS.—Sonora: near San Bernardo on Río Mayo on Sonora side of Sonora–Chihuahua boundary; *Alamos;* Guirocoba. Sinaloa: 1 km. S El Cajón, 1800 ft. (Armstrong and Jones, 1971:756). Chihuahua: Temoris, 440 ft. (Anderson, 1972:265). Sinaloa (Armstrong and Jones, 1971:756): 13 mi. ESE Badiraguato, 800 ft.; San Ignacio. Nayarit: Acaponeta. Sinaloa: Esquinapa; 6 mi. N El Dorado (Armstrong and Jones, 1971:756). Sonora: near Navajoa.

Lepus alleni tiburonensis Townsend

1912. *Lepus alleni tiburonensis* Townsend, Bull. Amer. Mus. Nat. Hist., 31:120, June 14, type from Tiburón Island, Gulf of California, Sonora. Known from Tiburón Island only.

Lepus capensis
Cape Hare

Total length, 640–700; tail, 70–100; hind foot, 130–150; ear from notch (dry), 79–100; weight 6⅝ to 10 pounds. Upper parts tawny, mixed with blackish hairs on back; underparts white including under side of tail; upper side of tail and terminal patch at distal end of outside of ears black; upper side of feet tawny like sides (not white or whitish). This is an introduced species.

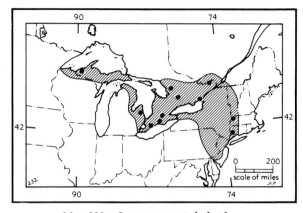

Map 232. *Lepus capensis hybridus.*

Lepus capensis hybridus Desmarest

1822. *Lepus hybridus* Desmarest, Mammalogie . . . , pt. 1, p. 349, *in* Encyclopédie méthodique . . . , name based on "Russac" of Pallas, Novae species quadrupedum e glirium ordine . . . , p. 5, 1778. Type locality, central Russia.

The name combination *Lepus capensis hybridus* is used above because Petter (1961:30–40) concluded that *Lepus europaeus* Pallas, 1778, including its subspecies *hybridus*, and the earlier named *Lepus capensis* Linnaeus, 1758 are conspecific.

MARGINAL RECORDS.—Michigan: Ontonagon County. Ontario: Parry Sound District, Burk's Falls; Lindsay; *Highway 38 near Hartington;* 5 mi. ENE Ottawa (Youngman, 1962:223); Pittsburg Township. Massachusetts: near North Adams. Connecticut: Washington. Ontario: Woodstock; St. Thomas; Highgate. Michigan: Sanilac County. See also Silver (1924:1134) for record of several introductions (possibly of subspecies other than *L. c. hybridus*) including one at Bethlehem in Pennsylvania, and one in 1888 at Jobstown in New Jersey.

ORDER **RODENTIA**—Rodents

The earliest known rodents (family Ischyromyidae) are from the late Paleocene of North America. When they first appear they are already clearly rodents, and there are no known fossils that are intermediate between rodents and the archaic mammalian groups. Furthermore, although the rodents probably exceed all other orders combined in variety and actual numbers, the fossil record is disproportionately poor. Because of this lack of fossils and because there are many instances of convergence, divergence, and parallelism, the current taxonomic arrangements of the higher taxonomic groups are correspondingly unsatisfactory.

Rodents are terrestrial, fossorial, arboreal, glissant, or semiaquatic mammals having a generalized type of brain and placentation. The feet are clawed, and the fibula does not articulate with the calcaneum. The masseter muscle is highly specialized, and its arrangement serves as a basis for the recognition of the principal suborders. Only 4 incisors, 2 above and 2 below, are present, and these grow continuously. The maximum dental formula is: i. $\frac{1}{1}$, c. $\frac{0}{0}$, p. $\frac{2}{1}$, m. $\frac{3}{3}$ = 22 permanent teeth, for instance in several squirrels. The lower jaw is somewhat loosely articulated to the skull and considerable rotatory motion is employed in chewing. Molariform teeth may be either hypsodont or brachydont. The crowns of the molariform series display several patterns of the enamel and dentine that seem to be basically different, but the homologies of these are not known.

Rodents are primarily herbivores, but many kinds show a marked predilection for animal matter, and some seem to require at least some of it in their diet. Food consists of grasses, stems, roots, leaves, nuts, fruits, bark, various small forms of animal life, both invertebrate and vertebrate; even cannibalism occurs occasionally in some groups.

The Order Rodentia comprises a group remarkably homogeneous with respect to the ordinal characters. Yet in other characters the variation is great. Included are the volant (flying) squirrels, the almost wholly subterranean pocket gophers, the saltatorial genera of various families, the semiaquatic beavers and muskrats, and various arboreal squirrels and mice, and the vast assemblage of terrestrial kinds. Some are quick, slender, and agile, and others are slow, heavy bodied, and clumsy. Collectively, rodents occupy nearly all available ecological situations except the air and the open seas. Some, like the sagebrush vole (*Lagurus*), are comparatively restricted ecologically; others (*e.g.*, the white-footed mice of the genus *Peromyscus*) are among the terrestrially ubiquitous animals in North America.

KEY TO SUBORDERS OF RODENTIA

1. Infraorbital canal not transmitting any part of medial masseter muscle (or at least not modified for transmission of the muscle). . . . . Sciuromorpha, p. 333
1'. Infraorbital canal transmitting medial masseter muscle and enlarged for that purpose.
  2. Infraorbital foramen greatly enlarged (see Fig. 501). . . . Hystricomorpha, p. 849
  2'. Infraorbital canal moderately enlarged (see Fig. 387) except in Zapodidae.
                Myomorpha, p. 605

SUBORDER **SCIUROMORPHA**

Infraorbital canal not transmitting any part of masseter medialis; dentition, i. $\frac{1}{1}$, c. $\frac{0}{0}$, p. $\frac{2-1}{1}$, m. $\frac{3}{3}$ (except in a few extinct genera); 4th premolar usually large and important. Tibia and fibula variably free or extensively fused.

The zygomasseteric structure of this suborder is of at least two distinct types. In the more primitive type the origin of the masseter is wholly beneath the infraorbital foramen and no part of the muscle extends onto the rostrum. In the advanced type the zygomatic plate is broadened and tilted upward and a part of the lateral masseter originates above the infraorbital foramen and on the rostrum.

This suborder includes the most primitive living rodents and the earliest known fossil rodents.

KEY TO NORTH AMERICAN FAMILIES OF SCIUROMORPHA

1. Skull broad, flat, and triangular in dorsal aspect; dentition, i. $\frac{1}{1}$, c. $\frac{0}{0}$, p. $\frac{2}{1}$, m. $\frac{3}{3}$; auditory bullae flask-shaped; angular process of mandible strongly inflected.
              Aplodontidae, p. 334
1'. Skull not broad and flat; dentition, i. $\frac{1}{1}$, c. $\frac{0}{0}$, p. $\frac{2 \text{ or } 1}{1}$, m. $\frac{3}{3}$; auditory bullae not flask-

shaped; angular process not strongly inflected.

2. Tail broad, flat, scaly; toes of hind feet webbed. . . . . Castoridae, p. 601

2'. Tail not broad, flat, scaly; toes of hind feet not webbed.

   3. External fur-lined cheek pouches present.

      4. Frontal bone backing postorbital process; skull robust and angular; auditory bullae not greatly inflated. . Geomyidae, p. 454

      4'. Frontal bone bearing postorbital process; skull weak and smooth; auditory bullae inflated. . . . . Heteromyidae, p. 526

   3'. External fur-lined cheek pouches absent. . . . . Sciuridae, p. 336

FAMILY APLODONTIDAE
Mountain Beaver

Origin of masseter muscle wholly beneath infraorbital foramen and no part of muscle reaches rostrum; zygomatic plate narrow and horizontal; tibia and fibula always separate; angular process of mandible markedly inflected; skull widened and flattened posteriorly.

Aplodontids are the only living Protrogomorphs, the group to which the earliest known rodents belong. *Aplodontia rufa* is the only living species.

Genus Aplodontia Richardson
Mountain Beaver

Revised by Taylor, Univ. California Publ. Zool., 17:435–504, May 29, 1918; see also Finley, Murrelet, 22:45–49, January 20, 1942, and Dalquest, Murrelet, 26:34–37, December 28, 1945.

1829. *Aplodontia* Richardson, Zool. Jour., 4:334, January. Type, *Aplodontia leporina* Richardson [= *Anisonyx rufa* Rafinesque]. Many variant spellings of *Aplodontia* are in the literature.

Skull unusually broad and flat, especially posteriorly; cheek-teeth, $\frac{5}{4}$, the first in upper series being a small simple peg; posterior cheek-teeth when slightly worn showing single simple basin; auditory bullae flask-shaped; palate extending posteriorly beyond tooth-rows; angular process of mandible greatly inflected; coronoid process high.

Aplodontia rufa
Mountain Beaver

External measurements: total length, 310–470; hind foot, 36–64. Body compact; legs short and stout; ears, eyes, and tail small; fur pinkish-cinnamon to brown, becoming grayish with age; relatively uniformly colored.

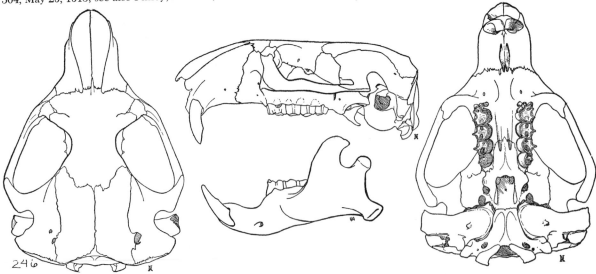

Fig. 246. *Aplodontia rufa californica*, ½ mi. S Marlette Lake, Nevada, No. 67066 M.V.Z., ♂, X 1.

Aplodontia rufa californica (Peters)

1864. *H[aplodon]. leporinus* var. *Californicus* Peters, Monatsb. preuss. Akad. Wiss., Berlin, p. 179, type assumed to be from Sierra Nevada of California (Grinnell, Proc. California Acad. Sci., ser. 4, 3:344, August 28, 1913; Hall, Murrelet, 22:50, January 20, 1942); specimens from Blue Canyon in central Sierra may be regarded as typical (Taylor, Univ. California Publ. Zool., 17:474, May 29, 1918).

1904. [*Aplodontia rufa*] *californica*, Trouessart, Catalogus mammalium . . . , Suppl., fasc. 2, p. 348.

1886. *Aplodontia major* Merriam, Ann. New York Acad. Sci., 3:316, May, type : from Sierra Nevada, Placer Co., California.

MARGINAL RECORDS.—California: Mt. Shasta; mts. 12 mi. W Susanville. Nevada: ⅞ mi. S, 2½ mi. W Lakeview. California: Mammoth; Clover Creek; Chinquapin; Lake Tahoe, Emerald Bay.

Aplodontia rufa humboldtiana Taylor

1916. *Aplodontia humboldtiana* Taylor, Proc. Biol. Soc. Washington, 29:21, February 24, type from Carlotta, Humboldt Co., California.

1918. *Aplodontia rufa humboldtiana* Taylor, Univ. California Publ. Zool., 17:470, May 29.

MARGINAL RECORDS.—California: Requa; 12 mi. N Hoopa, 3 mi. SW Weitzpek; Rio Dell, thence northward along coast to point of beginning.

Aplodontia rufa nigra Taylor

1914. *Aplodontia nigra* Taylor, Univ. California Publ. Zool., 12:297, April 15, type from Point Arena, Mendocino Co., California. Known only from type locality.

1918. *Aplodontia rufa nigra* Taylor, Univ. California Publ. Zool., 17:479, May 29.

Aplodontia rufa pacifica Merriam

1899. *Aplodontia pacifica* Merriam, Proc. Biol. Soc. Washington, 13:19, January 31, type from Newport, mouth Yaquina Bay, Lincoln Co., Oregon.

1918. *Aplodontia rufa pacifica*, Taylor, Univ. California Publ. Zool., 17:467, May 29.

MARGINAL RECORDS.—Oregon: Astoria; 11 mi. NW Linton; Eugene; Briggs Creek, 13 mi. SW Galice. California: 7 mi. ENE Smith River, thence northward along coast to Columbia Bay.

Aplodontia rufa phaea Merriam

1899. *Aplodontia phaea* Merriam, Proc. Biol. Soc. Washington, 13:20, January 31, type from Point Reyes, Marin Co., California.

1918. *Aplodontia rufa phaea*, Taylor, Univ. California Publ. Zool., 17:480, May 29.

MARGINAL RECORDS.—California: 5 mi. W Inverness; Lagunitas; 4 mi. S Olema.

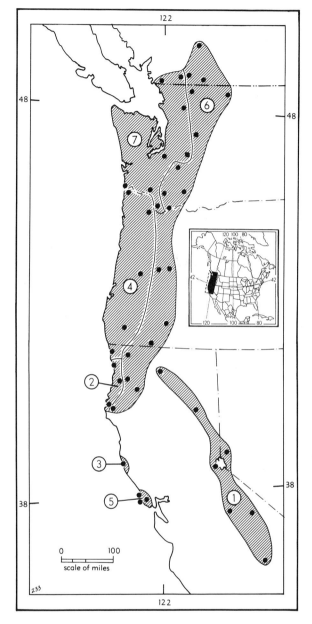

Map 233. *Aplodontia rufa.*

1. *A. r. californica*
2. *A. r. humboldtiana*
3. *A. r. nigra*
4. *A. r. pacifica*
5. *A. r. phaea*
6. *A. r. rainieri*
7. *A. r. rufa*

Aplodontia rufa rainieri Merriam

1899. *Aplodontia major rainieri* Merriam, Proc. Biol. Soc. Washington, 13:21, January 31, type from Paradise Creek, 5200 ft., S side Mt. Rainier, Pierce Co., Washington.

1904. [*Aplodontia rufa*] *rainieri*, Trouessart, Catalogus mammalium . . . , Suppl., fasc. 2, p. 348.

1916. *Aplodontia californica columbiana* Taylor, Univ. California Publ. Zool., 12:499, May 6, type from Roabs Ranch,

Hope, British Columbia. (Regarded as inseparable from *A. r. rainieri* by Dalquest, Univ. Kansas Publ., Mus. Nat. Hist., 2:369, April 9, 1948.)

MARGINAL RECORDS.—British Columbia: Nicola Valley near Merritt; Sterling Creek near Hedley. Washington: Loomis; Mt. Adams; Skamania; Mt. St. Helens; Mt. Rainier; Cascade Tunnel; Beaver Creek. British Columbia: Mile 14 on Hope–Princeton Highway.

Aplodontia rufa rufa (Rafinesque)

1817. *Anisonyx? rufa* Rafinesque, Amer. Month. Mag., 2:45, November. Type locality, neighborhood Columbia River, Oregon. Specimens from Marmot, Clackamas Co., regarded as typical (Taylor, Univ. California Publ. Zool., 17:455, May 29, 1918).
1886. *Aplodontia rufa*, Merriam, Ann. New York Acad. Sci., 3:316, May.
1899. *Aplodontia olympica* Merriam, Proc. Biol. Soc. Washington, 13:20, January 31, type from Quinault Lake, Grays Harbor Co., Washington.
1914. *Aplodontia chryseola* L. Kellogg, Univ. California Publ. Zool., 12:295, April 15, type from Jackson Lake, Siskiyou Co., California.
1916. *Aplodontia rufa grisea* Taylor, Univ. California Publ.

Zool., 12:497, May 6, type from Renton, near Seattle, King Co., Washington.

MARGINAL RECORDS.—British Columbia: Chilliwack. Washington: Sauk; Easton; Puyallup; 1½ mi. W Yaclot. Oregon: McKenzie Bridge; Fort Klamath, Anna Creek Canyon; N base Ashland Peak. California: Canyon Creek; Rio Dell; 10 mi. W forks of Salmon River; Poker Flat, 12 mi. NW Happy Camp. Oregon: Vida. Washington: 6 mi. NE Kelso; along N side Columbia River, to mouth Bear River, 5 mi. NE Ilwaco, and up coast to British Columbia: Aldergrove.

FAMILY SCIURIDAE
Squirrels and Relatives

Skull never of truly fossorial type; cheek-teeth $\frac{5}{4}$ or $\frac{4}{4}$, cuspidate, rooted, and usually not simplified; auditory bullae prominent, but usually not specially modified; well-developed postorbital processes; fibula and tibia not fused; tail completely haired. Geologic range: Oligocene to Recent.

The family is widely distributed and, within certain broad limitations imposed by structure, remarkably versatile ecologically; included are

KEY TO NORTH AMERICAN GENERA OF SCIURIDAE

1. Membrane present between foreleg and hind leg; modified for gliding; zygomatic plate low, slightly tilted upwards. . . . .*Glaucomys*, p. 447
1'. Membrane not present between foreleg and hind leg; not modified for gliding; zygomatic plate (usually) tilted strongly upward.
  2. No antorbital canal, the antorbital foramen piercing the zygomatic plate of the maxillary.
    3. Keel on tip of baculum on dorsal side; P3 present; tail more than 40% of total length. . . . .*Eutamias*, p. 340
    3'. Keel on tip of baculum on ventral side; P3 absent; tail less than 38% of total length. *Tamias*, p. 337
  2'. Antorbital canal present.
    4. Zygomatic breadth more than 48; anterior lower premolar with a paraconulid. . *Marmota*, p. 367
    4'. Zygomatic breadth less than 48; anterior lower premolar without paraconulid.
      5. Zygomata not parallel, but converging anteriorly with anterior part twisted toward a horizontal plane.
        6. Maxillary tooth-rows strongly convergent posteriorly. . . . .*Cynomys*, p. 410
        6'. Maxillary tooth-rows not strongly convergent posteriorly.
          7. Small masseteric tubercle directly below narrowly oval infraorbital foramen; cranium subrectangular from dorsal aspect; cranium flattened and slightly inclined downward to supernuchal line; clavobranchialis muscle absent.
               Ammospermophilus, p. 376
          7'. Medium to large masseteric tubercle ventral and slightly lateral to oval or subtriangular infraorbital foramen; cranium not subrectangular from dorsal aspect; cranium inflated and inclined downward to supernuchal line; clavobrachialis muscle present. . . . .*Spermophilus*, p. 381
      5'. Zygomata nearly parallel and nearly vertical throughout, that is, not twisted.
        8. Upper incisors projecting forward to or beyond tip of nasals.
          9. Postorbital process anterior to vertical line through posterior zygomatic root. . . . .*Syntheosciurus*, p. 438
          9'. Postorbital process almost directly above posterior zygomatic root.
              Microsciurus, p. 439
        8'. Upper incisors not projecting to or beyond tip of nasals.
          10. Baculum present; P3 usually well developed. . . . .*Sciurus*, p. 415
          10'. Baculum absent; P3 usually vestigial or absent. . . . .*Tamiasciurus*, p. 441

semifossorial, terrestrial, and volant species. Although primarily herbivorous, on occasion nearly all sciurids will accept animal food and some do so avidly.

As a whole, populations of the true squirrels seem to be less subject to periodic fluctuations in numbers of individuals than are many other kinds of rodents.

Genus Tamias Illiger—Eastern Chipmunk

Revised by A. H. Howell, N. Amer. Fauna, 52:11–23, November 30, 1929.

Concerning the arrangement of genera and subgenera of chipmunks, Howell (*op. cit.*) applied the name *Tamias* to the species (*T. striatus*) in eastern North America, and the generic name *Eutamias* to the several species in western North America and to the one Eurasian species. He (1929:26) divided the genus *Eutamias* into two subgenera: subgenus *Neotamias* for the several species of western North America and subgenus *Eutamias* for the one Eurasian species.—In 1953, J. A. White (Univ. Kansas Mus. Nat. Hist., 5:543–561) graphically listed characters of the genera and subgenera and arrived at the same arrangement as Howell (1929).—In 1940, Ellerman (The families and genera of living rodents, British Mus. Nat. Hist., pp. 426–437) recognized the genus *Tamias*, the subgenus *Tamias* as applying to the animals of eastern North America, the subgenus *Eutamias* as applying to the animals of Eurasia, and the subgenus *Neotamias* as applying to the animals of western North America. His arrangement has merit.—In 1965, Gromov [Fauna of the U.S.S.R., Mammals, 3(no. 2):70, 125, on the Marmotinae] recognized the tribe "Tamiini" to include two genera (*Sciurotamias* of China, and *Tamias* of Eurasia and North America). Under the genus *Tamias* he recognized two subgenera: subgenus *Tamias* for the animals of eastern North America, and subgenus *Eutamias* for those of both Eurasia and western North America. Consequently, *Tamias* was the generic name for all the chipmunks, and *Neotamias* disappeared as a junior synonym of the subgenus *Eutamias*. His arrangement, also, has much to recommend it.

The arrangement adopted here is that of Howell (1929) and White (1953).

1811. *Tamias* Illiger, Prodromus systematis mammalium et avium . . . , p. 83. Type, *Sciurus striatus* Linnaeus, 1758.

External measurements: 215–299; 78–113; 32–38. Greatest length of skull, 37.4–45.7. Skull lightly built, narrow; postorbital process small and weak; lachrymal not elongated; infraorbital foramen lacks canal, relatively larger than in most sciurids; P3 absent; head of malleus elongated; plane of manubrium of malleus forms 60° angle with plane of lamina; hypohyal and ceratohyal bones of hyoid apparatus fused in adults; conjoining tendon of anterior and posterior digastric muscles rounded in cross section; keel on ventral surface of tip of baculum, tip of baculum curved upward; tail less than 38 per cent of total length;

5 longitudinal dark and 4 longitudinal light stripes present but 2 dorsal light stripes at least twice as broad as other stripes; 4 lateral dark stripes short (after White, 1953:559).

Tamias striatus
Eastern Chipmunk

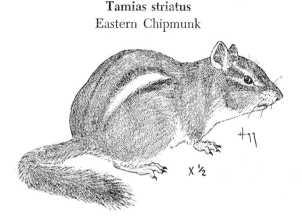

Characters as for the genus.

Tamias striatus doorsiensis C. A. Long

1971. *Tamias striatus doorsiensis* C. A. Long, Proc. Biol. Soc. Washington, 84:201, November 4, type from Peninsula State Park, Door Co., Wisconsin.

MARGINAL RECORDS (Long, 1971:202).—Wisconsin: type locality; *3 mi. N Baileys Harbor; 2 mi. N Jacksonport;* Sturgeon Bay; *6 mi. W Sturgeon Bay.*

Tamias striatus fisheri A. H. Howell

1925. *Tamias striatus fisheri* A. H. Howell, Jour. Mamm., 6:51, February 9, type from Merritts Corners, 3¾ mi. E, 1¼ mi. N Ossining, Westchester Co., New York (Paradiso, Jour. Mamm., 44:580, 1963).

MARGINAL RECORDS.—New York: Cohoes. Connecticut: Plainfield. Rhode Island: Providence. New Jersey: 1 mi. NW New Gretna, thence southward along coast to Virginia: Suffolk; Mountain Lake. West Virginia: Odd, 2900 ft.; 1½ mi. S Big Creek; 5 mi. E Huntington. Ohio: Carpenter; Athens; Maysville; Bowerston. Pennsylvania: New Castle; Butler County; Tyrone; Flowing Spring; Cresson; Laughlintown; Summit Mills. West Virginia: 7 mi. E Philippi; Berkeley Springs. Pennsylvania: Carlisle; Harveys Lake; Saylorsburg; Bushkill Creek, 7 mi. E Cresco. New York: Lanesville; Kiskatom.

Tamias striatus griseus Mearns

1891. *Tamias striatus griseus* Mearns, Bull. Amer. Mus. Nat. Hist., 3:231, June 5, type from Fort Snelling, Hennepin Co., Minnesota.

MARGINAL RECORDS.—Manitobia: Dauphin; Pine Falls, Winnipeg River. Ontario: Ingolf. Minnesota: Koochiching County; 2 mi. N, 2½ mi. E Grand

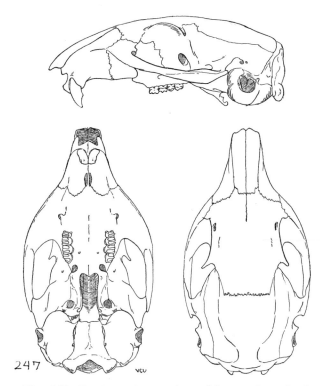

Fig. 247. *Tamias striatus griseus*, Missouri River bank, Doniphan Co., Kansas, No. 13701 K.U., ♀, X 1½.

Portage (Timm, 1975:16). Michigan: North Fox Island; South Fox Island. Wisconsin (Jackson, 1961: 144, May 29): Calumet County; *Green Lake County;* Marquette County; Buffalo County. Illinois: Mt. Carmel. Missouri: Williamsville; Independence. Kansas: Peoria; Onaga. Nebraska (Jones, 1964:119): Falls City; Nickerson; Engineer Cantonment. Iowa: Clay County (Bowles, 1975:62, Fig. 20). Minnesota: Lake Shetek State Park (Ernst and Ernst, 1972:377); Ortonville. South Dakota: Fort Sisseton. North Dakota: Kathryn; Larimore; Walhalla; Fish Lake, Birchwood P. O.; Turtle Mts. Manitoba: Turtle Mtn.

Tamias striatus lysteri (Richardson)

1829. *Sciurus (Tamias) lysteri* Richardson, Fauna Boreali-Americana, 1:181, pl. 15, June. Type locality, Pentanguishene, Ontario.
1886. *Tamias striatus lysteri*, Merriam, Amer. Nat., 20:242, March.

MARGINAL RECORDS.—New Brunswick: Bathurst. Nova Scotia: Cape Breton Island. Maine: Mount Desert Island (Manville, 1960:416, as *T. striatus* only), thence down coast to Massachusetts: Wareham. Connecticut: South Woodstock; Portland; Sharon Mtn. New York: Troy; Palensville [= Palenville]; *Kaaterskill Junction.* Pennsylvania: Mt. Pocono; Eaglesmere; Perry County; Adams County. Maryland: *Dans Mtn., 4 mi. NW Rawlings;* Accident; *Grantsville.* Pennsylvania: Summit; Tyronne; Venango County;

Erie. Ontario: Central Proton; Point Pelee; N end Georgian Bay; S shore Lake Nipissing; Ottawa. Quebec (Wrigley, 1969:207): St. Anne de Bellevue; *Mont Bruno;* Mont St. Hilaire; Mont Orford; *Ayers Cliff.* New Hampshire: Pinkham Notch. Quebec: St. Sebastien (Wrigley, 1969:207). New Brunswick: Edmundston.

Tamias striatus ohionensis Bole and Moulthrop

1942. *Tamias striatus ohionensis* Bole and Moulthrop, Sci. Publs., Cleveland Mus. Nat. Hist., 5(6):135, September 11, type from Cincinnati, Hamilton Co., Ohio.

MARGINAL RECORDS.—Wisconsin (Jackson, 1961:148): Fountain City; Fox Lake; Whitefish Bay. Indiana: Hebron. Ohio: Payne; Bettsville; *Overton;* Wooster; Loudonville; 7 mi. E Logan; Smoky Creek, Green Twp. Kentucky: Bath County. Indiana: Wheatland; Parke County; Mt. Ayr. Illinois: River Forest (Necker and Hatfield, 1941:49, as *T. s. griseus*). Wisconsin: Grant County (Jackson, 1961:144, Map 29).

Tamias striatus peninsulae Hooper

1942. *Tamias striatus peninsulae* Hooper, Occas. Pap. Mus. Zool., Univ. Michigan, 461:1, September 15, type from Barnhart Lake, 3 mi. SE Millersburg, Presque Isle Co., Michigan.

MARGINAL RECORDS.—Michigan: Beaver Island; Bay County; Bass Lake; "near" Lake Michigan, N of Muskegon State Park; South Manitou Island; North Manitou Island.

Tamias striatus pipilans Lowery

1943. *Tamias striatus pipilans* Lowery, Occas. Pap. Mus. Zool., Louisiana State Univ., 13:235, November 22, type from 5 mi. S Tunica, West Feliciana Parish, Louisiana.

MARGINAL RECORDS.—Mississippi (Kennedy, *et al.*, 1974:11, as *T. striatus* only): 2 mi. N Walls; 5 mi. SW Michigan City; Tishomingo County. Alabama: Woodville; Bucks Pocket; Talladega Mtn. Florida: 5 mi. SW Laurel Hill (Stevenson, 1962:110, as *T. striatus* only). Alabama: Greensboro. Mississippi: Jones County (Kennedy, *et al.*, 1974:11, as *T. striatus* only). Louisiana (Lowery, 1974:198): 2 mi. W Baton Rouge city limits on Mississippi River; Tunica. Mississippi (Kennedy, *et al.*, 1974:11, as *T. striatus* only): Natchez; 2 mi. S Vicksburg; Bolivar County.—See addenda.

Tamias striatus quebecensis Cameron

1950. *Tamias striatus quebecensis* Cameron, Jour. Mamm., 31:347, August 21, type from St. Félicien, Lake St. John Co., Quebec.

MARGINAL RECORDS.—Ontario: S end James Bay. Quebec: Lake Albanel; head Matamek River; Moisie Bay; Perce; Hatley; Parker. Ontario: Pancake Bay; Thunder Bay District; Kapuskasing.

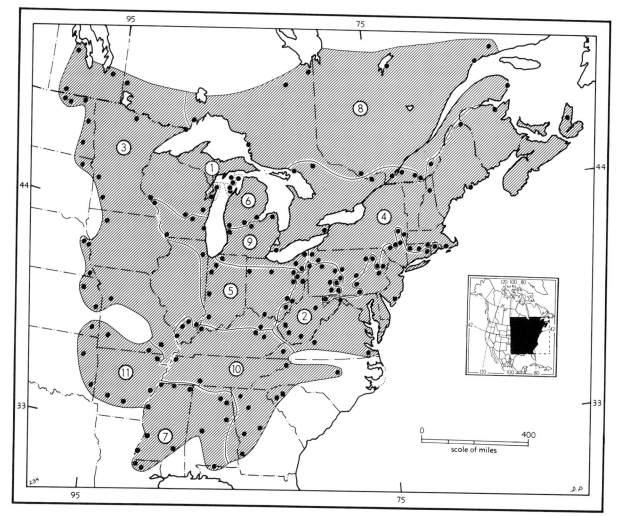

Map 234. *Tamias striatus.*

Guide to subspecies		
1. *T. s. doorsiensis*	3. *T. s. griseus*	6. *T. s. peninsulae*	9. *T. s. rufescens*
2. *T. s. fisheri*	4. *T. s. lysteri*	7. *T. s. pipilans*	10. *T. s. striatus*
	5. *T. s. ohionensis*	8. *T. s. quebecensis*	11. *T. s. venustus*

Tamias striatus rufescens Bole and Moulthrop

1942. *Tamias striatus rufescens* Bole and Moulthrop, Sci. Publs., Cleveland Mus. Nat. Hist., 5(6):130, September 11, type from Chesterland Caves, Chester Twp., Geauga Co., Ohio.

MARGINAL RECORDS.—Michigan: Huron County. Ohio: Lisbon; Ashtabula Township Park (Goodpaster and Hoffmeister, 1968:117); Evertt. Indiana: La Porte County. Michigan: Ottawa County; Gratiot County.

Tamias striatus striatus (Linnaeus)

1758. [*Sciurus*] *striatus* Linnaeus, Syst. nat., ed. 10, 1:64, type locality fixed by A. H. Howell (N. Amer. Fauna, 52:14, November 30, 1929) as upper Savannah River, South Carolina. A. H. Howell (*loc. cit.*) makes this restriction

primarily on the basis of Catesby's account, and points out that Merriam (Amer. Nat., 20:238, 1886) states that a specimen in his collection from Sylva, North Carolina, "may be regarded as the type of *striatus*," but it is obviously impossible to fix the type locality at a point outside the region where Catesby is known to have traveled.

1857. *Tamias striatus*, Baird, 11th Ann. Rept. Smiths. Inst., for 1856, p. 55.

1788. [*Sciurus striatus*] *americanus* Gmelin, Syst. nat., ed. 13, 1:150. Based on earlier citations, including Catesby's account.

MARGINAL RECORDS.—Illinois: Olney. Indiana: New Harmony. Kentucky: Lexington; Quicksand. Virginia: Cleveland. Tennessee: Roan Mtn. (Tuttle, 1968:133, as *T. striatus* only). North Carolina: Chapel Hill. South Carolina: Greenville; Abbeville (Golley, 1966:79). Georgia: Roswell; Clay

MAMMALS OF NORTH AMERICA

County (Golley, 1962:93); Wesleyan College Campus, Rivoli; 4 mi. N Geneva; Gordon County (Golley, 1962:93). Tennessee: 8 mi. NE Waynesboro; Reelfoot Lake, Samburg (Kellogg, 1939:271), but later referred to *T. s. ohionensis* by Goodpaster and Hoffmeister (1952:366). Illinois: Wolf Lake; Woodlawn.

Tamias striatus venustus Bangs

1896. *Tamias striatus venustus* Bangs, Proc. Biol. Soc. Washington, 10:137, December 28, type from Stilwell, Adair Co., Oklahoma.

MARGINAL RECORDS.—Kansas: Independence. Missouri: 8 mi. S Carthage; 20 mi. SE Alton. Arkansas: Clay County; Phillips County; Dallas County; *Hempstead County;* Delight. Oklahoma: 5 mi. W Smithville; Red Fork.—See addenda.

Genus Eutamias Trouessart
Western Chipmunks

The arrangement here followed is that of White, Univ. Kansas Publ., Mus. Nat. Hist., 5:543–561, 12 figs. in text, December 1, 1953. See remarks under genus *Tamias.*

1880. *Eutamias* Trouessart, Bull. Soc. d'Études Sci. d'Angers, 10(fasc. 1):86. Type, *Sciurus striatus asiaticus* Gmelin.

Skull lightly built, narrow; postorbital process light and weak; lachrymal not elongated; infraor-

bital foramen lacks canal, relatively larger than in most sciurids; P3 present; head of malleus not elongated; plane of manubrium of malleus at angle of 90° to plane of lamina; hypohyal and ceratohyal bones of hyoid apparatus fused in adults; conjoining tendon between anterior and posterior sets of digastric muscles ribbonlike; keel on dorsal side of tip of baculum; tail more than 40 per cent of total length; 5 longitudinal dark stripes evenly spaced and subequal in width; 2 lateral dark stripes short (after White, 1953:558).

Subgenus Neotamias A. H. Howell
Western Chipmunks

Revised by A. H. Howell, N. Amer. Fauna, 52:23–137, November 30, 1929.

1929. *Neotamias* A. H. Howell, N. Amer. Fauna, 52:26, November 30. Type, *Tamias asiaticus merriami* J. A. Allen.

Total length, 166–277; tail, 67–140. Greatest length of skull, 28.7–40.8. "Lambdoidal crest barely discernible; supraorbital notches even with, or posterior to, posterior notch of zygomatic plate; baculum with distinct keel on dorsal surface of tip which curves upward; pelage silky; ears long and pointed." (White, 1953:558.)

KEY TO SPECIES OF SUBGENUS NEOTAMIAS

1. Dorsal stripes (except median one) more or less indistinct.
  2. Occurring in California and Baja California.
    3. See Figs. 253*b, c.* . . . . *E. obscurus,* p. 355
    3'. See Figs. 253*b, c.* . . . . *E. merriami,* p. 354
  2'. Occurring east of the Californias. . . . . *E. dorsalis,* p. 356
1'. Dorsal stripes all distinctly marked.
  4. Greatest length of skull 37 or more.
    5. Occurring in Arizona, New Mexico, Texas, and México.
      6. Hind foot, 34–38; occurring in México, exclusive of Baja California. . . . . *E. bulleri,* p. 366
      6'. Hind foot, 32–36; occurring in Arizona, New Mexico, and Texas.
        7. Occurring in New Mexico and Texas; baculum shorter than 4 mm. . . *E. canipes,* p. 359
        7'. Occurring in Arizona and New Mexico; baculum longer than 4 mm.
           E. cinereicollis, p. 360
  5'. Occurring in British Columbia, Washington, Oregon, California, Baja California, and Nevada in Sierra Nevada.
    8. Tail less bushy; backs of ears distinctly bicolored in all pelages; tips of nasals not separated by median notch.
      9. Ears relatively short, not pointed; submalar dark stripe not black below ear. . . . . *townsendii*-subgroup, p. 350
      10. Tip of baculum longer than shaft (see Fig. 251*c*). . . . . *E. siskiyou,* p. 352
      10'. Tip of baculum less than ½ length of shaft (see Figs. 250*b, c,* 251*b*).
        11. Height of keel of baculum less than ¼ length of tip of baculum (see Fig. 250*c*). . . . . *E. ochrogenys,* p. 351
        11'. Height of keel of baculum more than ¼ length of tip of baculum (see Figs. 250*b,* 251*b*).
          12. Greatest depth of shaft of baculum more than 20% of length of shaft (see Fig. 251*b*). . . . . *E. senex,* p. 351

body, 106–127; incisive foramina small; inner pair of pale dorsal stripes relatively broad; tip of baculum 35% or less of length of shaft; angle between tip and shaft of baculum more than 115°. . . . .*E. amoenus*,   p. 347

    33′. Combination of following: length of skull, 33.5–36.7; length of head and body, 114–144; incisive foramina large; inner pair of pale dorsal stripes relatively narrow; tip of baculum 36% or more of length of shaft; angle between tip and shaft of baculum less than 115°.

        34. Outer pale stripes not broader than inner pale stripes; dark submalar stripe without black center below eye; tip of upper incisor anterior to posterior border of alveolus when skull is placed on horizontal surface; distal ½ of shaft of baculum laterally compressed. . . . .*E. umbrinus*,   p. 364

        34′. Outer pale stripes broader than inner pale stripes; dark submalar stripe with black center below eye; tip of upper incisor posterior to posterior border of alveolus when skull is placed on horizontal surface; distal ⅔ of shaft of baculum laterally compressed. . . . .*E. speciosus*,   p. 361

  31′. When outside of California and Nevada.

    35. Shaft of baculum thick, robust; keel of baculum at least ¼ as long as tip.

        36. Base of baculum markedly expanded; angle between tip and shaft of baculum less than 115°; upper parts of dark, somber hue. . . . .*E. umbrinus*,   p. 364

        36′. Base of baculum not markedly expanded; angle between tip and shaft of baculum more than 115°; upper parts more brightly colored. . . . .*E. quadrivittatus*,   p. 358

    35′. Shaft of baculum slender, light; keel of baculum less than ¼ as long as tip.

        37. Underside of tail grayish yellow; length of skull, 31.0–34.2; tip of baculum 28% or less of length of shaft. . . . .*E. minimus*,   p. 343

        37′. Underside of tail ochraceous; length of skull, 33.0–35.6; tip of baculum more than 28% of length of shaft. . . . .*E. amoenus*,   p. 347

Eutamias alpinus (Merriam)
Alpine Chipmunk

1893. *Tamias alpinus* Merriam, Proc. Biol. Soc. Washington, 8:137, December 28, type from Big Cottonwood Meadows, 10,000 ft., just S of Mount Whitney, Tulare Co., California.

1897. *Eutamias alpinus* (Merriam), Proc. Biol. Soc. Washington, 11:191, July 1.

External measurements: 166–195; 70–85; 28–31. Greatest length of skull, 28.9–31.7. Colors generally yellowish; light and dark stripes weakly contrasted. Skull broad and flattened. Smaller than other species of same area, except *E. minimus* from which *E. alpinus* differs in shorter tail that is more nearly flat in cross section and bright orange instead of dull grayish yellow beneath, longer and finer fur, paler dorsal coloration, more ochraceous in light dorsal stripes, larger and more flattened braincase, longer and blunter rostrum, wider palate and shorter incisors. Baculum: shaft thin; keel low, one-seventh of length of tip; tip 39 per cent of length of shaft; angle formed by tip and shaft 135°; distal third of shaft slightly compressed laterally; base slightly wider than shaft; shaft short, 2.17 mm.

MARGINAL RECORDS.—California: Mt. Conness; Warren Fork of Leevining Creek, 9700–10,000 ft.; Independence Creek, 10,000 ft.; Onion Valley; Olancha

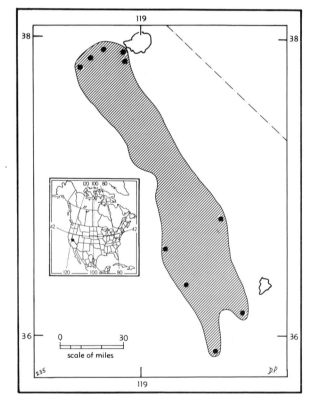

Map 235. *Eutamias alpinus*.

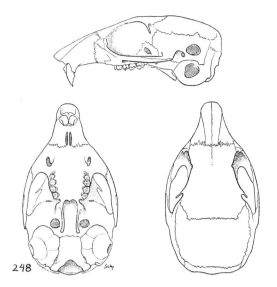

Fig. 248. *Eutamias alpinus*, Cottonwood Lakes, 11,000 ft., Inyo Co., California, No. 14942 M.V.Z., ♂, X 1½.

Peak, 9750–10,500 ft.; Horse Corral Meadows; Mineral King; Mt. Clark, 10,000 ft.; Mt. Hoffman, 10,200–10,700 ft.; Cold Canyon, 8000 ft.

Eutamias minimus
Least Chipmunk

X ½

External measurements: 167–225; 70–114; 26–35. Greatest length of skull, 28.7–34.2. Stripes well defined. Skull having high, narrow braincase. Differences from *E. alpinus* are mentioned in the account of that species. The only other species with which *E. minimus* is likely to be confused is *E. amoenus*. Certain subspecies of these two species (*e.g., E. m. operarius* and *E. a. amoenus*) closely resemble each other. As yet no specific diagnosis has been framed that will serve to distinguish the two species in all parts of their geographic ranges; but at any one place the two are distinguishable. For example, in Nevada and California, *E. minimus* differs from *E. amoenus* as

follows: smaller; paler; underside of tail yellowish instead of reddish; braincase less flattened; zygomatic arches less flattened; rostrum shorter; upper incisors less recurved. In Wyoming and southwestern Montana, and possibly in other areas, the tip of the baculum, in adult males, is less than 28 per cent of the length of the shaft (averages 24%), whereas the tip is more than 28 per cent in *E. amoenus* (averages 34%). For differences in other areas where the two species occur, reference should be made to A. H. Howell's (1929:36–77) "Revision of the American Chipmunks."

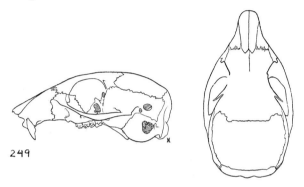

Fig. 249. *Eutamias minimus scrutator*, Wisconsin Creek, 7800 ft., Nevada, No. 45513 M.V.Z., ♀, X 1½.

E. minimus lives in sagebrush far distant from conifers as well as in sagebrush where conifers occur. In my experience, the habitat of *E. amoenus* always includes conifers.

Eutamias minimus atristriatus V. Bailey

1913. *Eutamias atristriatus* V. Bailey, Proc. Biol. Soc. Washington, 26:129, May 21, type from Penasco Creek, 7400 ft., 12 mi. E Cloudcroft, Sacramento Mts., New Mexico.
1922. *Eutamias minimus atristriatus*, A. H. Howell, Jour. Mamm., 3:178, August 4.

MARGINAL RECORDS.—New Mexico: 6 mi. E Cloudcroft; *type locality*; 4 mi. S, 14 mi. E Cloudcroft (Conley, 1970:701).

Eutamias minimus borealis (J. A. Allen)

1877. [*Tamias asiaticus*] var. *borealis* J. A. Allen, *in* Coues and Allen, Monog. N. Amer. Rodentia, p. 793, August, type from Fort Liard, Mackenzie.
1922. *Eutamias minimus borealis*, A. H. Howell, Jour. Mamm., 3:183, August 4.

MARGINAL RECORDS.—Yukon: 138 mi. N Watson Lake and 5 mi. E Little Hyland River, 6000 ft. (Youngman, 1968:74). Mackenzie: forks Nahanni and Liard rivers (Youngman, 1968:75); Fort Simpson; Hay River; Fort Resolution. Saskatchewan: Poplar Point, Athabasca Lake. Manitoba: The Pas; Wekusko Lake,

Eastside, District of The Pas; Oxford House. Ontario: Moose River; Minaki. North Dakota: Fort Pembina; Turtle Mts. Montana: Zortman; Big Snowy Range; 4 mi. W Tyler; Bear Paw Range. Saskatchewan: Fort Walsh. British Columbia (Cowan and Guiguet, 1965:136, 137, unless otherwise noted): Tornado Pass; Assiniboine; head Wapiti River; Tucheeda [= Tacheeda] Lake (Hall and Kelson, 1959:299); Nabesche River; Tocha [= Tochcha] Lake; Babine Mts.; Sustut River; Tatletuey [= Tatlatui] Lake; Minaker River. Yukon: Ida Lake [= McPherson Lake] (Youngman, 1968:74).

Eutamias minimus cacodemus Cary

1906. *Eutamias pallidus cacodemus* Cary, Proc. Biol. Soc. Washington, 19:89, June 4, type from head Corral Draw, Sheep Mtn., Big Badlands, South Dakota.
1922. *Eutamias minimus cacodemus*, A. H. Howell, Jour. Mamm., 3:183, August 4.

MARGINAL RECORDS.—South Dakota: *Cheyenne River Badlands; Badlands National Monument* (Sutton and Nadler, 1969:525); *Corral Draw;* type locality.

Eutamias minimus caniceps Osgood

1900. *Eutamias caniceps* Osgood, N. Amer. Fauna, 19:28, October 6, type from Lake Laberge, Yukon.
1922. *Eutamias minimus caniceps,* A. H. Howell, Jour. Mamm., 3:184, August 4.

MARGINAL RECORDS.—Yukon: Bonnet Plume Lake (Youngman, 1975:62, as *E. m. borealis*); Sheldon Mtn. Mackenzie: Nahanni River Mts. British Columbia (Cowan and Guiguet, 1965:137, unless otherwise noted): jct. Irons Creek and Liard River; Muncho Lake; *10 mi. S, 21 mi. E Muncho Lake;* 10 mi. S, 70 mi. W Fort Nelson; Groundhog Mtn.; Telegraph Creek; Atlin; Bennett City (Hall and Kelson, 1959:300); *Bennett Lake.* Yukon: 1½ mi. S, 3 mi. E Dalton Post; 6 mi. SW Kluane, 2550 ft.; Fort Selkirk; Dawson (Youngman, 1975:62, as *E. m. borealis* because he thought *caniceps* was a synonym of *borealis*).

Eutamias minimus caryi Merriam

1908. *Eutamias minimus caryi* Merriam, Proc. Biol. Soc. Washington, 21:143, June 9, type from Medano Ranch, San Luis Valley, Alamosa Co., Colorado.

MARGINAL RECORDS.—Colorado (Armstrong, 1972:98): 9 mi. E Center; 2 mi. S Great Sand Dunes National Monument; Mosca.

Eutamias minimus confinis A. H. Howell

1925. *Eutamias minimus confinis* A. H. Howell, Jour. Mamm., 6:52, February 9, type from head Trapper's Creek, 8500 ft., W slope Bighorn Mts., Wyoming.

MARGINAL RECORDS (Long, 1965a:556).—Wyoming: 38 mi. E Lovell, Bighorn National Forest; *1 mi. S, 5½ mi. W Buffalo; head N Fork Powder River;* 3 mi. SE Tensleep; *9 mi. N, 9 mi. E Tensleep; 2 mi. S, 2 mi. E Shell; Medicine Wheel Ranch, 28 mi. E Lovell.*

Eutamias minimus consobrinus (J. A. Allen)

1890. *Tamias minimus consobrinus* J. A. Allen, Bull. Amer. Mus. Nat. Hist., 3:112, June, type from Parleys Canyon, Wasatch Mts., near former site Barclay, Utah.
1901. *Eutamias minimus consobrinus,* Miller and Rehn, Proc. Boston Soc. Nat. Hist., 30(1):42, December 27.
1905. *Eutamias lectus* J. A. Allen, Mus. Brooklyn Inst. Arts and Sci., Sci. Bull., 1:117, March 31, type from Beaver Valley, Beaver Co., Utah.
1918. *Eutamias consobrinus clarus* V. Bailey, Proc. Biol. Soc. Washington, 31:31, May 16, type from Swan Lake Valley, Yellowstone National Park, Wyoming.

MARGINAL RECORDS.—Montana: head Sage Creek, Pryor Mts. Wyoming (Long, 1965a:557): *Beartooth Lake; Whirlwind Peak; Valley;* Needle Mtn.; *Lake Fork; Moccasin Lake, 4 mi. N, 19 mi. W Lander; 4 mi. S, 8½ mi. W Lander;* 3 mi. E South Pass City; *South Pass City; Big Sandy; Pinedale;* 2 mi. S, 19 mi. W Big Piney; *Kemmerer; 14 mi. S, 1 mi. W Kemmerer;* Fort Bridger; *Mountainview; Sage Creek, 10 mi. N Lonetree; 4½ mi. N, 4 mi. E Robertson.* Utah: Summit Springs. Colorado (Armstrong, 1972:101, 102): Pot Creek, near Pat's Hole; Axial Basin; 5 mi. W Craig; 43 mi. (by road) N Hayden, 8225 ft.; Pearl; Walden; 6 mi. N Grand Lake; between Pando and Mitchell, 9700 ft.; 18 mi. S Buford; Roan Plateau, 14 mi. SE Dragon, Utah. Utah: Nipple; SW slope Bald Peak, 10,500 ft., Unita Mts.; near Soldier Summit; Baldy Ranger Station; Elkhorn Guard Station, 9400 ft., Fishlake Plateau, 14 mi. N Torrey; Wildcat Ranger Station, 8700 ft., Boulder Mtn.; Bryce National Park; East Rim, just outside Zion National Park boundary; Cedar City; Beaver; 10 mi. E Sigurd; Ephraim; Butterfield Canyon, 3 mi. SW Butterfield Tunnel, 8000 ft. Idaho: Inkom; 12 mi. (via highway) NW Rea. Wyoming: Snow Pass, Mammoth Hot Springs (Long, 1965a:557). Also disjunct population in Arizona: *Tipover Spring* (Hoffmeister, 1971:169); *½ mi. S [Grand Canyon] park entrance* (ibid.); Bright Angel Spring, Kaibab Plateau.

Eutamias minimus grisescens A. H. Howell

1925. *Eutamias minimus grisescens* A. H. Howell, Jour. Mamm., 6:52, February 9, type from Farmer, Douglas Co., Washington.

MARGINAL RECORDS.—Washington: Douglas; Coulee City; Moses Coulee; Pasco; Vantage.

Eutamias minimus hudsonius Anderson and Rand

1944. *Eutamias minimus hudsonius* Anderson and Rand, Canadian Field-Nat., 57:133, January 24, type from Bird, Hudson Bay Railway, Mile 349, Manitoba.

MARGINAL RECORDS.—Manitoba: Herchmer, Hudson Bay Railway, Mile 412; type locality; Thicket Portage, Hudson Bay Railway, Mile 165; Alberta Lake near Flin Flon. Anderson (1947:114) states that *E. m. hudsonius* "probably occurs in extreme northwestern Ontario and northeastern Saskatchewan."

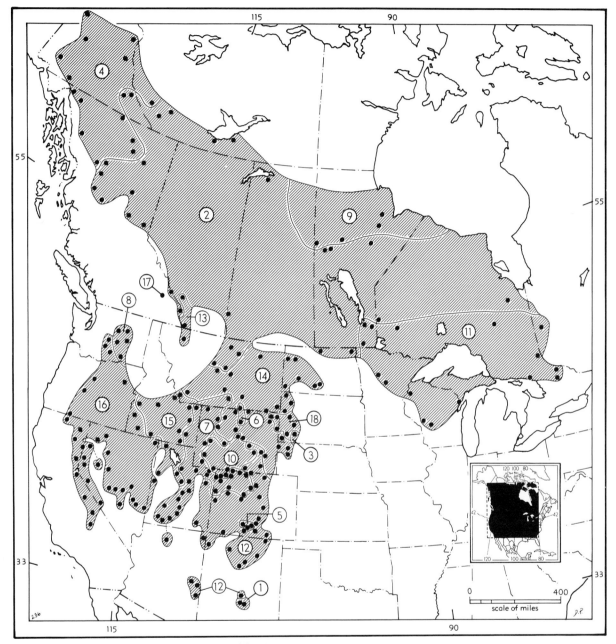

Map 236. *Eutamias minimus.*

1. *E. m. atristriatus*	7. *E. m. consobrinus*	13. *E. m. oreocetes*
2. *E. m. borealis*	8. *E. m. grisescens*	14. *E. m. pallidus*
3. *E. m. cacodemus*	9. *E. m. hudsonius*	15. *E. m. pictus*
4. *E. m. caniceps*	10. *E. m. minimus*	16. *E. m. scrutator*
5. *E. m. caryi*	11. *E. m. neglectus*	17. *E. m. selkirki*
6. *E. m. confinis*	12. *E. m. operarius*	18. *E. m. silvaticus*

Eutamias minimus minimus (Bachman)

1839. *Tamias minimus* Bachman, Jour. Acad. Nat. Sci. Philadelphia, 8:71, type from Green River, near mouth Big Sandy Creek, Sweetwater Co., Wyoming.

1901. *Eutamias minimus*, Miller and Rehn, Proc. Boston Soc. Nat. Hist., 30(1):42, December 27.

MARGINAL RECORDS.—Wyoming (Long, 1965a: 558, 559): *27 mi. N, 1 mi. E Powder River;* Bitter Creek, near Powder River; *Casper; 12 mi. S Alcova;* Spring Creek, 10 mi. W Marshall; *Saratoga.* Colorado: Snake River, 20 mi. W Baggs, Wyoming; *Lay;* Lily; Ladore. Utah: just N of Linwood. Wyoming (Long, 1965a:559): *mouth Burnt Fork, Henry's Fork; 15 mi. WSW Granger;*

Opal; *Fontenelle; Big Piney*; jct. Green River and New Fork; *2 mi. SE Big Sandy; 27 mi. N, 37 mi. E Rock Springs; 9 mi. S, 9 mi. W Waltman; 15 mi. N, 1 mi. W Waltman.*

Eutamias minimus neglectus (J. A. Allen)

1890. *Tamias quadrivittatus neglectus* J. A. Allen, Bull. Amer. Mus. Nat. Hist., 3:106, June, type from Montreal River, Ontario.

1922. *Eutamias minimus neglectus*, A. H. Howell, Jour. Mamm., 3:184, August 4.

1925. *Eutamias minimus jacksoni* A. H. Howell, Jour. Mamm., 6:53, February 9, type from Crescent Lake, Oneida Co., Wisconsin. Anderson and Rand, Canadian Field-Nat., 57:134, January 24, 1944, regard *E. m. jacksoni* as a synonym of *E. m. neglectus*. White, Univ. Kansas Publ., Mus. Nat. Hist., 5:618, December 1, 1953, accepts the subspecies as valid.

MARGINAL RECORDS.—Ontario: Lake Seul; Kapuskasing. Quebec: 6 mi. N Authier Nord. Ontario: Temagami; Lister Twp., Algonquin Park; Cache Lake, Algonquin Park; French River. Wisconsin: Marinette; *Waushara County* (Jackson, 1961:150, Map 30, pre-1900); Camp Douglas; Jackson County (Jackson, 1961:150, Map 30); mouth of Yellow River (*op. cit.*:154). Minnesota: Cass County; within 3 mi. Itasca State Park (Forbes, 1967:466); *Bear Paw Point, Itasca State Park* (Forbes, 1966:159). Manitoba: Caddy Lake, Sandilands Forest Reserve; Cedar Lake; Whiteshell Forest Reserve. Not found: Manitoba: Red Rock Lake. Ontario: Haveland Bay.

Eutamias minimus operarius Merriam

1905. *Eutamias amoenus operarius* Merriam, Proc. Biol. Soc. Washington, 18:164, June 29, type from Gold Hill, 7400 ft., Colorado.

1922. *Eutamias minimus operarius*, A. H. Howell, Jour. Mamm., 3:183, August 4.

1922. *Eutamias minimus arizonensis* A. H. Howell, Jour. Mamm., 3:178, August 4, type from Prieto Plateau, S end Blue Range, Greenlee Co., Arizona. Considered "synonymous" with *E. m. operarius* by Conley (Jour. Mamm., 51:700, November 30, 1970).

MARGINAL RECORDS.—Wyoming (Long, 1965a: 560): 6 mi. S, 2 mi. W Casper; Springhill; *27 mi. N, 7¹/₂ mi. E Laramie; 9 mi. N, 13 mi. E Laramie; 2 mi. W Horse Creek P.O.; 6 mi. W Islay.* Colorado: Livermore; Littleton; Elbert; Hardscrabble Canyon, 7 mi. above Wetmore; 5 mi. NE Great Sand Dunes National Monument (Armstrong, 1972:105); Blanca (*ibid.*); Fisher Peak. New Mexico: Raton Mts.; Halls Peak; 10 mi. NE Santa Fe; Sandia Crest, 9½ mi. E Alameda (Conley, 1970:701); Jemez Mts. Colorado (Armstrong, 1972:105): 5 mi. SW Alamosa; Florida. New Mexico: Chuska Mts. Arizona: Tunitcha Mts. Utah: Gooseberry Ranger Station, 8300 ft., Elk Ridge; Uncompahgre Indian Reservation. Colorado (Armstrong, 1972:103, 104, unless otherwise noted): ¼ mi. W Colorado National Monument, 6600 ft.; Middle Mamm Creek, near Rifle;

between Tennessee Pass and Leadville, 10,200 ft.; Mill Creek, Rocky Mountain National Park (Sutton and Nadler, 1969:526); 3 mi. SE Mountain Home, Wyoming. Wyoming (Long, 1965a:560): *5 mi. N, 5 mi. E Savory; S base Bridger Peak;* 6 mi. S, 10 mi. E Saratoga; *3 mi. SW Eagle Peak; 21¹/₂ mi. S, 24¹/₂ mi. W Douglas; 10 mi. S Casper; 7 mi. S, 2 mi. W Casper.* Also: New Mexico (Conley, 1970:701): R. 11 E, T. 10 S, ½ mi. N Sierra Blanca Peak, N ridge, 11,500 ft.; *R. 4 E, T. 11 S, E face Sierra Blanca Peak, 10,500–11,500 ft.* Arizona (*ibid.*): N fork White River, White Mts.; Alpine, 8000 ft.; Prieto Plateau, S end Blue Range, 9000 ft.

Eutamias minimus oreocetes Merriam

1897. *Eutamias oreocetes* Merriam, Proc. Biol. Soc. Washington, 11:207, July 1, type from Summit Mtn., at timberline, N of Summit Station, on Great Northern Railroad, Montana.

1922. *Eutamias minimus oreocetes*, A. H. Howell, Jour. Mamm., 3:183, August 4.

MARGINAL RECORDS.—Alberta: Forgetmenot Mtn.; Waterton Lakes National Park. Montana: type locality. British Columbia: *vic. Akamina Pass* (Cowan and Guiguet, 1965:137).

Eutamias minimus pallidus (J. A. Allen)

1874. *Tamias quadrivitatus* var. *pallidus* J. A. Allen, Proc. Boston Soc. Nat. Hist., 16:289. Type locality, Camp Thorne, near Glendive, Montana.

1922. *Eutamias minimus pallidus*, A. H. Howell, Jour. Mamm., 3:183, August 4.

MARGINAL RECORDS.—North Dakota: Williston; Goodall; Palace Buttes, 6 mi. N Cannon Ball; Parkin; 9 mi. E Dickinson (Genoways and Jones, 1972:11); Marmarth. South Dakota: NW ¼ sec. 15, R. 5 E, T. 22 N (Andersen and Jones, 1971:373); *2 mi. N, 5 mi. W Ludlow* (*ibid.*); 2 mi. S, 3¹/₄ mi. W Ludlow (*ibid.*); NE ¼ sec. 8, R. 8 E, T. 16 N (*ibid.*); Belle Fourche. Montana: Alzada. Wyoming (Long, 1965a:561, 562): 4 mi. S, 6 mi. W Rockypoint; *Moorcroft;* Thornton; *Upton; Newcastle.* South Dakota: *Wind Cave National Park* (Turner, 1974:64); Edgemont. Nebraska (Jones, 1964b:121): 10 mi. S Chadron; Hemingford; 6 mi. NW Harrison. Wyoming (Long, 1965a:561): *Laramie County;* 15 mi. SW Wheatland; *Douglas;* Powder River; head Bridger Creek; *2½ mi. W Shoshone; Meadow Creek; Fort Washakie;* 3 mi. S Dubois; 2 mi. S, 2 mi. E Meteetse. Montana: Little Bighorn River, 14 mi. S Crow Agency; Columbus; Dillon; Judith River; Jensen's Ranch; Darnell's Ranch, on Missouri River. Marginal records for *E. m. pallidus* in relation to *E. m. confinis* are, in Wyoming (Long, 1965a:561, 562): Sheridan; *Middle Butte, 38 mi. S, 19 mi. W Gillette; Otter Creek; 4 mi. N Hyattville;* Shell Creek, 1 mi. NW Shell.

For data on intergrades between *T. m. pallidus* and *T. m. silvaticus* of the Black Hills, and assignment of specimens to subspecies, see Turner (1974:63–68).

Eutamias minimus pictus (J. A. Allen)

1890. *Tamias minimus pictus* J. A. Allen, Bull. Amer. Mus. Nat. Hist., 3:115, June, type from Kelton, Utah.
1901. *Eutamias minimus pictus,* Miller and Rehn, Proc. Boston Soc. Nat. Hist., 30:42, December 27.
1890. *Tamias minimus melanurus* Merriam, N. Amer. Fauna, 4:22, October 8, type from W side Snake River, near Blackfoot, Idaho.

MARGINAL RECORDS.—Montana: Donovan. Idaho: Dubois; Idaho Falls; Pocatello; Malad City. Utah: Fairfield; Nephi; Ibapah; type locality. Idaho: 1 mi. N Idavada; Deer Flat; Salmon Valley, near Sawtooth City; Lemhi.

Eutamias minimus scrutator Hall and Hatfield

1934. *Eutamias minimus scrutator* Hall and Hatfield, Univ. California Publ. Zool., 40(6):321, February 12, type from near Blanco Mtn., 10,500 ft., White Mts., Mono Co., California.

MARGINAL RECORDS.—Washington: Ellensburg; Sunnyside. Oregon: Baker. Idaho: Silver City; 2½ mi. E Jordan Valley. Nevada: Goose Creek, 2 mi. W Utah boundary. Utah: 4 mi. S Gandy. Nevada: Eagle Valley, 3½ mi. N Ursine, 5900 ft.; Garden Valley, 8½ mi. NE Sharp; Burned Corral Canyon, Quinn Canyon Mts.; 3 mi. W Hamilton, 8400 ft.; Fish Spring Valley, ½ mi. N Fish Lake, 6500 ft.; 2½ mi. NE Silverbow, 7000 ft., Kawich Mts.; Toquima Mts., 2 mi. W Meadow Valley Ranger Station; Cloverdale Creek; Smiths Creek, 6800 ft.; Eldorado Canyon, Humboldt Range; Quinn River Crossing, 4100 ft.; 12 mi. N, 2 mi. E Gerlach, 4000 ft. California: Warm Spring, 9 mi. E Amedee. Nevada: Horse Canyon, Pahrum Peak, 5800 ft.; 2¾ mi. SW Pyramid; 6 mi. NE Virginia City, 6000 ft.; Cottonwood Canyon, Mt. Grant, 7400 ft.; N side Mt. Magruder, 7400 ft. California: 5 mi. SW Olancha Peak; Mazourka Canyon, Inyo Mts.; Mammoth; Diamond Valley, 1 mi. SE Woodfords; Vinton; Spaldings, Eagle Lake; 10 mi. SW Alturas; Mt. Hebron. Oregon: Tule Lake; Fremont; 2 mi. NE Prineville. Washington: Wiley City. Nevada: 1½ mi. NW Mountain Well, 6100 ft.

Eutamias minimus selkirki Cowan

1946. *Eutamias minimus selkirki* Cowan, Proc. Biol. Soc. Washington, 59:113, October 25, type from Paradise Mine, near Toby Creek, 19 mi. W Invermere, British Columbia. Known only from type locality.

Eutamias minimus silvaticus White

1952. *Eutamias minimus silvaticus* White, Univ. Kansas Publ., Mus. Nat. Hist., 5:261, April 10, type from 3 mi. NW Sundance, 5900 ft., Crook Co., Wyoming.

MARGINAL RECORDS.—Wyoming (Long, 1965a: 562): *15 mi. N Sundance, Black Hills National Forest;* 15 mi. ENE Sundance. South Dakota: Fort Meade; Rapid City; Buffalo Gap; 3 mi. SW Pringle (Turner,

1974:68). Wyoming (Long, 1965a:562): Newcastle; Devils Tower; *type locality.*

For data on intergrades between *T. m. silvaticus* and *T. m. pallidus* of Black Hills, and assignment of specimens to subspecies, see Turner (1974:63–68).

Eutamias amoenus
Yellow-pine Chipmunk

External measurements: 181–245; 73–108; 29–35. Greatest length of skull, 31.3–35.6. Some shade of reddish more prominent than in most subspecies of *E. minimus* (account of which see for comparison). Smaller, more tawny, and braincase more flattened than in *E. panamintinus.* Skull averaging smaller than in *E. umbrinus* or *E. speciosus.* In the Sierra Nevada of California where the three species occur in the same area, *E. amoenus* differs from *E. umbrinus* as follows: smaller; more reddish (less grayish) head and shoulders; broader inner light dorsal stripes; more ochraceous suffusion over light facial stripes and underparts; less massive skull with relatively broader braincase; less elevated rostrum; shorter upper incisors and less nearly parallel zygomatic arches; bent tip of baculum no more than, instead of more than, 35 per cent of length of shaft. *E. amoenus* differs from *E. speciosus,* as follows: smaller; shorter and broader-appearing ears; less sharply contrasting light and dark stripes; in *amoenus* the inner pair of light dorsal stripes is broader and the outer pair of light stripes, though not always narrower, is less conspicuous; light facial stripes more heavily washed with ochraceous; dark facial stripes less blackish; skull less massive; rostrum more pointed; incisive foramina smaller; bent tip of baculum no more than, instead of more than, 35 per cent of length of shaft (30–35% in *amoenus* and 47–55% in *speciosus*).

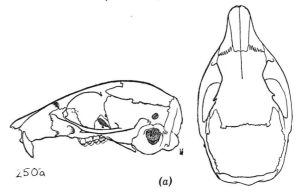

Fig. 250a. *Eutamias amoenus celeris,* Pine Forest Mtn., Nevada, No. 7931 M.V.Z., ♂, X 1½.

Eutamias amoenus affinis (J. A. Allen)

1890. *Tamias quadrivittatus affinis* J. A. Allen, Bull. Amer. Mus. Nat. Hist., 3:103, June, type from Ashcroft, British Columbia.
1922. *Eutamias amoenus affinis*, A. H. Howell, Jour. Mamm., 3:184, August 4.

MARGINAL RECORDS.—British Columbia (Cowan and Guiguet, 1965:140): Sorenson Lake; Clinton; Black Pines, 20 mi. N Kamloops; Kamloops; Glimpse Lake; Okanagan Mission; Osoyoos-Bridesville summit; Midway. Washington: Republic; Omak Lake; Chelan; Wenatchee; 10 mi. NW Ellensburg; Cleveland; 10 mi. N Grand Dalles; Mt. St. Helens; Easton; Wenatchee Lake; Hart Lake; Sheep Mts. British Columbia (Cowan and Guiguet, 1965:140): Ashnola River; Princeton; Lytton; Pavilion.

Eutamias amoenus albiventris Booth

1947. *Eutamias amoenus albiventris* Booth, Murrelet, 28(1):7, April 30, type from Wickiup Spring, 23 mi. W Anatone, Asotin–Garfield County boundary, Washington.

MARGINAL RECORDS.—Washington: Dayton; Anatone. Oregon: Cornucopia; Bourne; Meacham. Washington: Wallula (?); Prescott.

Eutamias amoenus amoenus (J. A. Allen)

1890. *Tamias amoenus* J. A. Allen, Bull. Amer. Mus. Nat. Hist., 3:90, June, type from Fort Klamath, Oregon.
1913. *Eutamias amoenus*, Merriam, Proc. Biol. Soc. Washington, 11:94, July 11.
1913. *Eutamias amoenus propinquus* Anthony, Bull. Amer. Mus. Nat. Hist., 32:6, March 7, type from Ironside, Malheur Co., Oregon.

MARGINAL RECORDS.—Idaho: vic. Riggins; Mill Creek, 14 mi. W Challis; Birch Creek, 10 mi. S Nicholia; mts. E of Birch Creek; Schutt's Mine; 8 mi. W Swan Lake; Malad. Utah: Pine Canyon, 6600 ft., Raft River Mts., 17 mi. NW Kelton; S Fork George Creek, 7000 ft., Raft River Mts., 5 mi. SE Yost. Nevada: summit between heads Copper and Coon creeks, Jarbidge Mts.; Cobb Creek, 6 mi. SW Mountain City. Oregon: Cedar Mts.; Steens Mts. Nevada: 12-Mile Creek, ½ mi. E California boundary. California: E face Warren Peak, 8700 ft., Warner Mts.; 13 mi. NE Termo (Sutton and Nadler, 1969:524); *8 mi. S Susanville;* Millford; 1½ mi. SE Sierraville; Lincoln Creek, 6200 ft.; Chaparral; *Sisson;* 3 mi. SW Weed; near head Little Shasta River, 4000 ft., N side Goosenest Mtn. Oregon: type locality; Diamond Lake; Sisters; Antelope; McEwen.

Eutamias amoenus canicaudus Merriam

1903. *Eutamias canicaudus* Merriam, Proc. Biol. Soc. Washington, 16:77, May 29, type from Spokane, Washington.
1922. *Eutamias amoenus canicaudus*, A. H. Howell, Jour. Mamm., 3:184, August 4.

MARGINAL RECORDS.—Washington: Marcus. Idaho: 5 mi. W Cocolalla. Montana: Prospect Creek, near Thompson Falls. Idaho: 2 mi. NE Weippe. Washington: Pullman; 8 mi. SW Waterville (Sutton and Nadler, 1969:525); ½ mi. E Devils Lake.

Eutamias amoenus caurinus Merriam

1898. *Eutamias caurinus* Merriam, Proc. Acad. Nat. Sci. Philadelphia, 50:352, October 4, type from Olympic Mts., Washington (timberline, near head Soleduck River).
1922. *Eutamias amoenus caurinus*, A. H. Howell, Jour. Mamm., 3:184, August 4.

MARGINAL RECORDS.—Washington: Happy Lake, Olympic Mts.; Deer Park; "near" head Dosewallips River; type locality.

Eutamias amoenus celeris Hall and Johnson

1940. *Eutamias amoenus celeris* Hall and Johnson, Proc. Biol. Soc. Washington, 53:155, December 19, type from near head Big Creek, 8000 ft., Pine Forest Mts., Humboldt Co., Nevada.

MARGINAL RECORDS.—Nevada: Alder Creek, 7000–8000 ft.; ridge "near" Pine Forest Mtn.; 30 mi. SW Denio (Sutton and Nadler, 1969:525).

Eutamias amoenus cratericus Blossom

1937. *Eutamias amoenus cratericus* Blossom, Occas. Pap. Mus. Zool., Univ. Michigan, 366:1, December 21, type from Grassy Cone, Craters of the Moon National Monument, 6000 ft., 26 mi. SW Arco, Butte Co., Idaho.

MARGINAL RECORDS.—Idaho: Sunset Cone; near Big Cinder Butte, Craters of the Moon; *30 mi. SW Arco, at S Base White Knob Mts.*

Eutamias amoenus felix (Rhoads)

1895. *Tamias quadrivittatus felix* Rhoads, Amer. Nat., 29:941, October, type from Church Mountain, New Westminster district, British Columbia, near international boundary.
1922. *Eutamias amoenus felix*, A. H. Howell, Jour. Mamm., 3:184, August 4.

MARGINAL RECORDS.—British Columbia: Purcell Point (Cowan and Guiguet, 1965:140); Alta Lake (*ibid.*); Tami Hy [= Tamihi] Creek. Washington: 19 mi. E Glacier (Sutton and Nadler, 1969:525); Mt. Baker. British Columbia (Cowan and Guiguet, 1965:140): North Vancouver; Gibsons Landing; Savary Island; *Lund;* Fawn Bluff.

Eutamias amoenus ludibundus Hollister

1911. *Eutamias ludibundus* Hollister, Smiths. Miscl. Coll., 56(26):1, December 5, type from Yellowhead Lake, 3700 ft., British Columbia.
1922. *Eutamias amoenus ludibundus*, A. H. Howell, Jour. Mamm., 3:184, August 4.

MARGINAL RECORDS.—Alberta: head Smoky River; Henry House; Astoria Creek. British Columbia: Clearwater Lake, Wells Gray Park (Cowan and Guiguet, 1965:142); Canim Lake; Horse Lake (Cowan and Guiguet, 1965:142); Lillooet; Coalmont. Washington: Hidden Lakes; Lyman Lake; Mt. Stuart; Lake Kachess; Boulder Cave; Bethel Ridge, 6000 ft., 30 mi. ESE Mt. Rainier, 3 mi. N Tieton Reservoir (Broadbooks, 1965:301, 302, as *E. amoenus* only); *Spirit Lake* (Sutton and Nadler, 1969:525); Mt. St. Helens. Oregon: Mt. Hood; Wapinitia; Three Sisters; O'Leary Mtn., 10 mi. S McKenzie Bridge. Washington: Glacier Basin, Mt. Rainier; *Winchester Mtn., Twin Lakes.* British Columbia: second summit, United States–Canada boundary, W of Skagit River; Hope; McGillivary Creek, Lillooet District; Quesnel; Indianpoint Lake (Cowan and Guiguet, 1965:142).

Cowan and Guiguet (1965:139–143) failed to account for continuity between the populations of the subspecies *E. a. ludibundus* in Washington and Oregon with the populations north of lat. 51° N in British Columbia. Because of that failure, the map herewith accords with A. H. Howell's (1929:74, 75) identification of specimens as subspecies *E. a. ludibundus* in British Columbia south of lat. 51° N to an extent that a corridor in British Columbia connects the northernmost and southernmost populations of *E. a. ludibundus.* The statement of Cowan and Guiguet (1965:142) that "specimens from the Cascade Range in British Columbia and Washington . . . are quite unlike topotypes" points up the need for further study in order to correctly assign a subspecific name (or names) to the populations south of lat. 51° N that Howell (*loc. cit.*) identified as *E. a. ludibundus* (in Washington and Oregon as well as in part of British Columbia).

Eutamias amoenus luteiventris (J. A. Allen)

1890. *Tamias quadrivittatus luteiventris* J. A. Allen, Bull. Amer. Mus. Nat. Hist., 3:101, June, type from "Chief Mountain Lake" [= Waterton Lake], Alberta (3½ mi. N United States–Canada boundary).

1922. *Eutamias amoenus luteiventris,* A. H. Howell, Jour. Mamm., 3:179, August 4.

MARGINAL RECORDS.—British Columbia: Kinbasket Lake. Alberta: Banff; foothills 40 mi. W Calgary; Burmis. Montana: St. Marys Lake; Highwood Mts.; Buffalo; Crazy Mts.; Reed Point; Red Lodge. Wyoming (Long, 1965a:563, 564): 16¼ mi. N, 17 mi. W Cody; Valley; 3¾ mi. E, 1 mi. S Moran; Merna; *Stanley;* LaBarge Creek. Idaho: head Crow Creek, Preuss Mts.; Warm River; Salmon River Mts.; Fiddle Creek; Seven Devils Mts.; Craig Mts. Montana: Superior; Thompson Falls. Washington: Newport; 15 mi. W Marcus. British Columbia (Cowan and Guiguet, 1965:142): Midway; Anarchist Mtn.; Rayleigh.

Eutamias amoenus monoensis Grinnell and Storer

1916. *Eutamias amoenus monoensis* Grinnell and Storer, Univ. California Publ. Zool., 17:3, August 23, type from Warren Fork of Leevining Creek, 9200 ft., Mono Co., California.

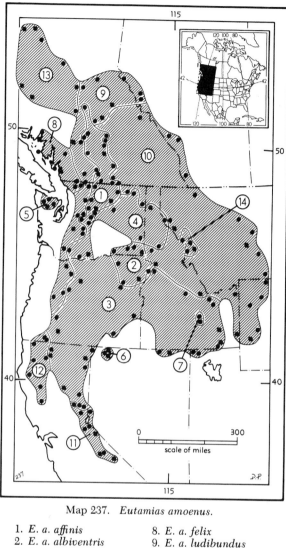

Map 237. *Eutamias amoenus.*

1. *E. a. affinis*
2. *E. a. albiventris*
3. *E. a. amoenus*
4. *E. a. canicaudus*
5. *E. a. caurinus*
6. *E. a. celeris*
7. *E. a. cratericus*
8. *E. a. felix*
9. *E. a. ludibundus*
10. *E. a. luteiventris*
11. *E. a. monoensis*
12. *E. a. ochraceus*
13. *E. a. septentrionalis*
14. *E. a. vallicola*

MARGINAL RECORDS.—California: Mohawk, 4400 ft. Nevada: W side Truckee River, 1 mi. W Verde, 4900 ft. California: Diamond Valley, 5500 ft., 1 mi. SE Woodfords; Swager Creek, 7600 ft., Sweetwater Range; Pine City, 8700 ft., near Mammoth; 5 mi. N Westgard Pass (Sutton and Nadler, 1969:525); Long Valley, 7300 ft., near Convict Creek; Silver Lake; 5 mi. SW Soda Springs (Sutton and Nadler, 1969:525).

Eutamias amoenus ochraceus A. H. Howell

1925. *Eutamias amoenus ochraceus* A. H. Howell, Jour. Mamm., 6:54, February 9, type from Studhorse Canyon, 6500 ft., Siskiyou Mts., California.

MARGINAL RECORDS.—Oregon: Ashland Peak; Siskiyou. California: head Deadfall Creek, W slope Mt. Eddy; Snow Mtn.; W side Thomas Creek, South Yolla Bolly Mtn.; near Blake Lookout, South Fork Mtn., 5700 ft.; head Redcap Creek, 5800 ft., 10 mi. E, 4 mi. N Hoopa; head E fork Dunn Creek, Siskiyou Mts.

Eutamias amoenus septentrionalis Cowan

1946. *Eutamias amoenus septentrionalis* Cowan, Proc. Biol. Soc. Washington, 59:110, October 25, type from Ootsa Lake P.O., on N shore Ootsa Lake, British Columbia.

MARGINAL RECORDS.—British Columbia: 8 mi. W Babine (Howell, 1929:74); Babine Lake; Vanderhoof (Howell, 1929:75); *Sinkut Mts.;* Puntchesakut Lake; *Stum Lake* (Cowan and Guiguet, 1965:143); Hanceville; *Cariboo Mtn. near Atnarko River* (Cowan and Guiguet, 1965:143); Stuie (*ibid.*); Kimsquit; *Rocher Déboulé;* Hazelton (Howell, 1929:75).

Eutamias amoenus vallicola A. H. Howell

1922. *Eutamias amoenus vallicola* A. H. Howell, Jour. Mamm., 3:179, August 4, type from Bass Creek, 3725 ft., near Stevensville, Montana.

MARGINAL RECORDS.—Montana: Lolo; Skalkaho Road; Canyon Creek; type locality; *6 mi. SW Lolo* (Sutton and Nadler, 1969:525).

townsendii-subgroup

C. H. Merriam in 1897 (194–199), A. H. Howell in 1929 (106–120), D. H. Johnson in 1943 (110–139), and Sutton and Nadler in 1974 (199–211), in four steps, have substantially increased knowledge of speciation and subspeciation of this subgroup. Additional studies, especially of differences in coloration in each of the two pelages (summer and winter), and determination of the form of the genitalia in certain populations, remain to be made before an ideally accurate map showing the geographic distribution of the taxa can be drawn.

I am indebted to Professor Dallas A. Sutton for the outlines of the ossa genitalia beyond of taxa of this species group (left sides of the baubella are shown).

Eutamias townsendii
Townsend's Chipmunk

External measurements: 235–263; 96–125; 34–38; 15–18. Greatest length of skull, 36.8–39.5. Length of the head and body approx. 139; average weight of adult males, 85 grams; coloration rich tawny to ochraceous tawny; top of head sayal brown or cinnamon to smoke gray, bordered on each side by fuscous black to fuscous; light facial stripes pinkish buff to cream white; underparts creamy white to grayish white. *E. t. townsendii* is the darker and more richly colored of the two subspecies. Baculum and os clitoris as shown in Fig. 250*b*. Skull resembling that of *E. senex.*

250b   **baculum**   **baubellum**

(*b*)

Fig. 250*b*. Ossa genitalia, *Eutamias t. townsendii*, lateral views, baculum right side, and baubella left side; after Sutton and Nadler (1974:208, 210), X 10.

Eutamias townsendii cooperi (Baird)

1855. *Tamias cooperi* Baird, Proc. Acad. Nat. Sci. Philadelphia, 7:334, type from Klickitat Pass, 4500 ft., Cascade Mts., Skamania Co., Washington (see Cooper, Amer. Nat., 2:531, December, 1968).
1919. *Eutamias townsendii cooperi,* Taylor, Proc. California Acad Sci., ser. 4, 9:110, July 12.

MARGINAL RECORDS.—British Columbia: Roab's Ranch, near Hope; Lightning Lakes near Allison Pass. Washington: Barron; Stehekin; Entiat River, 20 mi. from mouth; mts. near Wenatchee; McAllister Meadows, Tieton River; 9 mi. SW Fort Simcoe; White Salmon. Oregon: Parkdale; *3 mi. E Government Camp* (Sutton and Nadler, 1969:526); Wapinitia; O'Leary Mtn., 10 mi. S McKenzie Bridge; Fish Lake (Sutton and Nadler, 1969:525); halfway between Drew and Crater Lake (A. H. Howell, 1929:114, as *E. t. siskiyou*); Prospect (*op. cit.:* 117, as *E. t. senex*); Anchor; Glendale; Reston; Vida; Detroit. Washington: Skamania; Yacolt; White River, 3100 ft., Glacier National Park; Index Peak, 2700 ft.; Glacier. British Columbia: *Lihumitson Park* (Cowan and Guiguet, 1965:147); *Cultus Lake;* Chilliwack Lake. Washington: Chilliwack Creek, 30 mi. E Glacier.

Eutamias townsendii townsendii (Bachman)

1839. *Tamias Townsendii* Bachman, Jour. Acad. Nat. Sci. Philadelphia, 8(pt. 1):68, type from lower Columbia River, near lower mouth Willamette River, Oregon.
1897. *E[utamias]. townsendi,* Merriam, Proc. Biol. Soc. Washington, 11:192, July 1.
1842. *Tamias hindei* (typographical error for *hindsii*) Gray, Ann. Mag. Nat. Hist., ser. 1, 10:264, December, type locality not definitely known.
1903. *Tamias townsendii littoralis* Elliot, Field Columb. Mus., Publ. 74, Zool. Ser., 3(10):153, May 2, type from Marshfield, Oregon.

MARGINAL RECORDS.—British Columbia: New Westminster; Mt. Lehman; Chilliwack; Skagit. Washington: Chilliwack Creek, Whatcom Co. British Co-

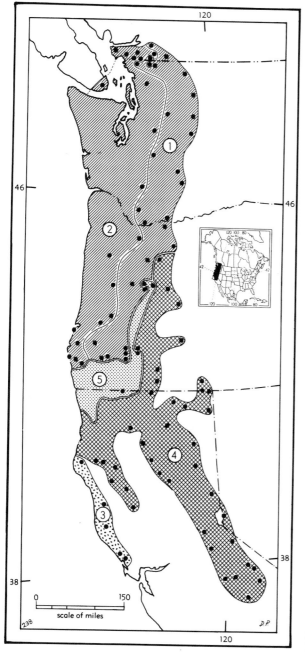

Map 238. *Eutamias townsendii* and related species.

1. *E. t. cooperi*   3. *E. ochrogenys*
2. *E. t. townsendii* 4. *E. senex*
        5. *E. siskiyou*

lumbia (Cowan and Guiguet, 1965:147): Church Mtn.; Sumas. Washington: Hamilton; North Bend; *Spirit Lake* (Sutton and Nadler, 1969:526); Mt. St. Helens. Oregon: Bissell; 10 mi. N Corvallis (Sutton and Nadler, 1969:526); Eugene; Elk Head; Oakland; Myrtle Point; Agness (A. H. Howell, 1929:114, as *E. t. siskiyou*); Rogue River Mts. (*op. cit.*:113, as *E. t. ochrogenys*);

Port Orford (*ibid.*, as *E. t. ochrogenys*); Empire, thence up coast to British Columbia: Esquimalt, Vancouver Island [introduced]; *South Westminster* (Cowan and Guiguet, 1965:147).

Eutamias ochrogenys Merriam
Redwood Chipmunk

1897. *Eutamias townsendii ochrogenys* Merriam, Proc. Biol. Soc. Washington, 11:195, 206, July 1, type from Mendocino, California.
1974. *Eutamias ochrogenys*, Sutton and Nadler, Southwestern Nat., 19:211, July 26.

External measurements: 252–277; 107–126; 37–39; 15–18. Greatest length of skull, 39–40.8. In length of head and body (approx. 159) largest member of subgenus *Neotamias*. Coloration in winter pelage dark tawny olive above; underparts grayish white strongly washed with pinkish buff or light pinkish cinnamon. In summer, pelage more tawny above than in winter, and underparts cinnamon buff (adapted from A. H. Howell, 1929:112, 113). This is the darkest member of the *townsendii*-subgroup. Skull averaging larger than in *E. senex* and *E. siskiyou*. Height of keel of baculum less than a quarter the length of tip of baculum (see Fig. 250c).

Sutton and Nadler (1974:205) offer a sophisticated mathematical procedure for separating this taxon from *E. senex* and *E. siskiyou*.

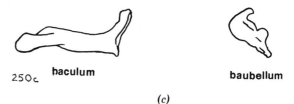

250c   baculum   baubellum

(c)

Fig. 250c. Ossa genitalia, *Eutamias ochrogenys*, lateral views, baculum right side, and baubella left side; after Sutton and Nadler (1974:208, 210), X 10.

MARGINAL RECORDS.—California: in redwood forests within *ca.* 25 mi. of Pacific Coast from Van Duzen River (Sutton and Nadler, 1974:203) southward to 5 mi. N Willits; Lake Leonard, 10 mi. NW Ukiah; Cazadero; *7 mi. SE Guerneville* (Sutton and Nadler, 1974:200); Freestone, thence up coast to point of beginning.

Eutamias senex (J. A. Allen)
California Chipmunk

1890. *Tamias senex* J. A. Allen, Bull. Amer. Mus. Nat. Hist., 3:83, June, type from summit of Donner Pass, Placer Co., California.
1897. *Eutamias senex*, Merriam, Proc. Biol. Soc. Washington, 11:196, July 1.

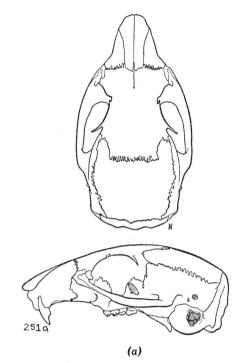

Fig. 251a. *Eutamias senex*, Blue Canyon, 5000 ft., Placer Co., California, No. 18870 M.V.Z., ♂, X 1½.

External measurements: 229–258; 95–112; 35–38. Greatest length of skull, 37.3–39.8. In Sierra Nevada and Cascades winter pelage paler than in *E. siskiyou;* top of head mixed pinkish cinnamon and fuscous spinkled with grayish white, bordered on each side with a stripe of fuscous; dark dorsal stripes fuscous black, more or less mixed with mikado brown, the median stripe usually darkest; light dorsal stripes grayish white, and the median pair sometimes faintly clouded with cinnamon; lateral stripes mikado brown; underparts creamy white. Summer pelage: sides darker (about sayal brown) and general tone of upper parts more ochraceous; median pair of light dorsal stripes often strongly mixed with pinkish buff (adapted from A. H. Howell, 1929:115). Sutton and Nadler (1974:205) offer a sophisticated mathematical procedure for separating this taxon from *E. ochrogenys* and *E. siskiyou.* Shape and great depth of shaft of baculum as well as great depth of os clitoris are distinctive (see Fig. 251b). California population west of Coast Range between Klamath and Van Duzen rivers resembling *E. ochrogenys* in being darker and larger than other populations of *E. senex* (*fide* Sutton and Nadler, 1974:203).

MARGINAL RECORDS.—Oregon: Mill Creek, 20 mi. W Warm Springs; Bend; Arnold Ice Cave; West Silver Creek, 4650 ft., Silver Lake; Naylox; Klamath

Fall. California: Picard; 20 mi. NW Canby, 4500 ft.; Lassen Creek. Oregon: Lakeview. California: Fort Bidwell; head N Fork Parker Creek, 5500 ft., Warner Mts.; 10 mi N Canby; Fort Crook; 5 mi. N Fredonyer Peak, 5700 ft.; 8 mi. S Susanville; Sierra Valley. Nevada: Glenbrook. California: Silver Creek; 1 mi. W Leevining (Sutton and Nadler, 1969:526); 10 mi. SE Leevining (Sutton and Nadler, 1974:200); Mammoth; Shaver Ranger Station, 5300 ft.; "vic." Chinquapin, 6200–7500 ft.; Crane Flat, 6300 ft.; N Fork Stanislaus River, 6700 ft.; Slipperyford (= Kyburz); [20 mi. SW] Quincy; 1 mi. S Inskip (Sutton and Nadler, 1974:200); Lyonsville; Baird; Castle Lake, 5434 ft.; Mt. Tomhead, 5000 ft.; Grindstone Creek, 6500 ft.; Snow Mtn.; 4 mi. S South Yolla Bolly Mtn.; Horse Ridge, 5500 ft., SE of Ruth (Johnson, 1943:115, as *E. t. siskiyou*); Van Duzen River, 12 mi. E Bridgeville (*ibid.,* as *E. t. siskiyou*), west along north bank Van Duzen River to Pacific Coast, thence up coast to mouth of Klamath River, thence eastward along south bank of that river into Oregon, thence northward to Crater Lake National Park (Sutton and Nadler, 1974:200). See Map 238.

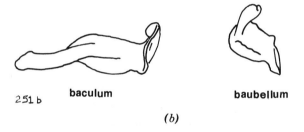

251 b   **baculum**   **baubellum**

(b)

Fig 251b. Ossa genitalia, *Eutamias senex*, lateral views, baculum right side, and baubella left side; after Sutton and Nadler (1974:208, 210), X 10.

Eutamias siskiyou A. H. Howell
Siskiyou Chipmunk

1922. *Eutamias townsendii siskiyou* A. H. Howell, Jour. Mamm., 3:180, August 4, type from near summit White Mtn., 6000 ft., Siskiyou Mts., California.
1974. *Eutamias siskiyou,* Sutton and Nadler, Southwestern Nat., 19:211, July 26.

External measurements: 250–268; 98–117; 35–38. Greatest length of skull, 39–40.8. Coloration: winter pelage darker than in *E. senex* but more grayish (less brownish) than in *E. ochrogenys;* top of head fuscous spinkled with pinkish cinnamon and grayish white; median dorsal stripes black; outer pair fuscous black, overlaid with sayal brown. General tone of upper parts in summer pelage more brownish than in winter pelage; in summer pelage outer pair of light dorsal stripes clear grayish white; inner pair much clouded with cinnamon; sides ochraceous tawny; otherwise, as in winter pelage (adapted from A. H. Howell, 1929:113, 114). Shape and shortness (less than 2 mm.) of shaft of baculum is distinctive in the *townsendii*-subgroup. See Fig. 251c.

Sutton and Nadler (1974:205) offer a sophisticated mathematical procedure for separating this taxon from *E. ochrogenys* and *E. senex*.

MARGINAL RECORDS AND BOUNDARY OF RANGE.—Oregon: McKenzie Bridge (Sutton and Nadler, 1974:200); W base Three Sisters, 5000 ft. (A. H. Howell, 1929:114, probably this species); Crater Lake National Park (Sutton and Nadler, 1974:200). California: W fork Cottonwood Creek, 4000 ft., 4½ mi. SW Hilt (Johnson, 1943:115), thence westward along N bank Klamath River to Pacific Coast, north up coast to mouth Rogue River, Oregon (*fide* Sutton and Nadler, 1974:200, 203), and eastward (along S bank Rogue River) almost to Crater Lake National Park, and thence north to point of beginning. See Map 238.

251c **baculum**   **baubellum**

(c)

Fig. 251c. Ossa genitalia, *Eutamias siskiyou*, lateral views, baculum right side, and baubella left side; after Sutton and Nadler (1974:208, 210), X 10.

Eutamias sonomae
Sonoma Chipmunk

External measurements: 220–272; 93–126; 32.0–39. Greatest length of skull, 36.6–39.7. Resembles *E. townsendii* but differs as follows: body paler; legs, tail, and ears longer; tail broader, more bushy; cheeks in winter, gray instead of brown; ears, in summer pelage, sparsely furred and unicolored instead of well furred and bicolored; central reddish area on underside of tail becoming paler rather than darker anteriorly; skull narrower; braincase relatively larger and more inflated; zygomatic arches more closely appressed to skull; anterior tips of nasals separated from one another by notch; incisive foramina shorter; posterior edge of palate thickened and having short spine instead of terminating in long, slender (thin) spine; upper incisors more recurved and angle of notch across occlusal surfaces more acute; cheek-teeth smaller.

Eutamias sonomae alleni A. H. Howell

1922. *Eutamias townsendii alleni* A. H. Howell, Jour. Mamm., 3:181, August 4, type from Inverness, Marin Co., California.

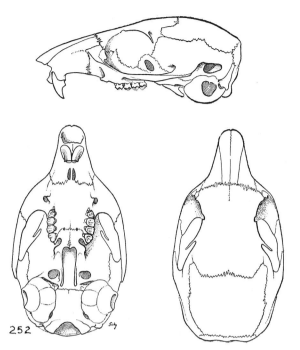

Fig. 252. *Eutamias sonomae sonomae*, Parks Creek, 1½ mi. SW Edgewood, 2900 ft., Siskiyou Co., California, No. 69197 M.V.Z., ♀, X 1½.

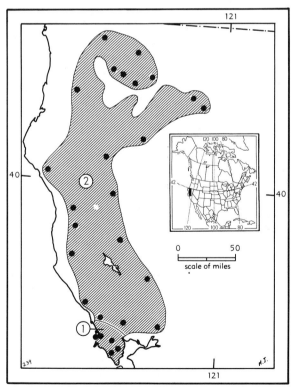

Map 239. *Eutamias sonomae.*

1. *E. s. alleni*   2. *E. s. sonomae*

MARGINAL RECORDS.—California: Point Reyes, 5 mi. W Inverness; type locality; Nicasio; Mailliard; Bolinas Ridge, 1350 ft., 2½ mi. S Lagunitas.

Eutamias sonomae sonomae Grinnell

1915. *Eutamias sonomae* Grinnell, Univ. California Publ. Zool., 12:321, January 20, type from 1 mi. W Guerneville, Sonoma Co., California.

MARGINAL RECORDS.—California: Seiad Valley, 1400 ft.; Forest House, 3000 ft., 3 mi. S Yreka; 1 mi. S Weed; Scott Mts., W of Gazelle; Scott River, 6 mi. NW Callahan; Salmon Mts., SW of Greenview; Dana; Fort Crook; Redding; 3 mi. W Knob; Coast Range, Tehama Co., 17 mi. W Paskenta; Fouts Springs; Rumsey, 500 ft.; Vacaville; Eldridge; Freestone, 300 ft.; 7 mi. W Cazadero; Christine; Sherwood; Laytonville; Briceland; Hoopa Valley. Californian localities not found are: Lime Gulch, 1 mi. E Castle Peak; Mt. Mill Hotel; Howell Mtn. (from Johnson, 1943:126), and Castle Peak; Hermitage; Kunz (from A. H. Howell, 1929:119).

Eutamias merriami
Merriam's Chipmunk

External measurements: 233–277; 84–140; 28–40. Greatest length of skull, 35.5–40.7. Tail long and bushy (75–97% of length of head and body); feet and ears long and slender; ears sparsely furred on convex surfaces in summer pelage; color grayish (notably ochraceous near coast); dorsal stripes all of approx. equal width; dark stripes gray or brown, seldom with black areas; light stripes grayish; cheeks and underparts white, more or less dulled by gray but in coastal areas usually suffused with ochraceous; tail-edging usually dull white but slightly buffy in some specimens; dorsal stripes more or less indistinct in winter pelage of *E. m. kernensis*, in this respect resembling *E. dorsalis* and *E. obscurus*; skull resembling that of *E. sonomae*. Large size distinguishes *E. merriami* from all species with which it shares its range except *E. quadrimaculatus* and *E. townsendii. E. merriami* differs from the two last named in longer and bushier tail, the edging of which is dull white or slightly buffy rather than pure white; narrower skull; more recurved incisors; presence of notch between anterior tips of nasals. It also differs from *E. townsendii* in grayish instead of brownish cheeks, and from *E. quadrimaculatus* in much paler submalar stripes. Compare ossa genitalia of *E. merriami* and *E. obscurus*.

Eutamias merriami kernensis Grinnell and Storer

1916. *Eutamias merriami kernensis* Grinnell and Storer, Univ. California Publ. Zool., 17:5, August 23, type from Fay Creek, 4100 ft., 6 mi. N Weldon, Kern Co., California.

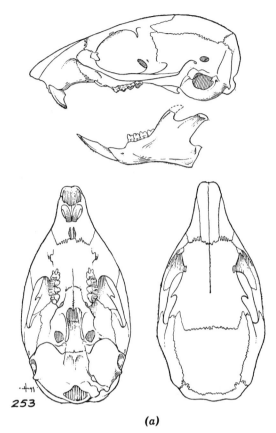

253

(a)

Fig. 253a. *Eutamias merriami pricei*, Santa Cruz, Santa Cruz Co., California, No. 233 K.U., ♂, X 1½.

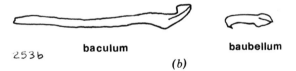

253b   **baculum**   **baubellum**

(b)

Fig. 253b. Ossa genitalia, *Eutamias m. merriami*, lateral views right side; after Callahan (1977:189), X 10.

MARGINAL RECORDS.—California: Onion Valley, 8500 ft., Sierra Nevada; W slope Walker Pass, 4600 ft.; 2 mi. N Sorell's Ranch, 4500 ft., Kelsoe Valley; French Gulch, 6700–7300 ft.; Kern River, 12 mi. below Bodfish; Kern River at Isabella; forks Big and Little Kern rivers. Not found: California: Claraville (Sutton and Nadler, 1969:525).

Eutamias merriami merriami (J. A. Allen)

1889. *Tamias asiaticus merriami* J. A. Allen, Bull. Amer. Mus. Nat. Hist., 2:176, October 21, type from San Bernardino Mts., 4500 ft., due north of San Bernardino, California.
1897. *E[utamias]. merriami*, Merriam, Proc. Biol. Soc. Washington, 11:191, July 1.
1916. *Eutamias merriami mariposae* Grinnell and Storer, Univ. California Publ. Zool., 17:4, August 23, type from El Portal, 2000 ft., Mariposa Co., California.

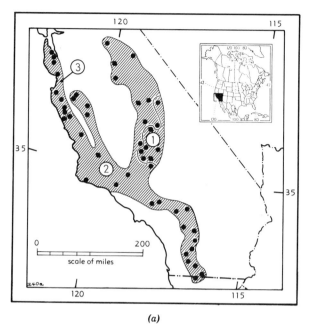

Map 240a. *Eutamias merriami.*

1. *E. m. kernensis*   2. *E. m. merriami*
                 3. *E. m. pricei*

MARGINAL RECORDS.—California: ¼ mi. E Columbia; "vic." Columbia Point, 5000 ft.; Kings River Canyon, 5000 ft.; Jordan Hot Springs, 6700 ft.; Doyle's Camp; Glenville; Tehachapi Peak; Pine Flats, N Fork San Gabriel River; E end Big Bear Lake (Callahan, 1977:199); Hidden Lake, 9000 ft., "near" Round Valley; Kenworthy (Callahan, 1977:199); Warner Pass; Laguna Mts.; Mountain Spring. Baja California: Nachogüero Valley. California: Cuyamaca Mts.; Witch Creek; Poppet Flat, 3700–4000 ft., San Jacinto Mts.; Santa Ana River, 5500–6000 ft.; Mt. Wilson, 5700 ft.; Matilija; Bulitos Creek, 7 mi. W Gaviota; Mission Creek, 2 mi. N San Antonio Mission; Paso Robles; head San Juan River, Carrizo Plains; Waltham Creek, 1850 ft., 4½ mi. SE Priest Valley; near Cook P.O., 1300 ft., Bear Valley; Butts Ranch, 3000 ft., 5 mi. NNE San Benito; 1 mi. SE summit San Benito Mtn., 4400 ft.; Old Fort Tejon, 3200 ft.; 1 mi. S Dunlap; Raymond, 940 ft.; 1 mi. W Coulterville, 1600 ft. Not found: California: Smith Mtn., San Diego Co. (A. H. Howell, 1929:126); Dark Canyon, Riverside Co. (Johnson, 1943:134); Sugar Loaf (Sutton and Nadler, 1969:525).

Eutamias merriami pricei (J. A. Allen)

1895. *Tamias pricei* J. A. Allen, Bull. Amer. Mus. Nat. Hist., 7:333, November 8, type from Portola, San Mateo Co., California.
1899. *(Eutamias Merriami) Pricei,* Trouessart, Catalogus mammalium . . . , fasc. 6 (appendix), p. 1312 (received June, 1899).

MARGINAL RECORDS.—California: Redwood City; Palo Alto; Arroyo Quito; Corralitos; San Francis-

quito Ranch; Chews Ridge; 2 mi. SW Abbotts; Santa Lucia Peak; vic. Chalk Peak.

Eutamias obscurus
Baja California Chipmunk

External measurements: 208–240; 75–120; 30–37. Greatest length of skull, 33.3–39.6. Adult summer pelage resembling that of *E. merriami kernensis* but paler, less yellowish, and dark dorsal stripes more reddish. Ossa genitalia distinctive.

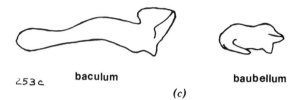

253c   **baculum**   **baubellum**
                              (c)

Fig. 253c. Ossa genitalia, *Eutamias o. obscurus,* lateral views, right side; after Callahan (1977:193), X 10.

Eutamias obscurus davisi Callahan

1977. *Eutamias obscurus davisi* Callahan, Jour. Mamm., 58:193, May 31, type from Barker's Reservoir, 1300 m., 16 km. SW Twentynine Palms, San Bernardino Co., California.

MARGINAL RECORDS (Callahan, 1977:200).—California: Northern Segment: Doble; type locality; Eagle Mts.; Pinyon Wells; 1½ mi. NW Big Bear City. Southern Segment: vic. Fuller's Mill; Toro Peak; Kenworthy; Black Mtn.

Eutamias obscurus meridionalis Nelson and Goldman

1909. *Eutamias merriami meridionalis* Nelson and Goldman, Proc. Biol. Soc. Washington, 22:23, March 10, type from Aguaje de San Esteban, approx. 1200 ft., about 25 mi. NW San Ignacio, Baja California.
1977. *Eutamias obscurus meridionalis,* Callahan, Jour. Mamm., 58:197, May 31.

MARGINAL RECORDS.—Baja California: San Pablo; *Rancho Las Calabasas, 17 mi. SE San Pablo* (Larson, 1964:634); type locality.

Eutamias obscurus obscurus (J. A. Allen)

1890. *Tamias obscurus* J. A. Allen, Bull. Amer. Mus. Nat. Hist., 3:70, June, type from Sierra San Pedro Mártir, near Vallecitos, Baja California.
1897. *E[utamias]. obscurus,* Merriam, Proc. Biol. Soc. Washington, 11:194, July 1.

MARGINAL RECORDS.—California (Callahan, 1977:200): Mountain Spring. Baja California: Hanson Laguna, Hanson Laguna Mts.; Rosarito Divide, San Pedro Mártir Mts.; El Rayo, Hanson Laguna Mts.

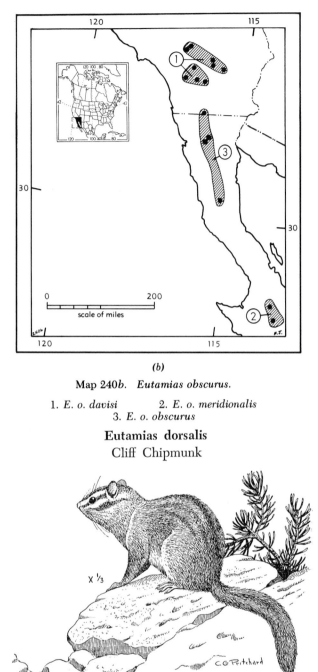

(b)

Map 240*b*. *Eutamias obscurus.*

1. *E. o. davisi*   2. *E. o. meridionalis*
3. *E. o. obscurus*

Eutamias dorsalis
Cliff Chipmunk

X ⅓

External measurements: 208–277; 89–140; 32.5–39.0. Greatest length of skull, 35.5–40.1. Upper parts smoke gray or neutral gray; dorsal stripes indistinct and in some stages of pelage obsolete; median stripe more pronounced than lateral stripes; postauricular patches grayish-white and poorly defined; upper side of tail fuscous-black overlain with tilleul buff; tail ochraceous-tawny, cinnamon or pinkish buff be-

low, bordered with fuscous black and edged with tilleul buff or grayish white; underparts creamy white, in some specimens tinged with buff. Top of braincase flattened, but slightly less so than in *E. panamintinus;* incisive foramina diverging posteriorly rather than parallel.

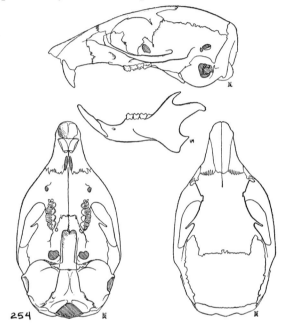

Fig. 254. *Eutamias dorsalis grinnelli,* SW base Groom Baldy, 7200 ft., Nevada, No. 47949 M.V.Z., ♂, X 1½.

Eutamias dorsalis carminis Goldman

1938. *Eutamias dorsalis carminis* Goldman, Proc. Biol. Soc. Washington, 51:56, March 18, type from Carmen Mts., 7400 ft., Coahuila.

MARGINAL RECORDS.—Coahuila: type locality; Sierra de la Madera, 23 mi. S, 5 mi. W Ocampo, 7500 ft.

Eutamias dorsalis dorsalis (Baird)

1855. *Tamias dorsalis* Baird, Proc. Acad. Nat. Sci. Philadelphia, 7:332, April, type from Fort Webster, copper-mines of the Mimbres, near present site of Santa Rita, Grant Co., New Mexico.
1897. *E[utamias]. dorsalis,* Merriam, Proc. Biol. Soc. Washington, 11:210, July 1.
1904. *Eutamias canescens,* J. A. Allen, Bull. Amer. Mus. Nat. Hist., 20:208, May 28, type from Guanaceví, Durango.

MARGINAL RECORDS.—Arizona: Supai Village, Cataract Creek, thence eastward along S bank Colorado River to Desert View (Hoffmeister, 1971:169); 1¾ mi. E Desert View (ibid.); SE [Grand Canyon] park boundary, route 64 (ibid.); Walnut, near Winona; Springerville. New Mexico: Datil Mts. Arizona: Fort Defiance. New Mexico (Findley, *et al.,* 1975:107): 1 mi. S Toadlena, Chuska Mts.; 15 mi. NE San Mateo

(Ranger tank); Armijo Lake, Sandia Mts.; Water Canyon, Magdalena Mts.; *2 mi. E South Baldy*; Kingston. Chihuahua (Anderson, 1972:272, 273): E side San Luis Mts., near Monument 63; 5 mi. NE Pacheco; 10 mi. WSW San Buenaventura; 34 mi. SSE Cuauhtémoc, 7000 ft.; Sierra Madre, *ca.* 10 km. ENE Guasarachi, 6000 ft. Durango: Guanaceví; 18 mi. S Tepehuanes, 8100 ft. (Baker, 1966:345). Chihuahua (Anderson, 1972:273): Sierra Madre, near Guadalupe y Calvo; El Cajon, 5 mi. W Churo, 6350 ft. Sonora: above Santa María Mine, near El Tigre. Arizona: Chiricahua Mts.; Rincon Mts.; *Santa Catalina Mts.* (Sutton and Nadler, 1969:525); Oracle; Fish Creek, Tonto National Forest; Weaver Mts.; Hualpai Mts.; Peach Springs; Pine Spring, Hualpai Indian Reservation.

Eutamias dorsalis grinnelli Burt

1931. *Eutamias dorsalis grinnelli* Burt, Jour. Mamm., 12:300, August 24, type from Mormon Well, Sheep Mts., 6500 ft., Clark Co., Nevada.

MARGINAL RECORDS.—Nevada: Goose Creek, 2 mi. W Utah line, 5000 ft. Utah: Pilot Range (Egoscue,

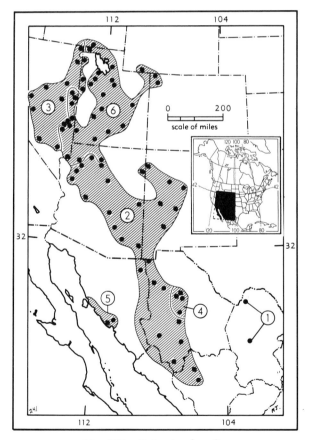

Map 241. *Eutamias dorsalis.*

1. *E. d. carminis*   4. *E. d. nidoensis*
2. *E. d. dorsalis*   5. *E. d. sonoriensis*
3. *E. d. grinnelli*   6. *E. d. utahensis*

1968:145); 1½ mi. N Gold Hill (Egoscue, 1968:146); Deep Creek Mts. (*ibid.*). Nevada: Eagle Valley, 3½ mi. N Ursine, 5600 ft.; 2 mi. SE Pioche, 6000 ft.; type locality; *Hidden Forest, Sheep Mts., 8500 ft.*; Belted Range, 5 mi. W White Rock Spring, 6950–7050 ft.; Manhattan; Wisconsin Creek, 7600 and 7800 ft.; 8 mi. W Eureka; Cherry Creek, 6800 ft.; Water Canyon, 8–10 mi. N Lund; Silver Zone Pass, 20 mi. NW Wendover (Egoscue, 1968:146).

Eutamias dorsalis nidoensis Lidicker

1960. *Eutamias dorsalis nidoensis* Lidicker, Proc. Biol. Soc. Washington, 73:267, December 30, type from 5 mi. N Cerro Campana, 5600 ft., Chihuahua, México.

MARGINAL RECORDS (Lidicker, 1960:272, unless otherwise noted).—Chihuahua: Arroyo del Nido, 30 mi. SW Gallego, 7000 ft.; type locality; 8 mi. NE Laguna, 7250 ft. (Anderson, 1972:273); Cañón del Alamo, Sierra del Nido, 7300 ft.; *Arroyo Mesteño, Sierra del Nido, 7600 ft.*

Eutamias dorsalis sonoriensis Callahan and Davis

1977. *Eutamias dorsalis sonoriensis* Callahan and Davis, Southwestern Nat., 22:71, March 1, type from 24 km. NE Guaymas, 100 m., Sonora.

MARGINAL RECORDS (Callahan and Davis, 1977:72).—Sonora: type locality; S end San Carlos Bay. Differences between *E. d. sonoriensis* and *E. d. dorsalis* in shape of ossa genitalia are shown (*op. cit.:*71).

Eutamias dorsalis utahensis Merriam

1897. *Eutamias dorsalis utahensis* Merriam, Proc. Biol. Soc. Washington, 11:210, July 1, type from Ogden, Utah.

MARGINAL RECORDS.—Idaho: near Bridge. Utah: type locality; Draper; Provo. Wyoming (Long, 1965a:564): *W side Green River, 1 mi. N Utah border*; Green River, 4 mi. NE Linwood, *in* Utah. Colorado (Armstrong, 1972:97): Colorado Highway 318, 8 mi. N jct. Moffat Co.; between Little Snake River Bridge and Lily. Utah: 7 mi. N Greenriver, 4100 ft.; Henry Mts. (Egoscue, 1968:146); 8 mi. S Escalante, 5200 ft.; 2 mi. SW Cave Lake Canyon, 5 mi. NW Kanab. Arizona: Kwagunt Creek, 2 mi. W Colorado River (Hoffmeister, 1971:169); *Shiva Temple* (*ibid.*); Green Spring, 6000 ft., 1 mi. S, 4 mi. E Mt. Dellenbaugh (Hoffmeister and Durham, 1971:29). Nevada: Cedar Basin, 30 mi. SE St. Thomas, 3500 ft.; Meadow Valley Wash, 7 mi. S Caliente, 4000 ft.; 11 mi. E Panaca; Lehman Cave, 7500 ft.; Smith Creek, Mt. Moriah, 6600 ft.; Hendry Creek, 7½ mi. SE Mt. Moriah, 6800 ft.; ¼ mi. W Utah boundary, lat. 38° 17′ N, 7300 ft. Utah: Hebron; Beaver; Tushar Mts. (Egoscue, 1968:146); 4 mi. E Oak City (*ibid.*); N end Granite Peak (Egoscue, 1964:389, as *E. dorsalis* only); N Cane Springs, Cedar Mts. (*ibid.*, as *E. dorsalis* only); Lakeside Mts. (Egoscue, 1968:146); George Creek Road jct., 5 mi. SE Yost, Raft River Mts.

Eutamias quadrivittatus
Colorado Chipmunk

External measurements: 197–235; 80–110; 28.4–35.0. Greatest length of skull, 33.4–36.8. Head smoke gray shaded with reddish; dorsal stripes distinct; underside of tail reddish; bent tip of baculum amounting to 30–44 per cent of length of shaft of baculum; cranial breadth averaging between 16.0 and 16.8 mm.

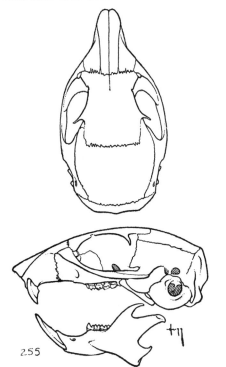

Fig. 255. *Eutamias quadrivittatus quadrivittatus*, 2 mi. S, 4 mi. W Coyote, 8100 ft., Rio Arriba Co., New Mexico, No. 41802 K.U., ♀, X 1½.

Eutamias quadrivittatus hopiensis Merriam

1905. *Eutamias hopiensis* Merriam, Proc. Biol. Soc. Washington, 18:165, June 29, type from Keam Canyon, Painted Desert, Arizona.
1922. *Eutamias quadrivittatus hopiensis*, A. H. Howell, Jour. Mamm., 3:184, August 4.

MARGINAL RECORDS.—Colorado (Armstrong, 1972:108): Ladder [Ladore?] Canyon, Dinosaur National Monument; 4 mi. NNW Cross Mtn.; White River, 20 mi. E Rangely; Atchee, 6600 ft.; 8 mi. W Rifle; McCoy; between Eagle and Wolcott, 6800 ft.; Somerset; 10 mi. SW Delta, 7025 ft.; Coventry; 1 mi. N Cahone, 6900 ft.; headquarters, Mesa Verde National Park. Arizona: type locality. Utah: Rainbow Bridge, 4000 ft. (Durrant, 1952:149); Fruita; 2 mi. W Orangeville; E side confluence Green and White rivers, 1 mi. SE Ouray, 4700 ft.

Eutamias quadrivittatus quadrivittatus (Say)

1823. *Sciurus quadrivittatus* Say, *in* Long, Account of an exped. . . . to the Rocky Mts. . . . , 2:45. Type locality, Arkansas River, Colorado, about 26 mi. below Cañon City.
1901. *Eutamias quadrivittatus*, Miller and Rehn, Proc. Boston Soc. Nat. Hist., 30(1):43, December 27.
1890. *Tamias quadrivittatus gracilis* J. A. Allen, Bull. Amer. Mus. Nat. Hist. 3:99, June, type from San Pedro, Socorro Co., New Mexico.
1909. *Eutamias quadrivittatus animosus* Warren, Proc. Biol. Soc. Washington, 22:105, June 25, type from Irwin Ranch, Las Animas Co., Colorado.

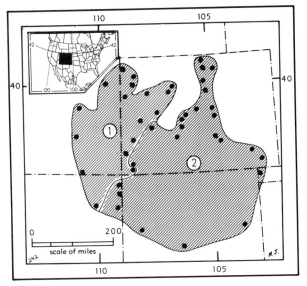

Map 242. *Eutamias quadrivittatus.*

1. *E. q. hopiensis*   2. *E. q. quadrivittatus*

MARGINAL RECORDS.—Colorado: Elkhorn; Spring Canyon, 7 mi. SE Fort Collins; Palmer Lake; Colorado Springs; Wetmore, Hardscrabble Canyon; 25 mi. SE Pueblo (Armstrong, 1972:111); Guame's Ranch; Baca County. Oklahoma: Kenton. New Mexico (Findley, *et al.*, 1975:110): 4 mi. NW Tucumucari; Manzano Peak; Agua Fria Creek, sec. 28, T. 10 N, R. 12 W. Arizona: 6 mi. S, 3 mi. W Sawmill (Hoffmeister and Nader, 1963:93); *4 mi. S, 3 mi. W Sawmill* (*ibid.*); Wheatfield Creek, W slope Tunitcha Mts., 7000 ft. (*ibid.*); summit Lukachukai Mts., 8000 ft., 15 mi. E Lukachukai Navajo School (*ibid.*). Colorado (Armstrong, 1972:111, 112): 1 mi. NW Dolares; 8 mi. W Sapinero; 5 mi. N, 22 mi. W Saguache, 10,000 ft.; 1½ mi. S Monarch; 5 mi. W Buena Vista; Cottonwood Springs; Tarryall Creek Camp; Lookout Mtn., near Golden; Boulder; Arkins.

Eutamias ruficaudus
Red-tailed Chipmunk

External measurements: 223–248; 101–121; 32–36. Greatest length of skull, 33.9–36.2. Upper

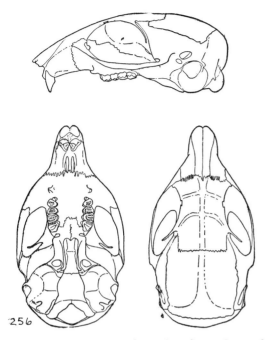

Fig. 256. *Eutamias ruficaudus ruficaudus*, Lolo Creek, 6½ mi. W Lolo, 3470 ft., Montana, No. 93358 M.V.Z., ♂, X 1½.

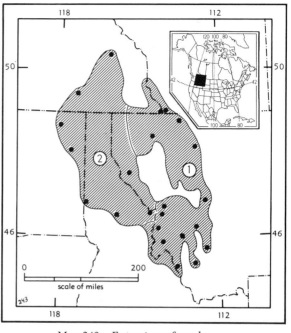

Map 243. *Eutamias ruficaudus*.

1. *E. r. ruficaudus*   2. *E. r. simulans*

parts and sides deep tawny; underside of tail ochraceous tawny to Sanford Brown; light dorsal stripes grayish brown, often mixed with ochraceous tawny; skull resembling that of *E. umbrinus umbrinus* but averaging smaller, with narrower rostrum and interorbital region; baculum 4 mm. or longer.

Eutamias ruficaudus ruficaudus A. H. Howell

1920. *Eutamias ruficaudus* A. H. Howell, Proc. Biol. Soc. Washington, 33:91, December 30, type from Upper St. Marys Lake, Montana.

MARGINAL RECORDS.—Alberta: Waterton Lake Park. Montana: type locality; summit Great Northern Railroad; Rogers Pass, Lewis and Clark Co. (Hoffmann, *et al.*, 1969:585, as *E. ruficaudus* only); *Pipestone Pass, 14 mi. SE Butte* (*ibid.*); Highland Mts., Silver Bow Co. (*ibid.*); Deer Lodge; head Birch Creek, Beaverhead Co. (Hoffmann, *et al.*, 1969:585, as *E. ruficaudus* only); Cutaway Pass, Granite Co. (*ibid.*); Upper Miner Lake, near Jackson (*ibid.*); Deer Mtn., Ravalli Co. (*ibid.*); Blodgett Creek, Ravalli Co. (*ibid.*); Bass Creek, Bitterroot Mts., NW Stevensville; 6½ mi. W Lolo; Upper Stillwater Lake. British Columbia: *Akamina Pass; Sage Pass.*

Eutamias ruficaudus simulans A. H. Howell

1922. *Eutamias ruficaudus simulans* A. H. Howell, Jour. Mamm., 3:179, August 4, type from Coeur d'Alene, Kootenai Co., Idaho.

MARGINAL RECORDS.—British Columbia: Invermere. Montana: Prospect Creek, near Thompson Falls. Idaho: Packers Meadow; 12 mi. E Weippe; Moscow. Washington: Loon Lake; Marcus. British Columbia: Nelson. White (1953:623) supposes that *E. r. simulans* does not intergrade with *E. r. ruficaudus* and is a distinct species.

Eutamias canipes
Gray-footed Chipmunk

Revised by Fleharty, Jour. Mamm., 41:235–242, May 20, 1960.

External measurements: 227–264; 91–115; 32–36. Greatest length of skull, 36.1–39.1. Resembles *E. cinereicollis* but paler; length of shaft of baculum, 3.1–3.49 mm. (4.7 in *E. cinereicollis*); angle formed by shaft and tip of baculum, 112–120° instead of 138–148°.

Eutamias canipes canipes V. Bailey

1902. *Eutamias cinereicollis canipes* V. Bailey, Proc. Biol. Soc. Washington, 15:117, June 2, type from Dog Canyon, 7000 ft., Guadalupe Mts., Texas.
1960. *E*[*utamias*]. *canipes canipes*, Fleharty, Jour. Mamm., 41:241, May 20.

MARGINAL RECORDS (A. H. Howell, 1929:102, unless otherwise noted).—New Mexico: 3 mi. S, 9 mi. W Corona (Fleharty, 1960:241); Jicarilla Mts.; Capitan Mts.; Ruidosa. Texas: McKittrick Canyon, 5900 ft.

(Davis and Robertson, 1944:267); Sierra Diablo (Davis, 1966:142, as *E. canipes* only). New Mexico: Mescalero.

Eutamias canipes sacramentoensis Fleharty

1960. *Eutamias canipes sacramentoensis* Fleharty, Jour. Mamm., 41:240, May 20, type from Sacramento Mountains, 1 mi. S Cloudcroft, 9000 ft., Otero Co., New Mexico.

MARGINAL RECORDS.—New Mexico: Cloudcroft (Sutton and Nadler, 1969:525); *type locality*.

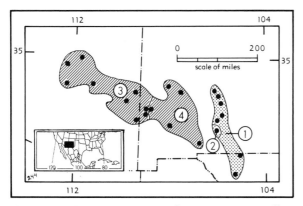

Map 244. *Eutamias canipes* and *Eutamias cinereicollis*.

1. *E. canipes canipes*
2. *E. canipes sacramentoensis*
3. *E. cinereicollis cinereicollis*
4. *E. cinereicollis cinereus*

Eutamias cinereicollis
Gray-collared Chipmunk

External measurements: 208–250; 90–115; 32–36. Greatest length of skull, 35.1–38.4. Resembles *E. umbrinus* but more grayish (less tawny), especially on shoulders; skull averaging longer.

Eutamias cinereicollis cinereicollis (J. A. Allen)

1890. *Tamias cinereicollis* J. A. Allen, Bull. Amer. Mus. Nat Hist., 3:94, June, type from San Francisco Mtn., Arizona.
1901. *Eutamias cinereicollis*, Miller and Rehn, Proc. Boston Soc. Nat. Hist., 30(1):40, December 27.

MARGINAL RECORDS.—Arizona: Little Spring, NW base San Francisco Mtn.; Springerville. New Mexico: 10 mi. NE Mogollon; 8 mi. SE Mogollon. Arizona: Blue River; Mt. Thomas, White Mts.; Mogollon Rim (Clothier, 1969:642, as *E. cinereicollis* only); *Baker Butte;* Mayer; Bill Williams Mtn.

Eutamias cinereicollis cinereus V. Bailey

1913. *Eutamias cinereicollis cinereus* V. Bailey, Proc. Biol. Soc. Washington, 26:130, May 21, type from Copper Canyon, 8200 ft., Magdalena Mts., New Mexico.

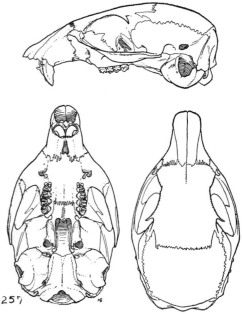

Fig. 257. *Eutamias cinereicollis cinereicollis*, Hannagan Meadows, 9500 ft., Arizona, No. 55373 M.V.Z., ♂, X 1½.

MARGINAL RECORDS.—New Mexico: Datil Range, 22 mi. NW Fort Tularosa; type locality; Organ Mts.; 6 mi. N, 15 mi. E Santa Rita (Fleharty, 1960:242); Mogollon Mts.

Eutamias quadrimaculatus (Gray)
Long-eared Chipmunk

1867. *Tamias quadrimaculatus* Gray, Ann. Mag. Nat. Hist., ser. 3, 20:435, December, type from Michigan Bluff, Placer Co., California.
1897. *E[utamias]. quadrimaculatus*, Merriam, Proc. Biol. Soc. Washington, 11:191, July 1.
1886. *Tamias macrorhabdotes* Merriam, Proc. Biol. Soc. Washington, 3:25, January 27, type from Blue Canyon, Placer Co., California.

External measurements: 200–250; 85–118; 34–37. Greatest length of skull, 36.3–38.5. Resembles *E. townsendii*, but differs as follows: smaller; longer ears; more brightly colored; generally more reddish and less grayish with less grayish dulling of white areas; bushier tail with more conspicuous white edging; less massive skull; longer nasals; narrower and shallower rostrum; shorter incisive foramina; smaller molariform teeth.

MARGINAL RECORDS.—California: 1 mi. SSE Prattville; 8 mi. NW Greenville; Grizzly Mtn. Nevada: 3 mi. S Mt. Rose, 8500 ft.; 10 mi. NW Minden. California: Markleeville; near jct. Sunrise Trail and Cloud's Rest Trail, 7000 ft.; Bass Lake; near Gentry's, Big Oak Flat Road; Placerville; Merrimac.

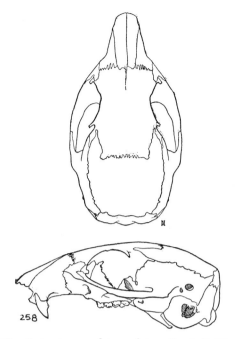

Fig. 258. *Eutamias quadrimaculatus*, 3 mi. S Mt. Rose, 8500 ft., Nevada, No. 88265 M.V.Z., ♂, X 1½.

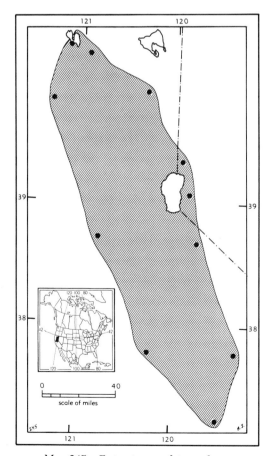

Map 245. *Eutamias quadrimaculatus*.

Eutamias speciosus
Lodgepole Chipmunk

External measurements: 197–241; 67–114; 30–36. Greatest length of skull, 33.5–36.1. This chipmunk lives in or near dense stands of lodgepole pine in the damper and more sheltered basins of the Sierra Nevada, whereas *E. umbrinus* there lives in the open forest of stunted limber pine and white bark pine on exposed and well-drained ridges and slopes near timber line. Differences between the two species are: in *E. speciosus*, size slightly smaller, ears longer, tail shorter, color darker, facial stripes darker, submalar stripe having (instead of lacking) black center below eye, subterminal area on underside of tail 20 (vs. 10) mm. long, skull shorter and broader, upper incisors shorter and more recurved, their outer borders forming an arc of a circle having a radius of 5 instead of 5.7 mm. When the skull rests on a horizontal surface and is viewed from the side, a perpendicular line through the posterior border of the alveolus falls anterior, rather than posterior, to the tip of the incisor. In *E. speciosus* the auditory bullae are smaller and the zygomatic arches converge anteriorly instead of being approx. parallel.

The late J. Grinnell and Tracy I. Storer in their "Animal life in the Yosemite," Univ. California Press, 1924, published the definitive account of this species. They wrote that it is perhaps the most abundant of the seven species of chipmunks inhabiting the Yosemite section and state that it "is the only one of the local chipmunks which habitually takes refuge well up in trees."

Eutamias speciosus callipeplus (Merriam)

1893. *Tamias callipeplus* Merriam, Proc. Biol. Soc. Washington, 8:136, December 28, type from Mt. Piños, 8800 ft., Ventura Co., California.
1897. *Eutamias speciosus callipeplus* (Merriam), Proc. Biol. Soc. Washington, 11:202, July 1.

MARGINAL RECORDS.—California: 1 mi. NE Mt. Piños, 8000 ft.; 3 mi. NW Frazier Borax Mine, 8100 ft.; *type locality*. Not found: Cañon de las Uvas (A. H. Howell, 1929:91).

Eutamias speciosus frater (J. A. Allen)

1890. *Tamias frater* J. A. Allen, Bull. Amer. Mus. Nat. Hist., 3:88, June, type from Donner, California.
1897. *Eutamias speciosus frater*, Merriam, Proc. Biol. Soc. Washington, 11:194, July 1.

MARGINAL RECORDS.—California: Eagle Lake; Campbell's Hot Springs, 5200 ft. Nevada: 3 mi. S Mt. Rose, 8500 ft.; Genoa (probably at higher elevation). California: Swager Creek, 7600 ft., Sweetwater Range;

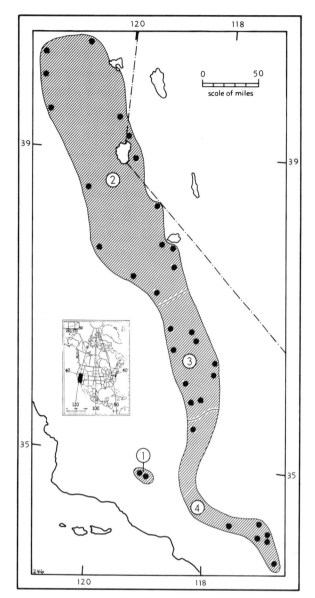

Map 246. *Eutamias speciosus.*

1. *E. s. callipeplus*　　3. *E. s. sequoiensis*
2. *E. s. frater*　　4. *E. s. speciosus*

Tioga Pass (Sutton and Nadler, 1969:526); Mono Craters, 8000 ft.; near Mammoth; Huntington Lake, 7000 ft.; Devils Peak, 6500 ft., near Fish Camp; vic. Crane Flat; Wrights Lake; Chaparral; Summit Creek, 5200 ft., 2 mi. E Mineral; Upper Lost Creek.

Eutamias speciosus sequoiensis A. H. Howell

1922. *Eutamias speciosus sequoiensis* A. H. Howell, Jour. Mamm., 3:180, August 4, type from Mineralking, 7300 ft., E fork Kaweah River, California.

MARGINAL RECORDS.—California: Hume, 5300 ft.; Kings River Canyon, 5000 ft.; Bubbs Creek; Little

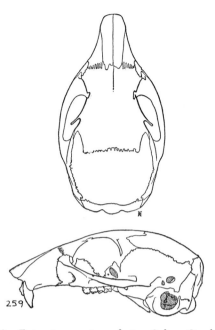

Fig. 259. *Eutamias speciosus frater*, Galena Creek, 8950 ft., Nevada, No. 88261 M.V.Z., ♂, X 1½.

Cottonwood Creek, 9500 ft.; Little Brush Meadow, 9700 ft., E slope Olancha Peak; Taylor Meadow, 7000 ft.; Cannell Meadow, 7500 ft.; headwaters N. Tule River; Alta Trail, Giant Forest.

Eutamias speciosus speciosus (Merriam)

1890. *Tamias speciosus* Merriam, in J. A. Allen, Bull. Amer. Mus. Nat. Hist., 3:86, June, type from Whitewater Creek, 7500 ft., San Bernardino Mts., California.
1897. *Eutamias speciosus* (Merriam), Proc. Biol. Soc. Washington, 11:194, July 1.

MARGINAL RECORDS.—California: French Gulch, Piute Mts.; Fawnskin Park; Sugarloaf; Dry Lake, 9000 ft.; Taquitz Valley, 8000 ft.; *Camp Angelus* (Sutton and Nadler, 1969:526); Mt. San Bernardino, 9000 ft.; Alpine City.

Eutamias panamintinus
Panamint Chipmunk

External measurements: 192–220; 80–102; 28.4–32.5. Greatest length of skull, 33.1–34.8. The bright, tawny colors with conspicuously gray rump, and flattened braincase are outstanding characteristics. Differences from *E. minimus* of the same geographic areas are: size larger; color more reddish, especially in summer pelage; central area of underside of tail more reddish and wider; skull larger and braincase more flattened. Differences from *E. amoenus* are: slightly shorter feet and ears; paler; gray rather than brown crown; narrower and paler facial stripes; less conspicuous dark dorsal stripes of which only

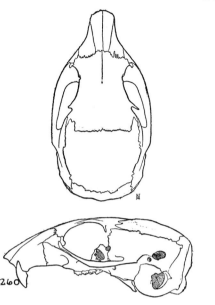

Fig. 260. *Eutamias panamintinus panamintinus*, ½ mi. W Wheeler Well, Nevada, No. 52149 M.V.Z., ♂, X 1½.

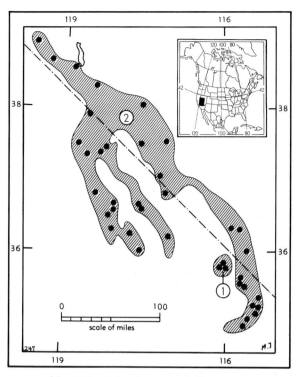

Map 247. *Eutamias panamintinus*.

1. *E. p. acrus*   2. *E. p. panamintinus*

part of the median one is black and the lateral pair almost obsolete; relatively narrower inner and broader outer stripes; skull, in general, broader; roof of braincase more nearly flat; nasals less prolonged anteriorly; incisive foramina longer; upper incisors less recurved. Differences from *E. quadrivittatus* are: lesser size; more reddish (less grayish) shoulders; less solidly black and less solidly white dorsal stripes; broader and more flattened braincase; shorter upper incisors; smaller cheek-teeth; less nearly parallel zygomatic arches. From *E. palmeri, E. panamintinus* differs in essentially the same fashion as from *E. umbrinus*. From *E. speciosus, E. panamintinus* differs in color in approx. the same way as from *E. umbrinus* but more pronouncedly. Braincase approx. same length but more flattened in *E. panamintinus*, which has less divergent upper tooth-rows and more pointed rostrum. From *E. quadrimaculatus, E. panamintinus* can be distinguished by lesser size alone; also the tail has an edging of buff rather than white. Differences from *E. dorsalis* are: reddish instead of grayish upper parts; reddish and distinct, instead of grayish and indistinct, submedian dark dorsal stripes; more flattened braincase; parallel, instead of anteriorly convergent, incisive foramina.

Eutamias panamintinus acrus Johnson

1943. *Eutamias panamintinus acrus* Johnson, Univ. California Publ. Zool., 48:94, December 24, type from 1¼ mi. SE Horse Spring, 5000 ft., Kingston Range, northeastern San Bernardino Co., California.

MARGINAL RECORDS.—California: Horse Spring, 4700 ft., Kingston Range; type locality; 2 mi. SW Horse Spring.

Eutamias panamintinus panamintinus (Merriam)

1893. *Tamias panamintinus* Merriam, Proc. Biol. Soc. Washington, 8:134, December 28, type from Johnson Canyon, near lower edge Piñon belt at approx. 5000 ft., vic. Hungry Bill's Ranch, Panamint Mts., California.
1897. *Eutamias panamintinus* Merriam, Proc. Biol. Soc. Washington, 11:194, July 1.
1931. *Eutamias panamintinus juniperus* Burt, Jour. Mamm., 12:298, August 24, type from ½ mi. W Wheeler Well, W slope Charleston Mts., Clark Co., Nevada.

MARGINAL RECORDS.—Nevada: Andersons Ranch; 2 mi. SW Pine Grove, 7250 ft.; Cottonwood Creek, Mt. Grant; Endowment Mine, Excelsior Mts.; Springdale Canyon, 6650 ft., Lone Mtn.; W side Stonewall Mtn., 6000 ft.; *Clark Canyon* (Deacon, *et al.*, 1964:402); Kyle Canyon, Charleston Peak; N side Potosi Mtn., 5800–8000 ft. California: 5 mi. SW Ivanpah, 4500 ft.; Government Holes; 5 mi. NE Granite Well, 5400 ft.; pass between Granite Mts. and Providence Mts.; Mitchell's; Cedar Canyon, 5000–5300 ft.; Mescal Cave; SE side Clark Mtn., 6300 ft.; N side Clark Mtn., 5400 ft. Nevada: W slope Charleston Peak, Wheeler Well. California: Fall Canyon, 5600 ft., Grapevine Mts. Nevada: 1 mi. S, 2½ mi. E Grapevine

Peak, 6700 ft.; Mt. Magruder. California: Roberts Ranch, 8250 ft., Wyman Creek; Hanaupah Canyon; 3 mi. E Jackass Spring; 3 mi. N Jackass Spring, 6000 ft.; 2½ mi. SE head Black Canyon, 8000 ft., White Mts.; Lone Pine; Hockett Trail, vic. Carrol Creek; Canyon, 5 mi. SW Olancha; Coso Mts.; Mountain Spring, Argus Mts.; Little Cottonwood Creek, 9000 ft., Sierra Nevada; Onion Valley; Bishop Creek, 7000 ft.; Rock Creek, 6200 ft., near Sherwin Hill, 21 mi. NW Bishop; near Antelope Peak, 6500 ft., 5 mi. N Benton.

Eutamias umbrinus
Uinta Chipmunk

External measurements: 196–243; 73–115; 30–35. Greatest length of skull, 33.5–36.8; ". . . pelage dark; sides dark; cranium narrow; baculum distinguishable from that of any other species (*E. palmeri* excepted) by combination of width of base more than ⅓ of length of shaft, distal ½ of shaft laterally compressed, and keel ¼ of length of tip." (White, 1953:571.)

This species has long been confused with *E. quadrivittatus* from which it can be distinguished externally only by the most subtle differences. Bacula of the two species differ much. The geographic range probably is more extensive than is now known; White (1953:620) remarks that bacula from as far west as Lardo, Valley Co., Idaho, seem to be of this species.

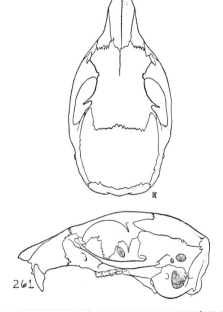

Fig. 261. *Eutamias umbrinus inyoensis*, mouth Pole Canyon, S side Baker Creek, E side Snake Mts., Nevada, No. 41574 M.V.Z., ♂, X 1½.

Eutamias umbrinus adsitus J. A. Allen

1905. *Eutamias adsitus* J. A. Allen, Mus. Brooklyn Inst. Arts and Sci., Sci. Bull., 1:118, March 31, type from Brigg's [= Britt's] Meadow, 10,000 ft., Beaver Mts., Utah.
1953. *Eutamias umbrinus adsitus*, White, Univ. Kansas Publ., Mus. Nat. Hist., 5:572, December 1.

MARGINAL RECORDS.—Utah: Great Basin Experiment Station; Carcass Creek, Grover, 7255 ft.; 18 mi. N Escalante, 9500 ft.; Bryce Canyon, 8200 ft.; *Cedar Breaks, 10,000 ft.*; West Rim, Zion National Park, 6500 ft.; Pine Valley Mts.; type locality. Also in Arizona: Jacob Lake (Sutton and Nadler, 1969:526); De Motte Park; *Point Imperial, 8800 ft.* (Hoffmeister, 1971:169); *Greenland Lake, Walhalla Plateau (ibid.)*; Bright Angel Spring; Powell Plateau (Hoffmeister, 1971:169).

Eutamias umbrinus fremonti White

1953. *Eutamias umbrinus fremonti* White, Univ. Kansas Publ., Mus. Nat. Hist., 5:576, December 1, type from 31 mi. N Pinedale, 8025 ft., Sublette Co., Wyoming.

MARGINAL RECORDS.—Montana (Hoffmann, *et al.*, 1969:586, as *E. umbrinus* only): 3 mi. NNE Cooke; Quad Creek, Beartooth Plateau. Wyoming (Long, 1965a:566): 16¼ mi. N, 17 mi. W Cody; Valley; 12 mi. N, 3 mi. W Shoshone; 17 mi. S, 6½ mi. W Lander; *Big Sandy*; LaBarge Creek. Idaho: Big Hole Mts., near Irwin. Wyoming: Tower Falls (Sutton and Nadler, 1969:527).

Eutamias umbrinus inyoensis Merriam

1897. *Eutamias speciosus inyoensis* Merriam, Proc. Biol. Soc. Washington, 11:202, 208, July 1, type from Black Canyon, 8200 ft., White Mts., Inyo Co., California.
1953. *Eutamias umbrinus inyoensis*, White, Univ. Kansas Publ., Mus. Nat. Hist., 5:573, December 1.

MARGINAL RECORDS.—Utah: head George and Clear creeks, 8500 ft., 5 mi. S Stanrod, Raft River Mts.; *Queen of Sheba Canyon, W side Deep Creek Mts., 8000 ft.*; Granite Creek Canyon, Deep Creek Mts., 6370 ft. (Egoscue, 1964:389, as *E. umbrinus* only). Nevada: Eagle Valley, 3½ mi. N Ursine, 5800 ft.; E and N slopes Irish Mtn.; Kawich Mts., 2⅝ mi. E Silverbow, 7300 ft. California: S slope Cirque Peak, 10,500 ft.; Mammoth Pass, 9800 ft. Nevada: Chiatovich Creek, 8200 ft.; Smiths Creek, 7100 ft.; 4 mi. S Tonkin, Denay Creek, Roberts Mts.; head Ackler Creek.

Eutamias umbrinus montanus White

1953. *Eutamias umbrinus montanus* White, Univ. Kansas Publ., Mus. Nat. Hist., 5:576, December 1, type from ½ mi. E and 3 mi. S Ward, 9400 ft., Boulder Co., Colorado.

MARGINAL RECORDS.—Wyoming (Long, 1965a: 567): Corner Mtn.; *3 mi. ESE Browns Peak*; 3½ mi. S Woods Landing. Colorado (Armstrong, 1972:113): 2 mi.

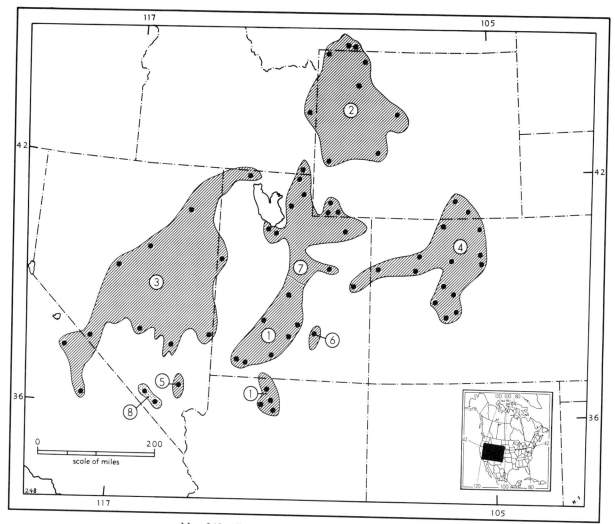

Map 248. *Eutamias umbrinus* and *Eutamias palmeri*.

Guide to kinds
1. *E. umbrinus adsitus*   3. *E. umbrinus inyoensis*   6. *E. umbrinus sedulus*
2. *E. umbrinus fremonti*   4. *E. umbrinus montanus*   7. *E. umbrinus umbrinus*
                             5. *E. umbrinus nevadensis*   8. *E. palmeri*

E Log Cabin, 7450 ft.; type locality; 3 mi. E Pinecliff;
Halfmoon Creek, 8 mi. SW Leadville, 10,000 ft.; St.
Elmo; Tomichi Dome; Crested Butte Peak, 9500 ft.;
Thomasville; McCoy; 10 mi. N New Castle. Utah: PR
Springs, 43 mi. S Ouray, 7950 ft. Colorado (Armstrong,
1972:112, 113): 29 mi. S Rangely; 1 mi. NW Pagoda
Peak, 10,400 ft.; Mt. Zirkel, 9775 ft.

Eutamias umbrinus nevadensis Burt

1931. *Eutamias quadrivittatus nevadensis* Burt, Jour.
   Mamm., 12:299, August 24, type from Hidden Forest, 8500
   ft., Sheep Mts., Clark Co., Nevada. Known only from type
   locality.
1953. *Eutamias umbrinus nevadensis*, White, Univ. Kansas
   Publ., Mus. Nat. Hist., 5:574, December 1.

Eutamias umbrinus sedulus White

1953. *Eutamias umbrinus sedulus* White, Univ. Kansas Publ.,
   Mus. Nat. Hist., 5:573, December 1, type from Mt. Ellen,
   Henry Mts., Garfield Co., Utah. Known only from type
   locality.

Eutamias umbrinus umbrinus (J. A. Allen)

1890. *Tamias umbrinus* J. A. Allen, Bull. Amer. Mus. Nat.
   Hist., 3:96, June, type from Blacks Fork, approx. 8000 ft.,
   Uinta Mts., Utah.
1901. *Eutamias umbrinus*, Miller and Rehn, Proc. Boston
   Soc. Nat. Hist., 30(1):45, December 27.

MARGINAL   RECORDS.—Idaho: ¼ mi. W
Copenhagen Basin. Utah: Monte Cristo, 18 mi. W

Woodruff, 8000 ft. Wyoming (Long, 1965a:568): 10 mi. S, 1 mi. W Robertson; Ft. Bridger; Beaver Creek, 4 mi. S Lonetree. Utah: Paradise Park, 15 mi. N, 21 mi. W Vernal, 10,050 ft.; jct. Argyle and Minnie Maud creeks; South Willow Creek Canyon, 7000 ft. (Egoscue, 1965:685); Butterfield Canyon, 3 mi. SW Butterfield Tunnel, 8000 ft.; Wasatch Mts., near Ogden; Spring Hollow, Logan Canyon.

Eutamias palmeri Merriam
Palmer's Chipmunk

1897. *Eutamias palmeri* Merriam, Proc. Biol. Soc. Washington, 11:208, July 1, type from Charleston Peak, 8000 ft., Clark Co., Nevada.

External measurements: 204–233; 74–101; 31–35. Greatest length of skull, 34.9–36.5. Dorsal stripes indistinct in winter pelage. From geographically adjoining populations of its closest relative, *E. umbrinus*, *E. palmeri* differs in browner (more reddish) dark dorsal stripes, more tawny color on underside of tail, shorter rostrum, and shorter upper incisors.

MARGINAL RECORDS.—Nevada: Deer Creek, 8250 ft.; type locality. See Map 248.

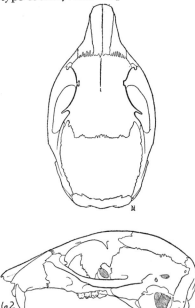

Fig. 262. *Eutamias palmeri*, 6 mi. N Charleston Park Resort, 7800 ft., Nevada, No. 86850 M.V.Z., ♂, X 1½.

Eutamias bulleri
Buller's Chipmunk

External measurements: 222–248; 93–113; 34–38. Greatest length of skull, 35.7–39.6. Facial

markings broader and more blackish than in *E. cinereicollis*; grayish collar present but faint in some specimens; dark dorsal stripes shaded with Mikado Brown; light dorsal stripes white or shaded with reddish; sides with wash of cinnamon or brownish; ventral face of tail reddish; skull resembling that of *E. cinereicollis* but larger.

Eutamias bulleri bulleri (J. A. Allen)

1889. *Tamias asiaticus bulleri* J. A. Allen, Bull. Amer. Mus. Nat. Hist., 2:173, October 21, type from Sierra de Valparaíso, Zacatecas.
1901. *Eutamias bulleri*, Miller and Rehn, Proc. Boston Soc. Nat. Hist., 30(1):40, December 27.

MARGINAL RECORDS.—Zacatecas: Sierra Madre, SW Sombrerete; type locality. Jalisco: 10 mi. NE Huejuquilla, 6800 ft. (Genoways and Jones, 1973:6). Durango: 28 mi. S, 17 mi. W Vicente Guerrero, 8350 ft. (Baker and Greer, 1962:83).

Eutamias bulleri durangae J. A. Allen

1903. *Eutamias durangae* J. A. Allen, Bull. Amer. Mus. Nat. Hist., 19:594, November 12, type from Arroyo de Bucy, approx. 7000 ft., Sierra de Candella, Durango.
1922. *Eutamias bulleri durangae*, A. H. Howell, Jour. Mamm., 3:184, August 4.
1905. *Tamias nexus* Elliot, Proc. Biol. Soc. Washington, 18:233, December 9, type from Coyotes, Durango.

MARGINAL RECORDS.—Chihuahua: 7 mi. SW El Vergel [= Lagunita], 7800 ft. (Anderson, 1972:272). Durango: type locality; Durango; 24 mi. SSE Durango, 7200 ft. (Baker and Greer, 1962:83); 40 mi. SW Durango (*ibid.*); 11 mi. SW Las Adjuntas (*ibid.*); 12 mi. S Las Adjuntas (*ibid.*); Cerro Huehuento. Chihuahua: Sierra Madre, near Guadalupe y Calvo.

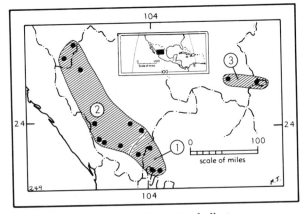

Map 249. *Eutamias bulleri*.

1. *E. b. bulleri*   2. *E. b. durangae*
3. *E. b. solivagus*

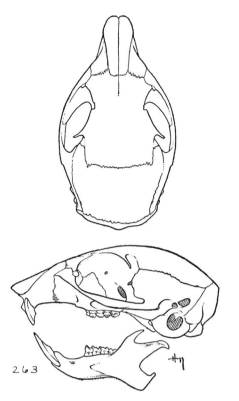

Fig. 263. *Eutamias bulleri durangae*, 9 mi. SW Las Adjuntas, 8900 ft., Durango, No. 54522 K.U., ♂, X 1½.

Eutamias bulleri solivagus A. H. Howell

1922. *Eutamias bulleri solivagus* A. H. Howell, Jour. Mamm., 3:179, August 4, type from Sierra Guadalupe, Coahuila.

MARGINAL RECORDS.—Coahuila: type locality; 12 mi. E San Antonio de Alazanas, 9000 ft.

Genus **Marmota** Blumenbach—Marmots

Revised by A. H. Howell, N. Amer. Fauna, 37:1–80, 15 pls., 3 figs. in text, April 7, 1915.

1777. *Glis* Erxleben, Systema regni animals . . . , 1:358. Type not designated; included *Glis marmota, G. monax, G. canadensis, G. cricetus, G. tscherkessicus, G. citellus, G. zemni, G. lemmus, G. migratorius, G. barabensis, G. arenarius, G. lagurus,* and *G. oeconomicus*. Not *Glis* Brisson, 1762, a genus of dormouse.

1779. *Marmota* Blumenbach, Handbuch der Naturgeschichte, 1:79. Type, *Mus marmorata* Linnaeus. Not *Marmota* Frisch, 1775; Frisch names are unavailable.

1780. *Arctomys* Schreber, Die Säugthiere . . . , pl. 208. Type, *Mus monax* Linnaeus.

1780. *Lagomys* Storr, Prodromus methodi mammalium . . . , p. 39. No type designated; 24 species included; ". . . typified by species of *Arctomys*" (Gill, Bull. Philos. Soc. Washington, 2:viii(App.), 1880).

1811. *Lipura* Illiger, Prodromus systematis mammalium et avium . . . , p. 95. Type, *Hyrax hudsonius* Schreber [= *Glis canadensis* Erxleben].

1923. *Marmotops* Pocock, Proc. Zool. Soc. London, p. 1200, February 13. Type, *Mus monax* Linnaeus; proposed as a subgenus.

External measurements: 418–820; 100–252; 68–113. Five digits, thumb rudimentary but bearing nail; palm with 5 pads (3 at bases of digits) and sole with 6 (4 at bases of digits). Pelage brown, reddish, or black; some species with white markings. Rostrum and cranium subequal; interorbital region much wider than postorbital region; P4 as large as or larger than M1; cheek-teeth high-crowned; metaloph complete on each upper molar, and on M3 turns posteriad and joins posterior cingulum; p4 molariform, its protolophid a transverse crest between protoconid and parametaconid; m1 and m2 parallelogram-shaped in occlusal outline; cheek pouch rudimentary and lacking retractor muscles.

KEY TO NORTH AMERICAN SPECIES OF MARMOTA

Marmota monax
Woodchuck

x ⅙

External measurements: 418–665; 100–155; 68–88. Eight mammae (2 pair pectoral, only 1 pair abdominal and 1 pair inguinal); posterior pad on sole of hind foot oval and situated near middle of sole. Color reddish or brownish; head without white except around nose; feet black or dark brown; interorbital region broad; postorbital processes projecting at right angles to long axis of skull or slightly forward of right angle; nasals, in posterior extent, usually noticeably wider than premaxillae; palate abruptly truncate at posterior border; incisive foramina widest posteriorly and narrowest anteriorly; maxillary tooth-rows approx. parallel instead of divergent anteriorly.

Marmota monax bunkeri Black

1935. *Marmota monax bunkeri* Black, Jour. Mamm., 16:319, November 15, type from 7 mi. SW of Lawrence, Douglas Co., Kansas.

MARGINAL RECORDS.—Nebraska: 6 mi. NW Wakefield (Jones, 1964c:123). Kansas: Doniphan Lake; 5½ mi. SE Fontana; Hamilton. Nebraska: near Nelson (Jones, 1964c:124).

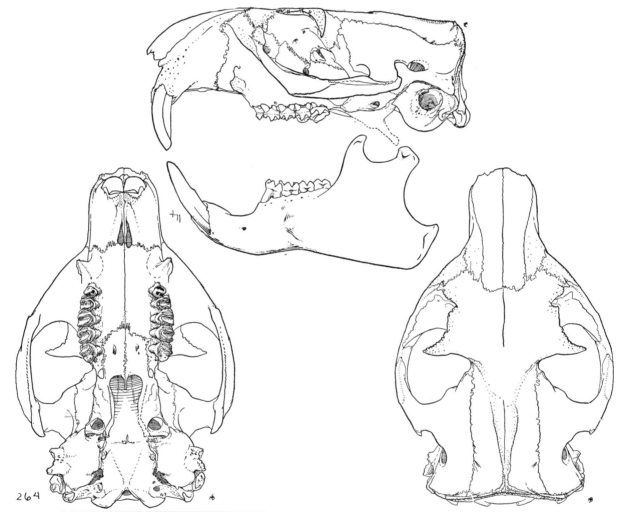

Fig. 264. *Marmota monax monax*, 3 mi. W, 3 mi. N Wilmington, Clinton, Co., Ohio, No. 81418 M.V.Z., ♀, X 1.

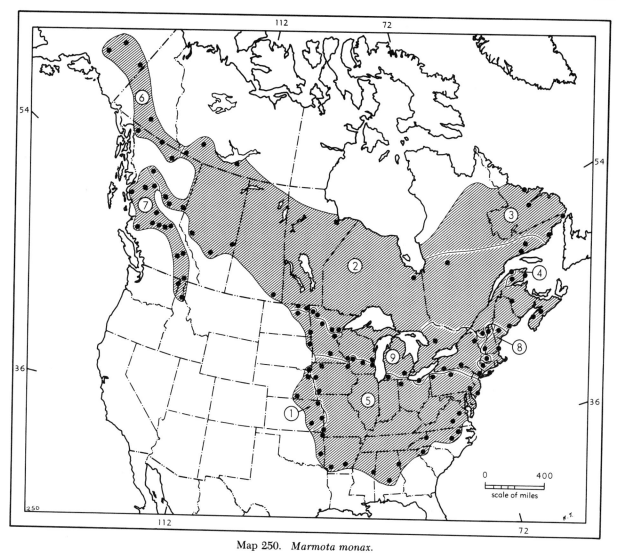

Map 250. *Marmota monax.*

1. *M. m. bunkeri*	4. *M. m. johnsoni*	7. *M. m. petrensis*
2. *M. m. canadensis*	5. *M. m. monax*	8. *M. m. preblorum*
3. *M. m. ignava*	6. *M. m. ochracea*	9. *M. m. rufescens*

Marmota monax canadensis (Erxleben)

1777. [*Glis*] *canadensis* Erxleben, Systema regni animalis . . . , 1:363, based primarily on the Quebec marmot of Pennant. From "Canada et ad fretum Hudsonis," but type locality fixed as Quebec, Quebec, Canada, by A. H. Howell, N. Amer. Fauna, 37:31, April 7, 1915.

1904. [*Marmota monax*] *canadensis*, Trouessart, Catalogus mammalium . . . , Suppl., p. 344.

1778. *Mus empetra* Pallas, Novae species quadrupedum e glirium ordine . . . , p. 75, based primarily on the Quebec marmot of Pennant.

1808. *Arctomys sibila* Wolf, Linne's Natursyst., 2:481, name proposed to include *Arctomys empetra* Pallas and *Arctomys pruinosa* Gmelin, 1788, supposed to be the same.

1820. *Arctomys melanopus* Kuhl, Beiträge zur Zoologie und vergleichenden Anatomie, p. 64. Type locality, Canada.

MARGINAL RECORDS.—Mackenzie: Simpson; mouth Buffalo River, Great Slave Lake. Manitoba: York Factory. Ontario: Moose River near Hudson Bay. Quebec: Eastmain River; Romaine River (Harper, 1961:40); Mingan (*ibid.*); Ste. Anne de Monts. New Brunswick: Arthurette. Nova Scotia: Newport; vic. Lake Kedgemakooge. Maine: Mount Desert Island. Vermont: Mt. Mansfield. Along N bank Ottawa River and W shores Lake Huron and Lake Superior to Minnesota: Carlton County; Aitkin; Thief River Falls. North Dakota: Pembina; Walhalla. Saskatchewan: Moose Mtn. Alberta: vic. Islay; Red Deer River; Entrance. British Columbia: Tupper Creek (Cowan and Guiguet, 1965:118); Misinchinka River (*ibid.*); Weston Creek (*ibid.*); near head Finlay River. Mackenzie: Liard.

Marmota monax ignava (Bangs)

1899. *Arctomys ignavus* Bangs, Proc. New England Zool. Club, 1:13, February 28, type from Black Bay, Strait of Belle Isle, Labrador.
1904. [*Marmota monax*] *ignavus*, Trouessart, Catalogus mammalium . . . , Suppl., p. 344.

MARGINAL RECORDS.—Quebec: *1½ mi. SE Lac Aulneau* (Harper, 1961:40, sight record only). Labrador: Ailik, Peters Cove; southward along coast to Point Armour. Quebec: 1 mi. E Cross River (Weaver, 1940:421). Labrador: head Hamilton Inlet [= Northwest River].

Marmota monax johnsoni Anderson

1943. *Marmota monax johnsoni* Anderson, Ann. Rept. Provancher Soc. Nat. Hist., Quebec, for 1942, p. 53, September 7, type from Percé, Gaspé Co., Quebec.

MARGINAL RECORDS.—Quebec: Gaspé Peninsula: *Federal Mine*; type locality; *Salmon Branch, Grand Cascapedia River*; Kelley's Camp, Berry Mountain Brook, near head of Grand Cascapedia River. Not found: near foot Mt. Lyall (Anderson, 1947:106).

Marmota monax monax (Linnaeus)

1758. [*Mus*] *monax* Linnaeus, Syst. nat., ed. 10, 1:60. Type locality, Maryland.
1904. [*Marmota*] *monax*, Trouessart, Catalogus mammalium . . . , Suppl., p. 344.

MARGINAL RECORDS.—Wisconsin (Jackson, 1961:124, Map 26): Monroe County; Columbia County; Ozaukee County. Michigan: Dowagiac. Ohio: Hicksville. Pennsylvania: Erie County; Sullivan County; Marple. New Jersey: Wading River. Virginia: Doswell; Brunswick County. North Carolina: Tillery (Paul and Cordes, 1969:372, as *Marmota monax* only); Clayton (*ibid.*); Roan Mtn. South Carolina: Greenville. Alabama: Piedmont; 16 mi. NW Prattville. Mississippi: near Bigbee Valley (T. 16 N, R. 19 E, Lot 4) (Ferguson, 1962:107). Arkansas: Lincoln County; Hemstead County. Oklahoma: 3 mi. S Red Oak (McCarley and Free, 1962:271, as *M. monax* only); *1 mi. S Spavinaw* (119482 KU); Davis Farm, 2 mi. SW Tri-state Monument (Long, 1961:255). Kansas (Long, 1961:256): *west edge of Pittsburg*; 2 mi. W Frontenac. Iowa (Bowles, 1975:64, 65): Walnut Twp., Fremont Co.; sec. 6, Belvidere Twp., Monona Co. South Dakota: Union County Park, 1200 ft. (101696 KU). Iowa (Bowles, 1975:64, 65): *Little Sioux River, Okoboji Twp.*; East Okoboji Lake; 8 mi. SW New Albin. Wisconsin: *La Crosse County* (Jackson, 1961:129).

Marmota monax ochracea Swarth

1911. *Marmota ochracea* Swarth, Univ. California Publ. Zool., 7:203, February 18, type from Forty-mile Creek, Alaska.
1915. *Marmota monax ochracea*, A. H. Howell, N. Amer. Fauna, 37:34, April 7.

MARGINAL RECORDS.—Alaska: Moose Creek, 30 mi. above Fairbanks; type locality. Yukon: Nisutlin River [Mile 40, Canol Road]. British Columbia: Lower Liard Crossing [Mile 213, Alaska Highway]; near jct. Liard and Trout rivers; Atlin. Alaska: Healy River; *Fairbanks*.

Marmota monax petrensis A. H. Howell

1915. *Marmota monax petrensis* A. H. Howell, N. Amer. Fauna, 37:33, April 7, type from Revelstoke, British Columbia.

MARGINAL RECORDS.—British Columbia: Driftwood River (Cowan and Guiguet, 1965:118); Vanderhoof (*ibid.*); Lynx Creek, Isaacs Lake; Glacier; Creston. Idaho: Thompson Pass. Washington: Pend Oreille Mts. British Columbia: type locality; Barkerville (Cowan and Guiguet, 1965:118); Quesnel; *Tiltzarone Lake* (Cowan and Guiguet, 1965:118); Nazko (*ibid.*); Lonesome Lake; Salvus; Kispiox.

Marmota monax preblorum A. H. Howell

1914. *Marmota monax preblorum* A. H. Howell, Proc. Biol. Soc. Washington, 27:14, February 2, type from Wilmington, Middlesex Co., Massachusetts.

MARGINAL RECORDS.—Maine: Norway; Eliot. Connecticut: *East Wallingford*; Sharon Mtn. Vermont: Saxtons River.

Marmota monax rufescens A. H. Howell

1914. *Marmota monax rufescens* A. H. Howell, Proc. Biol. Soc. Washington, 27:13, February 2, type from Elk River, Sherburne Co., Minnesota.

MARGINAL RECORDS.—North Dakota: *Grafton*. Minnesota: Marshall County; Hubbard County; Princeton; *Pine County*, thence along S shore of Lake Superior, thence almost due east to Ontario: Lake of Bays, *thence along S bank Ottawa River*. New York: Elizabethtown. Vermont: *Ferrisburg*; Lunenburg. Massachusetts: Easthampton. New York: Hastings; Miller Place; Owego; Allegheny State Park. Ohio: *Geauga Lake*; Cleveland. Michigan: Ann Arbor. Wisconsin: Sheboygan County (Jackson, 1961:124, Map 26); *Adams County* (*ibid.*); *Juneau County* (*ibid.*); Trempealeau (*op. cit.*:130). Minnesota: Nicollet County. South Dakota: Hartford Beach Park, Roberts Co. (101697 KU). North Dakota: Wahpeton; *Leonard*; Devils Lake.

Marmota flaviventris
Yellow-bellied Marmot

External measurements: 470–700; 130–220; 70–92. Ten mammae (2 pair pectoral, 2 pair abdominal, and 1 pair inguinal); posterior pad on sole of hind foot oval and situated near middle of sole. Color tawny, often frosted with white; white markings between eyes usual; sides of neck with conspicuous buffy patches; feet varying from light buff to hazel or dark brown (never black); interorbital region narrow; postorbital processes project-ing back of line drawn across their bases and at right angles to long axis of skull; nasals no broader posteriorly than premaxillae; posterior border of palate beveled at obtuse angle; incisive foramina constricted posteriorly or of equal width throughout; maxillary tooth-rows slightly divergent anteriorly.

Marmota flaviventris avara (Bangs)

1899. *Arctomys flaviventer avarus* Bangs, Proc. New England Zool. Club, 1:68, July 31, type from Okanagan, British Columbia.

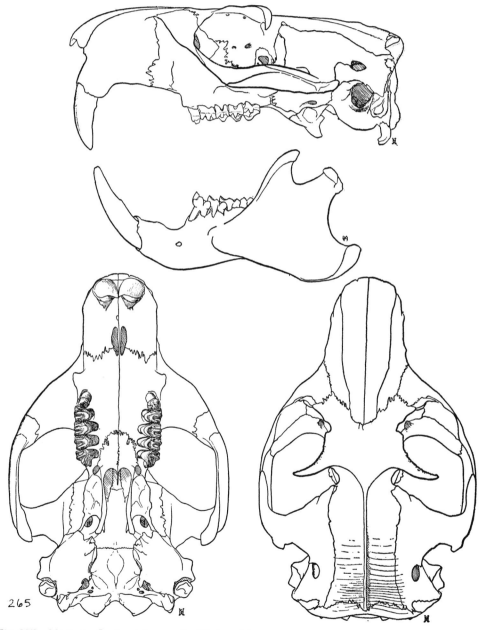

Fig. 265. *Marmota flaviventris parvula*, W slope Toquima Mtn., Nevada, No. 57476 M.V.Z., ♂, X 1.

1904. [*Marmota flaviventer*] *avarus*, Trouessart, Catalogus mammalium . . . , Suppl., p. 344.

MARGINAL RECORDS.—British Columbia: Williams Lake; Sicamous; Vernon; Trail. Washington: Spokane Bridge. Idaho: 10 mi. N St. Maries; 17 mi. N Prichard; Rapid River, near Riggins; Crane Creek, 15 mi. E Midvale; 1 mi. S Riddle. Nevada: Mountain City; head Ackler Creek, 6800 ft. Utah: Granite Creek, Deep Creek Mts., Juab Co., 8500 ft. (Shippee and Egoscue, 1958:276). Nevada: ½ mi. N Treasury Lake, 12,200 ft.; ¼ mi. W Hamilton; Granite Mtn.; Mt. Grant; Mt. Siegel; 3 mi. E Reno; Smoke Creek, 4300 ft.; 12-Mile Creek, 1 mi. E California boundary, 5300 ft. Oregon: Adel; Prineville. Washington: Bethel Ridge, 6000 ft., 30 mi. ESE Mt. Rainier, 3 mi. N Tieton Reservoir (Broadbooks, 1965:301, 302, as *M. flaviventris* only); 5 mi. N Entiat; Twisp; just outside Bellingham. British Columbia: Hope; *Spuzzum* (Cowan and Guiguet, 1965:120); *Boston Bar* (*ibid.*); Lytton; Ashcroft; *Sheep Creek Bridge on Fraser River* (Cowan and Guiguet, 1965:120).

Marmota flaviventris dacota (Merriam)

1889. *Arctomys dacota* Merriam, N. Amer. Fauna, 2:8, October 30, type from Custer, Custer Co., Black Hills, South Dakota.
1914. *M[armota]. f[laviventer]. dacota*, A. H. Howell, Proc. Biol. Soc. Washington, 27:15, February 2.

MARGINAL RECORDS.—Wyoming (Long, 1965a:570): 1 mi. S Warren Peak Lookout; *Bear Lodge Mtns.* South Dakota: Savoy; Tigerville, near Hill City; Custer. Wyoming: ½ mi. E Buckhorn, 6100 ft.

Marmota flaviventris engelhardti J. A. Allen

1905. *Marmota engelhardti* J. A. Allen, Mus. Brooklyn Inst. Arts and Sci., Sci. Bull., 1:120, March 31, type from Brigg's [= Britt's] Meadows, 10,000 ft., Beaver Range Mts., Beaver Co., Utah.
1915. *Marmota flaviventris engelhardti*, A. H. Howell, N. Amer. Fauna, 37:45, April 7.

MARGINAL RECORDS.—Utah: 5 mi. E Great Basin Experimental Station; head right fork Cottonwood Creek, Manti National Forest; Torrey, 6800 ft.; Long Valley, Markagunt Plateau; Gunlock (Stock, 1970:430); *Pine Valley* (*ibid.*); Dutch Creek Sink, Cedar Mtn., 8300 ft.; Cedar Breaks, 10,000 ft.; type locality; Fish Lake.

Lange (1956:289, 291) records bones of *Marmota flaviventris* from several Indian archaeological sites including the following: Arizona: Tse-an Olje Cave; Cylinder Cave; Tse-an Kaetan Cave; Tooth Cave; Government Cave, approx. 23 mi. W Flagstaff; Woodchuck Cave; Keet Seel. These occurrences are not shown on Map 251.

Marmota flaviventris flaviventris (Audubon and Bachman)

1841. *Arctomys flaviventer* Audubon and Bachman, Proc. Acad. Nat. Sci. Philadelphia, 1:99, type locality, "Mountains between Texas and California," but fixed as Mt. Hood, Oregon, by A. H. Howell (N. Amer. Fauna, 37:39, 40, April 7, 1915).
1904. [*Marmota*] *flaviventer*, Trouessart, Catalogus mammalium . . . , Suppl., p. 344.

MARGINAL RECORDS.—Oregon: type locality; Summer Lake. California: Steele Swamp; Petes Valley, 4500 ft.; Donner. Nevada: 3 mi. S Mt. Rose, 8500 ft. California: Hope Valley; Glen Alpine Springs; SW Black Butte, 6600 ft.; Hat Creek, 8000 ft.; Penoyar. Oregon: Klamath Lake; Crater Lake.

Marmota flaviventris fortirostris Grinnell

1921. *Marmota flaviventris fortirostris* Grinnell, Univ. California Publ. Zool., 21:242, November 7, type from 11,800 ft., McAfee Meadow, White Mts., Mono Co., California.

MARGINAL RECORDS.—California: near White Mtn. Peak; type locality; near Blanco Mtn.; Big Prospector Meadow, 10,300–10,700 ft.

Marmota flaviventris luteola A. H. Howell

1914. *Marmota flaviventer luteola* A. H. Howell, Proc. Biol. Soc. Washington, 27:15, February 2, type from Woods Post Office, 7500 ft., Medicine Bow Mts.; Albany Co., Wyoming.
1914. *Marmota flaviventer warreni* A. H. Howell, Proc. Biol. Soc. Washington, 27:16, February 2, type from Smith Trail, 2 mi. W Crested Butte, 10,000 ft., Gunnison Co., Colorado. Regarded as inseparable from *M. f. luteola* by Warren, Jour. Mamm., 17:392, November 16, 1936.
1915. *Marmota flaviventer campioni* Figgins, Proc. Biol. Soc. Washington, 28:147, September 21, type from 8 mi. NW Higho, Jackson Co., Colorado. Regarded as inseparable from *M. f. luteola* by Warren (*antea*).

MARGINAL RECORDS.—Wyoming (Long, 1965a:571): 21½ mi. S, 24½ mi. W Douglas; *3 mi. W Eagle Peak*; 17 mi. E Laramie. Colorado (Armstrong, 1972:116): Bingham Hill, 5 mi. NW Fort Collins; Colorado Springs; Monarch Park, 10,728 ft.; Osier. New Mexico: 5 mi. E Brazos Peak. Colorado: Florida; Cliff Palace (Anderson, 1961:42); *Weatherhill Mesa* (*ibid.*). Utah: 4 mi. W Geyser Pass, 10,000 ft., La Sal Mts.; Warner Ranger Station, 9750 ft., La Sal Mts. Colorado (Armstrong, 1972:116): Castle Park, ½ mi. from mouth Hells Canyon, Dinosaur National Monument; 16 mi. N Craig. Wyoming (Long, 1965a:571): 7½ mi. N, 18½ mi. E Savery; Bridgers Pass.

Marmota flaviventris nosophora A. H. Howell

1914. *Marmota flaviventer nosophora* A. H. Howell, Proc. Biol. Soc. Washington, 27:15, February 2, type from Willow Creek, 4000 ft., 7 mi. E Corvallis, Ravalli Co., Montana.

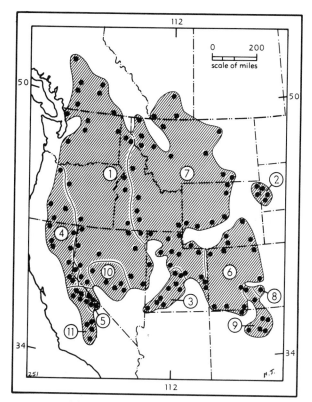

Map 251. *Marmota flaviventris.*

Guide to subspecies
1. *M. f. avara*
2. *M. f. dacota*
3. *M. f. engelhardti*
4. *M. f. flaviventris*
5. *M. f. fortirostris*
6. *M. f. luteola*
7. *M. f. nosophora*
8. *M. f. notioros*
9. *M. f. obscura*
10. *M. f. parvula*
11. *M. f. sierrae*

MARGINAL RECORDS.—Alberta: near Lake Newell (Soper, 1965:127); near Aden. Montana (Hoffmann, *et al.*, 1969:586, as *M. flaviventris* only, unless otherwise noted): 8 mi. N Nashua; 7 mi. W Tyler (Hall and Kelson, 1959:226); 8 mi. S Crow Agency; Lee Creek, 14 mi. SE Birney. Wyoming (Long, 1965a:571, 572): Sheridan; 17 mi. W Buffalo; *5 mi. N, 9 mi. E Tensleep;* 16 mi. S, 11 mi. W Waltman; 20 mi. S, 3½ mi. W Lander; Little Sandy Creek; *new fork Green River;* 10 mi. SE Afton. Idaho: Pegram. Utah: Laketown; Smith and Morehouse Creek. Wyoming: *9 mi. S Robertson;* 13 mi. S, 2 mi. E Robertson (Long, 1965a:572). Utah: Beaver Creek; Post Canyon, Book Cliffs, 75 mi. S Ouray; head right fork Cottonwood Creek, *ca.* 16 mi. NW Castledale; 6 mi. E Eureka; City Creek Canyon, 1½ mi. above forks, 4700 ft.; 6½ mi. W Brigham City, 4300 ft.; South Willow Canyon, 10,000 ft., base Deseret Peak; 7½ mi. SE Yost, Raft River Mts., 6500 ft. Nevada: 1 mi. S Contact, 4800 ft. Idaho: Salmon Creek, 8 mi. W Rogerson; 2 mi. S Hagerman; Lake Alturas; Elk City. Montana: Weeksville; [8 mi. E] Horse Plains; N side Pinkham Creek, *ca.* 4 mi. SW Rexford; Clearwater Junction (Hoffmann, *et al.*, 1969:586, as *M. flaviventris* only); Helena Golf Club

(*ibid.*); *Dearborn River* (*ibid.*); Browning (*ibid.*). Alberta: Waterton.

Marmota flaviventris notioros Warren

1934. *Marmota flaviventris notioros* Warren, Jour. Mamm., 15:62, February 15, type from 10,600 ft., near Marion Reservoir or Lake, Wet Mts., Custer Co., Colorado. Known only from type locality.

Marmota flaviventris obscura A. H. Howell

1914. *Marmota flaviventer obscura* A. H. Howell, Proc. Biol. Soc. Washington, 27:16, February 2, type from Wheeler Peak, 11,500 ft., 5 mi. S Twining, Taos Co., New Mexico.

MARGINAL RECORDS.—Colorado: head Raspberry Creek, N of Villa Grove; *Venable Lakes Trail* (Armstrong, 1972:117); Como Lake, 11,500 ft. (*ibid.*). New Mexico: Agua Fria Peak; 6 mi. W Rociada; 9 mi. NE Santa Fe; Panchuela, 8 mi. NW Pecos Baldy Lake; type locality.

Marmota flaviventris parvula A. H. Howell

1915. *Marmota flaviventer parvula* A. H. Howell, Proc. Biol. Soc. Washington, 27:14, February 2, type from Jefferson, Nye Co., Nevada.

MARGINAL RECORDS.—Nevada: Smiths Creek, 7500 ft.; Jefferson; Hot Creek Canyon, 6 mi. W Hot Creek P.O.; W slope Toquima Peak, 10,000 ft.; Arc Dome; Eastgate.

Marmota flaviventris sierrae A. H. Howell

1915. *Marmota flaventris sierrae* A. H. Howell, N. Amer. Fauna, 37:43, April 7, type from 9300 ft., head Kern River, Mt. Whitney, Tulare Co., California.

MARGINAL RECORDS.—California: Tuolumne Meadows; Silver Lake, 7300 ft.; vic. Mammoth; Bishop Creek, 8000 ft.; Cottonwood Lakes; Monache Meadows, near Olancha Peak; Cannell Meadows; E Fork Kaweah River; Big Meadows.

Marmota caligata
Hoary Marmot

External measurements: 620–820, 170–250, 90–113. Ten mammae (2 pair pectoral, 2 pair abdominal, and 1 pair inguinal); posterior pads on sole of hind foot subcircular and situated near edges of sole. Color mixed black and white, sometimes with brownish tinge especially on posterior half of upper parts; feet black or blackish brown, often with white markings on forefeet; interorbital region narrow; postorbital processes rarely projecting at right angles to long axis of skull and in most subspecies projecting back of line drawn across bases of processes and at right angle with long axis of skull; nasals narrowed posteriorly, averaging approx. same breadth as premaxillae; posterior border of palate beveled at obtuse angle; incisive foramina shaped differently depending on subspecies; maxillary tooth-rows divergent anteriorly.

When *Marmota c. broweri* was named, Hall was uncertain whether it was a subspecies or a species, and still is uncertain, although Bee and Hall reported results of study of some better specimens in 1956. The uncertainty probably can be dispelled when additional adult specimens are obtained and studied from localities that will show where gene-flow does and does not occur between named kinds of Alaskan marmots. Hoffmann and Nadler (1968:740–742), Rausch and Rausch (1971:86–101), and Vorontsov and Liapunova (1973:147) all mention the *2n* chromosome number (36, which is fewer than reported in any other taxon of *Marmota*) and refer to the animal as a species, *Marmota broweri*.

Marmota caligata broweri Hall and Gilmore

1934. *Marmota caligata broweri* Hall and Gilmore, Canadian Field-Nat., 48:57, April 2, type from Point Lay, Arctic Coast of Alaska; Rausch (Arctic, 6:117, July, 1953) states that type locality is probably actually head Kukpowruk River, about 69° N lat.

MARGINAL RECORDS.—Alaska: 50 mi. inland from Wainwright; 30–35 mi. N Tulugak Lake; *Hulahula River, in foothills S of Barter Island;* near Arctic Village, Chandalar River; Big Squaw Lake; mts. inland from Kotzbue Sound; Cape Thompson; type locality.

Marmota caligata caligata (Eschscholtz)

1829. *Arctomys caligatus* Eschscholtz, Zoologischer Atlas, pt. 2, p. 1, pl. 6. Type locality, near Bristol Bay, Alaska.
1903. *Marmotta* [sic] *caligata*, J. A. Allen, Bull. Amer. Mus. Nat. Hist., 19:539, October 10.
1788. *Arctomys pruinosa* Gmelin, Syst. nat., ed. 13, 1:144, based on hoary marmot of Pennant, may apply, at least in part, here but A. H. Howell (N. Amer. Fauna, 37:59, April 7, 1915), reviser of American marmots, rejects name as unidentifiable.

MARGINAL RECORDS.—Mackenzie: Black Mtn., SW of Aklavik; mountain W of Fort Goodhope. Yukon: Kalzas Creek. British Columbia: Atlin; Cheonee Mts. Alaska: Portland Canal; Port Snettisham; Yakutat Bay; Hinchinbrook Island; Cape Elizabeth; Kanatak, Portage Bay; Aleknagik Lake; Flat (226053 USNM); Toklat River near head; hills behind Nome; Fort Yukon.

Marmota caligata cascadensis A. H. Howell

1914. *Marmota caligata cascadensis* A. H. Howell, Proc. Biol. Soc. Washington, 27:17, February 2, type from Mt. Rainier, Pierce Co., Washington.

MARGINAL RECORDS.—British Columbia: Mc-Lean near Lillooet; Spences Bridge; Three Brothers Mtn. in Manning Park. Washington: Mt. Chopaka; near head Cascade River; mts. near Easton; Mt. Adams; type locality; Mt. Baker. British Columbia: Lihumitson Mts. near Chilliwack; Howe Sound; Alta Lake.

Marmota caligata nivaria A. H. Howell

1914. *Marmota caligata nivaria* A. H. Howell, Proc. Biol. Soc. Washington, 27:17, February 2, type from 6100 ft., mountains near Upper St. Marys Lake, Glacier Co., Montana.

MARGINAL RECORDS.—Alberta: Banff National Park; Mt. Forgetmenot. Montana (Hoffmann, *et al.*, 1969:587, as *M. caligata* only): type locality; Elk Calf Mtn.; Sugarloaf Mtn.; Goat Mtn. Lake; Upper Miner Lake, near Jackson. Idaho: Elk Summit, Salmon River Mts.; near Clarkia. Montana (Hoffmann, *et al.*, 1969:587, as *M. caligata* only): *Robinson Mtn., Lincoln Co.;* Northwest Peak, Lincoln Co.

Marmota caligata okanagana (King)

1836. *Arctomys okanaganus* King, Narrative of a journey to the shores of the Arctic Ocean . . . , 2:236. Type locality, region occupied by Okanagan Indians on borders Rocky Mts. between Columbia and Fraser rivers, subsequently fixed by A. H. Howell (Proc. Biol. Soc. Washington, 27:17, February 2, 1914) as Gold Range, British Columbia.
1915. *Marmota caligata okanagana,* A. H. Howell, N. Amer. Fauna, 37:64, April 7.

MARGINAL RECORDS.—Alberta: Henry House; *15 mi. S Henry House;* Healy Creek (Cowan and Guiguet, 1965:124); Carthew Lakes (*ibid.*). British Columbia: Rossland group of Monashee Range, near Rossland; Shuswap Range; Mt. Revelstoke (Cowan and Guiguet, 1965:124); Glacier (*ibid.*).

Marmota caligata oxytona Hollister

1912. *Marmota sibila* Hollister, Smiths. Miscl. Coll., 56(35):1, February 7, type from head Moose Pass branch of Smoky River, 7200 ft., Alberta. Not *Arctomys sibila* Wolf, 1808 [= *Glis canadensis* Erxleben].
1914. *Marmota oxytona* Hollister, Science, n.s., 39:251, February 13, a renaming of *M. sibila* Hollister.
1915. *Marmota caligata oxytona,* A. H. Howell, N. Amer. Fauna, 37:63, April 7.

MARGINAL RECORDS.—Yukon: S fork Macmillan River, Mile 268. Mackenzie: Nahanni River, near Glacier Lake (Youngman, 1968:74); Fort Liard. British Columbia: Laurier Pass; head Wapiti River. Alberta: type locality; Pobokton River. British Columbia: *Tonquin Pass;* Wells Gray Park; Barkerville; Stuart Lake (Cowan and Guiguet, 1965:125); mts. near Babine; Nine-mile Mtn., near Hazelton; mts. near Klappan River; Level Mtn.; *Sheslay River.* Yukon: Teslin Lake.

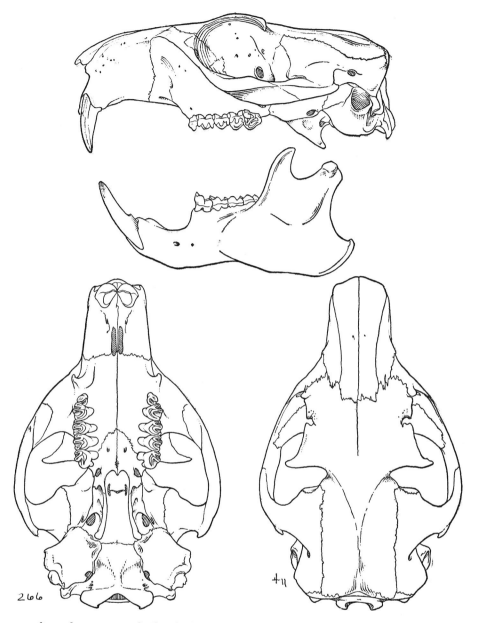

Fig. 266. *Marmota caligata broweri,* mouth Chamberlin Canyon, S end Lake Peters, 145° 08′ 34″, 69° 20′ 58″, 3690 ft., Brooks Range, Alaska, No. 50417 K.U., ♀, X 1.

Marmota caligata raceyi Anderson

1932. *Marmota caligata raceyi* Anderson, Ann. Rept. Canadian Nat. Mus. for 1931, 70:112, November 24, type from 6500 ft., Itcha Mts., Chilcotin Plateau, lat. 52° 45′ N, long. 125° W, British Columbia.

MARGINAL RECORDS.—British Columbia: mts. SW of Burns Lake (Cowan and Guiguet, 1965:125); type locality; 30 mi. E Bella Coola; *Mt. Brilliant, 5500 ft., Rainbow Mts.;* Wistaria.

Marmota caligata sheldoni A. H. Howell

1914. *Marmota caligata sheldoni* A. H. Howell, Proc. Biol. Soc. Washington, 27:18, February 2, type from Montague Island, Alaska. Known only from type locality.

Marmota caligata vigilis Heller

1909. *Marmota vigilis* Heller, Univ. California Publ. Zool., 5:248, February 18, type from W shore Glacier Bay, Alaska. Known only from type locality.
1915. *Marmota caligata vigilis,* A. H. Howell, N. Amer. Fauna, 37:61, April 7.

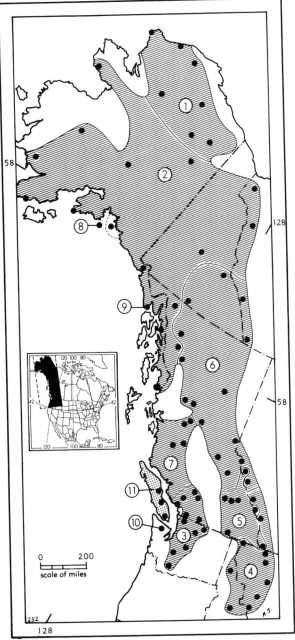

Map 252. *Marmota caligata* and related species.

Guide to kinds

1. *M. c. broweri*	6. *M. c. oxytona*
2. *M. c. caligata*	7. *M. c. raceyi*
3. *M. c. cascadensis*	8. *M. c. sheldoni*
4. *M. c. nivaria*	9. *M. c. vigilis*
5. *M. c. okanagana*	10. *M. olympus*
	11. *M. vancouverensis*

Marmota olympus (Merriam)
Olympic Mountain Marmot

1898. *Arctomys olympus* Merriam, Proc. Acad. Nat. Sci. Philadelphia, 50:352, October 4, type from timber line at head of Soleduc River, Olympic Mts., Clallam Co., Washington.

1904. [*Marmota*] *olympus*, Trouessart, Catalogus mammalium . . . , Suppl., p. 344.

External measurements: 680–785; 195–252; 94–112. Color brownish drab mixed with white; interorbital region and rostrum broad; otherwise closely resembling *M. caligata.*

Rausch (1953:120) considers *M. olympus,* as well as the several subspecies of *M. caligata,* to be only subspecifically differentiated from the Old World *Marmota marmota;* he employs for *olympus* the name *M[armota]. m[armota]. olympus.*

MARGINAL RECORDS.—Washington: Happy Lake; *Deer Park; Mt. Elinor; Mt. Steel; type locality.*

Marmota vancouverensis Swarth
Vancouver Marmot

1911. *Marmota vancouverensis* Swarth, Univ. California Publ. Zool., 7:201, February 18, type from Mt. Douglas, Vancouver Island, British Columbia.

External measurements: 670–750; 180–237; 91–110. Color uniformly dark brown; posterior border of nasals deeply V-shaped; otherwise closely resembles *M. caligata.*

Rausch (1953:120) thought *M. vancouverensis* should probably be regarded as a subspecies of *M. marmota* and used the name *M[armota].? m[armota]. vancouverensis.*

MARGINAL RECORDS.—British Columbia, Vancouver Island: Mt. Washington; Mt. Arrowsmith (Cowan and Guiguet, 1965:126); *Green Mtn.;* Jordan River; *Mt. Strata.* See Map 252.

Genus Ammospermophilus Merriam
Antelope Squirrels

Revised (under generic name *Citellus*) by A. H. Howell, N. Amer. Fauna, 56:166–183, May 18, 1938. Bryant, Amer. Midland Nat., 33:374–375, March, 1945, accords *Ammospermophilus* generic rank.

1892. *Ammospermophilus* Merriam, Proc. Biol. Soc. Washington, 7:27, April 13. Type, *Tamias leucurus* Merriam.

External measurements: 194–250; 54–94; 35–43. White stripe on each side of back extending from shoulder onto hip; upper parts otherwise grayish with ochraceous; tail usually curved over back exposing under surface. Small masseteric tubercle situated directly below oval infraorbital foramen; outer wall of this foramen inclined slightly mediad; interorbital region narrower than postorbital constriction; cranium

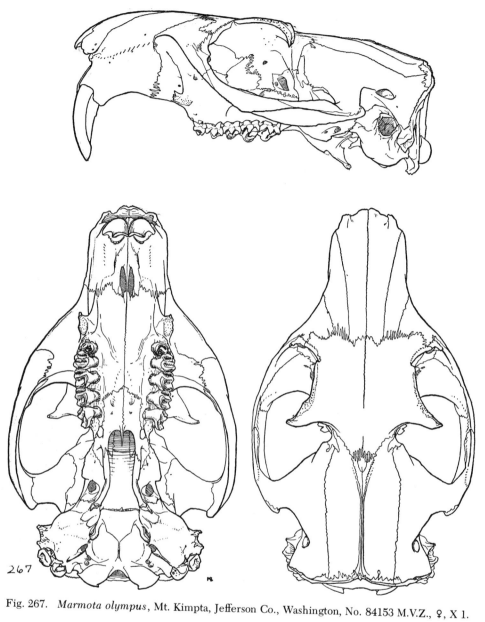

Fig. 267. *Marmota olympus*, Mt. Kimpta, Jefferson Co., Washington, No. 84153 M.V.Z., ♀, X 1.

nearly rectangular in dorsal outline; cheek-teeth low-crowned; metaloph on M1 and M2 not joining protocone; protolophid absent on p4; proximal end of baculum greatly enlarged and having a process on left side; clavobranchialis muscle absent.

KEY TO SPECIES OF AMMOSPERMOPHILUS

1. Underside of tail having median area white.
    2. Occurring in San Joaquin Valley of California. . . . .*A. nelsoni,* p. 381
    2'. Not occurring in San Joaquin Valley of California.

        3. Occurring on Espíritu Santo Island, Baja California; anterior upper premolar absent or rudimentary. . . . .*A. insularis,* p. 381
        3'. Not occurring on Espíritu Santo Island, Baja California; anterior premolar present and normal.
            4. Occurring in Coahuila, Texas, and E of Rio Grande in New Mexico. . . . .*A. interpres,* p. 380
            4'. Not occurring in Coahuila, Texas, or E of Rio Grande in New Mexico. . .*A. leucurus,* p. 378
1'. Underside of tail lacking median white area. . . . .*A. harrisii,* p. 378

Ammospermophilus harrisii
Harris' Antelope Squirrel

External measurements: 222–250; 74–94; 38–42. Greatest length of skull, 38.2–41.2. Upper parts in summer pinkish cinnamon more or less darkened with fuscous; in winter mouse gray; tail, above and below, mixed black and white and hence lacking clear white undersurface of the other species of the genus; in both seasons having two white stripes down back.

Ammospermophilus harrisii harrisii (Audubon and Bachman)

1854. *Spermophilus harrisii* Audubon and Bachman, The viviparous quadrupeds of North America, 3:267. Type locality restricted by A. H. Howell (N. Amer. Fauna, 56:167, May 18, 1938) to be in the Santa Cruz Valley, Arizona, at Mexican Boundary.
1907. *Ammospermophilus harrisii*, Mearns, Bull. U.S. Nat. Mus., 56:viii, 303, April 13.

MARGINAL RECORDS.—Arizona: 4 mi. SW Pierce Ferry (Cockrum, 1961:85); Peach Springs; Montezuma Well; Rice; Sheldon. New Mexico: 1 mi. E Redrock (Findley, *et al.*, 1975:115); 12 mi. NW Animas. Sonora: 32 km. SE Agua Prieta (Hoffmeister, 1977:150); Her-

mosillo; Ortiz; Agua Dulce. Arizona: Quitobaquita; 20 mi. SW Phoenix; Harquahala Mts.; Fort Mohave; Dolan Spring.

Ammospermophilus harrisii saxicola (Mearns)

1896. *Spermophilus harrisii saxicolus* Mearns, Preliminary diagnoses of new mammals from the Mexican border of the United States, p. 2, March 25 (preprint of Proc. U.S. Nat. Mus., 18:444, May 23, 1896), type from Tinajas Atlas, Yuma Co., Arizona.
1907. *Ammospermophilus harrisii saxicola*, Mearns, Bull. U.S. Nat. Mus., 56:viii, 306, April 13.
1937. *Ammospermophilus harrisii kinoensis* Huey, Trans. San Diego Soc. Nat. Hist., 8:352, June 15, type from Bahía Kino, Sonora.

MARGINAL RECORDS.—Arizona: Parker; Vicksburg; 10 mi. N Ajo; Quitovaquito [Quitobaquito] (Cockrum, 1961:87). Sonora: 45 mi. NE Puerto Libertad; Bahía Kino; Puerto Libertad. Arizona: type locality; Yuma; Ehrenburg (Cockrum, 1961:87).

Ammospermophilus leucurus
White-tailed Antelope Squirrel

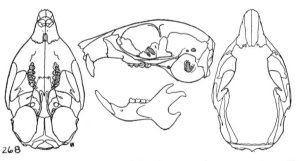

External measurements: 194–239; 54–87; 35–43. Upper parts brownish or cinnamon; two white stripes on back extending from sides to hips; tail broadly white or whitish below, bordered with fuscous black; in winter pelage more

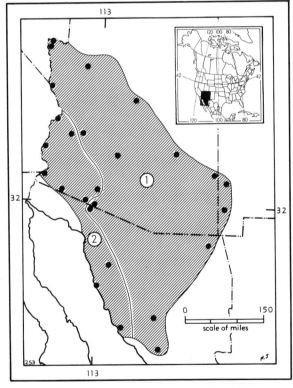

Map 253. *Ammospermophilus harrisii.*

1. *A. h. harrisii*   2. *A. h. saxicola*

Fig. 268. *Ammospermophilus leucurus leucurus*, 14 mi. E Searchlight, Nevada, No. 61472 M.V.Z., ♂, X 1.

grayish. Two molts per year, but hair of tail seems to be molted only in autumn.

Ammospermophilus leucurus canfieldae Huey

1929. *Ammospermophilus leucurus canfieldae* Huey, Trans. San Diego Soc. Nat. Hist., 5:243, February 27, type from Punta Prieta, Baja California.

MARGINAL RECORDS.—Baja California: Jaraguay, 58 mi. SE San Fernando; Calamahue; Yubay, 30 mi. SE Calamahue; Campo Los Angeles; Santo Domingo; San Andrés.

Ammospermophilus leucurus cinnamomeus (Merriam)

1890. *Tamias leucurus cinnamomeus* Merriam, N. Amer. Fauna, 3:52, September 11, type from Echo Cliffs, Painted Desert, Coconino Co., Arizona.
1907. *Ammospermophilus leucurus cinnamomeus,* Mearns, Bull. U.S. Nat. Mus., 56:299, April 13.

MARGINAL RECORDS.—Colorado: Rim Rock Drive near Coke Ovens, Colorado National Monument (McCoy and Miller, 1964:93); Coventry; Ashbaugh's Ranch, near McElmo. Arizona: Lukuchakai Indian School, 6500 ft. (Cockrum, 1961:89); Zuni River; Taylor; Winslow; Sunset Crater National Monument (Hoffmeister and Carothers, 1969:187, as *Spermophilus leucurus* only); Cedar Ridge, 6000 ft. [22 mi. S Grand Canyon Bridge] (Cockrum, 1961:88); Cedar Ranch Wash, Locket Tank; *Indian Garden, 3800 ft.* (Hoffmeister, 1971:168); Fredonia (Cockrum, 1961:88). Utah: Little Castle Valley; 20 mi. W Glade Park, Colorado. Colorado: *Sieber Ranch, Little Dolores River* (McCoy and Miller, 1964:93).

Ammospermophilus leucurus escalante (Hansen)

1955. *Citellus leucurus escalante* Hansen, Jour. Mamm., 36:274, May 26, type from 2 mi. SE Escalante, 5400 ft., Garfield Co., Utah.

MARGINAL RECORDS.—Utah: ½ mi. E Bicknell; Henry Mts., Kings Ranch; 8 mi. S Escalante, 5200 ft.; Kanab. Arizona (Hoffmeister and Durham, 1971:29, 30): Pipe Spring National Monument, 5000 ft.; mouth Fern Glen Canyon, Mile 168, 1750 ft.; Grand Canyon National Monument; *Dry Lake; Tasi Springs;* 1½ mi. NE Diamond Butte; 11 mi. S St. George. Utah: Beaverdam Wash; Leeds.

Ammospermophilus leucurus extimus Nelson and Goldman

1929. *Ammospermophilus leucurus extimus* Nelson and Goldman, Jour. Washington Acad. Sci., 19:281, July 19, type from Saccaton (15 mi. N Cape San Lucas), Baja California.

MARGINAL RECORDS.—Baja California: Santana; San Bruno; San Juanico Bay; Aguaje de San Esteban.

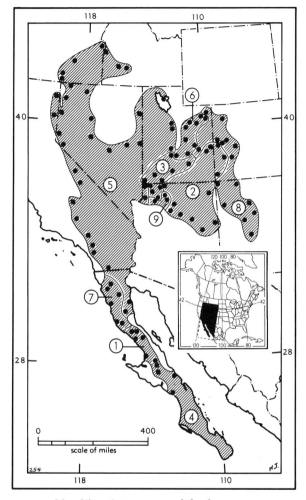

Map 254. *Ammospermophilus leucurus.*

Guide to subspecies
1. A. l. canfieldae
2. A. l. cinnamomeus
3. A. l. escalante
4. A. l. extimus
5. A. l. leucurus
6. A. l. notom
7. A. l. peninsulae
8. A. l. pennipes
9. A. l. tersus

Ammospermophilus leucurus leucurus (Merriam)

1889. *Tamias leucurus* Merriam, N. Amer. Fauna, 2:20, October 30, type from San Gorgonio Pass, Riverside Co., California.
1907. *Ammospermophilus leucurus* [*leucurus*], Mearns, Bull. U.S. Nat. Mus., 56:viii, 299, April 13.
1904. *Citellus l*[*eucurus*]. *vinnulus* Elliot, Field Columb. Mus., Publ. 79, Zool. Ser., 3:241, January 7, type from Keeler, Owens Lake, Inyo Co., California.

MARGINAL RECORDS.—Oregon: Vale. Idaho: Homedale; 5 mi. N Bruneau; Glenns Ferry. Nevada: Quinn River Crossing, 4100 ft.; S base Granite Peak, East Range; 8 mi. SE Eastgate; Hot Creek Range, 4 mi. N Hot Creek, 6400 ft.; 8 mi. N Lund; 1 mi. SE Smith Creek Cave, Mt. Moriah, 5800 ft.; 8 mi. S Wendover, 4700 ft. Utah: Promontory Point; Elberta; Nephi; Monroe. Arizona: Beaver Lodge, 1900 ft. (Hoffmeister and Durham, 1971:30); *6 mi. N Wolf Hole*; Grand Wash, 8 mi. S Pakoon Spring, thence along W side Colorado River (Hall, 1946:318). Baja California: San Felipe; Parral. California: Jacumba; Radec, 12 mi. E Temecula; Banning; Hesperia; Mojave; Kern River Valley, near Kernville; Independence; 8 mi. W Bishop. Nevada: East Walker River, 2 mi. N Morgan's Ranch, 5100 ft.; Washoe Lake. California: Amedee; Lower Alkali Lake. Nevada: Smoke Creek, 9 mi. E California boundary, 3900 ft.; mouth Little High Rock Canyon; Virgin Valley, 5000 ft. Oregon: Adel; South Warner Lake; S side Harney Lake (Olterman and Verts, 1972:26, as *S. leucurus* only).

Ammospermophilus leucurus notom (Hansen)

1955. *Citellus leucurus notom* Hansen, Jour. Mamm., 36:274, May 26, type from Notom, Wayne Co., Utah.

MARGINAL RECORDS.—Utah: 8 mi. S Vernal; 7 mi. SW Jensen, N bank Green River; Pariette Ranch, 8 mi. SW Ouray; 10 mi. S Greenriver; Hanksville; Torrey; Buckhorn Wash, San Rafael River; Wellington; Antelope Canyon, 10 mi. SW Myton.

Ammospermophilus leucurus peninsulae (J. A. Allen)

1893. *Tamias leucurus peninsulae* J. A. Allen, Bull. Amer. Mus. Nat. Hist., 5:197, August 18, type from San Telmo, Baja California.
1907. *Ammospermophilus leucurus peninsulae*, Mearns, Bull. U.S. Nat. Mus., 56:299, April 13.

MARGINAL RECORDS.—Baja California: Agua Escondido, near Hanson Laguna; Trinidad Valley; San Fernando; Rosario; type locality; San Rafael Valley, 20 mi. E Ojos Negros.

Ammospermophilus leucurus pennipes A. H. Howell

1931. *Ammospermophilus leucurus pennipes* A. H. Howell, Jour. Mamm., 12:162, May 14, type from Grand Junction, Colorado.

MARGINAL RECORDS.—Utah: Vernal. Colorado: White River, 20 mi. E Rangeley; Cameo (McCoy and Miller, 1964:93); Paonia (*ibid.*); *Hotchkiss*; Montrose (McCoy and Miller, 1964:93). New Mexico (Findley, *et al.*, 1975:117): ½ mi. S Archuleta; 1 mi. N La Ventana; Rio Rancho Estates; 1 mi. W Socorro; 35 mi. W Albuquerque; 4 mi. S, 8 mi. E Mexican Springs; ½ mi. ESE Four Corners boundary marker. Colorado (McCoy and Miller, 1964:93): type locality; *mouth Monument Canyon, Colorado National Monument; Fruita Reservoir, Colorado National Monument*; Mack. Utah: 1 mi. E Highway 160, 6 mi. S Valley City, 4500 ft.

Ammospermophilus leucurus tersus Goldman

1929. *Ammospermophilus leucurus tersus* Goldman, Jour. Washington Acad. Sci., 19:435, November 19, type from

lower end Prospect Valley, 4500 ft., Grand Canyon, Hualpai Indian Reservation, Arizona.

MARGINAL RECORDS.—Arizona: Great Thumb Point (Hoffmeister, 1971:96, as *A. leucurus* only); *2 mi. S Pasture Wash Ranger Station* (*op. cit.:*168); 3 mi. S Pasture Wash Ranger Station (*ibid.*); type locality; *Havasu Canyon* (Hoffmeister, 1971:96, as *A. leucurus* only).

Ammospermophilus interpres (Merriam)
Texas Antelope Squirrel

1890. *Tamias interpres* Merriam, N. Amer. Fauna, 4:21, October 8, type from El Paso, El Paso Co., Texas.
1905. *Ammospermophilus interpres*, V. Bailey, N. Amer. Fauna, 25:81, October 24.

External measurements: 220–235; 68–84; 36–40. Closely resembles *A. leucurus*; from the adjoining *A. leucurus pennipes*, *A. interpres* differs in having 2 black bands, instead of 1, on the hairs of the tail.

MARGINAL RECORDS.—New Mexico (Findley, *et al.*, 1975:118): 17 mi. N, 4½ mi. E Albuquerque; Carrizozo. Texas: 7 mi. N Pine Springs; *8 mi. W Big Lake* (Creel and Thornton, 1970:481, as *Citellus leucurus!*); Castle Mts.; High Bridge, mouth Pecos River. Coahuila: Jaral. Durango (Anderson, 1972:274): 3 mi. SW Lerdo; 6 mi. N Campana, 3750 ft. Chihuahua (Anderson, 1972:273, 274): 14 mi. SW Consolación, 5800 ft.; Sierra Rica, 29° 15′ N, 104° 05′ W, 4200 ft. Texas: El Paso. New Mexico: Organ Mts.; San Andres Mts.; 10 mi. NE Socorro.

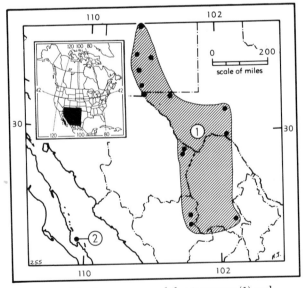

Map 255. *Ammospermophilus interpres* (1) and *Ammospermophilus insularis* (2).

Ammospermophilus insularis Nelson and Goldman
Espíritu Santo Island Antelope Squirrel

1909. *Ammospermophilus leucurus insularis* Nelson and Goldman, Proc. Biol. Soc. Washington, 22:24, March 10, type from Espíritu Santo Island, Baja California. Known only from Espíritu Santo Island.

External measurements: 210–240; 71–83; 36–40. Resembles *A. l. extimus* of adjoining mainland, but differs in larger size, darker flanks and legs, and absent or vestigial P3.

Ammospermophilus nelsoni (Merriam)
Nelson's Antelope Squirrel

1893. *Spermophilus nelsoni* Merriam, Proc. Biol. Soc. Washington, 8:129, December 28, type from Tipton, Tulare Co., California.

1909. *Ammospermophilus nelsoni*, Lyon and Osgood, Bull. U.S. Nat. Mus., 62:172.

1916. *Ammospermophilus nelsoni amplus* Taylor, Univ. California Publ. Zool., 17:15, October 3, type from 20 mi. S Los Baños, Merced Co., California.

External measurements: 218–240; 63–79; 40–43. Resembles *A. l. leucurus* but differs as follows: larger, more buffy (less grayish), more widely spreading zygomatic arches and larger auditory bullae.

MARGINAL RECORDS.—California: Los Baños, Dos Palos; Firebaugh; Mendota; Huron; type locality; 8 mi. NE Bakersfield; Rose Station, 4 mi. E Fort Tejon; Cuyama Valley; 8 mi. E Simmler, Carriso Plains; Alcalde; Panoche Pass.

Genus Spermophilus Cuvier
Spermophiles and Ground Squirrels

Revised by A. H. Howell, N. Amer. Fauna, 56:1–256, May 18, 1938, under the name *Citellus*. Subgenera treated by Bryant, Amer. Midland Nat., 33:257–390, March, 1945.

1816. *Citellus* Oken, Lehrbuch der Naturgeschichte, pt. 3, 2:842. Type, *Mus citellus* Linnaeus. (Oken names not available under the Regles. See Hershkovitz, Jour. Mamm., 30:289–301, August 17, 1949.)

1817. *Anisonyx* Rafinesque, Amer. Month. Mag., 2(1):45. Type, *Anisonyx brachiura* Rafinesque [= *Arctomys columbianus* Ord]. (Not *Anisonyx* Latreille, 1807, a coleopteran.)

1825. *Spermophilus* F. Cuvier, Des dents des mammifères, . . . , p. 255. Type, "*Mus citellus* Linn."

1830. *Citillus* Lichtenstein, Darstellung neuer oder wenig bekannter Säugethiere . . . , pl. 31, fig. 2 (not paged).

1844. *Colobotis* Brandt, Bull. Class. Phys.-Math. Acad. Imp. Sci. St. Pétersbourg, 2(23 and 24):366. Type, *Arctomys fulvus* Lichtenstein.

1844. *Otocolobus* Brandt, Bull. Class. Phys.-Math. Acad. Imp. Sci. St. Pétersbourg, 2(23 and 24)·382. Equals *Colobotis*.

1874?. *Colobates* Milne-Edwards, Recherches Hist. Nat. Mamm., 1:157.

1927. *Urocitellus* Obolenskij, Comp. Rend. Acad. Sci. U.S.S.R., p. 188. Type, *Spermophilus eversmanni* Brandt.

External measurements: 167–520; 32–252; 27–68. Brown, reddish, black, black and white, striped, spotted, variegated, or of solid color. Five digits on each foot; each digit with a claw, or nail (pollex has flat nail); palm with 5 pads (3 interdigital and 2 metacarpal); sole with 4 interdigital pads and no metatarsal pads. Four to seven pairs of mammae. Masseteric tubercle of medium or large size, situated ventral and slightly lateral to infraorbital foramen; infraorbital foramen oval or triangular, relatively broader than in *Ammospermophilus*; a bend present in dorsal profile of skull at junction of rostrum and cranium; zygomatic arches expanded posteriorly but not appressed so closely as in *Ammospermophilus, Eutamias,*

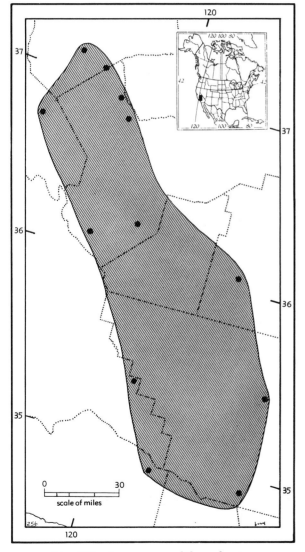

Map 256. *Ammospermophilus nelsoni.*

and *Tamias* and not spread so widely as in *Marmota*; upper tooth-rows parallel or diverging slightly posteriorly; M3 and m3 without complicated enamel folds in basins thus resembling *Ammospermophilus* and differing from *Cynomys*. Dentition, i. $\frac{1}{1}$, c. $\frac{0}{0}$, p. $\frac{1-2}{1}$, m. $\frac{3}{3}$. Clavobrachialis muscle present.

KEY TO SUBGENERA OF SPERMOPHILUS

1. Molars brachydont; parastyle ridge on M1 and M2 joining protocone without abrupt change of direction.
  2. Anterior upper premolar simple; less than ¼ size of P4.
    3. Upper incisors stout and distinctly recurved; tail of adults longer than 128; upper parts variegated black, white and buff. *Otospermophilus,*   p. 398
    3′. Upper incisors slender and not distinctly recurved; tail of adults shorter than 128; upper parts plain or with 1 white stripe on each side of back extending from shoulder all of way onto hip.
      4. Postorbital process long and slender; golden mantle on head and white stripe on each side of back from shoulder onto hip. *Callospermophilus,*   p. 406
      4′. Postorbital process short and thick; upper parts uniformly pinkish gray (not striped or spotted). *Xerospermophilus,*   p. 404
  2′. Anterior upper premolar bearing two cusps and a functional cutting edge; more than ¼ size of P4. *Poliocitellus,*   p. 397
1′. Molars hypsodont; parastyle ridge on M1 and M2 joining protocone with abrupt change of direction.
  5. Metaloph on P4 continuous; upper parts reddish or with fine white spots on gray background. *Spermophilus,*   p. 382
  5′. Metaloph on P4 not continuous; upper parts with nearly square white spots, in some species arranged in rows or upper parts clay color (*S. perotensis*). . . . . *Ictidomys,*   p. 391

Subgenus **Spermophilus** Cuvier

1825. *Spermophilus* F. Cuvier, Des dents des mammifères,
  . . . . p. 255. Type, "*Mus citellus* Linn."

External measurements: 167–495; 32–153; 29–68. Greatest length of skull, 32.4–65.8. Gray, reddish, or with fine white spots on gray background. Infraorbital foramen subtriangular, its lateral wall inclined ventrolaterad; masseteric tubercle situated more laterally than in *Otospermophilus;* interorbital and postorbital constric-

tions narrow and approx. equal in width; zygomatic arches expanded posteriorly; rostrum short and broad, expanded at tip and constricted at base; fossae anterolateral to incisive foramina shallow; anterior margin of alveolar border drops abruptly to join diastema instead of merging with diastema in a gradual curve as in other subgenera; cheek-teeth high-crowned; M1 and M2 narrowly triangular in occlusal outline; anterior cingulum joins protocone with abrupt change of direction on M1 and M2; metaloph joins protocone on each upper molar; M3 much larger than M2; posterior cingulum of M3 bends abruptly posteriad from protocone; p4 molariform, protolophid large and extends obliquely ventromediad from protoconid; protoconid of this tooth much larger than hypoconid; trigonid on lower cheek-teeth much higher than talonid; occlusal outline of m1 and m2 parallelogram-shaped; baculum having wider and deeper spoon than in *Callospermophilus* and rows of spines on margin more divergent; baculum otherwise as in *Otospermophilus;* cheek pouches of medium size; atlantoscapularis dorsalis muscle absent.

KEY TO NORTH AMERICAN SPECIES OF SUBGENUS SPERMOPHILUS

1. Upper parts unspotted and unmottled.
  2. Hind foot longer than 39.
    3. Underside of tail grayish.
                  S. armatus,   p. 386
    3′. Underside of tail buffy or reddish.
      4. Underside of tail buffy.
                 S. richardsonii,   p. 385
      4′. Underside of tail reddish.
                 S. beldingi,   p. 387
  2′. Hind foot less than 39. *S. townsendii,*   p. 382
1′. Upper parts spotted or mottled.
  5. Hind foot more than 43.
    6. Dorsal spots whitish. *S. parryii,*   p. 389
    6′. Dorsal spots buffy.
         S. columbianus,   p. 388
  5′. Hind foot less than 43.
    7. Upper parts grayish.
         S. washingtoni,   p. 384
    7′. Upper parts brownish.
         S. brunneus,   p. 385

Spermophilus townsendii
Townsend's Ground Squirrel

External measurements: 167–271; 32–72; 29–38. Greatest length of skull, 32.4–43.3. Upper parts plain, smoke gray shaded with pinkish buff or pinkish cinnamon; underparts whitish; underside of tail reddish. Rostrum stout, its sides nearly parallel; supraorbital borders of skull

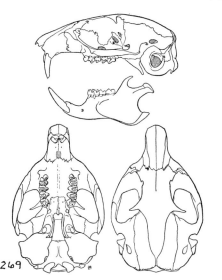

Fig. 269. *Spermophilus townsendii mollis*, 6 mi. E Stillwater, Nevada, No. 41525 M.V.Z., ♂, X 1.

slightly elevated; postorbital processes long, slender, decurved; temporal ridges lyrate, meeting posteriorly in old age and forming slight sagittal crest.

Spermophilus townsendii artemesiae (Merriam)

1913. *Citellus mollis artemesiae* Merriam, Proc. Biol. Soc. Washington, 26:137, May 21, type from Birch Creek, about 10 mi. S Nicholia, Idaho.
1938. *Citellus townsendii artemesiae*, A. H. Howell, N. Amer. Fauna, 56:65, May 18.
1913. *Citellus mollis* [*sic*] *pessimus* Merriam, Proc. Biol. Soc. Washington, 26:138, May 21, type from lower part of Big Lost River, Butte Co., Idaho.

MARGINAL RECORDS.—Idaho: Birch Creek, 2 mi. SE Kaufman; near Taber; Pingree, thence westward on N side Snake River to 3 mi. W Bliss; 3 mi. SE Arco.

Spermophilus townsendii canus Merriam

1898. *Spermophilus mollis canus* Merriam, Proc. Biol. Soc. Washington, 12:70, March 24, type from Antelope, Wasco Co., Oregon.
1938. *Citellus townsendii canus*, A. H. Howell, N. Amer. Fauna, 56:67, May 18.

MARGINAL RECORDS.—Oregon: 10 mi. N Baker; Cedar Mts. Nevada: Hot Spring; Virgin Valley; Long Valley Ranch. Oregon: Summer Lake; Fremont; Warmspring; 7 mi. E Antelope.

Spermophilus townsendii idahoensis (Merriam)

1913. *Citellus idahoensis* Merriam, Proc. Biol. Soc. Washington, 26:135, May 21, type from Payette, Idaho.
1939. *Citellus townsendii idahoensis*, Davis, Jour. Mamm., 20:182, May 15.

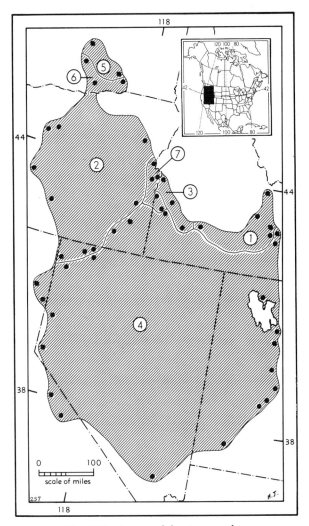

Map 257. *Spermophilus townsendii.*

Guide to subspecies
1. *S. t. artemesiae*
2. *S. t. canus*
3. *S. t. idahoensis*
4. *S. t. mollis*
5. *S. t. nancyae*
6. *S. t. townsendii*
7. *S. t. vigilis*

MARGINAL RECORDS.—Idaho: 2 mi. S Payette; 14 mi. SE Boise; Glenns Ferry, thence along N side Snake River to point of beginning.

Spermophilus townsendii mollis Kennicott

1863. *Spermophilus mollis* Kennicott, Proc. Acad. Nat. Sci. Philadelphia, 15:157. Type locality, Camp Floyd [= Fairfield], Utah Co., Utah.
1938. *Citellus townsendii mollis*, A. H. Howell, N. Amer. Fauna, 56:63, May 18.
1898. *Spermophilus mollis stephensi* Merriam, Proc. Biol. Soc. Washington, 12:69, March 24, type from Queen Station, near head Owens Valley, Mineral Co., Nevada.
1913. *Citellus leurodon* Merriam, Proc. Biol. Soc. Washington, 26:136, May 21, type from Murphy, in hills W of Snake River, Owyhee Co., Idaho.

1913. *Citellus mollis washoensis* Merriam, Proc. Biol. Soc. Washington, 26:138, May 21, type from Carson Valley [Douglas Co.?], Nevada.

MARGINAL RECORDS.—Idaho: Murphy, thence eastward along S side Snake River to *Blackfoot* (Davis, 1939:193); 3½ mi. E Wapello (*ibid.*); Ross Fork Creek. Utah: Promontory; Salt Lake City; 3 mi. N Saratoga Springs; Nephi; Manti; Salina; Richfield; 2 mi. W Cedar City. Nevada: Indian Springs. California: Long Valley; Mono Lake. Nevada: Carson City; Pyramid Lake. California: Honey Lake; Horse Lake. Nevada: mouth Little High Rock Canyon, 5000 ft.; Big Creek Ranch, Pine Forest Mts. Oregon: White Horse Sink; Rome.

Spermophilus townsendii nancyae Nadler

1968. *Spermophilus townsendi nancyae* Nadler, Cytogenetics, 7:153, April, type from 5 mi. N Richland, Benton Co., Washington.

MARGINAL RECORDS (Nadler, 1968:145, unless otherwise noted).—Washington: Ellensburg (Hall and Kelson, 1959:337); type locality; *Richland*, thence westward along N side Yakima River to point of beginning.

Spermophilus townsendii townsendii Bachman

1839. *Spermophilus townsendii* Bachman, Jour. Acad. Nat. Sci. Philadelphia, 8:61. Type locality, "On the Columbia River, about 300 miles above its mouth," Washington (but see T. H. Scheffer, Jour. Mamm., 27:395, November 25, 1946).
1898. *Spermophilus mollis yakimensis* Merriam, Proc. Biol. Soc. Washington, 12:70, March 24, type from Mabton, Yakima Co., Washington.

MARGINAL RECORDS.—Washington: Wiley City, thence eastward along S side Yakima River to *Kennewick*; 6³⁄₁₀ mi. E Badger Canyon (Nadler, 1966:581); Bickleton.

Spermophilus townsendii vigilis (Merriam)

1913. *Citellus canus vigilis* Merriam, Proc. Biol. Soc. Washington, 26:137, May 21, type from Vale, Oregon.
1938. *Citellus townsendii vigilis*, A. H. Howell, N. Amer. Fauna, 56:66, May 18.

MARGINAL RECORDS.—Oregon: Huntington; Ontario. Idaho: Homedale; 2 mi. W Reynolds Creek. Oregon: 1½ mi. SE Vale (Nadler, 1966:581); type locality.

Spermophilus washingtoni (A. H. Howell)
Washington Ground Squirrel

1938. *Citellus washingtoni washingtoni* A. H. Howell, N. Amer. Fauna, 56:69, May 18, type from Touchet, Walla Walla Co., Washington. (See T. H. Scheffer, Jour. Mamm., 27:395, November 26, 1946.)

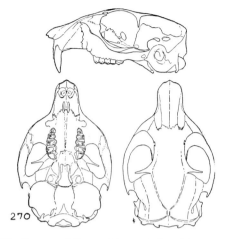

Fig. 270. *Spermophilus washingtoni*, 4 mi. W Pasco, Franklin Co., Washington, No. 93247 M.V.Z., ♂, X 1.

1938. *Citellus washingtoni loringi* A. H. Howell, N. Amer. Fauna, 56:71, May 18, type from Douglas, Washington. Dalquest (Univ. Kansas Publ., Mus. Nat. Hist., 2:271, April 9, 1948) regards *S. w. loringi* as inseparable from *S. washingtoni*.

External measurements: 185–245; 32–65; 30–38. Greatest length of skull, 35.0–41.4. Upper parts smoke gray flecked with white spots; tail

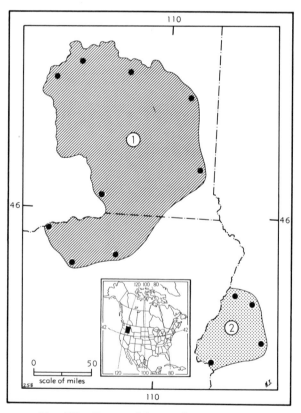

Map 258. *Spermophilus washingtoni* (1) and *Spermophilus brunneus* (2).

mixed fuscous and grayish white with blackish tip. Skull resembles that of *S. townsendii* but narrower in relation to length.

MARGINAL RECORDS.—Washington: Mansfield; Wilbur; Cheney; Wawawai. Oregon: Pilot Rock; Heppner; Willows. Washington: Pasco; Waterville.

Spermophilus brunneus (A. H. Howell)
Idaho Ground Squirrel

1928. *Citellus townsendii brunneus* A. H. Howell, Proc. Biol. Soc. Washington, 41:211, December 18, type from New Meadows, Adams Co., Idaho.
1938. *Citellus brunneus* A. H. Howell, N. Amer. Fauna, 56:72, May 18.

External measurements: 214–252; 46–61; 33–37. Greatest length of skull, 38.7–40.8. Upper parts dappled grayish brown, the brown predominating on lower part of back; nose, outer sides of hind legs, and ventral face of tail rusty brown; shoulders and forelegs ochraceous buff; underparts grayish fulvous; chin white; terminal hairs of tail with 5–8 alternating bands of black and white or fulvous. From *S. washingtoni* this species differs in ear larger (height from notch, 9–12), upper parts brown instead of gray and with smaller spots; pelage shorter and coarser; absence of distinct, white, ventrolateral stripe; auditory bullae smaller; rostrum shorter and broader; ratio of palatal to postpalatal length 82 as opposed to 70.5; sphenopalatine fissures smaller.

MARGINAL RECORDS.—Idaho: 1 mi. N Bear Creek R.S.; type locality; Van Wyck; Weiser.

Spermophilus richardsonii
Richardson's Ground Squirrel

External measurements: 253–337; 65–100; 39.5–49. Greatest length of skull, 42.0–48.6. Upper parts drab or smoke gray, more or less shaded with fuscous and dappled with cinnamon buff; under side of tail clay color, cinnamon buff, sayal brown, or ochraceous buff. Where the range of this species meets and overlaps that of *S. beldingi, S. richardsonii* can be recognized by relatively and actually longer tail, more intense cinnamon pigmentation on nose and underparts and ochraceous buff rather than reddish color on underside of tail. Where the ranges of *S. townsendii* and *S. richardsonii* overlap, the latter can be recognized by relatively and actually longer tail, cinnamon colored rather than whitish underparts, and tail with, rather than without, buffy white border. Where *S. armatus* and *S. richardsonii* occur together the latter can be rec-

ognized by the ochraceous buff instead of gray underside of the tail.

Robinson and Hoffmann (1975:79) by multivariate analysis of 36 cranial measurements found that *Spermophilus richardsonii richardsonii* differed more from any one of the three other subspecies (*S. r. aureus, S. r. elegans,* and *S. r. nevadensis*) than any one of the three last mentioned differed from each of the other two, and in conclusion (1975:87) wrote "If one were to . . . designate . . . different levels in the taxonomic hierarchy . . . *S. r. richardsonii* would . . . be differentiated . . . at the species level, and they [*S. r. aureus, S. r. elegans,* and *S. r. nevadensis*] would then be subspecies of *S. elegans,* the earliest named" of the three taxa. Robinson and Hoffmann (1975) did not present a revised classification but mentioned that chromosome divergence reported by Nadler, *et al.* (1971:298–305) were consistent with the analysis of cranial features.

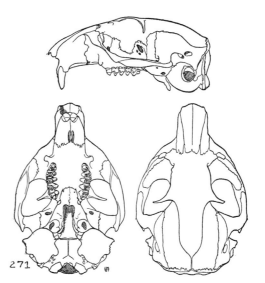

Fig. 271. *Spermophilus richardsonii nevadensis,* 4 mi. S Romano, Nevada, No. 70559 M.V.Z., ♂, X 1.

Spermophilus richardsonii aureus (Davis)

1939. *Citellus elegans aureus* Davis, The Recent mammals of Idaho, Caxton Printers, Caldwell, Idaho, p. 177, April 5, type from Double Springs, 16 mi. NE Dickey, Custer Co., Idaho.
1943. *Citellus richardsonii aureus,* Hall, Amer. Midland Nat., 29:378, March.

MARGINAL RECORDS.—Montana: 9 mi. N Wisdom; 3 mi. E Harrison (from map, Nadler, *et al.,* 1971:299); 5 mi. E Cameron (*ibid.*). Idaho: Henry Lake; near Teton Canyon; Patterson; Dickey; Forney. Montana: Big Hole Bench, W of Wisdom.

Spermophilus richardsonii elegans Kennicott

1863. *Spermophilus elegans* Kennicott, Proc. Acad. Nat. Sci. Philadelphia, 15:158, type from Fort Bridger, Wyoming.
1893. *Spermophilus r[ichardsoni]. elegans*, V. Bailey, Bull. Div. Ornith. and Mamm., U.S. Dept. Agric., 4:60.

MARGINAL RECORDS.—Wyoming (Long, 1965a:574, 575): 14 mi. N, 1 mi. W Waltman; 17 mi. S, 20 mi. W Douglas. Nebraska (Jones, 1964c:128): Bridgeport; Kimball. Colorado: T. 11, 12 N, R. 65–67 W, Weld Co. (Armstrong, 1972:123); 6 mi. W Boulder (*ibid.*); 2 mi. E Estabrook (Hansen, 1962:62); Colorado State Prison, near Cañon City (Armstrong, 1972:123); 21 mi. N Cotopaxi (Hansen, 1962:62); ½ mi. E Taylor Park Reservoir (*ibid.*); Rangeley. Utah: 2 mi. E Summit–Daggett county line, 2 mi. S Utah–Wyoming state line; Wasatch; 1 mi. N Randolph. Wyoming (Long, 1965a:574, 575): Cokeville; *2 mi. W Pinedale;* 1½ mi. NE Pinedale; *16 mi. S, 11 mi. W Waltman.*

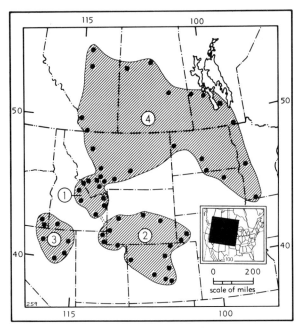

Map 259. *Spermophilus richardsonii.*

1. *S. r. aureus*   3. *S. r. nevadensis*
2. *S. r. elegans*   4. *S. r. richardsonii*

Spermophilus richardsonii nevadensis (A. H. Howell)

1928. *Citellus elegans nevadensis* A. H. Howell, Proc. Biol. Soc. Washington, 41:211, December 18, type from Paradise, Humboldt Co., Nevada.
1938. *Citellus richardsonii nevadensis* (A. H. Howell), N. Amer. Fauna, 56:77, May 18.

MARGINAL RECORDS.—Oregon: head Rattlesnake Creek. Idaho: 1 mi. S Riddle. Nevada: Metropolis; Ruby Valley; 4 mi. S Romono, Diamond Valley; 1 mi. N Winnemucca. Oregon: Malheur County, near McDermitt, Nevada.

Spermophilus richardsonii richardsonii (Sabine)

1822. *Arctomys richardsonii* Sabine, Trans. Linnean Soc. London, 13:589. Type locality, Carlton House, Saskatchewan.
1831. *Spermophilus richardsonii*, Cuvier, Supplément à l'histoire naturelle générale et particulière de Buffon, 1:323.

MARGINAL RECORDS.—Alberta: S end Baptiste Lake. Saskatchewan: Livelong; Prince Albert National Park (Beck, 1958:27); Touchwood Hills. Manitoba: Roblin (Soper, 1961:187); Lake Dauphin (*ibid.*); 1⅝ mi. NW St. Laurent (Tamsitt, 1962:74); 11 mi. E Emerson, 2 mi. N border. Minnesota (Heaney and Birney, 1975:31): E of Fergus Falls; Round Lake Waterfowl Station, Jackson Co. South Dakota: Aberdeen; *Sand Lake Wildlife Refuge* (Dickerman, 1960:403, as *Citellus richardsoni* only). North Dakota: Emmons County (Hewston, 1962:270, as "Richardson ground squirrel"); Bismark; Buford. Montana: N Bar Ranch, Flatwillow Creek (Hoffmann, *et al.*, 1969:588, as *S. richardsonii* only); Livingston; 25 mi. SW Bozeman (from map, Nadler, *et al.*, 1971:299); *7 mi. NE Harrison (ibid.);* Three Forks (*ibid.*); Toston; Birch Creek. Alberta: Cardston Road, Waterton Lakes National Park; 6 mi. E Maycroft; Edmonton.

Spermophilus armatus Kennicott
Uinta Ground Squirrel

1863. *Spermophilus armatus* Kennicott, Proc. Acad. Nat. Sci. Philadelphia, 15:158. Type locality, foothills Uinta Mts., near Fort Bridger, Wyoming.

External measurements: 280–303; 63–81; 42–45.5. Greatest length of skull, 46.3–48.5. Head, front of face, and ears cinnamon, sprinkled on crown with gray; sides of face and neck pale smoke gray; eye ring cartridge buff; front legs cinnamon buff, shading to pinkish buff on feet; general tone of dorsal area sayal brown or cinnamon buff, hairs tipped with pinkish buff, bases of hairs fuscous; sides paler than back, mixed cartridge buff and fuscous; thighs cinnamon; hind feet pinkish buff; tail, above and below, fuscous black, mixed with pale buff or buffy white, and edged with pinkish buff; underparts pinkish buff, shaded with buffy white. Underside of tail gray instead of ochraceous buff as in *S. richardsonii elegans*. Skull resembling that of *S. richardsonii elegans* but larger.

MARGINAL RECORDS.—Montana: West Gallatin River; West Boulder Creek, 18 mi. SE Livingston; Cooke. Wyoming (Long, 1965a:575–577): *29 mi. N, 32 mi. W Cody;* 16¼ mi. N, 17 mi. W Cody; 12 mi. N, 8 mi. W Cody; Valley; 1 mi. S, 3 mi. E Dubois; *Jackeys Creek;* W end Half Moon Lake; The Green River, 5 mi. W Green River; Opal; ¹/₁₀ mi. S Mountainview; 8 mi. S, 2½ mi. E Robertson. Utah: Fruitland; Mt. Pleasant; 20 mi. E Salina; Fish Lake; Maple Canyon; Mt. Timpanogos; Bountiful; Mantua; Pine Canyon, 17 mi. NE

Kelton. Idaho: Wickel Ranch, Elba; 4 mi. S Albion; Blackfoot; Arco; head Pahsimeroi River. Montana: Donovan.

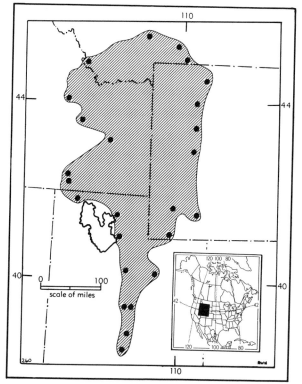

Map 260. *Spermophilus armatus.*

Spermophilus beldingi
Belding's Ground Squirrel

X ⅓

External measurements: 254–300; 55–76; 40–47. Greatest length of skull, 41.3–46.3. Upper parts smoke gray mixed with reddish brown or pinkish cinnamon; mid-dorsal area darkened with Sayal Brown; forehead pinkish cinnamon;

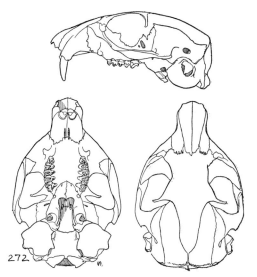

Fig. 272. *Spermophilus beldingi creber*, 7 mi. N Austin, Nevada, No. 70542 M.V.Z., ♂, X 1.

underparts grayish washed with pinkish cinnamon that is most pronounced on pectoral region, forelegs, forefeet, and hind feet; tail slightly darker than back on upper side but distinctly reddish, often hazel, beneath. Skull resembling that of *S. richardsonii*, but interorbital breadth greater relative to length of skull; in populations of *S. beldingi* geographically overlapping those of *S. richardsonii*, nasals averaging shorter relative to breadth of cranium. From *S. richardsonii*, *S. beldingi* is distinguished by shorter tail that is reddish rather than ochraceous buff beneath. Also, *S. beldingi* has more reddish in the pelage. From *S. townsendii*, *beldingi* is distinguished by larger size, cinnamon-colored rather than whitish underparts, maxillary tooth-row more, rather than less, than 8.9 mm.

Spermophilus beldingi beldingi Merriam

1888. *Spermophilus beldingi* Merriam, Ann. New York Acad. Sci., 4:317, December 28, type from Donner, Placer Co., California.

MARGINAL RECORDS.—California: Independence Lake. Nevada: 3 mi. S Mt. Rose, 8500–8600 ft.; ¼ mi. E Zephyr Cove, 6300 ft.; 6 mi. S Minden, 4850 ft. California: Walker Lake; Mono Lake; Bishop Creek; Little Pete Meadows, near head Middle Fork, Kings River; Sand Meadow; Bear Valley; head South Fork, American River, near Silver Lake.

Spermophilus beldingi creber (Hall)

1940. *Citellus beldingi crebrus* Hall, Murrelet, 21:59, December 20, type from Reese River Valley, 7 mi. N Austin, Lander Co., Nevada.

MARGINAL RECORDS.—Idaho: S bank Snake River, Homedale; S bank Snake River, 6 mi. S Rupert; 2 mi. S Malta; Bridge. Utah: Park Valley; *Grouse Creek Mts.; Grouse Creek; 12 mi. NW Grouse Creek.* Nevada: Goose Creek, 2 mi. W Utah boundary; Cedar Creek, 10 mi. NE San Jacinto, 6000 ft.; Steels Creek, N end Ruby Mts., 7000 ft.; Jerry Creek, 6700 ft.; W side Ruby Lake, 3 mi. N White Pine County line, 6200 ft.; Antone Creek, 1½ mi. SE Meadow Valley Ranger Station; Toquima Range, Meadow Creek (Valley) Ranger Station; Bell's Ranch, Reese River, 6890 ft.; Calico Mtn.; Big Creek, Pine Forest Mts., 6000–7000 ft.; head Leonard Creek, Pine Forest Mts., 6500 ft.; Alder Creek, Pine Forest Mts., 5000 ft. Oregon: 2 mi. W Jordan Valley.

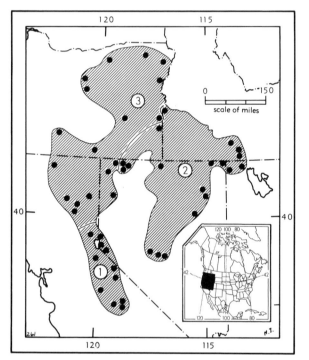

Map 261. *Spermophilus beldingi.*

1. *S. b. beldingi*   2. *S. b. creber*
3. *S. b. oregonus*

Spermophilus beldingi oregonus Merriam

1898. *Spermophilus oregonus* Merriam, Proc. Biol. Soc. Washington, 12:69, March 24, type from Swan Lake Valley, Klamath Basin, Oregon.
1938. *Citellus beldingi oregonus*, A. H. Howell, N. Amer. Fauna, 56:83, May 18.

MARGINAL RECORDS.—Oregon: Elgin; Joseph; Home; Rockville. Nevada: Badger; Massacre Creek, 5800 ft.; Rock Creek, 7000–7425 ft. California: Horse Lake; 12 mi. NE Prattville; North Fork, Feather River; Mt. Lassen, probably near Black Butte; Grass Lake. Oregon: Fort Klamath; Camas Prairie, E of Lakeview; Narrows; Prineville; Hay Creek; Heppner.

Spermophilus columbianus
Columbian Ground Squirrel

External measurements: 327–410; 80–116; 48–58. Greatest length of skull, 49.5–57. Nose and face tawny or hazel; occiput, nape, and sides of neck smoke gray; upper parts cinnamon buff or sayal brown, shaded with fuscous and in winter with smoke gray; hind legs and feet tawny or hazel; front feet ochraceous buff; tail gray or tawny; underparts ochraceous buff or tawny. Skull resembling that of *S. richardsonii* but longer and zygomatic arches less expanded posteriorly; dorsal outline more nearly flat; supraorbital margins of frontals not elevated or thickened; rostrum and nasals longer; upper tooth-rows nearly parallel instead of appreciably divergent anteriorly.

Spermophilus columbianus columbianus (Ord)

1815. *Arctomys Columbianus* Ord, *in* Guthrie, A new geog., hist., coml. grammar . . . , Philadelphia, Amer. ed. 2, 2:292 (described on p. 303). Based on Lewis and Clark's description of animals taken on a camas prairie between forks of Clearwater and Kooskooskie rivers, Idaho.
1891. *Spermophilus columbianus*, Merriam, N. Amer. Fauna, 5:39, July 30.
1817. *Anisonyx brachiura* Rafinesque, Amer. Month. Mag., 2:45. Based on same source as *Arctomys Columbianus* Ord.
1829. *Arctomys parryi* var. β., *erythrogluteia* Richardson,

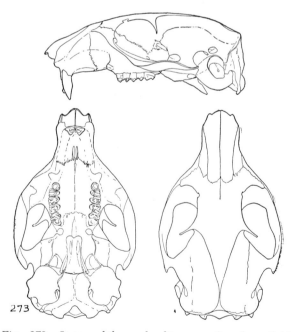

Fig. 273. *Spermophilus columbianus ruficaudus,* Cold Spring, 8 mi. E Austin, Grant Co., Oregon, No. 78467 M.V.Z., ♂, X 1.

Fauna Boreali-Americana, p. 161. Type locality, "Rocky Mountains, near sources of Elk River" [= Wolf Plain, 30 mi. W Rock Lake, Alberta].

1884. *Cynomys columbianus* True, Proc. U.S. Nat. Mus., 7(App., Circ. 29):593, November 29. (Apparently based on *Arctomys Colmbianus* Ord, 1815.)

1903. *Citellus columbianus albertae* J. A. Allen, Bull. Amer. Mus. Nat. Hist., 19:537, October 10, type from Canadian National Park, Banff, Alberta.

MARGINAL RECORDS.—British Columbia: mountains on E side lower Parsnip River. Alberta: Two Lakes, near Hat Mtn.; Smoky River Valley, 50 mi. N Jasper; Banff; Mt. Forgetmenot; Pincher Creek; Waterton Lakes National Park. Montana (Hoffmann, *et al.*, 1969:588, as *S. columbianus* only); Duck Lake; *12 mi. W Browning*; McClellan Creek E of Helena; Big Hole–Grasshopper Creek Divide; *McCaleb Ranch on Birch Creek*. Idaho: Mill Creek, 14 mi. W Challis; Ketchum; Craters of the Moon; Bald Mountain Ranger Station, 10 mi. S Idaho City; Shafer Butte. Oregon: 10 mi. N Harney; Strawberry Mts.; Ironside. Washington: Pullman; Colfax; Williams Lake; Loon Lake; 15 mi. E Tonasket. British Columbia (Cowan and Guiguet,

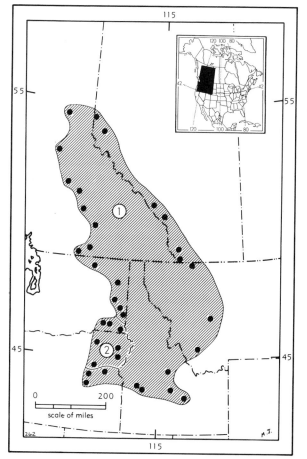

Map 262. *Spermophilus columbianus*.

1. *S. c. columbianus*   2. *S. c. ruficaudus*

1965:128): Ashnola River; Okanagan Falls; Okanagan Landing; Rayleigh; Bridge Lake; Lac la Hache; Cottonwood.

Spermophilus columbianus ruficaudus (A. H. Howell)

1928. *Citellus columbianus ruficaudus* A. H. Howell, Proc. Biol. Soc. Washington, 41:212, December 18, type from Wallowa Lake, Wallowa Co., Oregon.

MARGINAL RECORDS.—Washington: Dayton; Anatone. Oregon: type locality; Anthony, 6 mi. SW Cornucopia; Dixie Butte; Meacham. Washington: Prescott.

Spermophilus parryii
Arctic Ground Squirrel

External measurements: 332–495; 77–153; 50–68. Greatest length of skull, 50.7–65.8. Head tawny or cinnamon; rest of upper parts reddish brown, cinnamon or fuscous, more or less abundantly flecked with whitish spots; underparts ochraceous tawny to cinnamon buff in summer but ochraceous buff or grayish white in winter; tail above, ochraceous tawny, cinnamon or cinnamon buff, mixed with fuscous black; tail beneath, russet or tawny. Skull angular; highest at plane of postorbital processes; zygomata strongly twisted from the vertical plane; P3 a third to half size of P4.

Gromov, *et al.* (1963:310–314) concluded that (1) *S. undulatus* and *S. parryii* are two species, (2) both occur in Siberia, and (3) only *S. parryii* occurs in North America.

Modifications of subspecific boundaries as shown here have been suggested by Nadler and Youngman (1969:1051–1057) and by Nadler, *et al.* (1973:33–40), taking into account transferrins, ectoparasites, geographic origins of existing stocks, and even macromorphology (the last not described but referred to as in press by Youngman and by Hoffmann, Lyapunova, Nadler, and Vorontsov). But as of July 3, 1974, no formal systematic rearrangement of the several available names had been seen. A reclassification of the geographic variants of the Nearctic ground squirrels seems to be warranted. Chernyavsky (1972:199–214) has treated the closely related taxa of Asia.

Spermophilus parryii ablusus (Osgood)

1903. *Citellus plesius ablusus* Osgood, Proc. Biol. Soc. Washington, 16:25, March 19, type from Nushagak, Alaska.

1903. *Citellus stonei* J. A. Allen, Bull. Amer. Mus. Nat. Hist., 19:537, October 10, type from Stevana Flats, near Port Müller, Alaska Peninsula; not Wrangell, Alaska (see J. A. Allen, *op. cit.*:xvii).

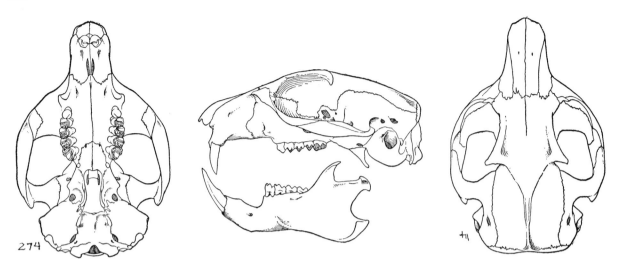

Fig. 274. *Spermophilus parryii kennicottii*, Bearpaw Creek, $1\frac{7}{10}$ mi. N, $1\frac{7}{10}$ mi. E Umiat, 152° 04′ 50″, 69° 23′ 30″, 550 ft., Alaska, No. 50437 K.U., ♂, X 1.

MARGINAL RECORDS.—Alaska: Eschscholtz Bay; Savage River, Mt. McKinley National Park; Jennie Creek; Talkeetna Mts., 60 mi. N Anchorage (Nadler, 1966:581); Anchorage; Ushagat Island; Unimak Island; Unalakleet; Golofnin Bay; Cooper Gulch, thence westward into Siberia, and eastward to Alaska: Cottonwood Creek; *Trail Creek*. Introduced also on Unalaska, Umnak, and Kavalga islands of the Aleutian group, but these occurrences are not shown on Map 263.

Spermophilus parryii kennicottii (Ross)

1861. A[*rctomys*]. *kennicottii* Ross, Canadian Nat. and Geol., 6:434, type from Fort Good Hope, Mackenzie.
1900. *Spermophilus barrowensis* Merriam, Proc. Washington Acad. Sci., 2:20, March 14, type from Point Barrow, Alaska.
1900. *Spermophilus beringensis* Merriam, Proc. Washington Acad. Sci., 2:20, March 14, type from Cape Lisburne, Alaska.

MARGINAL RECORDS.—Alaska: Point Barrow. Yukon: near mouth Firth River (Youngman, 1975:70). Mackenzie: Fort Anderson. Yukon: Richardson Mts., 13 mi. NE Lapierre House (Youngman, 1971:71). Alaska: Porcupine River, 12 mi. below Coleen River and near Salmon Trout River; Anaktuvuk Pass; Chandler Lake; Cape Thompson.

Spermophilus parryii kodiacensis J. A. Allen

1874. *Spermophilus parryi* var. *kodiacensis* J. A. Allen, Proc. Boston Soc. Nat. Hist., 16:292, lectotype from Kodiak Island, Alaska. Known only from Kodiak Island.

Spermophilus parryii lyratus (Hall and Gilmore)

1932. *Citellus lyratus* Hall and Gilmore, Univ. California Publ. Zool., 38:396, September 17, type from Iviktook Lagoon, about 35 mi. NW Northeast Cape, St. Lawrence Island, Bering Sea, Alaska.

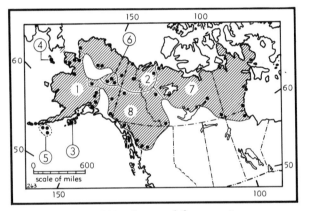

Map 263. *Spermophilus parryii.*

1. *S. p. ablusus*	5. *S. p. nebulicola*
2. *S. p. kennicottii*	6. *S. p. osgoodi*
3. *S. p. kodiacensis*	7. *S. p. parryii*
4. *S. p. lyratus*	8. *S. p. plesius*

MARGINAL RECORDS.—Alaska: St. Lawrence Island; type locality; *Kukuliak*.

Spermophilus parryii nebulicola (Osgood)

1903. *Citellus nebulicola* Osgood, Proc. Biol. Soc. Washington, 16:26, March 19, type from Nagai Island, Shumagin Islands, Alaska.

MARGINAL RECORDS.—Alaska: Shumagin Islands; Koniuji Island; Simeonof Island; type locality.

Spermophilus parryii osgoodi Merriam

1900. *Spermophilus osgoodi* Merriam, Proc. Washington Acad. Sci., 2:18, March 14, type from Fort Yukon, Alaska.

MARGINAL RECORDS.—Alaska: type locality; 20 mi. above Circle; 10 mi. above Hess Creek.

Spermophilus parryii parryii (Richardson)

1825. *Arctomys Parryii* Richardson, *in* Parry, Journal of a second voyage . . . , p. 316. Type locality, Five Hawser Bay, Lyon Inlet, Melville Peninsula, Hudson Bay, Canada.

1829. *Arctomys parryi* var. *phaeognatha* Richardson, Fauna Boreali-Americana, 1:161, type from Hudson Bay.

MARGINAL RECORDS.—Keewatin: *Melville Peninsula;* Marble Island; *ca.* 25 mi. S Cape Eskimo, Hudson Bay. Manitoba: Seal River Camp, 5 mi. N Seal River, W shore Hudson Bay (Nadler and Hoffmann, 1977:750). Mackenzie: Kasba Lake; Artillery Lake; Langton Bay, thence along Arctic Coast to point of beginning.—A specimen (18103 NMC) from *Mile 174 E Canol Road*, Mackenzie, mistakenly listed as S. *p. parryi* by Anderson (1947:110) was earlier listed as S. *p. plesius* by Rand (1945:37) and in 1976 was identified as S. *p. plesius* by R. S. Hoffmann (*in verbis*). It is not marginal for S. *p. plesius*.

Spermophilus parryii plesius Osgood

1900. *Spermophilus empetra plesius* Osgood, N. Amer. Fauna, 19:29, October 6, type from Bennett City, head of Lake Bennett, British Columbia.

MARGINAL RECORDS.—Alaska: Tanana Hills. Yukon: head Coal Creek. Mackenzie: mts. W Fort Norman; Fort Liard. British Columbia: Tatletuey Lake; head Klappan River; Glenora Mtn.; Stonehouse Creek near jct. with Kelsall River. Alaska: Chitina River Glacier; Tanana Crossing.

Subgenus Ictidomys J. A. Allen

1877. *Ictidomys* J. A. Allen, *in* Coues and Allen, Monog. N. Amer. Rodentia, p. 821, August. Type, *Sciurus tridecemlineatus* Mitchill.

1907. *Ictidomoides* Mearns, Bull. U.S. Nat. Mus., 56:328, April 13. Type, *Sciurus mexicanus* Erxleben, 1777.

External measurements: 170–380; 55–166; 27–51. Greatest length of skull, 34.0–52.5. Upper parts with nearly square white spots, in some species arranged in rows, or upper parts (in S. *perotensis*) clay color. Anterolateral walls of cranium nearly vertical; least width of postorbital region slightly greater than width of interorbital region; ventral mandibular incisure deep and acutely arched as in *Xerospermophilus;* crowns of cheek-teeth higher than those of *Otospermophilus* and lower than those of subgenus *Spermophilus;* trigon of P4, M1, and M2 narrowly V-shaped; anterior cingulum of M1 and M2 joins protocone with abrupt change of direction; metaloph on M1 and M2 separated from protocone by sulcus; tendency toward fusion of metaconule and metacone on M1 and M2. Metaloph indistinct or absent on M3; M3 much larger than M2; posterior cingulum on M3 bends abruptly posteriad from protocone; p4 not molariform, protolophid small; occlusal outline of m1 and m2 rhomboidal; cheek pouches of medium size; atlantoscapularis dorsalis muscle absent.

KEY TO SPECIES OF SUBGENUS ICTIDOMYS

1. Dorsal area striped. *S. tridecemlineatus,* p. 391
1'. Dorsal area spotted or plain.
   2. Dorsal spots in linear series.
               S. mexicanus, p. 394
   2'. Dorsal spots not in linear series or upper parts plain.
      3. Underparts buffy; upper parts plain or spots buffy. *S. perotensis,* p. 397
      3'. Underparts and dorsal spots white. . . . . . . *S. spilosoma,* p. 395

Spermophilus tridecemlineatus
13-lined Ground Squirrel

External measurements: 170–297; 60–132; 27–41. Greatest length of skull, 34.0–45.8. Upper parts marked with a series of alternating dark (brownish or blackish) and light longitudinal stripes; a row of nearly square white spots in each of the dark dorsal stripes; lowermost stripes on sides less well defined than on back; in some subspecies some of the light dorsal stripes are broken into spots. Skull long, narrow, and lightly built in comparison with that of S. *townsendii;* molariform tooth-rows only slightly convergent posteriorly.

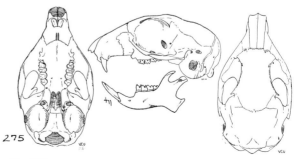

Fig. 275. *Spermophilus tridecemlineatus arenicola,* 6 mi. W Bird City, Cheyenne Co., Kansas, No. 12079 K.U., ♂, X 1.

Spermophilus tridecemlineatus alleni Merriam

1898. *Spermophilus tridecemlineatus alleni* Merriam, Proc. Biol. Soc. Washington, 12:71, March 24, type from near head of Canyon Creek, 8000 ft., west slope Bighorn Mts., Wyoming.

MARGINAL RECORDS (Long, 1965a:579).— Wyoming: W slope of Bighorn Mts., near head Canyon Creek; *head Kirby Creek*; Miners Delight, near head Twin Creek; New Fork Green River (Lander Road).

Spermophilus tridecemlineatus arenicola (A. H. Howell)

1928. *Citellus tridecemlineatus arenicola* A. H. Howell, Proc. Biol. Soc. Washington, 41:213, December 18, type from Pendennis, Kansas.
1955. *Spermophilus tridecemlineatus arenicola*, Hall, Univ. Kansas Mus. Nat. Hist., Miscl. Publ., 7:92, December 13.

MARGINAL RECORDS.—Kansas: Long Island; 7 mi. SW Sylvan Grove. Oklahoma: 3 mi. E Cherokee; 3 mi. W Orienta. Texas: Mobeetie; Briscoe County (Davis, 1966:132, as *Citellus tridecemlineatus* only); Lubbock; Dawson County (Davis, 1966:132, as *Citellus tridecemlineatus* only); Gaines County (*ibid.*, as *Citellus tridecemlineatus* only). New Mexico: Hobbs [2 mi. N] (Zimmerman and Cothran, 1976:706, as species only); Roswell; Loveless Lake, 10 mi. NW Capitan Mts.; 6 mi. N Cabra Spring (Findley, *et al.*, 1975:119, as species only); 4 mi. SW Cimarron. Colorado (Armstrong, 1972:124): La Veta, 7012 ft.; 4 mi. W Westcliffe; Eureka Hill, 4850 ft.

Spermophilus tridecemlineatus blanca Armstrong

1971. *Spermophilus tridecemlineatus blanca* Armstrong, Jour. Mamm., 52:533, August 26, type from 5 mi. W Antonito, Conejos Co., Colorado.

MARGINAL RECORDS.—Colorado (Armstrong, 1972:125): Moffat; Fort Garland; 2 mi. S San Acacio; type locality. New Mexico (Findley, *et al.*, 1975:118, 119, as *S. tridecemlineatus* only): 12 mi. NW Datil; San Augustin Plains near Monica Spring; Railroad Canyon, 10 mi. S Beaverhead. Arizona: Springerville. Assignment of material from Springerville and that from Catron and Socorro counties of New Mexico to this subspecies is provisional. Armstrong (1971:535), who studied some of this material, was uncertain about applying the name *S. t. blanca* to it.

Spermophilus tridecemlineatus hollisteri (V. Bailey)

1913. *Citellus tridecemlineatus hollisteri* V. Bailey, Proc. Biol. Soc. Washington, 26:131, May 21, type from Elk Valley, 8000 ft., Mescalero Indian Reservation, Sacramento Mts., New Mexico.

MARGINAL RECORDS.—New Mexico: 35 mi. NW Cimarron; Maxwell; 12 mi. N Las Vegas; *Las Vegas Golf Course, Las Vegas* (Studier and Baca, 1968:402, as *Citellus tridecemlineatus* only); type locality.

Spermophilus tridecemlineatus monticola (A. H. Howell)

1928. *Citellus tridecemlineatus monticola* A. H. Howell, Proc. Biol. Soc. Washington, 41:214, December 18, type from Marsh Lake, 9000 ft., White Mts., Arizona. Known only from type locality.

Spermophilus tridecemlineatus olivaceus J. A. Allen

1895. *Spermophilus tridecemlineatus olivaceus* J. A. Allen, Bull. Amer. Mus. Nat. Hist., 7:337, November 8, type from Custer, Custer Co., Black Hills, South Dakota.

MARGINAL RECORDS.—Wyoming: 15 mi. N Sundance (Long, 1965a:581). South Dakota: 1½ mi. S, ½ mi. E Spearfish (112545 KU); type locality. Wyoming (Long, 1965a:581): Newcastle; *1½ mi. NW Sundance*.

Spermophilus tridecemlineatus pallidus J. A. Allen

1874. [*Spermophilus tridecemlineatus*] var. *pallidus* J. A. Allen, Proc. Boston Soc. Nat. Hist., 16:291, February 4, a *nomen nudum*. "Hab. The dry plains and deserts of the interior westward to the Great Basin."
1877. [*Spermophilus tridecemlineatus*] var. *pallidus* J. A. Allen, *in* Coues and Allen, Monog. N. Amer. Rodentia, p. 872, August, 1877. Type locality, "Plains of lower Yellowstone River," Montana. (Specimen from mouth Yellowstone River designated as lectotype by A. H. Howell, N. Amer. Fauna No. 56:112, footnote, May 18, 1938.)

MARGINAL RECORDS.—Alberta: near Elkwater Lake. Saskatchewan: vic. Kealy Springs, Cypress Hills, 14 mi. N Ravenscrag; Salt Lake. Montana: type locality. North Dakota: Oakdale; Fort Clark; Mandan. Nebraska (Jones, 1964c:134): 6 mi. N Midway; head [? Little] Blue River; Beaver City; Benkelman. Colorado (Armstrong, 1972:126, 127): 4 mi. E Flagler; 15 mi. E Colorado Springs; Twin Lakes; [near] head Arkansas River; Como, 9800 ft.; Golden; Elkhorn, 7000 ft.; Canadian Creek. Wyoming (Long, 1965a:582): Centennial; Casper; 42 mi. S, 16 mi. W Gillette. Montana: Pryor Mts.; Billings; 20 mi. NE Roy.

Spermophilus tridecemlineatus parvus J. A. Allen

1895. *Spermophilus tridecemlineatus parvus* J. A. Allen, Bull. Amer. Mus. Nat. Hist., 7:337, November 8, type from Kennedys Hole, Uncompahgre Indian Reservation, 20 mi. northeast of Ouray, Uintah Co., Utah.

MARGINAL RECORDS.—Wyoming (Long, 1965a:582): Myersville; Independence Rock. Colorado (Armstrong, 1972:127): Routt County; W fork Elk Creek, 6–8 mi. above New Castle; 5 mi. W Rangely, 5600 ft. Utah: Pariette Bench, 12 mi. S Ouray; Red Creek, 2 mi. N Fruitland; Diamond Mtn. Wyoming (Long, 1965a:582): Green River; Big Sandy.

Spermophilus tridecemlineatus texensis Merriam

1898. *Spermophilus tridecemlineatus texensis* Merriam, Proc. Biol. Soc. Washington, 12:71, March 24, type from Gainesville, Cooke Co., Texas.

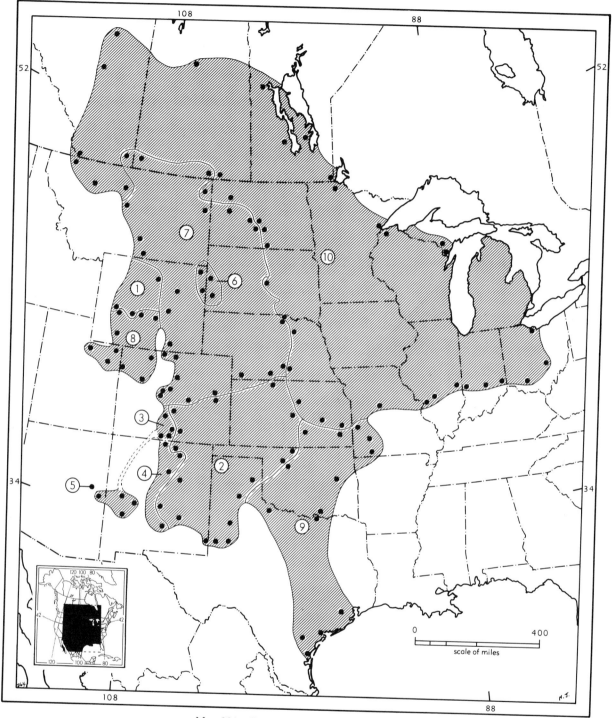

Map 264. *Spermophilus tridecemlineatus.*

Guide to
subspecies
1. *S. t. alleni*
2. *S. t. arenicola*

3. *S. t. blanca*
4. *S. t. hollisteri*
5. *S. t. monticola*
6. *S. t. olivaceus*

7. *S. t. pallidus*
8. *S. t. parvus*
9. *S. t. texensis*
10. *S. t. tridecemlineatus*

1899. *Spermophilus* (*Ictidomys*) *tridecemlineatus badius* Bangs, Proc. New England Zool. Club, 1:1, February 8, type from Stotesbury, Vernon Co., Missouri.

MARGINAL RECORDS.—Missouri: Stotesbury; Golden City; Washburn. Oklahoma: 3 mi. E Wainwright; 5 mi. N Colbert. Texas: Woodlawn Country Club Golf Course, 6 mi. N Sherman (McCarley, 1966:295); Richmond; Port Lavaca; Corpus Christi; Bee County; Vernon. Oklahoma: 5 mi. SW Canton. Kansas: Garden Plains; Neosho Falls.

Spermophilus tridecemlineatus tridecemlineatus (Mitchill)

1821. *Sciurus tridecem-lineatus* Mitchill, Med. Repos. (n.s.) [New York], 6(21):248. Type locality, central Minnesota (see J. A. Allen, Bull. Amer. Mus. Nat. Hist., 7:338, November 8, 1895).

1849. *Spermophilus tridecem lineatus*, Audubon and Bachman, The viviparous quadrupeds of North America, 1:294.

1822. *Arctomys hoodii* Sabine, Trans. Linnean Soc. London, 13:590, type from Carlton House, Saskatchewan. (Regarded by A. H. Howell, N. Amer. Fauna, 56:107, May 18, 1938, as inseparable from *S. t. tridecemlineatus* but arranged by Anderson, Bull. Nat. Mus. Canada, 102:111, January 24, 1947, as a tenable subspecies.)

MARGINAL RECORDS.—Alberta: near NE angle Baptiste Lake. Saskatchewan: S end Crean Lake (Soper, 1961:31, as *Citellus tridecemlineatus hoodii*). Manitoba: Bell River, a few mi. E Mafeking; McCreary; Geyser; near W end Whitemouth Lake. Minnesota: Williams; Duluth. Wisconsin: Herbster. Michigan: Dickinson County; Menominee County. Ohio: Sandusky County; Muskingum County; Bainbridge. Indiana: Franklin County (Mumford, 1969:59); Bartholomew County (*ibid.*); 5 mi. S Terre Haute. Illinois: Kansas; Hillsboro; Madison County. Missouri: Saline County. Kansas: Anderson County; 3 mi. SW Burdick; 3½ mi. W., ½ mi. S Beloit. Nebraska (Jones, 1964c:135, 136): Logan Township; Neligh; Spencer. South Dakota: Pierre. North Dakota: Zeeland; Bismark; Washburn; Fort Berthold. Montana: Johnson Lake. Saskatchewan: Estevan. Montana: Bear Paw Mts., 20 mi. SE Fort Assiniboine; Choteau; St. Mary Lake. Alberta: Chief Mountain Lake; Red Deer.—Introduced in Pennsylvania at Polk and persisting in vicinity (Richmond and Roslund, 1949:42; Doutt, *et al.*, 1973:112).

Spermophilus mexicanus
Mexican Ground Squirrel

External measurements: 280–380; 110–166; 38–51. Greatest length of skull, 41.0–52.5. Upper parts wood brown or buffy brown to sayal brown or snuff brown, with nearly square white spots arranged in longitudinal rows, usually 9, the spots in some animals partly confluent and in oth-

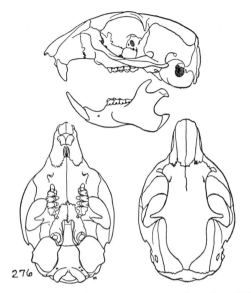

Fig. 276. *Spermophilus mexicanus parvidens*, ½ mi. NE Del Rio, Val Verde Co., Texas, No. 93789 M.V.Z., ♀, X 1.

ers faint; head buffy brown or wood brown, sprinkled with white; nose clay color or cinnamon buff; feet, sides, and underparts white to pinkish buff; tail above, mixed fuscous and grayish or buffy white; tail beneath, avellaneous

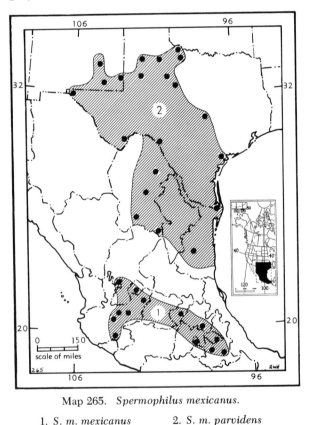

Map 265. *Spermophilus mexicanus.*

1. *S. m. mexicanus*   2. *S. m. parvidens*

to cinnamon buff, more or less obscured by grayish or buffy white. Skull resembling that of S. tridecemlineatus but larger, braincase less elongate, and zygomata more widely expanded.

Spermophilus mexicanus mexicanus (Erxleben)

1777. [*Sciurus*] *mexicanus* Erxleben, Systema regni animalis . . . , 1:428. Type locality, Toluca, México, by restriction (see Mearns, Preliminary diagnoses of new mammals from the Mexican border of the United States, p. 1, March 25, 1896).

1843. *Spermophilus mexicanus*, Wagner, *in* Schreber, Die Säugthiere . . . , Suppl., 3:250.

MARGINAL RECORDS.—Jalisco: La Mesa María de León (Genoways and Jones, 1973:10). Aguascalientes: 3 mi. S Aguascalientes (Ingles, 1959:388). Jalisco: Lagos. Querétaro: Tequisquiapan. Hidalgo: Irolo. Tlaxcala: Huamantla. Puebla: San Andrés, Chalchicomula; Atlixco. Distrito Federal: Tlalpam. Jalisco: 10 mi. SW Tepatitlán; Zapotlán; 21 mi. SW Guadalajara; Atemajac.

Spermophilus mexicanus parvidens Mearns

1896. *Spermophilus mexicanus parvidens* Mearns, Preliminary diagnoses of new mammals from the Mexican border of the United States, p. 1, March 25 (preprint of Proc. U.S. Nat. Mus., 18:443, May 23, 1896), type from Fort Clark, Kinney Co., Texas.

MARGINAL RECORDS.—Texas (Davis, 1966:133, as *Citellus mexicanus* only, unless otherwise noted): Lockett (Dalquest, 1968:16); Seymour (*ibid.*); Fisher County; Taylor County; Williamson County; Refugio County; *Rockport* (Hall and Kelson, 1959:348); Port Isabel (*ibid.*). Tamaulipas: Victoria. Coahuila: 8 mi. N LaVentura; 1 mi. N Parras; 5 mi. N, 2 mi. W Monclova; Sabinas. Texas: Del Rio; Black Gap Wildlife Management Area; El Paso. New Mexico: Roswell (Findley, *et al.*, 1975:121, as *S. mexicanus* only); 2 mi. NE Carlsbad (*ibid.*); Hobbs [2 mi. N] (Zimmerman and Cothran, 1976:706, as *S. mexicanus* only). Texas: Lamesa; vic. Lubbock County (Packard and Judd, 1968:536); Dickens County.

Spermophilus spilosoma
Spotted Ground Squirrel

X ½

External measurements: 185–253; 55–92; 28–38. Greatest length of skull, 34.1–42.7. Upper parts drab, cinnamon drab, avellaneous, smoke

gray, fawn, wood brown, snuff brown, or verona brown, more or less spotted with "squarish" white spots; tail above usually resembling back but having fuscous black at tip; tail beneath some shade of cinnamon. Skull resembles that of S. tridecemlineatus but relatively broader, especially in rostrum and interorbital region; auditory bullae much larger.

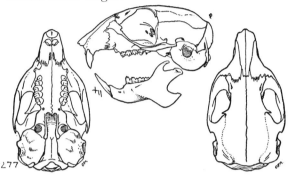

Fig. 277. *Spermophilus spilosoma marginatus*, N end Mariscal Mtn., 2300 ft., Brewster Co., Texas, No. 80346 M.V.Z., ♀, X 1.

Spermophilus spilosoma altiplanensis Anderson

1972. *Spermophilus spilosoma altiplanensis* Anderson, Bull. Amer. Mus. Nat. Hist., 148:275, September 8, type from 3 mi. ESE La Junta, Chihuahua, 6900 ft.

MARGINAL RECORDS (Anderson, 1972:276).—Chihuahua: 2 mi. N Tejolocachic, 6900 ft.; NE side Laguna de Bustillos, 6750 ft.; 24 mi. S Cuauhtémoc, 7700 ft.; 5 mi. W La Junta, 6900 ft.

Spermophilus spilosoma ammophilus Hoffmeister

1959. *Spermophilus spilosoma ammophilus* Hoffmeister, Proc. Biol. Soc. Washington, 72:37, May 1, type from 5½ mi. N Samalayuca, Chihuahua.

MARGINAL RECORDS (Hoffmeister, 1959:38, unless otherwise noted).—Chihuahua: 6 mi. S Ciudad Juárez, 3800 ft. (Anderson, 1972:276); *type locality*; 4⅘ mi. S Samalayuca.

Spermophilus spilosoma annectens Merriam

1893. *Spermophilus spilosoma annectens* Merriam, Proc. Biol. Soc. Washington, 8:132, December 28, type from "The Tanks," 12 miles from Point Isabel, Padre Island, Texas.

MARGINAL RECORDS.—Texas: mouth Pecos River; Mustang Island; Cameron County (Davis, 1966:135, as *Citellus spilosoma* only); 10 mi. from Edinburg (James and Hayse, 1963:574, as *Citellus spilosoma* only).

Spermophilus spilosoma bavicorensis Anderson

1972. *Spermophilus spilosoma bavicorensis* Anderson, Bull. Amer. Mus. Nat. Hist., 148:276, September 8, type from 2 mi. SW S. J. Babícora, Chihuahua, 7450 ft.

MARGINAL RECORDS (Anderson, 1972:276).—Chihuahua: La Varas, 7800 ft.; *10 mi. NNW Gómez Farías, 7800 ft.*; type locality.

Spermophilus spilosoma cabrerai (Dalquest)

1851. *Citellus spilosoma cabrerai* Dalquest, Proc. Biol. Soc. Washington, 64:106, August 24, type from 10 km. NNW Nuñez, San Luis Potosí.

MARGINAL RECORDS.—San Luis Potosí: 6 km. S Matehuala; 2 mi. NW Tepeyac, 3400 ft., 14 mi. N, 29 mi. W C. del Maíz; 6 km. SSW Nuñez.

Spermophilus spilosoma canescens Merriam

1890. *Spermophilus canescens* Merriam, N. Amer. Fauna, 4:38, October 8, type from Willcox, Cochise Co., Arizona.
1932. *Citellus spilosoma canescens,* V. Bailey, N. Amer. Fauna, 53:109, March 1.
1890. *Spermophilus spilosoma macrospilotus* Merriam, N. Amer. Fauna, 4:38, October 8, type from Oracle, Pinal Co., Arizona.
1901. [*Spermophilus spilosoma*] *microspilotus* Elliot, Field Columb. Mus., Publ. 45, Zool. Ser., 2:96, an accidental renaming of *macrospilotus*.
1902. *Spermophilus spilosoma arens* V. Bailey, Proc. Biol. Soc. Washington, 15:118, June 2, type from El Paso, El Paso Co., Texas.

MARGINAL RECORDS.—Arizona: Pima. New Mexico: Apache. Chihuahua: Monument 15, Mexican boundary line (Anderson, 1972:277). Texas: El Paso; Fort Hancock. Chihuahua (Anderson, 1972:277): 4 mi. S Gallegos, 5300 ft.; Aldama, 30 mi. NE Chihuahua, 4500 ft.; 1 mi. N, 3 mi. E Camargo, 4150 ft.; 1½ mi. W Jiménez; 5 mi. E Parral, 5700 ft. Durango: Río Ocampo (Hall and Kelson, 1959:350, as *S. s. pallescens*). Chihuahua (Anderson, 1972:277): 1½ mi. E General Trias, 1800 m.; 4 mi. SW San Buenaventura; San Diego; 28 mi. S Barenda. Sonora: La Noria. Arizona: Buenos Aires; Oracle.

Spermophilus spilosoma cryptospilotus Merriam

1890. *Spermophilus cryptospilotus* Merriam, N. Amer. Fauna, 3:57, September 11, type from "Tenebito" [= Dinnebito] Wash, Painted Desert, Coconino Co., Arizona.
1938. *Citellus spilosoma cryptospilotus,* A. H. Howell, N. Amer. Fauna, 56:130, May 18.

MARGINAL RECORDS.—Utah: Monticello; *Lockerby.* Colorado: McElmo Creek, S of Cortez. New Mexico: Thoreau. Arizona: Holbrook; Winslow; Moa Ave. Utah: 5 mi. S summit Navajo Mtn.

Spermophilus spilosoma marginatus V. Bailey

1890. *Spermophilus spilosoma major* Merriam, N. Amer. Fauna, 4:39, October 8, type from Albuquerque, Bernalillo Co., New Mexico. Not *Mus citellus* var. *major* Pallas, 1779 [= *Spermophilus major*], from near Samara, Russia.

1902. *Spermophilus spilosoma marginatus* V. Bailey, Proc. Biol. Soc. Washington, 15:118, June 2, type from Alpine, Brewster Co., Texas.

MARGINAL RECORDS.—Colorado (Armstrong, 1972:128): between Simla and Matheson; *halfway between Matheson and Resolis;* few mi. NW Cheyenne Wells. Kansas: Kinsley. Oklahoma: 5 mi. SW Canton; Wichita Mts. National Wildlife Refuge (Glass and Halloran, 1961:237). Texas: Burkburnett (Dalquest, 1968:16); Colorado; E base mts. at Fort Stockton; Brewster County (Davis, 1966:135, as *Citellus spilosoma* only); area of La Mota. Chihuahua (Anderson, 1972:278): 2 mi. S Ojinaga, 2600 ft.; *Mezquite, 3000 ft.*; Villa Ahumada; 6 mi. NW Hueso, 5000 ft. Texas: Van Horn. New Mexico: Mesilla; 8 mi. E Deming; St. Augustine Plains, 12 mi. N Monica Spring; Espanola; 4 mi. SW Cimarron. Colorado (Armstrong, 1972:128): Carrizo Creek; Pueblo.

Spermophilus spilosoma obsoletus Kennicott

1863. *Spermophilus obsoletus* Kennicott, Proc. Acad. Nat. Sci. Philadelphia, 15:157. Type locality, restricted by A. H. Howell (N. Amer. Fauna, 56:131, May 18, 1938) to "50 miles west of Fort Kearney, Nebraska."
1893. *Spermophilus spilosoma obsoletus,* V. Bailey, Bull. Div. Ornith. and Mamm., U.S. Dept. Agric., 4:580.

MARGINAL RECORDS.—South Dakota: S. Fork White River; Bennett County. Nebraska: Neligh; type locality. Kansas: 9 mi. NW St. Francis. Colorado (Armstrong, 1972:129): Tuttle; Sand Creek; White Rocks. Wyoming (Long, 1965a:577): Little Bear Creek [almost] 20 mi. SE Chugwater; *3 mi. E Wheatland;* 3 mi. W Guernsey; *6 mi. SW Spoon Butte.* Nebraska: Crawford (Jones, 1964c:130).

Spermophilus spilosoma oricolus Alvarez

1962. *Spermophilus spilosoma oricolus* Alvarez, Univ. Kansas Publ., Mus. Nat. Hist., 14:123, March 7, type from 1 mi. E La Pesca, Tamaulipas.

MARGINAL RECORDS (Alvarez, 1963:422).—Tamaulipas: 33 mi. S Washington Beach; 88 mi. S, 10 mi. W Matamores; *89 mi. S, 10 mi. W Matamores;* type locality.

Spermophilus spilosoma pallescens (A. H. Howell)

1928. *Citellus spilosoma pallescens* A. H. Howell, Proc. Biol. Soc. Washington, 41:212, December 18, type from La Ventura, Coahuila.
1956. *Spermophilus spilosoma pallescens,* Baker, Univ. Kansas Publ., Mus. Nat. Hist., 9:205, June 15.

MARGINAL RECORDS.—Chihuahua: Consolación, 5100 ft. (Anderson, 1972:278). Coahuila: 2 mi. S, 3 mi. E Hechicero, 4450 ft.; 11 mi. E Acebuches; 3 mi. NW Cuatro Ciénegas, 2450 ft.; 1 mi. N Saltillo, 5000 ft. Nuevo León: Doctor Arroyo. San Luis Potosí: Salado. Zacatecas: Cañitas. Durango: 4 mi. NNE Boquilla,

6300 ft. (Baker and Greer, 1962:81). Chihuahua (Anderson, 1972:278): Escalón, 60 mi. SE Jiménez; boca de Río San Pedro, 3850 ft.; 1 mi. NW Lázaro Cárdenas, 3850 ft.

Spermophilus spilosoma pratensis Merriam

1890. *Spermophilus spilosoma pratensis* Merriam, N. Amer. Fauna, 3:55, September 11, type from pine plateau at N foot San Francisco Mtn., Coconino Co., Arizona.
1890. *Spermophilus spilosoma obsidianus* Merriam, N. Amer. Fauna, 3:56, September 11, type from cedar belt, NE of San Francisco Mtn., Coconino Co., Arizona.

MARGINAL RECORDS.—Arizona: Grand Canyon; Deadmans Flat, NE San Francisco Mtn.; Walnut Canyon, Coconino National Forest; Seligman; Aubrey Valley, Hualpai Indian Reservation.

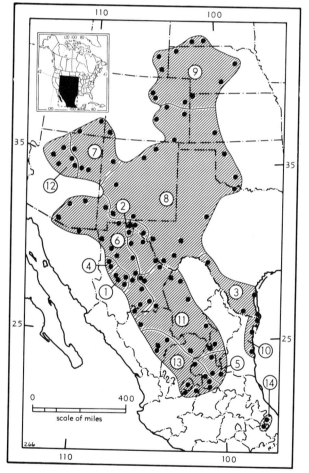

Map 266. *Spermophilus spilosoma* and
Spermophilus perotensis.

1. *S. s. altiplanensis*	8. *S. s. marginatus*
2. *S. s. ammophilus*	9. *S. s. obsoletus*
3. *S. s. annectens*	10. *S. s. oricolus*
4. *S. s. bavicorensis*	11. *S. s. pallescens*
5. *S. s. cabrerai*	12. *S. s. pratensis*
6. *S. s. canescens*	13. *S. s. spilosoma*
7. *S. s. cryptospilotus*	14. *S. perotensis*

Spermophilus spilosoma spilosoma Bennett

1833. *Spermophilus spilosoma* Bennett, Proc. Zool. Soc. London, p. 40, May 17. Type locality, Durango, México (see A. H. Howell, N. Amer. Fauna, 56:122, May 18, 1938).

MARGINAL RECORDS.—Durango: Hda. Atotonilco, 6680 ft. (Baker and Greer, 1962:81). San Luis Potosí: 2 km. E Illescas; Hda. la Parada; 10 mi. NE San Luis Potosí; Jesús María. Jalisco (Genoways and Jones, 1973:11): 2 mi. SW Matanzas, 7550 ft.; 8 mi. W Encarnación de Díaz, 6000 ft. Aguascalientes: Chicalote. Durango (Baker and Greer, 1962:81): Ciudad Durango; 5 mi. N, 12 mi. E Villa Madero.

Spermophilus perotensis Merriam
Perote Ground Squirrel

1893. *Spermophilus perotensis* Merriam, Proc. Biol. Soc. Washington, 8:131, December 28, type from Perote, Veracruz.

External measurements: 243–261; 57–78; 38–40. Greatest length of skull, 42.2–44.5. Upper parts clay color to wood brown or drab and indistinctly sprinkled with pinkish buff on hinder back; sides of body pinkish buff or cartridge buff; underparts and feet similar or slightly paler; tail above like back but pinkish buff below and bordered at distal end with blackish. Braincase high, narrow.

MARGINAL RECORDS.—Veracruz: type locality; *2 km. E Perote; 2 km. W Limón.* Puebla: 4 km. N San Salvador del Seco, 8000 ft. Veracruz: *Guadalupe Victoria* (Hall and Dalquest, 1963:269).

Subgenus **Poliocitellus** A. H. Howell

1938. *Poliocitellus* A. H. Howell, N. Amer. Fauna, 56:42, May 18. Type, *Arctomys franklinii* Sabine, 1822.

External measurements: 381–397; 136–153; 53.0–57.5. Greatest length of skull, 52.1–55.1. Head grayish; dorsum tawny olive or clay color; tail above and below blackish mixed with buff, overlaid and bordered with creamy white; underparts pinkish buff or buffy white. Skull long and narrow; postorbital and interorbital constrictions of approx. equal width; molars higher crowned than in *Otospermophilus* and lower crowned than in subgenus *Spermophilus*; M1 and M2 subquadrate in occlusal outline; trigon on P4, M1, and M2 broadly V-shaped; anterior cingulum usually joining protocone with abrupt change of direction; metalophs complete and mesostyles present on P4, M1, and M2; M3 slightly larger than M2; occlusal outline of m1 and m2 rhomboidal; baculum resembling that in subgenus *Spermophilus* but spoon relatively larger; cheek pouches large; atlantoscapularis dorsalis muscle absent.

Spermophilus franklinii (Sabine)
Franklin's Ground Squirrel

X ¼

1822. *Arctomys franklinii* Sabine, Trans. Linnean Soc. London, 13:587, type from Carlton House, Saskatchewan.
1827. *Spermophilus Franklini*, Lesson, Manuel de Mammalogie . . . , p. 244.

Characters as for the subgenus.

MARGINAL RECORDS.—Alberta: Athabasca Landing. Saskatchewan: Prince Albert National Park. Manitoba: The Pas (Soper, 1961:188); 7 mi. E Delta (Tamsitt, 1962:74); *Lake Winnipeg; Whitemouth Lake.* Ontario: Emo. Minnesota (Robins, 1971:31): Hibbing; Duluth. Big Horseshoe Lake (Jackson, 1961:142); Portage County (*op. cit.:* 140, Map 28); Racine (*op. cit.:* 142). Illinois: Grayslake. Indiana: north Liberty; near Rochester (Mumford, 1969:60); Tippecanoe County (*ibid.*). Illinois: Charleston; St. Clair County. Kansas: 5½ mi. N Moran; Neosho Falls; Hamilton; 1 mi. S, ½ mi. W Lindsborg; 9 mi. W Wakeeney. Nebraska

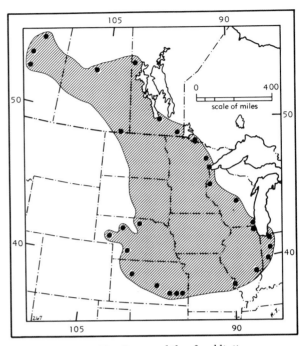

Map 267.   *Spermophilus franklinii.*

(Jones, 1964c:126): Curtis; 2 mi. NW Lisco; Kelso; near Bassett, thence northward in *Missouri Valley of North Dakota, and South Dakota.* Saskatchewan: Oxbow. Alberta: Pigeon Lake (Soper, 1965:142); Sandy Lake (*ibid.*); *near Rochester* (Keith and Meslow, 1966:541).

Subgenus Otospermophilus Brandt
Rock Squirrels

1844. *Otospermophilus* Brandt, Bull. Class. Phys.-Math. Acad. Imp. Sci. St. Pétersbourg, 2:379, March. Type, *Sciurus grammurus* Say.
1938. *Notocitellus* A. H. Howell, N. Amer. Fauna, 56:44, May 18. Type, *Spermophilus annulatus* Audubon and Bachman. *Notocitellus* arranged as a synonym of *Otospermophilus* by Bryant (1945:377).

External measurements: 315–540; 138–252; 43–65. Variegated pattern of black, white, and buff; infraorbital foramen oval; masseteric tubercle situated nearly ventral to infraorbital foramen; width at postorbital constriction slightly more than interorbital width; fossae anterolateral to incisive foramina deep; upper cheek-teeth low-crowned; M1 and M2 subquadrate in occlusal outline; metaloph on P4–M2 separated from protocone by sulcus; M3 slightly larger than M2; metaloph on M3 absent; protolophid on p4 absent and protoconid slightly larger than hypoconid; baculum with proximal end enlarged into a knob; cheek pouches large; atlantoscapularis dorsalis muscle present. For additional characters see Bryant (1945:376).

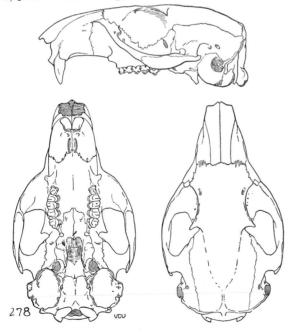

278

Fig. 278.   *Spermophilus franklinii,* 5 mi. N Moran, Allen Co., Kansas, No. 8640 K.U., ♀, X 1.

KEY TO SPECIES OF SUBGENUS OTOSPERMOPHILUS

1. Supraorbital foramen open; sides of head grayish (or mixed black and white); not in Tropical Life-zone.
    2. Nape and shoulders with dark median area; in United States occurring west of line from Toyabe Mts. (central Nevada) to Providence Mts. (SE California) and in México confined to Baja California.
        3. Dark median area on shoulders united with equally dark area on top of head; occurring south of 28° in Baja California. *S. atricapillus,* p. 402
        3'. Dark median area on shoulders darker than top of head; occurring north of 28° . . . . . *S. beecheyi,* p. 401
    2'. Nape and shoulders without dark median area; in United States occurring east of a line passing through westernmost parts of Toyabe Mts. (central Nevada) and Providence Mts. (SE California); in México not occurring in Baja California. . . . . *S. variegatus,* p. 399
1'. Supraorbital foramen closed; sides of head tawny or buffy; confined to Tropical Life-zone (Nayarit south into Guerrero).
    4. Cheeks tawny; tail ringed; nasals longer than 15.3. . . . *S. annulatus,* p. 403
    4'. Cheeks buffy; tail not ringed; nasals shorter than 15.3. . . . *S. adocetus,* p. 404

Spermophilus variegatus
Rock Squirrel

External measurements: 430–525; 172–252; 53–65. Greatest length of skull, 56.0–67.7. Upper parts variegated black and white, often with buff; head and forepart of back black in many subspecies; tail mixed black or brown and buffy white.

Spermophilus variegatus buckleyi Slack

1861. *Spermophilus buckleyi* Slack, Proc. Acad. Nat. Sci. Philadelphia, 13:314, type from Packsaddle Mountain, Llano Co., Texas.
1905. *Citellus variegatus buckleyi,* V. Bailey, N. Amer. Fauna, 25:84, October 24.

MARGINAL RECORDS.—Texas: San Saba River; Bull Creek; head Nueces River; Rocksprings.

Spermophilus variegatus couchii Baird

1855. *Spermophilus couchii* Baird, Proc. Acad. Nat. Sci. Philadelphia, 7:332, April, type from Santa Catarina, few mi. W Monterrey, Nuevo León.
1955. *Spermophilus variegatus couchii,* Baker, Univ. Kansas Publ., Mus. Nat. Hist., 9:207, June 15.

MARGINAL RECORDS.—Texas: Green Gulch, 5200 ft., Chisos Mts.; Boquillas. Nuevo León: type lo-

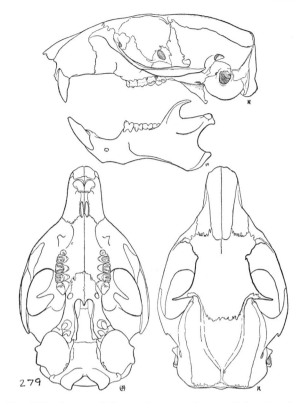

Fig. 279. *Spermophilus variegatus robustus,* Baker Creek, 7300 ft., Nevada, No. 41517 M.V.Z., ♂, X 1.

cality. Tamaulipas: Victoria. Coahuila: Sierra Encaranación; Sierra Guadalupe; 23 mi. S, 5 mi. W Ocampo (Baker, 1956:208). Chihuahua (Anderson, 1972:279): 29° 12′ N, 104° 16′ W; 12 mi. SSW Consolación, 5600 ft.

Spermophilus variegatus grammurus (Say)

1823. S[*ciurus*]. *grammurus* Say, *in* Long, Account of an exped. . . . to the Rocky Mts. . . . , 2:72, type from Purgatory River, near mouth Chacuaco Creek, Las Animas Co., Colorado.
1952. *Spermophilus variegatus grammurus,* Hall and Kelson, Univ. Kansas Publ., Mus. Nat. Hist., 5:346, December 15.
1913. *Citellus variegatus juglans* V. Bailey, Proc. Biol. Soc. Washington, 26:131, May 21, type from Glenwood, 5000 ft., Rio San Francisco, Catron Co., New Mexico.

MARGINAL RECORDS.—Colorado (Armstrong, 1972:130): 9 mi. NW Fort Collins; 3½ mi. E Fort Collins; Franktown; 20 mi. N Pueblo; JJ Ranch, 18 mi. S La Junta; Regnier. Oklahoma: Beaver County. New Mexico: Mosquero; between Cuervo and Montoya (Findley, *et al.,* 1975:128, as *S. variegatus* only); Carlsbad. Texas: Castle Mts.; Devils River; Presidio County (Davis, 1966:138, as *Citellus variegatus* only). Chihuahua (Anderson, 1972:279): 8 mi. NE Laguna, 7250 ft.; 18 mi. NNE San Juanito, 7250 ft. Sonora: Providencia Mines; Camoa; Tiburón Island. Arizona (Cockrum, 1961:83): 7 mi. E Papago Well; *Harquahala Mts., 3000 ft.;* Vicksburg, 1600 ft. California: Mitchell's; NW side Clark Mtn. Arizona (Cockrum, 1961:83): Beales Spring; 4 mi. SW Pierce Ferry, 2500 ft., thence to and eastward and northward along S bank Colorado River to Rainbow Lodge, 6400 ft. Utah (Durrant and Hansen, 1954:269): Moab, 4500 ft.; *mouth Nigger Bill Canyon, E side Colorado R., 4 mi. above Moab Bridge;* Castle Valley, 18 mi. NE Moab, 1600 ft. Colorado (Armstrong, 1972:130, 131): above Glenwood Springs; Dry Creek, ¾ mi. N Gunnison River, 7460 ft.; Coventry; Allison; Pagosa Springs; Trinidad; Crestone; Buena Vista, 8200 ft.; Cañon City; Dome Rock; Sugar Loaf.

Spermophilus variegatus robustus (Durrant and Hansen)

1954. *Citellus variegatus robustus* Durrant and Hansen, Proc. Biol. Soc. Washington, 67:264, November 15, type from Pass Creek, Deep Creek Mountains, 8000 ft., Juab Co., Utah.

MARGINAL RECORDS (Durrant and Hansen, 1954:268, unless otherwise noted).—Nevada: Cherry Creek, 6800 ft. Utah: Birch Creek, 7600 ft. Nevada: Baker Creek; Eagle Valley, 3½ mi. N Ursine, 5600–6000 ft. (Hall, 1946:313, as *C. v. grammurus*); *4 mi. NE Panaca (ibid.,* as *C. v. grammurus*); Meadow Valley Wash, 7 mi. S Caliente, 4000 ft. (*ibid.,* as *C. v. grammurus*); SW base Groom Baldy, 7200 ft. (*ibid.,* as *C. v. grammurus*); 5 mi. W White Rock Spring, Belted Range; *2 mi. E Silverbow, 6900 ft., Kawich Mts.;* Haws Canyon, 7000 ft.; Kingston Canyon, 6350 ft. Not found: type locality.

Spermophilus variegatus rupestris (J. A. Allen)

1903. *Citellus (Otospermophilus) grammurus rupestris* J. A. Allen, Bull. Amer. Mus. Nat. Hist., 19:595, November 12, type from Río Sestín, northwestern Durango.
1904. [*Citellus variegatus*] *rupestris,* Elliot, Field Columb. Mus., Publ. 95, Zool. Ser., 4:150.

MARGINAL RECORDS.—Sonora: Oposura. Chihuahua (Anderson, 1972:279, 280): Mojarachic; San Francisco de Borja, 5700 ft.; Cañón Gotera, 9 mi. NW Chihuahua, 5500 ft.; Santa Rosalía. Durango: Guazamota. Sinaloa: Sierra de Choix, 50 mi. NE Choix.

Spermophilus variegatus tularosae (Benson)

1932. *Citellus grammurus tularosae* Benson, Univ. California Publ. Zool., 38:336, April 14, type from French's Ranch, 5400 ft., 12 mi. NW Carrizozo, Lincoln Co., New Mexico.
1938. *Citellus variegatus tularosae,* A. H. Howell, N. Amer. Fauna, 56:145, May 18.

MARGINAL RECORDS.—New Mexico: type locality; Malpais Lava Beds.

Spermophilus variegatus utah (Merriam)

1903. *Citellus grammurus utah* Merriam, Proc. Biol. Soc. Washington, 16:77, May 29, type from foot Wasatch Mts., near Ogden, Weber Co., Utah.
1905. *Citellus variegatus utah,* Elliot, Field Columb. Mus., Publ. 105, Zool. Ser., 6:115, December 6.

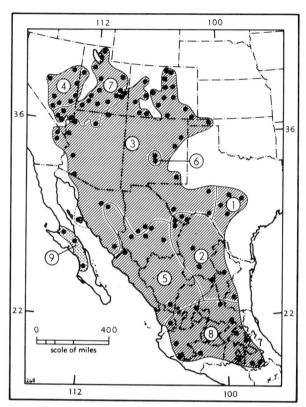

Map 268. *Spermophilus variegatus* and *Spermophilus atricapillus.*

Guide to kinds	
1. *S. v. buckleyi*	5. *S. v. rupestris*
2. *S. v. couchii*	6. *S. v. tularosae*
3. *S. v. grammurus*	7. *S. v. utah*
4. *S. v. robustus*	8. *S. v. variegatus*
	9. *S. atricapillus*

MARGINAL RECORDS.—Utah (Durrant and Hansen, 1954:269, unless otherwise noted): Logan (Durrant, 1952:120); along Green River, 15 mi. S Ouray, 4500 ft. (Durrant and Hansen, 1954:268); 13 mi. SE Thompsons, Arches National Monument; Ekkers Ranch, Robbers Roost, 25 mi. E Hanksville, 6000 ft.; Moki Tanks, Circle Cliffs; Henrieville. Arizona: 2 mi. N Cape Royal (Hoffmeister, 1971:168), thence to and westward along N bank Colorado River in appropriate habitats to Nevada (Durrant and Hansen, 1954:268): Cedar Basin, 3500 ft.; Charleston Park, 8000 ft.; *Charleston Mts.;* Willow Creek, 6000 ft. (Hall, 1946:313, as *C. v. grammurus*); Sheep Mts., 8500 ft. Utah (Durrant, 1952:120, unless otherwise noted): Pine Valley (Durrant, 1952:121, as *C. v. grammurus*); Parowan; Beaver, 6000 ft.; *Salina Canyon;* Salina; *1½ mi. SW Nephi, 5721 ft.;* 7 mi. N Nephi, 5730 ft.; Willow Springs; Ogden, 4293 ft.

Cockrum's (1961:83) cursorial referral of specimens from north of the Colorado River in northeastern Arizona to *S. v. grammurus* probably resulted from an oversight of Durrant and Hansen's (1954:263–271) careful study of subspecific characters. See also the later publication by Hoffmeister and Durham (1971:30).

Spermophilus variegatus variegatus (Erxleben)

1777. [*Sciurus*] *variegatus* Erxleben, Systema regni animalis . . . , 1:421. Type locality, Valley of México, near City of México (see Nelson, Science, n.s., 8:897, December 23, 1898).
1898. *Spermophilus variegatus*, Nelson, Science, n.s., 8:898, December 23.
1830. *Sciurus buccatus* Lichtenstein, Abh. k. Akad. Wiss., Berlin, 1827, p. 117.
1833. *Spermophilus macrourus* Bennett, Proc. Zool. Soc. London, p. 41, May 17, type from west México.

MARGINAL RECORDS.—San Luis Potosí: Puerto de Lobos; Hda. Capulín. Hidalgo: Encarnación; Tulancingo. Puebla: Tehuacán; 33 km. W Tehuacán, 8000 ft. Morelos: Tetela del Volcán. Michoacán: Tancítaro. Colima: Hda. San Antonio, base Volcán de Colima. Jalisco: Plantinar. Nayarit: Tepic. Zacatecas: Berriozábal.

Spermophilus beecheyi
California Ground Squirrel

External measurements: 357–500; 145–200; 50–64. Greatest length of skull, 51.6–62.4. Head cinnamon or brown; upper parts brown flecked with whitish or buffy; sides of neck and shoulders white or whitish, this color extending backward in two divergent stripes separated by triangular area of dark color. Skull averaging smaller than in *S. variegatus* but otherwise indistinguishable.

Spermophilus beecheyi beecheyi (Richardson)

1829. *Arctomys (Spermophilus) beecheyi* Richardson, Fauna Boreali-Americana, 1:170, type from "neighborhood of San Francisco and Monterey, in California"; restricted to Monterey, Monterey Co., California, by Grinnell, Univ. California Publ. Zool., 40:120, Sept. 26, 1933.
1831. *Spermophilus beecheyi*, Cuvier, Supplément à l'histoire naturelle générale et particulière de Buffon, 1:331.

MARGINAL RECORDS.—California: Walnut Creek; Corral Hollow, 8 mi. SW Tracy; Bitterwater; Idria Mines; Pozo; San Rafael Mts.; Heninger Flats, San Gabriel Mts.; Lytle Creek; Temescal [Canyon]; Point Firmin, San Pedro.

Spermophilus beecheyi douglasii (Richardson)

1829. *Arctomys ? (Spermophilus?) douglasii* Richardson, Fauna Boreali-Americana, 1:172. Type locality, bank Columbia River, Oregon.
1913. *Citellus beecheyi douglasi*, Grinnell, Proc. California Acad. Sci., ser. 4, 3:345, August 28.

MARGINAL RECORDS.—Washington: Tieton River Canyon (Broadbooks, 1961:257); 3 mi. NW Tampico (Broadbooks, 1961:258); Satus Creek, 6 mi. N Yakima County border; 21 mi. E Goldendale. Oregon: Warm Springs River; 10 mi. E McKenzie Bridge; Fort Klamath. California: Goose Lake; Lake City; Merrillville; Chico; Cherokee; Davis; Elmira; Fairfield, thence northward along coast to Oregon: Tillamook; Portland. Washington: White Salmon River.

Spermophilus beecheyi fisheri Merriam

1893. *Spermophilus beecheyi fisheri* Merriam, Proc. Biol. Soc. Washington, 8:133, December 28, type from S. Fork Kern River, 3 mi. above Onyx, California.

MARGINAL RECORDS.—California: Susanville. Nevada: 1 mi. SW Pyramid Lake; 5 mi. WSW Fallon; 1 mi. N Virginia City; Carson River, 5 mi. SE Minden, 5900 ft.; 6½ mi. S Minden, 4900 ft. California: Greenville; Nevada City; Coulterville; Soquel Mill, head N. Fork San Joaquin River; Mt. Whitney; Walker Pass; Tehachapi Peak; Mono Flats; Cuyama Valley; Dos Palos; Pacheco Pass; Modesto; Tracy; 6 mi. E Colusa; Yankee Hill; Oroville.

Spermophilus beecheyi nesioticus (Elliot)

1904. *Citellus nesioticus* Elliot, Field Columb. Mus., Publ. 90, Zool. Ser., 3:263, March 7, type from [near Avalon] Santa Catalina Island, California. Known from Santa Catalina Island only.
1913. *Citellus beecheyi nesioticus*, Grinnell, Proc. California Acad. Sci., ser. 4, 3:346, August 28.

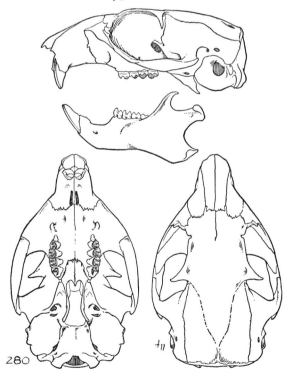

Fig. 280. *Spermophilus beecheyi nudipes*, Escondido, San Diego Co., California, No. 45956 K.U., ♂, × 1.

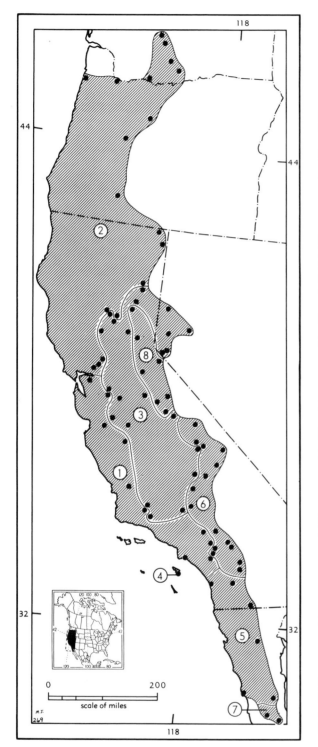

Map 269. *Spermophilus beecheyi.*

1. *S. b. beecheyi*
2. *S. b. douglasii*
3. *S. b. fisheri*
4. *S. b. nesioticus*
5. *S. b. nudipes*
6. *S. b. parvulus*
7. *S. b. rupinarum*
8. *S. b. sierrae*

Spermophilus beecheyi nudipes (Huey)

1931. *Citellus beecheyi nudipes* Huey, Trans. San Diego Soc Nat. Hist., 7(2):18, October 6, type from Hanson Laguna, Sierra Juárez, 5200 ft., Baja California.

MARGINAL RECORDS.—California: Oceanside; Grapevine Spring; Mountain Spring, 4 mi. N Monument 231, Mexican boundary line. Baja California: type locality; Mattomi [= Matomi]; San Quentín.

Spermophilus beecheyi parvulus (A. H. Howell)

1931. *Citellus beecheyi parvulus* A. H. Howell, Jour. Mamm., 12:160, May 14, type from Shepherd Canyon, Argus Mts., California.

MARGINAL RECORDS.—California: Owens Valley; Jackass Spring, Panamint Range; Argus Range; Little Lake; Victorville; Doble; Fish Creek; Palm Springs; Thomas Mtn.; Riverside; San Bernardino; Cameron; Piute Mts.; Lone Pine; NE base Mt. Williamson.

Spermophilus beecheyi rupinarum (Huey)

1931. *Citellus beecheyi rupinarum* Huey, Trans. San Diego Soc. Nat. Hist., 7(2):17, October 6, type from Cataviña, Baja California.

MARGINAL RECORDS.—Baja California: San Fernando; type locality.

Spermophilus beecheyi sierrae (A. H. Howell)

1938. *Citellus beecheyi sierrae* A. H. Howell, N. Amer. Fauna, 56:153, May 18, type from Emerald Bay, Lake Tahoe, El Dorado Co., California.

MARGINAL RECORDS.—California: Quincy; Markleeville; Little Yosemite; Wawona; near head Merced River; Big Trees; Blue Canyon.

Spermophilus atricapillus W. E. Bryant
Baja California Rock Squirrel

1889. *Spermophilus grammurus atricapillus* W. E. Bryant, Proc. California Acad., 2(ser. 2):26, June 20. Type locality, Comondú, Baja California.

External measurements: 410–465; 185–210; 55–60. Greatest length of skull, 54.8–58.5. Closely resembles *S. beecheyi* but darker, especially on head and anterior half of back; tail averages longer and skull averages smaller.

The geographic range of *S. atricapillus* seems to be separated from that of *S. beecheyi* by an area 40 mi. wide, uninhabited by any species of subgenus *Otospermophilus*.

MARGINAL RECORDS.—Baja California: San Pablo; type locality; San Ignacio. See Map 268.

Spermophilus annulatus
Ring-tailed Ground Squirrel

External measurements: 383–470; 186–238; 50–64. Greatest length of skull, 51.6–57.0. Upper parts nearly uniform mixed fuscous black and cinnamon buff or pale pinkish buff, blackish color often predominating on head and in some specimens on back; chin, throat, and sides of nose and face ochraceous buff; sides of neck, shoulders, and forelimbs hazel; ears and hind legs hazel or tawny; underparts warm buff or pinkish buff; tail mixed pinkish buff and black above, hazel be-

neath; tail approx. same length as head and body, distichous, narrow, not bushy, and with approx. 15 blackish annulations. Interorbital breadth more than 42 per cent of zygomatic breadth; anteroposterior diameter of upper incisors more than in the larger-skulled S. *beecheyi*.

Spermophilus annulatus annulatus Audubon and Bachman

1842. *Spermophilus annulatus* Audubon and Bachman, Jour. Acad. Nat. Sci. Philadelphia, 8:319, type from an unknown locality; subsequently designated as Manzanillo, Colima, by A. H. Howell (N. Amer. Fauna, 56:163, May 18, 1938).

MARGINAL RECORDS.—Jalisco (Genoways and Jones, 1973:7): La Cuesta, 1900 ft.; 6 mi. E Limón, 2700 ft.; Tolimán, 2200 ft. Colima: Hda. San Antonio, base Volcán de Colima. Guerrero: El Naranjo; La Unión, thence up coast to Colima: type locality. Jalisco: 25 mi. SW La Resolana (Genoways and Jones, 1973:7).

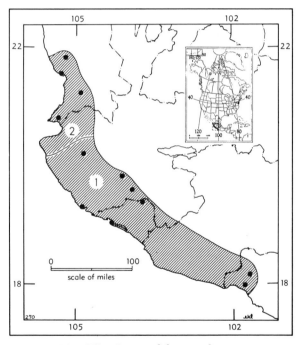

Map 270. *Spermophilus annulatus.*

1. S. a. annulatus 2. S. a. goldmani

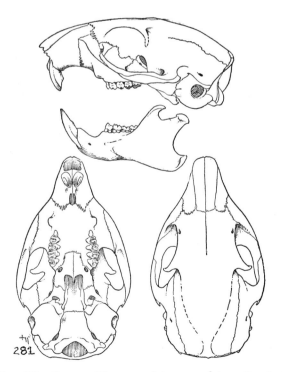

Fig. 281. *Spermophilus annulatus annulatus*, 5 mi. S Purificación, Jalisco, No. 33438 K.U., ♀, X 1.

Spermophilus annulatus goldmani Merriam

1902. *Spermophilus annulatus goldmani* Merriam, Proc. Biol. Soc. Washington, 15:69, March 22, type from Santiago, Nayarit.

MARGINAL RECORDS.—Nayarit: type locality; Compostela; Arroyo de San Juan Sanches, about 40 mi. SW Compostela; San Blas.

Spermophilus adocetus
Lesser Tropical Ground Squirrel

External measurements: 315–353; 138–168; 43–48. Greatest length of skull, 41.6–46.2. Resembles *S. annulatus* but smaller, paler (less reddish); tail without annulations; rostrum shorter and broader; interorbital region averaging 49 instead of 45 per cent of zygomatic breadth.

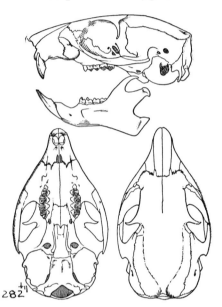

Fig. 282. *Spermophilus adocetus adocetus*, 9 mi. S Lombardia, 1500 ft., Michoacán, No. 38346 K.U., ♂, X 1.

Spermophilus adocetus adocetus (Merriam)

1903. *Citellus adocetus* Merriam, Proc. Biol. Soc. Washington, 16:79, May 29, type from La Salada, 40 mi. S Uruapan, Michoacán.

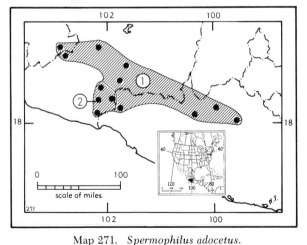

Map 271. *Spermophilus adocetus.*

1. *S. a. adocetus*   2. *S. a. infernatus*

1942. *Citellus adocetus arceliae* Villa-R., Anal. Inst. Biol., Univ. Nac. Autó. México, 13(1):357, October. *C. a. arceliae* synonymous with *S. a. adocetus* (Alvarez and Ramírez-P., Rev. Soc. Mexicana Hist. Nat., 29:182, December, 1968).

MARGINAL RECORDS.—Michoacán: near Tancítaro; Volcán Jorullo. Guerrero: 6 km. W Teloloapan (Alvarez and Ramírez-P., 1968:185); San Agustín Huapan (*ibid.*); 14 km. SW Arcelia (*ibid.*), thence along N side Río Balsas to La Escondida, *ca.* 20 mi. SE Balsas. Michoacán: type locality; Tepalcatepec (Alvarez and Ramírez-P., 1968:185). Jalisco: 8 mi. E Jilotlán de los Dolores, 2000 ft. (Genoways and Jones, 1973:7).

Spermophilus adocetus infernatus Alvarez and Ramírez-P.

1968. *Spermophilus adocetus infernatus* Alvarez and Ramírez-P., Rev. Soc. Mexicana Hist. Nat., 29:183, December, type from Pizandarán, 14 km. N [31 km. N, 19 km. E] El Infiernillo, Michoacán.

MARGINAL RECORDS (Alvarez and Ramírez-P., 1968:183).—Michoacán: 40 km. S Nueva Italia; type locality; 3 km. NE El Infiernillo; 30 km. N El Infiernillo.

Subgenus **Xerospermophilus** Merriam

1892. *Xerospermophilus* Merriam, Proc. Biol. Soc. Washington, 7:27, April 13. Type, *Spermophilus mohavensis* Merriam.

External measurements: 210–266; 57–107; 32–40. Greatest length of skull, 34.9–40.0. Upper parts pinkish gray (in *S. mohavensis*) and under surface of tail white, or (in *S. tereticaudus*) upper parts vinaceous cinnamon, pinkish cinnamon, light drab, cinnamon drab, or ecru drab; tail beneath drab or buff; upper parts without mottling. Skull short and broad; postorbital region wider than interorbital region; rostrum not constricted at base or expanded at tip; coronoid process of mandible and P3 smaller than in any other subgenus of the genus; molar teeth low-crowned; M1 and M2 subquadrate in occlusal outline; trigon on P4, M1, and M2 broadly V-shaped; anterior cingulum usually joins protocone with abrupt change of direction on M1 and M2; metaloph on P4, M1, and M2 separated from protocone by sulcus as in *Otospermophilus* and *Ictidomys*; M3 slightly larger than M2; posterior cingulum of M3 does not bend sharply posteriad from protocone as in subgenera *Spermophilus* and *Ictidomys*; P4 not molariform; trigonid on lower cheek-teeth slightly higher than talonid; occlusal outline of m1 and m2 rhomboidal as in *Otospermophilus*; baculum resembles that of subgenus *Spermophilus* but spoon relatively larger and entire baculum "stouter"; cheek pouches large; atlantoscapularis dorsalis muscle absent.

KEY TO SPECIES OF SUBGENUS XEROSPERMOPHILUS

1. Underside of tail white. . *S. mohavensis,* p. 405
1'. Underside of tail not white (usually cinnamon). . . . . *S. tereticaudus,* p. 405

Spermophilus mohavensis Merriam
Mohave Ground Squirrel

1889. *Spermophilus mohavensis* Merriam, N. Amer. Fauna, 2:15, October 30, type from near Rabbit Springs, *ca.* 15 mi. E Hesperia, San Bernardino Co., California (Grinnell and Dixon, Month. Bull. California Comm. Hort., 7:667, January 27, 1919).

External measurements: 210–230; 57–72; 32–38. Greatest length of skull, 38.1–40.0. In coloration resembles *S. townsendii mollis,* but upper parts more pinkish without trace of mottling; tail beneath whitish. Cranial characters as given for the subgenus.

MARGINAL RECORDS.—California: Haiwee Meadow, 10 mi. S Owens Lake; Salt Wells Valley, N end Mohave Desert; Oro Grande; Mohave River (= Rabbit Springs, 15 mi. E Hesperia); Hesperia; Palmdale; Mojave; Little Lake.

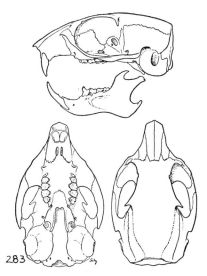

Fig. 283. *Spermophilus mohavensis,* 1 mi. NW Lovejoy Springs, 2300 ft., California, No. 44285 M.V.Z., ♀, X 1.

Spermophilus tereticaudus
Round-tailed Ground Squirrel

External measurements: 204–266; 60–107; 32–40. Greatest length of skull, 34.9–39.3. In winter, upper parts pinkish cinnamon, underparts white; in summer, upper parts brighter pinkish cinnamon and pelage shorter and harsher (there are two annual molts); tail long and slender and not broadly haired. Skull resembles that of *S. mohavensis* but smaller.

Spermophilus tereticaudus apricus (Huey)

1927. *Citellus tereticaudus apricus* Huey, Trans. San Diego Soc. Nat. Hist., 5(7):85, October 10, type from Valle de la Trinidad (lat. 31° 20' N; long. 115° 40' W), Baja California. Known only from type locality.

Spermophilus tereticaudus chlorus (Elliot)

1904. *Citellus chlorus* Elliot, Field Columb. Mus., Publ. 87, Zool. Ser., 3:242, January 7, type from Palm Springs, Riverside Co., California.
1913. *Citellus tereticaudus chlorus,* Grinnell, Proc. California Acad. Sci., ser. 4, 3:347, August 28.

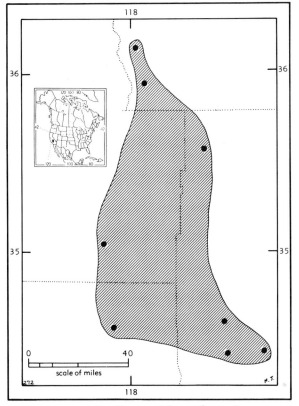

Map 272. *Spermophilus mohavensis.*

MARGINAL RECORDS.—California: Cabazon; Whitewater Station; Coachella; Mecca; Agua Caliente.

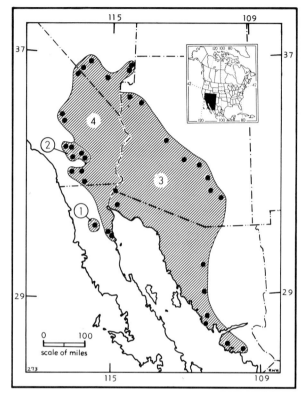

Map 273. *Spermophilus tereticaudus.*

1. *S. t. apricus*　　3. *S. t. neglectus*
2. *S. t. chlorus*　　4. *S. t. tereticaudus*

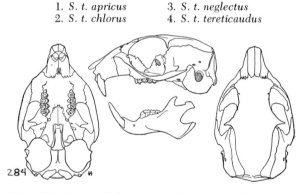

Fig. 284. *Spermophilus tereticaudus tereticaudus,* Furnace Creek Ranch, Death Valley, 178 ft. below sea level, Inyo Co., California, No. 27381 M.V.Z., ♀, X 1.

Spermophilus tereticaudus neglectus Merriam

1889. *Spermophilus neglectus* Merriam, N. Amer. Fauna, 2:17, October 30, type from Dolans Spring, 12 mi. NW Chloride, Mohave Co., Arizona.
1938. *Citellus tereticaudus neglectus,* A. H. Howell, N. Amer. Fauna, 56:187, May 18.
1891. *Spermophilus sonoriensis* Ward, Amer. Nat., 25:158, February, type from Hermosillo, Sonora.

1918. *Citellus tereticaudus arizonae* Grinnell, Proc. Biol. Soc. Washington, 31:105, November 29, type from Tempe, Maricopa Co., Arizona.

MARGINAL RECORDS.—Arizona: type locality; Hackberry; 2 mi. N Wickenburg; Tempe; 15 mi. N Florence (Cockrum, 1961:89); Tanque Verde (Cockrum, 1961:90); Rillito Creek, 5 mi. N Tucson; 30 mi. SE Tucson (Cockrum, 1961:90). Sonora: Querobabi; Hermosillo; Ortiz; Camoa, Río Mayo; Obregon; Guaymas; Ciénega Well, 30 mi. S Monument 204. Arizona: Colorado River at Monument 204, thence northward on E side Colorado River to at least *Fort Mohave.*

Spermophilus tereticaudus tereticaudus Baird

1858. *Spermophilus tereticaudus* Baird, Mammals, *in* Repts. Expl. Surv. . . . , 8(1):315, July 14, type from Old Fort Yuma, Imperial Co., California.
1904. *Citellus eremonomus* Elliot, Field Columb. Mus., Publ. 87, Zool. Ser., 3:243, January 7, type from Furnace Creek, Death Valley, Inyo Co., California.
1926. *Citellus tereticaudus vociferans* Huey, Proc. Biol. Soc. Washington, 39:29, July 30, type from San Felipe, Baja California.

MARGINAL RECORDS.—Nevada: Bunkerville; ½ mi. E St. Thomas, 2100 ft., thence southward on W side Colorado River to Baja California: San Felipe Bay; San Felipe. California: Coyote Well; La Puerta; Baregas Spring (= Borego); Daggett; Barstow; type locality. Nevada: Ash Meadows; Nevada Test Site (Jorgensen and Hayward, 1965:6); 4 mi. NW Las Vegas, 2100 ft.

Subgenus **Callospermophilus** Merriam
Golden-mantled Ground Squirrels

1897. *Callospermophilus* Merriam, Proc. Biol. Soc. Washington, 11:189, July 1. Type, *Sciurus lateralis* Say.

External measurements: 215–315; 52–118; 35–49. Greatest length of skull, 39.6–48.3. White stripe from shoulder to hip on each side of back; white stripe bordered below, and in most subspecies above, by black stripe; back gray, buff, cinnamon, or fawn; "mantle" on head and shoulders varying from cinnamon buff to tawny or russet; except in color and smaller size, closely resembling subgenus *Otospermophilus* but differing as follows: interorbital region relatively narrower in comparison with width at postorbital constriction; angular process of mandible relatively shorter; metaloph on M1 and M2 joins protocone; small protolophid present, instead of absent, on p4; atlantoscapularis dorsalis muscle present.

KEY TO SPECIES OF SUBGENUS CALLOSPERMOPHILUS

1. Occurring S of United States–Mexican
   boundary. . . . . S. *madrensis*, p. 410
1'. Occurring N of United States–Mexican
   boundary.
   2. Occurring in Cascade Mts. of southern
      British Columbia and Washington.
      S. *saturatus*, p. 409
   2'. Not occurring in Cascade Mts. of
      British Columbia and Washington.
      S. *lateralis*, p. 407

Spermophilus lateralis
Golden-mantled Ground Squirrel

X ⅓

External measurements: 230–308; 63–118;
35–46. Greatest length of skull, 39.6–45.6. Color
as described for the subgenus.

Spermophilus lateralis arizonensis (V. Bailey)

1913. *Callospermophilus lateralis arizonensis* V. Bailey, Proc.
Biol. Soc. Washington, 26:130, May 21, type from Little
Spring, 8250 ft., San Francisco Mtn., Arizona.

MARGINAL RECORDS.—Arizona: type locality;
Springerville. New Mexico (Findley, *et al.*, 1975:129,
130): Tularosa Mts.; Big Rocky Creek; 6 mi. N, 15 mi. E
Santa Rita. Arizona: Prieto Plateau; Montezuma Well;
Williams.

Spermophilus lateralis bernardinus Merriam

1893. *Spermophilus chrysodeirus brevicaudus* Merriam,
Proc. Biol. Soc. Washington, 8:134, December 28, type from
San Bernardino Peak, San Bernardino Co., California. Not
Spermophilus brevicauda Brandt, 1844, type from southern
Altai.
1898. *Spermophilus (Callospermophilus) bernardinus* Mer-
riam, Science, n.s., 8:782, December 2, a renaming of *S. c.
brevicaudus* Merriam.

1938. *Citellus lateralis bernardinus*, A. H. Howell, N. Amer.
Fauna, 56:209, May 18.

MARGINAL RECORDS.—California: Holcomb Valley; Sugarloaf; San Gorgonio Peak; head S. Fork Santa
Ana River.

Spermophilus lateralis castanurus (Merriam)

1890. *Tamias castanurus* Merriam, N. Amer. Fauna, 4:19, October 8, type from Park City, Wasatch Mts., Summit Co.,
Utah.
1938. *Citellus lateralis castanurus*, Howell, N. Amer. Fauna,
56:201, May 18.
1917. *Callospermophilus lateralis caryi* A. H. Howell, Proc.
Biol. Soc. Washington, 30:105, May 23, type from 7 mi. S
Fremont Peak, 10,400 ft., Wind River Mts., Fremont Co.,
Wyoming. (Arranged as a synonym of *S. l. castanurus* by
Long, Univ. Kansas Publ., Mus. Nat. Hist., 14:584–586, July
6, 1965.)

MARGINAL RECORDS.—Idaho: Big Hole Mts.
Wyoming (Long, 1965a:586): *Brooks Mtn.*; Jackeys
Creek, 4 mi. S Dubois; *Bull Lake*; ½ mi. S, 17½ mi. W
Lander; *LaBarge Creek*; Cokeville. Utah: type local-
ity; Santaquin Canyon, upper sawmill, N of Mt. Nebo;
Barclay; Tony Grove, Logan Canyon. Idaho: 8 mi. NE
Inkom.

Spermophilus lateralis certus (Goldman)

1921. *Callospermophilus lateralis certus* Goldman, Jour.
Mamm., 2:232, November 29, type from N base Charleston
Peak, Clark Co., Nevada.

MARGINAL RECORDS.—Nevada: Wheeler Well;
Charleston Peak; N side Potosi Mtn., 7000 ft., Spring
Mts.

Spermophilus lateralis chrysodeirus (Merriam)

1890. *Tamias chrysodeirus* Merriam, N. Amer. Fauna, 4:19,
October 8, type from Fort Klamath, Klamath Co., Oregon.
1936. *Citellus lateralis chrysodeirus*, A. H. Howell, N. Amer.
Fauna, 56:203, May 18.

MARGINAL RECORDS.—Oregon: Willows;
Meacham; Rock Creek; 10 mi. NW Silver Lake. Cali-
fornia: Goose Lake; Sierra Valley. Nevada: 2 mi. W Mt.
Rose summit; *1½ mi. N, 3 mi. E Edgewood*. California:
Mono Lake; near head San Joaquin River; Mt. Whit-
ney; Mulkey Meadows, 15 mi. S Mt. Whitney; E. Fork
Kaweah River, Sequoia National Park; Dinkey Creek,
N. Fork Kings River; Merced River, Fish Camp, S.
Fork; El Dorado County (Jameson and Mead,
1964:359, as *Citellus lateralis*); Shingletown; Mt.
Shasta; Beswick. Oregon: Four-mile Lake; Diamond
Lake; McKenzie Bridge; Mt. Hood; Miller, mouth of
Deschutes River.

Spermophilus lateralis cinerascens (Merriam)

1890. *Tamias cinerascens* Merriam, N. Amer. Fauna, 4:20, October 8, type from Helena, 4500 ft., Lewis and Clark Co., Montana.
1938. *Citellus lateralis cinerascens*, A. H. Howell, N. Amer. Fauna, 56:198, May 18.

MARGINAL RECORDS.—Montana (Hoffmann, *et al.*, 1969:589, unless otherwise noted, as *S. lateralis* only): Helena (Hall and Kelson, 1959:362); Mt. Edith; head Big Timber Creek; 20 mi. SW Red Lodge. Wyoming (Long, 1965a:587): Pahaska Tepee (Whirlwind Peak); *Yellowstone Lake, Yellowstone National Park.* Idaho: Henrys Lake. Montana: Deer Lodge County.

Spermophilus lateralis connectens (A. H. Howell)

1931. *Callospermophilus chrysodeirus connectens* A. H. Howell, Jour. Mamm., 12:161, May 14, type from Homestead, Oregon.
1938. *Citellus lateralis connectens* (A. H. Howell), N. Amer. Fauna, 56:205, May 18.

MARGINAL RECORDS.—Washington: Dayton; Anatone. Idaho: ½ mi. E Black Lake. Oregon: type locality; Anthony.

Spermophilus lateralis lateralis (Say)

1823. S[*ciurus*]. *lateralis* Say, *in* Long, Account of an exped. . . . to the Rocky Mts. . . . , 2:46. Type locality, Arkansas River, near Canyon City, Colorado (ca. 26 mi. below Canyon City, Merriam, Proc. Biol. Soc. Washington, 18:163, June 29, 1905).
1831. *Spermophilus lateralis*, Cuvier, Supplément à l'histoire naturelle générale et particulière de Buffon, 1:335.

MARGINAL RECORDS.—Isolated northern segment in Wyoming (Long, 1965a:589): *14 mi. S, 8½ mi. W Lander; 17 mi. S, 6½ mi. W Lander;* Miners Delight; *22 mi. S, 5½ mi. W Lander; 25 mi. S, 3 mi. W Lander; ½ mi. N, 3 mi. E South Pass City; South Pass City;* Big Sandy. Main segment of range, Wyoming (Long, 1965a:588, 589): Springhill; 2⅖ mi. W Horsecreek P.O.; *6 mi. W Islay.* Colorado (Armstrong, 1972:132–135): 12 mi. S Fort Collins; 2 mi. W Palmer Lake; East Spanish Peak, 11,000 ft. New Mexico (Findley, *et al.*, 1975:129, 130, as *S. lateralis* only): Long Canyon, 3 mi. N Catskill; Cimarron; Halls Peak; "Las Vegas Evergreen Valley"; 2 mi. N, 10 mi. E Jemez Springs; 1 mi. N La Ventana; 4 mi. N, 5 mi. W Dulce; Chuska Mts. Arizona: 12 mi. NW Fort Defiance, 7800 ft; Lukachukai Mts. Colorado: Mesa Verde National Park (Armstrong, 1972:134); Piñon Mesa, 8000 ft. (*op. cit.*:133); 28 mi. N, 5 mi. W Mack, 7250 ft. Utah: PR Springs, 7950 ft., 43 mi. S Ouray; 18 mi. N Escalante; Bryce National Park; Duck Creek, 9000 ft., Cedar Mtn.; *Blue Springs, 7975 ft.* (Stock, 1970:370); *2½ mi. SW Kolob Reservoir* (*ibid.*); Brian Head, 11,000 ft.; Britts Meadow, Beaver Range, 8500 ft.; Maple Canyon; Currant Creek, Uinta National Forest; Henrys Fork. Wyoming (Long,

1965a:589): 9 mi. S Robertson; 5 mi. SW Maxon. Colorado: Escalante Hills, 20 mi. SE Ladore; 5 mi. S Pagoda Peak, 9100 ft. Wyoming (Long, 1965a:588, 589): 5 mi. N, 5 mi. E Savery; Bridgers Pass; *North Fork Camp Ground, sec. 28, T. 16 N, R. 78 W.* Isolated southern segment in Arizona: Jacob Lake; *Point Imperial* (Hoffmeister, 1971:168); Greenland Spring; *Bright Angel Point* (Hoffmeister, 1971:168); Swamp Lake (*ibid.*).

Spermophilus lateralis mitratus (A. H. Howell)

1931. *Callospermophilus chrysodeirus mitratus* A. H. Howell, Jour. Mamm., 12:161, May 14, type from South Yolla Bolly Mtn., California.
1938. *Citellus lateralis mitratus* (A. H. Howell), N. Amer. Fauna, 56:210, May 18.

MARGINAL RECORDS.—California: Salmon Mts.; Castle Lake; Coast Range, 17 mi. W Paskenta; type locality; Salmon River, South Fork, Siskiyou Mts.

Spermophilus lateralis tescorum (Hollister)

1911. *Callospermophilus lateralis tescorum* Hollister, Smiths. Miscl. Coll., 56(26):2, December 5, type from head Moose Pass, branch Smoky River, 7000 ft., Alberta (near Moose Pass, British Columbia).

MARGINAL RECORDS.—British Columbia: Mt. Selwyn (Cowan and Guiguet, 1965:132); Sukunka River. Alberta: Wapiti River; Grand Cache River, 70 mi. N Jasper House; Canmore; Maycroft. Montana: Bear Creek; Bass Creek, near Stevensville. Idaho: Birch Creek; Craters of the Moon National Monument; Edna; Warren; 5 mi. W Cocolalla. British Columbia: Rossland; Barkerville.

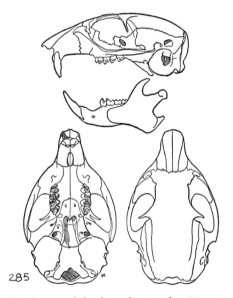

Fig. 285. *Spermophilus lateralis trepidus*, Kingston Ranger Station, 7500 ft., Nevada, No. 45473 M.V.Z., ♀, X 1.

Spermophilus lateralis trepidus (Taylor)

1910. *Callospermophilus trepidus* Taylor, Univ. California Publ. Zool., 5(6):283, February 12, type from head Big Creek, 8000 ft., Pine Forest Mts., Humboldt Co., Nevada.

1938. *Citellus lateralis trepidus*, A. H. Howell, N. Amer. Fauna, 56:206, May 18.

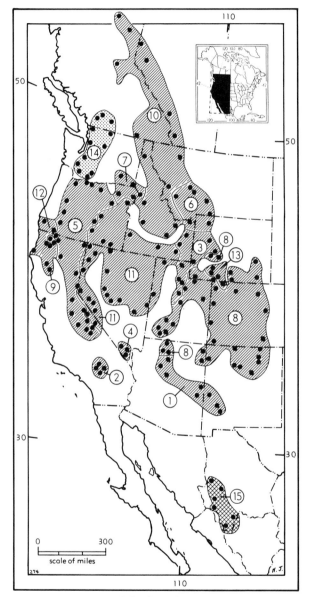

Map 274. *Spermophilus lateralis* and related species.

Guide to kinds
1. *S. l. arizonensis*
2. *S. l. bernardinus*
3. *S. l. castanurus*
4. *S. l. certus*
5. *S. l. chrysodeirus*
6. *S. l. cinerascens*
7. *S. l. connectens*
8. *S. l. lateralis*
9. *S. l. mitratus*
10. *S. l. tescorum*
11. *S. l. trepidus*
12. *S. l. trinitatis*
13. *S. l. wortmani*
14. *S. saturatus*
15. *S. madrensis*

1918. *Callospermophilus chrysodeirus perpallidus* Grinnell, Univ. California Publ. Zool., 17:429, April 25, type from near Big Prospector Meadow, 10,300 ft., White Mts., Mono Co., California.

MARGINAL RECORDS.—Oregon: Home. Idaho: Silver City; Albion; Bannock Mts., 8 mi. W Swan Lake. Utah: Clear Creek, 5 mi. SW Nafton; Deep Creek Mts. Nevada: Baker Creek, 8675 ft.; Quinn Canyon Mts., Burned Corral Canyon, 6700 ft.; Greenmonster Canyon, Monitor Range, 7500–9000 ft.; N slope Toquima Mtn., 9000 ft.; Shoshone Mts., 2 mi. W Indian Valley, 9000 ft.; 3 mi. W Carroll Summit, E of Eastgate; El Dorado Canyon, 8000 ft., Humboldt Range; 13 mi. N Paradise Valley, 6700 ft. Oregon: McDermitt Creek, 8 mi. NE McDermitt, Nevada; Steens Mts. Nevada: type locality; ½ mi. S Rock Creek, 6000 ft.; Horse Canyon, Pahrum Peak, 5800 ft.; 12-Mile Creek, ½ mi. E California boundary, 5300 ft. Oregon: Mt. Warner [= Hart Mtn.]; Burns. Isolated southwestern segment of range, Nevada: Edgewood; Lapon Canyon, Mt. Grant, 8900 ft.; Endowment Mine, 6500 ft., Excelsior Mts.; Chiatovich Creek, 8200 ft. California: Inyo Mts.; Mammoth Lakes.

Spermophilus lateralis trinitatis (Merriam)

1901. *Callospermophilus chrysodeirus trinitatis* Merriam, Proc. Biol. Soc. Washington, 14:126, July 19, type from E of Hoopa Valley, 5700 ft., Trinity Mts., California.

1938. *Citellus lateralis trinitatis*, A. H. Howell, N. Amer. Fauna, 56:211, May 18.

MARGINAL RECORDS.—Oregon: Briggs Creek, 13 mi. SW Galice; Siskiyou. California: type locality; Kneeland Hill, few mi. E Eureka (Yocom and Eley, 1972:29, as *Callospermophilus lateralis*).

Spermophilus lateralis wortmani (J. A. Allen)

1895. *Tamias wortmani* J. A. Allen, Bull. Amer. Mus. Nat. Hist., 7:335, November 8, type from Kinney Ranch, Bitter Creek, Sweetwater Co., Wyoming.

1911. *Callospermophilus lateralis wortmani*, Cary, N. Amer. Fauna, 33:84, August 17.

MARGINAL RECORDS.—Wyoming (Long, 1965a: 589): Superior; 22 mi. SSW Bitter Creek; *Kinney Ranch*. Colorado: Snake River, 20 mi. below Baggs, Wyoming. Wyoming (Long, 1965a:589): 15 mi. S, 2 mi. E Rock Springs; *Green River*.

Spermophilus saturatus (Rhoads)
Cascade Golden-mantled Ground Squirrel

1895. *Tamias lateralis saturatus* Rhoads, Proc. Acad. Nat. Philadelphia, 47:43, April 9, type from Lake Kichelos [= Keechelus], 8000 ft., Kittitas Co., Washington.

1938. *Citellus saturatus*, A. H. Howell, N. Amer. Fauna, 56:212, May 18.

External measurements: 286–315; 92–118; 43–49. Greatest length of skull, 44.0–48.3.

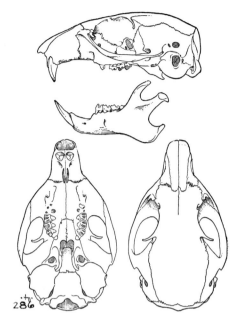

Fig. 286. *Spermophilus saturatus*, Stephens Pass, King Co., Washington, No. 22150 K.U., ♂, X 1.

Closely resembles S. *lateralis* but larger. Mantle poorly defined; inner pair of dark stripes obsolete or much reduced; outer pair of dark stripes reduced in length and obscurely defined.

MARGINAL RECORDS.—British Columbia: Tulameen; Hedley. Washington: Bauerman Ridge; Wenatchee; Cleveland; Goldendale; Trout Lake; Mt. St. Helens; Spray Park; type locality; Hannegan Pass. British Columbia: Skagit River. See Map 274.

Spermophilus madrensis (Merriam) Sierra Madre Mantled Ground Squirrel

1901. *Callospermophilus madrensis* Merriam, Proc. Washington Acad. Sci., 3:563, November 29, type from Sierra Madre, 7000 ft., near Guadalupe y Calvo, Chihuahua.

External measurements: 215–243; 52–66; 37–40. Greatest length of skull, 44.1–44.4. Closely resembles S. *lateralis* but smaller, tail shorter, colors much duller with scarcely a trace of mantle, black stripes short and poorly defined (tending to become obsolete), white stripes reaching nearly to base of tail; skull narrower and braincase more highly arched.

MARGINAL RECORDS.—Chihuahua (Anderson, 1972:274, 275, unless otherwise noted): Yaguirachic, 130 mi. W Chihuahua, 8500 ft.; 7 mi. NNE San Juanito, 8200 ft.; 7 mi. SW El Vergel [= Lagunita], 7800 ft.; Sierra Madre, near Guadalupe y Calvo; Batopilas (Hall and Kelson, 1959:363); 15 mi. S, 6 mi. E Creel, 7300 ft. See Map 274.

Genus **Cynomys** Rafinesque—Prairie Dogs

Revised by Hollister, N. Amer. Fauna, 40:1–36, 7 pls. 2, figs. in text, June 20, 1916.—**See** addenda.

1817. *Cynomys* Rafinesque, Amer. Month. Mag., 2:45, November. Type, *Cynomys socialis* Rafinesque [= *Arctomys ludivociana* Ord].
1819. *Monax* Warden, Statistical, political and historical account of the United States . . . , 1:226. Type, *Monax missouriensis* Warden [= *Arctomys ludoviciana* Ord].
1827. *Cynomis* Lesson, Manuel de mammalogie . . . , p. 244, a *lapsus*.
1894. *Cynomomus* Osborn, Science, 23:103, February 23, a *lapsus plumae* for *Cynomys*.

External measurements: 305–430; 30–115; 52–67. Condylobasal length, 51.9–64.0. Wrist and heel furred; manus with 5 distinct claws; mammae, 8–12. Skull broad and robust; occipital crest well developed; sagittal crest moderate anteriorly but well developed posteriorly; squamosal root of zygoma widespreading; antorbital foramen subtriangular, with prominent tubercle; tooth-rows strongly convergent posteriorly; individual cheek-teeth large and expanded laterally; 1st premolar large, nearly equalling 2nd; cheek-teeth more hypsodont than in *Spermophilus*, especially protocone; M3 with additional transverse ridge. Color pattern not sharply bicolored, usually grayish, brownish, or buffy.

KEY TO SPECIES OF CYNOMYS

1. Tail tipped with black; jugal heavy, thick, the outer surface at angle of ascending ramus presenting a broad, triangular surface.
  2. Black on tail covering most of distal half; posterior border of inflected angle of mandible nearly at right angles to axis of jaw. . . *C. mexicanus*,   p. 412
  2′. Black on tail confined to distal ⅓; posterior border of inflected angle of mandible at angle of approx. 45° to axis of jaw. . . . . *C. ludovicianus*,   p. 411
1′. Tail tipped with white; jugal weak, thin, flat, the outer surface at angle of ascending ramus only slightly thickened, the margin rounded, not distinctly triangular.
  3. Terminal half of tail white, without dark center.
    4. Color in summer reddish (Tawny-Olive to Clay Color); postorbital breadth *ca.* 14.2 in adult males; supraorbital notch prominent, almost foramenlike.
               C. parvidens,   p. 414
    4′. Color in summer grayish (Pinkish Buff to Pale Smoke Gray); postor-

bital breadth *ca.* 13.2 in adult males; supraorbital notch obsolete.

C. *leucurus*, p. 413

3'. Terminal half of tail with gray center.

C. *gunnisoni*, p. 414

Subgenus **Cynomys** Rafinesque

1817. *Cynomys* Rafinesque, Amer. Month. Mag., 2:45. Type, *Cynomys socialis* Rafinesque [= *Arctomys ludoviciana* Ord].

Tail comparatively long, averaging more than a fifth of total length; tipped with black. Skull massive; occipital region ovoid from posterior aspect; jugal thick and heavy, its outer surface at angle of ascending branch broad, triangular, with inferior vertex produced downward; maxillary root of zygoma correspondingly strengthened, shelf and suprajugal arm much thickened. Cheek-teeth larger, more expanded laterally than in subgenus *Leucocrossuromys*.

Cynomys ludovicianus
Black-tailed Prairie Dog

X ⅓

External measurements: 355–415; 72–115; 57–67. Condylobasal length, 56.6–64. Upper parts (summer) approx. dark pinkish cinnamon finely lined with black and buff; upper lip, sides of nose, and eye ring buffy or whitish; tail above like back for proximal two-thirds, black or black-ish brown distally; tail below vinaceous-cinnamon proximally, with distal third blackish or dark brown; underparts whitish or buffy white. See account of subgenus for diagnostic cranial characters.

Cynomys ludovicianus arizonensis Mearns

1890. *Cynomys arizonensis* Mearns, Bull. Amer. Mus. Nat. Hist., 2:305, February 21, type from Point of Mountain, near Willcox. Cochise Co., Arizona.—See addenda.
1892. *C[ynomys]. ludovicianus arizonensis*, Merriam, Proc. Biol. Soc. Washington, 7:158, July 27.

MARGINAL RECORDS.—New Mexico: San Pedro; Santa Rosa; Roswell; 8 mi. SW Carlsbad. Texas: 2 mi. E Sheffield; [30 mi. S] Sheffield; Brewster County (Davis, 1966:140, as *C. ludovicianus* only); Presidio County (*ibid.*); El Paso County (*ibid.*). Chihuahua (Anderson, 1972:280): Corralitos; San Diego; Llano de Las Carretas, 21 mi. W Cuervo. Sonora: Río San Pedro. Arizona: Fort Huachuca; Bonita. New Mexico: Cactus Flat, 20 mi. N Cliff; Chloride (Findley, *et al.*, 1975:132, as *C. ludovicianus* only); east foothills, Manzano Mts.

Cynomys ludovicianus ludovicianus (Ord)

1815. *Arctomys ludoviciana* Ord, *in* Guthrie, A new geog., hist., coml. grammar . . . , Philadelphia, Amer. ed. 2, 2:292 (description on p. 302). Type locality, "Upper Missouri River" ("vicinity of the Missouri, and throughout the greater part of Louisiana").
1858. *Cynomys ludovicianus*, Baird, Mammals, *in* Repts. Expl. and Surv. . . . , 8(1):xxxix, 331, July 14.
1817. *Cynomys socialis* Rafinesque, Amer. Month. Mag., 2:45, November. Type locality, "Plains of the Missouri."
1817. *Cynomys ?grisea* Rafinesque, Amer. Month. Mag., 2:45, November. Type locality, "On the Missouri."
1819. *Monax missouriensis* Warden, Statistical, political and historical account of the United States . . . , 1:226. Type locality, "The Missouri country." Based largely on the account of "Major Pike, in his expedition through Louisiana."

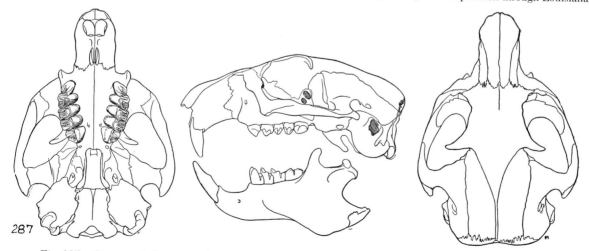

287

Fig. 287. *Cynomys ludovicianus ludovicianus*, Higgens, Lipscomb Co., Texas, No. 44365 M.V.Z., ♀, X 1.

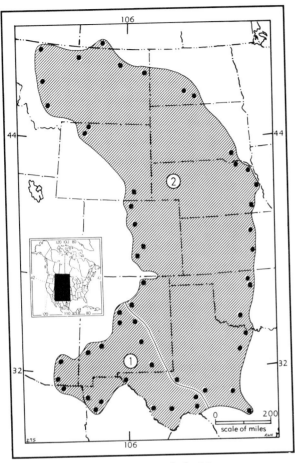

Map 275. *Cynomys ludovicianus.*

1. *C. l. arizonensis*   2. *C. l. ludovicianus*

1825. *Arctomys latrans* Harlan, Fauna Americana, p. 306. Type locality, "Plains of the Missouri."
1829. *C[ynomys]. cinereus* Richardson, Fauna Boreali-Americana, pt. 1, p. 155 (pro *grisea* Rafinesque, 1817).
1905. *Cynomys pyrrotrichus* Elliot, Proc. Biol. Soc. Washington, 18:139, April 18, type from White Horse Spring, Woods Co., Oklahoma.

MARGINAL RECORDS.—Saskatchewan: approx. 7½ mi. SE Val Marie. Montana: near mouth Milk River; Boxelder Creek. North Dakota: Glenullin; 4 mi. N Cannonball. South Dakota: Armour. Nebraska (Jones, 1964c:141): near Ponca Creek; 2 mi. N Wayne; Clark Creek, near Fontanelle; S of Wilber. Kansas: Clay County; Marion County. Oklahoma: Ponca Agency; 1 mi. N, 7 mi. W Stillwater; Wichita Mts. National Wildlife Refuge (Glass and Halloran, 1961:237). Texas: Henrietta; Fort Belknap; Mason; Bexar County (Davis, 1966:140, as *C. ludovicianus* only); head Devils River; Monahans. New Mexico: Pecos; 6 mi. NW Dawson, Van Bremmer Canyon (Findley, *et al.*, 1975:132, as *C. ludovicianus* only). Colorado: Soda Springs; near Colorado Springs; Boulder; Livermore. Wyoming: 8½ mi.

W Horse Creek P.O. (Long, 1965a:590); Ishawooa; Sage Creek. Montana: 3 mi. NW Three Forks; 10 mi. NW Craig; Shelby Junction; Fort Assiniboine.

Cynomys mexicanus Merriam
Mexican Prairie Dog

1892. *Cynomys mexicanus* Merriam, Proc. Biol. Soc. Washington, 7:157, July 27, type from La Ventura, Coahuila.

External measurements: 390–430; 89–115; 59.0–68.5. Condylobasal length, 58.4–60.5. Upper parts as in *C. ludovicianus,* but less reddish and more grayish and vinaceous-buff; black hairs more numerous, giving more grizzled effect; tail above like back in proximal half, black distally, the black extending proximally as a border; tail below as in *C. ludovicianus,* except for greater extent of black; underparts whitish. Selected characters in which the skull differs from that of *C. ludovicianus* are: auditory bullae more inflated; cheek-teeth triangular; nasals broad and usually posteriorly truncate; triangular plate of jugal especially well developed and greatly produced at downward point.

MARGINAL RECORDS.—Coahuila: Saltillo; 3 mi. N, 4 mi. W San Antonio de las Alzanas; type locality. San Luis Potosí: Vanegas. Coahuila: 3 mi. N Gómez Farías, 7000 ft.

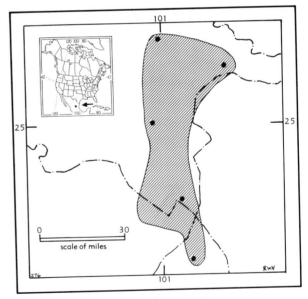

Map 276. *Cynomys mexicanus.*

Subgenus **Leucocrossuromys** Hollister

1916. *Leucocrossuromys* Hollister, N. Amer. Fauna, 40:23, June 20. Type, *Spermophilus gunnisoni* Baird.

Tail comparatively short, averaging less than a fifth total length; tipped with white. Skull not so massive as in subgenus *Cynomys;* occipital region elliptical-ovoid from posterior aspect; jugal weak, thin, flat, its outer surface at angle of ascending ramus only slightly thickened, and margin rounded, not triangular; maxillary root of zygoma correspondingly weak, the shelf and suprajugal arm not markedly thickened. Cheek-teeth smaller, less expanded laterally than in *Cynomys.*

Cynomys leucurus Merriam
White-tailed Prairie Dog

1890. *Cynomys leucurus* Merriam, N. Amer. Fauna, 3:59, September 11, type from Fort Bridger, Uinta Co., Wyoming.

External measurements: 340–370; 40–60; 60–65. Condylobasal length, 56.0–61.3. Upper parts pinkish buff mixed with black, which is most prominent on mid-dorsal region and rump; spots above eyes and cheeks brownish black; nose yellowish; upper side of tail proximally like rump but distal half white; underparts pale pinkish buff, slightly darker in axillary and inguinal regions. Skull large, robust; zygomatic arches widespread; postorbital processes robust and sharply decurved; tympanic bullae large and much inflated; cheek-teeth large, especially in transverse dimension.

MARGINAL RECORDS.—Northern segment: Montana: Clarks Fork; Sage Creek. Wyoming: Spring Creek; 17 mi. NNW Casper, 5650 ft.; Garrett; W of Cheyenne. Colorado (Armstrong, 1972:138): T. 12 N, R. 75 W, Chimney Rock Ranch, SW of Tie Siding, Wyom-

ing; Canadian Creek; Wolcott; Buford; Rangely. Utah: Duchesne; 1½ mi. N Tridell, 6700 ft. Colorado: Escalante Hills. Utah: Linwood; Uinta Mts.; eastern Morgan County. Wyoming: Fossil; 2 mi. N Big Piney, 6900 ft.; Dubois; Ishawooa. Southern segment: Utah: Emma Park, head Price River, near Colton; 8 mi. E Sunnyside. Colorado (Armstrong, 1972:138): 14 mi. N, 5 mi. W Mack, 5600 ft.; U.S. highways 6–24, 3 mi. N Mesa County line; Paonia; Ridgway; Montrose; Grand Junction. Utah: N side Colorado River, 4500 ft.; Ferron.

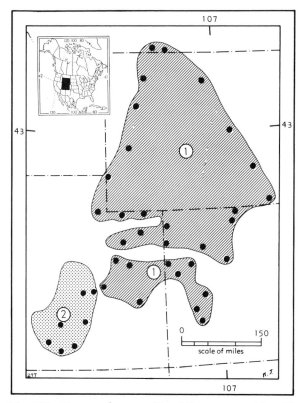

Map 277. *Cynomys leucurus* (1) and *Cynomys parvidens* (2).

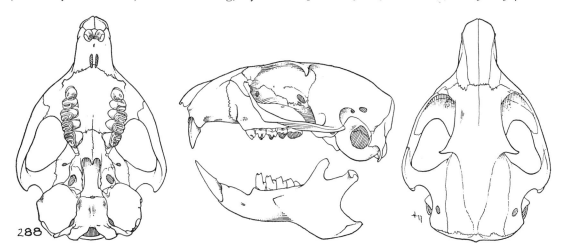

Fig. 288. *Cynomys leucurus,* W side Green River, 1 mi. N Utah border, Sweetwater Co., Wyoming, No. 16802 K.U., ♂, X 1.

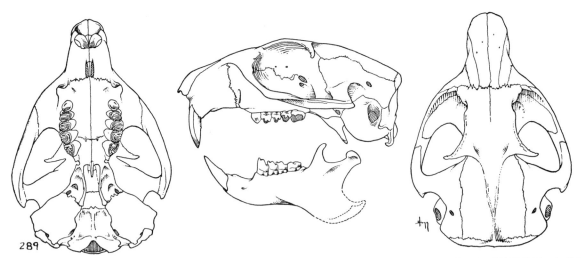

Fig. 289.   *Cynomys parvidens*, 1 mi. W Cedar City, 5800 ft., Iron Co., Utah, No. 15962 K.U., ♂, X 1.

Cynomys parvidens J. A. Allen
Utah Prairie Dog

1905. *Cynomys parvidens* J. A. Allen, Mus. Brooklyn Inst. Arts and Sci., Sci. Bull., 1:119, March 31, type from Buckskin Valley, Iron Co., Utah.

External measurements: 305–360; 30–57; 55–61. Condylobasal length of skull, 53.0–57.9. Upper parts cinnamon or clay color with mixture of buff and black-tipped hairs, usually slightly darker on rump; chin, sides of nose, and lips buffy white; upper side of tail proximally like rump, distal half white; underparts Cinnamon-Buff, grading to Cinnamon on pectoral and inguinal regions. Skull angular and robust; interorbital region broad; postorbital processes not abruptly projecting.

MARGINAL RECORDS (Durrant, 1952:109, unless otherwise noted).—Utah: Coal Mine Flat, 20 mi. E Salina, 4875 ft.; NE ¼ sec. 3, T. 30 S, R. 1 E, 8400 ft., of Wayne Co. (Pizzimenti and Nadler, 1972:281); Bryce Canyon; head Sevir River, 7200 ft. in Kane Co., W of Cedar City; head Buckthorn Flat, 1 mi. S Beaver–Iron County line, 6000 ft.; Salina (Ivie Farm), head Salina Canyon.

Cynomys gunnisoni
Gunnison's Prairie Dog

External measurements: 309–373; 39–68; 52.0–62.5. Condylobasal length of skull, 51.9–57.8. Upper parts approx. Cinnamon-Buff, darkest in mid-dorsal region and rump, lightly to moderately overlaid with black hairs; upper side of tail proximally like rump, distally mixed gray and white, bordered and tipped with white; underparts near Pale Cinnamon, often somewhat

buffy. Skull differs from that of *C. leucurus* and *C. parvidens* in certain selected characters as follows: maxillary arm of zygomatic arch more broadly spreading; mastoids smaller and more obliquely placed; auditory bullae smaller.

The breeding season is April and May. There are 3 to 6 young (mean 4.8).

Armstrong (1972:138–141) relates that in central Colorado this species is typically a mammal of the mountain parks, occurring in sites ranging in elevation from 6000 to 12,000 feet. In southwestern Colorado and adjacent areas, lower, more xeric habitats are utilized. Those habitats are comparable to those inhabited by *Cynomys leucurus* farther north.

Armstrong (*loc. cit.*) notes that *Cynomys gunnisoni* is the smallest of Coloradan prairie dogs, superficially resembles the white-tailed prairie dog, differs in smaller size, cranial details, habitat preferences, and details of social organization. He cites published literature providing details supporting his statements.

Cynomys gunnisoni gunnisoni (Baird)

1855. *Spermophilus gunnisoni* Baird, Proc. Acad. Nat. Sci. Philadelphia, 7:334, April, type from Cochetopa Pass, Saguache Co., Colorado.
1858. *Cynomys gunnisonii*, Baird, Mammals, *in* Repts. Expl. Surv. . . . , 8(1):xxxix, July 14.

MARGINAL RECORDS.—Colorado (Armstrong, 1972:140): Twin Creek [Twin Forks?]; sec. 31, T. 10 S, R. 66 W, Douglas Co.; Cascade, 7500 ft.; Greenwood; ½ mi. E Farisita; parks and plateaus of Trinidad region. New Mexico: Costilla Pass; 16 mi. SW Cimmaron, 9500–10,500 ft.; Coyote Creek; Jemez Mts.; Gallinas Mts.; La Jara Lake. Colorado (Armstrong, 1972:140): 12 mi. NE Cumbres, 8800 ft.; head Rio Grande; Ridgway;

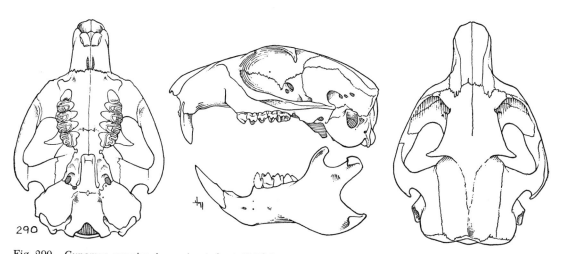

Fig. 290. *Cynomys gunnisoni gunnisoni*, 8 mi. N El Rito, Rio Arriba Co., New Mexico, No. 6236 K.U., ♀, X 1.

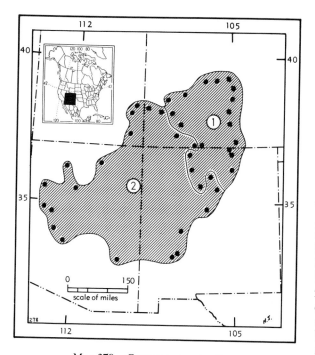

Map 278. *Cynomys gunnisoni*.

1. *C. g. gunnisoni*   2. *C. g. zuniensis*

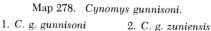

5 mi. W Olathe; near Crested Butte; Leadville; Jefferson.

Cynomys gunnisoni zuniensis Hollister

1916. *C[ynomys]. g[unnisoni]. zuniensis* Hollister, N. Amer. Fauna, 40:29, June 20, type from Wingate, McKinley Co., New Mexico.—See addenda.

MARGINAL RECORDS.—Utah: 25 mi. SE Moab. Colorado (Armstrong, 1972:141): 5 mi. W Uravan; Coventry; Vallecito; 5 mi. NW Chama, New Mexico. New Mexico: Espanola; Pecos; Manzano Mts.; Mag-

dalena; Fairview; Chloride. Arizona: 9-mile Spring, Ash Flat; Agua Fria Valley; 6 mi. S Yarnell, Peeples Valley; 6 mi. SE Kirkland (Cockrum, 1961:78); Simmons; Aubrey Valley; Pasture Wash (Hoffmeister, 1971:169); Kendrick Peak, 8200 ft., 20 mi. NW Flagstaff (Cockrum, 1961:78); Tonalea. Utah: 2 mi. S Blanding, Galbraith Ranch; Hewit Ranch, East Canyon.

Genus **Sciurus** Linnaeus—Tree Squirrels

1758. *Sciurus* Linnaeus, Syst. nat., ed. 10, 1:63. Type, *Sciurus vulgaris* Linnaeus.

Braincase strongly depressed posteriorly and without prominent ridges; palate broad, almost square posteriorly and terminating immediately behind tooth-rows; infraorbital foramen always forming canal; masseteric tubercle weak; M1 and M2 with 4 transverse crests, the 2nd and 3rd being most prominent and terminating in well-marked cusps; M3 with 1 prominent transverse crest; P3 present or absent and, when present, usually vestigial; size moderate; tail always bushy; ears prominent and tufted in some species.

Whoever undertakes to evaluate relationships of American species of tree squirrels and especially subgenera so far proposed should consult Nelson (1899), A. H. Howell (1938:48, 49), and Moore (1959, 1960). A satisfactory key to even the species of the subgenus *Sciurus* based on zoological characters remains to be made.

Owing to melanism and partial albinism some individuals of the two kinds can be difficult to identify to species by color. Two upper premolars instead of one premolar, and nasal bones not extending so far posteriorly in relation to the premaxillary bones differentiate *S. carolinensis* from *S. niger* (see Figs. 291 and 296).

Subgenus **Sciurus** Linnaeus

1758. *Sciurus* Linnaeus, Syst. nat., ed. 10, 1:63. Type, *Sciurus vulgaris* Linnaeus.
1880. *Neosciurus* Trouessart, Le Naturaliste, 2:292, October. Type, *Sciurus carolinensis* Gmelin.
1880. *Parasciurus* Trouessart, Le Naturaliste, 2(37):292, October. Type, *Sciurus niger* Linnaeus.
1880. *Echinosciurus* Trouessart, Le Naturaliste, 2:292, October. Type, *Sciurus hypopyrrhus* Wagler [= *Sciurus aureogaster* Cuvier].
1893. *Aphrontis* Schulze, Zeitschr. Naturwiss., Leipzig, 5te Folge, IV, p. 165. Type, *Sciurus vulgaris* Linnaeus.
1899. *Araeosciurus* Nelson, Proc. Washington Acad. Sci., 1:29, May 9. Type, *Sciurus oculatus* Peters.
1899. *Baiosciurus* Nelson, Proc. Washington Acad. Sci., 1:31, May 9. Type, *Sciurus deppei* Peters.
1909. *Tenes* Thomas, Ann. Mag. Nat. Hist., ser. 8, 3:468 (footnote), June. Type, *Sciurus persicus* (not of Erxleben, a *Glis*) = *Sciurus anomalus* Güldenstaedt.
1935. *Oreosciurus* Ognev, Abstr. zool. Inst. Moscow Univ., 2:50. Type, *Sciurus anomalus* Güldenstaedt. Antedated by *Tenes* Thomas.

Maxillary notch opposite M1; back part of jugal twisted; baculum approx. 10.5 mm. long.

Sciurus carolinensis
Gray Squirrel

External measurements: 430–500; 210–240; 60–70. Weight, 400–710 grams. Upper parts normally grizzled dark to light gray, usually with buffy under fur, buffy tone more pronounced on head, back, feet, and shoulders; underparts dark to light gray, usually with strong buffy suffusion. Melanism or partial albinism is common.

Sciurus carolinensis carolinensis Gmelin

1788. [*Sciurus*] *carolinensis* Gmelin, Syst. nat., ed. 13, 1:148. Type locality, Carolina.

MARGINAL RECORDS.—Indiana: Denver. Ohio: Smoky Creek. Kentucky: Carter Caves; Quicksand. Tennessee: Poor Valley Ridge; Walden Ridge. South Carolina: Richland. North Carolina: Statesville. Virginia: Charlotte Court House; Suffolk; Eastville. Florida: Micco; St. Petersburg. Alabama: Point Clear; Carlton. Louisiana (Lowery, 1974:181, as intergrade

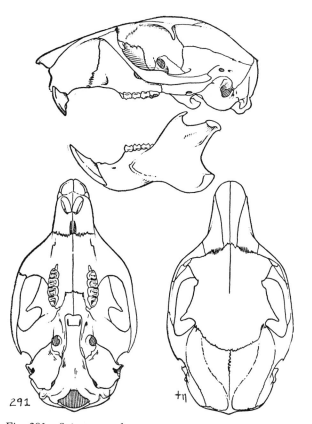

Fig. 291. *Sciurus carolinensis carolinensis*, 5 mi. NW Fall River, Greenwood Co., Kansas, No. 9637 K.U., ♂, X 1.

with *S. c. fuliginosus*): *2 mi. S Sicily Island; Fort Polk Game Reservation*; 2 mi. S Fort Polk. Texas: mouth of Colorado River; Cuero; Austin; Brazos. Oklahoma: Dougherty; Mohawk Park. Kansas: Cedar Vale; Baxter Springs (Jones and Cortner, 1961:288). Illinois: Union County. Indiana: *Sullivan County; Vigo County; Parke County; Montgomery County*; Tippecanoe County; *Carroll County; Cass County.*

Sciurus carolinensis extimus Bangs

1896. *Sciurus carolinensis extimus* Bangs, Proc. Biol. Soc. Washington, 10:158, December 28, type from Miami, Dade Co., Florida.
1937. *Sciurus carolinensis minutus* H. H. Bailey, Bailey Mus. Libr. Nat. Hist., Bull. 12 [p. 4], January 15, type from Key Largo, Monroe Co., Florida. Not *Sciurus minutus* du Chaillu, 1861.
1937. *Sciurus carolinensis matecumbei* H. H. Bailey, Jour. Mamm., 18:516, November 22, a renaming of *Sciurus carolinensis minutus* H. H. Bailey. *S. c. matecumbei* treated as a synonym of *S. c. extimus* by Hubbard and Banks, Proc. Biol. Soc. Washington, 83:329, September 25, 1970.

MARGINAL RECORDS (Layne, 1974:391, unless otherwise noted).—Florida: Eau Gallie (Sherman, 1937:114); type locality; *Cutler*; Key Largo; *Plantation*

Key; Upper Matecumbe Key; Willy Willy Mound; Everglades City; Fort Myers; *LaBelle; Immokalee; Deep Lake; Royal Palm Hammock.*

Sciurus carolinensis fuliginosus Bachman

1839. *Sciurus fuliginosus* Bachman, Proc. Zool. Soc. London, 1838, p. 97, February 7, type from near New Orleans, Louisiana.
1895. *Sciurus carolinensis fuliginosus*, Bangs, Proc. Boston Soc. Nat. Hist., 26:543, July 31.

MARGINAL RECORDS.—Louisiana: Jonesville (Lowery, 1974:181, as intergrade with *S. c. carolinensis*). Alabama: Stiggens Lake; *Chuckvee Bay;* Bayou Labatre. Louisiana (Lowery, 1974:180, 181): Port Sulphur; *Grand Caillou;* Cypremort Point; Interstate 10, W of Sulphur; *2 mi. N, 4 mi. W Longville; 8 mi. W Lecompte* (as intergrade with S. c. carolinensis).

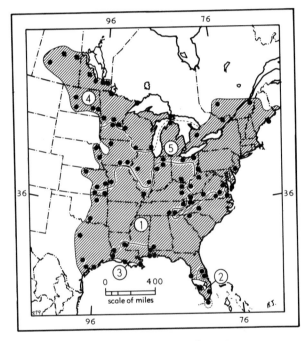

Map 279. *Sciurus carolinensis.*

1. *S. c. carolinensis*	3. *S. c. fuliginosus*
2. *S. c. extimus*	4. *S. c. hypophaeus*
	5. *S. c. pennsylvanicus*

Sciurus carolinensis hypophaeus Merriam

1886. *Sciurus carolinensis hypophaeus* Merriam, Science, 7:351, April 16, type from Elk River, Sherburne Co., Minnesota.

MARGINAL RECORDS.—Saskatchewan: Rose Valley, 13 mi. S Tisdale (Beck, 1958:30); 10 mi. NE Weekes (D. Hooper, 1973:238). Manitoba (Tamsitt, 1960:149): 3 mi. NW east entrance Riding Mountain National Park; 3 mi. S Delta; East Selkirk; Lake Jessie; Winnipeg River. Minnesota: Williams; Grand Marais (Timm, 1975:18, sight record only). Michigan: Chip-

pewa County (Burt, 1946:191, Map 42). Wisconsin (Jackson, 1961:164): Racine; Rutledge; Prescott. Minnesota: Anoka County; 2 mi. N Sartell; Grant County. North Dakota: Fargo; 1 mi. N Kathryn; 2 mi. S Bismarck; 4 mi. SW Velva. Saskatchewan: Saskatoon (Beck, 1958:30).

Sciurus carolinensis pennsylvanicus Ord

1815. *Sciurus Pennsylvanica* Ord, *in* Guthrie, A new geogr., hist., coml. grammar . . . , Philadelphia, Amer. ed. 2, 2:292. Type locality, Pennsylvania W of Allegany Ridge.
1894. *Sciurus carolinensis pennsylvanicus*, Rhoads, Appendix of reprint of Ord (*supra*), p. 19.
1815. *Sciurus hiemalis* Ord, *in* Guthrie, A new geogr., hist., coml. grammar . . . , Philadelphia, Amer. ed. 2, 2:292. Type locality, near Tuckerton, near Little Egg Harbor, New Jersey.
1830. *Sciurus leucotis* Gapper, Zool. Jour., 5:206, type from region between York (Toronto) and Lake Simcoe, Ontario.
1849. *Sciurus migratorius* Audubon and Bachman, The viviparous quadrupeds of North America, 1:265. Based on *S. leucotis* Gapper, 1830.

MARGINAL RECORDS.—Quebec: La Verendrye Prov. Park (Pirlot, 1962:135); St. Lambert (Wrigley, 1969:206, as *S. carolinensis* only). Maine: Mount Desert Island. New Jersey: Wading River. Virginia: Hampstead; Amelia County; Blacksburg. Tennessee: Sheeds Creek; 4 mi. NE Shady Valley. Virginia: head Moccasin Creek. West Virginia: 4 mi. E Huntington; Jacksons Mill; Clinton. Ohio: North Chagrin Metropolitan Park; Mud Creek. Indiana: Marshall County. Illinois: Newton; Peoria. Iowa: 3 mi. N, ½ mi. E Keosauqua (109700 KU); 5½ mi. E Moravia (109698 KU). Missouri: 2 mi. E, 7 mi. S Harrisonville (Jones and Cortner, 1961:286). Kansas: 5½ mi. SE Fontana (Jones and Cortner, 1961:288); 1 mi. SW Hamilton; Osborne (Black, 1937:179, based on mounted specimen, 159 KU, Osborne, Kansas, dated November 25, 1903, donated by F. L. Ward; specimen not found on September 10, 1971—E.R.H.). Nebraska (Jones, 1964c:144): Paddock's Grove (Omaha); Cedar County. Iowa: Clay County. Minnesota: *McLeod County;* near St. Cloud. Michigan (Burt, 1946:191, Map 42): Emmet County; *Cheboygan County.* Ontario: Mount Forest; Ottawa.—See addenda.

Sciurus aureogaster
Mexican Gray Squirrel

Revised by Musser, Miscl. Publ. Mus. Zool., Univ. Michigan, 137:1–112, 34 figs., 6 tables, November 1, 1968.

External measurements: 418–573; 206–315; 57–72. Upper parts light to dark grizzled gray with frosting of whitish, broken up by either nape and rump patches as well as shoulder and costal patches, or various combinations of these patterns (the patches vary in size and color); underparts varying from white through orange to deep chestnut; tail usually variegated grayish-white or grayish-buff in individuals having whitish underparts, and usually orange-red or chestnut in individuals having deep orange underparts; melanism, ranging from all black to partly blackish, common. Two upper premolars.

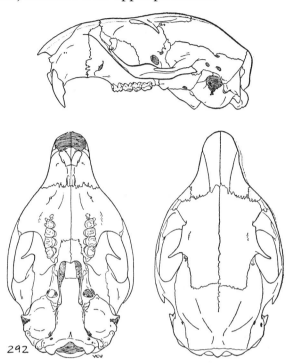

Fig. 292. *Sciurus aureogaster aureogaster*, Río Atoyac, 8 km. NW Potrero, Veracruz, No. 17910 K.U., ♂, X 1.

Of squirrels of the Western Hemisphere, individuals of this species are the most variable in color and pattern. Although individual variation in color is considerable in some populations, the variation in general is geographic. As a conse-

quence, two dozen or more names have been proposed for these squirrels (see synonymy under *Sciurus aureogaster aureogaster* and *S. a. nigrescens*). Musser (1968:Figs. 3, 4, 21, 22, 28, 29) portrays the geographic variation in color on which the taxonomic names were based and (*op. cit.*:94, 95) considered treating it by recognizing (1) a monotypic species, (2) two subspecies, (3) four subspecies, and (4) 22 subspecies. He chose to recognize two subspecies, and his choice is the one presented here.

Sciurus aureogaster aureogaster Cuvier

1829. [*Sciurus?*] *aureogaster* Cuvier, *in* É. Geoffroy St.-Hilaire and F. Cuvier, Hist. Nat. Mamm., 6, livr. 59, pl. with text, September (binomial published only at end of work, tableau génerale et méthodique, 7:4, 1842), type from "California"; restricted to Altamira, Tamaulipas, by Nelson (Proc. Washington Acad. Sci., 1:38, May 9, 1899), and accepted by Musser (Amer. Mus. Novit., 2438:8, November 12, 1970).

1830. *Sciurus rafiventer* Lichtenstein, Abh. k. Akad. Wiss., Berlin, p. 116 (1827). Based on a red-bellied, North American *Sciurus*.

1831. *Sciurus leucogaster* F. Cuvier, Supplément à l'histoire naturelle générale et particulière de Buffon, 1:300.

1831. *Sciurus hypopyrrhus* Wagler, Oken's Isis, p. 510, type from "Mexico"; restricted to Minatitlán, Veracruz, by Nelson (*op. cit.*:42).

1841. *Sciurus mustelinus* Audubon and Bachman, Proc. Acad. Nat. Sci. Philadelphia, p. 100. Based on a melanistic individual "received from California."

1841. *Sciurus ferruginiventris* Audubon and Bachman, Proc. Acad. Nat. Sci. Philadelphia, p. 101. Allegedly from California.

1845. *Sciurus ferrugineiventris* Schinz, Synopsis mammalium, 2:14, a variant spelling of *ferruginiventris*.

1855. *Sciurus hypoxanthus* (Lichtenstein MS) I. Geoffroy St.-Hilaire, Voyage de la Vénus, Zool. (text) p. 158 (on labels of squirrels from Berlin Museum, *fide* Nelson, *op. cit.*:38).

1855. *Sciurus chrysogaster* Giebel, *in* Schreber, Die Säugthiere, p. 650, an etymological substitute for *aureogaster*.

1867. *Sciurus hypopyrrhous* Gray, Ann. Mag. Nat. Hist., ser. 3, 20:424, a variant spelling of *hypopyrrhus*.

1867. *Macroxus morio* Gray, Ann. Mag. Nat. Hist., ser. 3, 2:424. Type locality unknown.

1867. *Macroxus maurus* Gray, Ann. Mag. Nat. Hist., ser. 3, 2:425. Type locality, Oaxaca (Sallé), Mexico.

1877. *Sciurus aureigaster* J. A. Allen, Rept. U.S. Geol. Surv. Territories, 9:750, a variant spelling of *aureogaster*.

1887. *Sciurus rufiventris* ? Rovirosa, La Naturaleza, 7:360 (1885–1886).

MARGINAL RECORDS (Musser, 1968:101, 102, unless otherwise noted).—Nuevo León: 20 km. NW General Terán, 900 ft.; *La Unión, 20 km. NE General Terán, 1000 ft.* Tamaulipas: San Fernando, 180 ft.; 7 mi. W La Pesca, 25 ft., thence down coast to Veracruz: Coatzacoalcos (Hall and Kelson, 1959:373). Tabasco: La Venta; 10 mi. E, 19 mi. N Macuspana; Monte Cristo

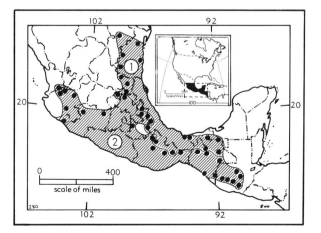

Map 280. *Sciurus aureogaster.*

1. *S. a. aureogaster*   2. *S. a. nigrescens*

Emiliano Zapata. Chiapas: *Laguna Ocotal, 950 m.;* La Florida, 50 km. E Altamirano, 525 m.; Simojovel, 660 m. Oaxaca: Santo Domingo; *mts. near Santo Domingo, 1600 ft.;* 11 mi. (Oaxaca de Juárez–Tuxtepec road) SW San Juan Bautista Valle Nacionál, 760 m. Veracruz: Xuchil. Puebla: 30 mi. E Huauchinango, 1200 ft.; *Apapantilla, 1000 ft.* Querétaro: Pinal de Amoles, 8000 ft. San Luis Potosí: 3½ mi. by road SW Xilitla, 740 m.; Ciudad de Valles; Rancho Martínez on Río Salto, 30 km. E Ciudad Maíz, 1600 ft.; *vic. El Salto Falls (Salto del Agua) on Río Salto.* Tamaulipas: Ejido Santa Isabel, 2 km. W Panamerican Hwy., 2000 ft. [6 km. N, 5 km. W El Carrizo]; Rancho Santa Rosa, 25 km. N, 13 km. W Ciudad Victoria, 260 m.

Sciurus aureogaster nigrescens Bennett

1833. *Sciurus nigrescens* Bennett, Proc. Zool. Soc. London, 1833:41, type specimen listed as from "that part of California which adjoins to Mexico" but shown by Musser (Amer. Mus. Novit., 2438:14, 15, November 12, 1970) to be from "the high mountains south and southeast of the town of Aquixtla, in northern Puebla."

1970. S[*ciurus*]. a[*ureogaster*]. *nigrescens*, Musser, Amer. Mus. Novit., 2438:16, November 12, 1970.

1837. *Sciurus albipes* Wagner, Abh. math.-phys., 101, bayerisch. Akad. Wiss. München, 2:501, Not *Sciurus albipes*, Kerr, 1792.

1837. *Sciurus socialis* Wagner, Abh. math.-phys., 101, bayerisch. Akad. Wiss. München, 2:504, type from near Tehuantepec, Oaxaca.

1843. *Sciurus varius* Wagner, *in* Schreber, Die Säugthiere . . . , Suppl., 3:168. Not *Sciurus varius* Pallas, 1831. A renaming of material that Wagner had earlier (1837) called *S. albipes.*

1867. [*Sciurus variegatus*] *poliopus* Fitzinger, Sitzungsb. k. Akad. Wiss., Wien, 55:478, March. Based on *Sciurus varius* var. β Wagner, 1843. Type locality by subsequent designation, Cerro San Felipe, Oaxaca.

1867. *Sciurus variegatus rufipes* Fitzinger, Sitzungsb. k. Akad. Wiss., Wien, 55:478, March. Based on *S. varius* var. γ Wagner, 1843.

1867. *Macroxus griseoflavus* Gray, Ann. Mag. Nat. Hist., ser. 3, 20:427, December, type from Guatemala; restricted to Dueñas by Nelson, Proc. Washington Acad. Sci., 1:67, May 9, 1899.

1867. *Macroxus leucops* Gray, Ann. Mag. Nat. Hist., ser. 3, 20:427, December, type from Oaxaca, México.

1878. Sc[*iurus*]. *affinis* (Reinh[ardt]., MS) Alston, Proc. Zool. Soc. London, June 18, in account of *S. griseoflavus.* Alston referred to a Mexican specimen, in the Copenhagen Museum, labeled "*Sc. affinis.*"

1890. *Sciurus cervicalis* J. A. Allen, Bull. Amer. Mus. Nat. Hist., 3:183, December 10, type from Hda. San Marcos, Tonila, Jalisco.

1893. *Sciurus nelsoni* Merriam, Proc. Biol. Soc. Washington, 8:144, December 29, type from Huitzilac, Morelos.

1898. *Sciurus albipes quercinus* Nelson, Proc. Biol. Soc. Washington, 12:150, June 3, type from mts. 15 mi. W Oaxaca, Oaxaca. Not *Sciurus quercinus* Erxleben, 1777.

1898. *Sciurus albipes nemoralis* Nelson, Proc. Biol. Soc. Washington, 12:151, June 3, type from Pátzcuaro, Michoacán.

1898. *Sciurus albipes colimensis* Nelson, Proc. Biol. Soc. Washington, 12:152, June 3, type from Hda. Magdalena, Colima.

1898. *Sciurus albipes effugius* Nelson, Proc. Biol. Soc. Washington, 12:152, June 3, type from high mts. [15 mi.] W Chilpancingo, Guerrero. Known only from type locality.

1898. *Sciurus nelsoni hirtus* Nelson, Proc. Biol. Soc. Washington, 12:153, June 3, type from Tochimilco, Puebla.

1898. *Sciurus aureogaster frumentor* Nelson, Proc. Biol. Soc. Washington, 12:154, June 3, type from Las Vigas, Veracruz.

1898. *Sciurus socialis cocos* Nelson, Proc. Biol. Soc. Washington, 12:155, June 3, type from Acapulco, Guerrero.

1898. *Sciurus wagneri* J. A. Allen, Bull. Amer. Mus. Nat. Hist., 10:453, November 10, a substitute name for the preoccupied *albipes* Wagner and *varius* Wagner. Allen overlooked the available name *poliopus* of Fitzinger 1867.

1898. [*Sciurus albipes*] *hernandezi* Nelson, Science, n.s., 8:783, December 2, a· renaming of *Sciurus albipes quercinus* Nelson.

1899. *Sciurus griseoflavus chiapensis* Nelson, Proc. Washington Acad. Sci., 1:69, May 9, type from San Cristóbal, Chiapas.

1904. *Sciurus poliopus senex* Nelson, Proc. Biol. Soc. Washington, 17:148, October 6, type from La Salada, 40 mi. S Uruapan, Michoacán.

1904. *Sciurus poliopus perigrinator* Nelson, Proc. Biol. Soc. Washington, 17:149, October 6, type from Piaxtla, Puebla. Known only from type locality.

1906. *Sciurus poliopus tepicanus* J. A. Allen, Bull. Amer. Mus. Nat. Hist., 22:243, July 25, type from Rancho Palo Amarillo, near Amatlán de Cañas, Nayarit.

1907. *Sciurus socialis littoralis* Nelson, Proc. Biol. Soc. Washington, 20:87, December 11, type from Puerto Angel, Oaxaca. Known only from type locality.

MARGINAL RECORDS (Musser, 1968:102–106, unless otherwise noted).—Nayarit: 6 mi. S Ixtlán del Río, 6800 ft. Jalisco: Cerro Viejo de Magdalena, 3 mi. NE Magdalena, 6500 ft.; *2 mi. N Tequila, 3700 ft.;* Barranca Ibarra, 3000 ft. Michoacán: near Patambán, 9000 ft.; 12 mi. W Ciudad Hidalgo, 9150 ft. Hidalgo: 10 mi. NNE Zimapán, 7000 ft.; *13 mi. NE Metepec (Hwy. 53), 6600 ft.* Puebla: *Apulco, 5½ road mi. NW Zacapoaxtla, 1375 m.;* Scapa, 3 mi. NE Huauchinango, 4000–4600 ft. Veracruz: ½ mi. NE Las Minas, 1400 m.; *3 km. E Las Vigas, 8000 ft.;* 5 mi. N Jalapa; Xico. Puebla: Tochimilco, 7500 ft. Oaxaca: Reyes Pápalo, 6700 and 9200 ft.; near La Parada, 8500 ft.; Cerro Zempoaltepec; *3½ mi. E Matías Romero;* Chimalapa (Goodwin, 1969:133). Chiapas: El Suspiro, 5 mi. N Berriozábal, 4500 ft.; *3 mi. S San Fernando, Las Vistas;* near San Cristóbal de las Casas, 8200–8500 ft.; *San José, 28 mi. ESE Comitán, 4900 ft.;* Lagos Montebello, 4350 ft. Guatemala: La Primavera; Sierra de las Minas, La Montañita; San Lorenzo, 4 mi. NE Volcán de Jumway, 5900–6000 ft.; Antigua; San Lucas; Volcán de Santa María. Chiapas: 2 mi. N Unión Juárez, 400 ft.; *Distrito Soconusco, Finca Juárez (ca. 14 km. E Escuintla), 1200 m.;* Mapastepec, thence up coast to Colima: 23 mi. (Hwy. 80) SE Manzanillo, 50 m.; *Pueblo Juárez (Hda. La Magdalena), 1500 ft.* Jalisco: 1 mi. N Tapalpa, 7800 ft.; 13 mi. WSW Ameca, 5100 ft.; *Rancho Palo Amarillo, near Amatlán de Cañas, 5000 ft.*

Sciurus colliaei
Collie's Squirrel

External measurements: 440–578; 203–305; 57–73. Upper parts yellowish gray coarsely grizzled and heavily overlaid with black; sides paler than back; nape, shoulders, legs, and ears variable, but usually either dark gray or of some shade approximating rufous; underparts white; base of tail like back; remainder black above with wash of white, below grizzled grayish or blackish yellow edged with white. Two upper premolars except that in northern populations P3 is absent on one or both sides in more than half the specimens.

Collie's squirrels live on the lowlands along the west coast of México, seldom above 2500 ft.

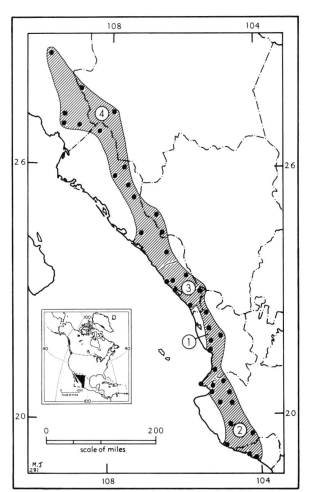

Map 281. *Sciurus colliaei.*

1. *S. c. colliaei*   3. *S. c. sinaloensis*
2. *S. c. nuchalis*   4. *S. c. truei*

Sciurus colliaei colliaei Richardson

1839. *Sciurus colliaei* Richardson, *in* Beechey, The zoology of Capt. Beechey's voyage . . . , p. 8, type from San Blas, Nayarit. (*Sciurus colliaei* Bachman, 1838, is a *nomen nudum.*)

MARGINAL RECORDS (Musser, 1968:106, 107, unless otherwise noted, as *S. colliaei* only).—Nayarit: Acaponeta, 200 ft.; *Platanares, 10 mi. E Ruiz;* 2 mi. E El Veando, 200 ft.; 5 mi. S Las Varas, 150 ft.; Banderas Bay, Mita Point; vic. San Blas; Santiago Ixcuintia, 200 ft.; 2 mi. SW Rosa Morada (Anderson, 1962:9). Not found: Sinaloa: Juan Lisiarraga Mountain; Los Peiles; ½ mi. by highway S Revolcaderos, 6600 ft.

Sciurus colliaei nuchalis Nelson

1899. *Sciurus colliaei nuchalis* Nelson, Proc. Washington Acad. Sci., 1:59, May 9, type from Manzanillo, Colima.

MARGINAL RECORDS (Musser, 1968:106, 107, unless otherwise noted, as *S. colliaei* only).—Jalisco: San Sebastián, 3000–7500 ft.; Mascota, 4700 ft.; 10 mi. SE Talpa de Allende, 5350 ft.; Sierra de Autlán, 20 mi. SSE Autlán, 6500 ft. Colima: 17 mi. (Hwy. 80) SE Manzanillo; *13 mi. (Hwy. 80) SE Manzanillo;* type locality. Jalisco: Tenacatita Bay; 20 km. WNW Purificación; Sierra de Cuale, 6500, 7300 ft.; Ixtapa, 300 ft.; Las Palmas, 1000 ft.

Sciurus colliaei sinaloensis Nelson

1899. *Sciurus sinaloensis* Nelson, Proc. Washington Acad. Sci., 1:60, May 9, type from Mazatlán, Sinaloa.
1962. *Sciurus colliaei sinaloensis,* Anderson, Amer. Mus. Novit., 2093:9, June 13.

MARGINAL RECORDS (Musser, 1968:107, as *S. colliaei* only).—Sinaloa: 5 km. SW Santa Lucía, 2150 ft.; *7 mi. ENE Plomosas, 6000 ft.;* 3 mi. SE Plomosas; Palmito; *Isla Palmito del Verde* (middle); 20 mi. SE Mazatlán; type locality; 8 km. N Villa Unión, 450 ft.

Sciurus colliaei truei Nelson

1899. *Sciurus truei* Nelson, Proc. Washington Acad. Sci., 1:61, May 9, type from Camoa, Río Mayo, Sonora.
1962. *Sciurus colliaei truei,* Anderson, Amer. Mus. Novit., 2093:12, June 13.

MARGINAL RECORDS (Musser, 1968:106–107, as *S. colliaei* only).—Sonora: San Javier. Chihuahua: Carimechi, Río Mayo; Barranca de Cobre, 4300 ft. Sinaloa: 15 km. N, 65 km. E Sinaloa, 4700 ft.; 20 km. N, 5 km. E Badiraguato; 13 mi. ESE Badiraguato. Durango: Chacala; *Santa Ana;* 12 km. E Cosalá, 1300 ft. Sinaloa: San Ignacio; 32 mi. SSE Culiacán; 10 km. S, 38 km. E Sinaloa, 800 ft.; 13 km. NNE Vaca, 1300 ft. Sonora: 8 mi. SE Alamos; Chinobampo; type locality.

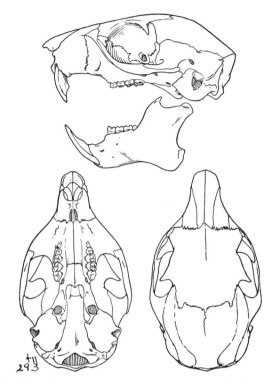

Fig. 293. *Sciurus yucatanensis yucatanensis*, Chichén-Itzá, Yucatán, No. 108179 U.S.N.M., ♂, X 1.

Sciurus yucatanensis
Yucatán Squirrel

External measurements: 450–500; 220–258; 53–65. Upper parts grizzled black and gray with suffusion of yellow to ochraceous buff; underparts varying from dirty white through a grizzled yellowish-gray to black; tail above black with wash of white, below with median stripe of dull gray fulvous or black and gray bordered with black and edged with white; ear tufts (sometimes present) dingy white; pelage thin, coarse, and stiff. Skull averages smaller (basal length *ca.* 45) than in neighboring species. Two upper premolars.

Dr. E. W. Nelson (1899:70,71) wrote that the distribution of this species was the arid tropical forests of the peninsula of Yucatan, and that the whitish ear tufts in certain pelages separated this species from the other Mexican and Central American squirrels of the subgenus *Echinosciurus*.

Sciurus yucatanensis baliolus Nelson

1901. *Sciurus yucatanensis baliolus* Nelson, Proc. Biol. Soc. Washington, 14:131, August 9, type from Apazote, Campeche.

MARGINAL RECORDS (Musser, 1968:108, unless otherwise noted, as *S. yucatanensis* only).—Campeche: S Campeche. Belize: Belize (Hall and Kelson, 1959:378); El Cayo. Guatemala: Remate. Tabasco: Monte Cristo Emiliano Zapata, 200 ft.; *12 mi. NW Balancán;* 22 mi. N Balancán. Campeche: 5 km. S Champotón.

Sciurus yucatanensis phaeopus Goodwin

1932. *Sciurus yucatanensis phaeopus* Goodwin, Amer. Mus. Novit., 574:1, October 22, type from Secanquim, 1600 ft., Alta Vera Paz, Guatemala.

MARGINAL RECORDS (Musser, 1968:108, as *S. yucatanensis* only).—Guatemala: Sayaxché; Secanquim; Finca Chama. Chiapas: 4 km. SW Sabana de San Quintín, 215 m.

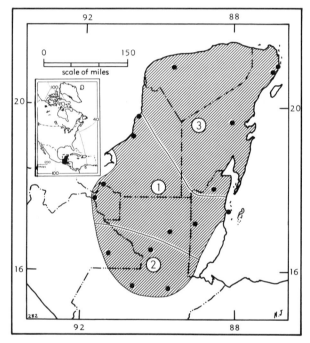

Map 282. *Sciurus yucatanensis.*

1. *S. y. baliolus*   2. *S. y. phaeopus*
              3. *S. y. yucatanensis*

Sciurus yucatanensis yucatanensis J. A. Allen

1877. [*Sciurus carolinensis*] var. *yucatanensis* J. A. Allen, *in* Coues and Allen, Monogr. N. Amer. Rodentia, p. 705, August, cotypes from Mérida, Yucatán.
1897. *Sciurus yucatanensis* J. A. Allen, Bull. Amer. Mus, Nat. Hist., 9:5, February 23.

MARGINAL RECORDS (Musser, 1968:108, unless otherwise noted, as *S. yucatanensis* only).—Quintana Roo: La Vega; Puerto Morelos; Felipe Carrillo Puerto, 30 m. Belize: Orange Walk (Hall and Kelson, 1959:378). Yucatán: Mérida. Not found: Guatemala: Pacomon.

Sciurus variegatoides
Variegated Squirrel

Revised by Harris, Miscl. Publ. Mus. Zool., Univ. Michigan, 38:7–39, September 4, 1937.

External measurements: 510–560; 240–305; 60–70. Weight, 450–520 grams. Upper parts highly variable, from blackish to grizzled yellowish gray; tail black above with heavy wash of white, sometimes appearing faintly annulated; tail below with median area of tawny to dark rufous, bordered with black and edged with white; underparts white to rich cinnamon-buff; when underparts not entirely white, white inguinal, axillary, and gular patches usually present; rump patch absent; dorsal patch on shoulder usually absent, and when present faintly expressed; pelage shiny, coarse, and bristly. S. variegatoides where its range meets that of S. aureogaster (the ranges meet only in Chiapas and Guatemala) is distinguished by feet whitish slightly speckled with black (not frosted buffy black or blackish as in aureogaster), and underparts white, rarely pigmented (instead of orange or chestnut, often black-flecked). S. variegatoides differs from the adjacent S. yucatanensis in larger skull and lack of black postauricular patches. The white underparts or white patches thereon distinguish S. variegatoides from S. aureogaster. Two upper premolars.

Sciurus variegatoides adolphei (Lesson)

1842. Macroxus adolphei Lesson, Nouveau tableau du régne animal . . . mammiferes, p. 112, type from Realejo, Nicaragua.
1920. Sciurus variegatoides adolphei, Goldman, Smiths. Miscl. Coll., 69(5):136, April 24.

?1905. Sciurus boothiae annalium Thomas, Ann. Mag. Nat. Hist., ser. 7, 16:309, September, type from Honduras.

MARGINAL RECORDS.—Nicaragua: Volcán Chinandega [= Volcán El Viejo]; Corinto; type locality.

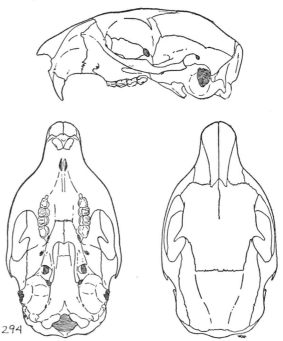

Fig. 294. Sciurus variegatoides rigidus, Agua Caliente, Cartago, Costa Rica, No. 16533 K.U., ♂, X 1.

Sciurus variegatoides atrirufus Harris

1930. Sciurus adolphei atrirufus Harris, Occas. Pap. Mus. Zool., Univ. Michigan, 219:2, October 15, type from Tambor, Nicoya Peninsula, Costa Rica. Known only from type locality.
1937. Sciurus variegatoides atrirufus Harris, Miscl. Publ. Mus. Zool., Univ. Michigan, 38:19, September 4.

Sciurus variegatoides bangsi Dickey

1928. Sciurus variegatoides bangsi Dickey, Proc. Biol. Soc. Washington, 41:7, February 1, type from Barra de Santiago, Dept. Ahuachapán, El Salvador.

MARGINAL RECORDS (Musser, 1968:108, unless otherwise noted, as S. variegatoides only).—Guatemala: Chiquimula, Linda Vista, 200 ft. El Salvador (Hall and Kelson, 1959:379): San José del Sacare, 3600 ft.; Hda. Chilata; type locality. Guatemala: Astillero (42 mi. SE Esquintla, on coast), 25 ft.; Finca El Cacahuito, Taxisco, 1300 ft.; Amayito.

Sciurus variegatoides belti Nelson

1899. Sciurus boothiae belti Nelson, Proc. Washington Acad. Sci., 1:78, May 9, type from Escondido River, 50 mi. from Bluefields, Nicaragua.

1937. *Sciurus variegatoides belti,* Harris, Miscl. Publ. Mus. Zool., Univ. Michigan, 38:13, September 4.

MARGINAL RECORDS.—Honduras: Yoro. Nicaragua: Greytown; Los Sábalos, Río San Juan; Chontales [Dept. of?]; Matagalpa; Río Coco (Harris, 1937:16). Not found: Honduras: Lancetilla; Ca[r]melina.

Sciurus variegatoides boothiae Gray

1843. *Sciurus boothiae* Gray, List of the . . . Mammalia in the . . . British Museum, p. 139, type from Honduras (restricted to San Pedro Sula by Nelson, Proc. Washington Acad. Sci., 1:77, May 9, 1899).
1937. *Sciurus variegatoides boothiae,* Harris, Miscl. Publ. Mus. Zool., Univ. Michigan, 38:12, September 4.
1842. *Sciurus richardsoni* Gray, Ann. Mag. Nat. Hist., ser. 1, 10:264. Not *Sciurus richardsoni* Bachman, 1838.
1845. *Sciurus fuscovariegatus* Schinz, Synopsis mammalium, 2:15.

MARGINAL RECORDS.—Honduras: type locality. Nicaragua: Jalapa; San Juan [de Murra]; Jícaro. Honduras: El Jaral. Guatemala: *along border of Honduras, on Caribbean Coast.* Not found: Laguna, Honduras.

Sciurus variegatoides dorsalis Gray

1849. *Sciurus dorsalis* Gray, Proc. Zool. Soc. London, for 1848, p. 138, June 1, type erroneously assumed to be from

Caracas, Venezuela. Nelson (Proc. Washington Acad. Sci., 1:7, May 9, 1899) regards specimens from Liberia, Costa Rica, as typical.
1920. *Sciurus variegatoides dorsalis,* Goldman, Smiths. Miscl. Coll., 69(5):136, April 24.

MARGINAL RECORDS.—Nicaragua: Tipitapa. Costa Rica: Miravalles; Chomes; Las Huacas; San Juanillo; La Cruz.

Sciurus variegatoides goldmani Nelson

1898. *Sciurus goldmani* Nelson, Proc. Biol. Soc. Washington, 12:149, June 3, type from Huehuetán, Chiapas.
1928. *Sciurus variegatoides goldmani,* Dickey, Proc. Biol. Soc. Washington, 41:8, February 1.

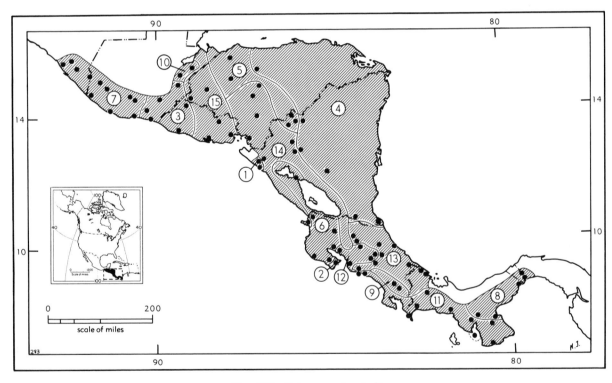

Map 283. *Sciurus variegatodies.*

Subspecies	4. S. v. belti	8. S. v. helveolus	12. S. v. rigidus
1. S. v. adolphei	5. S. v. boothiae	9. S. v. loweryi	13. S. v. thomasi
2. S. v. atrirufus	6. S. v. dorsalis	10. S. v. managuensis	14. S. v. underwoodi
3. S. v. bangsi	7. S. v. goldmani	11. S. v. melania	15. S. v. variegatoides

MARGINAL RECORDS (Musser, 1968:107, 108, unless otherwise noted, as *S. variegatoides* only).— Chiapas: Finca Esperanza (*ca.* 4 mi. NE Escuintla); 8 km. NE Huixtla, 160 m.; Cacahuatán. Guatemala: Finca Carolina, 3500 ft.; Finca Ciprés, 2000 ft.; San Pedro Yepocapa, 4900 ft; Finca El Zapote, 3200 ft.; San José (25 mi. S Esquintla, on coast); Concepción del Mar; Hda. California, sea level. Chiapas: 19 km. S Mapastepec, 25 ft.

Sciurus variegatoides helveolus Goldman

1912. *Sciurus variegatoides helveolus* Goldman, Smiths. Miscl. Coll., 56(36):3, February 19, type from Corozal, Canal Zone, Panamá.

MARGINAL RECORDS.—Panamá (Handley, 1966:777): Summit; Panamá; La Chorrera; Parita; Santiago.

Sciurus variegatoides loweryi McPherson

1972. *Sciurus variegatoides loweryi* McPherson, Rev. Biol. Trop., 19:191, February 22, type from 1 road mile S Paso Real, Río Escuadra, 150 m., Costa Rica.

MARGINAL RECORDS.—Costa Rica (McPherson, 1972:191, 192): 1 km. WNW Buenos Aires, 365 m.; type locality.

Sciurus variegatoides managuensis Nelson

1898. *Sciurus boothiae managuensis* Nelson, Proc. Biol. Soc. Washington, 12:150, June 3, type from Río Managua, Guatemala.
1937. *Sciurus variegatoides managuensis*, Harris, Miscl. Publ. Mus. Zool., Univ. Michigan, 38:17, September 4.

MARGINAL RECORDS.—Guatemala: Quiragua; "probably" Zacapa. Not found: Bobós.

Sciurus variegatoides melania (Gray)

1867. *Macroxus melania* Gray, Ann. Mag. Nat. Hist., ser. 3, 20:425, December, type from Point Burica, Costa Rica.
1920. *Sciurus variegatoides melania*, Goldman, Smiths. Miscl. Coll., 69(5):136, April 24.

MARGINAL RECORDS.—Costa Rica (McPherson, 1972:193): *Playa Río Parrita; Sardinal; Surubres*; Parrita, Finca La Ligia; Quepos. Panamá (Handley, 1966:777): Boquete; Remedios; Soná; Pesé; Guánico; Isla Cébaco; Divalá. Costa Rica: type locality.

Sciurus variegatoides rigidus Peters

1863. *Sciurus rigidus* Peters, Monatsb. preuss. Akad. Wiss., Berlin, p. 652, type from San José, Costa Rica.
1937. *Sciurus variegatoides rigidus*, Harris, Miscl. Publ. Mus. Zool., Univ. Michigan, 38:22, September 4.
1867. *Sciurus intermedius* Gray, Ann. Mag. Nat. Hist., ser. 3, 20:421, type from Guatemala.
1867. *Macroxus nicoyana* Gray, *op. cit.*:423, type from Nicoya, Costa Rica.
1933. *Sciurus variegatoides austini* Harris, Occas. Pap. Mus. Zool., Univ. Michigan, 266:1, June 28, type from Las Agujas, Puntarenas, Costa Rica. Regarded as inseparable from *S. v. rigidus* by Hall and Kelson, Univ. Kansas Publ., Mus. Nat. Hist., 5:356, December 15, 1952.

MARGINAL RECORDS.—Costa Rica: Zarcéro; Juan Viñas; San Isidro; Las Agujas; Chomes.

Sciurus variegatoides thomasi Nelson

1899. *Sciurus thomasi* Nelson, Proc. Washington Acad. Sci., 1:71, May 9, type from Talamanca, Costa Rica.
1937. *Sciurus variegatoides thomasi*, Harris, Miscl. Publ. Mus. Zool., Univ. Michigan, 38:24, September 4.

MARGINAL RECORDS.—Costa Rica: San Carlos; Guápiles; Pacuare; Cuábre. Panamá (Handley, 1966:777): 7 km. SSW Changuinola; Almirante. Costa Rica: Pozo Azul; La Carpintera; Perálta; Villa Quesada.

Sciurus variegatoides underwoodi Goldman

1932. *Sciurus boothiae underwoodi* Goldman, Jour. Washington Acad. Sci., 22(10):275, May 19, type from Monte Redondo, 5100 ft., "about" 30 mi. NW Tegucigalpa, Honduras.
1937. *Sciurus variegatoides underwoodi*, Harris, Miscl. Publ. Mus. Zool., Univ. Michigan, 38:9, September 4.

MARGINAL RECORDS.—Honduras: El Caliche Cedros. Nicaragua: San Rafael del Norte; Matagalpa. Costa Rica: Port Parker Bay. Nicaragua: Volcán El Viejo. Honduras: Cantarranas; Sabana Grande; type locality.

Sciurus variegatoides variegatoides Ogilby

1839. *Sciurus variegatoides* Ogilby, Proc. Zool. Soc. London, p. 117, December, type from El Salvador.
1842. *Macroxus pyladei* Lesson, Nouveau tableau du régne animal . . . mammifères, p. 112, type from San Carlos, El Salvador.
1843. *Sciurus griseocaudatus* Gray, in The zoology of the voyage of H.M.S. Sulphur . . . , 2(Mamm.):34, type from west coast of Central America.

MARGINAL RECORDS.—Guatemala: *highlands along El Salvador–Honduras border*. Honduras: Las Flores. El Salvador: Monte Cacaguatique; La Unión; Puerto del Triunfo. Honduras: Plan del Rancho.

Sciurus deppei
Deppe's Squirrel

External measurements: 343–387; 160–190; 54–58. Upper parts varying from grizzled dark rusty brown to yellowish brown and even grayish brown; outside of legs and feet usually dark gray; tail above black thinly washed with white, below varying from ochraceous to rich ferruginous, bordered with black and edged with white or pale yellow; underparts usually white or yellowish to dull rufous, but buffy patches of inguinal, axillary, and·gular regions extend well onto venter of some specimens. Small size distinguishes this squirrel from other species of subgenus. Two upper premolars.

Sciurus deppei usually lives in dense vegetation in humid lowlands and in dry more open forests of higher elevations.

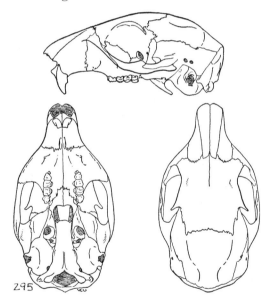

Fig. 295. *Sciurus deppei deppei*, 9 km. E Papantla, 300 ft., Veracruz, No. 23952 K.U., ♂, × 1.

Sciurus deppei deppei Peters

1863. *Sciurus deppei* Peters, Monatsb. preuss. Akad. Wiss., Berlin, p. 654, type from Papantla, Veracruz.
1867. *Macroxus tephrogaster* Gray, Ann. Mag. Nat. Hist., ser. 3, 20:431, type from México.
1867. *Macroxus taeniurus* Gray, *ibid.*, type from Guatemala.

MARGINAL RECORDS.—Veracruz: 12½ mi. N Tihuatlán, 300 ft. (Hall and Dalquest, 1963:271); 3 km.

W Guttierez [= Gutiérrez] Zamora, 300 ft. (*ibid.*); Catemaco. Chiapas: Tumbalá; *6 mi. SE Palenque*. Guatemala: La Perla; Secanquim. Honduras: Catacombas; Las Peinitas; Rancho Quemado; Comayaguela. El Salvador: Cerro Verde (Burt and Stirton, 1961:52). Guatemala: Volcán San Lucas; Finca Carolina. Chiapas: Cerro Ovando, 1800–2100 m.; Triunfo, 1950 m.; Ocuilapa. Oaxaca: mts. near Santo Domingo; Reyes. Veracruz: Córdoba; Jico; Las Vigas. Puebla: Huachinango. Veracruz: Zacualpan, 6000 ft. (Hall and Dalquest, 1963:271). Puebla: Metlaltoyuca.

Sciurus deppei matagalpae J. A. Allen

1908. *Sciurus deppei matagalpae* J. A. Allen, Bull. Amer. Mus. Nat. Hist., 24:660, October 13, type from San Rafael del Norte, Nicaragua.

MARGINAL RECORDS.—Nicaragua: Río Coco; Peña Blanca; type locality.

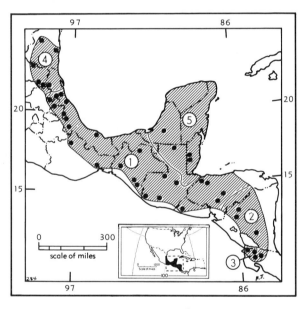

Map 284. *Sciurus deppei.*

1. *S. d. deppei*   3. *S. d. miravallensis*
2. *S. d. matagalpae*   4. *S. d. negligens*
5. *S. d. vivax*

Sciurus deppei miravallensis Harris

1931. *Sciurus miravallensis* Harris, Occas. Pap. Mus. Zool., Univ. Michigan, 277:1, June 4, type from Volcán de Miravalles, 1500 ft., Cordillera de Guanacaste, Costa Rica.
1943. *Sciurus deppei miravallensis* Harris, Occas. Pap. Mus. Zool., Univ. Michigan, 476:10, October 8.

MARGINAL RECORDS.—Costa Rica: Hda. Alemania; Hda. Santa María; Vijagua; type locality; *Volcán Orosi*.

Sciurus deppei negligens Nelson

1898. *Sciurus negligens* Nelson, Proc. Biol. Soc. Washington, 12:147, June 3, type from Altamira, Tamaulipas.
1953. *Sciurus deppei negligens*, Hooper, Occas. Pap. Mus. Zool., Univ. Michigan, 544:4, March 25.

MARGINAL RECORDS.—Tamaulipas (Alvarez, 1963:424): 9½ mi. SW Padilla; Mesa de Llera, 10 mi. NE Zamorina. Veracruz: Platón Sánchez, 800 ft. (Hall and Dalquest, 1963:272). San Luis Potosí: *12 mi. E Tamazunchale;* 1 mi. S Tamazunchale; Cerro Conejo; 10 km. E Platanito. Tamaulipas (Alvarez, 1963:424): *Rancho Santa Rosa; 3 mi. NE Guemes.*

Sciurus deppei vivax Nelson

1901. *Sciurus deppei vivax* Nelson, Proc. Biol. Soc. Washington, 14:131, August 9, type from Apazote, Campeche.

MARGINAL RECORDS.—Peninsula of Yucatán southward to Campeche: type locality. Belize: near El Cayo; transition area bordering Mountain Pine Ridge. Guatemala: Uaxactún; *Petén.*

Sciurus niger
Fox Squirrel

Subspecies occurring in United States reviewed by Osgood, Proc. Biol. Soc. Washington, 20:44–47, April 18, 1907; southern subspecies reviewed by Lowery and Davis, Occas. Pap. Mus. Zool., Louisiana State Univ., 9:153–172, March 4, 1942.

External measurements: 454–698; 200–330; 51–82. Highly variable in color both individually and geographically. There are three basic color phases, red or buff, gray, and black. Although these are individual color phases, one of them usually predominates locally. In one phase upper parts may be buffy or one of several shades of gray, but usually pale, sometimes with strong suffusion of black; toes, nose, and sometimes ears and tip of tail marked with white; tail above and below essentially same as upper parts and under-

parts of body, respectively; underparts like back but seldom with pronounced black suffusion. In the second color phase, upper parts yellowish or yellowish brown suffused with varying amounts of black; toes, nose, ears and tail usually not marked with white; underparts ordinarily yellowish to rufous, but in some individuals darker and in others whitish. Melanism, both partial and complete, is frequent in each of the two more nearly "normal" color phases. One upper premolar.

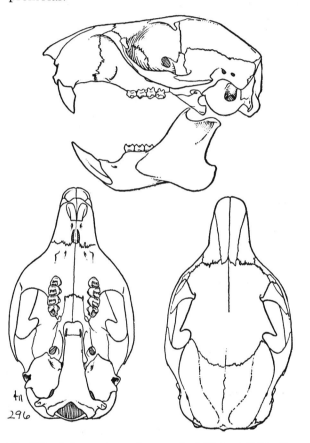

Fig. 296. *Sciurus niger rufiventer*, 6 mi. W, 6 mi. S Oberlin, Decatur Co., Kansas, No. 19027 K.U., ♂ X 1.

Sciurus niger avicennia A. H. Howell

1919. *Sciurus niger avicennia* A. H. Howell, Jour. Mamm., 1:37, November 28, type from Everglade[s], Collier Co., Florida.

MARGINAL RECORDS.—Florida: Fort Myers; Immokalee; The Big Cypress, Collier Co., few mi. from borders of Hendry and Broward counties; Miami (Moore, 1956:61); Florida City; Chokoloskee; type locality.—Layne (1974:392 remarks that *S. n. avicennia* "occupies the area west and south of the Everglades, typical *shermani* occurs south of the St. Lucie River and east of Lake Okeechobee, and the population of the Miami pinelands represents intergrades" between the two subspecies.

Sciurus niger bachmani Lowery and Davis

1942. *Sciurus niger bachmani* Lowery and Davis, Occas. Pap. Mus. Zool., Louisiana State Univ., 9:156, March 4, type from 10 mi. NW Enon, Washington Parish, Louisiana.

MARGINAL RECORDS.—Mississippi: Michigan City. Alabama: Sand Mtn., near Carpenter; Piedmont; Castleberry; Orange Beach. Louisiana: Nott; *7 mi. S Walker* (Lowery, 1974:187); *10 mi. E Baton Rouge; 2½ mi. NE Weyanoke (loc. cit.)*. Mississippi: 5 mi. N Woodville; Minter City.

Sciurus niger cinereus Linnaeus

1758. [*Sciurus*] *cinereus* Linnaeus, Syst. nat., ed. 10, p. 64. Type locality restricted to Cambridge, Dorchester Co., Maryland, by Barkalow (Proc. Biol. Soc. Washington, 69:13, May 21, 1956).
1877. *Sciurus niger* Var. *cinereus*, J. A. Allen, *in* Coues and Allen, Monog. N. Amer. Rodentia, p. 717, August.
1867. *Macroxus neglectus* Gray, Ann. Mag. Nat. Hist., ser. 3, 20:425, December, type from Wilmington, Newcastle Co., Delaware.
1920. *Sciurus niger bryanti* H. H. Bailey, Bull. Bailey Mus. and Lib. Nat. Hist., 1:1, August 1(?), type from Dorchester County, Maryland.

MARGINAL RECORDS.—Delaware: Wilmington (type locality of *Macroxus neglectus*). Maryland: type locality.

Sciurus niger limitis Baird

1855. *Sciurus limitis* Baird, Proc. Acad. Nat. Sci. Philadelphia, 7:331, April, type from Devils River, Valverde Co., Texas.
1907. S[*ciurus*]. n[*iger*]. *limitis*, Osgood, Proc. Biol. Soc. Washington, 20:45, April 18.

MARGINAL RECORDS.—Texas: Vernon; Montague County (Dalquest, 1968:16); Giddings; 30 mi. SW Eagle Lake; Aransas Refuge; mouth Nueces River. Coahuila: 2 mi. S, 3 mi. E San Juan de Sabinas, 1160 ft.; 8 mi. N, 4 mi. W Muzquiz, 1800 ft.; Fortín, 330 ft., 33 mi. N, 1 mi. E San Gerónimo. Texas: Independence Creek; 4 mi. N Bomarton (Baccus, 1971:181).

Sciurus niger ludovicianus Custis

1806. *Sciurus ludovicianus* Custis, Philadelphia Med. Phys. Jour., 2:47. Type locality, Red River of Louisiana; restricted to Natchitoches Parish by Lowery and Davis (Occas. Pap. Mus. Zool., Louisiana State Univ., 9:164, March 4, 1942).
1839. *Sciurus texianus* Bachman, Proc. Zool. Soc. London, p. 86, for 1838, February 7, type from "Mexico."
1877. [*Sciurus niger*] var. *ludovicianus*, J. A. Allen, *in* Coues and Allen, Monog. N. Amer. Rodentia, p. 718, August.

MARGINAL RECORDS.—Arkansas: Delight. Louisiana (Lowery, 1974:189, as intergrades with *S. n. subauratus*): 7 mi. W Beekman; *5 mi. N Monroe; 5 mi. S Monroe;* Jena; 4 mi. N Bunkie; *Thistlethwaite Game Refuge; Lafayette Parish;* Perry. Texas: near Beaumont; 2 mi. E Sheldon; Matagorda County (Davis, 1966:147, as *S. niger* only); Brenham; Milano; Groesbeck; Lamar County (Davis, 1966:147, as *S. niger* only); Texarkana.

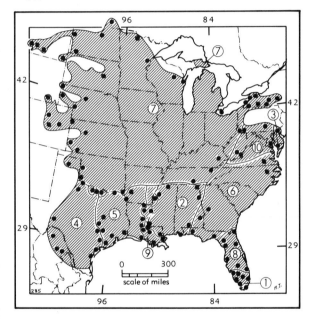

Map 285. *Sciurus niger.*

1. *S. n. avicennia*
2. *S. n. bachmani*
3. *S. n. cinereus*
4. *S. n. limitis*
5. *S. n. ludovicianus*
6. *S. n. niger*
7. *S. n. rufiventer*
8. *S. n. shermani*
9. *S. n. subauratus*
10. *S. n. vulpinus*

Sciurus niger niger Linnaeus

1758. [*Sciurus*] *niger* Linnaeus, Syst. nat., ed. 10, 1:64 (based on Catesby's Black Fox Squirrel). Type locality, probably southern South Carolina.
1758. [*Sciurus*] *cinereus* Linnaeus, Syst. nat., ed. 10, 1:64. Type locality, "America septentrionali"—may have applied in part to this subspecies, *niger*. Because of the early uncertainty of the application of the name, *cinereus* has been employed by one author or another as the name of almost every subspecies of both *S. niger* and *S. carolinensis*, although probably none of the post-Linnaean authors intended that *his* usage was to be regarded as a *new* name. Erroneous employment of the name arose through both misunderstanding of the original application and misidentification of specimens. No effort is made herein to allocate properly the many different usages of the name since several of them are of uncertain application. The name has been used in 12 or more senses. Now, of course, the name [*Sciurus*] *cinereus* Linnaeus is applied to the subspecies of *Sciurus niger* occurring at Cambridge, Dorchester Co., Maryland.

1802. *Sciurus capistratus* Bosc, Bull. Soc. Philom. Paris, No. 67, 3(7):145, October, type from Charleston, South Carolina.

MARGINAL RECORDS.—Virginia: Prince George County. North Carolina: Tarboro. Florida: Jefferson County; Okaloosa County. Alabama: Abbeville. Georgia: 1 mi. W Junction City.

Sciurus niger rufiventer É. Geoffroy St.-Hilaire

1803. *Sciurus rufiventer* É. Geoffroy St.-Hilaire, Catalogue des mammifères du Muséum National d'Histoire Naturelle, Paris, p. 176. Type locality, Mississippi Valley probably between southern Illinois and central Tennessee (Osgood, Proc. Biol. Soc. Washington, 20:44, April 18, 1907). For availability of this name of 1803 see remarks on page 72 concerning *Scalopus*.
1907. *Sciurus niger rufiventer*, Osgood, Proc. Biol. Soc. Washington, 20:44, April 18.
1820. *Sciurus ruber* Rafinesque, Ann. Nat., p. 4. Type locality, Missouri Territory.
1823. *Sciurus macroura* Say, *in* Long, Account of an exped. . . . to the Rocky Mts. . . . , 1:115. Type locality, northeastern Kansas. Not *Sciurus macrourus* Pennant, 1769 [= *Ratufa macroura macroura* from highlands of Ceylon].
1825. *Sciurus magnicaudatus* Harlan, Fauna Americana, p. 178, a renaming of *S. macroura* Say.
1851. *Sciurus rubicaudatus* Audubon and Bachman, The viviparous quadrupeds of North America, 2:pl. 55; text in letterpress, 2:30. Type locality, Kentucky.
1851. *Sciurus sayii* Audubon and Bachman, The viviparous quadrupeds of North America, 2:pl. 89, text in letterpress, 2:274. Type locality, somewhere in bottomlands of Wabash, Illinois, or Missouri rivers, or in Michigan.

MARGINAL RECORDS.—Manitoba: 5½ mi. W St. Claude (Wrigley, *et al.*, 1973:782). Minnesota: Koochiching County; 1 mi. S Barnum (Timm, 1975:38). Wisconsin: Menominee Indian Reservation; Sauble [= Sable ?] Point (Jackson, 1961:169), along southern shores of Great Lakes to Lake Erie. Ontario: Pelée Island (Banfield, 1962b:120, introduced). New York (Hamilton, 1963:124, 125): Dunkirk; Orchard Park; 6 mi. S Holley; Morris; Brooktondale; Waverly; Wellsville; Jamestown. Pennsylvania: Greene County. Kentucky: Jackson. Tennessee: Highcliff, 3 mi. E Jellico; 3 mi. S Fayetteville. Oklahoma: 10 mi. SE Broken Bow. Texas: Gainesville; Clay County (Davis, 1966:147, as *S. niger* only). Oklahoma: 10 mi. S, ½ mi. E Gould (Martin and Preston, 1970:50); *10 mi. N, ½ mi. W Hollis* (*ibid.*, sight record); 3 mi. W Cheyenne. Texas: Stinnett. Kansas: 14 mi. SW Meade; sec. 27, R. 39 W, T. 25 S, Hamilton Co.; Trego County; 23 mi. NW St. Francis. Colorado: Limon (Armstrong, 1972:145); Colorado Springs (*ibid.*); Denver (Borell, 1961:101, as *S. niger* only). Wyoming: Laramie (Long, 1965a:598). Nebraska (Jones, 1964c:148, 149): 5 mi. WSW Ogallala; Scotts Bluff National Monument; Thomas County; Monroe Canyon. South Dakota: 2 mi. S, 3 mi. E Fort Thompson, 1370 ft. Montana: sec. 16, T. 5 S, R. 62 E (Lampe, *et al.*, 1974:13). North Dakota: 1½ mi. W

Reeder. Montana (Hoffmann, *et al.*, 1969:589, as *S. niger* only): 7 mi. S Hardin, Big Horn River; Reed Point; Billings. North Dakota: 2 mi. S Dore; Pleasant Lake. Introduced in Upper Peninsula of Michigan (Burt, *in verbis*). Range extended into northeastern Colorado through natural invasion and introduction.

Sciurus niger shermani Moore

1956. *Sciurus niger shermani* Moore, Amer. Midland Nat., 55:56, January, type from 2 mi. E Univ. Florida Conservation Reserve, Welaka, Putnam Co., Florida.

MARGINAL RECORDS.—Florida: Nassau County; type locality; Jupiter; Highlands County; Hillsborough County; Levy County; Gilchrist County.— See terminal sentence under account of *S. n. avicennia* for Layne's (1974:392) statement on southern margin of range of *S. n. shermani*.

Sciurus niger subauratus Bachman

1839. *Sciurus subauratus* Bachman, Proc. Zool. Soc. London, p. 87, for 1838, February 7. Type locality restricted to Iberville Parish, Louisiana, by Lowery and Davis, Occas. Pap. Mus. Zool., Louisiana State Univ., 9:166, March 4, 1942.
1942. *Sciurus niger subauratus*, Lowery and Davis, Occas. Pap. Mus. Zool., Louisiana State Univ., 9:166, March 4.
1839. *Sciurus auduboni* Bachman, *ibid.*, p. 97, February 7. Type obtained at market, New Orleans, Lousiana; melanistic specimen of *S. n. subauratus*.

MARGINAL RECORDS.—Mississippi: Moorehead. Louisiana (Lowery, 1974:187–189): Tallulah; *5 mi. E Angola* (as intergrade with *S. n. bachmani*); *Bains* (as intergrade with *S. n. bachmani*); University [LSU, Baton Rouge]; *4 mi. S Whitehall*; Yscloskey; *Lake Hermitage*; 3 mi. SE Houma; *near Weeks Island*; *5 mi. W New Iberia, 2 mi. S Louisiana Hwy. 14*; 5 mi. N Port Barre; *5 mi. S Cottonport* (as intergrade with *S. n. ludovicianus*); *3 mi. W Marksville* (as intergrade with *S. n. ludovicianus*); 7 mi. SE Monroe (as intergrade with *S. n. ludovicianus*); *Swartz*; *3 mi. SE Bastrop*.

Sciurus niger vulpinus Gmelin

1788. [*Sciurus*] *vulpinus* Gmelin, Syst. nat., ed. 13, p. 147, based on specimens from the eastern United States including Blue Mountains (of Pennsylvania).
1954. *Sciurus niger vulpinus*, Barkalow, Jour. Elisha Mitchell Sci. Soc., 70:25, June.
1896. *Sciurus ludovicianus vicinus* Bangs, Proc. Biol. Soc. Washington, 10:150, December 28, type from White Sulphur Springs, West Virginia.

MARGINAL RECORDS.—Pennsylvania: Carlisle. Maryland: Calvert County. West Virginia: White Sulphur Springs; Lewisburg. Pennsylvania: Rothruck.

Sciurus oculatus
Peters' Squirrel

External measurements: 530–560; 254–269; 68–73. Upper parts either uniform grizzled gray or with median band or strong suffusion of black; ears and orbital ring dull white to buffy; tail above black with heavy white suffusion; tail below grizzled gray to yellowish brown bordered with black and edged with white; underparts varying from white with pale yellowish suffusion to rich ochraceous buff. Superficially this species resembles *S. carolinensis,* especially the more southern subspecies of the eastern gray squirrel. The most obvious difference is the absence of P3 in *S. oculatus.* Actual relationships genetically are thought to be more with *S. niger* than with *S. carolinensis. S. alleni* and *S. oculatus* closely resemble each other, but *oculatus* can be distinguished as follows: larger; venter usually buffy, rather than white; and postauricular patch prominent (buffy in *S. o. shawi* and dirty white in *S. o. tolucae*) instead of wanting or only faintly marked in pale grayish in *S. alleni.*

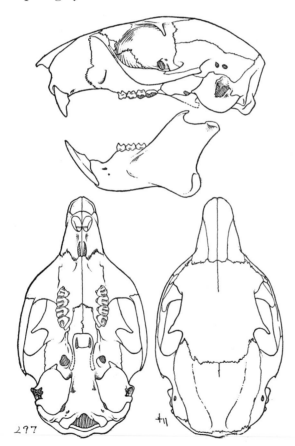

Fig. 297. *Sciurus oculatus oculatus,* Tulancingo, Hidalgo, No. 55609 U.S.N.M., ♂, × 1.

S. oculatus is most abundant in the pine forests of the Transition and Boreal life-zones, but occurs at lower elevations (5000 ft.) in oak forests.

Sciurus oculatus oculatus Peters

1863. *Sciurus oculatus* Peters, Monatsb. preuss. Akad. Wiss., Berlin, p. 653, type from México, probably near Las Vigas, Veracruz.
1830. *Sciurus capistratus* Lichtenstein, Abh. k. Akad. Wiss., Berlin, p. 116. Based on a North American *Sciurus* lacking a red belly. Not *Sciurus capistratus* Bosc, 1802.
1890. *Sciurus niger melanonotus* Thomas, Proc. Zool. Soc. London, p. 73, June, type from Las Vigas, Veracruz.

MARGINAL RECORDS.—Querétaro: Pinal de Amoles. Veracruz: Las Vigas. Puebla: Mt. Orizaba. Hidalgo: Tulancingo; Real del Monte; Encarnación.

Sciurus oculatus shawi Dalquest

1950. *Sciurus oculatus shawi* Dalquest, Occas. Pap. Mus. Zool., Louisiana State Univ., 23:4, July 10, type from Rancho San Francisco, 38 km. ESE city of San Luis Potosí, San Luis Potosí.

MARGINAL RECORDS.—San Luis Potosí: Villar; vic. Cerro Conejo at Llano de Garzas; Cerro Campanario; type locality.

Sciurus oculatus tolucae Nelson

1898. *Sciurus oculatus tolucae* Nelson, Proc. Biol. Soc. Washington, 12:148, June 3, type from N slope Volcáno de Toluca, State of México.

MARGINAL RECORDS.—Guanajuato: Guanajuato City. Querétaro: Tequisquiapan. Distrito Federal: Parres. México: type locality.

Sciurus alleni Nelson
Allen's Squirrel

1898. *Sciurus alleni* Nelson, Proc. Biol. Soc. Washington, 12:147, June 3, type from Monterrey, Nuevo León.

External measurements: 415–493; 175–235; 60–68. Upper parts yellowish brown, lightly grizzled, and darker medially than on sides; postauricular patch wanting or faintly grayish; tail above black washed with white, sometimes with buffy undertone; tail below, except for basal band like upper parts, yellowish brown to yellowish gray bordered with black and edged with white; underparts white or nearly so. See remarks under *S. oculatus.*

Allen's squirrels inhabit forests of oak, pine, and madroña mainly between 3000 and 5000 ft. but are reported at altitudes of 2000–8200 ft.

MARGINAL RECORDS.—Nuevo León: type locality; Río de San Juan; Linares, 1200 ft. Tamaulipas (Al-

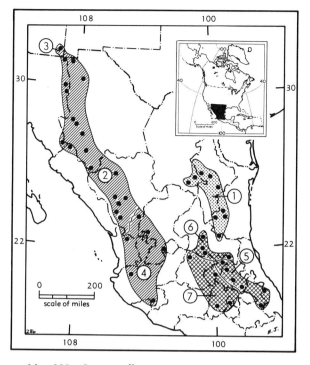

Map 286. *Sciurus alleni*, *Sciurus nayaritensis*, and *Sciurus oculatus*.

Guide to kinds
1. *S. alleni*
2. *S. n. apache*
3. *S. n. chiricahuae*
4. *S. n. nayaritensis*
5. *S. o. oculatus*
6. *S. o. shawi*
7. *S. o. tolucae*

varez, 1963:425): near Victoria; Joya Verde, 35 km. SW Ciudad Victoria. San Luis Potosí: 2 mi. S Pendencia. Tamaulipas: near Miquihuana. Coahuila: 10 mi. S General Cepeda; Diamante Pass, 8200 ft., Sierra Guadalupe, 4 mi. E, 3 mi. S Saltillo.

Sciurus nayaritensis
Nayarit Squirrel

Revised by Lee and Hoffmeister, Proc. Biol. Soc. Washington, 76:181–190, August 2, 1963.

External measurements: 530–575; 237–298; 70–84. Upper parts grayish washed with buff or yellow, and in southern populations washed with white. Underparts ochraceous (whitish in southern populations). One upper premolar. The close resemblance between this species and *S. niger* has caused some mammalogists to consider treating the two as a single species. *S. nayaritensis* in the western Sierra Madre of the Republic of México, lives mostly at higher elevations than does *Sciurus colliaei*. See Map 286.

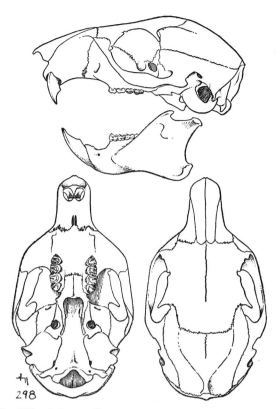

Fig. 298. *Sciurus alleni*, Diamante Pass, 4 mi. E, 3 mi. S Saltillo, 8200 ft., Coahuila, No. 35741 K.U., ♀, X 1.

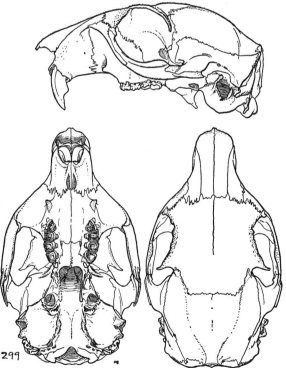

Fig. 299. *Sciurus nayaritensis chiricahuae*, Chiricahua Mts., 8200 ft., Cochise Co., Arizona, No. 64959 M.V.Z., ♀, X 1.

Sciurus nayaritensis apache J. A. Allen

1893. *Sciurus apache* J. A. Allen, Bull. Amer. Mus. Nat. Hist., 5:29, March 16, type from W slope Sierra de Nacori, 6300 ft., in eastern Sonora (van Rossem, Jour. Mamm., 17:417, November 16, 1936).
1963. *Sciurus nayaritensis apache*, Lee and Hoffmeister, Proc. Biol. Soc. Washington, 76:188, August 2.

MARGINAL RECORDS.—Sonora: *ca.* 3 mi. from New Mexico line, SE of Cloverdale (Lee and Hoffmeister, 1963:189). Chihuahua: Turkey Canyon, San Luis Mts.; Arroyo de Tapiecitas, Nuevo Casas Grandes (Alvarez and Aviña, 1963:34); near Ocampo (Anderson, 1972:284); Mojarachic (*ibid.*); Sierra Tarahumara (*ibid.*). Durango (Baker and Greer, 1962:85): Arroyo de Bucy; La Laguna; Pueblo Nuevo, 5000 ft.; 8 km. SW Las Adjuntas, 8650 ft.; 5 mi. SW El Salto; 25 mi. NNW El Salto; *E slope Cerro Huehueto*. Chihuahua (Anderson, 1972:284): 10 mi. SW Guadalupe y Calvo; Baborigame. Sinaloa: Sierra de Choix. Sonora: Baromico; type locality; 25 mi. W Colonia García (Anderson, 1972:284).

Sciurus nayaritensis chiricahuae Goldman

1933. *Sciurus chiricahuae* Goldman, Proc. Biol. Soc. Washington, 46:71, April 27, type from Cave Creek, Chiricahua Mts., Cochise Co., Arizona, 5200 ft. Known only from Chiricahua Mts., Arizona.
1963. *Sciurus nayaritensis chiricahuae*, Lee and Hoffmeister, Proc. Biol. Soc. Washington, 76:189, August 2.

MARGINAL RECORDS (Lee and Hoffmeister, 1963:189).—Arizona: *Pinery Canyon;* Cave Creek, 5200 ft.; *Green House Canyon, 1¼ mi. E Fly Park*.

Sciurus nayaritensis nayaritensis J. A. Allen

1889. *Sciurus alstoni* J. A. Allen, Bull. Amer. Mus. Nat. Hist., 2:167, October 21, type from Sierra Valparaíso, Zacatecas. Not *Sciurus alstoni* Anderson, 1879 [= *Callosciurus alstoni*], type from ?Borneo.
1890. [*Sciurus*] *nayaritensis* J. A. Allen, Bull. Amer. Mus. Nat. Hist., 2:vii, footnote, February, a renaming of *S. alstoni* J. A. Allen, 1889.

MARGINAL RECORDS.—Durango: 28 mi. S, 17 mi. W Vicente Guerrero, 8350 ft. (Baker and Greer, 1962:87). Zacatecas: type locality. Aguascalientes: Sierra Fría, 7500–8200 ft. (Lee and Hoffmeister, 1963:188). Jalisco: Barranca Beltram, E base Sierra Nevada de Colima; **Sierra de Juanacatlán**. Nayarit: Santa Teresa, 6800 ft. (Lee and Hoffmeister, 1963:188).

Sciurus arizonensis
Arizona Gray Squirrel

External measurements: 506–568; 240–310; 66–77. Upper parts uniform grizzled gray with tawny suffusion (in winter with pronounced ochraceous or fulvous dorsal stripe); postauricular

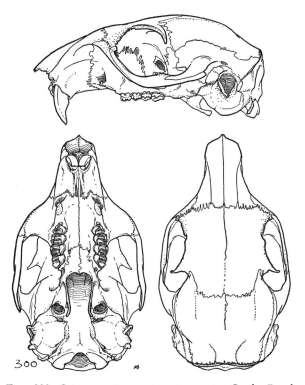

Fig. 300. *Sciurus arizonensis arizonensis*, Carr's Ranch, 5410 ft., Sierra Ancha, Arizona, No. 25500 M.V.Z., ♀, X 1.

patches of rusty yellow sometimes present; tail above black washed with white; tail below grizzled yellow-orange to rusty brown medially, bordered with black and edged with white; underparts white or nearly so. For tree squirrels, these animals differ much seasonally in color, being grayer in winter. Some authors have mistaken the striking seasonal difference of molting animals for geographic variation. Furthermore, it is known (Mearns, 1907:279) that walnuts, a preferred food, stain the pelage, especially the feet and underparts—a fact that has led at times to erroneous descriptions of color of the species. There is only one upper premolar.

Sciurus arizonensis arizonensis Coues

1867. *Sciurus arizonensis* Coues, Amer. Nat., 1:357, September, type from Fort Whipple, Yavapai Co., Arizona.

MARGINAL RECORDS.—Arizona: type locality; Fossil Creek; Christiphor [= Christopher] Creek, 27 [= 20½] mi. NE Payson (Cockrum, 1961:101); near Fort Apache. New Mexico (Findley, *et al.*, 1975:138, as *S. arizonensis* only): Frisco; *Willow Creek, 2³/₁₀ mi. E Mogollon;* 1 mi. NE Glenwood. Arizona: Carr's Ranch, Sierra Ancha.

Sciurus arizonensis catalinae Doutt

1931. *Sciurus arizonensis catalinae* Doutt, Ann. Carnegie Mus., 20:271, June 6, type from near Soldier Camp, 8000 ft., Santa Catalina Mts., Pima Co., Arizona.

MARGINAL RECORDS (Cockrum, 1961:101).—Arizona: Summerhaven, Santa Catalina Mts.; *type locality; Mt. Bigelow, 8000 ft.;* Spud Rock Ranger Station, Rincon Mts.

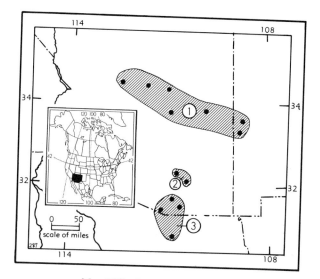

Map 287. *Sciurus arizonensis.*

1. *S. a. arizonensis*   2. *S. a. catalinae*
           3. *S. a. huachuca*

Sciurus arizonensis huachuca J. A. Allen

1894. *Sciurus arizonensis huachuca* J. A. Allen, Bull. Amer. Mus. Nat. Hist., 6:349, December 7, type from Huachuca Mts., southern Arizona.

MARGINAL RECORDS.—Arizona: Florida Canyon (Cockrum, 1961:103); type locality. Sonora: 32 mi. S Nogales. Arizona (Cockrum, 1961:103): Sycamore Canyon, Pajarito Mts.; *Madera Canyon, 6500 ft.*

Subgenus Hesperosciurus Nelson

1899. *Hesperosciurus* Nelson, Proc. Washington Acad. Sci., 1:27, May 9. Type, *Sciurus griseus* Ord.

Large squirrel; jugals relatively "weak" and relatively little twisted from vertical plane; skull broad, especially across parietals; nasals terminating posteriorly subequally with posterior tongues of premaxillae; molars massive.

Sciurus griseus
Western Gray Squirrel

External measurements: 510–570; 265–290; 74–80. Upper parts varying from dark gray to

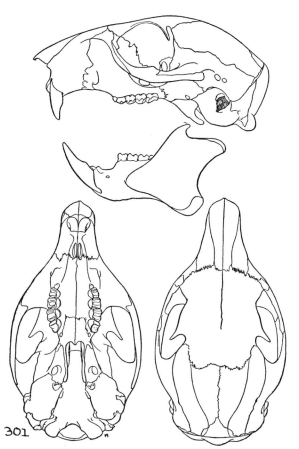

Fig. 301. *Sciurus griseus griseus,* French Gulch, Kern Co., California, No. 60026 M.V.Z., ♂, X 1.

light gray with yellowish or ochraceous wash; underparts varying from white to gray with tawny suffusion; tail gray but often with blackish or tawny suffusion. Two upper premolars.

Sciurus griseus anthonyi Mearns

1897. *Sciurus fossor anthonyi* Mearns, Preliminary diagnoses of new mammals of the genera *Sciurus, Castor, Neotoma,* and *Sigmodon,* from the Mexican border of the United States, p. 1, March 5 (preprint of Proc. U.S. Nat. Mus., 20:501, January 19, 1898), type from Campbells Ranch, Laguna Mts., San Diego Co., California.
1907. *Sciurus griseus anthonyi* Mearns, Bull. U.S. Nat. Mus., 56:264, April 13.

MARGINAL RECORDS.—California: Tehachapi Peak; Little Bear Valley; San Bernardino Peak; Santa Rosa Peak; Laguna Mts.; type locality; Smith Mtn.; San Gorgonio Pass; Oak Knoll, near Pasadena; Gaviota Pass; 7 mi. NE Santa Ynez; vic. Mt. Piños.

Sciurus griseus griseus Ord

1818. *Sciurus griseus* Ord, Jour. de Phys., Chim., Hist. Nat., et des Arts, 87:152. Type locality, The Dalles, Columbia River, Wasco Co., Oregon.

1841. *Sciurus leporinus* Audubon and Bachman, Proc. Acad. Nat. Sci. Philadelphia, p. 101, having habitat in "Northern parts of California."

1848. *Sciurus fossor* Peale, Mammalia and ornithology, *in* U.S. Expl. Exped. . . . , 8:55, from "southern parts of Oregon [where] there is a species of pine (*Pinus Lambertii*, Douglass,) which produces a cone about fifteen inches long, and eighteen in circumference."

1852. *Sciurus heermanni* Le Conte, Proc. Acad. Nat. Sci. Philadelphia, 6:149, September, type from California, probably central Sierran foothills in vic. Calaveras River, Calaveras Co.

MARGINAL RECORDS.—Washington: Lake Chelan; Cleveland. Oregon: Wapinitia; Mill Creek, 20 mi. W Warm Springs; Anchor. California: Beswick; Etna Mills; Bear Creek Valley, W of Dana; Battle Creek Meadows; Quincy. Nevada: Verdi; 3 mi. S, 5 mi. W Carson City. California: S. Fork Kings River Canyon;

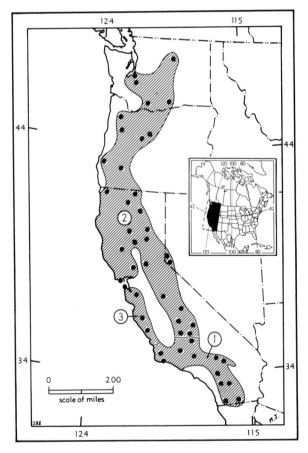

Map 288. *Sciurus griseus.*

1. S. g. anthonyi   2. S. g. griseus
3. S. g. nigripes

Walker Pass; Piute Mts.; Greenhorn Mts.; headwaters Tule River; Raymond; Placerville; 8 mi. E Chico; Tehama; Bartlett Mtn.; near Vacaville; Mt. Tamalpais; Shelley Creek. Oregon: Coquille; Salem; 3½ mi. E Newberg. Washington: Trout Lake; Roy; "near" Tacoma.

Sciurus griseus nigripes Bryant

1889. *Sciurus fossor nigripes* Bryant, Proc. California Acad. Sci., ser. 2, 2:25, June 20, type from coast region of San Mateo Co., California.

1894. [*Sciurus*] *griseus nigripes*, Rhoads, Amer. Nat., 28:525, June.

MARGINAL RECORDS.—California: near San Francisco; Scott Creek, Mt. Hamilton Range; Tassajara Creek, 6 mi. below Tassajara Springs; San Simeon.

Subgenus **Otosciurus** Nelson

1899. *Otosciurus* Nelson, Proc. Washington Acad. Sci., 1:28, May 9. Type, *Sciurus aberti* Woodhouse. **See** addenda.

Size large; ears long, broad with pronounced tufts (reduced in summer pelage); feet large; tail comparatively short and broad; upper parts mainly gray; lateral line usually black and distinct; skull short and broad; frontal area flattened; braincase wide and depressed; rostrum narrow and laterally compressed; nasals long (equaling interorbital breadth); premolars $\frac{2}{1}$.

The species of this subgenus are known collectively as the tassel-eared squirrels. They live between 6000 and 10,000 ft. elevation and almost never venture beyond the yellow-pine forest.

Sciurus aberti
Abert's Squirrel

External measurements: 463–584; 195–255; 65–80. Back dark grizzled iron gray; sides usually black; indistinct median dorsal stripe varying from rufous to chocolate brown; tail above same as back but overlaid with wash of white; tail below white (plumbeous underfur sometimes shows through guard-hairs) except for grizzled gray basal band. Geographic variation in this species is in color and extent of median dorsal stripe, and to a lesser extent in proportions of the skull.

Sciurus aberti aberti Woodhouse

1853. *Sciurus dorsalis* Woodhouse, Proc. Acad. Nat. Sci. Philadelphia, 6:110. Not *Sciurus dorsalis* Gray, 1848.

1853. *Sciurus aberti* Woodhouse, Proc. Acad. Nat. Sci. Philadelphia, 6(1852):220, type from San Francisco Mtn., Coconino Co., Arizona, a renaming of *S. dorsalis* Woodhouse.

1855. *Sciurus castanotus* Baird, Proc. Acad. Nat. Sci. Philadelphia, 7:332, type from "the Mimbres" [= "Coppermines" according to Baird, Mammals, *in* Repts. Expl. Surv. . . . , 8(1):266, July 14, 1858], New Mexico.

1858. *Sciurus castanonotus* Baird, Mammals, *in* Repts. Expl. Surv. . . . , 8(1):xxxvii, July 14. (An emendation; *castanotus* stated to have been a typographical error for *castanonotus*.)

1867. *Sciurus alberti* Gray, Ann. Mag. Nat. Hist., ser. 3, 20:417, a *lapsus* for *aberti*.

MARGINAL RECORDS.—Arizona: Yavapai Point (Hoffmeister, 1971:169); *Hull Tank* (*ibid.*); type locality; *50 mi. SE Flagstaff* (Cockrum, 1961:99); Springerville. New Mexico: Fort Wingate (Findley, *et al.*, 1975:137, as *S. aberti* only); SE "corner" Sandoval County (*op. cit.:* 136, Fig. 58); Red Canyon, 1 mi. S, 4 mi. W Manzano (*op. cit.:* 137); Magdalena Mts.; *ca.* 4 mi. SE Kingston, 9000 ft.; head Meadow Creek, *ca.* 8500 ft., Piños Altos Mts.; *tributary of Twin Sisters Creek* (Reynolds, 1966:550, as *S. aberti* only). Arizona: Prieto Plateau, S end Blue Range; Apache Maid Mtn.; Williams; *Grand Canyon Village* (Hoffmeister, 1971:169).

In 1940 and 1941 introduced into Santa Catalina Mts., Arizona (Lange, 1960:445). Possibly introduced in some places in New Mexico (Findley, *et al.*, 1975:135).

Sciurus aberti barberi J. A. Allen

1904. *Sciurus aberti barberi* J. A. Allen, Bull. Amer. Mus. Nat. Hist., 20:207, May 28, type from Colonia García, Chihuahua.

MARGINAL RECORDS.—Chihuahua (Anderson, 1972:281, 282): Pacheco; *Río Gavilán, 9 mi. SW Pacheco; type locality; 8 mi. S Colonia García, 7200 ft.;* Madera.

Sciurus aberti chuscensis Goldman

1931. *Sciurus aberti chuscensis* Goldman, Proc. Biol. Soc. Washington, 44:133, October 17, type from Chusca Mts., 9000 ft., New Mexico.

MARGINAL RECORDS.—New Mexico (Findley, *et al.*, 1975:137, as *S. aberti* only): Chuska Mts.; Tohatchi Mtn. Arizona: 12 mi. N Fort Defiance; Tunitcha Mts.

Sciurus aberti durangi Thomas

1893. *Sciurus aberti durangi* Thomas, Ann. Mag. Nat. Hist., ser. 6, 11:50, January, type from Ciudad Ranch, 100 mi. W Durango City, Durango.

MARGINAL RECORDS.—Durango (Baker and Greer, 1962:84): La Laguna; 24 mi. SSE Durango, 7200 ft.; 1½ mi. SW La Ciudad; 14 mi. W Las Adjuntas; type locality; Cerro Huehueto. The statement by Brand (1937:53) that *Sciurus aberti durangi* occurs in northwestern Chihuahua is in error; he was probably referring to *S. a. barberi*.

Sciurus aberti ferreus True

1894. *Sciurus aberti concolor* True, Diagnosis of new North American mammals, p. 1, April 26 (preprint of Proc. U.S. Nat. Mus., 17:241, November 15, 1894), type from Loveland, Larimer Co., Colorado. Not *Sc[iurus]. concolor* Blyth, 1855 [= *Callosciurus caniceps concolor*], type from Malacca.

1900. [*Sciurus aberti*] *ferreus* True, Proc. Biol. Soc. Washington, 13:183, November 30, a renaming of *S. a. concolor* True, 1894.

MARGINAL RECORDS.—Wyoming: ¼ mi. E Harriman (Brown, 1965c:516, as *S. aberti* only). Colorado (Armstrong, 1972:142, 143): Bellvue; type locality; Elbert; *3 mi. SW Eastonville*; 3 mi. N Falcon; Cañon City; around Bradford; Mosca Pass; Westcliffe; ½ mi. E Trout Creek Pass, near Antero Junction; Breckenridge; Montezuma; Horseshoe Park; 40 mi. by road up Poudre Canyon.

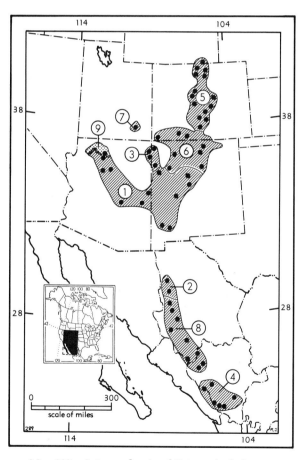

Map 289. *Sciurus aberti* and *Sciurus kaibabensis*.

1. *S. a. aberti*	5. *S. a. ferreus*
2. *S. a. barberi*	6. *S. a. mimus*
3. *S. a. chuscensis*	7. *S. a. navajo*
4. *S. a. durangi*	8. *S. a. phaeurus*
9. *S. kaibabensis*	

Sciurus aberti mimus Merriam

1904. *Sciurus aberti mimus* Merriam, Proc. Biol. Soc. Washington, 17:130, June 9, type from Hall Peak, S end Cimarron Mts., Mora Co., New Mexico.—See addenda.

MARGINAL RECORDS.—Colorado: Pagosa Springs; San Juan Mts., 10 mi. W Antonito. New Mexico: 8 mi. W Cimarron, 7800–8500 ft.; type locality; Las Vegas; Ranchos Vallejos; Jemez Mts. Colorado: "near" Park Well (Anderson, 1961:40).

Sciurus aberti navajo Durrant and Kelson

1947. *Sciurus aberti navajo* Durrant and Kelson, Proc. Biol. Soc. Washington, 60:79, July 2, type from 1 mi. E Kigalia Ranger Station, 30 mi. W Blanding, Natural Bridges National Monument Road, 8000 ft., San Juan Co., Utah.

MARGINAL RECORDS.—Utah: *Verdure;* type locality.—See addenda.

Sciurus aberti phaeurus J. A. Allen

1904. *Sciurus aberti phaeurus* J. A. Allen, Bull. Amer. Mus. Nat. Hist., 20:205, May 28, type from La Ciénega, 7500 ft., Durango.—See addenda.

MARGINAL RECORDS.—Chihuahua (Anderson, 1972:283): Yaguirachic, 130 mi. W Chihuahua, 8500 ft.; 30 mi. W Minyaca [= Miñaca]; 2 mi. NE San Juanito; La Unión, 10 km. N Guachochic, 8400 ft.; Sierra Madre near Guadalupe y Calvo [probably near San Julián]. Durango: type locality; Ciénega Corrales, 7000 ft. Chihuahua (Anderson, 1972:283): 10 mi. SW Guadalupe y Calvo, 7000 ft.; 5 mi. W Churo La Cueva, *ca.* 2700 m.

Sciurus kaibabensis Merriam
Kaibab Squirrel

1904. *Sciurus kaibabensis* Merriam, Proc. Biol. Soc. Washington, 17:129, June 9, type from Bright Angel Creek, top of Kaibab Plateau, N side Grand Canyon of Colorado River, Coconino Co., Arizona. Known only from Kaibab Plateau.

Similar in size to *Sciurus aberti.* Upper parts dark grizzled gray, usually with rufous or rusty median dorsal area; tail white above and below, sometimes with few blackish hairs; underparts dark gray to black.

The Kaibab squirrel is confined to the Kaibab Plateau of the northern side of the Grand Canyon of Arizona (including Walhalla Plateau and possibly Powell plateau). The area is approx. 40 mi. long and 20 mi. wide. Efforts to introduce this squirrel elsewhere—as in southern Utah—happily have been unsuccessful.—See addenda.

Subgenus Guerlinguetus Gray

1821. *Guerlinguetus* Gray, London Med. Repos., 15:304, April. Type, *Sciurus guerlinguetus* Gray [= *Sciurus aestuans* Linnaeus].
1823. *Macroxus* F. Cuvier, Des dents des mammifères, . . . , p. 161. Type by subsequent designation (Thomas, Proc. Zool. Soc. London, p. 933 for 1897, April, 1898), *Sciurus aestuans* Linnaeus.
1915. *Leptosciurus* J. A. Allen, Bull. Amer. Mus. Nat. Hist., 34:199, May 17. Type, *Sciurus rufoniger* Pucheran, 1845 [= *Macroxus pucheranii* Fitzinger, 1867, a name substituted for *rufoniger* Pucheran, *rufoniger* Pucheran being preoccupied by *Sciurus rufoniger* Gray, 1842].
1915. *Mesosciurus* J. A. Allen, *op. cit.:*212. Type, *Sciurus aestuans* var. *hoffmanni* Peters, 1863.
1915. *Histriosciurus* J. A. Allen, *op. cit.:*236. Type, *Sciurus gerrardi* Gray, 1861.
1915. *Simosciurus* J. A. Allen, *op. cit.:*280. Type, *Sciurus stramineus* Eydoux and Souleyet, 1841.

Resembles *Tamiasciurus hudsonicus* in shape and size of skull but cranium deeper and more highly arched; rostrum short and strongly "pinched in"; zygomata nearly parallel to axis of skull; postorbital processes short and slender; frontals swollen posteriorly; notch in maxillary plate opposite hinder part of P4 or division between P4 and M1; P3 absent; P4 subcircular or quadrate, instead of subtriangular as in *Parasciurus* and *Tamiasciurus;* differs further from *Parasciurus* in shorter rostrum, more swollen braincase, and position of notch in maxillary plate of zygoma (after A. H. Howell, 1938:50, 51).

Sciurus granatensis
Tropical Red Squirrel

External measurements: 382–440; 180–190; 50–56. Color varies so much individually and geographically that it is difficult to frame a description that includes all variants. The following

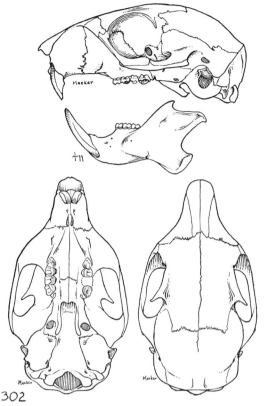

Fig. 302. *Sciurus granatensis morulus*, Barro Colorado Island, Canal Zone, Panamá, No. 45101 K.U., ♀, X 1.

description will serve for the great majority of specimens: upper parts mixed yellowish or rusty brown and black, with median area sometimes darker brown or even black; tail above varying from (1) color of upper parts but heavily overlaid with bright ferruginous (2) through intense rusty red to (3) black overlaid with white, with or without pronounced black tip; tail below with median area of dark yellowish brown but sometimes distinctly light yellowish or dark reddish brown; underparts dull rusty buff to deep ferruginous, often marked with white, especially on axial region and inguinal region.

These squirrels occur from sea level to more than 5000 ft. elevation mostly in tropical jungles, in comparatively localized populations.

Sciurus granatensis chiriquensis Bangs

1902. *Sciurus (Guerlinguetus) aestuans chiriquensis* Bangs, Bull. Mus. Comp. Zool., 39(2):22, April, type from Divalá, Chiriquí, Panamá.
1947. [*Sciurus granatensis*] *chiriquensis*, Hershkovitz, Proc. U.S. Nat. Mus., 97:7, August 25.

MARGINAL RECORDS.—Costa Rica: Cataratos San Carlos; Limón. Panamá: Almirante; 6 mi. E El

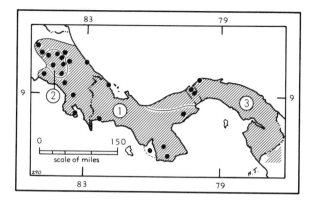

Map 290. *Sciurus granatensis*.

1. *S. g. chiriquensis*   2. *S. g. hoffmanni*
3. *S. g. morulus*

Valle (Handley, 1966:777); Cerro Viejo, Azuero Peninsula; Guanico (Handley, 1966:777); Cébaco Island, Montijo Bay; type locality. Costa Rica: Punta Jiménez; Palmar; El General; Pozo Azul Pirrís; *San Gerónimo Pirrís;* Peralta; Santa Clara; Zarcéro.

Sciurus granatensis hoffmanni Peters

1863. *Sciurus aestuans* var. *hoffmanni* Peters, Monatsb. preuss. Akad. Wiss., Berlin, p. 654, type from Costa Rica, subsequently restricted to Agua Caliente by Harris (Occas. Pap. Mus. Zool., Univ. Michiagn, 476:9, October 8, 1943).
1947. [*Sciurus granatensis*] *hoffmanni*, Hershkovitz, Proc. U.S. Nat. Mus., 97:7, August 25.
1867. *Macroxus xanthotus* Gray, Ann. Mag. Nat. Hist., ser. 3, 20:429, type from Cordillera de Tale, Veragua, Costa Rica.

MARGINAL RECORDS.—Costa Rica: Santa Clara; Rancho de Río Jiménez; W side Turrialba; Buena Vista; Escazú Heights; Volcán Poás.

Sciurus granatensis morulus Bangs

1900. *Sciurus variabilis morulus* Bangs, Proc. New England Zool. Club, 2:43, September 20, type from Loma del León, Panamá.
1955. *Sciurus granatensis morulus*, Miller and Kellogg, Bull. U.S. Nat. Mus., 205:257, March 3.
1913. *Sciurus variabilis choco* Goldman, Smiths. Miscl. Coll., 60(22):4, February 28, type from Cana, 3500 ft., Pirri Mts., eastern Panamá. Regarded as subspecifically inseparable from *morulus* by Handley, Checklist mamms. Panamá . . . , p. 777, November 22, 1966.

MARGINAL RECORDS.—Panamá: Portobello; Tabernilla; Gatún; eastward into South America.

Sciurus richmondi Nelson
Richmond's Squirrel

1898. *Sciurus richmondi* Nelson, Proc. Biol. Soc. Washington, 12:146, June 3, type from Escondido River, 50 mi. above Bluefields, Nicaragua.

Average external measurements of five adult topotypes: 361; 169; 52. Upper parts, including nose and base of tail, ochraceous or rusty brown, darker medially, purest on sides and lateral surfaces of legs; tail above black faintly suffused with tawny or rufous; tail below grizzled yellow-brown, bordered with black, and edged with ochraceous or yellow; underparts varying from dull yellow to dull buffy or ferruginous; yellowish postauricular patch sometimes present. Differs from S. granatensis hoffmanni in smaller average size. Probably this is only a subspecies of Sciurus granatensis.

Sciurus richmondi inhabits dense humid jungles.

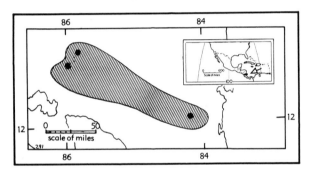

Map 291. Sciurus richmondi.

MARGINAL RECORDS.—Nicaragua: Río Tuma; type locality; Vijagua.

Genus Syntheosciurus Bangs—Montane Squirrels

1902. Syntheosciurus Bangs, Bull. Mus. Comp. Zool., 39:25, April. Type, Syntheosciurus brochus Bangs.
1904. Synthetosciurus Elliot, Field Columb. Mus., Publ. 95, Zool. Ser., 4:91, August 2, a renaming of Syntheosciurus Bangs.

Skull resembles that of Microsciurus but cranium more highly arched; frontals swollen; upper incisors projected forward; molariform teeth relatively large, P3 reaching crown of P4; auditory bullae small; postorbital processes slender (after A. H. Howell, 1938:52); skull thin and papery.

Various authors (Goodwin, 1946:365, for example) do not accord Syntheosciurus generic rank; some retain it as a subgenus. Certainly the characters allegedly diagnostic of the genus scarcely seem to indicate generic rank. This group, regardless of its nomenclatural status, seems to occupy a position intermediate between genus Sciurus and genus Microsciurus.

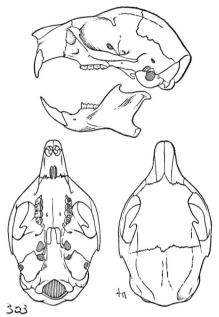

Fig. 303. Syntheosciurus brochus, Boquete, 7000 ft., Chiriquí, Panamá, No. 10402 M.C.Z., ♂, X 1. (After Bangs, Bull. Mus. Comp. Zool., 39:26, April, 1902.)

Syntheosciurus brochus
Bangs' Mountain Squirrel

Total length, 320; tail, 150; hind foot, 46. Upper parts finely mixed olivaceous bistre and dull tawny olive; orbital ring, sides of nose, and chin tawny olive; color of tail closely resembling that of back, fringed along sides with pale rusty and slightly more reddish, less olivaceous below; underparts, especially along middle line, strongly suffused with orange rufous (after Bangs, 1902:25, for S. brochus). Goodwin (1946:365) adds for the holotype of S. poasensis, upper parts, including top of head, ears, sides of body, and legs and feet, finely mixed Cinnamon-Buff and black, darkest medially; tail similar to upper parts and fringed with reddish Cinnamon-Buff. Measurements (in mm.) for the male of poasensis, the specimen (sex ?, 112029 KU) from Cerro Pando, and the male type and female topotype of S. brochus, in that order, are: total length, 295, 273, 320, 315; tail, 140, 120, 150, 145; hind foot, 44, 41, 46, 46; length of diastema I1–P4, 10.7, 11.2, ca. 10.6, 10.6. In the same order the upper incisors reportedly are smooth, barely grooved, grooved, grooved. More specimens may show that there is a single monotypic species, confirm the existence of two species, or (as suggested here) warrant the recognition of two subspecies.

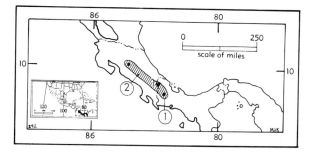

Map 292. *Syntheosciurus brochus* (1) and *Syntheosciurus poasensis* (2).

Syntheosciurus brochus brochus Bangs

1902. *Syntheosciurus brochus* Bangs, Bull. Mus. Comp. Zool., 39:25, April, type from Boquete, 7000 ft., Chiriquí, Panamá.

MARGINAL RECORDS.—Panamá: N slope Cerro Pando, 1920 m., Boca del Toro (112029 KU, skull only); type locality.

Syntheosciurus brochus poasensis (Goodwin)

1942. *Sciurus poasensis* Goodwin, Amer. Mus. Novit., 1218:1, February 11, type from Volcán Poás, 6700 ft., Alajuela, Costa Rica. Known only from type locality.

Genus Microsciurus J. A. Allen—Pygmy Squirrels

1895. *Microsciurus* J. A. Allen, Bull. Amer. Mus. Nat. Hist., 7:332, November 8. Type, *Sciurus (Microsciurus) alfari* J. A. Allen.

Smallest of the North and Central American tree squirrels; skull strongly depressed posteriorly, and postorbital process situated nearly or exactly over posterior zygomatic root; frontals broad; nasals short; jugal broad; upper incisors pro-odont, usually extending beyond plane of tip of nasals; palate normal; bullae small; cheekteeth as in *Sciurus*, but small outer (3rd) cusp of upper molars is often barely traceable; p3 present, and rather well developed (after Ellerman, 1940:319).

This group was originally described as a subgenus of *Sciurus* and was so treated by authors until Goldman (1912a:4) accorded it generic rank. Since J. A. Allen's (1915:188) review of the South American squirrels, the generic rank has not been seriously questioned.

KEY TO NORTH AMERICAN SPECIES OF MICROSCIURUS

1. Underparts slightly buffy to fulvous.
   M. alfari, p. 439
1'. Underparts ochraceous rufous.
   M. mimulus, p. 440

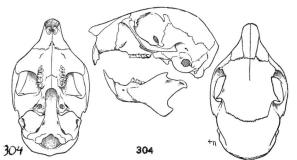

Fig. 304. *Microsciurus alfari browni*, Pozo Azul, Costa Rica, No. 140948 U.S.N.M., ♀, X 1.

Microsciurus alfari
Alfaro's Pygmy Squirrel

External measurements: 232–260; 75(?)–120; 33–40. There are two fairly distinct "color phases." So far as known, all specimens of *M. a. alticola* and *M. a. septentrionalis* have upper parts that are dull olive-brown or olive-black; specimens of other subspecies have upper parts that are finely mixed ochraceous-tawny on black. Tail above varies from ochraceous-tawny to cinnamon rufous, sometimes with grayish wash, usually black-tipped; tail below distinctly buffy; underparts varying from dull ochraceous buff to dull gray with yellowish wash.

The texture of the hair varies with the altitudinal occurrence of the subspecies. In those occurring at higher elevations the pelage is long, dense, and soft; the reverse characterizes those occurring at or near sea level.

Microsciurus alfari alfari (J. A. Allen)

1895. *Sciurus (Microsciurus) alfari* J. A. Allen, Bull. Amer. Mus. Nat. Hist., 7:333, November 8, type from Jiménez, Costa Rica.
1912. [*Microsciurus*] *alfari*, Goldman, Smiths. Miscl. Coll., 56(36):4, February 19.

MARGINAL RECORDS.—Costa Rica: type locality; Siquirres; Carrillo.

Microsciurus alfari alticola Goodwin

1943. *Microsciurus alfari alticola* Goodwin, Amer. Mus. Novit., 1218:2, February 11, type from Lajos [*sic*], Villa Quesada, 5000 ft., Alajuela, Costa Rica.

MARGINAL RECORDS.—Costa Rica: type locality; La Hondura.

Microsciurus alfari browni (Bangs)

1902. *Sciurus (Microsciurus) browni* Bangs, Bull. Mus. Comp. Zool., 39:24, April, type from Bugaba, 600 ft., Chiriquí, Panamá.

1914. *Microsciurus alfari browni*, J. A. Allen, Bull. Amer. Mus. Nat. Hist., 33:151, February 26.

MARGINAL RECORDS.—Costa Rica: San Gerónimo; El General; Agua Buena. Panamá (Handley, 1966:777): Quebrada Santa Clara, 3600–4200 ft.; *Río Gariché, 5300 ft.*; type locality. Costa Rica: Coto; Puerta Uvita; *Pozo Azul.*

Microsciurus alfari fusculus (Thomas)

1910. *Sciurus* (*Microsciurus*) *similis fusculus* Thomas, Ann. Mag. Nat. Hist., ser. 8, 6:503, November, type from Juntas, Río San Juan, Chocó District, Colombia, 400 ft.
1966. *M*[*icrosciurus*]. *a*[*lfari*]. *fusculus*, Handley, Ectoparasites of Panamá, Field Mus. Nat. Hist., p. 777, November 22.
1961. *Microsciurus flaviventer isthmius*, Cabrera, Rev. Mus. Argentino de Cienc. Nat., 4:357, August 25 (part).

MARGINAL RECORDS.—Panamá (Handley, 1966:777): Cerro Sapo; *Cana, 2000 ft.*; Río Setegantí, 2600 ft., thence into South America.

Microsciurus alfari septentrionalis Anthony

1920. *Microsciurus septentrionalis* Anthony, Jour. Mamm., 1:81, March 2, type from Sábalos, on Río San Juan, jct. Río Sábalos, Nicaragua.
1946. *Microsciurus alfari septentrionalis*, Goodwin, Bull. Amer. Mus. Nat. Hist., 87:367, December 31.

MARGINAL RECORDS.—Nicaragua: type locality. Costa Rica: La Vieja de San Carlos; La Vijagua.

Microsciurus alfari subsp. novum

1966. *Microsciurus alfari* ("undescribed subspecies") Handley, Ectoparasites of Panamá, Field Mus. Nat. Hist., p. 777, November 22.

MARGINAL RECORDS.—Panamá (Handley, 1966:777): Armila; Tacarcuna Casita, 1500 ft.; *Tacarcuna Village, 1950 ft.*

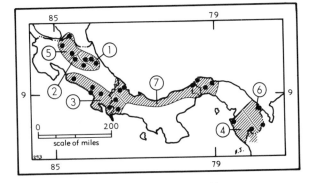
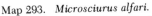

Map 293. *Microsciurus alfari.*

Guide to subspecies
1. *M. a. alfari*
2. *M. a. alticola*
3. *M. a. browni*
4. *M. a. fusculus*
5. *M. a. septentrionalis*
6. *M. alfari* subsp. novum
7. *M. a. venustulus*

Microsciurus alfari venustulus Goldman

1912. *Microsciurus alfari venustulus* Goldman, Smiths. Miscl. Coll., 56(36):4, February 19, type from Gatún, Canal Zone, Panamá.
1920. *M*[*icrosciurus*]. *i*[*sthmicus*]. *venustulus* Goldman, Smiths. Miscl. Coll., 69(5):143, April 24, a *lapsus.*

MARGINAL RECORDS.—Panamá (Handley, 1966:777): Almirante; type locality; Portobelo; Mandinga; Cerro Azul, 2000–2500 ft.; upper Río Changena, 5000 ft.

Microsciurus mimulus
Cloud-forest Pygmy Squirrel

External measurements of five specimens from Panamá: 236–268; 100–116; 35–38. Upper parts coarsely grizzled black and pale orange buff or buffy yellow, purest on cheeks and sides; tail above coarsely grizzled black and pale buff; tail below tawny ochraceous bordered with black and edged with grayish buff, becoming pure black at tip; underparts orange buffy, clearest on throat, chest, and inner surfaces of legs and lighter medially. Description based on that of *M. isthmius vivatus* Goldman [= *M. m. isthmius*].

This species is distinguished from *M. alfari*, which occurs in the same regions but usually at lower elevations, by: "much paler, more coarsely grizzled color of upper parts . . . , the maxillae encroach farther on the frontals between the lachrymals and premaxillae, and the interparietal is subtriangular instead of rectangular." (Goldman, 1912b:5.)

Microsciurus mimulus boquetensis (Nelson)

1903. *Sciurus* (*Microsciurus*) *boquetensis* Nelson, Proc. Biol. Soc. Washington, 16:121, September 30, type from Boquete, Chiriquí, Panamá.
1966. *M*[*icrosciurus*]. *m*[*imulus*]. *boquetensis*, Handley, Ectoparasites of Panamá, Field Mus. Nat. Hist., p. 778, November 22.

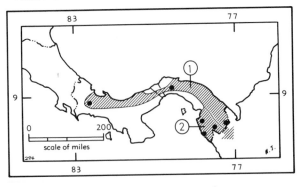

Map 294. *Microsciurus mimulus.*

1. *M. m. boquetensis*   2. *M. m. isthmius*

MARGINAL RECORDS.—Panamá (Handley, 1966:778): high ridges above Boquete; Cerro Azul, 2500–3000 ft.; *Cerro Malí, 4900 ft.;* Tacarcuna Village, 1950 ft., thence into South America.

Microsciurus mimulus isthmius (Nelson)

1899. *Sciurus (Microsciurus) isthmius* Nelson, Bull. Amer. Mus. Nat. Hist., 12:77, April 14, type from Truandó River, Colombia.
1966. *M[icrosciurus]. m[imulus]. isthmius,* Handley, Ectoparasites of Panamá, Field Mus. Nat. Hist., p. 778, November 22.
1912. *Microsciurus isthmius vivatus* Goldman, Smiths. Miscl. Coll., 60(2):4, September 20, type from near Cana, 3500 ft., Panamá.
1961. *Microsciurus flaviventer isthmius,* Cabrera, Rev. Mus. Argentino de Cienc. Nat., 4:357, August 25 (part).

MARGINAL RECORDS.—Panamá (Handley, 1966:778): Cerro Sapo, 3000 ft.; Cana, 3500 ft.; *20 mi. S Cana, 5000 ft.;* Río Jaqué, thence into South America.

Genus Tamiasciurus Trouessart—Red Squirrels

Revised by J. A. Allen, Bull. Amer. Mus. Nat. Hist., 10:249–298, July 22, 1898.

1880. *Tamiasciurus* Trouessart, Le Naturaliste, 2(37):292, October. Type, *Sciurus hudsonicus* Erxleben.

Notch in zygomatic plate opposite anterior margin of M1 or posterior part of P4 being farther forward than in *Sciurus.* Baculum vestigial and confined to glans penis.

Some of the subspecific names in *Tamiasciurus* may be incorrectly applied, and certain named kinds may not merit nomenclatural recognition. Essentially the arrangement used by R. M. Anderson (1947) is employed here.

KEY TO SPECIES OF TAMIASCIURUS

1. Underparts white or nearly so.
   T. hudsonicus, p. 441
1'. Underparts more or less rust colored.
   T. douglasii, p. 446

Tamiasciurus hudsonicus
Red Squirrel

External measurements: 270–385; 92–158; 35–57. Upper parts rusty reddish, brownish, or olivaceous gray, usually purest on sides; black or blackish lateral line usually present; tail above yellowish-rusty to rufous, usually lightly (but sometimes heavily) washed or grizzled with black and white; tail below yellowish-rufous or yellowish-gray to rusty red, edged with white- to fulvous-tipped hairs; underparts varying from white to grayish-white, or faintly washed with yellow.

Tamiasciurus hudsonicus subsp.?, according to Payne (1976:60–64) was introduced in 1963 on Newfoundland, is established, and as of 1976 was extending its geographic range on Newfoundland.

Tamiasciurus hudsonicus abieticola (A. H. Howell)

1929. *Sciurus hudsonicus abieticola* A. H. Howell, Jour. Mamm., 10:75, February 11, type from Highlands, North Carolina.
1937. *Tamiasciurus hudsonicus abieticola,* Kellogg, Proc. U.S. Nat. Mus., 84:459, October 7.

MARGINAL RECORDS.—West Virginia: Cheat Mtn., 3 mi. W Cheat Bridge, 3900 ft. North Carolina: Roan Mtn. South Carolina: Oconee. North Carolina: type locality. Tennessee: Buck Fork, Little Pigeon River.

Tamiasciurus hudsonicus baileyi (J. A. Allen)

1898. *Sciurus hudsonicus baileyi* J. A. Allen, Bull. Amer. Mus. Nat. Hist., 10:261, July 22, type from Bighorn Mountains, near head Kirby Creek, 8400 ft., Washakie Co., Wyoming.
1940. *Tamiasciurus hudsonicus baileyi,* Hayman and Holt, *in* Ellerman, The families and genera of living rodents, British Mus., 1:346, June 8.

MARGINAL RECORDS.—Montana: Bear Creek, Bear Paw Mts.; Pryor Mts. Wyoming (Long, 1965a:595): "near" Sheridan; Storey; *Casper Mts., 7 mi. S Casper;* Springhill; *2 mi. ESE Pole Mtn.; Sherman;* 8 mi. E Laramie; *Shirley Mts.; Ferris Mts.;* 8 mi. E Rongis, Green Mts.; hd. Kirby Creek, Bighorn Mts.; *Medicine Wheel Ranch, 28 mi. E Lovell.* Montana: Big Snowy Mts.

Tamiasciurus hudsonicus columbiensis A. H. Howell

1936. *Tamiasciurus hudsonicus columbiensis* A. H. Howell, Proc. Biol. Soc. Washington, 49:135, August 22, type from Raspberry Creek, about 30 mi. SE Telegraph Creek, northern British Columbia.

MARGINAL RECORDS.—Yukon: S fork Macmillan River, Canol Road. British Columbia: 1 mi. NW jct. Irons Creek and Liard River; jct. Trout and Liard rivers (Cowan and Guiguet, 1965:152); *Muncho Lake; 10 mi. S, 21 mi. E Muncho Lake* (Cowan and Guiguet, 1965:152); *S and 70 mi. W Fort Nelson (ibid.);* Summit Pass; Fort Grahame. Alberta: Folding Mountain-Roche; Canmore. British Columbia: Lake La Hache; Chilcotin River; *Stuie* (Cowan and Guiguet, 1965:152); Hagensborg; 10 mi. from Glenora. Yukon: McIntyre Creek, 2250 ft., 3 mi. NW Whitehorse; vic. Lake Leberge.

Tamiasciurus hudsonicus dakotensis (J. A. Allen)

1894. *Sciurus hudsonicus dakotensis* J. A. Allen, Bull. Amer. Mus. Nat. Hist., 6:325, November 7, type from Squaw Creek, Black Hills, Custer Co., South Dakota.

1940. *Tamiasciurus hudsonicus dakotensis*, Hayman and Holt, *in* Ellerman, The families and genera of living rodents, British Mus., 1:346, June 8.

MARGINAL RECORDS.—Montana: 5½ mi. S, 1 mi. E Ekalaka (Lampe, *et al.*, 1974:14); Long Pine Hills (Andersen and Jones, 1971:375). South Dakota: *Long Pine Hills (ibid.)*; type locality. Wyoming: Casper Mts., 7 mi. S Casper (Long, 1965a:595, as reidentified by Hoffmann and Jones, 1970:374); Devils Tower (Long, 1965a:596). Not found: Wyoming: J. A. Allen's (1898:260) record of occurrence as Belle Fourche and Long's (1965a:596) reference to the same perhaps refer to specimens from about 11 mi. to the east in South Dakota.

Tamiasciurus hudsonicus dixiensis Hardy

1942. *Tamiasciurus fremonti dixiensis* Hardy, Proc. Biol. Soc. Washington, 55:87, June 25, type from near Further Water, 9500 ft., Dixie National Forest, Pine Valley Mts., Washington Co., Utah.

1950. *T[amiasciurus]. h[udsonicus]. dixiensis* Hardy, Proc. Biol. Soc. Washington, 63:14, April 26.

MARGINAL RECORDS.—Utah: Elkhorn Guard Station, 9400 ft., 14 mi. N Torrey, Fishlake Plateau; Donkey Lake, 10,000 ft., Boulder Mtn.; Cyclone Lake, Aquarius Plateau; *Duck Creek, 9000 ft., Cedar Mtn.;* 2 mi. W Navajo Lake; *1 mi. up Middle Fork from Pine Valley Camp, Dixie National Forest;* type locality; *Parowan Canyon;* Puffer Lake, 9500 ft.

Tamiasciurus hudsonicus fremonti (Audubon and Bachman)

1853. *Sciurus fremonti* Audubon and Bachman, The viviparous quadrupeds of North America, 3(30):pl. 149, fig. 2; text, 3:237. Type locality, "Rocky Mountains," probably in the park region of central Colorado.

1950. *T[amiasciurus]. hudsonicus fremonti*, Hardy, Proc. Biol. Soc. Washington, 63:14, April 26.

1950. *Tamiasciurus hudsonicus wasatchensis* Hardy, Proc. Biol. Soc. Washington, 63:13, April 26, type from about 10,000 ft., in spruce-fir area along Skyline Drive E of Mt. Nebo, Juab Co., near Juab–Utah County line, Utah. Arranged as a synonym of *fremonti* by Durrant and Hansen, Jour. Mamm., 35:91, February 10, 1954.

MARGINAL RECORDS.—Wyoming (Long, 1965a:596): Bridgers Pass; sec. 28, T. 16 N, R. 78 W, N Fork Little Laramie River; *3 mi. above Woods Landing.* Colorado (Armstrong, 1972:147–149): 2¼ mi. W Fort Collins; U.S. Air Force Academy; Tercio. New Mexico: Chama. Colorado (Anderson, 1961:41): ¼ mi. NNW Middle Well, Prater Canyon, 7600 ft.; *Chickaree Draw.* Utah: *1 mi. W Geyser Pass, 9700 ft., La Sal Mts.; 3 mi. W Geyser Pass, 10,000 ft., La Sal Mts.;* Mt. Tomaski;

Clark Lake, La Sal Mts. Colorado: Baxter Pass, 8500 ft. (Armstrong, 1972:148). Utah: Hacking Lake, 15 mi. due NW Vernal, 9000 ft.; 2 mi. E Duchesne; Strawberry Reservoir; 5 mi. E Ferron Reservoir; *Bear Canyon Picnic Grounds, SE of Mt. Nebo;* Tinnys Flat, Santaquin Canyon; Smith and Morehouse Creek, Weber River. Wyoming (Long, 1965a:597): Fort Bridger; *4 mi. S Lonetree.* Utah: jct. Deep and Carter creeks, 7900 ft. Wyoming: Maxon (Long, 1965a:597). Also, Utah: 1 mi. E Jackson Camp, 21 mi. N Blanding, Abajo Mts., 10,200 ft.

Tamiasciurus hudsonicus grahamensis (J. A. Allen)

1894. *Sciurus hudsonicus grahamensis* J. A. Allen, Bull. Amer. Mus. Nat. Hist., 6:350, December 7, type from Graham Mountains, Graham Co., Arizona. Known only from Graham Mts.

1951. *T[amiasciurus]. h[udsonicus]. grahamensis*, Kelson, Univ. Utah Biol. Ser., 11(3):17, February 15.

Tamiasciurus hudsonicus gymnicus (Bangs)

1899. *Sciurus hudsonicus gymnicus* Bangs, Proc. New England Zool. Club, 1:28, March 31, type from Greenville, near Moosehead Lake, Piscataquis Co., Maine.

1938. *Tamiasciurus hudsonicus gymnicus*, F. L. Osgood, Jour. Mamm., 19:438, November 14.

MARGINAL RECORDS.—Nova Scotia: Cape North (Cameron, 1959:20). New Hampshire: Amherst (Hatt, 1929:16). Quebec: North Hatley (Anderson, 1942:33). Prince Edward Island: Fortune (Cameron, 1959:45). Not found: Prince Edward Island: Mermaid (Cameron, 1959:45).

Tamiasciurus hudsonicus hudsonicus (Erxleben)

1777. [*Sciurus vulgaris*] *hudsonicus* Erxleben, Systema regni animalis . . . , 1:416. Based on the Hudson Bay Squirrel of Pennant, 1771. Type locality, mouth Severn River, Ontario (see A. H. Howell, Proc. Biol. Soc. Washington, 49:134, August 22, 1936).

1923. *Tamisciurus hudsonicus*, Pocock, Proc. Zool. Soc. London, p. 213, July 6.

1788. [*Sciurus*] *hudsonius*, Gmelin, Syst. nat., ed. 13, 1:147. Based on several earlier citations, among them the Hudson Bay Squirrel of Pennant, 1771.

1822. *Sciurus rubrolineatus* Desmarest, Mammalogie . . . , 2:333, *in* Encyclopédie méthodique. . . . Based on "Écureuil rouge, Warden, Description des États-Unis, tom. 5, p. 630."

MARGINAL RECORDS.—Manitoba: Churchill (Soper, 1961:190). Ontario: type locality; Georgian Bay. Minnesota: Itasca County. Manitoba: Marchand (Soper, 1961:190); *Riding Mtn. (ibid.)*; Edwards Creek (Tamsitt, 1960:148); Riding Mts. (Soper, 1961:190); vic. Cormorant Lake, Mile 42 from The Pas.

Tamiasciurus hudsonicus kenaiensis A. H. Howell

1936. *Tamiasciurus hudsonicus kenaiensis* A. H. Howell, Proc. Biol. Soc. Washington, 49:136, August 22, type from

Hope, Cook Inlet, Alaska. Known only from Kenai Peninsula.

Tamiasciurus hudsonicus lanuginosus (Bachman)

1839. *Sciurus lanuginosus* Bachman, Proc. Zool. Soc. London, p. 101, for 1838, February 7, type from Fort McLoughlin, Campbell Island, British Columbia (see McCabe and Cowan, Trans. Royal Canadian Inst., Toronto, 25:164, February, 1945).

1956. *Tamiasciurus hudsonicus lanuginosus,* Cowan and Guiguet, British Columbia Prov. Mus., Handbook, 11:151, July 15.

1890. *Sciurus hudsonius* [sic] *vancouverensis* J. A. Allen, Bull. Amer. Mus. Nat. Hist., 3:165, November 14, type from Duncan Station, Vancouver Island, British Columbia. Not taxonomically distinct from *T. h. lanuginosus* according to McCabe and Cowan, Trans. Royal Canadian Inst., pp. 164, 165, February, 1945.

MARGINAL RECORDS.—British Columbia: mouth Skeena River; Hagensborg; Owikeno Lake; Calvert Island; Arran Rapids (Cowan and Guiguet, 1965:152); all of Vancouver Island, thence up coast, including several coastal islands, to Porcher Island. Not found: British Columbia: Little Gillard Island. Has been introduced on the Queen Charlotte Islands (Cowan and Guiguet, 1965:153).

Tamiasciurus hudsonicus laurentianus Anderson

1942. *Tamiasciurus hudsonicus laurentianus* Anderson, Ann. Rept. Provancher Soc. Nat. Hist., Quebec, for 1941, p. 31, July 14, type from Lac Marchant, near Moisie Bay, Saguenay Co., N shore Gulf of St. Lawrence, Quebec.

MARGINAL RECORDS.—Quebec: Strait of Belle Isle, thence southwestward along N shore of Gulf of St. Lawrence and St. Lawrence River to St. Maurice River. Not found: Newfoundland: *Camel Island* (Threlfall, 1969:198, as *Tamiasciurus hudsonicus* only; probably introduced).

Tamiasciurus hudsonicus loquax (Bangs)

1896. *Sciurus hudsonicus loquax* Bangs, Proc. Biol. Soc. Washington, 10:161, December 28, type from Liberty Hill, New London Co., Connecticut.

1936. *Tamiasciurus hudsonicus loquax,* A. H. Howell, Occas. Pap. Mus. Zool., Univ. Michigan, 338:1, July 7.

MARGINAL RECORDS.—Quebec (Wrigley, 1969:206, as *T. hudsonicus* only): Camp de la Roche; St. Sebastien; *Ayers Cliff.* New York: Heart Lake. Massachusetts: Worcester. New Jersey: Wading River, thence down coast to Virginia: Henrico County; Buckingham County. West Virginia: Berkeley Springs; Ogelbay Park. Ohio: 1 mi. E Ansonia. Indiana: Jefferson County; Vanderburg County. Michigan: Keeweenaw County. Ontario: Lake Nipissing. Quebec: Parker; Kamika Lake.

Tamiasciurus hudsonicus lychnuchus (Stone and Rehn)

1903. *Sciurus fremonti lychnuchus* Stone and Rehn, Proc. Acad. Nat. Sci. Philadelphia, 55:18, May 5, type from Forks of Ruidoso, Lincoln Co., New Mexico.

1955. *Tamiasciurus hudsonicus lychnuchus,* Miller and Kellogg, Bull. U.S. Nat. Mus., 205:262, March 3.

1929. *S*[*ciurus*]. *f*[*remonti*]. *ruidoso* Hatt, Roosevelt Wildlife Ann., 2(1): map facing p. 16, March. Name appears only on a map (not in text) and is a *lapsus* and *nomen nudum.*

MARGINAL RECORDS.—New Mexico: Capitan Mts.; type locality. Findley (1961:321) concluded that V. Bailey's (1932:80) record from Guadalupe Mts. was an error.

Tamiasciurus hudsonicus minnesota (J. A. Allen)

1899. *Sciurus hudsonicus minnesota* J. A. Allen, Amer. Nat., 33:640, August, type from Fort Snelling, Hennepin Co., Minnesota.

1940. *Tamiasciurus hudsonicus minnesota,* Hayman and Holt, *in* Ellerman, The families and genera of living rodents, British Mus., 1:346, June 8.

1943. *Tamiasciurus hudsonicus murii* A. H. Howell, Proc. Biol. Soc. Washington, 56:67, June 16, type from Moorehead, Clay Co., Minnesota.

MARGINAL RECORDS.—Manitoba: Carberry; Red River. Minnesota: S end Ten Mile Lake. Wisconsin: Outer Island (Jackson, 1961:171, Map 33); Detroit Harbor (*op. cit.:* 175). Illinois (Necker and Hatfield, 1941:50, as *T. h. loquax*): Onarga; Lawn Ridge. Iowa: Iowa City (Bowles, 1975:76); Dickinson County. North Dakota: Lisbon.—The alleged occurrence of this species in Iowa at Atlantic in Cass Co., Knoxville in Marion Co., and Pammel State Park in Madison Co., have been discounted with reason by Bowles (1975:72, 74, 75) and here are not accepted as valid.

Tamiasciurus hudsonicus mogollonensis (Mearns)

1890. *S*[*ciurus*]. *hudsonius* [sic] *mogollonensis* Mearns, Auk, 7:49, January, type from Quaking Asp Settlement, near summit Mogollon Mts., Coconino Co., Arizona.

1951. *T*[*amiasciurus*]. *h*[*udsonicus*]. *mogollonensis,* Kelson, Univ. Utah Biol. Ser., 11(3):17, February 15.

1898. *Sciurus fremonti neomexicanus* J. A. Allen, Bull. Amer. Mus. Nat. Hist., 10:291, July 22, type from Rayado Canyon, Colfax Co., New Mexico. (Regarded as inseparable from *mogollonensis* by V. Bailey, N. Amer. Fauna, 53:75, March 1, 1932.)

MARGINAL RECORDS.—New Mexico: Raton Mts.; Manzano Mts.; Beartrap Canyon, San Mateo Mts., 20 mi. S, 19 mi. W Magdalena (Findley, 1961:316); Iron Canyon, Black Range, 6 mi. N, 15 mi. E Santa Rita (*ibid.*); 10 mi. S Mogollon. Arizona: Prieto Plateau, S end Blue Range (Cockrum, 1961:105); Bakers Butte; Mt. Sitgreaves (Cockrum, 1961:104, 105); *Point Sublime* (Hoffmeister, 1971:170); *Swamp Lake (ibid.)*; *3 mi. S Dry Park Ranger Station (Cockrum,*

1961:104); De Motte [= V.T.] Park (*ibid.*); *Kaibab Plateau, E rim* (*ibid.*); *San Francisco Mtn.; 8 mi. SE Mormon Lake* (Cockrum, 1961:104); *18 mi. W Springerville* (Cockrum, 1961:105); Springerville. New Mexico: Mt. Taylor. Arizona (Cockrum, 1961:105): Spruce Creek, Tunitcha Mts.; *Lukachukai Pass, Lukachukai Mts.; near Roof Butte, Lukachukai Mts.; Lukachukai Mts., 15 mi. E Indian School.* New Mexico: Washington Pass, 6 mi. E Crystal, Chuska Mts. (Findley, 1961:316); San Juan Mts.

Tamiasciurus hudsonicus pallescens A. H. Howell

1942. *Tamiasciurus hudsonicus pallescens* A. H. Howell, Proc. Biol. Soc. Washington, 55:13, May 12, type from 8 mi. E Upham, McHenry Co., North Dakota.

MARGINAL RECORDS.—Saskatchewan: *Moose Mtn.* (Soper, 1961:31, as *T. h. hudsonicus*); Carlyle Lake (Beck, 1958:29, as *T. hudsonicus* only). Manitoba: Max Lake, Turtle Mts. North Dakota: Towner.

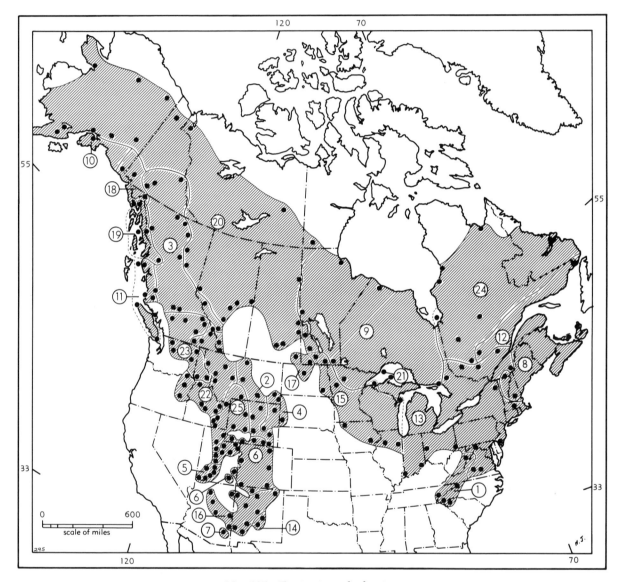

Map 295. *Tamiasciurus hudsonicus.*

Guide to subspecies
1. *T. h. abieticola*
2. *T. h. baileyi*
3. *T. h. columbiensis*
4. *T. h. dakotensis*
5. *T. h. dixiensis*
6. *T. h. fremonti*
7. *T. h. grahamensis*
8. *T. h. gymnicus*
9. *T. h. hudsonicus*
10. *T. h. kenaiensis*
11. *T. h. lanuginosus*
12. *T. h. laurentianus*
13. *T. h. loquax*
14. *T. h. lychnuchus*
15. *T. h. minnesota*
16. *T. h. mogollonensis*
17. *T. h. pallescens*
18. *T. h. petulans*
19. *T. h. picatus*
20. *T. h. preblei*
21. *T. h. regalis*
22. *T. h. richardsoni*
23. *T. h. streatori*
24. *T. h. ungavensis*
25. *T. h. ventorum*

Tamiasciurus hudsonicus petulans (Osgood)

1900. *Sciurus hudsonicus petulans* Osgood, N. Amer. Fauna, 19:27, October 6, type from Glacier, 1870 ft., White Pass, southern Alaska.

1936. *Tamiasciurus hudsonicus petulans*, A. H. Howell, Proc. Biol. Soc. Washington, 49:136, August 22.

MARGINAL RECORDS.—Yukon: Burwash Landing, Kluane Lake. Alaska: type locality; *Chilkat Valley;* Glacier Bay; head Chitina River.

Tamiasciurus hudsonicus picatus (Swarth)

1921. *Sciurus hudsonicus picatus* Swarth, Jour. Mamm., 2:92, May 2, type from Kupreanof Island, 25 mi. S Kake Village, at southern end of Keku Straits, SE Alaska.

1947. *Tamiasciurus hudsonicus picatus*, Anderson, Bull. Nat. Mus. Canada, Biol. Ser., 102:120, January 24.

MARGINAL RECORDS.—Alaska: Lynn Canal. British Columbia: Flood Glacier; Hazelton. Alaska: type locality; *Kulu* [= *Kuiu*] *Island.*

Tamiasciurus hudsonicus preblei A. H. Howell

1936. *Tamiasciurus hudsonicus preblei* A. H. Howell, Proc. Biol. Soc. Washington, 49:133, August 22, type from Fort Simpson, Mackenzie District, Northwest Territories.

MARGINAL RECORDS.—Alaska: vic. Arctic Village. Yukon: Porcupine River, 20 mi. NE Old Crow (Youngman, 1975:73). Mackenzie: Aklavik; Artillery Lake. Keewatin: mouth Windy River (Harper, 1956:19). Saskatchewan (Beck, 1958:29, as *T. hudsonicus* only): Reindeer Lake; Otosquen; *Madge Lake;* Kamsack; Fort Qu'Appelle; *Muscow;* Craven. Alberta (Soper, 1965:155, unless otherwise noted): Beaver River; White Earth Creek; Edmonton; Red Deer; Torrens Mtn. (Hall and Kelson, 1959:402). British Columbia: Aylard Creek; Redfern Lake, *Lower Liard Crossing.* Alaska: Yerrick Creek, 4 mi. N, 21 mi. W Tok Junction; *Swede Lake, Little Tok River*, various points between Paxson and Valdez; Tyonek; Lake Iliamna; Naknek Lake (Schiller and Rausch, 1956:197, as *T. h. kenaiensis*); Kowak River; near Anaktuvuk Pass.

Tamiasciurus hudsonicus regalis A. H. Howell

1936. *Tamiasciurus regalis* A. H. Howell, Occas. Pap. Mus. Zool., Univ. Michigan, 338:1, July 7, type from Belle Isle, Isle Royale, Michigan. Known only from Isle Royale.

1943. *Tamiasciurus hudsonicus regalis*, Burt, Occas. Pap. Mus. Zool., Univ. Michigan, 481:6, November 10.

Tamiasciurus hudsonicus richardsoni (Bachman)

1839. *Sciurus richardsoni* Bachman, Proc. Zool. Soc. London, for 1838, p. 100, February 7, type from head Big Lost River, Custer Co., Idaho (restricted by Merriam, N. Amer. Fauna, 5:50, July 30, 1891).

1939. *Tamiasciurus hudsonicus richardsoni*, Davis, The Recent mammals of Idaho, Caxton Printers, Caldwell, Idaho, p. 227, April 5.

MARGINAL RECORDS.—British Columbia: Assiniboine. Alberta: few mi. NW Crowsnest Lake; Waterton Lakes National Park. Montana: Summit; 9 mi. S Monarch. Idaho: Birch Creek; Lost River Mts.; Baker Creek, 12 mi. N Ketchum; Alturas Lake. Oregon: East Camp Creek, Unity; Meacham. Washington: Blue Mts. Idaho: Castle Creek Ranger Station. Montana: Bitterroot Valley; Missoula; Thompson Pass. British Columbia: *Rossland;* Deer Park; Nelson; Invermere.

Tamiasciurus hudsonicus streatori (J. A. Allen)

1898. *Sciurus hudsonicus streatori* J. A. Allen, Bull. Amer. Mus. Nat. Hist., 10:267, July 22, type from Ducks, British Columbia.

1936. *T[amiasciurus]. h[udsonicus]. streatori*, A. H. Howell, Proc. Biol. Soc. Washington, 49:135, August 22.

MARGINAL RECORDS.—British Columbia: 30 mi. N Mt. Revelstoke; Glacier; Monashee Pass. Washington: Marcus. Idaho: NE Idaho County; Lewis County. Washington: Fort Spokane; head Lake Chelan; Ruby Creek. British Columbia: Bridge River; *Pavilion;* Black Pines; *Mt. Todd* [= *Tod*].

Tamiasciurus hudsonicus ungavensis Anderson

1942. *Tamiasciurus hudsonicus ungavensis* Anderson, Ann. Rept. Provancher Soc. Nat. Hist., Quebec, for 1941, p. 33, July 14, type from Lake Waswanipi ("Woswonaby Post," Hudson's Bay Company), Abitibi District, Quebec, about 180 mi. SE intersection Quebec–Ontario interprovincial boundary with James Bay.

MARGINAL RECORDS.—Quebec: *Finger Lake* (not found), S *of Leaf Bay* (Harper, 1961:46): Chimo. Labrador: Rigoulette [*sic*]. Quebec: Lake Albanel; 47° 50′ N, 75° 35′ W, S of Clova (MacLeod and Cameron, 1961:282); type locality; Charlton Island, James Bay; Great Whale River; Sculpin Cove, Richmond Gulf (Edwards, 1963:6, recorded only as *T. hudsonicus*).

T. h. ungavensis introduced on Newfoundland in 1963 and on Camel Island in 1964 (N. F. Payne, MS, 1973; see also Payne, 1976:60–64).

Tamiasciurus hudsonicus ventorum (J. A. Allen)

1898. *Sciurus hudsonicus ventorum* J. A. Allen, Bull. Amer. Mus. Nat. Hist., 10:263, July 22, type from South Pass City, Wind River Mts., Fremont Co., Wyoming.

1939. *Tamiasciurus hudsonicus ventorum*, Davis, The Recent mammals of Idaho, Caxton Printers, Caldwell, Idaho, p. 229, April 5.

MARGINAL RECORDS.—Montana: Mystic Lake. Wyoming (Long, 1965a:597): 15½ mi. S, 3 mi. W Meeteetse; ½ mi. N, 3 mi. E South Pass City; type locality; LaBarge Creek. Utah: Old Canyon, 6 mi. W Randolph; Aspen Grove, Mt. Timpanogos; The Dam, Parleys Canyon; *Mt. Willard, Weber–Boxelder County line.* Idaho: Malad City; Indian Creek, 4 mi. S Pocatello; N fork Pocatello Creek, 6½ mi. NE Pocatello; Henrys Lake.

Tamiasciurus douglasii
Douglas' Squirrel

X ¼

External measurements: 270–348; 102–156; 45–55. Weights of five males and five females from eastern Oregon are, respectively, 256 and 255 grams. The color varies individually, geographically, and seasonally. Upper parts dark olivaceous brown to brownish gray, usually with broad median band of dark ferruginous to chestnut; tail above usually like median part of back for proximal two-thirds, then becoming black or blackish; tail below approx. grizzled rusty bordered with black and edged with buffy- or white-tipped hairs; underparts strong buffy gray through ochraceous tones to reddish orange with strong black wash.

The dark *Tamiasciurus douglasii albolimbatus* of the western parts of Washington and Oregon may meet and intergrade at one or more places with the paler *T. hudsonicus richardsoni* of the eastern parts of those states, a region drier than the humid coastal region. L. E. Hatton (MS., 1975) notes that one such place may be the coniferous forest of the Blue Mountains of northeastern Oregon, and states that the requisite forest habitat extends from there west, north, and northeast into the geographic ranges of the two subspecies mentioned above. If investigation reveals that intergradation occurs in the zone of contact of the two taxa, as seems likely, the four subspecies of *T. douglasii* (Bachman, 1839) of course will take the older specific name, *T. hudsonicus* (Erxleben, 1777), making 29 subspecies in all of a monotypic species of a monotypic genus.

Tamiasciurus douglasii albolimbatus (J. A. Allen)

1890. *Sciurus hudsonius* [sic] *californicus* J. A. Allen, Bull. Amer. Mus. Nat. Hist., 3:165, November 14, type from Blue

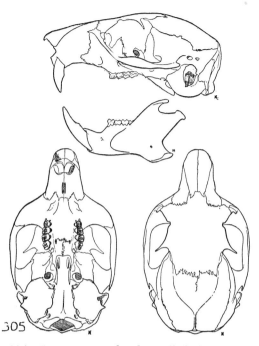

Fig. 305. *Tamiasciurus douglasi albolimbatus*, Incline Creek, 6500 ft., Nevada, No. 65215 M.V.Z., ♀, X 1.

Canyon, Placer Co., California. Not *Sciurus californicus* Lesson, 1847 [= ?*Citellus beecheyi*], type locality, "California."

1898. *Sciurus douglasii albolimbatus* J. A. Allen, Bull. Amer. Mus. Nat. Hist., 10:453, November 10, a renaming of *Sciurus hudsonius* [sic] *californicus* J. A. Allen, 1890.

1940. *Tamiasciurus douglasii albolimbatus*, Hayman and Holt, *in* Ellerman, The families and genera of living rodents, British Mus., 1:347, June 8.

MARGINAL RECORDS.—Oregon: 10 mi. W Wapineta; Strawberry Butte; 10 mi. N Harney; Warner Mts. California: Camp Bidwell; Big Valley Mts.; Honey Lake. Nevada: 3 mi. S Mt. Rose, 8500 ft. California: Markleeville; vic. Mammoth; Cottonwood Lakes; Kern River; E fork Kaweah River; S fork Merced River; Snow Mtn.; Sanhedrin Mtn.; Siskiyou Mts. Oregon: Prospect; Diamond Lake.

Tamiasciurus douglasii douglasii (Bachman)

1836. *Sciurus douglasii* Gray, Proc. Zool. Soc. London, p. 88, a *nomen nudum*.

1839. *Sciurus douglasii* Bachman, Proc. Zool. Soc. London, p. 99, for 1838, February 7, type from "shores of the Columbia River" (Audubon and Bachman, The viviparous quadrupeds of North America, 1:371, 1841), subsequently restricted by J. A. Allen (Bull. Amer. Mus. Nat. Hist., 10:284, July 22, 1898) to mouth Columbia River.

1940. *Tamiasciurus douglasii douglasii*, Hayman and Holt, *in* Ellerman, The families and genera of living rodents, British Mus., 1:347, June 8.

1842. *Sciurus belcheri* Gray, Ann. Mag. Nat. Hist., ser. 1, 10:263, type from mouth Columbia River.

1855. *Sciurus suckleyi* Baird, Proc. Acad. Nat. Sci. Philadelphia, 7:333, April, type from Steilacoom, Puget Sound, Washington.

MARGINAL RECORDS.—Washington: Simiahmoo [Bay]; Sauk; Tacoma; Kalama. Oregon: Eagle Creek; Scottsburgh; Marshfield; thence northward along coast to point of beginning.

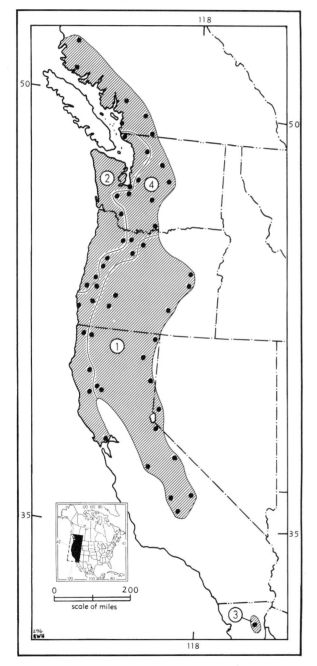

Map 296. *Tamiasciurus douglasii.*

1. *T. d. albolimbatus*
2. *T. d. douglasii*
3. *T. d. mearnsi*
4. *T. d. mollipilosus*

Tamiasciurus douglasii mearnsi (Townsend)

1897. *Sciurus hudsonius mearnsi* Townsend, Proc. Biol. Soc. Washington, 11:146, June 9, type from San Pedro Mártir Mts., about 7000 ft., Baja California.
1940. *Tamiasciurus douglasii mearnsi*, Hayman and Holt, *in* Ellerman, The families and genera of living rodents, British Mus., 1:348, June 8.

MARGINAL RECORDS.—Baja California: Vallecitos, Sierra San Pedro Mártir; *La Grulla.*

Tamiasciurus douglasii mollipilosus (Audubon and Bachman)

1841. *Sciurus molli-pilosus* Audubon and Bachman, Proc. Acad. Nat. Sci. Philadelphia, 1:102, October. Type locality, coast of northern California ("Most probably somewhere in Oregon" according to Grinnell, Univ. California Publ. Zool., 40:133, September 26, 1933).
1940. *Tamiasciurus douglasii mollipilosus*, Hayman and Holt, *in* Ellerman, The families and genera of living rodents, British Mus., 1:347, June 8.
1897. *Sciurus hudsonicus orarius* Bangs, Proc. Biol. Soc. Washington, 11:281, December 30, type from Philo, Mendocino Co., California.
1898. *Sciurus douglasii cascadensis* J. A. Allen, Bull. Amer. Mus. Nat. Hist., 10:277, July 22, type from Mt. Hood, Oregon (for status see V. Bailey, N. Amer. Fauna, 55:122, August 29, 1936, and Anderson, Bull. Nat. Mus. Canada, 102:121, January 24, 1937).

MARGINAL RECORDS.—British Columbia: S side Owikeno Lake (Cowan and Guiguet, 1965:156); Alta Lake (*ibid.*); Alexandra Lodge (*ibid.*); 1 mi. E Pinewoods. Washington: head Lake Chelan; Wenatchee; Natchez River; Goldendale. Oregon: Bald Mtn., head Clackamas River; Glendale. California: Gasduct [= Gasquet]; South Fork Mtn.; Willets, 1700–2000 ft.; Petaluma, thence northward along coast to Oregon: Port Orford; Elk Head; Eugene; Sweet Home; Mt. Hood. Washington: Tenino; Roy; Snoqualmie Falls. British Columbia: Vancouver, thence along coast, including *Bowen, Gambier,* and *Stuart* islands, to Kingcome Inlet (Cowan and Guiguet, 1965:156).

Genus **Glaucomys** Thomas
American Flying Squirrels

Revised by A. H. Howell, N. Amer. Fauna, 44:1–64, June 13, 1918.

1908. *Glaucomys* Thomas, Ann. Mag. Nat. Hist., ser. 8, 1:5, January. Type, *Mus volans* Linnaeus.

Fore- and hind-limbs connected from wrists to ankles by loose fold of fully haired skin; tail broad, flattened, almost parallel-sided, and tip rounded. Skull lightly constructed and somewhat flattened; nasals abruptly depressed at tip; dorsal profile of skull from nasals to postfrontal region nearly straight, abruptly depressed to occiput;

frontals long and narrow; interorbital region narrow; infraorbital foramen oval, vertical, and situated near P3; zygomatic plate tilted upward at angle of approx. 65° to basicranial axis; zygomata appressed, slightly convergent anteriorly, with nearly vertical sides; postorbital process broad at base, tapering abruptly to a point, and depressed at tip. Dentition, i. $\frac{1}{1}$, c. $\frac{0}{0}$, p. $\frac{2}{1}$, m. $\frac{3}{3}$.

Key to Species of Glaucomys

1. Size larger (total length more than 260; greatest length of skull more than 36); hairs on underparts gray basally.
G. sabrinus,   p. 450
1'. Size smaller (total length less than 260; greatest length of skull less than 36); hairs on underparts white or almost so basally. . . . . *G. volans*,   p. 448

Glaucomys volans
Southern Flying Squirrel

x½

External measurements: 210–253; 81–120; 28–33; greatest length of skull, 32.0–36.0; interorbital breadth, 6.5–7.1; zygomatic breadth, 20.0–21.3; maxillary tooth-row, 6.2–7.0. Upper parts drab, pinkish cinnamon, sayal brown, snuff brown, hair-brown, or yellowish wood brown according to season and subspecies; sides of face smoke gray, often washed with fuscous or buff; tail above hair-brown, snuff brown, verona brown, fuscous, or drab; tail below pinkish cinnamon, vinaceous-cinnamon, or pinkish buff; underparts white or creamy-white, often with cinnamon or pinkish buff sides; hairs of underparts white or whitish to base, except on glissant membranes and hind legs where plumbeous basally.

In some areas (e.g., in New York and Virginia) where *G. volans* and *G. sabrinus* occur together, certain specimens are identifiable to species only with difficulty. In every case, however, a consideration of *all* characters that usually serve to identify the species will permit recognition, whereas reliance on one or a few characters may not.

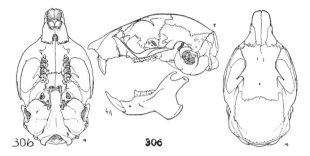

Fig. 306.   *Glaucomys volans saturatus*, 4 mi. SE Bergman, Boone Co., Arkansas, No. 95369 M.V.Z., ♀, X 1.

Glaucomys volans chontali Goodwin

1961. *Glaucomys volans chontali* Goodwin, Amer. Mus. Novit., 2059:3, November 29, type from Santo Domingo, Chontecomatlán, District of Yautepec, Oaxaca, México, *ca.* 7000 ft. Known only from type locality.

Glaucomys volans goldmani (Nelson)

1904. *Sciuropterus volans goldmani* Nelson, Proc. Biol. Soc. Washington, 17:148, October 6, type from 20 mi. SE Teopisca, Chiapas.
1918. *Glaucomys volans goldmani*, A. H. Howell, N. Amer. Fauna, 44:28, June 13.

MARGINAL RECORDS.—Chiapas: Palma Real, 6 mi. N Ocozocoautla, 1800 m. (Goodwin, 1961a:7); Ococingo [Ocosingo](*ibid.*); *9 mi. WNW Comitán* (*ibid.*); type locality.

Glaucomys volans herreranus Goldman

1936. *Glaucomys volans herreranus* Goldman, Jour. Washington Acad. Sci., 26:463, November 15, type from mts. of Veracruz.

MARGINAL RECORDS.—Tamaulipas: Aserradero del Infiernillo. Veracruz: Los Pescados, Cofre de Perote. Oaxaca: Sierra Madre, N of Zanatepec (Goodwin, 1969:135). Michoacán: 9 km. by highway S Pátzcuaro (sight record only). Querétaro: Pinal de Amoles. San Luis Potosí: 8 mi. by road E Santa Barbarita.

Glaucomys volans madrensis Goldman

1936. *Glaucomys volans madrensis* Goldman, Jour. Washington Acad. Sci., 26:463, November 15, type from Sierra Madre, Chihuahua. *Known only from type locality.*

Glaucomys volans oaxacensis Goodwin

1961. *Glaucomys volans oaxacensis* Goodwin, Amer. Mus. Novit., 2059:11, November 29, type from San Pedro Jilotepec, District of Tehuantepec, Oaxaca, México, *ca.* 5000 ft.

MARGINAL RECORDS (Goodwin, 1961a:12, unless otherwise noted).—Guerrero: Acahuizotla. Oaxaca: Cerro Yucuyácua, 7000 ft.; Cerro San Felipe, 2900–3000 m.; Santo Domingo Nejapa; type locality; *Tenango* (Goodwin, 1969:134).

Glaucomys volans querceti (Bangs)

1896. *Sciuropterus volans querceti* Bangs, Proc. Biol. Soc. Washington, 10:166, December 28, type from Citronelle, Citrus Co., Florida.
1918. *Glaucomys volans querceti*, A. H. Howell, N. Amer. Fauna, 44:26, June 13.

MARGINAL RECORDS.—Georgia: Montgomery. Florida: Lake Harney; Shell Hammock; *2 mi. S Miami* (Layne; 1974:392); *Goulds (ibid.); 4³⁄₁₀ mi. W Homestead (ibid.); 2 mi. E Royal Palm Hammock (ibid.);* Fort Myers; type locality. Georgia: Okefinokee Swamp.

Glaucomys volans saturatus A. H. Howell

1915. *Glaucomys volans saturatus* A. H. Howell, Proc. Biol. Soc. Washington, 28:110, May 27, type from Dothan, Houston Co., Alabama.

MARGINAL RECORDS.—Kentucky: Big Black Mtn. Tennessee: 7 mi. SW Crossville. Alabama: Sand Mtn., near Carpenter. North Carolina: Magnetic City, foot Roan Mtn.; Cranberry. South Carolina: Greenville; Plantersville. Georgia: Reidsville. Alabama: type locality. Florida: Milton. Louisiana (Lowery, 1974:204, unless otherwise noted): Lake Hermitage; *near Gibson* (Hall and Kelson, 1959:407); 3 mi. N Abbeville; *4 mi. N Basile; Bryceland;* 5 mi. N Homer. Arkansas: Delight. Oklahoma: 5 mi. SW Colbert; Wichita Mts. Wildlife Refuge (Glass and Halloran, 1961:237); Oklahoma City; Stillwater; Twin Lakes; Scraper. Arkansas: 1 mi. N Winslow; Sylamore Experimental Forest, Calico Rock (Ward and Leonard, 1968:530, as *G. volans* only). Kentucky: Hickman. Tennessee: 6 mi. SW Frankewing.

Glaucomys volans texensis A. H. Howell

1915. *Glaucomys volans texensis* A. H. Howell, Proc. Biol. Soc. Washington, 28:110, May 27, type from 7 mi. NE Sour Lake, Hardin Co., Texas.

MARGINAL RECORDS.—Texas: Gainesville; Texarkana. Louisiana (Lowery, 1974:204): Natchitoches; *1³⁄₅ mi. NW Flatwoods, along Hwy. 119;* Maplewood; *7 mi. W Lake Charles.* Texas: type locality; Cuero; Guadalupe River, 40 mi. E San Antonio; Aledo.

Glaucomys volans underwoodi Goodwin

1936. *Glaucomys volans underwoodi* Goodwin, Amer. Mus. Novit., 898:1, December 31, type from Zambrano, 4500 ft., Tegucigalpa, Honduras.

MARGINAL RECORDS (Goodwin, 1961a:15).—Guatemala: Tecpan. Honduras: Gracias; type locality. Guatemala: Dueñas.

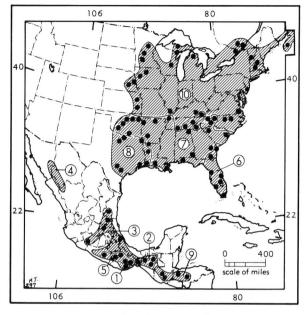

Map 297. *Glaucomys volans.*

1. G. v. chontali
2. G. v. goldmani
3. G. v. herreranus
4. G. v. madrensis
5. G. v. oaxacensis
6. G. v. querceti
7. G. v. saturatus
8. G. v. texensis
9. G. v. underwoodi
10. G. v. volans

Glaucomys volans volans (Linnaeus)

1758. [*Mus*] *volans* Linnaeus, Syst. nat., ed. 10, 1:63. Type locality fixed by Elliot (Field Columb. Mus., Zool. Ser., 2:109, 1901) as Virginia.
1915. [*Glaucomys*] *volans*, A. H. Howell, Proc. Biol. Soc. Washington, 28:109, May 27.
1778. *Sciurus volucella* Pallas, Novae species quadrupedum e glirium ordine . . . , p. 351, seemingly a renaming of *Mus volans* Linnaeus.
1808. *Pteromys virginianus* Tiedemann, Zoologie . . . , p. 451, a renaming of *Mus volans* Linnaeus.
1816. *Pteromys americana* Oken, Lehrbuch der Naturgeschichte, 2:865, a renaming of *Sciurus volucella* Pallas. Oken 1816 names are, in any case, non-Linnean and not available under the *Regles.*
1829. ? *Pteromys cucullatus* Fischer, Synopsis mammalium, p. 365. Type locality, Virginia (?). Based on "*Sciurus, Virginianus, volans*" Seba.
1896. *Sciuropterus silus* Bangs, Proc. Biol. Soc. Washington, 10:163, December 28, type from top of Katis Mtn., 3200 ft., near White Sulphur Springs, West Virginia.
1915. *Pteromys volans nebrascensis* Swenk, Univ. Nebraska Studies, 15:151, September 25, type from Nebraska City, Otoe Co., Nebraska.

MARGINAL RECORDS.—Minnesota: Aitkin. Wisconsin: Fond du Lac County (Jackson, 1961:177, Map 34). Michigan: Menominee County; Marquette (Haveman and Robinson, 1976:40, as *G. volans* only); Otsego County. Ontario: *ca.* 45° N Clayton. Quebec:

Fortune Lake (Youngman and Gill, 1968:227); Hudson (Wrigley, 1969:207, as *G. volans* only). Vermont: Rutland County. Nova Scotia (Wood and Tessier, 1974:83, as *G. volans* only): near Grafton Lake; *near Pebbleloggitch Lake.* New Hampshire: Hancock. Massachusetts: vic. Amherst (Muul, 1969:542, as *G. volans* only). New Jersey: 4 mi. NW New Gretna, thence down coast to Virginia: Suffolk. North Carolina: Raleigh; Apex. Tennessee: *Wautauga Valley;* Roan Mtn., 4100 ft.; Big Frog Mtn., 12 mi. W Copperhill, 2000 ft. Kentucky: 3⅗ km. E Burnside (Fassler, 1974:41, as *G. volans* only). Tennessee: 8 mi. N Waynesboro. Illinois: Olive Branch. Kansas: Woodson County; Topeka. Nebraska (Jones, 1964c:150, 151): Beaver Crossing; Florence. Iowa: Wall Lake; Humboldt. Minnesota (Heaney and Birney, 1975:32): 2 mi. N, ½ mi. E Morton; Stearns County; *Upper Long Lake, Crow Wing Co.* The alleged occurrence in Clearwater County, Minnesota (Gunderson and Beer, 1953:88), is not shown on Map 297 because Heaney and Birney (1975:32) suspect that the specimen was misidentified.

Glaucomys sabrinus
Northern Flying Squirrel

External measurements: 263–368; 115–180; 34–45. Greatest length of skull, 37.3–44.2. Upper parts varying from cinnamon to pecan brown according to subspecies; tail above from cinnamon to fuscous or even blackish, usually darkest near tip; tail below varying from pinkish cinnamon to nearly black; underparts white or creamy white, often washed with some shade of buffy or yellowish; sides of head and sometimes face gray, often with wash of buff, cinnamon, or fuscous.

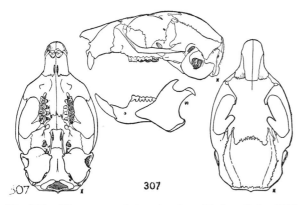

Fig. 307. *Glaucomys sabrinus lascivus*, Marlette Lake, 8000 ft., Nevada, No. 69640 M.V.Z., ♂, X 1.

Glaucomys sabrinus alpinus (Richardson)

1828. *Pteromys alpinus* Richardson, Zool. Jour., 3:519. Type locality, Jasper House, Alberta. (See A. H. Howell, N. Amer. Fauna, 44:40, June 13, 1918.)
1918. *Glaucomys sabrinus alpinus*, A. H. Howell, N. Amer. Fauna, 44:40, June 13.

MARGINAL RECORDS.—Yukon (Youngman, 1975:77, as *G. s. sabrinus*): Kathleen River, Haines Road; Louise Lake, 7½ mi. W Whitehorse. British Columbia: Lower Liard Crossing; Peace River Canyon, near Hudson Hope; Tupper Creek (Cowan and Guiguet, 1965:158). Alberta: Wildhay River country; Banff (Banfield, 1958:18); type locality. British Columbia: Henry Creek; Ten-mile Lake; Nulki Lake (Cowan and Guiguet, 1965:158); Babine Lake; Salvus; *Warm Springs, 15 mi. S Atlin;* Atlin (Cowan and Guiguet, 1965:158).

Glaucomys sabrinus bangsi (Rhoads)

1897. *Sciuropterus alpinus bangsi* Rhoads, Proc. Acad. Nat. Sci. Philadelphia, 49:321, July 19, type from near Raymond, Idaho Co., Idaho.
1918. *Glaucomys sabrinus bangsi*, A. H. Howell, N. Amer. Fauna, 44:38, June 13.
1915. *Glaucomys bullatus* A. H. Howell, Proc. Biol. Soc. Washington, 28:113, May 27, type from Sawtooth (= Alturas) Lake, E base Sawtooth Mts., Idaho. Regarded as inseparable from *G. s. bangsi* by Mayer (Murrelet, 22:31, September 15, 1941).

MARGINAL RECORDS.—Montana: Paola; Willow Creek. Idaho: 7 mi. W Yellowstone, 7000 ft. Wyoming (Long, 1965a:600): 31½ mi. N, 36 mi. W Cody; *middle fork Hay Creek, Bear Lodge Mts.;* ½ mi. N, 3 mi. E Buckhorn. South Dakota: Black Hills. Wyoming: 10 mi. NE Pinedale (Long, 1965a:600). Idaho: Justice Park; Ketchum. Oregon: Beech Creek. Washington: Wildcat Spring. Idaho: Golden.

Glaucomys sabrinus californicus (Rhoads)

1897. *Sciuropterus alpinus californicus* Rhoads, Proc. Acad. Nat. Sci. Philadelphia, 49:323, July 19, type from near Squirrel Inn, 5200 ft., San Bernardino Mts., San Bernardino Co., California.
1918. *Glaucomys sabrinus californicus*, A. H. Howell, N. Amer. Fauna, 44:56, June 13.

MARGINAL RECORDS.—California: vic. Red Ant Creek, Big Bear Lake; Idyllwild, Strawberry Valley, San Jacinto Mts.; type locality.

Glaucomys sabrinus canescens A. H. Howell

1915. *Glaucomys sabrinus canescens* A. H. Howell, Proc. Biol. Soc. Washington, 28:111, May 27, type from Portage la Prairie, Manitoba.

MARGINAL RECORDS.—Manitoba: Poplar Point, S end Lake Winnipeg. Minnesota: Breckinridge. North Dakota: Portland. Manitoba: Treesbank; Carberry; type locality.

Glaucomys sabrinus coloratus Handley

1953. *Glaucomys sabrinus coloratus* Handley, Proc. Biol. Soc. Washington, 66:191, December 2, type from Bald Knob, 5000 ft., 3½ mi. S summit Mt. Mitchell, Yancey Co., North Carolina.

MARGINAL RECORDS.—Tennessee: Roan Mtn., 5500 ft. North Carolina: Mt. Mitchell, 5000 ft. Tennessee: Blanket Mtn., 4000 ft.

Glaucomys sabrinus columbiensis A. H. Howell

1915. *Glaucomys sabrinus columbiensis* A. H. Howell, Proc. Biol. Soc. Washington, 28:111, May 27, type from Okanagan, British Columbia.

MARGINAL RECORDS.—British Columbia: Salmon River; Lumby; Canyon Creek in Kettle Valley; *Bridesville*. Washington: Molson; Lake Chelan; Stehekin. British Columbia: eastern Manning Park; Hedley; *Falkland*.

Glaucomys sabrinus flaviventris A. H. Howell

1915. *Glaucomys sabrinus flaviventris* A. H. Howell, Proc. Biol. Soc. Washington, 28:112, May 27, type from head Bear Creek, 6400 ft., Trinity Co., California.

MARGINAL RECORDS.—California: head Parker Creek, 7300 ft., Warner Mts.; Trinity Mts.; Grizzly Creek, 6000 ft.; Salmon Mts.

Glaucomys sabrinus fuliginosus (Rhoads)

1897. *Sciuropterus alpinus fuliginosus* Rhoads, Proc. Acad. Nat. Sci. Philadelphia, 49:321, July 19, type from Cascade Mts., near Martin Station, about 8000 ft., Kittitas Co., Washington.
1918. *Glaucomys sabrinus fuliginosus*, A. H. Howell, N. Amer. Fauna, 44:47, June 13.

MARGINAL RECORDS.—British Columbia: Blackwater Lake near Lillooet; Manning Park. Washington: Entiat River; Easton; Potato Hill, 15 mi. N Goldendale. Oregon: Belknap Springs; Crater Lake. California: Preston Peak, Siskiyou Mts. Oregon: Vida. Washington: Carson; Glacier Basin; Mt. Baker. British Columbia: North Vancouver (Cowan and Guiguet, 1965:160); *Mt. Sproatt (ibid.)*; Alta Lake; *Seton Lake*.

Glaucomys sabrinus fuscus Miller

1936. *Glaucomys sabrinus fuscus* Miller, Proc. Biol. Soc. Washington, 49:143, August 22, type from Cranberry Glades, 3300 ft., Pocahontas Co., West Virginia.

MARGINAL RECORDS.—West Virginia: Bickle Knob, 3900 ft., 7⁹⁄₁₀ mi. NE Elkins; *Cheat Bridge, 3900–4000 ft.*; type locality; *Mill Pt., Cranberry River, 3450 ft.*

Glaucomys sabrinus goodwini Anderson

1943. *Glaucomys sabrinus goodwini* Anderson, Ann. Rept. Provancher Soc. Nat. Hist., Quebec, for 1942, p. 55, September 7, type from Berry Mtn. Camp, jct. Berry Mtn. Brook and Grand Cascapedia River, about 1500 ft., Matane Co., Quebec.

MARGINAL RECORDS.—Quebec: Mt. Albert; type locality. Not found: Quebec: Tracadie.

Glaucomys sabrinus gouldi Anderson

1943. *Glaucomys sabrinus gouldi* Anderson, Ann. Rept. Provancher Soc., Nat. Hist., Quebec, for 1942, p. 56, September 7, type from Frizzleton, Inverness Co., Cape Breton Island, Nova Scotia.

MARGINAL RECORDS.—Nova Scotia: type locality. Prince Edward Island: Rollo Bay (Cameron, 1959:45).

Glaucomys sabrinus griseifrons A. H. Howell

1934. *Glaucomys sabrinus griseifrons* A. H. Howell, Jour. Mamm., 15:64, February 15, type from Lake Bay, Prince of Wales Island, Alaska. Known only from type locality.

Glaucomys sabrinus klamathensis (Merriam)

1897. *Sciuropterus alpinus klamathensis* Merriam, Proc. Biol. Soc. Washington, 11:225, July 15, type from Fort Klamath, 4200 ft., Klamath Co., Oregon.
1918. *Glaucomys sabrinus klamathensis*, A. H. Howell, N. Amer. Fauna, 44:52, June 13.

MARGINAL RECORDS.—Oregon: Paulina Lake; Upper Klamath Lake; *type locality; Crater Peak, 4 mi. S Crater Lake*.

Glaucomys sabrinus lascivus (Bangs)

1899. *Sciuropterus alpinus lascivus* Bangs, Proc. New England Zool. Club, 1:69, July 31, type from Tallac, El Dorado Co., California.
1918. *Glaucomys sabrinus lascivus*, A. H. Howell, N. Amer. Fauna, 44:55, June 13.

MARGINAL RECORDS.—California: Dana; Blacks Mtn. (McKeever, 1960:270); Willard Creek (*ibid.*). Nevada: Marlette Lake, 8000 ft. California: Kings River Canyon; Sherman Creek, Sequoia National Park; Fresno (mts. near?); Dudley, Smith Creek, 3000 ft.; Mill Creek, S base Mt. Lassen.

Glaucomys sabrinus latipes A. H. Howell

1915. *Glaucomys sabrinus latipes* A. H. Howell, Proc. Biol. Soc. Washington, 28:112, May 27, type from Glacier, British Columbia.

MARGINAL RECORDS.—British Columbia: Field; Tornado Pass (Cowan and Guiguet, 1965:161). Alberta: Waterton Lakes National Park. Montana: Nyack; Stanton Lake. Idaho: Orofino; Cedar Mtn. Washington: Loon Lake. British Columbia: *Westbridge; Shuswap*; Mt. Revelstoke (Cowan and Guiguet, 1965:161); *Glacier (ibid.)*.

Glaucomys sabrinus lucifugus Hall

1934. *Glaucomys sabrinus lucifugus* Hall, Occas. Pap. Mus. Zool., Univ. Michigan, 296:1, November 2, type from 12 mi. E Kamas, Summit Co., Utah.

MARGINAL RECORDS.—Utah: near Peterson; Henrys Fork, 8000 ft.; jct. Deep and Carter creeks, 7900 ft.; Paradise Park, 10,000 ft., Uinta Mts.; 15 mi. N Mountain Home; Ephraim [mts. near?]; S point, top Boulder Mtn., 11,000 ft.; Bryce National Park; Kolob Reservoir (Stock, 1970:430); Mt. Timpanogos.

Glaucomys sabrinus macrotis (Mearns)

1898. *Sciuropterus sabrinus macrotis* Mearns, Proc. U.S. Nat. Mus., 21:353, November 4, type from Hunter Mtn., 3300 ft., Catskill Mts., Greene Co., New York.

1915. *G[laucomys]. s[abrinus]. macrotis*, A. H. Howell, Proc. Biol. Soc. Washington, 28:111, May 27.

MARGINAL RECORDS.—New Brunswick: Edmundston; Miramichi Road, Gloucester Co., thence down coast to Massachusetts: Wilmington. Connecticut (Choate and Dubos, 1971:19): Union; Barkhamstead. New York: type locality. Pennsylvania: Monroe County (Grimm and Whitebread, 1952:60); McGees Mills; Horse Creek, 6 mi. E Oil City (Richmond and Roslund, 1949:45); Erie (mts. near ?). Michigan: Montcalm County. Wisconsin: Worden Twp., Clark Co. Minnesota: Elk River. Wisconsin (Jackson, 1961:181, 184): Burnett County; Douglas County, thence eastward on S shores lakes Superior and Huron, and through peninsular Ontario, to Quebec: Blue Sea Lake.

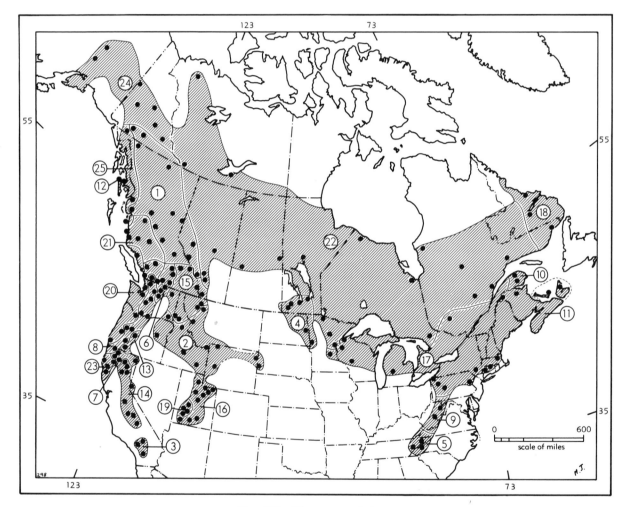

Map 298. *Glaucomys sabrinus*.

Guide to subspecies			
1. *G. s. alpinus*	5. *G. s. coloratus*	12. *G. s. griseifrons*	19. *G. s. murinauralis*
2. *G. s. bangsi*	6. *G. s. columbiensis*	13. *G. s. klamathensis*	20. *G. s. oregonensis*
3. *G. s. californicus*	7. *G. s. flaviventris*	14. *G. s. lascivus*	21. *G. s. reductus*
4. *G. s. canescens*	8. *G. s. fuliginosus*	15. *G. s. latipes*	22. *G. s. sabrinus*
	9. *G. s. fuscus*	16. *G. s. lucifugus*	23. *G. s. stephensi*
	10. *G. s. goodwini*	17. *G. s. macrotis*	24. *G. s. yukonensis*
	11. *G. s. gouldi*	18. *G. s. makkovikensis*	25. *G. s. zaphaeus*

Glaucomys sabrinus makkovikensis (Sornborger)

1900. *Sciuropterus sabrinus makkovikensis* Sornborger, Ottawa Naturalist, 14:48, June 6, type from Makkovik, Labrador.
1918. *Glaucomys sabrinus makkovikensis*, A. H. Howell, N. Amer. Fauna, 44:34, June 13.

MARGINAL RECORDS.—Labrador: type locality, thence down coast to Quebec: St. Augustine (Harper, 1961:48). Labrador: Northwest River.

Glaucomys sabrinus murinauralis Musser

1961. *Glaucomys sabrinus murinauralis* Musser, Proc. Biol. Soc. Washington, 74:120, August 11, type from Timid Springs, 10,300 ft., 1 mi. N Big Flat Guard Station, Tushar Mountains, Beaver Co., Utah.

MARGINAL RECORDS (Musser, 1961:125).—Utah: Fillmore Canyon, 8 mi. E Fillmore, 6800 ft.; type locality; Kents Lake, 8800 ft.; Indian Creek Guard Station, 9 mi. E Hwy. 91, 7875 ft.

Glaucomys sabrinus oregonensis (Bachman)

1839. *Pteromys oregonensis* Bachman, Jour. Acad. Nat. Sci. Philadelphia, 8:101, type from pine woods of the Columbia, near the sea. Probably near St. Helens, Columbia Co., Oregon. (See Rhoads, Proc. Acad. Nat. Sci. Philadelphia, 49:324, June, 1897.)
1918. *Glaucomys sabrinus oregonensis*, A. H. Howell, N. Amer. Fauna, 44:44, June 13.
1899. *Sciuropterus alpinus olympicus* Elliot, Field Columb. Mus., Publ. 30, Zool. Ser., 1:225, February 2, type from Happy Lake, Clallam Co., Washington. Regarded as inseparable from *G. s. oregonensis* by Dalquest (Univ. Kansas Publ., Mus. Nat. Hist., 2:295, April 9, 1948).

MARGINAL RECORDS.—British Columbia: Loughborough Inlet; *Bute Inlet;* North Vancouver (Cowan and Guiguet, 1965:162); *Stave Lake (ibid.);* Agassiz; *Upper Skagit River.* Washington: Camp Skagit; Cottage Lake. Oregon: Marmot; Elk Head; Gold Beach; thence northward along coast to British Columbia (Cowan and Guiguet, 1965:162): *Cortez Island; Quadra Island.*

Glaucomys sabrinus reductus Cowan

1937. *Glaucomys sabrinus reductus* Cowan, Proc. Biol. Soc. Washington, 50:79, June 22, type from Lonesome Lake, Atnarko River, approx. 52° 10′ N and 125° 45′ W, British Columbia.

MARGINAL RECORDS.—British Columbia: Wistaria; *Ootsa Lake; Quesnel;* Chezacut; type locality; Koeye River (Cowan and Guiguet, 1965:162); Ormidale Harbour (*ibid.*).

Glaucomys sabrinus sabrinus (Shaw)

1788. [*Sciurus*] *hudsonius* Gmelin, Syst. nat., ed. 13, 1:153. Type locality, mouth Severn River, Ontario. A homonym of

Sciurus hudsonius Gmelin, *antea*, p. 147, and possibly of *Sciurum hudsonium* Pallas, 1778.
1801. *Sciurus Sabrinus* Shaw, General zoology, 2:157, a renaming of *Sciurus hudsonius* Gmelin.
1915. [*Glaucomys*] *sabrinus*, A. H. Howell, Proc. Biol. Soc. Washington, 28:111, May 27.
1803. ?*Pteromys canadensis* É. Geoffroy St.-Hilaire, Catalogue des mammifères du Muséum d'Histoire Naturelle, Paris, p. 170, type from North America, probably Quebec (*fide* A. H. Howell, N. Amer. Fauna, 44:31, June 13, 1918).

MARGINAL RECORDS.—Mackenzie, Northwest Territories: Fort Anderson; Fort Resolution. Ontario: type locality; Moose Factory. Quebec: interior from Fort George (Harper, 1961:47); Kallio Lake; Matamek River; Tadousac; Lake Edward. Ontario: Trout Creek. Minnesota: Hinckley; Itasca County; Lake of the Woods County (46253 KU). Manitoba: Norway House. Saskatchewan: Cumberland House; 5 mi. S Battleford (Sealey, 1961:184, as *G. sabrinus* only). Alberta: Calgary; Didsbury; Edmonton. Mackenzie: Fort Liard.

Glaucomys sabrinus stephensi (Merriam)

1900. *Sciuropterus oregonensis stephensi* Merriam, Proc. Biol. Soc. Washington, 13:151, June 13, type from Sherwood, 2500 ft., Mendocino Co., California.
1918. *Glaucomys sabrinus stephensi*, A. H. Howell, N. Amer. Fauna, 44:57, June 13.

MARGINAL RECORDS.—California: Cecilville; Dos Rios; type locality; Eureka.

Glaucomys sabrinus yukonensis (Osgood)

1900. *Sciuropterus yukonensis* Osgood, N. Amer. Fauna, 19:25, October 6, type from Camp Davidson, Yukon River, near Alaska–Canada boundary, Yukon.
1918. *Glaucomys sabrinus yukonensis*, A. H. Howell, N. Amer. Fauna, 44:41, June 13.

MARGINAL RECORDS.—Alaska: 8 mi. N Tanana. Yukon, Northwest Territories: type locality; Mayo Lake, near head Stewart River; Frances Lake; Lapie River (Canol Road, Mile 132, near jct. Pelly and Ross rivers); Selkirk, jct. Pelly and Lewes rivers. Alaska: *Cook Inlet district;* head Toklat River.

Glaucomys sabrinus zaphaeus (Osgood)

1905. *Sciuropterus alpinus zaphaeus* Osgood, Proc. Biol. Soc. Washington, 18:133, April 18, type from Helm Bay, Cleveland Peninsula, southeastern Alaska.
1918. *Glaucomys sabrinus zaphaeus*, A. H. Howell, N. Amer. Fauna, 44:43, June 13.

MARGINAL RECORDS.—Yukon: 1½ mi. S, 3 mi. E Dalton Post, 2500 ft. British Columbia: Nass River; Princess Royal Island. Alaska: *type locality;* Etolin Island.

Family GEOMYIDAE—Pocket Gophers

Genera revised by Russell, Univ. Kansas Publ., Mus. Nat. Hist., 16:473–579, 9 figs., August 5, 1968.

Large fur-lined cheek pouches, one on each side of the face, opening outside the mouth is one of several characters shared by this family with the family Heteromyidae.

Medium-sized rodents (132–400 mm.) with thickset bodies and little if any external evidence of a neck; strongly modified for fossorial life by having, for instance, small eyes, small ears, and stout, strong-clawed forefeet; color relatively uniform in an individual, but varying from black to almost white depending on the kind. Skull massive and rugose; occiput large; auditory bullae relatively small; squamosals usually having pronounced ridges often united into a sagittal crest; zygomatic arches strong and widely flaring; interorbital constriction narrower than rostrum; anterior projection of nasals only slightly exceeding that of upper incisors; squamosals large and comprising much of braincase at expense of frontals and parietals; occipitals broad, and paroccipital processes pronounced; mandible heavy and provided with large coronoid process; root of lower incisor forming prominent process between condyle and angular process. Dentition, i. $\frac{1}{1}$, c. $\frac{0}{0}$, p. $\frac{1}{1}$, m. $\frac{3}{3}$. Cheek-teeth ever-growing and having enamel greatly reduced; anterior surface of incisors broad and flat, always smooth on lower teeth, but either smooth or grooved on upper teeth depending on the taxon.

Pocket gophers are widely distributed in North America and Central America. The temporal range is Upper Oligocene or Lower Miocene to Recent.

Pocket gophers lead an almost completely subterranean existence and are only rarely seen above ground and then only momentarily. The burrow systems are often extensive and usually marked by a series of mounds of earth. The mounds are not conical like those of the moles because the excavated earth is brought to the surface through an inclined lateral tunnel rather than a vertical shaft. Digging is accomplished mainly by means of the claws of the forefeet, but the incisor teeth also are used. The loose earth is compacted against the gopher's chest and pushed to the surface.

Food consists mainly of the underground parts of plants, especially the succulent portions. Forbs, however, are often cut back above ground around the mouth of a burrow. Stems are cut in short lengths and transported in the cheek pouches to storage chambers, that are lateral pockets of the burrow system, for later use. Pocket gophers do not hibernate.

They are decidedly unsocial; except for a short time when the young are in the mother's burrow

Key to Genera of Geomyidae

1. Anterior surface of upper incisors smooth, occasionally with a fine, indistinct groove near inner margin; form of 3rd upper molar elliptical and same (monoprismatic) as that in M1 and M2; basitemporal fossa (situated between m3 and lingual base of coronoid process) absent, except for shallow depression in largest individuals of *T. umbrinus* of Great Basin, forefoot small and narrow with small claws (see Fig. 329). . . . . *Thomomys*,   p. 455

1′. Anterior surface of upper incisors grooved (1 or 2 grooves depending on genus); form of 3rd upper molar never elliptical (instead biprismatic or partly so, except in some individuals of *Geomys* and *Pappogeomys*) and differing from the form of M1 and M2; basitemporal fossa (situated between m3 and lingual base of coronoid process) present; forefeet large and broad with large claws (see Fig. 329).

   2. Posterior wall of P4 having enamel plate, but usually restricted to lingual "end" of tooth (usually absent in subgenus *Orthogeomys* of genus *Orthogeomys*); M3 conspicuously bicolumnar, longer than wide owing to elongation of posterior loph.

      3. Upper incisors bisulcate; rostrum relatively narrow (see Fig. 316); length of enamel plate of M3 decidedly less than length of lingual plate; pelage soft and thick. . . . . *Zygogeomys*,   p. 496

      3′. Upper incisors unisulcate; rostrum relatively broad (see Fig. 325); length of lingual enamel plate of M3 approx. equal to, or more than, length of labial plate; pelage coarse, often hispid, and scant. . . . . *Orthogeomys*,   p. 506

   2′. Posterior wall of P4 lacking any trace of enamel; M3 not strongly bicolumnar, having shallow reentrant fold on labial side, and crown no longer than wide owing to shortness of posterior loph.

      4. Upper incisor bisulcate; both anterior and posterior walls of M1 and M2 having complete enamel plates. . . . . *Geomys*,   p. 497

      4′. Upper incisor unisulcate; enamel plate on posterior wall of M1 usually reduced to lingual side or absent (complete only in one species, *Pappogeomys bulleri*); enamel plate on posterior wall of M2 also absent in advanced species (subgenus *Cratogeomys*).   *Pappogeomys*,   p. 514

and when copulation takes place, any burrow system is occupied by only one animal. When two adults are placed together they fight viciously.

Pocket gophers have many natural enemies, and to protect themselves the species of small and medium size keep all entrances to a burrow closed with earthen plugs. It is thus easy for the collector to ascertain whether a given burrow is occupied: he need merely remove a plug; if there is an occupant, a new plug will be in place soon—sometimes, in fact, immediately.

Pocket gophers are relatively sedentary and live where there is suitable soil for burrowing. In many parts of their range, they are found in localized, isolated areas. Possibly because of their discontinuous distribution and great genetic plasticity, there are a great many kinds. Indeed, in many areas, each individual population possesses some degree of uniqueness. Which of these populations are to be recognized nomenclaturally is often a highly subjective matter. Because of their remarkably great variability, more than 300 kinds of pocket gophers have been formally named.

Genus **Thomomys** Wied-Neuwied
Smooth-toothed Pocket Gophers

1839. *Thomomys* Wied-Neuwied, Nova Acta Phys.-Med. Acad. Caesar. Leop.-Carol., 19(pt. 1):377. Type, *Thomomys rufescens* Wied-Neuwied.

1836. *Oryctomys* Eydoux and Gervais (in part), Mag. de Zool., 6:20, pl. 21. Type, *Oryctomys (Saccophorus) bottae* Eydoux and Gervais.

1903. *Megascapheus* Elliot, Field Columb. Mus., Publ. 76, Zool. Ser., 3(11):190, July 25. Type, *Diplostoma bulbivorum* Richardson.

1933. *Pleisothomomys* Gidley and Gazin, Jour. Mamm., 14:354, November 13. Type, *Pleisothomomys potomacensis* Gidley and Gazin.

External measurements: average of five adult males of *T. bulbivorus*, a large species, 300, 90, 42; holotype, adult male, of *T. talpoides pygmaeus*, the smallest subspecies, 177, 46, 22. Basilar length approx. 24–50, including both sexes. Females are smaller than males. Black to cream, in most taxa underparts essentially same color as upper parts. Upper incisors without grooving, excepting fine, indistinct sulcus rarely near inner margin (grooving more common in *T. monticola* than in other Recent species); crowns of cheek-teeth high, rooted and ever-growing; all molars, including M3, monoprismatic and anteroposteriorly compressed, sometimes (especially in subadults) having slight inflection on labial side in upper teeth and lingual side in lower teeth; molars bicolumnar in pre-final stages of wear

(seen in juveniles); enamel pattern interrupted in all cheek-teeth, loss occurring only at sides of each column; transverse enamel blade completely covering posterior face of both P4 and p4; all upper and lower molars with two transverse enamel blades, one on anterior surface and one on posterior surface, of each tooth, including M3; small 3rd plate sometimes persistent on broad side of tooth, labial side in upper molars and lingual side in lower molars (*T. bulbivorous*); skull generalized, neither unusually narrow and deep or broad and flat; usually without marked cresting or rugosity; masseteric ridge well developed and massive; basitemporal fossa absent (except for shallow depression in some individuals from the northern part of the Great Basin, *T. u. townsendii* and closely related subspecies); pelage soft, never harsh or hispid, covering body with thick coat of hair; forefoot exceptionally small for fossorial mammal, claws not especially long; body form remarkably fossorial.

The above characterization of the genus is adapted from Russell (1968:518) who interposed between the subfamily Geomyinae and its contained genera two tribes, the better, he thought, to show likenesses between the genera. He placed genus *Thomomys* in one tribe and the other Recent genera in a second tribe. His arrangement has merit, especially when considering extinct genera, but for the sake of simplicity is

KEY TO THE SPECIES OF THOMOMYS

not adopted here in listing only the Recent genera.

See also information beyond in the account of the species *Thomomys umbrinus*, page 469.

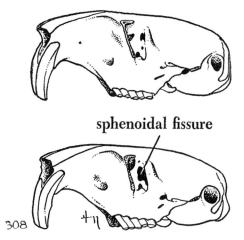

sphenoidal fissure

Fig. 308. Anterolateral views × 1½ of orbits of two skulls. Upper skull shows a small sphenoidal foramen in *Thomomys talpoides bridgeri* (Pocatello, Idaho, 6731 K.U., ♀), more often absent than present in *T. talpoides*. Lower skull shows instead a sphenoidal fissure in *Thomomys umbrinus ruidosae* (Ruidosa, New Mexico, 35157 K.U., ♀), always present in *T. umbrinus*.

pterygoid

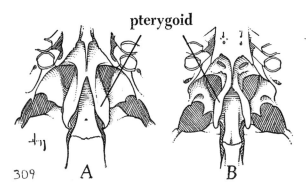

Fig. 309. Pterygoid regions, X 3, in: *Thomomys bulbivorus*, Benton Co., Oregon, No. 50385 K.U., ♀; *Thomomys umbrinus elkoensis*, Elburz, Nevada, No. 46460 K.U., ♀.

Subgenus **Thomomys** Wied-Neuwied

1839. *Thomomys* Wied-Neuwied, Nova Acta Phys.-Med. Acad. Caesar. Leop.-Carol., 19(pt. 1):377. Type, *Thomomys rufescens* Wied-Neuwied.
For other synonyms, see synonomy under the genus.

Differs from the extinct subgenus *Pleisothomomys* Gidley and Gazin primarily in that molars in cross section are pear-shaped instead of subcrescentic or ovate, and in molar crowns being abruptly narrowed at one "end" of each tooth instead of not narrowed. All Recent species belong in subgenus *Thomomys*.

Thomomys talpoides
Northern Pocket Gopher

External measurements: 165–253; 40–75; 20–31. Ear from notch, 5–6 mm. Color highly variable according to subspecies; most subspecies yellowish brown; some grayish brown; fewer pale grayish to plumbeous. Underparts paler usually washed with buffy or ochraceous. Skull robust; zygomatic arches usually widely spreading; sphenoidal fissure absent, occasionally a foramen present; incisive foramina anterior to infraorbital canal; anterior prism of P4 triangular; occlusal face of anterior prism of P4 with distinct anteromedial notch and area of occlusal face of anterior prism three-fourths or more that of posterior prism. Baculum 12–17 mm. long. For comparison with *T. mazama* and *T. monticola*, see accounts of those two species.

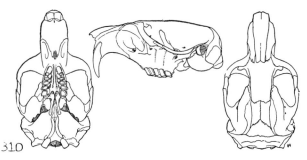

Fig. 310. *Thomomys talpoides monoensis*, Lapon Canyon, Mount Grant, Nevada, No. 63697 M.V.Z., ♂, X 1.

Thomomys talpoides aequalidens Dalquest

1942. *Thomomys talpoides aequalidens* Dalquest, Murrelet, 23:3, May 14, type from Abel Place, 2200 ft., 6 mi. SSE Dayton, Columbia Co., Washington.

MARGINAL RECORDS.—Washington: `Clarkston; Asotin; *Jim Creek, 3200 ft., 11 mi. SE Dayton;* 6½ mi. SE Walla Walla; *1⁹⁄₁₀ mi. SE Walla Walla;* Dayton.

Thomomys talpoides agrestis Merriam

1908. *Thomomys talpoides agrestis* Merriam, Proc. Biol. Soc. Washington, 21:144, June 9, type from Medano Ranch, San Luis Valley, Colorado.

MARGINAL RECORDS.—Colorado (Armstrong, 1972:158): 12 mi. NW Saguache; Crestone; Medano Pass, 9300 ft.; 22 mi. E Mosca; Trinchera River, 10 mi. E Fort Garland; Culebra Cañon, 8950 ft.; San Acacio; Alamosa.

Thomomys talpoides andersoni Goldman

1939. *Thomomys talpoides andersoni* Goldman, Jour. Mamm., 20:235, May 15, type from Medicine Hat, South Saskatchewan River, Alberta.

MARGINAL RECORDS.—Alberta: type locality; Milk River [Ranch, sec. 30, T. 2, R. 9 W of 4th meridian].

Thomomys talpoides attenuatus Hall and Montague

1951. *Thomomys talpoides attenuatus* Hall and Montague, Univ. Kansas Publ., Mus. Nat. Hist., 5:29, February 28, type from 3½ mi. W Horse Creek P.O., 7000 ft., Laramie Co., Wyoming.

MARGINAL RECORDS.—Wyoming (Long, 1965a:602): 27 mi. N, 2 mi. E Powder River; 12 mi. N, 6 mi. W Bill; 10 mi. N Hatcreek; *Rawhide Buttes; 2 mi. S, 9½ mi. E Cheyenne.* Colorado (Armstrong, 1972:158): 2½ mi. S, 12 mi. W Peetz; *Horsetail Creek, 17 mi. NW Stoneham;* Pawnee Buttes, 5300 ft.; Nunn; Goodwin Ranch, 25 mi. N Ft. Collins. Wyoming (Long, 1965a:602): Sherman; *1 mi. N, 5 mi. W Horse Creek; 15 mi. SW Wheatland;* 3 mi. SW Eagle Peak; 10 mi. S Casper.

Thomomys talpoides bridgeri Merriam

1901. *Thomomys bridgeri* Merriam, Proc. Biol. Soc. Washington, 14:113, July 19, type from Harvey's Ranch, Smith Fork, 6 mi. SW Old Fort Bridger, Uinta Co., Wyoming.
1939. *Thomomys talpoides bridgeri,* Goldman, Jour. Mamm., 20:234, May 14.

MARGINAL RECORDS.—Idaho: 3 mi. SW Victor; 10 mi. SE Irwin. Wyoming (Long, 1965a:603, 604): 3 mi. N, 11 mi. E Alpine; 34 mi. N, 4 mi. W Pinedale; 2 mi. S, 19 mi. W Big Piney; Fort Bridger; *4 mi. S Lonetree.* Idaho: Strawberry Creek (Canyon), 20 mi. NE Preston; Bridge; Indian Springs, 4 mi. S American Falls; Shelley; S side South Fork, 3 mi. W Swan Valley.

Thomomys talpoides bullatus V. Bailey

1914. *Thomomys talpoides bullatus* V. Bailey, Proc. Biol. Soc. Washington, 27:115, July 10, type from Powderville, Custer Co., Montana.

MARGINAL RECORDS.—Saskatchewan: Cypress Hills; Dollard. Montana: Johnson Lake. North Dakota: Buford. South Dakota (Anderson and Jones, 1971:376): 10 mi. S, 4 mi. W Reva; *10 mi. S, 5 mi. W Reva;* Crow Buttes. Wyoming (Long, 1965:605): Devils Tower; Buckhorn; *Newcastle;* 23 mi. SW Newcastle; *43½ mi. S, 13 mi. W Gillette; Belle Fourche River;* 43½ mi. S, 5½ mi. W Gillette; 2 mi. S, 6½ mi. W Buffalo; *Dayton.* Montana: Fort Custer; Red Lodge. Wyoming (Long, 1965a:605): Myersville; Miners Delight; *17 mi. S, 6¼ mi. W Lander; 2½ mi. N, 17½ mi. W Lander;* Meadow Creek on Wind River; Ishawooa Creek. Montana: Piney Buttes; Fort Assiniboine.

Thomomys talpoides caryi V. Bailey

1914. *Thomomys talpoides caryi* V. Bailey, Proc. Biol. Soc. Washington, 27:115, July 10, type from head Trapper Creek, 9500 ft., Bighorn Mts., Bighorn Co., Wyoming.

MARGINAL RECORDS.—Wyoming (Long, 1965a:606): 38 mi. E Lovell; 17½ mi. E Shell; *head Trapper Creek; 4 mi. N, 9 mi. E Tensleep;* Red Bank; 12 mi. E Shell; *Medicine Wheel Ranch, 28 mi. E Lovell.* Probably also in Bighorn Mts., Montana.

Thomomys talpoides cheyennensis Swenk

1941. *Thomomys talpoides cheyennensis* Swenk, Missouri Valley Fauna, 4:5, March 1, type from 2 mi. S Dalton, Cheyenne Co., Nebraska.

MARGINAL RECORDS.—Nebraska (Jones, 1964c:154, unless otherwise noted): Crescent Lake National Wildlife Refuge (Rickart, 1972:36, owl pellets); 4 mi. S Sidney; 9 mi. S Kimball; *Smeed.* Wyoming (Long, 1965a:606): *1 mi. W Pine Bluffs;* 12 mi. N, ½ mi. W Pine Bluffs. Nebraska (Jones, 1964c:154): Banner County; type locality.

Thomomys talpoides clusius Coues

1875. *Thomomys clusius* Coues, Proc. Acad. Nat. Sci. Philadelphia, 27:138, June 15, type from Bridger Pass, 18 mi. SW Rawlins, Carbon Co., Wyoming.
1915. *Thomomys talpoides clusius,* V. Bailey, N. Amer. Fauna, 39:100, November 15.

MARGINAL RECORDS.—Wyoming (Long, 1965a:607): 4 mi. N, 1 mi. W Shoshone; *Rattlesnake Mts.;* 40 mi. SW Casper; *Shirley Mts.; Fort Steele;* Bridger Pass; 8 mi. E Rongis, Green Mts.

Thomomys talpoides cognatus Johnstone

1955. *Thomomys talpoides cognatus* Johnstone, Canadian Field-Nat., 68:163, September 17, type from Crowsnest Pass, British Columbia.

MARGINAL RECORDS.—British Columbia: type locality; Flathead River, 20 mi. N U.S. boundary; East Newgate; Baynes Lake.

Thomomys talpoides columbianus V. Bailey

1914. *Thomomys fuscus columbianus* V. Bailey, Proc. Biol. Soc. Washington, 27:117, July 10, type from Touchet, Walla Walla Co., Washington.
1939. *Thomomys talpoides columbianus,* Goldman, Jour. Mamm., 20:234, May 15.

MARGINAL RECORDS.—Washington: Prescott; 7 mi. S Dixie, 4300 ft.; *2½ mi. S College Place, S side Walla Walla River; 4 mi. S Walla Walla.* Oregon: 1⅗ mi. S Milton; 7 mi. SW Milton; Pendleton; Willows; Umatilla. Washington: Fort Walla Walla; 2 mi. E jct. Snake and Columbia rivers, 350 ft.; 8 mi. E Eureka.

Thomomys talpoides confinis Davis

1937. *Thomomys talpoides confinis* Davis, Murrelet, 18:25, September 4, type from Gird Creek, near Hamilton, Ravalli Co., Montana.

MARGINAL RECORDS.—Montana: Hamilton; *type locality.*—See addenda.

Thomomys talpoides devexus Hall and Dalquest

1939. *Thomomys talpoides devexus* Hall and Dalquest, Murrelet, 20:3, April 30, type from 1 mi. WSW Neppel, Grant Co., Washington.
1939. *Thomomys talpoides ericaeus* Goldman, Jour. Mamm., 20:243, May 15, type from Badger Mts., 8 mi. SW Waterville, Douglas Co., Washington. (Regarded as inseparable from *devexus* by Dalquest and Scheffer, Amer. Nat., 78:434, September, 1944.)

MARGINAL RECORDS.—Washington: Badger Mts.; type locality; 2½ mi. E, 1½ mi. S Pasco, 400 ft.; 4 mi. W Pasco, 350 ft.

Thomomys talpoides douglasii (Richardson)

1829. *Geomys douglasii* Richardson, Fauna Boreali-Americana, 1:200. Type locality, near mouth Columbia River, probably near Vancouver, Washington.
1939. *Thomomys talpoides douglasii,* Goldman, Jour. Mamm., 20:234, May 15.

MARGINAL RECORDS.—Washington: Brush Prairie; type locality; 5 mi. E Vancouver.

Thomomys talpoides durranti Kelson

1949. *Thomomys talpoides durranti* Kelson, Proc. Biol. Soc. Washington, 62:143, August 23, type from Johnson Creek, 14 mi. N Blanding, 7500 ft., San Juan Co., Utah.

MARGINAL RECORDS.—Colorado (Armstrong, 1972:159): 22 mi. SW Meeker, 6200 ft.; 1 mi. S Rulison, 5200 ft.; 28 mi. N, 5 mi. W Mack, 7250 ft. Utah: 1 mi. SE Mesa Ranger Station, 9200 ft.; Warner Ranger Station, La Sal Mts., 9750 ft., and discontinuously in the Abajo Mt. area at Gooseberry Ranger Station, Elk Ridge, 8300 ft.; Dalton Spring, 5 mi. W Monticello, Abajo (= Blue) Mts., 8300 ft.; type locality; Oak Spring, middle fork Willow Creek, 15 mi. N Thompson. Colorado (Armstrong, 1972:159): W fork Douglas Creek, 35 mi. S Rangely, 8000 ft.

Thomomys talpoides falcifer Grinnell

1926. *Thomomys falcifer* Grinnell, Univ. California Publ. Zool., 30:180, December 10, type from Bells Ranch, 6890 ft., Reese River Valley, Nye Co., Nevada.
1939. *Thomomys talpoides falcifer,* Goldman, Jour. Mamm., 20:234, May 15.

MARGINAL RECORDS.—Nevada: 2 mi. E Unionville, 4500 ft.; Malloy Ranch, 5 mi. W Austin, 5500 ft., Reese River; Silver Creek, N Austin; Birch Creek Ranch, Big Smoky Valley, 5650 ft.; Kingston Ranger Station, 7500 ft.; Arc Dome; Slys Ranch, 7400 ft., Indian Valley, Shoshone Mts.; Cherry Valley (Meadows), 6450 ft.

Thomomys talpoides fisheri Merriam

1901. *Thomomys fuscus fisheri* Merriam, Proc. Biol. Soc. Washington, 14:111, July 19, type from Beckwith, Sierra Valley, Plumas Co., California.
1939. *Thomomys talpoides fisheri,* Goldman, Jour. Mamm., 20:234, May 15.

MARGINAL RECORDS.—Nevada: Cottonwood Creek, Virginia Mts., 4400 ft.; 6 mi. NE Virginia City, 6000 ft.; 8 mi. S Reno. California: type locality.

Thomomys talpoides fossor J. A. Allen

1893. *Thomomys fossor* J. A. Allen, Bull. Amer. Mus. Nat. Hist., 5:51, April 28, type from Florida, 7200 ft., La Plata Co., Colorado.
1939. *Thomomys talpoides fossor,* Goldman, Jour. Mamm., 20:234, May 15.

MARGINAL RECORDS.—Colorado (Armstrong, 1972:160): Middle Mamm Creek, near Rifle; 12 mi. SE Rifle; Gothic, 9500 ft.; 2 mi. S, 9 mi. E Crested Butte; 1½ mi. W Cochetopa Pass; 8 mi. S Monte Vista; 16 mi. SW Alamosa; Osier. New Mexico: Hopewell; head Santa Clara Creek, Jemez Mts.; Gallinas Mts.; Horse Lake. Colorado (Armstrong, 1972:160, unless otherwise noted): Chromo; Devil's Creek, near Dyche; type locality; La Plata City (Hall and Kelson, 1959:441); Goat Creek, Lone Cone Peak; SW¼ sec. 11, T. 48 N, R. 14 W, 9000 ft.; Piñon Mesa; 9 mi. S, 1 mi. W Glade Park P.O., 8800 ft.; Grand Mesa, 28 mi. E Grand Junction; 4 mi. S, 3 mi. E Collbran, 6800 ft. Isolated segment of range, Arizona: Lukachukai Mts. New Mexico: Chusca Mts. Arizona: Tunitcha Mts.

Specimens from 2nd isolated segment of *T. talpoides* here referred to *T. t. fossor* from the Sangre de Christo range of northern New Mexico deserve re-examination to learn whether they are referable instead to *T. t. agrestis,* as Fig. 55 of Armstrong (1972:155) suggests may be the case. Marginal localities for this segment are: Costilla Pass; Halls Peak; Pecos River Mts.; Pecos Baldy.

Thomomys talpoides fuscus Merriam

1891. *Thomomys clusius fuscus* Merriam, N. Amer. Fauna, 5:69, July 30, type from Summit Creek in mountains, head Big Lost River, Custer Co., Idaho.
1939. *Thomomys talpoides fuscus,* Davis, The Recent mammals of Idaho, Caxton Printers, Caldwell, Idaho, p. 253, April 5.
1901. *Thomomys myops* Merriam, Proc. Biol. Soc. Washington, 14:112, July 19, type from Conconully, E base Cascade Range, Okanogan Co., Washington.

MARGINAL RECORDS.—British Columbia: Westbridge; Eholt (Cowan and Guiguet, 1965:165); Trail. Washington: Round Top Mtn. Idaho: Hoodoo Valley; Mission. Montana: Corvallis; Lolo; Thompson Falls; Tobacco Plains; summit W of Blackfoot Station; W of Benton; Livingston; Fort Ellis; Lakeview. Idaho: Taylor Creek, 5 mi. S Montana line at Sheridan Mtn.; 3

mi. SW Victor; 5 mi. W St. Anthony. Montana (Hoffmann, et al., 1969:589): Birch Creek, 15 mi. N Dillon; Sawmill Creek–Twin Lakes, 14 mi. NW Jackson. Idaho: Salmon River Mts.; type locality; 16 mi. N Shoshone; 2 mi. E Acequia; 2 mi. S Hagerman; Crane Creek, 15 mi. E Midvale; Lewiston. Washington: Wawawai; Cheney; Loon Lake; Sherman Creek Pass; Duley Lake; Douglas; Wenatchee; 5 mi. NE Cle Elum; Easton; Mt. Stuart; Lucerne; Barron. British Columbia: Anarchist Mtn. (Cowan and Guiguet, 1965:165).

Thomomys talpoides gracilis Durrant

1939. *Thomomys quadratus gracilis* Durrant, Bull. Univ. Utah, 29(6):3, February 28, type from Pine Canyon, 6600 ft., 17 mi. NW Kelton, Boxelder Co., Utah.
1939. *Thomomys talpoides gracilis* Durrant, Bull. Univ. Utah, 30(5):6, October 24.

MARGINAL RECORDS.—Idaho: S fork Owyhee River, 12 mi. N Nevada line. Nevada: ½ mi. N Jarbidge. Utah: Yost; Park Valley; Etna. Nevada: Cleve Creek, 8100–8600 ft., Shell Creek Range; E side Shelbourne Pass, 6800 ft.; 1 mi. E Illipah, 6100 ft.; 3 mi. SW Hamilton, 8400 ft.; Nay Ranch, 7200 ft., Monitor Valley; Greenmonster Canyon, 7500–8200 ft., Monitor Range; Eureka; 5 mi. E Raines, Sulphur Spring Mts.; head Ackler Creek; Marys River, 5800 ft., 25 mi. N Deeth; 13 mi. N Paradise Valley, 6700 ft.; 5 mi. W Paradise Valley P.O.; Cottonwood Range.

Thomomys talpoides idahoensis Merriam

1901. *Thomomys idahoensis* Merriam, Proc. Biol. Soc. Washington, 14:114, July 19, type from Brich Creek, Clark Co., Idaho.
1939. *Thomomys talpoides idahoensis*, Davis, The Recent mammals of Idaho, Caxton Printers, Caldwell, Idaho, p. 251, April 5.

MARGINAL RECORDS (Thaeler, 1972:425).—Montana: Hagenbarth's, 13 mi. N Dillon (¾ mi. W highway); N Crazy Springs Creek crossing, Blacktail Valley. Idaho: Dubois; Mennan; ⅔ mi. N, 1 mi. W Swan Valley, 5400 ft., sec. 35, T. 2 N, R. 43 E; Alridge; Blackfoot; Big Butte; Big Lost River, near sink; Birch Creek Sink, 5100 ft.; Birch Creek, 10 mi. S Nicholia, 6400 ft. Montana: hill S of Bannock–Jackson Road (SW of Dillon).—See terminal remarks under *T. t. pygmaeus;* the literature cited there applies also to the uncertain status of *T. t. idahoensis.*—See also addenda.

Thomomys talpoides immunis Hall and Dalquest

1939. *Thomomys talpoides immunis* Hall and Dalquest, Murrelet, 20:4, April 30, type from 5 mi. S Trout Lake, Klickitat Co., Washington.

MARGINAL RECORDS.—Washington: Mt. St. Helens: Morrison Springs Ranger Station; Gotchen Creek Ranger Station; type locality.

Thomomys talpoides incensus Goldman

1939. *Thomomys talpoides incensus* Goldman, Jour. Mamm., 20:240, May 15, type from Shuswap, Yale District, British Columbia.

MARGINAL RECORDS.—British Columbia: type locality; Salmon Arm; *Grindrod* (Cowan and Guiguet, 1965:165); Monashee Pass (*ibid.*); Keremeos; Manning Park; Ashcroft; Savona; Kamloops.

Thomomys talpoides kaibabensis Goldman

1938. *Thomomys fossor kaibabensis* Goldman, Jour. Washington Acad. Sci., 28:333, July 15, type from De Motte Park, 9000 ft., Kaibab Plateau, Coconino Co., Arizona.
1939. *Thomomys talpoides kaibabensis* Goldman, Jour. Mamm., 20:234, May 15.

MARGINAL RECORDS (Hoffmeister, 1971:170, 171).—Arizona: type locality; *Greenland Lake, Walhalla Plateau;* Snowshoe Cabin, 8400 ft.; *Bright Angel Point;* Swamp Lake, 7700 ft.

Thomomys talpoides kelloggi Goldman

1939. *Thomomys talpoides kelloggi* Goldman, Jour. Mamm., 20:237, May 15, type from West Boulder Creek, Absaroka Mts., 18 mi. SE Livingston, Park Co., Montana.

MARGINAL RECORDS.—Montana: Boulder Creek, 8 mi. S Bigtimber; type locality; 12 mi. E, 7 mi. S Livingston, 4500 ft., West Boulder River; 15 mi. E, 4 mi. S Livingston, 4400 ft., West Boulder River.

Thomomys talpoides levis Goldman

1938. *Thomomys fossor levis* Goldman, Jour. Washington Acad. Sci., 28:336, July 15, type from Seven Mile Flat, 10,000 ft., 5 mi. N Fish Lake, Fish Lake Plateau, Sevier Co., Utah.
1939. *Thomomys talpoides levis* Goldman, Jour. Mamm., 20:234, May 15.

MARGINAL RECORDS.—Utah: type locality; Elkhorn Guard Station, 14 mi. N Torrey, 9400 ft., Fish Lake Plateau; Grover; 18 mi. N Escalante; summit Birch Creek, Escalante Mts.; Potato Hollow, Horse Pasture Plateau (Stock, 1970:430); *¼ mi. SW Kolob Reservoir (ibid.).*

Thomomys talpoides limosus Merriam

1901. *Thomomys limosus* Merriam, Proc. Biol. Soc. Washington, 14:116, July 19, type from White Salmon, Gorge of the Columbia, Klickitat Co., Washington.
1939. *Thomomys talpoides limosus,* Goldman, Jour. Mamm., 20:235, May 15.

MARGINAL RECORDS.—Washington: Paterson, 250 ft., thence westward along N side Columbia River to 5 mi. W White Salmon.

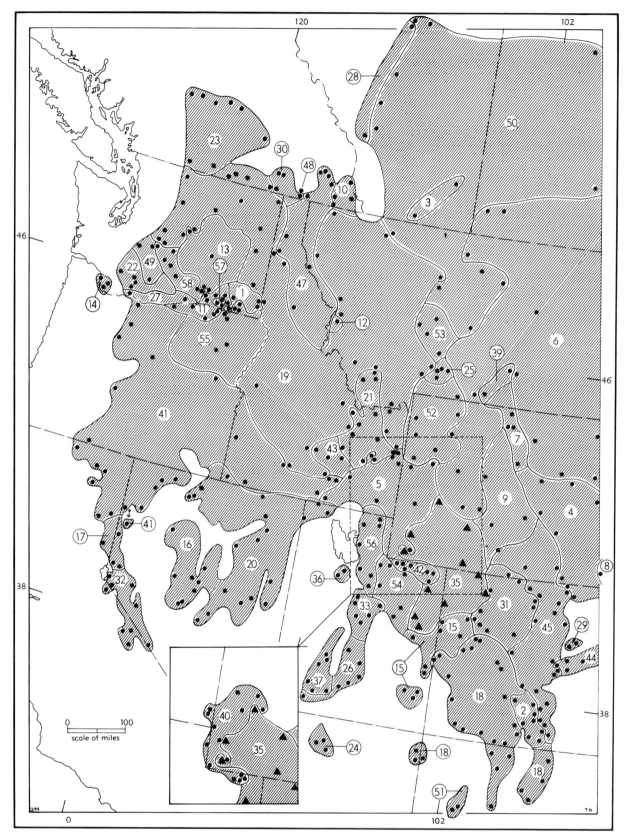

Map 299. Some subspecies of *Thomomys talpoides*. (See facing page for guide.)

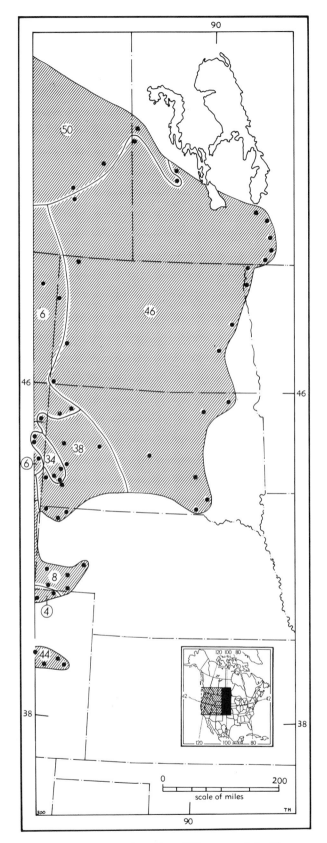

Map 300. Some subspecies of *Thomomys talpoides*.

461

Thomomys talpoides loringi V. Bailey

1914. *Thomomys fuscus loringi* V. Bailey, Proc. Biol. Soc. Washington, 27:118, July 10, type from South Edmonton, Alberta.
1939. *Thomomys talpoides loringi*, Goldman, Jour. Mamm., 20:234, May 14.

MARGINAL RECORDS.—Alberta: type locality; Moose Mtn.

Thomomys talpoides macrotis F. W. Miller

1930. *Thomomys talpoides macrotis* F. W. Miller, Proc. Colorado Mus. Nat. Hist., 9(3):41, December 14, type from D'Arcy Ranch, 2 mi. N Parker, Douglas Co., Colorado.

MARGINAL RECORDS (Armstrong, 1972:160).—Colorado: 4 mi. N, 10 mi. E Parker; type locality.

Thomomys talpoides medius Goldman

1939. *Thomomys talpoides medius* Goldman, Jour. Mamm., 20:241, May 15, type from Silver King Mine, summit Toad Mtn., 6 mi. S Nelson, West Kootenay District, British Columbia.

MARGINAL RECORDS.—British Columbia: Nelson; West Creston; Fruitvale; Ward's Ferry (Bonnington on Kootenay River, *ca.* 10 mi. below Nelson).

Thomomys talpoides meritus Hall

1951. *Thomomys talpoides meritus* Hall, Univ. Kansas Publ., Mus. Nat. Hist., 5:221, December 15, type from 8 mi. N, 19½ mi. E Savery, 8800 ft., Carbon Co., Wyoming.

MARGINAL RECORDS.—Wyoming (Long, 1965a:608): type locality; *7 mi. N, 17 mi. E Savery.* Colorado (Armstrong, 1972:161): near Pearl; 3 mi. below Cameron Pass; Willow Creek Pass, 9200 ft.; Allenton; W fork Elk Creek, 8 mi. above New Castle; Mud Spring, 8850 ft. [county line, due N Silt]; between Flag Creek and Grand Hogback, 9 mi. S Meeker; 7 mi. NE Meeker; 8 mi. NE Craig. Wyoming (Long, 1965a:608): *4 mi. N, 10 mi. E Savery; 5 mi. N, 3 mi. E Savery.*

Thomomys talpoides monoensis Huey

1934. *Thomomys quadratus monoensis* Huey, Trans. San Diego Soc. Nat. Hist., 7:373, May 31, type from Dexter Creek Meadow, 6800 ft., at confluence Dexter and Wet creeks, Mono Co., California.
1939. *Thomomys talpoides monoensis*, Goldman, Jour. Mamm., 20:234, May 15.

MARGINAL RECORDS.—Nevada: N side Carson River, 4700 ft., 4½ mi. NE Genoa; 2 mi. SW Pine Grove, 7250 ft.; Cottonwood Creek, Mt. Grant; Chiatovich Creek, 8200 ft. California: type locality; Woodfords; Leevining; Convict Creek. Nevada: 6 mi. S Minden. The three localities last mentioned from California stand in the literature as places where *T. t. fisheri* oc-

curs but the specimens need restudy in the light of the later naming of *T. t. monoensis.*

Thomomys talpoides moorei Goldman

1938. *Thomomys fossor moorei* Goldman, Jour. Washington Acad. Sci., 28:335, July 15, type from 1 mi. S Fairview, 6000 ft., Sanpete Co., Utah.
1939. *Thomomys talpoides moorei* Goldman, Jour. Mamm., 20:234, May 15.

MARGINAL RECORDS.—Utah: near Payson Lake; Colton; Lake Creek, 11 mi. E Mt. Pleasant; Ephraim; mouth Reddicks Canyon, Wales (= San Pitch) Mts., 7500 ft.; Mt. Nebo, 10,000 ft., 25 mi. SE Payson.

Thomomys talpoides nebulosus V. Bailey

1914. *Thomomys talpoides nebulosus* V. Bailey, Proc. Biol. Soc. Washington, 27:116, July 10, type from Jack Boyden's Ranch, 3750 ft., Sand Creek Canyon, 15 mi. NE Sundance, Crook Co., Wyoming.

MARGINAL RECORDS.—Wyoming: Bear Lodge Mts. (Long, 1965a:608). South Dakota: Spring Creek; Beaver Creek; Custer. Wyoming (Long, 1965a:608): *Rattlesnake Creek;* 1½ mi. NW Sundance.

Thomomys talpoides ocius Merriam

1901. *Thomomys clusius ocius* Merriam, Proc. Biol. Soc. Washington, 14:114, July 19, type from Mountainview, Smiths Fork, 4 mi. (by airline) SE Fort Bridger, Uinta Co., Wyoming.
1946. *Thomomys talpoides ocius*, Durrant, Univ. Kansas Publ., Mus. Nat. Hist., 1:17, August 15.

MARGINAL RECORDS.—Wyoming (Long, 1965a:610): jct. Green River and New Fork; 27 mi. N, 37 mi. E Rock Springs; Kinney Ranch, 21 mi. S Bitter Creek; *33 mi. S Bitter Creek.* Colorado (Armstrong, 1972:162): Snake River, 20 mi. W Bagg's Crossing; Craig; 5 mi. W Rangely. Utah: PR Springs, 43 mi. S Ouray, 7950 ft.; Brown Corral, 20 mi. S Ouray, 6250 ft.; Vernal. Wyoming (Long, 1965a:610): 3 mi. WSW Fort Bridger; *2½ mi. W Fort Bridger;* Cumberland; *Fontenelle.* Long (1964:754; 1965a:608) and Thaeler (1968:180) published information important to anyone reappraising the systematic position of *ocius.*

Thomomys talpoides oquirrhensis Durrant

1939. *Thomomys talpoides oquirrhensis* Durrant, Bull. Univ. Utah, 30(5):3, October 24, type from Settlement Creek, 6500 ft., Oquirrh Mts., Tooele Co., Utah.

MARGINAL RECORDS.—Utah: Rose Canyon, Oquirrh Mts., 5650 ft.; type locality.

Thomomys talpoides parowanensis Goldman

1938. *Thomomys fossor parowanensis* Goldman, Jour. Washington Acad. Sci., 28:334, July 15, type from Brian Head, 11,000 ft., Parowan Mts., Iron Co., Utah.

1939. *Thomomys talpoides parowanensis* Goldman, Jour. Mamm., 20:234, May 15.

MARGINAL RECORDS.—Utah: Britts Meadows, Beaver Mts., 8500 ft.; Puffer Lake, Beaver Mts.; $\frac{1}{4}$ mi. W Sunset Point, Bryce National Park, 8000 ft.; Duck Creek, Cedar Mts., 9000 ft.; type locality; Beaver Mts., Iron Co.

Thomomys talpoides pierreicolus Swenk

1941. *Thomomys talpoides pierreicolus* Swenk, Missouri Valley Fauna, 4:2, March 1, type from Wayside, Dawes Co., Nebraska.

MARGINAL RECORDS.—Montana: Alzada. South Dakota: Smithville. Nebraska: 5 mi. NW Chadron; Sand Creek Valley, near Horn; Indian Creek, N of Story. South Dakota: Elk Mtn.; Buffalo Gap; Rapid City; Fort Meade.

Thomomys talpoides pryori V. Bailey

1914. *Thomomys pryori* V. Bailey, Proc. Biol. Soc. Washington, 27:116, July 10, type from head Sage Creek, 6000 ft., Pryor Mts., Carbon Co., Montana.
1915. *Thomomys talpoides pryori* V. Bailey, N. Amer. Fauna, 39:104, November 15.

MARGINAL RECORDS.—Montana: Bighorn River, W side, near Fort Custer; type locality.

Thomomys talpoides pygmaeus Merriam

1901. *Thomomys pygmaeus* Merriam, Proc. Biol. Soc. Washington, 14:115, July 19, type from Montpelier Creek, 6700 ft., about 10 mi. NE Montpelier, Bear Lake Co., Idaho.
1939. *Thomomys talpoides pygmaeus*, Davis, The Recent mammals of Idaho, Caxton Printers, Caldwell, Idaho, p. 252, April 5.

MARGINAL RECORDS (Thaeler, 1972:426).—Wyoming: Surveyors Park, 8000 ft., 12 mi. NE Pinedale; Big Sandy; Big Piney; Fossil; Fort Bridger. Utah: Birch Creek, 2 mi. S Utah–Wyoming line; Hope Creek; 1 mi. W Summit Springs, 8000 ft.; Granite Park, 24 mi. S Manila. Wyoming: $14\frac{1}{2}$ mi. S, 1 mi. E Evanston, 8100 ft. Utah: 1 mi. SW Little Creek Reservoir. Wyoming: Cokeville, 6400 ft. Idaho: Montpelier Creek, $4\frac{3}{5}$ mi. N, $6\frac{1}{5}$ mi. E Montpelier, 6500 ft., sec. 10 T. 12 S, R. 45 E; Montpelier Creek, 12 mi. NE Montpelier, 6700 ft. Wyoming: Merna, 7800–8000 ft.—After Davis (1939:252) arranged *pygmaeus* as a subspecies of *T. talpoides*, Long (1964:754; 1965a:610) and Thaeler (1968:180; 1972:417–428) published information important to anyone reappraising the systematic position of *pygmaeus*.

Thomomys talpoides quadratus Merriam

1897. *Thomomys quadratus* Merriam, Proc. Biol. Soc. Washington, 11:214, July 15, type from The Dalles, Wasco Co., Oregon.
1939. *Thomomys talpoides quadratus*, Goldman, Jour. Mamm., 20:234, May 15.

MARGINAL RECORDS.—Oregon: type locality; Ochoco Ranger Station, 4000 ft., in Crook County; Ironside. Idaho: Reynolds Creek, 12 mi. S Snake River; S side Snake River, 2 mi. S Hagerman; Raft River, 2 mi. S Snake River; S side Snake River, 19 mi. SW American Falls; Elba; 1 mi. S Riddle. Nevada: meadow, Big Creek, 7000 ft., Pine Forest Mts.; Summit Lake; Soldier Meadows, 4600 ft.; 5 mi. E, $3\frac{1}{2}$ mi. N Granite Peak, 3900 ft.; Smoke Creek, California–Nevada boundary, 4300 ft.; Pahrum Peak (Hall, 1946:447). California (Thaeler, 1968:24, 25): $7\frac{1}{10}$ mi. N, $6\frac{1}{10}$ mi. E Litchfield; $3\frac{1}{2}$ mi. S, 1 mi. E? Susanville; *3 mi. N Merrillville; 1/10 mi. N, 1/5 mi. E Boyd Hill;* Bieber (S of town); $\frac{1}{2}$ mi. S Kelley Hot Springs; $10\frac{1}{2}$ mi. N, $13\frac{1}{10}$ mi. W Canby; *6 1/5 mi. S, 3/5 mi. W Macdoel;* $\frac{2}{5}$ mi. S, $\frac{9}{10}$ mi. W Macdoel. Oregon: Merrill; Fremont; Matoleus River; Wapinitia. Specimens from Oregon, for example, from Huntington, that stand in the older literature (see V. Bailey, 1915:127) as *T.* [*t.*] *fuscus* need re-examination to determine if they are *T. t. wallowa* or *T. t. quadratus*.

Thomomys talpoides ravus Durrant

1946. *Thomomys talpoides ravus* Durrant, Univ. Kansas Publ., Mus. Nat. Hist., 1:15, August 15, type from 19 mi. N Vernal, 8000 ft., Uintah Co., Utah.

MARGINAL RECORDS.—Utah: Henrys Fork, 8300 ft.; Hoop Lake, 8000 ft., Ashley National Forest; jct. Deep and Carter creeks, 7900 ft.; Vernal–Manila Road, 4 mi. W Greens Lake, 7500 ft.; type locality; Taylor Peak, 17 mi. N Vernal; E fork Blacks Fork, 31 mi. SSW Fort Bridger.

Thomomys talpoides relicinus Goldman

1939. *Thomomys talpoides relicinus* Goldman, Jour. Mamm., 20:239, May 15, type from Twin Springs, 20 mi. N Minidoka, Snake River Desert, Minidoka Co., Idaho.

MARGINAL RECORDS.—Idaho: Laidlaw Park, 35 mi. NW Minidoka; type locality; Sparks Well, 23 mi. NE Minidoka.

Thomomys talpoides retrorsus Hall

1951. *Thomomys talpoides retrorsus* Hall, Univ. Kansas Publ., Mus. Nat. Hist., 5:83, October 1, type from Flagler, Kit Carson Co., Colorado.

MARGINAL RECORDS.—Colorado: 8 mi. NE Elbert; type locality; 8 mi. S Seibert; Limon; Bijou Creek, near El Paso Co. line; Colorado Springs; Divide (Armstrong, 1972:162); Monument (*ibid.*).

Thomomys talpoides rostralis Hall and Montague

1951. *Thomomys talpoides rostralis* Hall and Montague, Univ. Kansas Publ., Mus. Nat. Hist., 5:27, February 28, type from 1 mi. E Laramie, 7164 ft., Albany Co., Wyoming.

MARGINAL RECORDS.—Wyoming (Long, 1965a: 611): Bridger Pass; 5 mi. N Laramie; 6 mi. S, 8 mi. E Laramie; *1 mi. SSE Pole Mtn.* Colorado (Armstrong, 1972:163): *3 mi. N Owl Canyon;* Waverly; Rawah Primitive Area, 10,500 ft.; Boulder; Denver; 10 mi. N Florissant, 8900 ft.; 4 mi. above Manitou; Cripple Creek; ½ mi. S Coaldale; 5 mi. SW Wetmore; 15 mi. NW Gardner; 5 mi. WSW Redwing; La Veta Pass; *La Veta;* 5 mi. S, 1 mi. W Cucharas Camps; near Poncha Pass, 8750 ft.; *6 mi. W Twin Lakes;* Independence Pass; *Summit County, 10,500 ft.;* Coulter; *4 mi. N Stillwater;* Estes Park; sand hills E of Canadian Creek.

Thomomys talpoides rufescens Wied-Neuwied

1839. *Thomomys rufescens* Wied-Neuwied, Nova Acta Phys.-Med. Acad. Caesar. Leop.-Carol., 19(pt. 1):378, type from Minnetaree Village, now Old Fort Clark, about 6 mi. S Stanton, Mercer Co., North Dakota.
1915. *Thomomys talpoides rufescens,* V. Bailey, N. Amer. Fauna, 39:98, November 15.

MARGINAL RECORDS.—Manitoba: Benito; Lake Dauphin; Selkirk Settlement; 2½ mi. S, 3 mi. E Beausejour (Wrigley and Dubois, 1973:169); 5 mi. E Richer (*ibid.*); *5 mi. E Marchand* (*ibid.*); Sandilands (*ibid.*); Caliento (*ibid.*). Minnesota: Humboldt; Robin. North Dakota: Portland; Valley City. South Dakota: Fort Sisseton; Aberdeen; White Lake; Armor; Fort Randall; Pierre. North Dakota: 13 mi. S, 26 mi. W Bowman (Genoways and Jones, 1972:13); *11 mi. S, 26 mi. W Bowman* (*ibid.*); 1 mi. S, 1 mi. W Medora (*ibid.*); Crosby. Saskatchewan: Red Fox Lake, NE Kendal.

Thomomys talpoides saturatus V. Bailey

1914. *Thomomys fuscus saturatus* V. Bailey, Proc. Biol. Soc. Washington, 27:117, July 10, type from Silver, near Saltese, Coeur d'Alene Mts., Missoula Co., Montana.
1939. *Thomomys talpoides saturatus,* Davis, The Recent mammals of Idaho, Caxton Printers, Caldwell, Idaho, p. 253, April 5.

MARGINAL RECORDS.—British Columbia (Cowan and Guiguet, 1965:167): Wasa; Fort Steele; Bull River; *Wardner;* West Newgate. Montana: Prospect Creek. Idaho: 2 mi. NE Weippe; Coeur d'Alene; 5 mi. W Cocolalla. British Columbia: Goatfell; St. Mary's Prairie.

Thomomys talpoides segregatus Johnstone

1955. *Thomomys talpoides segregatus* Johnstone, Canadian Field-Nat., 68:161, September 17, type from Goat Mtn., E side Kootenay River, near Wynndel, British Columbia.

MARGINAL RECORDS.—British Columbia: type locality; *Creston.*

Thomomys talpoides shawi Taylor

1921. *Thomomys douglasii shawi* Taylor, Proc. Biol. Soc. Washington, 34:121, June 30, type from Owyhigh Lake, 5100 ft., Mt. Rainier, Pierce Co., Washington.

1939. *Thomomys talpoides shawi,* Hall and Dalquest, Murelet, 20:4, April 30.

MARGINAL RECORDS.—Washington: Bumping Lake, 3 mi. NE Goose Prairie; Signal Peak, 4000 ft.; Glacier Basin, 5935 ft., Mt. Rainier.

Thomomys talpoides talpoides (Richardson)

1828. *Cricetus talpoides* Richardson, Zool. Jour., 3:518. Type locality fixed at near Fort Carlton (Carlton House), Saskatchewan River, Saskatchewan.
1858. *Thomomys talpoides,* Baird, Mammals, *in* Repts. Expl. Surv. . . . , 8(1):403, July 14.
1837. *Geomys borealis* Richardson, Sixth Ann. Rept. British Assn. for 1836, 5:150. Type locality, plains of the Saskatchewan.
1843. *Geomys unisulcatus* Gray, List of the . . . Mammalia in the . . . British Museum, p. 149, a *nomen nudum.*

MARGINAL RECORDS.—Alberta: St. Albert; Elk Island National Park. Saskatchewan: Prince Albert National Park. Manitoba: Swan River; Riding Mtn. Saskatchewan: Yorkton; Indian Head; Moose Jaw. Montana: Bearpaw Mts.; Zortman; Big Snowy Mts.; Highwood; Blackfoot. Alberta: Calgary; Didsbury; Red Deer.

Thomomys talpoides taylori Hooper

1940. *Thomomys talpoides taylori* Hooper, Occas. Pap. Mus. Zool., Univ. Michigan, 422:11, November 14, type from 6 mi. NE summit Mt. Taylor, about 8900 ft., near Fernandez summer camp, Valencia Co., New Mexico.

MARGINAL RECORDS.—New Mexico: type locality; Mirabel Spring, [SW slope] Mt. Taylor.

Thomomys talpoides tenellus Goldman

1939. *Thomomys talpoides tenellus* Goldman, Jour. Mamm., 20:238, May 15, type from Whirlwind Peak, 10,500 ft., Absaroka Range, Park Co., Wyoming.

MARGINAL RECORDS.—Montana: Beartooth Mts. Wyoming (Long, 1965a:612): Black Mtn.; *2 mi. S, 42 mi. W Cody;* Togwotee Pass; Jackson; *Teton Pass;* 18 mi. N, 9 mi. W Moran; *2 mi. S, 1 mi. W Snake River, 7000 ft.;* Old Faithful.

Thomomys talpoides trivialis Goldman

1939. *Thomomys talpoides trivialis* Goldman, Jour. Mamm., 20:236, May 15, type from near head Big Timber Creek, 5200 ft., about 15 mi. NW Big Timber, Crazy Mts., Sweetgrass Co., Montana.

MARGINAL RECORDS.—Montana: various local stations, Little Belt Mts.; type locality; 4 mi. S White Sulphur Springs, Castle Mts.; Camas Creek, 4 mi. S Fort Logan, Big Belt Mts.

Thomomys talpoides uinta Merriam

1901. *Thomomys uinta* Merriam, Proc. Biol. Soc. Washington, 14:112, July 19, type from Blacks Fork, 10,000 ft., N base Gilbert Peak, Uinta Mts., Summit Co., Utah.
1939. *Thomomys talpoides uinta*, Goldman, Jour. Mamm., 20:234, May 15.

MARGINAL RECORDS.—Utah: Smith and Morehouse Creek; 2 mi. S jct. Bear River and Haydens Fork; Petty Mtn., 15 mi. N Mountain Home, 9500 ft.; Christensen Ranch, Nine Mile Canyon, 10 mi. E Summit, 6300 ft.; Forks, Sunnyside, 9000 ft.; Currant Creek, Uinta Mts.

Thomomys talpoides wallowa Hall and Orr

1933. *Thomomys quadratus wallowa* Hall and Orr, Proc. Biol. Soc. Washington, 46:41, March 24, type from Catherine Creek, 3500 ft., 7 mi. E Telocaset, Union Co., Oregon.
1939. *Thomomys talpoides wallowa*, Goldman, Jour. Mamm., 20:234, May 15.

MARGINAL RECORDS.—Washington: Mountain Top, 4500 ft.; Twin Buttes Ranger Station, 25 mi. SE Dayton. Oregon: type locality; Anthony; Tollgate; Little Meadows, 20 mi. SE Walla Walla. Washington: *Blue Creek Ridge, 20 mi. E Walla Walla;* Hompeg Falls. Specimens from northeastern Oregon, for example, from Bingham Prairie, Elgin, and Huntington that stand in the literature as *T.* [*t.*] *fuscus* (V. Bailey, 1915:127), need re-examination because they probably are *T. t. wallowa.* (Those from Huntington possibly are *T. t. quadratus.*)

Thomomys talpoides wasatchensis Durrant

1946. *Thomomys talpoides wasatchensis* Durrant, Univ. Kansas Publ., Mus. Nat. Hist., 1:8, August 15, type from Midway, 5500 ft., Wasatch Co., Utah.

MARGINAL RECORDS.—Utah: Logan Canyon, Beaver Basin, Utah–Idaho line; Logan Mts., 20 mi. E Logan; head Grove Creek; Mt. Timpanogos, 1 mi. N Aspen Grove; mouth Big Cottonwood Canyon; Ogden; Avon.

Thomomys talpoides whitmani Drake and Booth

1952. *Thomomys talpoides whitmani* Drake and Booth, Walla Walla College Publ. Dept. Biol. Sci. Biol. Sta., 1(3):52, November 25, type from Whitman National Monument, 750 ft., 6 mi. W Walla Walla, Walla Walla Co., Washington.

MARGINAL RECORDS.—Washington: 7½ mi. N Walla Walla; *1¹⁄₅ mi. E Walla Walla;* 1 mi. S College Place; *2¹⁄₂ mi. SW College Place; 4 mi. W College Place;* 1 mi. W Lowden.

Thomomys talpoides yakimensis Hall and Dalquest

1939. *Thomomys talpoides yakimensis* Hall and Dalquest, Murrelet, 20:4, April 30, type from Selah, Yakima Co., Washington.

1939. *Thomomys talpoides badius* Goldman, Jour. Mamm., 20:242, May 15, type from Wenatchee, Chelan Co., Washington.

MARGINAL RECORDS.—Washington: Natches River, 40 mi. above mouth; Ellensburg; 2 mi. NW Richland, 400 ft.; Kennewick; Zillah; type locality; Cowiche.

Thomomys mazama
Mazama Pocket Gopher

External measurements: 175–273; 45–85; 21–35; ear from notch, 6.5–8.5. Reddish brown to black above; pelage of underparts plumbeous tipped with buffy; lips, nose, and postauricular patches black; wrists white. Mammae usually eight, two pairs pectoral and two pairs inguinal. Anterior openings of infraorbital canals posterior to anterior palatine foramina; occlusal surface of anterior prism of p4 having anteromedial notch and area of occlusal surface three-fourths or more of that of posterior prism.

From *T. monticola, T. mazama* (see Thaeler, 1968:19) differs, where the geographic ranges of the two species are no more than 10 mi. apart, as follows: dorsum generally bright reddish brown instead of Mummy Brown; ear from notch shorter (7.0–8.5 instead of 8.0–9.0 mm.); premaxillary tongues extending 1.5–2.5 mm. (instead of at most .5 mm.) posterior to nasals; nasals often truncate instead of V-shaped posteriorly; occipital region rounded instead of truncate and nearly vertical; base of rostrum broader on dorsal surface; anterior bases of pterygoids having, instead of lacking, broadened ventral shelves. The characters listed above, according to Thaeler (1968:19), vary enough to prevent any one character alone from being diagnostic, although in combination they serve to differentiate *mazama* from *monticola.* The baculum is long (22–31 mm. in Pacific Northwest) "always exceeding" that "of *monticola*" of comparable age."

From *T. talpoides,* according to Johnson and Benson (1960:20) *T. mazama* differs as follows: baculum longer (22–31 vs. 12–17 mm.); ear from notch longer (7.0–8.5 vs. 5.0–6.0 mm.); pelage reddish brown to black instead of yellowish brown and grayish brown; postauricular black patches more prominent; interparietal longer relative to its width.

Thomomys mazama couchi Goldman

1939. *Thomomys talpoides couchi* Goldman, Jour. Mamm., 20:243, May 15, type from 4 mi. N Shelton, Mason Co., Washington.

MARGINAL RECORDS.—Washington: type locality; Lost Lake Prairie, near Satsop.

Thomomys mazama glacialis Dalquest and Scheffer

1942. *Thomomys talpoides glacialis* Dalquest and Scheffer, Proc. Biol. Soc. Washington, 55:97, August 13, type from prairie 2 mi. S Roy, Pierce Co., Washington. Known only from Roy Prairie, Washington.

Thomomys mazama helleri Elliot

1903. *Thomomys helleri* Elliot, Field Columb. Mus., Publ. 74, Zool. Ser., 3:165, May 7, type from Goldbeach, mouth of Rogue River, Curry Co., Oregon.

MARGINAL RECORDS.—Oregon: *Wedderburn;* type locality.

Thomomys mazama hesperus Merriam

1901. *Thomomys hesperus* Merriam, Proc. Biol. Soc. Washington, 14:116, July 19, type from Tillamook, Tillamook Co., Oregon.

MARGINAL RECORDS.—Oregon: Elsie; Forest Grove; Grande Ronde; Philomath, W of Corvallis; Alsea; Oretown; type locality.

Thomomys mazama louiei Gardner

1950. *Thomomys talpoides louiei* Gardner, Jour. Mamm., 31:92, February 21, type from 12 mi. NNE Cathlamet (Crown-Zellerbach's Cathlamet Tree Farm), 2500 ft., Wahkiakum Co., Washington. Known only from type locality.

Thomomys mazama mazama Merriam

1897. *Thomomys mazama* Merriam, Proc. Biol. Soc. Washington, 11:214, July 15, type from Anna Creek, 6000 ft., near Crater Lake, Mt. Mazama, Klamath Co., Oregon.

MARGINAL RECORDS.—Oregon: W slope Mt. Hood; S base Mt. Hood, Summit House; Three Sisters; Diamond Lake; type locality; Elgin Ranch. California (Thaeler, 1968:20, 21, unless otherwise noted): 6 mi. SW Beswick (Hall and Kelson, 1959:446); *4 mi. W Bray;* $1\frac{7}{10}$ mi. N, $2\frac{9}{10}$ mi. W Tennant; Butte Creek, 5000 ft.; S side Grass Lake; $3\frac{9}{10}$ mi. W Ball Mtn.; *Spanish Spring;* Donomore Meadow, 15 mi. W Hilt; Deer Camp, head Doggett Creek; Poker Flat. Oregon: Siskiyou Mts.; Prospect; Pengra. Southern segment: California (Thaeler, 1968:20, 21, unless otherwise noted): $2\frac{1}{2}$ mi. E Etna; Weed; $3\frac{1}{10}$ mi. W Pondosa; Dixon Flat; Red Mtn., 5 mi. N, $5\frac{1}{10}$ mi. E Big Bend; Castle Lake; head Bear Creek (Hall and Kelson, 1959:446); S. Fork Salmon River, 5000 ft.; head Rush Creek, 6400 ft.

Thomomys mazama melanops Merriam

1899. *Thomomys melanops* Merriam, Proc. Biol. Soc. Washington, 13:21, January 31, type from timberline at head Soleduc River, Olympic Mts., Clallam Co., Washington.

MARGINAL RECORDS.—Washington: type locality; Happy Lake Ridge; Cat Creek, 4500 ft.; Canyon Creek, divide at head Bogachiel River.

Thomomys mazama nasicus Merriam

1897. *Thomomys nasicus* Merriam, Proc. Biol. Soc. Washington, 11:216, July 15, type from Farewell Bend, Deschutes River, Deschutes Co., Oregon.

MARGINAL RECORDS.—Oregon: type locality; Paulina Lake; Yamsey Mts.; Fort Klamath; mouth Davis Creek, Deschutes River; Bend.

Thomomys mazama niger Merriam

1901. *Thomomys niger* Merriam, Proc. Biol. Soc. Washington, 14:117, July 19, type from Seaton (= Mapleton), near mouth Umpqua River (= head tidewater, Siuslaw River), Lane Co., Oregon. (See Miller and Kellogg, Bull. U.S. Nat. Mus., 205:325, March 3, 1955.)

MARGINAL RECORDS.—Oregon: Benton–Lane County line; 2 mi. E Scottsburg; Seton; Mercer.

Thomomys mazama oregonus Merriam

1901. *Thomomys douglasi oregonus* Merriam, Proc. Biol. Soc. Washington, 14:115, July 19, type from Ely, near Oregon City, Willamette Valley, Clackamas Co., Oregon.

MARGINAL RECORDS.—Oregon: Scappoose (V. Bailey, 1936:254, as *T. douglasii douglasii,* but re-examined and identified as *T. mazama oregonus* by M. L. Johnson, *in Litt.,* July 27, 1971); Summit; 2 mi. W Parkdale; near Canby; Pedee; Black Rock; Forest Grove.

Thomomys mazama premaxillaris Grinnell

1914. *Thomomys monticola premaxillaris* Grinnell, Univ. California Publ. Zool., 12:312, November 21, type fom 2 mi. S South Yolla Bolly Mtn., 7500 ft., Tehama Co., California.

MARGINAL RECORDS.—California: 12 mi. N North Yolla Bolly Mtn.; type locality.

Thomomys mazama pugetensis Dalquest and Scheffer

1942. *Thomomys talpoides pugetensis* Dalquest and Scheffer, Proc. Biol. Soc. Washington, 55:96, August 13, type from 3 mi. S Olympia, Thurston Co., Washington. Known only from 3–5 mi. S Olympia, Washington (Lidicker, 1971:13).

Thomomys mazama tacomensis Taylor

1919. *Thomomys douglasii tacomensis* Taylor, Proc. Biol. Soc. Washington, 32:169, September 30, type from 6 mi. S Tacoma, Pierce Co., Washington.

MARGINAL RECORDS.—Washington: Tacoma; type locality.

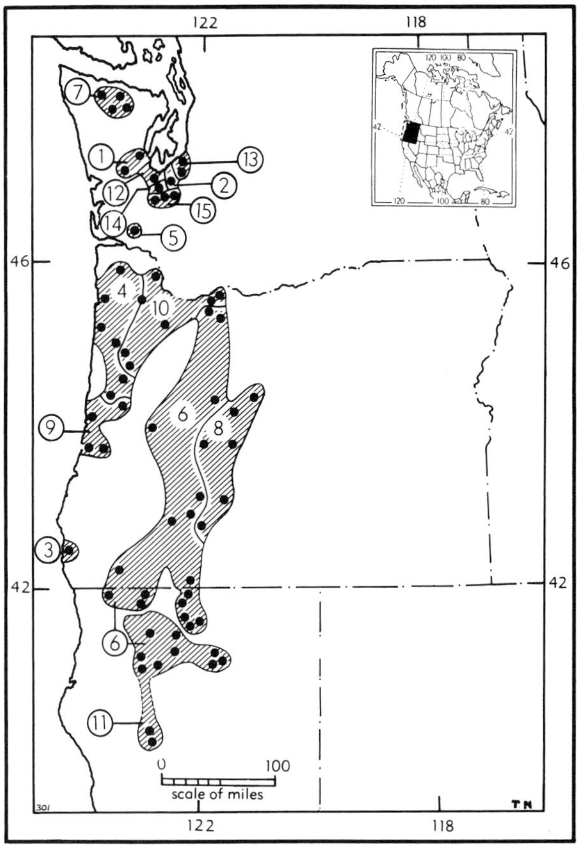

Map 301. *Thomomys mazama*.

Guide to subspecies
1. *T. m. couchi*
2. *T. m. glacialis*
3. *T. m. helleri*
4. *T. m. hesperus*
5. *T. m. louiei*
6. *T. m. mazama*
7. *T. m. melanops*
8. *T. m. nasicus*
9. *T. m. niger*
10. *T. m. oregonus*
11. *T. m. premaxillaris*
12. *T. m. pugetensis*
13. *T. m. tacomensis*
14. *T. m. tumuli*
15. *T. m. yelmensis*

Thomomys mazama tumuli Dalquest and Scheffer

1942. *Thomomys talpoides tumuli* Dalquest and Scheffer, Proc. Biol. Soc. Washington, 55:96, August 13, type from 7 mi. N Tenino, Thurston Co., Washington.

MARGINAL RECORDS.—Washington: type locality; *5 mi. N Tenino.*

Thomomys mazama yelmensis Merriam

1899. *Thomomys douglasi yelmensis* Merriam, Proc. Biol. Soc. Washington, 13:21, January 31, type from Tenino, Yelm Prairie, Thurston Co., Washington.

MARGINAL RECORDS.—Washington: type locality; 1 mi. W Vail; 2 mi. N Rochester.

Thomomys monticola J. A. Allen
Mountain Pocket Gopher

1893. *Thomomys monticolus* J. A. Allen, Bull. Amer. Mus. Nat. Hist., 5:48, April 28, type from Mt. Tallac, 7500 ft., El Dorado Co., California.

1899. *Thomomys monticola pinetorum* Merriam, N. Amer. Fauna, 16:97, October 28, type from Sisson, Siskiyou Co., California.

External measurements: 190–220; 55–80; 26–30; ear from notch, 8–9. Mummy Brown above; otherwise answers to description of *T. mazama*. From adjoining subspecies of *T. talpoides*, *T. monticola* differs as follows: color darker; ears longer (8–9 vs. 5–6 mm.) and pointed rather than rounded; two rather than three pairs of pectoral mammae; rostrum longer relative to length or breadth of braincase; zygomatic breadth less relative to length of skull; temporal ridges diverging posteriorly rather than nearly parallel; interparietal longer relative to its breadth; tympanic bullae larger; baculum longer (12–17 vs. 10.4–14.5 mm.).

311

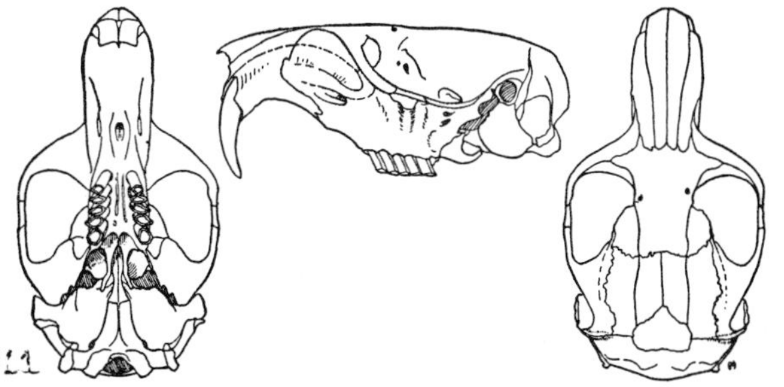

Fig. 311. *Thomomys monticola*, 3 mi. S Mount Rose, 8600 ft., Nevada, No. 65219 M.V.Z., ♂, X 1.

MARGINAL RECORDS.—California (Thaeler, 1968:15–19): Northern segment: N side Medicine Lake; $8\frac{1}{2}$ mi. N, 13 mi. W Canby; $\frac{1}{5}$ mi. N, $\frac{1}{10}$ mi. W Grouse Spring; $1\frac{1}{2}$ mi. N, $\frac{1}{5}$ mi. W Whitehorse; Dead Horse Summit; *$1\frac{3}{5}$ mi. S, $3\frac{7}{10}$ mi. W Bartle; Wagon Camp, Mt. Shasta;* Sisson [= Mt. Shasta, city]; *Spring Creek Ranch, $\frac{2}{5}$ mi. W Big Spring;* 2 mi. S Weed; *near Weed (W base "Cinder Cone");* $\frac{1}{2}$ mi. S, $\frac{1}{2}$ mi. E Tennant. Southern segment: 12 mi. W Burney; *$\frac{1}{10}$ mi. S, $\frac{3}{10}$ mi. W Lava Peak;* $1\frac{3}{10}$ mi. S, $1\frac{1}{2}$ mi. E Lava Peak; $9\frac{9}{10}$ mi. S, $4\frac{3}{10}$ mi. W Fredonyer Peak; *Gold Run Creek, $4\frac{1}{5}$ mi. S, $1\frac{7}{10}$ mi. W Susanville; $\frac{3}{10}$ mi. W Peter Lassen Grave; 5 mi. S, $\frac{1}{10}$ mi. E Johnstonville;* Rowland Ranch, $3\frac{1}{10}$ mi. S, 1 mi. E Doyle; Carmichael Ranch, $2\frac{1}{10}$ mi. N. $\frac{3}{10}$ mi. E Beckwourth; $1\frac{7}{10}$ mi. N, $1\frac{3}{10}$ mi. E Calpine; *$2\frac{1}{5}$ mi. S, $1\frac{2}{5}$ mi. E Loyalon.* Nevada: W side Truckee River, $\frac{1}{2}$ mi. W Verdi; 3 mi. S Mt. Rose; Carson (W of city?). California: Markleeville; Sonora Pass; Mt. Dana; vic. Mammoth; Mono Pass; Huntington Lake; Tenaya Lake; Big Trees; head South Fork American River; Heather Lake; Blue Canyon; 20 mi. SW Quincy (Thaler, 1968:18); Mineral

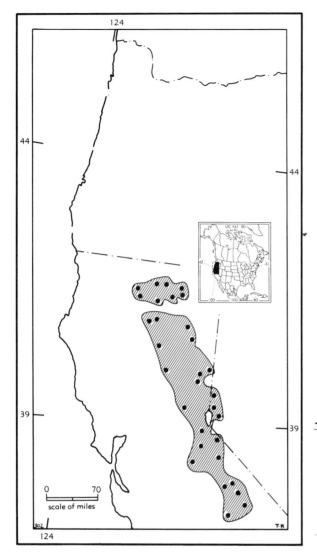

Map 302. *Thomomys monticola*.

(Thaeler, 1968:19); Supan Ranch, ⅕ mi. S, 3¼ mi. E Montgomery Creek P.O. (Thaeler, 1968:18).

Thomomys bulbivorus (Richardson)
Camas Pocket Gopher

1829. *Diplostoma bulbivorum* Richardson, Fauna Boreali-Americana, 1:206, type from "Banks of the Columbia River, Oregon," probably Portland, the only place near the Columbia River where it has been taken since. The type was reported as in the Hudson Bay Museum but has not been found (*fide* V. Bailey, N. Amer. Fauna, 39:40, November 15, 1915).

1855. *Thomomys bulbivorus*, Brandt, Beiträge zur nähern Kenntniss der Säugethiere Russland's, p. 188.

Average external measurements of five males: 300, 90, 42. Size large for genus. Color dark sooty brown, the plumbeous basal hairs showing through on underparts. Skull short and wide;

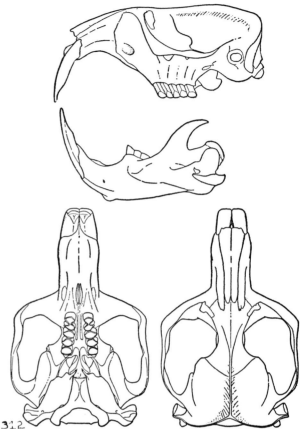

Fig. 312. *Thomomys bulbivorus*, Hillsboro, Washington Co., Oregon, No. 63479 M.V.Z., ♂, X 1.

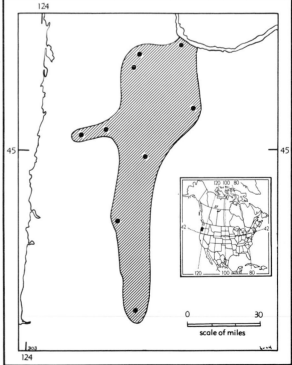

Map 303. *Thomomys bulbivorus*.

zygomatic arches usually widest posteriorly; nasals short; pterygoids convexly inflated and divided by narrow interpterygoid space; tympanic bullae relatively small; auditory meatus comparatively large; upper incisors markedly procumbent; molariform dentition weak.

MARGINAL RECORDS.—Oregon: Portland; Mulino; Salem; Eugene; Corvallis; Grande Ronde; Sheridan; Gaston; Forest Grove.

Thomomys umbrinus
Southern Pocket Gopher

External measurements: 132–340; 43–113; 22–45. Color varies from black to almost white according to subspecies (and in *T. u. albicaudatus*, individually); not sharply bicolored, but underparts paler than upper parts. Sphenoidal fissure present; incisive foramina posterior to anterior opening of infraorbital canal; interparietal relatively small; lamboidal suture usually approx. straight in region of interparietal.

When only a few study specimens of *Thomomys* were available from widely separated localities, systematic zoologists named most of the kinds as different species. In 1915 the number was reduced by V. Bailey to 40 (N. Amer. Fauna No. 39). In 1959 Hall and Kelson (The Mammals of North America) recognized six species. In the present work, only five species are recognized.

Many of the kinds of pocket gophers arranged as subspecies of the species *Thomomys umbrinus* (of Hall and Kelson, 1959) do not corssbreed at all places where the geographic ranges of adjacent kinds meet or closely approach one another. Since 1959, several mammalogists, who have studied pairs of subspecies, concluded that there was no intergradation between some of them, therefore that *T. umbrinus* (of Hall and Kelson, 1959) was a composite of two or three or many species. For example, failure to find intergradation (crossbreeding), at many places where looked for, between taxa of the southwestern United States and those of the tableland of México caused a number of mammalogists to recognize two species, *T. bottae* to the north and *T. umbrinus* to the south. Anderson (1966:197, Fig.

6), for one, did that. Hoffmeister (1969:89), however, later found "intermingling of genetical material" between the two "species" in Sycamore Canyon, Arizona. Also, study of specimens in the University of Kansas Museum of Natural History from eight localities in the state of Nuevo León and some from adjoining parts of Coahuila (although less minutely investigated than those from Sycamore Canyon) indicates an intermingling of genetic material in that region too. The "intermingling" in Arizona and in Nuevo León is regarded by this writer (Hall) as intergradation. Consequently the several taxa heretofore arranged as subspecies of *T. bottae* and *T. umbrinus* (in the sense of Anderson and others) are here arranged as subspecies of the latter because it was named first.

The five subspecies of *Thomomys baileyi* (of Hall and Kelson, 1959:435) were linked by intergradation of one of them (see Anderson, 1966:195) with *T. umbrinus*. *Thomomys townsendii* and *T. umbrinus* were tentatively retained as separate species by Hall and Kelson, 1959 because Hall (1946:451–455) had failed to find intergradation between them at the two places he looked for it. Later, at two other places Thaeler (1968:543) found a subspecies of *T. townsendii* intergrading (he termed it "hybridizing") with two subspecies of *T. umbrinus* (= *bottae* of Thaeler).

Consequently, *T. baileyi*, *T. townsendii*, and *T. bottae* of recent authors are here arranged as members of the species *Thomomys umbrinus*.

The several subspecies of *T. monticola* (of Hall and Kelson, 1959:446, 447) along with several subspecies formerly assigned to *T. talpoides* have been arranged as subspecies of *T. mazama* by Johnson and Benson (1960) leaving *T. monticola* as a monotypic species, and *T. talpoides* with fewer subspecies than formerly. *T. bulbivorus*, a monotypic species, remains as such.

Therefore, only five living species of *Thomomys* are recognized here. They are as follows: *T. talpoides*, *T. mazama*, *T. monticola*, *T. bulbivorus*, and *T. umbrinus*.

Thomomys umbrinus abbotti Huey

1928. *Thomomys bottae abbotti* Huey, Trans. San Diego Soc. Nat. Hist., 5:89, January 18, type from 1 mi. E El Rosario, Baja California (river bottom association), lat. 30° 03′ N, long. 115° 48′ W. Known only from type locality.
1959. *Thomomys umbrinus abbotti*, Hall and Kelson, Mammals of North America, Ronald Press, p. 416, Marcy 31.

Thomomys umbrinus absonus Goldman

1931. *Thomomys perpallidus absonus* Goldman, Jour. Washington Acad. Sci., 21:425, October 19, type from Jacobs Pools, 4000 ft., Houserock Valley, Coconino Co., Arizona.
1959. *Thomomys umbrinus absonus*, Hall and Kelson, Mammals of North America. Ronald Press, p. 418, March 31.

MARGINAL RECORDS.—Utah: Escalante, 5258 ft. Arizona: type locality; near Kanab Wash, southern boundary Kaibab Indian Reservation. Utah: Kanab, 4925 ft.

Thomomys umbrinus abstrusus Hall and Davis

1935. *Thomomys bottae abstrusus* Hall and Davis, Univ. California Publ. Zool., 40:391, March 13, type from Fish Spring Valley, 2 mi. SE Tulle Peak, 7000 ft., Nye Co., Nevada. Known only from type locality.

1959. *Thomomys umbrinus abstrusus,* Hall and Kelson, Mammals of North America, Ronald Press, p. 418, March 31.

Thomomys umbrinus acrirostratus Grinnell

1935. *Thomomys bottae acrirostratus* Grinnell, Univ. California Publ. Zool., 40:408, November 14, type from valley of Mad River, 2700 ft., 7 mi. above Ruth, Trinity Co., California.

1959. *Thomomys umbrinus acrirostratus,* Hall and Kelson, Mammals of North America, Ronald Press, p. 418, March 31.

MARGINAL RECORDS.—California: Helena, 1400 ft.; Deer Lick Springs, 3000 ft.; near Mad River bridge, SW base South Fork Mtn., 3000–3200 ft.; type locality.

Thomomys umbrinus actuosus Kelson

1951. *Thomomys bottae actuosus* Kelson, Univ. Kansas Publ., Mus. Nat. Hist., 5:67, October 1, type from Corona, Lincoln Co., New Mexico.

1959. *Thomomys umbrinus actuosus,* Hall and Kelson, Mammals of North America, Ronald Press, p. 418, March 31.

MARGINAL RECORDS.—New Mexico: 10 mi. S Mora, 7300 ft.; Las Vegas; type locality; NE slope Capitan Mts.; *SW foothills Capitan Mts.; near W end summit of ridge, Capitan Mts.;* Bear Canyon, San Andres Mts.; *E slope San Andres Mts.; E slope near S end Manzano Mts.;* E slope middle distance of range, Manzano Mts.; *E foothills near N end, Manzano Mts.;* San Pedro; *Pecos, 6800 ft.*

Thomomys umbrinus affinis Huey

1945. *Thomomys bottae affinis* Huey, Trans. San Diego Soc. Nat. Hist., 10:254, August 31, type from Jacumba, San Diego Co., California.

1959. *Thomomys umbrinus affinis,* Hall and Kelson, Mammals of North America, Ronald Press, p. 419, March 31.

MARGINAL RECORDS.—California: type locality; *Mountain Spring* (I have not been able to locate the specimen from Mountain Spring assigned by V. Bailey, 1915:57, to *T. b. nigricans;* it is almost surely better assigned to *T. u. affinis*).

Thomomys umbrinus agricolaris Grinnell

1935. *Thomomys bottae agricolaris* Grinnell, Univ. California Publ. Zool., 40:409, November 14, type from Stralock Farm, 3 mi. W Davis, Yolo Co., California.

1959. *Thomomys umbrinus agricolaris,* Hall and Kelson, Mammals of North America, Ronald Press, p. 419, March 31.

MARGINAL RECORDS.—California: Rumsey; Knights Landing; type locality; Rio Vista Junction; Denverton; 3 mi. W Vacaville.

Thomomys umbrinus albatus Grinnell

1912. *Thomomys albatus* Grinnell, Univ. California Publ. Zool., 10:172, June 7, type from W side Colorado River, at old Hanlon Ranch, near Pilot Knob, Imperial Co., California.

1959. *Thomomys umbrinus albatus,* Hall and Kelson, Mammals of North America, Ronald Press, p. 419, March 31.

MARGINAL RECORDS.—California: Alamo Duck Preserve, 8 mi. NW Calipatria; 10–12 mi. N Brawley (Bongardt, *et al.,* 1968:545); 2 mi. SW Potholes. Arizona: Yuma; 1 mi. N San Luis. Baja California: Colorado River, 20 mi. SW Pilot Knob; 5 mi. E Cerro Prieto. California: El Centro.

Thomomys umbrinus albicaudatus Hall

1930. *Thomomys perpallidus albicaudatus* Hall, Univ. California Publ. Zool., 32:444, July 8, type from Provo, 4510 ft., Utah Co., Utah.

1959. *Thomomys umbrinus albicaudatus,* Hall and Kelson, Mammals of North America, Ronald Press, p. 419, March 31.

MARGINAL RECORDS.—Utah: Bountiful, 4500 ft.; Draper, 4500 ft.; type locality; 7 mi. SW Nephi, 6000 ft.; 2 mi. W Murray, 4300 ft.; Rose Canyon, Oquirrh Mts., 5650 ft.; Vernon, 4300 ft.; Little Valley, Sheeprock Mts., 5500 ft.; Clover Creek, Onaqui Mts., 5500 ft.; Bauer, 4500 ft.; thence northward, E of Great Salt Lake to point of beginning.

Thomomys umbrinus albigularis Nelson and Goldman

1934. *Thomomys umbrinus albigularis* Nelson and Goldman, Jour. Mamm., 15:106, May 15, type from El Chico, 9800 ft., Sierra de Pachuca, Hidalgo.

MARGINAL RECORDS.—Hidalgo: type locality; 2 km. N Los Jacales, 7500 ft. (Hall and Dalquest, 1963:275); Tulancingo. Puebla: Rancho Ocotal Colorado, 5 mi. S, 2 mi. E Aquixtla, 8800 ft. (Musser, 1964:4, 7, as *T. umbrinus* only). Hidalgo: Real del Monte.

Thomomys umbrinus alexandrae Goldman

1933. *Thomomys alexandrae* Goldman, Jour. Washington, Acad. Sci., 23:464, October 15, type from plain 5 mi. SW Rainbow Lodge, near Navajo Mtn., 6200 ft., Coconino Co., Arizona.

1959. *Thomomys umbrinus alexandrae,* Hall and Kelson, Mammals of North America, Ronald Press, p. 419, March 31.

MARGINAL RECORDS.—Utah: Soldier Spring, Navajo Mtn., 8600 ft. Arizona: type locality.

Thomomys umbrinus alienus Goldman

1938. *Thomomys bottae alienus* Goldman, Jour. Washington Acad. Sci., 28:338, July 15, type from Mammoth, 2400 ft., San Pedro River, Pinal Co., Arizona.
1959. *Thomomys umbrinus alienus,* Hall and Kelson, Mammals of North America, Ronald Press, p. 419, March 31.

MARGINAL RECORDS.—Arizona (Cockrum, 1961:108): Juniper [Tank ?]; Safford, 2900 ft.; Duncan, 3500 ft. New Mexico: Redrock. Arizona (Cockrum, 1961:108): Fairbank, San Pedro River, 3700 ft.; 5 mi. SE Casabel; *5 mi. SE Redington;* San Pedro River, Redington; Oracle, 4500 ft.; Camp Grant [San Pedro River–Arivaipa Creek jct.]; Rice [San Carlos], 2700 ft.

Thomomys umbrinus alpinus Merriam

1897. *Thomomys alpinus* Merriam, Proc. Biol. Soc. Washington, 11:216, July 15, type from Big Cottonwood Meadows, 10,000 ft., 8 mi. SE Mt. Whitney peak, High Sierra, Inyo Co., California.
1959. *Thomomys umbrinus alpinus,* Hall and Kelson, Mammals of North America, Ronald Press, p. 419, March 31.

MARGINAL RECORDS.—California: Whitney Creek, 10,650 ft.; Cottonwood Creek, 9500 ft.; Olancha Peak; Siretta Meadows, 9000 ft.; Jordan Hot Springs.

Thomomys umbrinus alticolus J. A. Allen

1899. *Thomomys fulvus alticolus* J. A. Allen, Bull. Amer. Mus. Nat. Hist., 12:13, March 4, type from Sierra Laguna, 7000 ft., Baja California.
1959. *Thomomys umbrinus alticolus,* Hall and Kelson, Mammals of North America, Ronald Press, p. 419, March 31.

MARGINAL RECORDS.—Baja California: 7 mi. NW San Bartolo; type locality.

Thomomys umbrinus altivallis Rhoads

1895. *Thomomys altivallis* Rhoads, Proc. Acad. Nat. Sci. Philadelphia, 47:34, February 21, type from San Bernardino Mts., 5000 ft., California.
1959. *Thomomys umbrinus altivallis,* Hall and Kelson, Mammals of North America, Ronald Press, p. 419, March 31.

MARGINAL RECORDS.—California: *Doble; Fish Creek; Seven Oaks;* Fawnskin Valley.

Thomomys umbrinus amargosae Grinnell

1921. *Thomomys perpallidus amargosae* Grinnell, Univ. California Publ. Zool., 21:239, November 7, type from Shoshone, 1560 ft., Amargosa River, Inyo Co., California. Known only from type locality.
1959. *Thomomys umbrinus amargosae,* Hall and Kelson, Mammals of North America, Ronald Press, p. 149, March 31.

Thomomys umbrinus analogus Goldman

1938. *Thomomys umbrinus analogus* Goldman, Proc. Biol. Soc. Washington, 51:59, March 18, type from Sierra Guadalupe, about 12 mi. S General Cepeda, Coahuila.

MARGINAL RECORDS.—Coahuila: Jaral; 12 mi. E San Antonio de las Alazanas, 9000 ft.; Sierra Encarnación; *Sierra de Guadalupe, 7000 ft.*

Thomomys umbrinus angularis Merriam

1897. *Thomomys angularis* Merriam, Proc. Biol. Soc. Washington, 11:214, July 15, type from Los Baños, Merced Co., California.
1959. *Thomomys umbrinus angularis,* Hall and Kelson, Mammals of North America, Ronald Press, p. 419, March 31.

MARGINAL RECORDS.—California: 8 mi. S Tracy; type locality; Coalinga; Paso Robles; Kings City; Salinas; San Benito.

Thomomys umbrinus angustidens Baker

1953. *Thomomys bottae angustidens* Baker, Univ. Kansas Publ., Mus. Nat. Hist., 5:508, June 1, type from Sierra del Pino, 5250 ft., 6 mi. N, 6 mi. W Acebuches, Coahuila. Known only from vic. type locality.
1959. *Thomomys umbrinus angustidens,* Hall and Kelson, Mammals of North America, Ronald Press, p. 420, March 31.

Thomomys umbrinus anitae J. A. Allen

1898. *Thomomys fulvus anitae* J. A. Allen, Bull. Amer. Mus. Nat. Hist., 10:146, April 12, type from Santa Anita, Baja California.
1959. *Thomomys umbrinus anitae,* Hall and Kelson, Mammals of North America, Ronald Press, p. 420, March 31.

MARGINAL RECORDS.—Baja California: Tres Pachitas; Triunfo; type locality; Cape San Lucas.

Thomomys umbrinus aphrastus Elliot

1903. *Thomomys aphrastus* Elliot, Field Columb. Mus., Publ. 79, Zool. Ser., 3:219, August 15, type from San[to] Tomás, Baja California.
1959. *Thomomys umbrinus aphrastus,* Hall and Kelson, Mammals of North America, Ronald Press, p. 420, March 31.

MARGINAL RECORDS.—Baja California: type locality; extreme W end El Valle de la Trinidad; San Antonio; Rosarito; Socorro.

Thomomys umbrinus argusensis Huey

1931. *Thomomys argusensis* Huey, Trans. San Diego Soc. Nat. Hist., 7:43, December 19, type from Junction Ranch, Argus Mts., Inyo Co., California.
1959. *Thomomys umbrinus argusensis,* Hall and Kelson, Mammals of North America, Ronald Press, p. 420, March 31.

MARGINAL RECORDS.—California: type locality; *Mountain Spring; Orando (Arando) Mine.*

Thomomys umbrinus aridicola Huey

1937. *Thomomys bottae aridicola* Huey, Trans. San Diego Soc. Nat. Hist., 8:354, June 15, type from Ajo Railroad

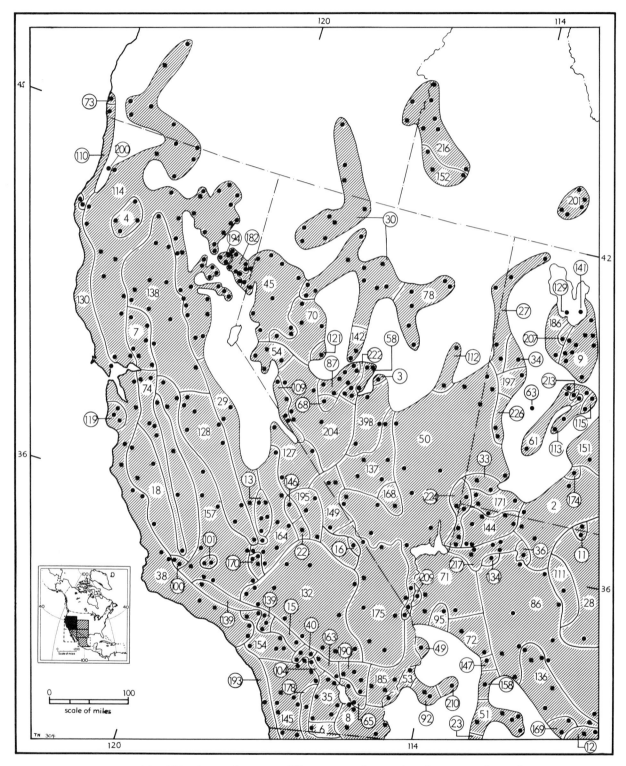

Map 304. Some subspecies of *Thomomys umbrinus*. (See facing page for guide.)

472

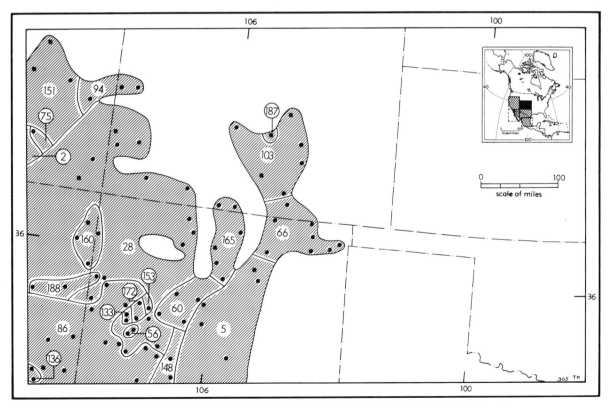

Map 305.   Some subspecies of *Thomomys umbrinus*.

473

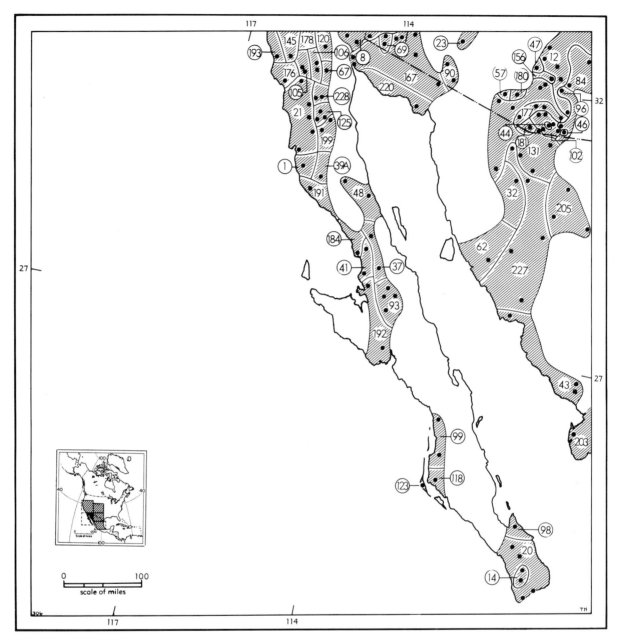

Map 306.   Some subspecies of *Thomomys umbrinus*. (See facing page for guide.)

474

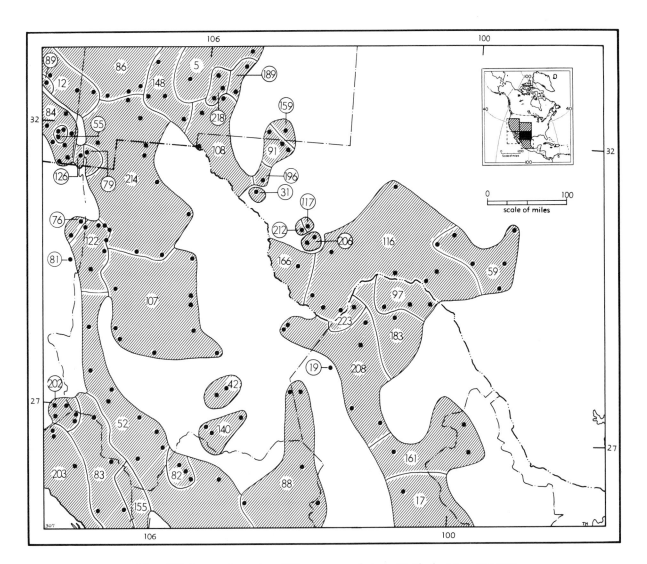

Map 307. Some subspecies of *Thomomys umbrinus*. Guide for Maps 306–307

1. *T. u. abbotti*	57. *T. u. comobabiensis*	106. *T. u. juarezensis*	178. *T. u. puertae*
5. *T. u. actuosus*	59. *T. u. confinalis*	107. *T. u. juntae*	180. *T. u. pusillus*
8. *T. u. albatus*	62. *T. u. convergens*	108. *T. u. lachuguilla*	181. *T. u. quercinus*
12. *T. u. alienus*	67. *T. u. cunicularius*	116. *T. u. limitaris*	183. *T. u. retractus*
14. *T. u. alticolus*	69. *T. u. depauperatus*	117. *T. u. limpiae*	184. *T. u. rhizophagus*
17. *T. u. analogus*	76. *T. u. divergens*	118. *T. u. litoris*	189. *T. u. ruidosae*
19. *T. u. angustidens*	79. *T. u. emotus*	120. *T. u. lucidus*	191. *T. u. ruricola*
20. *T. u. anitae*	81. *T. u. estanciae*	122. *T. u. madrensis*	192. *T. u. russeolus*
21. *T. u. aphrastus*	82. *T. u. evexus*	123. *T. u. magdalenae*	193. *T. u. sanctidiegi*
23. *T. u. aridicola*	83. *T. u. eximius*	125. *T. u. martirensis*	196. *T. u. scotophilus*
31. *T. u. baileyi*	84. *T. u. extenuatus*	126. *T. u. mearnsi*	199. *T. u. siccovallis*
32. *T. u. basilicae*	86. *T. u. fulvus*	131. *T. u. modicus*	202. *T. u. simulus*
37. *T. u. borjasensis*	88. *T. u. goldmani*	140. *T. u. nelsoni*	203. *T. u. sinaloae*
39A. *T. u. brazierhowelli*	89. *T. u. grahamensis*	145. *T. u. nigricans*	205. *T. u. sonoriensis*
41. *T. u. cactophilus*	90. *T. u. growlerensis*	148. *T. u. opulentus*	206. *T. u. spatiosus*
42. *T. u. camargensis*	91. *T. u. guadalupensis*	155. *T. u. parviceps*	208. *T. u. sturgisi*
43. *T. u. camoae*	93. *T. u. homorus*	156. *T. u. parvulus*	212. *T. u. texensis*
44. *T. u. caneloensis*	96. *T. u. hueyi*	159. *T. u. pectoralis*	214. *T. u. toltecus*
46. *T. u. carri*	97. *T. u. humilis*	161. *T. u. perditus*	218. *T. u. tularosae*
47. *T. u. catalinae*	98. *T. u. imitabilis*	166. *T. u. pervarius*	220. *T. u. vanrossemi*
48. *T. u. catavinensis*	99. *T. u. incomptus*	167. *T. u. phasma*	223. *T. u. villai*
52. *T. u. chihuahuae*	102. *T. u. intermedius*	176. *T. u. proximarinus*	227. *T. u. winthropi*
55. *T. u. collinus*	105. *T. u. jojobae*	177. *T. u. proximus*	228. *T. u. xerophilus*

475

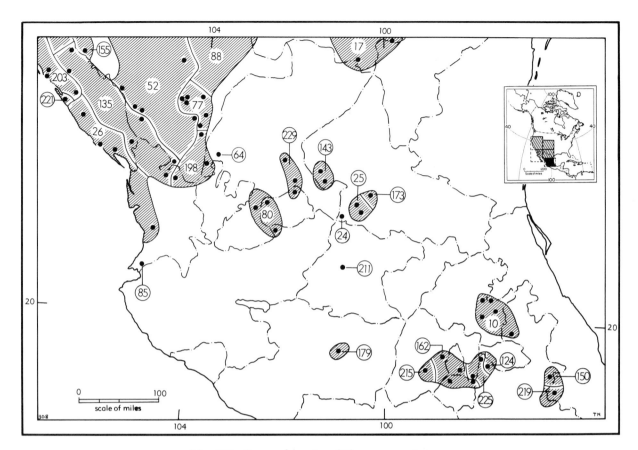

Map 308. Some subspecies of *Thomomys umbrinus*.

Guide to subspecies	64. *T. u. crassidens*	143. *T. u. newmani*	203. *T. u. sinaloae*
10. *T. u. albigularis*	77. *T. u. durangi*	150. *T. u. orizabae*	211. *T. u. supernus*
17. *T. u. analogus*	80. *T. u. enixus*	155. *T. u. parviceps*	215. *T. u. tolucae*
24. *T. u. arriagensis*	85. *T. u. extimus*	162. *T. u. peregrinus*	219. *T. u. umbrinus*
25. *T. u. atrodorsalis*	88. *T. u. goldmani*	173. *T. u. potosinus*	221. *T. u. varus*
26. *T. u. atrovarius*	124. *T. u. martinensis*	179. *T. u. pullus*	225. *T. u. vulcanius*
52. *T. u. chihuahuae*	135. *T. u. musculus*	198. *T. u. sheldoni*	229. *T. u. zacatecae*

right-of-way, about 2 mi. N Black Gap, 10 mi. S Gila Bend, Maricopa Co., Arizona. Known only from type locality.

1959. *Thomomys umbrinus aridicola*, Hall and Kelson, Mammals of North America, Ronald Press, p. 420, March 31.

Thomomys umbrinus arriagensis Dalquest

1951. *Thomomys umbrinus arriagensis* Dalquest, Jour. Washington Acad. Sci., 41:361, November 14, type from 1 km. S Arriaga, San Luis Potosí. Known only from type locality.

Thomomys umbrinus atrodorsalis Nelson and Goldman

1934. *Thomomys umbrinus atrodorsalis* Nelson and Goldman, Jour. Mamm., 15:111, May 15, type from Alvarez, 8000 ft., San Luis Potosí.

MARGINAL RECORDS.—San Luis Potosí: San Luis Potosí; *1 km. N Arenal*; type locality.

Thomomys umbrinus atrovarius J. A. Allen

1898. *Thomomys atrovarius* J. A. Allen, Bull. Amer. Mus. Nat. Hist., 10:148, April 12, type from Tatemales (near Rosario), Sinaloa.

1934. *Thomomys umbrinus atrovarius*, Nelson and Goldman, Jour. Mamm., 15:119, May 15.

MARGINAL RECORDS.—Sinaloa: 15 mi. NW Elota (Dunnigan, 1967:154); Rosario. Nayarit: Navarete, thence northward along coast to Sinaloa: Mazatlán.

Thomomys umbrinus aureiventris Hall

1930. *Thomomys perpallidus aureiventris* Hall, Univ. California Publ. Zool., 32:444, July 8, type from Fehlman Ranch, 3 mi. N Kelton, 4225 ft., Boxelder Co., Utah.

1959. *Thomomys umbrinus aureiventris*, Hall and Kelson, Mammals of North America, Ronald Press, p. 420, March 31.

MARGINAL RECORDS.—Utah: type locality; Utah–Nevada boundary, E side Tacoma Range, 4300 ft. Nevada: 15 mi. S Montello, 4 mi. W Pilot Peak. Utah: Ibapah, 5000 ft.; *Queen of Sheba Canyon, W side Deep Creek Mts., 5600 ft.;* Trout Creek.

Thomomys umbrinus aureus J. A. Allen

1893. *Thomomys aureus* J. A. Allen, Bull. Amer. Mus. Nat. Hist., 5:49, April 28, type from Bluff City, San Juan Co., Utah.

1959. *Thomomys umbrinus aureus*, Hall and Kelson, Mammals of North America, Ronald Press, p. 420, March 31.

1910. *Thomomys apache* V. Bailey, Proc. Biol. Soc. Washington, 23:79, May 4, type from Lake La Jara, 7500 ft., Jicarilla Apache Indian Reservation, New Mexico. Regarded as inseparable from *T. aureus* by Youngman (Univ. Kansas Publ., Mus. Nat. Hist., 9:370, February 21, 1958).

1936. *Thomomys bottae optabilis* Goldman, Jour. Washington Acad. Sci., 26:116, March 15, type from Coventry, 6500 ft., Montrose Co., Colorado. Regarded as inseparable from *T. aureus* by Youngman (Univ. Kansas Publ., Mus. Nat. Hist., 9:370, February 21, 1958).

MARGINAL RECORDS.—Colorado (Youngman, 1958:372): Bedrock, 5150 ft.; Coventry, 6800 ft.; 19 mi. N Dove Creek, 6100 ft.; 1 mi. N La Plata; 12 mi. W Pagosa Springs, 6700 ft. New Mexico: Horse Lake; near El Vado; Chama River, Gallina; Riley; St. Augustine Plains; Acoma; Wingate; Pueblo of Zuni. Arizona: Winslow. Utah: type locality; Monticello.

Thomomys umbrinus awahnee Merriam

1908. *Thomomys alpinus awahnee* Merriam, Proc. Biol. Soc. Washington, 21:146, June 9, type from Yosemite Valley, 4000 ft., near old Sentinel Hotel, Mariposa Co., California.

1959. *Thomomys umbrinus awahnee*, Hall and Kelson, Mammals of North America, Ronald Press, p. 420, March 31.

MARGINAL RECORDS.—California: type locality; Mineral King; *Kern River Lakes;* Taylor Meadow; *Tehachapi Peak; E fork Kaweah River, 5600 ft.;* Wawona; *S fork Merced River, 4000 ft.*

Thomomys umbrinus bachmani Davis

1937. *Thomomys townsendii bachmani* Davis, Jour. Mamm., 18:150, May 12, type from Quinn River Crossing, 4100 ft., Humboldt Co., Nevada.

MARGINAL RECORDS.—Oregon: 5 mi. SW Narrows. Nevada: McDermitt; type locality; 17½ mi. S, 5 mi. W Quinn River Crossing; 22 mi. N, 11½ mi. E Gerlach; Flowing Springs, 3½ mi. N, 7 mi. E Division Peak, 4200 ft.; Big Creek Ranch. Oregon: Tum Tum Lake; Lake Alvord. Also Nevada: Paradise; 1 mi. E Golconda, 4000 ft.; 18 mi. NE Iron Point, 4600 ft.; Argenta; Battle Mtn.; 18 mi. W Battle Mtn.; Lovelock; Toulon; *3 mi. N Lovelock* (Wentworth and Sutton, 1969:160); 1 mi. N Winnemucca; 13 mi. N Winnemucca (Bongardt, *et al.*, 1968:545).

Thomomys umbrinus baileyi Merriam

1901. *Thomomys baileyi* Merriam, Proc. Biol. Soc. Washington, 14:109, July 19, type from Sierra Blanca, Hudspeth Co., Texas. Known only from type locality.

Thomomys umbrinus basilicae Benson and Tillotson

1939. *Thomomys bottae occipitalis* Benson and Tillotson, Proc. Biol. Soc. Washington, 52:151, October 11, type from La Misión, 2 mi. W Magdalena, Sonora. Not *Thomomys bottae occipitalis* Dice, 1925, type a fossil, from Rancho La Brea deposits, California.

1940. *Thomomys bottae basilicae* Benson and Tillotson, Proc. Biol. Soc. Washington, 53:93, June 28, a renaming of *T. b. occipitalis*. Known only from type locality.

1959. *Thomomys umbrinus basilicae*, Hall and Kelson, Mammals of North America, Ronald Press, p. 420, March 31.

Thomomys umbrinus birdseyei Goldman

1937. *Thomomys bottae birdseyei* Goldman, Proc. Biol. Soc. Washington, 50:134, September 10, type from Pine Valley Mts., 5 mi. E Pine Valley, 8300 ft., Washington Co., Utah.

1959. *Thomomys umbrinus birdseyei*, Hall and Kelson, Mammals of North America, Ronald Press, p. 421, March 31.

MARGINAL RECORDS.—Utah: Hebron; type locality; *¾ mi. E town of Pine Valley, 6500 ft.; Pine Valley Campground, 6800 ft.*

Thomomys umbrinus bonnevillei Durrant

1946. *Thomomys bottae bonnevillei* Durrant, Univ. Kansas Publ., Mus. Nat. Hist., 1:41, August 15, type from Fish Springs, 4400 ft., Juab Co., Utah. Known only from type locality.

1959. *Thomomys umbrinus bonnevillei*, Hall and Kelson, Mammals of North America, Ronald Press, p. 421, March 31.

Thomomys umbrinus boregoensis Huey

1939. *Thomomys bottae boregoensis* Huey, Trans. San Diego Soc. Nat. Hist., 9:70, December 8, type from Beatty Ranch, Borego Valley, San Diego Co., California.

1959. *Thomomys umbrinus boregoensis*, Hall and Kelson, Mammals of North America, Ronald Press, p. 421, March 31.

1939. *Thomomys bottae aderrans* Huey, Trans. San Diego Soc. Nat. Hist., 9:71, December 8, type from Carrizo Creek, San Diego Co., California.

MARGINAL RECORDS.—California: 2 mi. N Oasis, near Thermal; Mecca; Harpers Well; Coyote Wells; Carrizo Creek; E side San Felipe Narrows.

Thomomys umbrinus boreorarius Durham

1952. *Thomomys bottae boreorarius* Durham, Jour. Mamm., 33:498, November 19, type from Swamp Point, 7522 ft., 18½ mi. NW Bright Angel Point, North Rim of Grand Canyon, Coconino Co., Arizona.

1959. *Thomomys umbrinus boreorarius*, Hall and Kelson, Mammals of North America, Ronald Press, p. 421, March 31.

MARGINAL RECORDS.—Arizona: *Powell Spring, 6209 ft.; NE end Powell Plateau, 7650 ft.*; type locality; *Swamp Lake* (Hoffmeister, 1971:170); *corral near Muav Trail, Powell Plateau (ibid.).*

Thomomys umbrinus borjasensis Huey

1945. *Thomomys bottae borjasensis* Huey, Trans. San Diego Soc. Nat. Hist., 10:262, August 31, type from San Borjas Mission, lat. 28° 52′ N, long. 113° 53′ W, Baja California.
1959. *Thomomys umbrinus borjasensis*, Hall and Kelson, Mammals of North America, Ronald Press, p. 421, March 31.

MARGINAL RECORDS.—Baja California: Yubay; type locality.

Thomomys umbrinus bottae (Eydoux and Gervais)

1836. *Oryctomys (Saccophorus) bottae* Eydoux and Gervais, Mag. de Zool., Paris, 6:23. Type locality, coast of California; name applied by Baird (Proc. Acad. Nat. Sci. Philadelphia, 7:335, April 1855) to the gopher occurring in vic. Monterey.
1959. *Thomomys umbrinus bottae*, Hall and Kelson, Mammals of North America, Ronald Press, p. 421, March 31.

MARGINAL RECORDS.—California: Walnut Creek; San Jose; Salinas Valley; Jamesburg; Jolon; Pleyto; Santa Margarita; Big Pine Mts.; Matilija; Santa Paula; Santa Monica; Alhambra, thence northward along coast (excepting range of *T. u. lorenzi*) to point of beginning.

Thomomys umbrinus brazierhowelli Huey

1960. *Thomomys umbrinus brazierhowelli* Huey, Trans. San Diego Soc. Nat. Hist., 12:407, February 1, type from San Fernando Mission, Baja California, México. Known only from type locality.

Thomomys umbrinus brevidens Hall

1932. *Thomomys bottae brevidens* Hall, Univ. California Publ. Zool., 38:330, February 27, type from Breen Creek, 7000 ft., Kawich Range, Nye Co., Nevada.
1959. *Thomomys umbrinus brevidens*, Hall and Kelson, Mammals of North America, Ronald Press, p. 421, March 31.

MARGINAL RECORDS.—Nevada: 1 mi. N Fish Lake, 6500 ft., Fish Spring Valley; 5 mi. E Nyala, 6000 ft.; Cactus Flat, 5700 ft., 7½ mi. SW Silverbow.

Thomomys umbrinus cabezonae Merriam

1901. *Thomomys cabezònae* Merriam, Proc. Biol. Soc. Washington, 14:110, July 19, type from Cabezon, San Grogonio Pass, Riverside Co., California.
1959. *Thomomys umbrinus cabezonae*, Hall and Kelson, Mammals of North America, Ronald Press, p. 421, March 31.

MARGINAL RECORDS.—California: Banning; Whitewater; Schains Ranch; *Vallevista, 1800 ft.*

Thomomys umbrinus cactophilus Huey

1929. *Thomomys bottae cactophilus* Huey, Trans. San Diego Soc. Nat. Hist., 5:241, February 27, type from Punta Prieta, lat. 28° 56′ N, long. 114° 12′ W, Baja: California.
1959. *Thomomys umbrinus cactophilus*, Hall and Kelson, Mammals of North America, Ronald Press, p. 421, March 31.

MARGINAL RECORDS.—Baja California: type locality; Santa Rosalía Bay.

Thomomys umbrinus camargensis Anderson

1972. *Thomomys umbrinus camargensis* Anderson, Bull. Amer. Mus. Nat. Hist., 148:288, September 8, type from 1 mi. S Camargo, 3950 ft., Chihuahua.

MARGINAL RECORDS (Anderson, 1972:290).—Chihuahua: 1 mi. NW Camargo, 4000 ft.; *type locality;* 1½ mi. N Boquilla de Conchos, 14 mi. SW Ciudad Camargo.

Thomomys umbrinus camoae Burt

1937. *Thomomys bottae camoae* Burt, Occas. Pap. Mus. Zool., Univ. Michigan, 344:1, January 5, type from Camoa, Río Mayo, Sonora.
1959. *Thomomys umbrinus camoae*, Hall and Kelson, Mammals of North America, Ronald Press, p. 421, March 31.

MARGINAL RECORDS.—Sonora: San José de Guaymas; type locality; Tesia.

Thomomys umbrinus caneloensis Lange

1959. *Thomomys bottae caneloensis* Lange, Proc. Biol. Soc. Washington, 72:131, November 4, type from Huachuca Mountains, western foothills, Canelo, 10 mi. S Elgin, 5100 ft., Santa Cruz Co., Arizona. Known only from type locality.

Thomomys umbrinus canus V. Bailey

1910. *Thomomys canus* V. Bailey, Proc. Biol. Soc. Washington, 23:79, May 4, type from Deep Hole, N end Smoke Creek Desert, Washoe Co., Nevada.
1959. *Thomomys umbrinus canus*, Hall and Kelson, Mammals of North America, Ronald Press, p. 421, March 31.

MARGINAL RECORDS.—Nevada: Granite Creek; type locality; sec. 7, T. 27 N, R. 28 E, Pershing Co. (Ghiselin, 1965:525); 2 mi. N Nixon; Fallon; 7 mi. S, 3½ mi. E Fallon; N side Carson River, 4300 ft., 1 mi. E Dayton; N side Truckee River, 9½ mi. E Reno, 4500 ft.; 4½ mi. S Flanagan, 4100 ft. California (Thaeler, 1968:7–9): 2⁹⁄₁₀ mi. S, 4⁹⁄₁₀ mi. E Herlong; 4¹⁄₅ mi. S Herlong; 3 mi. S, 1⁷⁄₁₀ mi. W Herlong; 2¹⁄₅ mi. S, ⁴⁄₅ mi. W Herlong; 2¹⁄₁₀ mi. S, ⁴⁄₅ mi. W Herlong; ²⁄₅ mi. S, 1³⁄₅ mi. W Herlong; 1³⁄₁₀ mi. W Herlong; High Rock Ranch. Nevada: Smoke Creek.

Thomomys umbrinus carri Lange

1959. *Thomomys bottae carri* Lange, Proc. Biol. Soc. Washington, 72:130, November 4, type from Huachuca Mts., NW

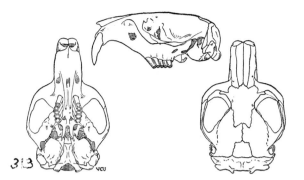

Fig. 313. *Thomomys umbrinus pullus*, 5 mi. S Pátzcuaro, Michoacán, No. 100515 M.V.Z., ♂, X 1.

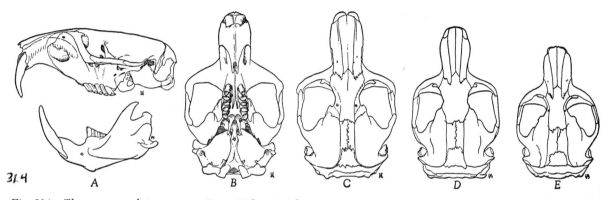

Fig. 314. *Thomomys umbrinus canus*, Deep Hole, Nevada, A, B, C, No. 41652 M.V.Z., ♂, X 1, and D No. 41660, ♀, X 1. E is *Thomomys umbrinus phelleoecus*, Hidden Forest, Nevada, No. 93119 M.V.Z., ♂, X 1. Note secondary sexual difference between C and D, and geographic (subspecific) difference between C and E.

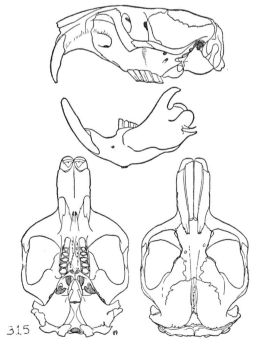

Fig. 315. *Thomomys umbrinus nevadensis*, Malloy Ranch, 5 mi. W Austin, Nevada, No. 37073 M.V.Z., ♂, X 1.

slope, Carr Peak, 8400 ft., Cochise Co., Arizona. Known only from type locality.

Thomomys umbrinus catalinae Goldman

1931. *Thomomys fulvus catalinae* Goldman, Jour. Washington Acad. Sci., 21:419, October 19, type from Summerhaven, Santa Catalina Mts., 7500 ft., Pima Co., Arizona.
1959. *Thomomys umbrinus catalinae*, Hall and Kelson, Mammals of North America, Ronald Press, p. 422, March 31.

MARGINAL RECORDS.—Arizona: type locality; *Sabino Canyon near Summerhaven.*

Thomomys umbrinus catavinensis Huey

1931. *Thomomys bottae catavinensis* Huey, Trans. San Diego Soc. Nat. Hist., 7:45, December 19, type from Cataviña, lat. 29° 54′ N, long. 114° 57′ W, Baja California.
1959. *Thomomys umbrinus catavinensis*, Hall and Kelson, Mammals of North America, Ronald Press, p. 422, March 31.

MARGINAL RECORDS.—Baja California: type locality; San Francisquito.

Thomomys umbrinus cedrinus Huey

1955. *Thomomys bottae cedrinus* Huey, Trans. San Diego Soc. Nat. Hist., 12:100, February 10, type from summit of Crossman Peak (Juniper–Piñon Belt), Chemehuevis Mts., Mohave Co., Arizona. Known only from type locality and Lucky Star Mine lower on mountain range.
1959. *Thomomys umbrinus cedrinus*, Hall and Kelson, Mammals of North America, Ronald Press, p. 422, March 31.

Thomomys umbrinus centralis Hall

1930. *Thomomys perpallidus centralis* Hall, Univ. California Publ. Zool., 32:445, July 8, type from 2½ mi. E Baker (1¼ mi. W Nevada–Utah boundary on 39th parallel), 5700 ft., White Pine Co., Nevada.

1959. *Thomomys umbrinus centralis,* Hall and Kelson, Mammals of North America, Ronald Press, p. 422, March 31.

MARGINAL RECORDS.—Utah: 1 mi. SE Gandy, 5000 ft.; White Valley (Tule Spring), 60 mi. W Delta; 5 mi. S Garrison, 5400 ft.; Cedar City. Nevada: Meadow Valley Wash, 7 mi. S Caliente; Black Canyon; Boulder City; Durban Ranch, Colorado River, 14 mi. E Searchlight; ½ mi. N California–Nevada Monument, Colorado River, 500 ft. California: Colorado River, 500 ft. Nevada: Trout Canyon, 6500 ft.; Amargosa River, 3½ mi. NE Beatty, 3400 ft.; Springdale, N end Oasis Valley; Indian Springs; Coyote Spring, 2800 ft.; Crystal Spring to ¾ mi. S thereof, 4000 ft., Pahranagat Valley; Grant Mts., 19½ mi. WSW Sunnyside; Cleveland Ranch, 6000 ft., Spring Valley.

Thomomys umbrinus cervinus J. A. Allen

1895. *Thomomys cervinus* J. A. Allen, Bull. Amer. Mus. Nat. Hist., 7:203, June 29, type from Phoenix, Maricopa Co., Arizona.
1959. *Thomomys umbrinus cervinus,* Hall and Kelson, Mammals of North America, Ronald Press, p. 422, March 31.

MARGINAL RECORDS.—Arizona: type locality; Tempe; jct. Salt and Verde rivers (877 ASU); Sacaton; Gila Bend.

Thomomys umbrinus chihuahuae Nelson and Goldman

1934. *Thomomys umbrinus chihuahuae* Nelson and Goldman, Jour. Mamm., 15:114, May 15, type from Sierra Madre, 7000 ft., about 65 mi. E Batopilas, Chihuahua.

MARGINAL RECORDS.—Chihuahua (Anderson, 1972:290): Yaguirachic, 130 mi. W Chihuahua, 8500 ft.; 15 mi. S, 6 mi. E Creel, 7300 ft.; type locality; 10 mi. SW Agostadero, 102 mi. (by road) W Parral, 8400 ft.; 7 mi. W El Vergel, 7800 ft. Durango: Río Sestín (Hall and Kelson, 1959:433, as *T. u. sheldoni*); La Boca *(ibid.)*; 24 mi. SSE Durango, 7200 ft. (Baker and Greer, 1962:94); Cueva, 8500 ft. *(ibid.)*; 7 mi. SW Las Adjuntas *(ibid.)*; E slope Cerro Huehueto *(ibid.).* Chihuahua (Anderson, 1972:290): near Guadalupe y Calvo; 2 mi. W Samachique, 7000 ft.; Mojarachic.

Thomomys umbrinus chrysonotus Grinnell

1912. *Thomomys chrysonotus* Grinnell, Univ. California Publ. Zool., 10:174, June 7, type from Ehrenberg, Yuma Co., Arizona.
1959. *Thomomys umbrinus chrysonotus,* Hall and Kelson, Mammals of North America, Ronald Press, p. 422, March 31.
1931. *Thomomys fulvus flavidus* Goldman, Jour. Washington Acad. Sci., 21:417, October 19, type from Parker, 350 ft., Yuma Co., Arizona.

MARGINAL RECORDS.—Arizona: Parker; 3 mi. S Ehrenberg.

Thomomys umbrinus cinereus Hall

1932. *Thomomys bottae cinereus* Hall, Univ. California Publ. Zool., 38:327, February 27, type from West Walker River, Smiths Valley, 4700 ft., Lyon Co., Nevada.
1959. *Thomomys umbrinus cinereus,* Hall and Kelson, Mammals of North America, Ronald Press, p. 422, March 31.

MARGINAL RECORDS.—Nevada: ½ mi. N Yerington; 3 mi. S Schurz, 4100 ft.; 9 mi. NE Wellington, 4800 ft.; East Walker River, 5000 ft., 5 mi. NW Morgans Ranch.

Thomomys umbrinus collinus Goldman

1931. *Thomomys fulvus collinus* Goldman, Jour. Washington Acad. Sci., 21:421, October 19, type from Fly Park, 9000 ft., Chiricahua Mts., Cochise Co., Arizona.
1959. *Thomomys umbrinus collinus,* Hall and Kelson, Mammals of North America, Ronald Press, p. 423, March 31.
1934. *Thomomys umbrinus chiricahuae,* Nelson and Goldman, Jour. Mamm., 15:117, May 15, type from Pinery Canyon, 7500 ft., Chiricahua Mts., Arizona.

MARGINAL RECORDS.—Arizona (Patton and Dingman, 1968:12, unless otherwise noted): 1 mi. below Rustlers Park; Rucker Canyon, *ca.* 5600 ft.; *mouth Turkey Creek, Chiricahua Mts.* (Hall and Kelson, 1959:423); El Coronado Ranch, West Turkey Canyon, Chiricahua Mts.; *1⁷⁄₁₀ mi. E El Coronado Ranch.*

Thomomys umbrinus collis Hooper

1940. *Thomomys bottae collis* Hooper, Occas. Pap. Mus. Zool., Univ. Michigan, 422:7, November 14, type from Shuman's Ranch, 30 mi. S Grants, sec. 30, T. 6 N, R. 10 W, Valencia Co., New Mexico.
1959. *Thomomys umbrinus collis,* Hall and Kelson, Mammals of North America, Ronald Press, p. 423, March 31.

MARGINAL RECORDS.—New Mexico: type locality; Point of Malpais, 12 mi. E Trechado.

Thomomys umbrinus comobabiensis Huey

1937. *Thomomys bottae comobabiensis* Huey, Trans. San Diego Soc. Nat. Hist., 8:354, June 15, type from 5 mi. NW Sells, 2400 ft., Pima Co., Arizona. Known only from type locality.
1959. *Thomomys umbrinus comobabiensis,* Hall and Kelson, Mammals of North America, Ronald Press, p. 423, March 31.

Thomomys umbrinus concisor Hall and Davis

1935. *Thomomys bottae concisor* Hall and Davis, Univ. California Publ. Zool., 40:390, March 13, type from Pott's Ranch, 6900 ft., Monitor Valley, Nye Co., Nevada.
1959. *Thomomys umbrinus concisor,* Hall and Kelson, Mammals of North America, Ronald Press, p. 423, March 31.

MARGINAL RECORDS.—Nevada: Wilson Creek, 7200 ft.; type locality; 8 mi. N Pine Creek Ranch, 7050 ft.

Thomomys umbrinus confinalis Goldman

1936. *Thomomys lachuguilla confinalis* Goldman, Jour. Washington Acad. Sci., 26:119, March 15, type from 35 mi. E Rock Springs, 2450 ft., Texas.

1959. *Thomomys umbrinus confinalis,* Hall and Kelson, Mammals of North America, Ronald Press, p. 423, March 31.

MARGINAL RECORDS (Dalquest and Kilpatrick, 1973:1–7, unless otherwise noted).—Texas: 3 mi. SE London; type locality; Uvalde County (Davis, 1966:150, as *T. bottae* only); 2 mi. S Rocksprings; 5 mi. W jct. U.S. 272 and FM (=Farm Road) 189, in Sutton Co.; 20 mi. W Sonora.

Thomomys umbrinus connectens Hall

1936. *Thomomys bottae connectens* Hall, Jour. Washington Acad. Sci., 26:296, July 15, type from Clawson Dairy, 5 mi. N Albuquerque, 4943 ft., Bernalillo Co., New Mexico.

1959. *Thomomys umbrinus connectens,* Hall and Kelson, Mammals of North America, Ronald Press, p. 423, March 31.

MARGINAL RECORDS.—New Mexico: Juan Tafoya; Bernalillo; Belen; Laguna.

Thomomys umbrinus contractus Durrant

1946. *Thomomys bottae contractus* Durrant, Univ. Kansas Publ., Mus. Nat. Hist., 1:50, August 15, type from Scipio, 5315 ft., Millard Co., Utah.

1959. *Thomomys umbrinus contractus,* Hall and Kelson, Mammals of North America, Ronald Press, p. 423, March 31.

MARGINAL RECORDS.—Utah: Oak City, 5000 ft.; type locality; Beaver, 6000 ft.

Thomomys umbrinus convergens Nelson and Goldman

1934. *Thomomys bottae convergens* Nelson and Goldman, Jour. Mamm., 15:123, May 15, type from Costa Rica Ranch, delta Sonora River, SW of Hermosillo, Sonora. Known only from type locality.

1959. *Thomomys umbrinus convergens,* Hall and Kelson, Mammals of North America, Ronald Press, p. 423, March 31.

Thomomys umbrinus convexus Durrant

1939. *Thomomys bottae convexus* Durrant, Proc. Biol. Soc. Washington, 52:159, October 11, type from E side Clear Lake, 4600 ft., Millard Co., Utah. Known only from type locality.

1959. *Thomomys umbrinus convexus,* Hall and Kelson, Mammals of North America, Ronald Press, p. 423, March 31.

Thomomys umbrinus crassidens Nelson and Goldman

1934. *Thomomys umbrinus crassidens* Nelson and Goldman, Jour. Mamm., 15:113, May 15, type from Sierra de Valparaíso, 8700 ft., Zacatecas. Known only from type locality.

Thomomys umbrinus crassus Chattin

1941. *Thomomys bottae crassus* Chattin, Trans. San Diego Soc. Nat. Hist., 9:274, April 30, type from 1½ mi. W Niland, −180 ft., Imperial Co., California.

1959. *Thomomys umbrinus crassus,* Hall and Kelson, Mammals of North America, Ronald Press, p. 423, March 31.

MARGINAL RECORDS.—California: Salt Creek; type locality; *5 mi. SW Niland, −212 ft.*

Thomomys umbrinus cultellus Kelson

1951. *Thomomys bottae cúltellus* Kelson, Univ. Kansas Publ., Mus. Nat. Hist., 5:64, ·October 1, type from Halls Peak, Mora Co., New Mexico.

1959. *Thomomys umbrinus cultellus,* Hall and Kelson, Mammals of North America, Ronald Press, p. 423, March 31.

MARGINAL RECORDS.—Colorado: Fisher Peak. New Mexico: Oak Canyon; Sierra Grande; $5\frac{9}{10}$ mi. N, $5\frac{1}{10}$ mi. W Clayton (Best, 1973:1318); $2\frac{2}{5}$ *mi. N, $7\frac{3}{5}$ mi. W Clayton (ibid.);* $4\frac{3}{10}$ mi. S Mt. Dora *(ibid.);* $6\frac{1}{10}$ mi. N Gladstone *(ibid.);* type locality; Long Canyon, 3 mi. N Catskill.

Thomomys umbrinus cunicularius Huey

1945. *Thomomys bottae cunicularius* Huey, Trans. San Diego Soc. Nat. Hist., 10:252, August 31, type from Los Palmitos (western end Pattie Basin), on southeastern basin of Sierra Juárez (desert slope), lat. 31° 44′ N, long. 115° 36′ W, Baja California. Known only from type locality.

1959. *Thomomys umbrinus cunicularius,* Hall and Kelson, Mammals of North America, Ronald Press, p. 423, March 31.

Thomomys umbrinus curtatus Hall

1932. *Thomomys bottae curtatus* Hall, Univ. California Publ. Zool., 38:329, February 27, type from San Antonio, 5400 ft., Nye Co., Nevada. Known only from type locality.

1959. *Thomomys umbrinus curtatus,* Hall and Kelson, Mammals of North America, Ronald Press, p. 423, March 31.

Thomomys umbrinus depauperatus Grinnell and Hill

1936. *Thomomys perpallidus depauperatus* Grinnell and Hill, Jour. Mamm., 17:4, February 17, type from E base Tinajas Altas Mts., 7 mi. S Raven Butte, 1150 ft., Yuma Co., Arizona.

1959. *Thomomys umbrinus depauperatus,* Hall and Kelson, Mammals of North America, Ronald Press, p. 423, March 31.

MARGINAL RECORDS.—Arizona: type locality; proximity of Tinajas Altas.

Thomomys umbrinus depressus Hall

1932. *Thomomys bottae depressus* Hall, Univ. California Publ. Zool., 38:326, February 27, type from Dixie Meadows (at S end Humboldt Salt Marsh), 3500 ft., Churchill Co., Nevada.

1959. *Thomomys umbrinus depressus,* Hall and Kelson, Mammals of North America, Ronald Press, p. 423, March 31.

MARGINAL RECORDS.—Nevada: 1⅔ mi. NE Ocala, 3900 ft.; type locality; 15 mi. SW Fallon, 5000 ft.; 1 mi. W Mountain Well.

Thomomys umbrinus desertorum Merriam

1901. *Thomomys desertorum* Merriam, Proc. Biol. Soc. Washington, 14:114, July 19, type from Mud Spring, Detrital Valley, Mohave Co., Arizona.
1959. *Thomomys umbrinus desertorum*, Hall and Kelson, Mammals of North America, Ronald Press, p. 424, March 31.

MARGINAL RECORDS.—Arizona: Mineral Park, Detrital Valley; Big Sandy Creek, 2000 ft.; type locality.

Thomomys umbrinus desitus Goldman

1936. *Thomomys bottae desitus* Goldman, Jour. Washington Acad. Sci., 26:113, March 15, type from Big Sandy River, 2000 ft., near Owen, Mohave Co., Arizona.
1959. *Thomomys umbrinus desitus*, Hall and Kelson, Mammals of North America, Ronald Press, p. 424, March 31.

MARGINAL RECORDS.—Arizona: type locality; Kirkland.

Thomomys umbrinus detumidus Grinnell

1935. *Thomomys bottae detumidus* Grinnell, Univ. California Publ. Zool., 40:405, November 14, type from 1½ mi. S (town of) Pistol River, 250 ft., Curry Co., Oregon. Known only from type locality.
1959. *Thomomys umbrinus detumidus*, Hall and Kelson, Mammals of North America, Ronald Press, p. 424, March 31.

Thomomys umbrinus diaboli Grinnell

1914. *Thomomys diaboli* Grinnell, Univ. California Publ. Zool., 12:313, November 21, type from Sweeney's Ranch, 22 mi. [by road] SW Los Baños, Diablo Range, Merced Co., California.
1959. *Thomomys umbrinus diaboli*, Hall and Kelson, Mammals of North America, Ronald Press, p. 424, March 31.

MARGINAL RECORDS.—California: Bryon; Pacheco Pass; type locality; divide W of McKittrick, 3000 ft.; Gilroy; Mt. Diablo.

Thomomys umbrinus dissimilis Goldman

1931. *Thomomys perpallidus dissimilis* Goldman, Jour. Washington Acad. Sci., 21:425, October 19, type from E slope Mt. Ellen, 8000 ft., Henry Mts., Garfield Co., Utah. Known only from type locality.
1959. *Thomomys umbrinus dissimilis*, Hall and Kelson, Mammals of North America, Ronald Press, p. 424, March 31.

Thomomys umbrinus divergens Nelson and Goldman

1934. *Thomomys bottae divergens* Nelson and Goldman, Jour. Mamm., 15:122, May 15, type from 4 mi. W Huachinera, 4000 ft., Río Bavispe, Sonora.
1959. *Thomomys umbrinus divergens*, Hall and Kelson, Mammals of North America, Ronald Press, p. 424, March 31.

MARGINAL RECORDS.—Sonora: Bacerac; type locality.

Thomomys umbrinus durangi Nelson and Goldman

1934. *Thomomys umbrinus durangi* Nelson and Goldman, Jour. Mamm., 15:114, May 15, type from Durango, Durango.

MARGINAL RECORDS (Baker and Greer, 1962:94).—Durango: 6 mi. NW La Pila, 6150 ft.; Nombre de Dios; 16 mi. S, 20 mi. W Vicente Guerrero, 6675 ft.; 5 mi. S Durango, 6200 ft.; 8 mi. NW Durango, 6200 ft.; 4 mi. S Morcillo, 6450 ft.

Thomomys umbrinus elkoensis Davis

1937. *Thomomys townsendii elkoensis* Davis, Jour. Mamm., 18:151, May 12, type from Evans, Eureka Co., Nevada.

MARGINAL RECORDS.—Nevada: 2 mi. W Halleck, 5200 ft.; *1 mi. E Halleck* (Wentworth and Sutton, 1969:160); 3 mi. S Halleck, 5200 ft.; type locality; 4 mi. S Romano; Winzell; Independence Valley.

Thomomys umbrinus emotus Goldman

1933. *Thomomys fulvus emotus* Goldman, Proc. Biol. Soc. Washington, 46:76, April 27, type from Animas Peak, 8000 ft., Animas Mts., New Mexico. Known only from upper slopes Animas Mts., New Mexico.
1934. *Thomomys umbrinus emotus*, Nelson and Goldman, Jour. Mamm., 15:116, May 15.

Thomomys umbrinus enixus Nelson and Goldman

1934. *Thomomys umbrinus enixus* Nelson and Goldman, Jour. Mamm., 15:112, May 15, type from Sierra Moroni, 8500 ft., near Plateado, Zacatecas.

MARGINAL RECORDS.—Aguascalientes: 12 mi. N Rincón de Romos, 6500 ft. (39790 KU). Jalisco: 1 mi. NE Villa Hidalgo, 6500 ft. (39794 KU). Zacatecas: type locality.

Thomomys umbrinus estanciae Benson and Tillotson

1939. *Thomomys bottae estanciae* Benson and Tillotson, Proc. Biol. Soc. Washington, 52:152, October 11, type from La Estancia, 6 mi. N Nacori, Sonora. Known only from type locality.
1959. *Thomomys umbrinus estanciae*, Hall and Kelson, Mammals of North America, Ronald Press, p. 424, March 31.

Thomomys umbrinus evexus Nelson and Goldman

1934. *Thomomys umbrinus evexus* Nelson and Goldman, Jour. Mamm., 15:115, May 15, type from Mt. San Gabriel, vic. Rosario, 10 mi. NW Villa Ocampo, Río Florida between 7000 and 9000 ft., Durango.

MARGINAL RECORDS.—Durango: type locality; Villa Ocampo, 4575 ft. (Baker and Greer, 1962:94, 95); 3 mi. E Las Nieves, 5400 ft. (*ibid.*).

Thomomys umbrinus eximius Nelson and Goldman

1934. *Thomomys umbrinus eximius* Nelson and Goldman, Jour. Mamm., 15:118, May 15, type from Sierra de Choix, about 20 mi. NE Choix, Sinaloa.

MARGINAL RECORDS (Dunnigan, 1967:157).—Sinaloa: *1 mi. E El Cajón;* type locality; 15 km. N, 65 km. E Sinaloa; 13 mi. ESE Badiraguato; 1 mi. S Pericos.

Thomomys umbrinus extenuatus Goldman

1935. *Thomomys bottae extenuatus* Goldman, Proc. Biol. Soc. Washington, 48:149, October 31, type from Willcox, 4000 ft., Cochise Co., Arizona.
1959. *Thomomys umbrinus extenuatus*, Hall and Kelson, Mammals of North America, Ronald Press, p. 424, March 31.

MARGINAL RECORDS.—Arizona (Cockrum, 1961: 112): Fort (Camp) Grant, 5200 ft.; San Simon Valley, San Simon Cattle Co. Ranch, 3800 ft.; 3 mi. E Portal; Pinery Canyon, 5000 ft., Chiricahua Mts.; Black Draw, 3900 ft., San Bernardino Ranch; San Bernardino Ranch, 15 mi. E Douglas; Sulphur Springs Valley, 30 mi. N Douglas; 12 mi. SE Dos Cabezos [Dos Cabezas], 4500 ft.; 15 mi. NE Benson, 4000 ft.

Thomomys umbrinus extimus Nelson and Goldman

1934. *Thomomys umbrinus extimus* Nelson and Goldman, Jour. Mamm., 15:119, May 15, type from Colomo, 600 ft., Nayarit. Known only from type locality.

Thomomys umbrinus fulvus (Woodhouse)

1852. *Geomys fulvus* Woodhouse, Proc. Acad. Nat. Sci. Philadelphia, 6:201, type from San Francisco Mtn., Coconino Co., Arizona.
1959. *Thomomys umbrinus fulvus*, Hall and Kelson, Mammals of North America, Ronald Press, p. 424, March 31.
1932. *Thomomys bottae nasutus* Hall, Proc. Biol. Soc. Washington, 45:96, June 21, type from W fork Black River, 7550 ft., Apache Co., Arizona. Regarded as inseparable from *T. u. fulvus* by Goldman (Proc. Biol. Soc. Washington, 48:156, October 31, 1935).

MARGINAL RECORDS.—Arizona (Cockrum, 1961: 112, 113, unless otherwise noted): Pasture Wash, jct. roads W-9 and W-9A (Hoffmeister, 1971:170); *N side Cedar Mtn. (ibid.);* E side Cedar Mtn. *(ibid.);* Wupatki Ruins; Sawmill Spring, 7300 ft., 8 mi. SE Mormon Lake; Bog Creek, 7 mi. E McNary; Springerville, 6500–7000 ft.; *16 mi. E Springerville.* New Mexico: Zuni River; Fort Wingate; Mt. Sedgwick; Largo Canyon, 10 mi. SW Quemado; Gallina Mts.; Copper Canyon, 8200 ft., Magdalena Mts.; Water Canyon, 6500 ft., Magdalena Mts.; 10 mi. E Chloride; *ca.* 4 mi. NW Kingston, 9500 ft.; Rio Mimbres; Burro Mts. Arizona (Cockrum, 1961:112, 113, unless otherwise noted): Rose Peak, 8700 ft.; W fork Black River, 7800 ft.; Sawmill, 5800 ft., 25 mi. NE Rice; N slope Baker Butte; Youngs Ranch, 7500 ft., Mingus Mtn., 6 mi. SE

Jerome; Bradshaw City; Prescott, 5300–5346 ft.; 5 mi. N Pine Springs; *2 mi. S Pasture Wash Ranger Station* (Hoffmeister, 1971:170); *Pasture Wash, 6300 ft. (ibid.).*

Thomomys umbrinus fumosus Hall

1932. *Thomomys bottae fumosus* Hall, Univ. California Publ. Zool., 38:329, February 27, type from Milman Ranch, Moores Creek, 19 mi. SE Millett P.O., Nye Co., Nevada.
1959. *Thomomys umbrinus fumosus*, Hall and Kelson, Mammals of North America, Ronald Press, p. 425, March 31.

MARGINAL RECORDS.—Nevada: Kingston Ranch, 16 mi. N Millett P.O.; Daniels Ranch, 12 mi. NE Millett P.O.; type locality; Peavine Ranch, 7 mi. N San Antonio, 6000 ft.; Cloverdale Ranch; South Twin River, 7000 ft.; Millett P.O.

Thomomys umbrinus goldmani Merriam

1901. *Thomomys goldmani* Merriam, Proc. Biol. Soc. Washington, 14:108, July 19, type from Mapimí, 3800 ft., Durango.
1934. *Thomomys umbrinus goldmani*, Nelson and Goldman, Jour. Mamm., 15:115, May 15.

MARGINAL RECORDS.—Coahuila: 3 mi. NE Sierra Mojada, 4100 ft.; 3 mi. SE Torreón, 3800 ft. Durango: *4 mi. WSW Lerdo, 3800 ft.;* 2 mi. S El Palmito (Baker and Greer, 1962:95); *1 mi. WSW Mapimi, 3800 ft.;* type locality. Chihuahua: SW slope Sierra Almagre, 19 mi. S, 4 mi. E Jaco, 5500 ft. (Anderson, 1972:290).

Thomomys umbrinus grahamensis Goldman

1931. *Thomomys fulvus grahamensis* Goldman, Jour. Washington Acad. Sci., 21:420, October 19, type from Graham Mts. (Pinaleno Mts. on some maps), 9200 ft., Graham Co., Arizona. Known only from upper slopes Graham Mts.
1959. *Thomomys umbrinus grahamensis*, Hall and Kelson, Mammals of North America, Ronald Press, p. 425, March 31.

Thomomys umbrinus growlerensis Huey

1937. *Thomomys bottae growlerensis* Huey, Trans. San Diego Soc. Nat. Hist., 8:353, June 15, type from 7 mi. E Papago Well, Pima Co., Arizona.
1959. *Thomomys umbrinus growlerensis*, Hall and Kelson, Mammals of North America, Ronald Press, p. 425, March 31.

MARGINAL RECORDS.—Arizona: Bates Well, Growler Pass; S end Puerto Blanco Mts.

Thomomys umbrinus guadalupensis Goldman

1936. *Thomomys bottae guadalupensis* Goldman, Jour. Washington Acad. Sci., 26:117, March 15, type from McKittrick Canyon, 7800 ft., Guadalupe Mts., Texas.
1959. *Thomomys umbrinus guadalupensis*, Hall and Kelson, Mammals of North America, Ronald Press, p. 425, March 31.

MARGINAL RECORDS.—New Mexico: Dog Canyon, Guadalupe Mts., 6800 ft. Texas: Mouth Pine

Springs Canyon, 5800 ft.; Burned Cabin, head McKit-trick Canyon, 7500 ft.

Thomomys umbrinus harquahalae Grinnell and Hill

1936. *Thomomys harquahalae* Grinnell and Hill, Jour. Mamm., 17:7, February 17, type from Ranegras Plain, 10 mi. W Hope, Yuma Co., Arizona.
1959. *Thomomys umbrinus harquahalae*, Hall and Kelson, Mammals of North America, Ronald Press, p. 425, March 31.

MARGINAL RECORDS.—Arizona: type locality; base Harquahala Mts., 5 mi. S Salome.

Thomomys umbrinus homorus Huey

1949. *Thomomys bottae homorus* Huey, Trans. San Diego Soc. Nat. Hist., 11:55, January 31, type from 1 mi. E Rancho Lagunitas, lat. 28° 20′ N, long. 113° 15′ W, Baja California.
1959. *Thomomys umbrinus homorus*, Hall and Kelson, Mammals of North America, Ronald Press, p. 425, March 31.

MARGINAL RECORDS.—Baja California: type locality; Rancho Unión, 15 mi. E Calmallí; Santana; Calmallí.

Thomomys umbrinus howelli Goldman

1936. *Thomomys bottae howelli* Goldman, Jour. Washington Acad. Sci., 26:116, March 15, type from Grand Junction, 4600 ft., Mesa Co., Colorado.
1959. *Thomomys umbrinus howelli*, Hall and Kelson, Mammals of North America, Ronald Press, p. 425, March 31.

MARGINAL RECORDS.—Colorado: type locality; Sieber Ranch, Little Dolores River (Youngman, 1958:373). Utah: 10 mi. N Moab.

Thomomys umbrinus hualpaiensis Goldman

1936. *Thomomys bottae hualpaiensis* Goldman, Jour. Washington Acad. Sci., 26:114, March 15, type from Hualpai Peak, Hualpai Mts., 7000 ft., Mohave Co., Arizona. Known only from type locality.
1959. *Thomomys umbrinus hualpaiensis*, Hall and Kelson, Mammals of North America, Ronald Press, p. 425, March 31.

Thomomys umbrinus hueyi Goldman

1938. *Thomomys bottae hueyi* Goldman, Jour. Washington Acad. Sci., 28:340, July 15, type from Spud Rock Ranger Station, 7400 ft., Rincon Mts., Pima Co., Arizona.
1959. *Thomomys umbrinus hueyi*, Hall and Kelson, Mammals of North America, Ronald Press, p. 425, March 31.

MARGINAL RECORDS.—Arizona: type locality; *head Miller Canyon, Huachuca Mts.*; Ramsey Canyon, 7000 ft., Huachuca Mts.

Thomomys umbrinus humilis Baker

1953. *Thomomys bottae humilis* Baker, Univ. Kansas Publ., Mus. Nat. Hist., 5:503, June 1, type from 3 mi. W Hda. San Miguel, 2200 ft., Coahuila.

1959. *Thomomys umbrinus humilis*, Hall and Kelson, Mammals of North America, Ronald Press, p. 425, March 31.

MARGINAL RECORDS.—Coahuila: 1 mi. S, 9 mi. W Villa Acuña; type locality; Cañón del Cochino, 3200 ft., 16 mi. N, 21 mi. E Piedra Blanca.

Thomomys umbrinus imitabilis Goldman

1939. *Thomomys bottae imitabilis* Goldman, Proc. Biol. Soc. Washington, 52:30, March 11, type from La Paz, Baja California. Known only from type locality.
1959. *Thomomys umbrinus imitabilis*, Hall and Kelson, Mammals of North America, Ronald Press, p. 426, March 31.

Thomomys umbrinus incomptus Goldman

1939. *Thomomys bottae incomptus* Goldman, Proc. Biol. Soc. Washington, 52:29, March 11, type from San Jorge, near Pacific Coast W of Pozo Grande, about 25 mi. SW Comondú, Baja California.
1959. *Thomomys umbrinus incomptus*, Hall and Kelson, Mammals of North America, Ronald Press, p. 426, March 31.

MARGINAL RECORDS.—Baja California: type locality; Matancita.

Thomomys umbrinus infrapallidus Grinnell

1914. *Thomomys infrapallidus* Grinnell, Univ. California Publ. Zool., 12:314, November 21, type from 7 mi. SE Simmler, Carrizo Plain, San Luis Obispo Co., California.
1959. *Thomomys umbrinus infrapallidus*, Hall and Kelson, Mammals of North America, Ronald Press, p. 426, March 31.

MARGINAL RECORDS.—California: type locality; 5 mi. N Painted Rock, Carrizo Plain.

Thomomys umbrinus ingens Grinnell

1932. *Thomomys bottae ingens* Grinnell, Univ. California Publ. Zool., 38:405, September 20, type from E side levee, 290 ft., 2 mi. due W of Millux, Buena Vista Lake, Kern Co., California.
1959. *Thomomys umbrinus ingens*, Hall and Kelson, Mammals of North America, Ronald Press, p. 426, March 31.

MARGINAL RECORDS.—California: 12 mi. S, 8 mi. W Bakersfield; type locality.

Thomomys umbrinus intermedius Mearns

1897. *Thomomys fulvus intermedius* Mearns, Proc. U.S. Nat. Mus., 19:719, July 30, type from summit Huachuca Mts., 9000 ft., Arizona.
1934. *Thomomys umbrinus intermedius*, Nelson and Goldman, Jour. Mamm., 15:117, May 15.
1932. *Thomomys burti* Huey, Trans. San Diego Soc. Nat. Hist., 7:158, July 28. Type from Madera Canyon, 6000 ft., Santa Rita Mts., Arizona. Regarded as inseparable from *Thomomys fulvus intermedius* Mearns by Lange, Proc. Biol. Soc. Washington, 72:127, November 4, 1959.

MARGINAL RECORDS.—Arizona (Patton and Dingman, 1968:12, unless otherwise noted): Gardner Canyon; 4 mi. W, 1 mi. N Fort [Huachuca] (Hoffmeister and Goodpaster, 1954:98, as *T. b. proximus*, but referred to *T. u. intermedius* by Lange, 1959:128); type locality; *mouth Italian Canyon;* Sycamore Canyon (west to locality 7 of Hoffmeister, 1969:80); Madera Canyon, 6000 ft., Santa Rita Mts. (Huey, 1932:158).

Thomomys umbrinus internatus Goldman

1936. *Thomomys bottae internatus* Goldman, Jour. Washington Acad. Sci., 26:115, March 15, type from Salida, 7000 ft., Chaffee Co., Colorado.
1959. *Thomomys umbrinus internatus*, Hall and Kelson, Mammals of North America, Ronald Press, p. 426, March 31.

MARGINAL RECORDS (Youngman, 1958:375).— Colorado: 1¼ mi. S Colorado Springs; St. Charles Mesa, 5600 ft.; fork Huerfano and Cuchara rivers; 5 mi. SE La Veta; 1½ mi. S Redwing; 2 mi. NNW Salida, 7100 ft.

Thomomys umbrinus jacinteus Grinnell and Swarth

1914. *Thomomys jacinteus* Grinnell and Swarth, Proc. California Acad. Sci., ser. 4, 4:154, December 30, type from Round Valley, 9000 ft., San Jacinto Mts., Riverside Co., California.
1959. *Thomomys umbrinus jacinteus*, Hall and Kelson, Mammals of North America, Ronald Press, p. 426, March 31.

MARGINAL RECORDS.—California: San Jacinto Peak, 10,200 ft.; *type locality; Tahquitz Valley, 8000 ft.; Tamarack Valley, 9400 ft.*

Thomomys umbrinus jojobae Huey

1945. *Thomomys bottae jojobae* Huey, Trans. San Diego Soc. Nat. Hist., 10:256, August 31, type from Sangre de Cristo, lat. 31° 52' N, long. 116° 06' W, Baja California.
1959. *Thomomys umbrinus jojobae*, Hall and Kelson, Mammals of North America, Ronald Press, p. 426, March 31.

MARGINAL RECORDS.—Baja California: La Huerta; type locality.

Thomomys umbrinus juarezensis Huey

1945. *Thomomys bottae juarezensis* Huey, Trans. San Diego Soc. Nat. Hist., 10:255, August 31, type from Laguna Hanson, Sierra Juárez, Baja California.
1959. *Thomomys umbrinus juarezensis*, Hall and Kelson, Mammals of North America, Ronald Press, p. 426, March 31.

MARGINAL RECORDS.—Baja California: type locality; El Rayo, Sierra Juárez.

Thomomys umbrinus juntae Anderson

1972. *Thomomys umbrinus juntae* Anderson, Bull. Amer. Mus. Nat. Hist., 148:291, September 8, type from Rancho San Ignacio, 4 mi. S, 1 mi. W Santo Tomás, Chihuahua.

MARGINAL RECORDS (Anderson, 1972:291).— Chihuahua: Gallego; 3 km. N, 10 km. W Est. Pinalé, 4900 ft.; Cañón de Potrero, 7 mi. W El Sauz; *5 mi. N Chihuahua, 4700 ft.;* Ciudad Chihuahua; jct. Río San Pedro and Río Conchos, 5 mi. N, 5 mi. E Meoqui, 3550 ft.; San Bernabé, near Cusihuiriachic; 2 mi. W Miñaca, 6900 ft.; type locality; 2 mi. SW Babícora, 7450 ft.

Thomomys umbrinus lachuguilla V. Bailey

1902. *Thomomys aureus lachuguilla* V. Bailey, Proc. Biol. Soc. Washington, 15:120, June 2, type from arid foothills near El Paso, El Paso Co., Texas.
1959. *Thomomys umbrinus lachuguilla*, Hall and Kelson, Mammals of North America, Ronald Press, p. 426, March 31.

MARGINAL RECORDS.—New Mexico: Alamagordo. Texas: El Paso. New Mexico: Organ.

Thomomys umbrinus lacrymalis Hall

1932. *Thomomys bottae lacrymalis* Hall, Univ. California Publ. Zool., 38:328, February 27, type from Arlemont [Chiatovich Ranch, Fish Lake Valley], 4900 ft., Esmeralda Co., Nevada.
1959. *Thomomys umbrinus lacrymalis*, Hall and Kelson, Mammals of North America, Ronald Press, p. 426, March 31.

MARGINAL RECORDS.—Nevada: Cat Creek, 4 mi. W Hawthorne; 1 mi. W Candelaria Junction, 5500 ft.; type locality; (upper) McNett Ranch, 5 mi. SW Arlemont, 5600 ft.

Thomomys umbrinus laticeps Baird

1855. *Thomomys laticeps* Baird, Proc. Acad. Nat. Sci. Philadelphia, 7:335, April, type from Humboldt Bay, Humboldt Co., California.
1959. *Thomomys umbrinus laticeps*, Hall and Kelson, Mammals of North America, Ronald Press, p. 426, March 31.

MARGINAL RECORDS.—Oregon: near Chetco, thence southward along coast to California: Rio Dell, on Eel River.

Thomomys umbrinus latirostris Merriam

1901. *Thomomys latirostris* Merriam, Proc. Biol. Soc. Washington, 14:107, July 19, type from Tanner Crossing, about 3 mi. above Cameron, Little Colorado River, Coconino Co., Arizona. (For status see Hoffmeister, Jour. Washington Acad. Sci., 45:127, April 25, 1955.) Known only from 3, 4½, and 5 mi. N Cameron.
1959. *Thomomys umbrinus latirostris*, Hall and Kelson, Mammals of North America, Ronald Press, p. 426, March 31.

Thomomys umbrinus latus Hall and Davis

1935. *Thomomys bottae latus* Hall and Davis, Univ. California Publ. Zool., 40:393, March 13, type from Cherry Creek, 6500 ft., White Pine Co., Nevada.
1959. *Thomomys umbrinus latus*, Hall and Kelson, Mammals of North America, Ronald Press, p. 426, March 31.

MARGINAL RECORDS.—Nevada: type locality; $6\frac{1}{2}$ mi. SE Ely, 6400 ft.

Thomomys umbrinus lenis Goldman

1942. *Thomomys townsendii lenis* Goldman, Proc. Biol. Soc. Washington, 55:75, June 25, type from Richfield, 5308 ft., Sevier Co., Utah.

1959. *Thomomys umbrinus lenis,* Hall and Kelson, Mammals of North America, Ronald Press, p. 427, March 31.

MARGINAL RECORDS.—Utah: Lynndyl, 4796 ft.; U.B. (= Yuba) Dam, 5000 ft.; Salina, 4375 ft.; type locality.

Thomomys umbrinus leucodon Merriam

1897. *Thomomys leucodon* Merriam, Proc. Biol. Soc. Washington, 11:215, July 15, type from Grant Pass, Rogue River Valley, Oregon.

1959. *Thomomys umbrinus leucodon,* Hall and Kelson, Mammals of North America, p. 427, March 31.

MARGINAL RECORDS.—Oregon: Cottage Grove; type locality; W slope Grizzly Peak; Picard (Thaeler, 1968:13). California (Thaeler, 1968:7, 9–13, unless otherwise noted): $\frac{1}{10}$ mi. S, $2\frac{1}{5}$ mi. W Macdoel; Edgewood; *Shasta River, 3 mi. S Edgewood; Gazelle;* Fort Jones; Castella; Dana; 8 mi. NW Day; $\frac{1}{5}$ mi. W Mud Spring; $\frac{1}{2}$ mi. S, $\frac{1}{2}$ mi. W Browns Well; 2 mi. E Canby; Rush Creek, 6 mi. NE Adin; $1\frac{4}{5}$ mi. N, $9\frac{4}{5}$ mi. W Madeline; 3 mi. S, $4\frac{4}{5}$ mi. E Hayden Hill L.O.; Coyote Flat; *Feather Lake;* McCoy Flat; *Norvell Flat; $4\frac{2}{5}$ mi. S, $3\frac{1}{10}$ mi. E Bogard R.S.;* $\frac{7}{10}$ mi. S, $1\frac{1}{2}$ mi. E Bogard Ranger Station; Butler Spring, Halls Flat; 3 mi. W Burney; Montgomery Creek P.O.; *Round Mtn.;* Turner's, Lyonsville P.O.; Rich Gulch, 8 mi. N, 11 mi. W Quincy; Prattville; Greenville, thence up Indian Creek to Genesse; Quincy; Mountain House, $13\frac{3}{10}$ mi. N, $12\frac{2}{5}$ mi. W Oroville; Crombergs, Middle Fork Feather River, thence up river to $\frac{1}{5}$ mi. S, $\frac{3}{10}$ mi. W Delleker; $4\frac{2}{5}$ mi. N, $2\frac{4}{5}$ mi. E Calpine; *$1\frac{3}{5}$ mi. N, $1\frac{1}{5}$ mi. E Calpine;* Downieville; Blue Canyon (Hall and Kelson, 1959:427); Fyffe [= Fyfee] (*ibid.*); Placerville (*ibid.*); 8 mi. W Grass Valley; Oregon House; 12 mi. W Woodleaf; *6 mi. S Berry Creek;* 3 mi. E Yankee Hill; Lyman's, 4 mi. NW Lyonsville; Long's, $4\frac{3}{10}$ mi. N, $1\frac{1}{2}$ mi. W Inskip Hill; 3 mi. W summit Mt. Sanhedrin (Hall and Kelson, 1959:427); Lower Lake (*ibid.*); Calistoga (*ibid.*); 8 mi. W Vacaville (Bailey, 1915:49); Novato (Hall and Kelson, 1959:427); Ukiah (*ibid.*); Laytonville (*ibid.*); Briceland (*ibid.*); Cuddeback (= Carlotta) (*ibid.*); Hoopa Valley (*ibid.*); Hilt. Oregon: Cave Junction (Olterman and Verts, 1972:27, as *T. bottae*); 13 mi. SW Galice (*ibid.*); Umpqua Valley, near Roseburg. Not found: California: Bartlett Mtn.; Post Creek; Slippery Ford (all from V. Bailey, 1915:48, 49).

Thomomys umbrinus levidensis Goldman

1942. *Thomomys bottae levidensis* Goldman, Proc. Biol. Soc. Washington, 55:76, June 25, type from Manti, about 5500 ft., Sanpete Co., Utah.

1959. *Thomomys umbrinus levidensis,* Hall and Kelson, Mammals of North America, Ronald Press, p. 427, March 31.

MARGINAL RECORDS.—Utah: Spring City; type locality.

Thomomys umbrinus limitaris Goldman

1936. *Thomomys lachuguilla limitaris* Goldman, Jour. Washington Acad. Sci., 26:118, March 15, type from 4 mi. W Boquillas, Brewster Co., Texas.

1959. *Thomomys umbrinus limitaris,* Hall and Kelson, Mammals of North America, Ronald Press, p. 427, March 31.

MARGINAL RECORDS.—Texas: Castle Mts.; Devils River, 13 mi. below Juno; Comstock; Samuels, 17 mi. W Langtry; Boquillas; Glenn Spring, 2606 ft.; E base Burro Mesa, 3500 ft.; 15 mi. S Marathon.

Thomomys umbrinus limpiae Blair

1939. *Thomomys bottae limpiae* Blair, Occas. Pap. Mus. Zool., Univ. Michigan, 403:2, June 16, type from Limpia Canyon, 1 mi. N Fort Davis, 4700 ft., Jeff Davis Co., Texas.

1959. *Thomomys umbrinus limpiae,* Hall and Kelson, Mammals of North America, Ronald Press, p. 427, March 31.

MARGINAL RECORDS.—Texas: *9 km. N, $9\frac{1}{2}$ km. E Fort Davis* (Williams and Baker, 1976:304); type locality; *2 mi. NW Fort Davis, 4800 ft.; Limpia Canyon.*

Thomomys umbrinus litoris Burt

1940. *Thomomys bottae litoris* Burt, Occas. Pap. Mus. Zool., Univ. Michigan, 424:1, November 29, type from Stearns Point, Magdalena Bay, Baja California. Known only from type locality.

1959. *Thomomys umbrinus litoris,* Hall and Kelson, Mammals of North America, Ronald Press, p. 427, March 31.

Thomomys umbrinus lorenzi Huey

1940. *Thomomys bottae lorenzi* Huey, Trans. San Diego Soc. Nat. Hist., 9:219, July 31, type from 7 mi. N Boulder Creek, Santa Cruz Co., California.

1959. *Thomomys umbrinus lorenzi,* Hall and Kelson, Mammals of North America, Ronald Press, p. 427, March 31.

MARGINAL RECORDS.—California: type locality; Scott Valley; Ben Lomond.

Thomomys umbrinus lucidus Hall

1932. *Thomomys bottae lucidus* Hall, Proc. Biol. Soc. Washington, 45:67, April 2, type from Las Palmas Canyon, 200 ft., W side Laguna Salada (N of 32° N lat.), Baja California. Known only from type locality.

1959. *Thomomys umbrinus lucidus,* Hall and Kelson, Mammals of North America, Ronald Press, p. 427, March 31.

Thomomys umbrinus lucrificus Hall and Durham

1938. *Thomomys bottae lucrificus* Hall and Durham, Proc. Biol. Soc. Washington, 51:15, February 18, type from Eastgate, Churchill Co., Nevada.

1959. *Thomomys umbrinus lucrificus*, Hall and Kelson, Mammals of North America, Ronald Press, p. 427, March 31.

MARGINAL RECORDS.—Nevada: type locality; *along creek flowing from Desatoya Mts. to Eastgate, 5025 ft.*

Thomomys umbrinus madrensis Nelson and Goldman

1934. *Thomomys umbrinus madrensis* Nelson and Goldman, Jour. Mamm., 15:115, May 15, type from Pilares Canyon, 6400 ft., 10 mi. NE Colonia García, and about 25 mi. SW Casas Grandes, Chihuahua.
1934. *Thomomys umbrinus caliginosus* Nelson and Goldman, Jour. Mamm., 15:116, May 15, type from 8 mi. W Altamirano, 8000 ft., Sierra Madre, Chihuahua.

MARGINAL RECORDS (Anderson, 1972:291, 292).—Chihuahua: 3 mi. NE Los Valles, 6700 ft.; Arroyo de la Tinaja, 5900 ft.; Colonia Juárez; 1½ mi. E Casa de Madera; *1 mi. E Cañón Pilares, 12 mi. E Pacheco, 7300 ft.;* 9 mi. SE Colonia García, 8200 ft.; *1 mi. E Chuhuichupa;* Chuhuichupa; *1 mi. N Chuhuichupa; 2 mi. N Chuhuichupa, 7000 ft.;* 8 mi. W Altamirano, 8000 ft.

Thomomys umbrinus magdalenae Nelson and Goldman

1909. *Thomomys magdalenae* Nelson and Goldman, Proc. Biol. Soc. Washington, 22:24, March 10, type from Magdalena Island, Baja California. Known only from type locality.
1959. *Thomomys umbrinus magdalenae*, Hall and Kelson, Mammals of North America, Ronald Press, p. 427, March 31.

Thomomys umbrinus martinensis Nelson and Goldman

1934. *Thomomys umbrinus martinensis* Nelson and Goldman, Jour. Mamm., 15:108, May 15, type from San Martín Texmelucán, 7400 ft., Puebla. Known only from type locality.

Thomomys umbrinus martirensis J. A. Allen

1898. *Thomomys fulvus martirensis* J. A. Allen, Bull. Amer. Mus. Nat. Hist., 10:147, April 12, type from La Grulla Meadow, Sierra San Pedro Mártir, 7400 ft., Baja California (see Huey, Trans. San Diego Soc. Nat. Hist., 5:89, January 18, 1928).
1959. *Thomomys umbrinus martirensis*, Hall and Kelson, Mammals of North America, Ronald Press, p. 427, March 31.

MARGINAL RECORDS.—Baja California: Piñón, W slope San Pedro Mártir Mts.; La Grulla, Sierra San Pedro Mártir; Valladares Creek, Sierra San Pedro Mártir.

Thomomys umbrinus mearnsi V. Bailey

1914. *Thomomys mearnsi* V. Bailey, Proc. Biol. Soc. Washington, 27:117, July 10, type from Grays Ranch, 5000 ft., Animas Valley, Grant Co., New Mexico. Known from type locality only.

Thomomys umbrinus melanotis Grinnell

1918. *Thomomys melanotis* Grinnell, Univ. California Publ. Zool., 17:425, April 25, type from Big Prospector Meadow, 10,500 ft., White Mts., California.
1959. *Thomomys umbrinus melanotis*, Hall and Kelson, Mammals of North America, Ronald Press, p. 427, March 31.

MARGINAL RECORDS.—California: Benton. Nevada: Lida, 6100 ft. California: Independence.

Thomomys umbrinus mewa Merriam

1908. *Thomomys mewa* Merriam, Proc. Biol. Soc. Washington, 21:146, June 9, type from Raymond, Madera Co., California.
1959. *Thomomys umbrinus mewa*, Hall and Kelson, Mammals of North America, Ronald Press, p. 428, March 31.

MARGINAL RECORDS.—California: Chinese; 3 mi. NE Coulterville; Wawona; Shaver Ranger Station; Three Rivers; Kernville; 8 mi. E Porterville; Merced; Lagrange.

Thomomys umbrinus minimus Durrant

1939. *Thomomys bottae minimus* Durrant, Proc. Biol. Soc. Washington, 52:161, October 11, type from Stansbury Island, Great Salt Lake, Tooele Co., Utah. Known only from type locality.
1959. *Thomomys umbrinus minimus*, Hall and Kelson, Mammals of North America, Ronald Press, p. 428, March 31.

Thomomys umbrinus minor V. Bailey

1914. *Thomomys bottae minor* V. Bailey, Proc. Biol. Soc. Washington, 27:116, July 10, type from Fort Bragg, Mendocino Co., California.
1959. *Thomomys umbrinus minor*, Hall and Kelson, Mammals of North America, Ronald Press, p. 428, March 31.

MARGINAL RECORDS.—California: Ferndale, thence down coast to near San Rafael.

Thomomys umbrinus modicus Goldman

1931. *Thomomys fulvus modicus* Goldman, Jour. Washington Acad. Sci., 21:418, October 19, type from La Osa (near Mexican boundary), southern end of Altar Valley, Pima Co., Arizona.
1959. *Thomomys umbrinus modicus*, Hall and Kelson, Mammals of North America, Ronald Press, p. 428, March 31.

MARGINAL RECORDS.—Arizona (Patton and Dingman, 1968:12, unless otherwise noted): Molino Basin, Santa Catalina Mts.; Fort Huachuca (Hall and Kelson, 1959:428); Sycamore Canyon (E to locality 6 of Hoffmeister, 1969:80). Sonora: Río Santa Cruz; Cerro Blanco; 35 mi. NW Magdalena; 5 mi. E Piquito. Arizona: Sells; Fresnall Canyon, Baboquivari Canyon, just below Allison Dam; Tucson.

Thomomys umbrinus mohavensis Grinnell

1918. *Thomomys perpallidus mohavensis* Grinnell, Univ. California Publ. Zool., 17:427, April 25, type from Mohave River bottom near Victorville, 2700 ft., San Bernardino Co., California.

1959. *Thomomys umbrinus mohavensis,* Hall and Kelson, Mammals of North America, Ronald Press, p. 428, March 31.

MARGINAL RECORDS.—California: Lone Willow Spring; Riggs Wash, 5½ mi. NE Silver Lake, 1900 ft.; 2½ mi. E Ludlow, 1842 ft.; Sheep Hole Mts., 21½ mi. S Amboy, 2300 ft.; Barker's Reservoir, 10 mi. S Twentynine Palms; Quail Spring, 17 mi. E Morongo Valley; Cushenbury Springs; 8 mi. SSE Hesperia; Granite Wells, 3950 ft.; Fairmont; Grapevine Ranch; 33 mi. E Mohave; 21 mi. SW Trona, 3050 ft.

Thomomys umbrinus morulus Hooper

1940. *Thomomys fulvus morulus* Hooper, Occas. Pap. Mus. Zool., Univ. Michigan, 422:9, November 14, type from Bill Porter's Ranch, 8 mi. SE Paxton, Valencia Co., New Mexico.

1959. *Thomomys umbrinus morulus,* Hall and Kelson, Mammals of North America, Ronald Press, p. 428, March 31.

MARGINAL RECORDS.—New Mexico: type locality; NW side Flagpole Crater.

Thomomys umbrinus muralis Goldman

1936. *Thomomys muralis* Goldman, Jour. Washington Acad. Sci., 26:112, March 15, type from lower end Prospect Valley, 4500 ft., Hualpai Indian Rservation, Grand Canyon, Arizona. Known only from type locality.

1959. *Thomomys umbrinus muralis,* Hall and Kelson, Mammals of North America, Ronald Press, p. 428, March 31.

Thomomys umbrinus musculus Nelson and Goldman

1934. *Thomomys umbrinus musculus* Nelson and Goldman, Jour. Mamm., 15:119, May 15, type from Pedro Pablo, 3500 ft., about 22 mi. E Acaponeta, Sierra de Teponahuaxtla, Nayarit.

MARGINAL RECORDS.—Sinaloa: 12 mi. NE Presa Sanalona (Dunnigan, 1967:159). Durango: 2 mi. N Pueblo Nuevo, 6000 ft. (Baker and Greer, 1962:95); *6 mi. S Pueblo Nuevo (ibid.);* Paso de Sihuacori (Crossin, *et al.,* 1973:199). Nayarit: type locality. Sinaloa: Plomosas; 18 mi. SE Culicán (Dunnigan, 1967:160).

Thomomys umbrinus mutabilis Goldman

1933. *Thomomys fulvus mutabilis* Goldman, Proc. Biol. Soc. Washington, 46:75, April 27, type from Camp Verde, Yavapai Co., Arizona.

1959. *Thomomys umbrinus mutabilis,* Hall and Kelson, Mammals of North America, Ronald Press, p. 428, March 31.

MARGINAL RECORDS.—Arizona: Clarkdale (Cockrum, 1961:117); Montezuma Well, near Camp Verde; H-bar Ranch, 10 mi. S Payson; Cazador Spring, S base Nanton Plateau; 20 mi. NE Calva (Cockrum,

1961:117); Gila Mts.; 9 mi. NW Roosevelt (Cockrum, 1961:117); type locality.

Thomomys umbrinus nanus Hall

1932. *Thomomys bottae nanus* Hall, Univ. California Publ. Zool., 38:331, February 27, type from S end Belted Range, 5½ mi. NW White Rock Spring, 7200 ft., Nye Co., Nevada.

1959. *Thomomys umbrinus nanus,* Hall and Kelson, Mammals of North America, p. 428, March 31.

MARGINAL RECORDS.—Nevada: Quinn Canyon Mts., Burned Corral Canyon, 6700–8700 ft.; N slope Irish Mtn., 7000–8000 ft.; Summit Spring, 4800 ft.; type locality; Kawich Range, 6000 ft., 1½ mi. E Kawich P.O.

Thomomys umbrinus navus Merriam

1901. *Thomomys leucodon navus* Merriam, Proc. Biol. Soc. Washington, 14:112, July 19, type from Red Bluff, Tehama Co., California.

1959. *Thomomys umbrinus navus,* Hall and Kelson, Mammals of North America, p. 428, March 31.

MARGINAL RECORDS.—California: 4 mi. SE Redding (Bongardt, *et al.,* 1968:545); Payne; Chico; 6 mi. E Oroville; Wheatland; Mokelumne Hill; Davis (Bongardt, *et al.,* 1968:545); Colusa; 5 mi. W Leesville; Sites; Willows; Battle Creek.

Thomomys umbrinus neglectus V. Bailey

1914. *Thomomys neglectus* V. Bailey, Proc. Biol. Soc. Washington, 27:117, July 10, type from Bear Flat Meadows, 6400 ft., San Antonio Peak, San Gabriel Mts., Los Angeles Co., California.

1959. *Thomomys umbrinus neglectus,* Hall and Kelson, Mammals of North America, p. 429, March 31.

MARGINAL RECORDS.—California: Mt. Piños; Bouquet Canyon; Mt. Islip, 7500 ft.; type locality.

Thomomys umbrinus nelsoni Merriam

1901. *Thomomys nelsoni* Merriam, Proc. Biol. Soc. Washington, 14:109, July 19, type from Parral, Chihuahua.

MARGINAL RECORDS (Anderson, 1972:292).—Chihuahua: Jiménez; 10 mi. SE Parral, 6000 ft.; type locality.

Thomomys umbrinus nesophilus Durrant

1936. *Thomomys bottae nesophilus* Durrant, Bull. Univ. Utah, 27(2):2, October 3, type from Antelope Island, Great Salt Lake, Davis Co., Utah. Known only from type locality.

1959. *Thomomys umbrinus nesophilus,* Hall and Kelson, Mammals of North America, Ronald Press, p. 429, March 31.

Thomomys umbrinus nevadensis Merriam

1897. *Thomomys nevadensis* Merriam, Proc. Biol. Soc. Washington, 11:213, July 15, type from Reese River Valley, 5 mi. W Austin, Lander Co., Nevada.

MARGINAL RECORDS.—Nevada: type locality; E side Reese River, 8 mi. W, 5 mi. S Austin.

Thomomys umbrinus newmani Dalquest

1951. *Thomomys umbrinus newmani* Dalquest, Jour. Washington Acad. Sci., 41:361, November 14, type from 7 km. NW Palma (village 12 km. NW Salinas), San Luis Potosí.

MARGINAL RECORDS.—San Luis Potosí: type locality; Cerro Peñón Blanco.

Thomomys umbrinus nicholi Goldman

1938. *Thomomys bottae nicholi* Goldman, Jour. Washington Acad. Sci., 28:337, July 15, type from 20 mi. S Wolf Hole (road to Parashonts), 5000 ft., Shivwits Plateau, Mohave Co., Arizona. Regarded as inseparable from *T. u. trumbullensis* by Durrant, Univ. Kansas Publ., Mus. Nat. Hist., 6:224, August 10, 1952.
1959. *Thomomys umbrinus nicholi*, Hall and Kelson, Mammals of North America, Ronald Press, p. 429, March 31.

MARGINAL RECORDS.—Arizona (Hoffmeister and Durham, 1971:31): 1 mi. NE Short Creek, 5040 ft.; Pipe Spring National Monument, 5000 ft.; S boundary Gardner Ranch, 5100 ft., 14 mi. N, 4 mi. W Mt. Dellenbaugh, Shivwits Plateau; 3 mi. W Lower Pigeon Spring, 4400 ft.; *type locality;* 10 mi. WSW Wolf Hole; 6 mi. N Wolf Hole.

Thomomys umbrinus nigricans Rhoads

1895. *Thomomys fulvus nigricans* Rhoads, Proc. Acad. Nat. Sci. Philadelphia, 47:36, February 21, type from Witch Creek, 2753 ft., 7 mi. W Julian, San Diego Co., California.
1959. *Thomomys umbrinus nigricans,* Hall and Kelson, Mammals of North America, Ronald Press, p. 429, March 31.

MARGINAL RECORDS.—California: Schains Ranch, W base San Jacinto Mts.; Kenworthy; Santa Rosa Peak; Julian; Laguna Mts. Baja California: Nachogüero Valley; Las Cruces; S end Valles de las Palmas. California: Jamul Creek; Poway; Escondido; Hemet Valley.

Thomomys umbrinus operarius Merriam

1897. *Thomomys operarius* Merriam, Proc. Biol. Soc. Washington, 11:215, July 15, type from Keeler, E side Owens Lake, Inyo Co., California. Known only from type locality.
1959. *Thomomys umbrinus operarius,* Hall and Kelson, Mammals of North America, Ronald Press, p. 429, March 31.

Thomomys umbrinus operosus Hatfield

1942. *Thomomys bottae operosus* Hatfield, Bull. Chicago Acad. Sci., 6:151, January 12, type from Peeples Valley, 4400 ft., 6 mi. N Yarnell, Yavapai Co., Arizona. Known only from type locality.
1959. *Thomomys umbrinus operosus,* Hall and Kelson, Mammals of North America, Ronald Press, p. 429, March 31.

Thomomys umbrinus opulentus Goldman

1935. *Thomomys bottae opulentus* Goldman, Proc. Biol. Soc. Washington, 48:150, October 31, type from Las Palomas, on the Rio Grande, Sierra Co., New Mexico.
1959. *Thomomys umbrinus opulentus,* Hall and Kelson, Mammals of North America, Ronald Press, p. 429, March 31.

MARGINAL RECORDS.—New Mexico: Socorro; San Marcial; Las Cruces; Garfield; Lake Valley; Cuchillo.

Thomomys umbrinus oreoecus Burt

1932. *Thomomys oreoecus* Burt, Trans. San Diego Soc. Nat. Hist., 7:154, July 28, type from Greenwater [Black Mts., 8 mi. SW Ryan], 4300 ft., Inyo Co., California.
1959. *Thomomys umbrinus oreoecus,* Hall and Kelson, Mammals of North America, Ronald Press, p. 429, March 31.

MARGINAL RECORDS.—Nevada: Thorps Mill. California: type locality. Nevada: 4¾ mi. E California boundary, 4200 ft.

Thomomys umbrinus orizabae Merriam

1893. *Thomomys orizabae* Merriam, Proc. Biol. Soc. Washington, 8:145, December 29, type from Mt. Orizaba, 9500 ft., Puebla.
1915. *Thomomys umbrinus orizabae,* V. Bailey, N. Amer. Fauna, 39:90, November 15.

MARGINAL RECORDS.—Puebla: type locality; *10–16 km. NNE San Andrés, W slope Mt. Orizaba, 10,000–11,000 ft.*

Thomomys umbrinus osgoodi Goldman

1931. *Thomomys perpallidus osgoodi* Goldman, Jour. Washington Acad. Sci., 21:424, October 19, type from Hanksville, Wayne Co., Utah.
1959. *Thomomys umbrinus osgoodi,* Hall and Kelson, Mammals of North America, Ronald Press, p. 429, March 31.

MARGINAL RECORDS.—Utah: ½ mi. N Spring Glen, 6150 ft.; Price River, 2 mi. SE Woodside, 4600 ft.; Notom, 6200 ft.; 5 mi. S Castle Dale, 5600 ft.

Thomomys umbrinus owyhensis Davis

1937. *Thomomys townsendii owyhensis* Davis, Jour. Mamm., 18:154, May 12, type from Castle Creek, 8 mi. S Oreana, Owyhee Co., Idaho.

MARGINAL RECORDS.—Idaho: 5 mi. NE Murphy (Wentworth and Sutton, 1969:160); *Hwy. 45, ½ mi. S Snake R.* (Wentworth and Sutton, 1969:160, 161); *Sinker Creek, 7 mi. SE Murphy;* Indian Cove; type locality.

Thomomys umbrinus paguatae Hooper

1940. *Thomomys bottae paguatae* Hooper, Occas. Pap. Mus. Zool., Univ. Michigan, 422:4, November 14, type from ½ mi.

N Cebolleta (Seboyeta P.O.), Valencia Co., New Mexico. Known only from type locality.

1959. *Thomomys umbrinus paguatae*, Hall and Kelson, Mammals of North America, Ronald Press, p. 429, March 31.

Thomomys umbrinus pallescens Rhoads

1895. *Thomomys bottae pallescens* Rhoads, Proc. Acad. Nat. Sci. Philadelphia, 47:36, February 21, type from Grapeland, San Bernardino Valley, San Bernardino Co., California.
1959. *Thomomys umbrinus pallescens*, Hall and Kelson, Mammals of North America, Ronald Press, p. 429, March 31.

MARGINAL RECORDS.—California: Palmdale; San Bernardino; El Casco; Riverside; Los Angeles; Saugus.

Thomomys umbrinus parviceps Nelson and Goldman

1934. *Thomomys simulus parviceps* Nelson and Goldman, Jour. Mamm., 15:121, May 15, type from Chacala, 3000 ft., Durango. Known only from type locality. (*T. u. parviceps* regarded as subspecifically inseparable from *T. u. musculus* by Dunnigan, Radford Review, 21:157, September 29, 1967.)
1959. *Thomomys umbrinus parviceps*, Hall and Kelson, Mammals of North America, Ronald Press, p. 430, March 31.

Thomomys umbrinus parvulus Goldman

1938. *Thomomys bottae parvulus* Goldman, Jour. Washington Acad. Sci., 28:339, July 15, type from pass between Santa Catalina and Rincon mts., 4500 ft., Pima Co., Arizona. Known only from type locality.
1959. *Thomomys umbrinus parvulus*, Hall and Kelson, Mammals of North America, Ronald Press, p. 430, March 31.

Thomomys umbrinus pascalis Merriam

1901. *Thomomys angularis pascalis* Merriam, Proc. Biol. Soc. Washington, 14:111, July 19, type from Fresno, San Joaquin Valley, Fresno Co., California.
1959. *Thomomys umbrinus pascalis*, Hall and Kelson, Mammals of North America, Ronald Press, p. 430, March 31.

MARGINAL RECORDS.—California: vic. Stockton; Oakdale; type locality; Tulare; *2–4 mi. W Porterville* (Bongardt, *et al.*, 1968:545); Bodfish; Tehachapi; Tejon Pass; N flank Mt. Piños; Cuyama Valley; Buttonwillow; Lemoore; Modesto.

Thomomys umbrinus patulus Goldman

1938. *Thomomys bottae patulus* Goldman, Jour. Washington Acad. Sci., 28:341, July 15, type from bottomland along Hassayampa River, 2000 ft., 2 mi. below Wickenburg, Maricopa Co., Arizona. Known only from type locality.
1959. *Thomomys umbrinus patulus*, Hall and Kelson, Mammals of North America, Ronald Press, p. 430, March 31.

Thomomys umbrinus pectoralis Goldman

1936. *Thomomys pectoralis* Goldman, Jour. Washington Acad. Sci., 26:120, March 15, type from vic. Carlsbad Cave, Carlsbad Cave National Monument, Eddy Co., New Mexico.
1959. *Thomomys umbrinus pectoralis*, Hall and Kelson, Mammals of North America, Ronald Press, p. 430, March 31.

MARGINAL RECORDS.—New Mexico: type locality. Texas: *Bell Canyon, 1 mi. N, 1 mi. E Nickel.*

Thomomys umbrinus peramplus Goldman

1931. *Thomomys fulvus peramplus* Goldman, Jour. Washington Acad. Sci., 21:423, October 19, type from Wheatfields Creek, 7000 ft. [about 27 mi. E Chin Lee], W slope Tunitcha Mts., Apache Co., Arizona.
1959. *Thomomys umbrinus peramplus*, Hall and Kelson, Mammals of North America, Ronald Press, p. 430, March 31.

MARGINAL RECORDS.—Arizona: type locality. New Mexico: Chusca Mts. Arizona: St. Michaels, E side Defiance Plateau; Canyon de Chelly, 7 mi. above mouth.

Thomomys umbrinus perditus Merriam

1901. *Thomomys perditus* Merriam, Proc. Biol. Soc. Washington, 14:108, July 19, type from Lampazos, Nuevo León.
1934. *Thomomys umbrinus perditus*, Nelson and Goldman, Jour. Mamm., 15:115, May 15.

MARGINAL RECORDS.—Nuevo León: type locality; Villadama. Coahuila: 2 mi. N, 18 mi. W Santa Teresa, 7500 ft.

Thomomys umbrinus peregrinus Merriam

1893. *Thomomys peregrinus* Merriam, Proc. Biol. Soc. Washington, 8:146, December 29, type from Salazar, 10,300 ft., México.
1915. *Thomomys umbrinus peregrinus*, V. Bailey, N. Amer. Fauna, 39:91, November 15.

MARGINAL RECORDS.—México: type locality. Distrito Federal: 2 mi. SSW Parres, 9000 ft. México: Volcán Popocatépetl. Morelos: 5 km. N Tres Cumbres. México: *1 mi. ESE Salazar, 9500 ft.*

Thomomys umbrinus perpallidus Merriam

1886. *Thomomys talpoides perpallidus* Merriam, Science, 8:588, December 24, type from Palm Springs, Riverside Co., California.
1959. *Thomomys umbrinus perpallidus*, Hall and Kelson, Mammals of North America, Ronald Press, p. 430, March 31.

MARGINAL RECORDS.—California: Whitewater; ½ mi. SE Thermal.

Thomomys umbrinus perpes Merriam

1901. *Thomomys aureus perpes* Merriam, Proc. Biol. Soc. Washington, 14:111, July 19, type from Lone Pine, Owens Valley, Inyo Co., California.
1959. *Thomomys umbrinus perpes*, Hall and Kelson, Mammals of North America, Ronald Press, p. 430, March 31.

MARGINAL RECORDS.—California: Bishop; type locality; Coso; vic. Freeman; Kelso Pass; Isbella; S. Fork Kern River, near Onyx; Haway Meadows, S of Owens Lake.

Thomomys umbrinus pervagus Merriam

1901. *Thomomys aureus pervagus* Merriam, Proc. Biol. Soc. Washington, 14:110, July 19, type from Española, Rio Arriba Co., New Mexico.
1959. *Thomomys umbrinus pervagus*, Hall and Kelson, Mammals of North America, Ronald Press, p. 430, March 31.

MARGINAL RECORDS.—Colorado (Youngman, 1958:373): Conejos River, 6 mi. W Antonito, 8300 ft.; 12 mi. E Antonito. New Mexico: Questa; 10 mi. N Santa Fe; Santa Clara Canyon; Chama Canyon, 6100 ft.

Thomomys umbrinus pervarius Goldman

1938. *Thomomys bottae pervarius* Goldman, Proc. Biol. Soc. Washington, 51:57, March 18, type from Lloyd Ranch, 35 mi. S Marfa, 4200 ft., Presidio Co., Texas. Known only from type locality.
1959. *Thomomys umbrinus pervarius*, Hall and Kelson, Mammals of North America, Ronald Press, p. 430, March 31.

Thomomys umbrinus phasma Goldman

1933. *Thomomys fulvus phasma* Goldman, Proc. Biol. Soc. Washington, 46:72, April 27, type from 2 mi. S Tule Tank, Tule Desert, near Mexican boundary, Yuma Co., Arizona.
1959. *Thomomys umbrinus phasma*, Hall and Kelson, Mammals of North America, Ronald Press, p. 430, March 31.

MARGINAL RECORDS.—Arizona: Dateland (Cockrum, 1961:118); Tule Well. Sonora: Quitobaquita; Cienega Well, 30 mi. S Monument 204; Colorado River, 20 mi. S Mexican boundary. Arizona: Welton; Tacna.

Thomomys umbrinus phelleoecus Burt

1933. *Thomomys phelleoecus* Burt, Jour. Mamm., 14:56, February 14, type from Hidden Forest, 8500 ft., Sheep Mts., Clark Co., Nevada.
1959. *Thomomys umbrinus phelleoecus*, Hall and Kelson, Mammals of North America. Ronald Press, p. 431, March 31.

MARGINAL RECORDS.—Nevada: Hidden Forest, 7700 ft.; *ridge N Wiregrass (= Wire) Spring, 8250 ft.; Hidden Forest road, 5300 and 5450 ft.*

Thomomys umbrinus pinalensis Goldman

1938. *Thomomys bottae pinalensis* Goldman, Jour. Washington Acad. Sci., 28:342, July 15, type from Oak Flat, 5 mi. E Superior, Pinal Mts., Pinal Co., Arizona. *T. b. pinalensis* was considered to be inseparable from *T. u. mutabilis* by Hall and Kelson (Univ. Kansas Publ., Mus. Nat. Hist., 5:360, December 15, 1952, and Mammals of North America, p. 428, March 31, 1959). In December 1972, at Arizona State University, T. Michael Young showed E. R. Hall six adult topotypes of *T. b. pinalensis* and 10 adults (five males, five

females) from 10 mi. N Globe, 3050 ft. that resembled the topotypes. These 16 specimens were compared with six adult topotypes (one male, five females) of *T. b. mutabilis* and seven adults from Tempe (three males, four females) of *T. cervinus*. *T. b. pinalensis* resembles *mutabilis* in color (*cervinus* is paler) and is significantly smaller than either *cervinus* or *mutabilis*. *T. u. pinalensis*, therefore, is here recognized as a valid taxon.

MARGINAL RECORDS.—Arizona: 10 mi. N Globe, 3050 ft. (869 ASU); type locality.

Thomomys umbrinus piutensis Grinnell and Hill

1936. *Thomomys bottae piutensis* Grinnell and Hill, Proc. Biol. Soc. Washington, 49:103, August 22, type from French Gulch, 6700 ft., Piute Mts., 2½ mi. NE Claraville, Kern Co., California.
1959. *Thomomys umbrinus piutensis*, Hall and Kelson, Mammals of North America, Ronald Press, p. 431, March 31.

MARGINAL RECORDS.—California: type locality; Kelso Valley; Walker Basin.

Thomomys umbrinus planirostris Burt

1931. *Thomomys perpallidus planirostris* Burt, Proc. Biol. Soc. Washington, 44:38, May 8, type from Zion National Park, Washington Co., Utah.
1959. *Thomomys umbrinus planirostris*, Hall and Kelson, Mammals of North America, Ronald Press, p. 431, March 31.

MARGINAL RECORDS.—Utah: East Entrance, 5725 ft., Zion National Park. Arizona: Fredonia. Utah: 6 mi. S St. George, 2700 ft.; Santa Clara Creek.

Thomomys umbrinus planorum Hooper

1940. *Thomomys bottae planorum* Hooper, Occas. Pap. Mus. Zool., Univ. Michigan, 422:5, November 14, type from 1½ mi. SW San Mateo, Valencia Co., New Mexico.
1959. *Thomomys umbrinus planorum*, Hall and Kelson, Mammals of North America, Ronald Press, p. 431, March 31.

MARGINAL RECORDS.—New Mexico: Horace Mesa, 1½ mi. S Canyon Lobo Ranger Station; 11 mi. SSE Grants.

Thomomys umbrinus potosinus Nelson and Goldman

1934. *Thomomys umbrinus potosinus* Nelson and Goldman, Jour. Mamm., 15:111, May 15, type from La Tinaja, 6000 ft. (at or near present town of Ventura), about 20 mi. NE San Luis Potosí, San Luis Potosí. Known only from type locality.

Thomomys umbrinus powelli Durrant

1955. *Thomomys bottae powelli* Durrant, Proc. Biol. Soc. Washington, 68:79, August 3, type from Hall Ranch, Salt Gulch, 8 mi. W Boulder, 6000 ft., Garfield Co., Utah. Known only from type locality.
1959. *Thomomys umbrinus powelli*, Hall and Kelson, Mammals of North America, Ronald Press, p. 431, March 31.

Thomomys umbrinus providentialis Grinnell

1931. *Thomomys providentialis* Grinnell, Univ. California Publ. Zool., 38:1, October 17, type from Purdy, 4500 ft., 6 mi. SE New York Mtn., Providence Range, San Bernardino Co., California.
1959. *Thomomys umbrinus providentialis,* Hall and Kelson, Mammals of North America, Ronald Press, p. 431, March 31.

MARGINAL RECORDS.—California: Twelve Mile Spring. Nevada: N side Potosi Mtn., 5800 ft.; 8 mi. SE Dead Mtn., 1900 ft. California: jct. Piute Springs and Searchlight roads, 2130 ft.; Turtle Mts., 12 mi. NE Sablon, 2492 ft.; $7\frac{1}{2}$ mi. NE Chubbuck, 2320 ft.; Pass between Granite and Providence mts., 4000 ft.; 5 mi. N Kelso Peak.

Thomomys umbrinus proximarinus Huey

1945. *Thomomys bottae proximarinus* Huey, Trans. San Diego Soc. Nat. Hist., 10:261, August 31, type from Boca la Playa, mesa bordering the sea, lat. 31° 32′ N, long. 116° 38′ W, 16 mi. W Santo Tomás, Baja California. Known only from type locality.
1959. *Thomomys umbrinus proximarinus,* Hall and Kelson, Mammals of North America, Ronald Press, p. 431, March 31.

Thomomys umbrinus proximus Burt and Campbell

1934. *Thomomys burti proximus* Burt and Campbell, Jour. Mamm., 15:151, May 15, type from Old Parker Ranch (Pickett's Ranch on U.S. Geological Survey Topographic Sheet, Patagonia Quadrangle, edition of August, 1905), 4800 ft., W slope Santa Rita Mts., Pima Co., Arizona.
1943. *Thomomys umbrinus proximus,* Goldman, Jour. Washington Acad. Sci., 33:147, May 15.

MARGINAL RECORDS.—Arizona: Empire Ranch, E of Santa Rita Mts., Fort Huachuca; *Carr Canyon Ranch, Huachuca Mts.* (Patton and Dingman, 1968:12); $\frac{1}{2}$ mi. N Clark Spring, Carr Canyon, Huachuca Mts. (*ibid.*); Arivaca; type locality.

Thomomys umbrinus puertae Grinnell

1914. *Thomomys nigricans puertae* Grinnell, Univ. California Publ. Zool., 12:315, November 21, type from La Puerta (Mason's Ranch), 5 mi. W Vallecitos, at lower end La Puerta Valley, San Diego Co., California. Known only from type locality.
1959. *Thomomys umbrinus puertae,* Hall and Kelson, Mammals of North America, Ronald Press, p. 431, March 31.

Thomomys umbrinus pullus Hall and Villa

1948. *Thomomys umbrinus pullus* Hall and Villa, Univ. Kansas Publ., Mus. Nat. Hist., 1:251, July 26, type from 5 mi. S Pátzcuaro, 7800 ft., Michoacán.

MARGINAL RECORDS.—Michoacán: *3 mi. S Pátzcuaro, 7800 ft.;* type locality.

Thomomys umbrinus pusillus Goldman

1931. *Thomomys fulvus pusillus* Goldman, Jour. Washington Acad. Sci., 21:422, October 19, type from Coyote Mts., 3000 ft., Pima Co., Arizona.
1959. *Thomomys umbrinus pusillus,* Hall and Kelson, Mammals of North America, Ronald Press, p. 432, March 31.

MARGINAL RECORDS.—Arizona: *ca. $1\frac{1}{2}$ mi. below Kit Peak National Observatory, Quinlan Mts.* (Patton and Dingman, 1968:12); type locality.

Thomomys umbrinus quercinus Burt and Campbell

1934. *Thomomys burti quercinus* Burt and Campbell, Jour. Mamm., 15:150, May 15, type from Peña Blanca Spring, Pajarito Mts., Arizona. Known only from type locality.
1943. *Thomomys umbrinus quercinus,* Goldman, Jour. Washington Acad. Sci., 33:147, May 15.

Thomomys umbrinus relictus Grinnell

1926. *Thomomys relictus* Grinnell, Univ. California Publ. Zool., 30:2, August 18, type from 2 mi. S Susanville, valley of Susan River, Lassen Co., California.

MARGINAL RECORDS (Thaeler, 1968:14).—California: $4\frac{1}{2}$ mi. ENE Susanville; $4\frac{3}{10}$ mi. N, $3\frac{2}{5}$ mi. W Litchfield; 1 mi. N, $\frac{3}{10}$ mi. W Wendel; 4 mi. ESE Amedee; *Garnier Ranch, $2\frac{2}{5}$ mi. S, $1\frac{1}{2}$ mi. W Herlong;* $3\frac{2}{5}$ mi. N, $2\frac{3}{5}$ mi. W Doyle; $3\frac{1}{2}$ mi. N, $3\frac{1}{5}$ mi. W Milford; Elysian Valley, $5\frac{3}{10}$ mi. S, $\frac{3}{10}$ mi. E Johnstonville; *$3\frac{1}{10}$ mi. S Susanville.*

Thomomys umbrinus retractus Baker

1953. *Thomomys bottae retractus* Baker, Univ. Kansas Publ., Mus. Nat. Hist., 5:507, June 1, type from Fortín, 3300 ft., 20 mi. N, 2 mi. E San Gerónimo, Coahuila (see Baker, Univ. Kansas Publ., Mus. Nat. Hist., 9:220, June 15, 1956). Known only from vic. type locality (see Baker, *loc. cit.*).
1959. *Thomomys umbrinus retractus,* Hall and Kelson, Mammals of North America, Ronald Press, p. 432, March 31.

Thomomys umbrinus rhizophagus Huey

1949. *Thomomys bottae rhizophagus* Huey, Trans. San Diego Soc. Nat. Hist., 11:54, January 31, type from Las Flores, lat. 28° 50′ N, long. 113° 32′ W, 7 mi. S Bahía de Los Angeles, Baja California. Known only from type locality.
1959. *Thomomys umbrinus rhizophagus,* Hall and Kelson, Mammals of North America, Ronald Press, p. 432, March 31.

Thomomys umbrinus riparius Grinnell and Hill

1936. *Thomomys perpallidus riparius* Grinnell and Hill, Jour. Mamm., 17:4, February 17, type from Blythe, Riverside Co., California.
1959. *Thomomys umbrinus riparius,* Hall and Kelson, Mammals of North America, Ronald Press, p. 432, March 31.

MARGINAL RECORDS.—California: 13 mi. NE Blythe; type locality; Ford Dry Lake, $21\frac{1}{2}$ mi. W Blythe.

Thomomys umbrinus robustus Durrant

1946. *Thomomys bottae robustus* Durrant, Univ. Kansas Publ., Mus. Nat. Hist., 1:30, August 15, type from Orr's Ranch, 4300 ft., Skull Valley, Tooele Co., Utah. Known only from type locality.
1959. *Thomomys umbrinus robustus*, Hall and Kelson, Mammals of North America, Ronald Press, p. 432, March 31.

Thomomys umbrinus rubidus Youngman

1958. *Thomomys bottae rubidus* Youngman, Univ. Kansas Publ., Mus. Nat. Hist., 9:376, February 21, type from 2⁹⁄₁₀ mi. E Cañon City, 5344 ft., Fremont Co., Colorado.
1959. *Thomomys umbrinus rubidus*, Hall and Kelson, Mammals of North America, Ronald Press, p. 1080, March 31.

MARGINAL RECORDS (Youngman, 1958:377).— Colorado: *Garden Park, Cañon City, 5344 ft.;* type locality.

Thomomys umbrinus rufidulus Hoffmeister

1955. *Thomomys bottae rufidulus* Hoffmeister, Jour. Washington Acad. Sci., 45:126, April 25, type from 2 mi. E Joseph City, Navajo Co., Arizona.
1959. *Thomomys umbrinus rufidulus*, Hall and Kelson, Mammals of North America, Ronald Press, p. 432, March 31.

MARGINAL RECORDS.—New Mexico: Gallup. Arizona: Navajo; type locality.

Thomomys umbrinus ruidosae Hall

1932. *Thomomys bottae ruidosae* Hall, Proc. Biol. Soc. Washington, 45:96, June 21, type from Ruidoso, 6700 ft., Lincoln Co., New Mexico.
1959. *Thomomys umbrinus ruidosae*, Hall and Kelson, Mammals of North America, Ronald Press, p. 432, March 31.

MARGINAL RECORDS.—New Mexico: type locality; Cloudcroft, 9000 ft.; *8 mi. NW Cloudcroft, 7000 ft.; Mescalero.*

Thomomys umbrinus rupestris Chattin

1941. *Thomomys bottae rupestris* Chattin, Trans. San Diego Soc. Nat. Hist., 9:272, April 30, type from 2 mi. E Clemens Well, 1131 ft., Riverside Co., California.
1959. *Thomomys umbrinus rupestris*, Hall and Kelson, Mammals of North America, Ronald Press, p. 432, March 31.

MARGINAL RECORDS.—California: 4 mi. SW Desert Center, 1020 ft.; Corn Spring, Chuckwalla Mts., 1573 ft.; Chocolate Mts., 11 mi. NE Niland; 3 mi. S Cottonwood Spring, 2400 ft.

Thomomys umbrinus ruricola Huey

1949. *Thomomys bottae ruricola* Huey, Trans. San Diego Soc. Nat. Hist., 11:53, January 31, type from 4 mi. N Santa Catarina Landing, lat. 29° 35′ N, long. 115° 17′ W, Baja California. Known only from type locality.

1959. *Thomomys umbrinus ruricola*, Hall and Kelson, Mammals of North America, Ronald Press, p. 432, March 31.

Thomomys umbrinus russeolus Nelson and Goldman

1909. *Thomomys bottae russeolus* Nelson and Goldman, Proc. Biol. Soc. Washington, 22:25, March 10, type from San Angel, WSW San Ignacio, Baja California.
1959. *Thomomys umbrinus russeolus*, Hall and Kelson, Mammals of North America, Ronald Press, p. 432, March 31.

MARGINAL RECORDS.—Baja California: Campo Los Angeles, Viscaino Desert; type locality.

Thomomys umbrinus sanctidiegi Huey

1945. *Thomomys bottae sanctidiegi* Huey, Trans. San Diego Soc. Nat. Hist., 10:258, August 31, type from Balboa Park, San Diego, California.
1959. *Thomomys umbrinus sanctidiegi*, Hall and Kelson, Mammals of North America, Ronald Press, p. 432, March 31.

MARGINAL RECORDS.—California: La Jolla; southward along coast to Baja California: Ensenada.

Thomomys umbrinus saxatilis Grinnell

1934. *Thomomys bottae saxatilis* Grinnell, Proc. Biol. Soc. Washington, 47:193, October 2, type from 1 mi. N Susanville, 4400 ft., Lassen Co., California.
1959. *Thomomys umbrinus saxatilis*, Hall and Kelson, Mammals of North America, Ronald Press, p. 432, March 31.

MARGINAL RECORDS (Thaeler, 1968:13).— California, with distance and direction from Susanville: 1½ mi. N, 1⁷⁄₁₀ mi. E; 3¹⁄₁₀ mi. S; 3⁷⁄₁₀ mi. S, ⁹⁄₁₀ mi. W; 4 mi. S, 1³⁄₁₀ mi. W; 4³⁄₁₀ mi. S, 1⅘ mi. W; 2⅘ mi. S, 7⅘ mi. W; *1 mi. NW; 1 mi. N.*

Thomomys umbrinus scapterus Elliot

1904. *Thomomys scapterus* Elliot, Field Columb. Mus., Publ. 87, Zool. Ser., 3:248, January 7, type from Hannopee Canyon, 7500 ft., Panamint Mts., Inyo Co., California.
1959. *Thomomys umbrinus scapterus*, Hall and Kelson, Mammals of North America, Ronald Press, p. 432, March 31.

MARGINAL RECORDS.—California: Jackass Spring, Panamint Mts.; Johnson Canyon, Panamint Mts.; near Lee Mine, 12 mi. N Darwin.

Thomomys umbrinus scotophilus Davis

1940. *Thomomys bottae scotophilus* Davis, Jour. Mamm., 21:204, May 16, type from 1½ mi. W Bat Cave, Sierra Diablo, Hudspeth Co., Texas.
1959. *Thomomys umbrinus scotophilus*, Hall and Kelson, Mammals of North America, Ronald Press, p. 432, March 31.

MARGINAL RECORDS.—Texas: *Bat Cave, ca. 2 mi. S head Victoria Canyon;* type locality.

Thomomys umbrinus sevieri Durrant

1946. *Thomomys bottae sevieri* Durrant, Univ. Kansas Publ., Mus. Nat. Hist., 1:45, August 15, type from Swasey Spring, 6500 ft., House Mtn., Millard Co., Utah. Known only from type locality.
1959. *Thomomys umbrinus sevieri*, Hall and Kelson, Mammals of North America, Ronald Press, p. 433, March 31.

Thomomys umbrinus sheldoni V. Bailey

1915. *Thomomys sheldoni* V. Bailey, N. Amer. Fauna, 39:93, November 15, type from Santa Teresa, 6800 ft., Sierra del Nayarit, Nayarit.
1934. *Thomomys umbrinus sheldoni*, Nelson and Goldman, Jour. Mamm., 15:113, May 15.

MARGINAL RECORDS.—Durango: 28 mi. S, 17 mi. W Vicente Guerrero (Baker and Greer, 1962:95). Zacatecas: Sierra Madre, 8500 ft. Nayarit: type locality.

Thomomys umbrinus siccovallis Huey

1945. *Thomomys bottae siccovallis* Huey, Trans. San Diego Soc. Nat. Hist., 10:258, August 31, type from El Cajón Canyon, 3200 ft., E base Sierra San Pedro Mátir, lat. 30° 54′ N, long. 115° 10′ W, Baja California.
1959. *Thomomys umbrinus siccovallis*, Hall and Kelson, Mammals of North America, Ronald Press, p. 433, March 31.

MARGINAL RECORDS.—Baja California: type locality; Mattomi [= Matomi].

Thomomys umbrinus silvifugus Grinnell

1935. *Thomomys bottae silvifugus* Grinnell, Univ. California Publ. Zool., 40:406, November 14, type from 16 mi. due E Patricks Point, near Coyote Peak, 3000 ft., Humboldt Co., California. Known only from type locality.
1959. *Thomomys umbrinus silvifugus*, Hall and Kelson, Mammals of North America, Ronald Press, p. 433, March 31.

Thomomys umbrinus similis Davis

1937. *Thomomys townsendii similis* Davis, Jour. Mamm., 18:155, May 12, type from Pocatello, Bannock Co., Idaho.

MARGINAL RECORDS.—Idaho: 1 mi. E Pingree; Fort Hall Indian School, 10 mi. N Pocatello; type locality; American Falls; 4 mi. NW American Falls.

Thomomys umbrinus simulus Nelson and Goldman

1934. *Thomomys simulus simulus* Nelson and Goldman, Jour. Mamm., 15:120, May 15, type from Alamos, 1200 ft., Sonora.
1959. *Thomomys umbrinus simulus*, Hall and Kelson, Mammals of North America, Ronald Press, p. 433, March 31.

MARGINAL RECORDS.—Sonora: type locality; Baromico. Sinaloa (Dunnigan, 1967:149): ¾ mi. ENE El Cajón; Tasajera. Sonora: 7 mi. SW Guirocoba.

Thomomys umbrinus sinaloae Merriam

1901. *Thomomys sinaloae* Merriam, Proc. Biol. Soc. Washington, 14:108, July 19, type from Altata, Sinaloa.
1959. *Thomomys umbrinus sinaloae*, Hall and Kelson, Mammals of North America, Ronald Press, p. 433, March 31.

MARGINAL RECORDS (Dunnigan, 1967:152).— Sinaloa: 3 mi. N El Fuerte; 1 mi. E Sinaloa; 4 km. SW Navolato; 2 mi. E San Lorenzo; Altata; thence northward along coast to 3 mi. N Higueras.

Thomomys umbrinus solitarius Grinnell

1926. *Thomomys solitarius* Grinnell, Univ. California Publ. Zool., 39:177, December 10, type from Fingerrock Wash, 5400 ft., Stewart Valley, Mineral Co., Nevada.
1959. *Thomomys umbrinus solitarius*, Hall and Kelson, Mammals of North America, Ronald Press, p. 433, March 31.

MARGINAL RECORDS.—Nevada: type locality; Lone Mtn., 6600 ft., 2½ mi. N, 12½ mi. W Tonopah; 4 mi. NE Arlemont, 4800 ft.; 7 mi. N Arlemont, 5500 ft.; S end Walker Lake, 4100 ft.

Thomomys umbrinus sonoriensis Nelson and Goldman

1934. *Thomomys umbrinus sonoriensis* Nelson and Goldman, Jour. Mamm., 15:118, May 15, type from 10 mi. E Chinapa, 3000 ft., Sonora.

MARGINAL RECORDS.—Sonora: type locality; Huásabas; Providencia Mines.

Thomomys umbrinus spatiosus Goldman

1938. *Thomomys baileyi spatiosus* Goldman, Proc. Biol. Soc. Washington, 51:58, March 18, type from Alpine, 4500 ft., Brewster Co., Texas.

MARGINAL RECORDS.—Texas: *Jeff Davis County* (Davis, 1966:151, as *T. baileyi* only); type locality; Paisano.

Thomomys umbrinus stansburyi Durrant

1946. *Thomomys bottae stansburyi* Durrant, Univ. Kansas Publ., Mus. Nat. Hist., 1:36, August 15, type from South Willow Creek, Stansbury Mts., 7500 ft., Tooele Co., Utah. Known only from type locality.
1959. *Thomomys umbrinus stansburyi*, Hall and Kelson, Mammals of North America, Ronald Press, p. 433, March 31.

Thomomys umbrinus sturgisi Goldman

1938. *Thomomys sturgisi* Goldman, Proc. Biol. Soc. Washington, 51:56, March 18, type from Sierra del Carmen, 6000 ft., Coahuila.
1959. *Thomomys umbrinus sturgisi*, Hall and Kelson, Mammals of North America, Ronald Press, p. 433, March 31.

MARGINAL RECORDS.—Coahuila: type locality; Sierra de la Encantada, 4100 ft., 37 mi. S, 21 mi. E Boquillas; 3 mi. NW Cuatro Ciénegas, 2450 ft.; 2 mi. N,

1 mi. W Ocampo, 4050 ft. Chihuahua (Anderson, 1972:288, as *T. bottae* ?subspecies): 1 mi. W Pyrámide, 5400 ft.; 4 mi. WNW Escobillas. The two Chihuahuan specimens are here assigned to this subspecies because Anderson (*loc. cit.*) stated they "can be matched in color with some specimens of *Thomomys bottae sturgis*" (= *T. umbrinus sturgisi*).

Thomomys umbrinus suboles Goldman

1928. *Thomomys fulvus suboles* Goldman, Proc. Biol. Soc. Washington, 41:203, December 18, type from Old Searchlight Ferry, Colorado River (Northwest of Kingman), Arizona. Known only from type locality.
1959. *Thomomys umbrinus suboles,* Hall and Kelson, Mammals of North America, Ronald Press, p. 433, March 31.

Thomomys umbrinus subsimilis Goldman

1933. *Thomomys fulvus subsimilis* Goldman, Proc. Biol. Soc. Washington, 46:74, April 27, type from Harquahala Mts., 3000 ft., Yuma Co., Arizona. Known only from type locality.
1959. *Thomomys umbrinus subsimilis,* Hall and Kelson, Mammals of North America, Ronald Press, p. 433, March 31.

Thomomys umbrinus supernus Nelson and Goldman

1934. *Thomomys umbrinus supernus* Nelson and Goldman, Jour. Mamm., 15:110, May 15, type from Santa Rosa, between 9500 and 10,000 ft., about 7 mi. NE Guanajuato, Guanajuato. Known only from type locality.

Thomomys umbrinus texensis V. Bailey

1902. *Thomomys fulvus texensis* V. Bailey, Proc. Biol. Soc. Washington, 15:119, June 2, type from head of Limpia Creek, 5500 ft., Davis Mts., Jeff Davis Co., Texas. Known only from type locality.
1939. *Thomomys umbrinus texensis,* Blair, Occas. Pap. Mus. Zool., Univ. Michigan, 403:2, June 16.

Thomomys umbrinus tivius Durrant

1937. *Thomomys bottae tivius* Durrant, Bull. Univ. Utah, 28(4):5, August 18, type from Oak Creek Canyon, 6 mi. E Oak City, 6000 ft., Millard Co., Utah. Known only from type locality.
1959. *Thomomys umbrinus tivius,* Hall and Kelson, Mammals of North America, Ronald Press, p. 434, March 31.

Thomomys umbrinus toltecus J. A. Allen

1893. *Thomomys toltecus* J. A. Allen, Bull. Amer. Mus. Nat. Hist., 5:52, April 28, type from Colonia Juárez, 4500 ft., Casas Grandes River, Chihuahua.
1959. *Thomomys umbrinus toltecus,* Hall and Kelson, Mammals of North America, Ronald Press, p. 434, March 31.

MARGINAL RECORDS.—New Mexico: Rio Mimbres; Deming. Chihuahua (Anderson, 1972:287, 288): Ojo Palomo Viejo, 4000 ft.; Vado de Fusiles, 4000 ft.; Laguna de Santa María; 3 mi. N Villa Ahumada; *Villa Ahumada, 1202 m.; 5 mi. N El Carmen;* 1 mi. W El Carmen; 11 mi. NNW San Buenaventura; *5 mi. W*

Mata Ortiz, 5900 ft.; 6 mi. W Mata Ortiz, 6300 ft.; Colonia Juárez, Casas Grandes River, 4500 ft.; Arroyo de Tinaja, 5600 ft. New Mexico: Adobe Ranch, N base Animas Mts.

Thomomys umbrinus tolucae Nelson and Goldman

1934. *Thomomys umbrinus tolucae* Nelson and Goldman, Jour. Mamm., 15:109, May 15, type from N slope Volcán de Toluca, 9500 ft., México. Known only from type locality.

Thomomys umbrinus townsendii (Bachman)

1839. *Geomys townsendii* Bachman, Jour. Acad. Nat. Sci. Philadelphia, 8:105. Type locality, erroneously given as "Columbia River," but probably near Nampa, Canyon Co., Idaho, where Townsend's party camped to trade with Indians, August 22, 1834 (V. Bailey, N. Amer. Fauna, 39:42, November 15, 1915).
1914. *Thomomys nevadensis atrogriseus* V. Bailey, Proc. Biol. Soc. Washington, 27:118, July 10, type from Nampa, Idaho.

MARGINAL RECORDS.—Idaho: Weiser; 2 mi. S Payette; Caldwell; Nampa; Hammett; Homedale. Oregon: Owyhee; Vale.

Thomomys umbrinus trumbullensis Hall and Davis

1934. *Thomomys bottae trumbullensis* Hall and Davis, Proc. Biol. Soc. Washington, 47:51, February 9, type from 3 mi. S Nixon Spring, Mt. Trumbull, Mohave Co., Arizona.
1959. *Thomomys umbrinus trumbullensis,* Hall and Kelson, Mammals of North America, Ronald Press, p. 434, March 31.

MARGINAL RECORDS.—Arizona (Hoffmeister and Durham, 1971:31): head Toroweap Valley; 2 mi. N Altar Point, 5800 ft.; *4 mi. NNE Vulcans Throne, Toroweap Valley; 3 mi. N Mt. Dellenbaugh, 6200 ft.;* Oak Grove, 5900 ft.; Waring Ranch (house), 5300 ft., Parashant Wash, Shivwits Plateau; Nixon Spring.

Thomomys umbrinus tularosae Hall

1932. *Thomomys baileyi tularosae* Hall, Univ. California Publ. Zool., 38:411, September 20, type from Cook Ranch, ½ mi. W Tularosa, Otero Co., New Mexico.

MARGINAL RECORDS.—New Mexico: type locality; *ca.* 2 mi. N Alamagordo; 11 mi. SW Alamagordo.

Thomomys umbrinus umbrinus (Richardson)

1829. *Geomys umbrinus* Richardson, Fauna Boreali-Americana, 1:202. Type locality, southern México; probably vic. Boca del Monte, Veracruz; type said to have come from "Cadadaguois, a town in southwestern Louisiana"; see V. Bailey, Proc. Biol. Soc. Washington, 19:3–6, January 29, 1906.
1855. *Thomomys umbrinus,* Baird, Proc. Acad. Nat. Sci. Philadelphia, 7:332.

MARGINAL RECORDS.—Veracruz: type locality; *Xuchil.*

Thomomys umbrinus vanrossemi Huey

1934. *Thomomys bottae vanrossemi* Huey, Trans. San Diego Soc. Nat. Hist., 8:1, August 10, type from Punta Peñascosa, Sonora. Known only from type locality.
1959. *Thomomys umbrinus vanrossemi*, Hall and Kelson, Mammals of North America, Ronald Press, p. 434, March 31.

Thomomys umbrinus varus Hall and Long

1960. *Thomomys umbrinus varus* Hall and Long, Proc. Biol. Soc. Washington, 73:35, August 10, type from 1 mi. S El Dorado, Sinaloa. Regarded as subspecifically inseparable from *Thomomys bottae sinaloae* by Dunnigan, Radford Review, 21:149, September 29, 1967. Known only from type locality and 1 mi. S thereof.

Thomomys umbrinus vescus Hall and Davis

1935. *Thomomys bottae vescus* Hall and Davis, Univ. California Publ. Zool., 40:389, March 13, type from S slope Mt. Jefferson, Toquima Range, 9000 ft., Nye Co., Nevada.
1959. *Thomomys umbrinus vescus*, Hall and Kelson, Mammals of North America, Ronald Press, p. 434, March 31.

MARGINAL RECORDS.—Nevada: 1 mi. E Jefferson, 7600 ft.; 5 mi. E Meadow Canyon Ranger Station; Meadow Canyon Ranger Station.

Thomomys umbrinus villai Baker

1953. *Thomomys bottae villai* Baker, Univ. Kansas Publ., Mus. Nat. Hist., 5:505, June 1, type from 7 mi. S, 2 mi. E Boquillas, 1800 ft., Coahuila. Known only from type locality.
1959. *Thomomys umbrinus villai*, Hall and Kelson, Mammals of North America, Ronald Press, p. 434, March 31.

Thomomys umbrinus virgineus Goldman

1937. *Thomomys bottae virgineus* Goldman, Proc. Biol. Soc. Washington, 50:133, September 10, type from Beaverdam Creek, 1500 ft., near confluence with Virgin River at Littlefield, Arizona.
1959. *Thomomys umbrinus virgineus*, Hall and Kelson, Mammals of North America, Ronald Press, p. 434, March 31.

MARGINAL RECORDS.—Utah: Beaverdam Wash, 8 mi. N Utah–Arizona border. Arizona (Hoffmeister and Durham, 1971:31): Littlefield; Middle Spring, 4500 ft.; Pakoon Springs, 2500 ft.; *Seven Springs, 1600 ft.*; Tasi Springs, 1600 ft. Nevada: Mesquite, 1750 ft.

Thomomys umbrinus vulcanius Nelson and Goldman

1934. *Thomomys umbrinus vulcanius* Nelson and Goldman, Jour. Mamm., 15:109, May 15, type from Volcán de Popocatépetl, 12,900 ft., México.

MARGINAL RECORDS.—Puebla: Río Otlatí, 8700 ft., 15 km. NW San Martín [Texmelucán]. México: N slope Mt. Popocatépetl, 13,500 ft.

Thomomys umbrinus wahwahensis Durrant

1937. *Thomomys bottae wahwahensis* Durrant, Bull. Univ. Utah, 28(4):3, August 18, type from Wah Wah Springs, 30 mi. W Milford, 6500 ft., Beaver Co., Utah.
1959. *Thomomys umbrinus wahwahensis*, Hall and Kelson, Mammals of North America, Ronald Press, p. 434, March 31.

MARGINAL RECORDS.—Utah: Desert Range Experiment Station, U.S. Forest Service, sec. 9, T. 25 S, R. 17 W, Salt Lake base meridian; type locality.

Thomomys umbrinus winthropi Nelson and Goldman

1934. *Thomomys bottae winthropi* Nelson and Goldman, Jour. Mamm., 15:122, May 15, type from Hermosillo, Sonora.
1959. *Thomomys umbrinus winthropi*, Hall and Kelson, Mammals of North America, Ronald Press, p. 434, March 31.

MARGINAL RECORDS.—Sonora: Saric; Magdalena; Ures; Ortiz; type locality.

Thomomys umbrinus xerophilus Huey

1945. *Thomomys bottae xerophilus* Huey, Trans. San Diego Soc. Nat. Hist., 10:257, August 31, type from near Diablito Spring, summit San Matías Pass between Sierra Juárez and Sierra San Pedro Mátir, Baja California.
1959. *Thomomys umbrinus xerophilus*, Hall and Kelson, Mammals of North America, Ronald Press, p. 434, March 31.

MARGINAL RECORDS.—Baja California: type locality; Aguajita Spring, El Valle de la Trinidad.

Thomomys umbrinus zacatecae Nelson and Goldman

1934. *Thomomys umbrinus zacatecae* Nelson and Goldman, Jour. Mamm., 15:112, May 15, type from Berriozábal, 6600 ft., Zacatecas.

MARGINAL RECORDS.—Zacatecas: 5 mi. NW Zacatecas, 7600 ft.; type locality. Aguascalientes: 12 mi. N Rincón de Romos, 6500 ft.

Genus Zygogeomys Merriam

Revised by Merriam, N. Amer. Fauna, 8:195–198, January 31, 1895; see also Russell, Univ. Kansas Publ., Mus. Nat. Hist., 16:523–525; August 5, 1968.

1895. *Zygogeomys* Merriam, N. Amer. Fauna, 8:195, January 31. Type, *Zygogeomys trichopus* Merriam.

Upper incisors bisulcate, major sulcus on inner side of median line and minor sulcus on inner convexity; M3 conspicuously bicolumnar, longer than wide owing to elongation of posterior loph; rostrum narrow in relation to its length (see Fig. 316); skull long and narrow; zygomata not widely spreading, slender, and anteroexternal angles rounded rather than expanded; maxillary and squamosal roots of zygomatic arch in contact above jugal; sagittal crest short, but well developed.

Zygogeomys trichopus
Michoacán Pocket Gopher

External measurements: in males, 343–346, 111–115, 46; in females, 292–322, 92–106, 38–43. Enamel plate on posterior wall of P4 restricted to lingual half of tooth; M3 partly biprismatic, the two lophs being separated by distinct outer sulcus; posterior part of M3 protected by two enamel plates, lingual plate especially long and extending to end of heel; mental foramen well anterior to anterior end of masseteric ridge; basitemporal fossa well developed and deep; pelage soft and thick.

Zygogeomys trichopus tarascensis Goldman

1938. *Zygogeomys trichopus tarascensis* Goldman, Proc. Biol. Soc. Washington, 51:211, December 23, type from 6 mi. SE Pátzcuaro, 8000 ft., Michoacán. Known only from type locality.

Zygogeomys trichopus trichopus Merriam

1895. *Zygogeomys trichopus* Merriam, N. Amer. Fauna, 8:196, January 31, type from Nahuatzen, Michoacán.

MARGINAL RECORDS.—Michoacán: type locality; Mt. Tancítaro, 6000–10,500 ft.

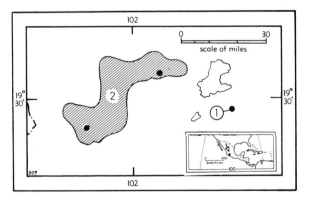

Map 309. *Zygogeomys trichopus.*

1. Z. t. tarascensis   2. Z. t. trichopus

Genus Geomys Rafinesque
Eastern Pocket Gophers

Revised: Merriam, N. Amer. Fauna, 8:109–145, January 31, 1895.

1817. *Geomys* Rafinesque, Amer. Month. Mag., 2:45, November. Type, *Geomys pinetis* Rafinesque.
1817. *Diplostoma* Rafinesque, Amer. Month. Mag., 2:44, November. Included *Diplostoma fusca* Rafinesque [= *Mus*

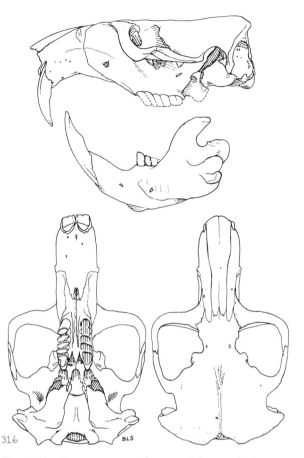

Fig. 316. *Zygogeomys trichopus trichopus*, Nahuatzen, Michoacán, No. 50100 U.S.N.M., ♂, X 1.

bursarius Shaw] and *D. alba* Rafinesque [=*Mus bursarius* Shaw] from Missouri River region.
1820. *Saccophorus* Kuhl, Beiträge zur Zoologie und vergleichenden Anatomie, pp. 65, 66. Type, *Mus bursarius* Shaw.
1823. *Pseudostoma* Say, *in* Long, Account of an exped. . . . to the Rocky Mts. . . . , 1:406. Type, *Mus bursarius* Shaw.
1825. *Ascomys* Lichtenstein, Abh. k. Akad. Wiss., Berlin, for 1822, p. 20, fig. 2. Type, *Ascomys canadensis* Lichtenstein [= *Mus busarius* Shaw] nominally from Canada.
1944. *Parageomys* Hibbard, Bull. Geol. Soc. Amer., 55:735, June. Type, *Parageomys tobinensis* Hibbard, an extinct (fossil) species. Valid as a subgenus.

External measurements in males, 217–357, 57–121, 27–43; in females, 187–316, 51–109, 23–39. Upper incisors bisulcate, major sulcus near median line and minor sulcus near inner border of tooth; M3 not strongly bicolumnar, crown no longer than wide owing to shortness of posterior loph, shallow re-entrant fold on labial side; posterior wall of P4 lacking any trace of enamel; in Recent species, P4 decidedly larger than p4 instead of subequal as in other genera;

basitemporal fossa (between coronoid process and m3) well developed; skull broad (see Figs. 317–324); zygomata in most species ordinarily wider across maxillary roots than across squamosal roots; pelage pale brown to black, usually paler below than above.

KEY TO SPECIES OF GEOMYS

1. Nasals hourglass-shaped (strongly constricted near middle); occurring in Alabama, Georgia, and Florida.
  2. Occurring on Cumberland Island, Georgia. . . . . . *G. cumberlandius,* p. 505
  2'. Not occurring on Cumberland Island.
    3. Fontanel on each side of skull between squamosal and parietal bones. . . . . . *G. fontanelus,* p. 505
    3'. No fontanel between squamosal and parietal bones.
      4. Interpterygoid space broadly U-shaped; nasals little constricted at middle. *G. colonus,* p. 505
      4'. Interpterygoid space broadly V-shaped; nasals much constricted at middle. *G. pinetis,* p. 504
1'. Nasals not hourglass-shaped; not in Alabama, Georgia, or Florida.
  5. Squamosal arm of zygoma ending in prominent knob over middle of jugal.
    6. Interparietal subquadrate; border of premaxilla at incisive foramina wedge-shaped. . . *G. arenarius,* p. 501
    6'. Interparetal triangular; border of premaxilla at incisive foramina subquadrate. . . . . *G. tropicalis,* p. 504
  5'. Squamosal arm of zygoma not ending in prominent knob over middle of jugal.
    7. Zygomatic breadth more anteriorly than posteriorly in most specimens; jugal longer than basioccipital including exoccipital condyle; posterior end of zygomatic arm of maxilla V-shaped at union with jugal. . . . . . . *G. bursarius,* p. 498
    7'. Zygomatic breadth approx. equal anteriorly and posteriorly in most specimens; jugal shorter than, or equal to, basioccipital including exoccipital condyle; posterior end of zygomatic arm of maxilla U-shaped at union with jugal.
                      G. personatus, p. 502

Geomys bursarius
Plains Pocket Gopher

X ¼

External measurements: in males, 217–357, 57–107, 27–43; in females, 187–316, 51–102, 23–39. Largest in north; smallest in south. Brown or black. Rostrum wider than basioccipital is long.

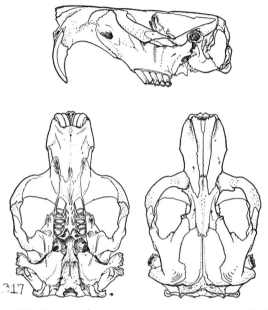

Fig. 317. *Geomys bursarius jugossicularis,* 4 mi. W Las Animas, 4100 ft., Bent Co., Colorado, No. 60784 M.V.Z., ♂, X 1.

Geomys bursarius ammophilus Davis

1940. *Geomys breviceps ammophilus* Davis, Texas Agric. Exp. Station Bull., 590:16, October 23, type from Cuero, De Witt Co., Texas.
1951. *Geomys bursarius ammophilus,* Baker and Glass, Proc. Biol. Soc. Washington, 64:57, April 13.

MARGINAL RECORDS.—Texas: Lavaca County (Davis, 1966:153, as *G. busarius* only); Inez; *Victoria;* type locality.

Geomys bursarius attwateri Merriam

1895. *Geomys breviceps attwateri* Merriam, N. Amer. Fauna, 8:135, January 31, type from Rockport, Texas.
1951. *Geomys bursarius attwateri,* Baker and Glass, Proc. Biol. Soc. Washington, 64:57, April 13.

MARGINAL RECORDS.—Texas: Caldwell County (Davis, 1966:153, as *G. bursarius* only); 4 mi. SE Luling; 3 mi. SW Victoria; Matagordà; 8 mi. SW Rockport; *Sinton* (Kennerly, 1959:253); 3 mi. S Skidmore (*ibid.,* but *4 mi. S Skidmore,* Kennerly, 1958:145); 11 mi. E Beeville; 5 mi. NW Berclaire (Kennerly, 1959:252); *3 mi. S Riverdale* (*ibid.*), along E bank San Antonio River (Kennerly, 1959:255), to ½ mi. S Falls City (Kennerly, 1959:254); 2 mi. NW Campbellton; 1 mi. N Moore; 18 mi. S San Antonio; Guadalupe County (Davis, 1966:153, as *G. bursarius* only).

Geomys bursarius brazensis Davis

1938. *Geomys breviceps brazensis* Davis, Jour. Mamm., 19:489, November 14, type from 5 mi. E Kurten, *in* Grimes Co., Texas.—See addenda.
1951. *Geomys bursarius brazensis,* Baker and Glass, Proc. Biol. Soc. Washington, 64:57, April 13.

MARGINAL RECORDS.—Texas: Terrell; Mineola; *Rusk County* (Davis, 1966:153, as *G. bursarius* only); 4 mi. NE Carthage; *Shelby County* (Davis, 1966:153, as *G. bursarius* only); San Augustine County (*ibid.*, as *G. bursarius* only); *Angelina County* (*ibid.*, as *G. bursarius* only); 1 mi. N Trinity; 4 mi. N Huffman; Fort Bend County (Davis, 1966:153, as *G. bursarius* only); Eagle Lake; Travis County (Davis, 1966:153, as *G. bursarius* only); 5 mi. E Bastrop; Milano; 1 mi. SE Regan.

Geomys bursarius breviceps Baird

1855. *Geomys breviceps* Baird, Proc. Acad. Nat. Sci. Philadelphia, 7:335, type from Prairie Mer Rouge, Lousiana. Known only from Morehouse Parish.
1951. *Geomys bursarius breviceps,* Baker and Glass, Proc. Biol. Soc. Washington, 64:57, April 13.

Geomys bursarius bursarius (Shaw)

1800. *Mus bursarius* Shaw, Trans. Linnean Soc. London, 5:227. Type locality, somewhere in upper Mississippi Valley. (Restricted to Elk River, Sherburne Co., Minnesota, by Swenk, Missouri Valley Fauna, 1:6, December 5.)
1829. *Geomys bursarius,* Richardson, Fauna Boreali-Americana, 1:203.
1817. *Diplostoma fusca* Rafinesque, Amer. Month. Mag., 2:44, November. Type locality, Missouri River Region.
1817. *Diplostoma alba* Rafinesque (*ibid.*). Type locality same.
1821. *Mus saccatus* Mitchill, Med. Repos. (n.s.) [New York], 6(21):249. Type locality "from area bordering on Lake Superior" [from west of Lake Superior according to Baird (1858:372)].
1825. *Ascomys canadensis* Lichtenstein, Abh. k. Akad. Wiss., Berlin, for 1822, p. 20, type from "Canada."

MARGINAL RECORDS.—Manitoba (Wrigley and Dubois, 1973:167): 3 mi. W Roseau River; *1 mi. S Gardenton.* Minnesota: eastern Kittson County; Clearwater County; Cass County; Carlton County. Wisconsin: Bayfield County (Jackson, 1961:186, Map 36); Prescott (*op. cit.:* 190). South Dakota: Scotland; Flandreau; Fort Sisseton. North Dakota: Ludden; Valley City; 10 mi. W Portland; Manvel. Manitoba (Wrigley and Dubois, 1973:167): Emerson; *2 mi. N, 6 1/2 mi. E Emerson; 7 1/2 mi. E Dominion City.*

Geomys bursarius dutcheri Davis

1940. *Geomys breviceps dutcheri* Davis, Texas Agric. Exp. Station Bull., 590:12, Oct. 23, type from Fort Gibson, Okla.
1951. *Geomys bursarius dutcheri,* Baker and Glass, Proc. Biol. Soc. Washington, 64:57, April 13.—See addenda.

MARGINAL RECORDS.—Oklahoma: near Garnett; type locality. Arkansas: Mulberry; *Ozark;* 10 mi. N Little Rock; Pine Bluff; El Dorado. Louisiana: Marion (Lowery, 1974:212); West Monroe (*ibid.*); Fishville; Pineville; Colfax; Keithville. Texas: Jefferson; Longview; vic. Arlington (Boley and Kennerly, 1969:348, as *G. bursarius* only); Decatur; Stoneburg (Dalquest, 1968:17). Oklahoma: 3½ mi. E Norman; 8 mi. W Red Fork; Tulsa. According to Sealander (1956:272), employees of the Arkansas Game and Fish Commission report pocket gophers from the following counties in Arkansas: *Washington; Cross; Phillips; Chicot; Ashley; Bradley.*

Geomys bursarius illinoensis Komarek and Spencer

1931. *Geomys bursarius illinoensis* Komarek and Spencer, Jour. Mamm., 12:405, November 11, type from 1 mi. S Momence, Kankakee Co., Illinois.

MARGINAL RECORDS.—Illinois: Ottawa [= South Ottawa]; Custer Park. Indiana: Jasper County; [N Wabash River] Tippecanoe Co., Illinois: Clinton; Boody; Collinsville; Havana; Oglesby.

Geomys bursarius industrius Villa and Hall

1947. *Geomys bursarius industrius* Villa and Hall, Univ. Kansas Publ., Mus. Nat. Hist., 1:226, November 29, type from 1½ mi. N Fowler, Meade Co., Kansas.

MARGINAL RECORDS.—Kansas: Larned, jct. Pawnee and Arkansas rivers; Pratt; Rezeau Ranch, 5 mi. N Belvidere; 7 mi. SW Kingsdown, Stephenson Ranch; State Lake and Park (Meade Co.); Cudahy Ash Pit, 7 mi. N Meade; 1 mi. W and 3½ mi. S Kinsley.

Geomys bursarius jugossicularis Hooper

1940. *Geomys lutescens jugossicularis* Hooper, Occas. Pap. Mus. Zool., Univ. Michigan, 420:1, June 28, type from Lamar, Prowers Co., Colorado.
1947. *Geomys bursarius jugossicularis,* Villa and Hall, Univ. Kansas Publ., Mus. Nat. Hist., 1:226, November 29.

MARGINAL RECORDS.—Colorado (Armstrong, 1972:166): Colorado Springs; Chivington. Kansas: 1 mi. E Coolidge; 1 mi. E Arkalon. Texas: 15 mi. E Texline. New Mexico (Best, 1973:1317): 1½ mi. N, 5⅕ mi. E Pasamonte; 7 3/10 mi. N, 1 mi. W Grenville. Colorado (Armstrong, 1972:166): 8 mi. S Campo; 8 mi. S Pritchett; 14 mi. N, 4 mi. E Springfield; type locality; Las Animas; Pueblo; 4 mi. SSE Cañon City.

Geomys bursarius knoxjonesi Baker and Genoways

1975. *Geomys bursarius knoxjonesi* Baker and Genoways, Occas. Pap. Mus. Texas Tech Univ., 29:1, April 25, type from 4 1/10 mi. N, 5 1/10 mi. E Kermit, Winkler Co., Texas.

MARGINAL RECORDS (Baker and Genoways, 1975:17, unless otherwise noted): New Mexico: 7⅕ mi. N, 11 3/10 mi. E Elkins. Texas: 4½ mi. SSW Morton; *3 2/5 mi. N, 3 3/10 mi. W Whiteface;* 1 7/10 mi. S, ½ mi. W Meadow; 2 9/10 mi. S Patricia; 3½ mi. E Monahans; *6 1/2 mi. SE Kermit.* New Mexico: 5 7/10 mi. E Loco Hills; 7/10 mi. N, 12⅗ mi. W Caprock; 15 mi. NE Roswell, Pecos River (Findley, *et al.*, 1975:152, as *G. bursarius* only).

Geomys bursarius llanensis V. Bailey

1905. *Geomys breviceps llanensis* V. Bailey, N. Amer. Fauna, 25:129, October 24, type from Llano, Texas.
1947. *Geomys bursarius llanensis*, Villa and Hall, Univ. Kansas Publ., Mus. Nat. Hist., 1:234, November 29.

MARGINAL RECORDS.—Texas: 1 mi. N Pontotoc; 3 mi. W, 1 mi. N Kingsland; 12 mi. S, 8 mi. W Llano; Castell.

Geomys bursarius ludemani Davis

1940. *Geomys breviceps ludemani* Davis, Texas Agric. Exp. Station Bull., 590:19, October 23, type from 7 mi. SW Fannett, Jefferson Co., Texas.—See addenda.
1951. *Geomys bursarius ludemani*, Baker and Glass, Proc. Biol. Soc. Washington, 64:58, April 13.

MARGINAL RECORDS.—Texas: type locality; Double Bayou, 10 mi. S Anahuac.

Geomys bursarius lutescens Merriam

1890. *Geomys bursarius lutescens* Merriam, N. Amer. Fauna, 4:51, October 8, type from sandhills on Birdwood Creek, Lincoln Co., Nebraska.
1938. *Geomys lutescens hylaeus* Blossom, Occas. Pap. Mus. Zool., Univ. Michigan, 368:1, April 6, type from 10 mi. S Chadron, Dawes Co., Nebraska. (Arranged as a synonym of *G. b. lutescens* by Jones, Univ. Kansas Publ., Mus. Nat. Hist., 16:156, October 1, 1964.)
1940. *Geomys lutescens levisagittalis* Swenk, Missouri Valley Fauna, 2:4, February 1, type from Spencer, Boyd Co., Nebraska. (Arranged as a synonym of *G. b. lutescens* by Jones, Univ. Kansas Publ., Mus. Nat. Hist., 16:156, October 1, 1964.)
1940. *Geomys lutescens vinaceus* Swenk, Missouri Valley Fauna, 2:7, February 1, type from Scottsbluff, Scotts Bluff Co., Nebraska (considered a synonym of *G. b. lutescens* by Russell and Jones, Trans. Kansas Acad. Sci., 58:513, January 23, 1956).

MARGINAL RECORDS.—Wyoming: 23 mi. SW Newcastle (Long, 1965a:613). South Dakota: Pine Ridge Agency; Rosebud Agency; Dog Ear Lake. Nebraska (Jones, 1964c:159, 160): *5 mi. WNW Spencer; 1 mi. WNW Spencer*; Spencer; Neligh; *1 mi. SW Neligh*; Bladen. Kansas: ½ mi. S, 1½ mi. E Alton (125039 KU); Hays State College Campus; 4 mi. S Scott City. Colorado (Armstrong, 1972:166): Kit Carson; Kiowa; Rose [D'Arcy] Ranch, 2 mi. N Parker; Wheatridge; Boulder; Big Thompson River, Loveland; *½ mi. WSW Mason-ville*; 14 mi. W Fort Collins; *13 1/5 mi. N, 3 3/5 mi. W Fort Collins*. Wyoming (Long, 1965a:613): Uva; 3 mi. N, 5 mi. E Orin.

Geomys bursarius major Davis

1940. *Geomys lutescens major* Davis, Texas Agric. Exp. Station Bull., 590:32, October 23, type from 8 mi. W Clarendon, Donley Co., Texas.
1947. *Geomys bursarius major*, Villa and Hall, Univ. Kansas Publ., Mus. Nat. Hist., 1:229, November 29.

MARGINAL RECORDS.—Kansas: 2 mi. S Ellsworth; Smoky Hill River, 1 mi. S, ½ mi. W Lindsborg; ½ mi. E McPherson; 8 mi. W Rosalia; 3 mi. SE Arkansas City. Oklahoma: Ponca Agency; Stillwater; 2 mi. E Norman; Apache; 12 mi. S Temple. Texas: Clay County (Dalquest, 1968:17); Brazos; 6 mi. S Waco; Colorado; Midland (Baker and Genoways, 1975:18); 2 1/10 mi. NE Big Spring (*ibid.*); 4½ mi. NW Post (*ibid.*); Lubbock (*ibid.*); *1 mi. W Morton* (*op. cit.*:17); 5 mi. W Morton (*ibid.*). New Mexico: 1 4/5 mi. S, 1 1/10 mi. E Lingo (*ibid.*); 1 mi. E Elida (*ibid.*); Mesa Jumanes, near Progresso; Santa Rosa; Bell Ranch (Findley, *et al.*, 1975:152, as *G. bursarius* only).

Geomys bursarius majusculus Swenk

1939. *Geomys bursarius majusculus* Swenk, Missouri Valley Fauna, 1:6, December 5, type from Lincoln, Nebraska.

MARGINAL RECORDS.—Iowa (Bowles, 1975:79, 80): New Albin, thence southward along W bank Mississippi River, to Burlington. Kansas: Fort Leavenworth; 11 mi. SW Lawrence; 8½ mi. SW Toronto; 4 mi. S, 14 mi. W Hamilton; 6 mi. S Lincolnville; ½ mi. S, 8 mi. E Osborne (125043 KU). Nebraska (Jones, 1964c:163): Hastings; *5 mi. W Tilden*; Oakdale; mouth Niobrara River. South Dakota: *southeastern part*. Iowa: 1½ mi. S, 1½ mi. E Granite (Bowles, 1975:80).

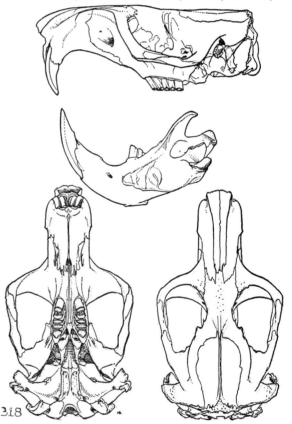

Fig. 318. *Geomys bursarius majusculus*, Lincoln, Lancaster Co., Nebraska, No. 97913 M.V.Z., holotype, ♂, X 1.

Geomys bursarius missouriensis McLaughlin

1958. *Geomys bursarius missouriensis* McLaughlin, Los Angeles Co. Mus. Contrib. Sci., 19:1, January 29, type from 2 mi. N Manchester, St. Louis Co., Missouri.

MARGINAL RECORDS.—Missouri (McLaughlin, 1958:4): St. Louis Co., vic. St. Louis; Williamsville; Hunter; *type locality*.

Geomys bursarius pratincola Davis

1940. *Geomys breviceps pratincolus* Davis, Texas Agric. Exp. Station Bull., 590:18, October 23, type from 2 mi. E Liberty, Liberty Co., Texas.—See addenda.
1951. *Geomys bursarius pratincolus*, Baker and Glass, Proc. Biol. Soc. Washington, 64:57, April 13.

MARGINAL RECORDS.—Louisiana (Lowery, 1974:212, as *G. b. dutcheri*, unless otherwise noted): Natchitoches; *3 mi. SW Boyce*; 5 mi. WSW Lecompte; 8 mi. SW Kinder; near edge Sweet Lake, 17 mi. S, 5 mi. E Lake Charles; Gum Cove, 15 mi. S Vinton (Davis, 1940:19). Texas: 13 mi. NE Sour Lake; 2 mi. E Liberty; 3 mi. W Livingston; Sabine County (Davis, 1966:153, as *G. bursarius* only).—Lowery (1974:211) assigned specimens from central-western and south-western Louisiana to *G. b. dutcheri*, instead of considering whether they are intergrades between *G. b. dutcheri* and *G. b. pratincola* and possibly more nearly like topotypes of the latter. This possibility needs to be investigated before the name *pratincola*, previously applied to the Louisiana specimens, is changed to *dutcheri*.

Geomys bursarius sagittalis Merriam

1895. *Geomys breviceps sagittalis* Merriam, N. Amer. Fauna, 8:134, January 31, type from Clear Creek, Galveston Bay, Galveston Co., Texas.
1951. *Geomys bursarius sagittalis*, Baker and Glass, Proc. Biol. Soc. Washington, 64:57, April 13.—See addenda.

MARGINAL RECORDS.—Texas: 3 mi. N La Porte; 3 mi. NE Webster; Arcadia; 4 mi. S Altoloma.

Geomys bursarius terricolus Davis

1940. *Geomys breviceps terricolus* Davis, Texas Agric. Exp. Station Bull., 590:17, October 23, type from 1 mi. N Texas City, Galveston Co., Texas. Known only from type locality.
1951. *Geomys bursarius terricolus*, Baker and Glass, Proc. Biol. Soc. Washington, 64:57, April 13.—See addenda.

Geomys bursarius texensis Merriam

1895. *Geomys texensis* Merriam, N. Amer. Fauna, 8:137, January 31, type from Mason, Mason Co., Texas.
1950. *Geomys bursarius texensis*, Baker, Jour. Mamm., 31:349, August 21.

MARGINAL RECORDS.—Texas: 10 mi. S Brady (Dalquest and Kilpatrick, 1973:5); ½ mi. W Castell; Gil-

lespie County (Davis, 1966:153, as *G. bursarius* only); *11 mi. SW Mason*; 10 mi. W Mason.

Geomys bursarius wisconsinensis Jackson

1957. *Geomys bursarius wisconsinensis* Jackson, Proc. Biol. Soc. Washington, 70:33, June 28, type from Lone Rock, Richland Co., Wisconsin.

MARGINAL RECORDS.—Wisconsin: Price County (Jackson, 1961:186, Map 36); type locality; Prairie du Chien (*op. cit.*:191); Pepin County (*op. cit.*:186, Map 36); Barron County (*ibid.*).

Geomys arenarius
Desert Pocket Gopher

External measurements: in males, 244–280, 74–95, 30–34; in females, 221–250, 58–84, 27–35. Brown; width of rostrum not exceeding length of basioccipital; squamosal arm of zygoma ending in prominent knob over middle of jugal.

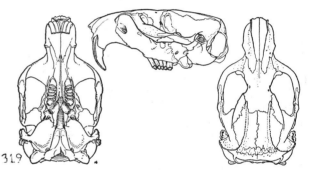

Fig. 319. *Geomys arenarius brevirostris*, E edge [white] sand [9 mi. W Tularosa], Tularosa Hot Springs Road, Otero Co., New Mexico, No. 50460 M.V.Z., holotype, ♀, X 1.

Geomys arenarius arenarius Merriam

1895. *Geomys arenarius* Merriam, N. Amer. Fauna, 8:139, January 31, type from El Paso Co., Texas.

MARGINAL RECORDS.—New Mexico: Deming; Las Cruces. Texas: Hudspeth County (Davis, 1966:154, as *G. arenarius* only). Chihuahua (Anderson, 1972:293): 1½ mi. NE Porvenir; 8 mi. S Samalayuca.

Geomys arenarius brevirostris Hall

1932. *Geomys arenarius brevirostris* Hall, Proc. Biol. Soc. Washington, 45:97, June 21, type from E edge [white] sand [9 mi. W Tularosa], Tularosa–Hot Springs Road, Otero Co., New Mexico.

MARGINAL RECORDS (Findley, *et al.*, 1975:153–154).—New Mexico: 5 mi. S, 11¼ mi. E San Antonio; type locality; *10 mi. SW Tularosa*; 18 mi. W Alamogordo.

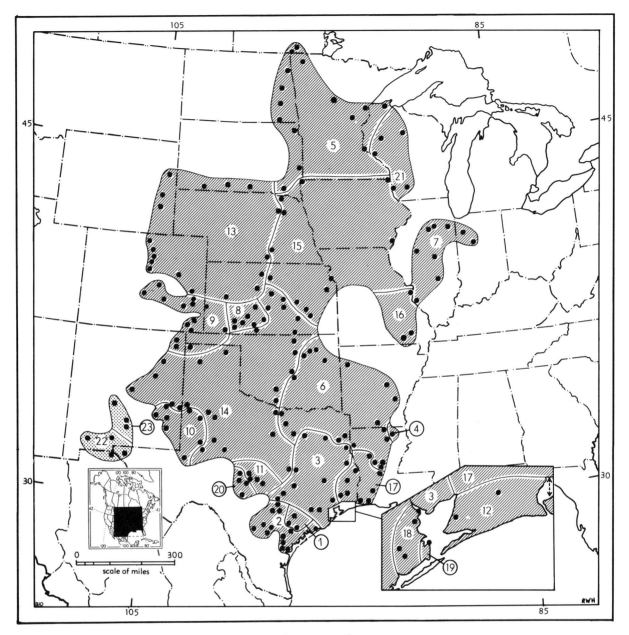

Map 310. *Geomys bursarius* and *Geomys arenarius*.

Guide to kinds			
1. *G. b. ammophilus*	6. *G. b. dutcheri*	12. *G. b. ludemani*	18. *G. b. sagittalis*
2. *G. b. attwateri*	7. *G. b. illinoensis*	13. *G. b. lutescens*	19. *G. b. terricolus*
3. *G. b. brazensis*	8. *G. b. industrius*	14. *G. b. major*	20. *G. b. texensis*
4. *G. b. breviceps*	9. *G. b. jugossicularis*	15. *G. b. majusculus*	21. *G. b. wisconsinensis*
5. *G. b. bursarius*	10. *G. b. knoxjonesi*	16. *G. b. missouriensis*	22. *G. a. arenarius*
	11. *G. b. llanensis*	17. *G. b. pratincola*	23. *G. a. brevirostris*

Geomys personatus
Texas Pocket Gopher

External measurements: in males, 248–326, 72–121, 30–42; in females, 225–305, 59–109, 30–39. Brown; squamosal arm of zygoma lacking prominent knob over middle of jugal.

Geomys personatus fallax Merriam

1895. *Geomys personatus fallax* Merriam, N. Amer. Fauna, 8:144, January 31, type from S side Nueces Bay, Nueces Co., Texas.

MARGINAL RECORDS.—Texas: ½ mi. S Falls City (Kennerly, 1959:254, but *½ mi. SE Falls City*, Kennerly,

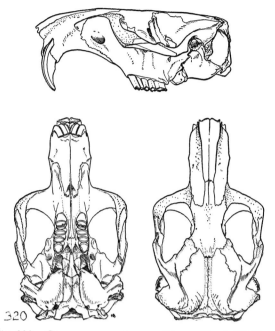

Fig. 320. *Geomys personatus maritimus*, 11 mi. SE Corpus Christi, Nueces Co., Texas, No. 84159 M.V.Z., ♂, X 1.

1958:145), along W bank San Antonio River (Kennerly, 1959:255), to 5 mi. W Riverdale (Kennerly, 1959:252); 9½ mi. N Berclaire (*ibid.*); 12 mi. N Beeville; Beeville; 1 mi. S Skidmore (Kennerly, 1959:253); 2½ mi. S Odem (*ibid.*); 6 mi. W Corpus Christi, to point approx. 7 mi. SE Corpus Christi (Kennerly, 1959:248); Sandia; 1½ mi. SW George West; 5 mi. S Three Rivers; 1½ mi. S Cambellton (Kennerly, 1959:252).

Geomys personatus fuscus Davis

1940. *Geomys personatus fuscus* Davis, Texas Agric. Exp. Station Bull., 590:30, October 23, type from Fort Clark [Bracketville], Kinney Co., Texas.

MARGINAL RECORDS.—Texas: Rio Grande at Del Rio; mouth Sycamore Creek, boundary between Val Verde and Kinney counties; type locality.

Geomys personatus maritimus Davis

1940. *Geomys personatus maritimus* Davis, Texas Agric. Exp. Station Bull., 590:26, October 23, type from Flour Bluff, 11 mi. SE Corpus Christi, Nueces Co., Texas. Known only from type locality.

Geomys personatus megapotamus Davis

1940. *Geomys personatus megapotamus* Davis, Texas Agric. Exp. Station Bull., 590:27, October 23, type from 4 mi. SE Oilton, Webb Co., Texas.

MARGINAL RECORDS.—Texas: S side Nueces River, 6 mi. W Cotulla; type locality; 3½ mi. SW Realitos; Falfurrias; Sauz Rancho, near Santa Monica; near Santa Rosa; 1 mi. SW Santa Elena; Carrizo [= Zapata].

Geomys personatus personatus True

1889. *Geomys personatus* True, Proc. U.S. Nat. Mus., 11:159 for 1888, January 5, type from Padre Island, Cameron Co., Texas.

MARGINAL RECORDS.—Texas: 14 mi. SW Port Aransas, Mustang Island; *type locality*. Tamaulipas (Alvarez, 1963:425): 33 mi. S Washington Beach; 35 mi. SSE Matamoros; 73 mi. S Washington Beach.

Geomys personatus streckeri Davis

1940. *Geomys personatus minor* Davis, Texas Agric. Exp. Station Bull., 590:29, October 23, type from Carrizo Springs, Dimmit Co., Texas. Not *Geomys minor* Gidley, 1922, a fossil. Known only from type locality.

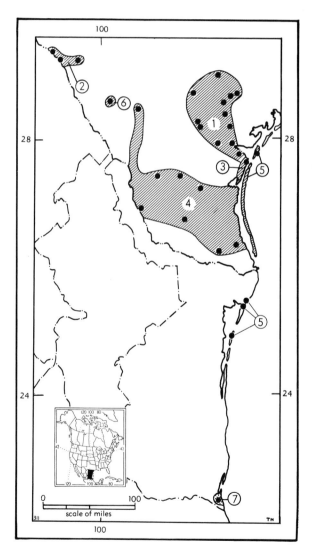

Map 311. *Geomys personatus* and *Geomys tropicalis*.

1. *G. p. fallax*   4. *G. p. megapotamus*
2. *G. p. fuscus*   5. *G. p. personatus*
3. *G. p. maritimus*   6. *G. p. streckeri*
          7. *G. tropicalis*

1943. *Geomys personatus streckeri* Davis, Jour. Mamm., 24:508, November 20, a renaming of *Geomys personatus minor* Davis.

Geomys tropicalis Goldman
Tropical Pocket Gopher

1915. *Geomys personatus tropicalis* Goldman, Proc. Biol. Soc. Washington, 28:134, June 29, type from Altamira, Tamaulipas.
1963. *Geomys tropicalis*, Alvarez, Univ. Kansas Publ., Mus. Nat. Hist., 14:426, May 20.

External measurements: in males, 260–265, 87–93, 33–35; in females, 235–250, 78–85, 31–33. Resembles *T. personatus* but differs in lower sagittal crest, presence of squamosal knob on zygomatic process of squamosal, and skull narrower (relative to its length) across zygomata and across mastoids. Differences from *T. arenarius* are as follows: narrower across zygomatic arches posteriorly than anteriorly, presence instead of absence of sagittal crest, triangular instead of subquadrate interparietal, and subquadrate instead of wedge-shaped border of premaxilla at incisive foramina.

MARGINAL RECORDS (Alvarez, 1963:427).— Tamaulipas: type locality; *1 mi. S Altamira; 10 mi. N Tampico.*

Geomys pinetis
Southeastern Pocket Gopher

External measurements: in males, 250–305, 81–96, 33–37; in females, 229–335, 76–82, 30.5–36. Dark brown to black. Width of rostrum more than length of basioccipital; nasals strongly constricted near middle and shaped like an hourglass.

Geomys pinetis austrinus Bangs

1898. *Geomys floridanus austrinus* Bangs, Proc. Boston Soc. Nat. Hist., 28:177, March, type from Belleair, Pinellas Co., Florida.
1952. *Geomys pinetis austrinus*, Harper, Proc. Biol. Soc. Washington, 65:37, January 29.

MARGINAL RECORDS.—Florida: Tarpon Springs; Arcadia, thence westward to coast and northward along coast to point of beginning.

Geomys pinetis floridanus (Audubon and Bachman)

1853. *Pseudostoma floridanus* Audubon and Bachman, The viviparous quadrupeds of North America, 3:242. Type locality, St. Augustine, St. Johns Co., Florida.
1952. *Geomys pinetis floridanus*, Harper, Proc. Biol. Soc. Washington, 65:37, January 29.

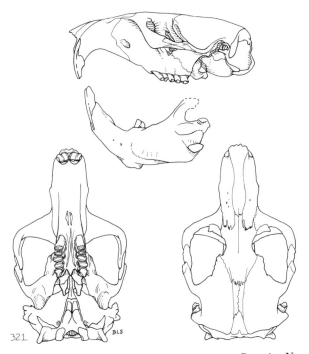

Fig. 321. *Geomys pinetis pinetis*, Augusta, Georgia, No. 63842 U.S.N.M., ♂, X 1.

MARGINAL RECORDS.—Florida: near Boulogne; Rose Bluff; Orlando; 1 mi. W Silver Springs; Gainesville; Chattahoochee. Georgia: Springhill Plantation, *ca.* 10 mi. SSW Thomasville. Florida: N of Macclenny.

Geomys pinetis goffi Sherman

1944. *Geomys tuza goffi* Sherman, Proc. New England Zool. Club, 23:38, August 30, type from Eau Gallie, Brevard Co., Florida.
1952. *Geomys pinetis goffi*, Harper, Proc. Biol. Soc. Washington, 65:37, January 29.

MARGINAL RECORDS.—Florida: type locality; $\frac{1}{2}$ mi. W Eau Gallie.

Geomys pinetis mobilensis Merriam

1895. *Geomys tuza mobilensis* Merriam, N. Amer. Fauna, 8:119, January 31, type from Point Clear, Mobile Bay, Baldwin Co., Alabama.
1952. *Geomys pinetis mobilensis*, Harper, Proc. Biol. Soc. Washington, 65:37, January 29.

MARGINAL RECORDS.—Alabama: Warrior River, near Lock 14; Seale. Florida: 6 mi. S, $\frac{1}{2}$ mi. W Wausaw; Milton. Alabama: 1 mi. N Fairhope.

Geomys pinetis pinetis Rafinesque

1806. *Mus tuza* Barton, Mag. für den neuesten Zustand der Naturkunde (ed. J. H. Voight), 12(6):488, November, *et auct.* (Type locality restricted to pine barrens near Augusta,

Georgia, by Bangs, Proc. Boston Soc. Nat. Hist., 28:175, March, 1898. According to Harper, Proc. Biol. Soc. Washington, 65:36, January 29, 1952, *tuza* of Barton is of uncertain application and is regarded as not available.)

1817. *Geomys pinetis* Rafinesque, Amer. Month. Mag., 2(1): 45, November. Type locality, Georgia in the region of the pines. (More restrictedly, Screven County according to Harper, Proc. Biol. Soc. Washington, 65:36, January 29, 1952.) Regarded as identical with *tuza* by Merriam, N. Amer. Fauna, 8:113, January 31, 1895.

MARGINAL RECORDS.—Georgia: Hollywood, 12 mi. S Augusta; Savannah; Sterling; 3 mi. SE Kingsland; St. Marys River, near Camp Pinckney; Hebardsville; Butler.

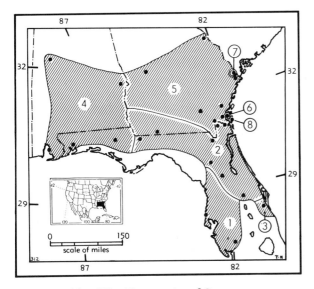

Map 312.   Four species of *Geomys*.

1. *G. pinetis austrinus*
2. *G. pinetis floridanus*
3. *G. pinetis goffi*
4. *G. pinetis mobilensis*
5. *G. pinetis pinetis*
6. *G. colonus*
7. *G. fontanelus*
8. *G. cumberlandius*

Geomys colonus Bangs
Colonial Pocket Gopher

1898. *Geomys colonus* Bangs, Proc. Boston Soc. Nat. Hist., 28:178, March, type from Arnot Plantation, about 4 mi. W St. Marys, Camden Co., Georgia.

External measurements: in males, 280–288, 89–100, 34–36; in females (average of six), 250, 78, 32. Rsembles *G. pinetis floridanus* but darker, upper parts being between seal-brown and sepia; palate wider between posterior molars; interpterygoid space broadly U-shaped, nasals shorter and shaped less like an hourglass.

MARGINAL RECORDS.—Georgia: Arnot (= Arnow) Plantation, approx. 4 mi. W St. Marys; *W of Shingle Swamp, between St. Marys and Kingsland.*

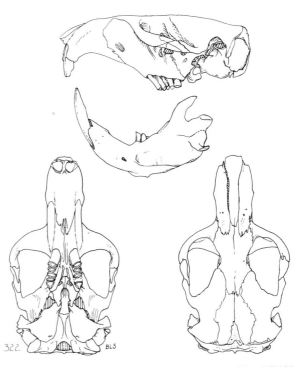

Fig. 322.   *Geomys colonus*, St. Marys, Georgia, No. 159183 U.S.N.M., ♂, X 1.

Geomys fontanelus Sherman
Sherman's Pocket Gopher

1940. *Geomys fontanelus* Sherman, Jour. Mamm., 21:341, August 13, type from 7 mi. NW Savannah, Chatham Co., Georgia. Known only from type locality.

External measurements: in males, 222–276, 70–105, 30–34; in females, 232–250, 61–95, 30–34. Described as more nearly related to *G. p. pinetis* than any other species and differing in slightly darker color, auditory bullae larger, presence of a fontanel on each side of skull between parietal and squamosal bones, and in more nearly triangular scapula with larger area for insertion of teres major muscle.

Geomys cumberlandius Bangs
Cumberland Island Pocket Gopher

1898. *Geomys cumberlandius* Bangs, Proc. Boston Soc. Nat. Hist., 28:180, March, type from Stafford Place, Cumberland Island, Camden Co., Georgia. Known only from Cumberland Island.

External measurements: Largest male in original series, 324, 114, 35.5; largest female, 283, 96, 34. Resembles *G. p. floridanus*, but tail longer, ascending arms of maxillae narrower, and zygomata less angled at posterior union with skull.

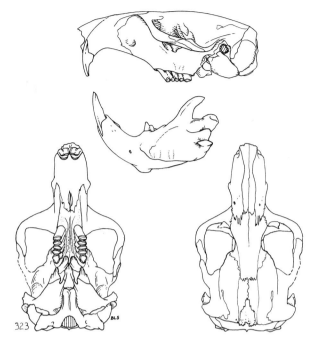

Fig. 323. *Geomys fontanelus*, 7 mi. N Savannah, Georgia,
No. 159498 U.S.N.M., ♂, X 1.

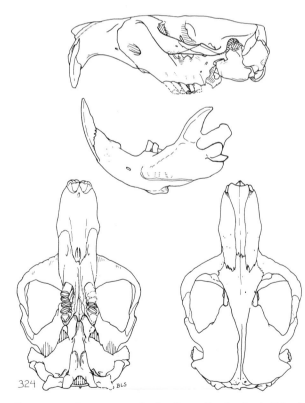

Fig. 324. *Geomys cumberlandius*, Cumberland Island,
Georgia, No. 159206 U.S.N.M., ♂, X 1.

Genus **Orthogeomys** Merriam
Giant Pocket Gophers

Revised by Merriam, N. Amer. Fauna, 18: 172–179, and
other pages, January 3, 1895. See also Russell, Univ. Kansas
Publ., Mus. Nat. Hist., 16:528–532, August 15, 1968.

1895. *Orthogeomys* Merriam, N. Amer. Fauna, 8:172, January
31. Type, *Geomys scalops* Thomas.
1895. *Heterogeomys* Merriam, N. Amer. Fauna, 8:179, January 31. Type, *Geomys hispidus* Le Conte. Valid as a
subgenus.
1895. *Macrogeomys* Merriam, N. Amer. Fauna, 8:185, January
31. Type, *Geomys heterodus* Peters. Valid as a subgenus.

External measurements: in males, 177–435,
70–140, 35–55; in females, 170–390, 70–128,
35–55. Upper incisors unisulcate; M3 partly
bilophodont, posterior loph long and tooth therefore longer than wide; enamel plates cover anterior surface and border re-entrant angles, of
upper- and lower-premolar; enamel plate on posterior border of p4; enamel plate on posterior
border of P4 limited to narrow blade on lingual
side (except in subgenus *Orthogeomys* where ordinarily absent); basitemporal fossa deep and
well defined except in subgenus *Orthogeomys*
where shallow; lingual, enamel plate of M3 approx. equal to, or more than, length of labial
plate; in relation to length of skull, rostrum broad
and posterior part of skull narrow (see Figs. 325–
328); angular process short in relation to length of
mandible; pelage coarse, in many instances hispid, and hairs so few in some individuals that
they appear to be naked. As now known the
genus occurs only in tropical life-zones.

KEY TO SUBGENERA OF ORTHOGEOMYS

1. Frontal wide and much inflated; no interorbital constriction; enamel plate on
   posterior wall of P4 usually absent, although small plate restricted to lingual
   end of wall rarely present.

   Orthogeomys, p. 507
1'. Frontal narrow and not much inflated; interorbital region decidedly constricted;
   enamel plate on posterior wall of P4 always present but short and restricted to
   lingual end of wall.
   2. Anterior margin of mesopterygoid
      fossa even with plane of posterior wall
      of M3; postorbital bar (process) weakly
      developed; anteroposterior occlusal
      length of M3 equal to, or less than,
      combined lengths of M1 and M2.

      Heterogeomys, p. 509
   2'. Anterior margin of mesoptergoid fossa
      decidedly behind plane of posterior
      wall of M3; postorbital bar (process)

strongly developed; anteroposterior occlusal length of M3 more than combined lengths of M1 and M2.

Macrogeomys, p. 512

Subgenus **Orthogeomys** Merriam

1895. *Orthogeomys* Merriam, N. Amer. Fauna, 8:172, January 31. Type, *Geomys scalops* Thomas.

External measurements: in males, 330–435, 95–140, 44–55; in females, 314–390, 91–125, 38–45. Pelage coarse, hispid or setose; sulcus on upper incisor slightly medial to mid-line but in some specimens touching middle; zygomatic breadth little more than mastoid breadth and in many skulls less. See additional characters in key to subgenera.

KEY TO SPECIES OF SUBGENUS ORTHOGEOMYS

1. Total length less than 345. *O. cuniculus*, p. 507
1'. Total length more than 345. *O. grandis*, p. 507

Orthogeomys cuniculus Elliot
Oaxacan Pocket Gopher

1905. *Orthogeomys cuniculus* Elliot, Proc. Biol. Soc. Washington, 18:234, December 9, type from Zanatepec, Oaxaca.

(Elliot, Field Columb. Mus., Publ. 115, Zool. Ser., 8:312, March 4, 1907, corrects type locality from Yautepec, Oaxaca, to Zanatepec.) Known only from type locality.

External measurements: in male, 330, 95, 44. According to Nelson and Goldman (1930:317), differs specifically from *O. g. scalops* in smaller size. In adult males: *O. g. scalops:* condylobasal length, 67.4; zygomatic breadth, 41.6; *O. cuniculus:* condylobasal length, 58.3; zygomatic breadth, 36.6.

Orthogeomys grandis
Large Pocket Gopher

External measurements: in males, 366–435, 97–140, 49–55; in females, 314–390, 91–121, 38–55. Pelage coarse, hispid, or setose.

Orthogeomys grandis alleni Nelson and Goldman

1930. *Orthogeomys grandis alleni* Nelson and Goldman, Jour. Mamm., 11:156, May 9, type from 2000 ft., near Acapulco, Guerrero.

MARGINAL RECORDS.—Jalisco: 8 mi. E Jilotlán de los Dolores, 2000 ft. (Genoways and Jones, 1969:755). Guerrero: Río Aguacatillo, 1000 ft., 30 km. N Acapulco. Oaxaca: Pinotepa; Zapotitlán (Goodwin, 1969:139); Puerto Angel; thence westward along coast to Guer-

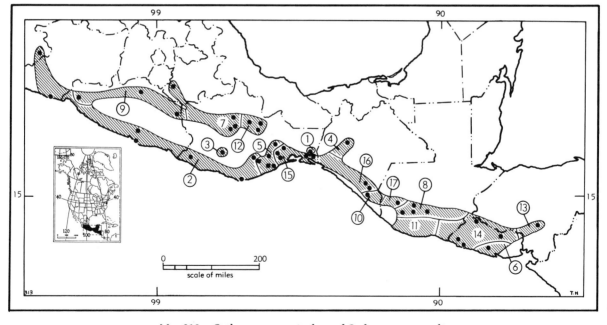

Map 313. *Orthogeomys cuniculus* and *Orthogeomys grandis*.

Guide to kinds		
1. *O. cuniculus*	6. *O. g. engelhardi*	12. *O. g. nelsoni*
2. *O. g. alleni*	7. *O. g. felipensis*	13. *O. g. pluto*
3. *O. g. alvarezi*	8. *O. g. grandis*	14. *O. g. pygacanthus*
4. *O. g. annexus*	9. *O. g. guerrerensis*	15. *O. g. scalops*
5. *O. g. carbo*	10. *O. g. huixtlae*	16. *O. g. soconuscensis*
	11. *O. g. latifrons*	17. *O. g. vulcani*

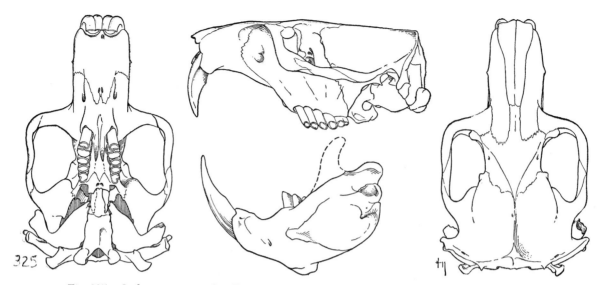

Fig. 325. *Orthogeomys grandis alleni*, ½ mi. E La Mira, 300 ft., Michoacán, No. 39807 K.U., ♀, X 1.

rero: type locality. Michoacán: ½ mi. E La Mira, 300 ft. (Genoways and Jones, 1969:755).

Orthogeomys grandis alvarezi Schaldach

1966. *Orthogeomys grandis alvarezi* Schaldach, Säugetrerkund. Mitteil., 14:292, October, type from ridge above Lachao, approx. 1700 m., *ca.* 40 km. N San Gabriel Mixtepec, Oaxaca.

MARGINAL RECORDS (Schaldach, 1966:292, 293).—Oaxaca: type locality; *ca. 30 km. N San Gabriel Mixtepec.*

Orthogeomys grandis annexus Nelson and Goldman

1933. *Orthogeomys grandis annexus* Nelson and Goldman, Proc. Biol. Soc. Washington, 46:195, October 26, type from Tuxtla Gutiérrez, 2600 ft., Chiapas. Known only from type locality.

Orthogeomys grandis carbo Goodwin

1956. *Orthogeomys grandis carbo* Goodwin, Amer. Mus. Novit., 1757:5, March 8, type from Escurano, 2500 ft., Cerro de San Pedro, 20 km. W Mixtequilla, Oaxaca.

MARGINAL RECORDS (Goodwin, 1969:139).—Oaxaca: Coatlán; *type locality;* Caja de Agua; *Morro Mazatlán;* Limón; Chontecomatlán; Santa Lucía.

Orthogeomys grandis engelhardi Felten

1957. *Orthogeomys grandis engelhardi* Felten, Senckenbergiana Biol., 38:151, April 1, type from Finca El Carmen, Volcán de San Vicente, El Salvador. Known only from type locality.

Orthogeomys grandis felipensis Nelson and Goldman

1930. *Orthogeomys grandis felipensis* Nelson and Goldman, Jour. Mamm., 11:157, May 9, type from Cerro San Felipe, 10 mi. N Oaxaca, Oaxaca.

MARGINAL RECORDS.—Puebla: 1 mi. SSW Tilapa, 3700 ft. (Genoways and Jones, 1969:755). Oaxaca: type locality; San Felipe del Agua (Goodwin, 1969:139); mts. 15 mi. SW Oaxaca.

Orthogeomys grandis grandis (Thomas)

1893. *Geomys grandis* Thomas, Ann. Mag. Nat. Hist., ser. 6, 12:270, October, type from Dueñas, Guatemala.
1895. *Orthogeomys grandis*, Merriam, N. Amer. Fauna, 8:175, January 31.

MARGINAL RECORDS.—Guatemala: San Lucas [= Tolimán]; type locality; near Pochuta; Finca Ciprés.

Orthogeomys grandis guerrerensis Nelson and Goldman

1930. *Orthogeomys grandis guerrerensis* Nelson and Goldman, Jour. Mamm., 11:158, May 9, type from El Limón, in valley of Río de las Balsas approx. 20 mi. NW La Unión, Guerrero.

MARGINAL RECORDS.—Guerrero: Río de las Balsas near Mexcala; Tlalistaquilla near Tlapa; type locality.

Orthogeomys grandis huixtlae Villa

1944. *Orthogeomys grandis huixtlae* Villa, Anal. Inst. Biol. Univ. Nac. Autó. México, 15(1):319, type from Finca Lubeca, 12 km. NE Huixtla, 850 m., Chiapas. Known only from type locality.

Orthogeomys grandis latifrons Merriam

1895. *Orthogeomys latifrons* Merriam, N. Amer. Fauna, 8:178, January 31, type from Guatemala; exact locality unknown, probably lowlands of southern part. Known only from Guatemala.
1930. *Orthogeomys grandis latifrons*, Nelson and Goldman, Jour. Mamm., 11:156, May 9.

Orthogeomys grandis nelsoni Merriam

1895. *Orthogeomys nelsoni* Merriam, N. Amer. Fauna, 8:176, January 31, type from Mt. Zempoaltepec, 8000 ft., Oaxaca.
1930. *Orthogeomys grandis nelsoni*, Nelson and Goldman, Jour. Mamm., 11:156, May 9.

MARGINAL RECORDS.—Oaxaca: Comaltepec; type locality; near Totontepec.

Orthogeomys grandis pluto Lawrence

1933. *Orthogeomys grandis pluto* Lawrence, Proc. New England Zool. Club, 13:66, May 8, type from Cerro Cantoral, N of Tegucigalpa, Honduras. Known only from type locality.

Orthogeomys grandis pygacanthus Dickey

1928. *Orthogeomys pygacanthus* Dickey, Proc. Biol. Soc. Washington, 41:9, February 1, type from Mt. Cacaguatique, 3500 ft., Dept. San Miguel, El Salvador. "No more than a subspecies of *grandis*" according to Burt and Stirton, Miscl. Publ. Mus Zool., Univ. Michigan, 117:53, September 22, 1961.

MARGINAL RECORDS.—El Salvador (Burt and Stirton, 1961:52): Los Esesmiles; type locality; Chilata; Cerro de los Naranjos.

Orthogeomys grandis scalops (Thomas)

1894. *Geomys scalops* Thomas, Ann. Mag. Nat. Hist., ser. 6, 13:437, May, type from Tehuantepec, Oaxaca.
1930. *Orthogeomys grandis scalops*, Nelson and Goldman, Jour. Mamm., 11:156, May 9.

MARGINAL RECORDS (Goodwin, 1969:139).—Oaxaca: Ixtepec; Zanatepec; Chahuites; type locality; Guigovéo.

Orthogeomys grandis soconuscensis Villa

1949. *Orthogeomys grandis soconuscensis* Villa, Anal. Inst. Biol. Univ. Nac. Autó. México, 19(1):267, April 8, type from Finca Esperanza, 710 m., 45 km. (by road) NW Huixtla, Chiapas.

MARGINAL RECORDS.—Chiapas: type locality; Finca Liquidámbar, 1210 m.

Orthogeomys grandis vulcani Nelson and Goldman

1931. *Orthogeomys grandis vulcani* Nelson and Goldman, Proc. Biol. Soc. Washington, 44:105, October 17, type from

Volcán Santa María, 9000 ft., Quezaltenango, Guatemala. Known only from type locality.

Subgenus **Heterogeomys** Merriam

List published by Nelson and Goldman, Proc. Biol. Soc. Washington, 42:147–152, March 30, 1929.

1895. *Heterogeomys* Merriam, N. Amer. Fauna, 8:179, January 31. Type, *Geomys hispidus* Le Conte.

External measurements: in males, 309–343, 74–101, 46–53; in females, 292–335, 75–98, 44–52. Dark brown. I1 unisulcate; sulcus wholly on inner side of median line and in some specimens on inner third; sulcus deep and abrupt; zygomata more widely spreading than in subgenus *Orthogeomys*. See additional characters in key to subgenera and Figs. 326 and 327.

KEY TO SPECIES OF SUBGENUS HETEROGEOMYS

1. Occurring at Xuchil, Veracruz; female with total length up to 361 and hind foot up to 54; pelage soft and wooly.
   O. lanius, p. 512
1'. Not occurring at Xuchil, Veracruz; female with total length less than 361 and hind foot shorter than 54; pelage harsh and coarse. . . . . *O. hispidus*, p. 509

Orthogeomys hispidus
Hispid Pocket Gopher

External measurements: in males, 309–343, 74–101, 46–53; in females, 292–335, 75–95, 44–52. Dark brown, pelage harsh and stiff; other characters as given for subgenus.

Orthogeomys hispidus cayoensis (Burt)

1937. *Heterogeomys hispidus cayoensis* Burt, Occas. Pap. Mus. Zool., Univ. Michigan, 365:1, December 16, type from Mountain Pine Ridge, 12 mi. S El Cayo, Belize.
1968. *Orthogeomys hispidus cayoensis*, Russell, Univ. Kansas Publ., Mus. Nat. Hist., 16:531, August 5.

MARGINAL RECORDS.—Belize: El Cayo; type locality.

Orthogeomys hispidus chiapensis (Nelson and Goldman)

1929. *Heterogeomys hispidus chiapensis* Nelson and Goldman, Proc. Biol. Soc. Washington, 42:151, March 30, type from Tenejapa, 16 mi. NE San Cristóbal, Chiapas.
1968. *Orthogeomys hispidus chiapensis*, Russell, Univ. Kansas Publ., Mus. Nat. Hist., 16:531, August 5.

MARGINAL RECORDS.—Tabasco: Montecristo. Guatemala: Chipoc; Guatemala City; Dueñas. Chiapas: type locality; Ocuilapa.

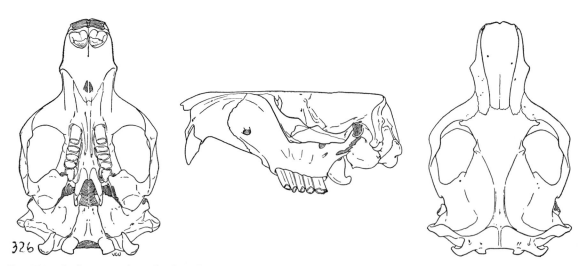

Fig. 326. *Orthogeomys hispidus hispidus*, Potrero Viejo, 7 km. W Potrero, 1700 ft., Veracruz, No. 19355 K.U., ♀, X 1.

Orthogeomys hispidus concavus (Nelson and Goldman)

1929. *Heterogeomys hispidus concavus* Nelson and Goldman, Proc. Biol. Soc. Washington, 42:148, March 30, type from Pinal de Amoles, Querétaro.
1968. *Orthogeomys hispidus concavus*, Russell, Univ. Kansas Publ., Mus. Nat. Hist., 16:531, August 5.

MARGINAL RECORDS.—San Luis Potosí: Valles; Tancanhuitz. Querétaro: type locality.

Orthogeomys hispidus hispidus (Le Conte)

1852. *G[eomys]. hispidus* Le Conte, Proc. Acad. Nat. Sci. Philadelphia, 6:158. Type locality, near Jalapa, Veracruz.
1968. *Orthogeomys hispidus hispidus*, Russell, Univ. Kansas Publ., Mus. Nat. Hist., 16:531, August 5.

MARGINAL RECORDS (Hall and Dalquest, 1963:277, unless otherwise noted).—Veracruz: 4 km. W Tlapacoyan, 1700 ft.; 5 km. N Jalapa, 4500 ft.; 2 km. N Motzorongo, 1500 ft.; *Montzorongo* (referred to *H. h.*

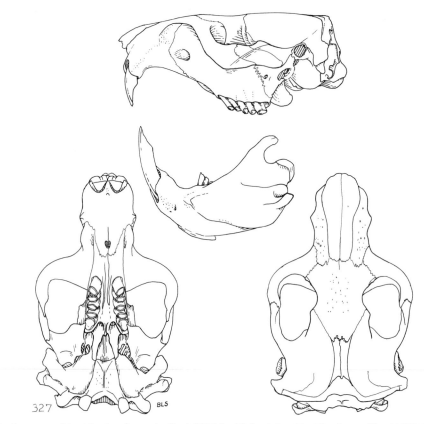

Fig. 327. *Orthogeomys hispidus hondurensis*, 8 mi. W Tela, 10 ft., Atlantida, Honduras, No. 12572 T.C.W.C., ♀, X 1.

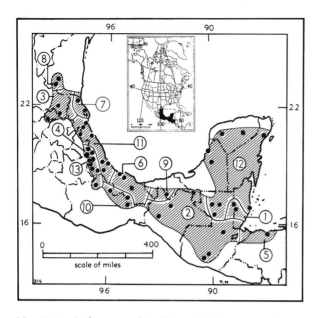

Map 314. *Orthogeomys hispidus* and *Orthogeomys lanius*

Guide to kinds
1. *O. h. cayoensis*
2. *O. h. chiapensis*
3. *O. h. concavus*
4. *O. h. hispidus*
5. *O. h. hondurensis*
6. *O. h. isthmicus*
7. *O. h. latirostris*
8. *O. h. negatus*
9. *O. h. teapensis*
10. *O. h. tehuantepecus*
11. *O. h. torridus*
12. *O. h. yucatanensis*
13. *O. lanius*

torridus by Goodwin, 1969:141). Oaxaca: Reyes [= Pápalo Santos Reyes] (Goodwin, 1969:141). Veracruz: *3 km. N Presidio, 1500 ft.;* Necostla [= Necoxtla]; *4 km. WNW Fortín, 3200 ft.;* Huatusco; Jico.

Orthogeomys hispidus hondurensis (Davis)

1966. *Heterogeomys hondurensis* Davis, Proc. Biol. Soc. Washington, 79:175, August 15, type from 8 mi. W Tela, Honduras. Known only from type locality. Here arranged as a subspecies of *O. hispidus* because an occasional specimen of *O. h. hispidus* (e.g., 1369 KU from Motzorongo, Veracruz) is morphologically intermediate between the specimens relied on by Davis, *op. cit. supra*, Fig. 1, in differentiating *O. hondurensis* from *O. hispidus*.

Orthogeomys hispidus isthmicus (Nelson and Goldman)

1929. *Heterogeomys hispidus isthmicus* Nelson and Goldman, Proc. Biol. Soc. Washington, 42:149, March 30, type from Jaltipan, Veracruz.

MARGINAL RECORDS.—Veracruz: 5 km. SE Lerdo de Tejada (Hall and Dalquest, 1963:278); *Tula* (*ibid.*); *3 km. E San Andrés Tuxtla, 1000 ft.;* Catemaco; 14 km. SW Coatzacoalcos, 100 ft.; Jesús Carranza, 250 ft.

Orthogeomys hispidus latirostris (Hall and Alvarez)

1961. *Heterogeomys hispidus latirostris* Hall and Alvarez, Anal. Escuela Nac. Cienc. Biol., Distrito Federal, México, 10:121, December 20, type from Hda. Tamiahua, Cobo Rojo, Veracruz.

1968. *Orthogeomys hispidus latirostris,* Russell, Univ. Kansas Publ., Mus. Nat. Hist., 16:531, August 5.

MARGINAL RECORDS (Hall and Dalquest, 1963:278).—Veracruz: Hda. El Carocol, Tamós; type locality.

Orthogeomys hispidus negatus (Goodwin)

1953. *Heterogeomys hispidus negatus* Goodwin, Amer. Mus. Novit., 1620:1, May 4, type from Gómez Feras [= Farías], 1300 ft., about 45 mi. S Ciudad Victoria, 10 mi. W Pan American Highway, Tamaulipas.
1968. *Orthogeomys hispidus negatus,* Russell, Univ. Kansas Publ., Mus. Nat. Hist., 16:531, August 5.

MARGINAL RECORDS (Alvarez, 1963:428).—Tamaulipas: *5 km. W El Carrizo; 2 km. W El Carrizo;* Ejido Santa Isabel; *Rancho Pano Ayuctle;* type locality.

Orthogeomys hispidus teapensis (Goldman)

1939. *Heterogeomys hispidus teapensis* Goldman, Jour. Washington Acad. Sci., 29:176, April 15, type from Teapa, Tabasco. Known only from type locality.

Orthogeomys hispidus tehuantepecus (Goldman)

1939. *Heterogeomys hispidus tehuantepecus* Goldman, Jour. Washington Acad. Sci., 29:175, April 15, type from mountains 12 mi. NW Santo Domingo and about 60 mi. N Tehuantepec, 1600 ft., Oaxaca.
1968. *Orthogeomys hispidus tehuantepecus,* Russell, Univ. Kansas Publ., Mus. Nat. Hist., 16:531, August 5.

MARGINAL RECORDS.—Oaxaca: La Gloria; type locality.

Orthogeomys hispidus torridus (Merriam)

1895. *Heterogeomys torridus* Merriam, N. Amer. Fauna, 8:183, January 31, type from Chichicaxtle, Veracruz.
1968. *Orthogeomys hispidus torridus,* Russell, Univ. Kansas Publ., Mus. Nat. Hist., 16:531, August 5.

MARGINAL RECORDS.—Veracruz: 4 km. N Tuxpan (Hall and Dalquest, 1963:278); 3 km. W Gutiérrez Zamora, 300 ft.; Boca del Río, 10 ft.; 15 km. W Piedras Negras, 400 ft.; type locality. Puebla: Rancho El Ajenjibre (Warner and Beer, 1957:17). Veracruz: 12½ mi. N Tihuatlán, 300 ft. (Hall and Dalquest, 1963:278).

Orthogeomys hispidus yucatanensis (Nelson and Goldman)

1929. *Heterogeomys hispidus yucatanensis* Nelson and Goldman, Proc. Biol. Soc. Washington, 42:150, March 30, type from Campeche, Campeche.
1968. *Orthogeomys hispidus yucatanensis,* Russell, Univ. Kansas Publ., Mus. Nat. Hist., 16:531, August 5.

MARGINAL RECORDS.—Yucatán: 2 km. E Chichén-Itzá (Birney, *et al.,* 1974:12). Quintana Roo: 6½ km. NE Playa del Carmen (*op. cit.:* 13). Belize: Stann Creek Valley. Guatemala: Uaxactún; Libertad; Chuntuqui. Campeche: Apazote; type locality. Yucatán: Calcehtok.

Orthogeomys lanius (Elliot)
Big Pocket Gopher

1905. *Heterogeomys lanius* Elliot, Proc. Biol. Soc. Washington, 18:235, December 9, type from Xuchil, Veracruz. Known only from type locality.
1968. *Orthogeomys lanius*, Russell, Univ. Kansas Publ., Mus. Nat. Hist., 16:531, August 5.

External measurements of the type, a female,: 361, 90, 54. In original description, distinguished from *H. h. hispidus* by larger size and soft wooly coat instead of harsh pelage. A re-examination of the specific versus subspecific status of *H. lanius* and *H. hispidus* is desirable. See Map 314.

Subgenus Macrogeomys Merriam
Central American Pocket Gophers

1895. *Macrogeomys* Merriam, N. Amer. Fauna, 8:185, January 31. Type, *Geomys heterodus* Peters.

External measurements: in males, 177–392, 70–120, 36–53; in females, 170–388, 70–128, 35–52.5. Dark brown to blackish; one species having white transverse band on lumbar region, some other species with white on rump or top of head; pelage wooly in some individuals, harsh in others, but hair never so sparse as in subgenus *Orthogeomys;* I1 unisulcate; sulcus on inner third of tooth, narrow and deep; surface of incisor flat on both sides of sulcus. See additional characters in key to subgenera.

KEY TO SPECIES OF SUBGENUS MACROGEOMYS

1. Occurring in extreme eastern Panamá .
         O. dariensis,   p. 513
1'. Not occurring in extreme eastern Panamá.
  2. Occurring in Nicaragua .
         O. matagalpae,   p. 514
  2'. Not occurring in Nicaragua.
    3. Pelage relatively soft; underparts clearly paler than upper parts
         O. heterodus,   p. 512
    3'. Pelage harsh and wooly; underparts not (or only slightly) paler than upper parts.
      4. Size large (320–390); color uniformly blackish (dorsal white markings absent). . *O. cavator*,   p. 512
      4'. Size small (less than 335); dorsal white markings present.
        5. Total length more than 210; white markings present on head and/or rump.
         O. cherriei,   p. 513

5'. Total length less than 210; white markings present on lumbar region and less distinct across abdomen.
         O. underwoodi,   p. 513

Orthogeomys heterodus
Variable Pocket Gopher

External measurements: in males, 335–392, 72–95, 48–52; in females, 315–380, 70–97, 47–51. Blackish, soft-haired, underparts paler than upper parts.

Orthogeomys heterodus cartagoensis (Goodwin)

1943. *Macrogeomys heterodus cartagoensis* Goodwin, Amer. Mus. Novit., 1227:2, April 22, type from Paso Ancho, Prov. Cartago, Costa Rica.
1968. *Orthogeomys heterodus cartagoensis*, Russell, Univ. Kansas Publ., Mus. Nat. Hist., 16:532, August 5.

MARGINAL RECORDS.—Costa Rica: Rancho Redondo, Volcán Irazú; El Sauce Peralta; Pozo Ancho; Cervantes; San Ramón Tres Ríos.

Orthogeomys heterodus dolichocephalus (Merriam)

1895. *Macrogeomys dolichocephalus* Merriam, N. Amer. Fauna, 8:189, January 31, type from San José, Costa Rica (probably Zarcero or Palmira according to Goodwin, Bull. Amer. Mus. Nat. Hist., 87:377, December 31, 1946).
1968. *Orthogeomys heterodus dolichocephalus*, Russell, Univ. Kansas Publ., Mus. Nat. Hist., 16:532, August 5.

MARGINAL RECORDS.—Costa Rica: Lajas Villa Quesada; Tapesco [= Tapezco].

Orthogeomys heterodus heterodus (Peters)

1865. *Geomys heterodus* Peters, Monatsb. preuss. Akad. Wiss., Berlin, p. 177, type from Costa Rica; exact locality unknown.
1968. *Orthogeomys heterodus heterodus*, Russell, Univ. Kansas Publ., Mus. Nat. Hist., 16:532, August 5.

MARGINAL RECORDS.—Costa Rica: Escazú; Escazú Heights; Sabanilla.

Orthogeomys cavator
Chiriquí Pocket Gopher

External measurements: in males, 370–390, 100–120, 52–53; in females, 320–380, 104–110, 48–51. Dark seal brown, almost blackish; uniformly colored; pelage harsh.

Orthogeomys cavator cavator (Bangs)

1902. *Macrogeomys cavator* Bangs, Bull. Mus. Comp. Zool., 39:42, April, type from Boquete, 4800 ft., Chiriquí, Panamá.

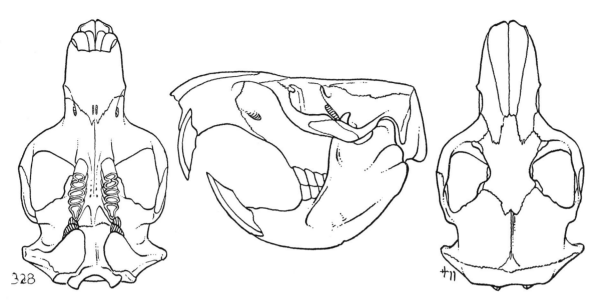

Fig. 328. *Orthogeomys heterodus dolichocephalus*, San José, Costa Rica, No. 36295 U.S.N.M., ♂, X 1. [After Merriam, N. Amer. Fauna, No. 8, Fig. 49, p. 104 and pl. 5, p. 231, January 31, 1895.]

MARGINAL RECORDS.—Panamá: Cerro Punta (Handley, 1966:778); type locality.

Orthogeomys cavator nigrescens (Goodwin)

1943. *Macrogeomys cavator nigrescens* Goodwin, Amer. Mus. Novit., 1227:3, April 22, type from El Muñeco (Río Navarro), 10 mi. S Cartago, Prov. Cartago, 4000 ft., Costa Rica. Known only from type locality.
1968. *Orthogeomys cavator nigrescens*, Russell, Univ. Kansas Publ., Mus. Nat. Hist., 16:532, August 5.

Orthogeomys cavator pansa (Bangs)

1902. *Macrogeomys pansa* Bangs, Bull. Mus. Comp. Zool., 39:44, April, type from Bogava [= Bugaba], 600 ft., Chiriquí, Panamá. Known only from type locality.
1968. *Orthogeomys cavator pansa*, Russell, Univ. Kansas Publ., Mus. Nat. Hist., 16:532, August 5.

Orthogeomys dariensis (Goldman)
Darién Pocket Gopher

1912. *Macrogeomys dariensis* Goldman, Smiths. Miscl. Coll., 60(2):8, September 20, type from Cana, 2000 ft., mts. of eastern Panamá.
1968. *Orthogeomys dariensis*, Russell, Univ. Kansas Publ., Mus. Nat. Hist., 16:532, August 5.

External measurements: in males, 358–401, 122–135, 51–53.5; in females, 348–388, 118–128, 48–52.5. Dull brown or black. Resembles *M. c. cavator* but color dull brown or black instead of rich seal brown and skull narrower posteriorly with low and nearly straight (or slightly convex), instead of high and sinuous, lambdoidal crest.

MARGINAL RECORDS.—Panamá: Tacarcuna; Tapalisa; type locality; Boca de Cupe. See Map 316.

Orthogeomys underwoodi (Osgood)
Underwood's Pocket Gopher

1931. *Macrogeomys underwoodi* Osgood, Field Mus. Nat. Hist., Publ. 295, Zool. Ser., 18(5):143, August 3, type from Alto de Jabillo Pirris, between San Gerónimo and Pozo Azul, western Costa Rica.
1968. *Orthogeomys underwoodi*, Russell, Univ. Kansas Publ., Mus. Nat. Hist., 16:532, August 5.

External measurements: in males, 277–299, 85–100, 36–38; in females, 280, 92, 35. Dark brown, almost blackish, on upper parts but paler below; transverse band, 13–22 mm. wide, of all white hair on lumbar region and extending less distinctly across abdomen. Resembles *M. cherriei* but lacks white on head, and rostrum narrower (11–13).

MARGINAL RECORDS.—Costa Rica: San Gerónimo; type locality.

Orthogeomys cherriei
Cherrie's Pocket Gopher

External measurements: in male, 323, 88, 41; in females, 263–280, 70–75, 38.5–44. Mummy brown to blackish brown with white markings on head.

Orthogeomys cherriei carlosensis (Goodwin)

1943. *Macrogeomys cherriei carlosensis* Goodwin, Amer. Mus. Novit., 1227:3, April 22, type from Cataratos, San Carlos, Alajuela, Costa Rica.
1968. *Orthogeomys cherriei carlosensis*, Russell, Univ. Kansas Publ., Mus. Nat. Hist., 16:532, August 5.

MARGINAL RECORDS.—Costa Rica: type locality; Villa Quesada.

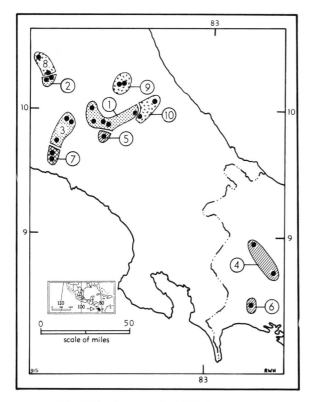

Map 315.   Four species of *Orthogeomys*.

1. *O. heterodus cartagoensis*
2. *O. heterodus dolichocephalus*
3. *O. heterodus heterodus*
4. *O. cavator cavator*
5. *O. cavator nigrescens*
6. *O. cavator pansa*
7. *O. underwoodi*
8. *O. cherriei carlosensis*
9. *O. cherriei cherriei*
10. *O. cherriei costaricensis*

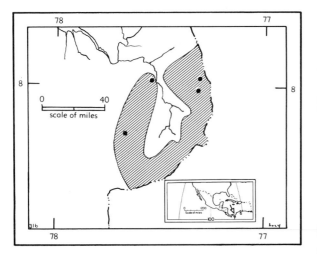

Map 316.   *Orthogeomys dariensis.*

Orthogeomys cherriei cherriei (J. A. Allen)

1893. *Geomys cherriei* J. A. Allen, Bull. Amer. Mus. Nat. Hist., 5:337, December 16, type from Santa Clara, Costa Rica.
1968. *Orthogeomys cherriei cherriei*, Russell, Univ. Kansas Publ., Mus. Nat. Hist., 16:532, August 5.

MARGINAL RECORDS.—Costa Rica: type locality; Jiménez.

Orthogeomys cherriei costaricensis (Merriam)

1895. *Macrogeomys costaricensis* Merriam, N. Amer. Fauna, 8:192, January 31, type from Pacuare, Costa Rica.
1968. *Orthogeomys cherriei costaricensis*, Russell, Univ. Kansas Publ., Mus. Nat. Hist., 16:532, August 5.

MARGINAL RECORDS.—Costa Rica: type locality; Santa Teresa Peralta.

Orthogeomys matagalpae (J. A. Allen)
Nicaraguan Pocket Gopher

1910. *Macrogeomys matagalpae* J. A. Allen, Bull. Amer. Mus. Nat. Hist., 28:97, April 30, type from Peña Blanca, Matagalpa, Nicaragua.
1968. *Orthogeomys matagalpae*, Russell, Univ. Kansas Publ., Mus. Nat. Hist., 16:532, August 5.

External measurements of male: 320, 80, 40. Dark brown with white crown-patch. Resembles *M. c. cherriei* but darker, smaller and with narrower (14 in adult male) rostrum.

MARGINAL RECORDS.—Nicaragua: Matagalpa; type locality.

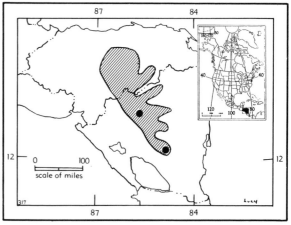

Map 317.   *Orthogeomys matagalpae.*

Genus **Pappogeomys** Merriam

Revised by Russell, Univ. Kansas Publ., Mus. Nat. Hist., 16:581–776, 10 figs., August 5, 1968.

1895. *Pappogeomys* Merriam, N. Amer. Fauna, 8:145, January 31. Type, *Geomys bulleri* Thomas.

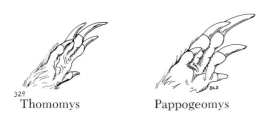

Thomomys   Pappogeomys

Fig. 329. Ventrolateral view of right forefoot: *Thomomys talpoides oquirrhensis*, Settlement Canyon, Tooele Co., Utah, No. 14737 K.U., ♂ ad., X 1; *Pappogeomys bulleri albinasus*, 2 mi. N, ½ mi. W Guadalajara, Jalisco, No. 31018 K.U., ♂ ad., X 1.

1895. *Cratogeomys* Merriam, N. Amer. Fauna, 8:150, January 31. Type, *Geomys merriami* Thomas.
1895. *Platygeomys* Merriam, N. Amer. Fauna, 8:162, January 31. Type, *Geomys gymnurus* Merriam. Regarded as inseparable from *Cratogeomys* by Hooper, Jour. Mamm., 27:397–399, November 25, 1946.

Outer surface of upper incisor having single median sulcus slightly displaced to inner side of tooth; posterior surface of P4 lacking enamel (exceptions are two individuals of subgenus *Pappogeomys*); basitemporal fossa deep; masseteric ridge on lower jaw well developed; pelage soft and dense in living species except in *P. fumosus* of Colima, in which the hairs are coarse and hispid.

KEY TO SUBGENERA OF PAPPOGEOMYS

1. Condylobasal length 35.0–45.5 in males and 31.3–42.2 in females; sagittal crest lacking; anterior angles of zygomata without lateral platelike expansions (see Fig. 330). . . . . *Pappogeomys*, p. 515
1'. Condylobasal length 43.7–74.5 in males and 38.9–65.9 in females; sagittal crest present; anterior angles of zygomata enlarged into platelike expansions (see Fig. 331). . . . . *Cratogeomys*, p. 517

Subgenus **Pappogeomys** Merriam

1895. *Pappogeomys* Merriam, N. Amer. Fauna, 8:145, January 31. Type, *Geomys bulleri* Thomas.

Head and body 150–188 in males and 142–183 in females; tail 63–88, 53–82; hind foot 28–34, 28–35. Sexual dimorphism less than in subgenus *Cratogeomys*. Claws on forefeet larger in relation to size of animal than in subgenus *Cratogeomys* and about as in *Geomys bursarius*.

Pelage long (approx. 10 mm.) and soft and covering all of body, except in *P. bulleri burti* in which pelage is a half shorter and three-fourths sparser than in other taxa of subgenus. Additional

characters listed above in key to subgenera of genus *Pappogeomys*.

KEY TO SPECIES OF SUBGENUS PAPPOGEOMYS

1. Enamel plate on posterior surface of M1 thin, usually extending across entire posterior wall (sometimes reduced, rarely absent); nasals emarginate posteriorly, forming V-shaped notch; nasal patch white or pale buffy, often absent. . . . *P. bulleri*, p. 515
1'. Enamel plate on posterior surface of M1 thick, restricted to lingual fourth of posterior wall; nasals truncate posteriorly; nasal patch bright ochraceous or buffy. *P. alcorni*, p. 516

Pappogeomys bulleri
Buller's Pocket Gopher

Characters as stated above in account of subgenus. Differs from *P. alcorni* as set forth above in key to species, and as described beyond in account of *alcorni*. Pelage of dorsum bicolored, basally pale gray to dark gray (black in melanistic individuals), apically black to ochraceous, tawny, and cinnamon depending on subspecies.

Pappogeomys bulleri albinasus Merriam

1895. *Pappogeomys albinasus* Merriam, N. Amer. Fauna, 8:149, January 31, type from Atemajac, a suburb of Guadalajara, Jalisco.
1939. *Pappogeomys bulleri albinasus*, Goldman, Jour. Mamm., 20:94, February 15.

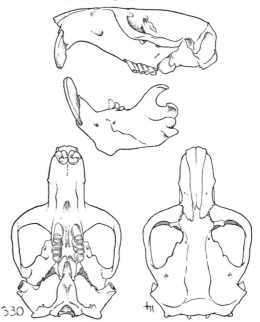

Fig. 330. *Pappogeomys bulleri albinasus*, 4 mi. W Guadalajara, Jalisco, No. 31031 K.U., ♂, X 1.

MARGINAL RECORDS (Russell, 1968:602).—
Jalisco: 4 mi. N, 13 mi. W Guadalajara; 2 mi. N, ½ mi. W
Guadalajara; *Ciudad Granja;* 10 mi. S, 8 mi. W
Guadalajara.

Pappogeomys bulleri amecensis Goldman

1939. *Pappogeomys bulleri amecensis* Goldman, Jour.
Mamm., 20:97, February 15, type from mountains near
Ameca, 6500 ft., Jalisco.

MARGINAL RECORDS (Russell, 1968:604).—
Jalisco: 5 mi. NNW Ameca; type locality; 13 mi. WSW
Ameca; *5 mi. NW Ameca.*

Pappogeomys bulleri bulleri (Thomas)

1892. *Geomys bulleri* Thomas, Ann. Mag. Nat. Hist., ser. 6,
10:196, August, type from near Talpa, W slope Sierra de
Mascota, 8500 (probably about 5000) ft., Jalisco.
1895. *Pappogeomys bulleri*, Merriam, N. Amer. Fauna, 8:159,
January 31.
1895. *Geomys nelsoni* Merriam, Proc. Biol. Soc. Washington,
7:164, September 29, type from N slope Sierra Nevada de
Colima, 6500 ft., Colima.
1939. *Pappogeomys bulleri flammeus* Goldman, Jour.
Mamm., 20:95, February 15, type from Milpillas, 5 mi. SW
San Sebastián, Jalisco.
1939. *Pappogeomys bulleri lagunensis* Goldman, Jour.
Mamm., 20:96, February 15, type from La Laguna, 6500 ft.,
Sierra de Juancatlán, Jalisco.

MARGINAL RECORDS (Russell, 1968:608, unless
otherwise noted).—Jalisco: Milpillas, 5 mi. SW San
Sebastián; La Laguna, Sierra de Juanacatlán; 4 mi. E
Atemajac de Brizuela, 8000 ft. (Genoways and Jones,
1969:750); 7 mi. S Tapalpa, 6800 ft. (*ibid.*); N slope
Sierra Nevada de Colima; *Volcán de Nieve, N slope
Sierra Nevada de Colima;* Sierra de Autlán; 2 mi. S La
Cuesta (Genoways and Jones, 1969:749, 750).

Pappogeomys bulleri burti Goldman

1939. *Pappogeomys bulleri burti* Goldman, Jour. Mamm.,
20:97, February 15, type from Tenacatita Bay, southwestern
coast of Jalisco.

MARGINAL RECORDS (Russell, 1968:610, unless
otherwise noted).—Jalisco: 10 mi. WSW La Huerta
(Genoways and Jones, 1969:751). Colima: 1 mi. S, 4 mi.
W Santiago; 34 mi. SE Manzanillo (Ingles, 1959:391); 3
mi. NE Cuyutlán. Jalisco: 10 mi. NNW Barrada
Navidad.

Pappogeomys bulleri infuscus Russell

1968. *Pappogeomys bulleri infuscus* Russell, Univ. Kansas
Publ., Mus. Nat. Hist., 16:610, August 5, type from Cerro
Tequila, 10,000 ft., 7 mi. S, 2 mi. W Tequila, Jalisco.

MARGINAL RECORDS (Russell, 1968:611).—
Jalisco: type locality; *7 mi. SSW Tequila.*

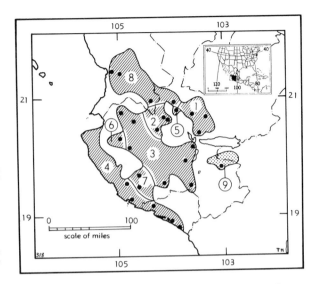

Map 318. *Pappogeomys bulleri* and *Pappogeomys alcorni.*

1. *P. b. albinasus*   5. *P. b. infuscus*
2. *P. b. amecensis*   6. *P. b. lutulentus*
3. *P. b. bulleri*   7. *P. b. melanurus*
4. *P. b. burti*   8. *P. b. nayaritensis*
              9. *P. alcorni*

Pappogeomys bulleri lutulentus Russell

1968. *Pappogeomys bulleri lutulentus* Russell, Univ. Kansas
Publ., Mus. Nat. Hist., 16:612, August 5, type from Sierra de
Cuale, 7300 ft., 9 km. N El Teosinte (= Desmoronado),
Jalisco. Known only from type locality.

Pappogeomys bulleri melanurus Genoways and Jones

1969. *Pappogeomys bulleri melanurus* Genoways and Jones,
Jour. Mamm., 50:748, November 28, type from 7½ mi. SE
Tecomate, 1500 ft., Jalisco.

MARGINAL RECORDS (Genoways and Jones,
1969:749).—Jalisco: 5 mi. S Purificación, *ca.* 1500 ft.;
type locality.

Pappogeomys bulleri nayaritensis Goldman

1939. *Pappogeomys bulleri nayaritensis* Goldman, Jour.
Mamm., 20:94, February 15, type from about 10 mi. S
Tepic, 5000 ft., Nayarit.

MARGINAL RECORDS (Russell, 1968:615).—
Nayarit: 2 mi. WNW Jalocotán; type locality. Jalisco: 3
mi. NE Magdalena. Nayarit: 6 mi. S Ixtlán del Río.

Pappogeomys alcorni Russell
Alcorn's Pocket Gopher

1957. *Pappogeomys alcorni* Russell, Univ. Kansas Publ., Mus.
Nat. Hist., 9:359, January 21, type from 4 mi. W Mazamitla,
6600 ft., Jalisco.

External measurements of 61328 KU, thought
by describer to be a male, are: 243, 78, and 30.

Corresponding measurements of two adult females from the type locality are: 210, 210; 61, 63; 29, 28. Condylobasal length of skull, 38.0, 36.9. Pelage Plumbeous basally and Orange-Cinnamon apically. Nasal patch Cinnamon Buff or Pinkish Buff instead of whitish or less slightly tinged with buff; anterior palatine foramina small and rounded instead of slitlike. On posterior wall of M1, enamel confined to inner fourth of wall whereas enamel in *P. bulleri* ordinarily covers entire posterior wall.

In the original description of *P. alcorni* the greater extent of enamel on the posterior wall of M1 of *P. bulleri* was stated to be diagnostic of every individual of *P. bulleri*. In Russell's (1968:596) later account of *P. bulleri* some specimens of *P. bulleri* are said to differ in this feature only slightly, if at all, from *P. alcorni*. If more specimens of *P. alcorni* are obtained, a study of them would be warranted to learn if the taxon differs specifically or only subspecifically from *P. bulleri*.

MARGINAL RECORDS (Russell, 1968:617).— Jalisco: type locality; *3 mi. WSW Mazamitla.*

Subgenus **Cratogeomys** Merriam

1895. *Cratogeomys* Merriam, N. Amer. Fauna, 8:150, January 31. Type, *Geomys merriami* Thomas.

Head and body 161–285 in males and 144–279 in females; tail 67–119, 50–105; hind foot 30–54, 27–54. Sexual dimorphism more than in subgenus *Pappogeomys*. Claws on forefoot smaller in relation to size of animal than in subgenus *Pappogeomys*.

Pelage typically long (approx. 8 mm.) and soft and covering entire body, except in *P. fumosus*, which has sparse, hispid, bristly pelage. Additional characters listed in key to subgenera of genus *Pappogeomys*.

KEY TO SPECIES OF SUBGENUS CRATOGEOMYS

1. Dorsal outline of lambdoidal crest convex posteriorly, never sinuous. Paroccipital processes small, not enlarged into flangelike knobs. Angular processes short, breadth across angular processes less than greatest length of mandible.
   2. Occlusal surface of M3 quadriform, posterior loph not elongated; squamosals not overlapping parietals; basioccipital parallel-sided or hourglass-shaped. . . . . *P. castanops,* p. 517
   2'. Occlusal surface of M3 obcordate, posterior loph elongated and displaced toward labial side; squamosals expanded medially, with increasing age progressively overlapping parietals and completely covering parietals in old adults; basioccipital strongly wedge-shaped, its anterior end distinctly narrower than posterior end. *P. merriami,* p. 521
1'. Dorsal outline of lambdoidal crest sinuous. Paroccipital processes enlarged into flangelike knobs. Angular processes long, breadth across angular processes more than greatest length of mandible.
   3. Females: Condylobasal length 45.6–46.7; rostrum relatively short (37.1–38.1% of condylobasal length) and broad (61.8–64.5% of length); adult male unknown. . . . . *P. neglectus,* p. 523
   3'. Condylobasal length 50.4–65.9 in females and 56.5–71.3 in males; rostrum relatively long (39.4–46.4% of condylobasal length in females, except for 38.2–38.8% in females of *P. t. brevirostris;* 39.4–45.5% of condylobasal length in males) and narrow (47.6–61.3% of length in females and 47.8–61.6% in males).
      4. Pelage harsh and bristly. Occurring only on Pacific Coastal Plain. . . . . . . *P. fumosus,* p. 523
      4'. Pelage soft and lax. Not occurring on Pacific Coastal Plain.
         5. Skull larger (condylobasal length 66.7–71.3 in males and 60.1–64.6 in females) and broader (squamosal breadth 43.1–52.3 in males and 41.6–47.8 in females).
                               P. gymnurus, p. 524
         5'. Skull smaller (condylobasal length 56.5–65.0 in males and 50.4–59.1 in females) and narrower (squamosal breadth 36.7–41.1 in males and 32.0–41.6 in females).
           6. Zygomata expanded laterally, zygomatic breadth 41.6–43.2 in females (adult male unknown). . . . . *P. zinseri,* p. 524
           6'. Zygomata not expanded laterally, zygomatic breadth 31.3–39.4 in females and 37.6–45.0 in males. . . . . *P. tylorhinus,* p. 523

castanops species-group
Pappogeomys castanops
Yellow-faced Pocket Gopher

Head and body, 161–255 in males and 144–212 in females; tail 67–105, 50–105; hind foot 30–42, 27–42. Color of pelage varies in tones of pale yellowish to dark reddish-brown. Dark-tipped hairs mixed in on back and top of head. Underparts varying from whitish to bright ochraceous-buff. Skull without strong platycephalic specializations; breadth across zygomata exceeding breadth across squamosals; posterior enamel plate of M1 and M2 absent; M3 quadriform or obcordate, lat-

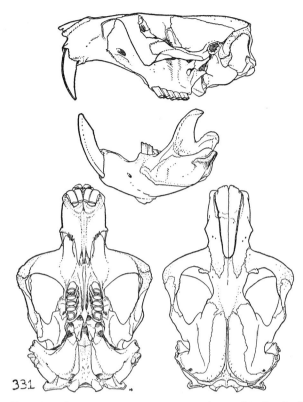

331

Fig. 331. *Pappogeomys castanops perplanus*, Big Bend of Rio Grande, 2000 ft., Brewster Co., Texas, No. 80360 M.V.Z., ♂, X 1.

eral plates reduced or absent. Additional characters listed in key to species of subgenus *Cratogeomys*.

Morphologically, *P. castanops* most closely resembles *P. merriami*. The relationship of the two is indicated by uniting them in the same species group (*castanops* group). *P. castanops* differs from its larger, paler relative in lack of cranial and dental specialization and less massive skull.

The 25 subspecies of *P. castanops* have been divided into two well-defined groups, the *excelsus* group and the *subnubilus* group. These groups are joined in at least one place by interbreeding populations (involving subspecies *P. c. goldmani*). Even though in most places where their geographic ranges meet or interdigitate, or even overlap, the two do not crossbreed, and therefore behave toward each other as species. The *subnubilus* subspecies-group consists of *P. c. consitus, elibatus, parviceps, perexiguus, peridoneus, planifrons, rubellus, subnubilus,* and *surculus*. Marginal records for each of these subspecies are shown on Map 319 by means of open circles. Marginal records for the 16 other subspecies, all included in the *excelsus* subspecies-

group, are shown on Map 319 by means of solid black circles.

The 25 named kinds of *Pappogeomys* here arranged as subspecies of the species *P. castanops* have been described in such detail by Russell (1968:581–776; 1969:337–371) as concerns sympatric and allopatric occurrence and intergradation that *P. castanops* now is one of the most attractive species for any investigator wishing to determine degree of evolution as correlated with different kinds of soils.

Pappogeomys castanops angusticeps (Nelson and Goldman)

1934. *Cratogeomys castanops angusticeps* Nelson and Goldman, Proc. Biol. Soc. Washington, 47:139, June 13, type from Eagle Pass, Maverick Co., Texas.
1968. *Pappogeomys castanops angusticeps,* Russell, Univ. Kansas Publ., Mus. Nat. Hist., 16:630, August 5.

MARGINAL RECORDS (Russell, 1968:632).— Texas: Fort Lancaster; *5 mi. S Howard Springs;* 20 mi. E Juno; type locality; 8 mi. S Langtry; Samuels; Black Gap, 50 mi. SSE Marathon; 3 mi. W Dryden; 15 mi. S Sheffield.

Pappogeomys castanops bullatus (Russell and Baker)

1955. *Cratogeomys castanops bullatus* Russell and Baker, Univ. Kansas Publ., Mus. Nat. Hist., 7:597, March 15, type from 2 mi. S, 6½ mi. E Nava, 810 ft., Coahuila.
1968. *Pappogeomys castanops bullatus,* Russell, Univ. Kansas Publ., Mus. Nat. Hist., 16:632, August 5.

MARGINAL RECORDS (Russell, 1968:635).— Coahuila: 2 mi. S, 12 mi. E Nava; *3 mi. S, 12 mi. E Nava.* Nuevo León: *1 mi. N Vallecillo;* Vallecillo; 3 mi. N Lampazos. Coahuila: 9 mi. S, 11 mi. E Sabinas; 8 mi. S, 8 mi. E Hda. La Mariposa; 10 mi. E Hda. La Mariposa; 29 mi. N, 6 mi. E Sabinas; *2 mi. S, 6½ mi. E Nava.*

Pappogeomys castanops castanops (Baird)

1852. *Pseudostoma castanops* Baird, *in* Rept. Stanbury's Expl. Surv. . . . Great Salt Lake of Utah . . . , App. C, p. 313, June, type from "prairie road to Bent's Fort," near present town of Las Animas, Bent Co., Colorado.
1968. *Pappogeomys castanops castanops,* Russell, Univ. Kansas Publ., Mus. Nat. Hist., 16:635, August 5.

MARGINAL RECORDS.—Kansas (Birney, *et al.,* 1971:374): 14 mi. S, 6 mi. E Dighton; *15 mi. S, 7½ mi. E Dighton;* 4 mi. S, ½ mi. W Jetmore; *9 mi. S Jetmore; 8½ mi. N, 6 mi. E Dodge City;* 7½ mi. N, 6 mi. E Dodge City; *5 mi. N, 2½ mi. E Dodge City;* 5 mi. N, 1 mi. E Dodge City; 10 mi. N, 4½ mi. E Cimarron. Colorado (Armstrong, 1972:168): Monon; Furnace [Furnish] Canyon. New Mexico: Clayton; 2⅕ mi. E Gladstone (Best, 1973:1317, as *P. castanops* only); Chico Springs. Colorado (Armstrong, 1972:167, 168): JJ Ranch, Higbee,

18 mi. S La Junta; Arkansas River, 26 mi. below Cañon City; 3 mi. W Pueblo on Beulah Road; Olney Springs; 4 mi. W Rocky Ford; near reservoirs several miles N of Lamar. Kansas: 5½ mi. N Syracuse (Birney, et al., 1971:374).

Pappogeomys castanops clarkii (Baird)

1855. *Geomys clarkii* Baird, Proc. Acad. Nat. Sci. Philadelphia, 7:322, April, type from Presidio del Norte, on the Río Grande, at or near present town of Ojinaja, Chihuahua.
1968. *Pappogeomys castanops clarkii*, Russell, Univ. Kansas Publ., Mus. Nat. Hist., 16:638, August 5.
1934. *Cratogeomys castanops convexus* Nelson and Goldman, Proc. Biol. Soc. Washington, 47:142, June 13, type from 7 mi. E Las Vacas (= Villa Acuña), Coahuila.

MARGINAL RECORDS (Russell, 1968:641).— Coahuila: Río Grande, 17 mi. S Dryden, Texas; Río Grande, opposite Samuels, Texas; Villa Acuña; 11 mi. W Hda. San Miguel; Cañón del Conchino, 16 mi. N, 21 mi. E Piedro Blanco. Texas: *1 mi. SW Boquillas;* Boquillas; Johnson's Ranch, on Big Bend of Río Grande; Castalon, Big Bend National Park; *mouth Sanata Elena Canyon, Big Bend National Park;* Lajitas. Chihuahua: *1½ mi. SE Ojinaga;* Ojinaga; *1½ mi. WNW Ojinaga.* Texas: *3 mi. NW Presidio.*

Pappogeomys castanops consitus (Nelson and Goldman)

1934. *Cratogeomys castanops consitus* Nelson and Goldman, Proc. Biol. Soc. Washington, 47:140, June 13, type from Gallego, 5500 ft., Chihuahua.
1968. *Pappogeomys castanops consitus*, Russell, Univ. Kansas Publ., Mus. Nat. Hist., 16:669, August 5.

MARGINAL RECORDS (Anderson, 1972:295).— Chihuahua: Samalyuca, 4200 ft.; 3½ mi. ESE Los Lamentos, 1420 m.; 1 mi. S Ojinaga; Santa Rosalía, 4100 ft.; 12 km. W Estación Encinillas, 5000 ft.; *Station Arados; type locality;* 3 mi. WNW El Sueco.

Pappogeomys castanops elibatus Russell

1968. *Pappogeomys castanops elibatus* Russell, Univ. Kansas Publ., Mus. Nat. Hist., 16:672, August 5, type from 12 mi. W San Antonio de las Alazanos, about 7500 ft., Coahuila.

MARGINAL RECORDS (Russell, 1968:673).— Coahuila: 4 mi. S, 6 mi. E Saltillo; *7 mi. S, 4 mi. E Bella Unión;* 2 mi. N, 2 mi. E San Antonio de las Alazanos; *type locality; 12 mi. S, 2 mi. E Artegia.*

Pappogeomys castanops excelsus (Nelson and Goldman)

1934. *Cratogeomys castanops excelsus* Nelson and Goldman, Proc. Biol. Soc. Washington, 47:143, June 13, type from 10 mi. W Laguna de Mayrán, Coahuila.
1968. *Pappogeomys castanops excelsus*, Russell, Univ. Kansas Publ., Mus. Nat. Hist., 16:641, August 5.

MARGINAL RECORDS (Russell, 1968:643).— Coahuila: 2 mi. S, 8 mi. E Americanos; 4 mi. N Acatita; 3 mi. N, 5 mi. W La Rosa; 11 mi. N, 10 mi. W San Lorenzo; *2 mi. E Torreón.* Durango: 4 mi. WSW Lerdo; Nuevo Mundo, 33 mi. NW Torreón, Coahuila; Tlahualilo.

Pappogeomys castanops goldmani (Merriam)

1895. *Cratogeomys castanops goldmani* Merriam, N. Amer. Fauna, 8:160, January 31, type from Cañitas, Zacatecas.
1968. *Pappogeomys castanops goldmani*, Russell, Univ. Kansas Publ., Mus. Nat. Hist., 16:643, August 5.

MARGINAL RECORDS (Russell, 1968:646).— Durango: 4 mi. NNE Boquilla. Coahuila: 1 mi. S Jimulco. Zacatecas: type locality. Durango: Hda. de Atotonilco; Río Nazos, 6 mi. NW Rodeo.

Pappogeomys castanops hirtus (Nelson and Goldman)

1934. *Cratogeomys castanops hirtus* Nelson and Goldman, Proc. Biol. Soc. Washington, 47:138, June 13, type from Albuquerque, Bernalillo Co., New Mexico.
1968. *Pappogeomys castanops hirtus*, Russell, Univ. Kansas Publ., Mus. Nat. Hist., 16:646, August 5.

MARGINAL RECORDS (Russell, 1968:648).—New Mexico: type locality; *South Valley, Albuquerque;* Rhodes Pass, San Andres Mts.; Parker Lake, E of Organ Mts. Texas: Municipal Golf Course, El Paso.

Pappogeomys castanops jucundus (Russell and Baker)

1955. *Cratogeomys castanops jucundus* Russell and Baker, Univ. Kansas Publ., Mus. Nat. Hist., 7:599, March 15, type from Hermanas, 1205 ft., Coahuila.
1968. *Pappogeomys castanops jucundus*, Russell, Univ. Kansas Publ., Mus. Nat. Hist., 16:648, August 5.

MARGINAL RECORDS (Russell, 1968:650).— Coahuila: Hermanas; *1 mi. S Hermanas; 2 mi. N, 1 mi. E Monclova;* Monclova; 16 km. S Cuatro Ciénegas; 6 mi. W Cuatro Ciénegas.

Pappogeomys castanops parviceps Russell

1968. *Pappogeomys castanops parviceps* Russell, Univ. Kansas Publ., Mus. Nat. Hist., 16:673, August 5, type from 18 mi. SW Alamogordo, 4400 ft., Otero Co., New Mexico.

MARGINAL RECORDS (Russell, 1968:676).—New Mexico: Carasal; Ancho; 3 mi. S Picacho. Texas: *7 mi. N Pine Springs;* foot Pine Canyon [= Pine Spring Canyon], Guadalupe Mts.; *mouth Pine Spring Canyon, Guadalupe Mts.* New Mexico: type locality; 9 mi. S Tularosa.

Pappogeomys castanops perexiguus Russell

1968. *Pappogeomys castanops perexiguus* Russell, Univ. Kansas Publ., Mus. Nat. Hist., 16:676, August 5, type from 6 mi. E Jaco, Chihuahua, 4500 ft., in Coahuila.

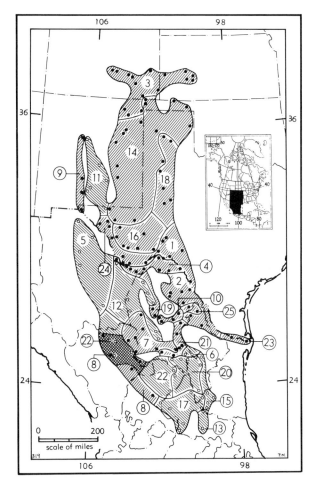

Map 319. *Pappogeomys castanops.*

Guide to subspecies
1. *P. c. angusticeps*	13. *P. c. peridoneus*
2. *P. c. bullatus*	14. *P. c. perplanus*
3. *P. c. castanops*	15. *P. c. planifrons*
4. *P. c. clarkii*	16. *P. c. pratensis*
5. *P. c. consitus*	17. *P. c. rubellus*
6. *P. c. elibatus*	18. *P. c. simulans*
7. *P. c. excelsus*	19. *P. c. sordidulus*
8. *P. c. goldmani*	20. *P. c. subnubilus*
9. *P. c. hirtus*	21. *P. c. subsimus*
10. *P. c. jucundus*	22. *P. c. surculus*
11. *P. c. parviceps*	23. *P. c. tamaulipensis*
12. *P. c. perexiguus*	24. *P. c. torridus*
	25. *P. c. ustulatus*

MARGINAL RECORDS (Russell, 1968:679).—Coahuila: 3 mi. N, 9 mi. E El Pino; 50 mi. N, 20 mi. W Ocampo; 18 mi. S, 14 mi. E Tanque Alvarez; 20 mi. S Hundido; 21 mi. S, 11 mi. E Australia; 3 mi. NE Sierra Majada; type locality.

Pappogeomys castanops peridoneus (Nelson and Goldman)

1934. *Cratogeomys castanops peridoneus* Nelson and Goldman, Proc. Biol. Soc. Washington, 47:148, June 13, type from Río Verde, 3000 ft., San Luis Potosí.

1968. *Pappogeomys castanops peridoneus*, Russell, Univ. Kansas Publ., Mus. Nat. Hist., 16:679, August 5.

MARGINAL RECORDS (Russell, 1968:680).—San Luis Potosí: 2 mi. NW Tepeyac; type locality.

Pappogeomys castanops perplanus (Nelson and Goldman)

1934. *Cratogeomys castanops perplanus* Nelson and Goldman, Proc. Biol. Soc. Washington, 47:136, June 13, type from Tascosa, Oldham Co., Texas.

1968. *Pappogeomys castanops perplanus*, Russell, Univ. Kansas Publ., Mus. Nat. Hist., 16:650, August 5.

1934. *Cratogeomys castanops lacrimalis* Nelson and Goldman, Proc. Biol. Soc. Washington, 47:137, June 13, type from Roswell, 3500 ft., Chaves Co., New Mexico.

MARGINAL RECORDS (Russell, 1968:653, unless otherwise noted).—Oklahoma: 7 mi. N Kenton; Hooker; 4 mi. E Elmwood P.O. Texas: Hemphill County (Davis, 1966:157, as *Cratogeomys castanops* only); Tascosa; Hale Center; Pecos; Scott Canyon, Delaware Mts.; foot Pine Canyon, Guadalupe Mts. New Mexico: Roswell; 35 mi. N Roswell; Santa Rosa; 1 mi. S, 2 mi. W Conchas Dam; Bell Ranch; ⅗ mi. N, ½ mi. W Amistad (Best, 1973:1317, as *P. castanops* only). Oklahoma: head Tesequite Canyon, 5 mi. S Kenton; *N side Black Mesa, 6 mi. N Kenton.*

Pappogeomys castanops planifrons (Nelson and Goldman)

1934. *Cratogeomys castanops planifrons* Nelson and Goldman, Proc. Biol. Soc. Washington, 47:146, June 13, type from Miquihuana, Tamaulipas.

1968. *Pappogeomys castanops planifrons*, Russell, Univ. Kansas Publ., Mus. Nat. Hist., 16:680, August 5.

MARGINAL RECORDS (Russell, 1968:682).—Nuevo León: 15 mi. W Montemorelos. Tamaulipas: type locality; 4 mi. N Jaumave; 8 mi. N Tula; 9 mi. SW Tula. Nuevo León: Dr. [Doctor] Arroyo.

Pappogeomys castanops pratensis Russell

1968. *Pappogeomys castanops pratensis* Russell, Univ. Kansas Publ., Mus. Nat. Hist., 16:653, August 5, type from 3 mi. S, 8 mi. W Alpine, 5100 ft., Brewster Co., Texas.

MARGINAL RECORDS (Russell, 1968:655).—Texas: Kent; Fort Stockton; 2 mi. E Sanderson; Marathon; 2 mi. S Paisano; Marfa; Upper Limpia Canyon, 5 mi. W Mt. Livermore.

Pappogeomys castanops rubellus (Nelson and Goldman)

1934. *Cratogeomys castanops rubellus* Nelson and Goldman, Proc. Biol. Soc. Washington, 47:147, June 13, type from Soledad, near San Luis Potosí, 6400 ft., San Luis Potosí.

1968. *Pappogeomys castanops rubellus*, Russell, Univ. Kansas Publ., Mus. Nat. Hist., 16:682, August 5.

MARGINAL RECORDS (Russell, 1968:685).—San Luis Potosí: 6 km. S Matehuala. Tamaulipas: Nicolás, 56 km. NW Tula. San Luis Potosí: Presa de Guadalupe; *7 km. W Presa de Guadalupe;* City of San Luis Potosí; type locality; 4½ mi. SW Herradura. Zacatecas: Villa de Cos.

Pappogeomys castanops simulans Russell

1968. *Pappogeomys castanops simulans* Russell, Univ. Kansas Publ., Mus. Nat. Hist., 16:656, August, 5, type from 17 mi. SE Washburn, Armstrong Co., Texas.

MARGINAL RECORDS (Russell, 1968:657).— Texas: 2 mi. E Amarillo; type locality; Big Springs; Stanton; *10 mi. E Lamesa;* 8 mi. NE Lamesa; *6½ mi. W Lubbock;* 2 mi. E Reese Air Force Base; *9 mi. NW Lubbock.*

Pappogeomys castanops sordidulus (Russell and Baker)

1955. *Cratogeomys castanops sordidulus* Russell and Baker, Univ. Kansas Publ., Mus. Nat. Hist., 7:600, March 15, type from 1½ mi. NW Ocampo, 3300 ft., Coahuila.
1968. *Pappogeomys castanops sordidulus,* Russell, Univ. Kansas Publ., Mus. Nat. Hist., 16:658, August 5.

MARGINAL RECORDS (Russell, 1968:660).— Coahuila: type locality; 5 mi. N, 19 mi. W Cuatro Ciénegas.

Pappogeomys castanops subnubilus (Nelson and Goldman)

1934. *Cratogeomys castanops subnubilus* Nelson and Goldman, Proc. Biol. Soc. Washington, 47:145, June 13, type from Carneros, Coahuila.
1968. *Pappogeomys castanops subnubilus,* Russell, Univ. Kansas Publ., Mus. Nat. Hist., 16:685, August 5.

MARGINAL RECORDS (Russell, 1968:688).— Coahuila: 11 mi. S, 4 mi. W General Cepeda; 1 mi. N Agua Nueva. Nuevo León: 7 mi. NW Providencia; Laguna; 1 mi. W Dr. [Doctor] Arroyo. Zacatecas: 15 mi. S Concepción del Oro.

Pappogeomys castanops subsimus (Nelson and Goldman)

1934. *Cratogeomys castanops subsimus* Nelson and Goldman, Proc. Biol. Soc. Washington, 47:144, June 13, type from Jaral [= San Antonio de Jaral], Coahuila.
1968. *Pappogeomys castanops subsimus,* Russell, Univ. Kansas Publ., Mus. Nat. Hist., 16:660, August 5.

MARGINAL RECORDS (Russell, 1968:663).— Coahuila: Hisachalo [= Huisachalo]; 2 mi. N Santa Cruz; 17 mi. N, 18 mi. W Saltillo; *N foot Sierra Guadalupe, 9 mi. S, 5 mi. W General Cepeda;* N foot Sierra Guadalupe, 10 mi. S, 5 mi. W General Cepeda; 1½ mi. N Parras; *12 mi. N, 10 mi. E Parras;* Hda. El Tulillo, *5 km. S Hipólito;* 3 mi. S, 3 mi. E Muralla.

Pappogeomys castanops surculus Russell

1968. *Pappogeomys castanops surculus* Russell, Univ. Kansas Publ., Mus. Nat. Hist., 16:688, August 5, type from La Zarca, Durango.

MARGINAL RECORDS (Russell, 1968:691).— Durango: 7 mi. NW La Zarca; 12 mi. E La Zarca; San Juan, 12 mi. W Lerdo. Zacatecas: Concepción del Oro; 8 mi. S Majoma. Durango: Río Nazas, 6 mi. NW Rodeo.

Pappogeomys castanops tamaulipensis (Nelson and Goldman)

1934. *Cratogeomys castanops tamaulipensis* Nelson and Goldman, Proc. Biol. Soc. Washington, 47:141, June 13, type from Matamoros, Tamaulipas.
1968. *Pappogeomys castanops tamaulipensis,* Russell, Univ. Kansas Publ., Mus. Nat. Hist., 16:663, August 5.

MARGINAL RECORDS (Russell, 1968:665).— Tamaulipas: 3 mi. SE Reynosa; type locality.

Pappogeomys castanops torridus Russell

1968. *Pappogeomys castanops torridus* Russell, Univ. Kansas Publ., Mus. Nat. Hist., 16:665, type from 3 mi. E Sierra Blanca, about 4000 ft., Hudspeth Co., Texas.

MARGINAL RECORDS (Russell, 1968:667).— Texas: Bat Cave, Diablo Mts.; Van Horn; Valentine; Harper Ranch, 37 mi. S Marfa; 8 mi. N Terlingua; *4 mi. E Terlingua;* 6 mi. S Terlingua; La Mota Rancho, 53 mi. S Marfa; *11 mi. W Valentine;* 3 mi. W Sierra Blanca; *1 mi. N, ½ mi. E Sierra Blanca.*

Pappogeomys castanops ustulatus (Russell and Baker)

1955. *Cratogeomys castanops ustulatus* Russell and Baker, Univ. Kansas Publ., Mus. Nat. Hist., 7:598, March 15, type from Don Martín, 800 ft., Coahuila.
1968. *Pappogeomys castanops ustulatus,* Russell, Univ. Kansas Publ., Mus. Nat. Hist., 16:667, August 5.

MARGINAL RECORDS (Russell, 1968:668).— Coahuila: type locality. Nuevo León: *9 mi. N, 2 mi. W Anahuac (= Rodrígues);* 4 mi. N, 1 mi. W Anahuac (= Rodrígues). Coahuila: *5 mi. SE Don Martín; 2 mi. SE Don Martín Dam.*

Pappogeomys merriami
Merriam's Pocket Gopher

Head and body, 200–285 in males and 180–253 in females; tail 74–126, 71–119; hind foot 38–53, 36–49. Color of pelage varies in tones of pale yellowish-buff to glossy black. Back and top of head darker than sides and face. Underparts varying from pale buff to bright ochraceous or rufous. Melanism common in *P. m. merriami* but rare or absent in other subspecies. Skull broader across zygomata than across squamosals; occlusal sur-

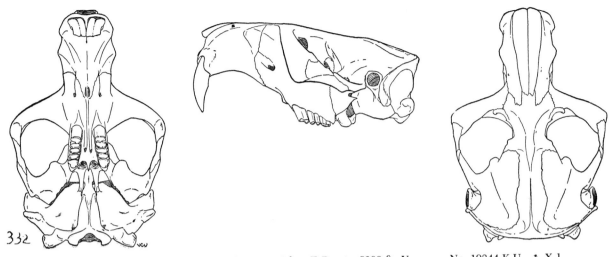

332

Fig. 332. *Pappogeomys merriami fulvescens*, 2 km. E Perote, 8300 ft., Veracruz, No. 19344 K.U., ♂, X 1.

face of M3 obcordate with short posterior loph; loph twisted to side; posterior apex of loph forming labial border of loph; lingual plate of loph lying transversely along posterior border of tooth; lower incisor distinctly beveled behind labial edge of enamel plate; additional characters listed in key to species of subgenus *Cratogeomys*.

Pappogeomys merriami estor (Merriam)

1895. *Cratogeomys estor* Merriam, N. Amer. Fauna, 8:155, January 31, type from Las Vigas, 8000 ft., Veracruz.
1968. *Pappogeomys merriami estor*, Russell, Univ. Kansas Publ., Mus. Nat. Hist., 16:698, August 5.
1934. *Cratogeomys perotensis estor*, Nelson and Goldman, Proc. Biol. Soc. Washington, 47:151, June 13.

MARGINAL RECORDS (Russell, 1968:700).— Veracruz: 7 km. SSE Jalacingo; *6 km. SE Altotonga;* 5 km. E Las Vigas; *2 km. W Las Vigas.*

Pappogeomys merriami fulvescens (Merriam)

1895. *Cratogeomys fulvescens* Merriam, N. Amer. Fauna, 8:161, January 31, type from Chalchicomula, 8200 ft., Puebla.
1968. *Pappogeomys merriami fulvescens*, Russell, Univ. Kansas Publ., Mus. Nat. Hist., 16:700, August 5.
1934. *Cratogeomys fulvescens subluteus* Nelson and Goldman, Proc. Biol. Soc. Washington, 47:152, June 13, type from Perote, 7800 ft., Veracruz.

MARGINAL RECORDS (Russell, 1968:703).— Veracruz: 2 km. N Perote; *2 km. E Perote.* Puebla: type locality. Veracruz: *6 mi. SW Perote.*

Pappogeomys merriami irolonis (Nelson and Goldman)

1934. *Cratogeomys merriami irolonis* Nelson and Goldman, Proc. Biol. Soc. Washington, 47:150, June 13, type from Irolo, 7600 ft., Hidalgo.

1968. *Pappogeomys merriami irolonis*, Russell, Univ. Kansas Publ., Mus. Nat. Hist., 16:703, August 5.

MARGINAL RECORDS (Russell, 1968:706).— Hidalgo: 10 mi. NW Apam. Puebla: Crus Alta; *2 mi. E Crus Alta.* Hidalgo: *type locality.*

Pappogeomys merriami merriami (Thomas)

1893. *Geomys merriami* Thomas, Ann. Mag. Nat. Hist., ser. 6, 12:271, October, type from southern México. ("Probably the Valley of Mexico" according to Merriam, N. Amer. Fauna, 8:152, January 31, 1895.)
1968. *Pappogeomys merriami merriami*, Russell, Univ. Kansas Publ., Mus. Nat. Hist., 16:706, August 5.
1895. *Cratogeomys oreocetes* Merriam, N. Amer. Fauna, 8:156, January 31, type from Mt. Popocatépetl, 11,000 ft., México.
1895. *Cratogeomys peregrinus* Merriam, N. Amer. Fauna, 8:158, January 31, type from Mt. Iztaccíhuatl, 11,500 ft., México.

MARGINAL RECORDS (Russell, 1968:709, unless otherwise noted).—México: 6 mi. S, 1 mi. W Texcoco; Monte Río Frío, 45 km. ESE Mexico City. Puebla: San Martín Texmelucan. México: 55 km. SE Mexico City, near Paso de Cortez; *N slope Mt. Popocatépetl.* Morelos: *1½ mi. SE Huitzilac;* 4 km. W Huitzilac (Ramírez-P., 1969:271). México: Lerma; *Salazar.* Distrito Federal: Coyoacan-Churubusco; *Ixtapalapa.*

Pappogeomys merriami peraltus (Goldman)

1937. *Cratogeomys perotensis peraltus* Goldman, Jour. Washington Acad. Sci., 27:403, September 15, type from Mount Orizaba, about 12,500 ft., Puebla [not Veracruz; see Hall and Dalquest, Univ. Kansas Publ., Mus. Nat. Hist., 14:280, May 20, 1963]. Known only from type locality.
1968. *Pappogeomys merriami peraltus*, Russell, Univ. Kansas Publ., Mus. Nat. Hist., 16:709, August 5.

Pappogeomys merriami perotensis (Merriam)

1895. *Cratogeomys perotensis* Merriam, N. Amer. Fauna, 8:154, January 31, type from Cofre de Perote, 9500 ft., Veracruz.

1968. *Pappogeomys merriami perotensis,* Russell, Univ. Kansas Publ., Mus. Nat. Hist., 16:712, August 5.

MARGINAL RECORDS (Russell, 1968:714).— Veracruz: *1 km. NW Pescados;* N slope Cofre de Perote.

Pappogeomys merriami saccharalis (Nelson and Goldman)

1934. *Cratogeomys merriami saccharalis* Nelson and Goldman, Proc. Biol. Soc. Washington, 47:149, June 13, type from Atlixco, 5400 ft., Puebla.

1968. *Pappogeomys merriami saccharalis,* Russell, Univ. Kansas Publ., Mus. Nat. Hist., 16:714, August 5.

MARGINAL RECORDS (Russell, 1968:716).— Puebla: type locality; *2 mi. S Atlixco.*

gymnurus species-group
Pappogeomys neglectus (Merriam)
Querétaro Pocket Gopher

1902. *Platygeomys neglectus* Merriam, Proc. Biol. Soc. Washington, 15:68, March 22, type from Cerro de la Calentura, 9500 ft., about 8 mi. NW Pinal de Amoles, Querétaro. Known only from type locality.

1968. *Pappogeomys neglectus,* Russell, Univ. Kansas Publ., Mus. Nat. Hist., 16:717, August 5.

Head and body 202– —in two males and 194–194 in two females; tail 90–96, 73–87; hind foot 42–42, 37–39. Hairs of upper parts Dark Mouse Gray basally and Ochraceous-Tawny apically, with some Bay-tipped hairs on back and top of head; pure Ochraceous-Tawny on sides and face; underparts Light Mouse Gray overlaid with Ochraceous-Buff; chin buffy; throat whitish; auricular patch small, blackish; hind foot whitish; tail sparsely clothed with ochraceous hairs. Skull smoothly rounded; squamosal breadth greater than zygomatic breadth; lambdoidal crest weakly developed, only slightly sinuous; upper incisors procumbent. Additional characters listed in key to species of subgenus *Cratogeomys.*

Pappogeomys fumosus (Merriam)
Smoky Pocket Gopher

1892. *Geomys fumosus* Merriam, Proc. Biol. Soc. Washington, 7:165, September 29, type from 3 mi. W Colima, 1700 ft., Colima.

1968. *Pappogeomys fumosus,* Russell, Univ. Kansas Publ., Mus. Nat. Hist., 16:719, August 5.

Head and body 222–242 in males and 194–213 in females; tail 74–98, 71–75; hind foot 40–44,

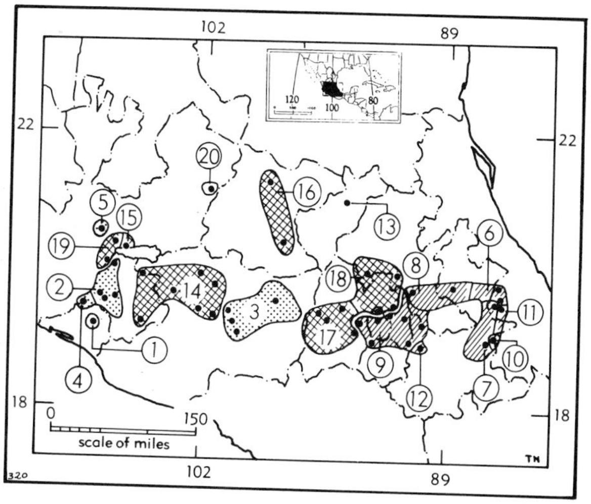

Map 320. Six species of *Pappogeomys.*

1. *P. fumosus*
2. *P. gymnurus gymnurus*
3. *P. gymnurus imparilis*
4. *P. gymnurus russelli*
5. *P. gymnurus tellus*
6. *P. merriami estor*
7. *P. merriami fulvescens*
8. *P. merriami irolonis*
9. *P. merriami merriami*
10. *P. merriami peraltus*
11. *P. merriami perotensis*
12. *P. merriami saccharalis*
13. *P. neglectus*
14. *P. tylorhinus angustirostris*
15. *P. tylorhinus atratus*
16. *P. tylorhinus brevirostris*
17. *P. tylorhinus planiceps*
18. *P. tylorhinus tylorhinus*
19. *p. tylorhinus zodius*
20. *P. zinseri*

39–41. Pelage coarse and bristly; upper parts Dark Mouse Gray basally and Mars Brown apically; underparts including chin and throat Gull Gray basally overlaid with Pale Ochraceous-Buff. Zygomata nearly parallel. Squamosal breadth less than zygomatic breadth. Additional characters listed in key to species of subgenus *Cratogeomys.*

MARGINAL RECORDS (Russell, 1968:721).— Colima: *type locality;* Colima City; *4 mi. SW Colima City.*

Pappogeomys tylorhinus
Naked-nosed Pocket Gopher

Head and body 212–276 in males and 193–248 in females; tail 79–106, 65–106; hind foot 37–47, 37–46. Pelage soft and lax; upper parts pale ochraceous-buff to glossy black; in brownish phases, top of head and back darker than sides and face; underparts white to bright shades of ochraceous (underparts never black even in melanistic individuals); throat gray or buffy; feet whitish or brownish. Skull rugose and angular. Angular processes elongated, breadth across processes greater than length of lower jaw; rostrum relatively narrow and lightly constructed; angle of maxillary arm of zygoma enlarged into platelike expansion; lower incisor lacking lateral bevel behind enamel plate. Additional characters listed in key to species of subgenus *Cratogeomys*. See Map 320.

Pappogeomys tylorhinus angustirostris (Merriam)

1903. *Platygeomys tylorhinus angustirostris* Merriam, Proc. Biol. Soc. Washington, 16:81, May 29, type from Cerro Patambán, 10,000 ft., Michoacán.
1968. *Pappogeomys tylorhinus angustirostris*, Russell, Univ. Kansas Publ., Mus. Nat. Hist., 16:727, August 5.
1939. *Platygeomys varius* Goldman, Jour. Mamm., 20:90, type from Uruapan, about 6000 ft., Michoacán.

MARGINAL RECORDS (Russell, 1968:731, unless otherwise noted).—Michoacán: Tangancícuaro; 2 mi. SE Zacapu, near village of Tacumbo; Nuevo San Juan (= Los Conejos); on road to Tzurarama Falls; 1 mi. N Tinquindin. Jalisco: *8 mi. E Jilotlán de los Dolores, 2400 ft.* (Genoways and Jones, 1969:754); Jilotlán de los Dolores, 2000 ft. (*ibid.*); *6 mi. S Mazamitla* (*ibid.*); *3 mi. WSW Mazamilta* (*ibid.*); 4 mi. W Mazamitla (Sierra del Tigre).

Pappogeomys tylorhinus atratus (Russell)

1953. *Cratogeomys gymnurus atratus* Russell, Univ. Kansas Publ., Mus. Nat. Hist., 5:539, October 15, type from top Cerro Viejo de Cuyutlán, 9700 ft., 19 mi. S, 9 mi. W Guadalajara, Jalisco. Known only from type locality.
1968. *Pappogeomys tylorhinus atratus*, Russell, Univ. Kansas Publ., Mus. Nat. Hist., 16:731, August 5.

Pappogeomys tylorhinus brevirostris Russell

1968. *Pappogeomys tylorhinus brevirostris* Russell, Univ. Kansas Publ., Mus. Nat. Hist., 16:733, August 5, type from 2 mi. E Celaya, 5800 ft., Guanajuato.

MARGINAL RECORDS (Russell, 1968:735).— Guanajuato: San Diego de la Unión; 5 mi. E Celaya.

Pappogeomys tylorhinus planiceps (Merriam)

1895. *Platygeomys planiceps* Merriam, N. Amer. Fauna, 8:168, January 31, type from N slope Volcán de Toluca, 9000 ft., México.
1968. *Pappogeomys tylorhinus planiceps*, Russell, Univ. Kansas Publ., Mus. Nat. Hist., 16:735, August 5.

MARGINAL RECORDS (Russell, 1968:738).— México: El Río (= San Bernabe), 14 mi. NW Toluca; type locality; *Isla, 3 mi. NW Tenango del Valle;* 10 mi. N, 6 mi. E Valle de Bravo; 3 mi. N, 7 mi. W San José Allende.

Pappogeomys tylorhinus tylorhinus (Merriam)

1895. *Platygeomys tylorhinus* Merriam, N. Amer. Fauna, 8:167, January 31, type from Tula, 6800 ft., Hidalgo.
1968. *Pappogeomys tylorhinus tylorhinus*, Russell, Univ. Kansas Publ., Mus. Nat. Hist., 16:739, August 5.

1947. *Cratogeomys tylorhinus arvalis* Hooper, Jour. Mamm., 28:45, February 15, type from Colonia del Valle, 2275 m., Mexico City, Distrito Federal.

MARGINAL RECORDS (Russell, 1968:742).— Hidalgo: type locality; 9 km. S Pachuca. México: Tempo del Sol, Piramida de San Juan Taotihuacán; 5 km. N Texcoco. Distrito Federal: *Colonia del Valle, Mexico City;* Coyoacán.

Pappogeomys tylorhinus zodius (Russell)

1953. *Cratogeomys zinseri zodius* Russell, Univ. Kansas Publ., Mus. Nat. Hist., 5:540, October 15, type from 13 mi. S, 15 mi. W Guadalajara, about 4500 ft., Jalisco.
1968. *Pappogeomys tylorhinus zodius* (Russell), Univ. Kansas Publ., Mus. Nat. Hist., 16:742, August 5.

MARGINAL RECORDS.—Jalisco: type locality; Mesa de Tapalpa, 4 mi. E Atemajac de Brizuela, 8000 ft. (Genoways and Jones, 1969:755).

Pappogeomys zinseri (Goldman)
Zinser's Pocket Gopher

1939. *Platygeomys zinseri* Goldman, Jour. Mamm., 20:91, February 15, type from Lagos, 6150 ft., Jalisco.
1968. *Pappogeomys zinseri*, Russell, Univ. Kansas Publ., Mus. Nat. Hist., 16:744, August 5.

Head and body 223 in one male (not fully adult) and 212–236 in five females; tail 100, 93–106; hind foot 43, 44–45. Pelage on upper parts Mouse Gray basally and Ochraceous-Tawny apically, becoming bright Ochraceous-Buff on sides and face; underparts Light Mouse Gray basally overlaid with Ochraceous-Buff; throat grayish; chin buffy; auricular area blackish; hind foot dark brownish above with white hairs about base of toes; zygomata widely spreading, averaging 73.9 per cent of condylobasal length; breadth across squamosals averaging 69.3 per cent of condylobasal length; rostrum long and moderately broad, averaging 55.1 per cent of its length. Additional characters listed above in key to species of subgenus *Cratogeomys*. See Map 320.

MARGINAL RECORDS (Russell, 1968:748).— Jalisco: ½ mi. NE Lagos de Moreno; *type locality*.

Pappogeomys gymnurus
Llano Pocket Gopher

Head and body 234–279 in males and 230–267 in females; tail 87–112, 75–105; hind foot 47–59, 44–54. Pelage soft and lax; upper parts pale brownish-buff to glossy brownish-black; top of head and back darker than sides or face; underparts creamy-white to bright ochraceous-tawny, or black like dorsum in most melanistic individ-

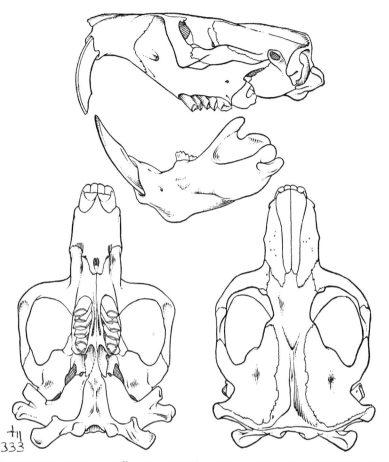

Fig. 333. *Pappogeomys gymnurus tellus*, 3 mi. W Tala, 4300 ft., Jalisco, No. 33454 K.U., holotype, ♀, X 1.

uals; throat usually grayish (overlaid with whitish or tinged with pale buff in *tellus*); dark auricular patches developed in lighter color phases, but not visible in melanistic pelages. Skull largest of the *gymnurus* group; breadth across squamosals (averaging 70.4–73.2% of condylobasal length) usually equaling or exceeding zygomatic breadth; rostrum long, 24.6–29.1. Additional characters listed above in key to species of subgenus *Cratogeomys*.

The strongly developed platycephalic specializations of the skull—broad and flat cranium, widely spreading rami of lower jaw and their elongated angular processes—separate *P. gymnurus* from members of the *castanops* species-group. See Map 320.

Pappogeomys gymnurus gymnurus (Merriam)

1892. *Geomys gymnurus* Merriam, Proc. Biol. Soc. Washington, 7:166, September 29, type from Zapotlán (= Ciudad Guzmán), 4000 ft., Jalisco.
1968. *Pappogeomys gymnurus gymnurus*, Russell, Univ. Kansas Publ., Mus. Nat. Hist., 16:751, August 5.

1939. *Platygeomys gymnurus inclarus* Goldman, Jour. Mamm., 20:88, February 15, type from N slope Sierra Nevada de Colima, 10,000 ft., Jalisco.
1953. *Cratogeomys zinseri morulus* Russell, Univ. Kansas Publ., Mus. Nat. Hist., 5:541, October 15, type from N end Lago de Sayula, 4400 ft., 9 mi. N, 2 mi. E Atoyac, Jalisco.

MARGINAL RECORDS (Russell, 1968:754, unless otherwise noted).—Jalisco: N end Lago de Sayula; 2 mi. N Ciudad Guzmán; 3½ mi. WNW Zapoltitic; 2½ mi. ENE Jazmín (109220 KU).

Pappogeomys gymnurus imparilis (Goldman)

1939. *Platygeomys gymnurus imparilis* Goldman, Jour. Mamm., 20:89, February 15, type from Pátzcuaro, Michoacán.
1968. *Pappogeomys gymnurus imparilis*, Russell, Univ. Kansas Publ., Mus. Nat. Hist., 16:754, August 5.

MARGINAL RECORDS (Russell, 1968:756).—Michoacán: 3 mi. S, 15 mi. W Ciudad Hidalgo; 1¾ mi. S Tacámbaro; 2 mi. E San Gregorio; 2 mi. W Pátzcuaro.

Pappogeomys gymnurus russelli Genoways and Jones

1969. *Pappogeomys gymnurus russelli* Genoways and Jones, Jour. Mamm., 50:751, November 28, type from 12 mi. S Tolimán, 7700 ft., Jalisco. Known only from type locality

Pappogeomys gymnurus tellus (Russell)

1953. *Cratogeomys gymnurus tellus* Russell, Univ. Kansas Publ., Mus. Nat. Hist., 5:537, October 15, type from 3 mi. W Tala, 4300 ft., Jalisco.
1968. *Pappogeomys gymnurus tellus* (Russell), Univ. Kansas Publ., Mus. Nat. Hist., 16:756, August 5.

MARGINAL RECORDS (Russell, 1968:757).— Jalisco: 1 mi. NE Tala; *1 mi. S El Refugio; type locality.*

FAMILY HETEROMYIDAE—Heteromyids

Skull thin, papery, and not strongly modified for fossorial existence; interorbital breadth greater than rostral breadth; nasals produced distally beyond incisors; zygomata slender and threadlike; interparietal reduced; tympanic bullae much inflated, the genera *Microdipodops* and *Dipodomys* constituting the extreme in this respect among American mammals; mastoids inflated and forming part of dorsal surface of skull; occipitals reduced; jaw small and weak; cheek-teeth rooted except in *Dipodomys*, and markedly simplified; molars quadritubercular; incisors thin and compressed.

In spite of the many obvious superficial differences, heteromyids and geomyids are closely related as shown by Hill (1937). This relationship can be shown taxonomically by arranging them as subfamilies of the family Heteromyidae or by according each of them family rank and uniting them in one superfamily. Both arrangements are in current use, the latter being perhaps the more common.

Some genera are somewhat murine in appearance (*Liomys* and *Heteromys*), whereas others are highly modified for a saltatorial mode of locomotion (*Dipodomys*). In the last-named genus, the hind limbs are long and powerful and the forelegs reduced.

KEY TO SUBFAMILIES AND GENERA OF HETEROMYIDAE

SUBFAMILY PEROGNATHINAE

Lophs of upper premolars unite first at or near center of tooth; protoloph usually single-cusped; lophs of upper molars unite progressively from lingual to buccal margins; those of lower premolars unite at center of tooth, giving an X-pattern; lophs of lower molars unite primitively at buccal margin, progressively at center of tooth, forming an H-pattern; cheek-teeth brachydont to hypsodont but always rooted; enamel pattern lost early in life; enamel always complete; upper incisor smooth or grooved; center of palate between premolars not ridged; ethmoid foramen in frontal; auditory region variable as to degree of inflation; ventral surface of tympanic bullae below level of grinding surface of upper cheek-teeth; no median ventral foramina in caudal vertebrae; astragalus articulating with cuboid (after Wood, 1935:88, 89).

Genus **Perognathus** Wied-Neuwied
Pocket Mice

Revised by Merriam, N. Amer. Fauna, 1:vii + 36 pp., 4 pls., October 25, 1889, and Osgood, N. Amer. Fauna, 18:1–72, 4 pls. 15 figs., September 20, 1900.

1839. *Perognathus* Wied-Neuwied, Nova Acta Phys.-Med. Acad. Caesar. Leop.-Carol., 19(pt. 1):368. Type, *Perognathus fasciatus* Wied-Neuwied.
1848. *Cricetodipus* Peale, Mammalia and ornithology, *in* U.S. Expl. Exped. . . . , 8:52. Type, *Cricetodipus parvus* Peale.
1868. *Abromys* Gray, Proc. Zool. Soc. London, May, p. 202. Type, *Abromys lordi* Gray.
1875. *Otognosis* Coues, Proc. Acad. Nat. Sci. Philadelphia, p. 305. Type, *Otognosis longimembris* Coues.

External measurements: 100–230; 44–143; 15–29. Small, slender; posterior limbs appreciably longer than anterior; skull delicate and somewhat flattened; mastoids large; auditory bullae large and well inflated; frontals little constricted; rostrum lightly constructed; nasals long and becoming semitubular anteriorly; infraorbital foramen reduced to a lateral opening in maxillae. Molars rooted and tuberculate; upper incisors strongly grooved. Dentition, i. $\frac{1}{1}$, c. $\frac{0}{0}$, p. $\frac{1}{1}$, m. $\frac{3}{3}$.

Genus *Perognathus* includes widely different species, which might be placed in different genera were it not for the many structurally intermediate species.

KEY TO SUBGENERA OF PEROGNATHUS

1. Pelage soft; soles of hind feet somewhat hairy; mstoids greatly developed, projecting beyond occipital plane; breadth of interparietal less than breadth of interorbital region (rarely equal in *P. longimembris*); auditory bullae meeting or nearly so anteriorly. . . . .*Perognathus*, p. 527
1'. Pelage harsh, sometimes with spiny bristles; soles of hind feet naked; mastoids relatively small, not projecting beyond occipital plane; breadth of interparietal equal to or greater than breadth of interorbital region; auditory bullae separated by almost entire width of basisphenoid. *Chaetodipus*, p. 543

Subgenus **Perognathus** Wied-Neuwied

1839. *Perognathus* Wied-Neuwied, Nova Acta Phys.-Med. Acad. Caesar. Leop.-Carol., 19(pt. 1):368. Type, *Perognathus fasciatus* Wied-Nieuwied.

External measurements: 110–198; 44–107; 15–27. Pelage throughout soft, no spines or bristles. Soles of hind feet more or less hairy. Mastoids greatly developed, projecting beyond plane of occiput; mastoid side of parietal longest. Breadth of interparietal (except rarely) less than breadth of interorbital region. Auditory bullae meeting, or almost so, anteriorly. Supraoccipital without lateral indentations by mastoids; ascending branches of supraoccipital slender and threadlike.

KEY TO NOMINAL SPECIES OF SUBGENUS PEROGNATHUS

1. Hind foot more than 20; antitragus lobed; occipitonasal length more than 24.
   2. Ears clothed with white hairs; tail pale.
      3. Dorsal color dark, black predominating. . . . .*P. alticola*, p. 533
      3'. Dorsal color ochraceous buff. . . . .*P. xanthonotus*, p. 533
   2'. Ears not clothed with white hairs; tail dark.
      4. Tail neither crested nor markedly tufted; olivaceous lateral line present; supraoccipital without lateral indentations by mastoid. . . . .*P. parvus*, p. 531
      4'. Tail crested and tufted; olivaceous lateral line absent; supraoccipital with lateral indentations by mastoid. . . . .*P. formosus*, p. 542
1'. Hind foot less than 20 (rarely more in *amplus* and *apache*); antitragus not lobed; occipitonasal length usually less than 24.
   5. Lower premolar distinctly larger than last molar.
      6. Occurs in San Joaquin Valley, California; mastoidal bullae relatively large. . . .*P. inornatus*, p. 541
      6'. Not occurring in San Joaquin Valley, California; mastoidal bullae moderate.*P. longimembris*, p. 536
   5'. Lower premolar equal to or smaller than last molar (in some subspecies barely larger).
      7. Total length more than 130 (total length of *P. amplus* rarely less than 130).
         8. Upper parts pinkish buff to ochraceous salmon; tail more than 70% of head-body. *P. amplus*, p. 540
         8'. Upper parts buffy to ochraceous; tail less than 70% of head-body. . . . . .*P. fasciatus*, p. 528
      7'. Total length less than 130.

9. Tail 57 or more.
    10. Interparietal breadth less than 4.0; dorsal coloration fulvous; occurring south of Colorado (subspecies *gilvus* and *merriami* may key out here; tail in some other subspecies of *P. flavus* rarely more than 56). . . . . *P. flavus*,   p. 533
    10'. Interparietal breadth more than 4.0; dorsal coloration not fulvous; occurring in Colorado and north and east thereof.
        11. Lower premolar distinctly smaller than last molar; dorsal coloration buffy, often strongly overlaid with black hairs to olive gray with yellowish spots on ears.
                                                          P. flavescens,   p. 529
        11'. Lower premolar equal to or but slightly smaller than last molar; dorsal coloration strongly olivaceous. *P. fasciatus*,   p. 528
9'. Tail 56 or less.
    12. Interparietal less than 4.0 wide; mastoidal breadth more than 80% of basilar length; dorsal coloration yellowish or buffy with large yellow postauricular patches. *P. flavus*,   p. 533
    12'. Interparietal more than 4.0 wide; mastoidal breadth less than 80% of basilar length; dorsal coloration olive gray with yellowish spots on ears. . . . .*P. flavescens*,   p. 529

Perognathus fasciatus
Olive-backed Pocket Mouse

External measurements: 128–135; 59–64; 16–18. Upper parts grayish- or buffy-olivaceous with black hairs intermingled; sides with bright, buffy lateral line; underparts white to buff depending on subspecies. Buffy postauricular spot present. Skull small with vaulted braincase; interparietal variable in shape, but usually approx. pentagonal and of moderate width; mastoids well developed and slightly projecting; auditory bullae barely meeting anteriorly; coronoid processes of mandible long and slender; lower premolar subequal to last molar or slightly smaller.

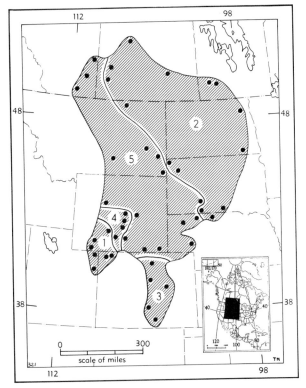

Map 321. *Perognathus fasciatus.*

Guide to subspecies
1. *P. f. callistus*
2. *P. f. fasciatus*
3. *P. f. infraluteus*
4. *P. f. litus*
5. *P. f. olivaceogriseus*

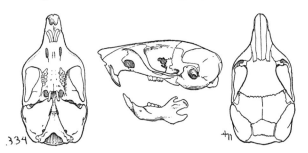

Fig. 334. *Perognathus fasciatus fasciatus*, Buford, North Dakota, No. 168599 U.S.N.M., ♂, X 1½.

Perognathus fasciatus callistus Osgood

1900. *Perognathus callistus* Osgood, N. Amer. Fauna, 18:28, September 20, type from Kinney Ranch, near Bitter Creek, Sweetwater Co., Wyoming.
1953. *Perognathus fasciatus callistus*, Jones, Univ. Kansas Publ., Mus. Nat. Hist., 5:524, August 1.

MARGINAL RECORDS.—Wyoming (Long, 1965a: 615): 27 mi. N, 37 mi. E Rock Springs; *25 mi. N, 38 mi. E Rock Springs.* Colorado: Sunny Peak; near Snake River, 7 mi. above jct. with Bear [River]. Utah: 15 mi.

N Bonanza; Bridgeport. Wyoming (Long, 1965a:615): Green River, 4 mi. ENE Linwood, Utah; Green River.

Perognathus fasciatus fasciatus Wied-Neuwied

1839. *Perognathus fasciatus* Wied-Neuwied, Nova Acta Phys.-Med. Acad. Caesar. Leop.-Carol., 191:369, type from upper Missouri River near jct. with the Yellowstone, near Buford, Williams Co., North Dakota.

MARGINAL RECORDS.—Saskatchewan: 2 mi. NE Grandora (Nero, 1958:176, as *P. fasciatus* only); Strawberry Lakes (Nero, 1965:37, as *P. fasciatus* only). Manitoba: Oak Lake; Aweme. North Dakota: 3 mi. W Park River (115918 KU); 1⅕ mi. S, 7⅕ mi. E Oakes. South Dakota: Colome. Nebraska (Jones, 1964c:165): *Sparks;* Valentine. South Dakota: Lacreek National Wildlife Refuge, 4 mi. S, 8 mi. E Martin (113089 KU); 15 mi. S, 4 mi. W Reva (Andersen and Jones, 1971:376); 4 mi. S, 7 mi. W Ladner (*ibid.*). Montana: Frenchman River. Saskatchewan (Beck, 1958:32): Skull Creek; Sceptre.

Perognathus fasciatus infraluteus Thomas

1893. *Perognathus infraluteus* Thomas, Ann. Mag. Nat. Hist., ser. 6, 11:406, May, type from Loveland, Larimer Co., Colorado.
1900. *Perognathus fasciatus infraluteus*, Osgood, N. Amer. Fauna, 18:19, September 20.

MARGINAL RECORDS.—Colorado: type locality; 7 mi. N Ramah; 4 mi. S La Veta (Turner, 1968:524); 1⅕ mi. (by road) NE Silver Cliff, 7900 ft. (Armstrong, 1972:172); Green Mtn., 5 mi. W Denver (Armstrong, 1972:171).   **See** addenda.

Perognathus fasciatus litus Cary

1911. *Perognathus fasciatus litus* Cary, Proc. Biol. Soc. Washington, 24:61, March 22, type from Sun, Sweetwater Valley, Natrona Co., Wyoming.

MARGINAL RECORDS.—Wyoming (Long, 1965a: 615): 16 mi. S, 11 mi. W Waltman; *type locality;* 8 mi. SE Lost Soldier; 2½ mi. N Wamsutter; *27 mi. N Table Rock;* Granite Mts., Fremont Co.,   **See** addenda.

Perognathus fasciatus olivaceogriseus Swenk

1940. *Perognathus flavescens olivaceogriseus* Swenk, Missouri Valley Fauna, 3:6, June 5, type from Little Bordeaux Creek, sec. 14, T. 33 N, R. 48 W, 3 mi. E Chadron, Dawes Co., Nebraska; see Jones, Univ. Kansas Publ., Mus. Nat. Hist., 5:520–522, August 1, 1953.
1953. *Perognathus fasciatus olivaceogriseus*, Jones, Univ. Kansas Publ., Mus. Nat. Hist., 5:520, August 1.

MARGINAL RECORDS.—Alberta: Empress (Smith 1969:227, as *P. fasciatus* only). Montana: 13 mi. E Miles City; 11½ mi. N, 3 mi. E Ekalaka (Lampe, *et al.*, 1974:16); Little Missouri River, 8 mi. NE Albion. South Dakota: White River flood plain, 7 mi. S Kadoka. Nebraska (Jones, 1964c:167): 12 mi. ESE Gordon; *10 mi. E Gordon;* Mirage Township; 2 mi. W Oskosh. Wyoming (Long, 1965a:616): 15 mi. ESE Cheyenne; Tie Siding Picnic Ground; Fort Steele; Casper; 40 mi. E Dubois. Montana: Lake Basin. Alberta: Medicine Hat; Foremost.   **See** addenda.

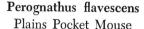

Perognathus flavescens
Plains Pocket Mouse

External measurements: 113–154; 47–73; 15–21. Upper parts varying, according to subspecies, from pale grayish buff to ochraceous buff and overlaid with darker wash of blackish or black; underparts white; subauricular spot small; postauricular spot, lateral line, and eye ring clear buff owing to absence of dark wash of blackish on these areas. Skull short, broad; nasals long; angular process of mandible shorter and wider than in *P. fasciatus;* tympanic bullae meeting anteriorly in northeastern subspecies; interparietal nearly as long as wide in southwestern subspecies but relatively wider in northeastern subspecies; lower premolar smaller than last molar in southwestern subspecies.

All six of the taxa (*apache, caryi, cleomophila, gypsi, melanotis,* and *relictus*) listed as subspecies of species *Perognathus apache* by Hall and Kelson (1959:481–483) are here listed as subspecies of species *Perognathus flavescens* (which has page priority) because specimens from eastern Colorado and eastern New Mexico so closely resemble those from farther west in Colorado and New Mexico that relationship at the subspecific level is indicated.

Perognathus flavescens apache Merriam

1889. *Perognathus apache* Merriam, N. Amer. Fauna, 1:14, October 25, type from Keams Canyon, Apache Co., Arizona.

MARGINAL RECORDS.—Utah: 1 mi. N Bluff, 4400 ft. Colorado (Armstrong, 1972:172): Morfield Ridge, Mesa Verde National Park, 7525 ft.; 8 mi. S Ignacio, 6100 ft. New Mexico: Chama River near Abiquiu; Espanola; San Pedro. Texas: 7½ mi. E El Paso City Hall (Jones and Lee, 1962:78). Chihuahua (Anderson, 1972:300): *10 mi. SE Zaragosa; 1 mi. E Samalayuca, 4500 ft.;* 2½ mi. S, 2 mi. W Samalyuca. New Mexico: Deming; Gallina Mts.; Fort Wingate. Arizona: Holbrook; Winslow; 3 mi. above Cedar Ranch Wash, 4500 ft. (Cockrum, 1961:126); Cedar Ridge, 6000 ft. (*ibid.*). Utah: Navajo Mountain Trading Post.

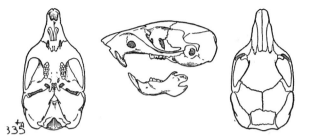

Fig. 335. *Perognathus flavescens flavescens*, Kennedy, Cherry Co., Nebraska, No. 66883, M.V.Z., ♀, X 1½.

Perognathus flavescens caryi Goldman

1918. *Perognathus apache caryi* Goldman, Proc. Biol. Soc. Washington, 31:24, May 16, type from 8 mi. W Rifle, Garfield Co., Colorado.

MARGINAL RECORDS.—Utah: Rainbow. Colorado: type locality; Sieber Ranch, Little Dolores Creek (Armstrong, 1972:172); Coventry. Utah: Johns Canyon, 5150 ft., San Juan River; river mile 119 [Hall Crossing], Glen Canyon, E bank (Durrant and Dean, 1959:83); 2 mi. E Highway 160, 6 mi. S Valley City, 4500 ft.; pump station, 4 mi. N Greenriver, 4100 ft.; Browns Corral, 20 mi. S Ouray, 6250 ft.

Perognathus flavescens cleomophila Goldman

1918. *Perognathus apache cleomophila* Goldman, Proc. Biol. Soc. Washington, 31:23, May 16, type from Winona, 6400 ft., Coconino Co., Arizona.—See addenda.

MARGINAL RECORDS (Cockrum, 1961:126).—Arizona: Navajo Indian Reservation boundary, 30 mi. NE Flagstaff; *1 mi. SE Grand Falls; type locality;* Walnut, 5 mi. from Turkey Tanks; *9 mi. NE Flagstaff, 6800 ft.*

Perognathus flavescens cockrumi Hall

1954. *Perognathus flavescens cockrumi* Hall, Univ. Kansas Publ., Mus. Nat. Hist., 7:589, November 15, type from 4½ mi. NE Danville, Harper Co., Kansas.

MARGINAL RECORDS.—Kansas: 1½ mi. S Wilson; Nickerson; 1¼ mi. N, 13¼ mi. W of K 15 & 24th street of North Newton (105231 KU); type locality. Oklahoma: ½ mi. S, 6 mi. W Canton. Kansas: Schwarz Canyon (Anderson and Nelson, 1958:305); 8 mi. N Ellinwood (*ibid.*).

Perognathus flavescens copei Rhoads

1894. *Perognathus copei* Rhoads, Proc. Acad. Nat. Sci. Philadelphia, for 1893, 46:404, January 27, type from near Mobeetie, Wheeler Co., Texas.
1905. *Perognathus flavescens copei*, V. Bailey, N. Amer. Fauna, 25:143, October 24.

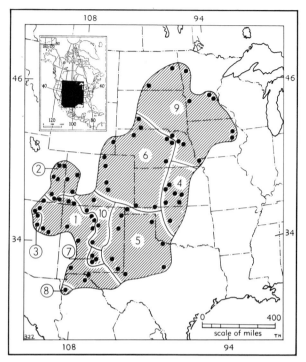

Map 322.   *Perognathus flavescens.*

1. *P. f. apache*	6. *P. f. flavescens*
2. *P. f. caryi*	7. *P. f. gypsi*
3. *P. f. cleomophila*	8. *P. f. melanotis*
4. *P. f. cockrumi*	9. *P. f. perniger*
5. *P. f. copei*	10. *P. f. relictus*

MARGINAL RECORDS.—Oklahoma: 3 mi. N Kenton. Texas (Davis, 1966:158, as *P. flavescens* only, unless otherwise noted): Hemphill County; near Vernon (Dalquest, 1968:17); Callahan County; 20 mi. N Monahans (Hall and Kelson, 1959:476). New Mexico (Williams and Findley, 1968:771): 15 mi. W, 7 mi. N Jal; 7 mi. E Hagerman; 2 mi. S, ½ mi. E Logan; 4 mi. S Clayton.

Perognathus flavescens flavescens Merriam

1889. *Perognathus fasicatus flavescens* Merriam, N. Amer. Fauna, 1:11, October 25, type from Kennedy, Cherry Co., Nebraska.
1900. *Perognathus flavescens*, Osgood, N. Amer. Fauna, 18:20, September 20.

MARGINAL RECORDS.—South Dakota: near Kadoka (Beer, 1961:103, as *P. flavescens* only); Rosebud Agency. Nebraska: Ewing; Neligh; Adams County. Kansas: Sand Creek, Cimarron River; 9 mi. N, 3 mi. E Elkhart. Colorado: Pueblo; Colorado Springs; Boulder County; Lindenmeier (Armstrong, 1972:173). Wyoming: 4½ mi. N, 2 mi. E Huntley (Brown and Metz, 1966:118). Nebraska: 11 mi. S Gordon (Jones, 1964c:170).

Perognathus flavescens gypsi Dice

1929. *Perognathus gypsi* Dice, Occas. Pap. Mus. Zool., Univ. Michigan, 203:1, June 19, type from White Sands, 12 mi. SW Alamogordo, Otero Co., New Mexico.

MARGINAL RECORDS.—New Mexico: 14 mi. W Tularosa; *10 mi. SW Tularosa;* type locality; 18 mi. SW Alamogordo.—**See** addenda.

Perognathus flavescens melanotis Osgood

1900. *Perognathus apache melanotis* Osgood, N. Amer. Fauna, 18:27, September 20, type from Casas Grandes, Chihuahua. Known only from type locality.—**See** addenda.

Perognathus flavescens perniger Osgood

1904. *Perognathus flavescens perniger* Osgood, Proc. Biol. Soc. Washington, 17:127, June 9, type from Vermillion, Clay Co., South Dakota.

MARGINAL RECORDS.—North Dakota: Finley. Minnesota: Polk County; Lac Qui Parle County; near Elk River; Anoka County. Iowa: Backbone Park, Delaware Co.; S of Center Point (Bowles, 1975:82). Missouri: 3⅜ mi. S Hamburg, Iowa (Easterla, 1967:479). Nebraska: Beemer; 1½ mi. S Pilger; Verdigris [= Verdigre]. North Dakota: Parkin [= Bremen].

Perognathus flavescens relictus Goldman

1938. *Perognathus apache relictus* Goldman, Jour. Mamm., 19:495, November 14, type from Medano Springs Ranch, 7600 ft., 15 mi. NE Mosca, San Luis Valley, Colorado.

MARGINAL RECORDS.—Colorado: type locality; *3 mi. S Great Sand Dunes National Monument* (Armstrong, 1972:173). New Mexico: 3 mi. S Pecos; Gran Quivira; *Santa Fe;* Lake Burford.

Perognathus parvus
Great Basin Pocket Mouse

External measurements: 148–198; 77–107; 19–27. Weights of 10 adult males and 10 adult females of *P. p. olivaceus* from Nevada (Hall, 1946:367) are, respectively: 25.4 (21.5–31.0), 20.5 (16.5–28.5). Upper parts approx. pinkish buff or ochraceous buff, thinly to heavily overlaid with blackish; underparts white to buffy; tail long, moderately penicillate, bicolored; antitragus lobed. Skull large, slightly rounded in dorsal profile; tympanic bullae well inflated, barely or nearly meeting anteriorly.

Perognathus parvus bullatus Durrant and Lee

1956. *Perognathus parvus bullatus* Durrant and Lee, Proc. Biol. Soc. Washington, 69:183, December 31, type from Ekker's Ranch, Robbers Roost, 25 mi. (by airline) E Hanksville, 6000 ft., Wayne Co., Utah.

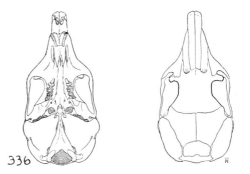

Fig. 336. *Perognathus parvus olivaceus,* Wisconsin Creek, 7800 ft., Nevada, No. 45599 M.V.Z., ♂, X 1½.

MARGINAL RECORDS.—Utah: Huntington (Hayward and Killpack, 1958:27); 1 mi. S San Rafael River, from Buckhorn Wash, 5200 ft.; Old Woman Wash, 23 mi. N Hanksville, 5200 ft.; type locality.

Perognathus parvus clarus Goldman

1917. *Perognathus parvus clarus* Goldman, Proc. Biol. Soc. Washington, 30:147, July 27, type from Cumberland, Lincoln Co., Wyoming.

MARGINAL RECORDS.—Montana (Hoffmann, *et al.,* 1969:589, 590, as *P. parvus* only): 10 mi. N Dillon; 30 mi. SE Dillon. Idaho: 5 mi. E Shelley. Wyoming (Long, 1965a:617): type locality; 26 mi. S, 21 mi. W Rock Springs. Utah: Linwood (Hayward and Killpack, 1958:27). Wyoming: Mountainview (Long, 1965a:617). Utah: Woodruff (Hayward and Killpack, 1958:27); Laketown. Idaho: 4 mi. N Rupert; Dickey; Lemhi; Birch Creek. Montana (Hoffmann, *et al.,* 1969:590, as *P. parvus* only); *10 mi. SW Dillon;* 5 mi. W Badger Pass.

Perognathus parvus columbianus Merriam

1894. *Perognathus columbianus* Merriam, Proc. Acad. Nat. Sci. Philadelphia, 46:263, September 27, type from Pasco, Franklin Co., Washington.
1948. *Perognathus parvus columbianus,* Dalquest, Univ. Kansas Publ., Mus. Nat. Hist., 2:299, April 9.

MARGINAL RECORDS.—Washington: Steamboat Rock; 2 mi. SW Coulee Dam; Sulphur Lake, 7 mi. E Connell; Lyon's Ferry, N side Snake River opposite Perry; type locality; Wenatchee; Waterville.

Perognathus parvus idahoensis Goldman

1922. *Perognathus parvus idahoensis* Goldman, Proc. Biol. Soc. Washington, 35:105, October 17, type from Echo Crater, 20 mi. SW Arco, Butte Co., Idaho.

MARGINAL RECORDS.—Idaho: mouth Little Cottonwood Creek Canyon, Craters of the Moon; type locality; Sparks Well, 23 mi. N Minidoka; Laidlaw Park, 20 mi. N Kimama.

Perognathus parvus laingi Anderson

1932. *Perognathus laingi* Anderson, Bull. Nat. Mus. Canada, 70:100, November 24, type from Anarchist Mtn., near Osoyoos-Bridesville summit, about 8 mi. E Osoyoos Lake, 3500 ft., lat. 49° 08' N, long. 119° 32' W, British Columbia.
1947. *Perognathus parvus laingi*, Anderson, Bull. Nat. Mus. Canada, 102:130, January 24.

MARGINAL RECORDS.—British Columbia: Ashcroft; Kamloops; Vernon; near Bridesville (Cowan and Guiguet, 1965:169); *type locality*.

Perognathus parvus lordi (Gray)

1868. *Abromys lordi* Gray, Proc. Zool. Soc. London, p. 202, May, type from southern British Columbia.
1939. *Perognathus parvus lordi*, Davis, The Recent mammals of Idaho, Caxton Printers, Caldwell, Idaho, p. 266, April 5.

MARGINAL RECORDS.—British Columbia: Keremeos; Oliver (Cowan and Guiguet, 1965:170); Osoyoos (*ibid.*); Midway. Washington: Marcus; Spokane Bridge; Pullman. Idaho: Lewiston. Washington: Asotin; Grande Ronde River, 6 mi. S Anatone; Washtucna; Okanogan; Chelan; Conconully.

Perognathus parvus mollipilosus Coues

1875. *P[erognathus]. mollipilosus* Coues, Proc. Acad. Nat. Sci. Philadelphia, 27:296, August 31, type from Fort Crook, Shasta Co., California.
1900. *Perognathus parvus mollipilosus*, Osgood, N. Amer. Fauna, 18:36, September 20.

MARGINAL RECORDS.—Oregon: W edge Wheeler Creek Canyon, 5500 ft., Crater Lake National Park; Summer Lake. California: Alturas; Likely; Amedee; Vinton; Beckwith; Susanville; Edgewood. Oregon: Lost River.

Perognathus parvus olivaceus Merriam

1889. *Perognathus olivaceus* Merriam, N. Amer. Fauna, 1:15, October 25, type from Kelton, near N end Great Salt Lake, Boxelder Co., Utah.
1900. *Perognathus parvus olivaceus*, Osgood, N. Amer. Fauna, 18:37, September 20.
1889. *Perognathus olivaceus amoenus* Merriam, N. Amer. Fauna, 1:16, October 25, type from Nephi, Juab Co., Utah.
1900. *Perognathus parvus magruderensis* Osgood, N. Amer. Fauna, 18:38, September 20, type from Mt. Magruder, 8000 ft., Esmerelda Co., Nevada. Regarded as inseparable from *olivaceus* by Hall, Mammals of Nevada, p. 367.
1939. *Perognathus parvus plerus* Goldman, Jour. Mamm., 20:352, August 14, type from N end Stansbury Island, Great Salt Lake, Utah. Regarded as inseparable from *olivaceus* by Durrant, Univ. Kansas Publ., Mus. Nat. Hist., 6:242, August 10.

MARGINAL RECORDS.—Idaho: Salmon Creek, 8 mi. W Rogerson. Utah: Clear Creek, 5 mi. SW Nafton 6500 ft.; Blacksmith Fork; Echo Junction (Hayward and Killpack, 1958:27); Millcreek Canyon, 5 mi. SE Salt Lake City, 4700 ft.; Bear Flat, head Slide Canyon, 3 mi.

E Provo; Nephi, 5059 ft.; Paradise Valley (Hayward and Killpack, 1958:27); Price (*ibid.*); Myton (*ibid.*); Roosevelt (*ibid.*); 7 mi. N Greenriver, 4100 ft.; Otter Creek; Minersville (Hayward and Killpack, 1958:27). Nevada: Eagle Valley, 3½ mi. N Ursine, 5900 ft.; N side Potosi Mtn., 5800 ft.; Clark Canyon (Deacon, *et al.*, 1964:403); 1 mi. S, 2½ mi. E Grapevine Peak, 6700 ft. California: Little Onion Valley; Bishop Creek; Fredericksburg, 5100 ft. Nevada: Smoke Creek, 9 mi. E California boundary, 3900 ft.; 17½ mi. W Deep Hole, 4750 ft.; Little High Rock Canyon, 5000 ft.; *1 mi. S Denio, Oregon, 4200 ft.*

Perognathus parvus parvus (Peale)

1848. *Cricetodipus parvus* Peale, Mammalia and ornithology, in U.S. Expl. Exped. . . . , 8:53, type from Oregon, probably near The Dalles, Wasco Co.
1858. *Perognathus parvus*, Cassin, Mammalia and ornithology, in U.S. Expl. Exped. . . . , 8:48.
1858. *Perognathus monticola* Baird, Mammals, in Repts. Expl. Surv. . . . , 8(1):422, July 14, type from west of Rocky

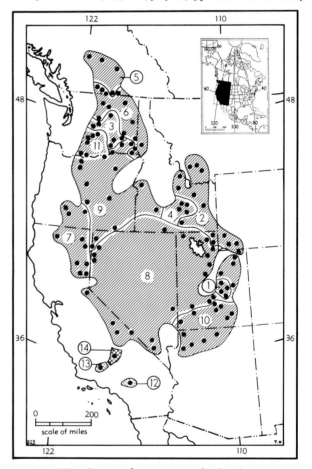

Map 323.   *Perognathus parvus* and related species.

1. *P. p. bullatus*	8. *P. p. olivaceus*
2. *P. p. clarus*	9. *P. p. parvus*
3. *P. p. columbianus*	10. *P. p. trumbullensis*
4. *P. p. idahoensis*	11. *P. p. yakimensis*
5. *P. p. laingi*	12. *P. alticola alticola*
6. *P. p. lordi*	13. *P. alticola inexpectatus*
7. *P. p. mollipilosus*	14. *P. xanthonotus*

Mts., St. Marys ? [= St. Marys Mission, Stevensville, Montana]; regarded by Osgood, N. Amer. Fauna, 18:36, September 20, 1900, as having been obtained at The Dalles, Oregon.

MARGINAL RECORDS.—Washington: eastern Garfield County. Oregon: Baker; Harney. Idaho: Crane Creek, 15 mi. E Midvale; 8 mi. N Hammett. Oregon: Denio. Nevada: 4½ mi. NE Painted Point, 5800 ft.; 1 mi. W Hausen, 4650 ft. Oregon: Fremont; Prineville; Maupin; The Dalles; Umatilla. Washington: Wallula; Prescott.

Perognathus parvus trumbullensis Benson

1937. *Perognathus parvus trumbullensis* Benson, Proc. Biol. Soc. Washington, 50:181, October 28, type from Nixon Spring, 6250 ft., Mt. Trumbull, Mohave Co., Arizona.

MARGINAL RECORDS.—Utah: Aquarius Guard Station; Bown's Reservoir; Hite (Durrant and Dean, 1959:85); river mile 113, Glen Canyon, west bank (*ibid.*); Hall Ranch. Arizona: Houserock Valley, North Canyon at edge juniper belt (Cockrum, 1961:131); Ryan (*ibid.*); type locality; slopes Mt. Dellenbaugh (Hoffmeister and Nader, 1963:93); 6 mi. N Wolf Hole, 4900 ft. Utah: 19 mi. W Enterprise; Parowan (Hayward and Killpack, 1958:27).

Perognathus parvus yakimensis Broadbooks

1954. *Perognathus parvus yakimensis* Broadbooks, Jour. Mamm., 35:96, February 10, type from 16 mi. NW Naches, Rocky Flat (or Rocky Prairie), 3800 ft., Yakima Co., Washington.

MARGINAL RECORDS.—Washington: Ellenburg; Vantage, thence southward and westward along Columbia River to Cliffs; Dallesport; 10 mi. W Wiley City; type locality.

Perognathus alticola
White-eared Pocket Mouse

External measurements: 160–181; 72–97; 21–23. Upper parts olivaceous-buff to near wood brown; underparts white; lateral line usually faintly expressed; tail bicolored or tricolored, above like upper parts proximally, but shading to dusky or black at tip, white below. Skull closely resembles that of *P. parvus* but smaller; ascending branches of supraoccipital exceedingly broad and heavy.

Perognathus alticola alticola Rhoads

1894. *Perognathus alticolus* Rhoads, Proc. Acad. Nat. Sci. Philadelphia, 45:412, January 27, type from Squirrel Inn, near Little Bear Valley, 5500 ft., San Bernardino Mts., San Bernardino Co., California.

MARGINAL RECORDS.—California: type locality; *1 mi. E Strawberry Peak, 5750 ft., San Bernardino Mts.*

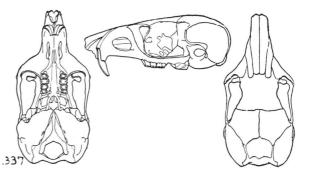

Fig. 337. *Perognathus alticola alticola*, 2 mi. E Strawberry Peak, 5750 ft., San Bernardino Co., California, No. 47408 M.V.Z., ♂, X 1½.

Perognathus alticola inexpectatus Huey

1926. *Perognathus alticola inexpectatus* Huey, Proc. Biol. Soc. Washington, 39:121, December 27, type from 14 mi. W Lebec, 600 ft., Kern Co., California. Known only from type locality.

Perognathus xanthonotus Grinnell
Yellow-eared Pocket Mouse

1912. *Perognathus xanthonotus* Grinnell, Proc. Biol. Soc. Washington, 25:128, July 31, type from Freeman Canyon, 4900 ft., E slope Walker Pass, Kern Co., California.

Measurements of the type, an adult male, are: 170, 85, 22.5. Upper parts "between ochraceous-buff and cream-buff, almost perfectly clear on sides of body and head, and but slightly obscured mid-dorsally with scanty dusky tippings to the hairs; feet and lower surface white; . . . tail well clothed with hairs, and distinctly penicillate, beneath white, above faint cream-buff with a slight dusky tinge on terminal fifth. "Skull.—Distinctly smaller than in *olivaceus*, mastoids and audital bullae notably so; closely similar to *alticola*." (Grinnell, 1912:128.)

MARGINAL RECORDS.—California: type locality; *W slope Walker Pass, 4600 ft.; head Kelso Valley, 5000 ft.*

Perognathus flavus
Silky Pocket Mouse

Mexican subspecies reviewed by Baker, Univ. Kansas Publ., Mus. Nat. Hist., 7:339–347, 1 fig. in text, February 15, 1954. See also Anderson, Bull. Amer. Mus. Nat. Hist.,

148:302–304, 308, 309, September 8, 1972, and Wilson, Proc. Biol. Soc. Washington, 86:175–192, May 31, 1973.

External measurements: 100–122; 44–60; 15–18. Upper parts finely lined with black on ochraceous buff and in some subspecies yellowish-buff and pinkish-buff; lateral line of buff faintly expressed; postauricular spot clear buff; subauricular spot white (in subspecies gilvus and merriami); underparts white, and in some specimens having a faint tawny suffusion; tail whitish below, dusky or buffy above. Skull: interparietal almost as long as wide; auditory bullae not meeting anteriorly; lower premolar no larger than last molar and in several subspecies smaller.

Geographic variation in this species is great. Until 1973 the taxa gilvus and merriami were considered to constitute a species separate from flavus.

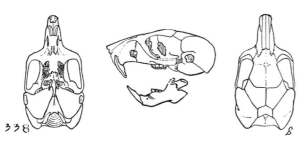

Fig. 338. Perognathus flavus bunkeri, 23 mi. (by road) NW St. Francis, Cheyenne Co., Kansas, No. 12092 K.U., ♀, X 1½.

In August 1966 Bahtiye Mursaloğlu and E. R. Hall observed that in University of Kansas materials from Coahuila, México, the morphological features of some specimens (e.g., 40300 from 1 mi. SW San Pedro de las Colonias, 58686 from ½ mi. E Las Margaritas, and 40913 from 4 mi. N Acatita) were intermediate between features characteristic of P. merriami and P. flavus pallescens. Mursaloğlu and Hall were uncertain about the cause of this intermediacy and did not publish their findings.

In 1972 (page 308) Sydney Anderson reported much the same condition in Chihuahuan material but recognized two species.

In 1973 (pages 175–191) Wilson reported morphological intermediacy in specimens from eastern New Mexico and interpreted it as intergradation between Perognathus merriami gilvus Osgood 1900 and Perognathus flavus flavus Baird 1885. If the two taxa intergrade, P. m. gilvus and P. m. merriami are subspecies of the earlier named P. flavus.

It is unusual but conceivable for essentially the same two "stocks" to intergrade in eastern New Mexico and to occur together without intergrad-

ing in Chihuahua. Perhaps such is the case; both Anderson and Wilson are careful observers. Even so, to clear up the matter a re-examination of specimens from the two critical areas, the eastern parts of New Mexico and Chihuahua, supported if possible by additional specimens, is warranted. Additional investigation of the P. merriami complex in Chihuahua is bound to be rewarding.

Because Wilson (op. cit.) published after Anderson (loc. cit.) did, Map 324 herewith shows only Perognathus flavus in the part of Chihuahua where Anderson thought both species (P. flavus, his Fig. 309, and P. merriami, his Fig. 313) occurred together.

Perognathus flavus bimaculatus Merriam

1889. Perognathus bimculatus Merriam, N. Amer. Fauna, 1:12, October 25, type from Fort Whipple, Yavapai Co., Arizona.
1900. Perognathus flavus bimaculatus, Osgood, N. Amer. Fauna, 18:24, September 20.

MARGINAL RECORDS.—Arizona: S rim Grand Canyon (Pasture Wash Ranger Station); 2 mi. NE Kirkland Junction, 4000 ft. (Cockrum, 1961:125); near Prescott; Aubrey Valley, 10 mi. S Pine Springs, 6000 ft. (Cockrum, 1961:125).

Perognathus flavus bunkeri Cockrum

1951. Perognathus flavus bunkeri Cockrum, Univ. Kansas Publ., Mus. Nat. Hist., 5:205, December 15, type from Conard Farm, 1 mi. E Coolidge, Hamilton Co., Kansas.

MARGINAL RECORDS.—Colorado (Armstrong, 1972:175): Horsetail Creek, 17 mi. NW Stoneham; Wray. Kansas: 23 mi. by road NW St. Francis; Vincent Ranch, 8 mi. N, 4 mi. W McAllaster, N. Fork Smoky [Hill] River; Wakeeney; Rezeau Ranch, 5 mi. N Belvidere. Oklahoma: 14 mi. S Olustee (Blair, 1939:115, as P. f. flavus; may be P. f. bunkeri); 2 mi. E Eva. Colorado (Armstrong, 1972:175): 1 mi. S, 7 mi. W Trinidad; 3 mi. W Salida; Loveland.

Perognathus flavus flavus Baird

1855. Perognatus [sic] flavus Baird, Proc. Acad. Nat. Sci. Philadelphia, 7:332, type from El Paso, El Paso Co., Texas.

MARGINAL RECORDS.—New Mexico: Chico Springs (Wilson, 1973:178). Texas (Davis, 1966:161, as P. flavus only): Moore County; Hutchinson County. New Mexico (Wilson, 1973:178): Santa Rosa; Carrizozo; Carlsbad. Texas (Wilson, 1973:178): Sierra Blanca; Valentine; Alpine. Chihuahua (Anderson, 1972:303): 15 mi. ESE Boquilla; 5 mi. N Parral. Durango: 7 mi. NNW La Zarca, 6000 ft. (Baker and Greer, 1962:96); Río Sestín (ibid.). Chihuahua (Anderson, 1972:303): 4 mi. NW San Francisco de Borja, 5700 ft.; 1 mi. N Arados, 1540 m.; 2½ mi. W El Carmen; 6 mi. SSW Nuevo Casas Grandes, 4800 ft.; 35 mi. NW Dublán, 5300 ft. Arizona: near Tumacacori Mission;

Baboquivari Mts.; 11 mi. W Casa Grande (Hoffmeister, 1959:16); SE Pinal County; central Graham County. New Mexico: O Bar O Canyon, 12 mi. S, 25 mi. E Reserve (Forbes, 1962:278, as *P. flavus* only); North Plains, 35 mi. S, 15 mi. W Grants (*ibid.*, as *P. flavus* only); Zuni Canyon; 1 mi. SE El Rito; Taos.

Perognathus flavus fuliginosus Merriam

1890. *Perognathus fuliginosus* Merriam. N. Amer. Fauna, 3:74, September 11, type from cedar belt NE of San Francisco Mtn., 7000 ft., Coconino Co., Arizona.

1900. *Perognathus flavus fulginosus,* Osgood, N. Amer. Fauna, 18:25, September 20.

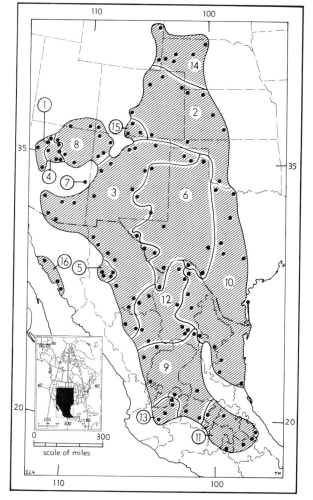

Map 324. *Perognathus flavus.*

1. *P. f. bimaculatus*	9. *P. f. medius*
2. *P. f. bunkeri*	10. *P. f. merriami*
3. *P. f. flavus*	11. *P. f. mexicanus*
4. *P. f. fuliginosus*	12. *P. f. pallescens*
5. *P. f. fuscus*	13. *P. f. parviceps*
6. *P. f. gilvus*	14. *P. f. piperi*
7. *P. f. goodpasteri*	15. *P. f. sanluisi*
8. *P. f. hopiensis*	16. *P. f. sonoriensis*

MARGINAL RECORDS.—Arizona: Red Horse Wash; Tanner Tank; 2⅖ mi. W Wupatki Ruins (Cockrum, 1961:126); Winona (Hoffmeister and Carothers, 1969:187, as *P. flavus* only); vic. Flagstaff; Bly.

Perognathus flavus fuscus Anderson

1972. *Perognathus flavus fuscus* Anderson, Bull. Amer. Mus. Nat. Hist., 148:304, September 8, type from 2 mi. W Miñaca, 6900 ft., Chihuahua.

MARGINAL RECORDS.—Chihuahua (Anderson, 1972:304): Rancho San Ignacio, 4 mi. S, 1 mi. W Santo Tomás; NE side Laguna de Bustillos, 6750 ft.; 20 mi. by road N Cuauhtémoc; type locality.

Perognathus flavus gilvus Osgood

1900. *Perognathus merriami gilvus* Osgood, N. Amer. Fauna, 18:22, September 20, type from Eddy, near Carlsbad, Eddy Co., New Mexico.

1973. *Perognathus flavus gilvus,* Wilson, Proc. Biol. Soc. Washington, 86:191, May 31.

MARGINAL RECORDS.—Texas: Mobeetie; 20 mi. E Rock Springs; Eagle Pass. Coahuila: 15 mi. SE Langtry, Texas. Texas: Boquillas (WIlson, 1973:177). Coahuila: 2 mi. SSE Castillión. Texas (Wilson, 1973:177): *15 mi. S Marathon; Marathon; Presidio County; Van Horn;* Kermit. New Mexico (Wilson, 1973:177): Carlsbad; 40 mi. W Roswell; 25 mi. W Tucumcari.—See also last paragraph of account above of *P. flavus* for records of occurrence in *Chihuahua* listed by Anderson (1972:308, 309).

Perognathus flavus goodpasteri Hoffmeister

1956. *Perognathus flavus goodpasteri* Hoffmeister, Proc. Biol. Soc. Washington, 69:55, May 21, type from 2¾ mi. NW Springerville, Apache Co., Arizona.

MARGINAL RECORDS.—Arizona: 3 mi. N Springerville; *type locality.*

Perognathus flavus hopiensis Goldman

1932. *Perognathus flavus hopiensis* Goldman, Proc. Biol. Soc. Washington, 45:89, June 21, type from Oraibi, 6000 ft., Hopi Indian Reservation, Navajo Co., Arizona.

MARGINAL RECORDS.—Colorado: Ashbaugh's Ranch. New Mexico: Fruitland; ¼ mi. SE base El Huerfano (Findley, *et al.,* 1975:163, as *P. flavus* only); 3 mi. N Crownpoint (Forbes, 1962:278, as *P. flavus* only); Wingate. Arizona: Holbrook; Winslow; entrance to Wupatki National Monument, along U.S. Highway 89; SE corner Grand Canyon National Park. Utah: Navajo Mountain Trading Post, 5 mi. SE Navajo Mtn.; ½ mi. NW Bluff, 4500 ft.

Perognathus flavus medius Baker

1954. *Perognathus flavus medius* Baker, Univ. Kansas Publ., Mus. Nat. Hist., 7:343, February 15, type from 1 mi. S, 6 mi. E Rincón de Romos, 6550 ft., Aguascalientes.

MARGINAL RECORDS.—Coahuila: 7 mi. S, 4 mi. E Bella Unión, 7200 ft. San Luis Potosí: Jesús María, 6000 ft., Guanajuato: 5 mi. E Celaya, 6000 ft.; 4 mi. N, 5 mi. W León, 7000 ft. Aguascalientes: 3 mi. SW Aguascalientes, 6100 ft. Jalisco: Huejugilla, 5400 ft. Durango: 4 mi. S Morcillo, 6450 ft. (Baker and Greer, 1962:96); 2 mi. N Cuencamé, 5200 ft. (*ibid.*).

Perognathus flavus merriami J. A. Allen

1892. *Perognathus merriami* J. A. Allen, Bull. Amer. Mus. Nat. Hist., 4:45, March 25, type from Brownsville, Cameron Co., Texas.
1973. *Perognathus flavus merriami*, Wilson, Proc. Biol. Soc. Washington, 86:191, May 31.
1896. *Perognathus mearnsi* J. A. Allen, Bull. Amer. Mus. Nat. Hist., 8:237, November 21, type from Watson's Ranch, 15 mi. SW San Antonio, Bexar Co., Texas.

MARGINAL RECORDS.—Texas: Baylor County (Baccus, 1971:181); Montague County (Dalquest, 1968:17); Palo Pinto County (Davis, 1966:160); Austin; Padre Island, thence down coast to Tamaulipas (Alvarez, 1963:429); 1 mi. S Alta Mira; 17 mi. SW Tula, 3900 ft.; Ciudad Victoria; *Hidalgo*. Nuevo León: Linares. Coahuila: Saltillo (Wilson, 1973:177); 5 mi. N, 2 mi. W Monclova; 29 mi. N, 6 mi. E Sabinas (Baker, 1956:236); 11 mi. W Hda. San Miguel (*ibid.*). Texas: Kerrville; Mason.

Perognathus flavus mexicanus Merriam

1894. *Perognathus flavus mexicanus* Merriam, Proc. Acad. Nat. Sci. Philadelphia, 46:265, September 27, type from Tlalpan, Distrito Federal, México.

MARGINAL RECORDS.—Querétaro: 6 mi. E Querétaro, 7400 ft. Hidalgo: Ixmiquilpan, 6000 ft. Veracruz: 2 km. W Perote, 8000 ft. Puebla: 10 km. W San Andrés, 8000 ft.; 7 mi. S, 3 mi. E Puebla, 6850 ft. Morelos: 10 mi. N Cuautla.

Perognathus flavus pallescens Baker

1954. *Perognathus flavus pallescens* Baker, Univ. Kansas Publ., Mus. Nat. Hist., 7:345, February 15, type from 1 mi. SW San Pedro de las Colonias, 3700 ft., Coahuila.

MARGINAL RECORDS.—Coahuila: Cañón del Cochino; ½ mi. E Las Margaritas; 3 mi. NW Cuatro Ciénegas; La Pastora Rancho, 15 mi. N, 41 mi. W Saltillo; 3 mi. N, 5 mi. W La Rosa; N foot Sierra Guadalupe, 6400 ft., 10 mi. S, 5 mi. W General Cepeda. Durango: 1½ mi. NW Nazas, 4100 ft. (Baker and Greer, 1962:96); 37 mi. W Mapimi, 5500 ft. (*ibid.*). Chihuahua (Anderson, 1972:304): 5 mi. W Jiménez; Sierra Almagre, 12 mi. S Jaco, 5300 ft.

Perognathus flavus parviceps Baker

1954. *Perognathus flavus parviceps* Baker, Univ. Kansas Publ., Mus. Nat. Hist., 7:344, February 15, type from 4 mi. W, 2 mi. S Guadalajara, 5100 ft., Jalisco.

MARGINAL RECORDS.—Jalisco: 1 mi. NE Villa Hidalgo, 6500 ft.; 3 mi. NW Yahualica; 2 mi. N, ½ mi. W Guadalajara; 21 mi. SW Guadalajara; *4 mi. W Guadalajara*.

Perognathus flavus piperi Goldman

1917. *Perognathus flavus piperi* Goldman, Proc. Biol. Soc. Washington, 30:148, July 27, type from 23 mi. SW Newcastle, Weston Co., Wyoming.

MARGINAL RECORDS.—Wyoming: type locality. Nebraska (Jones, 1964c:173): Alliance; Fort Niobrara National Widelife Refuge; Kelso; 5 mi. N Bridgeport; 6 mi. N Mitchell. Wyoming: between 5 mi. E and 18 mi. NW La Grange (Maxwell and Brown, 1968:144, as *P. flavus* only).

Perognathus flavus sanluisi Hill

1942. *Perognathus flavus sanluisi* Hill, Amer. Mus. Novit., 1212:1, December 7, type from 9 mi. E Center, Saguache Co. (or 20 mi. NW Alamosa, Alamosa Co.), 7580 ft., Colorado.

MARGINAL RECORDS.—Colorado (Armstrong, 1972:176): type locality; 5 mi. SSE Ft. Garland; Conejos River, 8300 ft.

Perognathus flavus sonoriensis Nelson and Goldman

1934. *Perognathus flavus sonoriensis* Nelson and Goldman, Jour. Washington Acad. Sci., 24:267, June 15, type from Costa Rica Ranch, lower Río Sonora, Sonora.

MARGINAL RECORDS.—Sonora: type locality; 3 mi. S Maytorena (Cockrum and Bradshaw, 1963:8); 10 mi. N Empalme (Bradshaw and Hayward, 1960:282, as *P. flavus* only).

Perognathus longimembris
Little Pocket Mouse

x 1

External measurements: 110–151; 53–86; 15–20. Upper parts approx. pinkish- or ochraceous-buff, the exact shades varying markedly geographically, overlaid with black or blackish hairs to a greater or lesser extent so that some subspecies appear quite dark and others are distinctly buffy in overall tone; underparts pale tawny to buffy or white, sometimes pectoral region alone is white and remainder tawny or buffy; tail usually bicolored.

Perognathus longimembris aestivus Huey

1928. *Perognathus longimembris aestivus* Huey, Trans. San Diego Soc. Nat. Hist., 5:87, January 18, type from Sangre de Cirsto, Valle San Rafael, western base Sierra Juárez, lat. 31° 52′ N, long. 116° 06′ W, Baja California.

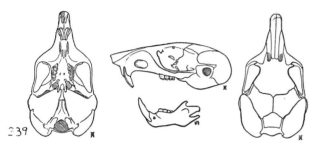

Fig. 339. *Perognathus longimembris panamintinus*, R. 38 E, T. 6 S, Esmeralda Co., Nevada, No. 38586 M.V.Z., ♀, X 1½.

MARGINAL RECORDS.—Baja California: type locality; El Valle de la Trinidad.

Perognathus longimembris arcus Benson

1935. *Perognathus longimembris arcus* Benson, Univ. California Publ. Zool., 40:451, December 31, type from Rainbow Bridge, San Juan Co., Utah.

MARGINAL RECORDS.—Utah: type locality; River Mile 43, Glen Canyon, E bank (Durrant and Dean, 1959:84).

Perognathus longimembris arizonensis Goldman

1931. *Perognathus longimembris arizonensis* Goldman, Proc. Biol. Soc. Washington, 44:134, October 17, type from 10 mi. S Jacobs Pools, 4000 ft., Houserock Valley, N side Marble Canyon of Colorado River, Arizona.

MARGINAL RECORDS.—Utah: Kaiparowits Plateau; River Mile 41, Glen Canyon, W bank (Durrant and Dean, 1959:84). Arizona (Cockrum, 1961:128): 7 mi. E Jacobs Pools; type locality; Fredonia. Utah: Kanab.

Perognathus longimembris bangsi Mearns

1898. *Perognathus longimembris bangsi* Mearns, Bull. Amer. Mus. Nat. Hist., 10:300, August 31, type from Palm Springs, Colorado Desert, Riverside Co., California.

1900. *Perognathus panamintinus arenicola* Stephens, Proc. Biol. Soc. Washington, 13:151, June 13, type from San Felipe Narrows, San Diego Co., California.

MARGINAL RECORDS.—California: Indian Cove, Joshua Tree National Monument (Chew and Butterworth, 1964:203, as *P. longimembris* only); E side San Felipe Narrows; *Borego Springs*; Whitewater Ranch.

Perognathus longimembris bombycinus Osgood

1907. *Perognathus bombycinus* Osgood, Proc. Biol. Soc. Washington, 20:19, February 23, type from Yuma, Yuma Co., Arizona.

1929. *Perognathus longimembris bombycinus*, Nelson and Goldman, Proc. Biol. Soc. Washington, 42:104, March 25.

MARGINAL RECORDS.—Arizona: 5½ mi. E Parker, Yuma Co. (Hoffmeister, 1959:16, as *P. longimembris* only); Wellton (Cockrum, 1961:128). Sonora: Pinacate Lava Flows (Patton, 1967:437); Colonia Lerdo. Baja California: San Felipe. California: 3 mi. W Pilot Knob. Arizona: near Ehrenberg.

Perognathus longimembris brevinasus Osgood

1900. *Perognathus panamintinus brevinasus* Osgood, N. Amer. Fauna, 18:30, September 20, type from San Bernardino, San Bernardino Co., California. According to Osgood (Proc. Biol. Soc. Washington, 31:96, June 29, 1918) this is perhaps a synonym of *P. longimembris longimembris*.

1928. *Perognathus longimembris brevinasus*, Huey, Trans. San Diego Soc. Nat. Hist., 5:88, January 18.

MARGINAL RECORDS.—California: San Fernando; type locality; Cabazon; *Aguanga*; 2½ mi. N Oak Grove; *Burbank*.

Perognathus longimembris gulosus Hall

1941. *Perognathus longimembris gulosus* Hall, Proc. Biol. Soc. Washington, 54:55, May 20, type from near [¼ mi. S] Smith Creek Cave, 5800 ft., Mt. Moriah, White Pine Co., Nevada.

MARGINAL RECORDS.—Utah: Kelton, 4225 ft.; 1 mi. S Callao (Shippee and Egoscue, 1958:276); 7 mi. NE Granite Peak (Granite Mtn.) (*ibid.*); ½ mi. E Wig Mtn. (*ibid.*); 6 mi. N Camel Back Mtn. (*ibid.*); 50 mi. W Milford; 55 mi. W of Milford, 5500 ft.; 5 mi. S Garrison, 5400 ft. Nevada: type locality; 8 mi. S Wendover, 4700 ft.; 13 mi. N Montello, 5000 ft.

Perognathus longimembris internationalis Huey

1939. *Perognathus longimembris internationalis* Huey, Trans. San Diego Soc. Nat. Hist., 9:47, August 31, type from Baja California side of international boundary at Jacumba, San Diego, California.

MARGINAL RECORDS.—California: La Puerta Valley; *San Felipe Valley*. Baja California: type locality.

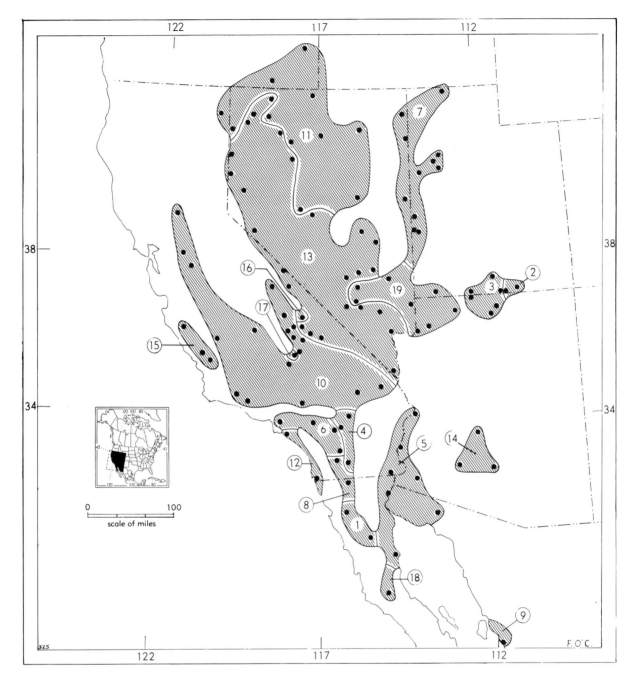

Map 325. *Perognathus longimembris.*

Guide to subspecies	5. *P. l. bombycinus*	10. *P. l. longimembris*	15. *P. l. psammophilus*
1. *P. l. aestivus*	6. *P. l. brevinasus*	11. *P. l. nevadensis*	16. *P. l. salinensis*
2. *P. l. arcus*	7. *P. l. gulosus*	12. *P. l. pacificus*	17. *P. l. tularensis*
3. *P. l. arizonensis*	8. *P. l. internationalis*	13. *P. l. panamintinus*	18. *P. l. venustus*
4. *P. l. bangsi*	9. *P. l. kinoensis*	14. *P. l. pimensis*	19. *P. l. virginis*

Perognathus longimembris kinoensis Huey

1935. *Perognathus longimembris kinoensis* Huey, Trans. San Diego Soc. Nat. Hist., 8:73, August 24, type from Bahía Kino (N end of sand-dune peninsula that borders bay and forms northern arm of estuary), Sonora. Known only from type locality.

Perognathus longimembris longimembris (Coues)

1875. *O[tognosis]. longimembris* Coues, Proc. Acad. Nat. Sci. Philadelphia, 27:305, August 31, type from Old Fort Tejon, Tehachapi Mountains, Kern Co., California.
1889. *Peognathus longimembris,* Merriam, N. Amer. Fauna, 1:13, October 25.

1904. *Perognathus elibatus* Elliot, Field Columb. Mus., Publ. 87, Zool. Ser., 3(14):252, January 7, type from Lockwood Valley, near Mt. Piños, Ventura Co., California.

1904. *Perognathus pericalles* Elliot, Field Columb. Mus., Publ. 87, Zool. Ser., 3(14):252, January 7, type from Keeler, Ownes Lake, Inyo Co., California.

MARGINAL RECORDS.—California: Marysville Buttes; Three Rivers; Onyx; Little Owen Lake; Olancha; W side Owens Lake; W of Independence toward Kearsarge Pass; Laws; 2 mi. SE Keeler; Lower Centennial Springs, Coso Hills; vic. Providence Mts.; Lavic; Pearblossom (Chew, *et al.*, 1965:478, as *P. longimembris* only); Lockwood Valley, 5500 ft., near Mt. Piños; San Emigdio; Huron; Ripon; Lodi.

Perognathus longimembris nevadensis Merriam

1894. *Perognathus nevadensis* Merriam, Proc. Acad. Nat. Sci. Philadelphia, 46:264, September 27, type from Halleck, East Humboldt Valley, Elko Co., Nevada.

1933. *Perognathus longimembris nevadensis*, Grinnell, Univ. California Publ. Zool., 40:147, September 26.

MARGINAL RECORDS.—Oregon: Rome. Nevada: 36 mi. NE Paradise Valley, 5500 ft.; 3 mi. S Izenhood; 2 mi. NW Halleck, 5200 ft.; 8 mi. E Eureka; Smiths Creek Valley, 3 mi. W Railroad Pass, 5500 ft.; 15 mi. SW Winnemucca; Jackson Creek Ranch, 17 mi. S, 5 mi. W Quinn River Crossing, 4000 ft.; 10 mi. SE Hausen, 4670 ft. California: near Eagleville. Oregon: Tumtum Lake.

Perognathus longimembris pacificus Mearns

1898. *Perognathus pacificus* Mearns, Bull. Amer. Mus. Nat. Hist., 10:299, August 31, type from Mexican boundary monument 258, shore of Pacific Ocean, San Diego Co., California.

1932. *Perognathus longimembris pacificus*, von Bloeker, Proc. Biol. Soc. Washington, 45:128, September 9.

1932. *Perognathus longimembris cantwelli* von Bloeker, Proc. Biol. Soc. Washington, 45:128, September 9, type from Hyperion, Los Angeles Co., California. Regarded as identical with *P. l. pacificus* by Huey, Trans. San Diego Soc. Nat. Hist., 9:48, 49, August 31, 1939.

MARGINAL RECORDS.—California: Palisades del Rey; type locality.

Perognathus longimembris panamintinus Merriam

1894. *Perognathus longimembris panamintinus* Merriam, Proc. Acad. Nat. Sci. Philadelphia, 46:265, September 27, type from Perognathus Flat, 5200 ft., Panamint Mts., Inyo Co., California.

MARGINAL RECORDS.—Nevada: 1½ mi. N Quinn River Crossing, 4100 ft.; 11 mi. E, 1 mi. N Jungo, 4200 ft.; 9½ mi. E, 1 mi. S Fanning, 4100 ft., Buena Vista Valley; 5 mi. SE Millett P.O., 5500 ft.; Railroad Valley, 2½ to 3¼ mi. S Lock's Ranch, 5000 ft.; 15 mi. WSW Sunnyside; 10 mi. N Seeman Pass, 4600 ft.; Penoyer Valley, 14 mi. NNW Groom Baldy; Belted Range, 2 to 4½ mi. NW Indian Spring, 5700–6300 ft.; Nevada Atomic Test

Site (Allred and Beck, 1963:190, as *P. longimembris* only); Indian Spring Valley, 14 mi. N Indian Springs, 3100 ft.; Mormon Well, 6500 ft.; Boulder City; 12 mi. S, 5 mi. E Searchlight, 2600 ft. California: Wild Rose Canyon, 6300 ft.; Willow Spring; Lee Flat; *Grapevine Canyon, 1 mi. W Jackass Springs*; Whitewater Canyon, Deep Springs Valley (Chew, *et al.*, 1965:478, as *P. longimembris* only); Morans, 5000 ft. Nevada: Mason Valley, 12 mi. E Wellington, 5000 ft.; 3 mi. E Reno; 3½ mi. NW Flanigan, 4200 ft.; Smoke Creek, 9 mi. E California boundary, 3900 ft.; mouth Little High Rock Canyon, 5000 ft.; Soldier Meadows, 4600 ft.

Perognathus longimembris pimensis Huey

1937. *Perognathus longimembris pimensis* Huey, Trans. San Diego Soc. Nat. Hist., 8:353, June 15, type from 11 mi. W Casa Grande, Pinal Co., Arizona.

MARGINAL RECORDS.—Arizona: Marinette; *ca.* 3 mi. W Casa Grande (Patton, 1967:437); *type locality; ca.* 8 mi. W Gila Bend (Patton, 1967:438).

Perognathus longimembris psammophilus von Bloeker

1937. *Perognathus longimembris psammophilus* von Bloeker, Proc. Biol. Soc. Washington, 50:153, September 10, type from W side Arroyo Seco, 150 ft., 4 mi. S Soledad, Monterey Co., California.

MARGINAL RECORDS.—California: type locality; Miguel; Sandiego Joe's (=Santiago Springs), 2600 ft.

Perognathus longimembris salinensis Bole

1937. *Perognathus longimembris salinensis* Bole, Sci. Publs., Cleveland Mus. Nat. Hist., 5(2):3, December 4, type from 1 mi. N Salt Camp, 1060 ft., W edge salt lake, Saline Valley, Inyo Co., California.

MARGINAL RECORDS.—California: *North Sand Dunes, Saline Valley*; type locality.

Perognathus longimembris tularensis Richardson

1937. *Perognathus longimembris tularensis* Richardson, Jour. Mamm., 18:510, November 22, type from 1 mi. W Kennedy Meadows, 6000 ft., S. Fork Kern River, Tulare Co., California.

MARGINAL RECORDS.—California: type locality; *Chimney Meadows, 6200 ft.*

Perognathus longimembris venustus Huey

1930. *Perognathus longimembris venustus* Huey, Trans. San Diego Soc. Nat. Hist., 6:233, December 24, type from San Agustín, lat. 30° N, long. 115° W, Baja California. Known only from type locality.

Perognathus longimembris virginis Huey

1939. *Perognathus longimembris virginis* Huey, Trans. San Diego Soc. Nat. Hist., 9:55, August 31, type from St. George, 2950 ft., Washington Co., Utah.

MARGINAL RECORDS.—Nevada: Desert Valley, 20 mi. SW Pioche, 5400–5700 ft. Utah: type locality.

Arizona (Hoffmeister and Durham, 1971:32, 33): *6 mi. S, 8 mi. W Fredonia;* Antelope Valley, 27 mi. SW Fredonia; 3 mi. W Lower Pigeon Spring, 4400 ft.; Hoover Dam Ferry [= Pierce Ferry?], 650 ft. (Cockrum, 1961:128). Nevada: S side Virgin River, ¾ mi. E Mesquite, 1750 ft.; Emigrant Valley, 9–9½ mi. S Oak Spring, 4400 ft.; 9 mi. E Wheelbarrow Peak.

Perognathus amplus
Arizona Pocket Mouse

External measurements: 123–170; 72–95; 17–22. Upper parts pinkish buff to pale ochraceous-salmon, overlaid with black that varies in degree so much that some subspecies appear distinctly blackish; underparts white or faintly washed with buff; tail more or less distinctly bicolored according to subspecies and season; lateral line buffy; orbital region usually distinctly paler than remainder of dorsum. Skull large; mastoids excessively developed, but auditory bullae not correspondingly large; interparietal small; nasals and rostrum long and slender; zygomatic arches narrower anteriorly than posteriorly.

This species, although possessing a unique combination of cranial characters, is not greatly unlike *P. longimembris* externally. In normal adult specimens *P. amplus* can usually be distinguished from *P. longimembris* by larger size, longer (actually and relatively) tail (more than 75% head-body length in *P. amplus*, less than 75% in *P. longimembris*).

Perognathus amplus ammodytes Benson

1933. *Perognathus amplus ammodytes* Benson, Proc. Biol. Soc. Washington, 46:110, April 27, type from 2 mi. S Cameron, Coconino Co., Arizona.

MARGINAL RECORDS.—Arizona: base of northern end Echo Cliffs, S of Grand Canyon Bridge; type locality.

Perognathus amplus amplus Osgood

1900. *Perognathus amplus* Osgood, N. Amer. Fauna, 18:32, September 20, type from Fort Verde, Yavapai Co., Arizona. Known only from type locality.

Perognathus amplus cineris Benson

1933. *Perognathus amplus cineris* Benson, Proc. Biol. Soc. Washington, 46:109, April 27, type from near Wupatki Ruins, Wupatki National Monument, 27 mi. NE Flagstaff, Coconino Co., Arizona. Known only from type locality.

Perognathus amplus jacksoni Goldman

1933. *Perognathus amplus jacksoni* Goldman, Jour. Washington Acad. Sci., 23:465, October 15, type from Congress Junction, 3000 ft., Yavapai Co., Arizona.

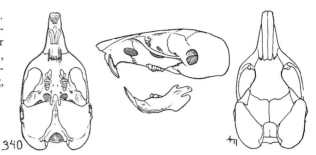

Fig. 340. *Perognathus amplus jacksoni*, 20 mi. SW Phoenix, Maricopa Co., Arizona, No. 7792 K.U., sex ?, X 1½.

MARGINAL RECORDS (Cockrum, 1961:129, 130).—Arizona: Kirkland, 3900 ft.; 30 mi. from Roosevelt on Young Rt. 1; Rice, 2700 ft.; 2 mi. N Florence; 20 mi. SW Phoenix; type locality.

Perognathus amplus pergracilis Goldman

1932. *Perognathus amplus pergracilis* Goldman, Jour. Washington Acad. Sci., 22:387, July 19, type from Hackberry, 3500 ft., Mohave Co., Arizona.

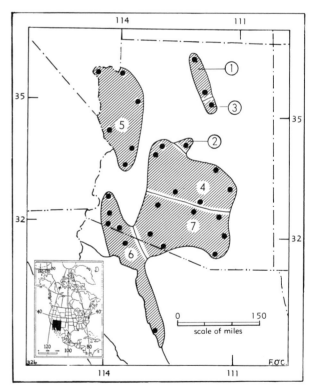

Map 326. *Perognathus amplus.*

Guide to subspecies
1. *P. a. ammodytes*
2. *P. a. amplus*
3. *P. a. cineris*
4. *P. a. jacksoni*
5. *P. a. pergracilis*
6. *P. a. rotundus*
7. *P. a. taylori*

MARGINAL RECORDS (Cockrum, 1961:130, 131, unless otherwise noted).—Arizona: Pierce Ferry, Colorado River; Peach Springs, 4000 ft.; Alamo Crossing (Patton, 1967:437); Salome; Lucky Star Mine, Chemehuevis Mts.; Hoover Dam Ferry.

Perognathus amplus rotundus Goldman

1932. *Perognathus amplus rotundus* Goldman, Jour. Washington Acad. Sci., 22:387, July 19, type from Wellton, Yuma Co., Arizona.

MARGINAL RECORDS.—Arizona (Cockrum, 1961:131): Castle Dome, 1200 ft.; Tule Well. Sonora: Pápago Tanks; 2 mi. WNW Puerto Libertad. Arizona: Tinajas Altas (Cockrum, 1961:131); type locality.

Perognathus amplus taylori Goldman

1932. *Perognathus amplus taylori* Goldman, Jour. Washington Acad. Sci., 22:488, October 19, type from Santa Rita Range Reserve (near NE station), 4000 ft., 35 mi. S Tucson, Pima Co., Arizona.

MARGINAL RECORDS.—Arizona: Gila Bend; 3 mi. SE Picacho (Cockrum, 1961:131); 5 mi. N Oracle; Saguaro National Monument (Patton, 1967:437); type locality; Sonoita Valley near Gray's Ranch; near Bates Well.

Perognathus inornatus
San Joaquin Pocket Mouse

For present status of subspecies of *Perognathus inornatus* see Osgood, Proc. Biol. Soc. Washington, 31:95–96, June 29, 1918.

External measurements: 128–160; 63–78; 18–21. Upper parts ochraceous-buff to pinkish overlaid with blackish hairs, extent of overlay changing overall tone in the several subspecies; lateral line moderately well marked; underparts white; tail faintly bicolored. Skull resembling that of *P. flavescens* and that of *P. longimembris;* mastoids large; auditory bullae apposed an-

teriorly; interparietal small, and approx. square; nasals short; coronoid process of mandible large.

This species is easily confused with *P. longimembris.* It is important to obtain fully adult animals on which to base the identification.

Perognathus inornatus inornatus Merriam

1889. *Perognathus inornatus* Merriam, N. Amer. Fauna, 1:15, October 25, type from Fresno, Fresno Co., California.

MARGINAL RECORDS.—California: Sites; Marysville Buttes; Lodi; type locality; Weldon; Walker Basin; Coalinga (Chew, *et al.,* 1965:478, as *P. inornatus* only); Panoche; Benicia.

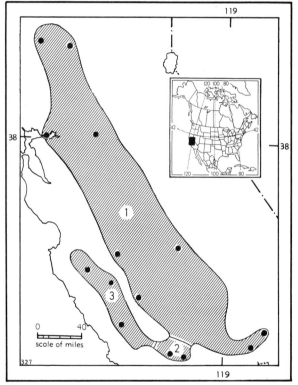

Map 327. *Perognathus inornatus.*

1. *P. i. inornatus* 2. *P. i. neglectus*
3. *P. i. sillimani*

Perognathus inornatus neglectus Taylor

1912. *Perognathus longimembris neglectus* Taylor, Univ. California Publ. Zool., 10:155, May 21, type from McKittrick, 1111 ft., Kern Co., California.
1918. *Perognathus inornatus neglectus,* Osgood, Proc. Biol. Soc. Washington, 31:96, June 29.

MARGINAL RECORDS.—California: Santiago Spring, edge of Carrizo Plain; type locality.

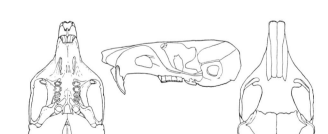

Fig. 341. *Perognathus inornatus inornatus,* B.M. 503, Panoche Creek, Fresno Co., California, No. 28371 M.V.Z., ♂, X 1½.

Perognathus inornatus sillimani von Bloeker

1937. *Perognathus inornatus sillimani* von Bloeker, Proc. Biol. Soc. Washington, 50:154, September 10, type from W side Arroyo Seco, 150 ft., 4 mi. S Soledad, Monterey Co., California.

MARGINAL RECORDS.—California: type locality; mouth Wild Horse Canyon, 500 ft.; Salinas Valley, 2 mi. S San Miguel.

Perognathus formosus
Long-tailed Pocket Mouse

External measurements: 172–211; 86–118; 22–26. Ten Nevadan males weighed 21.9 (19.8–24.7); 11 females, 20.2 (16.8–23.3) grams (Hall, 1946:371). Upper parts approx. wood brown but varying geographically; underparts white but sometimes faintly washed with buff; tail distinctly bicolored; tail long, sparsely haired, distinctly crested distally; interparietal markedly wider than long; mastoidal bullae projecting slightly beyond occiput; lower premolar larger than last molar.

Osgood (1900a:12) noted that *formosus* "is a *Perognathus* with strong inclination toward *Chaetodipus*." Here *P. formosus* is provisionally left in the subgenus *Perognathus*.

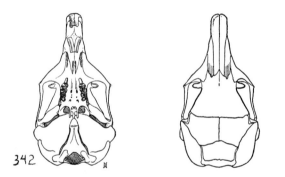

Fig. 342. *Perognathus formosus melanurus*, lat. 40° 28', 6 mi. E California boundary, Nevada, No. 73442 M.V.Z., ♂, X 1½.

Perognathus formosus cinerascens Nelson and Goldman

1929. *Perognathus formosus cinerascens* Nelson and Goldman, Proc. Biol. Soc. Washington, 42:105, March 25, type from San Felipe, northeastern Baja California.

MARGINAL RECORDS.—Baja California: type locality; NE El Mármol.

Perognathus formosus domisaxensis Cockrum

1956. *Perognathus formosus domisaxensis* Cockrum, Jour. Washington Acad. Sci., 46:131, May 7, type from Houserock Valley, 15 mi. W of [the Navajo] bridge, Coconino County, Arizona.

MARGINAL RECORDS.—Utah: Kane Creek, Mile 41, Glen Canyon (Durrant and Dean, 1959:85). Arizona: 2 mi. W Lees Ferry, 3250 ft.; Soap Creek, 15 mi. SW Lees Ferry; 6 mi. SE Fredonia.

Perognathus formosus formosus Merriam

1889. *Perognathus formosus* Merriam, N. Amer. Fauna, 1:17, October 25, type from St. George, Washington Co., Utah.

MARGINAL RECORDS.—Utah: Zion National Park. Arizona (Hoffmeister and Durham, 1971:33): *Pipe Spring National Monument, 5000 ft.; Antelope Valley, 29 mi. SW Fredonia;* Cutler Pockets; 4 mi. N Wolf Hole; *Larson's Ranch, 3 mi. E Black Rock Canyon, 3500 ft.* Utah: ½ mi. W St. George, 3000 ft.; 3 mi. SW St. George, 2800 ft.

Perognathus formosus incolatus Hall

1941. *Perognathus formosus incolatus* Hall, Proc. Biol. Soc. Washington, 54:56, May 20, type from 2 mi. W Smith Creek Cave, Mt. Moriah, White Pine Co., Nevada.

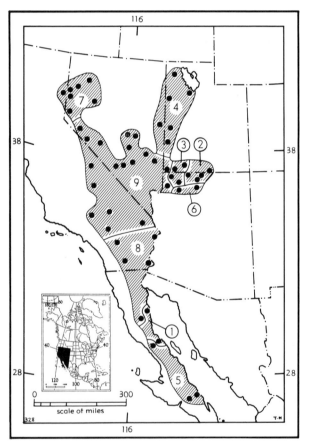

Map 328. *Perognathus formosus.*

Guide to subspecies
1. *P. f. cinerascens*
2. *P. f. domisaxensis*
3. *P. f. formosus*
4. *P. f. incolatus*

5. *P. f. infolatus*
6. *P. f. melanocaudus*
7. *P. f. melanurus*
8. *P. f. mesembrinus*
9. *P. f. mohavensis*

MARGINAL RECORDS.—Utah: Groome; 5 mi. S Timpie; White Valley, 65 mi. W Delta; Warm Cove, 55 mi. W Milford, 5500 ft. Nevada: type locality; *near (within ¼ mi. of) Smith Creek Cave, 5800 ft.* Utah: SE base Dutch Mtn. (Egoscue and Lewis, 1968:319); *N end Newfoundland Mts.*

Perognathus formosus infolatus Huey

1954. *Perognathus formosus infolatus* Huey, Trans. San Diego Soc. Nat. Hist., 12:1, March 1, type from 7 mi. W San Francisquito Bay, lat. 28° 30′ N, Gulf of California, Baja California.

MARGINAL RECORDS.—Baja California: 3 mi. W El Mármol; type locality; Barril.

Perognathus formosus melanocaudus Cockrum

1956. *Perognathus formosus melanocaudus* Cockrum, Jour. Washington Acad. Sci., 46:132, May 7, type from lower end Toroweap Valley (rim Grand Canyon), Mohave Co., Arizona.

MARGINAL RECORDS.—Arizona: Nankoweap Canyon, 3 mi. from Colorado River, 3600 ft. (Hoffmeister, 1971:171); *mouth Kwagunt Creek on Colorado River, 2800 ft.* (ibid.); *Deer Creek Falls, Mile 136, Kaibab National Forest, 1925 ft.* (ibid.); type locality; *½ mi. E Vulcans Throne* (Hoffmeister and Durham, 1971:33); *Toroweap Point, 6300 ft.* (ibid.).

Perognathus formosus melanurus Hall

1941. *Perognathus formosus melanurus* Hall, Proc. Biol. Soc. Washington, 54:57, May 20, type from lat. 40° 28′ N, 4000 ft., 6 mi. E California boundary, Washoe Co., Nevada.

MARGINAL RECORDS.—Nevada: 6 mi. N, 10½ mi. W Sulphur, 4000 ft.; 3½ mi. NE Toulon, 3950 ft.; N side Truckee River, 10 mi. E Reno, 4500 ft.; 2½ mi. E Flanigan, 4250 ft. California: 4½ mi. WNW Stacy. Nevada: type locality; 1 mi. NE Gerlach, 4000 ft.

Perognathus formosus mesembrinus Elliot

1904. *Perognathus mesembrinus* Elliot, Field Columb. Mus., Publ. 87, Zool. Ser., 3(14):251, January 7, type from Palm Springs, Riverside Co., California.

1929. *Perognathus formosus mesembrinus*, Nelson and Goldman, Proc. Biol. Soc. Washington, 42:106, March 25.

MARGINAL RECORDS.—California: Chemehuevis Valley; 1 mi. N Potholes. Baja California: Mattomi. California: San Felipe Canyon; type locality.

Perognathus formosus mohavensis Huey

1938. *Perognathus formosus mohavensis* Huey, Trans. San Diego Soc. Nat. Hist., 9:35, November 21, type from Bonanza King Mine, Providence Mts., San Bernardino Co., California.

MARGINAL RECORDS.—Nevada: West Walker River, 12 mi. S Yerington, 4600 ft. (= Wilson Canyon); 2½ mi. NW Blair Junction, 4950 ft.; NW base Timber Mtn., 4200 ft.; ½ mi. S Oak Spring, 5700 ft.; Old Mill, N end Reveille Valley, 6200 ft.; 6½ mi. N Hot Creek, 5900 ft.; 2½ mi. N and E Twin Springs, 5400 ft.; 9 mi. W Groom Baldy, 5500 ft.; Pahroc Spring; 5½ mi. N Elgin, 4000 ft. Utah: 1½ mi. E Beaverdam Wash, 8 mi. N Utah–Arizona border, 3200 ft. Arizona (Hoffmeister and Durham, 1971:33): *Beaver Dam (Lodge), 1900 ft.*; 3 mi. S, 8 mi. E Pakoon Springs; *Lake Mead, 1200 ft.*, thence southward W of Colorado River to California: type locality; Warren Station, near Mohave; Cushenberry Springs; 20 mi. S Trona; Lone Pine; Silver Canyon, White Mts. E of Laws. Nevada: Huntoon Valley, 5700 ft.

Subgenus **Chaetodipus** Merriam

1889. *Chaetodipus* Merriam, N. Amer. Fauna, 1:5, October 25. Type, *Perognathus spinatus* Merriam.

External measurements: 152–230; 83–143; 22–29. Pelage harsh with spiny bristles on rump; soles of hind feet naked. Mastoids relatively small, not projecting beyond plane of occiput; mastoidal side of parietal equal to or shorter than other sides; interparietal width equal to or greater than interorbital width. Auditory bullae separated by nearly full width of basisphenoid; supraoccipital with deep lateral indentations (except in *hispidus*); ascending branches of supraoccipital heavy and laminate.

KEY TO NOMINAL SPECIES OF SUBGENUS CHAETODIPUS

1. Rump with distinct spines or bristles.
    2. Lateral line well marked; pelage not markedly hispid; bristles moderate, usually on rump only.
        3. Ear less than 9.
            4. Spines on rump usually weakly developed; interparietal strap-shaped.
                5. Total length more than 180; rostrum and nasals broad and well developed. *P. nelsoni*, p. 553
                5′. Total length less than 180; rostrum and nasals narrow and weakly developed.
                                                                                P. intermedius, p. 551
            4′. Spines on rump strongly developed; interparietal pentagonal with a conspicuous anterior angle.
                6. Total length less than 170; cranium comparatively flattened; rostrum broad and well developed. . . . .*P. anthonyi*, p. 556

Perognathus baileyi
Bailey's Pocket Mouse

External measurements: 201–230; 110–125; 26–28. Upper parts grayish and washed with yellowish or tawny; underparts white or almost so; tail long, penicillate, and strongly crested, buffy above, whitish below. Skull large and heavily constructed; mastoid side of parietal approximately equal to other long sides; auditory bullae barely apposed anteriorly; interparietal relatively large; interparietal breadth approx. equal to least interorbital breadth; lower premolar equal to or slightly smaller than last molar.

Perognathus baileyi baileyi Merriam

1894. *Perognathus baileyi* Merriam, Proc. Acad. Nat. Sci. Philadelphia, 46:262, September 27, type from Magdalena, Sonora.

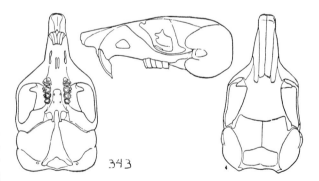

Fig. 343. *Perognathus baileyi hueyi,* Las Palmas Cañon, W side Languna Salada, U.S.–Mexican boundary, 300 ft., Baja California del Norte, No. 39199 M.V.Z., ♀, X 1½.

MARGINAL RECORDS.—Arizona: Congress Junction (Cockrum, 1961:134); New River; 3 mi. W Superior (Bateman, 1967:493); 6 mi. SSW Pima

(Hoffmeister, 1956:278). New Mexico (Findley, *et al.*, 1975:172, as *P. baileyi* only): Doubtful Canyon, 8 mi. N, 2 mi. E Steins; Guadalupe Canyon, 31 mi. S Rodeo. Sonora: 23 mi. S, 8 mi. E Nogales, 3200 ft.; Ures; Obregón; Navojoa (Patton and Jones, 1972:371). Sinaloa (*ibid.*): 1 mi. S, 6 mi. E El Carrizo; 42$\frac{1}{10}$ km. N Los Mochis. Sonora: Hermosillo; 5 mi. N Cornélio; Alamo Wash, 35 mi. NW Magdalena. Arizona: 7 mi. E Papago Well (Cockrum, 1961:135); Growler Mine, Organ Pipe Cactus National Monument.

Perognathus baileyi domensis Goldman

1928. *Perognathus baileyi domensis* Goldman, Proc. Biol. Soc. Washington, 41:204, December 18, type from Castle Dome, 1400 ft., at base Castle Dome Peak, Yuma Co., Arizona.

MARGINAL RECORDS.—Arizona: Harquahala Mts. (Cockrum, 1961:135); Cabeza Prieta Game Range. Sonora: Bahía San Carlos (Cockrum and Bradshaw, (1963:8); ½ mi. N Puerto Libertad. Arizona: Tinajas Altas, Gila Mts.; type locality; Kofa Mts., Willbank Ranch (Cockrum, 1961:135).

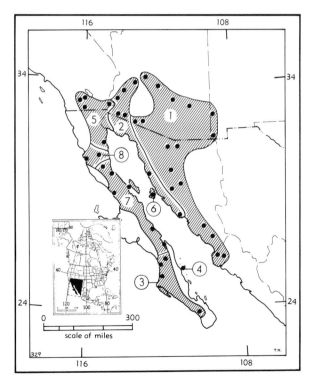

Map 329. *Perognathus baileyi.*

1. *P. b. baileyi*   5. *P. b. hueyi*
2. *P. b. domensis*   6. *P. b. insularis*
3. *P. b. extimus*   7. *P. b. mesidios*
4. *P. b. fornicatus*   8. *P. b. rudinoris*

Perognathus baileyi extimus Nelson and Goldman

1930. *Perognathus baileyi extimus* Nelson and Goldman, Jour. Washington Acad. Sci., 20:223, June 19, type from Tres Pachitas, 700 ft., 36 mi. S La Paz, Baja California.

MARGINAL RECORDS.—Baja California: Comondú; type locality; Matancita; San Jorge.

Perognathus baileyi fornicatus Burt

1932. *Perognathus baileyi fornicatus* Burt, Trans. San Diego Soc. Nat. Hist., 7(16): 164, October 31, type from Monserrate Island (lat. 25° 38′ N, long. 111° 02′ W), Gulf of California, Baja California. Known only from type locality.

Perognathus baileyi hueyi Nelson and Goldman

1929. *Perognathus baileyi hueyi* Nelson and Goldman, Proc. Biol. Soc. Washington, 42:106, March 25, type from San Felipe, northeastern Baja California.

MARGINAL RECORDS.—California: San Felipe Narrows; Bard. Baja California: type locality. California: Mountain Spring; Banner.

Perognathus baileyi insularis Townsend

1912. *Perognathus baileyi insularis* Townsend, Bull. Amer. Mus. Nat. Hist., 31:122, June 14, type from Tiburón Island, Gulf of California, Sonora. Known only from type locality.

Perognathus baileyi mesidios Huey

1964. *Perognathus baileyi mesidios* Huey, Trans. San Diego Soc. Nat. Hist., 13(7):112, January 15, type from San Borja Mission (*ca.* lat. 28° 45′ N), Baja California.

MARGINAL RECORDS (Nelson and Goldman, 1930:224, as *P. b. extimus*, here included in *P. b. mesidios* because Huey, 1964:112, states that *mesidios* ranges "South from Catavina to Conception Bay." Unless otherwise noted, records are from Nelson and Goldman).—Baja California: Onyx; Calamahue; Isla Smith (Lawlor, 1971:20, as *P. baileyi* only); San Bruno; Conception Bay (Huey, 1964:112); Punta Prieta.

Perognathus baileyi rudinoris Elliot

1903. *Perognathus baileyi rudinoris* Elliot, Field Columb. Mus., Publ. 74, Zool. Ser., 3(10):167, May 7, type from San Quintín, Baja California.
1903. *Perognathus knekus* Elliot, Field Columb. Mus., Publ. 74, Zool. Ser., 3(10):169, May 7, type from Rosarito, Sierra San Pedro Mártir, Baja California.

MARGINAL RECORDS.—Baja California: type locality; Rosarito.

Perognathus hispidus
Hispid Pocket Mouse

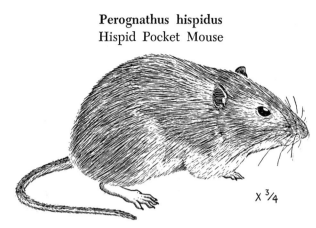

X ¾

Subspecies reviewed by Glass, Jour. Mamm., 28:174–179, June 1, 1947.

External measurements: 198–223; 90–113; 25–28. Upper parts ochraceous mixed with blackish hairs, sides usually only faintly paler than back; underparts white or nearly so; tail equal to or slightly shorter than head and body, tricolored, white below, buffy laterally, blackish above; lateral line distinct; pelage harsh. Skull large; rostrum robust; interorbital space wide; supraorbital bead conspicuous; mastoids relatively small, not bulging posteriorly; mastoid side of parietal short; interparietal wide; auditory bullae usually, but not always, well separated anteriorly; lower premolar approx. equal to last molar.

Perognathus hispidus hispidus Baird

1858. *Perognathus hispidus* Baird, Mammals, *in* Repts. Expl. Surv. . . . , 8(1):421, July 14, type from Charco Escondido, Tamaulipas.

MARGINAL RECORDS.—Texas: 10⅒ mi. N Vernon (Patton, 1967:36); Jefferson City. Louisiana: Vowells Mill; Glenmora (Lowery, 1974:216). Texas: 2½ mi N Hockley; Aransas Refuge; Brownsville. Tamaulipas: Sota la Marina (Alvarez, 1963:430); Victoria. Coahuila: Sabinas; 3 mi. N Cuatro Ciénegas; 6 mi. N, 2 mi. E La Babia; Cañón del Cochino, 3200 ft., 16 mi. N, 21 mi. E Piedra Blanca. Texas (Davis, 1966:164, as *P. hispidus* only): Val Verde County; Throckmorton County.

Perognathus hispidus paradoxus Merriam

1889. *Perognathus paradoxus* Merriam, N. Amer. Fauna, 1:24, October 25, type from Banner, Trego Co., Kansas.
1900. *Perognathus hispidus paradoxus*, Osgood, N. Amer. Fauna, 18:44, September 20.
1894. *Perognathus latirostris* Rhoads, Amer. Nat., 28:185, February, type from Rocky Mts.
1894. *Perognathus conditi* J. A. Allen, Bull. Amer. Mus. Nat. Hist., 6:318, November 7, type from San Bernardino Ranch, Cochise Co., Arizona. Regarded by Hoffmeister and Goodpaster (1954:103, 104) as a valid subspecies of *P. hispidus*.

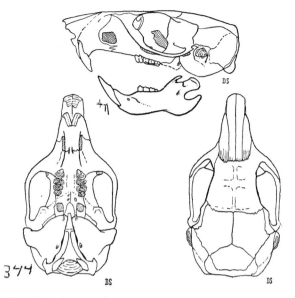

Fig. 344. *Perognathus hispidus paradoxus*, 6 mi. S Atwood, Rawlins Co., Kansas, No. 35092 K.U., ♂, X 1½.

MARGINAL RECORDS.—North Dakota: Wade. South Dakota: 2 mi. S, 3 mi. E Fort Thompson, 1470 ft. Nebraska: 2 mi. SE Niobrara; Red Cloud. Kansas: 4 mi. E Stockton; 3 mi. N, 2 mi. W Hoisington; Rezeau Ranch, 5 mi. N Belvidere. Oklahoma: 2 mi. W Edith. Texas: ¼ mi. S Chilicothe; Fisher County (Davis, 1966:164, as *P. hispidus* only); Mitchell County (*ibid.*, as *P. hispidus* only); Fort Stockton; Alpine. Chihuahua (Anderson, 1972:305): Station Arados; Santa Rosalía. Durango: 7 mi. NNE Boquilla, 6200 ft. (Baker, 1966:345). Chihuahua (Anderson, 1972:305): 2 mi. NE Hidalgo del Parral; Rancho San Ignacio, 4 mi. S, 1 mi. W Santo Tomás; Cases Grandes, 4300 ft.; 1½ mi. N San Francisco. Arizona: Santa Cruz R., W of Patagonia Mts. (Cockrum, 1961:136); 12 mi. NW Tucson. New Mexico (Findley, *et al.*, 1975:167, 168, as *P. hispidus* only): Dry Creek; Tularosa; Embudo Canyon, Sandia Mts.; ½ mi. N, ½ mi. E Algodones; Las Vegas; 3 mi. SE Cimarron. Colorado (Armstrong, 1972:176): 1 mi. S, 2 mi. W Walsenburg; Daniel's Park; 6 mi. S, 7 mi. W Fort Collins. Wyoming: 2½ mi. S Chugwater; 2 mi. N, 14 mi. W Hulett. Montana: 5 mi. N, 3½ mi. W Camp Crook, South Dakota (Pefaur and Hoffmann, 1971:247). South Dakota: 14 mi. S, 4 mi. W Reva (Andersen and Jones, 1971:377).

Perognathus hispidus spilotus Merriam

1889. *Perognathus paradoxus spilotus* Merriam, N. Amer. Fauna, 1:25, October 25, type from Gainesville, Cooke Co., Texas.
1937. *Perognathus hispidus spilotus*, Black, 30th Biennial Rept., Kansas State Bd. Agric., p. 183.
1904. *Perognathus hispidus maximus* Elliot, Field Columb. Mus., Publ. 87, Zool. Ser., 3(14):253, January 7, type from Noble, Cleveland Co., Oklahoma.

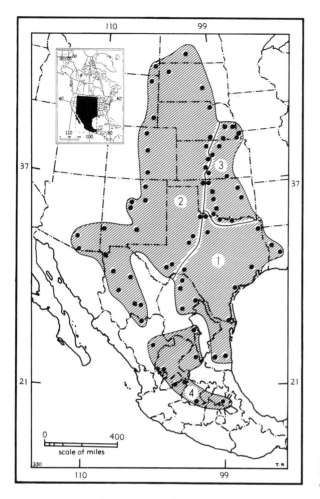

Map 330. *Perognathus hispidus.*

1. *P. h. hispidus*   3. *P. h. spilotus*
2. *P. h. paradoxus*   4. *P. h. zacatecae*

MARGINAL RECORDS.—Nebraska: 9 mi. N Lincoln (Jones, 1964c:176); ½ mi. W Manley (Genoways and Choate, 1970:121); Peru; 5 mi. SE Rulo. Kansas: Arkansas City. Oklahoma: Garnett; 3 mi. E Wainwright. Texas: type locality; *2 mi. S Marysville*; 3⅗ mi. N Henrietta (Patton, 1967:36). Oklahoma: Chattanooga; Wichita Mts. Wildlife Refuge (Glass and Halloran, 1961:238); 1 mi. N Hydro; 5 mi. W Canton; White Horse Springs. Kansas: 6 mi. N Aetna; 1½ mi. S Wilson; 4 mi. E Concordia. Nebraska: 1 mi. S Williams.

Perognathus hispidus zacatecae Osgood

1900. *Perognathus hispidus zacatecae* Osgood, N. Amer. Fauna, 18:45, September 20, type from Valparaíso, Zacatecas.

MARGINAL RECORDS.—Coahuila: 3 mi. S, 3 mi. E Bella Unión, 6750 ft. San Luis Potosí: 10 mi. S Matehuala (Gennaro and Salb, 1969:251). Jalisco: 2 mi. SW Matanzas, 7550 ft. (Genoways and Jones, 1973:12).

Hidalgo: 85 km. N Mexico City (9 km. S Pachuca), 8200 ft. Guanajuato: Celaya. Jalisco: Belén de Refugio, 5700 ft. (Genoways and Jones, 1973:12). Zacatecas: type locality. Durango: 8 km. SW Guadalupe Victoria (Hoffmeister, 1977:150).

penicillatus-group
Perognathus penicillatus
Desert Pocket Mouse

External measurements: 162–216; 83–129; 22–27. Upper parts yellowish-brown to yellowish-gray; underparts white to buffy; lateral line obscure or absent; tail long, markedly crested, penicillate, white below proximal to tuft; upper side of tail and tuft dusky. Skull of moderate size; rostrum robust and high; mastoid side of parietal equal to squamosal side, other sides much longer; interparietal pentagonal with all angles somewhat rounded; auditory bullae widely separated anteriorly.

Perognathus penicillatus angustirostris Osgood

1900. *Perognathus penicillatus angustirostris* Osgood, N. Amer. Fauna, 18:47, September 20, type from Carrizo Creek, Colorado Desert, Imperial Co., California.

MARGINAL RECORDS (Hoffmeister and Lee, 1967:373, 374, unless otherwise noted).—California: Barstow; Thousand Palms; *Indio*; Dos Palmas; 7 mi. NE Fort Yuma, 200 ft. Baja California: San Felipe (Hall and Kelson, 1959:498); Buena Vista Camp (*ibid.*). California: Coyote Wells (*ibid.*); Vallecitos (*ibid.*); Cathedral City; *2 mi. E Palm Springs*; Victorville; Hodge.

Perognathus penicillatus atrodorsalis Dalquest

1951. *Perognathus penicillatus atrodorsalis* Dalquest, Jour. Washington Acad. Sci., 41:362, November 14, type from 7 km. W Presa de Guadalupe, San Luis Potosí.

MARGINAL RECORDS (Hoffmeister and Lee, 1967:379, unless otherwise noted).—San Luis Potosí: 3 km. S Matehuala; *6 km. S Matehuala.* Nuevo León:

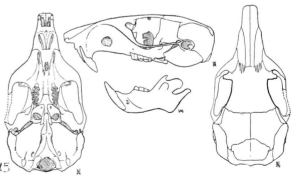

Fig. 345. *Perognathus penicillatus sobrinus*, 14 mi. E Searchlight, Nevada, No. 61569 M.V.Z., ♂, X 1½.

Doctor Arroyo. San Luis Potosí: 2 mi. NW Tepeyac, 3400 ft. (Hall and Kelson, 1959:498); Ciudad del Maíz region, 10³⁄₁₀ mi. NW on highway; type locality.

Perognathus penicillatus eremicus Mearns

1898. *Perognathus (Chaetodipus) eremicus* Mearns, Bull. Amer. Mus. Hist., 10:300, August 31, type from Fort Hancock, Hudspeth Co., Texas.
1900. *Perognathus penicillatus eremicus*, Osgood, N. Amer. Fauna, 18:48, September 20.

MARGINAL RECORDS (Hoffmeister and Lee, 1967:377, 378, unless otherwise noted).—New Mexico: Tularosa; Alamogordo; 3 mi. N, 3½ mi. E Loving (Findley, *et al.*, 1975:172, as *P. penicillatus* only); 3 mi. N, 2 mi. W Jal (*ibid.*). Texas: 6½ mi. SE Kermit (Packard and Garner, 1964:388); Monahans (Hall and Kelson, 1959:498); Val Verde County (Davis, 1966:165, as *P.*

penicillatus only); Boquillas. Coahuila: 50 mi. N, 20 mi. W Ocampo; 1 mi. S Hermanas; 17 mi. N, 8 mi. W Saltillo; *4 mi. NNW Saltillo*; La Ventura. San Luis Potosí: *31 km. S Santo Domingo*; Hernandez. Coahuila: Pico de Jimulco. Durango: 1½ mi. NW Nazas; 7 mi. NNW La Zarca. Chihuahua (Anderson, 1972:310): 20 mi. N, 7½ mi. E Parral; 3 mi. N Guadalupe Victoria; Mezquite, 3000 ft.; Ojo de Galeana, 4³⁄₁₀ mi. SE Galeana. New Mexico: Horse Canyon, Alamo Hueco Mts. (Findley, *et al.*, 1975:172, as *P. penicillatus* only); La Ciénega, Playas Valley; Deming; Garfield.

Perognathus penicillatus penicillatus Woodhouse

1852. *Perognathus penecillatus* [*sic*] Woodhouse, Proc. Acad. Nat. Sci. Philadelphia, 6:200, December. Type locality, by subsequent designation, 1 mi. SW Parker, Yuma Co., Arizona (see Hoffmeister and Lee, Jour. Mamm., 48:368, August 21, 1967).

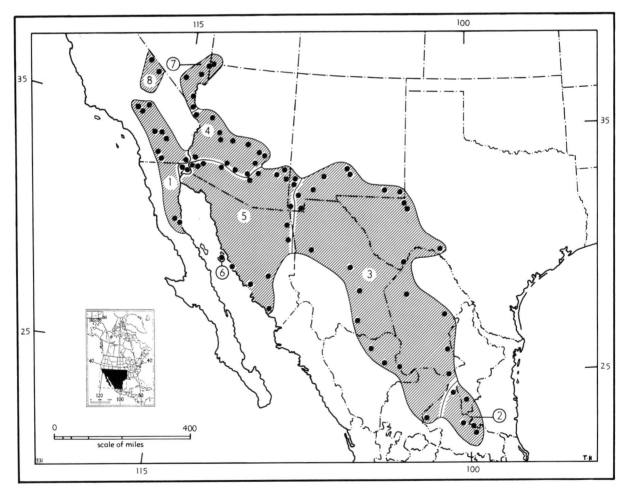

Map 331. *Perognathus penicillatus.*

Guide to subspecies	3. *P. p. eremicus*	6. *P. p. seri*
1. *P. p. angustirostris*	4. *P. p. penicillatus*	7. *P. p. sobrinus*
2. *P. p. atrodorsalis*	5. *P. p. pricei*	8. *P. p. stephensi*

1853. *Perognathus penicillatus* Woodhouse, *in* Sitgreaves, Rept. Exped. Zuni and Colorado rivers, U.S. Senate, 32 Cong., 2 Sess., Exec. No. 54, p. 49, pl. 3.

MARGINAL RECORDS (Hoffmeister and Lee, 1967:371).—Arizona: Topock; 2 mi. S, 1½ mi. E Wick-ieup; ½ mi. N Congress; 1 mi. N Wickenburg; New River; 3½ mi. N, 7 mi. W Roosevelt; Cutter, 6 mi. E Globe; Rice (San Carlos); 1½ mi. NW Hayden; 6¾ mi. S, 3 mi. E Picacho; 11 mi. W Casa Grande; 13 mi. S, 7½ mi. E Buckeye; Castle Dome Mine, Castle Dome Mts.; ½ mi. W San Luis; 4 mi. W Yuma, thence along E bank Colorado River to point of beginning.

Perognathus penicillatus pricei J. A. Allen

1894. *Perognathus pricei* J. A. Allen, Bull. Amer. Mus. Nat. Hist., 6:318, November 7, type from Oposura, Sonora.
1900. *Perognathus penicillatus pricei*, Osgood, N. Amer. Fauna, 18:47, September 20.

MARGINAL RECORDS (Hoffmeister and Lee, 1967:376, 377, unless otherwise noted).—Arizona: 8 mi. NNE Gila Bend; 10 mi. N Rillito; Mammoth (Hall and Kelson, 1959:498); Solomonsville, 3000 ft.; Clifton, 3500 ft.; 1 mi. S, ½ mi. E Franklin. New Mexico: 1 mi. E Redrock (Findley, *et al.*, 1975:172, as *P. penicillatus* only); 3½ mi. N, 1 mi. E Lordsburg; 31 mi. S Rodeo (Findley, *et al.*, 1975:172, as *P. penicillatus* only). Sonora: 8½ mi. NW Colonia Oaxaca; 1 mi. N Huachinera (Patton and Soule, 1967:267); 1 mi. S Tonichi (*ibid.*); Río Mayo, Navajoa (*ibid.*); Batamotal (Hall and Kelson, 1959:498); 2 mi. N Bahía Kino, thence northwestward along coast and north to Arizona: 11 mi. E Yuma; Wellton; Texas Hill.

Perognathus penicillatus seri Nelson

1912. *Perognathus penicillatus goldmani* Townsend, Bull. Amer. Mus. Nat. Hist., 31:122, June 14, type from Tiburón Island, Gulf of California, Sonora. Not *Perognathus goldmani* Osgood, 1900, type from Sinaloa, Sinaloa. Known only from type locality.
1912. *Perognathus penicillatus seri* Nelson, Proc. Biol. Soc. Washington, 25:116, June 29, a renaming of *goldmani* Townsend.

Perognathus penicillatus sobrinus Goldman

1939. *Perognathus penicillatus seorsus* Goldman, Proc. Biol. Soc. Washington, 52:34, March 11, type from sand flat along Virgin River, 7 mi. above Bunkerville, Clark Co., Nevada. [Regarded as Mohave County, Arizona, by Hardy, Jour. Mamm., 30:435, November 17, 1949.] Not *Perognathus spinatus seorsus* Burt, 1932, type from Danzante Island, Baja California.
1939. *Perognathus penicillatus sobrinus* Goldman, Jour. Mamm., 20:257, May 15, a renaming of *seorsus* Goldman.

MARGINAL RECORDS.—Utah: *"Terrys Ranch", Beaverdam Wash* (Stock, 1970:431). Arizona: Beaver Dam Lodge (Hoffmeister and Lee, 1967:372). Nevada (*ibid.*): type locality; 5 mi. SE Overton; Colorado River, 14 mi. E Searchlight, 500 ft.; opposite Fort Mohave; Vegas Valley.

Perognathus penicillatus stephensi Merriam

1894. *Perognathus (Chaetodipus) stephensi* Merriam, Proc. Acad. Nat. Sci. Philadelphia, 46:267, September 27, type from Mesquite Valley, NW arm Death Valley, Inyo Co., California.
1913. *Perognathus penicillatus stephensi*, Grinnell, Proc. California Acad. Sci., ser. 4, 3:333, August 28.

MARGINAL RECORDS (Hoffmeister and Lee, 1967:373).—California: Mesquite Valley; Stovepipe Wells Hotel.

Perognathus arenarius
Little Desert Pocket Mouse

External measurements: 136–182; 70–103; 20–23. Upper parts approx. buffy drab finely mixed with black; underparts white; lateral line usually absent; tail slightly exceeding head and body; skull small, relatively broad; braincase vaulted; nasals slender, zygomatic arches fragile; lower premolar larger than last molar; pelage soft and without bristles.

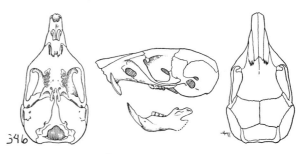

Fig. 346. *Perognathus arenarius arenarius*, San Jorge, Baja California, No. 146016 U.S.N.M., ♀, X 1½.

Perognathus arenarius albescens Huey

1926. *Perognathus arenarius albescens* Huey, Proc. Biol. Soc. Washington, 39:67, July 30, type from San Felipe, Baja California. Known only from type locality.

Perognathus arenarius albulus Nelson and Goldman

1923. *Perognathus penicillatus albulus* Nelson and Goldman, Proc. Biol. Soc. Washington, 36:159, May 1, type from Magdalena Island, Baja California.
1926. *Perognathus arenarius albulus*, Huey, Proc. Biol. Soc. Washington, 39:68, July 30.

MARGINAL RECORDS.—Baja California: type locality; *Estero Salinas* (Alvarez, 1958:3; this locality not found, but, according to Alvarez, "El Estero Salinas se encuentra en tierra peninsular, frente a la Isla Margarita").

Perognathus arenarius ambiguus Nelson and Goldman

1929. *Perognathus arenarius ambiguus* Nelson and Goldman, Proc. Biol. Soc. Washington, 42:108, March 25, type from Yubay, 30 mi. SE Calamahué, Baja California.

MARGINAL RECORDS.—Baja California: San Fernando; Pozo San Augustín, 20 mi. E San Fernando; mouth Calamahué Canyon; type locality; Pozo Altimirano; 20 mi. W San Ignacio; San Andrés; 25 mi. N Punta Prieta.

Perognathus arenarius ammophilus Osgood

1907. *Perognathus penicillatus ammophilus* Osgood, Proc. Biol. Soc. Washington, 20:20, February 23, type from Santa Margarita Island, Baja California. Known only from Santa Margarita Island.
1926. *Perognathus arenarius ammophilus*, Huey, Proc. Biol. Soc. Washington, 39:68, July 30.

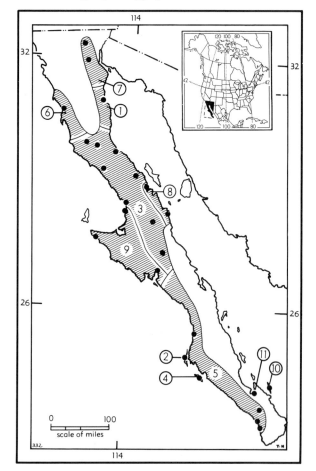

Map 332. *Perognathus arenarius.*

Guide to subspecies	
1. *P. a. albescens*	6. *P. a. helleri*
2. *P. a. albulus*	7. *P. a. mexicalis*
3. *P. a. ambiguus*	8. *P. a. paralios*
4. *P. a. ammophilus*	9. *P. a. sabulosus*
5. *P. a. arenarius*	10. *P. a. siccus*
	11. *P. a. sublucidus*

Perognathus arenarius arenarius Merriam

1894. *Perognathus arenarius* Merriam, Proc. California Acad. Sci., ser. 2, 4:461, September 25, type from San Jorge, near Comondú, Baja California.

MARGINAL RECORDS.—Baja California: type locality; Tres Pachitas (Banks, 1964:400); Todos Santos (*ibid.*); Pescadero (*ibid.*).

Perognathus arenarius helleri Elliot

1903. *Perognathus helleri* Elliot, Field Columb. Mus., Publ. 74, Zool. Ser., 3(10):166, May 7, type from San Quentín, Baja California. Known only from type locality.
1929. *Perognathus arenarius helleri*, Huey, Proc. Biol. Soc. Washington, 39:68, July 30.

Perognathus arenarius mexicalis Huey

1939. *Perognathus arenarius mexicalis* Huey, Trans. San Diego Soc. Nat. Hist., 9:57, August 31, type from Los Muertos Canyon fan, lat. 32° 27′ N, long. 115° 53′ W, Gaskill's Tank, near Laguna Salada, Baja California.

MARGINAL RECORDS.—Baja California: De Mara's Well, Laguna Salada; type locality.

Perognathus arenarius paralios Huey

1964. *Perognathus arenarius paralios* Huey, Trans. San Diego Soc. Nat. Hist., 13:113, January 15, type from Barril (lat. 28° 20′ N) on Gulf of California, Baja California.

MARGINAL RECORDS (Huey, 1964:114).—Baja California: Los Angeles Bay; type locality.

Perognathus arenarius sabulosus Huey

1964. *Perognathus arenarius sabulosus* Huey, Trans. San Diego Soc. Nat. Hist., 13:114, January 15, type from S side Scammon's Lagoon, Baja California.

MARGINAL RECORDS (Huey, 1964:114, unless otherwise noted).—Baja California: Santa Rosalía Bay; "*sandy areas about the lagoons on the western Vizcaíno Desert*"; northern part of San Ignacio Lagoon; Turtle (San Bartolomé) Bay (Nelson and Goldman, 1929:107).

Perognathus arenarius siccus Osgood

1907. *Perognathus penicillatus siccus* Osgood, Proc. Biol. Soc. Washington, 20:20, February 23, type from Ceralbo [= Cerralvo] Island, Baja California. Known only from Cerralvo Island.
1929. *Perognathus arenarius siccus*, Nelson and Goldman, Proc. Biol. Soc. Washington, 42:108, March 25.

Perognathus arenarius sublucidus Nelson and Goldman

1929. *Perognathus arenarius sublucidus* Nelson and Goldman, Proc. Biol. Soc. Washington, 41:109, March 25, type from La Paz, Baja California. Known only from type locality.

Perognathus dalquesti Roth
Dalquest's Pocket Mouse

1976. *Perognathus dalquesti* Roth, Jour. Mamm., 57:562, August 27, type from 4 mi. "SE Migriño (32 km. SSE Todos Santos)," Baja California Sur, México (23° 10′ N, 110° 07′ W).

External measurements: 148–180; 76–102; 20–25. Resembling *P. arenarius,* but averaging larger in all external measurements; tail averaging relatively longer; pelage darker; black margins present instead of lacking on ears; tail crested and with dark, almost black, dorsal stripe, instead of noncrested and almost uniformly pale; rump spines present instead of absent. See Map 334.

MARGINAL RECORDS (Roth, 1976:565).—Baja California Sur: 5 mi. N Cuñaño; Tres Pachitas; *Pescadero;* type locality.

Perognathus pernix
Sinaloan Pocket Mouse

External measurements: 162–175; 94–97; 22–24. Upper parts above lateral line hair brown; underparts white or almost so; lateral line buffy and well marked; tail long, thinly haired, faintly crested, brownish black above, whitish below; skull small, narrow, and elongate; mastoids small; interorbital breadth narrow; nasals comparatively broad and flattened; interparietal wide

Map 333. *Perognathus pernix.*

1. *P. p. pernix*   2. *P. p. rostratus*

and produced anteriorly; molariform teeth small; lower premolar larger than last molar; pelage slightly hispid.

Perognathus pernix pernix J. A. Allen

1898. *Perognathus pernix* J. A. Allen, Bull. Amer. Mus. Nat. Hist., 10:149, April 12, type from Rosario, Sinaloa.

MARGINAL RECORDS.—Sinaloa: 10 km. NE Pericos (Patton, 1967:36); Culiacán; type locality. Nayarit: Acaponeta. Sinaloa: Hacienda Island, a few miles W Escuinapa; Altata. Not found: Nayarit: Playa Novilleros.

Perognathus pernix rostratus Osgood

1900. *Perognathus pernix rostratus* Osgood, N. Amer. Fauna, 18:51, September 20, type from Camoa, Río Mayo, Sonora.

MARGINAL RECORDS.—Sonora: Tecoripa; Guiracoba. Sinaloa (Patton and Soule, 1967:267): San Rafael, 10 mi. N Guamuchil; San Miguel Zapote, Río Fuerte. Sonora (Patton and Soule, 1967:267): Masiaca; 1 mi. S, 7 mi. E Vicam. Not found: Sonora: Las Panelas.

intermedius-group
(*intermedius* through *anthonyi*)
Perognathus intermedius
Rock Pocket Mouse

External measurements: 152–180; 83–103; 19–24. Upper parts highly variable ranging from pale buffy gray (almost white) to nearly black, usually near drab; sides usually paler than back; underparts varying from buffy white to much darker; tail long, crested, and usually much darker distally than proximally, lighter below than above; skull with well-arched braincase; rostrum slender, depressed; interparietal wide, straplike; pelage hispid, weak spines often present on rump.

Perognathus intermedius ater Dice

1929. *Perognathus intermedius ater* Dice, Occas. Pap. Mus. Zool., Univ. Michigan, 203:2, June 19, type from Malpais Spring, 4150 ft., 15 mi. W Three Rivers, Otero Co., New Mexico.

MARGINAL RECORDS (Findley, *et al.,* 1975:170, as *P. penicillatus* only).—New Mexico: 12 mi. NW Carrizozo; near Carrizozo; type locality; 6½ mi. W *black lava, Oscura.*

Perognathus intermedius crinitus Benson

1934. *Perognathus intermedius crinitus* Benson, Proc. Biol. Soc. Washington, 47:199, October 2, type from 2⅜ mi. W Wupatki Ruins, Coconino Co., Arizona.

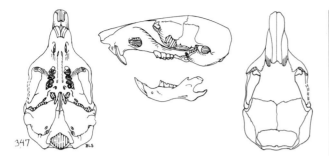

Fig. 347. *Perognathus intermedius intermedius*, 11 mi. NNW San Buenaventura, Chihuahua, No. 64012 K.U., ♀, X 1½.

MARGINAL RECORDS.—Utah: Aztec Creek, River Mile 68.5, Glen Canyon (Durrant and Dean, 1959:86); *Rainbow Bridge, 4000 ft.* Arizona (Cockrum, 1961:140, unless otherwise noted): Moa Ave; Walnut Tank, 10 mi. N Angell Augusta; Hance Rapids, Mile 76.5 (Hoffmeister, 1971:171); *Pipe Creek (ibid.); mouth Hermit Creek, Mile 95 (ibid.);* Hilltop *(ibid.);* Lower end Prospect Valley, 4500 ft., Hualpai Indian Reservation; mouth Diamond Creek, S side Colorado River. Utah: River Mile 28, Glen Canyon, near Warm Creek (Durrant and Dean, 1959:86). Not found: Arizona (Cockrum, 1961:140): Grand Canyon, Burro Spring; Burro Creek.

Perognathus intermedius intermedius Merriam

1889. *Perognathus intermedius* Merriam, N. Amer. Fauna, 1:18, October 25, type from Mud Spring, Mohave Co., Arizona.

1889. *Perognathus obscurus* Merriam, N. Amer. Fauna, 1:20, October 25, type from Camp Apache, Hidalgo Co., New Mexico.

MARGINAL RECORDS.—Arizona (Cockrum, 1961:139): Hoover Dam Ferry, 650 ft.; Peach Springs, 4000 ft.; Wickieup; 30 mi. N Phoenix; Wickenburg, 2000 ft.; Fish Creek, Tonto National Forest, 2000 ft. New Mexico (Gennaro, 1968:486, unless otherwise noted, as *P. intermedius* only): 3 mi. N, 2 mi. E White Water Canyon, Glenwood (Findley, *et al.*, 1975:170, as *P. intermedius* only); 1 mi. E Redrock; Lake Valley (V. Bailey, 1932:282); Las Palomas *(ibid.);* near Mush Mtn., 16¼ mi. S, 12¼ mi. W Correo; 6½ mi. S, 9 mi. W San Ysidro; 4 mi. S, 4 mi. W Penablanca; Mazano Mts., 1 mi. N, 17¼ mi. E Belen; 8⅛ mi. N, 1 mi. W Carrizozo. Texas: 7 mi. N Pine Springs, 5300 ft.; 8 mi. S Wink (Judd and Schmidly, 1969:382). Chihuahua (Anderson, 1972:307): 5 mi. WNW Ojinaga, 2500 ft.; 14 mi. SW Ciudad Camargo; 4 mi. NW Chihuahua, 4700 ft.; 2 mi. N, 9 mi. W Encinillas, 5500 ft.; 11 mi. NNW San Buenaventura; Casas Grandes, 4300 ft.; 2 mi. S, 5 mi. W San Francisco, 5500 ft. Arizona (Cockrum, 1961:140): Fort Huachuca; Pena Blanca Spring, 4500 ft.; Quitovaquito; 13 mi. N Ajo; 5 mi. S Wellton; 5 mi. NE Laguna; Parker, 400 ft.; type locality.

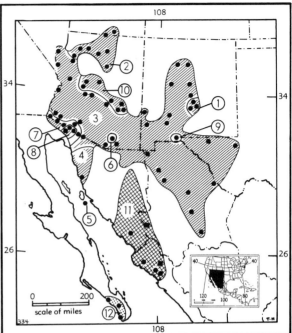

Map 334. *Perognathus intermedius, Perognathus goldmani,* and *Perognathus dalquesti.*

1. *P. i. ater*	7. *P. i. phasma*
2. *P. i. crinitus*	8. *P. i. pinacate*
3. *P. i. intermedius*	9. *P. i. rupestris*
4. *P. i. lithophilus*	10. *P. i. umbrosus*
5. *P. i. minimus*	11. *P. goldmani*
6. *P. i. nigrimontis*	12. *P. dalquesti*

Perognathus intermedius lithophilus Huey

1937. *Perognathus intermedius lithophilus* Huey, Trans. San Diego Soc. Nat. Hist., 8:355, June 15, type from Puerto Libertad (1½ mi. NW freshwater spring on beach), Sonora. Known only from type locality.

Perognathus intermedius minimus Burt

1932. *Perognathus penicillatus minimus* Burt, Trans. San Diego Soc. Nat. Hist., 7:164. October 31, type from Turners Island, lat. 28° 43′ N, long. 112° 19′ W, Gulf of California, Sonora. Known only from type locality.

1974. *Perognathus intermedius minimus,* Hoffmeister, Southwestern Nat., 19:213, July 26.

Perognathus intermedius nigrimontis Blossom

1933. *Perognathus intermedius nigrimontis* Blossom, Occas. Pap. Mus. Zool., Univ. Michigan, 265:1, June 21, type from Black Mtn., 10 mi. S Tucson, Pima Co., Arizona.

MARGINAL RECORDS.—Arizona: *Tumamoc Hill, Tucson* (Cockrum, 1961:142); type locality.

Perognathus intermedius phasma Goldman

1918. *Perognathus intermedius phasma* Goldman, Proc. Biol. Soc. Washington, 31:22, May 16, type from Tinajas Atlas, 1400 ft., Gila Mts., Yuma Co., Arizona.

MARGINAL RECORDS.—Arizona (Cockrum, 1961:142): Old Quarry near Yuma; 20 mi. S Wellton; Bates Well, 1385 ft.; E end Puerto Blanco Mts.; *7 mi. E Papago Well;* Tule Tank; *9 mi. E Tinajas Atlas.*

Perognathus intermedius pinacate Blossom

1933. *Perognathus intermedius pinacate* Blossom, Occas. Pap. Mus. Zool., Univ. Michigan, 273:4, October 31, type from Pápago Tanks, Pinacate Mts., Sonora.

MARGINAL RECORDS.—Arizona (Cockrum, 1961:142): 20 mi. E Tule Tank; Tule Desert, 3 mi. N Monument 182. Sonora: type locality.

Perognathus intermedius rupestris Benson

1932. *Perognathus intermedius rupestris* Benson, Univ. California Publ. Zool., 38:337, April 14, type from that part of the lava beds nearest to Kenzin, Dona Ana Co., New Mexico. Known only from type locality.

Perognathus intermedius umbrosus Benson

1934. *Perognathus intermedius umbrosus* Benson, Proc. Biol. Soc. Washington, 47:200, October 2, type from Camp Verde, Yavapai Co., Arizona

MARGINAL RECORDS.—Arizona (Cockrum, 1961:142): Cottonwood; Cazador Spring, S base Nantan Plateau, 4000 ft.; Gila Mts., 5800 ft.; Rice, 2800 ft.; Roosevelt, 2300 ft.; Congress Junction; Kirkland, 3900 ft.

Perognathus nelsoni
Nelson's Pocket Mouse

External measurements: 182–193; 104–117; 22–23. Upper parts dark brown; underparts white or almost so; lateral line fawn colored, well marked; tail heavily crested, blackish above, whitish below; pelage harsh; skull similar to that of *P. intermedius;* premaxillary tongues extending posterior to nasals.

On the Mexican tableland this is one of the commonest pocket mice; its relationships to *P. lineatus* are not well understood but seem to be close.

Perognathus nelsoni canescens Merriam

1894. *Perognathus (Chaetodipus) intermedius canescens* Merriam, Proc. Acad. Nat. Sci. Philadelphia, 46:267, September 27, type from Jaral, Coahuila.
1900. *Perognathus nelsoni canescens,* Osgood, N. Amer. Fauna, 18:54, September 20.

1938. *Perognathus collis* Blair, Occas. Pap. Mus. Zool., Univ. Michigan, 381:1, June 20, type from Limpia Canyon, 4800 ft., about 1 mi. NW Fort Davis, Texas. Regarded as synonym of *canescens* by Borell and Bryant, Univ. California Publ. Zool., 48:25, August 7, 1942.
1938. *Perognathus collis popei* Blair, Occas. Pap. Mus. Zool., Univ. Michigan, 381:3, June 20, type from Pinnacle Spring, Brewster Co., Texas. Regarded as inseparable from *canescens* by Borell and Bryant (*loc. cit.*).

MARGINAL RECORDS.—New Mexico: 4 mi. W White City. Texas: Sheffield; Comstock. Coahuila: Pánuco, 1220 ft.; 12 mi. N, 10 mi. E. Parras, 3850 ft. (Baker, 1956:238); 3 mi. SE Torreón, 3800 ft. (*ibid.*). Durango: 7 mi. N. Campaña, 3750 ft. (Baker and Greer, 1962:98); 12 mi. SSW Mapimí, 5000 ft. (*ibid.*); 37 mi. W Mapimí, 5500 ft. (*ibid.*). Chihuahua (Anderson, 1972:309): 1 mi. E Julimes; 13 mi. S, 28 mi. W Ojinaja, 3400 ft.; Consolación, 5100 ft. Texas: Pinnacle Spring; 1 mi. NW Fort Davis.

Perognathus nelsoni nelsoni Merriam

1894. *Perognathus (Chaetodipus) nelsoni* Merriam, Proc. Acad. Nat. Sci. Philadelphia, 46:266, September 27, type from Hda. La Parada, about 25 mi. NW Ciudad San Luis Potosí, San Luis Potosí.

MARGINAL RECORDS.—Chihuahua (Anderson, 1972:309): 2 mi. N, 6 mi. E Camargo, 4150 ft.; 5 mi. S Jiménez. Durango: 7 mi. NNW La Zarca, 6000 ft. (Baker and Greer, 1962:98); Mapimí. Coahuila (Baker, 1956:239): W foot Pico de Jimulco, 4600 ft.; 10 mi. S, 5 mi. W General Cepeda, 6500 ft.; 17 mi. N, 8 mi. W Saltillo. Nuevo León: 3 mi. SE Galeana, 5100 ft. (Dalby and Baker, 1967:195). Tamaulipas: Jaumave. San Luis Potosí: 2 mi. NW Tepeyac, 3400 ft., 14 mi. N, 29 mi. W Ciudad del Maíz; Jesús María. Jalisco: Lagos; Belén de Refugio, 5700 ft. (Genoways and Jones, 1973:12). Zacatecas: Hda. San Juan Capistrano. Durango: *2 km. NW Mezquital* (Crossin, *et al.,* 1973:197); 11 mi. N Mezquital, 5700 ft. (Baker and Greer, 1962:99); 4 mi. W Durango, 6200 ft.; Indé; 3 mi. E Las Nieves, 5400 ft. (*op. cit.*:98). Chihuahua: 2 mi. NE Hidalgo del Parral (Anderson, 1972:309).

Perognathus goldmani Osgood
Goldman's Pocket Mouse

1900. *Perognathus goldmani* Osgood, N. Amer. Fauna, 18:54, September 20, type from Sinaloa, Sinaloa.

External measurements of the type, an adult female: 202, 108, 28. Upper parts distinctly brown across shoulders and anterior part of dorsum, becoming darker—almost blackish—on rump; underparts whitish; lateral line pinkish buff, distinct; tail moderately long and heavily crested; sharply bicolored, blackish above, whitish below; skull large and especially robust; mastoids, compared to those of *P. nelsoni,* smaller

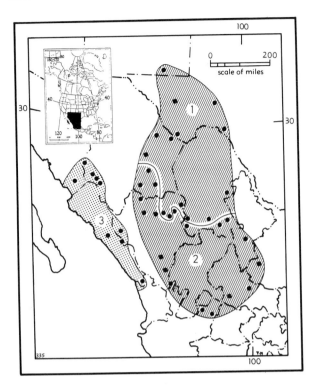

Map 335. *Perognathus nelsoni* and *Perognathus artus*.

1. *P. n. canescens*   2. *P. n. nelsoni*
3. *P. artus*

and more sculptured; nasals larger; braincase higher and narrower; pelage hispid with a few short bristles on rump; ears large and orbicular. See Map 334.

MARGINAL RECORDS (Anderson, 1964:22, 23, unless otherwise noted).—Chihuahua: Carimechi. Sinaloa: 8 km. N, 22 km. E Sinaloa, 400 ft.; 16 km. SE Topolobampo, 120 ft.; 1 mi. NW Topolobampo; 3 mi. N, 1 mi. E San Miguel. Sonora: 5³⁄₁₀ mi. E Navajoa (Patton, 1967:36).

Perognathus artus Osgood
Narrow-skulled Pocket Mouse

1900. *Perognathus artus* Osgood, N. Amer. Fauna, 18:55, September 20, type from Batopilas, Chihuahua.

Average external measurement of 5 adult topotypes: 191, 106, 24.6. Color as in *P. goldmani*. "The skins of *artus* are similar to those of *goldmani* except for slightly smaller size, less hairy tail, and broader dorsal tail stripe. The skulls differ in having a broader supraoccipital (least width 6.0, 6.4 mm. in two adult *artus*, and 5.7 mm. in one adult *goldmani*), smaller, more rugose and more prominently ridged mastoids,

and greater extensions of the premaxillae beyond the posterior borders of the nasals (less than 1 mm. in *goldmani*, more than 1 mm. in *artus*)." (Burt and Hooper, 1941:6.)

MARGINAL RECORDS (Anderson, 1964:23, 26, unless otherwise noted).—Chihuahua: Carimechi, Río Mayo; 3 mi. NE Temoris, 5600 ft.; Urique, 1700 ft.; type locality. Durango: Chacala, 564 m.; Santa Ana (Jones, 1964a:753). Nayarit: 21¾ mi. NW Acaponeta. Sinaloa: Culiacán. Sonora: 4 mi. NW Alamos.

Perognathus lineatus Dalquest
Lined Pocket Mouse

1951. *Perognathus lineatus* Dalquest, Jour. Washington Acad. Sci., 41:362, November 14, type from 1 km. S Arriaga, San Luis Potosí.

External measurements (average of eight males and seven females, respectively) are: 174, 174; 95, 98; 23, 23. Upper parts "dull gray, finely but distinctly lined with buffy, especially on head; general appearance of upper parts near Light Drab or Drab Gray; sides more grayish; underparts white separated from gray of sides by faint, indistinct line of pale buffy; tail dusky above and white beneath" (Dalquest, 1951:362). Skull larger and broader than in neighboring subspecies of *P. penicillatus*.

MARGINAL RECORDS.—San Luis Potosí: 6 km. S Matehuala; 10 km. NW Villar; Bledos; type locality; Cerro Peñón Blanco.

Perognathus fallax
San Diego Pocket Mouse

Revised by Huey, Trans. San Diego Soc. Nat. Hist., 12:415–419, February 1, 1960.

External measurements: 176–200; 88–118; 23. Upper parts rich brown, becoming blackish over rump; underparts white or whitish; lateral line buffy; tail crested, distinctly bicolored; braincase arched; interparietal wide, anterior angle obsolete; mastoids large; pelage harsh, having spines on rump.

Perognathus fallax fallax Merriam

1889. *Perognathus fallax* Merriam, N. Amer. Fauna, 1:19, October 25, type from Reche Canyon, 1250 ft., 3 mi. SE Colton, San Bernardino Co., California.

MARGINAL RECORDS.—California: 3 mi. N Claremont, 1600 ft.; San Bernardino (Huey, 1960:417); Banning, 2500 ft.; Jacumba. Baja California: San Matías Spring; Valle de la Trinidad (Huey, 1960:417); Ensenada (*ibid.*). California: San Onofre.

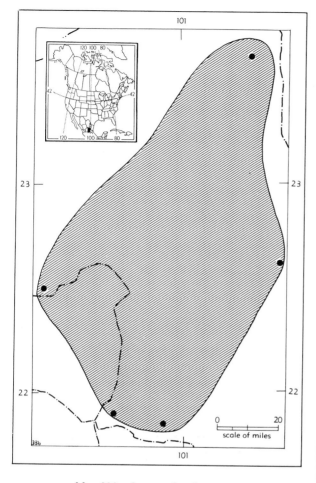

Map 336. *Perognathus lineatus.*

Perognathus fallax inopinus Nelson and Goldman

1929. *Perognathus fallax inopinus* Nelson and Goldman, Proc. Biol. Soc. Washington, 42:110, March 25, type from Turtle (San Bartolomé) Bay, Baja California.

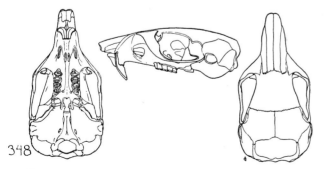

Fig. 348. *Perognathus fallax fallax,* 4¾ mi. N San Bernardino, 1600 ft., San Bernardino Co., California, No. 77119 M.V.Z., ♀, X 1½.

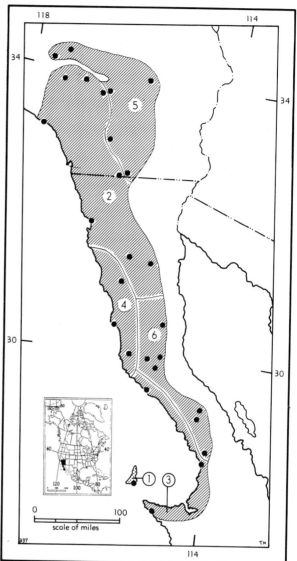

Map 337. *Perognathus anthonyi* and *Perognathus fallax.*

1. *P. anthonyi*	4. *P. f. majusculus*
2. *P. f. fallax*	5. *P. f. pallidus*
3. *P. f. inopinus*	6. *P. f. xerotrophicus*

MARGINAL RECORDS.—Baja California: Santa Catarina Landing (Huey, 1960:419); Santa Rosalía Bay, 28° 40′ N (*ibid.*); type locality.

Perognathus fallax majusculus Huey

1960. *Perognathus fallax majusculus* Huey, Trans. San Diego Soc. Nat. Hist., 12:418, February 1, type from San Quintín, Baja California.

MARGINAL RECORDS (Huey, 1960:418).—Baja California: Las Cabras; type locality; 10 mi. E El Rosario.

Perognathus fallax pallidus Mearns

1901. *Perognathus fallax pallidus* Mearns, Proc. Biol. Soc. Washington, 14:135, August 9, type from Mountain Spring, halfway up E slope Coast Range Mts., on Mexican boundary, Imperial Co., California.

MARGINAL RECORDS.—California: Oro Grande; Twenty-nine Palms; type locality; San Felipe Canyon; Cabezon; 2 mi. E Valyermo, 4500 ft.

Perognathus fallax xerotrophicus Huey

1960. *Perognathus fallax xerotrophicus* Huey, Trans. San Diego Soc. Nat. Hist., 12:419, February 1, type from Chapala, Baja California.

MARGINAL RECORDS (Huey, 1960:418, unless otherwise noted).—Baja California: Matomi (Hall and Kelson, 1959:503, as *P. f fallax*); San Agustín; Rancho Ramona, 7 mi. N Santa Catarina; type locality; San Andreas (*sic*); 25 mi. N Punta Prieta; San Fernando Mission.

Perognathus anthonyi Osgood
Anthony's Pocket Mouse

1900. *Perognathus anthonyi* Osgood, N. Amer. Fauna, 18:56, September 20, type from South Bay, Cerros (Cedros) Island, Baja California. Known only from type locality. See Map 337.

External measurements of type, adult female, are: 168, 92, 23.5. Upper parts grayish fawn mixed with black; underparts whitish; tail long, dusky above, whitish below; lateral line brownish fawn. "Skull.—Similar to *P. fallax*; cranium less arched; rostrum heavier; mastoids smaller; interparietal smaller and shorter; zygomatic breadth greater anteriorly." (Osgood, 1900a:56.)

Perognathus californicus
California Pocket Mouse

External measurements: 190–235; 103–143; 24–29. Upper parts brownish gray flecked with fulvous; underparts yellowish white; tail crested, bicolored; braincase markedly vaulted; mastoids especially small; mastoid breadth reduced; occiput greatly produced posteriorly; interparietal approx. twice as long as broad; lower premolar but little larger than last molar; pelage markedly hispid; strong spines present on rump and flanks; ears much elongated.

Perognathus californicus bensoni von Bloeker

1938. *Perognathus californicus bensoni* von Bloeker, Proc. Biol. Soc. Washington, 51:197, December 23, type from Stonewall Creek, 1300 ft., *ca.* 6⅗ mi. NE Soledad, Monterey Co., California.

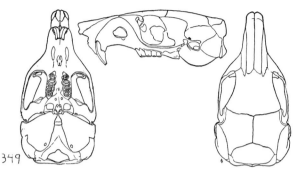

Fig. 349. *Perognathus californicus californicus*, N side Claremont Canyon, 600 ft., Alameda Co., California, No. 65120 M.V.Z., ♂, X 1½.

MARGINAL RECORDS.—California: 10 mi. W Gustine; Sweeney's Ranch, 1200 ft., Herrero Canyon, 22 mi. S Los Baños; Waltham Creek, 1850 ft., 4¼ mi. SE Priest Valley; Carrizo Plains, 2000 ft., 7 mi. SE Simmler; Santa Margarita, 996 ft.; Jolon, 1100 ft.; E base Sierra de Salinas, 250 ft., 2½ mi. W Soledad Mission.

Perognathus californicus bernardinus Benson

1930. *Perognathus californicus bernardinus* Benson, Univ. California Publ. Zool., 30:449, September 6, type from 2 mi. E Strawberry Peak, 5750 ft., San Bernardino Mts., San Bernardino Co., California.

MARGINAL RECORDS.—California: Big Pines, 6860 ft.; Seven Oaks, 5100 ft.; Strawberry Valley, 6000 ft.

Perognathus californicus californicus Merriam

1889. *Perognathus californicus* Merriam, N. Amer. Fauna, 1:26, October 25, type from Berkeley, Alameda Co., California.

1889. *Perognathus armatus* Merriam, N. Amer. Fauna, 1:27, October 25, type from Mt. Diablo, Contra Costa Co., California.

MARGINAL RECORDS.—California: E and S of San Francisco Bay from Sommersville; Gilroy; Portola; Redwood City, northward to point of beginning.

Perognathus californicus dispar Osgood

1900. *Perognathus californicus dispar* Osgood, N. Amer. Fauna, 18:58, September 20, type from Carpenteria, Santa Barbara Co., California.

MARGINAL RECORDS.—California: Eastern segment of range: Auburn; Hume; Dunlap. Western segment of range: San Luis Obispo; Santa Paula; San Fernando; Santa Monica, thence up coast to point of beginning. Note: the separated segments of the geographic range of this subspecies are undoubtedly the result of failure of some taxonomists to take cognizance of earlier published records of occurrence. The

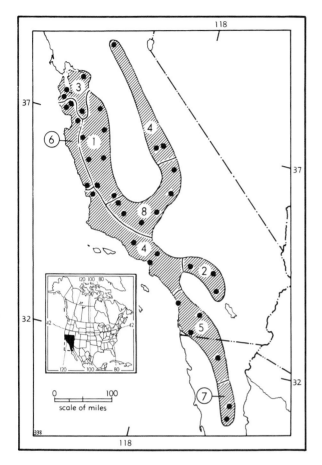

Map 338. *Perognathus californicus.*

1. *P. c. bensoni*   5. *P. c. femoralis*
2. *P. c. bernardinus*   6. *P. c. marinensis*
3. *P. c. californicus*   7. *P. c. mesopolius*
4. *P. c. dispar*   8. *P. c. ochrus*

eastern segment of the range is occupied by animals probably subspecifically distinct from those occurring along the coast.

Perognathus californicus femoralis J. A. Allen

1891. *Perognathus (Chaetodipus) femoralis* J. A. Allen, Bull. Amer. Mus. Nat. Hist., 3:281, June 30, type from Dulzura, San Diego Co., California.
1913. *Perognathus californicus femoralis*, Grinnell, Proc. California Acad. Sci., ser. 4, 3:335, August 28.

MARGINAL RECORDS.—California: mouth of Santa Margarita River; Banner. Baja California: Hanson Lagoon, Hanson Laguna Mts. California: type locality.

Perognathus californicus marinensis von Bloeker

1938. *Perognathus californicus marinensis* von Bloeker, Proc. Biol. Soc. Washington, 51:199, December 23, type from In-

dian Harbor, 50 ft., 1½ mi. S Marina, Monterey Co., California.

MARGINAL RECORDS.—California: Black Mountain; N end Gabilan Mts., 500 ft., 3½ mi. W San Juan Bautista; 4½ mi. S Morro; Bear Creek, 650 ft., 2–4½ mi. NE Boulder Creek.

Perognathus californicus mesopolius Elliot

1903. *Perognathus femoralis mesopolius* Elliot, Field Columb. Mus., Publ. 74, Zool. Ser., 3(10):168, May 7, type from Piñon, 5000 ft., Sierra San Pedro Mártir, Baja California.
1955. *Perognathus californicus mesopolius*, Miller and Kellogg, Bull. U.S. Nat. Mus., 205:280, March 3.

MARGINAL RECORDS.—Baja California: *type locality; Agua de las Fresas;* Santa Eulalia; Santa Rosa.

Perognathus californicus ochrus Osgood

1904. *Perognathus californicus ochrus* Osgood, Proc. Biol. Soc. Washington, 17:128, June 9, type from Santiago Spring, 16 mi. SW McKittrick, Kern Co., California.

MARGINAL RECORDS.—California: Jordan Hot Springs; Onyx; Tehachapi; Cuddy Canyon; Cayama Valley; Painted Rock, 25 mi. SE Simmler.

Perognathus spinatus
Spiny Pocket Mouse

External measurements: 164–225; 89–128; 20–28. Upper parts brownish to pale buffy yellow; underparts white or buffy-white; lateral line usually obsolete, and pale ecru when present; tail long, crested, brownish above, white below; skull comparatively slender and flattened; mastoids small (smaller, e.g., than in *P. fallax* and *P. intermedius*); interparietal broad and anterior angle faintly expressed; supraorbital ridge usually slightly trenchant; lower premolar and last molar approx. equal in size.

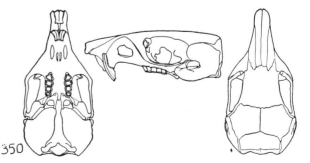

Fig. 350. *Perognathus spinatus spinatus,* Chocolate Mts., 11 mi. NE Niland, Imperial Co., California, No. 77285 M.V.Z., ♀, X 1½.

Perognathus spinatus broccus Huey

1960. *Perognathus spinatus broccus* Huey, Trans. San Diego Soc. Nat. Hist., 12:410, February 1, type from San Ignacio, lat. 27° 17′ N, Baja California.

MARGINAL RECORDS (Huey, 1960:412).—Baja California: type locality; Llano de San Bruno; 12 mi. S Mulegé, Concepción Bay; Comondú. Not found: Canipolé.

Perognathus spinatus bryanti Merriam

1894. *Perognathus bryanti* Merriam, Proc. California Acad. Sci., ser. 2, 4:458, September 25, type from San José Island, Gulf of California, Baja California. Known only from type locality.
1930. *P[erognathus]. s[pinatus]. bryanti*, Benson, Univ. California Publ. Zool., 30:452, September 6.

Perognathus spinatus evermanni Nelson and Goldman

1929. *Perognathus evermanni* Nelson and Goldman, Proc. Biol. Soc. Washington, 42:111, March 25, type from Mejía Island, near N end Angel de la Guarda Island, Baja California. Known only from Mejía Island.
1932. *Perognathus spinatus evermanni*, Burt, Trans. San Diego Soc. Nat. Hist., 7:165, October 31.

Perognathus spinatus guardiae Burt

1932. *Perognathus spinatus guardiae* Burt, Trans. San Diego Soc. Nat. Hist., 7:165, October 31, type from Puerto Refugio, 30 ft., N end Angel de la Guarda Island, Gulf of California, Baja California. Known only from type locality.

Perognathus spinatus lambi Benson

1930. *Perognathus spinatus lambi* Benson, Univ. California Publ. Zool., 32:452, September 6, type from San Gabriel, Espíritu Santo Island, Baja California. Known only from type locality.

Perognathus spinatus latijugularis Burt

1932. *Perognathus spinatus latijugularis* Burt, Trans. San Diego Soc. Nat. Hist., 7:168, October 31, type from San Francisco Island, lat. 24° 50′ N, long. 110° 34′ W, Gulf of California, Baja California. Known only from type locality.

Perognathus spinatus lorenzi Banks

1967. *Perognathus spinatus lorenzi* Banks, Proc. Biol. Soc. Washington, 80:101, July 28, type from South San Lorenzo Island (28° 36′ N lat., 112° 51′ W long.), Gulf of California, Baja California.

MARGINAL RECORDS.—Baja California (Banks, 1967b:101): North San Lorenzo Island; *South San Lorenzo Island.*

Perognathus spinatus magdalenae Osgood

1907. *Perognathus spinatus magdalenae* Osgood, Proc. Biol. Soc. Washington, 20:21, February 23, type from Magdalena Island, Baja California. Known only from type locality.

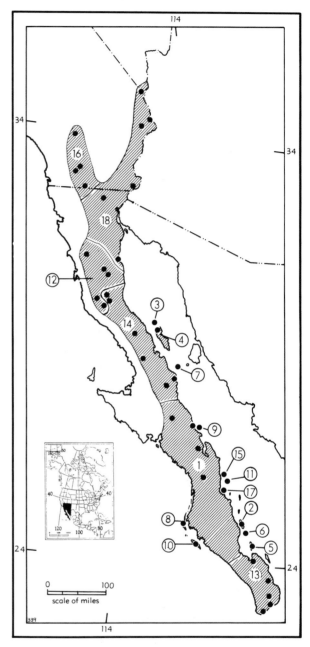

Map 339. *Perognathus spinatus.*

1. *P. s. broccus*
2. *P. s. bryanti*
3. *P. s. evermanni*
4. *P. s. guardiae*
5. *P. s. lambi*
6. *P. s. latijugularis*
7. *P. s. lorenzi*
8. *P. s. magdalenae*
9. *P. s. marcosensis*
10. *P. s. margaritae*
11. *P. s. occultus*
12. *P. s. oribates*
13. *P. s. peninsulae*
14. *P. s. prietae*
15. *P. s. pullus*
16. *P. s. rufescens*
17. *P. s. seorsus*
18. *P. s. spinatus*

Perognathus spinatus marcosensis Burt

1932. *Perognathus spinatus marcosensis* Burt, Trans. San Diego Soc. Nat. Hist., 7:166, October 31, type from San

Marcos Island, lat. 27° 13′ N, long. 112° 05′ W, Gulf of California, Baja California. Known only from the type locality.

Perognathus spinatus margaritae Merriam

1894. *Perognathus margaritae* Merriam, Proc. California Acad. Sci., ser. 2, 4:459, September 25, type from Margarita Island, Baja California. Known only from type locality.
1930. *P[erognathus]. s[pinatus]. margaritae*, Benson, Univ. California Publ. Zool., 32:452, September 6.

Perognathus spinatus occultus Nelson

1912. *Perognathus spinatus nelsoni* Townsend, Bull. Amer. Mus. Nat. Hist., 31:122, June 14, type from Carmen Island, Gulf of California, Baja California. Not *Perognathus (Chaetodipus) nelsoni* Merriam, 1894, type from Hda. La Parada, San Luis Potosí. Known only from type locality.
1912. *Perognathus spinatus occultus* Nelson, Proc. Biol. Soc. Washington, 25:116, June 29, a renaming of *nelsoni* Townsend.

Perognathus spinatus oribates Huey

1960. *Perognathus spinatus oribates* Huey, Trans. San Diego Soc. Nat. Hist., 12:409, February 1, type from San Fernando Mission, lat. 30° N, Baja California.

MARGINAL RECORDS (Huey, 1960:412, unless otherwise noted).—Baja California: Las Cabras; Parral (Hall and Kelson, 1959; 507, as *P. s. spinatus*); Matomi (*ibid.*); Rancho Ramona, 8 mi. N Santa Catarina; type locality.

Perognathus spinatus peninsulae Merriam

1894. *Perognathus spinatus peninsulae* Merriam, Proc. California Acad. Sci., ser. 2, 4:460, September 25, type from San José del Cabo, Baja California.

MARGINAL RECORDS (Huey, 1960:412).—Baja California: La Paz Mesa, 19 mi. N [= S?] La Paz; 7 mi. NW San Bartolo; Miraflores; San José del Cabo; Cape San Lucas.

Perognathus spinatus prietae Huey

1930. *Perognathus spinatus prietae* Huey, Trans. San Diego Soc. Nat. Hist., 6:232, December 24, type from 25 mi. N Punta Prieta, lat. 29° 24′ N, long. 114° 24′ W, Baja California.

MARGINAL RECORDS (Huey, 1960:412, unless otherwise noted).—Baja California: San Agustín (Hall and Kelson, 1959:507); Cataviña; type locality; [El] Barril; Santa Gertrudis Mission; San Borjas Mission. Not found: Not found: 12 mi. E El Arco.

Perognathus spinatus pullus Burt

1932. *Perognathus spinatus pullus* Burt, Trans. San Diego Soc. Nat. Hist., 7:166, October 31, type from Coronados Island, lat. 26° 06′ N, long. 111° 18′ W, Gulf of California, Baja California. Known only from type locality.

Perognathus spinatus rufescens Huey

1930. *Perognathus spinatus rufescens* Huey, Trans. San Diego Soc. Nat. Hist., 6:231, December 24, type from mouth Palm Canyon, Borego Valley, San Diego Co., California.

MARGINAL RECORDS.—California: Palm Springs; San Felipe Canyon; Mountain Spring; Vallecitos.

Perognathus spinatus seorsus Burt

1932. *Perognathus spinatus seorsus* Burt, Trans. San Diego Soc. Nat. Hist., 7:167, October 31, type from Danzante Island, lat. 25° 47′ N, long. 111° 11′ W, Gulf of California, Baja California. Known only from type locality.

Perognathus spinatus spinatus Merriam

1889. *Perognathus spinatus* Merriam, N. Amer. Fauna, 1:21, October 25, type from 25 mi. below The Needles, Colorado River, San Bernardino Co., California.

MARGINAL RECORDS.—Nevada: Granite Springs (Ryser, 1964:301). California: type locality; Pilot Knob, 200 ft. Baja California: San Felipe (Huey, 1960:412); Cocopah Mts. California: Horn Mine, 1000 ft., E base Turtle Mts.

Genus **Microdipodops** Merriam
Kangaroo Mice

Revised by Hall, Field Mus. Nat. Hist., Zool. Ser., 27:233–277, December 8, 1941.

1891. *Microdipodops* Merriam, N. Amer. Fauna, 5:115, July 30. Type, *Microdipodops megacephalus* Merriam.

External measurements: 130–180; 64–103; 22.3–27. Weight, 10.2–16.8 grams; basal length of skull (measured from anterior faces of upper incisors), 17.3–19.2; greatest breadth of skull, 17.7–20.8. Tail averaging slightly longer than head and body; tail short-haired, lacking terminal tuft and of greater diameter in middle than at base or tip; sole of hind foot densely covered with long hair; no dermal gland on back between shoulders; auditory bullae more highly inflated than in any other heteromyid, reaching below level of grinding surface of cheek-teeth and in many individuals extending anteriorly beyond glenoid fossae; bullae meeting in a symphysis across ventral face of basisphenoid; anterolateral face of zygomatic process of maxilla not much expanded, resulting in hamular process of lachrymal projecting free of maxilla; cheek-teeth hypsodont but each with more than one root, except M3 and m3; molars with H-pattern; P4 as in *Perognathus*, p4 with 5 or 6 cusps; cusps soon worn away with result that occlusal face of each cheek-tooth is an area of dentine completely surrounded by enamel; no pit behind m3; manus long and slender; tibia and fibula fused throughout almost three-fifths of their length; cervical vertebrae mostly fused; caudal vertebrae lacking median ventral foramen.

KEY TO SPECIES OF MICRODIPODOPS

1. Upper parts blackish; top of tail distally tipped with black; incisive foramina widest posteriorly or at middle; premaxillae extending but little posteriorly to nasals. . . . .*M. megacephalus,*   p. 560

1'. Upper parts near (*e*) Light Pinkish Cinnamon; top of tail not black distally but same color as base; incisive foramina parallelsided; premaxillae extending far behind nasals. . . . .*M. pallidus,*   p. 562

Microdipodops megacephalus
Dark Kangaroo Mouse

Upper parts brownish, blackish or grayish; top of distal sixth to half of tail ordinarily darker than back; hair of underparts plumbeous and white-tipped; hind foot, 23–25; anterior palatine foramina wide posteriorly and tapering to sharp point anteriorly; nasals extending posteriorly quite, or almost, as far as premaxillae. The few areas in which some of the above-mentioned characters do not hold are far removed from the geographic range of the other species, *M. pallidus.*

The area where the subspecies *M. m. leucotis* occurs is an example. Two specimens (25429 Univ. Utah, and 97308 KU) have been seen since the holotype was examined. These two resemble the species *M. pallidus* instead of the species *M. megacephalus* in several features pointed out in the original description of *M. m. leucotis,* and also in that the premaxillae extend far posteriorly to the nasals. Despite these resemblances to *M. pallidus,* the posteriorly wide anterior palatine foramina, short hind feet, and geographic near-

ness of *leucotis* to *M. m. megacephalus* influence me to let *leucotis* remain as a subspecies of *M. megacephalus.*

Possibly *leucotis* should be elevated to the rank of species. A straight, instead of convex, border on part of the lateral margin of each auditory bulla is a distinctive feature of the two skulls of *leucotis* in comparison with the few skulls of other named kinds of *Microdipodops* before me at this time (1976).

Microdipodops megacephalus albiventer Hall and Durrant

1937. *Microdipodops pallidus albiventer* Hall and Durrant, Jour. Mamm., 18:357, August 14, type from Desert Valley, 5300 ft., 21 mi. W Panaca, Lincoln Co., Nevada. Known only from type locality.

1941. *Microdipodops megacephalus albiventer,* Hall, Field Mus. Nat. Hist., Zool. Ser., 27:263, December 8.

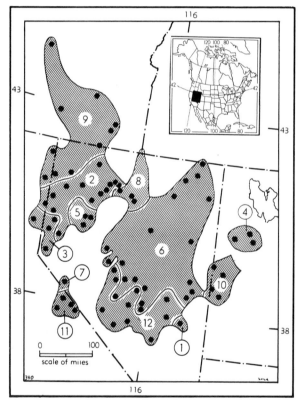

Map 340. *Microdipodops megacephalus.*

1. *M. m. albiventer*	7. *M. m. nasutus*
2. *M. m. ambiguus*	8. *M. m. nexus*
3. *M. m. californicus*	9. *M. m. oregonus*
4. *M. m. leucotis*	10. *M. m. paululus*
5. *M. m. medius*	11. *M. m. polionotus*
6. *M. m. megacephalus*	12. *M. m. sabulonis*

Fig. 351. *Microdipodops megacephalus leucotis,* N base Little Granite Mtn., 4700 ft., Tooele Co., Utah, No. 97308 K.U., ♂, X 1.

Microdipodops megacephalus ambiguus Hall

1941. *Microdipodops megacephalus ambiguus* Hall, Field Mus. Nat. Hist., Zool. Ser., 27:252, December 8, type from 1¼ mi. S Sulphur, 4050 ft., Humboldt Co., Nevada.

MARGINAL RECORDS.—Nevada: ½ mi. W Quinn River Crossing, 4100 ft.; 6 mi. N Golconda; 10 mi. NNW Golconda; 7 mi. N Winnemucca, 4400 ft.; 3 mi. SW Winnemucca, 4500 ft.; 15 mi. SW Winnemucca; 1 mi. W Humboldt, 4180 ft.; ¾ mi. S Sulphur, 4050 ft.; 1 mi. SE Wadsworth, 4200 ft.; 3½ mi. E Flanigan, 4200 ft. California: 1 mi. SW Warm Spring, 4000 ft. Nevada: Smoke Creek, 9 mi. E California boundary, 3900 ft.; 2½ mi. E, and 13 mi. N Gerlach, 4050 ft.

Microdipodops megacephalus californicus Merriam

1901. *Microdipodops californicus* Merriam, Proc. Biol. Soc. Washington, 14:128, July 19, type from Sierra Valley, near Vinton, Plumas Co., California.
1941. *Microdipodops megacephalus californicus*, Hall, Field Mus. Nat. Hist., Zool. Ser., 27:250, December 8.

MARGINAL RECORDS.—Nevada: 2¾ mi. SW Pyramid, 4300 ft.; 3½ mi. E Carson City, 4700 ft.; Junction House, 4500 ft. California: type locality.

Microdipodops megacephalus leucotis Hall and Durrant

1941. *Microdipodops megacephalus leucotis* Hall and Durrant, Murrelet, 22:6, April 30, type from 18 mi. SW Orr's Ranch, 4400 ft., Tooele Co., Utah.

MARGINAL RECORDS.—Utah (Shippee and Egoscue, 1958:277): 5 mi. SW Wig Mtn.; type locality; 1 mi. W Granite Mtn.; *7 mi. NE Granite Peak (Granite Mtn.)*.

Microdipodops megacephalus medius Hall

1941. *Microdipodops megacephalus medius* Hall, Field Mus. Nat. Hist., Zool. Ser., 27:256, December 8, type from 3 mi. S Vernon, 4250 ft., Pershing Co., Nevada.

MARGINAL RECORDS.—Nevada: 3 mi. SW Vernon, 4300 ft.; type locality; 21 mi. W, 2 mi. N Lovelock, 4000 ft.

Microdipodops megacephalus megacephalus Merriam

1891. *Microdipodops megacephalus* Merriam, N. Amer. Fauna, 5:116, July 30, type from Halleck, Elko Co., Nevada.

MARGINAL RECORDS.—Nevada: 9 mi. NE San Jacinto, 5300 ft.; Cobre, 6100 ft.; 5 mi. SE Greens Ranch, Steptoe Valley, 5900 ft.; 7 mi. SW Osceola, Spring Valley, 6275 ft.; 4 mi. S Shoshone, Spring Valley, 5900 ft.; 3 mi S Geyser Pass, Duck Valley, 6050 ft.; Old Mill, N end Reveille Valley, 6200 ft.; 2¼ mi. E, N Twin Spring, S end Hot Creek Valley, 5400 ft.; 3½ mi. E Hot Creek, Hot Creek Valley, 5650 ft.; 6½ mi. N Fish Lake, Fish Spring Valley, 6700 ft.; 15½ mi. NE Tonopah, Ralston Valley, 5800 ft.; 5 mi. N Belmont, Monitor Valley; 30 mi. N Belmont, Monitor Valley; Dutch Flat Schoolhouse, Reese River, 6715 ft.; 6 mi. ENE Smiths Creek Ranch, 5550 ft.; Winzell; 3–5 mi. W Halleck, 5200–5300 ft.; Marys River, 22 mi. N Deeth, 5800 ft.; 15 mi. S Contact, 5800 ft.

Microdipodops megacephalus nasutus Hall

1941. *Microdipodops megacephalus nasutus* Hall, Field Mus. Nat. Hist., Zool. Ser., 27:251, December 8, type from Fletcher, 6098 ft., Mineral Co., Nevada. Known only from type locality.

Microdipodops megacephalus nexus Hall

1941. *Microdipodops megacephalus nexus* Hall, Field Mus. Nat. Hist., Zool. Ser., 27:257, December 8, type from 3 mi. S Izenhood, Lander Co., Nevada.

MARGINAL RECORDS.—Nevada: 5 mi. NE Golconda; Izenhood; type locality.

Microdipodops megacephalus oregonus Merriam

1901. *Microdipodops megacephalus oregonus* Merriam, Proc. Biol. Soc. Washington, 14:127, July 19, type from Wild Horse Creek, 4 mi. NW Alvord Lake, Harney Co., Oregon.

MARGINAL RECORDS.—Oregon: Becker Ranch, Powell Butte; 1 mi. S Narrows, 4200 ft.; head Crooked Creek; White Horse Sink; 1½ mi. E Denio, 4200 ft. Nevada: 1 mi. E mouth Little High Rock Canyon, 5000 ft.; 10 mi. SE Hausen, 4675 ft. California: 6 mi. N Observation Peak, 5300 ft.; Sand Creek. Oregon: NE edge Alkali Lake, 4200 ft.

Microdipodops megacephalus paululus Hall and Durrant

1941. *Microdipodops megacephalus paululus* Hall and Durrant, Murrelet, 22:5, April 30, type from Pine Valley, ½ mi. E headquarters building, Desert Range Experimental Station, U.S. Forest Service, sec. 33, T. 25 S, R. 17 W, Salt Lake B.M., Millard Co., Utah.

MARGINAL RECORDS.—Utah: 4 mi. S Gandy, 5000 ft.; White Valley, 60 mi. W Delta; type locality; 5 mi. S Garrison, 5400 ft.

Microdipodops megacephalus polionotus Grinnell

1914. *Microdipodops polionotus* Grinnell, Univ. California Publ. Zool., 12:302, April 15, type from McKeever's Ranch, 2 mi. S Benton Station, 5200 ft., Mono Co., California.
1941. *Microdipodops megacephalus polionotus*, Hall, Field Mus. Nat. Hist., Zool. Ser., 27:251, December 8.

MARGINAL RECORDS.—California: E side Mono Lake; Pellisier Ranch, 5600 ft., 5 mi. N Benton; Taylor Ranch, 5300 ft., 2 mi. S Benton Station; Taylor Valley, 7000 ft., 25 mi. W Benton Station.

Microdipodops megacephalus sabulonis Hall

1941. *Microdipodops megacephalus sabulonis* Hall, Proc. Biol. Soc. Washington, 54:59, May 20, type from 5 mi. SE Kawich P.O., 5400 ft., Kawich Valley, Nye Co., Nevada.

MARGINAL RECORDS.—Nevada: 4 mi. SE Millet, 5500 ft.; 13 mi. NE San Antonio; 3 mi. S, 9 mi. W Tybo, 6200 ft.; 1 mi. N, 34 mi. E Tonopah, 5650 ft.; 16½ mi. WSW Sunnyside, White River Valley, 5500 ft.; 10 mi. N Seeman Pass, 4650 ft., Coal Valley; 15 mi. S Groom Baldy; 11½ mi. SW Silverbow, Cactus Flat, 5400 ft.; 14 mi. SE Goldfield, Stonewall Flat, 4700 ft.; 13½ mi. NW Goldfield, 4850 ft.

Microdipodops pallidus
Pale Kangaroo Mouse

Upper parts near Light Pinkish Cinnamon, lightly marked with buffy or blackish; upper parts of tail approx. same color as upper parts of body and lacking black tip; hair of underparts everywhere white to base; hind foot averaging more than 25 mm. long; anterior palatine foramina parallel-sided; premaxillae extending well behind nasals.

This species has been found only on fine sand that supports some plant growth.

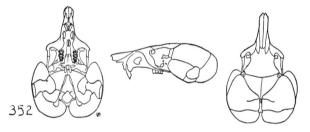

Fig. 352. *Microdipodops pallidus pallidus*, 8 mi. SE Blair, Nevada, No. 59344 M.V.Z., ♀, X 1.

Microdipodops pallidus ammophilus Hall

1941. *Microdipodops pallidus ammophilus* Hall, Field Mus. Nat. Hist., Zool. Ser., 27:273, December 8, type from Able Spring, 12½ mi. S Lock's Ranch, Railroad Valley, 5000 ft., Nye Co., Nevada.

MARGINAL RECORDS.—Nevada: Railroad Valley, 5000 ft., 2½ mi. S Lock's Ranch; 9½ mi. E New Reveille, Railroad Valley, 5100 ft.

Microdipodops pallidus pallidus Merriam

1901. *Microdipodops pallidus* Merriam, Proc. Biol. Soc. Washington, 14:127, July 19, type from Mountain Well, Churchill Co., Nevada.
1926. *Microdipodops megacephalus lucidus* Goldman, Proc. Biol. Soc. Washington, 39:127, December 27, type from 8 mi. SE Blair, 4500 ft., Esmeralda Co., Nevada.

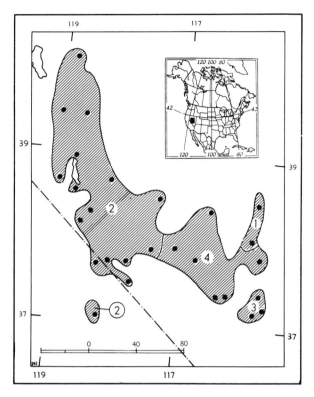

Map 341. *Microdipodops pallidus.*

1. *M. p. ammophilus*   3. *M. p. purus*
2. *M. p. pallidus*   4. *M. p. ruficollaris*

1927. *Microdipodops megacephalus dickeyi* Goldman, Proc. Biol. Soc. Washington, 40:115, September 26, type from 3 mi. SE Oasis, 5150 ft., Mono Co., California.

MARGINAL RECORDS.—Nevada: 21 mi. W, 2 mi. N Lovelock, 4000 ft.; type locality; Fingerrock Wash, Stewart Valley, 5400 ft.; 5½ mi. NE San Antonio, 5700 ft.; 13½ mi. NW Goldfield, 4850 ft.; 8 mi. SE Blair, 4500 ft.; 2 mi. SE Dyer, 4900–4950 ft.; mouth Palmetto Wash, 5350 ft. California: Deep Spring Valley, 4900–5000 ft.; 4½ to 5½ mi. SE Oasis, 5300 ft. Nevada: Huntoon Valley, 5700 ft.; Marietta, 4900 ft.; Cat Creek, 4 mi. W Hawthorne, 4500 ft.; 3 mi. S Schurz, 4100 ft.; 11¾ mi. S, 2¾ mi. E Yerrington, 4650 ft.; 9 mi. W Fallon, 4000 ft.

Microdipodops pallidus purus Hall

1941. *Microdipodops pallidus purus* Hall, Field Mus. Nat. Hist., Zool. Ser., 27:273, December 8, type from 14½ mi. S Groom Baldy, Lincoln Co., Nevada.

MARGINAL RECORDS.—Nevada: Desert Valley, 8 mi. SW Hancock Summit, 5300 ft.; 5½ mi. N Summit Spring, 4700 ft.; 15 mi. S Groom Baldy.

Microdipodops pallidus ruficollaris Hall

1941. *Microdipodos* [misspelling for *Microdipodops*] *pallidus ruficollaris* Hall, Proc. Biol. Soc. Washington, 54:60, May 20, type from 5 mi. SE Kawich P.O., 5400 ft., Kawich Valley, Nye Co., Nevada.

MARGINAL RECORDS.—Nevada: 3 mi. S, 9 mi. W Tybo, 6200 ft.; 17 mi. N Groom Baldy, Penoyer Valley; 5⁷⁄₁₀ mi. SE Kawich P.O., Kawich Valley, 5400 ft.; 6 mi. SW Kawich P.O., Gold Flat, 5100 ft.; 11½ mi. SW Silverbow, Cactus Flat, 5400 ft.; N shore Mud Lake, S end Ralston Valley, 5300 ft.

SUBFAMILY DIPODOMYINAE

Cheek-teeth as in Perognathinae, except that H-pattern is always present; cheek-teeth progressively hyposodont in geologically later kinds, which show progressive loss of enamel with result that cheek-teeth of *Dipodomys* are almost ever-growing; in adults, enamel limited to anterior and posterior plates; M$\frac{3}{3}$ small; increased height of crown not affecting pattern, which is rapidly destroyed, leaving only an oval of enamel; p4 never more than 5-cusped, the fifth appearing in center of metalophid; ventral surface of tympanic bullae rarely reaching level of grinding surface of cheek-teeth, never appreciably below that level; no ethmoid foramen in frontal; zygomatic root of maxilla expanded anteroposteriorly; center of palate between premolars ridged; pterygoid fossae double; caudal vertebrae with median ventral foramina; calcaneo-navicular or calcaneo-cuneiform articulation present (after Wood, 1935:117).

Genus Dipodomys Gray—Kangaroo Rats

1841. *Dipodomys* Gray, Ann. Mag. Nat. Hist., ser. 1, 7:521, August. Type, *Dipodomys phillipsii* Gray.
1867. *Perodipus* Fitzinger, Sitzungsb. k. Akad. Wiss., Wien, Math-Nat., Abth. 1, 56:126. Type, *Dipodomys agilis* Gambel. For taxonomic status, see Grinnell, Proc. Biol. Soc. Washington, 32:203, December 31, 1919.
1890. *Dipodops* Merriam, N. Amer. Fauna, 3:71, September 11. Type, *Dipodomys agilis* Gambel.

External measurements: 208–365; 100–212; 34–58; weight, 33.8–138.0 grams; basal length of skull (measured from front of upper incisors), 23.7–32.5; breadth of skull across auditory bullae, 21.3–31.9. Underparts, upper lips, spot above each eye, spot behind each ear, forelegs (except in some individuals, which have pigmented hairs on outer sides of forelegs), forefeet and antiplantar faces of hind feet white; white stripe extending from flank to base of tail and isolating patch of pigmented hair on each hind leg, this patch being some shade of buff or brownish like upper parts; tail white all around at base and having white stripes for entire length on each side; dark stripe extending down top of tail and in most species a second dark stripe on underside of tail; tail tufted in most species; fur silky, plumbeous basally on upper parts and white to base on underparts. Sleekness may result from secretion of dermal gland on back between shoulders. Hind legs large and fifth toe on large hind feet vestigial or wanting. Cervical vertebrae compressed, resulting in short neck.

Among American mammals, the huge auditory bullae are exceeded in size, relative to the remainder of the skull, only by members of the genus *Microdipodops*. The tail is longer than the head and body except that in *Dipodomys ordii* occasional individuals of the subspecies from southern Texas and northern Tamaulipas have the tail barely shorter than the head and body.

The linear arrangement of species, which follows, proceeds from the generalized to the specialized. Interspecific relationships are indicated to some extent by insertion of names of "groups," but the interspecific relationships could of course be shown best in a three-dimensional drawing. The relationships are suggested in two-dimensional drawings by Setzer (1949:Fig. 21) and Stock (1974:Fig. 6), who list also the detailed evidence for their one-dimensional (linear) arrangements of species. Protein variation in *Dipodomys* also reflects interspecific relationships (Johnson and Selander, 1971:377–405).

KEY TO SPECIES OF DIPODOMYS

1. Four toes on each hind foot.
    2. Occurring in northern Texas and southwestern Oklahoma. . . . . *D. elator*, p. 581
    2'. Not occurring in northern Texas and southwestern Oklahoma.
        3. Occurring in southern Oregon and northern California north of a line from Suisun Bay to Lake Tahoe. . . . . *D. heermanni*, p. 577
        3'. Occurring in California south of the mentioned line as well as in states east of California; also in México.
            4. Hind foot less than 42.
                5. Interorbital breadth more than half of basal length; tail-tip dusky or blackish brown.

ordii-group
Dipodomys ordii
Ord's Kangaroo Rat

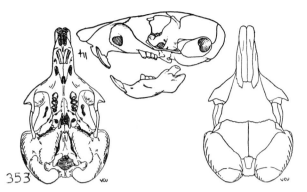

Fig. 353. *Dipodomys ordii richardsoni*, 1 mi. S Lamar, Prowers Co., Colorado, No. 15995 K.U., ♀, X 1.

Revised by Setzer, Univ. Kansas Publ., Mus. Nat. Hist., 1:473–573, December 27, 1949.

External measurements: 208–281; 100–163; 35–45. Five toes on each hind foot. The only other 5-toed kangaroo rats with which *Dipodomys ordii* shares parts of its range are *D. panamintinus* and *D. microps*. In *D. ordii* each of the lower incisors is awl-shaped instead of chisel-shaped as in *D. microps*. The hind foot of *D. ordii* is shorter than 44 mm. in the part of its range that is shared with the longer-footed, larger, *D. panamintinus*.

Schmidly and Hendricks (1976) arranged *D. compactus, D. o. largus, D. o. parvabullatus,* and *D. sennetti* (see nos. 7, 16, 30, and 35 of Map 342) as subspecies of species *Dipodomys compactus* True 1889 instead of as subspecies of species *Dipodomys ordii* Woodhouse 1853. As of 1977 the size and shape of the interparietal bone, small and pointed posteriorly in *ordii,* and larger and not pointed posteriorly in *compactus,* was thought to be the best single, simple character for separating the two species. This character seems to be of hardly more than subspecific grade, since essentially the same size and shape characteristic of *compactus* occurs elsewhere in the species *D. ordii,* for example, in some individuals of *D. o. luteolus, D. o. marshalli,* and *D. o. oklahomae.* Consequently the four taxa listed at the outset of this paragraph are here provisionally retained as subspecies of *Dipodomys ordii.*

Dipodomys ordii attenuatus Bryant

1939. *Dipodomys ordii attenuatus* Bryant, Occas. Pap. Mus. Zool., Louisiana State Univ., 5:65, November 10, type from mouth Santa Helena Canyon, 2146 ft., Big Bend of Rio Grande, Brewster Co., Texas.

MARGINAL RECORDS.—Texas: 6 mi. S Marathon; Cooper's Well, 47 mi. S Marathon; 10 mi. W San Vicente; mouth Santa Helena Canyon.

Dipodomys ordii celeripes Durrant and Hall

1939. *Dipodomys ordii celeripes* Durrant and Hall, Mammalia, 3:10, March, type from Trout Creek, 4600 ft., Juab Co., Utah.

MARGINAL RECORDS.—Nevada: 13 mi. N Montello, 5000 ft. Utah: Clifton Flat, 7 mi. SW Gold Hill, 6149 ft.; 35 mi. W Delta; 20 mi. SW Nephi; E side Clear Lake, 4600 ft.; Hendry Creek, 17 mi. S Gandy, 5000 ft. Nevada: Hendry Creek, 8 mi. SE Mt. Moriah, 6200 ft.; Cobre, 6100 ft.

Dipodomys ordii chapmani Mearns

1890. *Dipodomys chapmani* Mearns, Bull. Amer. Mus. Nat. Hist., 2:291, February 21, type from Fort Verde, Yavapai Co., Arizona.
1921. *Dipodomys ordii chapmani,* Grinnell, Jour. Mamm., 2:96, May 2.

MARGINAL RECORDS.—Arizona: lower end Prospect Valley, 4500 ft., Grand Canyon; Pasture Wash Ranger Station (Hoffmeister, 1971:171); Trash Tank (*ibid.*); Bill Williams Mtn.; ½ mi. S Camp Verde; Kirkland; Kingman.

Dipodomys ordii cinderensis Hardy

1944. *Dipodomys ordii cinderensis* Hardy, Proc. Biol. Soc. Washington, 57:53, October 31, type from approx. 4000 ft., immediately N of northern of two large cinder cones, Diamond Valley, 10 mi. N St. George, Washington Co., Utah.

MARGINAL RECORDS.—Utah: 11 mi. SE Lund; 4½ mi. NW Summit, 6 mi. W Parowan; Cedar City; type locality; N end Mountain Meadows.

Dipodomys ordii cineraceus Goldman

1939. *Dipodomys ordii cineraceus* Goldman, Jour. Mamm., 20:352, August 14, type from Dolphin Island, Great Salt Lake, 4250 ft., Boxelder Co., Utah. Known only from type locality.

Dipodomys ordii columbianus (Merriam)

1894. *Perodipus ordi columbianus* Merriam, Proc. Biol. Soc. Washington, 9:115, June 21, type from Umatilla, mouth Umatilla River, Plains of Columbia, Umatilla Co., Oregon.
1921. *Dipodomys ordii columbianus*, Grinnell, Jour. Mamm., 2:96, May 2.

MARGINAL RECORDS.—Washington: 6 mi. ENE Pasco (Broadbooks, 1958:300); 4 mi. E Burbank; Wallula. Idaho: Payette; Hammett; Arco; 5 mi. E Shelley; 3 mi. S Blackfoot; 5 mi. NW Michaud; 6 mi. SW American Falls; 8 mi. W Rogerson. Nevada: Marys River, 22 mi. N Deeth; 5 mi. SE Greens Ranch, Steptoe Valley; 4 mi. S Shoshone; Bells Ranch, Reese River, 6890 ft.; 2½ mi. NE Smiths Creek Ranch; 5 mi. N Beowawe; 1 mi. SE Tuscorara, 5900 ft.; 2 mi. SW Quinn River Crossing; 2½ mi. E, 11 mi. S Gerlach; Fox Canyon, 6 mi. S Pahrum Peak, 4800 ft.; 2¾ mi. SW Pyramid, 4300 ft. California: Beckwith; Dransfield, 6 mi. E Ravendale, 5300 ft.; 2 mi. W Red Rock P.O.; Eagleville. Oregon: 9 mi. S Adel, mouth 20 Mile Creek; Fort Rock; 2 mi. NE Prineville; 7 mi. E Madras; Arlington. Washington: Blalock Island (Broadbooks, 1958:300).

Dipodomys ordii compactus True

1889. *Dipodomys compactus* True, Proc. U.S. Nat. Mus., 11:160, January 5, type from Padre Island, Cameron Co., Texas.
1942. *Dipodomys ordii compactus*, Davis, Jour. Mamm., 23:332, August 13.

MARGINAL RECORDS.—Texas: *Padre Island, Nueces Co.* (Schmidly and Hendricks, 1976:236); type locality. Tamaulipas: Bagdad.

Dipodomys ordii cupidineus Goldman

1924. *Dipodomys ordii cupidineus* Goldman, Jour. Washington Acad. Sci., 14:372, September 19, type from Kanab Wash, southern boundary, Kaibab Indian Reservation, Mohave Co., Arizona.

MARGINAL RECORDS.—Utah: mouth Calf Creek, Escalante River; Willow Tank Springs. Arizona: 2 mi. W Lees Ferry; 10 mi. S Jacobs Pools, Houserock Valley; Dry Lake (Hoffmeister and Durham, 1971:21, 34); 20 mi. S Wolf Hole; 12 mi. N Jacob's Ranch, Cottonwood Canyon, 4250 ft. (Hoffmeister and Durham, 1971:19, 34); *10 mi. W Wolf Hole, 4850 ft.* (Hoffmeister and Durham, 1971:34). Utah: near Short Creek Road, S of town of Virgin; Cottonwood Canyon, 8 mi. NW Kanab, 4800 ft.

Dipodomys ordii durranti Setzer

1949. *Dipodomys ordii fuscus* Setzer, Univ. Kansas Publ., Mus. Nat. Hist., 1:555, December 27, type from Jaumave, Tamaulipas. Not *Dipodomys agilis fuscus* Boulware, 1943, type from 2½ mi. N La Purisima Mission, Santa Barbara Co., California.
1952. *Dipodomys ordii durranti* Setzer, Jour. Washington Acad. Sci., 42:391, December 17, a renaming of *Dipodomys ordii fuscus* Setzer, 1949.

MARGINAL RECORDS.—Coahuila: 11 mi. W Hda. San Miguel, 2200 ft. Texas (Schmidly and Hendricks, 1976:236): 2 mi. NE Carrizo Springs; 40 mi. SW Catarina, Rio Grande. Tamaulipas: Nuevo Laredo. Texas (Schmidly and Hendricks, 1976:236): 5 mi. N Zapata; *10 mi. NW Raymondville;* 17 mi. NW Edinburg. Tamaulipas: type locality; Tula. San Luis Potosí: 7⅗ mi. S Matehuala (Schmidly and Hendricks, 1976:236). Zacatecas: Lulú. Coahuila: San Juan Nepomuceno, 5 mi. N La Ventura; 17 mi. N, 8 mi. W Saltillo, 5200 ft.; 5 mi. N, 2 mi. W Monclova.

Dipodomys ordii evexus Goldman

1933. *Dipodomys ordii evexus* Goldman, Jour. Washington Acad. Sci., 23:468, October 15, type from Salida, 7000 ft., Chaffee Co., Colorado.

MARGINAL RECORDS.—Colorado (Armstrong, 1972:179): Browns Canyon, 7 mi. above Salida; *Salida; 21 mi. NW Cañon City;* 1 mi. N, 7 mi. E Cañon City.

Dipodomys ordii extractus Setzer

1949. *Dipodomys ordii extractus* Setzer, Univ. Kansas Publ., Mus. Nat. Hist., 1:534, December 27, type from 1 mi. E Samalayuca, 4500 ft., Chihuahua.

MARGINAL RECORDS.—Chihuahua (Anderson, 1972:315): 8 mi. NE Samalayuca, 4300 ft.; *type locality; 2½ mi. S, 2 mi. W Samalayuca, 1300 m.;* 1 mi. S Kilo, 4185 ft.

Dipodomys ordii fetosus Durrant and Hall

1939. *Dipodomys ordii fetosus* Durrant and Hall, Mammalia, 3:14, March, type from 2 mi. N Panaca, 4800 ft., Lincoln Co., Nevada.

MARGINAL RECORDS.—Utah: *5 mi. S Garrison, 5400 ft.;* Desert Range Experiment Station, 50 mi. W Milford, 5252 ft. Nevada: type locality; 10 mi. E Crystal Spring, 5000 ft.; 15 mi. S Groom Baldy; *Garden Valley;* 8½ mi. NE Sharp; White River Valley, 15 mi. WSW Sunnyside, 5500 ft.

Dipodomys ordii fremonti Durrant and Setzer

1945. *Dipodomys ordii fremonti* Durrant and Setzer, Bull. Univ. Utah, 35(26):21, June 30, type from Torrey, 7000 ft., Wayne Co., Utah.

MARGINAL RECORDS.—Utah: type locality; $1/4$ mi. N Grover.

Dipodomys ordii idoneus Setzer

1949. *Dipodomys ordii idoneus* Setzer, Univ. Kansas Publ., Mus. Nat. Hist., 1:546, December 27, type from San Juan, 12 mi. W Lerdo, 3800 ft., Durango.

MARGINAL RECORDS.—Coahuila: 3 mi. NE Sierra Mojada (Baker, 1956:248); 4 mi. N Acatita, 3600 ft.; ½ mi. S, 1 mi. E Jaral, 4 mi. S, 2 mi. W Hipolite. Durango: 5 mi. SE Lerdo, 3800 ft.; *type locality;* 1½ mi. NW Nazas, 4100 ft. (Baker and Greer, 1962:102); 37 mi. W Mapimí, 5500 ft. (*ibid.*); 6 mi. E Zavalza, 4150 ft. (*ibid.*).

Dipodomys ordii inaquosus Hall

1941. *Dipodomys ordii inaquosus* Hall, Proc. Biol. Soc. Washington, 54:58, May 20, type from 11 mi. E and 1 mi. N Jungo, 4200 ft., Humboldt Co., Nevada.

MARGINAL RECORDS.—Nevada: 18 mi. NE Iron Point, 4600 ft.; Izenhood; 23 mi. NW Battle Mountain; 15 mi. SW Winnemucca; 1 mi. N, 8 mi. E Jungo, 4200 ft.; 7 mi. N Winnemucca, 4400 ft.

Dipodomys ordii largus Hall

1951. *Dipodomys ordii largus* Hall, Univ. Kansas Publ., Mus. Nat. Hist., 5:40, October 1, type from Mustang Island, 14 mi. SW Port Aransas, Aransas Co., Texas.

MARGINAL RECORDS (Schmidly and Hendricks, 1976:236).—Texas: *5 mi. SW Port Aransas;* type locality; "*23 mi. SW Port Aransas.*"

Dipodomys ordii longipes (Merriam)

1890. *Dipodops longipes* Merriam, N. Amer. Fauna, 3:72, September 11, type from foot Echo Cliffs, Painted Desert, Coconino Co., Arizona.
1921. *Dipodomys ordii longipes,* Grinnell, Jour. Mamm., 2:96, May 2.
1933. *Dipodomys ordii cleomophila* Goldman, Jour. Washington Acad. Sci., 23:469, October 15, type from 5 mi. NE Winona, 6200 ft., Coconino Co., Arizona.

MARGINAL RECORDS.—Utah: Hatch Trading Post, Montezuma Creek, 25 mi. SE Blanding, 4500 ft. Colorado: Mesa Verde Park Entrance (Anderson, 1961:48). New Mexico: Ship Rock; Blanco; Chama Canyon; 1 mi. S Bernardo; Riley; 10 mi. S Quemado. Arizona: 3 mi. SE Springerville; Holbrook; Winslow; Flagstaff (Cockrum, 1961:150); Deadmans Flat, 6400 ft., NE San Francisco Mtn.; 20 mi. NE Lees Ferry. Utah: Johns Canyon, San Juan River, 5150 ft.

Dipodomys ordii luteolus (Goldman)

1917. *Perodipus ordii luteolus* Goldman, Proc. Biol. Soc. Washington, 30:112, May 23, type from Casper, Natrona Co., Wyoming.

1921. *Dipodomys ordii luteolus,* Grinnell, Jour. Mamm., 2:96, May 2.

MARGINAL RECORDS.—South Dakota: 9 mi. N Bison; 20 mi. SSE Philip. Nebraska (Jones, 1964c:180, 181): 1½ mi. SE Niobrara; Oakdale; 6 mi. S Monroe; 2–3 mi. SE Brady; *vic. North Platte;* 1–4 mi. N, 7 mi. E Ogallala; 1 mi. N, 2 mi. W Chappell. Colorado (Armstrong, 1972:180, 181): Julesburg; 4 mi. NE Burlington; Parker; Denver; Loveland; Goodwin Ranch, Box Elder Creek. Wyoming (Long, 1965a:619, 620): Fort Steele; *18 mi. NNE Sinclair;* 3 mi. W Splitrock; 2½ mi. W Shoshoni; *5 mi. N, 1 mi. W Shoshoni;* 27 mi. N, 1 mi. E Powder River; 12 mi. N, 6 mi. W Bill. South Dakota: Elk Mtn.

Dipodomys ordii marshalli Goldman

1937. *Dipodomys ordii marshalli* Goldman, Proc. Biol. Soc. Washington, 50:223, December 28, type from Bird Island, Great Salt Lake, 4300 ft., Utah.

MARGINAL RECORDS.—Utah: Kelton, 4300 ft.; type locality; *Carrington Island, Great Salt Lake, 4300 ft.;* Stansbury Island, Great Salt Lake, 4300 ft.; 14 mi. W Salt Lake City, 4300 ft.; 2 mi. W Grantsville.

Dipodomys ordii medius Setzer

1949. *Dipodomys ordii medius* Setzer, Univ. Kansas Publ., Mus. Nat. Hist., 1:519, December 27, type from Santa Rosa, Guadalupe Co., New Mexico.

MARGINAL RECORDS.—New Mexico: Rio Alamosa, 15 mi. N Ojo Caliente; Rinconada; Santa Rosa; 4 mi. W, 2¾ mi. N Clovis. Texas: 9 mi. SW Muleshoe; 7 mi. E Post; 2 mi. E Swenson (Ramsey and Carley, 1970:351); Colorado; Monahans. New Mexico: 15 mi. NE Roswell; 44 mi. NW Roswell; Pajarito; 12 mi. NW Alameda, 5500 ft.; 2 mi. SE El Rito.

Dipodomys ordii monoensis (Grinnell)

1919. *Perodipus monoensis* Grinnell, Univ. California Publ. Zool., 21:46, March 29, type from Pellisier Ranch, 5600 ft., 5 mi. N Benton Station, Mono Co., California.
1921. *Dipodomys ordii monoensis* Grinnell, Jour. Mamm., 2:96, May 2.

MARGINAL RECORDS.—Nevada: 21 mi. W, 2 mi. N Lovelock, 4000 ft.; ½ mi. NE Toulon; 1 mi. W Mountain Well, 5350 ft.; Eastgate, 4400 ft.; Fingerrock Wash, Stewart Valley, 5400 ft.; 2 mi. S Millett P.O., 5500 ft.; Fish Spring Valley, ½ mi. N Fish Lake, 6500 ft.; Railroad Valley, 2½ mi. S Lock's Ranch, 5000 ft.; Big Creek, Quinn Canyon Mts.; 5 mi. W White Rock Spring, 6950 ft., Belted Range; 1 mi. N Beatty; 2 mi. NW Palmetto. California: Deep Springs Valley; type locality. Nevada: West Walker River, Smiths Valley, 4700 ft.; ½ mi. S Pyramid Lake, 3950 ft.

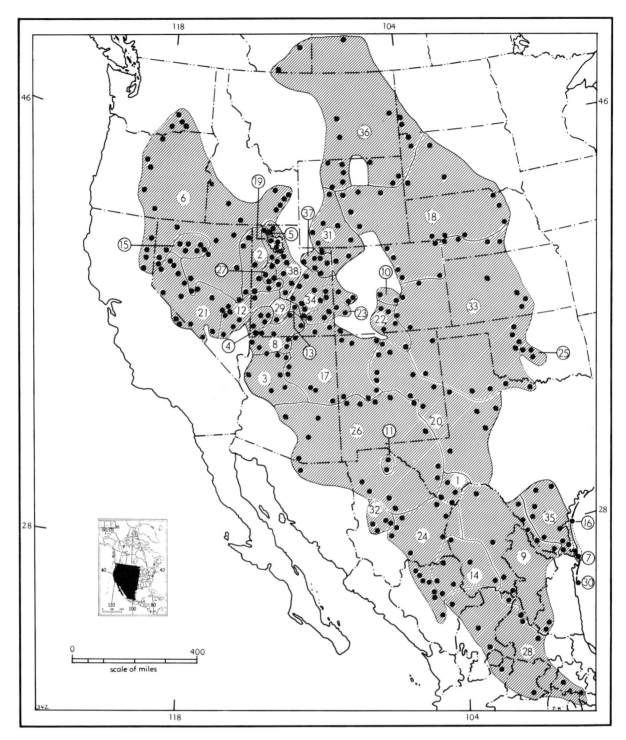

Map 342. *Dipodomys ordii.*

1. *D. o. attenuatus*
2. *D. o. celeripes*
3. *D. o. chapmani*
4. *D. o. cinderensis*
5. *D. o. cineraceus*
6. *D. o. columbianus*
7. *D. o. compactus*
8. *D. o. cupidineus*
9. *D. o. durranti*
10. *D. o. evexus*

11. *D. o. extractus*
12. *D. o. fetosus*
13. *D. o. fremonti*
14. *D. o. idoneus*
15. *D. o. inaquosus*
16. *D. o. largus*
17. *D. o. longipes*
18. *D. o. luteolus*
19. *D. o. marshalli*

20. *D. o. medius*
21. *D. o. monoensis*
22. *D. o. montanus*
23. *D. o. nexilis*
24. *D. o. obscurus*
25. *D. o. oklahomae*
26. *D. o. ordii*
27. *D. o. pallidus*
28. *D. o. palmeri*

29. *D. o. panguitchensis*
30. *D. o. parvabullatus*
31. *D. o. priscus*
32. *D. o. pullus*
33. *D. o. richardsoni*
34. *D. o. sanrafaeli*
35. *D. o. sennetti*
36. *D. o. terrosus*
37. *D. o. uintensis*
38. *D. o. utahensis*

Dipodomys ordii montanus Baird

1855. *Dipodomys montanus* Baird, Proc. Acad. Nat. Sci. Philadelphia, 7:334, April, type from Fort Massachusetts (now Fort Garland), Costilla Co., Colorado.
1921. *Dipodomys ordii montanus*, Grinnell, Jour. Mamm., 2:96, May 2.

MARGINAL RECORDS (Armstrong, 1972:181, unless otherwise noted).—Colorado: 1 mi. N Saguache, 8000 ft.; Crestone, 7851 ft.; 8 mi. S, 2 mi. E Fort Garland, 8000 ft. New Mexico: 4 mi. SW Cimmaron (Hall and Kelson, 1959:516). Colorado: Antonito.

Dipodomys ordii nexilis Goldman

1933. *Dipodomys ordii nexilis* Goldman, Jour. Washington Acad. Sci., 23:470, October 15, type from 5 mi. W Naturita, Montrose Co., Colorado.

MARGINAL RECORDS.—Utah: Cisco. Colorado: Bedrock, 5750 ft. (Armstrong, 1972:182); Coventry. Utah: Blanding; River Mile 113, Glen Canyon, E bank (Durrant and Dean, 1959:86); Snyders Pond, 22 mi. ENE Monticello; Castle Valley, 18 mi. NE Moab, 6000 ft.

Dipodomys ordii obscurus (J. A. Allen)

1903. *Perodipus obscurus* J. A. Allen, Bull. Amer. Mus. Nat. Hist., 19:603, November 12, type from Río Sestín, northwest Durango.
1921. *Dipodomys ordii obscurus*, Grinnell, Jour. Mamm., 2:96, May 2.

MARGINAL RECORDS.—Chihuahua (Anderson, 1972:317): Piñon Camp, Sierra Rica, 4900 ft.; 2 mi. SE Hechiceros, 4450 ft.; Sierra Almagre, 12 mi. S Jaco, 5300 ft. Durango (Baker and Greer, 1962:102, 103): ⅘ mi. S Alamillo, 5500 ft.; Río del Bocas; *type locality*; Mt. San Gabriel; 7 mi. NNW La Zarca, 6000 ft.; 4 mi. NNE Boquilla, 6300 ft.; 3 mi. E Las Nieves, 5400 ft.; Rosario. Chihuahua (Anderson, 1972:317): 7 mi. N Parral; 4 mi. NW San Francisco de Borja, 5700 ft.; General Trais, 1797 m.; 6 mi. NW Ciudad Chihuahua, 4750 ft.

Dipodomys ordii oklahomae Trowbridge and Whitaker

1940. *Dipodomys oklahomae* Trowbridge and Whitaker, Jour. Mamm., 21:343, August 13, type from N bank, South Canadian River, 2¼ mi. S Norman, Cleveland Co., Oklahoma.
1942. *Dipodomys ordii oklahomae*, Davis, Jour. Mamm., 23:332, August 13.

MARGINAL RECORDS.—Oklahoma: 4 mi. N Minco; type locality; T. 6 N, R. 1 W, Cleveland Co. (Thompson and Greer, 1969:108, as *D. ordii* only).

Dipodomys ordii ordii Woodhouse

1853. *D[ipodomys]. ordii* Woodhouse, Proc. Acad. Nat. Sci. Philadelphia, 6:224. Type locality, El Paso, El Paso Co., Texas.

MARGINAL RECORDS.—New Mexico: Gran Quivira, Mesa Jumanes; 40 mi. N Roswell; 2 mi. E Carlsbad. Texas: 5 mi. E Toyahvale; 14½ mi. S Fort Davis; *2 mi. S Paisano* (Schmidly and Hendricks, 1976:236). Chihuahua (Anderson, 1972:317): 12 km. W Est. Encinillas, 5000 ft.; Gallego; 11 mi. NNW San Buenaventura; 5 mi. SSW Nuevo Casas Grandes, 4800 ft. Sonora: Alamo Wash, 35 mi. NW Magdalena. Arizona: La Osa; Oracle (Cockrum, 1961:151); Marinette; 20 mi. NE Calva. New Mexico: Mangos Valley; Gallina Mts.; 10 mi. NE Socorro.

Dipodomys ordii pallidus Durrant and Setzer

1945. *Dipodomys ordii pallidus* Durrant and Setzer, Bull. Univ. Utah, 35(26):24, June 30, type from Old Lincoln Highway, 18 mi. SW Orr's Ranch, Skull Valley, 4400 ft., Tooele Co., Utah.

MARGINAL RECORDS.—Utah: type locality; 1 mi. N Lynndyl, 4768 ft.; Hinckley, 4600 ft.; *7 mi. S Fish Springs, 4400 ft.;* Fish Springs, 4400 ft.

Dipodomys ordii palmeri (J. A. Allen)

1881. *Dipodops ordii palmeri* J. A. Allen, Bull. Mus. Comp. Zool., 8(9):187, March, type from San Luis Potosí, State of San Luis Potosí.
1921. *Dipodomys ordii palmeri*, Grinnell, Jour. Mamm., 2:96, May 2.

MARGINAL RECORDS.—Durango: Hda. Atotonilco, 6680 ft. (Baker and Greer, 1962:103). Zacatecas: Cañitas. San Luis Potosí: Potrero Santa Ana, 7⅗ mi. S Matehuala; Tepeyac. Hidalgo: Ixmiquilpan; Irolo. Guanajuato: Celaya. Jalisco: 9 mi. N Encarnación, 1900 m. Zacatecas: Berriozábal.

Dipodomys ordii panguitchensis Hardy

1942. *Dipodomys ordii panguitchensis* Hardy, Proc. Biol. Soc. Washington, 55:90, June 25, type from 1 mi. S Panguitch, 6666 ft., Garfield Co., Utah. Known only from type locality.

Dipodomys ordii parvabullatus Hall

1951. *Dipodomys ordii parvabullatus* Hall, Univ. Kansas Publ., Mus. Nat. Hist., 5:38, October 1, type from 88 mi. S and 10 mi. W Matamoros, Tamaulipas.

MARGINAL RECORDS.—Tamaulipas: type locality; *90 mi. S, 10 mi. W Matamoros.*

Dipodomys ordii priscus Hoffmeister

1942. *Dipodomys ordii priscus* Hoffmeister, Proc. Biol. Soc. Washington, 55:167, December 31, type from Kinney Ranch, 21 mi. S Bitter Creek, 7100 ft., Sweetwater Co., Wyoming.

MARGINAL RECORDS.—Wyoming (Long, 1965a: 620): Wind River; *7 mi. N Fort Washakie;* 25 mi. N, 38 mi. E Rock Springs; 10½ to 11½ mi. N Baggs. Colorado: Bear River; W side White River, 1 mi. N Rangeley. Utah: N(E) side Green River, 1 mi. E Hideout

Trail Bridge, Hideout Canyon, 6400 ft. Wyoming (Long, 1965a:620, unless otherwise noted): *W side Green River;* 32 mi. S, 22 mi. W Rock Springs; *26 mi. S, 21 mi. W Rock Springs;* 10 mi. SW Granger (Setzer, 1949:548); Eden; *Ft. Washakie.*

Dipodomys ordii pullus Anderson

1972. *Dipodomys ordii pullus* Anderson, Bull. Amer. Mus. Nat. Hist., 148:317, September 8, type from El Rosario, Chihuahua, 6700 ft.

MARGINAL RECORDS.—Chihuahua (Anderson, 1972:318): 20 mi. by road N Cuauhtémoc; NE side Laguna de Bustillos, 6750 ft.; 1 mi. N, 4 mi. W Ciudad Guerrero; 4 mi. S, 4 mi. W Santo Tomás.

Dipodomys ordii richardsoni (J. A. Allen)

1891. *Dipodops richardsoni* J. A. Allen, Bull. Amer. Mus. Nat. Hist., 3:277, June 30, type from one of the sources of Beaver River, probably Harper Co., Oklahoma.—See addenda.
1921. *Dipodomys ordii richardsoni,* Grinnell, Jour. Mamm., 2:96, May 2.

MARGINAL RECORDS.—Nebraska (Jones, 1964c: 181, 182): 6 mi. NE Dickens; 3¾ mi. S Kearney; Hastings. Kansas: Ellis; Medora; Wichita; 4½ mi. NE Danville. Oklahoma: 4 mi. SE Cherokee; 3 mi. S Cleo Springs; T. 17 N, R. 5 W, Kingfisher Co. (Thompson and Greer, 1969:108, as *D. ordii* only); T. 12 N, R. 10 W, Canadian Co. (*ibid.*). Texas: Ringgold (Dalquest, 1968:17); Vernon; Cottle County (Davis, 1966:172, as *D. ordi* [*sic*] only); Floyd County (*ibid.*). New Mexico: 1 mi. S, 2 mi. W Conchas Dam, 4250 ft.; Clayton. Colorado (Armstrong, 1972:183): 1 mi. S, 7 mi. E Trinidad; 8 mi. W Gardner; Colorado Springs; Limon. Nebraska: Venango (Jones, 1964c:182).

Dipodomys ordii sanrafaeli Durrant and Setzer

1945. *Dipodomys ordii sanrafaeli* Durrant and Setzer, Bull. Univ. Utah, 35(26):26, June 30, type from 1½ mi. N Price, 5567 ft., Carbon Co., Utah.

MARGINAL RECORDS.—Utah: 12 mi. NE Price. Colorado (Armstrong, 1972:183): Grand Junction; Hotchkiss; 8 mi. W Olathe. Utah: 4 mi. N Greenriver, 4100 ft.; 3 mi. W Arches National Monument; River Mile 113, Glen Canyon, W bank (Durrant and Dean, 1959:86); King Ranch, 4800 ft.; Notom; 4 mi. E Mt. Alice, between Emery and Loa, 7450 ft.; 5 mi. S Castle Dale, 5600 ft.

Dipodomys ordii sennetti (J. A. Allen)

1891. *Dipodops sennetti* J. A. Allen, Bull. Amer. Mus. Nat. Hist., 3:226, April 29, type from Santa Rosa, 85 mi. SW Corpus Christi, Cameron Co., Texas. (See V. Bailey, N. Amer. Fauna, 25:146, October 24, 1905.)
1942. *Dipodomys ordii sennetti,* Davis, Jour. Mamm., 23:332, August 13.

MARGINAL RECORDS.—Texas: 2 mi. N Nixon; 2 mi. S Riviera (Schmidly and Hendricks, 1976:236); 1 mi. E Rudolf, Norias Division King Ranch (*ibid.*); 28 mi. E Raymondville; Santa Rosa [actually in Cameron Co.] (Schmidly and Hendricks, 1976:236); 20 mi. S Hebbronville (*ibid.*); 8 mi. E Encinal; 8 mi. NE Los Angeles; Somerset.

Dipodomys ordii terrosus Hoffmeister

1942. *Dipodomys ordii terrosus* Hoffmeister, Proc. Biol. Soc. Washington, 55:165, December 31, type from Yellowstone River, 5 mi. W Forsyth, 2750 ft., Rosebud Co., Montana.

MARGINAL RECORDS.—Saskatchewan: near Shackelton, 45–50 mi. NW Swift Current. Montana: Glendive. North Dakota (Genoways and Jones, 1972:18): 4½ mi. S, 1½ mi. W Golva; 6 mi. NE Marmarth. South Dakota (Andersen and Jones, 1971:377): 2 mi. N, 5 mi. W Ludlow; *14 mi. S, 4 mi. W Reva;* 15 mi. S, 4 mi. W Reva. Wyoming (Long, 1965a:621, unless otherwise noted): Newcastle (Setzer, 1949:524); 23 mi. SW Newcastle; Arvada; Greybull; 7 mi. S Basin; *7 mi. S, 5 mi. W Worland;* Kirby Creek; Sheep Creek, S base Owl Creek Mts.; 4 mi. N Garland. Montana: Billings; 24 mi. N Roundup, 8 mi. SW Flatwillow. Alberta: S side South Saskatchewan River, T. 19, R. 2, West 4 (H. C. Smith, 1972:53, as *D. ordii* only); near Medicine Hat.

Dipodomys ordii uintensis Durrant and Setzer

1945. *Dipodomys ordii uintensis* Durrant and Setzer, Bull. Univ. Utah, 35(26):27, June 30, type from Red Creek, 6700 ft., 2 mi. N Fruitland, Duchesne Co., Utah.

MARGINAL RECORDS.—Utah: 15 mi. W Vernal; Vernal; *E side Green River, 3 mi. S Jensen;* 20 mi. E Ouray; Brown Corral, 20 mi. S Ouray, 6250 ft.; 20 mi. S Myton; type locality.

Dipodomys ordii utahensis (Merriam)

1904. *Perodipus montanus utahensis* Merriam, Proc. Biol. Soc. Washington, 17:143, July 14, type from Ogden, Weber Co., Utah.
1921. *Dipodomys ordii utahensis,* Grinnell, Jour. Mamm., 2:96, May 2.

MARGINAL RECORDS.—Utah: 15 mi. E Park Valley, Raft River Mts., 5500 ft.; *Ogden, 4293 ft.;* 4 mi. N Draper, 4500 ft.; Provo, 4510 ft.; Spring City; 1 mi. W Aurora, 5190 ft.; 4 mi. W Nephi (Durrant, 1952:256); *Little Valley, Sheeprock Mts., 5500 ft.* (*ibid.*); *Clover Creek, Onaqui Mts., 5500 ft.* (*ibid.*); St. John, 4300 ft.; *Bauer, 4500 ft.* (Durrant, 1952:256); *Antelope Island, Great Salt Lake, 4250 ft.;* Promontory Point (*ibid.*).

Dipodomys microps
Chisel-toothed Kangaroo Rat

Revised by Hall and Dale, Occas. Pap. Mus. Zool., Louisiana State University, 4:47–63, November 10, 1939.

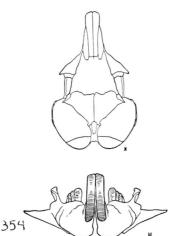

Fig. 354. *Dipodomys microps aquilonius*, 3½ mi. E Flanigan, Nevada, No. 73616 M.V.Z., ♂, X 1, but front view of lower incisors X 2.

External measurements: 254–297; 134–175; 39.0–46.4. Five toes on each hind foot. The only other 5-toed kangaroo rats with which *Dipodomys microps* shares its range are *Dipodomys panamintinus* and *Dipodomys ordii*. In *D. microps* the lower incisors are flat on their anterior faces and chisellike, whereas in *D. ordii* they are rounded and awllike. The hind foot of *D. microps* is shorter than 44 mm. in the part of its range that it shares with the longer-footed *D. panamintinus*.

Dipodomys microps alfredi Goldman

1937. *Dipodomys microps alfredi* Goldman, Proc. Biol. Soc. Washington, 50:221, December 28, type from Gunnison Island, 4300 ft., Great Salt Lake, Utah. Known only from Gunnison Island.

Dipodomys microps aquilonius Willett

1935. *Dipodomys microps aquilonius* Willett, Jour. Mamm., 16:63, February 14, type from 3 mi. E Eagleville, Modoc Co., California.

MARGINAL RECORDS.—California: 6 mi. E Cedarville, 4500 ft. Nevada: 1 mi. W Hausen, 5650 ft.; 12 mi. N, 2 mi. E Gerlach, 4000 ft.; 1 mi. NE Gerlach, 4000 ft.; ½ mi. S Pyramid Lake, 3950 ft. California: 9 mi. E Amedee; type locality.

Dipodomys microps bonnevillei Goldman

1937. *Dipodomys microps bonnevillei* Goldman, Proc. Biol. Soc. Washington, 50:222, December 28, type from Kelton, 4300 ft., Boxelder Co., Utah.

MARGINAL RECORDS.—Utah: Hardup; type locality; Old Lincoln Highway, 18 mi. SW Orr's Ranch, Skull Valley, 4400 ft.; Aurora; 2 mi. E Clear Lake, 4600 ft.; Pine Valley, sec. 33, T. 25 S, R. 17 W, Salt Lake

B.M., 50 mi. W Milford, 5500 ft. Nevada: 1 mi. N Baker; 2 mi. W Smith Creek Cave, Mt. Moriah, 6300 ft.; Cobre, 6100 ft.; 15 mi. S Contact, 5800 ft.

Dipodomys microps celsus Goldman

1924. *Dipodomys microps celsus* Goldman, Jour. Washington Acad. Sci., 14:372, September 19, type from 6 mi. N Wolf Hole, 3500 ft., Arizona.
1942. *Dipodomys microps woodburyi* Hardy, Proc. Biol. Soc. Washington, 55:89, June 25, type from Beaverdam Slope, approx. 3500 ft., W of Beaverdam Mts., Washington Co., Utah. Regarded as inseparable from *D. m. celsus* by Stock, Jour. Mamm., 51:431, May 20, 1970.

MARGINAL RECORDS.—Utah: 1½ mi. NW Diamond Valley; 5 mi. NW St. George; Gould's Ranch, Hurricane. Arizona: Kanab Wash, near S boundary

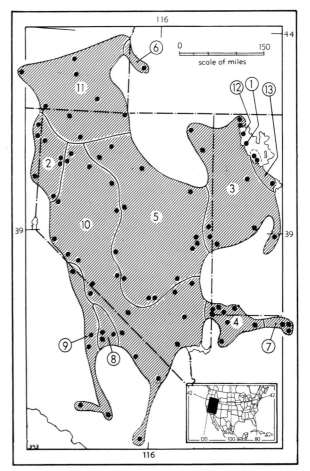

Map 343. *Dipodomys microps.*

Guide to subspecies	
1. *D. m. alfredi*	7. *D. m. leucotis*
2. *D. m. aquilonius*	8. *D. m. levipes*
3. *D. m. bonnevillei*	9. *D. m. microps*
4. *D. m. celsus*	10. *D. m. occidentalis*
5. *D. m. centralis*	11. *D. m. preblei*
6. *D. m. idahoensis*	12. *D. m. russeolus*
	13. *D. m. subtenuis*

Kaibab Indian Reservation; 3 mi. W Lower Pigeon Spring, 4400 ft. (Hoffmeister and Durham, 1971:34); type locality. Utah (as *D. m. woodburyi*): near Ed. Terry Ranch, 5000 ft.; Beaverdam Slope, *ca.* 3500 ft., W of Beaverdam Mts.

Dipodomys microps centralis Hall and Dale

1939. *Dipodomys microps centralis* Hall and Dale, Occas. Pap. Mus. Zool., Louisiana State Univ., 4:52, November 10, type from 4 mi. SE Romano, Diamond Valley, Eureka Co., Nevada.

MARGINAL RECORDS.—Nevada: 18 mi. NE Iron Point, 4600 ft.; ½ mi. S Beowawe; 5 mi. SE Greens Ranch, Steptoe Valley, 5900 ft.; 7 mi. SW Osceola, Spring Valley, 6100 ft.; 5½ mi. NW Shoshone P.O., 6100 ft.; Duck Valley, 3 mi. S Geyser, 6050 ft.; Coal Valley, 10 mi. N Seeman Pass, 4650 ft.; 9 mi. W Groom Baldy, 5500 ft.; 9 mi. E Wheelbarrow Peak; 1 mi. SW Cactus Spring, Cactus Range; San Antonio, 5400 ft.; Reese River Valley, 7 mi. N Austin; 15 mi. SW Winnemucca.

Dipodomys microps idahoensis Hall and Dale

1939. *Dipodomys microps idahoensis* Hall and Dale, Occas. Pap. Mus. Zool., Louisiana State Univ., 4:53, November 10, type from 5 mi. SE Murphy, Owyhee Co., Idaho.

MARGINAL RECORDS.—Idaho: *10 mi. E Murphy;* type locality.

Dipodomys microps leucotis Goldman

1931. *Dipodomys microps leucotis* Goldman, Proc. Biol. Soc. Washington, 44:135, October 17, type from 6 mi. W Colorado River Bridge, 3700 ft., Houserock Valley, N[= W] side Marble Canyon of Colorado River, Arizona.

MARGINAL RECORDS.—Arizona: type locality; 7 mi. E Jacobs Pool (Cockrum, 1961:153); ½ mi. E Navajo Bridge.

Dipodomys microps levipes (Merriam)

1904. *Perodipus microps levipes* Merriam, Proc. Biol. Soc. Washington, 17:145, July 14, type from Perognathus Flat, 5200 ft., Emigrant Gap, Panamint Mts., Inyo Co., California.
1931. *Dipodomys microps levipes*, Hall, Univ. California Publ. Zool., 37:5, April 10.

MARGINAL RECORDS.—California: 15 mi. N Darwin, 5200 to 5300 ft.; type locality; Darwin, 4800 ft.

Dipodomys microps microps (Merriam)

1904. *Perodipus microps* Merriam, Proc. Biol. Soc. Washington, 17:145, July 14, type from Lone Pine, Owens Valley, Inyo Co., California.
1921. *Dipodomys microps*, Grinnell, Jour. Mamm., 2:96, May 2.

MARGINAL RECORDS.—California: Pellisier Ranch, 5600 ft., 5 mi. N Benton Station; 2½ mi. NE Lone Pine; Victorville; 6 mi. N Lancaster (Csuti, 1971:50); 1½ mi. SW Olancha, 3900 ft.

Dipodomys microps occidentalis Hall and Dale

1939. *Dipodomys microps occidentalis* Hall and Dale, Occas. Pap. Mus. Zool., Louisiana State Univ., 4:56, November 10, type from 3 mi. S Schurz, 4100 ft., Mineral Co., Nevada.

MARGINAL RECORDS.—Nevada: 9½ mi. N Sulphur, 4050 ft.; 1 mi. W Humboldt, 4180 ft.; S slope Granite Peak, East Range; *Smiths Creek Valley, 5550 ft.;* 2 mi. W Railroad Pass; N shore Mud Lake, 5300 ft., S end Ralston Valley; 8½ mi. NE Springdale, 4250 ft.; 9½ mi. NW Crystal Spring, 4800 ft., Pahranagat Valley; Desert Valley, 5300 ft., 21 mi. W Panaca; Coyote Spring, 2800 ft.; 4 mi. NW Las Vegas. California: SE side Clark Mtn., 5100 ft.; Stubby Spring (Miller and Stebbins, 1964:306); 2 mi. E Greenwater, 3900 ft.; Salt Creek, NW arm Death Valley, −91 ft.; N end Deep Spring Valley, 5300 ft. Nevada: 2½ mi. N Dyer, 4850 ft.; Huntoon Valley, 5700 ft.; Fletcher, 6098 ft.; Smiths Valley, 7½ mi. NE Wellington, 4900 ft.; 1½ mi. N Wadsworth, 4100 ft.; 6 mi. N, 10 mi. W Sulphur, 4000 ft.

Dipodomys microps preblei (Goldman)

1921. *Perodipus microps preblei* Goldman, Jour. Mamm., 2:233, November 29, type from Narrows, Malheur Lake, Harney Co., Oregon.
1939. *Dipodomys microps preblei*, Hall and Dale, Occas. Pap. Mus. Zool., Louisiana State Univ., 4:54, November 10.

MARGINAL RECORDS.—Oregon: type locality; Buena Vista, 25 mi. S Narrows, 4300 ft.; White Horse Sink. Nevada: 36 mi. NE Paradise Valley, 5500 ft.; Jackson Creek Ranch, 4000 ft., 17½ mi. S, 5 mi. W Quinn River Crossing; Virgin Valley. Oregon: 8 mi. S Adel, E of mouth Twenty Mile Creek; Summer Lake.

Dipodomys microps russeolus Goldman

1939. *Dipodomys microps russeolus* Goldman, Jour. Mamm., 20:353, August 14, type from Dolphin Island, 4250 ft., Great Salt Lake, Utah. Known only from type locality.

Dipodomys microps subtenuis Goldman

1939. *Dipodomys microps subtenuis* Goldman, Jour. Mamm., 20:354, August 14, type from Carrington Island, 4250 ft., Great Salt Lake, Utah.

MARGINAL RECORDS.—Utah: type locality; *Badger Island;* Stansbury Island, 4250 ft.; Chimney Rock Pass, lat. 40° 04′ N, long. 111° 56′ W, Cedar Valley.

panamintinus-group
Dipodomys panamintinus
Panamint Kangaroo Rat

External measurements: 285–334; 156–202; 42–48. Five toes on each hind foot; pinna of ear and auditory bullae small; broad across maxillary processes of zygomatic arches.

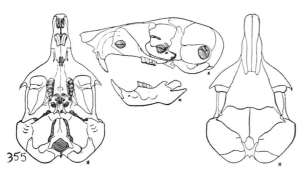

Fig. 355. *Dipodomys panamintinus leucogenys*, Junction House, Nevada, No. 49012 M.V.Z., ♂, X 1.

Dipodomys panamintinus argusensis Huey

1945. *Dipodomys mohavensis argusensis* Huey, Trans. San Diego Soc. Nat. Hist., 10(10):131, March 9, type from Junction Ranch, 5725 ft., Argus Mts., Inyo Co., California. Known only from type locality.
1955. *Dipodomys panamintinus argusensis*, Miller and Kellogg, Bull. U.S. Nat. Mus., 205:387, March 3.

Dipodomys panamintinus caudatus Hall

1946. *Dipodomys panamintinus caudatus* Hall, Mammals of Nevada, p. 409, July 1, type from 6 mi. S Granite Well, 3800 ft., Providence Mts., San Bernardino Co., California.

MARGINAL RECORDS.—Nevada: 3½ mi. S, 9 mi. W Searchlight, 4300 ft. California: *Purdy;* 2 mi. ESE Rock Spring; *type locality;* 5 mi. NE Granite Well.

Dipodomys panamintinus leucogenys (Grinnell)

1919. *Perodipus leucogenys* Grinnell, Univ. California Publ. Zool., 21:46, March 29, type from Pellisier Ranch, 5600 ft., 5 mi. N Benton Station, Mono Co., California.
1946. *Dipodomys panamintinus leucogenys*, Hall, Mammals of Nevada, p. 407, July 1.

MARGINAL RECORDS.—Nevada: ½ mi. S Pyramid Lake, 3950 ft.; 6 mi. NE Virginia City, 6000 ft.; 2 mi. SW Pine Grove, 7250 ft.; Lapon Canyon, Mt. Grant, 8900 ft.; *Huntoon Valley, 5700 ft.;* Endowment Mine, Excelsior Mts., 6930 ft. California: type locality; *Taylor Ranch, 5300 ft., 2 mi. S Benton Station;* Walters Ranch, 3900 ft., 2 mi. NNW Independence; 9 mi. W Bishop; Dry Creek, 6700–6900 ft., near Mono Lake. Nevada: N end Topaz Lake; 6½ mi. S Minden, 4900 ft.; 5½ mi. S Carson City, 4700 ft.; *3¹/₂ mi. E Carson City, 4700 ft.;* 10 mi. S Reno; 8 mi. NE Reno.

Dipodomys panamintinus mohavensis (Grinnell)

1918. *Perodipus mohavensis* Grinnell, Univ. California Publ. Zool., 17:428, April 25, type from ½ mi. E of Railway Station

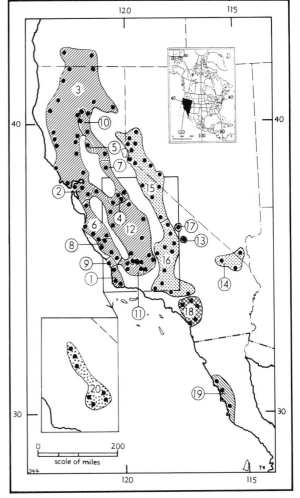

Map 344. Five species of *Dipodomys*.

1. *D. heermanni arenae*
2. *D. heermanni berkeleyensis*
3. *D. heermanni californicus*
4. *D. heermanni dixoni*
5. *D. heermanni eximius*
6. *D. heermanni goldmani*
7. *D. heermanni heermanni*
8. *D. heermanni jolonensis*
9. *D. heermanni morroensis*
10. *D. heermanni saxatilis*
11. *D. heermanni swarthi*
12. *D. heermanni tularensis*
13. *D. panamintinus argusensis*
14. *D. panamintinus caudatus*
15. *D. panamintinus leucogenys*
16. *D. panamintinus mohavensis*
17. *D. panamintinus panamintinus*
18. *D. stephensi*
19. *D. gravipes*
20. *D. ingens*

at Warren, 3275 ft., about 5 mi. N Mohave, Kern Co., California.

1946. *D[ipodomys]. panamintinus mohavensis,* Hall, Mammals of Nevada, p. 408, July 1.

MARGINAL RECORDS.—California: vic. Lone Pine; Olancha; Little Lake; Fay Creek; Hesperia; Rock Creek; Rye Canyon, near Castaic Junction (Csuti, 1971:50); Fairmont; 7 mi. W Mohave; Isabella; Walker Pass.

Dipodomys panamintinus panamintinus (Merriam)

1894. *Perodipus panamintinus* Merriam, Proc. Biol. Soc. Washington, 9:114, June 21, type from head Willow Creek, Panamint Mts., Inyo Co., California.

1921. *Dipodomys panamintinus,* Grinnell, Jour. Mamm., 2:95, May 2.

MARGINAL RECORDS.—California: type locality; *1 mi. S Lee Pump, 6100 ft., Panamint Mts.*

Dipodomys stephensi (Merriam)
Stephens' Kangaroo Rat

1907. *Perodipus stephensi* Merriam, Proc. Biol. Soc. Washington, 20:78, July 22, type from San Jacinto Valley [a little west of present town of Winchester toward Menifee], Riverside Co., California.

1921. *Dipodomys stephensi,* Grinnell, Jour. Mamm., 2:95, May 2.

1962. *Dipodomys cascus* Huey, Trans. San Diego Soc. Nat. Hist., 12:479, August 30, type from 1 mi. E Bonsall, 350 ft., San Diego Co., California. Regarded as a synonym of *D. stephensi* by Lackey (Trans. San Diego Soc. Nat. Hist., 14:315, December 13, 1967).

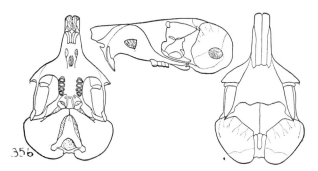

Fig. 356. *Dipodomys stephensi,* Perris, Riverside Co., California, No. 33570 M.V.Z., ♀, X 1.

External measurements: 277–300; 165–180; 41–43. Closely related to *Dipodomys panamintinus* and possibly only subspecifically distinct from it. Differs from *D. panamintinus* in larger auditory bullae and, on the average, in shorter hind foot.

MARGINAL RECORDS (Lackey, 1967:336, unless otherwise noted).—California: Reche Canyon, 4 mi. SE

Colton (Hall and Kelson, 1959:520); 6 mi. NW San Jacinto (*ibid.*); type locality; *4/5 mi. S, 5 1/5 mi. E Temecula;* 1 mi. S Pala; 2 *mi. N, 3 3/5 mi. E Bonsall;* 3/5 *mi. S, 2 mi. E San Luis Rey Mission;* 1 1/3 mi. NE San Luis Rey Mission; *SW corner of Fallbrook naval weapons annex* (Bleich and Schwartz, 1974:209); Temescal [Wash] (Hall and Kelson, 1959:520).

heermanni-group
Dipodomys elephantinus (Grinnell)
Big-eared Kangaroo Rat

1919. *Perodipus elephantinus* Grinnell, Univ. California Publ. Zool., 21:43, March 29, type from 1 mi. N Cook P.O., 1300 ft., Bear Valley, San Benito Co., California.

1921. *Dipodomys elephantinus* Grinnell, Jour. Mamm., 2:96, May 2.

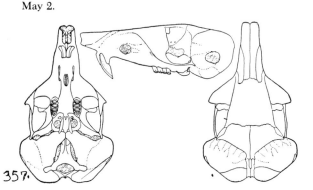

Fig. 357. *Dipodomys elephantinus,* Cook, 1300 ft., Bear Valley, San Benito Co., California, No. 28509 M.V.Z., ♀, X 1.

External measurements: 305–336; 183–210; 44–50. The big-eared kangaroo rat is closely related to, and possibly only subspecifically distinct from, *Dipodomys venustus* but differs from *D. venustus* in greater width (averaging 4.9) of rostrum near its end. See Map 345. In adults Grinnell (1922:101) found a unique character to be the greatly flared distal ends of the nasals—flared so much that a constriction appears in front of the premaxillae where the latter turn down to envelope the incisors.

MARGINAL RECORDS.—California: type locality; Stonewall Creek, 1300 ft., 6 mi. NE Soledad.

Dipodomys venustus
Narrow-faced Kangaroo Rat

External measurements: 293–332; 175–203; 44–47. This large, dark, 5-toed kangaroo rat with large external ears is narrow across the maxillary processes of the zygomatic arches of the skull. In this latter feature resemblance is shown to the species *Dipodomys agilis* from which *D. venustus* eventually may be found to be only subspecifically distinct.

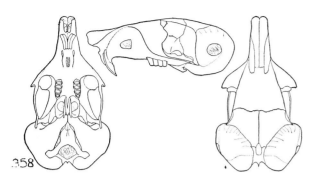

Fig. 358. *Dipodomys venustus venustus*, Doyle Gulch, 9 mi. E Santa Cruz, Santa Cruz Co., California, No. 46777 M.V.Z., ♂, X 1.

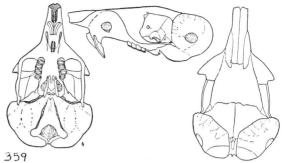

Fig. 359. *Dipodomys agilis agilis*, Mint Canyon, 2400 ft., Los Angeles Co., California, No. 47175 M.V.Z., ♂, X 1.

Dipodomys venustus sanctiluciae Grinnell

1919. *Dipodomys sanctiluciae* Grinnell, Proc. Biol. Soc. Washington, 32:204, December 31, type from 1 mi. SW Jolon, Monterey Co., California.
1921. *Dipodomys venustus sanctiluciae* Grinnell, Jour. Mamm., 2:96, May 2.

MARGINAL RECORDS.—California: Santa Lucia Peak, 5900 ft.; type locality; Santa Margarita; Chalk Peak, 3000 ft.

Dipodomys venustus venustus (Merriam)

1904. *Perodipus venustus* Merriam, Proc. Biol. Soc. Washington, 17:142, July 14, type from Santa Cruz, Santa Cruz Co., California.
1919. *Dipodomys venustus*, Grinnell, Proc. Biol. Soc. Washington, 32:204, December 31.

MARGINAL RECORDS.—California: Jasper Ridge, near Stanford University; Mt. Hamilton; Fremont Peak; type locality.

Dipodomys agilis
Agile Kangaroo Rat

External measurements: 265–319; 155–197; 40–46. Five toes on each hind foot; external ear medium to large; color dark; skull narrow across maxillary processes of zygomatic arches.

Dipodomys agilis agilis Gambel

1848. *Dipodomys agilis* Gambel, Proc. Acad. Nat. Sci. Philadelphia, 4:77, type from Los Angeles, Los Angeles Co., California.
1853. *D[ipodomys]. wagneri* Le Conte, Proc. Acad. Nat. Sci. Philadelphia, 6:224, January, type from an unknown locality.

MARGINAL RECORDS.—California: Schoolhouse Canyon; Matilija; Elizabeth Lake, 3400 ft.; N base Sugarloaf Mtn., 7500 ft., San Bernardino Mts.; *near Banning;* Schain's Ranch, 4900 ft., San Jacinto Mts.;

Kenworthy; ⅘ mi. S, 5⅓ mi. E Temecula (Lackey, 1967:336); *1³⁄₅ mi. N, ⁹⁄₅ mi. W Murrieta (ibid.);* mouth Trabuco Canyon, 1500 ft., thence northward along coast to point of beginning.

Dipodomys agilis cabezonae (Merriam)

1904. *Perodipus cabezonae* Merriam, Proc. Biol. Soc. Washington, 17:144, July 14, type from Cabezon, San Gorgonio Pass, Riverside Co., California.
1921. *Dipodomys agilis cabezonae,* Grinnell, Jour. Mamm., 2:96, May 2.

MARGINAL RECORDS.—California: type locality; near Dos Palmos Spring; vic. Mountain Spring. Baja California: *Tres Piños Mine, near Juárez;* 3 mi. NE Neji. California: *Banner, 2700 ft.;* Warner Pass, 3000 ft.

Dipodomys agilis fuscus Boulware

1943. *Dipodomys agilis fuscus* Boulware, Univ. California Publ. Zool., 46:393, September 16, type from 2½ mi. N La Purisima Mission, 600 ft., Santa Barbara Co., California.

MARGINAL RECORDS.—California: *C. A. Davis Ranch, 2 mi. NNW Lompoc, 400 ft.;* type locality.

Dipodomys agilis martirensis Huey

1927. *Dipodomys agilis martirensis* Huey, Trans. San Diego Soc. Nat. Hist., 5:7, February 20, type from La Grulla (E side of valley), 7500 ft., Sierra San Pedro Mártir, Baja California.

MARGINAL RECORDS.—Baja California: 14 mi. N Laguna Hanson (Alvarez, 1961:403); Laguna Hanson; *1 mi. E Laguna Hanson;* Diablito Spring, summit San Matías Pass; type locality; *Rosarito Divide;* Rosarito; Aquaito Spring, el Valle de la Trinidad; Sangre de Cristo.

Dipodomys agilis perplexus (Merriam)

1907. *Perodipus perplexus* Merriam, Proc. Biol. Soc. Washington, 20:79, July 22, type from Walker Basin, 3400 ft., Kern Co., California.

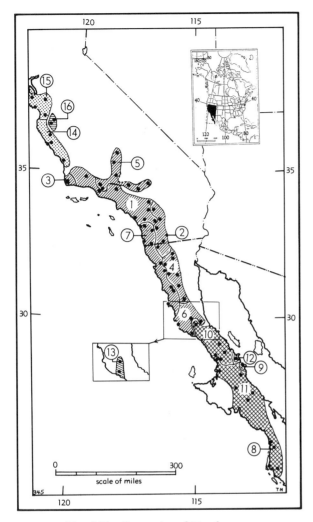

Map 345.   Six species of *Dipodomys*.

1. *D. agilis agilis*
2. *D. agilis cabezonae*
3. *D. agilis fuscus*
4. *D. agilis martirensis*
5. *D. agilis perplexus*
6. *D. agilis plectilis*
7. *D. agilis simulans*
8. *D. peninsularis australis*
9. *D. peninsularis eremoecus*
10. *D. peninsularis pedionomus*
11. *D. peninsularis peninsularis*
12. *D. antiquarius*
13. *D. paralius*
14. *D. venustus sanctiluciae*
15. *D. venustus venustus*
16. *D. elephantinus*

1921. *Dipodomys agilis perplexus*, Grinnell, Jour. Mamm.. 2:96, May 2.

MARGINAL RECORDS.—California: Trout Creek, toward head of S. Fork Kern River; *Rip Rap Mine, Piute Mts.*; 1½ mi. N Tehachapi; 10½ mi. S, 13½ mi. E Victor-

ville, 5200 ft., San Bernardino Mts.; Cajon Wash, ½ mi. SW Devore, 200 ft.; 1 mi. S, 2 mi. W Big Pines, 7400 ft.; head Piru Creek; S side Mt. Piños, 5500–6500 ft.; Kern River at Bodfish, 2400 ft.

Dipodomys agilis plectilis Huey

1951. *Dipodomys agilis plectilis* Huey, Trans. San Diego Soc. Nat. Hist., 11:240, April 30, type from mouth of canyon San Juan de Dios, lat. 30° 07′ N, Baja California.

MARGINAL RECORDS.—Baja California: type locality; 3 mi. W El Marmol; 4 mi. N Santa Catarina Landing; 1 mi. E El Rosario.

Dipodomys agilis simulans (Merriam)

1904. *Perodipus streatori simulans* Merriam, Proc. Biol. Soc. Washington, 17:144, July 14, type from Dulzura, San Diego Co., California.
1921. *Dipodomys agilis simulans*, Grinnell, Jour. Mamm., 2:96, May 2.
1925. *Dipodomys agilis latimaxillaris* Huey, Proc. Biol. Soc. Washington, 38:84, May 26, type from 2 mi. W Santo Domingo Mission, 30° 45′ N, 115° 58′ W, Baja California. Regarded as identical with *simulans* by Huey, Trans. San Diego Soc. Nat. Hist., 11:234, April 30, 1951.

MARGINAL RECORDS.—California: ½ mi. W Pala (Lackey, 1967:336); Santa Ysabel; type locality. Baja California: 3 mi. E Ojos Negros; Rancho San Pablo, 10 mi. SE Alamo; San José; Valledares; San Quintín, thence northward along coast to California: 1⅘ mi. NE San Luis Rey Mission (Lackey, 1967:336); 2⅖ mi. NE *San Luis Rey Mission (ibid.)*; ⁹/₁₀ mi. N, ¹/₁₀ mi E Bonsall *(ibid.)*.

Dipodomys paralius Huey
Santa Catarina Kangaroo Rat

1951. *Dipodomys paralius* Huey, Trans. San Diego Soc. Nat. Hist., 10:241, April 30, type from Santa Catarina Landing, lat. 29° 31′ N, Baja California.

External measurements: 255–280; 145–170; 38–41. Greatest length of skull, 37.2–39.2; breadth across bullae, 24.2–24.7. From its near relative, *Dipodomys agilis plectilis*, which occurs in the same geographic area, *D. paralius* differs in paler upper parts, lesser size, relatively more flattened auditory bullae, and more angular and more widely spreading zygomatic arches. The general outline of the skull is more nearly that of an equilateral triangle than that of an acute triangle, according to the original description.

MARGINAL RECORDS.—Baja California: Rancho La Ramona, 8 mi. N Santa Catarina; type locality. See addenda.

Dipodomys peninsularis
Baja California Kangaroo Rat

External measurements: 270–302; 158–189; 40–44. Greatest length of skull, 38.0–41.5; breadth across bullae, 24.1–26.9. Five toes on each hind foot. Closely resembles *Dipodomys agilis*, but differs in more inflated tympanic bullae, heavier-boned tail, and brighter, more buffy, coloration. For a taxonomic treatment of the subspecies of this species see Huey (1951:244–249), who deals similarly with all the other kinds of kangaroo rats known from Baja California (*op. cit.*). See Map 345, **and** addenda.

Dipodomys peninsularis australis Huey

1951. *Dipodomys peninsularis australis* Huey, Trans. San Diego Soc. Nat. Hist., 11:249, April 30, type from Santo Domingo, Magdalena Plain, lat. 25° 30′ N, Baja California.

MARGINAL RECORDS.—Baja California: San Jorgé; type locality; 9 mi. S El Refugio. **See** addenda.

Dipodomys peninsularis eremoecus Huey

1951. *Dipodomys peninsularis eremoecus* Huey, Trans. San Diego Soc. Nat. Hist., 11:248, April 30, type from 7 mi. W San Francisquito Bay, lat. 28° 30′ N, Baja California.

MARGINAL RECORDS.—Baja California: type locality; *Santa Teresa Bay.* **See** addenda.

Dipodomys peninsularis pedionomus Huey

1951. *Dipodomys peninsularis pedionomus* Huey, Trans. San Diego Soc. Nat. Hist., 11:247, April 30, type from 2 mi. N Chapala Dry Lake, on Llano de Santa Ana, lat. 29° 30′ N, long. 114° 35′ W, Baja California.

MARGINAL RECORDS.—Baja California: San Agustín; type locality; El Valle de Agua Amargá; San Borjas Mission; Punta Prieta. **See** addenda.

Dipodomys peninsularis peninsularis (Merriam)

1907. *Perodipus simulans peninsularis* Merriam, Proc. Biol. Soc. Washington, 20:79, July 22, type from Santo Domingo (Santo Domingo Landing), lat. 28° 51′ N, long. 114° W, Baja California.
1951. *Dipodomys peninsularis peninsularis*, Huey, Trans. San Diego Soc. Nat. Hist., 11:246, April 30.

MARGINAL RECORDS.—Baja California: 11 mi. S Punta Prieta; Santa Gertrudis Mission; Valle de Yaqui, 10 mi. W Santa Rosalía; San Ignacio; Poso Altimisano; La Lomita María. **See** addenda.

Dipodomys antiquarius Huey
Huey's Kangaroo Rat

1962. *Dipodomys antiquarius* Huey, Trans. San Diego Soc. Nat. Hist., 12:477, August 30, type from San Juan Mine,

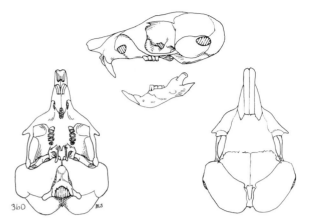

Fig. 360. *Dipodomys antiquarius*, San Juan Mine, Sierra San Borja, Baja California, No. 18903 S.D.S.N.H., ♂, X 1.

Sierra Borja, 400 ft., long. 113° 17′ W, lat. 28° 41′ N, Baja California del Norte. Known only from type locality.

External measurements: 247–285; 138–170; 40–42. Greatest length of skull, 385–410; breadth across bullae, 25.1–26.2. General coloration light buff; white side-stripes on tail about as wide as upper and lower black stripes; wider across maxillae and tympanic bullae than its near relative, *D. peninsularis*. See Map 345, **and** addenda.

Dipodomys heermanni
Heermann's Kangaroo Rat

External measurements: 250–340; 160–200; 38–47. Each hind foot has 4 or 5 toes, depending on the subspecies; external ear of moderate length; auditory bullae small to medium; skull broad across maxillary processes of zygomatic arches. (See Map 344.)

Dipodomys heermanni arenae Boulware

1943. *Dipodomys heermanni arenae* Boulware, Univ. California Publ. Zool., 46:392, September 16, type from 2 mi. NNW Lompoc, 400 ft., Santa Barbara Co., California.

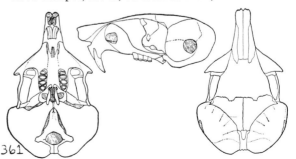

Fig. 361. *Dipodomys heermanni tularensis*, 6 mi. E Panoche, 1400 ft., Fresno Co., California, No. 72724 M.V.Z., ♀, X 1.

MARGINAL RECORDS.—California: $2\frac{3}{4}$ mi. S Oceano, 10–50 ft.; $2\frac{2}{5}$ mi. W Buellton, 350 ft.; type locality.

Dipodomys heermanni berkeleyensis Grinnell

1919. *Dipodomys berkeleyensis* Grinnell, Proc. Biol. Soc. Washington, 32:204, December 31, type from head Dwight Way, Berkeley, Alameda Co., California.
1921. *Dipodomys heermanni berkeleyensis* Grinnell, Jour. Mamm., 2:95, May 2.

MARGINAL RECORDS.—California: type locality; Mt. Diablo; vic. Livermore.

Dipodomys heermanni californicus Merriam

1890. *Dipodomys californicus* Merriam, N. Amer. Fauna, 4:49, October 8, type from Ukiah [W edge main road running S from Ukiah and about 1 mi. S of then the center of town], Mendocino Co., California.
1921. *Dipodomys heermanni californicus*, Grinnell, Jour. Mamm., 2:95, May 2.
1899. *Dipodomys californicus pallidulus* Bangs, Proc. New England Zool. Club, 1:65, July 31, type from Sites, Colusa Co., California.
1916. *Dipodomys californicus trinitatis* L. Kellogg, Univ. California Publ. Zool., 12:366, January 27, type from Helena, 1405 ft., Trinity Co., California.
1925. *Dipodomys heermanni gabrielsoni* Goldman, Proc. Biol. Soc. Washington, 38:33, March 12, type from Brownsboro, Jackson Co., Oregon. Regarded as inseparable from *californicus* by Grinnell and Linsdale, Univ. California Publ. Zool., 30:457, June 15, 1929.

MARGINAL RECORDS.—Oregon: Brownsboro; Swan Lake Valley. California: E side Tule Lake; Petes Valley; Hat Creek, 15 mi. S Cassel; Lyman's, 3300 ft.; 10 mi. N Red Bluff, 500 ft.; Willows; Sites; Rumsey; Vacaville; Lagunitas; 7 mi. W Cazadero; type locality; near Willets; 6 mi. SW Laytonville; Mad River; Helena; Scott Valley, 4 mi. S Fort Jones; Hornbrook.

Dipodomys heermanni dixoni (Grinnell)

1919. *Perodipus dixoni* Grinnell, Univ. California Publ. Zool., 21:45, March 29, type from Delhi, near Merced River, Merced Co., California.
1921. *Dipodomys heermanni dixoni* Grinnell, Jour. Mamm., 2:95, May 2.

MARGINAL RECORDS.—California: 3 mi. S Lagrange; [$1\frac{1}{2}$ mi. S] Merced Falls; type locality.

Dipodomys heermanni eximius Grinnell

1919. *Dipodomys californicus eximius* Grinnell, Proc. Biol. Soc. Washington, 32:205, December 31, type from Marysville Buttes, 300 ft., 3 mi. NW Sutter Co., California. Known only from type locality.
1921. *Dipodomys heermanni eximius* Grinnell, Jour. Mamm., 2:95, May 2.

Dipodomys heermanni goldmani (Merriam)

1904. *Perodipus goldmani* Merriam, Proc. Biol. Soc. Washington, 17:143, July 14, type from Salinas, mouth of Salinas Valley, Monterey Co., California.
1921. *Dipodomys heermanni goldmani*, Grinnell, Jour. Mamm., 2:95, May 2.

MARGINAL RECORDS.—California: vic. San Jose; Bear Valley, vic. Cook P.O.; Arroyo Seco, S of Paraiso Springs; Jamesburg, valley of Carmel River; Seaside.

Dipodomys heermanni heermanni Le Conte

1853. *D[ipodomys]. heermanni* Le Conte, Proc. Acad. Nat. Sci. Philadelphia, 6:224. Type locality, Sierra Nevada, California, probably in Upper Sonoran Life-zone on Calaveras River, Calaveras Co.; see Grinnell, Univ. California Publ. Zool., 24:47, June 17, 1922.
1894. *Perodipus streatori* Merriam, Proc. Biol. Soc. Washington, 9:113, June 21, type from Carbondale, Amador Co., California.

MARGINAL RECORDS.—California: Carbondale; [6 mi. E] Coulterville; *vic. Calaveras River, foothill district.*

Dipodomys heermanni jolonensis Grinnell

1919. *Dipodomys jolonensis* Grinnell, Proc. Biol. Soc. Washington, 32:203, December 31, type from floor of valley 1 mi. SW Jolon, Monterey Co., California.
1921. *Dipodomys heermanni jolonensis* Grinnell, Jour. Mamm., 2:95, May 2.

MARGINAL RECORDS.—California: vic. King City; *San Lucas;* Creston; 2 mi. S San Miguel; *Pleyto;* type locality.

Dipodomys heermanni morroensis (Merriam)

1907. *Perodipus morroensis* Merriam, Proc. Biol. Soc. Washington, 20:78, July 22, type from 4 mi. S Morro, San Luis Obispo Co., California. Known only from type locality.
1943. *Dipodomys heermanni morroensis*, Boulware, Univ. California Publ. Zool., 46:393, September 16.

Dipodomys heermanni saxatilis Grinnell and Linsdale

1929. *Dipodomys heermanni saxatilis* Grinnell and Linsdale, Univ. California Publ. Zool., 30:453, June 15, type from near Dale's, N side Paine's Creek, 700 ft., Tehama Co., California.

MARGINAL RECORDS.—California: near Longs, 1 mi. S jct. N and S forks Battle Creek; *Inskip Forebay, 6 mi. SW Manton;* Limekiln, 1200 ft.; *8 mi. SE Chico, 450 ft.;* 4 mi. SE Chico, 450 ft.; $2\frac{1}{2}$ mi. NE Tehama, 400 ft.; *type locality.*

Dipodomys heermanni swarthi (Grinnell)

1919. *Perodipus swarthi* Grinnell, Univ. California Publ. Zool., 21:44, March 29, type from 7 mi. SE Simmler, Carrizo Plain, San Luis Obispo Co., California.
1921. *Dipodomys heermanni swarthi* Grinnell, Jour. Mamm., 2:95, May 2.

MARGINAL RECORDS.—California: type locality; vic. McKittrick; San Emigdio; extreme northern Santa Barbara Co. [Cuyama Valley].

Dipodomys heermanni tularensis (Merriam)

1904. *Perodipus agilis tularensis* Merriam, Proc. Biol. Soc. Washington, 17:143, July 14, type from Alila, now Earlimart, Tulare Co., California.
1921. *Dipodomys heermanni tularensis*, Grinnell, Jour. Mamm., 2:95, May 2.

MARGINAL RECORDS.—California: Tracy; Raymond; Dunlap; Tipton; 8 mi. NE Bakersfield; mouth Caliente Creek Wash; *ca. 20 mi. S Bakersfield;* lower San Emigdio Creek Wash, 450 ft.; Taft; Temblor Mts. [12 mi.] W McKittrick; near Lost Hills; 1 mi. N, 4½ mi. E Panoche; *22 mi. S Los Baños;* Los Baños.

Dipodomys gravipes Huey
San Quintín Kangaroo Rat

1925. *Dipodomys gravipes* Huey, Proc. Biol. Soc. Washington, 38:83, May 26, type from 2 mi. W Santo Domingo Mission, Baja California, lat. 30° 45′ N, long. 115° 58′ W.

External measurements: 286–310; 157–180; 43–44. "A large-sized, heavy bodied, small-eared animal, with thick tail of medium length, belonging to the *heermanni* group" (Huey, 1925:83). See Map 344.

MARGINAL RECORDS.—Baja California: 3 mi. S San Telmo; type locality; 1 mi. S San Ramón.

Dipodomys ingens (Merriam)
Giant Kangaroo Rat

1904. *Perodipus ingens* Merriam, Proc. Biol. Soc. Washington, 17:141, July 14, type from Painted Rock, 20 [= 12] mi. SE Simmler, Carrizo Plain, San Luis Obispo Co., California.
1921. *Dipodomys ingens*, Grinnell, Jour. Mamm., 2:95, May 2.

External measurements: 311–348; 157–198; 46–55. Weight, 131–180 grams, hence the heaviest of all kangaroo rats. Five toes on each hind foot. Tail and ear short in relation to length of head and body. Skull broad across maxillary processes of zygomatic arches. (See Map 344.)

MARGINAL RECORDS.—California: near mouth of Laguna Seca Creek; Buttonwillow; near Buena Vista Lake; extreme northern Santa Barbara County;

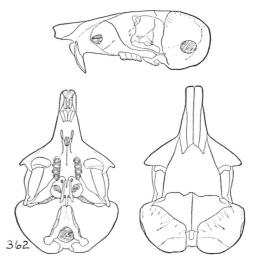

Fig. 362. *Dipodomys ingens*, Panoche Creek, 2 mi. SE Panoche, 1200 ft., Fresno Co., California, No. 72763 M.V.Z., ♀, X 1.

Carrizo Plain, near Santiago Spring; Panoche Creek, foothills of Coast Range, *ca.* 50 mi. W Fresno; *ca.* 12 mi. S Los Baños.

spectabilis-group
Dipodomys spectabilis
Banner-tailed Kangaroo Rat

External measurements: 310–349; 180–208; 47–51; weight 98–132 grams. Four toes on each hind foot. Distal 40 mm. of tail white; proximal to this white area, tail black all round; lateral white stripes present only on proximal half of tail.

Dipodomys spectabilis baileyi Goldman

1923. *Dipodomys spectabilis baileyi* Goldman, Proc. Biol. Soc. Washington, 36:140, May 1, type from 40 mi. W Roswell, Chaves Co., New Mexico.

MARGINAL RECORDS.—New Mexico (Findley, *et al.*, 1975:182, 183, as *D. spectabilis* only, unless otherwise noted): 6 mi. N of NE corner of [Santa Fe] Municipal Airport; *Lamy;* 3½ mi. S, 8½ mi. E Clines Corners; Capitan; type locality; Santa Rosa; 9 mi. N Floyd; 1 mi. N, 3 mi. W Caprock (Judd and Schmidly, 1969:382); 30 mi. E Carlsbad (V. Bailey, 1932: 259). Texas: Texas Highway 115, on Andrews–Winkler County line (Packard and Judd, 1968:537); *7 mi. W Tarzan* (Ramsey and Carley, 1970:351); 1 mi. S, 18 mi. W Stanton (*ibid.*); *5 mi. N, 8 mi. W Midland* (*ibid.*); *Odessa;* Pecos County (Davis, 1966:167 as *D. spectabilis* only); 4 mi. SW Fort Davis, 5000 ft.; 7 mi. NE Marfa; Valentine; Sierra Blanca; lower edge Franklin Mts. New Mexico (Findley, *et al.*, 1975:182, 183, as *D. spectabilis* only, unless otherwise noted): Rio Alamosa (Vorhies and Taylor, 1922:9); Datil Mts.; Juan Tofoya; 2½ mi. N, 15 mi. W San Ysidro.

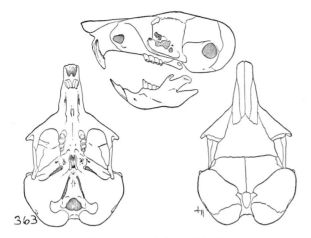

Fig. 363. *Dipodomys spectabilis perblandus*, 30 mi. S Tucson, Pima Co., Arizona, No. 7680 K.U., ♂, X 1.

Dipodomys spectabilis clarencei Goldman

1933. *Dipodomys spectabilis clarencei* Goldman, Jour. Washington Acad. Sci., 23:467, October 15, type from Blanco, San Juan Co., New Mexico.

MARGINAL RECORDS.—New Mexico: Fruitland; type locality; 57 mi. N, 3 mi. W Grants (Findley, *et al.*, 1975:182, as *D. spectabilis* only); 15 mi. NW Gallup. Arizona: 8 mi. S Chinle (Hoffmeister and Nader, 1963:92).—Hoffmeister (1977:150) lists two specimens of *D. spectabilis* labeled as from "5 mi. E Rainbow Lodge," which is 112 km. W and 80 km. N of the westernmost record of *D. s. clarencei*. The specimens are of uncertain subspecific identity, and the locality is not shown on Map 346.

Dipodomys spectabilis cratodon Merriam

1907. *Dipodomys spectabilis cratodon* Merriam, Proc. Biol. Soc. Washington, 20:75, July 22, type from Chicalote, Aguascalientes.

MARGINAL RECORDS.—San Luis Potosí: 2 km. E Illescas; 4 km. E Salinas. Aguascalientes: type locality. Alvarez (1961:408) recorded one specimen in the Instituto de Biología, Univ. Nac. Autó. México, from *3 mi. N Lulú, Zacatecas*, as *Dipodomys spectabilis cratodon*. John O. Matson, *in verbis*, identifies three other specimens, in the California Museum of Vertebrate Zoology, taken at the same time and place as *Dipodomys nelsoni*.

Dipodomys spectabilis intermedius Nader

1965. *Dipodomys spectabilis intermedius* Nader, Proc. Biol. Soc. Washington, 78:50, July 21, type from $16\frac{7}{10}$ mi. SW Bámori, 1900 ft., Sonora.

MARGINAL RECORDS.—Sonora (Nader, 1965:52): *5 mi. W Querobabi*; 45 mi. N Hermosillo, 2100 ft.; type locality.

Dipodomys spectabilis perblandus Goldman

1933. *Dipodomys spectabilis perblandus* Goldman, Jour. Washington Acad. Sci., 23:466, October 15, type from Calabasas, 3500 ft., Santa Cruz Co., Arizona.

MARGINAL RECORDS.—Arizona: Florence (Hoffmeister and Nader, 1963:92); Oracle; 3 mi. S, 14 mi. E Continental (Cockrum, 1961:144); type locality. Sonora: La Sauceda, 15 mi. NNE Cananea (Alvarez, 1961:407); Cerro Blanco; Noria; 2 mi. S Sasabe. Arizona: 75 mi. SW Tucson; Indian Oasis [= Sells, Pima Co.]; Organ Pipe Cactus National Monument, 30 mi. S Ajo (Cockrum, 1961:144); Ajo.

Dipodomys spectabilis spectabilis Merriam

1890. *Dipodomys spectabilis* Merriam, N. Amer. Fauna, 4:46, October 8, type from Dos Cabezos, Cochise Co., Arizona.

MARGINAL RECORDS.—New Mexico: Silver City; Lake Valley (Findley, *et al.*, 1975:182, as *D. spectabilis* only); 15 mi. W Las Cruces (*ibid.*). Chihuahua (Anderson, 1972:318): 4 mi. S, 1 mi. E Moctezuma, 4500 ft.; 6 mi. S, 15 mi. W Coyamé, 5500 ft.; 30 mi. S Chihuahua, 5200 ft.; $4\frac{2}{5}$ mi. SE San Buenaventura; 2 mi. S, 7 mi. W Casas Grandes; Rancho San Francisco, 5100 ft. Arizona: San Bernardino Ranch; mouth Montezuma Canyon; *8 mi. SE Fort Huachuca; 7 mi. ESE Fort Huachuca*; type locality; SW slopes Graham Mts., near Fort Grant. New Mexico: Lordsburg.

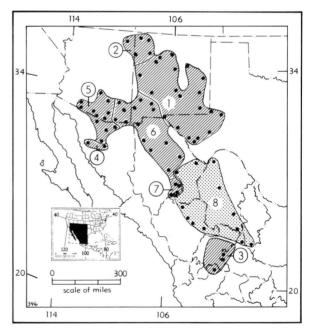

Map 346. *Dipodomys spectabilis* and *Dipodomys nelsoni*.

1. *D. s. baileyi*	5. *D. s. perblandus*
2. *D. s. clarencei*	6. *D. s. spectabilis*
3. *D. s. cratodon*	7. *D. s. zygomaticus*
4. *D. s. intermedius*	8. *D. nelsoni*

Dipodomys spectabilis zygomaticus Goldman

1923. *Dipodomys spectabilis zygomaticus* Goldman, Proc. Biol. Soc. Washington, 36:140, May 1, type from Parral, southern Chihuahua.

MARGINAL RECORDS.—Chihuahua (Anderson, 1972:318, 319, unless otherwise noted): 2 mi. SE Boquilla, 4700 ft.; 15 mi. ESE Boquilla, 4700 ft.; 21½ mi. WSW Jiménez, 4800 ft.; *1 mi. S, 3³/₁₀ mi. E Valle de Allende* (95062 KU); *1¹/₄ mi. S, ¹/₂ mi. E Valle de Allende* (96799 KU); 2 mi. W Parral, 6200 ft.

Dipodomys nelsoni Merriam
Nelson's Kangaroo Rat

1907. *Dipodomys nelsoni* Merriam, Proc. Biol. Soc. Washington, 20:75, July 22, type from La Ventura, Coahuila.

External measurements: 310–330; 177–204; 45–50. Four toes on each hind foot. Averages slightly smaller than *D. spectabilis* and has shorter (20 mm.) terminal white area on tail, or tip of tail black.

In a graph showing the relationship of maxillary breadth to total length of skull, Anderson (1972:314) easily distinguished all specimens available to him of *D. nelsoni* from those of *D. spectabilis*. Additional specimens from Chihuahua now are catalogued in the University of Kansas Museum of Natural History. These include No. 95042 from 3 mi. N, ½ mi. W Salaices, and Nos. 95053 and 95063 in a series of 39 saved in 1963 from southeast of the town of Valle de Allende. The three are less easily identified to taxon by means of the graph than are other specimens. Even so, I am not convinced that the three are intergrades and consequently provisionally here retain *D. nelsoni* as a species instead of arranging it as a subspecies of *D. spectabilis*.

Nevertheless, additional specimens from places where the two taxa meet are needed for study to determine the taxonomic status of *D. nelsoni*. One such place is, of course, on the south side of the Río Allende between a point 1 mi. S, 1³/₁₀ mi. E of the town of Valle de Allende and another point 1 mi. farther east. In this mile it was impossible to dislodge individuals of *Dipodomys* from their (easily located) burrows by means of a shovel because holes through the hard subsurface layer of rock enabled the rats to remain out of the collector's reach. West and east of this mile it was easy to dislodge the kangaroo rats because the subsurface layer of rocks lacked holes. Baited snap traps caught next to no rats (the time was mid-May) in either area. The collector speculated that the best means of obtaining specimens from the mile-wide area having holes in the rocky substratum might be to use a spotlight and shot pistol at night, when the rats are above ground.

MARGINAL RECORDS.—Coahuila: 7 mi. S, 2 mi. E Boquillas, 1800 ft.; 2½ mi. S, 21 mi. E Ocampo, 3500 ft.; Treviño; type locality. Nuevo León: Dr. Arroyo. San Luis Potosí: 6 km. S Matehuala. Durango: 2 mi. E Pedriceña, 4300 ft.; 4 mi. SE Casco, 5850 ft. (Baker and Greer, 1962:102); 7 mi. NNW La Zarca, 6000 ft. (*ibid.*); 12 mi. NW La Resolana, 6000 ft. (*ibid.*). Chihuahua (Anderson, 1972:313, unless otherwise noted): 1 mi. S, ¾ mi. E Valle de Allende (95053 KU); *1 mi. S, 4¹/₂ mi. W [= SW on S side Río Valle de Allende] Salaices* (95044 KU); *14¹/₂ mi. WSW Jiménez, 5000 ft.*; 1 mi. E Julimes; Piñón Camp, Sierra Rica, 4900 ft.

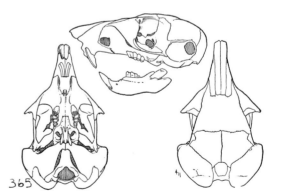

Fig. 365. *Dipodomys elator*, Henrietta, Texas, No. 107244 U.S.N.M., ♀, X 1.

merriami-group
Dipodomys elator Merriam
Texas Kangaroo Rat

1894. *Dipodomys elator* Merriam, Proc. Biol. Soc. Washington, 9:109, June 21, type from Henrietta, Clay Co., Texas.

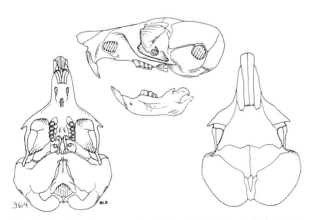

Fig. 364. *Dipodomys nelsoni*, 2½ mi. W, 21 mi. S Ocampo, 3500 ft., Coahuila, No. 56636 K.U., ♂, X 1.

External measurements: 317, 196, 46. Four toes on each hind foot. Superficially, in size and color pattern, resembling *D. spectabilis.* Skull narrow interorbitally; interparietal wide; orbit large.

Grinnell (1921:95) and Setzer (1949:495) consider this species to be more closely allied to *D. phillipsii* than to *D. spectabilis* but Davis (1942:329) takes the opposite view and points out that some features ally *D. elator* with *D. merriami.* All three authors seem to agree that *D. elator* is a species, and not subspecifically linked to any other named kind of *Dipodomys.* See Map 348.

MARGINAL RECORDS.—Oklahoma: Chattanooga. Texas (Martin and Matocha, 1972:876): 5 mi. E Henrietta; 22 mi. S Henrietta; 43 mi. SW Henrietta; 7 mi. E Matador; 3½ mi. E Lazare; 8 mi. NE Quanah.

The alleged occurrence (Blair, 1954:248) at 3 mi. W Gatesville is questioned by Dalquest and Collier (1964:149) and Martin and Matocha (1972:874, 875). The locality is indicated on Map 348 by a question mark.

Dipodomys phillipsii
Phillips' Kangaroo Rat

External measurements: 230–304; 149–192; 36–45. Four toes on each hind foot; arietiform facial markings black and extensive; dark stripes on tail uniting in distal third; tip of tail white or black; auditory bullae relatively small.

Dipodomys phillipsii oaxacae Hooper

1947. *Dipodomys phillipsii oaxacae* Hooper, Jour. Mamm., 28:48, February 17, type from Teotitlán, 950 m., Oaxaca.

MARGINAL RECORDS.—Puebla: 1½ mi. W Tehuitzingo, 3570 ft. (Genoways and Jones, 1971:283). Oaxaca: type locality.

Dipodomys phillipsii ornatus

1894. *Dipodomys ornatus* Merriam, Proc. Biol. Soc. Washington, 9:110, June 21, type from Berriozábal, Zacatecas.

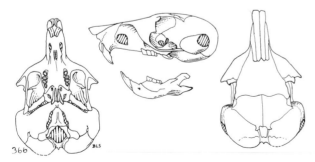

Fig. 366. *Dipodomys phillipsii ornatus*, 8 mi. SE Zacatecas, 7225 ft., Zacatecas, No. 48977 K.U., ♀, X 1.

1971. *Dipodomys phillipsii ornatus,* Genoways and Jones, Jour. Mamm., 52:281, May 28.

MARGINAL RECORDS (Genoways and Jones, 1971:282).—Durango: SE end Laguna de Santiaguillo, Santa Cruz; 6 mi. NW La Pila, 6150 ft. Zacatecas: 12 mi. N, 7 mi. E Fresnillo. San Luis Potosí: 1 km. N Arenal; Bledos. Querétaro: Tequisquiapam. Guanajuato: 4 mi. N, 5 mi. W León, 7000 ft. Jalisco: *8 mi. W Encarnación de Díaz, 6000 ft.;* 1 mi. NE Villa Hidalgo, 6550 ft. Zacatecas: Hda. San Juan Capistrano. Durango: Durango; *4 mi. S Morcillo, 6450 ft.; 9 mi. N Durango, 6200 ft.*

Dipodomys phillipsii perotensis Merriam

1894. *Dipodomys perotensis* Merriam, Proc. Biol. Soc. Washington, 9:111, June 21, type from Perote, Veracruz.
1944. *Dipodomys phillipsii perotensis,* Davis, Jour. Mamm., 25:391, December 12.

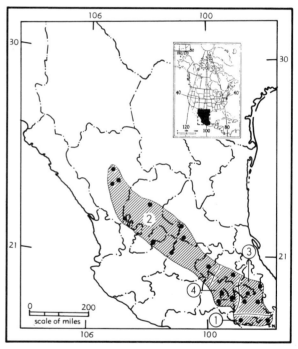

Map 347. *Dipodomys phillipsii.*

| 1. *D. p. oaxacae* | 3. *D. p. perotensis* |
| 2. *D. p. ornatus* | 4. *D. p. phillipsii* |

MARGINAL RECORDS (Genoways and Jones, 1971:283, unless otherwise noted).—Veracruz: 2 km. N Perote, 8000 ft.; *2 km. E Perote, 8300 ft.* Puebla: *W base Mt. Orizaba, 9000 ft.* (Hall and Kelson, 1959:529); Chalchicomula; 7 mi. S, 3 mi. E Puebla, 6850 ft. Tlaxcala: Huamantla.

Dipodomys phillipsii phillipsii Gray

1841. *Dipodomys philipii* [*sic*] Gray, Ann. Mag. Nat. Hist., ser. 1, 7:522, August, type from near Real del Monte, Valley of México, Hidalgo.
1893. *Dipodomys phillipsi*, Merriam, Proc. Biol. Soc. Washington, 8:91, July 18.

MARGINAL RECORDS.—Hidalgo: type locality. México: W base Mt. Popocatépetl, near Amecameca; near peak of Huitzilac, near Cruz del Marqués, 9000 ft. Distrito Federal: 17 km. ESE Mexico City (Genoways and Jones, 1971:281). Hidalgo: *85 km. N Mexico City, 9 km. S Pachuca, 8200 ft.*

Dipodomys merriami
Merriam's Kangaroo Rat

Revised by Lidicker, W. Z., Jr., Univ. California Publ. Zool., 67:125–218, August 4, 1960, and modified by Huey, L. M., Trans. San Diego Soc. Nat. Hist., 13:85–168, January 15, 1964.

External measurements: 234–259; 135–161; 36–41. Weight, 38.4–46.9 grams. Four toes on each hind foot; hind feet and tail slender; auditory bullae large; skull narrow across maxillary processes of zygomatic arches; interorbital breadth of skull more than half of basal length.

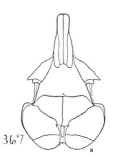

Fig. 367 . *Dipodomys merriami merriami*, 14 mi. E Searchlight, Nevada, No. 61580 M.V.Z., ♂, X 1.

Dipodomys merriami ambiguus Merriam

1890. *Dipodomys ambiguus* Merriam, N. Amer. Fauna, 4:42, October 8, type from El Paso, El Paso Co., Texas.
1901. [*Dipodomys merriami*] *ambiguus*, Elliot, Field Columb. Mus., Publ. Zool., ser. 45, 2:234, March 6.

MARGINAL RECORDS (Lidicker, 1960:181–183, unless otherwise noted).—New Mexico: Arroyo de las Lomatas Negras, 2 mi. S, 3 mi. W Bernalillo, 5150 ft. (Gennaro, 1968:488, as *P. merriami* only); ¾ mi. N, 8 mi. E Isleta (Gennaro, 1968:489, as *P. merriami* only); 2 mi. N, 5 mi. W Carrizozo (*ibid.*, as *P. merriami* only); Alamogordo. Texas: 7 mi. N Pine Springs, 5300 ft. New Mexico: Carlsbad Cave; Black River, near Carlsbad; *Carlsbad;* 40 mi. N El Paso Gap (Findley, *et al.*, 1975:185); 2⅖ mi. SW Dunlap (Best, 1971:210); 9 mi. N, 11½ mi. W Fort Sumner (Findley, *et al.*, 1975:186); 7 mi. E Roswell; 8⅕ mi. S, 9³/₁₀ mi. E Roswell; 4 mi. N

Caprock (Judd and Schmidly, 1969:383). Texas: 12 mi. S Flower Grove (Ramsey and Carley, 1970:351, as *D. m. merriami*); Langtry (Hall and Kelson, 1959:532). Coahuila: 15 mi. N, 8 mi. W Piedras Negras; 2 mi. S, 11 mi. E Nava, 810 ft. Nuevo León: 38 mi. S Nuevo Laredo, 400 ft.; Rancho 14 de Mayo, municipio de China; 10 km. W Monterrey, 480 m.; 14 km. N Monterrey; Lampazos. Coahuila: 3 mi. E Arteaga [= Artegia], 5300 ft.; 1 mi. S, 1 mi. E La Rosa, 5200 ft.; La Pastora Rancho, 15 mi. N, 41 mi. W Saltillo; 1 mi. N San Lorenzo, 4200 ft.; Jimulco. Durango (Baker and Greer, 1962:100): 2 mi. NW Chocolate, 4500 ft.; 1½ mi. NW Nazas, 4100 ft.; 6¹/₁₀ mi. S Alamillo, 4600 ft.; 7 mi. NNW La Zarca, 6000 ft. Chihuahua: 5 mi. E Parral, 5700 ft.; *2 mi. NE Hidalgo del Parral;* 9 mi. SE San Lucas, Río San Pedro, 5300 ft.; Chihuahua; *4 mi. N, 2 mi. W Chihuahua, 4750 ft.;* 2 mi. W Parrita (Anderson, 1972:312); 1 mi. S Kilo, 4185 ft.; Samalayuca; *5½ mi. N Samalayuca.* New Mexico: 1 mi. E Strauss; Fort Cummings; Cuchillo, 4700 ft.; San Mateo Canyon, San Mateo Mts.; near Mush Mtn., 16¼ mi. S, 17¼ mi. W Correo (Gennaro, 1968:488, as *P. merriami* only); 1 mi. E Suwanee; 18 mi. W Albuquerque.

Dipodomys merriami annulus Huey

1951. *Dipodomys merriami annulus* Huey, Trans. San Diego Soc. Nat. Hist., 11:224, April 30, type from Barril, Gulf of California, lat. 28° 20′ N, long. 112° 50′ W, Baja California.

MARGINAL RECORDS.—Baja California Norte: Las Flores near Los Angeles Bay; type locality; San Borjas Mission (Lidicker, 1960:201).

Dipodomys merriami arenivagus Elliot

1904. *Dipodomys m*[*erriami*]. *arenivagus* Elliot, Field Columb. Mus., Publ. 87, Zool. Ser., 3(14):249, January 7, type from San Felipe, Baja California.

MARGINAL RECORDS (Lidicker, 1960:193, 194).—California: Clark's Well; *4 mi. N, 4¹/₂ mi. W Truckhaven, 21 mi. NW Kane Spring*[*s*]; 12 mi. N Kane Springs; *New River, near Salton Sea;* Brawley. Baja California Norte: SE base Signal Mtn.; *Cerro Centinela, 13 mi. WSW Mexicali, 300 ft.;* 13 mi. N El Mayor; 40 mi. N San Felipe; Puerto de Calamajue; la bocana [boca?] de la de [*sic*] Cañón de Santa María, 10 mi. E Missión de Santa María; *15 mi. NW Calamajue Mission, 1600 ft.;* mouth El Cajón Canyon, base San Pedro Mártir Mts., 2300 ft.; *valley E base San Pedro Mártir Mts.;* Cañón Esperanza; *Gaskill's Tank, Los Muertos Canyon Fan, near Laguna Salada; De Mara's Well, W side Laguna Salada;* Las Palmas Canyon, W side Laguna Salada, 15 mi. S north end, 200 ft.; 6 mi. S, 3½ mi. E Mountain Springs, California. California: *Carrizo Canyon, 6⁹/₁₀ mi. S, ¹/₅ mi. E Mountain Palm Springs;* Vallecito; Yaqui Well; *5 mi. NW Borrego P.O., 700 ft.; mouth Palm Canyon, Borrego Valley.*

Dipodomys merriami atronasus Merriam

1894. *Dipodomys merriami atronasus* Merriam, Proc. Biol. Soc. Washington, 9:113, June 21, type from Hda. La Parada, about 25 mi. NW San Luis Potosí, State of San Luis Potosí.

MARGINAL RECORDS (Lidicker, 1960:178).— Coahuila: 5 mi. NW Saltillo; 26 mi. E Saltillo. Nuevo León: Doctor Arroyo, 5800 ft. Tamaulipas (Alvarez, 1963:432): *Nicolás; 8 mi. N Tula; Tula; 9 mi. SW Tula.* San Luis Potosí: 3 mi. NW Esperanza; Río Verde; 1 km. W Bledos. Aguascalientes: 8 mi. E Aguascalientes; *5 mi. E Aguascalientes; ½ mi. W Rincón de Romos.* Zacatecas: 2 mi. ESE Trancoso, 7000 ft.; *8 mi. SE Zacatecas, 7225 ft.;* 4½ mi. E Fresnillo. Durango: Hda. de Atotonilco (= 12 mi. SE Yerbanís), 6680 ft.; 9 mi. N Durango; 2 mi. E Pedriceña. Coahuila: *10 mi. W Saltillo.*

Dipodomys merriami brunensis Huey

1951. *Dipodomys merriami brunensis* Huey, Trans. San Diego Soc. Nat. Hist., 11:225, April 30, type from Llano de San Bruno, Baja California.

MARGINAL RECORDS.—Baja California Sur: El Valle de Yaqui, NW of Santa Rosalía; Canipolé.

Dipodomys merriami collinus Lidicker

1960. *Dipodomys merriami collinus* Lidicker, Univ. California Publ. Zool., 67:194, August 4, type from 3¼ mi. S, 2¼ mi. E Scissors Crossing, Earthquake Valley, San Diego Co., California.

MARGINAL RECORDS (Lidicker, 1960:196).— California: Radec, 2000 ft.; *1¼ mi. N, ⅓ mi. E Aguanga, 2300 ft.; 1 mi. SE Aguanga, 2150 ft.; 2½ mi. N Oak Grove;* San Felipe Valley; *Scissors Crossing, Earthquake Valley; type locality; Mason Valley* (= La Puerta Valley).

Dipodomys merriami frenatus Bole

1936. *Dipodomys merriami frenatus* Bole, Sci. Publs., Cleveland Mus. Nat. Hist., 5(1):1, January 17, type from Toquerville, Washington Co., Utah.

MARGINAL RECORDS (Lidicker, 1960:188, unless otherwise noted).—Utah: Veyo; type locality; *Coal Pit Wash, Zion Park; Springdale, 3900 ft.* Arizona: *25 mi. S Hurricane* (Hoffmeister and Durham, 1971:35); 6 mi. N Wolf Hole, 3500 ft.; *11 mi. SW St. George* (Hoffmeister and Durham, 1971:35). Utah: 5 mi. NW St. George; Chadburn's Ranch, 16 mi. NW St. George, 4000 ft.

Dipodomys merriami llanoensis Huey

1951. *Dipodomys merriami llanoensis* Huey, Trans. San Diego Soc. Nat. Hist., 11:226, April 30, type from Buena Vista, Magdalena Plain, lat. 24° 50′ N, long. 111° 50′ W, Baja California.

MARGINAL RECORDS (Lidicker, 1960:203, as *D. m. melanurus,* but see Huey, 1964:123).—Baja California Sur: San Jorge (25° 44′ N, 112° 07′ W); Santo Domingo (lat. 25° 31′ N); near El Refugio; 24³⁄₁₀ mi. by road SE El Refugio, 24° 33′ N, 111° 35′ W, 100 ft.

Dipodomys merriami mayensis Goldman

1928. *Dipodomys merriami mayensis* Goldman, Proc. Biol. Soc. Washington, 41:141, October 15, type from Alamos, Sonora.

MARGINAL RECORDS (Lidicker, 1960:185, 186).— Sonora: 6 mi. NNW Ciudad Obregón; *1 mi. N Ciudad Obregón, 200 ft.;* Camoa, Río Mayo; *12 mi. WNW Alamos;* type locality; Guirocoba. Sinaloa: 1 mi. NE El Fuerte. Sonora: 33 mi. SSE Navojoa; 3 mi. NNW Bacavachi [= Bacabachi]; Los Médanos, 30 mi. SE Obregón; 1 mi. S Vicam, 200 ft.

Dipodomys merriami melanurus Merriam

1893. *Dipodomys merriami melanurus* Merriam, Proc. California Acad. Sci., ser. 2, 3:345, June 5, type from San José del Cabo, Baja California.

MARGINAL RECORDS.—Baja California Sur: La Paz, and southward over rest of Baja California.

Dipodomys merriami merriami Mearns

1890. *Dipodomys merriami* Mearns, Bull. Amer. Mus. Nat. Hist., 2:290, February 21, type from New River, between Phoenix and Prescott, Maricopa Co., Arizona.
1894. *Dipodomys simiolus* Rhoads, Proc. Acad. Nat. Sci. Philadelphia, 45:410, January 27, type from Agua Caliente, now Palm Springs, Riverside Co., California.
1894. *Dipodomys similis* Rhoads, Proc. Acad. Nat. Sci. Philadelphia, 45:411, January 27, type from Whitewater, Riverside Co., California.
1894. *Dipodomys merriami nevadensis* Merriam, Proc. Biol. Soc. Washington, 9:111, June 21, type from Pyramid Lake, Washoe Co., Nevada.
1894. *Dipodomys merriami nitratus* Merriam, Proc. Biol. Soc. Washington, 9:112, June 21, type from Keeler, E side Owens Lake, Inyo Co., California.
1904. *Dipodomys merriami mortivallis* Elliot, Field Columb. Mus., Publ. 87, Zool. Ser., 3(14):250, January 7, type from Furnace Creek, Death Valley, Inyo Co., California.
1907. *Dipodomys merriami kernensis* Merriam, Proc. Biol. Soc. Washington, 20:77, July 22, type from Onyx, W end Walker Pass, Kern Co., California.
1937. *Dipodomys merriami regillus* Goldman, Proc. Biol. Soc. Washington, 50:75, June 22, type from Tule Well, Tule Desert between Cabeza Prieta Mts. and Tule Mts., Yuma Co., Arizona.

MARGINAL RECORDS (Lidicker, 1960:169–176, unless otherwise noted).—Nevada: Quinn River Crossing; 1¼ mi. N Sulphur, 4050 ft.; 3½ mi. NE Toulon, 4500 ft.; *1 mi. S, 9½ mi. E Fanning, Buena Vista Valley, 4100 ft.;* Eastgate, 4400 ft.; 2 mi. E Hawthorne, 4300 ft.; 8 mi. S Mina; 3½–4 mi. SE Coaldale, 4850 ft. (Hall,

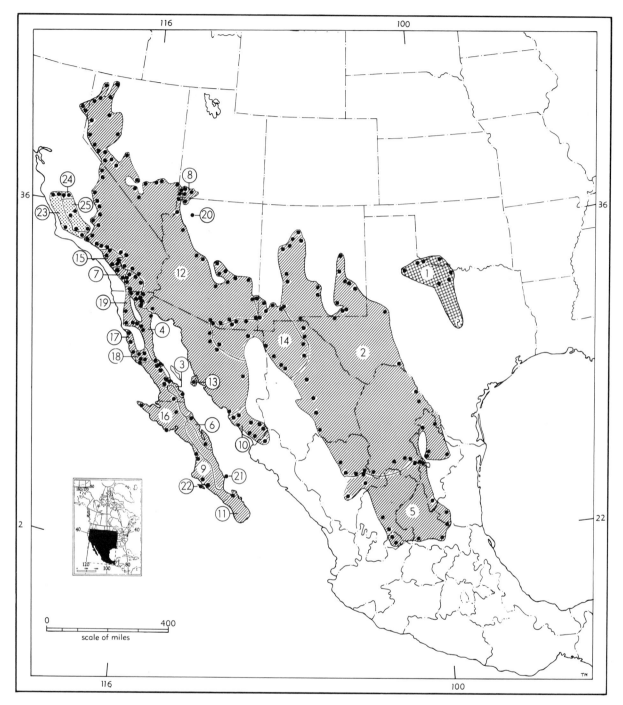

Map 348. *Dipodomys merriami* and related species.

Guide to kinds

1. *D. elator*
2. *D. merriami ambiguus*
3. *D. merriami annulus*
4. *D. merriami arenivagus*
5. *D. merriami atronasus*
6. *D. merriami brunensis*
7. *D. merriami collinus*
8. *D. merriami frenatus*
9. *D. merriami llanoensis*
10. *D. merriami mayensis*
11. *D. merriami melanurus*
12. *D. merriami merriami*
13. *D. merriami mitchelli*
14. *D. merriami olivaceus*
15. *D. merriami parvus*
16. *D. merriami platycephalus*
17. *D. merriami quintinensis*
18. *D. merriami semipallidus*
19. *D. merriami trinidadensis*
20. *D. merriami vulcani*
21. *D. insularis*
22. *D. margaritae*
23. *D. nitratoides brevinasus*
24. *D. nitratoides exilis*
25. *D. nitratoides nitratoides*

1946:428); 4½ to 5¾ mi. NE San Antonio, 5640–5700 ft. (*ibid.*); Gold Flat, 6 mi. W Kawich P.O., 5150 ft.; NW base Timber Mtn., 4200 ft.; *northern edge Jackass Flats, ca. 9½ mi. S Topopah Springs* (Rowland and Turner, 1964:57); NW base Skull Mtn., 3500 ft.; ½ *mi. N Oak Spring, 6500 ft.*; 9 mi. E Wheelbarrow Peak; 3 mi. N Crystal Spring, 4000 ft.; Pahroc Spring; Beaverdam Creek. Utah: *Beaverdam Wash, 8 mi. N Utah–Arizona border;* 1½ mi. E Beaverdam Wash, 8 mi. N Utah–Arizona border, 2800 ft.; *W slope Beaverdam Mts., 5 mi. N Utah–Arizona border.* Arizona: Beaver Dam Creek [= Beaverdam Creek], near mouth just above Littlefield, 1500 ft.; *13 mi. S, 4 mi. W Littlefield* (Hoffmeister and Durham, 1971:35); *3 mi. N, 1 mi. W Pakoon Springs* (*ibid.*); *Grand Wash, 2 mi. S, 3 mi. E Pakoon Springs* (*ibid.*); Tasi Springs (*ibid.*); Peach Springs, 4000 ft.; Kirkland Junction; Turkey Creek, 3400 ft.; 2 mi. S Florence Junction; 3 mi. N Globe; Roosevelt Lake, N end mouth Bumblebee Creek, 2300 ft.; *H Bar Ranch, 10 mi. S Payson, 3300 ft.*; 2 mi. S Payson; 35 mi. from Roosevelt, Young Rt. Road; Cassadore Springs, 23 mi. ENE Globe, 4050 ft.; 3 mi. S, 3 mi. E Maverick Mtn.; 2 mi. W Ash Peak; 2 mi. E Portal; Empire Ranch, E from Santa Rita Mts.; *Santa Rita Mts., Santa Cruz Co.*; Tubac; E slope Baboquivari Mts., 10 mi. N international boundary. Sonora: *5 mi. S Sasabe;* 26 mi. S Sasabe, 5 mi. N Los Molinos, 2000 ft.; 50 mi. S Sasabe, on new Altar Road, 1400 ft.; Llano; 1 mi. N Batuc, Río Moctezuma; Pitahaya, 40 km. SE Empalme; El Doctor. Baja California: Mt. Mayor; E side Cocopah Mts., 21 mi. SSE Mexicali, 100 ft.; Imperial Canal, 11 mi. E Mexicali. California: Frink; Mecca; 10 mi. S Palm Springs; *10 mi. W Palm Springs;* Cabazon, 1700 ft.; 1 mi. NW Warren's Well, Morongo Valley; *Burn's Canyon;* Doble, San Bernardino Mts., 7000 ft.; *Cushenbury Springs, San Bernardino Mts., 4000 ft.*; 15 mi. W Hesperia; 5 mi. NW Shoemaker; 2 mi. NW Palmdale; Sears cattle camp, W Antelope Valley; Bean Spring; *7 mi. W Mojave;* ½ *mi. E Warren (railroad station)*; Red Rock Canyon; *Kelso Creek Valley, 11 mi. SSE Weldon, 3500 ft.*; Weldon; *Onyx, S. Fork Kern River, 2750 ft.*; ½ mi. NW Little Lake; ½ mi. W Olancha; Lukken Ranch, 5 mi. SW Lone Pine; 11 mi. NW Independence; *Taboose Creek, 12 mi. SW Big Pine, 4000 ft.*; Fish Slough, Owens Valley, 4200 ft.; *Chalfant Siding, 10 mi. N, 2 mi. E Bishop;* 5 mi. S Benton Station; *1½ mi. N Benton Station; Pellisier Ranch, Owens Valley.* Nevada: Huntoon Valley, 5700–6000 ft.; *Cat Creek, 4 mi. W Hawthorne, 4500 ft.*; Wichman; 3 mi. S Hudson; 3½ mi. E Carson City, 4700 ft.; 8 mi. SSW Sutcliffe, Pyramid Lake. California: 4½ mi. WNW Stacy, 4000 ft.; 5 mi. E Litchfield, 4600 ft. Nevada: 10 mi. W Deep Hole, Smoke Creek Desert; 12 mi. N, 2 mi. E Gerlach, 4000 ft.; Soldier Meadows, 4600 ft. Not precisely located: Nevada: *Osobb* (Lidicker, 1960:174).

Dipodomys merriami mitchelli Mearns

1897. *Dipodomys mitchelli* Mearns, Proc. U.S. Nat. Mus., 19:719, July 30, type from Tiburón Island, Gulf of California, Sonora. Known only from Tiburón Island.

1938. *Dipodomys merriami mitchelli*, Burt, Miscl. Publ. Mus. Zool., Univ. Michigan, 39:48, February 15.

Dipodomys merriami olivaceus Swarth

1929. *Dipodomys merriami olivaceus* Swarth, Proc. California Acad. Sci., ser. 4, 18:356, April 26, type from Fairbank, Cochise Co., Arizona.

MARGINAL RECORDS (Lidicker, 1960:184, 185, unless otherwise noted).—Arizona: York, 3500 ft. New Mexico: Redrock; 6 mi. SSE Lordsburg, 4200 ft.; 14 mi. S Silver City; 6 mi. N Deming; *2 mi. E Deming.* Chihuahua: 8 mi. N Villa Ahumada, 1190 m.; *3 mi. N Villa Ahumada;* 4 mi. S, 1 mi. E Moctezuma; 4⅔ mi. SE San Buenaventura (Anderson, 1972:313); 11 mi. NNW San Buenaventura; Colonia Juárez; Llano de Carretas, 27 mi. W Cuervo. Sonora: 30 km. S Agua Prieta on Railroad; 2 mi. W Magdalena. Arizona: La Osa; *Tumacacori Mission;* Fort Huachuca; type locality; *Tombstone;* Moore Ranch, White Water Creek, 14 mi. S Light P.O., Sulphur Spring Valley. New Mexico: 3 mi. N Rodeo; *Steins Pass.*

Dipodomys merriami parvus Rhoads

1894. *Dipodomys parvus* Rhoads, Amer. Nat., 28:70, January, type from Reche Canyon, 4 mi. SE Colton, San Bernardino Co., California. (See Grinnell, Univ. California Publ. Zool., 24:82, June 17, 1922.)
1901. [*Dipodomys merriami*] *parvus*, Elliot, Field Columb. Mus., Publ. 45, Zool. Ser., 2:234, March 6.

MARGINAL RECORDS (Lidicker, 1960:190).— California: Cajon Wash, ½ mi. S Devore; *Redlands;* Vallevista, San Jacinto Valley; Menifee; *4 mi. SW Perris; 1 mi. S Riverside;* 5 mi. E Ontario.

Dipodomys merriami platycephalus Merriam

1907. *Dipodomys platycephalus* Merriam, Proc. Biol. Soc. Washington, 20:76, July 22, type from Calmallí, Baja California.
1927. *Dipodomys merriami platycephalus*, Huey, Trans. San Diego Soc. Nat. Hist., 5:66, July 6.

MARGINAL RECORDS (Lidicker, 1960:199).— Baja California Norte: mouth Calamajue Canyon; *San Francisquito;* Ubai [Yubay], 30 mi. SE Calamajue; Valle de Agua Amarga (15 mi. W Bahía Los Angeles); type locality; *4 mi. E El Arco.* Baja California Sur: Campo Los Angeles; Punta Abrejos; 1 mi. SE Cabo Tórtola. Baja California Norte: Santa Catarina Landing.

Dipodomys merriami quintinensis Huey

1951. *Dipodomys merriami quintinensis* Huey, Trans. San Diego Soc. Nat. Hist., 11:222, April 30, type from 5 mi. E San Quintín, Baja California.

MARGINAL RECORDS (Lidicker, 1960:200).— Baja California Norte: N end San Quintín Plain; *Santo Domingo;* 10 mi. E San Quintín; 10 mi. E El Rosario.

Dipodomys merriami semipallidus Huey

1927. *Dipodomys merriami semipallidus* Huey, Trans. San Diego Soc. Nat. Hist., 5:65, July 6, type from 7 mi. N Santa Catarina, Baja California, lat. 29° 45′ N.

MARGINAL RECORDS (Lidicker, 1960:199, as *D. m. platycephalus*, unless otherwise noted, but see Huey, 1964:122).—Baja California Norte: 10 mi. N Cataviña; *Onyx (El Marmol); 5 mi. S El Marmol; 12½ mi. by road S El Marmol;* 7 mi. S Cataviña; Santa Catarina (Huey, 1964:122); San Fernando.

Dipodomys merriami trinidadensis Huey

1951. *Dipodomys merriami trinidadensis* Huey, Trans. San Diego Soc. Nat. Hist., 11:220, April 30, type from Aquajito Spring, El Valle de la Trinidad, Baja California.

MARGINAL RECORDS (Lidicker, 1960:197, 198).—California: Mountain Springs (Imperial Co.). Baja California Norte: summit San Matías Pass, near Diablito Spring; type locality; San José; Sangre de Cristo (El Valle de la San Rafael); *La Huerta, W base Laguna Hanson Mts.; International boundary at Jacumba.* California: *Jacumba; Mountain Springs* (San Diego Co.).

Dipodomys merriami vulcani Benson

1934. *Dipodomys merriami vulcani* Benson, Proc. Biol. Soc. Washington, 47:181, October 2, type from lower end Toroweap Valley, about ½ mi. E Vulcan's Throne, Mohave Co., Arizona.

MARGINAL RECORDS (Lidicker, 1960:189, unless otherwise noted).—Arizona: *Sullivan Ranch, 3 mi. N Mt. Emma, 6247 ft.; 1 mi. NE Vulcan[']s Throne* (Hoffmeister and Durham, 1971:35); type locality; *½ mi. S Dry Lake, Grand Canyon National Monument, 4350 ft.; Dry Lake, Grand Canyon National Monument, 4500 ft.*

Dipodomys insularis Merriam
San José Island Kangaroo Rat

1907. *Dipodomys insularis* Merriam, Proc. Biol. Soc. Washington, 20:77, July 22, type from San José Island, Gulf of California, Baja California. Known only from San José Island.

External measurements of five specimens average 249, 146, 39.6. Closely allied to *D. merriami*, this insular rat is described as paler than its relatives of the nearby mainland. See Map 348.

Dipodomys margaritae Merriam
Margarita Island Kangaroo Rat

1907. *Dipodomys margaritae* Merriam, Proc. Biol. Soc. Washington, 20:76, July 22, type from Santa Margarita Island, Baja California. Known only from Santa Margarita Island.

External measurements of three specimens average: 240; 149; 38.2. Closely allied to *D. merriami*, this insular rat is described as differing in smaller auditory bullae and smaller size generally. See Map 348.

Dipodomys nitratoides
Fresno Kangaroo Rat

External measurements: 211–253; 120–152; 33–37. Rostrum short, narrow at base and its sides nearly parallel. The morphological differences between species *D. nitratoides* and *D. merriami* are slight; it is principally because intergradation does not occur between them that they are treated as distinct species. The Tehachapi Mountains isolate the small 4-toed kangaroo rats in the San Joaquin Valley from *D. merriami*, which occurs widely to the east and south. See Map 348.

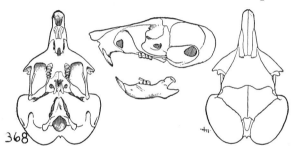

Fig. 368. *Dipodomys nitratoides brevinasus,* 7 mi. SE Simmler, San Luis Obispo Co., California, No. 10747 K.U., ♀, X 1.

Dipodomys nitratoides brevinasus Ginnell

1920. *Dipodomys merriami brevinasus* Grinnell, Jour. Mamm., 1:179, August 24, type from Hayes Station, 19 mi. SW Mendota, Fresno Co., California.
1921. *Dipodomys nitratoides brevinasus* Grinnell, Jour. Mamm., 2:96, May 2.

MARGINAL RECORDS.—California: near Mendota; *21½ mi. W Fresno;* near Wheeler Ridge, 600 ft.; 7 mi. SE Simmler; 11 mi. E Llanado; *type locality.*

Dipodomys nitratoides exilis Merriam

1894. *Dipodomys merriami exilis* Merriam, Proc. Biol. Soc. Washington, 9:113, June 21, type from Fresno, San Joaquin Valley, Fresno Co., California.
1921. *Dipodomys nitratoides exilis,* Grinnell, Jour. Mamm., 2:96, May 2.

MARGINAL RECORDS.—California: type locality; immediately S Kerman, 210 ft.

Dipodomys nitratoides nitratoides Merriam

1894. *Dipodomys merriami nitratoides* Merriam, Proc. Biol. Soc. Washington, 9:112, June 21, type from Tipton, San Joaquin Valley, Tulare Co., California.
1921. *Dipodomys nitratoides nitratoides,* Grinnell, Jour. Mamm., 2:96, May 2.

MARGINAL RECORDS.—California: type locality; Caliente Wash; N side Buena Vista Lake; 15 mi. S Corcoran.

deserti-group
Dipodomys deserti
Desert Kangaroo Rat

External measurements: 305–377; 180–215; 50–58. Weight, 83–138 grams. Upper parts pale ochraceous buff and remainder of body white; ventral dark stripe on tail lacking in most specimens; distal third of tail crested; long hair forming crest dusky, except that distal 25 mm. of tail white; 4 toes on hind foot; inflation of auditory bullae maximum for genus; enlargement of mastoidal bullae so restricting space for interparietal and supraoccipital that these bones are barely visible on dorsal surface of skull.

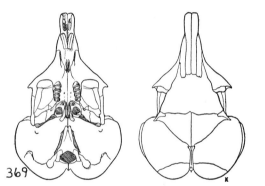

Fig. 369. *Dipodomys deserti deserti,* 21 mi. W, 2 mi. N Lovelock, 4000 ft., Nevada, No. 73694 M.V.Z., ♂, X 1.

Dipodomys deserti aquilus Nader

1965. *Dipodomys deserti aquilus* Nader, Proc. Biol. Soc. Washington, 78:52, July 21, type from 1½ mi. NW High Rock Ranch, sec. 26, T. 28 N, R. 17 E, *ca.* 12 mi. SE Wendel, 4080 ft., Lassen Co., California.

MARGINAL RECORDS (Nader, 1965:54).—Nevada: 1 mi. N, 8 mi. E Jungo, 4200 ft.; 2 mi. N, 21 mi. W Lovelock, 4000 ft.; 3¼ mi. NNE Toulon, 3900 ft.; *3 mi. E Toulon, 3900 ft.;* 4½ mi. N, 2½ mi. W Nixon; 4½ mi. N, 4½ mi. W Flanigan, 4000 ft. California: type locality.

Dipodomys deserti arizonae Huey

1955. *Dipodomys deserti arizonae* Huey, Trans. San Diego Soc. Nat. Hist., 12:99, February 10, type from 3 mi. SE Picacho, Pinal Co., Arizona.

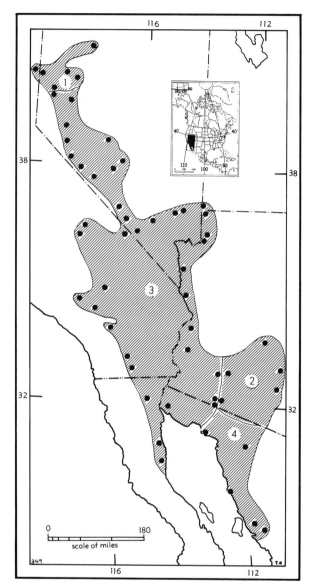

Map 349. *Dipodomys deserti.*

1. *D. d. aquilus*	3. *D. d. deserti*
2. *D. d. arizonae*	4. *D. d. sonoriensis*

MARGINAL RECORDS.—Arizona: 10 mi. N, 2 mi. E Scottsdale (Clothier, 1960:517, recorded as *D. spectabilis perblandus* [= *D. deserti arizonae*, Hoffmeister and Nader, 1963:92]); Florence (Hoffmeister and Nader, 1963:92, recorded as *D. deserti* only); type locality; 7 mi. E Papago Well (Huey, 1955:100); 10 mi. S Gila Bend (*ibid.*).

Dipodomys deserti deserti Stephens

1887. *Dipodomys deserti* Stephens, Amer. Nat., 21:42, January, type from Mohave River [3–4 mi. from, and opposite, Hesperia], San Bernardino Co., California.

1904. *Dipodomys deserti helleri* Elliot, Field Columb. Mus., Publ. 87, Zool. Ser., 3:249, January 7, type from Keeler, Owens Lake, Inyo Co., California. Regarded as identical with *deserti* by Grinnell, Univ. California Publ. Zool., 24:106, June 17, 1922.

MARGINAL RECORDS.—Nevada: ½ mi. SE Wadsworth, 4200 ft.; 4 mi. W Fallon, 4000 ft.; San Antonio, 5400 ft.; 1 mi. N, 34 mi. E Tonopah, Ralston Valley, 5650 ft.; N shore Mud Lake, 5300 ft.; 1 mi. N Beatty, 3400 ft.; Amargosa Desert, 20 mi. SE Beatty, 2500 ft.; 3½ mi. N Pahrump, 2667 ft.; Indian Spring Valley, 14 mi. N Indian Springs, 3100 ft.; Coyote Spring, 2800 ft.; Carp. Utah: 8 mi. N Utah–Arizona border, 2800 ft., Beaverdam Wash. Arizona: Beaver Dam (Lodge), 1900 ft. (Hoffmeister and Durham, 1971:35); Grand Wash, 2100 ft., 2 mi. S, 3 mi. W Pakoon Springs (*ibid.*); *Seven Springs, 1600 ft.* (*ibid.*); Lake Mead, 1200 ft., foot Grand Canyon Wash; 31 mi. N, 2½ mi. W Camp Mohave (Cockrum, 1961:154); Wm. Roberts Ranch, opposite Needles (*ibid.*); 11 mi. NW Bouse (*ibid.*); 10 mi. S Quartzsite (*ibid.*); Sentinel; Growler Valley, about 3 mi. SW Growler Mine; 1 mi. E Quitovaquita, Abra Valley. Sonora: El Doctor. Baja California: San Felipe, Gulf of California; 56 km. S San Felipe (Alvarez, 1961:422); De Mara's Well, W side Laguna Salada, 35 mi. below international boundary. California: Carrizo Creek; Borego Spring; Whitewater; type locality; 16 mi. NE Palmdale (Butterworth, 1964:213, as *D. deserti* only); Barstow; Olancha; Keeler; Kelley's Well, Amargosa River. Nevada: 8 mi. SE Blair; 7 mi. N Arlemont, 5500 ft.; Huntoon Valley, 5700 ft.; Cat Creek, 4 mi. W Hawthorne, 4500 ft.; 8 mi. SE Schurz, 4100 ft.

Dipodomys deserti sonoriensis Goldman

1923. *Dipodomys deserti sonoriensis* Goldman, Proc. Biol. Soc. Washington, 36:139, May 1, type from La Libertad Ranch, 30 mi. E Sierra Seri, Sonora.

MARGINAL RECORDS.—Sonora: near Puerto Peñasco (Cockrum and Bradshaw, 1963:8); Rancho Noche Buena Viejo, 120 km. SE Sonoyta (Alvarez, 1961:422, as *D. d. deserti*); type locality; Costa Rica Ranch; sand dunes, Puerto Libertad (Cockrum and Bradshaw, 1963:8).

SUBFAMILY HETEROMYINAE

Lophs of lower premolars first unite, when worn, at buccal side, next at lingual side; stylids progressively present on p4 in geologically later kinds, developing at any point on tooth; lophs of upper premolars unite first at lingual side, then at buccal side; protoloph of P4 formed of more than one cusp; lophs of upper molars always, and of lowers usually, unite at two ends, surrounding central basin; external cingulum of lower teeth migrates to anterior side, internal cingulum of upper teeth migrates to posterior side, developing secondary connection with middle of adjacent loph, forming Y-shaped crest; cheek-teeth rooted but progressively high-crowned; in later hypsodont kinds, entire crown increasing; and the pattern is long retained; 2 pairs of pits for pterygoid muscles; ethmoid foramen present in dorsal part of orbit; ventral surface of tympanic bullae never reaching level of grinding surface of upper cheek-teeth; median ventral foramina at anterior end of centra of caudal vertebrae; masseteric crest ending above, not behind, mental foramen; astragalo-cuboid articulation (after Wood, 1935:165).

Genus **Liomys** Merriam—Spiny Pocket Mice

Revised by Genoways, Special Publ. Mus. Texas Tech Univ., 5:1–368, December 7, 1973.

1902. *Liomys* Merriam, Proc. Biol. Soc. Washington, 15:44, March 5. Type *Heteromys alleni* Coues.

External measurements: 183–300; 81–169; 22–36. Greatest length of skull, 26.0–38.9. Pelage hispid, stiff spines mingled with soft slender hairs; tail usually well haired; soles of hind feet haired; upper parts brownish; underparts white; lateral line usually present in three species but absent in two (*salvini* and *adspersus*); auditory region of skull uninflated; interpterygoid fossa U-shaped; angle of mandible strongly everted; incisors asulcate; posterior molars decidedly narrower than premolars; 2 lophids on p4; entostyle close to hypocone causing Y-shaped median valley of P4 to be poorly formed.

Genoways (1973:45) notes that a simple key to species of *Liomys* is difficult to construct owing to much geographic variation within each of three of the five species but, that because the few zones of overlap between species are narrow, only a single species is likely to occur at a trapping station. He (*loc. cit.*) mentions that three species may occur in one small part of southeastern Jalisco and that two of these (*L. irroratus* and *L. pictus*) occur together in a zone extending from central Jalisco southward through Michoacán, and Guerrero into Oaxaca, but that in this zone *L. irroratus* favors the drier uplands whereas *L. pictus* favors the more moist lowlands. *L. pictus* and *L. salvini* may occur together from the vicinity of Reforma in Oaxaca eastward to Tonalá in Chiapas. Genoways' key, somewhat abbreviated, follows.

KEY TO SPECIES OF LIOMYS

1. Plantar tubercles 5; upper parts grayish brown; lateral stripe, usually present, pale pinkish to buffy; pterygoid processes broad (see figs.); shaft of baculum oval to tip. . . . . *L. irroratus,*   p. 590

1'. Plantar tubercles 6; upper parts reddish brown, chocolate brown, or paler; lateral stripe ochraceous or absent; pterygoid processes narrow (see figs.); shaft of baculum flattened dorsoventrally (also in some instances compressed laterally at some point).

     2. Upper parts reddish brown; lateral stripe ochraceous; hairs on back not curled upward and not visible above spines; interorbital region broad (relative to length of skull); distal end of shaft of baculum having ventral keel, shaft flattened immediately posterior to keel; not upturned at tip.

         3. Keel on baculum 0.85–1.25; in SE Jalisco greatest length of skull 28.9–32.0 and hind foot rarely more than 30. . . . . *L. pictus,*   p. 592

         3'. Keel on baculum 1.30; in SE Jalisco greatest length of skull 33.0–35.3 and hind foot rarely less than 30; occurring only in SE Jalisco. . . . . *L. spectabilis,*   p. 593

     2'. Upper parts chocolate brown or paler; lateral stripe absent; hairs on back curled up and visible above spines; interorbital region narrow (relative to length of skull); no ventral keel on baculum, shaft flattened almost throughout; upturned at tip.

         4. Small, greatest length of skull averaging less than 33.5; not occurring in Panamá (central Costa Rica north to SE Oaxaca). . . . . *L. salvini,*   p. 594

         4'. Large, greatest length of skull averaging 34.5; occurring only in Panamá. . . *L. adspersus,*   p. 595

Liomys irroratus
Mexican Spiny Pocket Mouse

External measurements: 194–300; 95–169; 22.0–36.0. Greatest length of skull, 27.3–36.9. See key to species for macroscopic features. Protoloph and metaloph of P4 each tricuspid; on metaloph, hypocone largest, metacone only slightly smaller, entostyle distinct but situated near hypocone; re-entrant angle on labial side of p4 separate from median valley; urethral lappets of glans penis trilobed; head of spermatozoon long, apex rounded, and no neck discernible between head and midpiece of spermatozoon; chromosomes, 2n = 60, FN = 62; only known host of the anopluran species *Fahrenholzia ehrlichi* and *F. texana.*

In and near areas where both *L. spectabilis* and *L. torridus* occur, the latter differs as follows: averaging smaller in external measurements and length of skull; cranium broader in relation to length of skull; baculum having ovoid (not dorsoventrally flattened) shaft, upturned (instead of

straight) tip, no ventral keel (instead of keel); broad (rather than narrow) pterygoid processes; 5 (not 6) plantar tubercles; brown (not reddish) upper parts; pale pinkish to buffy (not ochraceous) lateral line; and details (see Genoways, 1973:48) of shape of glans penis and sperm.

Liomys irroratus alleni (Coues)

1881. *Heteromys alleni* Coues, Bull. Mus. Comp. Zool., 8:187, March, type from Hda. Angostura, Río Verde, San Luis Potosí.

1911. *Liomys irroratus alleni,* Goldman, N. Amer. Fauna, 34:56, September 7.

1902. *Liomys canus* Merriam, Proc. Biol. Soc. Washington, 15:44, March 5, type from near Parral, Chihuahua.

1947. *Liomys irroratus pullus* Hooper, Jour. Mamm., 28:47, February 17, type from Tlalpan, 2250 m., Distrito Federal.

1948. *Liomys irroratus acutus* Hall and Villa-R., Univ. Kansas Publ., Mus. Nat. Hist., 1:253, July 26, type from 2 mi. W Pátzcuaro, 7700 ft., Michoacán.

MARGINAL RECORDS (Genoways, 1973:105, 106).—Chihuahua: Santa Rosalía; La Ciéneg[u]illa Springs, 46⁹⁄₁₀ mi. S Jiménez. Durango: Indé; Río Nazas, 10 mi. NNW Rodeo. Zacatecas: 5 mi. SE Río Grande, 6360 ft.; 10 mi. N Rancho Grande, 6200 ft.; Río Nieves, 1 mi. N Rancho Grande; 5 mi. NW Zacatecas, 7600 ft. San Luis Potosí: 8 mi. SW Ramos, 6700 ft.; Cerro Peñón Blanco; Ahualulco; Villar; San Carlos. Tamaulipas: Nicolás, 56 km. NW Tula, 5500 ft. Nuevo León: 5 mi. W Ascención, 6500 ft.; Pablillo, 28 km. S, 38 km. W Linares, 7400 ft.; San Francisco; *Ojo de Agua;* Iturbide, Sierra Madre Oriental, 5000 ft.; Ibarilla, 35 mi. S Linares, 3000 ft.; *Aramberri, 3600 ft.; 1¹⁄₂ mi. N Zaragoza, 4500 ft.;* 1 mi. S Zaragoza, 4600 ft. Tamaulipas: Jaumave. San Luis Potosí: 1½ mi. E Río Verde; Hda. Capulín. Querétaro: Jalpan. Hidalgo: Maguey Verde, 8½ mi. NE Zimapán, 7100 ft.; Zacualti-

Fig. 370. *Liomys irroratus irroratus,* 3 mi. W Mitla, Oaxaca, No. 68855 K.U., ♂, X 1.

pan. Veracruz: 10 km. SW Jacales, 6500 ft. Hidalgo: 1 km. N Epazoyucan, 2550 m. Tlaxcala: 5 mi. W Ciudad Tlaxcala. México: 4 km. SSE San Rafael, 2460 m. Distrito Federal: San Gregorio Altapulco; *200 m. N San Mateo Xalpa.* México: Tenancingo, 2050 m.; Temascaltepec. Michoacán: 5 mi. S Pátzcuaro, 7800 ft.; Tancítaro, 6000 ft.; 1 mi. N Tingüindín, 6300 ft. Jalisco: *3 mi. WSW Mazamitla;* 4 mi. W Mazamitla, 6600 ft. Michoacán: *2 mi. SE La Palma, SE side Lago de Chapala.* Jalisco: 1 mi. S Ocotlán, 5000 ft.; *2 mi. WNW Ocotlán, 5000 ft.;* 2 mi. E Zapotlanejo; 2 mi. S, ½ mi. W Tepatitlán; 10 mi. NE Yahualica. Zacatecas: 3 mi. SW Jalpa, 4600 ft. Jalisco: 3 mi. S Huejúcar, 5900 ft.; 5 mi. NE Huejuquilla, 6200 ft. Zacatecas: Hda. San Juan Capistrano; 4 mi. NNW Chalchihuites, 7000 ft. Durango: Durango; 8 mi. NW Durango; 3 mi. SE Tepehuanes, 5840 ft.; Navarro. Chihuahua: near Parral.

Liomys irroratus bulleri (Thomas)

1893. *Heteromys bulleri* Thomas, Ann. Mag. Nat. Hist., ser. 6, 11:330, April, type from La Laguna, Sierra de Juanacatlán, Jalisco.
1973. *Liomys irroratus bulleri,* Genoways, Special Publ. Mus. Texas Tech Univ., 5:106, December 7.

MARGINAL RECORDS (Genoways, 1973:108).— Jalisco: type locality; *3 km. N Soyatlán del Oro; Rancho de Colomo, 3 km. E Soyatlán del Oro;* 5 km. S Soyatlán del Oro; *5 km. W Soyatlán del Oro.*

Liomys irroratus guerrerensis Goldman

1911. *Liomys guerrerensis* Goldman, N. Amer. Fauna, 34:62, September 7, type from Omilteme, Guerrero.
1973. *Liomys irroratus guerrerensis,* Genoways, Special Publ. Mus. Texas Tech Univ., 5:108, December 7.

MARGINAL RECORDS (Genoways, 1973:110).— Guerrero: 1 mi. NW Omilteme; *2 mi. E Omilteme;* 15 km. SW Chilpancingo; *1 mi. SW Omilteme.*

Liomys irroratus irroratus (Gray)

1868. *Heteromys irroratus* Gray, Proc. Zool. Soc. London, p. 205, May, type from Oaxaca (restricted by Genoways, Special Publ. Mus. Texas Tech Univ., 5:111, December 7), Oaxaca. *H. irroratus* has line priority over *H. albolimbatus*.
1911. *Liomys irroratus,* Goldman, N. Amer. Fauna, 34:53, September 7.
1868. *Heteromys albolimbatus* Gray, Proc. Zool. Soc. London, 205, May, lectotype from La Parada, Oaxaca (Thomas, Ann. Mag. Nat. Hist., ser. 9, 19:552, May 1927).
1956. *Liomys irroratus yautepecus* Goodwin, Amer. Mus. Novit., 1757:7, March 8, type from Rancho Sauce, San Pedro Jilotepec, 5000 ft., Oaxaca.

MARGINAL RECORDS (Genoways, 1973:112).— Oaxaca: Sierra Juárez; Yalalag; Mt. Zempoaltepec; San Pedro Jilotepec; Zapotitlán; Santa María Candelaria; Río Jalatengo, Kilometer 178 Oaxaca–Puerto Angel Road, 4275 ft.; 8 mi. SSW Juchatengo, 6300 ft.; San Andrés Chicahuaxtla; 7 mi. NNE Oaxaca.

Liomys irroratus jaliscensis (J. A. Allen)

1906. *Heteromys jaliscensis* J. A. Allen, Bull. Amer. Mus. Nat. Hist., 22:251, July 25, type from Las Canoas, 7000 ft., Jalisco.
1911. *Liomys irroratus jalicensis* [*sic*], Goldman, N. Amer. Fauna, 34:60, September 7.

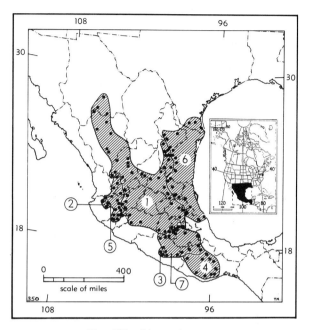

Map 350. *Liomys irroratus.*

Guide to subspecies
1. *L. i. alleni*
2. *L. i. bulleri*
3. *L. i. guerrerensis*
4. *L. i. irroratus*
5. *L. i. jaliscensis*
6. *L. i. texensis*
7. *L. i. torridus*

MARGINAL RECORDS (Genoways, 1973:115).— Jalisco: 1 mi. NW Mezquitic, 5000 ft.; 3 mi. N Villa Guerrero, 5600 ft. Zacatecas: 1 mi. N Santa Rosa; *8 mi. S Moyahua.* Jalisco: Atemajac; 7 km. SE Tonalá; 3 mi. NW Chapala; 8 mi. E Tizapán; 2 mi. N Ciudad Guzmán; *Zapotlán* [= *Ciudad Guzmán*]; *⁹/₁₀ mi. S Ciudad Guzmán; 2⁷/₁₀ mi. WNW Zapotiltic;* type locality; 5 mi. W Atenquique; *2½ mi. ENE Jazmín, 6800 ft.;* 10 mi. W Ciudad Guzmán, 6500 ft.; 27 mi. S, 12 mi. W Guadalajara; *7 km. S Acatlán; 2 mi. SW Ameca, 4000 ft.;* 7 mi. W Ameca, 4000 ft. Nayarit: Ojo de Agua. Jalisco: *6 mi. W San Marcos, 5400 ft.;* 2 mi. NW Magdalena, 4500 ft.; Bolaños, 2800 ft.

Liomys irroratus texensis Merriam

1902. *Liomys texensis* Merriam, Proc. Biol. Soc. Washington, 15:44, March 5, type from Brownsville, Texas.
1911. *Liomys irroratus texensis,* Goldman, N. Amer. Fauna, 34:59, September 7.
1911. *Liomys irroratus pretiosus* Goldman, N. Amer. Fauna, 34:58, September 7.

MARGINAL RECORDS (Genoways, 1973:119).— Texas: 10 mi. NW Raymondville; Laguna Atascosa, 10 mi. E Rio Hondo, thence southward along coast to Veracruz: Nautla; 4 km. W Tlapacoyan. Puebla: Pahuatlán, 1100 m.; Metlaltoyuca. Veracruz: Platón Sánchez. San Luis Potosí: 3 km. S Tamazunchale; Apetsco; 30 km. W Valles; Puerto del Lobos; El Salto. Tamaulipas: near headwaters Río Sabinas, 10 km. N, 8 km. W El Encino; Sierra Madre Oriental, 5 mi. S, 3 mi. W Ciudad Victoria, 1900 ft.; Rancho Santa Rosa, 25 km. N, 13 km. W Ciudad Victoria, 260 m.; 7 km. SW La Purísima; Villa Mainero, 1700 ft. Nuevo León: 7 mi. S, 16 mi. W Linares, 2200 ft.; Río Ramas, 4 mi. W Allende, 1700 ft.; 3 mi. SW Monterrey; *Monterrey;* China. Texas: 17 mi. NW Edinburg.

Liomys irroratus torridus Merriam

1902. *Liomys torridus* Merriam, Proc. Biol. Soc. Washington, 15:45, March 5, type from Cuicatlán, Oaxaca.

1911. *Liomys irroratus torridus*, Goldman, N. Amer. Fauna, 34:55, September 7.

1902. *Liomys irroratus minor* Merriam, Proc. Biol. Soc. Washington, 15:45, March 5, type from Huajuapa de León, Oaxaca.

1903. *Heteromys exiguus* Elliot, Field Columb. Mus. Publ., Zool. Ser., 3:146, March 20, type from Puente de Ixtla, Morelos.

MARGINAL RECORDS (Genoways, 1973:123).— Puebla: San Martín; 6 km. E Totimehuacán, 2150 m. Veracruz: Acultzingo. Oaxaca: Teotitlán, 950 m.; 3 km. NNE Cuicatlán, 600 m.; *1 km. S Cuicatlán, 590 m.;* Huajuapan de León, 1600 m.; Tlapancingo. Guerrero: Tlalixtaquilla; 5 km. N Agua del Obispo; 4 mi. W Chilpancingo, 5800 ft.; 2 mi. W Xochipala, 4400 ft.; *Texcalzintla, 6 km. NNW Teloloapan; Laguna Honda, 1¹/₂ km. SSW Yerbabuena, 1840 m.;* Yerbabuena, 1850 m.; 7 km. SW Cacahuamilpa. Morelos: *2 mi. SW Michapa, 5000 ft.;* Tetecala; *Cuernavaca;* 1½ mi. SE Huitzilac, 8000 ft.; 20 km. NE Cuernavaca, 2100 m.; 20 km. NE Cuautla.

Liomys pictus
Painted Spiny Pocket Mouse

External measurements: 183–294; 91–168; 22–36. Greatest length of skull, 26.0–36.7. See key to species for macroscopic features. Protoloph of P4 unicuspid; metaloph of P4 tricuspid but cusps so connected as not to form discrete cones, hypocone largest, entostyle connected to hypocone; re-entrant angle on labial margin of p4 not reaching median valley; urethral lappets of glans penis trilobed; head of spermatozoon long, apex pointed, and distinct neck between head and midpiece of spermatozoon; chromosomes, 2n = 48, FN = 66.

In and near areas where both *L. spectabilis* and *L. pictus* occur, the latter differs as follows: smaller (see measurements of *L. p. plantinaren-*

sis, L. p. pictus, and *L. spectabilis* in Genoways, 1973), coloration (*op. cit.*), and details of shape of spermatozoon and baculum (*op. cit.*).

In and near areas where both *L. salvini* and *L. pictus* occur, the latter differs as follows: shallower re-entrant angle on labial side of lower premolar, presence instead of absence of ventral keel on baculum, long (not short) head on spermatozoon, trilobed (not bilobed) urethral lappets on glans penis, reddish brown (not chocolate brown) upper parts, presence (not absence) of a lateral stripe, and hairs on back curled upward and visible above spines (instead of not curled upward and concealed beneath spines).

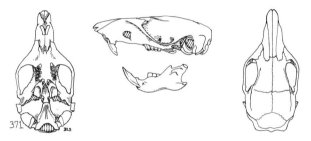

Fig. 371. *Liomys pictus pictus*, San Sebastián, 4400 ft., Jalisco, No. 112276 K.U., ♂, X 1.

Liomys pictus annectens (Merriam)

1902. *Heteromys annectens* Merriam, Proc. Biol. Soc. Washington, 15:43, March 5, type from Pluma Hidalgo, Oaxaca.

1973. *Liomys pictus annectens*, Genoways, Special Publ. Mus. Texas Tech Univ., 5:175, December 7.

MARGINAL RECORDS (Genoways, 1973:179).— Guerrero: 1 mi. NW Omilteme. Oaxaca: San Vincente; Kilometer 136 on Oaxaca–Puerto Escondido Road; Río Molino; Sierra de San Felipe Lachilló; type locality; *4 mi. S Candelaria;* ½ mi. SE San Gabriel Mixtepec; Zacatepec.

Liomys pictus hispidus (J. A. Allen)

1897. *Heteromys hispidus* J. A. Allen, Bull. Amer. Mus. Nat. Hist., 9:56, March 15, type from Rancho El Colomo, Compostela, Nayarit.

1973. *Liomys pictus hispidus*, Genoways, Special Publ. Mus. Texas Tech Univ., 5:179, December 7.

1902. *Liomys sonorana* Merriam, Proc. Biol. Soc. Washington, 15:47, March 5, type from Alamos, Sonora.

1906. *Heteromys pictus escuinapae* J. A. Allen, Bull. Amer. Mus. Nat. Hist., 22:211, July 25, type from Esquinapas [= Escuinapa], Sinaloa.

MARGINAL RECORDS (Genoways, 1973:185).— Sonora: 23 mi. S, 5 mi. E Nogales; E bank Río Yaqui, 1 mi. S El Novillo. Chihuahua: near Batopilas (approx. 20 mi. W Batopilas). Sinaloa: 44 km. ENE Sinaloa; 20 km. N, 5 km. E Badiraguato. Durango: Chacala. Sinaloa: 6 km. E Cosalá; 1 mi. E Santa Lucía. Durango:

2 mi. E Pueblo Nuevo; 2 mi. SW Mezquital (Crossin, *et al.*, 1973:197). Nayarit: La Cuchara, approx. 40 mi. E Acaponeta; Plantanares, 10 mi. E Ruiz; 20 mi. SE Tepic; ½ mi. W Jalisco–Nayarit border on Mexican Highway 15. Jalisco: 1 mi. N Tequila; 9 mi. NNE Guadalajara; El Zapote; 7 mi. W Ameca. Nayarit: Amatlán de Cañas; 1 mi. SW San José del Conde; type locality; 17 mi. E San Blas; 17 km. SE Tuxpan, thence northward along coast to Sinaloa: 24 km. S Gusave. Sonora: Chinobampo; Tecoripa; Ures.

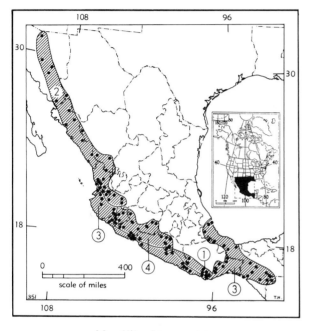

Map 351. *Liomys pictus.*

1. *L. p. annectens*   3. *L. p. pictus*
2. *L. p. hispidus*   4. *L. p. plantinarensis*

Liomys pictus pictus (Thomas)

1893. *Heteromys pictus* Thomas, Ann. Mag. Nat. Hist., ser. 6, 12:233, September, type from Mineral San Sebastián, 4300 ft., Jalisco.
1911. *Liomys pictus*, Goldman, N. Amer. Fauna, 34:33, September 7.
1902. *Liomys pictus rostratus* Merriam, Proc. Biol. Soc. Washington, 15:46, March 5, type from near Ometepec, Guerrero.
1902. *Liomys pictus isthmius* Merriam, Proc. Biol. Soc. Washington, 15:46, March 5, type from Tehuantepec, Oaxaca.
1902. *Liomys veraecrucis* Merriam, Proc. Biol. Soc. Washington, 15:47, March 5, type from San Andrés Tuxtla, Veracruz.
1902. *Liomys obscurus* Merriam, Proc. Biol. Soc. Washington, 15:48, March 5, type from Carrizal, Veracruz.
1902. *Liomys phaeura* Merriam, Proc. Biol. Soc. Washington, 15:48, March 5, type from Pinotepa, Oaxaca.
1902. *Liomys orbitalis* Merriam, Proc. Biol. Soc. Washington, 15:48, March 5, type from Catemaco, Veracruz.
1903. *Heteromys* (*Liomys*) *paralius* Elliot, Field Columb.

Mus. Publ., Zool. Ser., 3:233, September 3, type from San Carlos, Veracruz.
1956. *Liomys pinetorum* Goodwin, Amer. Mus. Novit., 1791:2, September 28, type from San Miguel, *ca.* 4000 ft., 24 km. NE Tonalá, Cerro Tres Picos, District of Tonalá, Chiapas.

MARGINAL RECORDS (Genoways, 1973:191).— Nayarit: Santiago; *Navarrete; 10 mi. E San Blas;* 5 mi. S Las Varas. Jalisco: type locality; 3 mi. N Mascota; La Cuesta; 7 mi. SE Tapalpa; 1 mi. N San Gabriel; *2 km. NE San Marcos.* Colima: Quesería; 4 mi. SE Colima; 23 km. SW Colima. Michoacán: 16 mi. S Arteaga; 8 km. N Melchor Ocampo. Guerrero: *Zacatula;* 5 mi. (by road) E Zacatula; 2 km. SE Zihuatanejo; 5½ mi. N Agua del Obispo; 4 km. NE Colotilpa; near Ometepec (approx. 9 mi. SE Ometepec). Oaxaca: Llano Grande; Pinotepa de Don Luis; Jamiltepec; Cycad Camp, Kilometer 233 on Oaxaca–Puerto Escondido Road; Chacalapa; San Bartolo Yautepec; Nejapa, 60 mi. NW Tehuantepec; Ixcuintepec; San José Lachiguirí; Guichicovi; 8 mi. S Veracruz–Oaxaca state line. Veracruz: Achotal; Jimba; Otatitlán. Oaxaca: Vincente. Veracruz: 2 km. N Paraje Nuevo; 3 mi. NW Plan del Río, San Carlos, thence S along coast to 14 km. SW Coatzacoalcos. Chiapas: 17 mi. W Bochil; El Chorreadero, 6 km. NE Chiapa de Corzo; San Bartolomé; San Vincente. Guatemala: Nentón. Chiapas: 35 mi. SE Comitán; Jaltenango; Villa Flores; 9 mi. SE, 8 mi. NE Tonalá (*sic*); *Madre Mía;* Puerto Arista, thence northward along coast to Nayarit: San Blas.

Liomys pictus plantinarensis Merriam

1902. *Liomys plantinarensis* Merriam, Proc. Biol. Soc. Washington, 15:46, type from Platanar [spelled "Plantinar" on specimen label], Jalisco.
1911. *Liomys pictus plantinarensis*, Goldman, N. Amer. Fauna, 34:37, September 7.
1904. *Liomys parviceps* Goldman, Proc. Biol. Soc. Washington, 17:82, March 21, type from La Salada, 40 mi. S Uruapan, Michoacán.

MARGINAL RECORDS (Genoways, 1973:194).— Jalisco: 3 mi. NE Contla. Michoacán: Los Reyes; Apatzingán; near La Huacana; 2½ mi. S, 1 mi. E Tacámbaro; 3⅜ mi. by road S Tzitzio; Agua Blanca. Guerrero: Arroyo Alcholoya, 7³⁄₁₀ km. N Teloloapan; 2 km. ENE Los Sabinos; *Iguala;* 3200 m. SSE Iguala; 1½ mi. SE Zumpango; Río Balsas; El Limón. Michoacán: El Atuto, 3 km. NE Infiernillo; 7 mi. S Tumbiscato; 11 mi. by road E Dos Aguas; ½ mi. W Coalcomán. Colima: Trapichillos. Jalisco: Piguamo [= Pihuamo]; 6 km. S Atenqueque; 3½ mi. WNW Zapoltitic.

Liomys spectabilis Genoways
Jaliscan Spiny Pocket Mouse

1971. *Liomys spectabilis* Genoways, Occas. Pap. Mus. Nat. Hist., Univ. Kansas, 5:1, June 18, type from 2⅕ mi. NE Contla, 3850 ft., Jalisco.

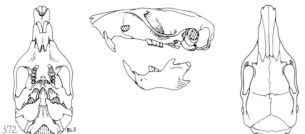

Fig. 372. *Liomys spectabilis*, 2⅛ mi. NE Contla, 3850 ft., Jalisco, No. 96051 K.U., ♂, X 1.

Fig. 373. *Liomys salvini salvini*, 2³⁄₁₀ mi. W, ¼ mi. N Iztapa, 5 ft., Guatemala, No. 65042 K.U., ♂, X 1.

External measurements: 242–280; 122–142; 29.5–32.0. Greatest length of skull, 33.0–35.3. See key to species for macroscopic features. No unworn premolars available but probably resemble those of *L. pictus;* urethral lappets of glans penis trilobed; head of spermatozoon long, apex pointed, and distinct neck between head and midpiece of spermatozoon; chromosomes, 2n = 48, FN = 64. For comparisons with *L. irroratus* and *L. pictus*, see accounts of those species. Known from only a restricted area in southern Jalisco.

MARGINAL RECORDS (Genoways, 1973:200).—Jalisco: 8½ mi. S Mazamitla; *6 km. NE Contla;* 12 mi. NE Pihuamo; 8 mi. SW Tecalitlán; type locality; 8 km. N Contla.

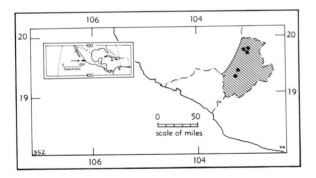

Map 352. *Liomys spectabilis.*

Liomys salvini
Salvin's Spiny Pocket Mouse

External measurements: 185–272; 81–155; 22–32. Greatest length of skull, 28.7–36.4. See key to species for macroscopic features. Protoloph of P4 unicuspid, metaloph having 3 or 4 cusps, metacone in some specimens larger than hypocone, entostyle distinctly separated from other cusps of metaloph; re-entrant angle on labial margin of p4 connects with median valley; ure-

thral lappets of glans penis bilobed; head of spermatozoon short, apex bluntly rounded, distinct neck between head and midpiece; chromosomes, 2n = 56, FN = 86. For comparison with *L. pictus* see account of that species.

Liomys salvini crispus Merriam

1902. *Liomys crispus* Merriam, Proc. Biol. Soc. Washington, 15:49, March 5, type from Tonalá, Chiapas.
1973. *Liomys salvini crispus*, Genoways, Special Publ. Mus. Texas Tech Univ., 5:234, December 7.
1902. *Liomys crispus setosus* Merriam, Proc. Biol. Soc. Washington, 15:49, March 5, type from Huehuetán, Chiapas.

MARGINAL RECORDS (Genoways, 1973:236).—Oaxaca: Reforma. Chiapas: 6 mi. NW Tonalá; *type locality;* approx. 12½ km. SE Tonalá; Pijijiapan; 20 km. SE Pijijiapan; Mapastepec; *11 mi. NW Escuintla;* 13½ km. (Huixtla–Motozintla Road) NE Huixtla; Guatimoc. Guatemala: 3 mi. W Mazatenago; Hda. California, thence northward along coast to point of beginning.

Liomys salvini salvini (Thomas)

1893. *Heteromys salvini* Thomas, Ann. Mag. Nat. Hist., ser. 6, 11:331, April, type from Dueñas, Sacatepéquez, Guatemala.
1911. *Liomys salvini*, Goldman, N. Amer. Fauna, 34:50, September 7.
1893. *Heteromys salvini nigrescens* Thomas, Ann. Mag. Nat. Hist., ser. 6, 12:234, September, type from Costa Rica, probably Escazú (see Goodwin, Bull. Amer. Mus. Nat. Hist., 87:374, December 31, 1946).
1902. *Liomys heterothrix* Merriam, Proc. Biol. Soc. Washington, 15:50, March 5, type from San Pedro Sula, Cortés, Honduras.
1932. *Liomys anthonyi* Goodwin, Amer. Mus. Novit., 528:2, May 23, type from Sacapulas, 4500 ft., El Quiché, Guatemala.
1938. *Liomys salvini aterrimus* Goodwin, Amer. Mus. Novit., 987:4, May 13, type from Sabanillas de Pirris, *ca.* 3730 ft., San José, Costa Rica.

MARGINAL RECORDS (Genoways, 1973:242).—Honduras: San Pedro Sula. Guatemala: 3 mi. E Jocotán. El Salvador: El Tablón; Lake Coatepeque; 10 mi. NW Santa Tecla; 1 mi. NW San Salvador; 6 mi. E San Salvador; Mt. Cacaguatique. Honduras: Monte Redondo; El Caliche Orica; Catacamas; Escuela Agricultura Panamericana. Nicaragua: Quilalí; Savala; Santa Rosa, 17 km. N, 15 km. E Boaco; 1 km. N, 2½ km. W Villa Somoza; 2 km. N, 3 km. E Mérida, Isla de Ometepe. Costa Rica: 9 km. N Liberia, 4 km. E Interamerican Hwy.; 1⁷⁄₁₀ mi. by road W Tilarán; 1 mi. N Santa Ana; *Los Higuerones;* Monte Rey, 22 km. S San José; Sabanilla de Pirrís; Boca del Barranca; San Juanillo, thence northward along coast to Nicaragua: 3 mi. SE San Pablo; Moyogalpa, NW end Isla de Ometepe, 40 m.; 14 km. S Boaco; 8 km. (by road) N Las Maderas; 9 mi. NNE Estelí. Honduras: La Piedra de Jesús. El Salvador: Río Goascorán, 13° 30′ N; Pine Peaks, 3 mi. W Volcán Conchagua, thence northward along coast to Guatemala: Tiquisate; San Lucas; Scapulas; La Primavera; ½ mi. N, 1 mi. E Salamá; *Río Santiago, 152 km. NE Guatemala City (17 mi. NE Río Hondo);* Gualán.

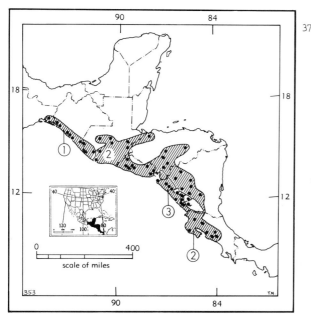

Map 353. *Liomys salvini.*

1. *L. s. crispus*   2. *L. s. salvini*
3. *L. s. vulcani*

Liomys salvini vulcani (J. A. Allen)

1908. *Heteromys vulcani* J. A. Allen, Bull. Amer. Mus. Nat. Hist., 24:652, October 13, type from Volcán de Chinandega, *ca.* 4000 ft., Chinandega, Nicaragua.
1946. *Liomys salvini vulcani,* Goodwin, Bull. Amer. Mus. Nat. Hist., 87:374, December 31.

MARGINAL RECORDS (Genoways, 1973:244).— Nicaragua: 4½ km. N Cosigüina; type locality; Hda. Bellavista, Volcán Casita; Hda. Las Colinas, 4 mi. WNW Puerto Momotambo; Hda. Corpus Christi; 2 mi. N Sabana Grande; Hda. Cutirre, Volcán Mombacho; NW side Isla de Zapatera; La Calera, Nandaime; 1 mi. SE Masachapa.

Liomys adspersus (Peters)
Panamanian Spiny Pocket Mouse

1874. *Heteromys adspersus* Peters, Monatsb. preuss. Akad. Wiss., Berlin, p. 357, May, type from Panamá, by restriction Panama City (Goldman, Smiths. Miscl. Coll., 69(5):118, April 26, 1920).
1911. *Liomys adspersus,* Goldman, N. Amer. Fauna, 34:51, September 7.

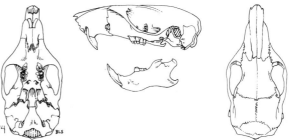

Fig. 374. *Liomys adspersus,* 2 km. NE Buenos Aires, Panamá, Panamá, No. 121528 K.U., ♂, X 1.

External measurements: 222–285; 107–148; 26–34. Greatest length of skull, 32.2–38.9. See key to species for macroscopic features. Occlusal surfaces of premolars, urethral lappets of glands penis, and spermatozoon as described for *Liomys salvini;* chromosomes, 2n = 56, FN probably 84. Known only from Panamá.

MARGINAL RECORDS (Genoways, 1973:254).— Panamá: Cerro Azul; Chepo. Canal Zone: Balboa; Fort Kobbe. Panamá: Río Hato; Guánico; Paracoté; Guabalá; 2 mi. NE Tolé; Río Santa María, Santa Fé. Canal Zone: Río Chágres; *Juan Mina.* Panamá: 2 km. NE Buenos Aires.

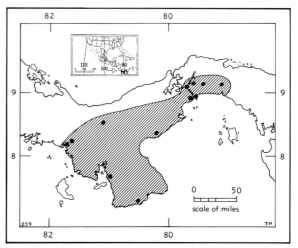

Map 354. *Liomys adspersus.*

Genus **Heteromys** Desmarest
Spiny Pocket Mice

Revised by Goldman, N. Amer. Fauna, 34:14–32, September 7, 1911.

1817. *Heteromys* Desmarest, Nouv. Dict. Hist. Nat., 14:181. Type, *Mus anomalus* Thompson.

External measurements: 272–357; 128–201; 31–44. Pelage composed of flattened, anteriorly grooved bristles or spines mingled with slender hairs; tail usually longer than head and body, not conspicuously crested or penciled at tip. Skull elongated; mastoids rather small, appearing externally entirely posterior to auditory meatus; audital bullae not overlapped by pterygoids; interpterygoid fossa V-shaped, narrowing gradually to a rather acute point anteriorly; 3 or 4 lophids on p4; molar crowns in early life completely divided by a transverse sulcus into two parallel enamel loops, the grinding surface becoming flat, and the loops uniting first at inner ends of upper row and outer ends of lower row, and continued wear obliterating sulcus or leaving a central enamel island (modified from Goldman, 1911:14, 15).

The record (see No. 20 on Map 355) of *Heteromys* in Guerrero is more than 200 mi. west of any other known occurrence of the genus. The record is provided by osseus remains found in owl pellets in the "Cueva del Cañón del Zopilote," situated 13 km. south of the Mexcala bridge at an elevation of 720 m. (Ramírez-P. and Sánchez-H., 1974:107–111). Identification of the remains to species likely will have to wait until a revisionary study is made of specimens from Central America and southern Mexico. Goldman's revision (1911:14–32) was as good as could be made on the basis of the few specimens then available, but not as good as could be made now by using also specimens collected since that time.

KEY TO SUBGENERA OF HETEROMYS

1. Posterior molars narrower than premolars; pelage with numerous stiff bristles or spines. . . . . *Heteromys*, p. 596
1'. Posterior molars equal to or broader than premolars; pelage harsh but bristles soft. . . . . *Xylomys*, p. 600

Subgenus **Heteromys** Desmarest

1817. *Heteromys* Desmarest, Nouv. Dict. Hist. Nat., 14:181. Type, *Mus anomalus* Thompson.

Characters are those of the genus.

KEY TO NOMINAL SPECIES OF SUBGENUS HETEROMYS

1. Soles of hind foot hairy from posterior tubercle to heel; occurring on Peninsula of Yucatán.
   H. gaumeri, p. 599
1'. Soles of hind foot naked posteriorly; not occurring on Peninsula of Yucatán.
   2. Parietals not extending laterally across temporal ridges. . . . . *H. temporalis*, p. 599
   2'. Parietals reaching laterally across temporal ridges.
      3. Zygomatic breadth usually less than 14; tail usually less than 140; occurring only in extreme eastern Panamá below 3000 ft. . . . . *H. australis*, p. 599
      3'. Zygomatic breadth usually more than 14; tail usually more than 140; not occurring as above.
         4. Nasals broadly expanded distally and abruptly narrowing posteriorly. *H. longicaudatus*, p. 599
         4'. Nasals neither broadly expanded distally nor abruptly narrowing posteriorly.
            5. Upper parts lacking a pronounced sprinkling of slender, ochraceous hairs.
               H. goldmani, p. 598
            5'. Upper parts with pronounced sprinkling of slender buffy or ochraceous hairs.
               6. Upper parts distinctly blackish; occurring in southern Veracruz and adjoining parts of north-central Oaxaca. . . . .*H. lepturus*, p. 599
               6'. Upper parts grayish to brownish or, if not grayish to brownish, not in Veracruz or Oaxaca.
                  7. In Sierra de Choapan and Sierra de los Mije. . . . .*H. nigricaudatus*, p. 600
                  7'. Not recorded from Sierra de Choapan and Sierra de los Mije.
                     H. desmarestianus, p. 596

desmarestianus-group
Heteromys desmarestianus
Desmarest's Spiny Pocket Mouse

External measurements: 255–345; 130–190; 31–42. Upper parts mouse gray to blackish, darker shades usually limited to mid-dorsal region, sprinkled with slender, ochraceous hairs; underparts white; lateral line sometimes present, but when present usually not pronounced, buffy or ochraceous; tail longer than head and body, sparsely haired; dusky above, white below; outer surface of forelegs usually washed with ochraceous. Skull large in most subspecies (greatest length *ca.* 35); cranium vaulted; interparietal variable; bullae small; interpterygoid fossa V-shaped. Pelage thin.

This species-group presents some perplexing taxonomic problems; the current nomenclatural arrangement is unsatisfactory and the keys are correspondingly unsatisfactory.

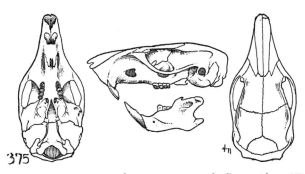

Fig. 375. *Heteromys desmarestianus subaffinis,* 5 km. SE Turrialba, 1950 ft., Prov. Cartago, Costa Rica, No. 26965 K.U., ♀, X 1.

Heteromys desmarestianus chiriquensis Enders

1938. *Heteromys desmarestianus chiriquensis* Enders, Proc. Acad. Nat. Sci. Philadelphia, 90:141, September 20, type from Cerro Pando, between Río Chiriquí Viejo and its tributary Río Colorado, about 10 mi. from Volcán de Chiriquí, Panamá.

MARGINAL RECORDS.—Panamá: type locality; Cerro Punta (Handley, 1966:779). Costa Rica: Auga Buena, Puntarenas.

Heteromys desmarestianus crassirostris Goldman

1912. *Heteromys crassirostris* Goldman, Smiths. Miscl. Coll., 60(2):10, September 20, type from near head Río Limón, 5000 ft., Mt. Pirri, eastern Panamá.
1920. *Heteromys desmarestianus crassirostris* Goldman, Smiths. Miscl. Coll., 69(5):117, April 24.

MARGINAL RECORDS.—Panamá: Mt. Tacarcuna; type locality, thence southward into South America.

Heteromys desmarestianus desmarestianus Gray

1868. *Heteromys desmarestianus* Gray, Proc. Zool. Soc. London, p. 204, May, type from Cobán, Guatemala.
1928. *Heteromys desmarestianus psakastus* Dickey, Proc. Biol. Soc. Washington, 41:10, February 1, type from Los Esesmiles, 8000 ft., Chalatenango, El Salvador. Stated by Goldman (Jour. Washington Acad. Sci., 27:419, October 15, 1937) and reaffirmed by Goodwin (Bull. Amer. Mus. Nat. Hist., 87:371, December 31, 1946) to be identical with *H. d. desmarestianus,* but not, by them, so arranged nomenclatorially.

MARGINAL RECORDS.—Yucatán: Chichén-Itzá; Tekom. Belize: Belize. Guatemala: Puebla. El Salvador (Burt and Stirton, 1961:53): Los Esesmiles; Mt. Cacaguatique. Guatemala: Volcán San Lucas. Chiapas: Ocuilapa. Tabasco: Teapa. Chiapas: 6 mi. SE Palenque. Guatemala: Chamá; Secanquim. Belize: Double Falls; Kate s Lagoon. Not found: X-Cala-Koop, Yucatán (Laurie, 1953:387). That part of range in Yucatán and Belize is based on record-stations from which the specimens were identified only to species by Laurie (*loc. cit.*).

Heteromys desmarestianus fuscatus J. A. Allen

1908. *Heteromys fuscatus* J. A. Allen, Bull. Amer. Mus. Nat. Hist., 24:652, October 13, type from Tuma, Nicaragua.
1920. *Heteromys desmarestianus fuscatus,* Goldman, Smiths. Miscl. Coll., 69(5):115, footnote, April 26.

MARGINAL RECORDS.—Honduras: La Mica; Cerro Cantoral. Nicaragua: type locality; Chontales; Matagalpa. Honduras: Monte Linderos; Cerro Pucca.

Heteromys desmarestianus griseus Merriam

1902. *Heteromys griseus* Merriam, Proc. Biol. Soc. Washington, 15:42, March 5, type from mountains near Tonalá, Chiapas.
1911. *Heteromys desmarestianus griseus,* Goldman, N. Amer. Fauna, 34:22, September 7.

MARGINAL RECORDS.—Oaxaca (Goodwin, 1969:149): 40 mi. S Acayucan, Veracruz; La Gloria; Río Negro. Chiapas: type locality. Oaxaca: *Guichicovi.*

Heteromys desmarestianus panamensis Goldman

1912. *Heteromys panamensis* Goldman, Smiths. Miscl. Coll., 56(36):9, February 19, type from Cerro Azul, near headwaters Río Chagres, 2800 ft., Panamá.
1920. *Heteromys desmarestianus panamensis* Goldman, Smiths. Miscl. Coll., 69(5):117, April 24.

MARGINAL RECORDS.—Panamá: Cerro Brujo; type locality.

Heteromys desmarestianus planifrons Goldman

1937. *Heteromys desmarestianus planifrons* Goldman, Jour. Washington Acad. Sci., 27:418, October 15, type from San Gerónimo Pirrís, Costa Rica.

MARGINAL RECORDS—Costa Rica: Hda. Santa María; Cataratos, San Carlos; El General; type locality.

Heteromys desmarestianus repens Bangs

1902. *Heteromys repens* Bangs, Bull. Mus. Comp. Zool., 39:45, April, type from Boquete, S slope Volcán de Chiriquí, 4000 ft., Panamá.
1920. *Heteromys desmarestianus repens,* Goldman, Smiths. Miscl. Coll., 69(5):115, April 24.

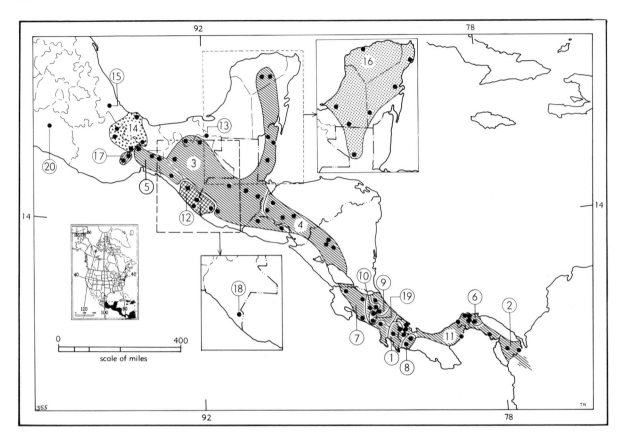

Map 355.   Some species of the genus *Heteromys*.

Guide to kinds
1. *H. desmarestianus chiriquensis*
2. *H. desmarestianus crassirostris*
3. *H. desmarestianus desmarestianus*
4. *H. desmarestianus fuscatus*
5. *H. desmarestianus griseus*
6. *H. desmarestianus panamensis*

7. *H. desmarestianus planifrons*
8. *H. desmarestianus repens*
9. *H. desmarestianus subaffinis*
10. *H. desmarestianus underwoodi*
11. *H. desmarestianus zonalis*
12. *H. goldmani*
13. *H. longicaudatus*

14. *H. lepturus*
15. *H. temporalis*
16. *H. gaumeri*
17. *H. nigricaudatus*
18. *H. nelsoni*
19. *H. oresterus*
20. *H. ? See p. 596*

MARGINAL RECORDS.—Panamá: type locality; Boquerón.

Heteromys desmarestianus subaffinis Goldman

1937. *Heteromys desmarestianus subaffinis* Goldman, Jour. Washington Acad. Sci., 27:420, October 15, type from Angostura, 1980 ft., S side Río Reventazón, opposite Turrialba, Costa Rica.

MARGINAL RECORDS.—Costa Rica: Suerre; type locality; 5 km. SE Turrialba.

Heteromys desmarestianus underwoodi Goodwin

1943. *Heteromys desmarestianus underwoodi* Goodwin, Amer. Mus. Novit., 1227:1, April 22, type from Escazú, about 3000 ft., about 7 mi. SW San José, San José, Costa Rica.

MARGINAL RECORDS.—Costa Rica: Isla Nievo Irazú; type locality.

Heteromys desmarestianus zonalis Goldman

1912. *Heteromys zonalis* Goldman, Smiths. Miscl. Coll., 56(36):9, February 19, type from Río Indio, near Gatún, Canal Zone, Panamá.
1920. *Heteromys desmarestianus zonalis* Goldman, Smiths. Miscl. Coll., 69(5):116, April 24.

MARGINAL RECORDS.—Panamá (Handley, 1966: 779): Fort Sherman; Madden Dam; Maxon Ranch, Río Trinidad; Cerro Campana, 6 mi. E El Valle; upper Río Changena, 2300–2600 ft.; Almirante, 7 km. SSW Changuinola; Salud. Handley (*ibid.*) reported specimens from the coast of San Blas, Panamá, as an undescribed subspecies.

Heteromys goldmani Merriam
Goldman's Spiny Pocket Mouse

1902. *Heteromys goldmani* Merriam, Proc. Biol. Soc. Washington, 15:41, March 5, type from Chicharras, Chiapas.

External measurements: 300–350; 170–201; 35–41. Goldman (1911:24, 25) characterized the skull as follows: "In general form similar to that of *H. desmarestianus*, but larger; rostrum less decurved, flatter above; premaxillae narrower and more tapering posteriorly . . . zygomata more spreading anteriorly."

MARGINAL RECORDS.—Chiapas: Catarina, 1300 m.; type locality. Guatemala: Zunil, 5000 ft. Chiapas: Huehuetán.

Heteromys longicaudatus Gray
Long-tailed Spiny Pocket Mouse

1868. *Heteromys longicaudatus* Gray, Proc. Zool. Soc. London, p. 204, May, type from México. Known precisely only from Montecristo, Tabasco (Goldman, 1911:24).

An adult female from Montecristo, Tabasco, measures 295, 170, 37. Upper parts approx. mouse gray, with sprinkling of buffy hairs; underparts and feet white; buffy lateral line present but faint; tail dusky above, white below. Skull large, broad, robust; rostrum short, robust; nasals, frontals, premaxillae broad; zygomata and interparietal narrow.

Heteromys lepturus Merriam
Santo Domingo Spiny Pocket Mouse

1902. *Heteromys goldmani lepturus* Merriam, Proc. Biol. Soc. Washington, 15:42, March 5, type from mountains near Santo Domingo (few miles W of Guichicovi), Oaxaca.
1911. *Heteromys lepturus*, Goldman, N. Amer. Fauna, 34:25, September 7.

The type measures: 340, 191, 39. This "species" is closely related to *H. desmarestianus* and *H. longicaudatus*. It is said (Goldman, 1911:26) to differ from *longicaudatus* in longer and more decurved rostrum, more highly arched braincase, broader and—with reference to the nasals—more posteriorly extended premaxillae, more widely spreading zygomata anteriorly; broader frontals, more strongly developed supraorbital shelves, wider interparietal, less sharply incurved temporal ridges; decidedly broader basioccipital.

MARGINAL RECORDS.—Veracruz: San Andrés Tuxtla; 25 km. SE Jesús Carranza. Oaxaca (Goodwin, 1969:151): *type locality;* Ixcuintepec; Tarabundí; Tuxtepec.

Heteromys temporalis Goldman
Motzorongo Spiny Pocket Mouse

1911. *Heteromys temporalis* Goldman, N. Amer. Fauna, 34:26, September 7, type from Motzorongo, Veracruz.

The type measures: 320, 180, 37. The species closely resembles *H. goldmani* in most aspects but is unique in the genus in that the temporal ridges follow the parietosquamosal suture rather than crossing the parietal.

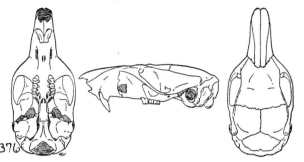

Fig. 376. *Heteromys temporalis*, 2 km. N Motzorongo, 1500 ft., Veracruz, No. 19363 K.U., ♂, X 1.

MARGINAL RECORDS.—Veracruz: 2 km. N Motzorongo, 1800 ft.; *type locality.*

Heteromys gaumeri J. A. Allen and Chapman
Gaumer's Spiny Pocket Mouse

1897. *Heteromys gaumeri* J. A. Allen and Chapman, Bull. Amer. Mus. Nat. Hist., 9:9, February 23, type from Chichén-Itzá, Yucatán.

Three adults from Tunkás, Yucatán, measured: 295–300; 160–166; 34–35. "Skull rather long and angular, with supraorbital ridges strongly developed laterally as overhanging shelves" (Goldman, 1911:30). The species resembles *H. longicaudatus* and *H. desmarestianus* but, unlike those species, has the soles of the hind feet hairy posteriorly.

MARGINAL RECORDS.—Yucatán: 66 km. NE Mérida (93638 KU). Quintana Roo: Puerto Morelos; 4 km. NNE Felipe Carrillo Puerto (92137 KU); 85 km. W Chetumal (93649 KU). Guatemala: 11 km. NE Flores (Ryan, 1960:11). Campeche: 103 km. SE Escárcega (93646 KU); 7½ km. W Escárcega (92153 KU).

anomalus-group
Heteromys australis
Southern Spiny Pocket Mouse

Measurements of the type of *H. australis conscius* are: 260, 133, 32. Two adult topotypes of *conscius* measure: 240, 251; 120, 131; 31, 33.5. Upper parts slaty black finely grizzled with gray hairs; underparts variable, but usually white; tail brownish above, lighter below, but not sharply bicolored. Skull short, broad; braincase well inflated; rostrum short, massive; auditory bullae small.

Heteromys australis conscius Goldman

1913. *Heteromys australis conscius* Goldman, Smiths. Miscl. Coll., 60(22):8, February 28, type from Cana, 2000 ft., mountains of eastern Panamá.

MARGINAL RECORDS.—Panamá (Handley, 1966: 778): Tacarcuna Village, 1950 ft.; Boca de Río Paya, 500 ft., thence into South America and northward to Panamá (*ibid.*): Río Setegantí, 2600 ft.; Cana, 1800–2000 ft.; Tacarcuna Casita, 1500–1700 ft.

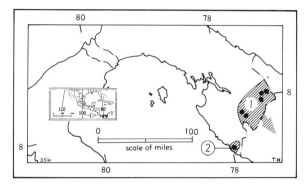

Map 356. *Heteromys australis*.

1. *H. a. conscius*   2. *H. a. pacificus*

Heteromys australis pacificus Pearson

1939. *Heteromys australis pacificus* Pearson, Not. Naturae, Acad. Nat. Sci. Philadelphia, 6:4, June 8, type from Amagal, 1000 ft., S of Guayabo Bay, Darién, Panamá. Known only from type locality.

Heteromys nigricaudatus Goodwin
Goodwin's Spiny Pocket Mouse

1956. *Heteromys nigricaudatus* Goodwin, Amer. Mus. Novit., 1791:4, September 28, type from Mazatlán, about 1500 ft., Oaxaca.

The type, an adult female, measured: 283; 153; 36.5. Upper parts approx. mummy brown; underparts white; no lateral line; tail short, blackish and nearly unicolored; skull short and broad (greatest length, 36; zygomatic breadth, 17).

MARGINAL RECORDS.—Oaxaca: type locality; Ixcuintepec.

Subgenus **Xylomys** Merriam

1902. *Xylomys* Merriam, Proc. Biol. Soc. Washington, 15:43, March 5. Type, *Heteromys nelsoni* Merriam.

Xylomys can readily be distinguished from subgenus *Heteromys* as follows: markedly greater development of posterior part of toothrow; more intricate enamel folds in posterior upper molars; greater lateral extension of pari-

etals along lambdoidal crest (they reach, or almost reach, mastoids); comparatively soft pelage.

KEY TO SPECIES OF SUBGENUS XYLOMYS

1. Ears edged with white; premaxillary tongues extending posterior to nasals.
   H. oresterus, p. 600
1'. Ears without white edges; premaxillary tongues coterminous with nasals.
   H. nelsoni, p. 600

Heteromys nelsoni Merriam
Nelson's Spiny Pocket Mouse

1902. *Heteromys* (*Xylomys*) *nelsoni* Merriam, Proc. Biol. Soc. Washington, 15:43, March 5, type from Pinabete, 8200 ft., Chiapas. Known only from type locality.

Measurements of type are: 356, 195, 43.5. Upper parts mouse gray to blackish; underparts white; tail dusky above, whitish below except for dusky tip. Skull long, slender; nasals markedly broadened, especially distally; premaxillary tongues and nasals crowded well forward by anterior extension of frontals.

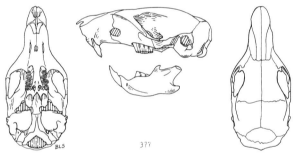

Fig. 377. *Heteromys nelsoni*, Pinabete, Chiapas, No. 77578 U.S.N.M., ♀, X 1.

Heteromys oresterus Harris
Mountain Spiny Pocket Mouse

1932. *Heteromys oresterus* Harris, Occas. Pap. Mus. Zool., Univ. Michigan, 248:4, August 4, type from El Copey de Dota, Cordillera de Talamanca, 6000 ft., Costa Rica.

Average external measurements of eight adult topotypes are: 340, 174, 40. Upper parts blackish gray with sprinkling of ochraceous or buffy hairs; underparts white; ears edged with white; tail blackish above, white below except for black basal area, white tipped. Skull resembling that of "*repens* but with longer rostrum more inflated anteriorly; premaxillae reaching posteriorly beyond the nasals; palate narrower; bullae less inflated" (Harris, 1932:4).

MARGINAL RECORDS.—Costa Rica: El Muñeco; type locality.

INDEX TO VERNACULAR NAMES

INDEX TO TECHNICAL NAMES

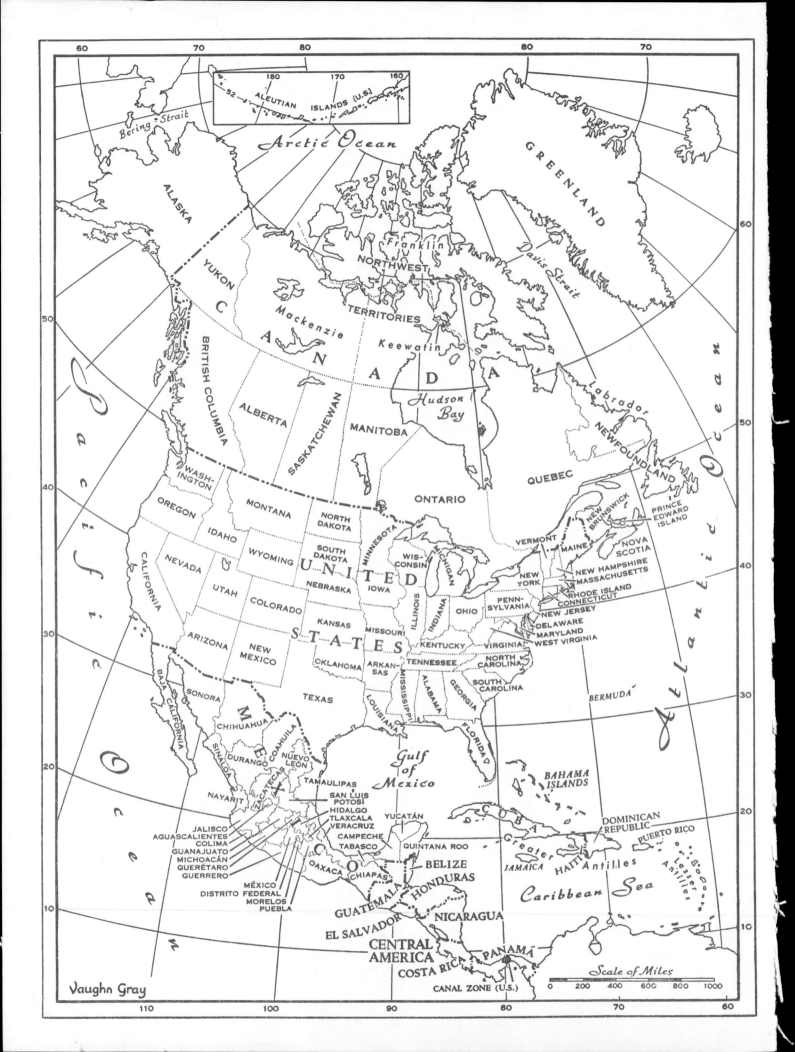